Fumio Hiroshima, József Lőrinczi

Feynman–Kac-Type Theorems and Gibbs Measures on Path Space

De Gruyter Studies in Mathematics

Volume 34/2

Fumio Hiroshima, József Lőrinczi

Feynman–Kac-Type Theorems and Gibbs Measures on Path Space

Volume 2:
Applications in Rigorous Quantum Field Theory

3rd, updated edition

DE GRUYTER

Mathematics Subject Classification 2020
Primary: 47D08, 81Q10, 35J10; Secondary: 47-01, 60-01

Authors
Prof. Dr. Fumio Hiroshima
Kyushu University
Faculty of Mathematics
744 Motooka
Nishiku
819-0395 Fukuoka
Japan
hiroshima@math.kyushu-u.ac.jp

Prof. Dr. József Lőrinczi
Alfréd Rényi Institute of Mathematics
Reáltanoda utca 13-15
1053 Budapest
Hungary
lorinczi@renyi.hu

ISSN 0179-0986
ISBN 978-3-11-223148-7
ISBN 978-3-11-223149-4 (PDF)
ISBN 978-3-11-223150-0 (E-PUB)
DOI https://doi.org/10.1515/9783112231494

Library of Congress Control Number: 2026943532

Bibliographic information published by the Deutsche Nationalbibliothek
The Deutsche Nationalbibliothek lists this publication in the Deutsche Nationalbibliografie; detailed bibliographic data are available on the Internet at https://dnb.dnb.de.

De Gruyter and Walter de Gruyter GmbH are part of De Gruyter Brill.
www.degruyterbrill.com

Questions about General Product Safety Regulation: productsafety@degruyterbrill.com

Printing: CPI books GmbH, Leck

Contents

Contents of Volume 1

Preface to the third edition

We are grateful to many colleagues who carefully read earlier versions of the manuscript and pointed out numerous misprints and errors, which in the present edition have been amended. In particular, we would like to thank Jinqiao Duan, Fritz Gesztesy, Haruto Inoue, Niels Jacob, DongYun Lee, Michael Pang, Michał Ryznar, Soichiro Sakamoto, Itaru Sasaki, Yuki Tsujimoto, Yuma Takabayashi, Kota Ujino, Vanja Wagner, and Yuki Yamamoto.

The first named author was supported in part by JSPS KAKENHI Grant Numbers 20K20886, 20H01808, and 25H00595. The first named author also gratefully acknowledges the kind hospitality of Technische Universität Darmstadt, Hanoi University of Science and Technology, Thammasat University, Paderborn University, the Hausdorff School at the University of Bonn, and the National University of Singapore. In particular, at the Hausdorff School at the University of Bonn, the first named author benefited from very insightful discussions on quantum field theory and path integrals, which were highly encouraging. Part of this work was carried out during the workshop "Quantum Field Theory and Microlocal Analysis" in Roscoff, as well as during the above-mentioned visits. The first named author expresses his sincere thanks to his collaborators, Zied Ammari, Marco Falconi, Benjamin Hinrichs, Oliver Matte, and Tomoyuki Shirai. The first named author is also grateful to Ms. Tsubura Imabayashi for her administrative assistance. Finally, he would like to thank his family, Yuki, Kotaro, and Yui for their warm and constant support.

The second named author is thankful to Giacomo Ascione, Anup Biswas, Atsuhide Ishida, Kamil Kaleta, Itaru Sasaki, Xiaochuan Yang, for recent collaborations which have added new developments in the field, and contributed to the present edition. He also gratefully thanks IHES Bures-sur-Yvette, Scuola Normale Superiore, Naples, Università Sapienza, Rome, Tokyo University of Science, and Great Bay University, Dongguan, for research and lecturing opportunities, providing an environment of sustained interest in this book, and interactions which inspired further work on these subjects and shaped the present edition. Special thanks are due to Isaac Newton Institute, Cambridge, for giving an opportunity to host the 4-month thematic programme "Fractional Differential Equations", having strong and motivating connections with these topics and promoting further research. Also, the yearly series of workshops "Fractional Calculus, Probability, and Non-Local Operators," which has been a joint project with Gianni Pagnini and Enrico Scalas began at BCAM Bilbao, and which has recently come to its 10th round, has provided an active platform of interactions across the fields which cluster around topics that also include and had an impact on the subject matter of this book.

We both thank Kristin Berber-Nerlinger and Melanie Götz of Walter de Gruyter/Brill for their professionalism and support through the process of the present edition.

June 2026

https://doi.org/10.1515/9783112231494-204

Preface to the second edition

In the present edition of this monograph, first published in 2011, the material covered has doubled and thus it was natural to split the presentation in two volumes. This second part is a companion volume to "Feynman–Kac-Type Formulae and Gibbs Measures", where we laid the foundations to the applications of Feynman–Kac representations to a rigorous study of models of quantum field theory presented in this volume.

Our goal in the present edition was to make the material more self-contained and up-to-date, reflecting recent work by the authors and their collaborators on developing new methods of functional integration to ground state problems of fundamental models of quantum field theory. After a detailed introduction to a mathematical treatment of the free Euclidean quantum field and infinite dimensional Ornstein–Uhlenbeck processes, we devote separate chapters to the Nelson, the Pauli–Fierz, and the spin-boson models. While our approach is via path integrals, there is much interaction between functional analysis, operator theory and stochastic analysis at work, which is presented systematically in the hope that newcomers to the field will find it useful, while it will be a resource also to experts. In the presentation we typically follow the pattern of first introducing and discussing a model Hamiltonian on Fock space (suitable for the operator language) and a unitary equivalent L^2 space (suitable for the stochastic language), then derive a functional integral representation, next describe the related path measure as a Gibbs measure, and finally study ground state properties of the initial Hamiltonian by making use of this Gibbs measure. We recommend to use this book with reference to the material of Volume 1, and we provide further commentary in Chapter 5 to a selection of the vast bibliography.

We are grateful to our collaborators Massimiliano Gubinelli, Masao Hirokawa, Oliver Matte, Toshimitsu Takaesu for recent joint work on the problems discussed in this book, and express our appreciation to Gonzalo A. Bley, Thomas Norman Dam, Jacob Schach Møller, Yuma Takabayashi, for comments on the previous edition.

The first named author thanks the kind hospitality of Aarhus University and the International Network Program of the Danish Agency for Science, Technology and Innovation, where part of this work has been done. His work was also financially supported by JSPS KAKENHI Grant Number JP16H03942 and JSPS KAKENHI Grant Number JP16K17612. The second named author thanks IHES, Bures-sur-Yvette, for visiting fellowships through several years, where part of this book has been written, and specifically Professor David Ruelle for sustained interest and discussions. Last but not least, we also express our appreciation to Sabina Dabrowski of Walter de Gruyter, and Ina Talandienė and her team at VTeX UAB, Lithuania, for their professionalism and support.

https://doi.org/10.1515/9783112231494-205

1 Free Euclidean quantum field and Ornstein–Uhlenbeck processes

1.1 Background

A free quantum field describes a large collection of quantum particles of the same type in which there is no interaction between any of these particles or with any other particles in the environment. In this section, we introduce basic notions of the mathematical theory of free quantum fields. There are various formulations of this theory of which we will discuss two. One uses field operators acting on a Hilbert space, providing a framework in which problems of spectral theory, scattering theory, and others are addressed directly. Another is the so-called Euclidean quantum field theory which uses an L^2 space-valued random process to address similar problems by using probabilistic methods. Our goal here is to explain the equivalence of the theory of free boson fields with the theory of infinite-dimensional Gaussian random processes. The essential link between the two descriptions is a Feynman–Kac-type formula.

First, we define the quantum field in boson Fock space. While there are many advantages to the Fock space picture, what is needed for an approach using a Feynman–Kac-type formula is that the Hamiltonian is given in Schrödinger representation, i. e., acting on an $L^2(\mathscr{Q}, d\mu)$ space. Fock space is not of this form, and thus the Gaussian space introduced in the previous section and the Wiener–Itô–Segal isomorphism are needed.

Secondly, we discuss the quantum field in terms of Euclidean quantum field theory. In order to obtain the integral representation of $(F, e^{-tH_f}G)$ for $F, G \in L^2(\mathscr{Q}, d\mu)$ with its Hamiltonian H_f, one way is to change e^{itH} to e^{-tH}, regarding it as the analytic continuation of t to it. In this procedure, the Minkowski metric turns into a Euclidean metric, $-t^2+\mathbf{x}^2 \to t^2+\mathbf{x}^2$, $(t, \mathbf{x}) \in \mathbb{R}\times\mathbb{R}^3$. In the Lorentz covariant formulation of scalar quantum field theory, the so-called *two-point function* is given by

$$(1, \phi(f)\phi(g)1) = \int_{\mathbb{R}^4\times\mathbb{R}^4} f(x)g(y)W(x-y)dxdy$$

with the *Wightman distribution* $W(x-y) = (1, \phi(x)\phi(y)1)$. Here, $x = (t, \mathbf{x})$ and $\phi(x) = e^{-itH}\phi((0,\mathbf{x}))e^{itH}$. The Wightman distribution of the free field Hamiltonian is

$$W((t,\mathbf{x})) = \frac{1}{2}\frac{1}{(2\pi)^3}\int_{\mathbb{R}^3} \frac{e^{it\sqrt{|k|^2+\nu^2}-ik\cdot\mathbf{x}}}{\sqrt{|k|^2+\nu^2}}dk.$$

The *field at time zero* is given by

$$\int_{\mathscr{Q}} \phi(f)\phi(g)d\mu = \frac{1}{2}\int_{\mathbb{R}^3} \overline{\hat{f}(k)}\hat{g}(k)\frac{dk}{(2\pi)^3\sqrt{|k|^2+\nu^2}}.$$

https://doi.org/10.1515/9783112231494-001

Due to the singularity of the so-called *Feynman propagator*

$$\frac{1}{k_0^2 - |k|^2 - \nu^2},$$

it is natural to make an analytic continuation into the region $k_0^2 < 0$. The depth and utility of this point of view has been long appreciated, and Nelson discovered a method of recovering a Minkowski regional field theory from Euclidean field theory. The constructive quantum field theory developed initially by Glimm and Jaffe made use of the Euclidean method.

The analytic continuation $t \mapsto it$ of $W((t, \mathbf{x}))$ yields the *Schwinger function*

$$W((it, \mathbf{x})) = \frac{1}{2}\frac{1}{(2\pi)^4}\int_{\mathbb{R}^4} \frac{e^{itk_0 - ik\cdot\mathbf{x}}}{|k|^2 + \nu^2 + k_0^2}d\mathbf{k}, \quad \mathbf{k} = (k_0, k) \in \mathbb{R} \times \mathbb{R}^3.$$

The Euclidean time-zero quantum field $\phi_{\mathrm{E}}(f)$ has covariance

$$\int_{\mathscr{Q}_{\mathrm{E}}} \phi_{\mathrm{E}}(f)\phi_{\mathrm{E}}(g)d\mu_{\mathrm{E}} = \frac{1}{2}\int_{\mathbb{R}^4} \overline{\hat{f}(\mathbf{k})}\hat{g}(\mathbf{k})\frac{d\mathbf{k}}{|k|^2 + \nu^2 + k_0^2}.$$

To establish the connection between $L^2(\mathscr{Q})$ and the Euclidean space $L^2(\mathscr{Q}_{\mathrm{E}}, d\mu_{\mathrm{E}})$, a family of operators $\{\mathrm{I}_t\}_{t\in\mathbb{R}}$ is introduced. The functional integral representation of $e^{-t(H_{\mathrm{f}}+H_{\mathrm{I}})}$ with some perturbation H_{I} can be also constructed by using the Trotter product formula, the Markov property of $\mathrm{I}_s\mathrm{I}_s^*$, and the identity $\mathrm{I}_t^*\mathrm{I}_s = e^{-|s-t|H_{\mathrm{f}}}$, and we have

$$(F, e^{-tH}G) = \int_{\mathscr{Q}_{\mathrm{E}}} \overline{F_0}G_t e^K d\mu_E, \tag{1.1.1}$$

where $F_0 = \mathrm{I}_0 F$, $G_t = \mathrm{I}_t G$ and e^K denotes an integral kernel. The right-hand side of (1.1.1) is the integral over $\mathscr{Q}_{\mathrm{E}}$, e. g., $\mathscr{S}'_{\mathrm{real}}(\mathbb{R}^{d+1})$, which we call *functional integral* in distinction to *path integral*.

An alternative approach is based on a path measure constructed on $C(\mathbb{R}; \mathscr{M}_{-2})$, i. e., the set of $\mathscr{M}_{-2}$-valued continuous paths, where $\mathscr{M}_{-2}$ is the dual of a Hilbert space $\mathscr{M}_{+2}$. For path integral representations of the heat semigroup $(f, e^{-tH}g)$ in quantum mechanics, a path measure is given on $C([0, \infty); \mathbb{R}^d)$. In quantum field theory, the same procedure yields $\mathscr{M}_{-2}$-valued paths. In the finite mode, $a_k = (1/\sqrt{2})(x_k + \partial_k)$ and $a_k^* = (1/\sqrt{2})(x_k - \partial_k)$, $k = 1, \ldots, n$, and the free field Hamiltonian H_{f} becomes

$$H_{\mathrm{f}} \to \sum_{k=1}^n \sqrt{|k|^2 + \nu^2}a_k^* a_k = \sum_{k=1}^n \frac{\sqrt{|k|^2 + \nu^2}}{2}(-\Delta_k + |x_k|^2 - 1).$$

In this representation, H_{f} has the form of an infinite-dimensional harmonic oscillator. It is known that the generator of the d-dimensional Ornstein–Uhlenbeck process on $C(\mathbb{R};\mathbb{R}^d)$ is the harmonic oscillator on $L^2(\mathbb{R}^d)$. This suggests that it might be possible to construct an infinite-dimensional Ornstein–Uhlenbeck process $(\xi_t)_{t\in\mathbb{R}}$ on $\mathfrak{Y} = C(\mathbb{R}, \mathscr{M}_{-2})$ whose generator is the free field Hamiltonian H_{f}. This can be done by way of Kolmogorov's consistency theorem through the finite-dimensional distributions, giving

$$(F, e^{-tH}G) = \int_{\mathfrak{Y}} \overline{F(\xi_0)}G(\xi_t)e^{K(\xi)}d\mathcal{G}$$

with a suitable kernel $K(\xi)$.

1.2 Boson Fock space

1.2.1 Second quantization

In this section we start the above outlined program by explaining the Fock space-based formulation of quantum field theory. Instead of physical space-time dimension, we work in general d dimension.

Let $\mathscr{W}$ be a separable Hilbert space over $\mathbb{C}$. Consider the operation $\otimes^n_{\mathrm{sym}}$ of n-fold symmetric tensor product defined through the symmetrization operator

$$S_n(f_1 \otimes \cdots \otimes f_n) = \frac{1}{n!}\sum_{\pi\in\wp_n} f_{\pi(1)} \otimes \cdots \otimes f_{\pi(n)}, \quad n \geq 1,$$

where $f_1, \ldots, f_n \in \mathscr{W}$ and $\wp_n$ denotes the permutation group of order n. Define

$$\mathcal{F}_{\mathrm{b}}^{(n)} = \mathcal{F}_{\mathrm{b}}^{(n)}(\mathscr{W}) = \otimes^n_{\mathrm{sym}}\mathscr{W} = S_n(\otimes^n\mathscr{W}),$$

where $\otimes^0_{\mathrm{sym}}\mathscr{W} = \mathbb{C}$. The space

$$\mathcal{F}_{\mathrm{b}} = \mathcal{F}_{\mathrm{b}}(\mathscr{W}) = \bigoplus_{n=0}^{\infty}\mathcal{F}_{\mathrm{b}}^{(n)}(\mathscr{W}),$$

where $\bigoplus_{n=0}^{\infty}$ is understood to be completed (rather than simple algebraic) direct sum, is called the *boson Fock space* over $\mathscr{W}$. The Fock space $\mathcal{F}_{\mathrm{b}}$ can be identified with the space of ℓ_2-sequences $(\Psi^{(n)})_{n\in\mathbb{N}}$ such that $\Psi^{(n)} \in \mathcal{F}_{\mathrm{b}}^{(n)}$ and

$$\|\Psi\|^2_{\mathcal{F}_{\mathrm{b}}} = \sum_{n=0}^{\infty}\|\Psi^{(n)}\|^2_{\mathcal{F}_{\mathrm{b}}^{(n)}} < \infty. \tag{1.2.1}$$

This $\mathcal{F}_{\mathrm{b}}$ is a Hilbert space endowed with the scalar product

$$(\Psi, \Phi)_{\mathcal{F}_{\mathrm{b}}} = \sum_{n=0}^{\infty}(\Psi^{(n)}, \Phi^{(n)})_{\mathcal{F}_{\mathrm{b}}^{(n)}}.$$

The vector

$$\Omega_{\rm b} = (1, 0, 0, \dots)$$

is called *Fock vacuum*. Write P_n for the projection from $\mathcal{F}_{\rm b}$ onto $\mathcal{F}_{\rm b}^{(n)}$. The subspace $P_n\mathcal{F}_{\rm b}$ can be interpreted as consisting of the states of the quantum field having exactly n boson particles. Functions in $P_n\mathcal{F}_{\rm b}$ then determine the exact behavior of these particles, while the permutation symmetry corresponds to the fact that the particles are indistinguishable. In a model of an electron coupled to its radiation field which will be studied in Chapter 3, the particles will be photons. If $\|\Phi\|_{\mathcal{F}_{\rm b}} = 1$, then $(\Phi, P_n\Phi)_{\mathcal{F}_{\rm b}}$ represents the probability of having exactly n bosons in the state described by Φ. Note that by (1.2.1) this probability must decay faster than $1/\sqrt{n}$ for large n in order to have $\Phi \in \mathcal{F}_{\rm b}$. The reason for this constraint is mathematical convenience rather than physical necessity, a fact that we will encounter when discussing the infrared divergence of a specific quantum field model.

In the description of the free quantum field, the following operators acting in $\mathcal{F}_{\rm b}$ are used. There are two fundamental boson particle operators, the *creation operator* denoted by $a^*(f), f \in \mathscr{W}$, and the *annihilation operator* by $a(f), f \in \mathscr{W}$, both acting on $\mathcal{F}_{\rm b}$, defined by

$$(a^*(f)\Psi)^{(n)} = \sqrt{n}S_n(f \otimes \Psi^{(n-1)}), \quad n \geq 1,$$
$$(a^*(f)\Psi)^{(0)} = 0$$

with domain

$$D(a^*(f)) = \left\{(\Psi^{(n)})_{n\geq 0} \in \mathcal{F}_{\rm b} \;\middle|\; \sum_{n=1}^{\infty} n\|S_n(f \otimes \Psi^{(n-1)})\|^2_{\mathcal{F}_{\rm b}^{(n)}} < \infty\right\},$$

and

$$a(f) = (a^*(\bar{f}))^*.$$

As the terminology suggests, the action of $a^*(f)$ increases the number of bosons by one, while $a(f)$ decreases it by one. Since one is the adjoint operator of the other, the relation $(\Phi, a(f)\Psi) = (a^*(\bar{f})\Phi, \Psi)$ holds. Furthermore, since both operators are closable by the dense definition of their adjoints, we will use and denote their closed extensions by the same symbols. Let $D \subset \mathscr{W}$ be a dense subset. It is known that

$$\text{L. H.}\left\{\Omega_{\rm b}, a^*(f_1)\cdots a^*(f_n)\Omega_{\rm b} \mid f_j \in D, j = 1, \dots, n, n \geq 1\right\}$$

is dense in $\mathcal{F}_{\rm b}(\mathscr{W})$, where L. H. is a shorthand for the linear hull. The space

$$\mathcal{F}_{\rm b,fin} = \left\{(\Psi^{(n)})_{n\geq 0} \in \mathcal{F}_{\rm b} \mid \Psi^{(m)} = 0 \text{ for all } m \geq M \text{ with some } M\right\}$$

is called the *finite boson subspace.* The field operators a, a^* leave $\mathcal{F}_{\mathrm{b,fin}}$ invariant and satisfy the *canonical commutation relations*

$$[a(f), a^*(g)] = (\bar{f}, g)1, \quad [a(f), a(g)] = 0, \quad [a^*(f), a^*(g)] = 0$$

on $\mathcal{F}_{\mathrm{b,fin}}$.

Given a bounded operator T on $\mathscr{W}$, the *second quantization* of T is the operator $\Gamma(T)$ on $\mathcal{F}_{\mathrm{b}}$ defined by

$$\Gamma(T) = \bigoplus_{n=0}^{\infty}(\otimes^n T).$$

Here it is understood that $\otimes^0 T = 1$. In most cases $\Gamma(T)$ is an unbounded operator resulting from the fact that it is given by a countable direct sum. However, for a contraction operator T, the second quantization $\Gamma(T)$ is also a contraction on $\mathcal{F}_{\mathrm{b}}$, or equivalently, Γ is a functor

$$\Gamma : \mathscr{C}(\mathscr{W} \to \mathscr{W}) \to \mathscr{C}(\mathcal{F}_{\mathrm{b}} \to \mathcal{F}_{\mathrm{b}}),$$

of the set $\mathscr{C}(X \to Y)$ of contraction operators from X to Y. The functor Γ has the semigroup property, while $\mathscr{C}(\mathscr{W} \to \mathscr{W})$ is a $*$-algebra with respect to operator multiplication and conjugation $*$ (i. e., taking adjoints). The map Γ pulls this structure over to $\mathcal{F}_{\mathrm{b}}$ so that

$$\Gamma(S)\Gamma(T) = \Gamma(ST), \quad \Gamma(S)^* = \Gamma(S^*), \quad \Gamma(1) = 1, \tag{1.2.2}$$

for $S, T \in \mathscr{C}(\mathscr{W} \to \mathscr{W})$. The above definitions imply the intertwining properties

$$\Gamma(T)a^*(f) = a^*(Tf)\Gamma(T), \tag{1.2.3}$$
$$a(f)\Gamma(T) = \Gamma(T)a(jT^*jf), \tag{1.2.4}$$

which can be checked directly. From this, the commutation relations

$$[\Gamma(T), a^*(f)] = a^*((T-1)f)\Gamma(T), \tag{1.2.5}$$
$$[\Gamma(T), a(f)] = \Gamma(T)a(j(1-T^*)jf), \tag{1.2.6}$$

follow, where $jf = \bar{f}$ denotes the complex conjugate of f. If, in particular, T satisfies $T^*T = 1$, then (1.2.3) and (1.2.4) yield

$$\Gamma(T)a^*(f)\Gamma(T^*) = a^*(Tf), \tag{1.2.7}$$
$$\Gamma(T)a(f)\Gamma(T^*) = a(jTjf). \tag{1.2.8}$$

For a self-adjoint operator h on $\mathscr{W}$, the structure relations (1.2.2) imply in particular that

$$\{\Gamma(e^{ith}) \ : \ t \in \mathbb{R}\}$$

is a strongly continuous one-parameter unitary group on $\mathcal{F}_b$. By Stone's theorem, there exists a unique self-adjoint operator $d\Gamma(h)$ on $\mathcal{F}_b$ such that

$$\Gamma(e^{ith}) = e^{itd\Gamma(h)}, \quad t \in \mathbb{R}.$$

The operator $d\Gamma(h)$ is called the *differential second quantization* of h, or simply second quantization of h. Thus we have

$$e^{itd\Gamma(h)} a^*(f) e^{-itd\Gamma(h)} = a^*(e^{ith} f), \tag{1.2.9}$$

$$e^{itd\Gamma(h)} a(f) e^{-itd\Gamma(h)} = a(je^{ith} jf). \tag{1.2.10}$$

Since

$$d\Gamma(h) = -i\frac{d}{dt}\Gamma(e^{ith})\lceil_{t=0}$$

on a suitable domain, we have

$$d\Gamma(h) = 0 \oplus \left\{ \bigoplus_{n=1}^{\infty} \left(\overline{\sum_{j=1}^{n} 1 \otimes \cdots \otimes \overset{j}{h} \otimes \cdots \otimes 1} \right) \right\}, \tag{1.2.11}$$

where the bar denotes closure, and j on top of h indicates its position in the product. Thus the action of $d\Gamma(h)$ on each $\mathcal{F}_b^{(n)}$ is given by

$$d\Gamma(h)\Omega_b = 0$$

and

$$d\Gamma(h) a^*(f_1) \cdots a^*(f_n)\Omega_b = \sum_{j=1}^{n} a^*(f_1) \cdots a^*(hf_j) \cdots a^*(f_n)\Omega_b.$$

It can be also seen by (1.2.11) that

$$\mathrm{Spec}(d\Gamma(h)) = \overline{\{\lambda_1 + \cdots + \lambda_n \mid \lambda_j \in \mathrm{Spec}(h), j = 1, \ldots, n, n \geq 1\} \cup \{0\}},$$

$$\mathrm{Spec}_p(d\Gamma(h)) = \{\lambda_1 + \cdots + \lambda_n \mid \lambda_j \in \mathrm{Spec}_p(h), j = 1, \ldots, n, n \geq 1\} \cup \{0\}.$$

If $0 \notin \mathrm{Spec}_p(h)$, the multiplicity of 0 in $\mathrm{Spec}_p(d\Gamma(h))$ is one.

A crucial operator in quantum field theory is the *boson number operator*

$$N = d\Gamma(1)$$

defined as the second quantization of the identity operator on $\mathscr{W}$. Since

$$N\Omega_b = 0,$$

$$Na^*(f_1) \cdots a^*(f_n)\Omega_b = na^*(f_1) \cdots a^*(f_n)\Omega_b,$$

it follows that

$$\mathrm{Spec}(N) = \mathrm{Spec}_{\mathrm{p}}(N) = \mathbb{N} \cup \{0\}.$$

Let ρ be a real-valued measurable function on $\mathbb{R}$. The operator $\rho(N)$ is self-adjoint on the dense domain

$$D(\rho(N)) = \left\{ (\Phi^{(n)})_{n\geq 0} \in \mathcal{F}_{\mathrm{b}} \,\middle|\, \sum_{n=0}^{\infty} \rho(n)^2 \|\Phi^{(n)}\|^2_{\mathcal{F}_{\mathrm{b}}^{(n)}} < \infty \right\}.$$

In the case of $\rho(x) = e^{\beta x}$ with $\beta > 0$ (note the positive exponent), $\Psi \in D(e^{\beta N})$ implies that the probability of finding n bosons decays exponentially as $n \to \infty$.

We will use the following facts below. The superscript in $a^\sharp$ indicates that any of the creation or annihilation operators is meant.

Proposition 1.1 (Relative bound for $a^\sharp(f)$). *Let h be a positive self-adjoint operator, $f \in D(h^{-1/2})$ and $\Psi \in D(\mathrm{d}\Gamma(h)^{1/2})$. Then $\Psi \in D(a^\sharp(f))$ and*

$$\|a(f)\Psi\| \leq \|h^{-1/2}f\|\|\mathrm{d}\Gamma(h)^{1/2}\Psi\|, \tag{1.2.12}$$

$$\|a^*(f)\Psi\| \leq \|h^{-1/2}f\|\|\mathrm{d}\Gamma(h)^{1/2}\Psi\| + \|f\|\|\Psi\|. \tag{1.2.13}$$

In particular, $D(\mathrm{d}\Gamma(h)^{1/2}) \subset D(a^\sharp(f))$, whenever $f \in D(h^{-1/2})$.

While we do not give a proof of this proposition, we will prove it for a special case below.

Example 1.2.

(1) A first application of the above estimates is that $a^*(f)$ and $a(f)$ are well defined on $D(N^{1/2})$ for all $f \in \mathscr{W}$. For $f \in \mathscr{W}$,

$$\|a(f)\Psi\| \leq \|f\|\|N^{1/2}\Psi\|, \tag{1.2.14}$$

$$\|a^*(f)\Psi\| \leq \|f\|(\|N^{1/2}\Psi\| + \|\Psi\|). \tag{1.2.15}$$

(2) Another application is considering a perturbation of $\mathrm{d}\Gamma(h)$ by adding the operator

$$\Phi(f) = \frac{1}{\sqrt{2}}(a^*(f) + a(\bar{f})).$$

Let $f \in D(h^{-1/2})$. Then $\mathrm{d}\Gamma(h) + \alpha\Phi(f)$ is self-adjoint on $D(\mathrm{d}\Gamma(h))$ for all $\alpha \in \mathbb{R}$. This follows from (1.2.12), (1.2.13), and the Kato–Rellich theorem.

To obtain the commutation relations between $a^\sharp(f)$ and $\mathrm{d}\Gamma(h)$, suppose that $f \in D(h^{-1/2}) \cap D(h)$. Then

$$[\mathrm{d}\Gamma(h), a^*(f)]\Psi = a^*(hf)\Psi, \quad [\mathrm{d}\Gamma(h), a(f)]\Psi = -a(\overline{h\bar{f}})\Psi, \tag{1.2.16}$$

for $\Psi \in D(\mathrm{d}\Gamma(h)^{3/2}) \cap \mathcal{F}_{\mathrm{b,fin}}$. By a limiting argument, (1.2.16) can be extended to $\Psi \in D(\mathrm{d}\Gamma(h)^{3/2})$, and it is seen that $a^\sharp(f)$ maps $D(\mathrm{d}\Gamma(h)^{3/2})$ into $D(\mathrm{d}\Gamma(h))$.

Lemma 1.3. *Let $f \in \bigcap_{n=1}^\infty \left(D(h^n) \cap D(h^{n-1/2})\right)$. Then*

$$a^\sharp(f) : D(\mathrm{d}\Gamma(h)^{n+1/2}) \to D(\mathrm{d}\Gamma(h)^n)$$

for all $n \geq 1$. In particular, $a^\sharp(f)$ maps $\bigcap_{n=1}^\infty D(\mathrm{d}\Gamma(h)^n)$ into itself.

Next we discuss the tensor product of boson Fock spaces.

Proposition 1.4.

(1) *Let $\mathscr{W}_1$ and $\mathscr{W}_2$ be Hilbert spaces. Then*

$$\mathcal{F}_{\mathrm{b}}(\mathscr{W}_1 \oplus \mathscr{W}_2) \cong \mathcal{F}_{\mathrm{b}}(\mathscr{W}_1) \otimes \mathcal{F}_{\mathrm{b}}(\mathscr{W}_2).$$

(2) *Let A and B be self-adjoint operator in $\mathscr{W}_1$ and $\mathscr{W}_2$, respectively. Then*

$$d\Gamma(A \oplus B) \cong d\Gamma(A) \otimes 1 + 1 \otimes d\Gamma(B).$$

Proof. Let Ω_1 (resp. Ω_2) be the Fock vacuum in $\mathcal{F}_{\mathrm{b}}(\mathscr{W}_1)$ (resp. $\mathcal{F}_{\mathrm{b}}(\mathscr{W}_2)$). Define the operator $U : \mathcal{F}_{\mathrm{b}}(\mathscr{W}_1 \oplus \mathscr{W}_2) \to \mathcal{F}_{\mathrm{b}}(\mathscr{W}_1) \otimes \mathcal{F}_{\mathrm{b}}(\mathscr{W}_2)$ by

$$U \prod_{j=1}^n a^*(f_j \oplus g_j)\Omega_{\mathrm{b}} = \prod_{j=1}^n (a^*(f_j) \otimes 1 + 1 \otimes a^*(g_j))\, \Omega_1 \otimes \Omega_2.$$

Here $f_j \in \mathscr{W}_1, j = 1, \ldots, n$ and $g_i \in \mathscr{W}_2, i = 1, \ldots, n$ We see that

$$\left\| U \prod_{j=1}^n a^*(f_j \oplus g_j)\Omega_{\mathrm{b}} \right\|_{\mathcal{F}_{\mathrm{b}}(\mathscr{W}_1 \oplus \mathscr{W}_2)} = \left\| \prod_{j=1}^n (a^*(f_j) \otimes 1 + 1 \otimes a^*(g_j))\, \Omega_1 \otimes \Omega_2 \right\|_{\mathcal{F}_{\mathrm{b}}(\mathscr{W}_1) \otimes \mathcal{F}_{\mathrm{b}}(\mathscr{W}_2)}$$

From this equality, U can be extended to the unitary operator from $\mathcal{F}_{\mathrm{b}}(\mathscr{W}_1 \oplus \mathscr{W}_2)$ to $\mathcal{F}_{\mathrm{b}}(\mathscr{W}_1) \otimes \mathcal{F}_{\mathrm{b}}(\mathscr{W}_2)$, thus statement (1) follows.

Next we prove statement (2). We have $\Gamma(e^{itA \oplus itB}) = \Gamma(e^{itA} \oplus e^{itB})$ and

$$\begin{aligned}
U\Gamma(e^{itA} \oplus e^{itB}) \prod_{j=1}^n a^*(f_j \oplus g_j)\Omega_{\mathrm{b}} &= U \prod_{j=1}^n a^*(e^{itA} f_j \oplus e^{itB} g_j)\Omega_{\mathrm{b}} \\
&= \prod_{j=1}^n \left(a^*(e^{itA} f_j) \otimes 1 + 1 \otimes a^*(e^{itB} g_j)\right) \Omega_1 \otimes \Omega_2 \\
&= \left(\Gamma(e^{itA}) \otimes 1 + 1 \otimes \Gamma(e^{itB})\right) \prod_{j=1}^n (a^*(f_j) \otimes 1 + 1 \otimes a^*(g_j))\, \Omega_1 \otimes \Omega_2 \\
&= \left(\Gamma(e^{itA}) \otimes 1 + 1 \otimes \Gamma(e^{itB})\right) U \prod_{j=1}^n a^*(f_j \oplus g_j)\Omega_{\mathrm{b}}.
\end{aligned}$$

Hence

$$U\Gamma(e^{itA} \oplus e^{itB})U^{-1} = \Gamma(e^{itA}) \otimes 1 + 1 \otimes \Gamma(e^{itB})$$

follows. Taking the derivative at $t = 0$, yields statement (2). □

Inductively, we can also see that

$$\mathcal{F}_{\mathrm{b}}\Big(\bigoplus_{j=1}^{n} \mathscr{W}_j\Big) \cong \bigotimes_{j=1}^{n} \mathcal{F}_{\mathrm{b}}(\mathscr{W}_j). \tag{1.2.17}$$

1.2.2 The case $\mathscr{W} = L^2(\mathbb{R}^d)$

Take now $\mathscr{W} = L^2(\mathbb{R}^d)$ and consider the boson Fock space $\mathcal{F}_{\mathrm{b}}(L^2(\mathbb{R}^d))$. In this case, for $n \in \mathbb{N}$ the space $\mathcal{F}_{\mathrm{b}}^{(n)}$ can be identified with the set of symmetric functions on $L^2(\mathbb{R}^{dn})$ through

$$\otimes_{\mathrm{sym}}^{n} L^2(\mathbb{R}^d) \cong \Big\{f \in L^2(\mathbb{R}^{dn}) \mid f(k_1,\ldots,k_n) = f(k_{\pi(1)},\ldots,k_{\pi(n)}), \forall \pi \in \wp_n\Big\}.$$

The creation and annihilation operators are realized as

$$(a(f)\Psi)^{(n)}(k_1,\ldots,k_n) = \sqrt{n+1}\int_{\mathbb{R}^d} f(k)\Psi^{(n+1)}(k,k_1,\ldots,k_n)dk, \quad n \geq 0,$$

$$(a^*(f)\Psi)^{(n)}(k_1,\ldots,k_n) = \begin{cases} \frac{1}{\sqrt{n}}\sum_{j=1}^{n} f(k_j)\Psi^{(n-1)}(k_1,\ldots,\hat{k}_j,\ldots,k_n), & n \geq 1, \\ 0, & n = 0. \end{cases}$$

Here $\Psi \in \mathcal{F}_{\mathrm{b}}^{(n)}$ is denoted as a pointwise defined function for convenience, however, all of these expressions are to be understood in the L^2-sense. Formally, we also use a notation of kernels $a(k)$ and $a^*(k)$ by writing

$$(a(k)\Psi)^{(n)}(k_1,\ldots,k_n) = \sqrt{n+1}\Psi^{(n+1)}(k,k_1,\ldots,k_n),$$

$$(a^*(k)\Psi)^{(n)}(k_1,\ldots,k_n) = \frac{1}{\sqrt{n}}\sum_{j=1}^{n} \delta(k-k_j)\Psi^{(n-1)}(k_1,\ldots,\hat{k}_j,\ldots,k_n),$$

and

$$[a(k), a^*(k')] = \delta(k-k'), \quad [a(k), a(k')] = 0 = [a^*(k), a^*(k')].$$

Let $\omega_\nu : L^2(\mathbb{R}^d) \to L^2(\mathbb{R}^d)$ be the multiplication operator called *dispersion relation* given by

$$\omega_\nu(k) = \sqrt{|k|^2 + \nu^2}, \quad k \in \mathbb{R}^d,$$

with $\nu \geq 0$. Here ν describes the *boson mass*. The second quantization of the dispersion relation is

$$(d\Gamma(\omega_\nu)\Psi)^{(n)}(k_1,\ldots,k_n) = \left(\sum_{j=1}^{n} \omega_\nu(k_j)\right)\Psi^{(n)}(k_1,\ldots,k_n).$$

Definition 1.5 (Free field Hamiltonian). The self-adjoint operator $d\Gamma(\omega_\nu)$ is called the *free field Hamiltonian* on $\mathcal{F}_b(L^2(\mathbb{R}^d))$ and we use the notation

$$H_f = d\Gamma(\omega_\nu).$$

The spectrum of the free field Hamiltonian is $\mathrm{Spec}(H_f) = [0,\infty)$, with component $\mathrm{Spec}_p(H_f) = \{0\}$, which is of single multiplicity with $H_f\Omega_b = 0$. Formally, we may write the free field Hamiltonian as

$$H_f = \int_{\mathbb{R}^d} \omega_\nu(k)a^*(k)a(k)dk.$$

Physically, this describes the total energy of the free field since $a^*(k)a(k)$ gives the number of bosons carrying momentum k, multiplied with the energy $\omega_\nu(k)$ of a single boson and integrated over all momenta. The commutation relations are

$$[H_f, a(f)] = -a(\omega_\nu f), \quad [H_f, a^*(f)] = a^*(\omega_\nu f),$$

and in formal expression they can be written as

$$[d\Gamma(\omega_\nu), a(k)] = -\omega_\nu(k)a(k), \quad [d\Gamma(\omega_\nu), a^*(k)] = \omega_\nu(k)a^*(k).$$

The relative bounds of $a^\sharp(f)$ with respect to the free Hamiltonian H_f can be seen from (1.2.18) and (1.2.19). If $f/\sqrt{\omega_\nu} \in L^2(\mathbb{R}^d)$, then

$$\|a(f)\Psi\| \leq \|f/\sqrt{\omega_\nu}\|\|H_f^{1/2}\Psi\|, \tag{1.2.18}$$

$$\|a^*(f)\Psi\| \leq \|f/\sqrt{\omega_\nu}\|\|H_f^{1/2}\Psi\| + \|f\|\|\Psi\| \tag{1.2.19}$$

hold, following directly from the lemma below.

Lemma 1.6. *Let $h : \mathbb{R}^d \to \mathbb{C}$ be a measurable function, and $g \in D(h)$. Then for every $\Psi \in D(d\Gamma(|h|^2)^{1/2})$, we have $\Psi \in D(a^\sharp(hg))$, and it follows that*

$$\|a(hg)\Psi\|^2 \leq \|g\|^2\|d\Gamma(|h|^2)^{1/2}\Psi\|^2, \tag{1.2.20}$$

$$\|a^*(hg)\Psi\|^2 \leq \|g\|^2\|d\Gamma(|h|^2)^{1/2}\Psi\|^2 + \|hg\|^2\|\Psi\|^2. \tag{1.2.21}$$

Proof. Let $\Psi \in D(d\Gamma(|h|^2))$. First, note that

$$(\Psi^{(n)}, (d\Gamma(|h|^2)\Psi)^{(n)}) = \sum_{i=1}^{n} \int_{\mathbb{R}^{nd}} |\Psi^{(n)}(k_1,\dots,k_n)|^2 |h(k_i)|^2 dk_1 \cdots dk_n$$

$$= n \int_{\mathbb{R}^{nd}} |\Psi^{(n)}(k_1,\dots,k_n)|^2 |h(k_1)|^2 dk_1 \cdots dk_n.$$

Thus

$$\|(a(hg)\Psi)^{(n-1)}\|^2 = n \int_{\mathbb{R}^{(n-1)d}} \left| \int_{\mathbb{R}^d} g(k_1)h(k_1)\Psi^{(n)}(k_1,\dots,k_n)dk_1 \right|^2 dk_2 \cdots dk_n$$

$$\leq n\|g\|^2 \int_{\mathbb{R}^{nd}} |h(k_1)|^2 |\Psi^{(n)}(k_1,\dots,k_n)|^2 dk_1 \cdots dk_n$$

$$= \|g\|^2 (\Psi^{(n)}, (d\Gamma(|h|^2)\Psi)^{(n)}),$$

and summing over n gives (1.2.20). Due to the closedness of both $d\Gamma(|h|^2)$ and $a(hg)$, we can extend to $\Psi \in D(d\Gamma(|h|^2)^{1/2})$. Equality (1.2.21) can be derived from the canonical commutation relation $[a(hg), a^*(hg)] = \|hg\|^2$ and (1.2.20). □

We can show the next result similarly to Lemma 1.6.

Lemma 1.7. *Let $h : \mathbb{R}^d \to \mathbb{C}$ be measurable, and $g_j \in D(h)$ for $j = 1,\dots,m$. Then for every $\Psi \in D(d\Gamma(|h|^2)^{m/2})$, we have $\Psi \in D(\prod_{j=1}^m a(hg_j))$, and it follows that*

$$\left\| \prod_{j=1}^{m} a(hg_j)\Psi \right\| \leq \left(\prod_{j=1}^{m} \|g_j\| \right) \|d\Gamma(|h|^2)^{m/2}\Psi\|. \tag{1.2.22}$$

Proof. Let $\Psi \in D(d\Gamma(|h|^2)^m)$. First, note that

$$(\Psi^{(n)}, (d\Gamma(|h|^2)^m\Psi)^{(n)}) = \int_{\mathbb{R}^{nd}} |\Psi^{(n)}(k_1,\dots,k_n)|^2 \left(\sum_{i=1}^{n} |h(k_i)|^2 \right)^m dk_1 \cdots dk_n.$$

Using permutation symmetry of $\Psi^{(n)}(k_1,\dots,k_n)$, we can replace

$$\left(\sum_{i=1}^{n} |h(k_i)|^2 \right)^m = \sum_{\{i_1,\dots,i_m\}\subset\{1,\dots,n\}} \prod_{j=1}^{m} |h(k_{i_j})|^2$$

with

$$\left(\sum_{i=1}^{n} |h(k_i)|^2 \right)^m = C(n,m) \prod_{j=1}^{m} |h(k_j)|^2.$$

Here $C(n,m) = n(n-1)\cdots(n-m+1)$. We have

$$(\Psi^{(n)}, (d\Gamma(|h|^2)^m\Psi)^{(n)}) = C(n,m) \int_{\mathbb{R}^{nd}} |\Psi^{(n)}(k_1,\ldots,k_n)|^2 \prod_{j=1}^m |h(k_j)|^2 dk_1\cdots dk_n.$$

On the other hand, by the definition of the annihilation operators, we have

$$\begin{aligned}
&\left\| \left(\prod_{j=1}^m a(hg_j)\Psi\right)^{(n-m)} \right\|^2 \\
&= C(n,m) \int_{\mathbb{R}^{(n-m)d}} \left| \int_{\mathbb{R}^{md}} \left(\prod_{j=1}^m g_j(k_j)h(k_j)\right) \Psi^{(n)}(k_1,\ldots,k_n) dk_1\cdots dk_m \right|^2 dk_{m+1}\cdots dk_n \\
&\leq C(n,m) \left(\prod_{j=1}^m \|g_j\|^2\right) \int_{\mathbb{R}^{nd}} \left(\prod_{j=1}^m |h(k_j)|^2\right) |\Psi^{(n)}(k_1,\ldots,k_n)|^2 dk_1\cdots dk_n \\
&= \left(\prod_{j=1}^m \|g_j\|^2\right) (\Psi^{(n)}, (d\Gamma(|h|^2)^m\Psi)^{(n)}),
\end{aligned}$$

and summing over n gives (1.2.22). Due to the closedness of the operators $d\Gamma(|h|^2)^{m/2}$ and $\prod_{j=1}^m a(hg_j)$, we can extend to $\Psi \in D(d\Gamma(|h|^2)^{m/2})$. □

Corollary 1.8. *If $f_j \in L^2(\mathbb{R}^d)$ and $g_j \in D(1/\sqrt{\omega_\nu})$ for $j = 1,\ldots,n$, then*

$$\left\|\prod_{j=1}^n a(f_j)\Psi\right\| \leq \left(\prod_{j=1}^n \|f_j\|\right) \|N^{n/2}\Psi\|, \quad \Psi \in D(N^{n/2}),$$

$$\left\|\prod_{j=1}^n a(g_j)\Phi\right\| \leq \left(\prod_{j=1}^n \|g_j/\sqrt{\omega_\nu}\|\right) \|H_f^{n/2}\Phi\|, \quad \Phi \in D(H_f^{n/2}).$$

Proof. This is obtained directly from Lemma 1.7 by putting $h = 1$ and $h = \sqrt{\omega_\nu}$. □

1.2.3 The case $\mathscr{W} = \mathbb{C}1$

Now we consider the boson Fock space over the one-dimensional vector space $\mathscr{W} = \mathbb{C}1$ over $\mathbb{C}$, and write for simplicity $\mathbb{C}$ for $\mathbb{C}1$. We have

$$\mathcal{F}_b(\mathbb{C}) = \overline{\{a^*(1)^n\Omega_b \mid n \in \dot{\mathbb{N}}\}}.$$

We see that $a(1)\Omega_b = 0$ and the canonical commutation relation $[a(1), a^*(1)] = 1$ is satisfied. On the other hand, we consider the operators

$$a = \frac{1}{\sqrt{2}}\left(x + \frac{d}{dx}\right), \quad a^* = \frac{1}{\sqrt{2}}\left(x - \frac{d}{dx}\right),$$

acting in $L^2(\mathbb{R})$, and $\Phi_0(x) = \pi^{-1/4}e^{-x^2/2}$. We see that $\|\Phi_0\| = 1$, $a\Phi_0 = 0$, and the canonical commutation relation $[a, a^*] = 1$ holds. It can be also seen that

$$\Phi_n = \frac{1}{\sqrt{n!}}(a^*)^n\Phi_0$$

is the nth Hermite function and $\{\Phi_n\}_{n\in\dot{\mathbb{N}}}$ is a complete orthonormal system of $L^2(\mathbb{R})$, i. e.,

$$L^2(\mathbb{R}) = \overline{\mathrm{L.H.}\{(a^*)^n\Omega_\mathrm{b} \mid n \in \dot{\mathbb{N}}\}}.$$

Now we construct a unitary operator between $\mathcal{F}_\mathrm{b}(\mathbb{C})$ and $L^2(\mathbb{R})$. Let $U : \mathcal{F}_\mathrm{b}(\mathscr{W}) \to L^2(\mathbb{R})$ be defined by

$$Ua^*(1)^n\Omega_\mathrm{b} = (a^*)^n\Phi_0, \quad n \in \dot{\mathbb{N}}.$$

The operator U maps a dense subspace to a dense subspace, and, furthermore, we see that $\|Ua^*(1)^n\Omega_\mathrm{b}\|_{\mathcal{F}_\mathrm{b}(\mathscr{W})} = \|(a^*)^n\Phi_0\|_{L^2(\mathbb{R})}$. From this equality, U can be extended to the unitary operator from $\mathcal{F}_\mathrm{b}(\mathscr{W})$ to $L^2(\mathbb{R})$. Then

$$\mathcal{F}_\mathrm{b}(\mathbb{C}) \cong L^2(\mathbb{R})$$

follows. By the discussion above, we also see that

$$\mathcal{F}_\mathrm{b}^{(n)}(\mathbb{C}) \cong \mathrm{L.H.}\{\Phi_n\}, \quad n \in \dot{\mathbb{N}}$$

and $\dim\mathcal{F}_\mathrm{b}^{(n)}(\mathbb{C}) = 1$ for $n \in \dot{\mathbb{N}}$. We summarize this in the proposition below.

Proposition 1.9. *Let $d \geq 1$. It follows that*

$$\mathcal{F}_\mathrm{b}(\mathbb{C}^d) \cong L^2(\mathbb{R}^d)$$

and

$$\mathcal{F}_\mathrm{b}^{(n)}(\mathbb{C}^d) \cong \mathrm{L.H.}\{\Phi_{n_1} \otimes \cdots \otimes \Phi_{n_k} \mid n_1 + \cdots + n_k = n, n_j \in \dot{\mathbb{N}}, k \geq 0\}.$$

Proof. For $d = 1$, we already showed it. In the case of $d \geq 2$, inductively we have

$$\mathcal{F}_\mathrm{b}(\mathbb{C}^d) = \mathcal{F}_\mathrm{b}(\oplus^d\mathbb{C}) \cong \otimes^d\mathcal{F}_\mathrm{b}(\mathbb{C}) \cong \otimes^d L^2(\mathbb{R}) \cong L^2(\mathbb{R}^d),$$

thus the first statement follows. Since

$$\mathcal{F}_\mathrm{b}(\mathbb{C}^d) \cong \otimes^d\mathcal{F}_\mathrm{b}(\mathbb{C}) = \otimes^d\bigoplus_{n=0}^\infty\mathcal{F}_\mathrm{b}^{(n)}(\mathbb{C}) = \bigoplus_{n=0}^\infty\mathcal{K}_\mathrm{b}^{(n)}$$

with

$$\mathcal{K}_{\mathrm{b}}^{(n)} = \bigoplus_{n_1+\cdots+n_k=n, k\geq 0} \mathcal{F}_{\mathrm{b}}^{(n_1)}(\mathbb{C}) \otimes \cdots \otimes \mathcal{F}_{\mathrm{b}}^{(n_k)}(\mathbb{C}),$$

the second statement also follows. □

The number operator N in $\mathcal{F}_{\mathrm{b}}(\mathbb{C})$ acts as

$$UNa^*(1)^n \Omega_{\mathrm{b}} = nUa^*(1)^n \Omega_{\mathrm{b}} = n(a^*)^n \Phi_0 = a^* a (a^*)^n \Phi_0.$$

Hence the number operator in $\mathcal{F}_{\mathrm{b}}(\mathbb{C})$ corresponds to the Hamiltonian of the harmonic oscillator

$$N \cong a^* a = -\frac{1}{2}\Delta + \frac{1}{2}x^2 - \frac{1}{2}$$

and $\mathrm{Spec}(N) = \dot{\mathbb{N}}$.

Example 1.10 (Mehler's formula). Let $0 < c < 1$. The second quantization $\Gamma(c)$ is the integral operator on $L^2(\mathbb{R})$ and its integral kernel is given by

$$\Gamma(c)(x,y) = \left(\frac{1}{\pi(1-c^2)}\right)^{1/2} \exp\left(-\frac{(1+c^2)(x^2+y^2) - 4xyc}{2(1-c^2)}\right).$$

Example 1.11.

(1) Let $\mathscr{W}$ be a Hilbert space. Then $\mathcal{F}_{\mathrm{b}}(\mathbb{C}^d \oplus \mathscr{W}) \cong L^2(\mathbb{R}^d) \otimes \mathcal{F}_{\mathrm{b}}(\mathscr{W})$ and the relation $d\Gamma(1_{\mathbb{C}^d} \oplus 1_{\mathscr{W}}) \cong a^* a \otimes 1 + 1 \otimes N$ holds.

(2) Let $\mathscr{W} = L^2(\mathbb{R}^d)$. Then $\mathcal{F}_{\mathrm{b}}(\mathbb{C}^d \oplus L^2(\mathbb{R}^d)) \cong L^2(\mathbb{R}^d) \otimes \mathcal{F}_{\mathrm{b}}(L^2(\mathbb{R}^d))$ and the relation $d\Gamma(1_{\mathbb{C}^d} \oplus \omega) \cong a^* a \otimes 1 + 1 \otimes H_{\mathrm{f}}$ holds.

1.2.4 Segal–Bargmann space

In Section 1.2.3 we showed that $L^2(\mathbb{R}) \cong \mathcal{F}_{\mathrm{b}}(\mathbb{C})$ and we established that $L^2(\mathbb{R})$ can be regarded as a special case of the boson Fock space. Here we discuss an alternative representation of $L^2(\mathbb{R})$ with an annihilation operator and a creation operator.

Definition 1.12 (Segal–Bargmann space). The *Segal–Bargmann space* $\mathcal{B}$ is the space of holomorphic functions f on $\mathbb{C}$ satisfying the condition

$$\|f\|_{\mathcal{B}}^2 = \frac{1}{\pi} \int_{\mathbb{C}} |f(z)|^2 \exp(-|z|^2)\, dz < \infty,$$

where $dz = dxdy$ under the identification $\mathbb{C} \cong \mathbb{R} \oplus \mathbb{R}$ by $z = x + iy$.

Space $\mathcal{B}$ is a Hilbert space with respect to the associated inner product

$$(f,g)_{\mathcal{B}} = \frac{1}{\pi}\int_{\mathbb{C}} \overline{f(z)}g(z)\exp(-|z|^2)\,dz.$$

Let $F_a(z) = \exp(\bar{a}z)$. The function F_a is called a *coherent state* with parameter a, and the function $\kappa(a,z) = \overline{F_a(z)}$ is called the *reproducing kernel* for the Segal–Bargmann space. Note that

$$f(a) = \frac{1}{\pi}\int_{\mathbb{C}} \kappa(a,z)f(z)\exp(-|z|^2)\,dz.$$

Any function $f \in \mathcal{B}$ can be represented as $f(z) = \sum_{n=0}^{\infty} c_n z^n$, and we see that $(z^n, z^m)_{\mathcal{B}} = 0$ unless $n = m$. Then

$$\{z^n\}_{n\in\dot{\mathbb{N}}}$$

comprises a complete orthonormal system of $\mathcal{B}$. We define $b^* = z$ and $b = \frac{d}{dz}$. We see that the adjoint of b is b^* under the Segal–Bargmann inner product, and it satisfies that $[b, b^*] = 1$. An important fact is that there exists a unitary operator U from $L^2(\mathbb{R})$ to $\mathcal{B}$ defined by

$$(Uf)(z) = \int_{\mathbb{R}} \exp\left(-\frac{1}{2}(z^2 - 2\sqrt{2}zx + x^2)\right)f(x)dx$$

and the inverse formula is given by

$$f(x) = \int_{\mathbb{C}} \exp\left(-\frac{1}{2}(\bar{z}^2 - 2\sqrt{2}\bar{z}x + x^2)\right)(Uf)(z)e^{-|z|^2}\,dz.$$

This leads to the identification $\mathcal{B} \cong L^2(\mathbb{R})$.

Proposition 1.13. *It follows that*

$$L^2(\mathbb{R}) \cong \mathcal{B} = \bigoplus_{n=0}^{\infty} \mathrm{L.H.}\{z^n\}.$$

Under this identification, we have

$$z \cong \frac{1}{\sqrt{2}}\left(x - \frac{d}{dx}\right), \quad \frac{d}{dz} \cong \frac{1}{\sqrt{2}}\left(x + \frac{d}{dx}\right),$$

and it follows that

$$z\frac{d}{dz} \cong -\frac{1}{2}\Delta + \frac{x^2}{2} - \frac{1}{2}.$$

Example 1.14. The eigenvalue equation

$$\left(-\frac{1}{2}\Delta + \frac{x^2}{2} - \frac{1}{2}\right)\Phi = \alpha\Phi$$

can be reduced to finding a holomorphic function Φ such that

$$z\frac{d}{dz}\Phi(z) = \alpha\Phi(z).$$

The solution is $\Phi(z) = z^\alpha$ which is holomorphic. Hence $\alpha \in \dot{\mathbb{N}}$ and

$$\Phi(z) = z^n, \quad n \in \mathbb{N},$$

that is, $z^n \in \mathcal{B}$ corresponds to the nth Hermite function $\left(\frac{1}{\sqrt{2}}(x - \frac{d}{dx})\right)^n \Phi_0$ in $L^2(\mathbb{R})$, where $\Phi_0(x) = \pi^{-1/4}e^{-|x|^2/2}$.

1.2.5 Segal fields

The creation and annihilation operators are not symmetric and do not commute. Roughly speaking, the creation operator corresponds to $\frac{1}{\sqrt{2}}(x - \partial_x)$ and the annihilation operator to $\frac{1}{\sqrt{2}}(x + \partial_x)$ in $L^2(\mathbb{R}^d)$. We can, however, construct symmetric and commutative operators by combining the two field operators leading to so-called *Segal fields*. As it will be seen below, Segal fields linearly span Fock space, which suggests that $\mathcal{F}_\mathrm{b}$ may be realized conveniently as a space $L^2(\mathcal{Q})$ with infinitely many variables on a suitable $\mathcal{Q}$ instead of necessarily using operators. This is due to the fact that Segal fields are commutative. In this representation, Segal fields translate into real-valued multiplication operators.

Definition 1.15 (Segal field). The Segal field $\Phi(f)$ on the boson Fock space $\mathcal{F}_\mathrm{b}(\mathscr{W})$ is defined by

$$\Phi(f) = \frac{1}{\sqrt{2}}(a^*(f) + a(\bar{f})), \quad f \in \mathscr{W},$$

and its *conjugate momentum* by

$$\Pi(f) = \frac{i}{\sqrt{2}}(a^*(f) - a(\bar{f})), \quad f \in \mathscr{W}.$$

As above, $\bar{f}$ denotes the complex conjugate of f.

By the above definition, both $\Phi(f)$ and $\Pi(g)$ are symmetric, however, they are not linear in f and g over $\mathbb{C}$. Note that, in contrast, they are linear operators over $\mathbb{R}$. It is straightforward to check that

$$[\Phi(f),\Pi(g)] = i\,\mathrm{Re}(f,g),$$

and

$$[\Phi(f),\Phi(g)] = i\,\mathrm{Im}(f,g), \quad [\Pi(f),\Pi(g)] = i\,\mathrm{Im}(f,g).$$

In particular, for real-valued f and g, the *canonical commutation relations* become

$$[\Phi(f),\Pi(g)] = i(f,g), \quad [\Phi(f),\Phi(g)] = [\Pi(f),\Pi(g)] = 0.$$

Applying the bounds (1.2.18)–(1.2.19) to $h = 1$, we see that $\mathcal{F}_{\mathrm{b,fin}}$ is the set of *analytic vectors* of $\Phi(f)$, i. e.,

$$\operatorname*{s-lim}_{m\to\infty}\sum_{n=0}^{m}\frac{\|\Phi(f)^n\Psi\|t^n}{n!} < \infty$$

for $\Psi \in \mathcal{F}_{\mathrm{b,fin}}$ and $t \geq 0$. The following is a general result.

Proposition 1.16 (Nelson's analytic vector theorem). *Let K be a symmetric operator on a Hilbert space. We assume that there exists a dense subspace $\mathcal{D} \subset D(K)$ such that $\sum_{n=0}^{\infty}\|K^n f\|t^n/n! < \infty$ for $f \in \mathcal{D}$ and $t > 0$. Then K is essentially self-adjoint on $\mathcal{D}$.*

By Nelson's analytic vector theorem, $\Phi(f)$ and $\Pi(g)$ are essentially self-adjoint on $\mathcal{F}_{\mathrm{b,fin}}$. We keep denoting the closures of $\Phi(f)\lceil_{\mathcal{F}_{\mathrm{b,fin}}}$ and $\Pi(g)\lceil_{\mathcal{F}_{\mathrm{b,fin}}}$ by the same symbols.

Next we review the algebraic calculus involving commutators, which are useful in a rigorous treatment of commutation relations. Suppose that $[A,B] = C$ and $[C,B] = D$, ignoring domain issues for simplicity. Let $n \in \mathbb{N}$, and suppose that $D = 0$, i. e., C and B are commutative. Then we have

$$[A,B^n] = nB^{n-1}C. \tag{1.2.23}$$

We define e^B by the formal series $\sum_{n=0}^{\infty}\frac{B^n}{n!}$. By (1.2.23), it can be directly seen that

$$[A,e^B] = e^B C.$$

In particular, $e^{-B}Ae^B = A + C$ and

$$e^{-B}A^n e^B = (A + C)^n$$

follow. Next consider the case where $[D,B] = 0$, giving

$$[A,B^n] = nB^{n-1}C + \frac{n(n-1)}{2}B^{n-2}D.$$

From this, it follows that

$$[A,e^B] = e^B\left(C + \frac{1}{2}D\right)$$

and, in particular, we have

$$e^{-B}A^n e^B = \left(A + C + \frac{1}{2}D\right)^n. \tag{1.2.24}$$

These relations can be generalized. Define $\mathrm{ad}_A B = [A,B]$. Hence $\mathrm{ad}_A^2 B = [[A,B],B]$ and $\mathrm{ad}_A^3 B = [[[A,B],B],B]$, etc., follow. Thus (1.2.24) can be written as

$$e^{-B}A^n e^B = \left(\mathrm{ad}_A^0 B + \mathrm{ad}_A^1 B + \frac{1}{2}\mathrm{ad}_A^2 B\right)^n \tag{1.2.25}$$

with $\mathrm{ad}_A^0 B = A$ and $\mathrm{ad}_A^1 B = \mathrm{ad}_A B$. In general, it can be seen that

$$e^{-B}A^m e^B = \left(\sum_{n=0}^{\infty} \frac{1}{n!}\mathrm{ad}_A^n B\right)^m. \tag{1.2.26}$$

These expressions are formal and require a check of the convergence of $\sum_{n=0}^{\infty} \frac{1}{n!}\mathrm{ad}_A^n B$ and specifying the domain where both sides of (1.2.26) are well defined.

Example 1.17.

(1) Since $[\Phi(f), i\Pi(g)] = -\operatorname{Re}(f,g)$, we have

$$\Phi(f)e^{i\Pi(g)} = e^{i\Pi(g)}(\Phi(f) - \operatorname{Re}(f,g))$$

and

$$e^{-i\Pi(g)}\Phi(f)e^{i\Pi(g)} = \Phi(f) - \operatorname{Re}(f,g). \tag{1.2.27}$$

(2) Let h be self-adjoint and $g \in D(h)$. It is seen that $[\mathrm{d}\Gamma(h), i\Pi(g)] = -\Phi(hg)$ and $[-\Phi(hg), i\Pi(g)] = (hg,g)$. Hence

$$\mathrm{d}\Gamma(h)e^{i\Pi(g)} = e^{i\Pi(g)}\left(\mathrm{d}\Gamma(h) - \Phi(hg) + \frac{1}{2}(hg,g)\right).$$

We also see that

$$e^{-i\Pi(g)}\mathrm{d}\Gamma(h)e^{i\Pi(g)} = \mathrm{d}\Gamma(h) - \Phi(hg) + \frac{1}{2}(hg,g). \tag{1.2.28}$$

Equalities (1.2.27)–(1.2.28) can rigorously be verified by a limiting argument as in the proposition below.

Proposition 1.18.

(1) *The operator $e^{-i\Pi(g)}$ maps $D(\Phi(f))$ onto itself, and* (1.2.27) *holds on $D(\Phi(f))$.*
(2) *If h is self-adjoint and $g \in D(h)$, then $e^{-i\Pi(g)}$ maps $D(\mathrm{d}\Gamma(h))$ onto itself and* (1.2.28) *is satisfied on $D(\mathrm{d}\Gamma(h))$.*

Proof. Let $F \in \mathcal{F}_{\mathrm{b,fin}}$. We have

$$\Phi(f) \sum_{m=0}^{n} \frac{(i\Pi(g))^m}{m!} F = \sum_{m=0}^{n} \frac{(i\Pi(g))^m}{m!} \Phi(f)F - \sum_{m=0}^{n-1} \frac{(i\Pi(g))^m}{m!} \operatorname{Re}(f,g)F.$$

By Nelson's analytic vector theorem, we see that

$$\sum_{m=0}^{n} \frac{(i\Pi(g))^m}{m!} F \to e^{i\Pi(g)}F,$$

$$\Phi(f) \sum_{m=0}^{n} \frac{(i\Pi(g))^m}{m!} F \to e^{i\Pi(g)}(\Phi(f) - \operatorname{Re}(f,g))F$$

as $n \to \infty$. Then $e^{i\Pi(g)}F \in D(\Phi(f))$ and (1.2.27) follows on $\mathcal{F}_{\mathrm{b,fin}}$ by the closedness of $\Phi(f)$. Let $F \in D(\Phi(f))$. Since $\mathcal{F}_{\mathrm{b,fin}}$ is a core of $\Phi(f)$, there exists a sequence $(F_n)_{n\in\mathbb{N}} \subset \mathcal{F}_{\mathrm{b,fin}}$ such that $F_n \to F$ and $\Phi(f)F_n \to \Phi(f)F$ as $n \to \infty$. It follows that

$$e^{-i\Pi(g)}\Phi(f)e^{i\Pi(g)}F_n = (\Phi(f) - \operatorname{Re}(f,g))F_n$$

and $e^{i\Pi(g)}F_n \to e^{i\Pi(g)}F$ as $n \to \infty$. Hence the closedness of $\Phi(f)$ yields again that $e^{i\Pi(g)}F \in D(\Phi(f))$ and (1.2.27) is true on $D(\Phi(f))$, and (1) is proven. Part (2) can be shown similarly. □

Corollary 1.19.

(1) *If $f \not\equiv 0$, then*

$$\operatorname{Spec}(\Phi(f)) = \mathbb{R}, \quad \operatorname{Spec}_{\mathrm{p}}(\Phi(f)) = \emptyset.$$

(2) *Let h be self-adjoint and $g \in D(h)$. Then*

$$\inf \operatorname{Spec}(\mathrm{d}\Gamma(h) + \Phi(hg)) = -\frac{1}{2}(hg,g),$$

$$(\mathrm{d}\Gamma(h) + \Phi(hg))e^{i\Pi(g)}\Omega_{\mathrm{b}} = -\frac{1}{2}(hg,g)e^{i\Pi(g)}\Omega_{\mathrm{b}}.$$

Proof. For every $\alpha \in \mathbb{R}$, the operator $e^{i\alpha\Pi(f)}$ is unitary, and we have

$$e^{-i\alpha\Pi(f)}\Phi(f)e^{i\alpha\Pi(f)} = \Phi(f) - \alpha\|f\|^2.$$

Since $\alpha \in \mathbb{R}$ is arbitrary, $\operatorname{Spec}(\Phi(f)) = \mathbb{R}$ follows. Let Ψ be an eigenvector of $\Phi(f)$ such that $\Phi(f)\Psi = E\Psi$. Then $\Phi(f)e^{i\alpha\Pi(f)}\Psi = (E - \alpha\|f\|^2)e^{i\alpha\Pi(f)}\Psi$. This yields that $e^{i\alpha\Pi(f)}\Psi$ is an eigenvector of $\Phi(f)$ associated with eigenvalue $E-\alpha\|f\|^2$. This contradicts the separability of $\mathcal{F}_{\mathrm{b}}$, and (1) follows. By

$$e^{-i\Pi(g)}(\mathrm{d}\Gamma(h) + \Phi(hg))e^{i\Pi(g)} = \mathrm{d}\Gamma(h) - \frac{1}{2}(hg,g)$$

and $\inf \operatorname{Spec}(\mathrm{d}\Gamma(h)) = 0$, implying (2). □

Example 1.20. Let h be self-adjoint. Choosing any $n \in \mathbb{N}$, it is important to identify which space is left by $e^{-\Phi(f)^{2n}}$ invariant. We have

$$\eta = [d\Gamma(h), -\Phi(f)^{2n}] = 2ni\Phi^{2n-1}(f)\Pi(hf) + \frac{2n(2n-1)}{2}\Phi^{2n-2}(f)(hf, f)$$

and

$$\xi = [\eta, -\Phi(f)^{2n}] = -(2n)^2\Phi(f)^{4n-2}(hf, f).$$

Thus we can derive that

$$[d\Gamma(h), e^{-\Phi(f)^{2n}}] = e^{-\Phi(f)^{2n}}\left(\eta + \frac{1}{2}\xi\right).$$

In particular,

$$d\Gamma(h)^m e^{-\Phi(f)^{2n}} = e^{-\Phi(f)^{2n}}\left(d\Gamma(h) + \eta + \frac{1}{2}\xi\right)^m.$$

Set $h = 1$, for which $d\Gamma(1) = N$ is the number operator. Let $C^\infty(N) = \bigcap_{m=0}^\infty D(N^m)$. For $F \in C^\infty(N)$, a computation gives that for arbitrary $m \in \mathbb{N}$,

$$N^m e^{-\Phi(f)^{2n}} F$$
$$= e^{-\Phi(f)^{2n}}\left(N + 2ni\Phi(f)^{2n-1}\Pi(f) + \frac{2n(2n-1)}{2}\Phi(f)^{2n-2}\|f\|^2 - \frac{(2n)^2}{2}\Phi(f)^{4n-2}\|f\|^2\right)^m F.$$

This implies that $e^{-\Phi(f)^{2n}}$ leaves $C^\infty(N)$ invariant.

1.2.6 Wick product

Loosely speaking, the so-called Wick product $:a^\sharp(f_1)\cdots a^\sharp(f_n):$ is defined in a product of creation and annihilation operators by moving the creation operators to the left and the annihilation operators to the right without taking the commutation relations into account. For example, $:a(f_1)a^*(f_2)a(f_3)a^*(f_4): = a^*(f_2)a^*(f_4)a(f_1)a(f_3)$.

Definition 1.21 (Wick product). The *Wick product* $:\prod_{i=1}^n \Phi(g_i):$ of $\prod_{i=1}^n \Phi(g_i)$ is recursively defined by the equalities

$$:\Phi(f): = \Phi(f),$$
$$:\Phi(f)\prod_{i=1}^n \Phi(f_i): = \Phi(f):\prod_{i=1}^n \Phi(f_i): - \frac{1}{2}\sum_{j=1}^n (f, f_j):\prod_{i\neq j} \Phi(f_i):.$$

By the above definition, we have

$$:\Phi(f)^n: = 2^{-n/2}\sum_{m=0}^{n}\binom{n}{m}a^*(f)^m a(f)^{n-m},$$

or

$$:\Phi(f)^n: = \sum_{k=0}^{[n/2]}\frac{n!}{k!(n-2k)!}\Phi(f)^{n-2k}\left(-\frac{1}{4}\|f\|^2\right)^k = n!\sum_{k=0}^{[n/2]}\sum_{m+2k=n}\frac{\Phi(f)^m}{m!}\frac{\left(-\frac{1}{4}\|f\|^2\right)^k}{k!}.$$

Here $[m]$ denotes the integer part of m. Note that $:\Phi(f_1)\cdots\Phi(f_n):\Omega_{\mathrm{b}} = 2^{-n/2}a^*(f_1)\cdots a^*(f_n)\Omega_{\mathrm{b}}$. From this, the orthogonality property

$$\left(:\prod_{i=1}^{n}\Phi(f_i):\Omega_{\mathrm{b}}, :\prod_{i=1}^{m}\Phi(g_i):\Omega_{\mathrm{b}}\right) = \delta_{nm}2^{-n/2}\sum_{\pi\in\wp_n}\prod_{i=1}^{n}(g_i, f_{\pi(i)})$$

follows. Hence

$$\overline{\mathrm{L.H.}\left\{:\prod_{i=1}^{n}\Phi(f_i):\Omega_{\mathrm{b}}\,\middle|\, f_j\in\mathscr{W}, j=1,\dots,n\right\}} = \mathcal{F}_{\mathrm{b}}^{(n)}.$$

The Wick product of the exponential can be computed directly to yield

$$:e^{\alpha\Phi(f)}:\Omega_{\mathrm{b}} = \operatorname*{s-lim}_{m\to\infty}\sum_{n=0}^{m}\frac{\alpha^n}{n!}:\Phi(f)^n:\Omega_{\mathrm{b}} = e^{-(\alpha^2/4)\|f\|^2}e^{\alpha\Phi(f)}\Omega_{\mathrm{b}}. \tag{1.2.29}$$

Hence for real-valued f and g,

$$(\Omega_{\mathrm{b}}, \Phi(f)\Omega_{\mathrm{b}}) = 0, \quad (\Omega_{\mathrm{b}}, \Phi(f)\Phi(g)\Omega_{\mathrm{b}}) = \frac{1}{2}(f, g), \tag{1.2.30}$$

and

$$(\Omega_{\mathrm{b}}, e^{\alpha\Phi(f)}\Omega_{\mathrm{b}}) = e^{(\alpha^2/4)\|f\|^2}. \tag{1.2.31}$$

Since the commutator $[\Phi(f), \Phi(g)] = 0$, equalities (1.2.30)–(1.2.31) suggest that with a real-valued test function f the expression $\Phi(f)$ can be realized as a Gaussian random variable, as it will be further explored in the next section.

1.2.7 Exponentials of creation and annihilation operators

The field operator $\Phi(f), f\in\mathscr{W}$, is a self-adjoint operator, while $a^*(f)$ and $a(f)$ are not. Nevertheless, we can define $e^{a^*(f)}$ and $e^{a(f)}$ by the series

$$e^{a^\sharp(f)} = \sum_{n=0}^{\infty} \frac{1}{n!} a^\sharp(f)^n,$$

which are closed operators and play an important role in quantum field theory. Furthermore, $e^{a^*(f)}\Omega_b$ becomes an eigenvector of the annihilation operator $a(g)$ with $g \neq 0$ such that $a(g)e^{a^*(f)}\Omega_b = (\bar{g}, f)e^{a^*(f)}\Omega_b$. Hence we see that the spectrum of the closed operator $a(g)$ coincides with $\mathbb{C}$ and $\mathrm{Spec}_p(a(g)) = \mathbb{C}$ follows. We discuss now these operators rigorously.

Let $f \in \mathscr{W}$ and define the *exponential of the creation operators* F_f by

$$F_f = \sum_{n=0}^{\infty} \frac{1}{n!} a^*(f)^n$$

with the domain

$$D(F_f) = \left\{ \Phi \in \bigcap_{n=1}^{\infty} D(a^*(f)^n) \,\middle|\, \sum_{n=0}^{\infty} \frac{1}{n!} \|a^*(f)^n \Phi\| < \infty \right\}.$$

Let $\Phi \in \mathcal{F}_b^{(m)}$. Then we have

$$\|F_f \Phi\| \leq \|\Phi\| + \sum_{n=1}^{\infty} \frac{\sqrt{m+n-1} \cdots \sqrt{m}}{n!} \|f\|^n \|\Phi\| < \infty,$$

and $\mathcal{F}_{b,\mathrm{fin}} \subset D(F_f)$ follows. We also define the *exponential of annihilation operators* by

$$G_f = \sum_{n=0}^{\infty} \frac{1}{n!} a(f)^n$$

with the domain

$$D(G_f) = \left\{ \Phi \in \bigcap_{n=1}^{\infty} D(a(f)^n) \,\middle|\, \sum_{n=0}^{\infty} \frac{1}{n!} \|a(f)^n \Phi\| < \infty \right\}.$$

For convenience, we write

$$F_f = e^{a^*(f)}, \quad G_f = e^{a(f)},$$

unless any confusion may arise. It is seen that $(e^{a^*(f)})^* \supset e^{a(\bar{f})}$, which implies that $e^{a^*(f)}$ is closable. We will denote the closure of $e^{a^*(f)}$ by the same symbol. Similarly, we denote the closure of $e^{a(f)}$ by the same symbol. The vector defined by

$$C(f) = e^{a^*(f)}\Omega_b$$

is called a *coherent vector*. By the definition of the creation operators, $C(f)$ is formally written as

$$C(f) = \sum_{n=0}^{\infty} \frac{\otimes_{\rm sym}^n f}{\sqrt{n!}}.$$

Let D be a subspace of $\mathscr{W}$ and define the linear hull of coherent vectors with test functions in D by

$$\mathscr{C}(D) = \mathrm{L.H.}\left\{C(g) \mid g \in D\right\}.$$

Proposition 1.22 (Denseness of coherent vectors).
(1) *If $f_1, \ldots, f_n \in \mathscr{W}$ are linearly independent vectors, then $C(f_1), \ldots, C(f_n)$ are also linearly independent.*
(2) *If $D \subset \mathscr{W}$ is a dense subset, then $\mathscr{C}(D)$ is also dense in $\mathcal{F}_{\rm b}(\mathscr{W})$.*

Proof. Suppose that $\sum_{j=1}^n a_j C(f_j) = 0$ with $a_j \in \mathbb{C}$ for $j = 1, \ldots, n$. In particular, we have $\sum_{j=1}^n a_j a^*(f_j)\Omega_{\rm b} = 0$, which implies that $\sum_{j=1}^n a_j f_j = 0$. Then $a_1 = \cdots = a_n = 0$ and statement (1) follows. In order to prove (2), it is enough to show that $\mathscr{C}(D)^\perp = \{0\}$. Let $\Phi \in \mathcal{F}_{\rm b}(\mathscr{W})$ be such that $(\Phi, C(f)) = 0$ for any $f \in D$. Hence $(\Phi^{(n)}, a^*(f)^n\Omega_{\rm b}) = 0$ for any n. Putting $f = \sum_{j=1}^m a_j f_j$ with $a_j \in \mathbb{C}$, we see that

$$0 = \left(\Phi^{(n)}, a^*\left(\sum_{j=1}^m a_j f_j\right)^n \Omega_{\rm b}\right) = n! \sum_{\substack{j_1+\cdots+j_m=n,\\ j_i\geq 0, i=1,\ldots,m}} \frac{a_1^{j_1}\cdots a_m^{j_m}}{j_1!\cdots j_m!}\left(\Phi^{(n)}, \prod_{i=1}^m a^*(f_i)^{j_i}\Omega_{\rm b}\right).$$

The right-hand side is a polynomial of degree n in $a_1, \ldots, a_n$, identically zero. In particular, $(\Phi^{(n)}, a^*(f_1)\cdots a^*(f_n)\Omega_{\rm b}) = 0$. Since $\mathrm{L.H.}\{a^*(f_1)\cdots a^*(f_n)\Omega_{\rm b} \mid f_j \in D\}$ is dense in $\mathcal{F}_{\rm b}{}^{(n)}(\mathscr{W})$, $\Phi^{(n)} = 0$ follows. As n is arbitrary, we conclude that $\Phi = 0$. □

It can be seen that $C(g) \in \bigcap_{n=0}^\infty D(a^*(f)^n)$ and

$$\left\|C(f+g) - \sum_{n=0}^M \frac{a^*(f)^n}{n!} C(g)\right\| \to 0$$

as $M \to \infty$. Also, $C(g) \in D(e^{a^*(f)})$ and

$$e^{a^*(f)}C(g) = C(f+g)$$

follows by the closedness of $e^{a^*(f)}$. We also see that

$$a(g)e^{a^*(f)}\Phi = e^{a^*(f)}a(g)\Phi + (\bar{g}, f)e^{a^*(f)}\Phi$$

for $\Phi \in \mathcal{F}_{\rm b,fin}$. In particular,

$$a(g)C(f) = (\bar{g}, f)C(f), \tag{1.2.32}$$

and recursively we get

$$P(a(g))C(f) = P((\bar{g}, f))C(f)$$

for any polynomial P. We summarize these facts in the proposition below.

Proposition 1.23 (Algebraic properties). *Let $f, g \in \mathscr{W}$, and P be a polynomial. Then*
(1) $e^{a^*(g)}e^{a^*(f)}\Omega_{\mathrm{b}} = e^{a^*(f+g)}\Omega_{\mathrm{b}}$;
(2) $P(a(g))e^{a^*(f)}\Omega_{\mathrm{b}} = P((\bar{g}, f))e^{a^*(f)}\Omega_{\mathrm{b}}$;
(3) $e^{a(g)}e^{a^*(f)}\Omega_{\mathrm{b}} = e^{(\bar{g}, f)}e^{a^*(f)}\Omega_{\mathrm{b}}$.

Proof. Claims (1) and (2) are already shown. Since

$$\sum_{n=0}^{M} \frac{a(g)^n}{n!} e^{a^*(f)}\Omega_{\mathrm{b}} = \sum_{n=0}^{M} \frac{(\bar{g}, f)^n}{n!} e^{a^*(f)}\Omega_{\mathrm{b}}$$

and the right-hand side converges to $e^{(\bar{g}, f)}e^{a^*(f)}\Omega_{\mathrm{b}}$ as $M \to \infty$, statement (3) follows by the closedness of $e^{a(g)}$. □

Corollary 1.24. *Let $f \in \mathscr{W}$ and $\alpha \in \mathbb{C}$. Then it follows that on $D(e^{a^\sharp(f)})$,*

$$\sum_{n=0}^{\infty} \frac{(a^\sharp(f) + \alpha)^n}{n!} = e^{\alpha} e^{a^\sharp(f)}. \tag{1.2.33}$$

Proof. Let $\Phi \in D(e^{a^\sharp(f)})$. We have

$$\sum_{n=0}^{m} \frac{(a^\sharp(f) + \alpha)^n}{n!} \Phi = \sum_{k=0}^{m} \frac{a^\sharp(f)^k}{k!} \sum_{l=0}^{m-k} \frac{\alpha^l}{l!} \Phi.$$

It follows that

$$\begin{aligned}
&\left\| \sum_{k=0}^{m} \frac{a^\sharp(f)^k}{k!} \sum_{l=0}^{m-k} \frac{\alpha^l}{l!} \Phi - e^{\alpha} e^{a^\sharp(f)} \Phi \right\| \\
&\leq \left\| \sum_{k=0}^{m} \frac{a^\sharp(f)^k}{k!} \sum_{l=0}^{m-k} \frac{\alpha^l}{l!} \Phi - \sum_{k=0}^{m} \frac{a^\sharp(f)^k}{k!} e^{\alpha} \Phi \right\| + e^{\alpha} \left\| \sum_{k=0}^{m} \frac{a^\sharp(f)^k}{k!} \Phi - e^{a^\sharp(f)} \Phi \right\|.
\end{aligned}$$

By a simple limiting argument, it is seen that the right-hand side above converges to zero as $m \to \infty$, which yields (1.2.33). □

We can also determine the spectrum of $a^\sharp(f)$ by using their algebraic properties. Recall that $\mathrm{Spec}_{\mathrm{p}}(A)$ and $\mathrm{Spec}_{\mathrm{r}}(A)$ describe the point spectrum of A and the residual spectrum of A, respectively.

Proposition 1.25 (Spectral properties). *Let $f \neq 0$. Then it follows that*

$$\mathrm{Spec}(a(f)) = \mathrm{Spec}_{\mathrm{p}}(a(f)) = \mathbb{C},$$
$$\mathrm{Spec}(a^*(f)) = \mathrm{Spec}_{\mathrm{r}}(a^*(f)) = \mathbb{C}.$$

In particular, $\mathrm{Spec}_{\mathrm{p}}(a^*(f)) = \emptyset$.

Proof. Since $a(f)C(g) = (\bar{f}, g)C(g)$, we have $a(f)C(af/\|f\|^2) = aC(af/\|f\|^2)$ for every $\alpha \in \mathbb{C}$. This implies that $C(af/\|f\|^2)$ is an eigenfunction associated with eigenvalue α. This shows the first statement. To prove the second, suppose that $a^*(f)\Phi = 0$. By the identity $\|a^*(f)\Phi\|^2 = \|a(f)\Phi\|^2 + \|f\|^2\|\Phi\|^2$, we have $\Phi = 0$, and hence $0 \notin \mathrm{Spec}_{\mathrm{p}}(a^*(f))$. Let $a^*(f)\Phi = z\Phi$ with $z \neq 0$. Since $z\Phi^{(0)} = 0$ and $\sqrt{n}S_n(f \otimes \Phi^{(n-1)}) = z\Phi^{(n)}$ for $n \geq 1$, by induction $\Phi^{(n)} = 0$ for all $n \geq 0$ and $\Phi = 0$. Thus $z \notin \mathrm{Spec}_{\mathrm{p}}(a^*(f))$ and $a^*(f) - z$ is injective for every $z \in \mathbb{C}$. Using the decomposition $\mathcal{F}_{\mathrm{b}} = \ker(a(\bar{f}) - z^*) \oplus \overline{\mathrm{Ran}(a^*(f) - z)}$ and $\dim\ker(a(\bar{f}) - z^*) \geq 1$, we see that $\overline{\mathrm{Ran}(a^*(f) - z)} \neq \mathcal{F}_{\mathrm{b}}$. This implies that the domain of $(a^*(f) - z)^{-1}$ is not dense, and the second statement follows. □

Proposition 1.26 (Continuity). *Let $\Phi \in \mathcal{F}_{\mathrm{b,fin}}$. The map $\mathcal{W} \ni f \mapsto e^{a^\sharp(f)}\Phi \in \mathcal{F}_{\mathrm{b}}$ is strongly continuous.*

Proof. Suppose that $f_m \to f$ strongly in $\mathcal{W}$ as $m \to \infty$. Let $\Phi \in \mathcal{F}_{\mathrm{b}}^{(N)}$. We see that $e^{a(f_m)}\Phi = \sum_{n=0}^N \frac{a(f_m)^n}{n!}\Phi$. Since $\sum_{n=0}^N \frac{a(f_m)^n}{n!}\Phi \to \sum_{n=0}^N \frac{a(f)^n}{n!}\Phi$ as $m \to \infty$,

$$\lim_{m\to\infty} e^{a(f_m)}\Phi = e^{a(f)}\Phi$$

follows. Next consider the continuity of $f \mapsto e^{a^*(f)}\Phi$. Let $\varepsilon > 0$ and $\|f_m - f\| < \varepsilon$ for a sufficiently large m. We fix $c > 0$ such that $\|f_m\| < c$ for all m. It is seen that

$$\begin{aligned}\left\|e^{a^*(f_m)}\Phi - e^{a^*(f)}\Phi\right\| &\leq \sum_{n=1}^{\infty} \frac{\sqrt{(N+n-1)}\cdots\sqrt{N}}{n!} \sum_{k=0}^{n-1} \|f_m\|^k \|f_m - f\| \|f\|^{n-k-1} \|\Phi\| \\ &\leq \varepsilon \sum_{n=1}^{\infty} \frac{c^{n-1}\sqrt{(N+n-1)}\cdots\sqrt{N}}{(n-1)!} \|\Phi\|.\end{aligned}$$

This implies that $\|e^{a^*(f_n)}\Phi - e^{a^*(f)}\Phi\| \to 0$ as $n \to \infty$. □

Proposition 1.27 (Differentiability). *Let h be a self-adjoint operator in $\mathcal{W}$, $f \in D(h)$, and $\Phi \in \mathcal{F}_{\mathrm{b,fin}}$. Then the map $\mathbb{R} \ni t \mapsto e^{a^\sharp(e^{ith}f)}\Phi \in \mathcal{F}_{\mathrm{b}}$ is strongly differentiable and*

$$\frac{d}{dt} e^{a^\sharp(e^{ith}f)}\Phi = a^\sharp(ihe^{ith}f)e^{a^\sharp(e^{ith}f)}\Phi.$$

Proof. The proof is straightforward. Let $\varepsilon \in \mathbb{R}$ and $\Phi \in \mathcal{F}_{\mathrm{b}}^{(N)}$. We show the statement only for $a^*(f)$, the proof for $a(f)$ is similar. We write $a^*(e^{ith}f) = a^*(0)$ and $a^*(e^{i(t+\varepsilon)h}f) = a^*(\varepsilon)$ for notational simplicity. We have

$$\frac{1}{\varepsilon}\left(e^{a^*(\varepsilon)} - e^{a^*(0)}\right)\Phi - a^*(ihe^{ith}f)e^{a^*(0)}\Phi = A_\varepsilon + B_\varepsilon,$$

where

$$A_\varepsilon = a^*\left(\left(\frac{e^{i\varepsilon h}-1}{\varepsilon} - ih\right)e^{ith}f\right)\sum_{n=1}^{\infty}\frac{1}{n!}\sum_{k=0}^{n-1}a^*(\varepsilon)^{n-k-1}a^*(0)^k\Phi,$$

$$B_\varepsilon = a^*(ihe^{ith}f)\left(\sum_{n=1}^{\infty}\frac{1}{n!}\sum_{k=0}^{n-1}a^*(\varepsilon)^{n-k-1}a^*(0)^k - \sum_{n=0}^{\infty}\frac{1}{n!}a^*(0)^n\right)\Phi.$$

We see that

$$\|A_\varepsilon\| \le \left\|\frac{e^{i\varepsilon h}-1}{\varepsilon}f - ihf\right\|\sum_{n=1}^{\infty}\frac{\sqrt{N+n-1}\cdots\sqrt{N}}{n!}\|f\|^{n-1}\|\Phi\|$$

and

$$\begin{aligned}\|B_\varepsilon\| &\le \sum_{n=1}^{\infty}\frac{1}{(n-1)!}\sum_{k=0}^{n-1}\|(a^*(\varepsilon)^{n-k-1} - a^*(0)^{n-k-1})a^*(0)^k a(ihe^{ith}f)\Phi\|\\ &\le \sum_{n=1}^{\infty}\frac{\sqrt{N+n-1}\cdots\sqrt{N}}{(n-1)!}\sum_{k=0}^{n-1}(n-k-1)\|f\|^{n-2}\|hf\|\|(e^{i\varepsilon h}-1)f\|\|\Phi\|\\ &\le \|hf\|\|(e^{i\varepsilon h}-1)f\|\sum_{n=1}^{\infty}\frac{\sqrt{N+n-1}\cdots\sqrt{N}}{(n-1)!}\frac{n(n-1)}{2}\|f\|^{n-2}\|\Phi\|.\end{aligned}$$

Hence $\lim_{\varepsilon\to 0}\|A_\varepsilon\| = 0$ and $\lim_{\varepsilon\to 0}\|B_\varepsilon\| = 0$. □

Next we discuss the relationship between $e^{a^*(f)}$ and the second quantization $\Gamma(T)$. Let h be a self-adjoint operator in $\mathscr{W}$ and define

$$\begin{aligned}\mathscr{D} &= \mathrm{L.H.}\{a^*(f_1)\cdots a^*(f_n)\Omega_\mathrm{b}, \Omega_\mathrm{b} \mid f_j \in \mathscr{W}, j = 1,\dots,n, n \ge 1\},\\ \mathscr{D}_h &= \mathrm{L.H.}\{a^*(f_1)\cdots a^*(f_n)\Omega_\mathrm{b}, \Omega_\mathrm{b} \mid f_j \in D(h), j = 1,\dots,n, n \ge 1\}.\end{aligned}$$

Recall that $jf = \bar{f}$ denotes the complex conjugate of f.

Proposition 1.28 (Intertwining properties).

(1) *Let T be a contraction operator on $\mathscr{W}$. Then the following hold on $\mathcal{F}_{\mathrm{b,fin}}$:*

$$\begin{aligned}\Gamma(T)e^{a^*(f)} &= e^{a^*(Tf)}\Gamma(T),\\ \Gamma(T)e^{a(jT^*jf)} &= e^{a(f)}\Gamma(T).\end{aligned}$$

(2) *Let h be self-adjoint in $\mathscr{W}$ and $f \in D(h)$. Then the following hold on $\mathscr{D}_h$:*

$$\begin{aligned}\mathrm{d}\Gamma(h)e^{a^*(f)} &= a^*(hf)e^{a^*(f)} + e^{a^*(f)}\mathrm{d}\Gamma(h),\\ \mathrm{d}\Gamma(h)e^{a(f)} &= -a(jhjf)e^{a(f)} + e^{a(f)}\mathrm{d}\Gamma(h).\end{aligned}$$

Proof. Let $\Phi = \prod_{j=1}^m a^*(g_j)\Omega_b \in \mathscr{D}$. It follows that $\Phi \in D(e^{a^*(f)})$ and $\Gamma(T)\mathscr{D} \subset \mathscr{D}$. We have

$$\Gamma(T)e^{a^*(f)}\Phi = \sum_{n=0}^{\infty} \frac{a^*(Tf)^n}{n!} \prod_{j=1}^{m} a^*(Tg_j)\Omega_b = e^{a^*(Tf)}\Gamma(T)\Phi,$$

and the first statement of (1) is proven on $\mathscr{D}$. Let $\Phi \in \mathcal{F}_b^{(N)}$. Then there exists $\Phi_n \in \mathscr{D}$ such that $\Phi_n \in \mathcal{F}_b^{(N)}$ and $\|\Phi_n - \Phi\| \to 0$ as $n \to \infty$. We can also see that $e^{a^*(f)}\Phi_n \to e^{a^*(f)}\Phi$ as $n \to \infty$. The limit of $\Gamma(T)e^{a^*(f)}\Phi_n$ exists, and the identity $\Gamma(T)e^{a^*(f)}\Phi_n = e^{a^*(Tf)}\Gamma(T)\Phi_n$ implies that $\Gamma(T)e^{a^*(f)}\Phi = e^{a^*(Tf)}\Gamma(T)\Phi$. Hence the first statement of (1) is proven on $\mathcal{F}_{b,fin}$. The second statement of (1) can be shown by taking the adjoint of both sides of the first statement. To show (2), let $\Phi \in \mathscr{D}_h$ and $T = e^{ith}$. It follows that

$$\Gamma(e^{ith})e^{a^*(f)}\Phi = e^{a^*(e^{ith}f)}\Gamma(e^{ith})\Phi. \tag{1.2.34}$$

Similarly as in Proposition 1.27, it can be seen that the right-hand side above is differentiable with respect to t at $t = 0$. The result is

$$\frac{d}{dt}e^{a^*(e^{ith}f)}\Gamma(e^{ith})\Phi = ia^*(hf)e^{a^*(f)}\Phi + ie^{a^*(f)}\mathrm{d}\Gamma(h)\Phi. \tag{1.2.35}$$

This implies that the left-hand side of (1.2.34) is also differentiable with respect to t, thus $e^{a^*(f)}\Phi \in D(\mathrm{d}\Gamma(h))$, and the derivative of the left-hand side at $t = 0$ is

$$\frac{d}{dt}\Gamma(e^{ith})e^{a^*(f)}\Phi = i\mathrm{d}\Gamma(h)e^{a^*(f)}\Phi. \tag{1.2.36}$$

Comparing (1.2.35) with (1.2.36), the first statement of (2) follows. The second statement can be shown by taking the adjoint of both sides in the first statement. □

Finally, we discuss the representation of $e^{\Phi(f)}$ in terms of $e^{a^*(f)}$ and $e^{a(f)}$ which is known as *Baker–Campbell–Hausdorff formula.* Define

$$\mathscr{D}_b = \mathrm{L.H.}\left\{C(g), \Phi \mid g \in \mathscr{W}, \Phi \in \mathcal{F}_{b,fin}\right\}.$$

Proposition 1.29 (Baker–Campbell–Hausdorff formula). *Let $f \in \mathscr{W}$ and $\alpha \in \mathbb{C}$. Then*

$$e^{\alpha(a^*(f)+a(\bar{f}))} = e^{\alpha a^*(f)}e^{\alpha a(\bar{f})}e^{\frac{1}{2}\alpha^2\|f\|^2} \tag{1.2.37}$$

holds on $\mathscr{D}_b$.

Proof. We show (1.2.37) on $C(g)$, the proof on $\mathcal{F}_{b,fin}$ is similar. We have

$$e^{\alpha a^*(f)}e^{\alpha a(\bar{f})}C(g) = e^{\alpha(f,g)}C(\alpha f + g). \tag{1.2.38}$$

Let $\psi(f) = a^*(f) + a(\bar{f})$, which is self-adjoint and we have

$$e^{\alpha\psi(f)} = \sum_{n=0}^{\infty} \frac{\alpha^n \psi(f)^n}{n!} \tag{1.2.39}$$

on $\mathcal{F}_{\mathrm{b,fin}}$. Let $C_m(g) = \sum_{n=0}^{m} \frac{a^*(g)^n}{n!}\Omega_{\mathrm{b}}$. By using the expansion (1.2.39), we have the expression

$$e^{\alpha\psi(f)} C_m(g) = \sum_{n=0}^{m} \frac{(a^*(g) + \alpha(f,g))^n}{n!} e^{\alpha\psi(f)}\Omega_{\mathrm{b}}.$$

Together with $e^{\alpha\psi(f)}\Omega_{\mathrm{b}} = e^{\frac{1}{2}\alpha^2\|f\|^2} e^{\alpha a^*(f)}\Omega_{\mathrm{b}}$, this implies

$$e^{\alpha\psi(f)} C_m(g) = \sum_{n=0}^{m} \frac{(a^*(g) + \alpha(f,g))^n}{n!} e^{\frac{1}{2}\alpha^2\|f\|^2} e^{\alpha a^*(f)}\Omega_{\mathrm{b}}.$$

By Corollary 1.24, we have

$$e^{\alpha\psi(f)} C(g) = e^{\alpha(f,g)} e^{\frac{1}{2}\alpha^2\|f\|^2} C(\alpha f + g). \tag{1.2.40}$$

Using (1.2.38) and (1.2.40), the claim follows. □

Example 1.30. Choosing $\alpha = i$ in the Baker–Campbell–Hausdorff formula, we obtain for the unitary operator $e^{i(a^*(f)+a(\bar{f}))}$ such that

$$e^{i(a^*(f)+a(\bar{f}))} = e^{ia^*(f)} e^{ia(\bar{f})} e^{-\frac{1}{2}\|f\|^2}.$$

Since $\|e^{i(a^*(f)+a(\bar{f}))}\| = 1$, the operator $e^{ia^*(f)}e^{ia(\bar{f})}$ is bounded and it follows that

$$\|e^{ia^*(f)} e^{ia(\bar{f})}\| = e^{\frac{1}{2}\|f\|^2}.$$

From the discussions above, we also obtain the following expressions of the action of the operators on coherent vectors:

$$\begin{aligned}
e^{-i\Phi(f)} C(g) &= e^{-\|f\|^2/4} e^{-i(f,g)/\sqrt{2}} C(g - if/\sqrt{2}), \\
e^{-i\Pi(f)} C(g) &= e^{-\|f\|^2/4} e^{-(f,g)/\sqrt{2}} C(g + f/\sqrt{2}), \\
e^{-i\mathrm{d}\Gamma(A)} C(g) &= C(e^{-iA} g).
\end{aligned}$$

1.2.8 The case $\mathscr{W} = L^2(\mathbb{R}^d)$

From now on, we consider the case when $\mathscr{W} = L^2(\mathbb{R}^d)$, and let ω_ν be the multiplication operator by $\omega_\nu = \omega_\nu(k) = \sqrt{|k|^2 + \nu^2}$, $\nu \geq 0$, in $L^2(\mathbb{R}^d)$. Thus $H_{\mathrm{f}} = \mathrm{d}\Gamma(\omega_\nu)$.

We estimate the norm of the product $\|\prod_{j=1}^m a^*(f_j)\Phi\|$ of creation operators. Since we already have a bound on the product of annihilation operators $\|\prod_{j=1}^m a(f_j)\Phi\|$ by Corollary 1.8 when $\mathscr{W} = L^2(\mathbb{R}^d)$, we may also estimate $\|\prod_{j=1}^m a^*(f_j)\Phi\|$, which is though technically more complicated than for annihilation operators. To obtain this, we use the fact

$$\Big\|\prod_{j=1}^m a^*(f_j)\Phi\Big\|^2 = \Big(\Phi, \prod_{j=1}^m a(\bar{f}_j)\prod_{j=1}^m a^*(f_j)\Phi\Big)$$

and compute the commutator

$$\Big[\prod_{j=1}^m a(\bar{f}_j), \prod_{j=1}^m a^*(f_j)\Big].$$

We find that

$$\Big\|\prod_{j=1}^m a^*(f_j)\Phi\Big\|^2 = \Big\|\prod_{j=1}^m a(\bar{f}_j)\Phi\Big\|^2 + \Big(\Phi, \Big[\prod_{j=1}^m a(\bar{f}_j), \prod_{j=1}^m a^*(f_j)\Big]\Phi\Big).$$

It remains to estimate the second term on the right-hand side.

Lemma 1.31. *Let $\Phi \in \mathcal{F}_{\mathrm{b,fin}}$ and $f_i, g_j \in \mathscr{W}$ for $i,j = 1,\ldots,n$. Then*

$$\prod_{j=1}^n a(\bar{g}_j)\prod_{j=1}^n a^*(f_j)\Phi = \sum_{m=0}^n \sum_{C_m \ni A} \sum_{C_{n-m}\ni B} \sum_{\substack{\sigma:A^c\to B\\ \text{bijection}}} \Big(\prod_{l\in A^c}(g_l, f_{\sigma(l)})\Big)\Big(\prod_{p\in B^c} a^*(f_p)\Big)\Big(\prod_{q\in A} a(\bar{g}_q)\Big)\Phi. \tag{1.2.41}$$

Here $C_k = \{A \subset \{1,\ldots,n\} \mid \#A = k\}$, $C_0 = \emptyset$, and in $\sum_{\sigma:A^c\to B,\ \text{bijection}}$ the summation is extended over all bijections from A^c to B. In particular,

$$\Big[\prod_{j=1}^n a(\bar{g}_j), \prod_{j=1}^n a^*(f_j)\Big] = \sum_{m=0}^{n-1} \sum_{C_m \ni A} \sum_{C_{n-m}\ni B} \sum_{\substack{\sigma:A^c\to B\\ \text{bijection}}} \Big(\prod_{l\in A^c}(g_l, f_{\sigma(l)})\Big)\Big(\prod_{p\in B^c} a^*(f_p)\Big)\prod_{q\in A} a(\bar{g}_q).$$

Proof. For $n = 1$, we have $a(\bar{g})a^*(f)\Phi = (g,f)\Phi + a^*(f)a(\bar{g})\Phi$. For $n \ge 2$, the proof can be completed by induction. □

From this lemma, we can estimate $\|\prod_{j=1}^n a^*(f_j)\Phi\|$.

Lemma 1.32. *Let $f_i, g_j \in D(1/\sqrt{\omega_\nu})$ for $i,j = 1,\ldots,n$, and $\Phi \in D(H_{\mathrm{f}}^{n/2})$. Then*

$$\Big|\Big(\prod_{j=1}^n a^*(g_j)\Phi, \prod_{j=1}^n a^*(f_j)\Phi\Big)\Big| \le n!2^n\Big(\prod_{l=1}^n \|f_l\|_\omega \|g_l\|_\omega\Big)\sum_{m=0}^n \frac{1}{m!}\|H_{\mathrm{f}}^{m/2}\Phi\|^2,$$

where $\|f\|_\omega = \|f\| + \|f/\sqrt{\omega_\nu}\|$.

Proof. By Lemma 1.31, we have

$$\Big(\prod_{j=1}^{n} a^*(g_j)\Phi, \prod_{j=1}^{n} a^*(f_j)\Phi\Big) = \Big(\Phi, \prod_{j=1}^{n} a(\bar{g}_j) \prod_{j=1}^{n} a^*(f_j)\Phi\Big)$$
$$= \sum_{m=0}^{n} \sum_{C_m \ni A} \sum_{C_{n-m} \ni B} \sum_{\substack{\sigma: A^c \to B \\ \text{bijection}}} \Big(\Phi, \prod_{l \in A^c} (g_l, f_{\sigma(l)}) \prod_{p \in B^c} a^*(f_p) \prod_{q \in A} a(\bar{g}_q)\Phi\Big). \tag{1.2.42}$$

Using $\| \prod_{j=1}^{n} a(h_j)\Phi\| \le \prod_{j=1}^{n} \|h_j/\sqrt{\omega_\nu}\| \|H_{\mathrm{f}}^{n/2}\Phi\|$ and $\#B^c = m = \#A$, the right-hand side of (1.2.42) can be estimated as

$$\Big|\Big(\Phi, \prod_{p \in B^c} a^*(f_p) \prod_{q \in A} a(\bar{g}_q)\Phi\Big)\Big| = \Big|\Big(\prod_{p \in B^c} a(\bar{f}_p)\Phi, \prod_{q \in A} a(\bar{g}_q)\Phi\Big)\Big|$$
$$\le \Big(\prod_{p \in B^c} \|f_p/\sqrt{\omega_\nu}\|\Big)\Big(\prod_{q \in A} \|g_q/\sqrt{\omega_\nu}\|\Big) \|H_{\mathrm{f}}^{m/2}\Phi\|^2.$$

Since $\|f\| \le \|f\|_\omega$ and $\|f/\sqrt{\omega}\| \le \|f\|_\omega$, we have

$$\Big|\Big(\Phi, \prod_{l \in A^c} (g_l, f_{\sigma(l)}) \prod_{p \in B^c} a^*(f_p) \prod_{q \in A} a(\bar{g}_q)\Phi\Big)\Big| \le \Big(\prod_{l=1}^{n} \|g_l\|_\omega \|f_l\|_\omega\Big) \|H_{\mathrm{f}}^{m/2}\Phi\|^2. \tag{1.2.43}$$

Hence by (1.2.42) and (1.2.43), we have

$$\Big|\Big(\Phi, \prod_{j=1}^{n} a(\bar{g}_j) \prod_{j=1}^{n} a^*(f_j)\Phi\Big)\Big| \le \sum_{m=0}^{n} \sum_{C_m \ni A} \sum_{C_{n-m} \ni B} \sum_{\substack{\sigma: A^c \to B \\ \text{bijection}}} \Big(\prod_{l=1}^{n} \|g_l\|_\omega \|f_l\|_\omega\Big) \|H_{\mathrm{f}}^{m/2}\Phi\|^2$$
$$= n! \sum_{m=0}^{n} \frac{n!}{(n-m)!m!} \frac{1}{m!} \Big(\prod_{l=1}^{n} \|g_l\|_\omega \|f_l\|_\omega\Big) \|H_{\mathrm{f}}^{m/2}\Phi\|^2$$
$$\le n! 2^n \sum_{m=0}^{n} \frac{1}{m!} \Big(\prod_{l=1}^{n} \|g_l\|_\omega \|f_l\|_\omega\Big) \|H_{\mathrm{f}}^{m/2}\Phi\|^2. \qquad \square$$

By Lemma 1.32 and Corollary 1.8, we have useful bounds for products of annihilation and creation operators. We summarize them below. Suppose that $f_j \in D(1/\sqrt{\omega_\nu})$ for $j = 1, \dots, n$. Then for $\Phi \in D(H_{\mathrm{f}}^{n/2})$, we have

$$\Big\|\prod_{j=1}^{n} a(f_j)\Phi\Big\| \le \Big(\prod_{j=1}^{n} \|f_j/\sqrt{\omega_\nu}\|\Big) \|H_{\mathrm{f}}^{n/2}\Phi\|, \tag{1.2.44}$$

$$\Big\|\prod_{j=1}^{n} a^*(f_j)\Phi\Big\| \le \sqrt{n!} 2^{n/2} \Big(\prod_{l=1}^{n} \|f_l\|_\omega\Big)\Big(\sum_{m=0}^{n} \frac{1}{m!} \|H_{\mathrm{f}}^{m/2}\Phi\|^2\Big)^{1/2}. \tag{1.2.45}$$

By introducing a scaling parameter $0 < s < 1$, we also have

$$\left\| \prod_{j=1}^{n} a(f_j)\Phi \right\| \le s^{-n/2} \left(\prod_{j=1}^{n} \|f_j/\sqrt{\omega_\nu}\| \right) \|(sH_{\mathrm{f}})^{n/2}\Phi\|, \tag{1.2.46}$$

$$\left\| \prod_{j=1}^{n} a^*(f_j)\Phi \right\| \le \sqrt{n!}2^{n/2}s^{-n/2} \left(\prod_{l=1}^{n} \|f_l\|_\omega \right) \left(\sum_{m=0}^{n} \frac{1}{m!} \|(sH_{\mathrm{f}})^{m/2}\Phi\|^2 \right)^{1/2}. \tag{1.2.47}$$

Although the exponential operator $e^{a^*(f)}$ is unbounded, it can be seen that $e^{a^*(f)}e^{-\frac{t}{2}H_{\mathrm{f}}}$ is bounded for every $t > 0$.

Proposition 1.33 (Boundedness). *Let $t > 0$.*

(1) *If $f \in D(1/\sqrt{\omega_\nu})$, then both $e^{a^*(f)}e^{-\frac{t}{2}H_{\mathrm{f}}}$ and $\overline{e^{-\frac{t}{2}H_{\mathrm{f}}}e^{a(f)}}$ are bounded.*

(2) *Let N be the number operator and $f \in L^2(\mathbb{R}^d)$. Then both $e^{a^*(f)}e^{-\frac{t}{2}N}$ and $\overline{e^{-\frac{t}{2}N}e^{a(f)}}$ are bounded.*

Proof. Choose $t < 1$. By (1.2.47) for any $s < t$, we have for $\Phi \in \mathcal{F}_{\mathrm{b}}$,

$$\begin{aligned}\left\| \sum_{n=0}^{m} \frac{1}{n!} a^*(f)^n e^{-\frac{t}{2}H_{\mathrm{f}}}\Phi \right\| &\le \sum_{n=0}^{m} \frac{1}{n!} \|a^*(f)^n e^{-\frac{t}{2}H_{\mathrm{f}}}\Phi\| \\ &\le \sum_{n=0}^{m} \frac{1}{\sqrt{n!}} 2^{n/2}s^{-n/2}\|f\|_\omega^n \left(\sum_{k=0}^{n} \frac{1}{k!} \|(sH_{\mathrm{f}})^{k/2}e^{-\frac{t}{2}H_{\mathrm{f}}}\Phi\|^2 \right)^{1/2}.\end{aligned}$$

We see that $\left(\sum_{n=0}^{m} \frac{1}{n!} a^*(f)^n e^{-\frac{t}{2}H_{\mathrm{f}}}\Phi\right)_{m\in\mathbb{N}}$ is a Cauchy sequence in $\mathcal{F}_{\mathrm{b}}$. Hence $e^{-\frac{t}{2}H_{\mathrm{f}}}\Phi \in D(e^{a^*(f)})$, and on both sides above we have

$$\|e^{a^*(f)}e^{-\frac{t}{2}H_{\mathrm{f}}}\Phi\| \le A(f,s)\|e^{-\frac{1}{2}(t-s)H_{\mathrm{f}}}\Phi\|,$$

as $m \to \infty$, where

$$A(f,s) = \sum_{n=0}^{\infty} \frac{1}{\sqrt{n!}} 2^{n/2}s^{-n/2}\|f\|_\omega^n. \tag{1.2.48}$$

Choosing s such that $s < t$, we see that $\|e^{-\frac{1}{2}(t-s)H_{\mathrm{f}}}\Phi\| \le \|\Phi\|$ and $e^{a^*(f)}e^{-\frac{t}{2}H_{\mathrm{f}}}$ for $t < 1$ is bounded. Let now $t \ge 1$. Choosing $s = 1$ above, we have

$$\|e^{a^*(f)}e^{-\frac{t}{2}H_{\mathrm{f}}}\Phi\| \le A(f,1)\|e^{-\frac{1}{2}(t-1)H_{\mathrm{f}}}\Phi\| \le A(f,1)\|\Phi\|.$$

Thus $e^{a^*(f)}e^{-\frac{t}{2}H_{\mathrm{f}}}$ for $t \ge 1$ is bounded. Finally, since $(e^{-\frac{t}{2}H_{\mathrm{f}}}e^{a(f)})^* \supset e^{a^*(\bar{f})}e^{-\frac{t}{2}H_{\mathrm{f}}}$, the second statement follows, and (1) is proven. Part (2) is similarly proven by replacing ω_ν with the identity 1. □

We can also estimate the norms of these operators, which can be obtained from the estimates in the proof of Proposition 1.33.

Corollary 1.34.

(1) *If* $f \in D(1/\sqrt{\omega_\nu})$, *then*

$$\|e^{a^*(f)}e^{-\frac{t}{2}H_f}\| \le \sqrt{2}e^{(2/s)\|f\|_\omega^2}\|e^{-\frac{1}{2}(t-s)H_f}\|, \quad 0 < s < t < 1,$$
$$\|e^{a^*(f)}e^{-\frac{t}{2}H_f}\| \le \sqrt{2}e^{2\|f\|_\omega^2}\|e^{-\frac{1}{2}(t-1)H_f}\|, \qquad t \ge 1.$$

In particular, we have

$$\|e^{a^*(f)}e^{-tH_f}e^{a(\bar{f})}\| \le 2e^{(4/s)\|f\|_\omega^2}, \quad 0 < s < t < 1,$$
$$\|e^{a^*(f)}e^{-tH_f}e^{a(\bar{f})}\| \le 2e^{4\|f\|_\omega^2}, \qquad t \ge 1.$$

(2) *If* $f \in L^2(\mathbb{R}^d)$, *then*

$$\|e^{a^*(f)}e^{-\frac{t}{2}N}\| \le \sqrt{2}e^{(2/s)(2\|f\|)^2}\|e^{-\frac{1}{2}(t-s)N}\|, \quad 0 < s < t < 1,$$
$$\|e^{a^*(f)}e^{-\frac{t}{2}N}\| \le \sqrt{2}e^{2(2\|f\|)^2}\|e^{-\frac{1}{2}(t-1)N}\|, \qquad t \ge 1.$$

In particular,

$$\|e^{a^*(f)}e^{-tN}e^{a(\bar{f})}\| \le 2e^{(16/s)\|f\|^2}, \quad 0 < s < t < 1,$$
$$\|e^{a^*(f)}e^{-tN}e^{a(\bar{f})}\| \le 2e^{16\|f\|^2}, \qquad t \ge 1.$$

Proof. We can estimate $A(f,s)$ defined in (1.2.48) as

$$\begin{aligned} A(f,s) &= \sum_{n=0}^\infty \frac{1}{\sqrt{n!}} 2^{n/2}s^{-n/2}\|f\|_\omega^n = \sum_{n=0}^\infty \frac{1}{\sqrt{n!}} 2^{n/2}\|f\|_\omega^n s^{-n/2}2^n 2^{-n} \\ &\le \left(\sum_{n=0}^\infty \frac{1}{n!}4^n\|f\|_\omega^{2n}s^{-n}\right)^{1/2}\left(\sum_{n=0}^\infty 2^{-n}\right)^{1/2} \le \sqrt{2}e^{(2/s)\|f\|_\omega^2}. \end{aligned}$$

Claim (1) follows from Proposition 1.33, and (2) can be derived by replacing ω_ν with 1. □

We have already obtained the strong continuity of the map

$$L^2(\mathbb{R}^d) \ni f \mapsto e^{a^\sharp(f)}\Phi.$$

We can furthermore prove the uniform continuity of the map

$$f \mapsto e^{a^*(f)}e^{-\frac{t}{2}H_f}, \quad t > 0.$$

Corollary 1.35 (Uniform continuity).

(1) *Let* $f, g \in D(1/\sqrt{\omega_\nu})$. *Then*

$$\begin{aligned} &\|e^{a^*(f)}e^{-\frac{t}{2}H_f} - e^{a^*(g)}e^{-\frac{t}{2}H_f}\| \le \sqrt{2}\|f-g\|_\omega e^{(2/s)(\|f\|_\omega+\|g\|_\omega+1)^2}, \quad 0 < s < t < 1, \\ &\|e^{a^*(f)}e^{-\frac{t}{2}H_f} - e^{a^*(g)}e^{-\frac{t}{2}H_f}\| \le \sqrt{2}\|f-g\|_\omega e^{2(\|f\|_\omega+\|g\|_\omega+1)^2}, \qquad t \ge 1. \end{aligned} \tag{1.2.49}$$

In particular, let $f, f_n \in D(1/\sqrt{\omega_v})$ for $n \geq 1$ be such that $\|f - f_n\|_\omega \to 0$ as $n \to \infty$. Then $e^{a^*(f_n)}e^{-\frac{t}{2}H_f}$ is uniformly convergent to $e^{a^*(f)}e^{-\frac{t}{2}H_f}$ as $n \to \infty$.

(2) *Let $f, g \in L^2(\mathbb{R}^d)$. Then*

$$\begin{aligned} \|e^{a^*(f)}e^{-\frac{t}{2}N} - e^{a^*(g)}e^{-\frac{t}{2}N}\| &\leq 2\sqrt{2}\|f - g\|e^{(2/s)(2\|f\|+2\|g\|+1)^2}, \quad 0 < s < t < 1, \\ \|e^{a^*(f)}e^{-\frac{t}{2}N} - e^{a^*(g)}e^{-\frac{t}{2}N}\| &\leq 2\sqrt{2}\|f - g\|e^{2(2\|f\|+2\|g\|+1)^2}, \quad t \geq 1. \end{aligned} \tag{1.2.50}$$

In particular, let $f, f_n \in L^2(\mathbb{R}^d)$ for $n \geq 1$ be such that $\|f - f_n\| \to 0$ as $n \to \infty$. Then $e^{a^*(f_n)}e^{-\frac{t}{2}N}$ is uniformly convergent to $e^{a^*(f)}e^{-\frac{t}{2}N}$ as $n \to \infty$.

Proof. It is seen by inspection that

$$\begin{aligned} \|(a^*(f)^n - a^*(g)^n)\Psi\| &\leq \sum_{j=1}^n \left\|a^*(f)^{j-1}a^*(f-g)a^*(g)^{n-j}\Psi\right\| \\ &\leq \sum_{j=1}^n \sqrt{n!}2^{n/2}s^{-n/2}\|f\|_\omega^{j-1}\|f-g\|_\omega\|g\|_\omega^{n-j}\Big(\sum_{m=0}^n \frac{1}{m!}\|(sH_f)^{m/2}\Psi\|^2\Big)^{1/2} \\ &\leq \sqrt{n!}2^{n/2}s^{-n/2}(\|f\|_\omega + \|g\|_\omega + 1)^n\Big(\sum_{m=0}^n \frac{1}{m!}\|(sH_f)^{m/2}\Psi\|^2\Big)^{1/2}\|f-g\|_\omega. \end{aligned}$$

Hence from

$$\|(e^{a^*(f)}e^{-\frac{t}{2}H_f} - e^{a^*(g)}e^{-\frac{t}{2}H_f})\Phi\| \leq \sum_{n=0}^\infty \frac{1}{n!}\|(a^*(f)^n - a^*(g)^n)e^{-\frac{t}{2}H_f}\Phi\|,$$

(1.2.49) follows. The bounds (1.2.50) can be derived replacing ω_v by the identity map 1. □

We can slightly improve Theorem 1.33 and Corollary 1.34 above. In the corollary, we gave bounds on $e^{a^*(f)}e^{-\frac{t}{2}H_f}$ and $e^{a^*(f)}e^{-\frac{t}{2}N}$ by estimating $A(f, s)$. This estimate, however, can be modified by introducing a parameter $q > 1$ as

$$A(f, s) \leq \left(\sum_{n=0}^\infty \frac{1}{n!}2^n q^{2n}\|f\|_\omega^{2n}s^{-n}q^{-n}\right)^{1/2}\left(\sum_{n=0}^\infty 1^2 \cdot q^{-n}\right)^{1/2} \leq \sqrt{\frac{q}{q-1}}e^{(q/s)\|f\|_\omega^2}.$$

An advantage of this estimate is to get the coefficient $\sqrt{\frac{q}{q-1}}$ as close to 1 as convenient. This will be used in Lemma 2.17 in the next chapter. Here we show a modified version of Corollary 1.34.

Corollary 1.36. *Let* $q > 1$.

(1) *Let* $f \in D(1/\sqrt{\omega_v})$. *Then*

$$\|e^{a^*(f)}e^{-\frac{t}{2}H_f}\| \le \sqrt{\frac{q}{q-1}}e^{(q/s)\|f\|_\omega^2}\|e^{-\frac{1}{2}(t-s)H_f}\|, \quad 0 < s < t < 1,$$

$$\|e^{a^*(f)}e^{-\frac{t}{2}H_f}\| \le \sqrt{\frac{q}{q-1}}e^{q\|f\|_\omega^2}\|e^{-\frac{1}{2}(t-1)H_f}\|, \quad t \ge 1.$$

In particular, we have

$$\|e^{a^*(f)}e^{-tH_f}e^{a(\bar{f})}\| \le \frac{q}{q-1}e^{(2q/s)\|f\|_\omega^2}, \quad 0 < s < t < 1,$$

$$\|e^{a^*(f)}e^{-tH_f}e^{a(\bar{f})}\| \le \frac{q}{q-1}e^{2q\|f\|_\omega^2}, \quad t \ge 1.$$

(2) *Let* $f \in L^2(\mathbb{R}^d)$. *Then*

$$\|e^{a^*(f)}e^{-\frac{t}{2}N}\| \le \sqrt{\frac{q}{q-1}}e^{(q/s)(2\|f\|)^2}\|e^{-\frac{1}{2}(t-s)N}\|, \quad 0 < s < t < 1,$$

$$\|e^{a^*(f)}e^{-\frac{t}{2}N}\| \le \sqrt{\frac{q}{q-1}}e^{q(2\|f\|)^2}\|e^{-\frac{1}{2}(t-1)N}\|, \quad t \ge 1.$$

In particular, we have

$$\|e^{a^*(f)}e^{-tN}e^{a(\bar{f})}\| \le \frac{q}{q-1}e^{(8q/s)\|f\|^2}, \quad 0 < s < t < 1,$$

$$\|e^{a^*(f)}e^{-tN}e^{a(\bar{f})}\| \le \frac{q}{q-1}e^{8q\|f\|^2}, \quad t \ge 1.$$

In later developments discussed in the chapters below, we often encounter exponentials of creation and annihilation operators, which will require some estimates. Next we discuss some technical estimates in this direction. Let $\Phi(f) = \frac{1}{\sqrt{2}}(a^*(f) + a(\bar{f}))$. Products such as

$$\prod_{j=1}^m \Phi(f_j)$$

can be represented in terms of sums of Wick-ordered operators. For simplicity, we apply the shorthand

$$c^m_{j_1,\dots,j_k} = :\Phi(f_1)\cdots\widehat{\Phi(f_{j_1})}\cdots\widehat{\Phi(f_{j_k})}\cdots\Phi(f_m): \tag{1.2.51}$$

below, where $\widehat{\Phi(f)}$ means skipping $\Phi(f)$. We also write

$$c^m = :\Phi(f_1)\cdots\Phi(f_m):. \tag{1.2.52}$$

Lemma 1.37 (Wick's theorem). *Let $f_j \in L^2(\mathbb{R}^d)$, $j = 1, \dots, n$. Then $\prod_{j=1}^m \Phi(f_j)$ can be represented in terms of Wick-ordered operators as*

$$\prod_{j=1}^{2n} \Phi(f_j) = c^{2n} + \sum_{k=1}^{n} \frac{1}{2^k} \sum_{j_1,\dots,j_{2k}}^{\text{pair}} \left(\prod_{i=1}^{k} (f_{j_{2i-1}}, f_{j_{2i}}) \right) c^{2n}_{j_1,\dots,j_{2k}}, \quad m = 2n, \tag{1.2.53}$$

where $\sum^{\text{pair}}_{j_1,\dots,j_{2k}}$ is run over all k-pairs formed of the elements of $\{1, \dots, 2n\}$, and

$$\prod_{j=1}^{2n+1} \Phi(f_j) = c^{2n+1} + \sum_{k=1}^{n} \frac{1}{2^k} \sum_{j_1,\dots,j_{2k}}^{\text{pair}} \left(\prod_{i=1}^{k} (f_{j_{2i-1}}, f_{j_{2i}}) \right) c^{2n+1}_{j_1,\dots,j_{2k}}, \quad m = 2n+1, \tag{1.2.54}$$

where $\sum^{\text{pair}}_{j_1,\dots,j_{2k}}$ is run over all k-pairs formed of the elements of $\{1, \dots, 2n+1\}$.

Proof. The lemma directly follows by the definition of Wick product. □

Lemma 1.38. *If $f \in L^2(\mathbb{R}^d)$, then*

$$\Phi(f)^n = n! \sum_{k=0}^{[n/2]} \sum_{l+m+2k=n} \frac{\left(\frac{\|f\|^2}{4}\right)^k}{k!} \frac{\left(\frac{1}{\sqrt{2}} a^*(f)\right)^l}{l!} \frac{\left(\frac{1}{\sqrt{2}} a(\bar{f})\right)^m}{m!}. \tag{1.2.55}$$

Here $[m]$ denotes the integer part of m, i. e., $[2n/2] = n$ and $[(2n+1)/2] = n$.

Proof. The number of all k-pairs chosen from $\{1, \dots, n\}$ is

$$\frac{1}{k!} \binom{n}{2k} \binom{2k}{2} \binom{2k-2}{2} \cdots \binom{2}{2} = \frac{1}{2^k} \frac{n!}{k!(n-2k)!}.$$

We then have from (1.2.53) that

$$\Phi(f)^n = \sum_{k=0}^{[n/2]} \frac{n!}{k!(n-2k)!} \left(\frac{\|f\|^2}{4} \right)^k :\Phi(f)^{n-2k}:,$$

where

$$:\Phi(f)^m: = 2^{-m/2} \sum_{k=0}^{m} \binom{m}{k} a^*(f)^k a(\bar{f})^{m-k}.$$

From here, we obtain

$$\begin{aligned} \Phi(f)^n &= \sum_{k=0}^{[n/2]} \frac{n!}{k!(n-2k)!} \left(\frac{\|f\|^2}{4} \right)^k 2^{-\frac{n-2k}{2}} \sum_{l=0}^{n-2k} \binom{n-2k}{l} a^*(f)^l a(\bar{f})^{n-2k-l} \\ &= n! \sum_{k=0}^{[n/2]} \sum_{l=0}^{n-2k} \frac{\left(\frac{\|f\|^2}{4}\right)^k}{k!} \frac{\left(\frac{1}{\sqrt{2}} a^*(f)\right)^l}{l!} \frac{\left(\frac{1}{\sqrt{2}} a(\bar{f})\right)^{n-2k-l}}{(n-2k-l)!} = (1.2.55). \end{aligned}$$

□

Remark 1.39 (Baker–Campbell–Hausdorff formula: formal derivation). We can derive the Baker–Campbell–Hausdorff formula from (1.2.55) by the formal expansion

$$e^{\Phi(f)} = \sum_{n=0}^{\infty} \frac{\Phi(f)^n}{n!} = \sum_{n=0}^{\infty} \sum_{l+m+2k=n} \frac{(\frac{\|f\|^2}{4})^k}{k!} \frac{(\frac{1}{\sqrt{2}}a^*(f))^l}{l!} \frac{(\frac{1}{\sqrt{2}}a(\bar{f}))^m}{m!} = e^{\frac{1}{\sqrt{2}}a^*(f)} e^{\frac{1}{\sqrt{2}}a(\bar{f})} e^{\frac{\|f\|^2}{4}}.$$

This is suggestive, however, a rigorous derivation will also control the operator domains.

Consider $X = e^{a^*(g)}e^{a(\bar{g})}$ for $g \in L^2(\mathbb{R}^d)$. We have already seen above that the operator $e^{a^*(g)}e^{-\frac{t}{2}H_f}e^{a(\bar{g})}$ is bounded for every $t > 0$. This can be extended to more general cases which will be useful later on. From equality (1.2.55), we have the explicit form of the Wick ordering of $\Phi(f)^n$, and see that

$$\begin{aligned} X\Phi(f)^n\Psi &= n! \sum_{k=0}^{[n/2]} \sum_{l+m+2k=n} \frac{\left(\frac{\|f\|^2}{4}\right)^k}{k!} X \frac{\left(\frac{1}{\sqrt{2}}a^*(f)\right)^l}{l!} \frac{\left(\frac{1}{\sqrt{2}}a(\bar{f})\right)^m}{m!} \Psi \\ &= n! \sum_{k=0}^{[n/2]} \sum_{l+m+2k=n} \frac{\left(\frac{\|f\|^2}{4}\right)^k}{k!} \frac{\left(\frac{1}{\sqrt{2}}a^*(f)+y\right)^l}{l!} X \frac{\left(\frac{1}{\sqrt{2}}a(\bar{f})\right)^m}{m!} \Psi. \end{aligned} \tag{1.2.56}$$

Here we write $y = (g,f)/\sqrt{2}$. The operator $X\Phi(f)^n$ is unbounded, but our interest is in estimating the operator $Z(t,n)$, which is defined by inserting e^{-tH_f} into $X\Phi(f)^n$ as

$$Z(t,n) = n! \sum_{k=0}^{[n/2]} \sum_{l+m+2k=n} \frac{\left(\frac{\|f\|^2}{4}\right)^k}{k!} \frac{\left(\frac{1}{\sqrt{2}}a^*(f)+y\right)^l}{l!} e^{a^*(\bar{g})}e^{-tH_f}e^{a(\bar{g})} \frac{\left(\frac{1}{\sqrt{2}}a(\bar{f})\right)^m}{m!}.$$

Furthermore, below we will be interested also in estimates of

$$Z(t) = \sum_{n=0}^{\infty} \frac{t^n}{n!} Z(t,n)$$

and its norm.

Remark 1.40. Similarly to the formal derivation of the Baker–Campbell–Hausdorff formula in Remark 1.39, we can proceed to obtain further expressions of interest. Note that for any real-valued functions f, g, we have $[\Phi(f), \Phi(g)] = 0$ and $e^{\Phi(g)}e^{\Phi(f)} = e^{\Phi(f+g)}$. The Baker–Campbell–Hausdorff formula gives

$$e^{\Phi(g)}e^{\Phi(f)} = e^{\frac{1}{4}\|f+g\|^2} e^{\frac{a^*(f+g)}{\sqrt{2}}} e^{\frac{a(f+g)}{\sqrt{2}}}. \tag{1.2.57}$$

This can also be obtained from (1.2.56) by a formal computation. First of all,

$$e^{\Phi(g)}e^{\Phi(f)} = \sum_{n=0}^{\infty} \frac{1}{n!} e^{\frac{1}{4}\|g\|^2} e^{\frac{a^*(g)}{\sqrt{2}}} e^{\frac{a(\bar{g})}{\sqrt{2}}} \Phi(f)^n.$$

Replacing g by $g/\sqrt{2}$ in equality (1.2.56), we then obtain

$$\sum_{n=0}^{\infty}\frac{1}{n!}e^{\frac{1}{4}\|g\|^2}e^{\frac{a^*(g)}{\sqrt{2}}}e^{\frac{a(\bar{g})}{\sqrt{2}}}\Phi(f)^n$$

$$= e^{\frac{1}{4}\|g\|^2}\sum_{k=0}^{[n/2]}\sum_{l+m+2k=n}\frac{\left(\frac{\|f\|^2}{4}\right)^k}{k!}\frac{\left(\frac{1}{\sqrt{2}}a^*(f)+\frac{1}{2}(g,f)\right)^l}{l!}e^{\frac{a^*(g)}{\sqrt{2}}}e^{\frac{a(\bar{g})}{\sqrt{2}}}\frac{\left(\frac{1}{\sqrt{2}}a(\bar{f})\right)^m}{m!}$$

$$= e^{\frac{1}{4}\|g\|^2}e^{\frac{1}{4}\|f\|^2}e^{\frac{a^*(f)}{\sqrt{2}}+\frac{1}{2}(f,g)}e^{\frac{a^*(g)}{\sqrt{2}}}e^{\frac{a(\bar{g})}{\sqrt{2}}}e^{\frac{a(\bar{f})}{\sqrt{2}}} = e^{\frac{1}{4}\|f+g\|^2}e^{\frac{a^*(f+g)}{\sqrt{2}}}e^{\frac{a(f+g)}{\sqrt{2}}}.$$

Here we used that f and g are real, thus $\|f\|^2+\|g\|^2+2(f,g)=\|f+g\|^2$.

Lemma 1.41. *Let $f \in D(1/\sqrt{\omega_V})$. Then $Z(t,n)$ is bounded for every $t>0$, with the following bounds:*

(1) *if $t \geq 1$, then*

$$\|Z(t,n)\Psi\| \leq z_t(n)\left\|e^{-\frac{1}{2}(t-1)H_f}\right\|^2\|\Psi\|,$$

where

$$z_t(n) = n!\sum_{k=0}^{[n/2]}\sum_{l+m+2k=n}\frac{\left(\frac{\|f\|^2}{4}\right)^k(\|f\|_\omega+|y|)^l\|f\|_\omega^m}{k!l!m!}\xi(l,m),$$

with $y=(g,f)/\sqrt{2}$ and

$$\xi(l,m) = \sum_{j,j'=0}^{\infty}\frac{(\sqrt{2}\|g\|_\omega)^{j+j''}\sqrt{(j+l)!}\sqrt{(j'+m)!}}{j!j'!};$$

(2) *if $t<1$, then for every $0<s<t$,*

$$\|Z(t,n)\Psi\| \leq z_t(s,n)\left\|e^{-\frac{1}{2}(t-s)H_f}\right\|^2\|\Psi\|,$$

where

$$z_t(s,n) = n!\sum_{k=0}^{[n/2]}\sum_{l+m+2k=n}\frac{\left(\frac{\|f\|^2}{4}\right)^k(\|f\|_\omega+|y|)^l\|f\|_\omega^m}{k!l!m!}\xi_s(l,m)$$

and

$$\xi_s(l,m) = \sum_{j,j'=0}^{\infty}\frac{(\sqrt{2}\|g\|_\omega)^{j+j''}\sqrt{(j+l)!}\sqrt{(j'+m)!}}{s^{(j+l)/2}s^{(j'+m)/2}j!j'!}.$$

Proof. In this proof, we set $s = 1$ for $t \geq 1$ and $0 < s < t$ for $t < 1$. Since

$$\Big\| \sum_{n=0}^{N} \frac{a^*(f)^n}{n!} \Phi \Big\| \leq \sum_{n=0}^{N} \frac{1}{n!} \|a^*(f)^n \Phi\| \leq \sum_{n=0}^{N} \frac{(\sqrt{2}\|f\|_\omega)^n s^{-n/2}}{\sqrt{n!}} \|e^{\frac{1}{2}sH_f} \Phi\|$$

and

$$\begin{aligned}
\|(a^*(f) + y)^m a^*(g)^n \Psi\| &\leq \sum_{k=0}^{m} \binom{m}{k} |y|^k \|a^*(f)^{m-k} a^*(g)^n \Psi\| \\
&\leq \sum_{k=0}^{m} \binom{m}{k} |y|^k (\sqrt{2}\|f\|)^{m-k} (\sqrt{2}\|g\|_\omega)^n \sqrt{(n+m-k)!} s^{-(n+m)/2} \|e^{\frac{1}{2}sH_f} \Psi\| \\
&\leq s^{-(n+m)/2} (\sqrt{2}\|f\| + |y|)^m (\sqrt{2}\|g\|_\omega)^n \sqrt{(n+m)!} \|e^{\frac{1}{2}sH_f} \Psi\|,
\end{aligned}$$

we have for nonnegative integers $m \geq 0$,

$$\begin{aligned}
\Big\|(a^*(g) + y)^m e^{a^*(f)} \Phi\Big\| &\leq \sum_{n=0}^{\infty} \frac{1}{n!} \|(a^*(g) + y)^m a^*(f)^n \Phi\| \\
&\leq \sum_{n=0}^{\infty} \frac{(\sqrt{2}\|f\|_\omega + |y|)^m (\sqrt{2}\|g\|_\omega)^n \sqrt{(n+m)!}}{s^{(n+m)/2} n!} \|e^{\frac{1}{2}sH_f} \Phi\|.
\end{aligned}$$

Also,

$$\begin{aligned}
\|Z(t,n)\Psi\| &\leq \sum_{k=0}^{[n/2]} \sum_{l+m+2k=n} \frac{n! \left(\frac{\|f\|^2}{4}\right)^k}{k!l!m!} \Big\| \Big(\frac{a^*(f)}{\sqrt{2}} + y\Big)^l e^{a^*(g)} e^{-\frac{t}{2}H_f} e^{a(\bar{g})} \Big(\frac{a(\bar{f})}{\sqrt{2}}\Big)^m \Psi \Big\| \\
&\leq \sum_{k=0}^{[n/2]} \sum_{l+m+2k=n} \frac{n! \left(\frac{\|f\|^2}{4}\right)^k}{k!l!m!} \Big\| \Big(\frac{a^*(f)}{\sqrt{2}} + y\Big)^l e^{a^*(g)} e^{-\frac{t}{2}H_f} \Big\| \Big\| e^{-\frac{t}{2}H_f} e^{a(\bar{g})} \Big(\frac{a(\bar{f})}{\sqrt{2}}\Big)^m \Big\| \|\Psi\|.
\end{aligned} \tag{1.2.58}$$

Applying the estimates

$$\begin{aligned}
\Big\| \Big(\frac{a^*(f)}{\sqrt{2}} + |y|\Big)^l e^{a^*(g)} e^{-\frac{t}{2}H_f} \Big\| &\leq A \sum_{j=0}^{\infty} \frac{(\sqrt{2}\|g\|_\omega)^j (\|f\|_\omega + |y|)^l \sqrt{(j+l)!}}{s^{(j+l)/2} j!}, \\
\Big\| e^{-\frac{t}{2}H_f} e^{a(\bar{g})} \Big(\frac{a(\bar{f})}{\sqrt{2}}\Big)^m \Big\| &\leq A \sum_{j=0}^{\infty} \frac{(\sqrt{2}\|g\|_\omega)^j \|f\|_\omega^m \sqrt{(j+m)!}}{s^{(j+m)/2} j!}
\end{aligned}$$

in (1.2.58) with $A = \|e^{-\frac{1}{2}(t-s)H_f}\|$, the claim follows. □

Next we estimate the operator $Z(t) = \sum_{n=0}^{\infty} \frac{t^n}{n!} Z(t,n)$.

Lemma 1.42. *If $f, g \in D(1/\sqrt{\omega_v})$, then $Z(t)$ is bounded for $t > 0$ with*

$$\|Z(t)\| \leq \frac{e^{\frac{\|f\|^2}{4}}\|e^{-\frac{t-s}{2}H_f}\|^2 C(h)^2}{\left(1 - \frac{h}{\sqrt{s}}(\sqrt{2}\|g\|_\omega + \|f\|_\omega + y)\right)\left(1 - \frac{h}{\sqrt{s}}(\sqrt{2}\|g\|_\omega + \|f\|_\omega)\right)}. \tag{1.2.59}$$

Here $s = 1$ for $t \geq 1$ and $0 < s < t$ for $t < 1$, and $C(h)$ is a constant dependent on $h > 0$ such that $1 - \frac{h}{\sqrt{s}}(\sqrt{2}\|g\|_\omega + \|f\|_\omega + y) > 0$ for $y = (g, f)/\sqrt{2}$.

Proof. We set again $s = 1$ for $t \geq 1$ and $0 < s < t$ for $t < 1$. By Lemma 1.41 and the definition of $Z(t)$, we have

$$\|Z(t)\Psi\| \leq z_t \left\| e^{-\frac{1}{2}(t-1)H_f} \right\|^2 \|\Psi\|,$$

where

$$\begin{aligned} z_t &= \sum_{n=0}^{\infty} t^n \sum_{l+m+2k=n} \frac{\left(\frac{\|f\|^2}{4}\right)^k (\|f\|_\omega + |y|)^l \|f\|_\omega^m}{k!l!m!} \xi_s(l,m) \\ &= \sum_{n=0}^{\infty} \sum_{l+m+2k=n} \frac{\left(\frac{\|tf\|^2}{4}\right)^k (\|tf\|_\omega + t|y|)^l \|tf\|_\omega^m}{k!l!m!} \xi_s(l,m). \end{aligned}$$

Replacing tf and $t|y|$ by f and $|y|$, respectively, we get

$$z_t = \sum_{n=0}^{\infty} \sum_{l+m+2k=n} \frac{\left(\frac{\|f\|^2}{4}\right)^k}{k!} \frac{(\|f\|_\omega + |y|)^l}{l!} \frac{\|f\|_\omega^m}{m!} \xi_s(l,m) = e^{\|f\|^2/4} \xi_s(y) \xi_s(0), \tag{1.2.60}$$

where

$$\xi_s(y) = \sum_{j=0}^{\infty} \sum_{l=0}^{\infty} \frac{(\sqrt{2}\|g\|_\omega)^j (\|f\|_\omega + y)^l \sqrt{(j+l)!}}{s^{(j+l)/2} l! j!}.$$

Let $a = \sqrt{2}\|g\|_\omega/\sqrt{s}$ and $b = (\|f\|_\omega + y)/\sqrt{s}$. We rewrite $\xi_s(y)$ as

$$\xi_s(y) = \sum_{j=0}^{\infty} \sum_{l=0}^{\infty} \frac{a^j b^l \sqrt{(j+l)!}}{l! j!}. \tag{1.2.61}$$

Let $0 < h$ be such that $h < 1/(a+b)$. Then

$$\xi_s(y) = \sum_{j,l=0}^{\infty} \frac{a^j b^l h^{j+l} (j+l)!}{l! j!} \frac{(1/h)^j (1/h)^l}{\sqrt{(j+l)!}} \leq \frac{C}{1 - h(a+b)} < \infty.$$

Here $C = C(h) = \sup_{(j,l) \in \mathbb{N} \times \mathbb{N}} \frac{(1/h)^j (1/h)^l}{\sqrt{(j+l)!}}$. It can be also shown that $\xi(0)$ is finite. Hence we have (1.2.59), and z_t is finite. □

In the next step we extend Lemma 1.42. The operator $Z(t,n)$ is defined through the commutator $[X,\Phi(f)^n]$, where $X = e^{a^*(g)}e^{a(\bar{g})}$. We consider the operator $W(t,n)$ defined through the commutator

$$\Big[X, \prod_{j=1}^{n} \Phi(f_j)\Big]. \tag{1.2.62}$$

To determine this commutator explicitly, we apply Wick's theorem to the product

$$\prod_{j=1}^{n} \Phi(f_j) = c^n + \sum_{k=1}^{[n/2]} \frac{1}{2^k} \sum_{j_1,\dots,j_{2k}}^{\text{pair}} \Big(\prod_{i=1}^{k} (f_{j_{2i-1}}, f_{j_{2i}})\Big) c^n_{j_1,\dots,j_{2k}}.$$

Recall that $\sum^{\text{pair}}_{j_1,\dots,j_{2k}}$ denotes summation over all k-pairs chosen from $\{1,\dots,n\}$, and c^n and $c^n_{j_1,\dots,j_{2k}}$ are defined by (1.2.52) and (1.2.51), respectively. Since

$$c^n = :\prod_{j=1}^{n} \Phi(f_j): = 2^{-n/2} \Big(\sum_{k=1}^{n} \sum_{\{j_1,\dots,j_k\}\subset\{1,\dots,n\}} \prod_{j\in\{j_1,\dots,j_k\}} a^*(f_j) \prod_{i\in\{j_1,\dots,j_k\}^c} a(\bar{f}_i) + \prod_{i=1}^{n} a(\bar{f}_i) \Big),$$

we can compute $c^n_{j_1,\dots,j_{2k}}$ as

$$c^n_{j_1,\dots,j_{2k}} = 2^{-(n-2k)/2} \Big(\sum_{l=1}^{n-2k} \sum_{\{i_1,\dots,i_l\}\subset\{j_1,\dots,j_{2k}\}^c} \prod_{j\in\{i_1,\dots,i_l\}} a^*(f_j) \prod_{i\in\{i_1,\dots,i_l\}^c\cap\{j_1,\dots,j_{2k}\}^c} a(\bar{f}_i) + a^n_{j_1,\dots,j_{2k}} \Big),$$

where

$$a^n_{j_1,\dots,j_{2k}} = a(\bar{f}_1)\cdots\widehat{a(\bar{f}_{j_1})}\cdots\widehat{a(\bar{f}_{j_{2k}})}\cdots a(\bar{f}_n),$$

and we note that $\{j_1,\dots,j_{2k}\}^c = \{1,\dots,n\} \setminus \{j_1,\dots,j_{2k}\}$. Hence

$$\begin{aligned} X\prod_{j=1}^{n} \Phi(f_j) &= Xc^n + \sum_{k=1}^{[n/2]} \frac{1}{2^k} \sum_{j_1,\dots,j_{2k}}^{\text{pair}} \Big(\prod_{i=1}^{k} (f_{j_{2i-1}}, f_{j_{2i}})\Big) Xc^n_{j_1,\dots,j_{2k}} \\ &= 2^{-n/2} X \prod_{i=1}^{n} a(\bar{f}_i) + \sum_{l=1}^{n} \sum_{\{i_1,\dots,i_l\}\subset\{1,\dots,n\}} \bar{A}X\bar{B} \\ &\quad + \sum_{k=1}^{[n/2]} \frac{1}{2^k} \sum_{j_1,\dots,j_{2k}}^{\text{pair}} \Big(\prod_{i=1}^{k} (f_{j_{2i-1}}, f_{j_{2i}})\Big) \Big(Xa^n_{j_1,\dots,j_{2k}} + \sum_{l=1}^{n-2k} \sum_{\{i_1,\dots,i_l\}\subset\{j_1,\dots,j_{2k}\}^c} AXB \Big), \end{aligned}$$

where

$$A = A(\{i_1,\dots,i_l\}) = \prod_{j\in\{i_1,\dots,i_l\}} \Big(\frac{a^*(f_j)}{\sqrt{2}} + y_j \Big), \tag{1.2.63}$$

$$B = B(\{i_1,\ldots,i_l\}^c \cap \{j_1,\ldots,j_{2k}\}^c) = \prod_{i\in\{i_1,\ldots,i_l\}^c\cap\{j_1,\ldots,j_{2k}\}^c} \frac{a(\bar{f}_i)}{\sqrt{2}}, \tag{1.2.64}$$
$$\bar{A} = A(\{i_1,\ldots,i_l\}),$$
$$\bar{B} = B(\{i_1,\ldots,i_l\}^c) = \prod_{i\in\{i_1,\ldots,i_l\}^c} \frac{a(\bar{f}_i)}{\sqrt{2}},$$

with $y_j = (g, f_j)/\sqrt{2}$. It is simply written as

$$X\prod_{j=1}^n \Phi(f_j) = \sum_{k=0}^{[n/2]} \frac{1}{2^k} \sum_{j_1,\ldots,j_{2k}}^{\text{pair}} \left(\prod_{i=1}^k (f_{j_{2i-1}}, f_{j_{2i}})\right) \sum_{l=0}^{n-2k} \sum_{\{i_1,\ldots,i_l\}\subset\{j_1,\ldots,j_{2k}\}^c} AXB.$$

Here and in what follows, it is understood that $\{i_1,\ldots,i_l\} = \emptyset$ if $l = 0$, $\{j_1,\ldots,j_{2k}\}^c = \{1,\ldots,n\}$ if $k = 0$ and

$$\sum_{j_1,\ldots,j_{2k}}^{\text{pair}} \left(\prod_{i=1}^k (f_{j_{2i-1}}, f_{j_{2i}})\right) \sum_{l=0}^{n-2k} \sum_{\{i_1,\ldots,i_l\}\subset\{j_1,\ldots,j_{2k}\}^c} AXB = \sum_{l=0}^{n} \sum_{\{i_1,\ldots,i_l\}\subset\{1,\ldots,n\}} AXB$$

if $k = 0$. Since $X\prod_{j=1}^m \Phi(f_j)$ is not bounded, we then define $W(t,n)$ by inserting e^{-tH_f} into $X\prod_{j=1}^m \Phi(f_j)$ as

$$W(t,n) = \sum_{k=0}^{[n/2]} \frac{1}{2^k} \sum_{j_1,\ldots,j_{2k}}^{\text{pair}} \sum_{l=0}^{n-2k} \sum_{\{i_1,\ldots,i_l\}\subset\{j_1,\ldots,j_{2k}\}^c} \left(\prod_{i=1}^k (f_{j_{2i-1}}, f_{j_{2i}})\right) Ae^{a^*(g)} e^{-tH_f} e^{a(\bar{g})} B. \tag{1.2.65}$$

Furthermore, consider

$$W(t) = \sum_{n=0}^{\infty} \frac{t^n}{n!} W(t,n). \tag{1.2.66}$$

The formula proved in Theorem 1.43 below will be used in estimating integral kernels of semigroups generated by the so-called Pauli–Fierz model with spin 1/2 discussed in Chapter 3 below, or the spin-boson model in Chapter 4. The integral kernels of both are of the form given by (1.2.66). Next we show that these integral kernels are bounded.

Theorem 1.43. *Let $f_j \in D(1/\sqrt{\omega_\nu})$, $j \in \mathbb{N}$. We suppose that there exists $a > 0$ such that $\|f_j\|_\omega \le a$ for every j. Then $W(t,n)$ and $W(t)$ are bounded for every $t > 0$, with norm*

$$\|W(t)\| \le \frac{e^{\frac{a^2}{4}} \|e^{-\frac{t-s}{2}H_f}\|^2 C(h)^2}{\left(1 - \frac{h}{\sqrt{s}}(\sqrt{2}\|g\|_\omega + a + y)\right)\left(1 - \frac{h}{\sqrt{s}}(\sqrt{2}\|g\|_\omega + a)\right)}.$$

Here $s = 1$ for $t \ge 1$, $0 < s < t$ for $t < 1$, $y = a\|g\|/\sqrt{2}$, and $C(h)$ is a constant dependent on $h > 0$ such that $1 - \frac{h}{\sqrt{s}}(\sqrt{2}\|g\|_\omega + a + y) > 0$.

Proof. We set $s = 1$ for $t \geq 1$ and $0 < s < t$ for $t < 1$. Let A and B be as given in (1.2.63) and (1.2.64), respectively. We estimate

$$\left\|Ae^{a^*(g)}e^{-tH_f}e^{a(\bar{g})}B\right\| \leq \left\|Ae^{a^*(g)}e^{-\frac{t}{2}H_f}\right\|\left\|e^{-\frac{t}{2}H_f}e^{a(\bar{g})}B\right\|,$$

with

$$\left\|Ae^{a^*(g)}e^{-\frac{t}{2}H_f}\Phi\right\| \leq \sum_{j=0}^{\infty}\frac{(\alpha+y)^l(\sqrt{2}\|g\|_\omega)^j\sqrt{(l+j)!}}{s^{(j+l)/2}j!}\left\|e^{-\frac{1}{2}(t-s)H_f}\right\|,$$

$$\left\|e^{-\frac{t}{2}H_f}e^{a(\bar{g})}B\right\| \leq \sum_{j=0}^{\infty}\frac{\alpha^{n-2k-l}(\sqrt{2}\|g\|_\omega)^j\sqrt{(n-2k-l+j)!}}{s^{(j+n-2k-l)/2}j!}\left\|e^{-\frac{1}{2}(t-s)H_f}\right\|.$$

Let $y = \alpha\|g\|/\sqrt{2}$, then $|y_j| \leq y$. Similarly to the proof of Lemma 1.41, we obtain the bounds

$$\|W(t,n)\Psi\| \leq w_t(n)\left\|e^{-\frac{1}{2}(t-1)H_f}\right\|^2\|\Psi\|, \quad t \geq 1,$$

where

$$w_t(n) = n!\sum_{k=0}^{[n/2]}\sum_{l+m+2k=n}\frac{\left(\frac{\alpha^2}{4}\right)^k(\alpha+y)^l\alpha^m}{k!l!m!}\xi(l,m),$$

and

$$\|W(t,n)\Psi\| \leq w_t(s,n)\left\|e^{-\frac{1}{2}(t-s)H_f}\right\|^2\|\Psi\|, \quad 0 < s < t < 1,$$

where

$$w_t(s,n) = n!\sum_{k=0}^{[n/2]}\sum_{l+m+2k=n}\frac{\left(\frac{\alpha^2}{4}\right)^k(\alpha+y)^l\alpha^m}{k!l!m!}\xi_s(l,m).$$

Here $\xi(l,m)$ and $\xi_s(l,m)$ are defined as in Lemma 1.41. This shows that $W(t,n)$ is bounded. To show boundedness of $W(t)$, we proceed to the proof of Lemma 1.42. We have

$$\|W(t)\Psi\| \leq w_t\left\|e^{-\frac{1}{2}(t-1)H_f}\right\|^2\|\Psi\|,$$

where

$$w_t = \sum_{n=0}^{\infty}\sum_{l+m+2k=n}\frac{\left(\frac{t^2\alpha^2}{4}\right)^k(t\alpha+ty)^l(t\alpha)^m}{k!l!m!}\xi_s(l,m).$$

We replace $t\alpha$ and ty by α and y, respectively, and get

$$W_t = \sum_{n=0}^{\infty} \sum_{l+m+2k=n} \frac{\left(\frac{\alpha^2}{4}\right)^k}{k!} \frac{(\alpha+y)^l}{l!} \frac{\alpha^m}{m!} \xi_s(l,m) = e^{\alpha^2/4} \xi_s(y)\xi_s(0),$$

where

$$\xi_s(y) = \sum_{j=0}^{\infty} \sum_{l=0}^{\infty} \frac{(\sqrt{2}\|g\|_\omega)^j (\alpha+y)^l \sqrt{(j+l)!}}{s^{(j+l)/2} l! j!}.$$

The right-hand side above is finite as shown in Lemma 1.42. □

Finally, we prove continuity of $W(t,n)$ and $W(t)$. Here $W(t,n)$ and $W(t)$ are defined by (1.2.65) and (1.2.66), respectively. Let $f_j, g \in L^2(\mathbb{R}^d)$ for $j \in \mathbb{N}$. Then $W(t,n)$ is a map from $L^2(\mathbb{R}^d) \times (\times^n L^2(\mathbb{R}^d)) \ni g \times (\times_{j=1}^n f_j) \mapsto W(t,n)$.

Lemma 1.44.
(1) *The map $L^2(\mathbb{R}^d) \times (\times^n L^2(\mathbb{R}^d)) \ni g \times (\times_{j=1}^n f_j) \mapsto W(t,n)$ is continuous in the uniform norm.*
(2) *Suppose that there exists $\alpha > 0$ such that $f_j^h \to f_j$ as $h \to 0$ is uniformly in j and $\alpha = \sup\{\|f_j^h\|_\omega \mid j \in \mathbb{N}, h > 0\}$. Then the map $\times^\infty L^2(\mathbb{R}^d) \ni \times_{j=1}^\infty f_j \mapsto W(t,n)$ is continuous in the uniform norm.*

Proof. We set $W(t,n) = W^0(t,n) + W^+(t,n)$, where

$$W^+(t,n) = \sum_{k=1}^{[n/2]} \frac{1}{2^k} \sum_{j_1,\dots,j_{2k}}^{\text{pair}} \sum_{l=0}^{n-2k} \sum_{\{i_1,\dots,i_l\}\subset\{j_1,\dots,j_{2k}\}^c} \left(\prod_{i=1}^k (f_{j_{2i-1}}, f_{j_{2i}})\right) A e^{a^*(g)} e^{-tH_f} e^{a(\bar{g})} B,$$

$$W^0(t,n) = \sum_{l=0}^{n} A e^{a^*(g)} e^{-tH_f} e^{a(\bar{g})} B.$$

We show the statements (1) and (2) for $W^0(t,n)$ and $W^+(t,n)$.

Suppose that $g^h \to g$ and $f_j^h \to f_j$ as $h \to \infty$ in $\|\cdot\|_\omega$. Let $W_h^+(t,n)$ be defined by $W^+(t,n)$ with f_j and g replaced by f_j^h for $j = 1,\dots,n$, and g^h, respectively. Define the finite constants $\alpha = \sup\{\|f_j^h\|_\omega \mid j = 1,\dots,n, h > 0\}$ and $\beta = \sup\{\|g^h\|_\omega \mid h > 0\}$. We estimate $\|W_h^+(t,n) - W^+(t,n)\|$. Let

$$\xi(l,m) = \sum_{j,j'=0}^{\infty} \frac{(\sqrt{2}\beta)^{j+j''} \sqrt{(j+l)!} \sqrt{(j'+m)!}}{j! j'!}.$$

We see that

$$W_h^+(t,n) - W^+(t,n) = \sum_{k=1}^{[n/2]} \frac{1}{2^k} \sum_{j_1,\dots,j_{2k}}^{\text{pair}} \sum_{l=0}^{n-2k} \sum_{\{i_1,\dots,i_l\}\subset\{j_1,\dots,j_{2k}\}^c} \sum_{p=1}^{5} J_p,$$

where $\sum_{p=1}^4 J_p^0$ and $\sum_{p=1}^5 J_p$ are obtained by telescoping $W_h^+(t,n) - W^+(t,n)$ defined by

$$J_1 = \left(\prod_{i=1}^k (f_{j_{2i-1}}^h, f_{j_{2i}}^h) - \prod_{i=1}^k (f_{j_{2i-1}}, f_{j_{2i}})\right) A_h e^{a^*(g^h)} e^{-tH_f} e^{a(\bar{g}^h)} B_h,$$

$$J_2 = \prod_{i=1}^k (f_{j_{2i-1}}, f_{j_{2i}})(A_h - A) e^{a^*(g^h)} e^{-tH_f} e^{a(\bar{g}^h)} B_h,$$

$$J_3 = \prod_{i=1}^k (f_{j_{2i-1}}, f_{j_{2i}}) A (e^{a^*(g^h)} - e^{a^*(g)}) e^{-tH_f} e^{a(\bar{g}^h)} B_h,$$

$$J_4 = \prod_{i=1}^k (f_{j_{2i-1}}, f_{j_{2i}}) A e^{a^*(g)} e^{-tH_f} (e^{a(\bar{g}^h)} - e^{a(\bar{g})}) B_h,$$

$$J_5 = \prod_{i=1}^k (f_{j_{2i-1}}, f_{j_{2i}}) A e^{a^*(g)} e^{-tH_f} e^{a(\bar{g})} (B_h - B).$$

Suppose that there exists $\delta > 0$ such that $\|f_j^h - f_j\|_\omega \le \varepsilon$ for any j and $\|g^h - g\| \le \varepsilon$ for every $h < \delta$. Then

$$\sum_{k=1}^{[n/2]} \frac{1}{2^k} \sum_{j_1,\dots,j_{2k}}^{\text{pair}} \sum_{l=0}^{n-2k} \sum_{\{i_1,\dots,i_l\}\subset\{j_1,\dots,j_{2k}\}^c} \|J_1\| \le \varepsilon n! \sum_{k=1}^{[n/2]} \sum_{l+m+2k=n} \frac{\frac{a}{2}\left(\frac{a^2}{4}\right)^{k-1}}{(k-1)!} \frac{(a+y)^l a^m}{l!m!} \xi(l,m).$$

Similar estimates can be obtained for J_2 and J_4, and the net result is

$$\begin{aligned} &\sum_{k=1}^{[n/2]} \frac{1}{2^k} \sum_{j_1,\dots,j_{2k}}^{\text{pair}} \sum_{l=0}^{n-2k} \sum_{\{i_1,\dots,i_l\}\subset\{j_1,\dots,j_{2k}\}^c} (\|J_2\| + \|J_4\|) \\ &\quad \le \varepsilon n! \sum_{k=1}^{[n/2]} \sum_{l+m+2k=n} \frac{\left(\frac{a^2}{4}\right)^k}{k!} \left(\frac{(a+y)^{l-1} a^m}{(l-1)!m!} + \frac{(a+y)^l a^{m-1}}{l!(m-1)!} \right) \xi(l,m). \end{aligned}$$

Finally, J_3 and J_5 are estimated as

$$\begin{aligned} &\sum_{k=1}^{[n/2]} \frac{1}{2^k} \sum_{j_1,\dots,j_{2k}}^{\text{pair}} \sum_{l=0}^{n-2k} \sum_{\{i_1,\dots,i_l\}\subset\{j_1,\dots,j_{2k}\}^c} (\|J_3\| + \|J_5\|) \\ &\quad \le \varepsilon n! \sum_{k=1}^{[n/2]} \sum_{l+m+2k=n} \frac{\left(\frac{a^2}{4}\right)^k}{k!} \frac{(a+y)^l a^m}{l!m!} (\xi(l+1,m) + \xi(l,m+1)). \end{aligned}$$

Together with them, (1) follows. By the estimates above, it is seen that

$$\|(W^+(t) - W_h^+(t))\Psi\| \le \varepsilon w_t^+ \left\| e^{-\frac{1}{2}(t-1)H_f} \right\|^2 \|\Psi\|,$$

where

$$w_t^+ = \sum_{n=1}^{\infty}\sum_{k=1}^{[n/2]}\sum_{l+m+2k=n}(a_{klm}+b_{klm}+c_{klm})$$

with

$$a_{klm} = \frac{\frac{a}{2}\left(\frac{a^2}{4}\right)^{k-1}}{(k-1)!}\frac{(a+y)^l a^m}{l!m!}\xi(l,m),$$

$$b_{klm} = \frac{\left(\frac{a^2}{4}\right)^k}{k!}\left(\frac{(a+y)^{l-1}a^m}{(l-1)!m!}+\frac{(a+y)^l a^{m-1}}{l!(m-1)!}\right)\xi(l,m),$$

$$c_{klm} = \frac{\left(\frac{a^2}{4}\right)^k}{k!}\frac{(a+y)^l a^m}{l!m!}(\xi(l+1,m)+\xi(l,m+1)).$$

Thus (2) follows. Using the decomposition

$$W_h^0(t,n) - W^0(t,n) = \sum_{l=0}^{n}\sum_{\{i_1,\dots,i_l\}\subset\{1,\dots,n\}}\sum_{p=1}^{4} J_p^0,$$

where

$$J_1^0 = (A_h - A)e^{a^*(g^h)}e^{-tH_\mathrm{f}}e^{a(\bar{g}^h)}B_h,$$

$$J_2^0 = A(e^{a^*(g^h)} - e^{a^*(g)})e^{-tH_\mathrm{f}}e^{a(\bar{g}^h)}B_h,$$

$$J_3^0 = Ae^{a^*(g)}e^{-tH_\mathrm{f}}(e^{a(\bar{g}^h)} - e^{a(\bar{g})})B_h,$$

$$J_4^0 = Ae^{a^*(g)}e^{-tH_\mathrm{f}}e^{a(\bar{g})}(B_h - B),$$

we can similarly prove the statements (1) and (2) for $W^0(t,n)$ and $W^0(t)$. □

1.3 $\mathscr{Q}$-spaces

1.3.1 Gaussian random processes

We define a family of Gaussian random variables indexed by a given real vector space $\mathfrak{h}$ on a probability space $(\mathscr{Q},\Sigma,\mu)$, and study $L^2(\mathscr{Q}) = L^2(\mathscr{Q},\Sigma,\mu)$. Note that $L^2(\mathscr{Q})$ is a vector space over $\mathbb{C}$. Our main goal here is to show unitary equivalence of $L^2(\mathscr{Q})$ and $\mathscr{F}_\mathrm{b}(\mathfrak{h}_\mathbb{C})$, where $\mathfrak{h}_\mathbb{C}$ is the complexification of $\mathfrak{h}$.

Definition 1.45 (Gaussian random variables indexed by $\mathfrak{h}$). We say that $(\phi(f), f \in \mathfrak{h})$ is a family of *Gaussian random variables on a probability space* $(\mathscr{Q},\Sigma,\mu)$ *indexed by a real inner product space* $\mathfrak{h}$ whenever

(1) $\phi : \mathfrak{h} \ni f \mapsto \phi(f)$ is a map from $\mathfrak{h}$ to a Gaussian random variable on $(\mathcal{Q}, \Sigma, \mu)$ with zero mean and covariance

$$\int_{\mathcal{Q}} \phi(f)\phi(g)d\mu = \frac{1}{2}(f,g)_{\mathfrak{h}},$$

(2) $\phi(\alpha f + \beta g) = \alpha\phi(f) + \beta\phi(g)$, $\alpha, \beta \in \mathbb{R}$,

(3) Σ is the completion of the minimal σ-field generated by $\{\phi(f) \mid f \in \mathfrak{h}\}$.

Property (3) in Definition 1.45 is expressed by saying that the σ-field Σ is *full*. The existence of such a family of Gaussian random variables will be discussed in Section 1.4. Here we first look at some properties. Define

$$\mathscr{S}(\mathcal{Q}) = \left\{F(\phi(f_1), \ldots, \phi(f_n)) \,\middle|\, F \in \mathscr{S}(\mathbb{R}^n), f_j \in \mathfrak{h}, j = 1, \ldots, n, n \geq 1\right\}.$$

Theorem 1.46. *Let $(\mathcal{Q}, \Sigma, \mu)$ be a probability space and $(\phi(f), f \in \mathfrak{h})$ a family of Gaussian random variables indexed by $\mathfrak{h}$ on it. Then the following hold:*

(1) *If Σ is a full σ-field, then $\mathscr{S}(\mathcal{Q})$ is dense in $L^2(\mathcal{Q})$.*

(2) *If Σ is a complete σ-field and $\mathscr{S}(\mathcal{Q})$ is dense in $L^2(\mathcal{Q})$, then Σ is a full σ-field.*

Proof. To prove (1), we regard $\mathscr{S}(\mathcal{Q})$ as a set of bounded multiplication operators on $L^2(\mathcal{Q})$. Hence the closure $\overline{\mathscr{S}(\mathcal{Q})}$ taken in strong topology is a von Neumann algebra. It follows that $\overline{\mathscr{S}(\mathcal{Q})} \subset L^\infty(\mathcal{Q})$, since a strongly convergent bounded operator is bounded. Let $\mathcal{P}$ be the set of all projections in $\overline{\mathscr{S}(\mathcal{Q})}$. Every projection is the characteristic function of a measurable subset in $\mathcal{Q}$. Let $\Sigma_{\mathcal{P}}$ denote the family of measurable subsets of $\mathcal{Q}$ associated with $\mathcal{P}$, satisfying $\Sigma_{\mathcal{P}} \subset \Sigma$. It is easily seen by a limiting argument that $\phi(f)$ is measurable with respect to $\Sigma_{\mathcal{P}}$. Hence $\Sigma = \Sigma_{\mathcal{P}}$, as Σ is the minimal σ-field generated by $\{\phi(f) \mid f \in \mathfrak{h}\}$. Since a function in $L^\infty(\mathcal{Q})$ can be approximated by a finite linear sum of characteristic functions and $\Sigma = \Sigma_{\mathcal{P}}$, $\overline{\mathscr{S}(\mathcal{Q})} = L^\infty(\mathcal{Q})$ follows as sets of bounded operators. It is clear that $L^\infty(\mathcal{Q}) = \overline{\mathscr{S}(\mathcal{Q})}1 \subset \overline{\mathscr{S}(\mathcal{Q})}^{\|\cdot\|_{L^2(\mathcal{Q})}}$. Since $L^\infty(\mathcal{Q})$ is dense in $L^2(\mathcal{Q})$, it follows that $\overline{\mathscr{S}(\mathcal{Q})}^{\|\cdot\|_{L^2(\mathcal{Q})}} = L^2(\mathcal{Q})$. This shows (1).

Next we prove (2). Let Σ_0 be the completion of the minimal σ-field generated by $\{\phi(f) \mid f \in \mathfrak{h}\}$, and then $\Sigma_0 \subset \Sigma$. We shall prove $\Sigma_0 \supset \Sigma$. Let M be the L^2-closure of the set of all Σ_0-measurable functions in $L^2(\mathcal{Q})$. Then for any $f \in M$ there exists a sequence of Σ_0-measurable functions f_n such that f_n converges to f a. e. Since Σ_0 is complete, f is also Σ_0-measurable. Thus any $f \in M$ is Σ_0-measurable. Let P_M be the projection on M. Let $A \in \Sigma$. By the assumption, there exists a sequence $\Phi_n \in \mathscr{S}(\mathcal{Q})$ such that $\|\Phi_n - 1_A\|_{L^2(\mathcal{Q})} \to 0$ as $n \to \infty$. Since $\Phi_n \in M$, we can see that

$$\|\Phi_n - 1_A\|^2 = \|1_A P_{M^c}\|^2 + \|(\Phi_n - 1_A)P_M\|^2$$

and $\|(\Phi_n - 1_A)P_M\|^2 \to 0$ as $n \to \infty$. This implies that $\|1_A P_{M^c}\| = 0$ and then $1_A P_M = 1_A$. Hence 1_A is a Σ_0-measurable function. Thus A is Σ_0 measurable and $\Sigma_0 \supset \Sigma$. This completes (2). □

The following is a useful consequence saying that every nonnegative function in $L^2(\mathscr{Q})$ can be approximated by nonnegative functions in $\mathscr{S}(\mathscr{Q})$.

Corollary 1.47. *Let $(\mathscr{Q}, \Sigma, \mu)$ be a probability space such that Σ is full. Let $F \in L^2(\mathscr{Q})$ be nonnegative. Then there exists a sequence $(F_n)_{n\in\mathbb{N}} \subset \mathscr{S}(\mathscr{Q})$ such that $F_n \geq 0$ for $n \in \mathbb{N}$ and $\|F_n - F\|_{L^2(\mathscr{Q})} \to 0$ as $n \to \infty$.*

Proof. Notice that $\mathscr{S}(\mathscr{Q})$ is dense in $L^2(\mathscr{Q})$. There exists $(G_n)_{n\in\mathbb{N}}^\infty \subset \mathscr{S}(\mathscr{Q})$ such that $G_n \to F$ as $n \to \infty$. Choosing a subsequence n', we see that $G_{n'} \to F$ as $n' \to \infty$ a. s. Thus $|G_{n'}| \to F$ a. s. We approximate $|G_n|(\phi(f_1), \ldots, \phi(f_m))$ by functions in $\mathscr{S}(\mathscr{Q})$. Let $\rho \in C_0^\infty(\mathbb{R}^d)$ such that

$$\rho(x) \begin{cases} \geq 0, & |x| < 1, \\ = 0, & |x| \geq 1, \end{cases} \quad \text{and} \quad \int_{\mathbb{R}^d} \rho(x)dx = 1.$$

Define $\rho_\varepsilon(x) = \varepsilon^{-d}\rho(x/\varepsilon)$ and $F_n^\varepsilon(x) = \int_{\mathbb{R}^d} \rho_\varepsilon(x-y)|G_n|(y)dy$. It then follows for every $1 \leq p < \infty$ that $\|F_n^\varepsilon - |G_n|\|_{L^p(\mathbb{R}^d)} \to 0$ as $\varepsilon \downarrow 0$. This yields

$$\begin{aligned} &\|F_n^\varepsilon(\phi(f_1), \ldots, \phi(f_m)) - |G_n|(\phi(f_1), \ldots, \phi(f_m))\|_{L^2(\mathscr{Q})} \\ &= \left\| \frac{1}{(2\pi)^{m/2}} \int_{\mathbb{R}^m} (\check{F}_n^\varepsilon(k) - |\check{G}_n|(k)) e^{-i\sum_{j=1}^m k_j \phi(f_j)} dk \right\|_{L^2(\mathscr{Q})} \\ &\leq \frac{1}{(2\pi)^{m/2}} \int_{\mathbb{R}^m} \left| \check{F}_n^\varepsilon(k) - |\check{G}_n|(k) \right| dk \to 0, \quad \text{as } \varepsilon \downarrow 0. \end{aligned}$$
□

We define Wick products in $L^2(\mathscr{Q})$ in the same way as for boson Fock space.

Definition 1.48 (Wick product). The *Wick product* of $\prod_{i=1}^n \phi(f_i)$ is recursively defined by

$$\begin{aligned} &{:}\phi(f){:} = \phi(f), \\ &{:}\phi(f)\prod_{i=1}^n \phi(f_i){:} = \phi(f){:}\prod_{i=1}^n \phi(f_i){:} - \frac{1}{2}\sum_{j=1}^n (f, f_j){:}\prod_{i\neq j} \phi(f_i){:}. \end{aligned} \tag{1.3.1}$$

Since μ is a Gaussian measure, any Wick-ordered polynomial satisfies

$${:}\prod_{i=1}^n \phi(f_i){:} \in L^2(\mathscr{Q}, d\mu).$$

Definition 1.48 has the virtue of allowing explicit calculations, and we will need especially equality (1.3.1) later on. In particular, for $f_i, g_j \in \mathfrak{h}$,

$$\left(:\prod_{i=1}^{n} \phi(f_i):, :\prod_{i=1}^{m} \phi(g_i): \right) = \delta_{mn} \sum_{\pi \in \wp_n} 2^{-n} \prod_{i=1}^{n} (f_i, g_{\pi(i)}).$$

Furthermore,

$$:e^{\alpha\phi(f)}: = \text{s-}\lim_{m\to\infty} \sum_{n=0}^{m} \frac{\alpha^n}{n!} :\phi(f)^n: = e^{\frac{\alpha^2\|f\|^2}{4}} e^{\alpha\phi(f)}. \tag{1.3.2}$$

Write

$$L_n^2(\mathscr{Q}) = \overline{\text{L.H.} \left\{ :\prod_{i=1}^{n} \phi(f_i): \middle| f_i \in \mathfrak{h}, i = 1, \dots, n \right\}}.$$

We see that $L_n^2(\mathscr{Q}) \subset L^2(\mathscr{Q})$ and $L_m^2(\mathscr{Q}) \perp L_n^2(\mathscr{Q})$ if $n \neq m$.

Proposition 1.49 (Wiener–Itô decomposition). *We have that*

$$L^2(\mathscr{Q}) = \bigoplus_{n=0}^{\infty} L_n^2(\mathscr{Q}). \tag{1.3.3}$$

Proof. Denote $\mathscr{G} = \bigoplus_{n=0}^{\infty} L_n^2(\mathscr{Q})$. It is clear that $\mathscr{G} \subset L^2(\mathscr{Q})$. By (1.3.2), $e^{i\phi(f)} \in \mathscr{G}$, which means that $F(\phi(f_1), \dots, \phi(f_n)) \in \mathscr{G}$ for $F \in \mathscr{S}(\mathbb{R}^n)$. Since $\mathscr{S}(\mathscr{Q})$ is dense in $L^2(\mathscr{Q})$ by Theorem 1.46, $L^2(\mathscr{Q}) = \mathscr{G}$ follows. □

Equality (1.3.3) is called the *Wiener–Itô decomposition*. The following result says that the family of Gaussian random variables $(\phi(f), f \in \mathfrak{h})$ is unique up to unitary equivalence.

Proposition 1.50 (Uniqueness of Gaussian random variables). *Suppose that $(\phi(f), f \in \mathfrak{h})$ and $(\phi'(f), f \in \mathfrak{h})$ are Gaussian random variables indexed by a real vector space $\mathfrak{h}$ on $(\mathscr{Q}, \Sigma, \mu)$ and $(\mathscr{Q}', \Sigma', \mu')$, respectively. Then there exists a unitary operator $U : L^2(\mathscr{Q}) \to L^2(\mathscr{Q}')$ such that $U1 = 1$ and $U\phi(f) = \phi'(f)$.*

Proof. Define $U : L^2(\mathscr{Q}) \to L^2(\mathscr{Q}')$ by $U:\phi(f_1)\cdots\phi(f_n): = :\phi'(f_1)\cdots\phi'(f_n):$. By Proposition 1.49, U can be extended to a unitary operator. □

1.3.2 Wiener–Itô–Segal isomorphism and positivity improving

Define the *complexification* of the real Hilbert space $\mathfrak{h}$ by

$$\mathfrak{h}_{\mathbb{C}} = \left\{ \{f, g\} \mid f, g \in \mathfrak{h} \right\}$$

such that for $\lambda \in \mathbb{R}$,

$$\lambda\{f, g\} = \{\lambda f, \lambda g\}, \quad i\lambda\{f, g\} = \{-\lambda g, \lambda f\}, \quad \{f, g\} + \{f', g'\} = \{f + f', g + g'\}.$$

The scalar product on $\mathfrak{h}_{\mathbb{C}}$ is defined by

$$(\{f,g\},\{f',g'\})_{\mathfrak{h}_{\mathbb{C}}} = (f,f') + (g,g') + i((f,g') - (g,f')).$$

With these definitions, $(\mathfrak{h}_{\mathbb{C}}, (\cdot,\cdot)_{\mathfrak{h}_{\mathbb{C}}})$ is a Hilbert space over $\mathbb{C}$.

Proposition 1.51 (Wiener–Itô–Segal isomorphism). *There exists a unitary operator* $\theta_{\mathrm{W}} : \mathcal{F}_{\mathrm{b}}(\mathfrak{h}_{\mathbb{C}}) \to L^2(\mathscr{Q})$ *such that*
(1) $\theta_{\mathrm{W}}\Omega_{\mathrm{b}} = 1$;
(2) $\theta_{\mathrm{W}}\mathcal{F}_{\mathrm{b}}^{(n)}(\mathfrak{h}_{\mathbb{C}}) = L_n^2(\mathscr{Q})$;
(3) $\theta_{\mathrm{W}}\Phi(f)\theta_{\mathrm{W}}^{-1} = \phi(f)$, *where* $\Phi(f), f \in \mathfrak{h}$, *is a Segal field.*

Proof. Define $\theta_{\mathrm{W}} : \mathcal{F}_{\mathrm{b}}(\mathfrak{h}_{\mathbb{C}}) \to L^2(\mathscr{Q})$ by

$$\theta_{\mathrm{W}}{:}\prod_{i=1}^{n}\Phi(f_i){:}\Omega_{\mathrm{b}} = {:}\prod_{i=1}^{n}\phi(f_i){:}, \quad f_1,\dots,f_n \in \mathfrak{h},$$
$$\theta_{\mathrm{W}}\Omega_{\mathrm{b}} = 1.$$

Notice that $\|{:}\prod_{i=1}^{n}\Phi(f_i){:}\Omega_{\mathrm{b}}\|_{\mathcal{F}_{\mathrm{b}}(\mathfrak{h}_{\mathbb{C}})} = \|{:}\prod_{i=1}^{n}\phi(f_i){:}\|_{L^2(\mathscr{Q})}$ can be checked and subspace $\mathrm{L.H.}\{{:}\prod_{i=1}^{n}\Phi(f_i){:}\Omega_{\mathrm{b}}, \Omega_{\mathrm{b}} \mid f_i \in \mathfrak{h}, i = 1,\dots,n, n \geq 1\}$ is dense in the Fock space $\mathcal{F}_{\mathrm{b}}(\mathfrak{h}_{\mathbb{C}})$. It is straightforward to check that θ_{W} can be extended to a unitary operator from $\mathcal{F}_{\mathrm{b}}(\mathfrak{h}_{\mathbb{C}})$ to $L^2(\mathscr{Q})$ by Proposition 1.49, and the proposition follows. □

The unitary operator θ_{W} is called the *Wiener–Itô–Segal isomorphism*, and it allows identifying $\mathcal{F}_{\mathrm{b}}(\mathfrak{h}_{\mathbb{C}})$ with $L^2(\mathscr{Q})$.

Let $T : \mathfrak{h} \to \mathfrak{h}$ be a contraction operator. This T can be linearly extended to a contraction on $\mathfrak{h}_{\mathbb{C}}$. The operator $\theta_{\mathrm{W}}\Gamma(T)\theta_{\mathrm{W}}^{-1} : L^2(\mathscr{Q}) \to L^2(\mathscr{Q})$ is called the *second quantization* of T, and we will denote it by the same symbol $\Gamma(T)$. It is seen directly that $\Gamma(T)$ acts as

$$\Gamma(T){:}\prod_{i=1}^{n}\phi(f_i){:} = {:}\prod_{i=1}^{n}\phi(Tf_i){:}$$

and

$$\Gamma(T)1 = 1.$$

For a self-adjoint operator h, the *differential second quantization* $\theta_{\mathrm{W}}\mathrm{d}\Gamma(h)\theta_{\mathrm{W}}^{-1}$ will also be denoted by the same symbol $\mathrm{d}\Gamma(h)$. We have

$$\mathrm{d}\Gamma(h){:}\phi(f_1)\cdots\phi(f_n){:} = \sum_{j=1}^{n}{:}\phi(f_1)\cdots\phi(hf_j)\cdots\phi(f_n){:},$$

with

$$\mathrm{d}\Gamma(h)1 = 0$$

for $f_j \in D(h), j = 1,\dots,n$.

Using Wiener–Itô–Segal isomorphism allows identifying $\mathcal{F}(\mathfrak{h}_{\mathbb{C}})$ with $L^2(\mathcal{Q})$ and discussing the fundamental question of positivity of eigenfunctions. In particular, one expects the same positivity properties of ground states of self-adjoint operators acting in $L^2(\mathcal{Q})$ as of operators acting on $L^2(\mathbb{R}^d)$. Now we turn to discussing the relationship between positivity preserving operators and second quantization.

Let (M,μ) be a σ-finite measure space. Recall that a bounded operator A on $L^2(M,d\mu)$ is called a positivity preserving operator if $(f,Ag) \geq 0$ for all nonnegative $f,g \in L^2(M,d\mu)$. Operator A is also called a positivity improving operator if $(f,Ag) > 0$ for all nonnegative $f,g \in L^2(M,d\mu)$.

Proposition 1.52 (Positivity preserving). *Let T be a contraction operator on a real Hilbert space $\mathfrak{h}$. Then $\Gamma(T)$ is positivity preserving on $L^2(\mathcal{Q})$.*

Proof. Note that $\Gamma(T){:}\exp(\alpha\phi(f)){:} = {:}\exp(\alpha\phi(Tf)){:}$ for every $\alpha \in \mathbb{C}$. Thus

$$\Gamma(T)e^{\alpha\phi(f)} = {:}e^{\alpha\phi(Tf)}{:}e^{\frac{1}{4}\alpha^2\|f\|^2} = e^{\alpha\phi(Tf)}e^{\frac{1}{4}\alpha^2(f,(1-T^*T)f)}.$$

Hence for $F(\phi(f_1),\dots,\phi(f_n)) \in \mathcal{S}(\mathcal{Q})$,

$$\begin{aligned}&\Gamma(T)F(\phi(f_1),\dots,\phi(f_n))\\ &= \frac{1}{(2\pi)^{n/2}}\int_{\mathbb{R}^n}\hat{F}(k_1,\dots,k_n)e^{-\frac{1}{4}\sum_{i,j=1}^n(f_i,(1-T^*T)f_j)k_ik_j}e^{i\sum_{j=1}^n k_j\phi(Tf_j)}dk_1\cdots dk_n.\end{aligned} \tag{1.3.4}$$

Since $\|T\| \leq 1$, $\{(f_i,(1-T^*T)f_j)\}_{i,j}$ is positive semidefinite. The right-hand side of (1.3.4) is expressed by the convolution of F and the Gaussian kernel D_T, i. e.,

$$\Gamma(T)F(\phi(f_1),\dots,\phi(f_n)) = (2\pi)^{-n/2}(F * D_T)(\phi(Tf_1),\dots,\phi(Tf_n)).$$

Thus $F \geq 0$ implies that $\Gamma(T)F \geq 0$. Let $\Psi \in L^2(\mathcal{Q})$ be a nonnegative function. By Corollary 1.47, we can construct a sequence $(F_n)_{n\in\mathbb{N}} \subset \mathcal{S}(\mathcal{Q})$ such that $F_n \to \Psi$ as $n \to \infty$ in $L^2(\mathcal{Q})$ and $F_n \geq 0$. Hence $\Gamma(T)\Psi \geq 0$ follows by a limiting argument. □

Next we show a sufficient condition under which a positivity preserving operator $\Gamma(T)$ becomes a positivity improving operator. First, we give an abstract necessary and sufficient condition for a semigroup to be positivity improving.

Proposition 1.53 (Positivity improving semigroup). *Let (M,μ) be a σ-finite measure space, and A a self-adjoint operator on $L^2(M)$, bounded from below. Suppose that e^{-tA} is positivity preserving for all $t > 0$ and $E = \inf \mathrm{Spec}(A)$ is an eigenvalue. Then the following properties are equivalent:*

(1) *E is a simple eigenvalue with a strictly positive eigenvector.*
(2) *e^{-tA} is positivity improving for all $t > 0$.*

Proof. (1) → (2) Fix $t > 0$. Let $U = e^{-tA}$ and write $S = U/\|U\|$. Let $E_S(\lambda)$ be the spectral projection of S. We have

$$\begin{aligned}\lim_{n\to\infty}(f, S^n g) &= \lim_{n\to\infty}\int_{[0,1]}\lambda^n d(f, E_S(\lambda)g)\\ &= \lim_{n\to\infty}\int_{[0,1)}\lambda^n d(f, E_S(\lambda)g) + (f, E_S(\{1\})g) = (f, E_S(\{1\})g).\end{aligned}$$

Thus S^n weakly converges to the projection $E_S(\{1\})$ and s-$\lim_{n\to\infty} S^n = E_S(\{1\})$ follows. Note that $E_S(\{1\}) = (v,\cdot)v$, where v is the unique and strictly positive eigenvector of T such that $Tv = \|T\|v$. Let $f, g \in L^2(M)$ and $f, g \geq 0$. Note that $\{t > 0 \mid (f, e^{-tA}g) \geq 0\} = [0,\infty)$. Hence we see that $\lim_{n\to\infty}(f, S^n g) = (f, v)(v, g) > 0$, and $(f, S^n g) > 0$ for some $n > 0$. This implies that

$$B = \{t > 0 \mid (f, e^{-tA}g) > 0\} \neq \emptyset.$$

We see that $F(z) = (f, e^{-zA}g)$ is an analytic function in a neighborhood of the positive axis. This implies that $(0,\infty) \setminus B$ is a discrete set, in particular, B contains arbitrarily small positive numbers. Fix $s \in B$. We see that

$$(f, e^{-sA}g) = \int_M f(m)(e^{-sA}g)(m)d\mu > 0$$

and $f(m)(e^{-sA}g)(m)$ is not identically zero. Define the function

$$\eta(m) = \min\{f(m), (e^{-sA}g)(m)\},$$

and note that it is not identically zero. Hence for arbitrary $t > 0$, we have

$$(f, e^{-(t+s)A}g) = (f, e^{-tA}(e^{-sA}g)) \geq (f, e^{-tA}\eta) = (e^{-tA}f, \eta) \geq (e^{-tA}\eta, \eta) > 0.$$

This implies that $e^{-(t+s)A}$ is positivity improving for arbitrary $t > 0$. Hence e^{-tA} is positivity improving for any $t \geq s$, and we can chose arbitrary small positive s. This implies that e^{-tA} is positivity improving for all $t > 0$.

(2) → (1) Fix $t > 0$ and write $U = e^{-tA}$. We prove that there is no nontrivial closed subspace invariant under U or any other bounded multiplication operator on $L^2(M)$. Suppose that $S \subset L^2(M)$ is a nontrivial closed subspace invariant under U and every bounded multiplication operator. Let $f \in S$ and write $h = \bar{f}/|f|$. The function h defines a bounded multiplication operator. Since S is invariant under every bounded multiplication operator, $|f| = hf \in S$. Similarly, we have $|g| \in S^\perp$ for $g \in S^\perp$. Let $f \in S$ and $g \in S^\perp$ be such that f and g are not identically zero. Since U leaves S invariant, $U|f| \in S$. Hence we have $(|g|, U|f|) = 0$, which is in contradiction to the positivity improving property of U. This implies that there is no nontrivial closed subspace invariant under U and any

bounded multiplication operator. By assumption, $\|U\|$ is an eigenvalue. Let f be an eigenvector of U associated with $\|U\|$ and suppose that f is real-valued. Since $|f| \pm f \geq 0$ and U is positivity improving, it follows that either $U(|f| \pm f) > 0$ or identically $U(|f| \pm f) = 0$. Thus either $U|f| > \pm Uf$ or identically $U|f| = \pm Uf$. We have

$$U|f| > |Uf| \quad \text{or} \quad U|f| = |Uf|, \tag{1.3.5}$$

however,

$$\|U\|\|f\|^2 \geq (|f|, U|f|) \geq (|f|, |Uf|) \geq (f, Uf) = \|U\|\|f\|^2.$$

Hence

$$0 = (|f|, U|f| - |Uf|) = \int_M |f(m)||(U|f| - |Uf|)(m)| d\mu.$$

From (1.3.5), it follows that $U|f| = |Uf|$, which implies that $U|f| = \|U\||f|$. In particular, $|f|$ is also an eigenvector of U. Let $S = \{g \in L^2(M) \mid gf = 0 \text{ a. e.}\}$. We see that S is a subspace of $L^2(M)$ invariant under every bounded multiplication operator. We show that $US \subset S$. Let $S_+ = \{g \in S \mid g \geq 0\}$. For $g \in S_+$, we have $(Ug, |f|) = (g, U|f|) = \|U\|(g, |f|) = 0$, and hence $Ug \in S_+$, i. e., U leaves S_+ invariant. Since $S = S_+ - S_+ + i(S_+ - S_+)$, U leaves S invariant. Thus S is invariant under U and every bounded multiplication operator, and thus $S = M$ or $S = \{0\}$, but $f \notin S$. We have $S = \{0\}$ and $f \neq 0$ a. e., which implies that $|f| > 0$ a. e. As a result, it follows that every real-valued eigenvector with eigenvalue $\|U\|$ is strictly positive and $U|f| = \|U\||f|$. Thus $|f| - f$ is a nonnegative eigenvector such that $U(|f| - f) = \|U\|(|f| - f)$. Since U is positivity improving, either $|f| - f > 0$ a. e. or $|f| - f = 0$ a. e. This means that f is either strictly positive or strictly negative. Let f and g be two real-valued eigenvectors such that $Uf = \|U\|f$ and $Ug = \|U\|g$. It is impossible to choose f and g such that $(f, g) = 0$. We see that a real-valued eigenvector is unique up to multiple constants. Now let $f = \Re f + i\Im f$ be an arbitrary eigenvector with eigenvalue $\|U\|$. Since $\Re(Uf) = U\Re f$, $\Re f$ and $\Im f$ are eigenvectors of U. The uniqueness of real-valued eigenvectors implies that $\Re f = a\Im f$ with some $a \in \mathbb{R}$. Hence f is the complex multiple of the unique real-valued eigenvector, i. e., $f = (1 + ia)\Re f$. Thus $g = f/(1 + ia)$ is the strictly positive eigenvectors with $Ug = \|U\|g$. □

Corollary 1.54 (Positivity improving). *Let* $\mathfrak{h}_{\mathbb{C}} = L^2(\mathbb{R}^d)$ *and* $f : \mathbb{R}^d \to \mathbb{R}$ *be a nonnegative multiplication operator on* $L^2(\mathbb{R}^d)$ *such that the Lebesgue measure of the set* $\{k \in \mathbb{R}^d \mid f(k) = 0\}$ *is zero. Let* $d\Gamma(f)$ *be the differential second quantization on* $L^2(\mathscr{Q})$. *Then the semigroup* $\{e^{-td\Gamma(f)} : t > 0\}$ *is positivity improving.*

Proof. The semigroup $e^{-td\Gamma(f)}$ is positivity preserving and 1 is a simple eigenvector of $d\Gamma(f)$ with eigenvalue zero, since $\mathrm{Spec_p}(f) \ni 0$. Hence the corollary follows from Proposition 1.53. □

Finally, we show in an abstract setting an expression of the ground state eigenvalue of a self-adjoint operator K acting in $L^2(M)$. Let

$$\kappa = \inf \mathrm{Spec}(K)$$

and suppose that K has a ground state ϕ, i. e., $K\phi = \kappa\phi$, and

$$(\phi, f)_{L^2(M)} \neq 0. \tag{1.3.6}$$

By spectral theory, we can directly see that

$$\kappa = -\lim_{t\to\infty} \frac{1}{t} \log(f, e^{-tK} f). \tag{1.3.7}$$

Lemma 1.55. *Let $(M, \mathscr{F}, \mu)$ be a measure space, and K a self-adjoint operator on $L^2(M)$, bounded from below. Suppose that K has a unique strictly positive ground state. We suppose that $f \in L^2(M)$ is a nonnegative function, not identically zero. Then* (1.3.7) *holds.*

Proof. Let E denote the spectral projection of K. Write $\nu_f(\cdot) = (f, E(\cdot)f)$. We have

$$(f, e^{-tK} f) = e^{-t\kappa}\Big(\int_{(\kappa,\infty)} e^{-t(\lambda-\kappa)} d\nu_f(\lambda) + \nu_f(\{\kappa\})\Big).$$

Note that $\int_{(\kappa,\infty)} e^{-t(\lambda-\kappa)} d\nu_f(\lambda) + \nu_f(\{\kappa\}) \to \nu_f(\{\kappa\}) > 0$ as $t \to \infty$. This implies

$$-\frac{1}{t}\log(f, e^{-tK} f) = \kappa - \frac{1}{t}\log\Big(\int_{(\kappa,\infty)} e^{-t(\lambda-\kappa)} d\nu_f(\lambda) + \nu_f(\{\kappa\})\Big) \to \kappa$$

as $t \to \infty$. □

As seen in Proposition 1.53, if e^{-tK} is positivity improving and K has a ground state, then the ground state is unique and strictly positive. Hence (1.3.7) follows for any nonnegative f, e. g., $f(x) = 1_{\{|x|\le 1\}}$. However, we can prove (1.3.7) under weaker conditions without assuming the existence of any ground state.

Lemma 1.56 (Ground state energy formula). *Let M be a metric space, $(M, \mathscr{F}, \mu)$ a σ-finite measure space, and K a self-adjoint operator on $L^2(M)$, bounded from below. Suppose that e^{-tK} is positivity preserving for $t \ge 0$, and $f \in L^2(M)$ satisfies $\inf_{m\in D} f(m) > 0$ for every compact set $D \subset M$. Then* (1.3.7) *holds, in particular, if $\mu(M) < \infty$, then*

$$\kappa = -\lim_{t\to\infty} \frac{1}{t} \log(1, e^{-tK} 1). \tag{1.3.8}$$

Proof. Denote by E the spectral projection of the self-adjoint operator K, and let E_f be the infimum of the support of measure $\nu_f(\cdot) = (f, E(\cdot)f)$. We have

$$
\begin{aligned}
-\frac{1}{t}\log(f, e^{-tK}f) &= -\frac{1}{t}\log\int_{E_f}^{\infty} e^{-\lambda t} d\nu_f(\lambda) \\
&= -\frac{1}{t}\log e^{-E_f t}\int_{E_f}^{\infty} e^{-(\lambda-E_f)t} d\nu_f(\lambda) = E_f - \frac{1}{t}\log\int_{E_f}^{\infty} e^{-(\lambda-E_f)t} d\nu_f(\lambda).
\end{aligned}
$$

Note that $\nu_f([E_f, E_f+\varepsilon)) \neq 0$ for every $\varepsilon > 0$ by the definition of E_f. We have

$$
\frac{1}{t}\log\int_{E_f}^{\infty} e^{-(\lambda-E_f)t} d\nu_f(\lambda) \leq \frac{1}{t}\log\left(\nu_f([E_f, E_f+\varepsilon)) + e^{-\varepsilon t}\right) \to 0,
$$

since $\nu_f([E_f, E_f+\varepsilon)) + e^{-\varepsilon t} \to \nu_f([E_f, E_f+\varepsilon)) \neq 0$. Thus

$$
-\lim_{t\to\infty}\frac{1}{t}\log(f, e^{-tK}f) = E_f.
$$

It remains to prove $\kappa = E_f$. Let $\mathscr{O} = \{g \in L^2(M) \mid \operatorname{supp} g \subset \bigcup_{N>0} B_N\}$, where B_N is a ball of radius N centered at the origin. Notice that $\mathscr{O}$ is dense in $L^2(M)$. We show that

$$
\inf_{g\in\mathscr{O}} E_g = \kappa. \tag{1.3.9}
$$

Since trivially $\inf_{g\in\mathscr{O}} E_g \geq \kappa$, it suffices to prove that $\inf_{g\in\mathscr{O}} E_g \leq \kappa$. Suppose, to the contrary, that $\inf_{g\in\mathscr{O}} E_g > \kappa$. Then there exists $\varepsilon > 0$ such that $E([\kappa, \kappa+\varepsilon))L^2(M) \neq \emptyset$. Take a nonzero $h \in E([\kappa, \kappa+\varepsilon))L^2(M)$. Thus for every $g \in \mathscr{O}$,

$$
(g, h) = (g, E([\kappa, \kappa+\varepsilon))h) = (E([\kappa, \kappa+\varepsilon))g, h) = (0, h) = 0
$$

holds. This is in contradiction to the denseness of $\mathscr{O}$, and hence (1.3.9) holds. Let $g \in \mathscr{O}$. We have $|g| \leq Cf$ on M, where

$$
C = \frac{\operatorname{ess\,sup}_{m\in M}|g(m)|}{\operatorname{ess\,inf}_{m\in \operatorname{supp} g} f(m)} < \infty,
$$

and notice that $\operatorname{ess\,inf}_{m\in\operatorname{supp} g} f(m) > 0$ since $\operatorname{supp} g$ is compact. Since e^{-tK} is positivity preserving, we see that

$$
(g, e^{-tK}g) \leq (|g|, e^{-tK}|g|) \leq C^2(f, e^{-tK}f),
$$

and hence

$$
-\lim_{t\to 0}\frac{1}{t}\log(g, e^{-tK}g) \geq -\lim_{t\to 0}\frac{1}{t}\log C^2(f, e^{-tK}f).
$$

This means that $E_g \geq E_f$ for every $g \in \mathscr{O}$, and $\kappa = \inf_{g\in\mathscr{O}} E_g \geq E_f$. Since $E_f \geq \kappa$ trivially holds, $E_f = \kappa$ and the lemma follows. □

1.3.3 Hypercontractivity

Let Δ be the d-dimensional Laplacian. It is known that $e^{t\Delta}$, $t > 0$, is a contraction from $L^p(\mathbb{R}^d)$ to $L^q(\mathbb{R}^d)$ for $1 \le p \le q \le \infty$. In this section we discuss the hypercontractivity of $\Gamma(T)$, as a counterpart of the L^p–L^q bound of $e^{t\Delta}$.

Let T be a contraction operator on a real Hilbert space $\mathfrak{h}$. It was shown that $\Gamma(T) : L^2(\mathscr{Q}) \to L^2(\mathscr{Q})$ is positivity preserving and $\Gamma(T)1 = \Gamma(T)^*1 = 1$. Operators with this property are called doubly Markovian. In the definition below, we use a more abstract setting.

Definition 1.57 (Doubly Markovian operator). Let $(M, \mathcal{B}_M, P)$ be a probability space. A bounded operator S on $L^2(M)$ is *doubly Markovian* if

(1) S is positivity preserving;
(2) $S1 = S^*1 = 1$.

Lemma 1.58. *Suppose that S is a doubly Markovian operator on $L^2(M)$. Then its adjoint S^* is also doubly Markovian on it.*

Proof. Saying that S is positivity preserving is equivalent to $(f, Sg) \ge 0$ for all positive $f, g \in L^2(M)$. Thus if S is positivity preserving, so is S^*. Hence S^* is also doubly Markovian. □

Every doubly Markovian operator has the following property.

Theorem 1.59 (L^p-contraction). *Let $(M, \mathcal{B}_M, P)$ be a probability space, and suppose that S is doubly Markovian on $L^2(M)$. Then S is a contraction on $L^p(M)$ for every $1 \le p \le \infty$, i. e.,*

$$\|Sf\|_p \le \|f\|_p$$

for $f \in L^2(M) \cap L^p(M)$.

Proof. Let $f \in L^1(M) \cap L^2(M)$ and $f \ge 0$. We have

$$\|Sf\|_{L^1} = (1, |Sf|)_{L^2} = (1, Sf)_{L^2} = (S^*1, f)_{L^2} = (1, f)_{L^2} = \|f\|_{L^1}.$$

For general $f \in L^1(M) \cap L^2(M)$, we have

$$\|Sf\|_{L^1} = (1, |Sf|)_{L^2} \le (1, S|f|)_{L^2} = \|S|f|\|_{L^1} = \|f\|_{L^1}.$$

Hence S is a contraction on $L^1(M)$. By Lemma 1.58, S^* is also doubly Markovian and thus a contraction on $L^1(M)$, i. e., $\|S^*f\|_1 \le \|f\|_1$ for $f \in L^1(M) \cap L^2(M)$. By duality, $\|Sf\|_\infty \le \|f\|_\infty$. The theorem follows by the Riesz–Thorin interpolation theorem. □

Now we investigate the second quantization $\Gamma(T)$ of a contraction T. We denote the norm on the Banach space $L^p(\mathscr{Q})$ by $\|\cdot\|_p$ for $1 \le p \le \infty$.

Lemma 1.60. *Let T be a contraction. Then $\Gamma(T)$ is doubly Markovian and a contraction on $L^p(\mathscr{Q})$, for $1 \le p \le \infty$.*

Proof. We have already seen that $\Gamma(T)$ is doubly Markovian. Hence $\Gamma(T)$ is a contraction on $L^p(\mathscr{Q})$ by Theorem 1.59. □

Lemma 1.61. *Let $\Phi \in L_n^2(\mathscr{Q})$. Then it follows that*

$$\|\Phi\|_4 \le 2^n \|\Phi\|_2.$$

Proof. Let $\mathbb{Z}_+^n$ denote the set of sequences of length n of nonnegative integers. Let $\Phi \in L_n^2(\mathscr{Q})$ and $\{e_i\}_{i\in\mathbb{N}}$ be an orthogonal system in $\mathfrak{h}$ such that $(e_i, e_j)_2 = 2\delta_{ij}$. Since

$$\Big\{:\prod_{l=1}^n \phi(e_{i_l}): \mid i_1 \le i_2 \le \cdots \le i_n\Big\}$$

form the orthogonal system in $L_n^2(\mathscr{Q})$, we have

$$\Phi = \sum_{i=(i_1,\ldots,i_n)\in\mathbb{Z}_+^n} a_i :\prod_{l=1}^n \phi(e_{i_l}):$$

and

$$\Big(:\prod_{l=1}^n \phi(e_{i_l}):, :\prod_{l=1}^n \phi(e_{j_l}):\Big)_2 = \sum_{\pi\in\wp_n} \delta_{i_1 j_{\pi(1)}} \cdots \delta_{i_n j_{\pi(n)}}.$$

Thus we see that

$$\begin{aligned}\|\Phi\|_2^2 &= \sum_{i,j\in\mathbb{Z}_+^n} \bar{a}_i a_j \Big(:\prod_{l=1}^n \phi(e_{i_l}):, :\prod_{l=1}^n \phi(e_{j_l}):\Big)_2 \\ &= \sum_{\pi\in\wp_n}\sum_{i\in\mathbb{Z}_+^n} \bar{a}_i a_{\pi(i)} = n! \sum_{i\in\mathbb{Z}_+^n} |a_i|^2. \end{aligned} \tag{1.3.10}$$

Here we used that $a_i = a_{\pi(i)}$ for every $\pi \in \wp_n$, where $\pi(i) = (i_{\pi(1)}, \ldots, i_{\pi(n)})$. Let P_n be the set of pairings of $4n$ distinct objects $i = \{i_1, \ldots, i_n\}$, $j = \{j_1, \ldots, j_n\}$, $k = \{k_1, \ldots, k_n\}$, $l = \{l_1, \ldots, l_n\} \in \mathbb{Z}_+^n$ in such a way that no two elements of i, no two elements of j, no two elements of k, and no two elements of l are paired together. In other words, let $P_n = \{X_1, \ldots, X_N\}$ with $X_m = \{\{a_1^m, b_1^m\}, \ldots, \{a_{2n}^m, b_{2n}^m\}\}$, $m = 1, \ldots, N$, where N denotes the number of elements of P_n, and $\{a_p^m, b_p^m\} \not\subset i$, $\{a_p^m, b_p^m\} \not\subset j$, $\{a_p^m, b_p^m\} \not\subset k$ and $\{a_p^m, b_p^m\} \not\subset l$ hold for all $1 \le m \le N$ and $1 \le p \le 2n$. Let $|P_n|$ denote the number of elements in P_n. We see that

$$\|\Phi\|_4^4 = \sum_{i,j,k,l\in\mathbb{Z}_+^n} \bar{a}_i \bar{a}_j a_k a_l \int_{\mathscr{Q}} :\prod_{l=1}^n \phi(e_{i_l})::\prod_{l=1}^n \phi(e_{j_l})::\prod_{l=1}^n \phi(e_{k_l})::\prod_{l=1}^n \phi(e_{m_l}): d\mu$$

is a sum of $|P_n|$ terms each of which is $\delta_{p_1q_1}\cdots\delta_{p_{2n}q_{2n}}$ with $\{\{p_1,q_1\},\dots,\{p_{2n},q_{2n}\}\}\in P_n$. After computations, we finally obtain

$$\|\Phi\|_4^4 \le |P_n|\Big(\sum_{i\in\mathbb{Z}_+^n}|a_i|^2\Big)^2.$$

From this and (1.3.10), we have

$$\|\Phi\|_4^4 \le \frac{|P_n|}{(n!)^2}\|\Phi\|_2^4.$$

Here $|P_n|$ is less than the number of ways of pairing $4n$ objects into $2n$ pairs without any restriction. Hence $|P_n| \le (4n)!/(2n)!2^{2n}$. Thus

$$\frac{|P_n|}{(n!)^2} \le \frac{1}{2^{2n}}\binom{4n}{2n}\binom{2n}{n} \le \frac{1}{2^{2n}}2^{4n}2^{2n} \le 2^{4n}.$$

Thus we have $\|\Phi\|_4^4 \le 2^{4n}\|\Phi\|_2^4$. □

Lemma 1.62. *Let $2\le p\le 4$ and $\Phi\in L_n^2(\mathscr{Q})$. Then*

$$\|\Phi\|_p \le (4^{1-\frac{2}{p}})^n\|\Phi\|_2.$$

Proof. By the Hölder inequality, we have

$$\|\Phi\|_p^p \le \|\Phi\|_4^{2p-4}\|\Phi\|_2^{4-p}.$$

Thus the claim follows from Lemma 1.61. □

Lemma 1.63. *Let $2 < p \le 4$ and $0 < c < \frac{1}{4^{1-\frac{2}{p}}}$. Then*

$$\|\Gamma(c)\Phi\|_p \le \frac{1}{1-c4^{1-\frac{2}{p}}}\|\Phi\|_2, \quad \Phi\in L^2(\mathscr{Q}).$$

Proof. Let $\Phi = \bigoplus_{n=0}^\infty \Phi^{(n)} \in \bigoplus_{n=0}^\infty L_n^2(\mathscr{Q})$. Then we have

$$\begin{aligned}\|\Gamma(c)\Phi\|_p &\le \sum_{n=0}^\infty \|\Gamma(c)\Phi^{(n)}\|_p \le \sum_{n=0}^\infty \|\Gamma(c)\Phi^{(n)}\|_2(4^{1-\frac{2}{p}})^n\\ &\le \sum_{n=0}^\infty (c4^{1-\frac{2}{p}})^n\|\Phi^{(n)}\|_2 \le \|\Phi\|_2\sum_{n=0}^\infty (c4^{1-\frac{2}{p}})^n.\end{aligned}$$ □

We can make this lemma sharper by using the identities $L^p(\mathscr{Q}\oplus\mathscr{Q}) \cong L^p(\mathscr{Q})\otimes L^p(\mathscr{Q})$ and $\Gamma(c\oplus c)\cong\Gamma(c)\otimes\Gamma(c)$.

Lemma 1.64. *Let* $2 < p \le 4$ *and* $0 < c < \frac{1}{4^{1-\frac{2}{p}}}$. *Then*

$$\|\Gamma(c)\Phi\|_p \le \|\Phi\|_2, \quad \Phi \in L^2(\mathscr{Q}).$$

Proof. By Lemma 1.63, we have $\|\Gamma(c)\Phi\|_p \le d\|\Phi\|_2$ with some d. Since $\|c \oplus c\| \le \frac{1}{4^{1-\frac{2}{p}}}$, we have

$$\|\Gamma(c \oplus c)\Psi\|_{L^p(\mathscr{Q}\oplus\mathscr{Q})} \le d\|\Psi\|_{L^2(\mathscr{Q}\oplus\mathscr{Q})}$$

for $\Psi \in L^2(\mathscr{Q} \oplus \mathscr{Q})$. On the other hand, for $\Psi = \Phi \otimes \Phi$ we have

$$\|\Gamma(c \oplus c)\Psi\|_{L^p(\mathscr{Q}\oplus\mathscr{Q})} = \|\Gamma(c)\Phi \otimes \Gamma(c)\Phi\|_{L^p(\mathscr{Q})\otimes L^p(\mathscr{Q})} = \|\Gamma(c)\Phi\|_p^2$$

and $\|\Psi\|_{L^2(\mathscr{Q}\oplus\mathscr{Q})} = \|\Phi\|_2^2$. Combining these expressions, we see that $\|\Gamma(c)\Phi\|_p^2 \le d\|\Phi\|_2^2$ and

$$\|\Gamma(c)\Phi\|_p \le \sqrt{d}\|\Phi\|_2.$$

Repeating this procedure, it follows that

$$\|\Gamma(c)\Phi\|_p \le d^{1/2n}\|\Phi\|_2$$

for every $n \ge 1$, and the lemma follows. □

Theorem 1.65 (Hypercontractivity). *Let* $2 < p \le 4$. *Suppose that* T *is a contraction such that* $\|T\| \le 4^{\frac{2}{p}-1}$. *Then*

$$\|\Gamma(T)\Phi\|_p \le \|\Phi\|_2, \quad \Phi \in L^2(\mathscr{Q}).$$

Proof. We have $\Gamma(T) = \Gamma(T/\|T\|)\Gamma(\|T\|)$. Since $\Gamma(T/\|T\|)$ is a contraction on L^p by Lemma 1.60 and $\Gamma(\|T\|)$ is a contraction from L^2 to L^p by Lemma 1.64, the theorem follows. □

Example 1.66. Let $2 < p \le 4$. Suppose that $e^{-t} \le 4^{\frac{2}{p}-1}$. Then

$$\|e^{-tN}\Phi\|_p \le \|\Phi\|_2.$$

In particular, $e^{-tN}\Phi \in \bigcap_{2\le p\le 4} L^p(\mathscr{Q})$ for $t > \log 4$.

A more refined result is available, which we present together with a concise proof.

Theorem 1.67 (Hypercontractivity). *Let* $1 < p \le q < \infty$ *and*

$$\|T\| \le \sqrt{(p-1)/(q-1)}.$$

Then $\Gamma(T)$ *is a contraction from* L^p *to* L^q, *i. e.,*

$$\|\Gamma(T)\Phi\|_q \le \|\Phi\|_p, \quad \Phi \in L^p(\mathscr{Q}).$$

Proof. We give an outline of the proof. Let $h \in \mathfrak{h}$ and consider the Wick exponential $\Phi_h = e^{\phi(h)-\frac{1}{2}\|h\|^2}$. It can be seen that

$$\|\Phi_h\|_p = \exp\Big(\frac{p-1}{2}\|h\|^2\Big).$$

Since $\Gamma(T)$ acts on Wick exponentials by $\Gamma(T)\Phi_h = e^{\phi(Th)-\frac{1}{2}\|Th\|^2}$, we have

$$\|\Gamma(T)\Phi_h\|_q = \exp\Big(\frac{q-1}{2}\|Th\|^2\Big).$$

Suppose that $\|T\| \le \sqrt{(p-1)/(q-1)}$, we obtain

$$\frac{q-1}{2}\|Th\|^2 \le \frac{q-1}{2}\|T\|^2\|h\|^2 \le \frac{p-1}{2}\|h\|^2.$$

Hence

$$\|\Gamma(T)\Phi_h\|_q \le \exp\Big(\frac{p-1}{2}\|h\|^2\Big) = \|\Phi_h\|_p.$$

This estimate can be extend from L^p to L^q. □

1.3.4 Lorentz covariant quantum fields

Following the conventions in physics, a quantum field is required to have the so-called Lorentz covariance property, i. e., to transform under the Lorentz group. The measure $\delta(k_0^2 - k^2 - \nu^2)d\mathbf{k}$, $\mathbf{k} = (k_0, k) \in \mathbb{R} \times \mathbb{R}^3$, on the 4-dimensional space-time $\mathbb{R}^4$ supported on the light cone with mass ν is Lorentz-covariant and

$$\int_{\mathbb{R}^3} \frac{1}{2\omega_\nu(k)}\frac{dk}{(2\pi)^3} = \int_{\mathbb{R}^4} (2\pi)\delta(k_0^2 - |k|^2 - \nu^2)\Gamma_{\{k_0>0\}}\frac{d\mathbf{k}}{(2\pi)^4}$$

is a Lorentz invariant integral. After normalization, the Lorentz covariant scalar field is formally given by

$$\phi(t,x) = \int_{\mathbb{R}^3} \frac{1}{\sqrt{2\omega_\nu(k)}}\left(a(k)e^{-it\omega_\nu(k)+ik\cdot x} + a^*(k)e^{it\omega_\nu(k)-ik\cdot x}\right)dk.$$

Instead of Segal fields, the conventional *time-zero scalar field* in $\mathcal{F}_b(L^2(\mathbb{R}^3))$ is obtained,

$$\frac{1}{\sqrt{2}}(a^*(\hat{f}/\sqrt{\omega_\nu}) + a(\tilde{\hat{f}}/\sqrt{\omega_\nu})). \tag{1.3.11}$$

Here $\tilde{\hat{f}}(k) = \hat{f}(-k)$. Note that $\tilde{\hat{f}} = \bar{\hat{f}}$ for real f, and (1.3.11) is linear in f. In Chapter 2, we will discuss the Nelson model of a scalar quantum field given by (1.3.11).

Note that (1.3.11) can be generalized by using a covariance operator C. Suppose that $(\phi(f), f \in \mathfrak{h})$ is a family of Gaussian random variables on $(\mathscr{Q}, \Sigma, \mu)$. Let $C : \mathfrak{h} \to \mathfrak{h}$ be self-adjoint and $\ker C = \{0\}$. Define $\mathfrak{h}'$ by the closure of $D(C)$ with respect to the norm $\|\cdot\|_{\mathfrak{h}'} = \|C\cdot\|_{\mathfrak{h}}$, i. e., $\mathfrak{h}' = \overline{D(C)}^{\|C\cdot\|_{\mathfrak{h}}}$ and $(f,g)_{\mathfrak{h}'} = (Cf, Cg)_{\mathfrak{h}}$. Then a quantum field $\phi_C(f)$ is defined by $\phi_C(f) = \phi(Cf)$ for $f \in D(C)$, and

$$(\phi(Cf), \phi(Cg))_{L^2(\mathscr{Q})} = \frac{1}{2}(Cf, Cg)_{\mathfrak{h}}, \quad f, g \in D(C).$$

Let $f \in \mathfrak{h}'$. Then there exists a sequence $f_n \in D(C)$ such that $\|f_n - f\|_{\mathfrak{h}'} \to 0$ as $n \to \infty$. Thus we can see that $\|\phi_C(f_n) - \phi_C(f_m)\|_{L^2(\mathscr{Q})} = \|f_n - f_m\|_{\mathfrak{h}'} \to 0$ as $n, m \to \infty$ and $(\phi_C(f_n))_{n\in\mathbb{N}}$ is a Cauchy sequence in $L^2(\mathscr{Q})$. Define $\phi_C(f)$ for $f \in \mathfrak{h}'$ by the limit $\text{s-lim}_{n\to\infty} \phi_C(f_n)$. For $\phi_C(f)$ the Wick product can be defined in the same way as for $\phi(f)$, i. e.,

$$\begin{aligned} &:\phi_C(f): = \phi_C(f), \\ &:\phi_C(f) \prod_{i=1}^{n} \phi_C(f_i): = \phi_C(f) :\prod_{i=1}^{n} \phi_C(f_i): - \frac{1}{2} \sum_{j=1}^{n} (f, f_j)_{\mathfrak{h}'} :\prod_{i\neq j} \phi_C(f_i):. \end{aligned}$$

The Wiener–Itô–Segal isomorphism $\theta_W : L^2(\mathscr{Q}, d\mu) \to \mathscr{F}_b(\mathfrak{h}'_C)$ acts as

$$\theta_W :\prod_{i=1}^{n} \phi_C(f_i): = a^*(f_1) \cdots a^*(f_n)\Omega_b, \quad \theta_W 1 = \Omega_b.$$

Here a^* and a are the creation and annihilation operators in $\mathscr{F}_b(\mathfrak{h}'_C)$, respectively, satisfying the canonical commutation relations

$$[a(f), a^*(g)] = (\bar{f}, g)_{\mathfrak{h}'}, \quad [a(f), a(g)] = 0 = [a^*(f), a^*(g)].$$

In particular, $[a(f), a^*(g)] = (\overline{Cf}, Cg)_{\mathfrak{h}}$ for $f, g \in D(C)$. The map θ_W can be extended to a unitary operator between $L^2(\mathscr{Q})$ and $\mathscr{F}_b(\mathfrak{h}'_C)$ by linearity and using the fact that L. H. $\{\prod_{j=1}^{n} a^*(f_j)\Omega_b, \Omega_b \mid f_j \in D(C), j = 1, \ldots, n, n \in \mathbb{N}\}$ is dense in $\mathscr{F}_b(\mathfrak{h}'_C)$. Thus the unitary equivalence $L^2(\mathscr{Q}) \cong \mathscr{F}_b(\mathfrak{h}')$ is implemented by θ_W.

1.4 Existence of $\mathscr{Q}$-spaces

1.4.1 Countable product spaces

In the previous section we have seen that a $\mathscr{Q}$-space is unique up to unitary equivalence. On the other hand, the $\mathscr{Q}$-space is not canonically given. In this section, we consider explicit choices of a measure space $(\mathscr{Q}, \Sigma, \mu)$ on which we will be able to construct a family of Gaussian random variables indexed by a real Hilbert space $\mathfrak{h}$.

Theorem 1.68. *Let $\mathfrak{h}$ be a real Hilbert space. Then there exists a probability space $(\mathscr{Q}, \Sigma, \mu)$ and a family of Gaussian random variables $(\phi(f), f \in \mathfrak{h})$ with mean zero and covariance*

$$\int_{\mathscr{Q}} \phi(f)\phi(g)\mu(d\phi) = \frac{1}{2}(f, g)_{\mathfrak{h}}.$$

Proof. We show that $\mathscr{Q}$ can be obtained as a countable product space and, roughly, the measure μ may be regarded as the infinite product of Gaussian distributions

$$d\mu = \prod_{i=1}^{\infty} \pi^{-1/2} e^{-x_i^2} dx_i \quad \text{on} \quad \prod_{i=1}^{\infty} \mathbb{R}.$$

We establish this next.

Let $\dot{\mathbb{R}} = \mathbb{R} \cup \{\infty\}$ be the one-point compactification of $\mathbb{R}$, and put $\mathscr{Q} = \times_{n=1}^{\infty} \dot{\mathbb{R}}$. By the Tychonoff theorem, $\mathscr{Q}$ is a compact Hausdorff space. Let $C(\mathscr{Q})$ be the set of continuous functions on $\mathscr{Q}$, and $C_{\mathrm{fin}}(\mathscr{Q}) \subset C(\mathscr{Q})$ the set of continuous functions on $\mathscr{Q}$ depending on a finite number of variables. For $F = F(x_1, \ldots, x_n) \in C_{\mathrm{fin}}(\mathscr{Q})$, define $\ell : C_{\mathrm{fin}}(\mathscr{Q}) \to \mathbb{C}$ by

$$\ell(F) = \frac{1}{\pi^{n/2}} \int_{\mathbb{R}^n} F(x_1, \ldots, x_n) e^{-\sum_{i=1}^n x_i^2} dx_1 \cdots dx_n.$$

Since $|\ell(F)| \le \|F\|_\infty$, ℓ is a bounded linear functional on $C_{\mathrm{fin}}(\mathscr{Q})$. On the other hand, $C_{\mathrm{fin}}(\mathscr{Q})$ is a dense subset in $C(\mathscr{Q})$ by the Stone–Weierstrass theorem, thus ℓ can be uniquely extended to a bounded linear functional $\tilde{\ell}$ on the Banach space $(C(\mathscr{Q}), \|\cdot\|_\infty)$. It is clear that $\tilde{\ell}(F) \ge 0$, for $F \ge 0$. The Riesz–Markov theorem furthermore implies that there exists a probability space $(\mathscr{Q}, \Sigma, \mu)$ such that $\tilde{\ell}(F) = \int_{\mathscr{Q}} F(q) d\mu(q)$. Let $\{e_n\}_{n=1}^{\infty}$ be a complete orthonormal system of a real Hilbert space $\mathfrak{h}$. For each e_n, we define $e_n(q) = x_n$ for $q = \{x_k\}_{k=1}^{\infty} \in \mathscr{Q}$. Note that

$$\int_{\mathscr{Q}} |e_n(q)|^2 d\mu(q) = \tilde{\ell}(e_n^2) = \frac{1}{\pi^{1/2}} \int_{\mathbb{R}} x^2 e^{-x^2} dx = \frac{1}{2}.$$

Since $C(\mathscr{Q})$ is dense in $L^2(\mathscr{Q})$ in the $\|\cdot\|_{L^2(\mathscr{Q})}$ norm, $C_{\mathrm{fin}}(\mathscr{Q})$ is dense in $L^2(\mathscr{Q})$. By Theorem 1.46, the minimal σ-field generated by $\{e_n(\cdot)\}_{n=1}^{\infty}$ is Σ. On expanding $f \in \mathfrak{h}$ as $f = \sum_{n=1}^{\infty} a_n e_n$ and writing $\phi_n(f) = \sum_{k=1}^{n} a_k e_k(\cdot)$, it is seen that

$$\|\phi_m(f) - \phi_n(f)\|_{L^2(\mathscr{Q})}^2 = \frac{1}{2} \sum_{k=m+1}^{n} a_k^2 \to 0, \quad m, n \to \infty.$$

Thus $\phi(f) = \text{s-lim}_{n\to\infty} \phi_n(f)$ exists in $L^2(\mathscr{Q})$ and $\int_{\mathscr{Q}} e^{i\phi(f)} d\mu = e^{-(1/4)\|f\|^2}$ easily follows. From this, we obtain $\int_{\mathscr{Q}} \phi(f) d\mu = 0$ and

$$\int_{\mathscr{Q}} \phi(f)\phi(g) d\mu = \frac{1}{2}(f, g),$$

and thereby we have constructed the desired family of Gaussian random variables $(\phi(f), f \in \mathfrak{h})$ indexed by $\mathfrak{h}$ on $(\mathscr{Q}, \Sigma, \mu)$. □

1.4.2 Bochner and Minlos theorems

Next we construct the $\mathscr{Q}$ space in Theorem 1.68. Like seen above, this is given by the countable product $\times_{n=1}^{\infty} \dot{\mathbb{R}}$, however, the family of Gaussian random variables $(\phi(f), f \in \mathfrak{h})$, depends on the base of the given Hilbert space $\mathfrak{h}$, therefore, the realization is not unique. The $\mathscr{Q}$-space introduced in this section is obtained through a generalization of Bochner's theorem, known as *Minlos' theorem*. In this construction, the measure on $\mathscr{Q}$ is given on an abstract nuclear space such as $\mathscr{S}'(\mathbb{R}^d)$.

Let X be a random variable on $(\Omega, \mathscr{F}, P)$, and consider its characteristic function

$$C(z) = \mathbb{E}_P[e^{iz\cdot X}], \quad z \in \mathbb{R}^d.$$

Clearly, $C(z)$ has the properties

$$(1) \sum_{i,j=1}^{n} a_i \bar{a}_j C(z_i - z_j) \geq 0, \quad (2)\ C \text{ is continuous}, \quad (3)\ C(0) = 1. \tag{1.4.1}$$

The converse statement is known as Bochner's theorem.

Theorem 1.69 (Bochner). *Suppose that a function $C : \mathbb{R}^d \to \mathbb{C}$ satisfies condition* (1.4.1). *Then there exists a probability space $(\mathbb{R}^d, \mathcal{B}(\mathbb{R}^d), \mu_{\mathrm{B}})$ such that*

$$C(z) = \int_{\mathbb{R}^d} e^{iz\cdot x} \mu_{\mathrm{B}}(dx).$$

Proof. Let $C_t(z) = e^{-t|z|^2/2}$. It is seen that

$$\rho_t(x) = \frac{1}{(2\pi)^{d/2}} \int_{\mathbb{R}^d} e^{-iz\cdot x} C_t(z) C(z) dz \geq 0$$

and

$$\int_{\mathbb{R}^d} \rho_t(x) dx = (2\pi)^{d/2} C_t(0) C(0) = (2\pi)^{d/2}.$$

Put $\mu_t(dx) = (2\pi)^{-d/2} \rho_t(x) dx$. Thus $\mu_t(dx)$ is a probability measure on $\mathbb{R}^d$ with characteristic function $C_t(z)C(z)$. Since $\lim_{t\downarrow 0} C_t(z)C(z) = C(z)$ pointwise and $C(z)$ is continuous, there exists a measure μ such that μ_t is weakly convergent to μ and its characteristic function is $C(z)$. Setting $\mu = \mu_B$ proves the claim. □

A function C satisfying (1) of (1.4.1) for all $z_i, z_j \in \mathbb{C}$, $\alpha_i, \alpha_j \in \mathbb{C}$ and $n \in \mathbb{N}$ is called *positive definite*. Note that any positive definite function which is continuous at the origin is automatically uniformly continuous by Lemma 1.113 below. Bochner's theorem can be extended to infinite dimension, however, as it will be illustrated in Remark 1.75 below, the Hilbert space $\mathfrak{h}$ is too small to support the corresponding measure μ. To replace $\mathfrak{h}$, we will introduce a suitable nuclear space.

Let $\mathfrak{h}$ be a complete and separable topological vector space with a countable family of norms $\{\|\cdot\|_n\}_{n=0}^{\infty}$ such that $\|\cdot\|_0 < \|\cdot\|_1 < \|\cdot\|_2 < \cdots$, and there exists a scalar product $(\cdot,\cdot)_n$ generating these norms so that $\|f\|_n^2 = (f,f)_n$ for each n. Define

$$\mathfrak{h} = \bigcap_{n=0}^{\infty} E_n,$$

where $E_n = \overline{\mathfrak{h}}^{\|\cdot\|_n}$ and E_n is a Hilbert space with the scalar product $(\cdot,\cdot)_n$. Identifying E_0 with E_0^* gives rise to the nested inclusions

$$\cdots \subset E_2 \subset E_1 \subset E_0 \cong E_0^* \subset E_1^* \subset E_2^* \subset \cdots.$$

We observe that $\mathfrak{h}^* = \bigcup_{n=0}^{\infty} E_n^*$. For $x \in E_n^*$, we denote its norm by $\|x\|_{-n} = \sup_{\xi \in E_n} |x(\xi)|$.

Definition 1.70 (Nuclear space). Space $\mathfrak{h}$ is a called *nuclear* whenever for every $m > 0$ there exists $n > m$ such that the injection $\iota_{nm} : E_n \to E_m$ is a Hilbert–Schmidt operator.

Example 1.71. Let $H_{\mathrm{osc}} = \frac{1}{2}(-\Delta + |x|^2 - d)$ be the *harmonic oscillator* on $L^2(\mathbb{R}^d)$ with a purely discrete spectrum $\{n_1 + \cdots + n_d \mid n_j \in \mathbb{N} \cup \{0\}, j = 1, \ldots, d\}$. Using a suitable orthonormal basis $\{g_{i_1,\ldots,i_d}\}_{(i_1,\ldots,i_d)\in(\mathbb{N}\cup\{0\})^d}$, H_{osc} is diagonalized and $(g_{i_1,\ldots,i_d}, Hg_{i_1,\ldots,i_d}) = \sum_{j=1}^d i_j$. It is easily seen that $(H_{\mathrm{osc}} + 1)^{-(d+\varepsilon)/2}$ is a Hilbert–Schmidt operator for every $\varepsilon > 0$ and

$$\mathrm{Tr}[(H_{\mathrm{osc}} + 1)^{-(d+\varepsilon)}] = \sum_{i_1=0}^{\infty} \cdots \sum_{i_d=0}^{\infty} \left(\sum_{\nu=1}^{d} (i_\nu + 1) \right)^{-(d+\varepsilon)}.$$

Let $(\Psi, \Phi)_n = (\Psi, (H_{\mathrm{osc}} + 1)^n \Phi)$. Then $\|\Psi\|_n < \|\Psi\|_{n+1}$ and denote the completion of $L^2(\mathbb{R}^d)$ with respect to $\|\cdot\|_n$ by $\mathcal{H}_n$. We have $\cdots \subset \mathcal{H}_{n+1} \subset \mathcal{H}_n \subset \cdots$, and it can be proven that

$$\mathscr{S} = \bigcap_{n=0}^{\infty} \mathcal{H}_n, \quad \mathscr{S}' = \bigcup_{n=0}^{\infty} \mathcal{H}_{-n}.$$

Here we identify the dual space $\mathcal{H}_n^*$ with $\mathcal{H}_{-n}$. Let $1_{n+1} = 1 : \mathcal{H}_{n+1} \to \mathcal{H}_n$ be the identity map. Thus $(H_{\mathrm{osc}}+1)^{1/2} 1_{n+1}$ is unitary and so $1 = 1_{n+1} \cdots 1_{n+D+1} : \mathcal{H}_{n+D+1} \to \mathcal{H}_n$ is a Hilbert–Schmidt operator for sufficiently large D since $1 = (H_{\mathrm{osc}}+1)^{-D/2}(H_{\mathrm{osc}}+1)^{D/2} 1$ is a product of a Hilbert–Schmidt operator with a unitary operator.

Theorem 1.72 (Minlos). *Let $\mathfrak{h}$ be a nuclear space, and suppose that $C : \mathfrak{h} \to \mathbb{C}$ satisfies*

$$(1) \sum_{i,j=1}^{n} a_i \bar{a}_j C(\xi_i - \xi_j) \geq 0, \quad (2)\ C \text{ is continuous}, \quad (3)\ C(0) = 1. \tag{1.4.2}$$

Then there exists a probability space $(\mathfrak{h}^*, \mathcal{B}, \mu)$ such that

$$C(\xi) = \int_{\mathfrak{h}^*} e^{i\langle\langle x,\xi\rangle\rangle} d\mu(x), \quad \xi \in \mathfrak{h}, \tag{1.4.3}$$

where $\langle\langle x, \xi\rangle\rangle$ denotes the pairing between $x \in \mathfrak{h}^*$ and $\xi \in \mathfrak{h}$.

Proof. Suppose that $\mathfrak{h} = \bigcap_{n=0}^{\infty} E_n$. Let $F \subset \mathfrak{h}$ be a finite-dimensional subspace, and for $\xi_1, \ldots, \xi_n \in F$ and $B \in \mathcal{B}(\mathbb{R}^n)$ define

$$A_{\xi_1,\ldots,\xi_n,B} = \{x \in \mathfrak{h}^* \mid (\langle\langle x, \xi_1\rangle\rangle, \ldots, \langle\langle x, \xi_n\rangle\rangle) \in B\}.$$

Let $\mathfrak{A}_F = \sigma(\{A_{\xi_1,\ldots,\xi_n,B} \mid \xi_1, \ldots, \xi_n \in F, B \in \mathcal{B}(\mathbb{R}^n), n \in \mathbb{N}\})$ be the minimal σ-field generated by $\{A_{\xi_1,\ldots,\xi_n,B} \mid \xi_1, \ldots, \xi_n \in F, B \in \mathcal{B}(\mathbb{R}^n), n \in \mathbb{N}\}$ and

$$\mathfrak{A}_\infty = \bigcup_{F\subset\mathfrak{h}, \dim F<\infty} \mathfrak{A}_F.$$

This $\mathfrak{A}_\infty$ is a finitely additive family of subsets of $\mathfrak{h}^*$. Taking $\mathfrak{A} = \sigma(\mathfrak{A}_\infty)$, we obtain the measurable space $(\mathfrak{h}^*, \mathfrak{A})$.

We prove the theorem in four steps. In Step 1, we construct a probability space $(\mathfrak{h}^*/F^\perp, \mathcal{B}_F, m_F)$, where $\mathfrak{h}^*/F^\perp$ is a finite-dimensional space. In Step 2, a probability space $(\mathfrak{h}^*, \mathfrak{A}_F, \tilde{m}_F)$ is constructed for each F and, varying F, we construct a measurable space $(\mathfrak{h}^*, \mathfrak{A}, m)$ with a finitely additive set function m. In Step 3, we show that m can be extended to a measure μ. In Step 4, we complete the proof.

Step 1. Suppose that $\dim F = n$. Define $F^\perp = \{x \in \mathfrak{h}^* \mid \langle\langle x, \xi\rangle\rangle = 0, \forall \xi \in F\}$ and consider the quotient space $\mathfrak{h}^*/F^\perp$. For $\bar{x} \in \mathfrak{h}^*/F^\perp$ with representing element $x \in \mathfrak{h}^*$, we define $\langle\langle \bar{x}, \xi\rangle\rangle_F = \langle\langle x, \xi\rangle\rangle$ for $\xi \in F$, which is well defined. Define the linear map

$$T : \mathfrak{h}^*/F^\perp \to F^*, \quad T : \bar{x} \mapsto \langle\langle \bar{x}, \cdot\rangle\rangle_F.$$

Note that T is injective. Let $\{\xi_m\}_{m=1}^n$ be a basis of F. There exists $x_j \in \mathfrak{h}^*$ such that $\langle\langle x_j, \xi_m\rangle\rangle = \delta_{jm}$. Hence $\{T\bar{x}_j\}_{j=1}^n = \{\langle\langle \bar{x}_j, \cdot\rangle\rangle_F\}_{j=1}^n$ forms a basis of F^* and, in particular, T is surjective. Hence $\mathfrak{h}^*/F^\perp \cong F^*$. Thus we can identify $\mathfrak{h}^*/F^\perp$ with F^* by T. In particular, $\dim \mathfrak{h}^*/F^\perp = n$. The restriction of $C(\xi)$ for $\xi \in F \cong \mathbb{R}^n$, denoted by $C_F(\xi)$, satisfies (1.4.1). The Bochner theorem yields that there exists a probability space

$$(\mathbb{R}^n, \mathcal{B}(\mathbb{R}^n), \Phi)$$

such that

$$C_F(\xi) = \int_{\mathbb{R}^n} e^{ix\cdot\xi}\Phi(dx), \quad \xi \in \mathbb{R}^n.$$

The identifications $F \cong \mathbb{R}^n$ and $\mathfrak{h}^*/F^\perp \cong F^* \cong (\mathbb{R}^n)^* \cong \mathbb{R}^n$ induce a measure m_F on $\mathfrak{h}^*/F^\perp$ such that

$$C_F(\xi) = \int_{\mathfrak{h}^*/F^\perp} e^{i\langle\langle \bar{x},\xi\rangle\rangle_F} dm_F(\bar{x}), \quad \xi \in F.$$

Thus we obtain the probability space

$$(\mathfrak{h}^*/F^\perp, \mathcal{B}_F, m_F),$$

where $\mathcal{B}_F = \mathcal{B}(\mathfrak{h}^*/F^\perp)$.

Step 2. Let $\pi_F : \mathfrak{h}^* \to \mathfrak{h}^*/F^\perp$ be the surjective map defined by $\pi_F(x) = \bar{x}$. Hence

$$\pi_F^{-1} : \mathfrak{h}^*/F^\perp \to \mathfrak{h}^*.$$

We have $\pi_F^{-1}(\mathcal{B}_F) = \mathfrak{A}_F$ and define a measure $\tilde{m}_F$ on $(\mathfrak{h}^*, \mathfrak{A}_F)$ by the image measure of m_F by π_F^{-1},

$$\tilde{m}_F(A) = m_F \circ \pi_F(A), \quad A \in \mathfrak{A}_F.$$

Thus we obtain the probability space

$$(\mathfrak{h}^*, \mathfrak{A}_F, \tilde{m}_F)$$

with

$$C_F(\xi) = \int_{\mathfrak{h}^*} e^{i\langle\langle x,\xi\rangle\rangle} d\tilde{m}_F(x), \quad \xi \in F. \tag{1.4.4}$$

It is seen that if $F \subset G \subset \mathfrak{h}$ with $\dim G < \infty$, then $(\mathfrak{h}^*, \mathfrak{A}_G, \tilde{m}_G) \supset (\mathfrak{h}^*, \mathfrak{A}_F, \tilde{m}_F)$, i. e., $\mathfrak{A}_F \subset \mathfrak{A}_G$ and $\tilde{m}_F(B) = \tilde{m}_G(B)$ for $B \in \mathfrak{A}_F$. Let $A \in \mathfrak{A}$. Then $A \in \mathcal{B}_F$ for a finite-dimensional subspace F. We define

$$m(A) = \tilde{m}_F(A), \quad A \in \mathfrak{A}.$$

This is well defined by the fact that $(\mathfrak{h}^*, \mathfrak{A}_G, \tilde{m}_G) \supset (\mathfrak{h}^*, \mathfrak{A}_F, \tilde{m}_F)$. We will see below that m is a finitely additive set function on $\mathfrak{A}$. Let $A_1, \ldots, A_n \in \mathfrak{A}$ be such that $A_i \cap A_j = \emptyset$ for $i \neq j$. Assume that $A_i \in \mathfrak{A}_{F_i}$ for a finite-dimensional subspace $F_i \subset \mathfrak{h}$. Let F be the subspace spanned by $F_1, \ldots, F_n$. Then $A_i \in \mathfrak{A}_F$ and $\tilde{m}_F(A_i) = \tilde{m}_{F_i}(A_i)$ for $i = 1, \ldots, n$. We have

$$m\Big(\sum_{i=1}^n A_i\Big) = \tilde{m}_F\Big(\sum_{i=1}^n A_i\Big) = \sum_{i=1}^n \tilde{m}_F(A_i) = \sum_{i=1}^n \tilde{m}_{F_i}(A_i) = \sum_{i=1}^n m(A_i).$$

We construct the measurable space

$$(\mathfrak{h}^*, \mathfrak{A}, m)$$

with the finitely additive set function m.

Step 3. In order to extend m to a measure on the measurable space $(\mathfrak{h}^*, \mathfrak{A})$, we need two lemmas below.

Lemma 1.73. *Let P be a probability measure on $(\mathbb{R}^n, \mathcal{B}(\mathbb{R}^n))$ such that its characteristic function ϕ has the property that $|\phi(z) - 1| < \varepsilon$ on $A = \{z \in \mathbb{R}^n \mid \sum_{i=1}^n a_i^2 z_i^2 < \gamma^2\}$ with some $a \in \mathbb{R}^n$ and $\gamma \in \mathbb{R}$. Let B_t be a ball with radius t and $\beta = (1 - e^{-1/2})^{-1}$. Then*

$$P(B_t^c) < \beta^2 \left(\varepsilon + \frac{2|a|^2}{\gamma^2 t^2} \right).$$

Proof. It is clear that

$$I = \int_{\mathbb{R}^n} (1 - e^{-|x|^2/(2t^2)}) dP(x) \geq (1 - e^{-1/2}) P(B_t^c). \tag{1.4.5}$$

On the other hand,

$$\begin{aligned} I &= \left(\frac{t^2}{2\pi}\right)^{n/2} \int_{\mathbb{R}^n} (1 - \phi(z)) e^{-(t^2/2)|z|^2} dz \\ &\leq \left(\frac{t^2}{2\pi}\right)^{n/2} \left| \int_A (1 - \phi(z)) e^{-(t^2/2)|z|^2} dz + \int_{A^c} (1 - \phi(z)) e^{-(t^2/2)|z|^2} dz \right| \\ &\leq \varepsilon + \left(\frac{t^2}{2\pi}\right)^{n/2} \frac{2}{\gamma^2} \int_{A^c} \left(\sum_{j=1}^n a_j^2 z_j^2 \right) e^{-(t^2/2)|z|^2} dz \leq \varepsilon + \frac{2|a|^2}{\gamma^2 t^2}. \end{aligned} \tag{1.4.6}$$

Combining (1.4.5)–(1.4.6), the statement follows. □

Lemma 1.74. *The finitely additive set function m defined on $(\mathfrak{h}^*, \mathfrak{A})$ can be extended to a measure on $(\mathfrak{h}^*, \mathfrak{A})$ if and only if for every $0 < \varepsilon < 1$ there exist $n \in \mathbb{N}$ and a set $S_n = \{x \in \mathfrak{h}^* \mid \|x\|_{-n} \leq \gamma_n\}$ with a suitable $\gamma_n > 0$, such that $m(A) < \varepsilon$ holds for every $A \in \mathfrak{A}$ with $A \cap S_n = \emptyset$.*

Proof. ($\Longrightarrow$) Suppose that m can be extended to a measure μ on $\mathfrak{h}^*$. Assume that $\gamma_n \to \infty$ as $n \to \infty$. Then $\bigcup_{n=1}^\infty S_n = \mathfrak{h}^*$, and $\mu(S_n^c) < \varepsilon$ for a sufficiently large n. Hence $(A) \leq \mu(S_n^c) < \varepsilon$.

($\Longleftarrow$) Assume that $\bigcup_{n=1}^\infty A_n = \mathfrak{h}^*$ and $A_m \cap A_n = \emptyset$ for $m \neq n$. Since m is a finitely additive set function, $\sum_{n=1}^\infty m(A_n) \leq 1$. Suppose that $\sum_{n=1}^\infty m(A_n) = 1 - 3\varepsilon < 1$. There exists an open set A_n' such that $A_n' \supset A_n$ and $m(A_n' \setminus A_n) < \varepsilon/2^n$. Since S_n is weakly compact, there exists k such that $A' = \bigcup_{n=1}^k A_n'$ is an open covering of S_n. Thus

$$1 = m(A' \cup A'^c) = m(A') + m(A'^c) \leq \sum_{n=1}^{k} m(A_n) + \varepsilon + \varepsilon \leq 1 - \varepsilon < 1,$$

which is a contradiction. Hence we have $\sum_{n=1}^{\infty} m(A_n) = 1$. This implies that m is completely additive, and the existence of the extension of m follows from Hopf's extension theorem. □

Using the two lemmas above, we can complete the proof of Theorem 1.72. By (2) of (1.4.1), $C(\cdot)$ is continuous for every $\|\cdot\|_p$. Fix a p. By (3) of (1.4.1), for every $\varepsilon > 0$ there exists $B_\gamma \subset E_\mathrm{p}$ such that

$$|C(\xi) - 1| \leq \frac{\varepsilon}{2\beta^2}, \quad \xi \in B_\gamma, \tag{1.4.7}$$

where $\beta = 1/(1 - e^{-1/2})$. There exists $n > p$ such that the injection $\iota_{np} : E_n \to E_\mathrm{p}$ is a Hilbert–Schmidt operator. Also, there exists a neighborhood V of 0 in E_n such that $\iota_{np} V \subset B_\gamma$. Take $B_t \subset E_n^* \subset \mathfrak{h}^*$, where

$$t = \frac{2\beta\|\iota_{np}\|_{\mathrm{HS}}}{\sqrt{\varepsilon\gamma^2}}.$$

Let $F \subset E_n$ be a finite-dimensional subspace and $A \subset \mathfrak{h}^* \setminus B_t$ with $A \in \mathfrak{A}_F \subset \mathfrak{A}$. Note that $\pi_F(B_t) \cap \pi_F(A) = \emptyset$. Since ι_{np} is a Hilbert–Schmidt operator, there exists a complete orthonormal system $\{e_m\}_{m=1}^{\infty} \subset E_n$ such that $\iota_{np} e_m = \lambda_m e_m$. Since $\dim F = n$, $y \in F \cap V$ can be expressed as $y = z_1 e_1 + \cdots + z_n e_n$ with some $e_1, \ldots, e_n$ and $z_1, \ldots, z_n \in \mathbb{R}$. For $y = z_1 e_1 + \cdots + z_n e_n \in F \cap V$, $\iota_{np} y \in B_\gamma$ implies that $\|\iota_{np} y\|_p^2 = z_1^2\lambda_1^2 + \cdots + z_n^2\lambda_n^2 \leq \gamma^2$. We identify $\iota_{np}(F \cup V)$ with a subspace of $\mathbb{R}^n$. Set $D = \iota_{np}(F \cap V) = \{z \in \mathbb{R}^n \mid \sum_{i=1}^n \lambda_i^2 z_i^2 \leq \gamma^2\} \subset E_\mathrm{p}$. The restriction of $C(\xi)$ to $\xi \in F$, $C_F(\xi)$, satisfies $|C_F(\xi) - 1| \leq \varepsilon/(2\beta^2)$ for $\xi \in D$ by (1.4.7). By Lemma 1.73, we estimate as

$$m_F([\pi_F(B_t)]^c) \leq \beta^2 \left(\frac{\varepsilon}{2\beta^2} + \frac{2}{\gamma^2 t^2} \sum_{i=1}^{n} \lambda_i^2 \right) \leq \frac{\varepsilon}{2} + \frac{2\beta^2 \|\iota_{np}\|_{\mathrm{HS}}^2}{\gamma^2 t^2} = \varepsilon.$$

Hence $m(A) = \tilde{m}_F(A) = m_F(\pi_F(A)) \leq m_F([\pi_F(B_t)]^c) < \varepsilon$. By a limiting argument, we conclude that $m(A) < \varepsilon$ for every $A \subset \mathfrak{h}^*$ such that $A \cap B_t = \emptyset$. Hence there exists an extended measure $\mu \subset m$ on $(\mathfrak{h}^*, \mathfrak{A})$ by Lemma 1.74. Thus we constructed the probability space

$$(\mathfrak{h}^*, \mathfrak{A}, \mu).$$

Step 4. By Step 3, we can write (1.4.4) as

$$C(\xi) = \int_{\mathfrak{h}^*} e^{i\langle\langle x, \xi \rangle\rangle} d\mu(x)$$

for $\xi \in \mathscr{E}$. □

Remark 1.75. Let $\mathfrak{h}$ be a Hilbert space and $C(\xi) = \exp(-(1/4)\|\xi\|^2)$ for $\xi \in \mathfrak{h}$. By Minlos' theorem, $C(\xi)$ satisfies conditions (1.4.2). Let $\{\xi_n\}_{n=1}^{\infty}$ be a complete orthonormal system of $\mathfrak{h}$. If a measure μ on $\mathfrak{h}$ satisfying (1.4.3) exists, then

$$\int_{\mathfrak{h}} e^{iz(x,\xi_n)} d\mu(x) = e^{-(1/4)z^2}. \tag{1.4.8}$$

Since $(x, \xi_n) \to 0$ as $n \to \infty$, the left-hand side of (1.4.8) converges to 1, in contradiction to the right-hand side. Moreover, from $C(\sum_{k=1}^n z_k \xi_k) = \exp\left(-(1/4)\sum_{k=1}^n z_k^2\right)$ and the law of large numbers, it would follow that

$$\lim_{n\to\infty} \frac{1}{n} \sum_{k=1}^{n} \langle\!\langle x, \xi_k \rangle\!\rangle^2 = 1,$$

which is also a contradiction. We further discuss Gaussian measures on Hilbert spaces in Section 1.6.

Let $\mathscr{S}_{\text{real}}(\mathbb{R}^d)$ denote the set of real-valued, rapidly decreasing, and infinitely differentiable functions on $\mathbb{R}^d$. It is known that $\mathscr{S}_{\text{real}}(\mathbb{R}^d)$ is a nuclear space.

Corollary 1.76. *Suppose that $\mathscr{S}_{\text{real}}(\mathbb{R}^d)$ is dense in a real Hilbert space $\mathcal{H}$. Then there exists a family of Gaussian random variables $(\phi(f), f \in \mathcal{H})$ indexed by $\mathcal{H}$ on a probability space $(\mathscr{S}'_{\text{real}}(\mathbb{R}^d), \mathcal{B}, \mu)$.*

Proof. Write $C(\xi) = \exp(-(1/4)\|\xi\|^2)$ for $\xi \in \mathscr{S}_{\text{real}}(\mathbb{R}^d)$. Since

$$C(\xi) = \int_{\mathscr{Q}} e^{i\phi(f)} \mu(d\phi)$$

by Theorem 1.68, C satisfies (1.4.2). Thus Minlos' theorem implies that a probability space $(\mathscr{S}'_{\text{real}}(\mathbb{R}^d), \mathcal{B}, \mu)$ exists such that

$$C(\xi) = \int_{\mathscr{S}'_{\text{real}}(\mathbb{R}^d)} e^{i\langle\!\langle \phi, \xi \rangle\!\rangle} d\mu(\phi).$$

Define $\phi(\xi) = \langle\!\langle \phi, \xi \rangle\!\rangle$ and $\phi \in \mathscr{S}'_{\text{real}}(\mathbb{R}^d)$ and $\xi \in \mathscr{S}_{\text{real}}(\mathbb{R}^d)$. We have

$$\int_{\mathscr{S}'_{\text{real}}(\mathbb{R}^d)} \phi(\xi) d\mu = 0, \qquad \int_{\mathscr{S}'_{\text{real}}(\mathbb{R}^d)} \phi(\xi)\phi(\eta) d\mu = \frac{1}{2}(\xi, \eta), \quad \xi, \eta \in \mathscr{S}_{\text{real}}(\mathbb{R}^d).$$

Since $\mathcal{B}$ is the minimal σ-field generated by cylinder sets, $\mathcal{B}$ is the minimal σ-field generated by $\{\phi(\xi), \xi \in \mathscr{S}_{\text{real}}(\mathbb{R}^d)\}$. For $f \in \mathcal{H}$, there is a sequence $(\xi_n)_{n\in\mathbb{N}} \subset \mathscr{S}_{\text{real}}(\mathbb{R}^d)$ such that $\xi_n \to f$ as $n \to \infty$ in $\mathcal{H}$. Thus

$$\int_{\mathscr{S}'_{\mathrm{real}}(\mathbb{R}^d)} |\phi(\xi_n) - \phi(\xi_m)|^2 d\mu = \frac{1}{2}\|\xi_n - \xi_m\|^2_{\mathcal{H}} \to 0$$

as $n, m \to \infty$. Hence $(\phi(\xi_n))_{n\in\mathbb{N}}$ is a Cauchy sequence in $L^2(\mathscr{S}'_{\mathrm{real}}(\mathbb{R}^d))$. Define $\phi(f) = \text{s-lim}_{n\to\infty} \phi(\xi_n)$. Then $\{\phi(f) \mid f \in \mathcal{H}\}$ is a family of Gaussian random variables on $(\mathscr{S}'_{\mathrm{real}}(\mathbb{R}^d), \mathcal{B}, \mu)$ indexed by $\mathcal{H}$. □

Example 1.77. Suppose that $B(f,g)$ is a positive semidefinite quadratic form on $\mathscr{S}_{\mathrm{real}}(\mathbb{R}^d)$. Denote the completion of $\mathscr{S}_{\mathrm{real}}(\mathbb{R}^d)$ with respect to the norm $\|\cdot\| = \sqrt{B(\cdot,\cdot)}$ by $\mathfrak{h}$. Thus $\mathfrak{h}$ is the real Hilbert space with inner product $B(\cdot,\cdot)$. Hence by Theorem 1.68, there exists a probability space $(\mathscr{Q}, \Sigma, \mu)$ with a Gaussian measure μ such that $\int_{\mathscr{Q}} e^{i\phi(f)}\mu(d\phi) = e^{-B(f,f)/4}$. From this, we see that $e^{-B(f,f)/4}$ satisfies (1.4.2). By Minlos' theorem, we can construct a measure μ on $\mathscr{S}'_{\mathrm{real}}(\mathbb{R}^d)$ such that

$$\int_{\mathscr{S}'_{\mathrm{real}}(\mathbb{R}^d)} e^{i\langle\langle f,x\rangle\rangle} d\mu(x) = e^{-B(f,f)/4}.$$

1.5 Functional integral representation of the Euclidean quantum field

1.5.1 Basic results in Euclidean quantum field theory

In this section we define the state space of a single boson and study Euclidean quantum fields. Let $\mathscr{S}'(\mathbb{R}^d)$ be the *space of tempered distributions* on $\mathbb{R}^d$. For $s \in \mathbb{R}$, consider the *inhomogeneous Sobolev space* $H^s(\mathbb{R}^d)$, i. e.,

$$H^s(\mathbb{R}^d) = \left\{u \in \mathscr{S}'(\mathbb{R}^d) \mid \hat{u} \in L^1_{\mathrm{loc}}(\mathbb{R}^d), (1+|k|^2)^{s/2}\hat{u} \in L^2(\mathbb{R}^d)\right\}.$$

Here $\hat{u}$ describes the Fourier transform of u on $\mathscr{S}'(\mathbb{R}^d)$. In particular, $H^0(\mathbb{R}^d) = L^2(\mathbb{R}^d)$ and $H^s(\mathbb{R}^d) \subset L^2(\mathbb{R}^d)$ for every $s > 0$. The scalar product on $H^s(\mathbb{R}^d)$ is defined by

$$(f,g)_{H^s(\mathbb{R}^d)} = ((1+|k|^2)^{s/2}\hat{f}, (1+|k|^2)^{s/2}\hat{g})_{L^2(\mathbb{R}^d)}.$$

Hence $H^s(\mathbb{R}^d)$ is a Hilbert space equipped with the scalar product defined above. Furthermore, we can regard $H^{-s}(\mathbb{R}^d)$ as the dual of $H^s(\mathbb{R}^d)$, for every $s \in \mathbb{R}$.

We can also define *homogeneous Sobolev spaces*. For $s \in \mathbb{R}$, the homogeneous Sobolev space $\dot{H}^s(\mathbb{R}^d)$ is defined by $H^s(\mathbb{R}^d)$ with $(1+|k|^2)^{s/2}$ replaced by $|k|^s$, i. e.,

$$\dot{H}^s(\mathbb{R}^d) = \left\{u \in \mathscr{S}'(\mathbb{R}^d) \mid \hat{u} \in L^1_{\mathrm{loc}}(\mathbb{R}^d), |k|^s\hat{u} \in L^2(\mathbb{R}^d)\right\}.$$

The scalar product on $\dot{H}^s(\mathbb{R}^d)$ is given by

$$(f,g)_{\dot{H}^s(\mathbb{R}^d)} = (|k|^s\hat{f}, |k|^s\hat{g})_{L^2(\mathbb{R}^d)}.$$

It is known that $\dot{H}^s(\mathbb{R}^d)$ is a Hilbert space if and only if $s < d/2$. Let $\mathscr{S}_0(\mathbb{R}^d)$ be the set of $u \in \mathscr{S}(\mathbb{R}^d)$ such that the Fourier transform $\hat{u}$ vanishes near $k = 0$. It is also known that $\mathscr{S}_0(\mathbb{R}^d)$ is dense in $\dot{H}^s(\mathbb{R}^d)$ for $s < d/2$. Moreover, $\dot{H}^{-s}(\mathbb{R}^d)$ can be regarded as the dual of $\dot{H}^s(\mathbb{R}^d)$ for $|s| < d/2$.

Example 1.78.

(1) Let δ_x be the delta distribution with mass at $x = 0$ in d-dimensions. Then

$$\hat{\delta}_x = \frac{1}{(2\pi)^{d/2}}$$

and $\delta_x \in H^\varepsilon(\mathbb{R}^d)$ for $\varepsilon < -d/2$.

(2) Let $f \in \dot{H}^{-1/2}(\mathbb{R}^d)$ and δ_x be the delta distribution with mass at $x = 0$ in one-dimension. Then $\delta_x \otimes f \in \dot{H}^{-1}(\mathbb{R}^{d+1})$. We can actually see that

$$\widehat{\delta_x \otimes f}(k_0, k) = \frac{1}{(2\pi)^{1/2}} \hat{f}(k)$$

and

$$\left\| \frac{\widehat{\delta_x \otimes f}}{(k_0^2 + |k|^2)^{1/2}} \right\|^2_{L^2(\mathbb{R}^{d+1})} = \frac{1}{2\pi} \int_{\mathbb{R}^{d+1}} \frac{|\hat{f}(k)|^2}{k_0^2 + |k|^2} dk_0 dk = \frac{1}{2} \int_{\mathbb{R}^d} \frac{|\hat{f}(k)|^2}{|k|} dk < \infty.$$

In view of applications to quantum field theory, we make some modifications to these spaces. Let $\omega_\nu(k) = \sqrt{|k|^2 + \nu^2}$, where $\nu \geq 0$. We define $H_\nu^s(\mathbb{R}^d)$ by $H^s(\mathbb{R}^d)$ with $(1 + |k|^2)^{1/2}$ replaced by ω_ν. Hence $H^s(\mathbb{R}^d)$ and $H_\nu^s(\mathbb{R}^d)$ are equivalent for $\nu > 0$, and $H_\nu^s(\mathbb{R}^d) = \dot{H}^s(\mathbb{R}^d)$ for $\nu = 0$. We write

$$\mathcal{H}_s(\mathbb{R}^d) = \begin{cases} H_\nu^s(\mathbb{R}^d), & \nu > 0, \\ \dot{H}^s(\mathbb{R}^d), & \nu = 0. \end{cases}$$

Two specific Hilbert spaces will play a special role, as singled out next.

Definition 1.79 ($\mathcal{H}_M$, $\hat{\mathcal{H}}_M$, $\mathcal{H}_E$, and $\hat{\mathcal{H}}_E$).

(1) We write $\mathcal{H}_M = \mathcal{H}_{-1/2}(\mathbb{R}^d)$ and $\mathcal{H}_E = \mathcal{H}_{-1}(\mathbb{R}^{d+1})$.

(2) We define the Fourier transform (in the sense of tempered distributions) of $\mathcal{H}_M$ and $\mathcal{H}_E$ by $\hat{\mathcal{H}}_M$ and $\hat{\mathcal{H}}_E$, respectively.

We note that $\hat{F}, \hat{G} \in \hat{\mathcal{H}}_E$ if and only if $\hat{F}/\omega_\nu, \hat{G}/\omega_\nu \in L^2(\mathbb{R}^{d+1})$ and define

$$(\hat{F}, \hat{G})_{\hat{\mathcal{H}}_E} = 2(\hat{F}/\omega_\nu, \hat{G}/\omega_\nu)_{L^2(\mathbb{R}^{d+1})}. \tag{1.5.1}$$

Here the coefficient 2 is added for a normalization which will be used in Lemma 1.81 below. We also see that $\hat{f}, \hat{g} \in \hat{\mathcal{H}}_M$ if and only if $\hat{f}/\sqrt{\omega_\nu}, \hat{g}/\sqrt{\omega_\nu} \in L^2(\mathbb{R}^d)$ and

$$(\hat{f},\hat{g})_{\hat{\mathcal{H}}_M} = (\hat{f}/\sqrt{\omega_\nu},\hat{g}/\sqrt{\omega_\nu})_{L^2(\mathbb{R}^d)}. \tag{1.5.2}$$

For notational simplicity, we write

$$(\hat{F},\hat{G})_{\hat{\mathcal{H}}_E} = (\hat{F},\hat{G})_{-1} = (F,G)_{-1},$$
$$(\hat{f},\hat{g})_{\hat{\mathcal{H}}_M} = (\hat{f},\hat{g})_{-1/2} = (f,g)_{-1/2}.$$

Although $\mathcal{H}_M$, $\hat{\mathcal{H}}_M$, $\mathcal{H}_E$, and $\hat{\mathcal{H}}_E$ depend on the space dimension and $\nu \geq 0$, we do not indicate this explicitly. We also define two specific real Hilbert spaces below.

Definition 1.80 ($\mathscr{M}$ and $\mathscr{E}$). We define
(1) $\mathscr{M} = \{f \in \mathcal{H}_M \mid f \text{ is real-valued}\}$,
(2) $\mathscr{E} = \{f \in \mathcal{H}_E \mid f \text{ is real-valued}\}$.

Both $\mathscr{M}$ and $\mathscr{E}$ are Hilbert spaces over $\mathbb{R}$. Note that $\mathscr{M}_\mathbb{C} \cong \mathcal{H}_M$ and $\mathscr{E}_\mathbb{C} \cong \mathcal{H}_E$. In summary, we have $\mathscr{M}_\mathbb{C} \cong \mathcal{H}_M \cong \hat{\mathcal{H}}_M$ and $\mathscr{E}_\mathbb{C} \cong \mathcal{H}_E \cong \hat{\mathcal{H}}_E$.

Let $(\phi(f), f \in \mathscr{M})$ be a family of Gaussian random variables indexed by $f \in \mathscr{M}$ on a probability space $(\mathscr{Q},\Sigma,\mu)$ and $(\phi_\mathrm{E}(F), F \in \mathscr{E})$ on $(\mathscr{Q}_\mathrm{E},\Sigma_\mathrm{E},\mu_\mathrm{E})$. The family of Gaussian random variables $(\phi_\mathrm{E}(F), F \in \mathscr{E})$ is called a *Euclidean field*. One possible choice of $\mathscr{Q}$ and $\mathscr{Q}_\mathrm{E}$ is $\mathscr{S}'_{\mathrm{real}}(\mathbb{R}^d)$ and $\mathscr{S}'_{\mathrm{real}}(\mathbb{R}^{d+1})$, respectively, however, we do not fix any particular space. Note that $\phi(f)$ and $\phi_\mathrm{E}(F)$ are Gaussian random variables with mean zero and covariance

$$\int_{\mathscr{Q}} \phi(f)\phi(g)d\mu = \frac{1}{2}(f,g)_{-1/2}, \quad \int_{\mathscr{Q}_\mathrm{E}} \phi_\mathrm{E}(F)\phi_\mathrm{E}(G)d\mu_\mathrm{E} = \frac{1}{2}(F,G)_{-1}.$$

Under the identification $\mathscr{M}_\mathbb{C} \cong \hat{\mathcal{H}}_M$, we have

$$L^2(\mathscr{Q}) \cong \mathcal{F}_\mathrm{b}(\hat{\mathcal{H}}_M) \tag{1.5.3}$$

by the Wiener–Itô–Segal isomorphism $\theta_W : \mathcal{F}_\mathrm{b}(\hat{\mathcal{H}}_M) \to L^2(\mathscr{Q})$, as well as

$$\theta_\mathrm{W}^{-1}\phi(f)\theta_\mathrm{W} = \frac{1}{\sqrt{2}}(a_M^*(\hat{f}) + a_M(\bar{\hat{f}})).$$

Here a_M^* and a_M denote the annihilation and creation operators, respectively, in $\mathcal{F}_\mathrm{b}(\hat{\mathcal{H}}_M)$, which satisfy the canonical commutation relations

$$[a_M(\hat{f}), a_M(\hat{g})] = 0 = [a_M^*(\hat{f}), a_M^*(\hat{g})], \quad [a_M(\hat{f}), a_M^*(\hat{g})] = (\bar{\hat{f}},\hat{g})_{-1/2}. \tag{1.5.4}$$

We introduce a family of transformations I_t from $L^2(\mathscr{Q})$ to $L^2(\mathscr{Q}_\mathrm{E})$ through the second quantization of a specific transformation τ_t from $\mathscr{M}$ to $\mathscr{E}$. Define

$$\tau_t : \mathscr{M} \to \mathscr{E}, \quad \tau_t : f \mapsto \delta_t \otimes f.$$

Here $\delta_t(x) = \delta(x-t)$ is the delta function with mass at t. Note that for $f \in \mathscr{M}$,

$$\overline{\widehat{\delta_t \otimes f}(k,k_0)} = \widehat{\delta_t \otimes f}(-k,-k_0) = \widehat{\delta_t \otimes f}(k,k_0).$$

Thus $\overline{\delta_t \otimes f} = \delta_t \otimes f$, which implies that τ_t preserves realness. Define

$$\hat{\omega}_\nu = \omega_\nu(-i\nabla) = \sqrt{-\Delta + \nu^2}.$$

Lemma 1.81. *It follows that*

$$\tau_t^* \tau_s = e^{-|t-s|\hat{\omega}_\nu}, \quad s,t \in \mathbb{R}.$$

In particular, τ_t is an isometry between $\mathscr{M}$ and $\mathscr{E}$ for every $t \in \mathbb{R}$.

Proof. We see that

$$\begin{aligned}(\delta_t \otimes f, \delta_s \otimes g)_{-1} &= \frac{1}{\pi} \int_{\mathbb{R}^d} \overline{\hat{f}(k)} \hat{g}(k) dk \int_{\mathbb{R}} e^{-i(s-t)k_0} \frac{1}{\omega_\nu(k)^2 + |k_0|^2} dk_0 \\ &= \int_{\mathbb{R}^d} \overline{\hat{f}(k)} \hat{g}(k) \frac{e^{-|s-t|\omega_\nu(k)}}{\omega_\nu(k)} dk = (f, e^{-|s-t|\hat{\omega}_\nu} g)_{-1/2}. \end{aligned} \tag{1.5.5}$$

□

Definition 1.82 (Second quantization of contraction operators).

(1) Let $T \in \mathscr{C}(\mathscr{M} \to \mathscr{M})$. Define $\Gamma(T) \in \mathscr{C}(L^2(\mathscr{Q}) \to L^2(\mathscr{Q}))$ by

$$\Gamma(T){:}\phi(f_1)\cdots\phi(f_n){:} = {:}\phi(Tf_1)\cdots\phi(Tf_n){:} \tag{1.5.6}$$

with $\Gamma(T)1_{\mathrm{M}} = 1_{\mathrm{M}}$, where 1_{M} denotes the identity function in $L^2(\mathscr{Q})$.

(2) Let $T \in \mathscr{C}(\mathscr{E} \to \mathscr{E})$. Define $\Gamma_{\mathrm{E}}(T) \in \mathscr{C}(L^2(\mathscr{Q}_{\mathrm{E}}) \to L^2(\mathscr{Q}_{\mathrm{E}}))$ by (1.5.6) with ϕ replaced by ϕ_{E}.

(3) Let $T \in \mathscr{C}(\mathscr{M} \to \mathscr{E})$. Define $\Gamma_{\mathrm{Int}}(T) \in \mathscr{C}(L^2(\mathscr{Q}) \to L^2(\mathscr{Q}_{\mathrm{E}}))$ by

$$\Gamma_{\mathrm{Int}}(T){:}\phi(f_1)\cdots\phi(f_n){:} = {:}\phi_{\mathrm{E}}(Tf_1)\cdots\phi_{\mathrm{E}}(Tf_n){:}$$

with $\Gamma_{\mathrm{Int}}(T)1_{\mathrm{M}} = 1_{\mathrm{E}}$, where 1_{E} denotes the identity function in $L^2(\mathscr{Q}_{\mathrm{E}})$.

Definition 1.83. Let $\mathrm{I}_t = \Gamma_{\mathrm{Int}}(\tau_t) : L^2(\mathscr{Q}) \to L^2(\mathscr{Q}_{\mathrm{E}})$, $t \in \mathbb{R}$, be a family of isometries, i. e.,

$$\mathrm{I}_t 1_{\mathrm{M}} = 1_{\mathrm{E}}, \quad \mathrm{I}_t{:}\phi(f_1)\cdots\phi(f_n){:} = {:}\phi_{\mathrm{E}}(\delta_t \otimes f_1)\cdots\phi_{\mathrm{E}}(\delta_t \otimes f_n){:}.$$

Definition 1.84 (Free field Hamiltonian). The self-adjoint operator

$$\hat{H}_{\mathrm{f}} = \mathrm{d}\Gamma(\hat{\omega}_\nu)$$

is called the *free field Hamiltonian* in $L^2(\mathscr{Q})$.

Note the following relationship between H_{f} on $\mathcal{F}_{\mathrm{b}}(\hat{\mathcal{H}}_M)$ and $\hat{H}_{\mathrm{f}}$. We have

$$\hat{H}_{\mathrm{f}} = \theta_{\mathrm{W}} H_{\mathrm{f}} \theta_{\mathrm{W}}^{-1},$$

where $\theta_{\mathrm{W}} : \mathcal{F}_{\mathrm{b}}(\hat{\mathcal{H}}_M) \to L^2(\mathscr{Q})$ is the Wiener–Itô–Segal isomorphism. From the identity $\tau_s^* \tau_t = e^{-|t-s|\hat{\omega}_\nu}$, it follows that

$$\mathrm{I}_t^* \mathrm{I}_s = e^{-|t-s|\hat{H}_{\mathrm{f}}}, \quad s, t \in \mathbb{R}. \tag{1.5.7}$$

From equality (1.5.7), we obtain the following representation.

Proposition 1.85 (Functional integral representation for free field Hamiltonian). *Let $F, G \in L^2(\mathscr{Q})$ and $t \geq 0$. Then*

$$(F, e^{-t\hat{H}_{\mathrm{f}}} G)_{L^2(\mathscr{Q})} = \int_{\mathscr{Q}_{\mathrm{E}}} \overline{F_0}(\phi) G_t(\phi) d\mu_{\mathrm{E}}(\phi), \tag{1.5.8}$$

where $F_0 = \mathrm{I}_0 F$ and $G_t = \mathrm{I}_t G$.

Proof. Noting the factorization property (1.5.7), we see directly that

$$(F, e^{-t\hat{H}_{\mathrm{f}}} G)_{L^2(\mathscr{Q})} = (F, \mathrm{I}_0^* \mathrm{I}_t G)_{L^2(\mathscr{Q})} = (\mathrm{I}_0 F, \mathrm{I}_t G)_{L^2(\mathscr{Q}_{\mathrm{E}})}.$$

Hence the proposition follows. □

In order to extend equality (1.5.8) to Hamiltonians of the form $\hat{H}_{\mathrm{f}}$ + perturbation, we will use the Markov property of projections

$$\mathrm{E}_t = \mathrm{I}_t \mathrm{I}_t^*, \quad t \in \mathbb{R}, \tag{1.5.9}$$

which will be proven below. The following argument shows why the Markov property is useful. Let

$$\hat{H} = \hat{H}_{\mathrm{f}} + H_{\mathrm{I}}$$

with interaction term H_{I}. By the Trotter product formula, we have

$$e^{-t\hat{H}} = \operatorname*{s-lim}_{n\to\infty} (e^{-(t/n)\hat{H}_{\mathrm{f}}} e^{-(t/n)H_{\mathrm{I}}})^n. \tag{1.5.10}$$

Substituting $e^{-|t-s|\hat{H}_{\mathrm{f}}} = \mathrm{I}_t^* \mathrm{I}_s$ into (1.5.10), we have

$$e^{-t\hat{H}} = \operatorname*{s-lim}_{n\to\infty} \mathrm{I}_0^* \left(\prod_{j=1}^n \mathrm{I}_{tj/n} e^{-(t/n)H_{\mathrm{I}}} \mathrm{I}_{tj/n}^* \right) \mathrm{I}_t,$$

where $\prod_{j=1}^n T_j = T_1 T_2 \cdots T_n$. Thus we obtain

$$e^{-t\hat{H}} = \operatorname*{s-lim}_{n\to\infty} \mathrm{I}_0^* \left(\prod_{j=1}^n \mathrm{E}_{tj/n} e^{-(t/n)H_{\mathrm{I}}(tj/n)} \mathrm{E}_{tj/n} \right) \mathrm{I}_t \tag{1.5.11}$$

by the definition of the projection E_s. Here $H_\mathrm{I}(t/n)$ denotes an operator acting on $L^2(\mathscr{Q}_\mathrm{E})$. Applying the Markov property of E_s implies that all E_s in equation (1.5.11) can be disregarded, thus the net result is

$$\begin{aligned}(F, e^{-t\hat{H}}G) &= \lim_{n\to\infty}\left(F_0, \left(\prod_{j=1}^n e^{-(t/n)H_\mathrm{I}(tj/n)}\right)G_t\right)\\ &= \lim_{n\to\infty}\left(F_0, e^{-\sum_{j=1}^\infty (t/n)H_\mathrm{I}(tj/n)}G_t\right)\\ &= \int_{\mathscr{Q}_\mathrm{E}} \overline{F_0}G_t e^{-\int_0^t H_\mathrm{I}(s)ds}d\mu_\mathrm{E}.\end{aligned}$$

We investigate the Markov property of E_t and establish the functional integral representation of $(F, e^{-t\hat{H}}G)$ in the next section.

We finally consider estimates of the operator $\mathrm{I}_a^* e^{\phi_\mathrm{E}(f)}\mathrm{I}_b$ in $L^2(\mathscr{Q})$. Formally,

$$(F, e^{-t\hat{H}}G) = \int_{\mathscr{Q}_\mathrm{E}} \overline{F_0}G_t e^{-\int_0^t H_\mathrm{I}(s)ds}d\mu_\mathrm{E}$$

and the right-hand side above can be rewritten as

$$\int_{\mathscr{Q}_\mathrm{E}} \overline{F_0}G_t e^{-\int_0^t H_\mathrm{I}(s)ds}d\mu_\mathrm{E} = \left(F, \mathrm{I}_0^* e^{-\int_0^t H_\mathrm{I}(s)ds}\mathrm{I}_t G\right).$$

We are interested in $\mathrm{I}_0^* e^{-\int_0^t H_\mathrm{I}(s)ds}\mathrm{I}_t$ as an operator in $L^2(\mathscr{Q})$, in particular, operators of the form $\mathrm{I}_a e^{\phi_\mathrm{E}(f)}\mathrm{I}_b$ are of special interest. As it will be seen below, $e^{\phi_\mathrm{E}(f)}$ is an unbounded operator but $\mathrm{I}_a e^{\phi_\mathrm{E}(f)}\mathrm{I}_b$ is bounded when $a \neq b$. In the massive case, when $\omega(k) = \sqrt{|k|^2 + \nu^2}$ and $\nu > 0$, it can be seen that $e^{\phi_\mathrm{E}(f)} \in \bigcap_{p\geq 1} L^p(\mathscr{Q}_\mathrm{E})$ and $\mathrm{I}_a e^{\phi_\mathrm{E}(f)}\mathrm{I}_b$ becomes a bounded operator by using the so-called hypercontractivity of e^{-tH_f}. This will be discussed in Theorem 1.87 below. In this section we discuss the boundedness of $\mathrm{I}_a^* e^{\phi_\mathrm{E}(f)}\mathrm{I}_b$ also for the massless case.

First, we consider the intertwining properties of I_t and τ_t. We can identify $\phi_\mathrm{E}(f)$ with $\frac{1}{\sqrt{2}}(a_E^*(\hat{f}) + a_E(\tilde{\hat{f}}))$, and $\phi(f)$ with $\frac{1}{\sqrt{2}}(a_M^*(\hat{f}) + a_M(\tilde{\hat{f}}))$, respectively. Under this correspondence, I_t can be identified as a map from $\mathcal{F}_\mathrm{b}(\hat{\mathcal{H}}_M)$ to $\mathcal{F}_\mathrm{b}(\hat{\mathcal{H}}_E)$, i. e.,

$$\mathrm{I}_t \prod_{j=1}^n a_M^*(\hat{f}_j)\Omega_\mathrm{b} = \prod_{j=1}^n a_E^*(\hat{\tau}_t f_j)\Omega_\mathrm{b}^\mathrm{E}$$

and I_t^* from $\mathcal{F}_\mathrm{b}(\hat{\mathcal{H}}_E)$ to $\mathcal{F}_\mathrm{b}(\hat{\mathcal{H}}_M)$. Here $\Omega_\mathrm{b}^\mathrm{E}$ denotes the Fock vacuum in $\mathcal{F}_\mathrm{b}(\hat{\mathcal{H}}_E)$.

Lemma 1.86 (Intertwining properties). *On the finite particle subspace, the relations*

$$\mathrm{I}_t a_M^*(\hat{f}) = a_E^*(\widehat{\tau_t f})\mathrm{I}_t, \tag{1.5.12}$$

$$\mathrm{I}_t a_M(\hat{f}) = a_E(\widehat{\tau_t f})\mathrm{I}_t, \tag{1.5.13}$$

$$\mathrm{I}_t^* a_E^*(\hat{f}) = a_M^*(\widehat{\tau_t^* f})\mathrm{I}_t^*, \tag{1.5.14}$$

$$\mathrm{I}_t^* a_E(\widehat{e_t f}) = a_M(\widehat{\tau_t^* f})\mathrm{I}_t^* \tag{1.5.15}$$

hold, where $e_t = \tau_t\tau_t^*$. In particular,

$$\mathrm{I}_t a_M^*(\widehat{\tau_t^* f}) = a_E^*(\widehat{e_t f})\mathrm{I}_t, \tag{1.5.16}$$

$$\mathrm{I}_t a_M(\widehat{\tau_t^* f}) = a_E(\widehat{e_t f})\mathrm{I}_t = a_E(\hat{f})\mathrm{I}_t, \tag{1.5.17}$$

$$\mathrm{I}_t^* a_E^*(\widehat{\tau_t f}) = a_M^*(\hat{f})\mathrm{I}_t^*, \tag{1.5.18}$$

$$\mathrm{I}_t^* a_E(\widehat{\tau_t f}) = a_M(\hat{f})\mathrm{I}_t^*. \tag{1.5.19}$$

Proof. These intertwining properties can be checked by applying vectors of the form $\prod_{j=1}^n a_M^*(\hat{f}_j)\Omega_\mathrm{b}$ for (1.5.12), (1.5.13), (1.5.16), and (1.5.17), and $\prod_{j=1}^n a_E^*(\hat{f}_j)\Omega_\mathrm{b}^\mathrm{E}$ for (1.5.14), (1.5.15), (1.5.18), and (1.5.19). Note that

$$\begin{aligned} a_E(\widehat{e_t f})\mathrm{I}_t \prod_{j=1}^n a_M^*(\hat{f}_j)\Omega_\mathrm{b} &= a_E(\widehat{e_t f}) \prod_{j=1}^n a_E^*(\widehat{\tau_t f_j})\Omega_\mathrm{b}^\mathrm{E} = \sum_{j=1}^n (e_t\bar{f}, \tau_t f_j) \prod_{i\neq j}^n a_E^*(\widehat{\tau_t f_i})\Omega_\mathrm{b}^\mathrm{E} \\ &= \sum_{j=1}^n (\bar{f}, \tau_t f_j) \prod_{i\neq j}^n a_E^*(\widehat{\tau_t f_i})\Omega_\mathrm{b}^\mathrm{E} = a_E(\hat{f})\mathrm{I}_t \prod_{j=1}^n a_M^*(\hat{f}_j)\Omega_\mathrm{b}. \end{aligned}$$

Then also the second equality of (1.5.17) follows. □

Theorem 1.87. *Suppose that $\hat{f} \in \hat{\mathcal{H}}_E$ and $\widehat{\tau_t^* f}/\sqrt{\omega} \in \hat{\mathcal{H}}_M$ for $t = a, b$ with $a \neq b$. Then*

$$\|\mathrm{I}_a^* e^{\phi_\mathrm{E}(f)}\mathrm{I}_b\| \le 2\exp\Big(\frac{1}{4}\|\hat{f}\|_{\hat{\mathcal{H}}_E} + 2\Big(1 \vee \frac{1}{|a-b|}\Big)(\|\widehat{\tau_a^* f}\|_\omega^2 + \|\widehat{\tau_b^* f}\|_\omega^2)\Big).$$

Here $x \vee y = \max\{x, y\}$ and

$$\|\widehat{\tau_a^* f}\|_\omega = \|\widehat{\tau_a^* f}\|_{\hat{\mathcal{H}}_M} + \|\widehat{\tau_a^* f}/\sqrt{\omega}\|_{\hat{\mathcal{H}}_M} = \|\widehat{\tau_a^* f}/\sqrt{\omega}\|_{L^2(\mathbb{R}^d)} + \|\widehat{\tau_a^* f}/\omega\|_{L^2(\mathbb{R}^d)}.$$

Proof. We identify $\phi_\mathrm{E}(f)$ with $\frac{1}{\sqrt{2}}(a_E^*(\hat{f}) + a_E(\bar{\hat{f}}))$, where $a_E^\sharp$ denotes the annihilation and creation operators in $\mathcal{F}_\mathrm{b}(\hat{\mathcal{H}}_E)$. By the Baker–Campbell–Hausdorff formula, we have

$$e^{\phi_\mathrm{E}(f)} = e^{\frac{1}{\sqrt{2}}a_E^*(\hat{f})} e^{\frac{1}{\sqrt{2}}a_E(\bar{\hat{f}})} e^{\frac{1}{4}\|\hat{f}\|_{\hat{\mathcal{H}}_E}^2}.$$

We compute τ_b^* as

$$(\tau_b f, G)_{-1} = \int_{\mathbb{R}^d} \frac{\bar{\hat{f}}(k)}{\sqrt{|k|}} \frac{\hat{g}(k)}{\sqrt{|k|}} dk = (\hat{f}, \widehat{\tau_b^* G})_{-1/2},$$

where

$$\hat{g}(k) = \sqrt{\frac{2}{\pi}} \int_{\mathbb{R}} \frac{|k|}{k_0^2 + |k|^2} e^{ibk_0}\hat{G}(k_0, k)dk_0.$$

We have $\widehat{\tau_b^* G} = \hat{g}$. The intertwining properties in Lemma 1.86 yield

$$\mathrm{I}_a^* e^{\phi_{\mathrm{E}}(f)} \mathrm{I}_b = e^{\frac{1}{\sqrt{2}} a_M^*(\widehat{\tau_a^* f})} e^{-|a-b| H_{\mathrm{f}}} e^{\frac{1}{\sqrt{2}} a_M(\widetilde{\widehat{\tau_b^* f}})} e^{\frac{1}{4}\|\hat{f}\|^2_{\hat{\mathcal{H}}_E}}.$$

The operators $e^{\frac{1}{\sqrt{2}} a_M^*(\widehat{\tau_a^* f})} e^{-\frac{|a-b|}{2} H_{\mathrm{f}}}$ and $e^{-\frac{|a-b|}{2} H_{\mathrm{f}}} e^{\frac{1}{\sqrt{2}} a_M(\widetilde{\widehat{\tau_b^* f}})}$ are bounded with

$$\|e^{\frac{1}{\sqrt{2}} a_M^*(\widehat{\tau_a^* f})} e^{-\frac{|a-b|}{2} H_{\mathrm{f}}}\| \le \sqrt{2} \exp\Big(2\Big(1 \vee \frac{1}{|a-b|}\Big)\big(\|\widehat{\tau_a^* f}\|^2_{\hat{\mathcal{H}}_M} + \|\widehat{\tau_a^* f}/\sqrt{\omega}\|^2_{\hat{\mathcal{H}}_M}\big)\Big),$$

$$\|e^{-\frac{|a-b|}{2} H_{\mathrm{f}}} e^{\frac{1}{\sqrt{2}} a_M(\widetilde{\widehat{\tau_b^* f}})}\| \le \sqrt{2} \exp\Big(2\Big(1 \vee \frac{1}{|a-b|}\Big)\big(\|\widehat{\tau_b^* f}\|^2_{\hat{\mathcal{H}}_M} + \|\widehat{\tau_b^* f}/\sqrt{\omega}\|^2_{\hat{\mathcal{H}}_M}\big)\Big)$$

by Corollary 1.34. Hence we have

$$\begin{aligned}
\|\mathrm{I}_a^* e^{\phi_{\mathrm{E}}(f)} \mathrm{I}_b\| &\le \|e^{\frac{1}{\sqrt{2}} a_M^*(\widehat{\tau_a^* f})} e^{-\frac{|a-b|}{2} H_{\mathrm{f}}}\| \|e^{-\frac{|a-b|}{2} H_{\mathrm{f}}} e^{\frac{1}{\sqrt{2}} a_M(\widetilde{\widehat{\tau_b^* f}})}\| e^{\frac{1}{4}\|\hat{f}\|^2_{\hat{\mathcal{H}}_E}} \\
&\le 2 \exp\Big(\frac{1}{4}\|\hat{f}\|^2_{\hat{\mathcal{H}}_E} + 2\Big(1 \vee \frac{1}{|a-b|}\Big)\big(\|\widehat{\tau_a^* f}\|^2_{\omega} + \|\widehat{\tau_b^* f}\|^2_{\omega}\big)\Big). \qquad \square
\end{aligned}$$

We single out some special cases of Theorem 1.87 which will be used below in the study of the Nelson Hamiltonian.

Corollary 1.88. *Let $T \ge 0$, and $X : \mathbb{R} \to \mathbb{R}^d$ be a measurable function. Let $f \in \mathcal{H}_M$, i. e., $\hat{f}/\sqrt{\omega} \in L^2(\mathbb{R}^d)$. Define*

$$\Phi_X = \phi_{\mathrm{E}}\Big(\int_0^T (\tau_s f)(\cdot - X(s)) ds\Big).$$

The following hold:
(1) *If $\hat{f}/\omega, \hat{f}/\sqrt{\omega^3} \in L^2(\mathbb{R}^d)$, then*

$$\|\mathrm{I}_0^* e^{\Phi_X} \mathrm{I}_T\| \le 2 \exp\Big(\frac{T}{2}\|\hat{f}/\omega\|^2 + 4(T \vee 1)(\|\hat{f}/\omega\|^2 + \|\hat{f}/\sqrt{\omega^3}\|^2)\Big). \tag{1.5.20}$$

(2) *If $\hat{f}/\omega \in L^2(\mathbb{R}^d)$, then*

$$\|\mathrm{I}_0^* e^{\Phi_X} \mathrm{I}_T\| \le 2 \exp\Big(\frac{T}{2}\|\hat{f}/\omega\|^2 + 4T(T \vee 1)(\|\hat{f}/\sqrt{\omega}\|^2 + \|\hat{f}/\omega\|^2)\Big). \tag{1.5.21}$$

Proof. We see that

$$\Big\|\int_0^T \tau_s f ds\Big\|^2_{\hat{\mathcal{H}}_E} = \int_0^T ds \int_0^T (e^{-|s-t|\omega}\hat{f}, \hat{f})_{\hat{\mathcal{H}}_M} dt = 2\int_0^T ((1 - e^{-s\omega})\hat{f}/\sqrt{\omega}, \hat{f}/\sqrt{\omega})_{\hat{\mathcal{H}}_M} ds \le 2T\|\hat{f}/\omega\|^2.$$

We also see that

$$\left\| \tau_T^* \int_0^T \tau_s f ds \right\|_{\mathcal{H}_M}^2 \le T\|\hat{f}/\omega\|^2, \quad \left\| \tau_T^* \int_0^T \tau_s f ds/\sqrt{\omega} \right\|_{\mathcal{H}_M}^2 \le T\|\hat{f}/\sqrt{\omega^3}\|^2,$$

$$\left\| \tau_0^* \int_0^T \tau_s f ds \right\|_{\mathcal{H}_M}^2 \le T\|\hat{f}/\omega\|^2, \quad \left\| \tau_0^* \int_0^T \tau_s f ds/\sqrt{\omega} \right\|_{\mathcal{H}_M}^2 \le T\|\hat{f}/\sqrt{\omega^3}\|^2.$$

Hence (1) follows by Theorem 1.87. For (2), we estimate as

$$\left\| \tau_T^* \int_0^T \tau_s f ds \right\|_{\mathcal{H}_M}^2 \le T^2\|\hat{f}/\sqrt{\omega}\|^2, \quad \left\| \tau_T^* \int_0^T \tau_s f ds/\sqrt{\omega} \right\|_{\mathcal{H}_M}^2 \le T^2\|\hat{f}/\omega\|^2,$$

$$\left\| \tau_0^* \int_0^T \tau_s f ds \right\|_{\mathcal{H}_M}^2 \le T^2\|\hat{f}/\sqrt{\omega}\|^2, \quad \left\| \tau_0^* \int_0^T \tau_s f ds/\sqrt{\omega} \right\|_{\mathcal{H}_M}^2 \le T^2\|\hat{f}/\omega\|^2. \qquad \square$$

Hence (2) follows.

Corollary 1.89. *Let Φ_X be as in Corollary 1.88. Suppose that $f \in \mathcal{H}_M$, i. e., $\hat{f}/\sqrt{\omega} \in L^2(\mathbb{R}^d)$, and $\hat{f}/\omega, \hat{f}/\sqrt{\omega^3} \in L^2(\mathbb{R}^d)$. Let*

$$E(\hat{f}) = \max \begin{Bmatrix} \frac{1}{2}\|\hat{f}/\omega\|^2 + 4(\|\hat{f}/\omega\|^2 + \|\hat{f}/\sqrt{\omega^3}\|^2) \\ \frac{1}{2}\|\hat{f}/\omega\|^2 + 4(\|\hat{f}/\sqrt{\omega}\|^2 + \|\hat{f}/\omega\|^2) \end{Bmatrix}. \tag{1.5.22}$$

Then

$$\|\mathrm{I}_0^* e^{\Phi_X} \mathrm{I}_T\| \le 2e^{TE(\hat{f})}, \quad T \ge 0. \tag{1.5.23}$$

Proof. We have

$$\|\mathrm{I}_0^* e^{\Phi_X} \mathrm{I}_T\| \le 2\exp\left(\frac{T}{2}\|\hat{f}/\omega\|^2 + 4T(\|\hat{f}/\omega\|^2 + \|\hat{f}/\sqrt{\omega^3}\|^2)\right)$$

for $T \ge 1$, and

$$\|\mathrm{I}_0^* e^{\Phi_X} \mathrm{I}_T\| \le 2\exp\left(\frac{T}{2}\|\hat{f}/\omega\|^2 + 4T(\|\hat{f}/\sqrt{\omega}\|^2 + \|\hat{f}/\omega\|^2)\right)$$

for $T \le 1$. $\square$

By Corollary 1.88 above, it can be seen that if $\hat{f}/\sqrt{\omega}, \hat{f}/\omega \in L^2(\mathbb{R}^d)$, then $\mathrm{I}_0^* e^{\Phi_X} \mathrm{I}_T$ is a bounded operator, and $\sup_{T\ge 0} \frac{1}{T^2} \log \|\mathrm{I}_0^* e^{\Phi_X} \mathrm{I}_T\| < \infty$. Furthermore, the extra condition $\hat{f}/\sqrt{\omega^3} \in L^2(\mathbb{R}^d)$ gives that $\sup_{T\ge 0} \frac{1}{T} \log \|\mathrm{I}_0^* e^{\Phi_X} \mathrm{I}_T\| \le E(\hat{f}) < \infty$.

In Corollary 1.36, we introduced a parameter $q > 1$ and obtained an alternative bound on $e^{a^*(f)}e^{-tH_f}$. Similarly, we can also modify Theorem 1.87 and Corollary 1.88.

Theorem 1.90. *Let $q > 1$. Suppose that $\hat{f} \in \hat{\mathcal{H}}_E$ and $\widehat{\tau_t^* f}/\sqrt{\omega} \in \hat{\mathcal{H}}_M$ for $t = a, b$ with $a \neq b$. Then*

$$\|\mathrm{I}_a^* e^{\Phi_{\mathrm{E}}(f)} \mathrm{I}_b\| \leq \frac{q}{q-1} \exp\left(\frac{1}{4}\|\hat{f}\|_{\hat{\mathcal{H}}_E}^2 + q\left(1 \vee \frac{1}{|a-b|}\right)(\|\widehat{\tau_a^* f}\|_\omega^2 + \|\widehat{\tau_b^* f}\|_\omega^2)\right).$$

Corollary 1.91. *Let $q > 1$ and Φ_X be as in Corollary 1.88. Suppose that $f \in \mathcal{H}_M$, i. e., $\hat{f}/\sqrt{\omega} \in L^2(\mathbb{R}^d)$. Then the following hold:*

(1) *If $\hat{f}/\omega, \hat{f}/\sqrt{\omega^3} \in L^2(\mathbb{R}^d)$, then*

$$\|\mathrm{I}_0^* e^{\Phi_X} \mathrm{I}_T\| \leq \frac{q}{q-1} \exp\left(\frac{T}{2}\|\hat{f}/\omega\|^2 + 2q(T \vee 1)(\|\hat{f}/\omega\|^2 + \|\hat{f}/\sqrt{\omega^3}\|^2)\right). \tag{1.5.24}$$

In particular, let

$$E_q(\hat{f}) = \max\left\{\begin{array}{l} \frac{1}{2}\|\hat{f}/\omega\|^2 + 2q(\|\hat{f}/\omega\|^2 + \|\hat{f}/\sqrt{\omega^3}\|^2) \\ \frac{1}{2}\|\hat{f}/\omega\|^2 + 2q(\|\hat{f}/\sqrt{\omega}\|^2 + \|\hat{f}/\omega\|^2) \end{array}\right\}. \tag{1.5.25}$$

Then it follows that

$$\|\mathrm{I}_0^* e^{\Phi_X} \mathrm{I}_T\| \leq \frac{q}{q-1} e^{TE_q(\hat{f})}, \quad T \geq 1. \tag{1.5.26}$$

(2) *If $\hat{f}/\omega \in L^2(\mathbb{R}^d)$, then*

$$\|\mathrm{I}_0^* e^{\Phi_X} \mathrm{I}_T\| \leq \frac{q}{q-1} \exp\left(\frac{T}{2}\|\hat{f}/\omega\|^2 + qT(T \vee 1)(\|\hat{f}/\sqrt{\omega}\|^2 + \|\hat{f}/\omega\|^2)\right). \tag{1.5.27}$$

1.5.2 Markov property of projections

Let $\mathcal{O} \subset \mathbb{R}$ and define

$$\mathscr{E}(\mathcal{O}) = \{f \in \mathscr{E} \mid \operatorname{supp} f \subset \mathcal{O} \times \mathbb{R}^d\}.$$

We denote the projection $\mathscr{E} \to \mathscr{E}(\mathcal{O})$ by $e_{\mathcal{O}}$. Let $\Sigma_{\mathcal{O}}$ be the minimal σ-field generated by $\{\phi_{\mathrm{E}}(f) \mid f \in \mathscr{E}(\mathcal{O})\}$, and define

$$\mathscr{E}_{\mathcal{O}} = \{\Phi \in L^2(\mathscr{Q}_{\mathrm{E}}) \mid \Phi \text{ is } \Sigma_{\mathcal{O}}\text{-measurable}\}.$$

Also, let $e_t = \tau_t \tau_t^*$, $t \in \mathbb{R}$. The collection $\{e_t\}_{t \in \mathbb{R}}$ is a family of projections from $\mathscr{E}$ to $\operatorname{Ran}(\tau_t)$. Let Σ_t, $t \in \mathbb{R}$, be the minimal σ-field generated by $\{\phi_{\mathrm{E}}(f) \mid f \in \operatorname{Ran}(e_t)\}$, and define

$$\mathscr{E}_t = \{\Phi \in L^2(\mathscr{Q}_E) \mid \Phi \text{ is } \Sigma_t\text{-measurable}\}, \quad t \in \mathbb{R}.$$

We will see below that $F \in \mathscr{E}([a,b])$ can be characterized by $\operatorname{supp} F \subset [a,b] \times \mathbb{R}^d$.

Lemma 1.92.
(1) *$\mathscr{E}(\{t\}) = \operatorname{Ran}(e_t)$ and every $f \in \operatorname{Ran}(e_t)$ can be expressed as $f = \delta_t \otimes g$ for some $g \in \mathscr{M}$. In particular, $e_{\{t\}} = e_t$.*
(2) *$\mathscr{E}([a,b]) = \overline{\text{L.H.}\{f \in \mathscr{E} \mid f \in \operatorname{Ran}(e_t), a \le t \le b\}}^{\|\cdot\|_{-1}}$ holds.*

Proof. (1) The key fact is that any $f \in \mathscr{S}'_{\text{real}}(\mathbb{R}^{d+1})$ with support $\{t\} \times \mathbb{R}^d$ can be expressed as

$$f(x_0, x) = \sum_{j=0}^{n} \delta^{(j)}(x_0 - t) \otimes g_j(x), \quad (x_0, x) \in \mathbb{R} \times \mathbb{R}^d,$$

where $g_j \in \mathscr{S}'_{\text{real}}(\mathbb{R}^d)$ and $\delta^{(j)}$ is the jth derivative of the delta distribution δ. We have

$$\hat{f}(k_0, k) = \frac{1}{\sqrt{2\pi}} \sum_{j=0}^{n} (ik_0)^j e^{-ik_0 t} \hat{g}_j(k), \quad (k_0, k) \in \mathbb{R} \times \mathbb{R}^d,$$

and hence $n = 0$ and $g_0 \in \mathscr{M}$ is necessary and sufficient for $f \in \mathscr{E}$. Then $f \in \mathscr{E}(\{t\})$ if and only if there exists $g \in \mathscr{M}$ such that $f = \delta_t \otimes g$, which gives that $f \in \operatorname{Ran}(e_t)$. In particular, $\mathscr{E}(\{t\}) \subset \operatorname{Ran}(e_t)$. The inclusion $\operatorname{Ran}(e_t) \subset \mathscr{E}(\{t\})$ is trivial.

(2) Let $A_1 = \mathscr{E}((a,b))$ and take $f \in \mathscr{E}([a,b])$. For $\lambda \in (1,\infty)$, we have $f_\lambda(x_0, x) = f(\lambda x_0 + (1-\lambda)(a+b)/2, x) \in A_1$, and $\|f_\lambda - f\|_{-1} \to 0$ as $\lambda \to 1$. Hence A_1 is dense in $\mathscr{E}([a,b])$.

Let $A_2 = \{f \in A_1 \mid f(x_0, \cdot) \in C^\infty(\mathbb{R}^d) \text{ for all } x_0 \in \mathbb{R}\}$, and define

$$f_\varepsilon(x_0, x) = \int_{\mathbb{R}^d} \rho_\varepsilon(x-y) f(x_0, y)\, dy,$$

where $\rho_\varepsilon(x) = \rho(x/\varepsilon)/\varepsilon^d$, $\rho \in C_0^\infty(\mathbb{R}^d)$, $\rho(x) \ge 0$, $\int_{\mathbb{R}^d} \rho(x) dx = 1$ and $\operatorname{supp}\rho \subset \{x \in \mathbb{R}^d \mid |x| \le 1\}$. Then $f_\varepsilon \in A_2$ for all sufficiently small ε and $\|f_\varepsilon - f\|_{-1} = 0$ as $\varepsilon \to 0$. Hence A_2 is dense in $\mathscr{E}([a,b])$.

Next we show that $A_3 = C_0^\infty(\mathbb{R}^{d+1}) \cap A_1$ is dense in $\mathscr{E}([a,b])$. For $\nu > 0$, we define

$$\chi_n(x) = \begin{cases} 1, & |x| \le n, \\ 0, & |x| > n+1, \end{cases} \qquad \chi_n \in C_0^\infty(\mathbb{R}^d) \text{ and } 0 \le \chi_n \le 1.$$

For $f \in A_2$, we have $f_n = \chi_n f \in A_2$ and $\|f_n - f\|_{-1} = 0$ as $n \to \infty$. Since A_2 is dense in A_3, A_3 is dense in $\mathscr{E}([a,b])$. If $\nu = 0$, let $A_4 = \{f \in C_0^\infty(\mathbb{R}^{d+1}) \cap A_1 \mid \operatorname{supp}\hat{f} \subset \mathbb{R}^{d+1} \setminus \{0\}\}$. We similarly show that A_4 is dense in $\mathscr{E}([a,b])$.

Finally, we show that any $f \in A_3$ can be represented as a limit of finite linear sums of vectors in $\operatorname{Ran}(e_t)$, $a < t < b$. Take $f \in A_3$ and define

$$f_n(x_0, x) = \sum_{j=-\infty}^{\infty} \frac{1}{n} \delta\left(x_0 - \frac{j}{n}\right) f\left(\frac{j}{n}, x\right).$$

We have

$$\hat{f}_n(k_0, k) = \sum_{j=-\infty}^{\infty} \frac{1}{n} g\left(\frac{j}{n}, k\right) \frac{e^{-ijk_0/n}}{\sqrt{2\pi}},$$

where $g(X, k) = (2\pi)^{-d/2} \int_{\mathbb{R}^d} f(X, x) e^{-ikx} dx$. Note that

$$|\hat{f}_n(k_0, k)| \le c_N / (|k|^2 + 1)^N$$

for some N and $\hat{f}_n(k_0, k) \to \hat{f}(k_0, k)$ as $n \to \infty$ for every k. Hence $\|f_n - f\|_{-1} \to 0$ as $n \to \infty$ by the Lebesgue dominated convergence theorem. This completes the proof of (2) with $\nu \neq 0$. In the case of $\nu = 0$, we use A_3 replaced by A_4. □

Lemma 1.93. *Let $a \le b \le t \le c \le d$. Then:*
(1) $e_a e_b e_c = e_a e_c$;
(2) $e_{[a,b]} e_t e_{[c,d]} = e_{[a,b]} e_{[c,d]}$;
(3) $e_c e_b e_a = e_c e_a$;
(4) $e_{[c,d]} e_t e_{[a,b]} = e_{[c,d]} e_{[a,b]}$.

Proof. Since

$$e_a e_b e_c = \tau_a \tau_a^* \tau_b \tau_b^* \tau_c \tau_c^* = \tau_a e^{-|a-b|\hat{\omega}_\nu} e^{-|b-c|\hat{\omega}_\nu} \tau_c^* = \tau_a e^{-|a-c|\hat{\omega}_\nu} \tau_c^* = \tau_a \tau_a^* \tau_c \tau_c^* = e_a e_c,$$

(1) readily follows. Take $f, g \in \mathscr{E}$. By Lemma 1.92 (2), it is clear that

$$e_{[c,d]} f = \underset{n\to\infty}{\text{s-lim}} \sum_{\alpha=1}^{N_n} f_{n_\alpha}, \quad f_{n_\alpha} \in \text{Ran}(e_{t_{n_\alpha}}), \quad t_{n_\alpha} \in [c, d],$$

$$e_{[a,b]} g = \underset{m\to\infty}{\text{s-lim}} \sum_{\beta=1}^{N_m} g_{m_\beta}, \quad g_{m_\beta} \in \text{Ran}(e_{t_{m_\beta}}), \quad t_{m_\beta} \in [a, b].$$

Hence by (1), we have

$$(e_{[a,b]} e_t e_{[c,d]} f, g) = \lim_{m,n\to\infty} \sum_{\alpha,\beta=1}^{N_n, M_m} (e_t f_{n_\alpha}, g_{m_\alpha}) = \lim_{m,n\to\infty} \sum_{\alpha,\beta=1}^{N_n, M_m} (f_{n_\alpha}, g_{m_\alpha}) = (e_{[a,b]} e_{[c,d]} f, g).$$

This gives (2). Statements (3) and (4) are similarly proven. □

Define

$$\mathrm{E}_t = \mathrm{I}_t \mathrm{I}_t^* = \Gamma_{\mathrm{E}}(e_t), \quad t \in \mathbb{R} \quad \text{and} \quad \mathrm{E}_{\mathscr{O}} = \Gamma_{\mathrm{E}}(e_{\mathscr{O}}), \quad \mathscr{O} \subset \mathbb{R}.$$

Notice that $\mathrm{E}_{\{t\}} = \mathrm{E}_t$ for $t \in \mathbb{R}$ by Lemma 1.92. The following relationships hold between the projections $\mathrm{E}_{[a,b]}$ and the set of $\Sigma_{[a,b]}$-measurable functions.

Proposition 1.94.
(1) $\mathrm{Ran}(\mathrm{E}_{[a,b]}) = \mathscr{E}_{[a,b]}$;
(2) $\mathrm{E}_{[a,b]}\mathrm{E}_t\mathrm{E}_{[c,d]} = \mathrm{E}_{[a,b]}\mathrm{E}_{[c,d]}$ *and* $\mathrm{E}_{[c,d]}\mathrm{E}_t\mathrm{E}_{[a,b]} = \mathrm{E}_{[c,d]}\mathrm{E}_{[a,b]}$ *hold for* a, b, c, d *such that* $a \le b \le t \le c \le d$;
(3) $\mathrm{E}_{[a,b]}\mathrm{E}_{[c,d]} = \mathrm{E}_{[c,d]}\mathrm{E}_{[a,b]} = \mathrm{E}_{[a,b]}$ *if* $[a,b] \subset [c,d]$.

Proof. Let $\{g_n\}_{n=1}^\infty$ be a complete orthonormal system of $\mathscr{E}([a,b])$. We have

$$\mathrm{Ran}(\mathrm{E}_{[a,b]}) = \overline{\mathrm{L.H.}\{:\phi_{\mathrm{E}}(g_1)^{n_1}\cdots\phi_{\mathrm{E}}(g_k)^{n_k}: \mid g_j \in \{g_n\}_{n=1}^\infty, n_j \ge 0, j = 1,\ldots,k, k \ge 0\}},$$

which implies that $\mathrm{Ran}(\mathrm{E}_{[a,b]}) \subset \mathscr{E}_{[a,b]}$. Noticing that $:\exp(i\phi_{\mathrm{E}}(f)): \in \mathrm{Ran}(\mathrm{E}_{[a,b]})$, we obtain $\exp(i\phi_{\mathrm{E}}(f)) \in \mathrm{Ran}(\mathrm{E}_{[a,b]})$. Thus $F(\phi_{\mathrm{E}}(f_1),\ldots,\phi_{\mathrm{E}}(f_n)) \in \mathrm{Ran}(\mathrm{E}_{[a,b]})$, for $F \in \mathscr{S}(\mathbb{R}^n)$ and $f_1,\ldots,f_n \in \mathscr{E}([a,b])$. Since

$$\{F(\phi_{\mathrm{E}}(f_1)\cdots\phi_{\mathrm{E}}(f_n)) \mid F \in \mathscr{S}(\mathbb{R}^n), f_1,\ldots,f_n \in \mathscr{E}([a,b])\}$$

is dense in $\mathscr{E}_{[a,b]}$, we conclude that $\mathscr{E}_{[a,b]} \subset \mathrm{Ran}(\mathrm{E}_{[a,b]})$. Hence (1) is obtained. Claim (2) follows from (2) and (4) of Lemma 1.93. Since a $\Sigma_{[a,b]}$-measurable function is also $\Sigma_{[c,d]}$-measurable, (3) also follows. □

Remark 1.95. Part (1) of Proposition 1.94 implies that $\mathrm{E}_{[a,b]}$ is a projection to the set of $\Sigma_{[a,b]}$-measurable functions in $L^2(\mathscr{Q}_{\mathrm{E}})$. This is a conditional expectation, and we can write $\mathrm{E}_{[a,b]}F$ as $\mathbb{E}[F|\Sigma_{[a,b]}]$. We also denote E_tF by $\mathbb{E}[F|\Sigma_t]$. Part (2) of Proposition 1.94 is called the *Markov property* of the family E_s, $s \in \mathbb{R}$.

Proposition 1.96 (Markov property). *Let* $F \in \mathscr{E}_{s+t}$. *Then*

$$\mathbb{E}[F|\Sigma_{(-\infty,s]}] = \mathbb{E}[F|\Sigma_s].$$

Proof. It follows that $\mathbb{E}[F|\Sigma_{(-\infty,s]}] = \mathrm{E}_{(-\infty,s]}\mathrm{E}_{s+t}F = \mathrm{E}_{(-\infty,s]}\mathrm{E}_s\mathrm{E}_{s+t}F$ from the Markov property. Since $\mathrm{E}_{(-\infty,s]}\mathrm{E}_s = \mathrm{E}_{(-\infty,s]}\mathrm{E}_{\{s\}} = \mathrm{E}_{\{s\}} = \mathrm{E}_s$ by Proposition 1.94, we see that $\mathrm{E}_{(-\infty,s]}\mathrm{E}_s\mathrm{E}_{s+t}F = \mathrm{E}_s\mathrm{E}_{s+t}F = \mathrm{E}_sF = \mathbb{E}[F|\Sigma_s]$. □

1.5.3 Feynman–Kac–Nelson formula

Consider the polynomial

$$P(X) = a_{2n}X^{2n} + a_{2n-1}X^{2n-1} + \cdots + a_1X + a_0$$

with $a_{2n} > 0$. For $f \in \mathscr{M}$, define

$$H_{\mathrm{I}} = {:}P(\phi(f)){:}.$$

Note that H_{I} is a polynomial of $\phi(f)$ of degree $2n$ and there exists $M > -\infty$ such that $H_{\mathrm{I}} > M$. Define the self-adjoint operator

$$H_P = \hat{H}_{\mathrm{f}} \dotplus H_{\mathrm{I}} \tag{1.5.28}$$

acting in $L^2(\mathscr{Q})$. As was already mentioned in this section, a key ingredient in constructing a functional integral representation of $(F, e^{-tH_P}G)$ is the Trotter product formula and the Markov property of the family E_s, $s \in \mathbb{R}$. Before going to state the functional integral representation of e^{-tH_P}, we give a corollary to Lemma 1.94.

Corollary 1.97. *Let $F \in \mathscr{E}_{[a,b]}$, $G \in \mathscr{E}_{[c,d]}$ with $a \le b \le t \le c \le d$. Then it follows that $(F, E_t G) = (F, G)$.*

Proof. By Proposition 1.94, it is immediate to see that

$$(F, E_t G) = (\mathrm{E}_{[a,b]}F, \mathrm{E}_t \mathrm{E}_{[c,d]} G) = (\mathrm{E}_{[a,b]}F, \mathrm{E}_{[c,d]} G) = (F, G). \qquad \square$$

Now we can prove the functional integral representation of H_P given by (1.5.28).

Theorem 1.98 (Feynman–Kac–Nelson formula). *Let $F, G \in L^2(\mathscr{Q})$. Then*

$$(F, e^{-tH_P}G) = \int_{\mathscr{Q}_{\mathrm{E}}} \overline{F_0} G_t \exp\left(- \int_0^t {:}P(\phi_{\mathrm{E}}(\delta_s \otimes f)){:}ds \right) d\mu_{\mathrm{E}},$$

where $F_0 = \mathrm{I}_t F$ and $G_t = \mathrm{I}_t G$.

Proof. By the Trotter product formula and the fact $e^{-|s-t|H_{\mathrm{f}}} = \mathrm{I}_t^* \mathrm{I}_s$, we have

$$\begin{aligned}
(F, e^{-tH_P}G) &= \lim_{n\to\infty} (F, (e^{-(t/n)H_{\mathrm{f}}} e^{-(t/n)H_{\mathrm{I}}})^n G) \\
&= \lim_{n\to\infty} \left(F_0, \prod_{i=1}^n (\mathrm{I}_{ti/n} e^{-(t/n)H_{\mathrm{I}}} \mathrm{I}_{ti/n}^*) G_t \right) \\
&= \lim_{n\to\infty} \left(F_0, \prod_{i=1}^n (\mathrm{E}_{ti/n} R_i \mathrm{E}_{ti/n}) G_t \right),
\end{aligned} \tag{1.5.29}$$

where $R_i = \exp(-(t/n){:}P(\phi_{\mathrm{E}}(\delta_{it/n} f)){:})$ and we used that

$$\mathrm{I}_s \exp(-tH_{\mathrm{I}}) \mathrm{I}_s^* = \mathrm{E}_s \exp(-t{:}P(\phi_{\mathrm{E}}(\delta_s \otimes f)){:}) \mathrm{E}_s \tag{1.5.30}$$

as operators. In fact, we can check that

$$\mathrm{I}_t \phi(f){:}\prod_{i=1}^n \phi(f_i){:} = \mathrm{I}_t \left({:}\phi(f) \prod_{i=1}^n \phi(f_i){:} + \frac{1}{2} \sum_{j=1}^n (f, f_j) {:}\prod_{i \ne j}^n \phi(f_i){:} \right)$$

$$= :\phi_{\mathrm{E}}(\delta_t \otimes f) \prod_{i=1}^{n} \phi_{\mathrm{E}}(\delta_t \otimes f_i): + \frac{1}{2} \sum_{j=1}^{n} (\delta_t \otimes f, \delta_t \otimes f_j) :\prod_{i \neq j}^{n} \phi(\delta_t \otimes f_i):$$

$$= \phi_{\mathrm{E}}(\delta_t \otimes f) \mathrm{I}_t :\prod_{i=1}^{n} \phi(f_i):.$$

Thus $\mathrm{I}_t \phi(f) \mathrm{I}_t^* = \phi_{\mathrm{E}}(\delta_t \otimes f)\mathrm{E}_t = \mathrm{E}_t \phi_{\mathrm{E}}(\delta_t \otimes f)\mathrm{E}_t$. Inductively, it follows that

$$\mathrm{I}_t \left(\prod_{j=1}^{n} \phi(f_j) \right) \mathrm{I}_t^* = \mathrm{E}_t \left(\prod_{j=1}^{n} \phi_{\mathrm{E}}(\delta_t \otimes f_j) \right) \mathrm{E}_t.$$

A limiting argument gives then (1.5.30). Notice that

$$\left(F_0, \prod_{i=1}^{n} (\mathrm{E}_{ti/n} R_i \mathrm{E}_{ti/n}) G_t \right) = \left(\underbrace{F_0}_{\in \mathscr{E}_{\{0\}}}, \mathrm{E}_{t/n} \underbrace{R_1 \mathrm{E}_{t/n} (\mathrm{E}_{2t/n} R_2 \mathrm{E}_{2t/n}) \cdots (\mathrm{E}_t R_n \mathrm{E}_t) G_t}_{\in \mathscr{E}_{[t/n,t]}} \right). \tag{1.5.31}$$

By Corollary 1.97, we can remove $\mathrm{E}_{t/n}$ in (1.5.31). Furthermore, note that

$$\left(F_0, \prod_{i=1}^{n} (\mathrm{E}_{ti/n} R_i \mathrm{E}_{ti/n}) G_t \right) = \left(\underbrace{R_1 F_0}_{\in \mathscr{E}_{[0,t/n]}}, \mathrm{E}_{t/n} \underbrace{(\mathrm{E}_{2t/n} R_2 \mathrm{E}_{2t/n}) \cdots (\mathrm{E}_t R_n \mathrm{E}_t) G_t}_{\in \mathscr{E}_{[2t/n,t]}} \right). \tag{1.5.32}$$

Similarly, $\mathrm{E}_{t/n}$ can be removed in (1.5.32), and since

$$\left(F_0, \prod_{i=1}^{n} (\mathrm{E}_{ti/n} R_i \mathrm{E}_{ti/n}) G_t \right) = \left(\underbrace{R_1 F_0}_{\in \mathscr{E}_{[0,t/n]}}, \mathrm{E}_{2t/n} \underbrace{R_2 \mathrm{E}_{2t/n} \cdots (\mathrm{E}_t R_n \mathrm{E}_t) G_t}_{\in \mathscr{E}_{[2t/n,t]}} \right),$$

also $\mathrm{E}_{2t/n}$ can be removed. Recursively, we can delete all E_s in the rightmost side of (1.5.29). Thus

$$\begin{aligned}
(F, e^{-tH_P} G) &= \lim_{n\to\infty} (F_0, R_1 \cdots R_n G_t) \\
&= \lim_{n\to\infty} \left(F_0, \exp\left(-\sum_{j=1}^{n} (t/n) :P(\phi_{\mathrm{E}}(\delta_{jt/n} \otimes f)): \right) G_t \right) \\
&= \int_{\mathscr{Q}_{\mathrm{E}}} \overline{F_0} G_t \exp\left(-\int_0^t :P(\phi_{\mathrm{E}}(\delta_s \otimes f)): ds \right) d\mu_{\mathrm{E}}.
\end{aligned}$$

□

1.5.4 van Hove Hamiltonian

We consider the so-called *van Hove Hamiltonian* H_{vH}, which is the simplest model in quantum field theory. It is given by H_P by choosing $P(x) = x$, i. e.,

$$H_{\mathrm{vH}} = \hat{H}_{\mathrm{f}} + \phi(f). \tag{1.5.33}$$

Alternatively, H_{vH} can be also defined by $H_{\mathrm{f}} + \Phi(\hat{f})$ on $\mathcal{F}_{\mathrm{b}}(\hat{\mathcal{H}}_M)$, where

$$\Phi(\hat{f}) = \frac{1}{\sqrt{2}}(a_M^*(\hat{f}) + a_M(\tilde{\hat{f}})), \quad \hat{f} \in \hat{\mathcal{H}}_M.$$

Here a_M and a_M^* are the annihilation operator and the creation operator in $\mathcal{F}_{\mathrm{b}}(\hat{\mathcal{H}}_M)$, respectively, and they satisfy canonical commutation relations (1.5.4). Suppose that $\hat{f}/\omega, \hat{f}/\sqrt{\omega} \in L^2(\mathbb{R}^d)$ and f is a real-valued function. Then H_{vH} is self-adjoint on $D(\hat{H}_{\mathrm{f}})$ and bounded from below by the Kato–Rellich theorem.

Corollary 1.99 (Functional integral representation of van Hove Hamiltonian). *We have*

$$(\Phi, e^{-tH_{\mathrm{vH}}}\Psi)_{L^2(\mathcal{Q})} = (\mathrm{I}_0\Phi, e^{-\phi_{\mathrm{E}}(\int_0^t \delta_s \otimes f ds)}\mathrm{I}_t\Psi)_{L^2(\mathcal{Q}_{\mathrm{E}})}. \tag{1.5.34}$$

Proof. Let $H_{\mathrm{vH}}^{\varepsilon} = \hat{H}_{\mathrm{f}} + y_{\varepsilon}(\phi(f))$, where $y_{\varepsilon}(x) = x + \varepsilon x^2$ for $\varepsilon \geq 0$. $H_{\mathrm{vH}}^{\varepsilon}$ is self-adjoint on $D(\hat{H}_{\mathrm{f}})$ and essentially self-adjoint on any core of $\hat{H}_{\mathrm{f}}$ for sufficiently small $0 \leq \varepsilon$ by the Kato–Rellich theorem. In particular, $e^{-tH_{\mathrm{vH}}^{\varepsilon}} \to e^{-tH_{\mathrm{vH}}}$ strongly as $\varepsilon \downarrow 0$. By the Feynman–Kac–Nelson formula, it can be seen that the semigroup generated by $H_{\mathrm{vH}}^{\varepsilon}$ is represented as

$$(\Phi, e^{-tH_{\mathrm{vH}}^{\varepsilon}}\Psi)_{L^2(\mathcal{Q})} = (\mathrm{I}_0\Phi, e^{-\int_0^t y_{\varepsilon}(\phi_{\mathrm{E}}(\int_0^t \delta_s \otimes f))ds}\mathrm{I}_t\Psi)_{L^2(\mathcal{Q}_{\mathrm{E}})}. \tag{1.5.35}$$

Taking $\varepsilon \downarrow 0$ on both sides, the result follows. □

In order to factorize $e^{-\phi_{\mathrm{E}}(\int_0^t \delta_s \otimes f ds)}$ in (1.5.34) in a suitable form, we recall that

$$\phi_{\mathrm{E}}(F) \cong \frac{1}{\sqrt{2}}(a_E^*(\hat{F}) + a_E(\tilde{\hat{F}})), \quad \hat{F} \in \hat{\mathcal{H}}_E, \tag{1.5.36}$$

$$\phi(f) \cong \frac{1}{\sqrt{2}}(a_M^*(\hat{f}) + a_M(\tilde{\hat{f}})), \quad \hat{f} \in \hat{\mathcal{H}}_M, \tag{1.5.37}$$

where $a_E^*(\hat{F})$ and $a_E(\hat{F})$ are the creation and annihilation operators, respectively, on boson Fock space $\mathcal{F}_{\mathrm{b}}(\hat{\mathcal{H}}_E)$, i. e., they satisfy

$$[a_E(\hat{F}), a_E(\hat{G})] = 0 = [a_E^*(\hat{F}), a_E^*(\hat{G})], \quad [a_E(\hat{F}), a_E^*(\hat{G})] = (\tilde{\hat{F}}, \hat{G})_{\hat{\mathcal{H}}_E}.$$

We use the identifications (1.5.36) and (1.5.37). By the Baker–Campbell–Hausdorff formula, we see that

$$e^{-\phi_{\mathrm{E}}(\int_0^t \delta_s \otimes f ds)} = e^{-\frac{1}{\sqrt{2}}a_E^*(\int_0^t \widehat{\delta_s \otimes f} ds)}e^{-\frac{1}{\sqrt{2}}a_E(\int_0^t \widetilde{\widehat{\delta_s \otimes f}} ds)}e^{\frac{1}{4}\|\int_0^t \delta_s \otimes f ds\|_{\mathcal{H}_E}^2}.$$

Recall that $\tau_s f = \delta_s \otimes f$ and $\tau_s^* \tau_t = e^{-|s-t|\hat{\omega}}$. The intertwining properties

$$\mathrm{I}_0^* e^{a_E^*(\hat{F})} = e^{a_M^*(\widehat{\tau_0^* F})}\mathrm{I}_0^*, \quad \mathrm{I}_t e^{a_M(\widetilde{\widehat{\tau_t^* F}})} = e^{a_E(\tilde{\hat{F}})}\mathrm{I}_t$$

are satisfied. Thus we have

$$\mathrm{I}_0^* e^{-\phi_{\mathrm{E}}(\int_0^t \delta_s \otimes f ds)} \mathrm{I}_t = e^{-\frac{1}{\sqrt{2}} a_M^*(\int_0^t e^{-s\omega} \hat{f} ds)} e^{-tH_{\mathrm{f}}} e^{-\frac{1}{\sqrt{2}} a_M(\int_0^t e^{-(t-s)\omega} \tilde{\hat{f}} ds)} e^{\frac{1}{2} W_t}, \tag{1.5.38}$$

where

$$W_t = \frac{1}{2} \left\| \int_0^t \delta_s \otimes f ds \right\|_{\mathcal{H}_E}^2 = \frac{1}{2} \int_0^t ds \int_0^t dr \int_{\mathbb{R}^d} e^{-|s-r|\omega(k)} \frac{|\hat{f}(k)|^2}{\omega(k)} dk.$$

The following result gives an expression of the van Hove Hamiltonian in terms of exponentials of the creation and annihilation operators.

Proposition 1.100. *Denote*

$$U(f,t) = -\frac{1}{\sqrt{2}} \frac{1}{\omega} (1 - e^{-t\omega}) \hat{f}, \quad \tilde{U}(f,t) = -\frac{1}{\sqrt{2}} \frac{1}{\omega} (1 - e^{-t\omega}) \tilde{\hat{f}}.$$

Then $U(f,t), \tilde{U}(f,t) \in \hat{\mathcal{H}}_M$ and it follows that

$$(\Phi, e^{-tH_{\mathrm{vH}}} \Psi)_{L^2(\mathcal{Q})} = \left(\Phi, e^{a_M^*(U(f,t))} e^{-t\hat{H}_{\mathrm{f}}} e^{a_M(\tilde{U}(f,t))} \Psi \right)_{L^2(\mathcal{Q})} e^{\frac{1}{2} W_t}. \tag{1.5.39}$$

Proof. From (1.5.34) and (1.5.38), it follows that

$$(\Phi, e^{-tH_{\mathrm{vH}}} \Psi)_{L^2(\mathcal{Q})} = \left(\Phi, e^{a_M^*(S(f,t))} e^{-t\hat{H}_{\mathrm{f}}} e^{a_M(\tilde{S}(f,t))} \Psi \right)_{L^2(\mathcal{Q})} e^{\frac{1}{2} W_t},$$

where

$$S(f,t) = -\frac{1}{\sqrt{2}} \int_0^t e^{-s\omega} \hat{f} ds, \quad \tilde{S}(f,t) = -\frac{1}{\sqrt{2}} \int_0^t e^{-s\omega} \tilde{\hat{f}} ds,$$

and $S(f,t) = U(f,t)$, $\tilde{S}(f,t) = \tilde{U}(f,t)$. □

We note that the operator $e^{a_M^*(U(f,t))} e^{-t\hat{H}_{\mathrm{f}}} e^{a_M(\tilde{U}(f,t))}$ given in (1.5.39) is bounded, since $e^{a_M^*(U(f,t))} e^{-(t/2)\hat{H}_{\mathrm{f}}}$ and $e^{-(t/2)H_{\mathrm{f}}} e^{a_M(\tilde{U}(f,t))}$ are bounded for all $t > 0$.

1.6 Infinite dimensional Ornstein–Uhlenbeck processes

1.6.1 Abstract theory of Gaussian measures on Hilbert spaces

As was explained in Remark 1.75, it is not possible to construct a Gaussian measure on any Hilbert space $\mathcal{K}$ such that the covariance form is defined by the scalar product on $\mathcal{K}$. Nevertheless, if the covariance is given in the form $(A^{-1}\cdot, A^{-1}\cdot)_{\mathcal{K}}$ with a suitable self-adjoint operator A having a Hilbert–Schmidt inverse, i. e., such that $\mathrm{Tr}(A^{-2}) < \infty$, then there exists a Gaussian measure on an enlarged Hilbert space.

We define $\mathbb{R}^\infty$ by the infinite direct product $\prod_{n=1}^\infty \mathbb{R}$, and define a topology on $\mathbb{R}^\infty$ by the product topology. Let $\mathcal{K}$ be a separable Hilbert space over $\mathbb{R}$, $L_1^+(\mathcal{K})$ be the set of symmetric positive trace class operators on $\mathcal{K}$, and consider

$$Q \in L_1^+(\mathcal{K}), \quad a \in \mathcal{K}.$$

Let $\{e_n\}_{n=1}^\infty$ be a complete orthonormal system of $\mathcal{K}$ such that $Qe_n = \lambda_n e_n$. Hence $\operatorname{Tr} Q = \sum_{j=1}^\infty \lambda_j$. Consider

$$\ell^2 = \left\{ x = (x_k)_{k=1}^\infty \in \mathbb{R}^\infty \;\middle|\; \sum_{k=1}^\infty |x_k|^2 < \infty \right\}.$$

Define the map $\gamma : \mathcal{K} \to \ell^2$ by $\gamma(x) = (x_k)_{k=1}^\infty$, where $x_k = (e_k, x)_\mathcal{K}$. Then γ is a unitary operator and we write $\gamma(a) = (a_k)_{k=1}^\infty$. Let $\mu_k = N_{a_k,\lambda_k}$ be a Gaussian measure on $\mathbb{R}$ with mean a_k and covariance λ_k. Also, let

$$\bar{\mu} = \prod_{k=1}^\infty \mu_k \quad \text{on } (\mathbb{R}^\infty, \mathcal{B}(\mathbb{R}^\infty)).$$

Due to the expressions

$$\int_{\mathbb{R}^n} \sum_{j=1}^n x_j^2 \prod_{j=1}^n \mu_j(dx_j) = \sum_{j=1}^n \lambda_j + \sum_{j=1}^n a_j^2,$$

$$\int_{\mathbb{R}^n} \sum_{j=1}^n x_j h_j \prod_{j=1}^n \mu_j(dx_j) = \sum_{j=1}^n a_j h_j,$$

$$\int_{\mathbb{R}^n} \left(\sum_{j=1}^n (x_j - a_j) h_j \right) \left(\sum_{i=1}^n (x_i - a_i) k_i \right) \prod_{j=1}^n \mu_j(dx_j) = \sum_{j=1}^n \lambda_j h_j k_j,$$

$$\int_{\mathbb{R}^n} e^{i \sum_{i=1}^n x_j h_j} \prod_{j=1}^n \mu_j(dx_j) = e^{i \sum_{i=1}^n a_j h_j - \frac{1}{2} \sum_{i=1}^n \lambda_j h_j^2},$$

we find that

$$\int_{\mathbb{R}^\infty} \|x\|_\mathcal{K}^2 \bar{\mu}(dx) = \operatorname{Tr} Q + \|a\|_\mathcal{K}^2, \quad \int_{\mathbb{R}^\infty} (x, h)_\mathcal{K} \bar{\mu}(dx) = (a, h)_\mathcal{K},$$

$$\int_{\mathbb{R}^\infty} (x - a, h)_\mathcal{K} (x - a, k)_\mathcal{K} \bar{\mu}(dx) = (Qh, k)_\mathcal{K},$$

$$\int_{\mathbb{R}^\infty} e^{i(x,h)_\mathcal{K}} \bar{\mu}(dx) = e^{i(a,h)_\mathcal{K} - \frac{1}{2} \|Q^{1/2} h\|_\mathcal{K}^2}.$$

Notice that $\ell^2 \subset \mathbb{R}^\infty$.

Lemma 1.101. *Let* $Q \in L_1^+(\mathcal{K})$ *with* $\operatorname{Ker} Q = \{0\}$ *and* $a \in \mathcal{K}$. *Then* ℓ^2 *is* $\bar{\mu}$*-measurable and* $\mu(\ell^2) = 1$.

Proof. Define the projection $P_n : \mathbb{R}^\infty \to \mathbb{R}^\infty$ by $P_n((x_k)_{k=1}^\infty) = (x_1, \ldots, x_n, 0, \ldots)$. The map $\mathbb{R}^\infty \ni x \mapsto f_n(x) = \|P_n x\|_{\ell^2}^2$ is continuous, and f_n is measurable. Thus $\lim_{n\to\infty} f_n(x) = f(x)$ is also measurable, i. e., $f(x) = \|x\|_{\ell^2}^2$ is a measurable function. This means that $\{x \in \mathbb{R}^\infty \mid f(x) < \infty\}$ is a measurable set. We also see that

$$\int_{\mathbb{R}^\infty} f(x) d\mu = \lim_{n\to\infty} \int_{\mathbb{R}^\infty} f_n(x) d\mu = \lim_{n\to\infty} \sum_{k=1}^n \int_{\mathbb{R}^\infty} \|x_k\|^2 d\mu$$

$$= \lim_{n\to\infty} \sum_{k=1}^n (\lambda_k + a_k^2) = \operatorname{Tr} Q + \|a\|^2 < \infty.$$

This implies that $\mu(\{x \in \mathbb{R}^\infty \mid f(x) < \infty\}) = 1$ and the lemma is proven. □

Theorem 1.102. *Let* $Q \in L_1^+(\mathcal{K})$ *and* $a \in \mathcal{K}$. *Suppose that* $\operatorname{Ker} Q = \{0\}$. *Then there exists a probability measure* μ_∞ *on* $(\mathcal{K}, \mathcal{B}(\mathcal{K}))$ *such that*

$$\int_{\mathcal{K}} \|x\|_{\mathcal{K}}^2 \mu_\infty(dx) = \operatorname{Tr} Q + \|a\|_{\mathcal{K}}^2, \tag{1.6.1}$$

$$\int_{\mathcal{K}} (x, h)_{\mathcal{K}} \mu_\infty(dx) = (a, h)_{\mathcal{K}}, \tag{1.6.2}$$

$$\int_{\mathcal{K}} (x - a, h)_{\mathcal{K}} (x - a, k)_{\mathcal{K}} \mu_\infty(dx) = (Qh, k)_{\mathcal{K}}, \tag{1.6.3}$$

$$\int_{\mathcal{K}} e^{i(x,h)_{\mathcal{K}}} \mu(dx) = e^{i(a,h)_{\mathcal{K}} - \frac{1}{2}\|Q^{1/2}h\|_{\mathcal{K}}^2}. \tag{1.6.4}$$

Proof. It is known that $\mathcal{K} \cong \ell^2$ since $\mathcal{K}$ is separable by the assumption. Then it suffices to construct a Gaussian measure on ℓ^2. By Lemma 1.101, it follows that $\mu_\infty = \bar{\mu}\lceil_{\ell^2}$ is a probability measure satisfying (1.6.1)–(1.6.4). □

Definition 1.103 (Gaussian measures on a Hilbert space). The probability measure μ_∞ on $(\mathcal{K}, \mathcal{B}(\mathcal{K}))$ defined in Theorem 1.102 is called a *Gaussian measure* with mean $a \in \mathcal{K}$ and covariance $Q \in L_1^+(\mathcal{K})$.

By Theorem 1.102, we can construct a Gaussian measure on a Hilbert space, which satisfies (1.6.4). Let $a = 0$. Then we have

$$\int_{\mathcal{K}} e^{i(Q^{-1/2}x,h)} \mu_\infty(dx) = e^{-\frac{1}{2}\|h\|^2}. \tag{1.6.5}$$

In (1.6.5), a requirement is that $x \in D(Q^{-1/2}) = Q^{1/2}\mathcal{K}$, thus one may expect

$$\int_{Q^{1/2}\mathcal{K}} e^{i(Q^{-1/2}x,h)} \mu_\infty(dx) = e^{-\frac{1}{2}\|h\|^2} \tag{1.6.6}$$

to hold. Putting $X_h(x) = (Q^{-1/2}x, h)$, the expression (1.6.6) would appear to give a Gaussian random variable $X_h(\cdot)$ with covariance $\|h\|^2_{\mathcal{K}}$. The next result, however, says that this cannot be the case.

Lemma 1.104. *Let $Q \in L_1^+(\mathcal{K})$ and $\operatorname{Ker} Q = \{0\}$. If μ_∞ is the probability measure on $(\mathcal{K}, \mathcal{B}(\mathcal{K}))$ given in Theorem 1.102, then $\mu(Q^{1/2}\mathcal{K}) = 0$.*

Proof. We identify $\mathcal{K}$ with ℓ^2. Notice that

$$Q^{1/2}\mathcal{K} = \bigcup_{n=1}^{\infty} \left\{ y = (y_k)_{k=1}^{\infty} \in \mathbb{R}^\infty \;\Big|\; \sum_{j=1}^{\infty} \frac{1}{\lambda_j} y_j^2 \le n^2 \right\} = \bigcup_{n=1}^{\infty} U_n.$$

Moreover, define

$$U_{nk} = \left\{ y = (y_k)_{k=1}^{\infty} \in \mathbb{R}^\infty \;\Big|\; \sum_{j=1}^{2k} \frac{1}{\lambda_j} y_j^2 \le n^2 \right\}.$$

Since $U_{nk} \downarrow U_n$ as $k \to \infty$, we have $\mu_\infty(U_n) = \lim_{k\to\infty} \mu_\infty(U_{nk})$. By a direct computation,

$$\mu_\infty(U_{nk}) = \frac{(\prod_{j=1}^{2n} \lambda_j)^{-1/2}}{(2\pi)^{n/2}} \int_{\mathbb{R}^{2k}} 1_{\{\sum_{j=1}^{2k} \frac{1}{\lambda_j} y_j^2 \le n^2\}} e^{-\frac{1}{2}\sum_{j=1}^{2k} \frac{1}{\lambda_j}|y_j|^2} dy = \frac{S_{2k-1}}{(2\pi)^k} \int_0^n e^{-\frac{1}{2}r^2} r^{2k-1} dr,$$

where $S_{2k-1} = \frac{2\pi^k}{(k-1)!}$. Hence we have

$$\mu(U_{nk}) \le \frac{S_{2k-1}}{(2\pi)^k} \int_0^n r^{2k-1} dr \le \frac{n^{2k}}{2^k k!} \to 0$$

as $k \to \infty$. Hence $\mu(U_n) = 0$ for each n and the proposition follows. □

The dense linear space $Q^{1/2}\mathcal{K}$ is called the *Cameron–Martin space*. This space is crucial in defining Gaussian measures on a Hilbert space. Recall that, in general, when a probability measure P is *absolutely continuous* with respect to a probability measure Q, the notation $P \ll Q$ is used and the *Radon–Nikodým* derivative is denoted by $\frac{dP}{dQ}$. Thus

$$P(A) = \int_A \frac{dP}{dQ}(x) dQ(x).$$

When this is not the case, P is said to be *singular* with respect to Q, denoted $P \perp Q$, i. e., in this case there exists a measurable set A such that $P(A) = 0$ and $Q(A) = 1$. If $P \ll Q$ and $Q \ll P$ hold simultaneously, denoted $P \sim Q$, then P and Q are called *equivalent*.

Let $(\Omega, \mathscr{F})$ be a measurable space, and μ, ν probability measures on this space. Take $\xi = \frac{1}{2}(\mu + \nu)$; then we have $\mu \ll \xi$ and $\nu \ll \xi$. The expression

$$H(\mu, \nu) = \int_{\Omega} \sqrt{\frac{d\mu}{d\xi}\frac{d\nu}{d\xi}}\, d\xi$$

is called the *Hellinger integral.*

Lemma 1.105. *Let μ and ν be probability measures on a measurable space $(\Omega, \mathscr{F})$. If $H(\mu, \nu) = 0$, then $\mu \perp \nu$.*

Proof. Denote $f = \frac{d\mu}{d\xi}$ and $g = \frac{d\nu}{d\xi}$. The equality $H(\mu, \nu) = \int_{\Omega} \sqrt{f(x)g(x)}\, d\xi = 0$ implies that $fg = 0$ a. e. Define the sets $A = \{x \in \Omega \mid f(x) = 0\}$, $B = \{x \in \Omega \mid g(x) = 0\}$ and $C = A \cup B = \{x \in \Omega \mid f(x)g(x) = 0\}$. Thus $\mu(A) = \int_A f(x)\, d\xi = 0$ and $\nu(B) = \int_B g(x)\, d\xi = 0$. Since $\mu(C) = \nu(C) = 1$, we can conclude that $\mu(B \setminus A) = 1$ and $\nu(A \setminus B) = 1$ and $(B \setminus A) \cap (A \setminus B) = \emptyset$. □

Lemma 1.106. *Let μ and ν be probability measures on a measurable space $(\Omega, \mathscr{F})$. If $\mu \sim \nu$, then $H(\mu, \nu) = \int_{\Omega} \sqrt{\frac{d\nu}{d\mu}}\, d\mu = \int_{\Omega} \sqrt{\frac{d\mu}{d\nu}}\, d\nu$.*

Proof. We have $\frac{d\mu}{d\xi}\frac{d\nu}{d\xi} = \frac{d\mu}{d\xi}\frac{d\nu}{d\mu}\frac{d\mu}{d\xi}$. Hence

$$H(\mu, \nu) = \int_{\Omega} \sqrt{\frac{d\mu}{d\xi}\frac{d\nu}{d\mu}\frac{d\mu}{d\xi}}\, d\xi = \int_{\Omega} \sqrt{\frac{d\nu}{d\mu}}\, d\mu.$$

The second expression is shown similarly. □

Example 1.107. Choose $\mu = N_{0,\lambda}$ and $\nu = N_{a,\lambda}$ to be Gaussian measures on $\mathbb{R}$. We have $\mu \sim \nu$ and

$$\frac{d\nu}{d\mu} = e^{-\frac{a^2}{\lambda} + \frac{ax}{\lambda}}.$$

In this case the Hellinger integral $H(\mu, \nu)$ is

$$H(\mu, \nu) = \int_{\mathbb{R}} \sqrt{e^{-\frac{a^2}{\lambda} + \frac{ax}{\lambda}}}\, d\mu = e^{-\frac{1}{8}\frac{a^2}{\lambda}}.$$

We extend this to n-dimensional case.

Example 1.108. Let $\mu_n = N_{0,(\lambda_1,\dots,\lambda_n)}$ and $\nu_n = N_{(a_1,\dots,a_n),(\lambda_1,\dots,\lambda_n)}$ be Gaussian measures on $\mathbb{R}^n$. By similar computations as above, it is seen that

$$H(\mu_n, \nu_n) = e^{-\frac{1}{8}\sum_{j=1}^{n}\frac{a_j^2}{\lambda_j}}.$$

From Example 1.108 above, formally we see that

$$\lim_{n\to\infty} H(\mu_n, \nu_n) = e^{-\frac{1}{8}\|Q^{-1/2}a\|^2}.$$

In particular, $a \notin Q^{1/2}\mathcal{K}$ implies that the right-hand side is zero, and then $\mu_\infty \perp \nu_\infty$ may be expected. In general, it can be seen that

$$H\Big(\prod_{j=1}^{n} \mu_j, \prod_{j=1}^{n} \nu_j\Big) = \prod_{j=1}^{n} H(\mu_j, \nu_j).$$

This identity can be extended to infinite product measures and applied to investigating the absolute continuity of infinite product measures.

Proposition 1.109 (Kakutani). *Let μ_k and ν_k be probability measures on $(\mathbb{R}, \mathcal{B}(\mathbb{R}))$ and suppose that $\mu_k \sim \nu_k$, for each $k \in \mathbb{N}$. Let $\mu = \prod_{k=1}^\infty \mu_k$ and $\nu = \prod_{k=1}^\infty \nu_k$. Then*

$$H(\mu, \nu) = \prod_{j=1}^{\infty} H(\mu_j, \nu_j). \tag{1.6.7}$$

Furthermore, if $H(\mu, \nu) > 0$, then $\mu \sim \nu$, and if $H(\mu, \nu) = 0$, then $\mu \perp \nu$. When $\mu \sim \nu$, the Radon–Nikodým derivative is given by

$$\frac{d\mu}{d\nu}(x) = \prod_{j=1}^{\infty} \frac{d\mu_j}{d\nu_j}(x_j). \tag{1.6.8}$$

Proof. Write

$$\Psi_n(x) = \Psi_n(x_1, \ldots, x_n) = \prod_{j=1}^{n} \sqrt{\frac{d\nu_j}{d\mu_j}(x_j)}$$

for $x \in \mathbb{R}^\infty$. We show that $(\Psi_n)_{n\in\mathbb{N}}$ is a Cauchy sequence in $L^2(\mathbb{R}^\infty, d\mu)$. It is straightforward to see that

$$\begin{aligned}\|\Psi_n - \Psi_m\|^2 &= \int_{\mathbb{R}^\infty} \prod_{j=1}^{n} \frac{d\nu_j}{d\mu_j}(x_j) \Big|1 - \prod_{j=n+1}^{m} \sqrt{\frac{d\nu_j}{d\mu_j}(x_j)}\Big|^2 \mu(dx) \\ &= 2\Big(1 - \prod_{j=n+1}^{m} \int_{\mathbb{R}} \sqrt{\frac{d\nu_j}{d\mu_j}(x_j)}\mu_j(dx_j)\Big) = 2\Big(1 - \prod_{j=n+1}^{m} H(\mu_j, \nu_j)\Big).\end{aligned} \tag{1.6.9}$$

We consider two cases. First, suppose that

$$\lim_{n\to\infty} \prod_{j=1}^{n} H(\mu_j, \nu_j) > 0.$$

We have $-\sum_{n=1}^{\infty} \log H(\mu_n, \nu_n) < \infty$ and $(-\log H(\mu_n, \nu_n))_{n\in\mathbb{N}}$ is a Cauchy sequence. Thus

$$\lim_{m,n\to\infty} \prod_{j=n+1}^{m} H(\mu_j, \nu_j) = 1,$$

which implies by (1.6.9) that $(\Psi_n)_{n\in\mathbb{N}}$ is a Cauchy sequence. Hence there exists $\Psi \in L^2(\mathbb{R}^\infty, d\mu)$ such that $\Psi_n \to \Psi$ as $n \to \infty$. We also see that $(\Psi_n^2)_{n\in\mathbb{N}}$ is a Cauchy sequence in $L^1(\mathbb{R}^\infty, d\mu)$ by the bound

$$\Big| \int_{\mathbb{R}^\infty} |\Psi_m^2(x) - \Psi_n^2(x)| d\mu(x) \Big|^2 \le 4 \int_{\mathbb{R}^\infty} |\Psi_m(x) - \Psi_n(x)|^2 d\mu(x).$$

Let $A \in \mathcal{B}(\mathbb{R}^\infty)$ and $\chi_n(x) = 1_A(P_n x)$, where $P_n x = (x_1, \dots, x_n, 0, \dots)$ for $x \in \mathbb{R}^\infty$. We have

$$\int_{\mathbb{R}^\infty} \chi_n(x)\nu(dx) = \int_{\mathbb{R}^n} \chi_n(x) \prod_{j=1}^{n} \frac{d\nu_j}{d\mu_j} \prod_{j=1}^{n} \mu_j(dx_j) = \int_{\mathbb{R}^\infty} \chi_n(x)\Psi_n^2(x)\mu(dx).$$

Hence

$$\nu(A) = \int_{\mathbb{R}^\infty} 1_A(x)\nu(dx) = \lim_{n\to\infty} \int_{\mathbb{R}^\infty} \chi_n(x)\Psi_n^2(x)\mu(dx).$$

Since

$$\begin{aligned}
&\Big| \int_{\mathbb{R}^\infty} 1_A(x)\Psi^2(x)\mu(dx) - \int_{\mathbb{R}^\infty} \chi_n(x)\Psi_n^2(x)\mu(dx) \Big| \\
&\quad \le \Big| \int_{\mathbb{R}^\infty} 1_A(x)\Psi^2(x)\mu(dx) - \int_{\mathbb{R}^\infty} \chi_n(x)\Psi^2(x)\mu(dx) \Big| \\
&\quad\quad + \Big| \int_{\mathbb{R}^\infty} \chi_n(x)\Psi^2(x)\mu(dx) - \int_{\mathbb{R}^\infty} \chi_n(x)\Psi_n^2(x)\mu(dx) \Big| \to 0
\end{aligned}$$

as $n \to \infty$ by the L^1-convergence of Ψ_n^2, we conclude that

$$\nu(A) = \lim_{n\to\infty} \int_{\mathbb{R}^\infty} \chi_n(x)\Psi_n^2(x)\mu(dx) = \int_{\mathbb{R}^\infty} 1_A \Psi^2(x)\mu(dx).$$

Hence $\nu \ll \mu$ and the Radon–Nikodým derivative is given by

$$\frac{d\nu}{d\mu}(x) = \Psi^2(x).$$

This implies (1.6.8). Furthermore, it follows that

$$
\begin{aligned}
H(\mu,\nu) &= \int_{\mathbb{R}^\infty} \sqrt{\frac{d\nu}{d\mu}(x)}\mu(dx) = \int_{\mathbb{R}^\infty} \Psi(x)\mu(dx) = \lim_{n\to\infty} \int_{\mathbb{R}^\infty} \Psi_n(x)\mu(dx) \\
&= \lim_{n\to\infty} \prod_{j=1}^n \int_{\mathbb{R}} \sqrt{\frac{d\nu_j}{d\mu_j}(x_j)}\mu_j(dx_j) = \lim_{n\to\infty} \prod_{j=1}^n H(\mu_j,\nu_j).
\end{aligned}
$$

Hence (1.6.7) is obtained. Next suppose that

$$
\lim_{n\to\infty} \prod_{j=1}^n H(\mu_j,\nu_j) = 0.
$$

Let $\varepsilon > 0$ and $A_n = \{x = (x_1,\dots,x_n) \in \mathbb{R}^n \mid \Psi_n^2(x,0,\dots) = \prod_{j=1}^n \frac{d\nu_j}{d\mu_j}(x_j) > 1\}$. We have

$$
\Big(\prod_{j=1}^n \mu_j\Big)(A_n) \le \int_{A_n} \Psi_n(x) \prod_{j=1}^n \mu_j(dx_j) \le \prod_{j=1}^n H(\mu_j,\nu_j) < \varepsilon
$$

for sufficiently large n. Since $\prod_{j=1}^n \frac{d\nu_j}{d\mu_j}(x_j) \le 1$ on $\mathbb{R}^n \setminus A_n$ and the identity

$$
\Big(\prod_{j=1}^n \frac{d\nu_j}{d\mu_j}\Big)\Big(\prod_{j=1}^n \frac{d\mu_j}{d\nu_j}\Big) = 1,
$$

we have $\prod_{j=1}^n \frac{d\mu_j}{d\nu_j}(x_j) > 1$ on $\mathbb{R}^n \setminus A_n$. In the same way as in the estimate above, we obtain that

$$
\Big(\prod_{j=1}^n \nu_j\Big)(\mathbb{R}^n \setminus A_n) \le \int_{\mathbb{R}^n\setminus A_n} \sqrt{\prod_{j=1}^n \frac{d\mu_j}{d\nu_j}} \prod_{j=1}^n \nu_j(dx_j) \le \prod_{j=1}^n H(\mu_j,\nu_j) < \varepsilon
$$

for sufficiently large n. Define $\tilde{A}_n = A_n \times \prod_{j=n+1}^\infty \mathbb{R}$. Then $\mu(\tilde{A}_n) \le \varepsilon$ and $\nu(\mathbb{R}^\infty \setminus \tilde{A}_n) \le \varepsilon$. Hence there exists B_n such that $\mu(B_n) \le \frac{\varepsilon}{2^n}$ and $\nu(\mathbb{R}^\infty \setminus B_n) \le \frac{\varepsilon}{2^n}$. Let $C = \bigcap_{n=1}^\infty B_n$. We have $\mu(C) \le \frac{\varepsilon}{2^n}$ for all $n \in \mathbb{N}$, and this implies that $\mu(C) = 0$. We also see that $\nu(\mathbb{R}^\infty \setminus C) \le \varepsilon$. Hence for each $n \in \mathbb{N}$ there exists C_n such that $\mu(C_n) = 0$ and $\nu(\mathbb{R}^\infty \setminus C_n) \le 1/n$. Let $D = \bigcup_{n=1}^\infty C_n$. We have

$$
\mu(D) = 0 \quad \text{and} \quad \nu(\mathbb{R}^\infty \setminus D) = 0.
$$

From this $\mu \perp \nu$ follows. Finally, we see that

$$
H(\mu,\nu) = \int_D \sqrt{\frac{d\mu}{d\xi}\frac{d\nu}{d\xi}}d\xi + \int_{D^c} \sqrt{\frac{d\mu}{d\xi}\frac{d\nu}{d\xi}}d\xi
$$

$$\le \left(\int_D \frac{d\mu}{d\xi}d\xi\right)^{1/2}\left(\int_D \frac{d\nu}{d\xi}d\xi\right)^{1/2} + \left(\int_{D^c} \frac{d\mu}{d\xi}d\xi\right)^{1/2}\left(\int_{D^c} \frac{d\nu}{d\xi}d\xi\right)^{1/2}$$
$$= (\mu(D)\nu(D))^{1/2} + (\mu(D^c)\nu(D^c))^{1/2} = 0.$$

Hence $H(\mu,\nu) = 0 = \lim_{n\to\infty}\prod_{j=1}^n H(\mu_j,\nu_j)$ follows. □

There are various applications of Proposition 1.109. One allows comparing Gaussian measures with zero and nonzero means.

Corollary 1.110. *Let* $Q \in L_1^+(\mathcal{K})$, $\operatorname{Ker} Q = \{0\}$ *and* $a \in \mathcal{K}$.
(1) *If* $a \notin Q^{1/2}\mathcal{K}$, *then* $N_{a,Q} \perp N_{0,Q}$.
(2) *If* $a \in Q^{1/2}\mathcal{K}$, *then* $N_{a,Q} \sim N_{0,Q}$ *and*

$$\frac{dN_{0,Q}}{dN_{a,Q}} = e^{-\frac{1}{2}\|Q^{-1/2}a\|^2 + (Q^{-1/2}a,x)}.$$

Proof. We have

$$H(\mu,\nu) = \lim_{n\to\infty}\prod_{j=1}^n H(\mu_j,\nu_j) = \lim_{n\to\infty}\prod_{j=1}^n e^{-\frac{1}{8}\frac{a_j^2}{\lambda_j}}.$$

By Proposition 1.109, if $a \in Q^{1/2}\mathcal{K}$, then $H(\mu,\nu) > 0$ and $\mu \sim \nu$, if $a \notin Q^{1/2}\mathcal{K}$, then $H(\mu,\nu) = 0$ and $\mu \perp \nu$. The Radon–Nikodým derivative is also given by

$$\frac{dN_{0,Q}}{dN_{a,Q}} = \lim_{n\to\infty} e^{-\frac{1}{2}\sum_{j=1}^n \frac{a_j^2}{\lambda_j} + \sum_{j=1}^n \frac{a_j x_j}{\lambda_j}} = e^{-\frac{1}{2}\|Q^{-1/2}a\|^2 + (Q^{-1/2}a,x)}.$$ □

From this corollary, it follows that either $N_{a,Q} \sim N_{0,Q}$ or $N_{a,Q} \perp N_{0,Q}$, and clearly $N_{a,Q} \sim N_{0,Q}$ if and only if a is in the Cameron–Martin space $Q^{1/2}\mathcal{K}$, and if a is not contained in the Cameron–Martin space, then $N_{a,Q} \perp N_{0,Q}$. In the next corollary, we compare the Gaussian measures $N_{0,Q}$ and $N_{0,aQ}$, $a \ge 0$.

Corollary 1.111. *Let* $Q, R \in L_1^+(\mathcal{K})$ *and* $\operatorname{Ker} Q = \operatorname{Ker} R = \{0\}$. *Suppose that* $[Q,R] = 0$. *Let* $\{e_n\}_{n=1}^\infty$ *be a complete orthonormal system such that* $Qe_n = \lambda_n e_n$ *and* $Re_n = r_n e_n$ *for each* $n \in \mathbb{N}$. *Then*
(1) $N_{0,Q} \sim N_{0,R}$ *if and only if* $\sum_{k=1}^\infty \left(\frac{\lambda_k - r_k}{\lambda_k + r_k}\right)^2 < \infty$.
(2) *If* $N_{0,Q} \nsim N_{0,R}$, *then* $N_{0,Q} \perp H_{0,R}$.

Proof. A computation gives

$$H(N_{0,Q}, N_{0,R})^4 = \prod_{j=1}^\infty \left(1 - \left(\frac{\lambda_k - r_k}{\lambda_k + r_k}\right)^2\right).$$

We see that $\sum_{k=1}^\infty \left(\frac{\lambda_k - r_k}{\lambda_k + r_k}\right)^2 < \infty$ if and only if $H(N_{0,Q}, N_{0,R}) > 0$. □

Corollary 1.112. *Let $a > 0$, $a \neq 1$, $Q \in L_1^+(\mathcal{K})$ and* $\operatorname{Ker} Q = \{0\}$. *Then* $N_{0,Q} \perp N_{0,aQ}$.

Proof. Let $r_j = a\lambda_j$. Since $\sum_{k=1}^{\infty}\left(\frac{\lambda_k - r_k}{\lambda_k + r_k}\right)^2 = \sum_{k=1}^{\infty}\left(\frac{1-a}{1+a}\right)^2 = \infty$, the claim follows by Corollary 1.111. □

Using the above corollary, we can construct an uncountable family of Gaussian measures $\{N_{0,aQ}\}_{a\geq 0}$ such that $N_{0,aQ} \perp N_{0,\beta Q}$ for any pair $a \neq \beta$. This indicates that Gaussian measures on an infinite-dimensional Hilbert space generally have a more complicated structure than in finite dimensions.

1.6.2 Abstract theory of Borel measures on Hilbert spaces

Consider the Borel measure space $(\mathcal{K}, \mathcal{B}(\mathcal{K}), \mu)$ on a separable Hilbert space $\mathcal{K}$ over $\mathbb{R}$, and define

$$\hat{\mu}(x) = \int_{\mathcal{K}} e^{i(x,y)} \mu(dy).$$

It is seen that
(1) $\hat{\mu}(0) = \mu(\mathcal{K})$;
(2) $x \mapsto \hat{\mu}(x)$ is continuous;
(3) $\hat{\mu}$ is positive definite.

A natural converse question is if it is possible to construct a Borel measure on $(\mathcal{K}, \mathcal{B}(\mathcal{K}))$ such that properties (1)–(3) are satisfied. We summarize the properties of positive definite functions next.

Lemma 1.113. *Let $\phi : \mathcal{K} \to \mathbb{C}$ be positive definite. Then the following hold:*
(1) $|\phi(x)| \leq |\phi(0)|$, $\overline{\phi(x)} = \phi(-x)$.
(2) $|\phi(x) - \phi(y)| \leq 2\sqrt{\phi(0)}\sqrt{\phi(0) - \phi(x-y)}$.
(3) $|\phi(0) - \phi(x)| \leq \sqrt{2\phi(0)}\sqrt{\phi(0) - \Re\phi(x)}$.

Proof. For $x, y \in \mathcal{K}$, define

$$A = \begin{pmatrix} \phi(0) & \phi(x) \\ \phi(-x) & \phi(0) \end{pmatrix} \quad \text{and} \quad B = \begin{pmatrix} \phi(0) & \phi(x) & \phi(y) \\ \phi(-x) & \phi(0) & \phi(y-x) \\ \phi(-y) & \phi(x-y) & \phi(0) \end{pmatrix}.$$

Since ϕ is positive definite, both A and B are positive definite matrices. We have $A^* = A$. Hence $\overline{\phi(x)} = \phi(-x)$ follows, and $\det A \geq 0$ implies that $|\phi(x)| \leq \phi(0)$. It is straightforward to see that

$$\begin{aligned}\det B &= \phi(0)^3 - \phi(0)|\phi(x-y)|^2 - \phi(0)|\phi(x) - \phi(y)|^2 \\ &\quad + 2\Re\left(\phi(y)\overline{\phi(x)}\phi(x-y) - \phi(0)\right).\end{aligned} \tag{1.6.10}$$

Using the inequality $a^3 - ab^2 \leq 2a^2|a-b|$ for $|b| < a$, we see that

$$\phi(0)^3 - \phi(0)|\phi(x-y)|^2 \leq 2\phi(0)^2|\phi(0) - \phi(x-y)|.$$

Inserting this into (1.6.10), we obtain

$$0 \leq \det B \leq 4\phi(0)^2|\phi(0) - \phi(x-y)| - \phi(0)|\phi(x) - \phi(y)|^2,$$

which proves (2). Part (3) follows from the estimate

$$|\phi(0) - \phi(x)|^2 \leq 2\phi(0)^2 - 2\phi(0)\Re\phi(x). \qquad \square$$

Proposition 1.114 (Minlos–Sazonov). *Let $\mathcal{K}$ be a separable Hilbert space over $\mathbb{R}$ and $\phi : \mathcal{K} \to \mathbb{C}$ be positive definite. The following properties are equivalent:*

(1) *There exists a finite Borel measure μ on $(\mathcal{K}, \mathcal{B}(\mathcal{K}))$ such that*

$$\phi(x) = \int_{\mathcal{K}} e^{i(x,y)}\mu(dy).$$

(2) *For every $\varepsilon > 0$, there is a symmetric positive trace class operator Q_ε such that $(Q_\varepsilon x, x) < 1$ implies that $\Re(\phi(0) - \phi(x)) < \varepsilon$.*

(3) *There exists $Q \in L_1^+(\mathcal{K})$ such that ϕ is continuous with respect to $\|\cdot\|_Q$, where $\|x\|_Q = \sqrt{(Qx,x)}$.*

Proof. (1) $\Longrightarrow$ (2) Suppose (1). For every $\gamma > 0$, we obtain

$$\begin{aligned}\Re(\phi(0) - \phi(x)) &= \int_{\|z\|\leq\gamma} (1 - \cos(x,z))\mu(dz) + \int_{\|z\|>\gamma} (1 - \cos(x,z))\mu(dz) \\ &\leq \frac{1}{2}\int_{\|z\|\leq\gamma} (z,x)^2\mu(dz) + 2\mu(\{z \in \mathcal{K} \mid \|z\| > \gamma\}).\end{aligned}$$

Define the sesquilinear form $B : \mathcal{K} \times \mathcal{K} \to \mathbb{C}$ by $B(f,g) = \int_{\|z\|\leq\gamma}(z,f)(z,g)\mu(dz)$. By the Riesz representation theorem, there exists a bounded operator $\bar{B}$ such that $B(f,g) = (f, \bar{B}g)$. Furthermore, for any complete orthonormal system $\{e_j\}_{j=1}^\infty$ of $\mathcal{K}$, we see that

$$\sum_{j=1}^{\infty}(e_j, \bar{B}e_j) = \int_{\|z\|\leq\gamma} \|z\|^2\mu(dz) \leq \gamma^2 < \infty,$$

which implies that $\bar{B}$ is a trace class operator. On the other hand, there exists γ such that $\mu(\{z \in \mathcal{K} \mid \|z\| > \gamma\}) \leq \varepsilon/4$. Thus we have

$$\Re(\phi(0) - \phi(x)) \leq \frac{\varepsilon}{2}(x, Q_\varepsilon x) + \frac{\varepsilon}{2},$$

where we set $Q_\varepsilon = \bar{B}/\varepsilon$. Hence (2) follows.

(2) $\Longrightarrow$ (1) Suppose (2). Let $\|x\|^2 \le 1/\|Q_\varepsilon\|$. We have $|(Q_\varepsilon x, x)| \le 1$, and furthermore $\Re(\phi(0) - \phi(x)) \le \varepsilon$, which implies that ϕ is continuous at $x = 0$ and then continuous at all $x \in \mathcal{K}$. Let $\{e_j\}_{j=1}^\infty$ be a complete orthonormal system in $\mathcal{K}$. Let $\Lambda = \mathbb{N}$ and for $\{i_1, \dots, i_n\} \subset \Lambda$ define

$$f_{i_1,\dots,i_n}(x_1, \dots, x_n) = \phi(x_1 e_{i_1} + \dots + x_n e_{i_n}), \quad f_{i_1,\dots,i_n} : \mathbb{R}^n \to \mathbb{C},$$

which is positive definite. Thus by the Bochner theorem, there exists a finite Borel measure $\mu_{i_1,\dots,i_n}$ on $(\mathbb{R}^n, \mathcal{B}(\mathbb{R}^n))$ such that

$$f_{i_1,\dots,i_n}(x) = \int_{\mathbb{R}^n} e^{ix\cdot y} \mu_{i_1,\dots,i_n}(dy). \tag{1.6.11}$$

We define the family of probability spaces $(\mathbb{R}^A, \mathcal{B}(\mathbb{R}^A), \mu_A)$, $A \subset \Lambda$, $\#A < \infty$. Let $F((x_0, x)) = f(x_0)$ for $x_0 \in \mathbb{R}^n$ and $x \in \mathbb{R}^m$ with $f \in \mathscr{S}(\mathbb{R}^n)$. We have

$$\begin{aligned}
&\int_{\mathbb{R}^{n+m}} F((x_0, x)) \mu_{i_1,\dots,i_n,i_{n+1},\dots,i_{n+m}}(dx_0 dx) \\
&= \int_{\mathbb{R}^{n+m}} f(x_0) \mu_{i_1,\dots,i_n,i_{n+1},\dots,i_{n+m}}(dx_0 dx) \\
&= \frac{1}{(2\pi)^{n/2}} \int_{\mathbb{R}^n} \check{f}(k_0) dk_0 \int_{\mathbb{R}^{n+m}} e^{-ik_0\cdot x_0} \mu_{i_1,\dots,i_n,i_{n+1},\dots,i_{n+m}}(dx_0 dx) \\
&= \frac{1}{(2\pi)^{n/2}} \int_{\mathbb{R}^n} \check{f}(k_0) dk_0 \int_{\mathbb{R}^n} e^{-ik_0\cdot x_0} \mu_{i_1,\dots,i_n}(dx_0) = \int_{\mathbb{R}^n} f(x_0) \mu_{i_1,\dots,i_n}(dx_0).
\end{aligned}$$

Let $E = E_{i_1} \times \dots \times E_{i_n}$ with $E_{i_j} \in \mathcal{B}(\mathbb{R})$. By a limiting argument we see that

$$\begin{aligned}
\mu_{i_1,\dots,i_n,i_{n+1},\dots,i_{n+m}}(E \times \underbrace{\mathbb{R} \times \dots \times \mathbb{R}}_{m}) &= \int_{\mathbb{R}^{n+m}} 1_{E \times \underbrace{\mathbb{R} \times \dots \times \mathbb{R}}_{m}}(x_0, x) \mu_{i_1,\dots,i_n,i_{n+1},\dots,i_{n+m}}(dx_0 dx) \\
&= \int_{\mathbb{R}^n} 1_E(x_0) \mu_{i_1,\dots,i_n}(dx_0) = \mu_{i_1,\dots,i_n}(E).
\end{aligned}$$

The Kolmogorov's consistency condition

$$\mu_{i_1,\dots,i_n,i_{n+1},\dots,i_{n+m}}(E \times \underbrace{\mathbb{R} \times \dots \times \mathbb{R}}_{m}) = \mu_{i_1,\dots,i_n}(E)$$

then follows. By the Kolmogorov's extension theorem, there exists a measure γ on a measurable space $(\mathbb{R}^\infty, \mathcal{B}(\mathbb{R}^\infty))$ such that

$$\gamma \circ \pi_{i_1,\dots,i_n}^{-1}(E_{i_1} \times \dots \times E_{i_n}) = \mu_{i_1,\dots,i_n}(E_{i_1} \times \dots \times E_{i_n}), \tag{1.6.12}$$

where $\pi_{i_1,\dots,i_n} : \mathbb{R}^\infty \to \mathbb{R}^n$ is the projection defined by $\pi_{i_1,\dots,i_n}(x) = (x_{i_1},\dots,x_{i_n})$. Define $X : \mathbb{R}^\infty \to \mathbb{R}$ by $X_i(x) = \pi_{\{i\}}(x) = x_i$, and $X : \mathbb{R}^\infty \to \mathcal{K}$ by $X(x) = \sum_{j=1}^\infty X_j(x)e_j$. We show that

$$\sum_{j=1}^{\infty} X_j^2(x) < \infty \tag{1.6.13}$$

for γ-a. e. in Lemma 1.115 below. Hence X is measurable and defines the image measure on $(\mathcal{K}, \mathcal{B}(\mathcal{K}))$ by $\mu = \gamma \circ X^{-1}$. For $x \in \mathcal{K}$, we define $\tilde{x} = ((e_j, x))_{j=1}^\infty \in \mathbb{R}^\infty$. We have $X(\tilde{x}) = x$ and

$$\begin{aligned}\int_{\mathcal{K}} e^{i(x,y)}\mu(dy) &= \int_{\mathbb{R}^\infty} e^{i(X(\tilde{x}),X(\tilde{y}))}\gamma(d\tilde{y}) = \lim_{n\to\infty}\int_{\mathbb{R}^\infty} e^{i\sum_{j=1}^n \tilde{x}_j\tilde{y}_j}\gamma(d\tilde{y}) \\ &= \lim_{n\to\infty}\int_{\mathbb{R}^n} e^{i\sum_{j=1}^n \tilde{x}_j\tilde{y}_j}\mu_{1,\dots,n}(d\tilde{y}) = \lim_{n\to\infty}\phi\Big(\sum_{j=1}^n \tilde{x}_j e_j\Big) = \phi(x).\end{aligned}$$

The first equality follows by a change of variables, the second by the Lebesgue dominated convergence theorem, the third by Kolmogorov's extension theorem (1.6.12), the fourth by the Bochner theorem (1.6.11), and the fifth is derived from the continuity of ϕ. This completes the proof of (1).

(2) ⇒ (3) Suppose (2) holds. Take $\varepsilon = 1/k$ in (2) and λ_k such that

$$\sum_{k=1}^{\infty} \lambda_k \operatorname{Tr} Q_{1/k} < \infty.$$

Define $Q = \sum_{k=1}^\infty \lambda_k Q_{1/k}$. This Q is a positive and symmetric trace class operator. Let $(Qx,x) < \lambda_k$. We have $(Q_{1/k}x,x) < 1$, which by the assumption implies that $\Re(\phi(0) - \phi(x)) < 1/k$. Hence $\phi(x)$ is continuous at $x = 0$ by $\|\cdot\|_Q$, and continuous at all $x \in \mathcal{K}$.

(3) ⇒ (2) Suppose (3) holds. For every $\varepsilon > 0$, there exists $\delta > 0$ such that $\|x\|_Q < \delta$ implies $\Re(\phi(0) - \phi(x)) < \varepsilon$. Set $Q_\varepsilon = Q/\delta$. We see that $\Re(\phi(0) - \phi(x)) < \varepsilon$ for x satisfying $(Q_\varepsilon x, x) < 1$ and Q_ε satisfies (2). □

Lemma 1.115. *The property* (1.6.13) *holds.*

Proof. Let N be a Gaussian measure on $\mathbb{R}^n$ such that $\int_{\mathbb{R}^n} e^{-ikx}N(dx) = e^{-\frac{1}{2}|k|^2}$. We have

$$\begin{aligned}\phi(0) - \int_{\mathbb{R}^\infty} e^{-\frac{1}{2}\sum_{j=1}^n X_{k+j}^2}\gamma(dx) &= \phi(0) - \int_{\mathbb{R}^\infty}\gamma(dx)\int_{\mathbb{R}^n} e^{-i\sum_{j=1}^n y_j X_{k+j}}N(dy) \\ &= \phi(0) - \int_{\mathbb{R}^n} N(dy)\int_{\mathbb{R}^\infty} e^{-i\sum_{j=1}^n y_j X_{k+j}}\gamma(dx) = \phi(0) - \int_{\mathbb{R}^n}\phi\Big(\sum_{j=1}^n y_j e_{k+j}\Big)N(dy) \\ &= \int_{\mathbb{R}^n}\Big(\phi(0) - \phi\Big(\sum_{j=1}^n y_j e_{k+j}\Big)\Big)N(dy).\end{aligned}$$

Since $(Q_\varepsilon x, x) < 1$ gives $\mathbb{R}(\phi(0) - \phi(x)) < \varepsilon$, we have

$$(\phi(0) - \mathbb{R}\phi(x)) \le \varepsilon + 2\phi(0)(Q_\varepsilon x, x)$$

for $x \in \mathcal{K}$. Hence

$$\begin{aligned}
\phi(0) - \int_{\mathbb{R}^\infty} e^{-\frac{1}{2}\sum_{j=1}^n X_{k+j}^2} \gamma(dx) &\le \varepsilon + 2\phi(0) \int_{\mathbb{R}^n} \Big(Q_\varepsilon \sum_{j=1}^n y_j e_{k+j}, \sum_{j=1}^n y_j e_{k+j} \Big) N(dy) \\
&= \varepsilon + 2\phi(0) \sum_{i,j=1}^n (Q_\varepsilon e_{k+j}, e_{k+i}) \int_{\mathbb{R}^n} y_i y_j N(dy) \\
&= \varepsilon + 2\phi(0) \sum_{j=1}^n (Q_\varepsilon e_{k+j}, e_{k+j}) \int_{\mathbb{R}^n} y_j^2 N(dy) = \varepsilon + 2\phi(0) \sum_{j=1}^n (Q_\varepsilon e_{k+j}, e_{k+j}).
\end{aligned}$$

We obtain

$$\Big| \phi(0) - \int_{\mathbb{R}^\infty} e^{-\frac{1}{2}\sum_{j=1}^n X_{k+j}^2} \gamma(dx) \Big| \le \varepsilon + 2\phi(0) \sum_{j=k+1}^\infty (Q_\varepsilon e_j, e_j).$$

Taking $k \to \infty$ and $\varepsilon \to 0$ gives

$$\lim_{k\to\infty} \int_{\mathbb{R}^\infty} e^{-\frac{1}{2}\sum_{j=k+1}^\infty X_j^2} \gamma(dx) = \phi(0).$$

This implies that $e^{-\frac{1}{2}\sum_{j=k+1}^\infty X_j^2}$ converges to the constant function 1 in $L^1(\mathbb{R}^\infty, \gamma)$. Hence there exists a subsequence k' such that $e^{-\frac{1}{2}\sum_{j=k'+1}^\infty X_j^2}$ converges to 1 γ-a. e., which shows $\sum_{j=1}^\infty X_j^2 < \infty$ γ-a. e. □

Definition 1.116 (Sazonov topology). Let $\mathcal{K}$ be a real separable Hilbert space. A nonnegative trace class operator B defines the seminorm

$$p_B(f) = \sqrt{(f, Bf)_{\mathcal{K}}}.$$

The *Sazonov topology* on $\mathcal{K}$ is the locally convex topology defined by the family of seminorms $\{p_B \mid B \text{ is a nonnegative trace class operator on } \mathcal{K}\}$.

Proposition 1.117 (Borel measures on a Hilbert space). *Let $\mathcal{K}$ be a separable Hilbert space over $\mathbb{R}$. Suppose that $C : \mathcal{K} \to \mathbb{C}$ is positive definite. There exists a finite Borel measure μ on $\mathcal{K}$ such that*

$$C(f) = \int_{\mathcal{K}} e^{i(\psi, f)_{\mathcal{K}}} d\mu(\psi)$$

if and only if $C(\cdot)$ is continuous in the Sazonov topology.

Proof. This follows from Proposition 1.114. □

Proposition 1.117 is called the Minlos–Sazonov theorem.

Example 1.118. Let $B \geq 0$ be a self-adjoint bounded operator and $C_B(f) = e^{-(f,Bf)}$. It is well known that C_B is continuous in the Sazonov topology if and only if B is of trace class. Therefore, whenever B is of trace class, there is a finite Borel measure μ_B on $\mathcal{K}$ such that $\int_{\mathcal{K}} e^{i(\psi,f)_{\mathcal{K}}} d\mu_B(\psi) = e^{-(f,Bf)_{\mathcal{K}}}$.

Let $A \geq 0$ be a self-adjoint operator on $\mathcal{K}$ such that A^{-1} is a Hilbert–Schmidt operator. Define

$$C^\infty(A) = \bigcap_{n=0}^{\infty} D(A^n)$$

and write $\|f\|_n = \|A^{n/2}f\|_{\mathcal{K}}$. Also, the completion

$$\mathcal{K}_n = \overline{C^\infty(A)}^{\|\cdot\|_n}$$

defines a Hilbert space with scalar product

$$(f,g)_n = (A^{n/2}f, A^{n/2}g)_{\mathcal{K}}, \quad f,g \in D(A^{n/2}).$$

We have the sequence of Hilbert spaces

$$\cdots \subset \mathcal{K}_{+2} \subset \mathcal{K}_{+1} \subset \mathcal{K}_0 \subset \mathcal{K}_{-1} \subset \mathcal{K}_{-2} \subset \cdots.$$

Under the identification $\mathcal{K}_0 \cong \mathcal{K}_0^*$, we also have

$$\mathcal{K}_n^* \cong \mathcal{K}_{-n}, \quad n \geq 1.$$

Denote the dual pair between $\mathcal{K}_{-2}$ and $\mathcal{K}_{+2}$ by $\langle\langle \psi, f\rangle\rangle$. An immediate corollary of the Minlos–Sazonov theorem is the following.

Lemma 1.119. *If $L \geq 0$ is a bounded self-adjoint operator on $\mathcal{K}$ such that $[A^{-1}, L] = 0$, then there exists a Borel probability measure m_L on $(\mathcal{K}_{-2}, \mathcal{B}(\mathcal{K}_{-2}))$ such that*

$$\int_{\mathcal{K}_{-2}} e^{i\langle\langle \psi,f\rangle\rangle} dm_L(\psi) = e^{-\frac{1}{2}(f,Lf)_0}. \tag{1.6.14}$$

In particular, if $L = 1$, then there exists a probability measure m_1 on $\mathcal{K}_{-2}$ such that

$$\int_{\mathcal{K}_{-2}} e^{i\langle\langle \psi,f\rangle\rangle} dm_1(\psi) = e^{-\frac{1}{2}\|f\|_0^2}.$$

Proof. We show that A^{-1} is a Hilbert–Schmidt operator also on $\mathcal{K}_{-2}$. To see this, note that for a complete orthonormal system $\{e_i\}_{i=1}^{\infty}$ of $\mathcal{K}$ consisting of eigenvectors of A, the set $\{Ae_i\}_{i=1}^{\infty}$ is also a complete orthonormal system of $\mathcal{K}_{-2}$, and we put $\tilde{e}_i = Ae_i$. We have

$$\sum_{i=1}^{\infty}(\tilde{e}_i, A^{-2}\tilde{e}_i)_{-2} = \sum_{i=1}^{\infty}(e_i, A^{-2}e_i)_0 = \mathrm{Tr}(A^{-2}) < \infty.$$

Hence A^{-1} is a Hilbert–Schmidt operator on $\mathcal{K}_{-2}$. Moreover, since

$$\|Lf\|_{-2} = \|A^{-1}Lf\|_0 = \|LA^{-1}f\|_0 \le \|L\|\|f\|_{-2},$$

L is bounded on $\mathcal{K}_{-2}$, so that LA^{-1} is a Hilbert–Schmidt operator on $\mathcal{K}_{-2}$. By Proposition 1.117 and Example 1.118, there exists a probability measure m_L on $\mathcal{K}_{-2}$ such that

$$\int_{\mathcal{K}_{-2}} e^{i(\psi,f)_{-2}} dm_L(\psi) = e^{-\frac{1}{2}(f,LA^{-2}f)_{-2}}, \quad f \in C^{\infty}(A).$$

Replacing f by $A^2 f$ on the right-hand side above, we have

$$\int_{\mathcal{K}_{-2}} e^{i\rho(\psi,f)} dm_L(\psi) = e^{-\frac{1}{2}(f,Lf)_0}, \tag{1.6.15}$$

where we put

$$\rho(\psi,f) = (\psi, A^2 f)_{-2}, \quad f \in C^{\infty}(A), \quad \psi \in \mathcal{K}_{-2}.$$

Thus we have the bound $|\rho(\psi,f)| \le \|\psi\|_{-2}\|f\|_2$. Using this, we can extend $\rho(\psi,f)$ to $f \in \mathcal{K}_{+2}$ by

$$\rho(\psi,f) = \lim_{n\to\infty} \rho(\psi,f_n),$$

where $f_n \to f$ in $\mathcal{K}_{+2}$ as $n \to \infty$. This $\rho(\psi,f)$ with $f \in \mathcal{K}_{+2}$ also satisfies $|\rho(\psi,f)| \le \|\psi\|_{-2}\|f\|_2$, hence $\rho(\psi,f) = \langle\langle \psi,f\rangle\rangle$. Moreover, for $f \in \mathcal{K}_2$ there exists a sequence $(f_n)_{n\in\mathbb{N}} \subset C^{\infty}(A)$ such that $f_n \to f$ in both $\mathcal{K}_0$ and $\mathcal{K}_2$ as $n \to \infty$, since A is closed and $C^{\infty}(A)$ is a core of A. Hence (1.6.15) can be extended to $f \in \mathcal{K}_{+2}$ by a limiting argument, and the lemma follows. □

We can extend Lemma 1.119 to $f \in \mathcal{K}$.

Theorem 1.120 (Gaussian measure m_L on $\mathcal{K}_{-2}$). *Suppose that $L \ge 0$ is a bounded self-adjoint operator on $\mathcal{K}$ and $[A^{-1}, L] = 0$. Then there exists a Borel probability measure m_L on $(\mathcal{K}_{-2}, \mathcal{B}(\mathcal{K}_{-2}))$ and a family of Gaussian random variables $\psi(f)$ indexed by $f \in \mathcal{K}$ such that*

(1) $\mathcal{K} \ni f \mapsto \psi(f) \in \mathbb{R}$ *is linear;*
(2) $\psi(f) = \langle\langle \psi, f \rangle\rangle$ *for* $f \in \mathcal{K}_{+2}$*;*
(3) *it satisfies*

$$\int_{\mathcal{K}_{-2}} e^{i\psi(f)} dm_L(\psi) = e^{-\frac{1}{2}(f,Lf)_0}. \tag{1.6.16}$$

Proof. For $f \in \mathcal{K}_{+2}$, we write $\psi(f) = \langle\langle \psi, f \rangle\rangle$. Equality (1.6.16) follows by Lemma 1.119. For $f \in \mathcal{K} \setminus \mathcal{K}_{+2}$, $\psi(f)$ can be defined through a limiting procedure. From (1.6.14), it can be seen that

$$\int_{\mathcal{K}_{-2}} \psi(f)^2 dm_L(\psi) = (f, Lf)_0, \quad f \in \mathcal{K}_{+2}.$$

Since for $f \in \mathcal{K}$ there exists a sequence $(f_n)_{n\in\mathbb{N}} \subset \mathcal{K}_{+2}$ such that $f_n \to f$ strongly in $\mathcal{K}$ as $n \to \infty$, we can define $\psi(f)$ by

$$\psi(f) = \operatorname*{s-lim}_{n\to\infty} \psi(f_n)$$

in $L^2(\mathcal{K}_{-2}, dm_L)$. The theorem follows by a limiting argument. □

Remark 1.121. The map $\mathcal{K}_{+2} \ni f \mapsto \psi(f) \in \mathbb{R}$ is continuous with $|\psi(f)| \le \|\psi\|_{-2}\|f\|_2$, but not as a map $\mathcal{K} \ni f \mapsto \psi(f) \in \mathbb{R}$.

Example 1.122. It is impossible to construct a Gaussian measure on a Hilbert space $\mathcal{K}_0$ with covariance $(\cdot,\cdot)_{\mathcal{K}_0}$. By Theorem 1.120, however, we can construct a Gaussian measure with covariance $(\cdot,\cdot)_{\mathcal{K}_0}$ on Banach space $\mathcal{K}_{-2}$ by taking the identity as L.

1.6.3 Fock space as a function space

In order to construct functional integral representation of quantum field Hamiltonians, it will be convenient to define boson Fock space as an L^2-space over a function space with a Gaussian measure. From now on, we will work in the following setup:

(1) *Dispersion relation.* We choose $\nu = 0$ and $\omega_0(k) = \omega(k) = |k|$.
(2) *Lorentz-covariant space.* The real Hilbert space $\mathscr{M}$ is the real part of $\mathcal{H}_M$ for ω_ν replaced by ω.
(3) *Hilbert–Schmidt operator.* A given positive self-adjoint operator $\mathcal{D}$ with Hilbert–Schmidt inverse on $\mathscr{M}$ and such that $\sqrt{\omega}\mathcal{D}^{-1}$ is bounded on $\mathscr{M}$. Let $\mathrm{Spec}(\mathcal{D}) = \{\lambda_i\}_{i=1}^\infty$ and $\{e_i\}_{i=1}^\infty$ be the normalized eigenvectors such that $\mathcal{D}e_i = \lambda_i e_i$. In this case $\mathrm{Tr}(\mathcal{D}^{-2}) = \sum_{i=1}^\infty \lambda_i^{-2} < \infty$. We define a real vector space by

$$\mathscr{M}_n = \overline{C^\infty(\mathcal{D})}^{\|\mathcal{D}^{n/2}\cdot\|_{\mathscr{M}}}.$$

Remark 1.123. The dispersion relation ω can be chosen to be more generally a measurable function $\overline{\omega} : \mathbb{R}^d \to \mathbb{R}$, with $0 \le \overline{\omega}(k) = \overline{\omega}(-k)$, and $\overline{\omega}(k) = 0$ only on a set of Lebesgue-measure zero. The results in this section will equally apply to $\overline{\omega}$.

Remark 1.124.
(1) The positive self-adjoint operator

$$T = \left(-\frac{1}{2}\Delta + \frac{1}{2}|k|^2 + \frac{1}{2}\right)^{(d+1)/2}$$

is a specific case of $\mathcal{D}$ above, see Example 1.71. Indeed, T^{-1} is a Hilbert–Schmidt operator and $\sqrt{\omega}T^{-1}$ is bounded due to $\| \sqrt{\omega}(-\Delta + |k|^2 + 1)^{-(d+1)/2} \| \le 1$.
(2) The condition that $\sqrt{\omega}\mathcal{D}^{-1}$ is bounded on $\mathcal{M}$ will be used in Lemma 1.134 below, where we will prove path continuity of the infinite-dimensional Ornstein–Uhlenbeck process. Moreover, this condition yields that the embedding

$$\iota : \mathcal{M}_{+2} \ni f \mapsto f \in L^2(\mathbb{R}^d)$$

is bounded. It can indeed be seen that

$$\|f\|_{L^2(\mathbb{R}^d)} = \|\sqrt{\omega}\mathcal{D}^{-1}\mathcal{D}f\|_{\mathcal{M}} \le \|\sqrt{\omega}\mathcal{D}^{-1}\|\|f\|_{\mathcal{M}_{+2}}. \tag{1.6.17}$$

We identify $\mathcal{M}_{+2}$ with the topological dual of $\mathcal{M}_{-2}$ by $\mathcal{M}_{-2}^* \cong \mathcal{M}_{+2}$, and denote the pairing between $\mathcal{M}_{+2}$ and $\mathcal{M}_{-2}$ by $\langle\langle \cdot, \cdot \rangle\rangle$.

Theorem 1.125 (Gaussian measure G on $\mathcal{M}_{-2}$). *In the above setup, there exists a probability measure G on $\mathcal{M}_{-2}$ and a family of Gaussian random variables $\xi(f)$ indexed by $f \in \mathcal{M}$ such that*
(1) $\mathcal{M} \ni f \mapsto \xi(f) \in \mathbb{R}$ *is linear;*
(2) $\xi(f) = \langle\langle \xi, f \rangle\rangle$ *when* $f \in \mathcal{M}_{+2}$;
(3) *the expression*

$$\int_{\mathcal{M}_{-2}} e^{i\xi(f)} d\mathsf{G} = e^{-\frac{1}{2}\|f\|^2_{\mathcal{H}_M}}$$

holds.

Proof. This is immediate by Theorem 1.120 with $A = \mathcal{D}, L = 1, \mathcal{K} = \mathcal{M}, \psi(f) = \xi(f)$, and $m_L = \mathsf{G}$. □

Now we discuss the unitary equivalence of the boson Fock space $\mathcal{F}_b(\hat{\mathcal{H}}_M)$ and $L^2(\mathcal{M}_{-2}, d\mathsf{G})$. Let

$$\Phi(f) = \frac{1}{\sqrt{2}}(a_M^*(\hat{f}) + a_M(\tilde{\hat{f}})), \quad \hat{f} \in \hat{\mathcal{H}}_M,$$

where a_M and a_M^* denote the creation and annihilation operators, respectively, in the boson Fock space $\mathcal{F}_b(\hat{\mathcal{H}}_M)$. The field operator $\Phi(f)$ is formally written as

$$\Phi(f) = \int_{\mathbb{R}^d} \frac{1}{\sqrt{2\omega(k)}} \left(a^*(k)\hat{f}(k) + a(k)\hat{f}(-k)\right) dk.$$

Recall that Ω_b denotes the Fock vacuum. For $f, g \in \mathscr{M}$, we see that

$$(\Phi(f)\Omega_b, \Phi(g)\Omega_b)_{\mathcal{F}_b(\hat{\mathcal{H}}_M)} = \frac{1}{2}(\hat{f}, \hat{g})_{\hat{\mathcal{H}}_M} = \mathbb{E}_G[\xi(f)\xi(g)]. \tag{1.6.18}$$

For $f \in \hat{\mathcal{H}}_M, f = g + ih$ with $g, h \in \mathscr{M}$, define

$$\xi(f) = \xi(g) + i\xi(h).$$

The Wick product of $\prod_{j=1}^n \xi(f_j)$ is inductively defined by

$$:\xi(f): = \xi(f),$$
$$:\xi(f)\prod_{j=1}^n \xi(f_j): = \xi(f):\prod_{j=1}^n \xi(f_j): - \frac{1}{2}\sum_{i=1}^n (f, f_i)_{\mathcal{H}_M} :\prod_{j\neq i}^n \xi(f_j):.$$

Proposition 1.126 (Unitary equivalence). *There exists a unitary operator $\theta_M : \mathcal{F}_b(\hat{\mathcal{H}}_M) \to L^2(\mathscr{M}_{-2}, dG)$ satisfying (1)–(3) below:*

(1) $\theta_M \Omega_b = 1$;
(2) $\theta_M :\prod_{i=1}^n \Phi(\hat{f}_i):\Omega_b = :\prod_{i=1}^n \xi(f_i):$;
(3) $\theta_M \Phi(\hat{f})\theta_M^{-1} = \xi(f)$ *for* $f \in \mathscr{M}$ *as an operator.*

In particular, it follows that

$$\mathcal{F}_b(\hat{\mathcal{H}}_M) \cong L^2(\mathscr{M}_{-2}, dG). \tag{1.6.19}$$

Proof. The proof is similar to that of Proposition 1.51. We define the linear operator $\theta_M : \mathcal{F}_b(\hat{\mathcal{H}}_M) \to L^2(\mathscr{M}_{-2}, dG)$ by

$$\theta_M \Omega_b = 1, \quad \theta_M :\prod_{i=1}^n \Phi(\hat{f}_i):\Omega_b = :\prod_{i=1}^n \xi(f_i):, \quad f_i \in \mathscr{M}. \tag{1.6.20}$$

From the commutation relations, it follows that

$$\left\|:\prod_{i=1}^n \Phi(\hat{f}_i):\Omega_b\right\|_{\mathcal{F}_b(\hat{\mathcal{H}}_M)} = \left\|:\prod_{i=1}^n \xi(f_i):\right\|_{L^2(\mathscr{M}_{-2}, dG)}$$

and the linear hull of $:\prod_{i=1}^{n}\Phi(\hat{f}_i):\Omega_{\mathrm{b}}$ and $:\prod_{i=1}^{n}\xi(f_i):$ is a dense subspace in $\mathcal{F}_{\mathrm{b}}(\hat{\mathcal{H}}_M)$ and $L^2(\mathcal{M}_{-2}, d\mathrm{G})$, respectively. Hence θ_M can be extended to the unitary operator from $\mathcal{F}_{\mathrm{b}}(\hat{\mathcal{H}}_M)$ to $L^2(\mathcal{M}_{-2}, d\mathrm{G})$ and the proposition follows. □

Definition 1.127 (Second quantization of contraction operators).
(1) For $T \in \mathscr{C}(\mathcal{M} \to \mathcal{M})$ its second quantization $\tilde{\Gamma}(T) \in \mathscr{C}(L^2(\mathcal{M}_{-2}, d\mathrm{G}) \to L^2(\mathcal{M}_{-2}, d\mathrm{G}))$ is defined by

$$\tilde{\Gamma}(T):\xi(f_1)\cdots\xi(f_n): = :\xi(Tf_1)\cdots\xi(Tf_n):, \tag{1.6.21}$$

with $\tilde{\Gamma}(T)1 = 1$, where 1 denotes the identity function in $L^2(\mathcal{M}_{-2}, d\mathrm{G})$.
(2) Let h be a self-adjoint operator in $\mathcal{M}_{\mathbb{C}}$. Its differential second quantization $d\tilde{\Gamma}(h)$ is defined by the self-adjoint operator such that $\tilde{\Gamma}(e^{-ith}) = e^{-itd\tilde{\Gamma}(h)}$ for $t \in \mathbb{R}$.

Definition 1.128 (Free field Hamiltonian). The free field Hamiltonian in $L^2(\mathcal{M}_{-2}, d\mathrm{G})$ is defined by

$$\tilde{H}_{\mathrm{f}} = d\tilde{\Gamma}(\omega(-i\nabla)). \tag{1.6.22}$$

From this definition, it can be seen that $\tilde{H}_{\mathrm{f}} = \theta_M H_{\mathrm{f}} \theta_M^{-1}$, where H_{f} is the free field Hamiltonian of $\mathcal{F}_{\mathrm{b}}(\hat{\mathcal{H}}_M)$. In the previous section, we defined the self-adjoint operator $H_P = \hat{H}_{\mathrm{f}} + :P(\phi(f)):$ in $L^2(\mathcal{Q})$ and have obtained a functional integral representation of the semigroup e^{-tH_P} by making use of the Euclidean field. The operator

$$\tilde{H}_P = \tilde{H}_{\mathrm{f}} + :P(\xi(f)): \tag{1.6.23}$$

is the operator associated with H_P in $L^2(\mathcal{M}_{-2}, d\mathrm{G})$ by $\theta_M H_P \theta_M^{-1} = \tilde{H}_P$. We can also construct a functional integral representation of $e^{-t\tilde{H}_P}$ in terms of an infinite-dimensional Ornstein–Uhlenbeck process which will be investigated next.

1.6.4 Infinite-dimensional Ornstein–Uhlenbeck process

In this section we define an infinite-dimensional Ornstein–Uhlenbeck process as a Hilbert space-valued random process $(\xi_s)_{s\in\mathbb{R}}$ on the set of continuous paths $C(\mathbb{R}; \mathcal{M}_{-2})$. The one-dimensional Ornstein–Uhlenbeck process $(X_t)_{t\in\mathbb{R}}$ has been discussed in Sections 2.4.5 and 5.1.5 in Volume 1, and we have seen that it is a stationary Gaussian Markov process defined on the probability space $\mathfrak{X} = C(\mathbb{R}; \mathbb{R})$ with a measure characterized by having mean and covariance given by

$$\mathbb{E}_\mu[X_t] = 0 \quad \text{and} \quad \mathbb{E}_\mu[X_s X_t] = \frac{\sigma^2}{2\omega} e^{-|s-t|\omega}.$$

The construction of an infinite-dimensional Ornstein–Uhlenbeck process is similar to this. Denote the set of $\mathscr{M}_{-2}$-valued continuous functions on the real line $\mathbb{R}$ by

$$\mathfrak{Y} = C(\mathbb{R}; \mathscr{M}_{-2}). \tag{1.6.24}$$

On $\mathfrak{Y}$ we introduce the topology derived from the metric

$$\rho_{\mathfrak{Y}}(f,g) = \sum_{k=0}^{\infty} \frac{1}{2^k}\left[\left(\sup_{0\le|t|\le k} \|f(t) - g(t)\|_{\mathscr{M}_{-2}}\right)\wedge 1\right]. \tag{1.6.25}$$

This metric induces the locally uniform topology, giving rise to the Borel σ-field $\mathcal{B}(\mathfrak{Y})$.

In order to construct $(\xi_s)_{s\in\mathbb{R}}$, we introduce the Euclidean version of $\mathscr{M}$, $\mathscr{E}$ and a self-adjoint operator D. Let $\mathscr{E}$ be given by Definition 1.79 with ω_ν replaced by ω. Let D be a positive self-adjoint operator with Hilbert–Schmidt inverse on $\mathscr{E}$, and define

$$\mathscr{E}_n = \overline{C^\infty(D)}^{\|D^{n/2}\cdot\|_{\mathscr{E}}}.$$

The main theorem in this section is stated below; it encapsulates the key conclusion of our analysis and will play a crucial role in the arguments that follow.

Theorem 1.129.

(1) *There exists a probability measure $\mathcal{G}$ on $(\mathfrak{Y}, \mathcal{B}(\mathfrak{Y}))$ and an $\mathscr{M}_{-2}$-valued continuous random process $\mathbb{R} \ni s \mapsto \xi_s \in \mathscr{M}_{-2}$ on the probability space $(\mathfrak{Y}, \mathcal{B}(\mathfrak{Y}), \mathcal{G})$ such that*

$$\mathbb{E}_{\mathcal{G}}[(\xi_s, \xi_t)_{\mathscr{M}_{-2}}] = \frac{1}{4}\sum_{i,j=1}^{\infty} \frac{1}{\lambda_i\lambda_j}(e_i, e^{-|s-t|\hat{\omega}} e_j)_{\mathcal{H}_M},$$

where $\mathrm{Spec}(\mathcal{D}) = \{\lambda_j\}_{j=1}^\infty$, *and $\{e_j\}_{j=1}^\infty$ are the normalized eigenvectors $\mathcal{D}e_j = \lambda_j e_j$.*

(2) *There exists a family of random processes $(\xi_s(f))_{s\in\mathbb{R}}$ indexed by $f \in \mathscr{M}$ on the probability space $(\mathfrak{Y}, \mathcal{B}(\mathfrak{Y}), \mathcal{G})$ such that the following properties hold:*
   (a) *(linearity) $\mathscr{M} \ni f \mapsto \xi_s(f) \in \mathbb{R}$ is a linear map;*
   (b) *(boundedness) $\xi_s(f) = \langle\langle \xi_s, f\rangle\rangle$ for $f \in \mathscr{M}_{+2}$, in particular, $|\xi_s(f)| \le \|\xi_s\|_{\mathscr{M}_{-2}}\|f\|_{\mathscr{M}_{+2}}$ for $f \in \mathscr{M}_{+2}$;*
   (c) *(path continuity)*
      (i) *if $f \in \mathscr{M}_{+2}$, then $\mathbb{R} \ni s \mapsto \xi_s(f) \in \mathbb{R}$ is almost surely continuous;*
      (ii) *if $f \in \mathscr{M}$, then $\mathbb{R} \ni s \mapsto \xi_s(f) \in L^2(\mathscr{E}_{-2}, d\mathcal{G})$ is strongly continuous;*
   (d) *(characteristic function) $\xi_s(f)$ is a* Gaussian random process *with respect to $\mathcal{G}$ with mean zero and*

$$\mathbb{E}_{\mathcal{G}}[e^{i\sum_{j=1}^n \xi_{s_j}(f_j)}] = \exp\left(-\frac{1}{4}\sum_{i,j=1}^{n}(f_i, e^{-|s_i-s_j|\hat{\omega}} f_j)_{\mathcal{H}_M}\right).$$

We will prove Theorem 1.129 below.

Definition 1.130 (Infinite-dimensional Ornstein–Uhlenbeck process).
The $\mathscr{M}_{-2}$-valued ran dom process $(\xi_s)_{s\in\mathbb{R}}$ on the probability space $(\mathfrak{Y}, \mathcal{B}(\mathfrak{Y}), \mathcal{G})$ given in Theorem 1.129 is called an *infinite-dimensional Ornstein–Uhlenbeck process.*

The next lemma follows from Theorem 1.120.

Lemma 1.131. *There exists a Borel probability measure γ on $(\mathscr{E}_{-2}, \mathcal{B}(\mathscr{E}_{-2}))$ and a family of Gaussian random variables $\tilde{\xi}(f)$ indexed by $f \in \mathscr{E}$ with mean zero such that*

(1) $\mathscr{E} \ni f \mapsto \tilde{\xi}(f) \in \mathbb{R}$ *is linear;*

(2) $\tilde{\xi}(f) = \langle\langle \tilde{\xi}, f \rangle\rangle_{\mathscr{E}}$ *when* $f \in \mathscr{E}_{+2}$, *where* $\langle\langle \tilde{\xi}, f \rangle\rangle_{\mathscr{E}}$ *denotes the pairing between* $\mathscr{E}_{+2}$ *and* $\mathscr{E}_{-2}$;

(3) *it satisfies*

$$\int_{\mathscr{E}_{-2}} e^{i\tilde{\xi}(f)} d\gamma(\tilde{\xi}) = e^{-\frac{1}{2}\|f\|^2_{\mathscr{E}}}.$$

Take $g \in \mathscr{M}$. We have already checked in (1.5.5) that $\delta_t \otimes g \in \mathscr{E}$ and

$$(\delta_s \otimes f, \delta_t \otimes g)_{\mathscr{E}} = (f, e^{-|s-t|\hat{\omega}} g)_{\mathcal{H}_M}$$

for all $f, g \in \mathscr{M}$. For every $t \in \mathbb{R}$ and $g \in \mathscr{M}$, put

$$\tilde{\xi}_t(g) = \tilde{\xi}(\delta_t \otimes g).$$

In particular, it follows that

$$\int_{\mathscr{E}_{-2}} \tilde{\xi}_s(f) d\gamma(\tilde{\xi}) = 0, \quad \int_{\mathscr{E}_{-2}} \tilde{\xi}_s(f)\tilde{\xi}_t(g) d\gamma(\tilde{\xi}) = \frac{1}{2}(f, e^{-|s-t|\hat{\omega}} g)_{\mathcal{H}_M}.$$

Lemma 1.132. *For every fixed $t_1 < t_2 < \cdots < t_n \in \mathbb{R}$, there exists a Gaussian measure $\mathcal{G}_{t_1,\ldots,t_n}$ on $\mathscr{M}^n_{-2} = \times_{k=1}^n \mathscr{M}_{-2}$ and a family of random variables $\tilde{\xi}_{t_1,\ldots,t_n}(g)$ indexed by $g \in \mathscr{M}^n$ such that*

(1) $\mathscr{M}^n \ni g \mapsto \tilde{\xi}_{t_1,\ldots,t_n}(g) \in \mathbb{R}$ *is linear;*

(2) $\tilde{\xi}_{t_1,\ldots,t_n}(g) = \langle\langle \tilde{\xi}_{t_1,\ldots,t_n}, g \rangle\rangle$ *when* $g \in \mathscr{M}^n_{+2}$, *where* $\langle\langle \cdot, \cdot \rangle\rangle$ *denotes the pairing between* $\mathscr{M}^n_{+2}$ *and* $\mathscr{M}^n_{-2}$;

(3)

$$\int_{\mathscr{M}^n_{-2}} e^{i\tilde{\xi}_{t_1,\ldots,t_n}(g)} d\mathcal{G}_{t_1,\ldots,t_n} = \int_{\mathscr{E}_{-2}} e^{i\sum_{j=1}^n \tilde{\xi}_{t_j}(g_j)} d\gamma(\tilde{\xi}) \tag{1.6.26}$$

for all $g = (g_1, \ldots, g_n) \in \mathscr{M}^n_{+2}$.

Proof. By the definition of $\tilde{\xi}_t$, we have

$$\int_{\mathscr{E}_{-2}} e^{i\sum_{j=1}^n \tilde{\xi}_{t_j}(g_j)} d\gamma(\tilde{\xi}) = \exp\Big(-\frac{1}{4}\sum_{j,l=1}^n (\delta_{t_j}\otimes g_j, \delta_{t_l}\otimes g_l)_{\mathscr{E}} \Big).$$

The quadratic form

$$(f,g) \mapsto Q_{t_1,\ldots,t_n}(f,g) = \sum_{j,l=1}^n (\delta_{t_j}\otimes f_j, \delta_{t_l}\otimes g_l)_{\mathscr{E}}$$

on $\mathscr{M}^n \times \mathscr{M}^n$ satisfies $|Q_{t_1,\ldots,t_n}(f,g)| \leq \|f\|_{\mathscr{M}^n}\|g\|_{\mathscr{M}^n}$. Thus there exists a bounded operator $L_{t_1,\ldots,t_n} = L$ on $\mathscr{M}^n$ such that $Q_{t_1,\ldots,t_n}(f,g) = (f, Lg)_{\mathscr{M}^n}$. Hence we can write

$$\int_{\mathscr{E}_{-2}} e^{i\sum_{j=1}^n \tilde{\xi}_{t_j}(g_j)} d\gamma(\tilde{\xi}) = e^{-\frac{1}{4}(g,Lg)_{\mathscr{M}^n}}.$$

Let $\mathcal{D}_n = \oplus^n \mathcal{D}$. Notice that

$$\begin{aligned}(f, L\mathcal{D}_n^{-1}g)_{\mathscr{M}^n} &= \sum_{i,j=1}^n (\delta_{t_j}\otimes f_j, \delta_{t_l}\otimes \mathcal{D}^{-1}g_l)_{\mathscr{E}} = \sum_{i,j=1}^n (\delta_{t_j}\otimes \mathcal{D}^{-1}f_j, \delta_{t_l}\otimes g_l)_{\mathscr{E}} \\ &= (\mathcal{D}_n^{-1}f, Lg)_{\mathscr{M}^n} = (f, \mathcal{D}_n^{-1}Lg)_{\mathscr{M}^n}.\end{aligned}$$

Thus $[L, \mathcal{D}_n^{-1}] = 0$. By Theorem 1.120, there exist a measure $\mathcal{G}_{t_1,\ldots,t_n}$ and a random variable $\tilde{\xi}_{t_1,\ldots,t_n}(g)$ such that

$$\int_{\mathscr{M}_{-2}^n} e^{i\tilde{\xi}_{t_1,\ldots,t_n}(g)} d\mathcal{G}_{t_1,\ldots,t_n} = e^{-\frac{1}{4}(g,Lg)_{\mathscr{M}^n}},$$

and hence (1.6.26) is obtained. □

Let $J = \mathbb{R}\times\mathbb{N}$ be the index set and $\Lambda \subset J$ be such that $|\Lambda| < \infty$, where $|\Lambda| = \#\Lambda$. Write $\mathscr{M}_{-2}^{\Lambda} = \{f : \Lambda \to \mathscr{M}_{-2}\}$, let $\pi_\Lambda : \mathscr{M}_{-2}^J \to \mathscr{M}_{-2}^{\Lambda}$ be the projection defined by $\pi_\Lambda f = f\lceil_\Lambda$, and define

$$\mathscr{A} = \Big\{\pi_\Lambda^{-1}(E) \in \mathscr{M}_{-2}^J \,|\, \Lambda \subset J, |\Lambda| < \infty, E \in \mathcal{B}(\mathscr{M}_{-2}^{|J|})\Big\}.$$

Now we construct a probability measure $\mathcal{G}$ on $(\mathscr{M}_{-2}^J, \sigma(\mathscr{A}))$ by using Kolmogorov's extension theorem.

Lemma 1.133. *There exists a probability measure $\mathcal{G}$ on $(\mathscr{M}_{-2}^J, \sigma(\mathscr{A}))$, an $\mathscr{M}_{-2}$-valued random process $(\xi_t)_{t\in\mathbb{R}}$, and an $\mathbb{R}$-valued random process $(\xi_t(f))_{t\in\mathbb{R}}$ with $f \in \mathscr{M}$ on $(\mathscr{M}_{-2}^J, \sigma(\mathscr{A}), \mathcal{G})$ such that the following hold:*

(1) *Its covariance is*

$$\mathbb{E}_{\mathcal{G}}\left[(\xi_s, \xi_t)_{\mathcal{M}_{-2}}\right] = \frac{1}{4}\sum_{j=1}^{\infty}\frac{1}{\lambda_j^2}(e_j, e^{-|s-t|\hat{\omega}}e_j)_{\mathcal{H}_M}, \tag{1.6.27}$$

where $\{\lambda_i\}_{i=1}^{\infty}$ *and* $\{e_i\}_{i=1}^{\infty}$ *are given in Section* 1.6.3. *In particular,*

$$\mathbb{E}_{\mathcal{G}}[\|\xi_t\|^2_{\mathcal{M}_{-2}}] = \frac{1}{4}\operatorname{Tr}(\mathcal{D}^{-2}).$$

(2) $\mathcal{M} \ni f \mapsto \xi_s(f) \in \mathbb{R}$ *is linear, and* $\xi_s(f) = \langle\langle \xi_s, f\rangle\rangle$ *whenever* $f \in \mathcal{M}_{-2}$.

(3) $(\xi_t(f))_{t\in\mathbb{R}}$ *is a Gaussian random process with zero mean and*

$$\mathbb{E}_{\mathcal{G}}[e^{i\sum_{j=1}^n \xi_{s_j}(f_j)}] = \exp\left(-\frac{1}{4}\sum_{i,j=1}^{n}(f_i, e^{-|s_i-s_j|\hat{\omega}}f_j)_{\mathcal{H}_M}\right).$$

Proof. Write $\tilde{e}_j = \mathcal{D}^{-1}e_j = \lambda_j^{-1}e_j$. The set $\{\tilde{e}_j\}_{j=1}^{\infty}$ forms a complete orthonormal system in $\mathcal{M}_{+2}$. For every $k_1, \ldots, k_n \in J$ with $k_j = (t_j, i_j)$, and $E_1 \times \cdots \times E_n \in \mathcal{B}(\mathbb{R}^n)$, put

$$\mu_{k_1,\ldots,k_n}(E_1 \times \cdots \times E_n) = \mathbb{E}_{\gamma}\Big[\prod_{j=1}^{n} 1_{E_j}(\tilde{\xi}_{t_j}(\tilde{e}_{i_j}))\Big]. \tag{1.6.28}$$

By Lemma 1.132, there exists a probability measure $\mathcal{G}_{t_1,\ldots,t_n}$ on $\mathcal{M}_{-2}^n$ such that

$$\mu_{k_1,\ldots,k_n}(E_1 \times \cdots \times E_n) = \int_{\mathcal{M}_{-2}^n}\prod_{j=1}^{n} 1_{E_j}(\tilde{\xi}_{i_j})d\mathcal{G}_{t_1,\ldots,t_n},$$

where $\tilde{\xi}_{i_j} = \tilde{\xi}_{t_1,\ldots,t_n}(\underbrace{0 \oplus \cdots \overset{j}{\tilde{e}_{i_j}} \cdots \oplus 0}_{n})$. Since the family of set functions $\{\mu_\Lambda\}_{\Lambda\subset J, |\Lambda|<\infty}$ satisfies the consistency condition (1.6.28), by the Kolmogorov's extension theorem there exists a probability measure $\mathcal{G}$ on $(\mathcal{M}_{-2}^J, \sigma(\mathscr{A}))$ such that

$$\mathcal{G}(\pi_\Lambda^{-1}(E_1 \times \cdots \times E_n)) = \mu_{k_1,\ldots,k_n}(E_1 \times \cdots \times E_n) \tag{1.6.29}$$

for $\Lambda = \{k_1, \ldots, k_n\} \subset J$. Let $(\xi_{s,j})_{(s,j)\in J} \in \mathbb{R}^J$. Equality (1.6.29) can be rewritten as

$$\mathbb{E}_{\mathcal{G}}\Big[\prod_{j=1}^{n} 1_{E_j}(\xi_{t_j,i_j})\Big] = \mathbb{E}_{\gamma}\Big[\prod_{j=1}^{n} 1_{E_j}(\tilde{\xi}_{t_j}(\tilde{e}_{i_j}))\Big].$$

By a limiting argument, we have for bounded Borel measurable functions F on $\mathbb{R}^n$ that

$$\mathbb{E}_{\mathcal{G}}[F(\xi_{s_1,i_1}, \ldots, \xi_{s_n,i_n})] = \mathbb{E}_{\gamma}[F(\tilde{\xi}_{s_1}(\tilde{e}_{i_1}), \ldots, \tilde{\xi}_{s_n}(\tilde{e}_{i_n}))]. \tag{1.6.30}$$

Similarly, by taking limits, we arrive at

$$\mathbb{E}_{\mathcal{G}}[\xi_{s,i}] = \mathbb{E}_{\gamma}[\tilde{\xi}_s(\tilde{e}_i)] = 0,$$
$$\mathbb{E}_{\mathcal{G}}[\xi_{s,i}\xi_{t,j}] = \mathbb{E}_{\gamma}[\tilde{\xi}_s(\tilde{e}_i)\tilde{\xi}_t(\tilde{e}_j)] = \frac{1}{4\lambda_i\lambda_j}(e_i, e^{-|s-t|\hat{\omega}}e_j)_{\mathcal{H}_M}.$$

Let $b_j = \mathcal{D}e_j$, so that $\{b_j\}_{j=1}^{\infty}$ is a complete orthonormal system in $\mathcal{M}_{-2}$. Define the vector

$$\xi_s = \sum_{j=1}^{\infty} \xi_{s,j} b_j \in \mathcal{M}_{-2}.$$

We can check that $\|\xi_s\|_{\mathcal{M}_{-2}} < \infty$ almost surely with respect to $\mathcal{G}$ for every s since

$$\mathbb{E}_{\mathcal{G}}[\|\xi_s\|^2_{\mathcal{M}_{-2}}] = \mathbb{E}_{\mathcal{G}}\Big[\sum_{j=1}^{\infty} \xi_{s,j}^2\Big] = \frac{1}{4}\operatorname{Tr}(\mathcal{D}^{-2}) < \infty.$$

Define $\xi_s(f) = \langle\langle \xi_s, f\rangle\rangle$ for $f \in \mathcal{M}_{+2}$. Thus

$$\xi_s(f) = \sum_{j=1}^{\infty} \xi_{s,j}\langle\langle b_j, f\rangle\rangle, \quad f \in \mathcal{M}_{+2},$$

and

$$\mathbb{E}_{\mathcal{G}}[\xi_s(f)] = \lim_{n\to\infty} \sum_{j=1}^{n} \mathbb{E}_{\mathcal{G}}[\xi_{s,j}\langle\langle b_j, f\rangle\rangle] = 0.$$

Next we show (3). By (1.6.30) and a limiting argument, we have for $f_j \in \mathcal{M}_{+2}$,

$$\mathbb{E}_{\mathcal{G}}[e^{i\sum_{j=1}^{n}\xi_{s_j}(f_j)}] = \mathbb{E}_{\gamma}[e^{i\sum_{j=1}^{n}\tilde{\xi}_{s_j}(f_j)}] = \exp\left(-\frac{1}{4}\sum_{i,j=1}^{n}(f_i, e^{-|s_i-s_j|\hat{\omega}}f_j)_{\mathcal{H}_M}\right). \tag{1.6.31}$$

Now we extend $\xi_s(f)$ to $f \in \mathcal{M}$, which is a routine argument. By (1.6.31),

$$\mathbb{E}_{\mathcal{G}}[\xi_s(f)\xi_t(g)] = \frac{1}{2}(f, e^{-|s-t|\hat{\omega}}g)_{\mathcal{H}_M}.$$

In particular, $\mathbb{E}_{\mathcal{G}}[\xi_s(f)^2] = \|f\|^2_{\mathcal{H}_M}$. Thus $\xi_s(f)$ for $f \in \mathcal{M}$ is defined by

$$\xi_s(f) = \operatorname*{s-lim}_{n\to\infty} \xi_s(f_n)$$

in $L^2(\mathcal{M}_{-2}^J, d\mathcal{G})$ for $f_n \in \mathcal{M}_{+2}$ such that $f_n \to f$ in $\mathcal{M}$. It is not difficult to check that (3) is satisfied for $\xi_s(f)$ with $f \in \mathcal{M}$. □

The last point is to establish the almost sure continuity of sample paths ξ_s for $\mathcal{G}$.

Lemma 1.134. *There exists a continuous version of* $(\xi_t)_{t\in\mathbb{R}}$.

Proof. Using the Kolmogorov–Čentsov theorem, we only need to show that

$$\mathbb{E}_{\mathcal{G}}[\|\xi_s - \xi_t\|^4_{\mathcal{M}_{-2}}] \leq C|s-t|^2 \tag{1.6.32}$$

for some $C > 0$. Let $g \in \mathcal{M}_{+2}$. A direct computation gives

$$\begin{aligned}\mathbb{E}_{\mathcal{G}}[|\xi_s(g) - \xi_t(g)|^4] &= \mathbb{E}_{\gamma}[|\tilde{\xi}_s(g) - \tilde{\xi}_t(g)|^4] = 3(g, (1 - e^{-|s-t|\hat{\omega}})g)^2_{\hat{\mathcal{H}}_M} \\ &= 3\left(\int_{\mathbb{R}^d} \frac{1 - e^{-|s-t|\omega(k)}}{\omega(k)} |\hat{g}(k)|^2 dk\right)^2 \leq 3|s-t|^2 \|g\|^4_{L^2(\mathbb{R}^d)}.\end{aligned}$$

Notice that the embedding $\iota : \mathcal{M}_{+2} \to L^2(\mathbb{R}^d)$ is bounded with the bound

$$\|\iota g\|_{L^2(\mathbb{R}^d)} \leq \|\sqrt{\omega}\mathcal{D}^{-1}\| \|g\|_{\mathcal{M}_{+2}}.$$

From this, it follows that

$$\mathbb{E}_{\mathcal{G}}[|\xi_s(g) - \xi_t(g)|^4] \leq C\|g\|^4_{\mathcal{M}_{+2}}$$

with a constant $C > 0$. Since

$$\|\xi_s\|_{\mathcal{M}_{-2}} = \sup_{f\neq 0, f\in\mathcal{M}_{+2}} \frac{|\xi_s(f)|}{\|f\|_{\mathcal{M}_{+2}}},$$

there exists $f_\varepsilon \in \mathcal{M}_{+2}$ such that

$$\|\xi_s - \xi_t\|_{\mathcal{M}_{-2}} - \varepsilon \leq \frac{|\xi_s(f_\varepsilon) - \xi_t(f_\varepsilon)|}{\|f_\varepsilon\|_{\mathcal{M}_{+2}}},$$

and we obtain that

$$\mathbb{E}_{\mathcal{G}}[\|\xi_s - \xi_t\|^4_{\mathcal{M}_{-2}}] \leq \varepsilon + C|s-t|^2$$

for every $\varepsilon > 0$. Hence (1.6.32) follows. □

Proof of Theorem 1.129. We denote the continuous version of $(\xi_t)_{t\in\mathbb{R}}$ by $\bar{\xi} = (\bar{\xi}_t)_{t\in\mathbb{R}}$, and the image measure $\mathcal{G} \circ \bar{\xi}^{-1}$ on $\mathfrak{Y}$ by the same symbol $\mathcal{G}$. Thus we have constructed above the probability space $(\mathfrak{Y}, \mathcal{B}(\mathfrak{Y}), \mathcal{G})$. Parts (1) and (2) of the theorem follow by Lemma 1.133 except for the path continuity, which we show now. Whenever $f \in \mathcal{M}_{+2}$, we see that $\xi_s(f) = \langle\langle \xi_s, f\rangle\rangle$. Hence $s \mapsto \xi_s(f)$ is continuous almost surely since $\mathbb{R} \ni s \mapsto \xi_s \in \mathcal{M}_{-2}$ is continuous. For $f \in \mathcal{M}$,

$$\lim_{s\to t} \mathbb{E}_{\mathcal{G}}[|\xi_s(f) - \xi_t(f)|^2] = \frac{1}{2}\lim_{s\to t}(f, (1 - e^{-|s-t|\hat{\omega}})f)_{\mathcal{H}_M} = 0.$$

This gives the desired result. □

1.6.5 Markov property

The random process $(\xi_t)_{t\in\mathbb{R}}$ is stationary due to the form of its covariance as given by (1.6.27). In Proposition 1.96, we have shown that the Euclidean quantum field has the Markov property. In this section we show that $(\xi_t)_{t\in\mathbb{R}}$ also is a Markov process.

Recall that $\delta_s \otimes f \in \mathscr{E}$ for $f \in \mathscr{M}$, and $(\phi_{\mathrm{E}}(F), F \in \mathscr{E})$ is a Euclidean field on $(\mathscr{Q}_{\mathrm{E}}, \Sigma_{\mathrm{E}}, \mu_{\mathrm{E}})$. First, we show the relationship of $\tilde{\xi}_s(f) = \tilde{\xi}(\delta_s \otimes f)$, $(\xi_s)_{s\in\mathbb{R}}$, and the Euclidean field $\phi_s(f) = \phi_{\mathrm{E}}(\delta_s \otimes f)$.

Proposition 1.135. *Let F_j be Borel measurable bounded functions on $\mathbb{R}$, and $f_j \in \mathscr{M}$. Then*

$$\mathbb{E}_{\mathcal{G}}\Big[\prod_{j=1}^n F_j(\xi_{t_j}(f_j))\Big] = \mathbb{E}_\gamma\Big[\prod_{j=1}^n F_j(\tilde{\xi}_{t_j}(f_j))\Big] = \mathbb{E}_{\mu_{\mathrm{E}}}\Big[\prod_{j=1}^n F_j(\phi_{t_j}(f_j))\Big].$$

Proof. Let $F \in \mathscr{S}(\mathbb{R}^n)$. From (1.6.31), it follows that

$$\begin{aligned}
&\mathbb{E}_{\mathcal{G}}[F(\xi_{s_1}(f_1), \dots, \xi_{s_n}(f_n))] \\
&= \frac{1}{(2\pi)^{n/2}} \int_{\mathbb{R}^{nd}} \check{F}(k_1, \dots, k_n) \mathbb{E}_{\mathcal{G}}[e^{i\sum_{j=1}^n k_j \xi_{s_j}(f_j)}] dk_1 \cdots dk_n \\
&= \frac{1}{(2\pi)^{n/2}} \int_{\mathbb{R}^{nd}} \check{F}(k_1, \dots, k_n) \mathbb{E}_\gamma[e^{i\sum_{j=1}^n k_j \tilde{\xi}_{s_j}(f_j)}] dk_1 \cdots dk_n \\
&= \mathbb{E}_\gamma[F(\tilde{\xi}_{s_1}(f_1), \dots, \tilde{\xi}_{s_n}(f_n))].
\end{aligned}$$

The first equality can be completed by a limiting argument, the second is proven similarly. □

Lemma 1.136. *Let $\mathscr{E}' \subset \mathscr{E}$ be a closed subspace, and $P : \mathscr{E} \to \mathscr{E}'$ be the corresponding orthogonal projection. Consider $\mathscr{G}_P = \sigma(\tilde{\xi}(f) \mid f \in \mathscr{E}')$. Then for every $\alpha \in \mathbb{C}$ and $f \in \mathscr{E}$, it follows that*

$$\mathbb{E}_\gamma[e^{\alpha\tilde{\xi}(f)} | \mathscr{G}_P] = e^{\alpha\tilde{\xi}(Pf)} e^{\frac{\alpha^2}{2}\|f - Pf\|^2_{\mathscr{E}}}.$$

Proof. We have that

$$\mathbb{E}_\gamma[e^{\alpha\tilde{\xi}(f)} | \mathscr{G}_P] = \mathbb{E}_\gamma[e^{\alpha\tilde{\xi}(Pf)} e^{\alpha\tilde{\xi}(f-Pf)} | \mathscr{G}_P] = e^{\alpha\tilde{\xi}(Pf)} \mathbb{E}_\gamma[e^{\alpha\tilde{\xi}(f-Pf)} | \mathscr{G}_P], \tag{1.6.33}$$

since $\tilde{\xi}(Pf)$ is $\mathscr{G}_P$-measurable. Moreover, since $\tilde{\xi}(f - Pf)$ is independent of $\mathscr{G}_P$, we have

$$(1.6.33) = e^{\alpha\tilde{\xi}(Pf)} \mathbb{E}_\gamma[e^{\alpha\tilde{\xi}(f-Pf)}] = e^{\alpha\tilde{\xi}(Pf)} e^{\frac{\alpha^2}{2}\|f-Pf\|^2_{\mathscr{E}}}.$$ □

For any interval $I \subset \mathbb{R}$, write

$$\mathscr{F}_I = \sigma(\mathrm{L.H.}\,\{\xi_t(f) \mid t \in I, f \in \mathscr{M}\}).$$

Theorem 1.137 (Markov property). *Let F be a bounded $\mathscr{F}_{[s,\infty)}$-measurable function. Then*

$$\mathbb{E}_{\mathcal{G}}[F|\mathscr{F}_{(-\infty,s]}] = \mathbb{E}_{\mathcal{G}}[F|\mathscr{F}_{\{s\}}]. \tag{1.6.34}$$

Proof. Denote by $\mathscr{E}_{(-\infty,s]}$ the closed subspace of $\mathscr{E}$ generated by the linear hull of the set $\{\delta_r \otimes f \in \mathscr{E} \mid r \leq s, f \in \mathscr{M}\}$, and similarly by $\mathscr{E}_{\{s\}}$ the closed subspace generated by $\{\delta_s \otimes f \in \mathscr{E} \mid f \in \mathscr{M}\}$. Write $P_{(-\infty,s]}$ and $P_{\{s\}}$ for the corresponding projections. We claim that for every $t \geq s$ and every $g \in \mathscr{M}$,

$$P_{(-\infty,s]}(\delta_t \otimes g) = \delta_s \otimes e^{-(t-s)\hat{\omega}}g = P_{\{s\}}(\delta_s \otimes e^{-(t-s)\hat{\omega}}g). \tag{1.6.35}$$

Indeed, $\delta_s \otimes e^{-(t-s)\hat{\omega}}g \in \mathscr{E}_{(-\infty,s]}$, and for all $r \leq s$,

$$(\delta_r \otimes f, \delta_t \otimes g)_{\mathscr{E}} = \int_{\mathbb{R}^d} e^{-(s-r)\omega(k)}\overline{\hat{f}(k)}(e^{-(t-s)\hat{\omega}}g)\hat{\ }(k)\frac{dk}{2\omega(k)} = (\delta_r \otimes f, \delta_s \otimes e^{-(t-s)\hat{\omega}}g)_{\mathscr{E}}.$$

From this, we get

$$(F, \delta_t \otimes g)_{\mathscr{E}} = (F, \delta_s \otimes e^{-(t-s)\hat{\omega}}g)_{\mathscr{E}}, \quad F \in \mathscr{E}_{(-\infty,s]},$$

by linearity and approximation, and thus the first equality in (1.6.35) is shown. The second equality now follows from the fact that $\delta_s \otimes e^{-(t-s)\hat{\omega}}g$ is not only in $\mathscr{E}_{(-\infty,s]}$ but also in $\mathscr{E}_{\{s\}}$. By Lemma 1.136, we furthermore have

$$\mathbb{E}_{\gamma}[e^{\alpha\tilde{\xi}_t(g)}|\mathscr{G}_{P_{(-\infty,s]}}] = e^{\alpha\tilde{\xi}(P_{(-\infty,s]}\delta_t\otimes g)}\exp\left(\frac{\alpha^2}{2}\|(1-P_{(-\infty,s]})\delta_t \otimes g\|^2_{\mathscr{E}}\right),$$

and Proposition 1.135 and (1.6.33) give

$$\mathbb{E}_{\mathcal{G}}[e^{\alpha\xi_t(g)}|\mathscr{F}_{(-\infty,s]}] = e^{\alpha\xi_s(e^{-(t-s)\hat{\omega}}g)}\exp\left(\frac{\alpha^2}{4}(g,(1-e^{-2(t-s)\hat{\omega}})g)_{\mathscr{M}}\right) \in \mathscr{F}_{\{s\}}$$

for $t \geq 0$. Again by approximation, we find $\mathbb{E}_{\mathcal{G}}[F|\mathscr{F}_{(-\infty,s]}] \in \mathscr{F}_{\{s\}}$ for all bounded $\mathscr{F}_{[s,\infty)}$-measurable functions. Thus (1.6.34) follows. □

1.6.6 Regular conditional Gaussian probability measures

The Gaussian measure $\mathcal{G}$ is a probability measure on $(\mathfrak{Y}, \mathcal{B}(\mathfrak{Y}))$ and $\xi_0 : \mathfrak{Y} \to \mathscr{M}_{-2}$ is a random variable. Since both $\mathfrak{Y} = C(\mathbb{R}; \mathscr{M}_{-2})$ and $\mathscr{M}_{-2}$ are Polish spaces, we can define the regular conditional probability measure $\mathcal{G}^{\xi}$ with $\xi \in \mathscr{M}_{-2}$ by

$$\mathcal{G}^{\xi}(\cdot) = \mathcal{G}(\cdot \mid \xi_0 = \xi).$$

In particular, we may assume that

$$\mathcal{G}^{\xi}(\xi_0 \in B) = 1_{\xi}(B), \quad B \in \mathcal{B}(\mathcal{M}_{-2}),$$

and

$$\mathcal{G}^{\xi}(\xi_0 = \xi) = 1, \quad \mathcal{G} \circ \xi_0^{-1}\text{-a. s. } \xi.$$

We compute $\mathbb{E}^{\xi}_{\mathcal{G}}[\xi_t(f)]$ and $\mathbb{E}^{\xi}_{\mathcal{G}}[\xi_t(f)\xi_s(g)]$ for later use. Note that

$$\mathbb{E}_{\mathcal{G}}[e^{\sum_{j=1}^{n}\xi_{t_j}(f_j)}|\mathcal{F}_{\{0\}}] = \mathbb{E}_{\mathcal{G}}[e^{\sum_{j=1}^{n}\xi_{t_j}(f_j)}|\sigma(\xi_0)]. \tag{1.6.36}$$

It is seen that the right-hand side of (1.6.36) is a function of ξ_0 and then we write $h(\xi_0)$. Then $\mathbb{E}^{\xi}_{\mathcal{G}}[e^{\sum_{j=1}^{n}\xi_{t_j}(f_j)}]$ is defined by $h(\xi_0)$ with ξ_0 replaced by ξ.

Theorem 1.138. *If $f, g \in \mathcal{M}$ and $\xi \in \mathcal{M}_{-2}$, then*

$$\mathbb{E}^{\xi}_{\mathcal{G}}[\xi_t(f)] = \xi(e^{-|t|\hat{\omega}}f), \tag{1.6.37}$$

$$\mathbb{E}^{\xi}_{\mathcal{G}}[\xi_t(f)\xi_s(g)] = \frac{1}{2}(f, e^{-|s-t|\hat{\omega}}g)_{\mathcal{H}_M} - \frac{1}{2}(f, e^{-(|t|+|s|)\hat{\omega}}g)_{\mathcal{H}_M} + \xi(e^{-|t|\hat{\omega}}f)\xi(e^{-|s|\hat{\omega}}g). \tag{1.6.38}$$

Proof. Similarly as in the proof of Theorem 1.137, we see that

$$\mathbb{E}_{\mathcal{G}}[e^{\sum_{j=1}^{n}\xi_{t_j}(f_j)}|\mathcal{F}_{\{r\}}] = e^{\sum_{j=1}^{n}\xi_r(e^{-|t_j-r|\hat{\omega}}f_j)}e^{Q/4}, \tag{1.6.39}$$

where

$$Q = \sum_{i,j=1}^{n}\left((f_i, e^{-|t_i-t_j|\hat{\omega}}f_j)_{\mathcal{H}_M} - (f_i, e^{-|t_i-r|\hat{\omega}}e^{-|t_j-r|\hat{\omega}}f_j)_{\mathcal{H}_M}\right).$$

Let $\alpha, \beta \in \mathbb{R}$. Inserting αf and βg into f_1 and f_2 in (1.6.39), respectively, and setting $n = 2$ and $r = 0$, we have

$$\mathbb{E}_{\mathcal{G}}[e^{\alpha\xi_t(f)+\beta\xi_s(g)}|\sigma(\xi_0)] = e^{\alpha\xi_0(e^{-|t|\hat{\omega}}f)+\beta\xi_0(e^{-|s|\hat{\omega}}g)}e^{Q/4},$$

where

$$\begin{aligned}Q = {} & \alpha^2(\|f\|^2_{\mathcal{H}_M} - (f, e^{-2|t|\hat{\omega}}f)_{\mathcal{H}_M}) + \beta^2(\|g\|^2_{\mathcal{H}_M} - (f, e^{-2|s|\hat{\omega}}g)_{\mathcal{H}_M}) \\ & + 2\alpha\beta((f, e^{-|s-t|\hat{\omega}}g)_{\mathcal{H}_M} - (f, e^{-(|t|+|s|)\hat{\omega}}g)_{\mathcal{H}_M}).\end{aligned}$$

Hence

$$\mathbb{E}^{\xi}_{\mathcal{G}}[e^{\alpha\xi_t(f)+\beta\xi_s(g)}] = e^{\alpha\xi(e^{-|t|\hat{\omega}}f)+\beta\xi(e^{-|s|\hat{\omega}}g)}e^{Q/4}.$$

Equality (1.6.37) is derived through $\partial_\alpha\mathbb{E}^{\xi}_{\mathcal{G}}[e^{\alpha\xi_t(f)+\beta\xi_s(g)}]_{\alpha=0}$, while formula (1.6.38) can be obtained via $\partial_\alpha\partial_\beta\mathbb{E}^{\xi}_{\mathcal{G}}[e^{\alpha\xi_t(f)+\beta\xi_s(g)}]_{\alpha=\beta=0}$. □

Using Theorem 1.138, we can also compute the covariance to be

$$\begin{aligned}\operatorname{cov}(\xi_t(f);\xi_t(g)) &= \mathbb{E}^{\xi}_{\mathcal{G}}[\xi_t(f)\xi_t(g)] - \mathbb{E}^{\xi}_{\mathcal{G}}[\xi_t(f)]\mathbb{E}^{\xi}_{\mathcal{G}}[\xi_t(g)] \\ &= \frac{1}{2}(f, e^{-|s-t|\hat{\omega}}g)_{\mathcal{H}_M} - \frac{1}{2}(f, e^{-(|t|+|s|)\hat{\omega}}g)_{\mathcal{H}_M}.\end{aligned}$$

In particular, $\operatorname{cov}(\xi_t(f);\xi_t(g))$ is independent of ξ.

1.6.7 Feynman–Kac–Nelson formula by path measures

Using $\mathcal{G}$, an alternative functional integral representation of $(F, e^{-tH_P}G)$ in Theorem 1.98 can be given. Define

$$\tilde{H}_P = \tilde{H}_{\mathrm{f}} \dotplus :P(\xi(f)):, \quad f \in \mathscr{M},$$

where we recall that $\tilde{H}_{\mathrm{f}}$ denotes the free field Hamiltonian on $L^2(\mathscr{M}_{-2}, d\mathrm{G})$.

Theorem 1.139 (Feynman–Kac–Nelson formula). *Let $F, G \in L^2(\mathscr{M}_{-2}, d\mathrm{G})$. Then*

$$(F, e^{-t\tilde{H}_P}G) = \mathbb{E}_{\mathcal{G}}\Big[\overline{F(\xi_0)}G(\xi_t)e^{-\int :P(\xi_s(f)):ds}\Big].$$

Proof. Let $F(\xi) = P_1(\xi(f_1),\dots,\xi(f_n))$ and $G(\xi) = P_2(\xi(f_1),\dots,\xi(f_m))$ with a polynomial P_j. In the proof of Theorem 1.98, we have

$$(\theta_{\mathrm{W}}^{-1}F, e^{-tH_P}\theta_{\mathrm{W}}^{-1}G)_{\mathcal{F}_{\mathrm{b}}} = \lim_{n\to\infty}\mathbb{E}_{\gamma}\Big[\overline{\theta_{\mathrm{W}}^{-1}F(\tilde{\xi}_0)}e^{-\sum_{j=1}^{n-1}(tj/n):P(\tilde{\xi}_{tj/n}(f)):}\theta_{\mathrm{W}}^{-1}G(\tilde{\xi}_t)\Big].$$

Employing the unitary equivalence $\theta_{\mathrm{W}}e^{-tH_P}\theta_{\mathrm{W}}^{-1} = e^{-t\tilde{H}_P}$ and Proposition 1.135, we see that

$$\begin{aligned}(F, e^{-t\tilde{H}_P}G)_{L^2(\mathscr{M}_{-2},d\mathrm{G})} &= \lim_{n\to\infty}\mathbb{E}_{\mathcal{G}}\Big[\overline{F(\xi_0)}e^{-\sum_{j=1}^{n-1}(tj/n):P(\xi_{tj/n}(f)):}G(\xi_t)\Big] \\ &= \mathbb{E}_{\mathcal{G}}\Big[\overline{F(\xi_0)}e^{-\int :P(\xi_s(f)):ds}G(\xi_t)\Big].\end{aligned}$$

Here we used the strong continuity of $s \mapsto \xi_s(f)$. By a limiting argument, F, G can be extended to $F, G \in L^2(\mathscr{M}_{-2}, d\mathrm{G})$. □

Corollary 1.140 (Generator and invariant measure of $(\xi_t)_{t\in\mathbb{R}}$). *The generator of $(\xi_t)_{t\in\mathbb{R}}$ is $-\tilde{H}_{\mathrm{f}}$ and its invariant measure is* G.

Proof. Let $F, G \in L^2(\mathscr{M}_{-2}, d\mathrm{G})$. By Theorem 1.139, we have

$$(F, e^{-t\tilde{H}_{\mathrm{f}}}G) = \mathbb{E}_{\mathcal{G}}\Big[\overline{F(\xi_0)}G(\xi_t)\Big].$$

Hence the generator of $(\xi_t)_{t\in\mathbb{R}}$ is $-\tilde{H}_{\mathrm{f}}$. Let $p(t,\xi,A)$ be the probability transition kernel of $(\xi_t)_{t\in\mathbb{R}}$. Since $\tilde{H}_{\mathrm{f}}1 = 0$, we have

$$\int_{\mathcal{M}_{-2}} p(t,\xi,A)d\mathrm{G} = (1, e^{-t\tilde{H}_f}1_A)_{L^2(\mathcal{M}_{-2})} = \mathrm{G}(A).$$

This implies that G is an invariant measure for $(\xi_t)_{t\in\mathbb{R}}$. □

Making use of the regular conditional probability measure $\mathcal{G}^\xi$, we can derive the functional integral representation of $e^{-t\tilde{H}_P}$.

Corollary 1.141. *Let* $G \in L^2(\mathcal{M}_{-2}, d\mathrm{G})$. *Then*

$$(e^{-t\tilde{H}_P}G)(\xi) = \mathbb{E}^\xi_{\mathcal{G}}[e^{-\int :P(\xi_s(f)):ds}G(\xi_t)], \quad a.e.\ \xi \in \mathcal{M}_{-2}.$$

Proof. The proof is straightforward. By Theorem 1.139, we have

$$(F, e^{-t\tilde{H}_P}G) = \mathbb{E}_{\mathrm{G}}\Big[\overline{F(\xi)}\mathbb{E}^\xi_{\mathcal{G}}\Big[G(\xi_t)e^{-\int :P(\xi_s(f)):ds}\Big]\Big].$$

Since F is arbitrary, the corollary follows. □

1.7 Connection with infinite-dimensional stochastic analysis and Malliavin calculus

1.7.1 Finite-dimensional case

While the subject of this brief section differs from the main theme of this book, here we point out some important connections with infinite-dimensional analysis, which is a language that some readers may find more familiar. Euclidean quantum field theory and commutative or noncommutative infinite-dimensional stochastic analysis have a deep relationship, which would deserve and require a book-length discussion, therefore we need to refer the interested reader to a well-established literature for details. Here we will only highlight some basic concepts and facts, and mostly will have to skip proofs.

First, we make our observations in one dimension. Let $(\mathbb{R}, \mathcal{B}(\mathbb{R}), P)$ be a probability space, and a standard Gaussian random variable be given on it. Denote by

$$P(E) = \frac{1}{\sqrt{2\pi}}\int_E e^{-\frac{x^2}{2}}dx, \quad E \in \mathcal{B}(\mathbb{R}),$$

the related standard Gaussian measure, and $\mathbb{E}[f] = \int_{\mathbb{R}} f(x)P(dx)$.

Denote $\dot{\mathbb{N}} = \{0\} \cup \mathbb{N}$. Consider $L^2(\mathbb{R}, dP)$ and its dense subspace

$$\mathcal{L} = \mathrm{L.H.}\{x^n \mid n \in \dot{\mathbb{N}}\}.$$

Also, consider the differentiation operator

$$\mathrm{d}f(x) = \frac{df(x)}{dx}$$

with the domain Dom(d) $= \mathcal{L}$. (To avoid confusion with operators, in this section we denote operator domain by Dom.) Its adjoint d^* in $L^2(\mathbb{R}, dP)$ is obtained by a straightforward calculation.

Proposition 1.142. *For every $f \in \mathcal{L}$, we have*

$$\mathrm{d}^* f(x) = -\mathrm{d}f(x) + xf(x), \quad x \in \mathbb{R}.$$

Proof. Integration by parts gives

$$\int_{\mathbb{R}} \mathrm{d}f(x) \cdot h(x) dP(x) = -\int_{\mathbb{R}} f(x)\mathrm{d}^* h(x) dP(x), \quad h \in \mathcal{L}.$$ □

A similar calculation and an extension argument give the corollary below.

Corollary 1.143. *On $\mathcal{L}$ we have*

$$[\mathrm{d}, \mathrm{d}^*] = 1,$$

where the right-hand side denotes the identity operator in $L^2(\mathbb{R}, dP)$.

For higher derivatives, we write d^n to mean d repeated n times, and similarly we write $(\mathrm{d}^*)^n$ to mean n iterations of d^*. It is understood that $\mathrm{d}^0 = 1 = (\mathrm{d}^*)^0$. Consider the operator

$$N = \mathrm{d}^* \mathrm{d} \tag{1.7.1}$$

on $L^2(\mathbb{R}, dP)$. On $\mathcal{L}$ we have

$$Nf(x) = -\frac{d^2 f(x)}{dx^2} + x\frac{df(x)}{dx}.$$

This means that N is the generator of an Ornstein–Uhlenbeck process with state space $\mathbb{R}$ and path measure given by P. As seen in Section 5.1.5 in Volume 1, its eigenfunctions are given by the Hermite polynomials

$$H_n = (-1)^n e^{\frac{x^2}{2}} \frac{d^n}{dx^n} e^{-\frac{x^2}{2}}, \quad n \in \dot{\mathbb{N}}, \tag{1.7.2}$$

so that $NH_n = nH_n$, $n \in \dot{\mathbb{N}}$, holds. It can be obtained directly from (1.7.2) that

$$H_n = (\mathrm{d}^*)^n 1, \quad n \in \dot{\mathbb{N}}, \tag{1.7.3}$$

and it is also direct to see that the set of functions

$$\left\{\frac{1}{\sqrt{n!}}H_n\right\}_{n\in\dot{\mathbb{N}}} \tag{1.7.4}$$

forms a complete orthonormal system in $L^2(\mathbb{R}, dP)$.

Remark 1.144 (Hermite polynomial). In Section 5.1.5 in Volume 1, we used Hermite polynomials defined as

$$\bar{H}_n = (-1)^n e^{x^2}\frac{d^n}{dx^n}e^{-x^2}, \quad n \in \dot{\mathbb{N}}. \tag{1.7.5}$$

See also (2.6.26) below. Although the two definitions (1.7.2) and (1.7.5) are not exactly identical, each is a rescaling of the other as

$$\bar{H}_n(x) = 2^{\frac{n}{2}}H_n\left(\sqrt{2}x\right) \quad \text{and} \quad H_n(x) = 2^{-\frac{n}{2}}\bar{H}_n\left(\frac{x}{\sqrt{2}}\right).$$

Hermite polynomials defined by (1.7.5) are orthogonal with respect to a Gaussian weight function so that

$$\frac{1}{\sqrt{\pi}}\int_{-\infty}^{\infty}\bar{H}_m(x)\bar{H}_n(x)e^{-x^2}\,dx = 2^n n!\delta_{mn}$$

holds, hence

$$\left\{\frac{1}{\sqrt{n!2^n}}\bar{H}_n\right\}_{n\in\dot{\mathbb{N}}} \tag{1.7.6}$$

forms a complete orthonormal system in $L^2(\mathbb{R}, e^{-x^2}/\sqrt{\pi}dx)$. They also satisfy

$$\left(-\frac{1}{2}\frac{d^2}{dx^2} + x\frac{d}{dx}\right)\bar{H}_n(x) = n\bar{H}_n(x).$$

The functions defined by (1.7.5) are sometimes called physicists' Hermite polynomials, and those defined by (1.7.2) are called probabilists' Hermite polynomials.

The above lead to the following series expansion formula.

Proposition 1.145. *For every $f \in C^\infty(\mathbb{R})$ with $\frac{d^n f}{dx^n} \in L^2(\mathbb{R}, dP)$, $n \in \dot{\mathbb{N}}$, we have*

$$f = \sum_{n=0}^{\infty}\frac{1}{n!}\mathbb{E}_P[d^n f]H_n. \tag{1.7.7}$$

Proof. Since the functions (1.7.4) also belong to $C^\infty(\mathbb{R})$, we have for every $f \in C^\infty(\mathbb{R})$ that there exists a sequence of numbers $c_0, c_1, \ldots \in \mathbb{R}$ such that

$$f(x) = \sum_{n=0}^{\infty} c_n H_n(x), \quad \text{a.e. } x \in \mathbb{R}.$$

Using orthogonality, it follows that

$$c_n = \frac{1}{n!} \int_{\mathbb{R}} f(x) H_n(x) dP(x) = \frac{1}{n!} \mathbb{E}_P[f H_n].$$

By Proposition 1.142 and equality (1.7.3), we furthermore have

$$\mathbb{E}_P[f H_n] = \mathbb{E}_P[f (\mathrm{d}^*)^n 1] = \mathbb{E}_P[\mathrm{d}^n f],$$

proving the claim. □

Next we consider $d \geq 1$ and an $\mathbb{R}^d$-valued Gaussian random variable with related probability measure

$$P(E) = \frac{1}{(2\pi)^{d/2}} \int_E e^{-\frac{|x|^2}{2}} dx, \quad E \in \mathcal{B}(\mathbb{R}^d),$$

and again write $\mathbb{E}_P[f] = \int_{\mathbb{R}^d} f(x) dP(x)$ to keep the notation simple. Recall that, in general, given an open subset $U \subset \mathbb{R}^n$, a function $f : U \to \mathbb{R}^m$ is said to have a directional derivative at $x \in U$ in direction $y \in \mathbb{R}^n$ if the limit

$$D_y f(x) = \lim_{\varepsilon \downarrow 0} \frac{f(x + \varepsilon y) - f(x)}{\varepsilon} \in \mathbb{R}^m \tag{1.7.8}$$

exists. With the canonical basis vectors $e_1, \dots, e_n$ of $\mathbb{R}^n$, the derivative $D_{e_j} f(x) = \partial_{x_j} f(x) = \partial_j f(x)$ is the partial derivative of f with respect to x_j. Furthermore, f is called differentiable at $x \in U$ whenever there exists an $m \times n$-matrix $D(x)$ such that

$$\lim_{\substack{z \to 0 \\ z \in \mathbb{R}^n}} \frac{|f(x + z) - f(x) - D(x) z|}{|z|} = 0. \tag{1.7.9}$$

Then the classical fact is that differentiability of f at $x \in U$ implies that $D_y f$ exists for all $y \in \mathbb{R}^n$ and $D_y f(x) = D(x) y$ holds, and conversely, if all $\partial_j f(x)$ for $x \in U$ exist and are continuous functions of x, then f is differentiable at x and $\partial_j f_i(x) = D_{ij}(x)$, where $f_i(x)$ is the ith component of $f(x)$.

Returning to our setting, with the canonical basis vectors $e_1, \dots, e_d$ of $\mathbb{R}^d$, we thus define the d-dimensional counterpart of d to be the partial derivatives ∂_j in direction e_j. Its adjoint in $L^2(\mathbb{R}^d, dP)$ is now

$$\partial_j^* f(x) = -\partial_j f(x) + x_j f(x).$$

Corollary 1.143 becomes

$$[\partial_j, \partial_k^*] = \delta_{jk} 1, \quad j, k = 1, \dots, d.$$

To define higher order derivatives and their adjoints, consider the map $\ell : \{1,\dots,d\} \to \dot{\mathbb{N}}$ and the set $\mathcal{M}_d = \left\{\ell \mid \ell : \{1,\dots,d\} \to \dot{\mathbb{N}}\right\}$. Then for every $\ell \in \mathcal{M}_d$, we define

$$\partial^\ell = \partial_1^{\ell(1)} \cdots \partial_d^{\ell(d)} \quad \text{and} \quad (\partial^*)^\ell = (\partial_1^*)^{\ell(1)} \cdots (\partial_d^*)^{\ell(d)}.$$

The operator corresponding to (1.7.1) is now

$$N = \sum_{j=1}^d \partial_j^* \partial_j,$$

and the eigenvalue equation

$$NH_\ell = |\ell| H_\ell, \quad |\ell| = \sum_{j=1}^d \ell(j), \quad \ell \in \mathcal{M}_d,$$

is satisfied by the d-dimensional Hermite polynomials

$$H_\ell(x) = \Pi_{j=1}^d H_{\ell(j)}(x_j) = (\partial^*)^\ell \, 1_{\mathbb{R}^d},$$

where $1_{\mathbb{R}^d}$ denotes the identity function on $\mathbb{R}^d$. A complete orthonormal system of $L^2(\mathbb{R}^d, dP)$ is given by

$$\left\{\frac{1}{\Pi_{j=1}^d \ell(j)!} H_\ell\right\}_{\mathcal{M}_d}.$$

Finally, Proposition 1.145 generalizes now as below.

Proposition 1.146. *For every $f \in C^\infty(\mathbb{R}^d)$ with $\frac{d^n f}{dx^n} \in L^2(\mathbb{R}^d, dP)$, $n \in \dot{\mathbb{N}}$, we have*

$$f = \sum_{\ell \in \mathcal{M}_d} \frac{1}{\Pi_{j=1}^d \ell(j)!} \mathbb{E}_P[\partial^\ell f] H_\ell. \tag{1.7.10}$$

From the above, it is clear that the operators ∂^*, ∂, and N are reminiscent of the creation, annihilation, and number operators, respectively. Next we discuss their counterparts on infinite-dimensional spaces, and show that the analogies observed in finite dimensions are indeed valid.

1.7.2 Stochastic derivative and Cameron–Martin space

We consider the counterparts of (1.7.8)–(1.7.9) in Banach spaces. Let $\mathfrak{B}$ be a general Banach space over $\mathbb{R}$ with norm $\|\cdot\|_{\mathfrak{B}}$, $U \subset \mathfrak{B}$ an open subset in the topology generated by the given norm, and $f : U \to \mathbb{R}^m$. Then the function f is said to have a Gâteaux (directional) derivative $D_y f(x)$ in direction $y \in \mathfrak{B}$ at $x \in U$ if a similar expression to that on the right-hand side of (1.7.8) exists, and f is called Fréchet differentiable at x if there exists a bounded linear map $D : \mathfrak{B} \to \mathbb{R}^m$ with components $D_j, j = 1, \ldots, m$, in the dual space $\mathfrak{B}^*$ of $\mathfrak{B}$ such that the expression on the right-hand side of (1.7.9) exists with $|z|$ replaced by $\|z\|_{\mathfrak{B}}$. Furthermore, Gâteaux and Fréchet differentiability are related similarly as in finite-dimensional Euclidean spaces.

In what follows, we restrict to a specific Banach space by choosing

$$\mathfrak{B} = \mathscr{X} = C(\mathbb{R}^+; \mathbb{R}),$$

and consider the standard Wiener space $(\mathscr{X}, \mathcal{B}(\mathscr{X}), \mathcal{W})$, where $\mathcal{W}$ is Wiener measure defined in Definition 2.95 in Volume 1, and used throughout in this book. We note that the constructions that we will present are possible to do for the more general class of the so-called isonormal Gaussian processes, but we only discuss it for one-dimensional Brownian motion $(B_t)_{t \geq 0}$ on a standard Wiener space. Also, we note that, since there exists no Lebesgue-type measure on $C(\mathbb{R}^+; \mathbb{R})$ or in other infinite-dimensional spaces, the role of a suitable (Gaussian) measure on the given space is now clear, which does have an infinite-dimensional extension. We denote the expectation with respect to $\mathcal{W}$ by $\mathbb{E}$.

First, we consider the case when Brownian motion is run in a bounded time-interval only. Let $T > 0$ be arbitrary, and take the interval $[0, T]$. We denote the projection of $\mathscr{X}$ to this interval by $\mathscr{X}_T = C([0, T]; \mathbb{R})$. From the construction of a standard Wiener space, we know that the Borel σ-field $\mathcal{B}(\mathscr{X}_T)$ coincides with the σ-field generated by cylinder sets. It will then suffice to consider derivatives of functions F in

$$\begin{aligned} C(\mathscr{X}_T) = \bigcup_{n \in \mathbb{N}} \Big\{ F : \mathscr{X}_T \to \mathbb{R} \mid \exists f \in C^\infty_{\text{poly}}(\mathbb{R}^n),\ \exists\, 0 < t_1 < \cdots < t_n < T \text{ such that} \\ F(\omega) = f(\omega(t_1), \ldots, \omega(t_n)),\ \omega \in \mathscr{X}_T \Big\} \end{aligned}$$

called smooth cylinder functions, where

$$\begin{aligned} C^\infty_{\text{poly}}(\mathbb{R}^n) = \Big\{ f \in C^\infty(\mathbb{R}^n) \mid \forall k = (k_1, \ldots, k_n) \in \dot{\mathbb{N}}^n,\ \exists \text{ polynomial } p_k \text{ with} \\ |f^{(k)}(x)| \leq p_k(\bar{x}), x = (x_1, \ldots, x_n), \bar{x} = (|x_1|, \ldots, |x_n|) \Big\}, \end{aligned}$$

and $f^{(k)}$ denote usual derivatives. It is known that $C(\mathscr{X}_T)$ is dense in $L^p(\mathscr{X}_T, d\mathcal{W})$ for every $p \geq 1$.

In order to construct a directional derivative, we need translations of paths $\omega \in \mathscr{X}_T$. Define the operator $\theta_\varpi : \mathscr{X}_T \to \mathscr{X}_T$ by $\theta_\varpi(\omega) = \omega + \varpi$, i. e., a pathwise translation by ϖ: $\theta_\varpi(\omega)(t) = \omega(t) + \varpi(t)$ for $0 \leq t \leq T$. The map θ_ϖ is measurable for every ϖ and it is an $\mathscr{X}_T$-valued random variable with induced probability measure $\mathcal{W}^\varpi = \mathcal{W} \circ \theta_\varpi^{-1}$. Consider the subset of paths

$$\mathscr{C}_T = \Big\{h \in \mathscr{X}_T \mid \exists g_h \in L^2([0,T],dt) \text{ such that } h(t) = \int_0^t g_h(s)ds,\ t \in [0,T]\Big\}.$$

Denoting g_h corresponding to h by $\dot{h}$, under the inner product

$$(h_1, h_2) = \int_0^T \dot{h}_1(t)\dot{h}_2(t)dt,$$

the set $\mathscr{C}_T$ becomes a Hilbert space, called the *Cameron–Martin space*. An application of the Girsanov theorem then implies that whenever $\varpi \in \mathscr{C}_T$, the measure $\mathcal{W}^{\varpi}$ is absolutely continuous with respect to Wiener measure, with Radon–Nikodým derivative

$$\frac{d\mathcal{W}^{\varpi}}{d\mathcal{W}} = \exp\Bigg(\int_0^T \dot{\varpi}(s)dB_s - \frac{1}{2}\int_0^T \dot{\varpi}(s)^2 ds\Bigg). \tag{1.7.11}$$

Furthermore, it is known that $\mathcal{W}^{\varpi}$ is absolutely continuous with respect to $\mathcal{W}$ if and only if $\varpi \in \mathscr{C}_T$. The implication is then that there is a natural limitation in taking Gâteaux derivatives of Brownian paths in arbitrary directions, and, to keep absolute continuity, the directions need to be selected from the Cameron–Martin space. Then for every $\varpi \in \mathscr{C}_T$, we can consider the directional derivative

$$D_{\varpi}F(\omega) = \lim_{\varepsilon\to 0} \frac{(F \circ \theta_{\varepsilon\varpi})(\omega) - F(\omega)}{\varepsilon}.$$

Definition 1.147 (Stochastic differentiability). Let $F : \mathscr{X}_T \to \mathbb{R}$, and suppose that $D_{\varpi}F$ exists as a limit in $L^2(\mathscr{X}_T)$, for every $\varpi \in \mathscr{C}_T$. Also, assume that there exists a random process $(u_t)_{t\geq 0}$ with the property $\int_0^T \mathbb{E}[u_t^2]dt < \infty$ such that

$$D_{\varpi}F(\omega) = \int_0^{\infty} u_t(\omega)\dot{\varpi}(t)dt, \quad \text{for every } \varpi \in \mathscr{C}_T.$$

Then F is said to be *stochastically differentiable*, and its *stochastic derivative* is

$$\mathrm{D}_t F(\omega) = u_t(\omega), \quad \mathcal{W}\text{-a.s.} \tag{1.7.12}$$

The set of stochastically differentiable random variables is denoted by $\mathcal{D}^{1,2}$.

Note that $\mathrm{D}F = \mathrm{D}_t F(\omega) \in L^2(\mathscr{X}_T \times [0,T])$. The above concept can be applied to all $F = F(\omega) = f(\omega(t_1), \dots, \omega(t_n)) \in C(\mathscr{X}_T)$, and we get

$$D_{\varpi}F(\omega) = \sum_{j=1}^{n} \frac{\partial f}{\partial x_j}(\omega(t_1), \dots, \omega(t_n))1_{[0,t_j]}. \tag{1.7.13}$$

To obtain the adjoint of the directional derivative D_ϖ, we use the following integration by parts formula.

Proposition 1.148. *For every $\varpi \in \mathscr{C}_T$ and $F, G \in C(\mathscr{X}_T)$, we have*

$$\mathbb{E}[D_\varpi F \cdot G] = \mathbb{E}[F D^*_\varpi G],$$

with

$$D^*_\varpi G = -D_\varpi G + \int_0^T \dot{\varpi}(t) dB_t.$$

Proof. By the Cameron–Martin formula, we have

$$\begin{aligned}
\mathbb{E}[F \circ \theta_{\varepsilon\varpi} G] - \mathbb{E}[FG] &= \int_{\mathscr{X}_T} F(\omega + \varepsilon\varpi) G(\omega) d\mathcal{W}(\omega) - \int_{\mathscr{X}_T} F(\omega) G(\omega) d\mathcal{W}(\omega) \\
&= \int_{\mathscr{X}_T} F(\omega) G(\omega - \varepsilon\varpi) d\mathcal{W}^{\varepsilon\varpi}(\omega) - \int_{\mathscr{X}_T} F(\omega) G(\omega) d\mathcal{W}(\omega) \\
&= -\int_{\mathscr{X}_T} F(\omega) \left(G(\omega) - G(\omega - \varepsilon\varpi) \frac{d\mathcal{W}^{\varepsilon\varpi}}{d\mathcal{W}}(\omega) \right) d\mathcal{W}(\omega).
\end{aligned}$$

Then after dividing by ε, taking the limit $\varepsilon \to 0$, using (1.7.11), dominated convergence, and the properties of $C(\mathscr{X}_T)$, the claim follows. □

1.7.3 Malliavin derivative and divergence operator on $L^2(\mathscr{X})$

In this section we take $\mathbb{R}^+$ instead of bounded intervals, and consider a Brownian motion on $(\mathscr{X}, \mathcal{B}(\mathscr{X}), \mathcal{W})$. The space $L^2(\mathbb{R}^+)$ is a Hilbert space with the inner product

$$\langle f, g \rangle = \int_0^\infty \bar{f}(t) g(t) dt.$$

For $f \in L^2(\mathbb{R}^+)$, consider the stochastic integral

$$I(f) = \int_0^\infty f(t) dB_t$$

of the nonrandom function f. Using the stochastic integral, Brownian motion B_t can be reexpressed by choosing $f(t) = 1_{[0,t]}$. Due to the Itô isometry, for each f the linear functional $I(f)$ is a Gaussian random variable with mean zero and variance $\|f\|^2_{L^2(\mathbb{R}^+)} = \langle f, f \rangle$. Also, consider

$$\mathcal{S} = \bigcup_{n\in\mathbb{N}} \Big\{F : \mathscr{X} \to \mathbb{R} \mid \exists h \in C^{\infty}_{\text{poly}}(\mathbb{R}^n),\ \exists 0 < t_1 < \cdots < t_n \text{ such that}$$
$$F = h(I(f_1), \dots, I(f_n)), f_1, \dots, f_n \in L^2(\mathbb{R}^+)\Big\}.$$

The set $\mathcal{S}$ is dense in $L^2(\mathbb{R}^+)$. Inspired by (1.7.13), we have the following concept.

Definition 1.149 (Malliavin derivative on $L^2(\mathscr{X})$). For $F = h(I(f_1), \dots, I(f_n)) \in \mathcal{S}$, the $L^2(\mathbb{R}^+)$-valued random variable

$$D_t F = \sum_{j=1}^{n} \frac{\partial h}{\partial x_j}(I(f_1), \dots, I(f_n)) f_j(t)$$

is called the *Malliavin derivative* of F.

Since $\mathbb{E}[\|D_t F\|^2_{L^2(\mathbb{R}^+)}] < \infty$, it is seen that D_t is a map

$$D_t : L^2(\mathscr{X}) \to L^2(\mathscr{X}; L^2(\mathbb{R}^+)).$$

We use the notation

$$D_g F = \langle DF, g\rangle = \int_0^{\infty} (D_t F)(t) g(t) dt.$$

Higher order derivatives can similarly be defined. For $F \in \mathcal{S}$, we have for $k \in \mathbb{N}$,

$$D^k_{t_1,\dots,t_k} F = D_{t_1} \cdots D_{t_k} F = \sum_{i=1}^{k} \sum_{j_i=1}^{n} \frac{\partial^k h}{\partial x_{j_1} \cdots \partial x_{j_k}} (I(f_1), \dots, I(f_n)) f_{j_1}(t_1) \cdots f_{j_k}(t_k)$$

and it is clear that

$$D^k : L^2(\mathscr{X}) \to L^2(\mathscr{X}; \otimes^k L^2(\mathbb{R}^+)).$$

We use the identifications

$$L^2(\mathscr{X}; \otimes^k L^2(\mathbb{R}^+)) \cong L^2(\mathscr{X}) \otimes (\otimes^k L^2(\mathbb{R}^+)) \cong L^2(\mathscr{X}) \otimes L^2(\mathbb{R}^k_+) \cong L^2(\mathscr{X} \times \mathbb{R}^k_+).$$

We also define the adjoint of the Malliavin derivative. Consider the family of smooth cylinder processes

$$\mathcal{S}' = \bigcup_{n\in\mathbb{N}} \Big\{u : \mathscr{X} \times \mathbb{R}^+ \to \mathbb{R} \mid \exists F_1, \dots, F_n \in \mathcal{S},\ \exists f_1, \dots, f_n \in L^2(\mathbb{R}^+)$$
$$\text{such that } u_t = \sum_{j=1}^{n} F_j f_j(t)\Big\}.$$

Definition 1.150 (Divergence operator on $L^2(\mathscr{X})$). The linear operator

$$\delta : L^2(\mathscr{X} \times \mathbb{R}^+) \to L^2(\mathscr{X}),$$

called the *divergence operator*, is defined by the equality

$$\mathbb{E}[F\delta(u)] = \mathbb{E}[\langle DF, u\rangle], \quad F \in \mathcal{S},\, u \in \mathcal{S}'. \tag{1.7.14}$$

Integration by parts applied to this definition gives the following expression.

Proposition 1.151. *If $F = h(I(f_1), \dots, I(f_n)) \in \mathcal{S}$ and $u = \sum_{j=1}^n F_j f_j(t) \in \mathcal{S}'$, then*

$$\delta(u) = \sum_{j=1}^n F_j I(f_j) - \sum_{j=1}^n \langle DF_j, f_j\rangle.$$

Example 1.152. We have $D_t F = f(t)$ for $F = I(f)$, and $\delta(f) = I(f)$ for $f \in L^2(\mathbb{R}^+)$.

Proposition 1.153. *Let $F \in \mathcal{S}$, $u \in \mathcal{S}'$, and $g \in L^2(\mathbb{R}^+)$. Then*
(1) $D_g\delta(u) = \delta(D_g u) + \langle g, u\rangle$;
(2) $\delta(Fu) = F\delta(u) - \langle DF, u\rangle$;
(3) *for every $v \in \mathcal{S}'$,*

$$\mathbb{E}[\delta(u)\delta(v)] = \mathbb{E}\Big[\int_0^\infty u_t v_t dt\Big] + \mathbb{E}\Big[\int_0^\infty\int_0^\infty (D_s u_t)(D_t v_s) ds dt\Big].$$

The following observation allows extending the Malliavin operators to more singular functions.

Lemma 1.154. *The Malliavin derivative $D^k\lceil_{\mathcal{S}}$ is a closable operator from $L^p(\mathscr{X})$ to $L^p(\mathscr{X}; \otimes^k L^2(\mathbb{R}^+))$, for every $p \geq 1$.*

We denote the closed extension $\overline{D^k\lceil_{\mathcal{S}}}$ by the same label D^k. Define the family of norms

$$\|F\|_{k,p} = \Bigg(\mathbb{E}[|F|^p] + \mathbb{E}\Big[\sum_{j=1}^k \Big(\int_{(R^+)^n} |D^j_{t_1,\dots,t_j} F|^2 dt_1 \cdots dt_j\Big)^{p/2}\Big]\Bigg)^{1/p}, \quad k \in \mathbb{N},\ p \geq 1,$$

and denote $\mathbb{D}^{k,p} = \overline{\mathcal{S}}^{\|\cdot\|_{k,p}}$. Then the domain of the extended kth order Malliavin derivative D^k is $\mathrm{Dom}(D^k) = \mathbb{D}^{k,p}$, in particular, since $\mathcal{S}$ is a core of D^k, we see that $F \in \mathbb{D}^{k,p}$ if and only if there exists $(F_n)_{n\in\mathbb{N}} \subset \mathcal{S}$ such that $F_n \overset{L^p(\mathscr{X})}{\longrightarrow} F$ and $D^k F_n \overset{L^p(\mathscr{X};\otimes^k L^2(\mathbb{R}^+))}{\longrightarrow} D^k F$ as $n \to \infty$. Let $p = 2$ and $k = 1$. Then $\mathbb{D}^{1,2}$ is a Hilbert space with the inner product

$$(F, G)_{\mathbb{D}^{1,2}} = \mathbb{E}[FG] + \mathbb{E}[\langle DF, DG\rangle].$$

A random variable $F \in \mathbb{D}^{1,2}$ is called *Malliavin differentiable.*

The divergence operator δ can also be extended by requiring (1.7.14) to hold for every $F \in \mathbb{D}^{1,2}$. Since δ is the adjoint of D, it is a closed operator so that if $(u_n)_{n\in\mathbb{N}} \subset \mathcal{S}'$ is such that $u_n \overset{L^2(\mathcal{X};L^2(\mathbb{R}^+))}{\longrightarrow} u$ and $\delta(u_n) \overset{L^2(\mathcal{X})}{\longrightarrow} G$ as $n \to \infty$, then $u \in \mathrm{Dom}(\delta)$ and $\delta(u) = G$. We also consider the closure $\widetilde{\mathbb{D}}^{k,p} = \overline{\mathcal{S}'}^{\|\cdot\|_{k,p}}$.

Proposition 1.155. *The formulae in Proposition* 1.153 *can be shown to hold also for the extended operators in the following sense:*

(1) *If* $u \in \widetilde{\mathbb{D}}^{2,2}$ *and* $g \in L^2(\mathbb{R}^+)$, *then* $D_g u \in \mathrm{Dom}(\delta)$, $\delta(u) \in \mathbb{D}^{1,2}$, *and*

$$D_g\delta(u) = \delta(D_g u) + \langle g, u\rangle$$

holds.

(2) *If* $F \in \mathbb{D}^{1,2}$ *and* $u \in \mathrm{Dom}(\delta)$ *are such that* $Fu \in L^2(\mathcal{X}; L^2(\mathbb{R}^+))$, *then*

$$\delta(Fu) = F\delta(u) - \langle DF, u\rangle$$

whenever the right-hand side is square-integrable.

(3) *If* $u, v \in \widetilde{\mathbb{D}}^{1,2}$, *then*

$$\mathbb{E}[\delta(u)\delta(v)] = \mathbb{E}\Big[\int_0^\infty u_t v_t dt\Big] + \mathbb{E}\Big[\int_0^\infty\int_0^\infty (D_s u_t)(D_t v_s) ds dt\Big].$$

Remark 1.156. The following chain rule applies for the Malliavin derivative. Consider $\phi \in C^1(\mathbb{R})$ with the property that there exist $\lambda \geq 0$, $C > 0$ such that $|\phi'(x)| \leq C(1 + |x|^\lambda)$. If $F \in \mathbb{D}^{1,p}$ with $p \geq \lambda + 1$, then $\phi(F) \in \mathbb{D}^{1,q}$ with $q = \frac{p}{\lambda+1}$ and $D\phi(F) = \phi'(F)DF$.

Remark 1.157 (Malliavin derivative and stochastic derivative). The Malliavin derivative on $\mathbb{D}^{1,2}$ is consistent with the stochastic derivative given in Definition 1.147, and on the space $\mathbb{D}^{1,2} \cap \mathcal{D}^{1,2}$ the two operators coincide.

Remark 1.158 (Divergence operator and Skorokhod integral). If $u \in \mathrm{Dom}(\delta)$ is a progressively measurable random process, then

$$\delta(u) = \int_0^\infty u_t dB_t,$$

i. e., it is a stochastic integral. When $(u_t)_{t\geq 0}$ is *not* adapted, $\delta(u)$ can be obtained as a more general stochastic integral called the *Skorokhod integral*. This can be expressed as

$$\delta(u) = \int_0^\infty u_t \delta B_t.$$

For details, we refer to the literature.

1.7.4 Wiener–Itô chaos expansion

Recall the class $M^2(S,T)$ of square integrable processes, adapted to the augmented filtration $(\mathscr{F}_t^{\mathrm{BM}})_{t\geq 0}$ of Brownian motion, introduced in Definition 2.94 in Volume 1. A basic property of Itô integrals with integrands in $M^2(S,T)$ is that they are martingales with respect to the same filtration (Theorem 2.132 in Volume 1). The following converse statements also hold.

Theorem 1.159 (Martingale representation theorem). *If $(M_t)_{t\geq 0}$ is a martingale with respect to $(\mathscr{F}_t^{\mathrm{BM}})_{t\geq 0}$ such that $M_t \in L^2(\mathscr{X})$, $t \geq 0$, then there exists a unique random process $(f_t)_{t\in[0,T]} \in M^2(0,T)$, for all $T > 0$, such that*

$$M_t = \mathbb{E}[M_t] + \int_0^t f_s dB_s.$$

In particular, for a random variable $X \in L^2(\mathscr{X})$, the random process $(M_t)_{t\geq 0}$ with $M_t = \mathbb{E}[X|\mathscr{F}_t^{\mathrm{BM}}]$, is a martingale, on which applying the martingale representation theorem gives the theorem below.

Theorem 1.160 (Itô representation theorem). *Let $T > 0$. If $F \in L^2(\mathscr{X})$ is an $\mathscr{F}_T^{\mathrm{BM}}$-measurable random variable, then there exists a unique random process $(f_t)_{t\in[0,T]} \in M^2(0,T)$ such that*

$$F = \mathbb{E}[F] + \int_0^T f_t dB_t. \tag{1.7.15}$$

At each fixed t, we can apply the Itô representation theorem to the random variable f_t in (1.7.15) again. This gives that for each $t \in [0,T]$ there exists a unique random process $(g_{s,t})_{s\in[0,t]} \in M^2(0,t)$ such that

$$f_t = \mathbb{E}[f_t] + \int_0^t g_{s,t} dB_s.$$

Inserting this into (1.7.15) furthermore gives

$$F = \mathbb{E}[F] + \int_0^T \mathbb{E}[f_t] dB_t + \int_0^T dB_t \int_0^t g_{s,t} dB_s.$$

Iterating this procedure indefinitely leads to a canonical representation of an arbitrary random variable in $L^2(\mathscr{X})$, which we discuss next.

Take an arbitrary $n \in \mathbb{N}$ and consider a function $f : (\mathbb{R}^+)^n \to \mathbb{R}$. The function f is called symmetric if

$$f(t_{\pi(1)},\dots,t_{\pi(n)}) = f(t_1,\dots,t_n), \quad \text{for all } \pi \in \wp_n, \ t_1,\dots,t_n \in \mathbb{R}^+,$$

where $\wp_n$ denotes the permutation group of order n. The symmetrization of a function f is defined by

$$\tilde{f}(t_1,\dots,t_n) = \frac{1}{n!}\sum_{\pi\in\wp_n} f(t_{\pi(1)},\dots,t_{\pi(n)}).$$

Clearly, f is symmetric if and only if $\tilde{f} = f$. We denote the space of square-integrable symmetric functions with respect to Lebesgue measure on $(\mathbb{R}^+)^n$ by $L^2_{\text{sym}}(\mathbb{R}^n_+)$. For $f \in L^2(\mathbb{R}^+)$, consider again the stochastic integral

$$I(f) = \int_0^\infty f(t)dB_t.$$

Definition 1.161 (Iterated stochastic integral). Let $n \in \mathbb{N}$ and $f \in L^2_{\text{sym}}(\mathbb{R}^n_+)$. The *n-fold iterated stochastic integral* of f is defined by

$$I_n(f) = n!\int_0^\infty dB_{t_n}\int_0^{t_n} dB_{t_{n-1}}\cdots\int_0^{t_2} f(t_1,\dots,t_n)dB_{t_1}.$$

Let $f \in L^2(\mathbb{R}^n_+)$. The n-fold iterated stochastic integral of f is defined by $I(f) = I(\tilde{f})$.

Note that for each iterate the integrands are adapted, and, due to Itô isometry, the iterated stochastic integrals are well defined to any order n. Also, from the properties of the stochastic integral, it is straightforward to deduce the orthogonality relation

$$\mathbb{E}[I_m(f)I_n(g)] = n!(\tilde{f},\tilde{g})_{L^2(\mathbb{R}^n_+)}\delta_{mn}, \quad m,n \in \mathbb{N}. \tag{1.7.16}$$

In the special case when $f(t_1,\dots,t_n) = h(t_1)\cdots h(t_n)$, with a suitable $h \in L^2(\mathbb{R}^+)$, there is the more explicit expression

$$I_n\Big(\prod_{j=1}^n h(t_j)\Big) = \|h\|^n_{L^2(\mathbb{R}^+)}H_n\Big(\frac{I(h)}{\|h\|_{L^2(\mathbb{R}^+)}}\Big), \tag{1.7.17}$$

where H_n is the nth Hermite polynomial defined by (1.7.2). From these properties, the following fundamental fact can be obtained.

Theorem 1.162 (Wiener–Itô chaos expansion on $L^2(\mathscr{X})$). *Let $T > 0$ and suppose that $F \in L^2(\mathscr{X})$ is $\mathscr{F}_T^{\text{BM}}$-measurable. Then there exists a unique sequence of nonrandom functions $(f_n)_{n\in\mathbb{N}}$ with $f_n \in L^2_{\text{sym}}(\mathbb{R}^n_+)$, $n \in \mathbb{N}$, such that*

$$F = \sum_{n=0}^\infty I_n(f_n)$$

in $L^2(\mathscr{X})$, *where* $I_0(f_0) = \mathbb{E}[F]$. *Furthermore, the isometry*

$$\|F\|^2_{L^2(\mathscr{X})} = \sum_{n=0}^{\infty} n!\|f_n\|^2_{L^2_{\mathrm{sym}}(\mathbb{R}^n_+)}$$

holds.

Define $\mathcal{H}_0 = \mathbb{R}$ and $\mathcal{H}_n = \overline{\mathrm{L.H.}\{H_n(I(f)) \mid f \in L^2(\mathbb{R}^n_+),\ \|f\|_{L^2(\mathbb{R}^n_+)} = 1\}}$ for $n \in \mathbb{N}$, where the closure is taken in the $L^2(\mathscr{X})$ norm. The subspace $\mathcal{H}_n$ is called the nth *Wiener chaos*. Due to (1.7.17), the subspaces $\mathcal{H}_m$ and $\mathcal{H}_n$ are orthogonal unless $m = n$. Theorem 1.162 then amounts to saying that the orthogonal decomposition

$$L^2(\mathscr{X}) = \bigoplus_{n=0}^{\infty} \mathcal{H}_n \tag{1.7.18}$$

holds.

1.7.5 Malliavin derivative and divergence operator on Wiener chaos

From Theorem 1.162, we know that a random variable included in $L^2(\mathscr{X})$ can be canonically represented through its Wiener–Itô chaos expansion, which makes it natural to see how the Malliavin operators D and δ act on functions in this representation.

Proposition 1.163. *Let* $F \in L^2(\mathscr{X})$ *and consider its nth Wiener chaos* $\mathcal{H}_n$, *for arbitrary* $n \in \dot{\mathbb{N}}$, *and* $I_n(f_n)$, *where* $f_n \in L^2_{\mathrm{sym}}(\mathbb{R}^n_+)$ *is uniquely determined. Then* $I_n(f_n) \in \mathbb{D}^{1,2}$ *and*

$$D_t I_n(f_n) = nI_{n-1}(f_n(\cdot, t)), \quad n \in \mathbb{N}, \tag{1.7.19}$$

where

$$I_{n-1}(f_n(\cdot, t)) = \int_0^{\infty} dB_{t_{n-1}} \int_0^{t_{n-1}} dB_{t_{n-2}} \cdots \int_0^{t_1} f(t_1, \dots, t_{n-1}, t) dB_{t_1}.$$

Proof. First, consider the case $f_n(t_1, \dots, t_n) = \prod_{j=1}^n h_n(t_j)$, with $h_n \in L^2(\mathbb{R}_+)$ such that $\|h\|_{L^2(\mathbb{R}_+)} = 1$. Then by using (1.7.17), we obtain

$$\begin{aligned} D_t I_n(f_n) &= D_t H_n(I(h_n)) = H_n'(I(h_n)) D_t(I(h_n)) \\ &= nH_{n-1}(I(h_n))h_n(t) = nh_n(t)I_{n-1}\Big(\prod_{j=1}^{n-1} h_n(t_j)\Big) = nI_{n-1}(f_n(\cdot, t)). \end{aligned}$$

The general case can be obtained by using that the Hermite polynomials make a complete orthonormal system in $L^2(\mathbb{R}^+)$. Furthermore, by using the isometry (1.7.16), we obtain

$$\mathbb{E}\Big[\int_0^\infty (D_t I_n(f_n))^2 dt\Big] = n^2 \int_0^\infty \mathbb{E}[I_{n-1}(f_n(\cdot,t))^2]dt = n^2(n-1)\int_0^\infty \|f_n(\cdot,t)\|^2_{L^2_{\text{sym}}(\mathbb{R}_+^{n-1})}dt$$
$$= nn!\|f_n\|^2_{L^2_{\text{sym}}(\mathbb{R}_+^n)} = n\mathbb{E}[I_n(f_n)^2],$$

which shows that $I_n(f_n) \in \mathbb{D}^{1,2}$. □

The above observation leads then to the following result.

Theorem 1.164 (Malliavin derivative on Wiener chaos in $L^2(\mathscr{X})$). *Let $F \in L^2(\mathscr{X})$ with Wiener–Itô chaos expansion $F = \sum_{n=0}^\infty I_n(f_n)$. We have that $F \in \mathbb{D}^{1,2}$ if and only if*

$$\mathbb{E}[\|DF\|^2_{L^2(\mathscr{X})}] = \sum_{n=0}^\infty nn!\|f_n\|^2_{L^2_{\text{sym}}(\mathbb{R}_+^n)} < \infty$$

and then

$$D_t F = \sum_{n=0}^\infty n I_{n-1}(f_n(\cdot,t)).$$

Higher-order Malliavin derivatives are similarly obtained. For $k \geq 2$, it can be shown that $F \in \mathbb{D}^{k,2}$ if and only if

$$\sum_{n=0}^\infty n^k n!\|f_n\|^2_{L^2_{\text{sym}}(\mathbb{R}_+^n)} < \infty,$$

and then

$$D^k_{t_1,\dots,t_k} F = \sum_{n=k}^\infty n(n-1)\cdots(n-k+1) I_{n-k}(f_n(\cdot,t_1,\dots,t_k)),$$

where

$$I_{n-k}(f_n(\cdot,t_1,\dots,t_k)) = \int_0^\infty dB_{s_{n-k}} \int_0^{s_{n-k}} dB_{s_{n-k-1}} \cdots \int_0^{s_1} f(s_1,\dots,s_{n-k},t_1,\dots,t_k)dB_{s_1}.$$

Furthermore, if $F = \bigcap_{k\geq 1}\mathbb{D}^{k,2}$, then

$$f_n = \frac{1}{n!}\mathbb{E}[D^n F]$$

and thus

$$F = \sum_{n=0}^\infty \frac{1}{n!} I_n(\mathbb{E}[D^n F]),$$

as a counterpart of (1.7.7) and (1.7.10).

Next we consider the adjoint δ of D and consider its action on Wiener chaos.

Theorem 1.165 (Divergence operator on Wiener chaos). *Consider $u \in L^2(\mathscr{X} \times \mathbb{R}^+)$ with Wiener–Itô chaos expansion $u_t = \sum_{n=0}^{\infty} I_n(f_n(\cdot, t))$. Then $\delta(u) \in \mathrm{Dom}(\delta)$ if and only if*

$$\delta(u) = \sum_{n=0}^{\infty} I_{n+1}(\tilde{f}_n)$$

is a convergent series in $L^2(\mathscr{X})$.

Proof. We present one direction of the implication only, and show that for arbitrary $n \in \mathbb{N}$ and with $G = I_n(g)$, $g \in L^2(\mathbb{R}^+)$, we have

$$\mathbb{E}[I_n(\tilde{f}_{n-1})G] = \mathbb{E}[\langle u, DG\rangle] = \mathbb{E}[\delta(u)G], \quad G \in \mathcal{H}_n,$$

i. e., identify $\delta(u)$ by its projection to the nth Wiener chaos. Let $u \in \mathrm{Dom}(\delta)$. Using that $D_t G = nI_{n-1}(g(\cdot, t))$ and (1.7.16), after interchanging operations we obtain

$$\begin{aligned}
\mathbb{E}[\langle u, DG\rangle] &= n \sum_{m=0}^{\infty} \int_0^{\infty} \mathbb{E}[I_m(f_m(\cdot, t))I_{n-1}(g(\cdot, t))]dt \\
&= n \sum_{m=0}^{\infty} m!\delta_{m\,n-1} \int_0^{\infty} (\tilde{f}_m(\cdot, t), \tilde{g}(\cdot, t))_{L^2(\mathbb{R}_+^m)} dt \\
&= n(n-1)! \int_0^{\infty} (\tilde{f}_{n-1}(\cdot, t), \tilde{g}(\cdot, t))_{L^2(\mathbb{R}_+^{n-1})} dt \\
&= n!(\tilde{f}_{n-1}, \tilde{g})_{L^2(\mathbb{R}_+^n)} = \mathbb{E}[I_n(\tilde{f}_{n-1})I_n(\tilde{g})] = \mathbb{E}[I_n(\tilde{f}_{n-1})G].
\end{aligned}$$

Since n is arbitrary and $u \in \mathrm{Dom}(\delta)$, the divergence is obtained as the series given the statement, which is convergent. □

An application of the above is the following celebrated formula, which identifies the integrand in the Itô representation theorem for Malliavin differentiable random variables.

Theorem 1.166 (Clark–Ocone formula). *Let $T > 0$ and $F \in \mathbb{D}^{1,2}$ be an $\mathscr{F}_T$-measurable random variable. Then*

$$F = \mathbb{E}[F] + \int_0^T \mathbb{E}[D_t F | \mathscr{F}_t] dB_t.$$

1.7.6 Infinite-dimensional Ornstein–Uhlenbeck semigroup

Similarly to Mehler's formula in the finite-dimensional case (see Section 5.1.5 in Volume 1), we can construct the following semigroup.

Definition 1.167 (Infinite-dimensional Ornstein–Uhlenbeck semigroup). Let $F \in L^2(\mathscr{X})$ with Wiener–Itô chaos expansion $F = \sum_{n=0}^{\infty} I_n(f_n)$. Also, let $(W_t)_{t\geq 0}$ be an independent Brownian motion $(B_t)_{t\geq 0}$ on $(\mathscr{X}, \mathcal{B}(\mathscr{X}), \mathcal{W})$, and define

$$T_t(F) = \mathbb{E}_W \left[F(e^{-t}B + \sqrt{1-e^{-2t}}W)\right], \quad t \geq 0,$$

where $\mathbb{E}_W$ denotes expectation with respect to the Brownian motion $(W_t)_{t\geq 0}$. The one-parameter semigroup $\{T_t : t \geq 0\}$ on $L^2(\mathscr{X})$ is called the *infinite-dimensional Ornstein–Uhlenbeck semigroup.*

The following representation of the infinite-dimensional Ornstein–Uhlenbeck semigroup on chaos expansions holds.

Proposition 1.168. *Let $F \in L^2(\mathscr{X})$ with Wiener–Itô chaos expansion $F = \sum_{n=0}^{\infty} I_n(f_n)$. Then*

$$T_t(F) = \sum_{n=0}^{\infty} e^{-nt} I_n(f_n).$$

The generator of the infinite-dimensional Ornstein–Uhlenbeck semigroup is defined by

$$LF = \lim_{\varepsilon \downarrow 0} \frac{1}{t}(T_t - 1)F,$$

for $F \in L^2(\mathscr{X})$ for which the limit exists in $L^2(\mathscr{X})$. The following property holds.

Proposition 1.169 (Generator of infinite-dimensional Ornstein–Uhlenbeck semigroup). *Let $F \in L^2(\mathscr{X})$ with Wiener–Itô chaos expansion $F = \sum_{n=0}^{\infty} I_n(f_n)$.*

(1) *The property $F \in \mathrm{Dom}(L)$ holds if and only if $\sum_{n=1}^{\infty} n^2 \|f_n\|_{L^2_{\mathrm{sym}}(\mathbb{R}^n_+)} < \infty$ and then*

$$LF = -\sum_{n=1}^{\infty} n I_n(f_n).$$

(2) *The property $F \in \mathrm{Dom}(L)$ holds if and only if $F \in \mathbb{D}^{1,2}$ and $DF \in \mathrm{Dom}(\delta)$, and then*

$$\delta DF = -LF.$$

From Theorem 1.169, we also obtain that

$$[D, \delta] = 1_{L^2(\mathscr{X})},$$

as a counterpart of Corollary 1.143.

1.7.7 Malliavin derivative on white noise space

In this section we briefly sketch an alternative way of introducing the Malliavin operators, when instead of Wiener space the space of tempered distributions is used.

For simplicity, we only consider the one-dimensional case, as before. Denote the duality product for $f \in \mathscr{S}'_{\mathrm{real}}(\mathbb{R})$ and $\phi \in \mathscr{S}_{\mathrm{real}}(\mathbb{R})$ by $\langle\!\langle f, \phi \rangle\!\rangle$. Consider the measurable space $(\mathscr{S}'_{\mathrm{real}}(\mathbb{R}), \mathscr{F})$, where $\mathscr{F}$ is its Borel σ-field. By the Bochner–Minlos–Sazonov theorem (Proposition 1.114 above), there exists a Gaussian probability measure μ on $\mathscr{S}'_{\mathrm{real}}(\mathbb{R})$ such that

$$\int_{\mathscr{S}'_{\mathrm{real}}(\mathbb{R})} e^{i\langle\!\langle \omega, \phi \rangle\!\rangle} d\mu(\omega) = e^{-\frac{1}{2}\|\phi\|^2}, \quad \phi \in \mathscr{S}_{\mathrm{real}}(\mathbb{R}).$$

We denote

$$(L^2) = L^2(\mathscr{S}'_{\mathrm{real}}(\mathbb{R}), d\mu). \tag{1.7.20}$$

Definition 1.170 (White noise process). The random process $(w_\phi)_{\phi \in \mathscr{S}_{\mathrm{real}}(\mathbb{R})}$ on the probability space $(\mathscr{S}'_{\mathrm{real}}(\mathbb{R}), \mathscr{F}, \mu)$ indexed by $\mathscr{S}_{\mathrm{real}}(\mathbb{R})$ and given by the measurable map $w : \mathscr{S}_{\mathrm{real}}(\mathbb{R}) \times \mathscr{S}'_{\mathrm{real}}(\mathbb{R}) \to \mathbb{R}$,

$$w_\phi(\omega) = \langle\!\langle \omega, \phi \rangle\!\rangle, \quad \omega \in \mathscr{S}'_{\mathrm{real}}(\mathbb{R}), \ \phi \in \mathscr{S}_{\mathrm{real}}(\mathbb{R}),$$

is called a *white noise process*.

Remark 1.171. By Wiener–Itô–Segal isomorphism (Proposition 1.51), we have the equivalence $(L^2) \cong \mathcal{F}_{\mathrm{b}}(L^2(\mathbb{R}))$. Under this equivalence, the Segal field $\Phi(\phi)$ is equivalent to w_ϕ for $\phi \in L^2_{\mathrm{real}}(\mathbb{R})$. Below we construct the triplet $(\mathcal{S}) \subset (L^2) \subset (\mathcal{S})^*$ in (1.7.21), which is a counterpart of the triplet $\mathscr{S}(\mathbb{R}) \subset L^2(\mathbb{R}) \subset \mathscr{S}'(\mathbb{R})$, and thereby we show that white noise analysis can be interpreted as a distribution theory on boson Fock space.

Using the isometry

$$\mathbb{E}_\mu[w_\phi^2] = \int_{\mathscr{S}'_{\mathrm{real}}(\mathbb{R})} \langle\!\langle \omega, \phi \rangle\!\rangle^2 d\mu(\omega) = \|\phi\|^2,$$

the white noise process can be extended to $\phi \in L^2_{\mathrm{real}}(\mathbb{R})$. Writing

$$B_t(\omega) = \begin{cases} \langle\!\langle \omega, 1_{[0,t)} \rangle\!\rangle, & t \geq 0, \\ \langle\!\langle \omega, 1_{(t,0]} \rangle\!\rangle, & t < 0, \end{cases}$$

we can define a two-sided Brownian motion $(B_t)_{t\in\mathbb{R}}$ indexed by $\mathbb{R}$, and by further using the Kolmogorov–Čentsov theorem, it is possible to show that it has a continuous version. We have that

$$w_\phi(\omega) = I(\phi)(\omega) = \int_{-\infty}^{\infty} \phi(t)dB_t(\omega)$$

holds as a stochastic integral, and then $\mathbb{E}_\mu[w_\phi^2] = \|\phi\|^2$ is usual Itô isometry.

The white noise variant of Wiener–Itô chaos expansion can be obtained in the following way. Recall the Hermite polynomials as given in (1.7.2) and consider the related family of Hermite functions given by

$$\mathrm{e}_n(t) = \frac{1}{\sqrt{\sqrt{\pi}(n-1)!}} H_{n-1}(\sqrt{2}t)e^{-\frac{t^2}{2}}, \quad n \in \mathbb{N},$$

which are Schwartz functions and make a complete orthonormal system in $L^2(\mathbb{R})$. One can check that

$$\begin{aligned}\int_{\mathbb{R}} \mathrm{e}_n(t)\mathrm{e}_m(t)dt &= \frac{1}{\sqrt{\pi(n-1)!(m-1)!}} \int_{\mathbb{R}} H_{n-1}(\sqrt{2}t)H_{m-1}(\sqrt{2}t)e^{-t^2}dt \\ &= \frac{1}{\sqrt{(n-1)!(m-1)!}} \int_{\mathbb{R}} H_{n-1}(t)H_{m-1}(t)\frac{e^{-t^2/2}}{\sqrt{2\pi}}dt = \delta_{nm}.\end{aligned}$$

Define

$$J_n(\omega) = w_{\mathrm{e}_n}(\omega).$$

Also, let $\mathcal{I} = \bigcup_{n=1}^{\infty} \left\{\ell = (\ell_1, \ldots, \ell_n) \mid \ell_1, \ldots, \ell_n \in \dot{\mathbb{N}}\right\}$ be the family of nonnegative integer multiindices, write $L_0 = 1$, and for $\ell \in \mathcal{I}$ define

$$L_\ell(\omega) = \prod_{k=1}^{n} H_{\ell_k}(J_k(\omega)), \quad \ell = (\ell_1, \ldots, \ell_n) \neq 0.$$

Theorem 1.172 (Wiener–Itô chaos expansion on white noise). *Choose $T \in \mathbb{R}$ and consider an $\mathcal{F}_T$-measurable random variable $F \in (L^2)$. There exists a unique sequence of numbers $(c_\ell)_{\ell \in \mathcal{I}} \subset \mathbb{R}$ such that*

$$F = \sum_{\ell \in \mathcal{I}} c_\ell L_\ell.$$

Moreover, $\|F\|^2_{(L^2)} = \sum_{\ell \in \mathcal{I}} \ell!\, c_\ell^2$ holds, where $\ell! = \ell_1! \cdots \ell_n!$ for $\ell = (\ell_1, \ldots, \ell_n) \in \mathcal{I}$.

We note that the equivalence with the expansion given in Theorem 1.162 is secured by the following relationships. We can show by using (1.7.17) that

$$I_n(\otimes^{\ell}_{\mathrm{sym}} \mathrm{e}_n) = \prod_{k=1}^{n} H_{\ell_k}(J_k) = L_\ell, \quad \ell = (\ell_1, \ldots, \ell_n) \in \mathcal{I},\ n = |\ell|.$$

On the other hand, f_n in the expansion in Theorem 1.162 can be expressed like

$$f_n = \sum_{\substack{\ell=(\ell_1,\dots,\ell_m)\in\mathcal{I}\\ |\ell|=n}} c_\ell\,(\otimes^{\ell_1} e_1)\otimes_{\mathrm{sym}}\cdots\otimes_{\mathrm{sym}}(\otimes^{\ell_m} e_m).$$

These relationships allow introducing the following spaces. For $m \in \mathbb{N}$ and $\ell = (\ell_1,\dots,\ell_m)$, denote $(m,\ell) = \prod_{k=1}^m (2k)^{\ell_k}$.

Definition 1.173 (Hida space). Consider $\phi = \sum_{\ell\in\mathcal{I}} a_\ell L_\ell$ and $F = \sum_{\ell\in\mathcal{I}} b_\ell L_\ell$.

(1) Let $p > 0$. Then ϕ is said to belong to the *Hida test function Hilbert space* $(\mathcal{S})_p$ if

$$\|\phi\|^2_{(\mathcal{S})_p} = \sum_{\ell\in\mathcal{I}} (m,\ell)^{p\ell}\,\ell!\,a_\ell^2 < \infty.$$

We define the *Hida test function space* by $(\mathcal{S}) = \bigcap_{p>0}(\mathcal{S})_p$.

(2) Let $q > 0$. Then F is said to belong to the *Hida distribution Hilbert space* $(\mathcal{S})_{-q}$ if

$$\|F\|^2_{(\mathcal{S})_{-q}} = \sum_{\ell\in\mathcal{I}} (m,\ell)^{-q\ell}\,\ell!\,b_\ell^2 < \infty.$$

We define the *Hida distribution space* by $(\mathcal{S})^* = \bigcup_{q>0}(\mathcal{S})_{-q}$.

We note that $(\mathcal{S})^*$ is the dual space of $(\mathcal{S})$, and for $F \in (\mathcal{S})^*$ and $\phi \in (\mathcal{S})$ of the form in the definition above, the formula

$$\langle\langle F,\phi\rangle\rangle = \sum_{\ell\in\mathcal{I}} \ell!\,a_\ell b_\ell$$

applies. Here we also used $\langle\langle\cdot,\cdot\rangle\rangle$ for the distribution product between $(\mathcal{S})^*$ and $(\mathcal{S})$. Also, the inclusions

$$(\mathcal{S}) \subset (\mathcal{S})_p \subset (L^2) \subset (\mathcal{S})_{-q} \subset (\mathcal{S})^* \tag{1.7.21}$$

hold for every $0 < p,q$. The topology on $(\mathcal{S})$ is given by requiring that $(\phi_n)_{n\in\mathbb{N}} \subset (\mathcal{S})$ is convergent to $\phi \in (\mathcal{S})$ if $\|\phi_n - \phi\|_{(\mathcal{S})_p} \to 0$ as $n \to \infty$, for every $p > 0$. Similarly, the topology on $(\mathcal{S})^*$ is defined by saying that $(F_n)_{n\in\mathbb{N}} \subset (\mathcal{S})^*$ is convergent to $F \in (\mathcal{S})^*$ if there exists $q > 0$ such that $\|F_n - F\|_{(\mathcal{S})_{-q}} \to 0$ as $n \to \infty$.

Using these spaces, now we can introduce Malliavin derivative on white noise.

Definition 1.174 (Malliavin derivative on white noise). Let $F \in (L^2)$, and $\varpi \in L^2(\mathbb{R})$ be a nonrandom function. The directional derivative of F in $(\mathcal{S})^*$ is given by

$$D_\varpi F(\omega) = \lim_{\varepsilon\to 0}\frac{F(\omega+\varepsilon\varpi) - F(\omega)}{\varepsilon}, \quad \varpi \in \mathscr{S}'(\mathbb{R}),$$

if the limit exists in $(\mathcal{S})^*$. If, moreover, there exists a function $u : \mathbb{R} \to (\mathcal{S})^*$ such that $\int_{-\infty}^{\infty} u(t)\varpi(t)dt$ is a convergent integral in $(\mathcal{S})^*$ and

$$D_{\varpi}F = \int_{-\infty}^{\infty} u(t)\varpi(t)dt, \quad \varpi \in L^2(\mathbb{R}),$$

then F is called *Malliavin differentiable* in $(\mathcal{S})^*$ and

$$u(t) = D_t F, \quad t \in \mathbb{R},$$

is called its *Malliavin derivative* in $(\mathcal{S})^*$ at t.

Remark 1.175 (Malliavin derivative on Wiener chaos in (L^2)).

(1) We can also express the Malliavin derivative on the Wiener chaos in white noise variables. A calculation gives

$$D_t L_\ell = \sum_{k=1}^{m} \Big(\prod_{\substack{j=1 \\ j \neq k}}^{m} H_{\ell_j}(J_j) \Big) \ell_k H_{\ell_k - 1}(J_k) \mathrm{e}_k(t), \quad \ell = (\ell_1, \ldots, \ell_m) \in \mathcal{I}.$$

In fact, this can be used for a definition of D_t on the expansion series. For $F = \sum_{\ell \in \mathcal{I}} c_\ell L_\ell \in (\mathcal{S})^*$, the Malliavin derivative at $t \in \mathbb{R}$ is then defined to be

$$D_t F = \sum_{\ell \in \mathcal{I}} c_\ell \sum_{k=1}^{\infty} \Big(\prod_{\substack{j=1 \\ j \neq k}}^{m} H_{\ell_j}(J_j) \Big) \ell_k H_{\ell_k - 1}(J_k) \mathrm{e}_k(t),$$

and $\mathrm{Dom}(D_t)$ contains all $F \in (\mathcal{S})^*$ for which this series is convergent in the topology of $(\mathcal{S})^*$. It can be shown that $(L^2) \subset \mathrm{Dom}(D_t) \subset (\mathcal{S})^*$.

(2) Higher order derivatives and a divergence operator as the adjoint of the Malliavin derivative can similarly be defined on white noise as previously seen; for details, we refer to the literature.

2 The Nelson model by path measures

2.1 Preliminaries

In this section we consider a model of an electrically charged, spinless, and nonrelativistic particle interacting with a scalar boson field, which we call *Nelson model.* The coupling between particle and field is assumed to be linear. First, we explain how the Nelson model can be derived from a Lagrangian, and then continue with a rigorous definition and analysis in the following sections.

The *Lagrangian density* $\mathscr{L}_{\mathrm{N}} = \mathscr{L}_{\mathrm{N}}(x,t)$, $(x,t) \in \mathbb{R}^3 \times \mathbb{R}$ of the Nelson model is described by

$$\mathscr{L}_{\mathrm{N}} = i\Psi^*\dot{\Psi} + \frac{1}{2m}\partial_j\Psi^*\partial_j\Psi + \frac{1}{2}\partial_\mu\phi\partial^\mu\phi - \frac{1}{2}\nu^2\phi^2 + \Psi^*\Psi\phi,$$

where $\Psi = \Psi(x,t)$ is a complex scalar field describing a nonrelativistic, spinless electron, and $\phi(x,t)$ is a neutral scalar field describing scalar bosons, $\nu \geq 0$ is the mass of bosons, and $m > 0$ is the mass of the electron. Here $\partial_\mu\phi\partial^\mu\phi = \dot{\phi}\dot{\phi} - \partial_j\phi\partial_j\phi$, $\partial_j = \partial_{x_j}$, the dots denote time derivative, and the stars denote complex conjugate. The dynamics of this system is given by the *Euler–Lagrange equation* leading to the system of nonlinear field equations

$$\begin{cases} (\Box + \nu^2)\phi(x,t) = \Psi^*(x,t)\Psi(x,t), \\ \left(i\partial_t + \frac{1}{2m}\Delta\right)\Psi(x,t) = \phi(x,t)\Psi(x,t). \end{cases}$$

The Hamiltonian density is derived from the *Legendre transform* of $\mathscr{L}_{\mathrm{N}}$. The conjugate momenta are defined by

$$\Phi = \frac{\partial \mathscr{L}_{\mathrm{N}}}{\partial \dot{\Psi}} = i\Psi^*, \tag{2.1.1}$$

$$\pi = \frac{\partial \mathscr{L}_{\mathrm{N}}}{\partial \dot{\phi}} = \dot{\phi}. \tag{2.1.2}$$

Thus the Hamiltonian density $H_{\mathrm{N}} = H_{\mathrm{N}}(x,t)$ is given by

$$H_{\mathrm{N}} = \Phi\dot{\Psi} + \pi\dot{\phi} - \mathscr{L}_{\mathrm{N}} = \frac{1}{2m}|\partial_x\Psi|^2 + \frac{1}{2}(\dot{\phi}^2 + (\partial_x\phi)^2 + \nu^2\phi^2) - \Psi^*\Psi\phi$$

and the Hamiltonian

$$H_{\mathrm{N}} = \int_{\mathbb{R}^3} \left[\Psi^*\left(-\frac{1}{2m}\Delta\right)\Psi + \frac{1}{2}(\dot{\phi}^2 + (\partial_x\phi)^2 + \nu^2\phi^2) - \Psi^*\Psi\phi\right] dx.$$

Since the Nelson model describes a charged particle interacting with a scalar field, and the particle is supposed to carry low energy, there is no annihilation and creation of

https://doi.org/10.1515/9783112231494-002

particles and the number of particles is conserved. In order to obtain the quantized model, the kinetic part is replaced by

$$\int_{\mathbb{R}^3} \Psi^* \left(-\frac{1}{2m}\Delta\right) \Psi dx \to -\frac{1}{2m}\Delta$$

and the interaction by

$$-\int_{\mathbb{R}^3} \Psi^* \Psi \phi dx \to \phi(x).$$

Adding an external potential V, the formal expression of the Nelson model is

$$-\frac{1}{2m}\Delta + V + H_{\mathrm{f}} + \phi(x),$$

where

$$H_{\mathrm{f}} = \frac{1}{2}\int_{\mathbb{R}^3} \left(\dot{\phi}^2 + (\partial_x \phi)^2 + \nu^2\phi^2\right) dx.$$

2.2 The Nelson model in Fock space

2.2.1 Definition of the Nelson model

We give a rigorous definition of the Nelson model. In the previous section, we introduced the Hilbert spaces $\mathcal{F}_{\mathrm{b}}(\hat{\mathcal{H}}_M)$, $L^2(\mathcal{Q})$, and $L^2(\mathcal{M}_{-2}, d\mathrm{G})$, serving as state spaces of the quantum field. We have seen the unitary equivalence

$$\mathcal{F}_{\mathrm{b}}(\hat{\mathcal{H}}_M) \cong L^2(\mathcal{Q}) \cong L^2(\mathcal{M}_{-2}, d\mathrm{G}) \tag{2.2.1}$$

in (1.5.3) and (1.6.19), following the identifications

$$\frac{1}{\sqrt{2}}(a_M^*(\hat{f}) + a_M(\bar{\hat{f}})) \cong \phi(f) \cong \xi(f), \quad f \in \mathcal{M}. \tag{2.2.2}$$

Here a_M^* and a_M denote the creation and annihilation operators, respectively, in the boson Fock space $\mathcal{F}_{\mathrm{b}}(\hat{\mathcal{H}}_M)$, satisfying the canonical commutation relations

$$[a_M(\hat{f}), a_M(\hat{g})] = 0 = [a_M^*(\hat{f}), a_M^*(\hat{g})], \quad [a_M(\hat{f}), a_M^*(\hat{g})] = (\bar{\hat{f}}, \hat{g})_{\hat{\mathcal{H}}_M}. \tag{2.2.3}$$

The Hilbert space $L^2(\mathbb{R}^d)$ describes the state space of the nonrelativistic particle, and the *scalar boson field* is defined by

$$\mathcal{F}_{\mathrm{N}} = \mathcal{F}_{\mathrm{b}}(\hat{\mathcal{H}}_M),$$

i. e., the boson Fock space over $\hat{\mathcal{H}}_M$. The joint state space is described by the tensor product

$$\mathcal{H}_{\mathrm{N}} = L^2(\mathbb{R}^d) \otimes \mathcal{F}_{\mathrm{N}}.$$

The *free particle Hamiltonian* is described by the Schrödinger operator

$$H_{\mathrm{p}} = -\frac{1}{2}\Delta + V$$

acting in $L^2(\mathbb{R}^d)$. We denote the ground state energy of H_{p} by

$$E_{\mathrm{p}} = \inf \mathrm{Spec}(H_{\mathrm{p}}).$$

We introduce a class of potentials.

Definition 2.1 (Definition of $\mathcal{R}_{\mathrm{Kato}}$). Let V be relatively bounded with respect to $-\frac{1}{2}\Delta$ with a relative bound strictly smaller than one, i. e., $D(V) \subset D(-(1/2)\Delta)$ and

$$\|Vf\| \le a\left\|-\frac{1}{2}\Delta f\right\| + b\|f\|$$

for $f \in D(V)$ with some $a < 1$ and $b \ge 0$. Then we say that $V \in \mathcal{R}_{\mathrm{Kato}}$.

The following are standing assumptions throughout Section 2.2.

Assumption 2.2. The following conditions hold:
(1) (Dispersion relation) $\omega(k) = \omega_\nu(k) = \sqrt{|k|^2 + \nu^2}$, $\nu \ge 0$.
(2) (Charge distribution) $\varphi \in \mathscr{S}'(\mathbb{R}^d)$, $\overline{\hat{\varphi}(k)} = \hat{\varphi}(-k)$ and $\hat{\varphi}/\omega, \hat{\varphi}/\sqrt{\omega} \in L^2(\mathbb{R}^d)$.
(3) (External potential) $V \in \mathcal{R}_{\mathrm{Kato}}$.

Note that we use the notation ω instead of ω_ν for notational simplicity unless any confusion may arise. We also note that condition $\overline{\hat{\varphi}(k)} = \hat{\varphi}(-k)$ is necessary and sufficient for φ being a real tempered distribution. The *free field Hamiltonian*

$$H_{\mathrm{f}} = \mathrm{d}\Gamma(\omega)$$

on $\mathcal{F}_{\mathrm{N}}$ accounts for the energy carried by the field configuration. The *particle–field interaction Hamiltonian* H_{I} acting on the Hilbert space $\mathcal{H}_{\mathrm{N}}$ describes then the interaction energy between the boson field and the particle. To give a definition of this operator, we identify $\mathcal{H}_{\mathrm{N}}$ as the space of $\mathcal{F}_{\mathrm{N}}$-valued L^2-functions on $\mathbb{R}^d$:

$$\mathcal{H}_{\mathrm{N}} \cong \int_{\mathbb{R}^d}^{\oplus} \mathcal{F}_{\mathrm{N}} dx = \left\{F : \mathbb{R}^d \to \mathcal{F}_{\mathrm{N}} \,\middle|\, \int_{\mathbb{R}^d} \|F(x)\|^2_{\mathcal{F}_{\mathrm{N}}} dx < \infty\right\}.$$

For every $x \in \mathbb{R}^d$, the operator $H_{\mathrm{I}}(x)$ is defined by the time-zero field

$$H_{\mathrm{I}}(x) = \frac{1}{\sqrt{2}}\left(a_M^*(\hat{\varphi}e^{-ik\cdot x}) + a_M(\tilde{\hat{\varphi}}e^{ik\cdot x})\right).$$

We will use the formal notation

$$H_{\mathrm{I}}(x) = \int_{\mathbb{R}^d} \frac{1}{\sqrt{2\omega(k)}}\left(\hat{\varphi}(k)e^{-ik\cdot x}a^*(k) + \hat{\varphi}(-k)e^{ik\cdot x}a(k)\right)dk$$

for convenience. Here we assume that $[a(k), a^*(k')] = \delta(k-k')$. Since $\overline{\hat{\varphi}(k)} = \hat{\varphi}(-k)$, $H_{\mathrm{I}}(x)$ is symmetric, and it can be shown by using Nelson's analytic vector theorem that $H_{\mathrm{I}}(x)$ is essentially self-adjoint on the finite particle subspace

$$\mathcal{F}_{\mathrm{N,fin}} = \mathcal{F}_{\mathrm{b,fin}}(\hat{\mathcal{H}}_M).$$

We denote the self-adjoint extension of $H_{\mathrm{I}}(x)$ by $\overline{H_{\mathrm{I}}(x)}$. The interaction H_{I} is then defined by the self-adjoint operator

$$H_{\mathrm{I}} = \int_{\mathbb{R}^d}^{\oplus} \overline{H_{\mathrm{I}}(x)}dx$$

acting as $(H_{\mathrm{I}}\Psi)(x) = \overline{H_{\mathrm{I}}(x)}\Psi(x)$ with domain

$$D(H_{\mathrm{I}}) = \left\{\Psi \in \mathcal{H}_{\mathrm{N}} \mid \Psi(x) \in D(\overline{H_{\mathrm{I}}(x)}) \text{ a. e. } x \text{ and } \int_{\mathbb{R}^d} \|\overline{H_{\mathrm{I}}(x)}\Psi(x)\|^2_{\mathcal{F}_{\mathrm{N}}}dx < \infty\right\}.$$

Definition 2.3 (Nelson Hamiltonian in Fock space). The operator

$$H_{\mathrm{N}} = H_{\mathrm{p}} \otimes 1 + 1 \otimes H_{\mathrm{f}} + H_{\mathrm{I}} \tag{2.2.4}$$

acting in $\mathcal{H}_{\mathrm{N}}$ is called the *Nelson Hamiltonian*.

Let

$$H_0 = H_{\mathrm{p}} \otimes 1 + 1 \otimes H_{\mathrm{f}}, \quad D(H_0) = D(H_{\mathrm{p}} \otimes 1) \cap D(1 \otimes H_{\mathrm{f}}).$$

The first natural question about the Nelson Hamiltonian is whether it is self-adjoint.

Proposition 2.4 (Self-adjointness). *We have the following properties:*

(1) *H_0 is self-adjoint on $D(H_0)$ and bounded from below.*

(2) *H_{N} is self-adjoint on $D(H_0)$ and bounded from below. Furthermore, it is essentially self-adjoint on any core of H_0.*

Proof. Operator H_0 is self-adjoint and bounded from below as it is the sum of two commuting self-adjoint operators bounded from below. By Lemma 1.6, $D(H_{\mathrm{I}}) \supset D(H_0)$, and Proposition 1.6 gives

$$\|\overline{H_\mathrm{I}(x)}\Psi\|_{\mathcal{F}_\mathrm{N}} \le (2\|\hat{\varphi}/\omega\| + \|\hat{\varphi}/\sqrt{\omega}\|)\|(H_\mathrm{f}+1)^{1/2}\Psi\|_{\mathcal{F}_\mathrm{N}}$$

for $\Psi \in D(H_\mathrm{f})$ and every $x \in \mathbb{R}^d$. Thus for $\Phi \in D(1 \otimes H_\mathrm{f})$,

$$\|H_\mathrm{I}\Phi\|_{\mathcal{H}_\mathrm{N}} \le (2\|\hat{\varphi}/\omega\| + \|\hat{\varphi}/\sqrt{\omega}\|)\|1 \otimes (H_\mathrm{f}+1)^{1/2}\Phi\|_{\mathcal{H}_\mathrm{N}}.$$

Since

$$\|1 \otimes (H_\mathrm{f}+1)^{1/2}\Psi\| \le \|(H_0 - E_\mathrm{p} + 1)^{1/2}\Psi\| \le \varepsilon\|H_0\Psi\| + b_\varepsilon\|\Psi\|$$

with some $b_\varepsilon \ge 0$, the Kato–Rellich theorem yields that H_N is self-adjoint on $D(H_0)$ and bounded from below. □

2.2.2 Infrared and ultraviolet divergences

It is a standard fact in quantum field theory that a rigorous description runs into mathematical difficulties caused by divergent integrals at the two ends of the spectrum. Since in quantum theory the electron is regarded as a point particle, the physical form factor corresponds to $\varphi_\mathrm{ph}(x) = \delta(x)$. Due to $\hat{\varphi}_\mathrm{ph}(k) = (2\pi)^{-d/2}$, this implies

$$\int_{\mathbb{R}^d} \frac{|\hat{\varphi}_\mathrm{ph}(k)|^2}{\omega(k)} dk = \infty.$$

This situation is called *ultraviolet divergence*. In a mathematical description therefore, a regularized form factor $\hat{\varphi}$ is imposed so that the above integral is finite and $H_\mathrm{I}(x)$ is well defined as an operator on $\mathcal{F}_\mathrm{N}$.

At the other end of the energy spectrum, an infrared divergence is encountered. Suppose that $\hat{\varphi}(k) = (2\pi)^{-d/2}$ for $|k| < \varepsilon$, with some $\varepsilon > 0$. Since for massless bosons the physical dispersion relation is $\omega(k) = |k|$, in this case

$$\int_{|k|<\varepsilon} \frac{|\hat{\varphi}(k)|^2}{\omega(k)^3} dk = \infty, \quad d \le 3. \tag{2.2.5}$$

This is called *infrared divergence*. In the physical understanding of the model, the ground state expectation of the number of bosons should be finite, i. e., for the ground state Ψ_g of the Nelson Hamiltonian H_N and the boson number operator N one should have $(\Psi_\mathrm{g}, N\Psi_\mathrm{g}) < \infty$. It will be seen in (2.6.19) below that, in fact,

$$C_1 \int_{\mathbb{R}^d} \frac{|\hat{\varphi}(k)|^2}{\omega(k)^3} dk - C_2 \le (\Psi_\mathrm{g}, N\Psi_\mathrm{g}) \le C_3 \int_{\mathbb{R}^d} \frac{|\hat{\varphi}(k)|^2}{\omega(k)^3} dk$$

with constants $C_1, C_2, C_3 > 0$. Moreover, under condition (2.2.5), it can be proven that there is no ground state in $\mathcal{H}_N$, which justifies the expectation that in the ground state only a finite number of bosons bind to the particle. To cope with this difficulty, we impose the *infrared regularity condition*

$$\int_{\mathbb{R}^d} \frac{|\hat{\varphi}(k)|^2}{\omega(k)^3} dk < \infty \tag{2.2.6}$$

so that whenever the above integral is infinite, we speak of *infrared singularity*. Note that, in contrast, for massive bosons where $\omega(k) = \sqrt{|k|^2 + \nu^2}$ with $\nu > 0$, no infrared divergence occurs. Although the cutoffs are mathematical artifacts and not intrinsically physical properties, we impose both an ultraviolet and an infrared cutoff in order that the model be well defined, and at a stage of discussing this theory we will ask whether in a proper limit these cutoffs can be removed.

Two specific choices of ω and $\hat{\varphi}$ will be particularly important, and we single them out. Choosing

$$\omega(k) = |k| \quad \text{and} \quad \hat{\varphi}(k) = g 1_{\{\kappa<|k|<\Lambda\}}$$

with $0 < \kappa < \Lambda$, we arrive at the *massless Nelson model* in which the mass of bosons is zero. Here κ and Λ are respectively the infrared and ultraviolet cut-off parameters, and g is the coupling constant. Choosing

$$\omega(k) = \sqrt{|k|^2 + \nu^2} \quad \text{and} \quad \hat{\varphi}(k) = g 1_{\{|k|<\Lambda\}},$$

with $\nu > 0$ and $\Lambda > 0$, gives the *massive Nelson model* with nonzero mass ν of bosons, ultraviolet cut-off Λ and coupling constant g. Both of these cases satisfy (2.2.6), except

$$\omega(k) = |k| \quad \text{and} \quad \hat{\varphi}(k) = g 1_{\{|k|<\Lambda\}}, \quad d \leq 3,$$

when an infrared singularity occurs.

2.2.3 Embedded eigenvalues

The Nelson Hamiltonian is of the form $H_N = H_0 + H_I$. The spectrum of H_0 is well known, in particular, when the boson field is massless, all the eigenvalues of H_0 are embedded in the continuous spectrum. It is of great interest to study the behavior of the embedded eigenvalues resulting from the interaction term H_I in a nonperturbative way. While there are few general results on the perturbation of embedded eigenvalues, the study of perturbations of the discrete spectrum can be done by Kato's theory for sufficiently weak couplings. From this theory, it can be seen that under a small perturbation the discrete spectrum remains discrete. In our present context, however, the investigation of embedded eigenvalues is crucial. Since even for an arbitrarily small coupling constant the bottom of the spectrum of H_0 is an embedded eigenvalue, the existence of a ground state of H_N is nontrivial for any value of the coupling constant. One of the goals of the mathematical analysis of the Nelson Hamiltonian in this chapter is to establish the behavior of embedded eigenvalues under perturbations by using functional integral methods. In particular, we will show the existence of a ground state for all values of the coupling constant. It is, however, worth noting that the proof of existence of a ground state is not constructive. One way of studying the qualitative properties of the ground state is to apply functional integrals. Using this, we get probabilistic expressions for ground state expectations $(\Psi_g, \mathcal{O}\Psi_g)$ for a range of observables $\mathcal{O}$.

2.3 The Nelson model in function space

2.3.1 Infinite-dimensional Ornstein–Uhlenbeck processes and $P(\phi)_1$-processes

We will make use of the infinite-dimensional Ornstein–Uhlenbeck process $(\xi_t)_{t\in\mathbb{R}}$ discussed in Section 1.6 to obtain a unitary equivalent representation of H_N. We assume a real Hilbert space $\mathscr{M}$, and a positive self-adjoint operator $\mathcal{D}$ in $\mathscr{M}$ with a Hilbert–Schmidt inverse to be given as in Section 1.6.3. A family of Gaussian random variables $\xi(f), f \in \mathscr{M}$, on the probability space $(\mathscr{M}_{-2}, \mathcal{B}(\mathscr{M}_{-2}), \mathrm{G})$ can be constructed as shown in the previous section. It has already been seen that $L^2(\mathscr{M}_{-2}, d\mathrm{G})$ and $\mathcal{F}_N$ are unitary equivalent under

$$\xi(f) \cong \frac{1}{\sqrt{2}}(a_M^*(\hat{f}) + a_M(\tilde{\hat{f}})), \quad f \in \mathscr{M}.$$

Recall that $(\xi_t(f))_{t\in\mathbb{R}}$ is an infinite-dimensional Ornstein–Uhlenbeck process on the probability space $(\mathfrak{Y}, \mathcal{B}(\mathfrak{Y}), \mathcal{G})$ with mean zero, such that $\mathscr{M} \ni f \mapsto \xi_t(f) \in \mathbb{R}$ is linear, $\xi_t(f) = \langle\langle \xi_t, f \rangle\rangle$ for $f \in \mathscr{M}_{+2}$, and

$$\mathbb{E}_{\mathcal{G}}[e^{i\sum_{j=1}^n \xi_{s_j}(f_j)}] = e^{-\frac{1}{4}\sum_{i,j=1}^n (f_i, e^{-|s_i-s_j|\hat{\omega}} f_j)_{\mathcal{H}_M}}.$$

Also, $(\xi_t(f))_{t\in\mathbb{R}}$ is a stationary Markov process. Using the regular conditional probability measure $\mathcal{G}^\xi(\cdot) = \mathcal{G}(\cdot \mid \xi_0 = \xi)$, we have $\mathbb{E}_{\mathcal{G}}[\ldots] = \int_{\mathscr{M}_{-2}} \mathbb{E}_{\mathcal{G}}^\xi[\ldots] d\mathrm{G}$, where $\mathbb{E}_{\mathcal{G}}^\xi = \mathbb{E}_{\mathcal{G}^\xi}$.

We define the Nelson Hamiltonian on function space instead of $\mathcal{H}_N$. Recall that $V : \mathbb{R}^d \to \mathbb{R}$ is called a Kato-class potential, denoted $V \in \mathcal{K}(\mathbb{R}^d)$, whenever

$$\lim_{r\to 0} \sup_{x\in\mathbb{R}^d} \int_{B_r(x)} |g(x-y)V(y)|\, dy = 0$$

holds, where $B_r(x)$ is the closed ball of radius r centered at x, and

$$g(x) = \begin{cases} |x|, & d = 1, \\ -\log|x|, & d = 2, \\ |x|^{2-d}, & d \geq 3. \end{cases}$$

If $1_K V \in \mathcal{K}(\mathbb{R}^d)$ for any compact set $K \subset \mathbb{R}^d$, then V is called a local Kato-class potential, denoted $V \in \mathcal{K}_{\mathrm{loc}}(\mathbb{R}^d)$. Potential V is Kato-decomposable whenever $V = V_+ - V_-$ with $V_+ \in \mathcal{K}_{\mathrm{loc}}(\mathbb{R}^d)$ and $V_- \in \mathcal{K}(\mathbb{R}^d)$. It is known that a nonnegative function V is in $\mathcal{K}(\mathbb{R}^d)$ if and only if

$$\lim_{t \downarrow 0} \sup_{x \in \mathbb{R}^d} \mathbb{E}^x \Big[\int_0^t V(B_s)\, ds \Big] = 0. \tag{2.3.1}$$

By Khasminskii's lemma and (2.3.1), $V \in \mathcal{K}(\mathbb{R}^d)$ satisfies

$$\sup_{x \in \mathbb{R}^d} \mathbb{E}^x [e^{\int_0^t V(B_s)\, ds}] < \infty. \tag{2.3.2}$$

The following are standing assumptions in Sections 2.3.1 and 2.3.2.

Assumption 2.5. The following conditions hold:
(1) (Dispersion relation) $\omega(k) = \omega_\nu(k) = \sqrt{|k|^2 + \nu^2}$, $\nu \geq 0$.
(2) (Charge distribution) $\varphi \in \mathscr{S}'(\mathbb{R}^d)$, $\overline{\hat{\varphi}(k)} = \hat{\varphi}(-k)$ and $\hat{\varphi}/\omega, \hat{\varphi}/\sqrt{\omega} \in L^2(\mathbb{R}^d)$.
(3) (External potential) V is Kato-decomposable and H_{p} has a ground state Ψ_{p} with $H_{\mathrm{p}}\Psi_{\mathrm{p}} = E_{\mathrm{p}}\Psi_{\mathrm{p}}$.

The difference between Assumption 2.2 in the previous section and Assumption 2.5 here is in the condition on the external potentials. In Assumption 2.5, the existence of a ground state of H_{p} is taken for granted, which is too strong to include the case $V = 0$, but it has an advantage in constructing functional integral representations, e. g., it allows defining the Nelson Hamiltonian on an L^2-space with a probability measure. Recall that the ground state Ψ_{p} of H_{p} is unique and it has a strictly positive version, which we will choose throughout. It is directly seen that

$$\theta_M H_{\mathrm{I}}(x) \theta_M^{-1} = \xi(\varphi(\cdot - x))$$

for every $x \in \mathbb{R}^d$, where θ_M is an isomorphism between $\mathcal{F}_{\mathrm{b}}(\hat{\mathcal{H}}_M)$ and $L^2(\mathscr{M}_{-2}, d\mathsf{G})$ given in Proposition 1.126. We need one more step. Since H_{N} is an operator in $\mathcal{H}_{\mathrm{N}}$, and since thus far we only have the isomorphism $\mathcal{F}_{\mathrm{N}} \cong L^2(\mathscr{M}_{-2}, d\mathsf{G})$, we still need to transform the first factor H_{p}. Instead of using identity, we choose the ground state transform instead as it has the benefit that the transformed Hamiltonian becomes a sum of the generator of a random process and a multiplication operator.

Let

$$d\mathsf{N}_0 = \Psi_{\mathrm{p}}^2 dx$$

and recall the *ground state transform*

$$U_{\Psi_p}: L^2(\mathbb{R}^d, d\mathsf{N}_0) \to L^2(\mathbb{R}^d, dx), \quad f \mapsto \Psi_p f.$$

As in Volume 1, we use the notations

$$\mathfrak{X} = C(\mathbb{R}; \mathbb{R}^d), \quad \mathscr{X} = C([0,\infty); \mathbb{R}^d).$$

Since V is Kato-decomposable and H_p has a strictly positive ground state Ψ_p, by Theorem 5.15 in Volume 1, there exists a $P(\phi)_1$-process $(X_t)_{t\geq 0}$ on $(\mathfrak{X}, \mathcal{B}(\mathfrak{X}))$, with Markov generator

$$L_p = U_{\Psi_p}(H_p - E_p)U_{\Psi_p}^{-1}.$$

This process can be realized as the coordinate process on $(\mathfrak{X}, \mathcal{B}(\mathfrak{X}))$, $X_t(\omega) = \omega(t)$, under the probability measure $\mathcal{N}_0^x$. Define the filtrations $\mathscr{F}_t^+ = \sigma(X_s, 0 \leq s \leq t)$ and $\mathscr{F}_t^- = \sigma(X_s, t \leq s \leq 0)$. The $P(\phi)_1$-process satisfies the following properties:

(1) (Initial distribution) $\mathcal{N}_0^x(X_0 = x) = 1$.

(2) (Reflection symmetry) $(X_t)_{t\geq 0}$ and $(X_t)_{t\leq 0}$ are independent, and $X_{-t} \stackrel{\mathrm{d}}{=} X_t$ for $t \in \mathbb{R}$.

(3) (Diffusion property) $(X_t)_{t\geq 0}$ and $(X_t)_{t\leq 0}$ are diffusion processes with respect to $(\mathscr{F}_t^+)_{t\geq 0}$ and $(\mathscr{F}_t^-)_{t\leq 0}$, respectively, i. e.,

$$\mathbb{E}_{\mathcal{N}_0^x}[X_{t+s}|\mathscr{F}_s^+] = \mathbb{E}_{\mathcal{N}_0^x}[X_{t+s}|\sigma(X_s)] = \mathbb{E}_{\mathcal{N}_0^{X_s}}[X_t],$$
$$\mathbb{E}_{\mathcal{N}_0^x}[X_{-t-s}|\mathscr{F}_{-s}^-] = \mathbb{E}_{\mathcal{N}_0^x}[X_{-t-s}|\sigma(X_{-s})] = \mathbb{E}_{\mathcal{N}_0^{X_{-s}}}[X_{-t}]$$

for $s, t \geq 0$, and $\mathbb{R} \ni t \mapsto X_t \in \mathbb{R}^d$ is a. s. continuous, where $\mathbb{E}_{\mathcal{N}_0^{X_s}}$ means $\mathbb{E}_{\mathcal{N}_0^y}$ evaluated at $y = X_s$.

(4) (Shift invariance) Let $-\infty < t_0 \leq t_1 \leq \cdots \leq t_n < \infty$. Then

$$\int_{\mathbb{R}^d} \mathbb{E}_{\mathcal{N}_0^x}\Big[\prod_{j=0}^n f_j(X_{t_j})\Big] d\mathsf{N}_0 = (f_0, e^{-(t_1-t_0)L}f_1 \cdots e^{-(t_n-t_{n-1})L}f_n)_{L^2(d\mathsf{N}_0)}$$

for $f_0, f_n \in L^2(d\mathsf{N}_0)$ and $f_j \in L^\infty(\mathbb{R}^d)$, $j = 1, \ldots, n-1$, and the finite-dimensional distributions of the process are shift invariant, i. e.,

$$\int_{\mathbb{R}^d} \mathbb{E}_{\mathcal{N}_0^x}\Big[\prod_{j=1}^n f_j(X_{t_j})\Big] d\mathsf{N}_0 = \int_{\mathbb{R}^d} \mathbb{E}_{\mathcal{N}_0^x}\Big[\prod_{j=1}^n f_j(X_{t_j+s})\Big] d\mathsf{N}_0, \quad s \in \mathbb{R}.$$

Write

$$\mathsf{P}_0 = \mathsf{N}_0 \otimes \mathsf{G},$$

which is a probability measure on $\mathbb{R}^d \otimes \mathscr{M}_{-2}$. The unitary operator

$$U_{\Psi_p} \otimes \theta_M : \mathcal{H}_N \to L^2(\mathbb{R}^d \otimes \mathscr{M}_{-2}, d\mathsf{P}_0) \tag{2.3.3}$$

implies then the unitary equivalence $L^2(\mathbb{R}^d \otimes \mathscr{M}_{-2}, d\mathsf{P}_0) \simeq \mathcal{H}_N$. For convenience, we write $L^2(\mathbb{R}^d \otimes \mathscr{M}_{-2}, \mathsf{P}_0)$ simply as $L^2(\mathsf{P}_0)$, and similarly, $L^2(\mathsf{N}_0)$ and $L^2(\mathsf{G})$ for $L^2(\mathbb{R}^d, d\mathsf{N}_0)$ and $L^2(\mathscr{M}_{-2}, d\mathsf{G})$, respectively. The space $L^2(\mathbb{R}^d) \otimes L^2(\mathsf{G})$ will be continued to be denoted by $\mathcal{H}_N$.

Definition 2.6 (Nelson Hamiltonian in $L^2(\mathsf{P}_0)$). The Nelson Hamiltonian in $L^2(\mathsf{P}_0)$ is defined by

$$L_N = L_p \otimes 1 + 1 \otimes \tilde{H}_f + \tilde{H}_I,$$

with

$$\tilde{H}_I : F(x, \xi) \mapsto \xi(\varphi(\cdot - x))F(x, \xi).$$

Equivalently,

$$\tilde{H}_I = \int_{\mathbb{R}^d}^{\oplus} \xi(\varphi(\cdot - x))\Psi_p(x)^2 dx$$

under the identification

$$L^2(\mathsf{P}_0) \cong \int_{\mathbb{R}^d}^{\oplus} L^2(\mathsf{G})\Psi_p(x)^2 dx.$$

Clearly, L_N and $H_N - E_p$ are unitary equivalent under (2.3.3), and L_N is self-adjoint on $D(L_p \otimes 1) \cap D(1 \otimes \tilde{H}_f)$, which is proven in the same way as in Proposition 2.4. In what follows, we also simplify the notation to H_f for $\tilde{H}_f$, and H_I for $\tilde{H}_I$.

Now we turn to deriving the functional integral representation for the Nelson model. Recall that

$$d\mathcal{N}_0 = d\mathcal{N}_0^x \otimes d\mathsf{N}_0$$

is the path measure on $\mathcal{B}(\mathfrak{X}) \otimes \mathcal{B}(\mathbb{R}^d)$ of the $P(\phi)_1$-process $(X_t)_{t\in\mathbb{R}}$ generated by L_p. We write

$$\mathcal{P}_0 = \mathcal{N}_0 \otimes \mathcal{G}.$$

Also, we already know that

$$(f, e^{-tL_p}g)_{L^2(\mathsf{N}_0)} = \mathbb{E}_{\mathcal{N}_0}[\overline{f(X_0)}g(X_t)] \tag{2.3.4}$$

and

$$(\Psi, e^{-t(H_f+\xi(f))}\Phi)_{L^2(\mathsf{G})} = \mathbb{E}_{\mathcal{G}}[\overline{\Psi(\xi_0)}\Phi(\xi_t)e^{-\int_0^t \xi_s(f)ds}]. \tag{2.3.5}$$

Theorem 2.7 (Functional integral representation for Nelson Hamiltonian). *For every $t \geq 0$ and $F, G \in L^2(\mathrm{P}_0)$, we have*

$$(F, e^{-tL_\mathrm{N}} G)_{L^2(\mathrm{P}_0)} = \mathbb{E}_{\mathcal{P}_0}[\overline{F(X_0, \xi_0)} e^{-\int_0^t \xi_s(\varphi_{X_s}) ds} G(X_t, \xi_t)], \tag{2.3.6}$$

or, equivalently,

$$(\Psi_\mathrm{p} F, e^{-tH_\mathrm{N}} \Psi_\mathrm{p} G)_{\mathcal{H}_\mathrm{N}} = e^{-tE_\mathrm{p}} \mathbb{E}_{\mathcal{P}_0}[\overline{F(X_0, \xi_0)} e^{-\int_0^t \xi_s(\varphi_{X_s}) ds} G(X_t, \xi_t)],$$

where $\varphi_{X_s}(\cdot) = \varphi(\cdot - X_s)$ and $(\Psi_\mathrm{p} F)(x, \xi) = \Psi_\mathrm{p}(x) F(x, \xi)$, $x \in \mathbb{R}^d$.

Proof. First, assume that

$$\begin{aligned} F &= F(x, \xi) = F_1(x) F_2(\xi(f_1), \dots, \xi(f_n)), \\ G &= G(x, \xi) = G_1(x) G_2(\xi(g_1), \dots, \xi(g_m)), \end{aligned}$$

with $F_1, G_1 \in \mathscr{S}(\mathbb{R}^d)$, $F_2 \in \mathscr{S}(\mathbb{R}^n)$, and $G_2 \in \mathscr{S}(\mathbb{R}^m)$. By the Trotter product formula, we have

$$\begin{aligned} (F, e^{-tL_\mathrm{N}} G) &= \lim_{n\to\infty} (F, (e^{-(t/n)L_\mathrm{p}} e^{-(t/n)H_\mathrm{I}} e^{-(t/n)H_\mathrm{f}})^n G) \\ &= \lim_{n\to\infty} \int_{\mathbb{R}^d} \mathbb{E}_{\mu_\mathrm{E}}[\overline{F(X_0, \phi_0)} G(X_t, \phi_t) e^{-\sum_{j=1}^n \phi_{jt/n}(\varphi_{X_{jt/n}})}] d\mathcal{N}_0, \end{aligned}$$

where $\phi_t(f) = \phi_\mathrm{E}(\delta_t \otimes f)$ is the Euclidean field; see (1.5.30). We also note that

$$\begin{aligned} &|\mathbb{E}_{\mu_\mathrm{E}}[\overline{F(X_0, \phi_0)} G(X_t, \phi_t) e^{-\sum_{j=1}^n \phi_{jt/n}(\varphi_{X_{jt/n}})}]| \\ &\quad \leq 2\|F(X_0, \cdot)\| \|G(X_t, \cdot)\| e^{\frac{t}{2}\|\hat{\varphi}/\sqrt{\omega}\|^2 + 2t(t\vee 1)(\|\hat{\varphi}/\sqrt{\omega}\|^2 + \|\hat{\varphi}/\omega\|^2)} \end{aligned}$$

where $\|\cdot\| = \|\cdot\|_{L^2(\mathscr{Q})}$. Here we used Corollary 1.88. Since $\phi_t(f)$ and $\xi_t(f)$ are identically distributed,

$$(F, e^{-tL_\mathrm{N}} G) = \lim_{n\to\infty} \mathbb{E}_{\mathcal{P}_0}[\overline{F(X_0, \xi_0)} G(X_t, \xi_t) e^{-\sum_{j=1}^n \xi_{jt/n}(\varphi_{X_{jt/n}})}].$$

Since $t \mapsto \xi_t(\varphi_{X_t})$ is almost surely continuous, the theorem follows for F, G above. The proof can be completed by a limiting argument. □

We also present a functional integral representation for further types of Green functions.

Theorem 2.8 (Euclidean Green functions). *Let Choose the arbitrary time-points $0 = t_0 < t_1 < \dots < t_n = t$, $n \in \mathbb{N}$. Suppose that $F_0, F_n \in L^2(\mathrm{P}_0)$ and $F_1, \dots, F_{n-1} \in L^\infty(\mathrm{P}_0)$. Then*

$$\Big(F_0, \prod_{j=1}^{n} e^{-(t_j-t_{j-1})L_{\mathrm{N}}} F_j\Big)_{L^2(\mathrm{P}_0)} = \mathbb{E}_{\mathcal{P}_0}\Big[\overline{F_0(X_0,\xi_0)} e^{-\int_0^t \xi_s(\varphi_{X_s})ds}\Big(\prod_{j=1}^{n-1} F_j(X_{t_j},\xi_{t_j})\Big)F_n(X_t,\xi_t)\Big]. \tag{2.3.7}$$

Proof. The proof is similar to that of Theorem 2.7, which is based on the Trotter product formula. We skip the details. □

As in the finite-dimensional setting, we will use the functional integral representation in Theorem 2.7 to study the eigenfunctions of H_{N}. If we can suppose that H_{N} has a unique ground state Ψ_{g}, then

$$\|e^{-T(H_{\mathrm{N}}-E_{\mathrm{N}})}F\|_{\mathcal{H}_{\mathrm{N}}}^{-1} e^{-T(H_{\mathrm{N}}-E_{\mathrm{N}})}F \to \frac{(\Psi_{\mathrm{g}},F)_{\mathcal{H}_{\mathrm{N}}}}{|(\Psi_{\mathrm{g}},F)_{\mathcal{H}_{\mathrm{N}}}|}\frac{1}{\|\Psi_{\mathrm{g}}\|_{\mathcal{H}_{\mathrm{N}}}}\Psi_{\mathrm{g}} \tag{2.3.8}$$

strongly as $T \to \infty$, where $E_{\mathrm{N}} = \inf \mathrm{Spec}(H_{\mathrm{N}})$. Thus the properties of Ψ_{g} can be studied by an analysis of the left-hand side of (2.3.8) via (2.3.6) and taking the limit. This is the strategy we will follow.

Corollary 2.9. *If $f, g \in L^2(\mathrm{N}_0)$, then for every $T > 0$,*

$$(f \otimes 1, e^{-TL_{\mathrm{N}}} g \otimes 1)_{L^2(\mathrm{P}_0)} = \mathbb{E}_{\mathcal{N}_0}[\overline{f(X_0)}g(X_T)e^{\frac{1}{2}\int_0^T ds \int_0^T W(X_s-X_t,s-t)dt}], \tag{2.3.9}$$

where

$$W(x,t) = \int_{\mathbb{R}^d} \frac{|\hat{\varphi}(k)|^2}{2\omega(k)} e^{-ik\cdot x} e^{-\omega(k)|t|} dk. \tag{2.3.10}$$

Proof. Let $I_T = \int_0^T \xi_s(\varphi_{X_s})ds$. We have

$$(f \otimes 1, e^{-TL_{\mathrm{N}}} g \otimes 1)_{L^2(\mathrm{P}_0)} = \mathbb{E}_{\mathcal{P}_0}[\overline{f(X_0)}g(X_t)e^{-I_T}] = \mathbb{E}_{\mathcal{N}_0}[\overline{f(X_0)}g(X_t)\mathbb{E}_{\mathcal{G}}[e^{-I_T}]].$$

Note that a finite linear sum of the form $\sum_{i=j}^n \xi_{s_j}(f_j)$ is a Gaussian random variable and

$$\mathbb{E}_{\mathcal{G}}[e^{ia\sum_{j=1}^n \xi_{s_j}(f_j)}] = \int_{\mathscr{M}_{-2}^n} e^{ia\phi(\sum_{j=1}^n \delta_{s_j}\otimes f_j)} dy = e^{-(a^2/4)\|\sum_{j=1}^n \delta_{s_j}\otimes f_j\|_{\mathcal{H}_M}^2}.$$

Also, I_T is a Gaussian random variable with covariance

$$\mathbb{E}_{\mathcal{G}}[I_T^2] = \int_0^T ds \int_0^T \mathbb{E}_{\mathcal{G}}[\xi_s(\varphi_{X_s})\xi_t(\varphi_{X_t})]dt = \int_0^T ds \int_0^T W(X_s - X_t, s-t)dt.$$

Hence it follows that

$$\mathbb{E}_G[e^{-I_T}] = \exp\left(\frac{1}{2}\int_0^T ds \int_0^T W(X_s - X_t, s-t)dt\right).$$

Fubini's theorem implies the claim. □

Definition 2.10 (Pair potential). We call $W(x,t)$ given by (2.3.10) the *pair potential* of the Nelson Hamiltonian and $\int_0^T ds \int_0^T W(X_s - X_t, s-t)dt$ the *pair interaction* of the Nelson Hamiltonian.

In this chapter we often use the following estimates:

$$\left|\int_0^T ds \int_0^T W(X_s - X_t, s-t)dt\right| \le T\int_{\mathbb{R}^d} \frac{|\hat\varphi(k)|^2}{\omega(k)^2}dk - \int_{\mathbb{R}^d} \frac{|\hat\varphi(k)|^2}{\omega(k)^3}(1-e^{-T\omega(k)})dk,$$

$$\left|\int_{-T}^T ds \int_{-T}^T W(X_s - X_t, s-t)dt\right| \le 2T\int_{\mathbb{R}^d} \frac{|\hat\varphi(k)|^2}{\omega(k)^2}dk - \int_{\mathbb{R}^d} \frac{|\hat\varphi(k)|^2}{\omega(k)^3}(1-e^{-2T\omega(k)})dk,$$

$$\left|\int_0^T ds \int_{-T}^0 W(X_s - X_t, s-t)dt\right| \le \frac{1}{2}\int_{\mathbb{R}^d} \frac{|\hat\varphi(k)|^2}{\omega(k)^3}(1-e^{-T\omega(k)})^2 dk.$$

In particular, we see that

$$\left|\int_0^T ds \int_0^T W(X_s - X_t, s-t)dt\right| \le T\int_{\mathbb{R}^d} \frac{|\hat\varphi(k)|^2}{\omega(k)^2}dk,$$

$$\left|\int_{-T}^T ds \int_{-T}^T W(X_s - X_t, s-t)dt\right| \le 2T\int_{\mathbb{R}^d} \frac{|\hat\varphi(k)|^2}{\omega(k)^2}dk,$$

$$\left|\int_0^T ds \int_{-T}^0 W(X_s - X_t, s-t)dt\right| \le \frac{1}{2}\int_{\mathbb{R}^d} \frac{|\hat\varphi(k)|^2}{\omega(k)^3}dk,$$

$$\left|\int_0^T ds \int_{-T}^0 W(X_s - X_t, s-t)dt\right| \le \frac{T}{2}\int_{\mathbb{R}^d} \frac{|\hat\varphi(k)|^2}{\omega(k)^2}dk,$$

$$\left|\int_0^T ds \int_{-T}^0 W(X_s - X_t, s-t)dt\right| \le \frac{T^2}{2}\int_{\mathbb{R}^d} \frac{|\hat\varphi(k)|^2}{\omega(k)}dk.$$

2.3.2 Euclidean field and Brownian motion

For later use, we give an alternative functional integral representation of the expression $(F, e^{-tH_N}G)$ in terms of the Euclidean field $\phi_E(f)$ and two-sided Brownian motion

$(B_t)_{t\in\mathbb{R}}$ instead of the infinite-dimensional Ornstein–Uhlenbeck process $(\xi_t(f))_{t\in\mathbb{R}}$ or the $P(\phi)_1$-process $(X_t)_{t\in\mathbb{R}}$. Let $(\phi(f), f \in \mathscr{M})$ be a family of Gaussian random variables on $(\mathscr{Q}, \Sigma, \mu)$. As it has been seen in (1.5.3) in Section 1.5, the spaces $L^2(\mathscr{Q})$ and $\mathcal{F}_\mathrm{N}$ are unitary equivalent.

Definition 2.11 (Nelson Hamiltonian in $L^2(\mathbb{R}^d)\otimes L^2(\mathscr{Q})$). The *Nelson Hamiltonian* in $L^2(\mathbb{R}^d)\otimes L^2(\mathscr{Q})$ is defined by

$$\hat{H}_\mathrm{N} = H_\mathrm{p}\otimes 1 + 1\otimes \hat{H}_\mathrm{f} + \hat{H}_\mathrm{I}. \tag{2.3.11}$$

Here $\hat{H}_\mathrm{f}$ denotes the free field Hamiltonian in $L^2(\mathscr{Q})$, and

$$\hat{H}_\mathrm{I} = \int_{\mathbb{R}^d}^{\oplus} \phi(\varphi(\cdot - x))dx,$$

under the identification $L^2(\mathbb{R}^d)\otimes L^2(\mathscr{Q}) \cong \int_{\mathbb{R}^d}^{\oplus} L^2(\mathscr{Q})dx$.

Note that $\varphi(\cdot - x) \in \mathscr{M}$ for every $x \in \mathbb{R}^d$ in Definition 2.11. It is seen that (2.3.11) is unitary equivalent to H_N. We also denote the self-adjoint operator $\hat{H}_\mathrm{N}$ and the total Hilbert space $L^2(\mathbb{R}^d)\otimes L^2(\mathscr{Q})$ by H_N and $\mathcal{H}_\mathrm{N}$, respectively.

Theorem 2.12 (Functional integral representation for Nelson Hamiltonian). *For every $t \ge 0$ and $F, G \in \mathcal{H}_\mathrm{N}$,*

$$(F, e^{-tH_\mathrm{N}}G)_{\mathcal{H}_\mathrm{N}} = \int_{\mathbb{R}^d} \mathbb{E}^x\left[e^{-\int_0^t V(B_s)ds}(\mathrm{I}_0F(B_0), e^{-\phi_\mathrm{E}(\int_0^t \delta_s\otimes\varphi(\cdot-B_s)ds)}\mathrm{I}_tG(B_t))_{L^2(\mathscr{Q}_\mathrm{E})}\right]dx, \tag{2.3.12}$$

where I_t, $t \in \mathbb{R}$, denotes the family of isometries given in Definition 1.83, and ϕ_E is the Euclidean field indexed by $\mathscr{E}$. Here $F, G \in \mathcal{H}_\mathrm{N}$ are regarded as $L^2(\mathscr{Q})$-valued L^2-functions on $\mathbb{R}^d$.

Proof. For $\varepsilon \ge 0$, let $\hat{H}_\mathrm{I}^\varepsilon = y_\varepsilon(\hat{H}_\mathrm{I})$, where $y_\varepsilon(X) = X + \varepsilon X^2$. Then $\hat{H}_\mathrm{I}^\varepsilon$ is bounded from below for $\varepsilon > 0$. Moreover, $H_\mathrm{N}^\varepsilon = H_\mathrm{p}\otimes 1 + 1\otimes\hat{H}_\mathrm{f} + \hat{H}_\mathrm{I}^\varepsilon$ is self-adjoint on $D(H_\mathrm{p}\otimes 1 + 1\otimes\hat{H}_\mathrm{f})$ and essentially self-adjoint on any core of $H_\mathrm{p}\otimes 1 + 1\otimes\hat{H}_\mathrm{f}$ for $0 \le \varepsilon < c$ with some c. In particular, $e^{-tH_\mathrm{N}^\varepsilon} \to e^{-tH_\mathrm{N}}$ strongly as $\varepsilon \downarrow 0$. First, we construct a functional integral representation of $(F, e^{-tH_\mathrm{N}^\varepsilon}G)$ for $\varepsilon > 0$ and, by a limiting argument, we obtain the extension $(F, e^{-tH_\mathrm{N}}G)$ to $\varepsilon = 0$. For simplicity, first we assume that $V \in C_0^\infty(\mathbb{R}^d)$. Let $h = -\frac{1}{2}\Delta$. By the Trotter product formula and the factorization formula $e^{-|s-t|\hat{H}_\mathrm{f}} = \mathrm{I}_s^*\mathrm{I}_t$, we have

$$e^{-tH_\mathrm{N}^\varepsilon} = \operatorname*{s-lim}_{n\to\infty}(e^{-\frac{t}{n}\hat{H}_\mathrm{I}^\varepsilon}e^{-\frac{t}{n}h}e^{-\frac{t}{n}V}e^{-\frac{t}{n}\hat{H}_\mathrm{f}})^n = \operatorname*{s-lim}_{n\to\infty}\mathrm{I}_0^*\left(\prod_{j=0}^{n-1}\mathrm{I}_{\frac{jt}{n}}e^{-\frac{t}{n}\hat{H}_\mathrm{I}^\varepsilon}e^{-\frac{t}{n}h}e^{-\frac{t}{n}V}\mathrm{I}_{\frac{jt}{n}}^*\right)\mathrm{I}_t. \tag{2.3.13}$$

Inserting (2.3.13) into $(F, e^{-tH_\mathrm{N}^\varepsilon}G)$ gives

$$(F, e^{-tH_N^\varepsilon} G) = \lim_{n\to\infty} \Big(\mathrm{I}_0 F, \Big(\prod_{j=0}^{n-1} \mathrm{I}_{\frac{jt}{n}} e^{-\frac{t}{n}\hat{H}_\mathrm{I}^\varepsilon} e^{-\frac{t}{n}h} e^{-\frac{t}{n}V} \mathrm{I}^*_{\frac{jt}{n}} \Big) \mathrm{I}_t G \Big).$$

Here $\prod_{j=1}^n T_j = T_1 \cdots T_n$. Using the identity $\mathrm{I}_s e^{-\hat{H}_\mathrm{I}^\varepsilon} \mathrm{I}^*{}_s = E_s e^{-\hat{H}_\mathrm{I}^\varepsilon(s,x)} E_s$ for $s \in \mathbb{R}$, where

$$\hat{H}_\mathrm{I}^\varepsilon(s,x) = \int_{\mathbb{R}^d}^{\oplus} y_\varepsilon\left(\phi_\mathrm{E}(\delta_s \otimes \varphi(\cdot - x))\right) dx$$

and $E_s = \mathrm{I}_s \mathrm{I}_s^*$ is a projection, we see that

$$(F, e^{-tH_N^\varepsilon} G) = \lim_{n\to\infty} \left(\mathrm{I}_0 F, \left(\prod_{j=0}^{n-1} E_{\frac{jt}{n}} e^{-\frac{t}{n}\hat{H}_\mathrm{I}^\varepsilon(\frac{jt}{n},x)} e^{-\frac{t}{n}h} e^{-\frac{t}{n}V} E_{\frac{jt}{n}} \right) \mathrm{I}_t G \right).$$

By the Markov property, we may drop E_s on the right-hand side above. We obtain

$$(F, e^{-tH_N^\varepsilon} G) = \lim_{n\to\infty} \left(\mathrm{I}_0 F, \left(\prod_{j=0}^{n-1} e^{-\frac{t}{n}\hat{H}_\mathrm{I}^\varepsilon(\frac{jt}{n},x)} e^{-\frac{t}{n}h} e^{-\frac{t}{n}V} \right) \mathrm{I}_t G \right).$$

The right-hand side above can be expressed in terms of Wiener measure. Since for any time-division $t_0 \le t_1 \le \cdots \le t_n$,

$$\Big(f_0, \prod_{j=1}^{n} e^{-(t_j - t_{j-1})h} f_j\Big) = \int_{\mathbb{R}^d} \mathbb{E}^x \Big[\overline{f_0(B_0)} \Big(\prod_{j=1}^{n} f_j(B_{t_j}) \Big)\Big] dx,$$

it follows that

$$(F, e^{-tH_N^\varepsilon} G) = \lim_{n\to\infty} \int_{\mathbb{R}^d} \mathbb{E}^x \Big[e^{-\sum_{j=0}^{n-1} \frac{t}{n} V(B_{\frac{jt}{n}})} \Big(\mathrm{I}_0 F(B_0), e^{-\sum_{j=0}^{n-1} \frac{t}{n} \hat{H}_\mathrm{I}^\varepsilon(\frac{jt}{n}, B_{\frac{jt}{n}})} \mathrm{I}_t G(B_t) \Big) \Big] dx.$$

Note that $s \mapsto \delta_s \otimes \varphi(\cdot - B_s)$ is strongly continuous as a map $\mathbb{R} \to \mathscr{E}$, almost surely. Hence $s \mapsto \phi_\mathrm{E}(\delta_s \otimes \varphi(\cdot - B_s))$ is also strongly continuous as $\mathbb{R} \to L^2(\mathscr{Q}_E)$. We can compute the limit and the result is

$$(F, e^{-tH_N^\varepsilon} G) = \int_{\mathbb{R}^d} \mathbb{E}^x \left[e^{-\int_0^t V(B_s) ds} \left(\mathrm{I}_0 F(B_0), e^{-\varepsilon Q_t - \phi_\mathrm{E}(\int_0^t \delta_s \otimes \varphi(\cdot - B_s) ds)} \mathrm{I}_t G(B_t) \right) \right] dx.$$

Here

$$Q_t = \int_0^t \phi_\mathrm{E}(\delta_s \otimes \varphi(\cdot - B_s))^2 ds.$$

We see that $H_N^\varepsilon \to H_N$ on $D(H_\mathrm{p} \otimes 1 + 1 \otimes H_\mathrm{rad})$, and $e^{-tH_N^\varepsilon} G \to e^{-tH_N} G$ follows as $\varepsilon \downarrow 0$. On the other hand, we have the bound

$$\left|\left(\mathrm{I}_0 F(B_0), e^{-\varepsilon Q_t - \phi_{\mathrm{E}}(\int_0^t \delta_s \otimes \varphi(\cdot - B_s) ds)} \mathrm{I}_t G(B_t)\right)\right|$$
$$\leq \|F(B_0)\| \|G(B_t)\| e^{\frac{t}{2}\|\hat{\varphi}/\sqrt{\omega}\|^2 + 2t(t\vee 1)(\|\hat{\varphi}/\sqrt{\omega}\|^2 + \|\hat{\varphi}/\omega\|^2)}$$

and

$$\int_{\mathbb{R}^d} \mathbb{E}^x \left[e^{-\int_0^t V(B_s)ds} \|F(B_0)\| \|G(B_t)\|\right] dx < \infty.$$

Hence the Lebesgue dominated convergence theorem yields

$$\lim_{\varepsilon \downarrow 0} \int_{\mathbb{R}^d} \mathbb{E}^x \left[e^{-\int_0^t V(B_s)ds} \left(\mathrm{I}_0 F(B_0), e^{-\varepsilon Q_t - \phi_{\mathrm{E}}(\int_0^t \delta_s \otimes \varphi(\cdot - B_s) ds)} \mathrm{I}_t G(B_t)\right)\right] dx$$
$$= \int_{\mathbb{R}^d} \mathbb{E}^x \left[e^{-\int_0^t V(B_s)ds} \left(\mathrm{I}_0 F(B_0), e^{-\phi_{\mathrm{E}}(\int_0^t \delta_s \otimes \varphi(\cdot - B_s) ds)} \mathrm{I}_t G(B_t)\right)\right] dx,$$

implying

$$(F, e^{-tH_{\mathrm{N}}} G) = \int_{\mathbb{R}^d} \mathbb{E}^x \left[e^{-\int_0^t V(B_s)ds} \left(\mathrm{I}_0 F(B_0), e^{-\phi_{\mathrm{E}}(\int_0^t \delta_s \otimes \varphi(\cdot - B_s) ds)} \mathrm{I}_t G(B_t)\right)\right] dx. \tag{2.3.14}$$

Thus the theorem follows for $V \in C_0^\infty(\mathbb{R}^d)$. By a limiting argument similar to the proof of Theorem 4.88 in Volume 1, we can extend (2.3.12) to $V \in \mathcal{R}_{\mathrm{Kato}}$. □

Remark 2.13. We can also derive the self-adjoint generator H_{N} from Theorem 2.12 by taking the derivative of the functional integral representation with respect to t at $t = 0$. Suppose that V and φ are sufficiently smooth and bounded. Also assume that F and G are sufficiently smooth and $F, G \in D(H_{\mathrm{N}})$ for simplicity. Let

$$R_t = \int_{\mathbb{R}^d} \mathbb{E}^x \left[e^{-\int_0^t V(B_s)ds} (\mathrm{I}_0 F(B_0), e^{-\phi_{\mathrm{E}}(\int_0^t \delta_s \otimes \varphi(\cdot - B_s) ds)} \mathrm{I}_t G(B_t))_{L^2(\mathscr{Q}_{\mathrm{E}})}\right] dx.$$

We have $R_t - R_0 = a_t + b_t + c_t$, where

$$a_t = \int_{\mathbb{R}^d} \mathbb{E}^x \left[(e^{-\int_0^t V(B_s)ds} - 1)(\mathrm{I}_0 F(B_0), e^{-\phi_{\mathrm{E}}(\int_0^t \delta_s \otimes \varphi(\cdot - B_s) ds)} \mathrm{I}_t G(B_t))_{L^2(\mathscr{Q}_{\mathrm{E}})}\right] dx,$$
$$b_t = \int_{\mathbb{R}^d} \mathbb{E}^x \left[(\mathrm{I}_0 F(B_0), (e^{-\phi_{\mathrm{E}}(\int_0^t \delta_s \otimes \varphi(\cdot - B_s) ds)} - 1) \mathrm{I}_t G(B_t))_{L^2(\mathscr{Q}_{\mathrm{E}})}\right] dx,$$
$$c_t = \int_{\mathbb{R}^d} \mathbb{E}^x \left[(\mathrm{I}_0 F(B_0), (\mathrm{I}_t G(B_t) - \mathrm{I}_0 G(B_0)))_{L^2(\mathscr{Q}_{\mathrm{E}})}\right] dx.$$

It is straightforward to see that

$$\lim_{t\to 0}\frac{a_t}{t} = \int_{\mathbb{R}^d} \mathbb{E}^x\left[-V(B_0)(\mathrm{I}_0F(B_0),\mathrm{I}_0G(B_0))_{L^2(\mathscr{Q}_\mathrm{E})}\right]dx = (F,-VG),$$

$$\lim_{t\to 0}\frac{b_t}{t} = \int_{\mathbb{R}^d} \mathbb{E}^x\left[(\mathrm{I}_0F(B_0), -\phi_\mathrm{E}(\delta_0\otimes\varphi(\cdot - B_0))\mathrm{I}_0G(B_0))_{L^2(\mathscr{Q}_\mathrm{E})}\right]dx = (F,-\phi(\varphi(\cdot - x))G).$$

Using $\mathrm{I}_0^*\mathrm{I}_t = e^{-tH_\mathrm{f}}$, we can also see that

$$\lim_{t\to 0}\frac{c_t}{t} = \lim_{t\to 0}\frac{1}{t}(F,(e^{-tH_\mathrm{f}}e^{t\Delta/2}-1)G)_{L^2(\mathscr{Q}_\mathrm{E})} = (F,(-H_\mathrm{f}+\frac{1}{2}\Delta)G).$$

Together we have

$$\lim_{t\to 0}\frac{1}{t}(R_t - R_0) = (F,-H_\mathrm{N}G).$$

Hence we can derive H_N as a generator. Conversely, in a similar way to the Schrödinger operator $-\frac{1}{2}\Delta + V$ with singular external potentials, we can also define self-adjoint operator H_N with singular external potentials V through the functional integral representation in Sections 2.3.3 and 2.4 below.

Suppose that $\int_{\mathbb{R}^d}|\hat{\varphi}(k)|^2/\omega(k)^3dk<\infty$, and define

$$E(\hat{\varphi}) = \max\left\{\begin{matrix}\frac{1}{2}\|\hat{\varphi}/\omega\|^2 + 4(\|\hat{\varphi}/\omega\|^2 + \|\hat{\varphi}/\sqrt{\omega^3}\|^2)\\ \frac{1}{2}\|\hat{\varphi}/\omega\|^2 + 4(\|\hat{\varphi}/\sqrt{\omega}\|^2 + \|\hat{\varphi}/\omega\|^2)\end{matrix}\right\}. \tag{2.3.15}$$

From (1.5.23), it follows that

$$\|\mathrm{I}_0^*e^{-\phi_\mathrm{E}(\int_0^t\delta_s\otimes\varphi(\cdot - B_s)ds)}\mathrm{I}_t\| \le 2e^{tE(\hat{\varphi})},\quad t\ge 0. \tag{2.3.16}$$

More generally, for $q>1$ define

$$E_q(\hat{\varphi}) = \max\left\{\begin{matrix}\frac{1}{2}\|\hat{\varphi}/\omega\|^2 + 2q(\|\hat{\varphi}/\omega\|^2 + \|\hat{\varphi}/\sqrt{\omega^3}\|^2)\\ \frac{1}{2}\|\hat{\varphi}/\omega\|^2 + 2q(\|\hat{\varphi}/\sqrt{\omega}\|^2 + \|\hat{\varphi}/\omega\|^2)\end{matrix}\right\}. \tag{2.3.17}$$

By (1.5.26), we then have

$$\|\mathrm{I}_0^*e^{-\phi_\mathrm{E}(\int_0^t\delta_s\otimes\varphi(\cdot - B_s)ds)}\mathrm{I}_t\| \le \frac{q}{q-1}e^{tE_q(\hat{\varphi})},\quad t\ge 0. \tag{2.3.18}$$

Corollary 2.14 (Diamagnetic inequality). *Assume $\int_{\mathbb{R}^d}\frac{|\hat{\varphi}(k)|^2}{\omega(k)^3}dk<\infty$, and take $q>1$. Then*

$$|(F,e^{-tH_\mathrm{N}}G)_{\mathcal{H}_\mathrm{N}}| \le \frac{q}{q-1}(\|F\|_{L^2(\mathscr{Q})}, e^{-t(H_\mathrm{p}-E_q(\hat{\varphi}))}\|G\|_{L^2(\mathscr{Q})})_{L^2(\mathbb{R}^d)}.$$

Proof. From the functional integral representation of $(F, e^{-tH_N}G)$ and the bound (2.3.18), we see that

$$|(F, e^{-tH_N}G)| \le \frac{q}{q-1} e^{tE_q(\hat{\phi})} \int_{\mathbb{R}^d} \mathbb{E}^x \left[e^{-\int_0^t V(B_s)ds} \|F(B_0)\| \cdot \|G(B_t)\| \right] dx. \qquad \square$$

Similarly to Theorem 2.8, we can also construct a functional integral representation for other types of Green functions.

Theorem 2.15 (Euclidean Green functions). *Let* $F_0, F_n \in \mathcal{H}_N$, *and consider* $F_j = f_j \otimes \Phi_j(\phi(g_1^j), \ldots, \phi(g_{n_j}^j))$, *where* $f_j \in L^\infty(\mathbb{R}^d)$, $\Phi_j \in L^\infty(\mathbb{R}^{n_j})$, $g_i^j \in \mathscr{M}$, $i = 1, \ldots, n_j$, $j = 1, \ldots, n-1$. *Take arbitrary* $0 = t_0 < t_1 < \cdots < t_n = t$, $n \in \mathbb{N}$. *Then*

$$\left(F_0, \left(\prod_{j=1}^n e^{-(t_j - t_{j-1})H_N} F_j\right)\right)_{\mathcal{H}_N}$$

$$= \int_{\mathbb{R}^d} \mathbb{E}^x \left[e^{-\int_0^t V(B_s)ds} \left(\mathrm{I}_0 F_0(B_0), e^{-\phi_E(\int_0^t \delta_s \otimes \varphi(\cdot - B_s)ds)} \left(\prod_{j=1}^{n-1} \tilde{F}_j(B_{t_j}) \right) \mathrm{I}_t F_n(B_t) \right) \right] dx, \qquad (2.3.19)$$

where $\tilde{F}_j(x) = f_j(x)\Phi_j(\phi_E(\delta_{t_j} \otimes g_1^j), \ldots, \phi_E(\delta_{t_j} \otimes g_{n_j}^j))$.

Proof. For smooth functions with compact support f_j and Φ_j, $j = 1, \ldots, n$, we have

$$\mathrm{I}_s F_j \mathrm{I}_s^* = f_j E_s \Phi_j \left(\phi_E(\delta_s \otimes g_{n_1}^j), \ldots, \phi_E(\delta_s \otimes g_{n_1}^j) \right) E_s$$

as a bounded operator. Similarly to the proof of Theorem 2.12, the Trotter product formula implies (2.3.19). For general F_j, a limiting argument can be used. $\square$

2.3.3 Extension to general external potential

In Assumption 2.2, we required the external potential to belong to $\mathcal{R}_{\mathrm{Kato}}$. In Section 2.3.3, we extend the functional integral representation to a wider class of external potentials, but we use Assumption 2.16 on dispersion relations and charge distributions below.

Assumption 2.16. The following conditions hold:
(1) (Dispersion relation) $\omega(k) = \omega_\nu(k) = \sqrt{|k|^2 + \nu^2}$, $\nu \ge 0$.
(2) (Charge distribution) $\varphi \in \mathscr{S}'(\mathbb{R}^d)$, $\overline{\hat{\varphi}(k)} = \hat{\varphi}(-k)$ and $\hat{\varphi}/\omega, \hat{\varphi}/\sqrt{\omega} \in L^2(\mathbb{R}^d)$.

For convenience, we denote H_N with $V \equiv 0$ by

$$K = -\frac{1}{2}\Delta \otimes 1 + 1 \otimes \hat{H}_f + \hat{H}_I.$$

Let $z > 0$ and suppose that

$$\| |V|(-(1/2)\Delta + z)^{-1} \| \le a \tag{2.3.20}$$

with an $a > 0$. We see that

$$\| |V|(K+z)^{-1}F\| \le a\| \left(1 + \hat{H}_{\rm I}(-(1/2)\Delta + \hat{H}_{\rm f} + z)^{-1}\right)^{-1} F\|$$

and we also know that $\|\hat{H}_{\rm I}(-(1/2)\Delta + \hat{H}_{\rm f} + z)^{-1}F\| \le b\|F\|$ with some b. Whenever $b < 1$, we furthermore have

$$\| \left(1 + \hat{H}_{\rm I}(-(1/2)\Delta + \hat{H}_{\rm f} + z)^{-1}\right)^{-1} \| \le \frac{1}{1-b}$$

and

$$\| |V|(K+z)^{-1}\| \le \frac{a}{1-b}.$$

It is, however, not trivial to see $b < 1$ in general, and the right-hand side above is $a/(1-b) > a$. Surprisingly, in the lemma below we can, nevertheless, show that (2.3.20) implies

$$\| |V|(K+z)^{-1}\| \le a$$

under the infrared regular condition $\int_{\mathbb{R}^d} |\hat{\varphi}(k)|^2/\omega(k)^3 dk < \infty$. Denote

$$\operatorname{sgn} F(x) = \begin{cases} F(x)/\|F(x)\|_{L^2(\mathscr{Q})} & \text{if } \|F(x)\|_{L^2(\mathscr{Q})} \ne 0, \\ 0 & \text{if } \|F(x)\|_{L^2(\mathscr{Q})} = 0. \end{cases}$$

Lemma 2.17. *Let $\int_{\mathbb{R}^d} |\hat{\varphi}(k)|^2/\omega(k)^3 dk < \infty$ hold.*
(1) *If V is $-\frac{1}{2}\Delta$-form bounded with a relative bound a, then $|V|$ is also K-form bounded with a relative bound smaller than a.*
(2) *If V is $-\frac{1}{2}\Delta$-bounded with a relative bound a, then $|V|$ is also K-bounded with a relative bound smaller than a.*

Proof. Let $q > 1$, and $\psi \in C_0^\infty(\mathbb{R}^d)$ be such that $\psi > 0$. Applying $F = \operatorname{sgn}((K+z)^{-1/2}G)\,\psi$ for $z < \inf \operatorname{Spec}(K)$ in the diamagnetic inequality, we obtain

$$|(F, e^{-tK}G)_{\mathcal{H}_{\rm N}}| \le \frac{q}{q-1} e^{tE_q(\hat{\varphi})}(\|F\|_{L^2(\mathscr{Q})}, e^{-t(-(1/2)\Delta)}\|G\|_{L^2(\mathscr{Q})}),$$

where $E_q(\hat{\varphi})$ is given by (2.3.17). By formula $(T+z)^{-1/2} = \frac{1}{\sqrt{\pi}}\int_0^\infty t^{-1/2}e^{-t(T+z)}dt$ for any self-adjoint operator T and $z < \inf \operatorname{Spec}(T)$, we furthermore get

$$(\psi, \|(K+z)^{-1/2}G\|_{L^2(\mathscr{Q})})_{L^2(\mathbb{R}^d)} \le \frac{q}{q-1}\left(\psi, (-\frac{1}{2}\Delta - E_q(\hat{\varphi}) + z)^{-1/2}\|G\|_{L^2(\mathscr{Q})}\right)_{L^2(\mathbb{R}^d)}$$

for $z < \inf \operatorname{Spec}(K)$ and $z < \inf \operatorname{Spec}(-\frac{1}{2}\Delta - E_q(\hat{\varphi}))$. Since $\psi \in C_0^\infty(\mathbb{R}^d)$ is arbitrary, we have

$$\|(K+z)^{-1/2}G(x)\|_{L^2(\mathscr{Q})} \leq \frac{q}{q-1}\Big(-\frac{1}{2}\Delta - E_q(\hat{\varphi}) + z\Big)^{-1/2} \|G(x)\|_{L^2(\mathscr{Q})}$$

for a. e. $x \in \mathbb{R}^d$. We obtain that

$$\frac{\||V|^{1/2}(K+z)^{-1/2}G\|_{\mathcal{H}_N}}{\|G\|_{\mathcal{H}_N}} \leq \frac{q}{q-1} \frac{\left\||V|^{1/2}(-\frac{1}{2}\Delta - E_q(\hat{\varphi}) + z)^{-1/2}\|G(\cdot)\|_{L^2(\mathscr{Q})}\right\|_{L^2(\mathbb{R}^d)}}{\|G\|_{\mathcal{H}_N}}.$$

Taking the supremum on both sides above with respect to $\|G\|_{\mathcal{H}_N} \neq 0$, we furthermore obtain

$$\||V|^{1/2}(K+z)^{-1/2}\| \leq \frac{q}{q-1}\left\||V|^{1/2}\Big(-\frac{1}{2}\Delta - E_q(\hat{\varphi}) + z\Big)^{-1/2}\right\|.$$

Since V is $-\frac{1}{2}\Delta$-form bounded with relative bound a, V is $|-(1/2)\Delta - E(\hat{\varphi})|$-form bounded with relative bound smaller than a. This gives $\||V|^{1/2}(-(1/2)\Delta - E(\hat{\varphi}) + z)^{-1/2}\| \leq a$ and

$$\||V|^{1/2}(K+z)^{-1/2}\| \leq \frac{q}{q-1}a,$$

where $q > 1$ is arbitrary. Then $\||V|^{1/2}(K+z)^{-1/2}\| \leq a$ follows and (1) is proven. Part (2) can be shown similarly. □

Theorem 2.18 (Functional integral representation of Nelson Hamiltonian with singular external potential). *Let* $\int_{\mathbb{R}^d} |\hat{\varphi}(k)|^2/\omega(k)^3 dk < \infty$ *hold. Take* $K \dotplus V_+ \dot{-} V_-$ *with* $V_+ \in L^1_{\mathrm{loc}}(\mathbb{R}^d)$ *and* $-\frac{1}{2}\Delta$*-relatively form bounded* V_- *with relative bound strictly less than* 1. *Then* (2.3.12) *holds for* $K \dotplus V_+ \dot{-} V_-$.

Proof. The proof is similar to the Feynman–Kac formula for Schrödinger operator with singular external potentials in Theorem 4.88 in Volume 1, with $h = -\frac{1}{2}\Delta$ replaced by K. If $V \in L^\infty(\mathbb{R}^d)$, then (2.3.12) holds. Assume that $V_+ \in L^1_{\mathrm{loc}}(\mathbb{R}^d)$ and V_- is relatively form bounded with respect to $-\frac{1}{2}\Delta$. Define

$$V_{+n}(x) = \begin{cases} V_+(x), & V_+(x) < n, \\ n, & V_+(x) \geq n, \end{cases} \qquad V_{-m}(x) = \begin{cases} V_-(x), & V_-(x) < m, \\ m, & V_-(x) \geq m, \end{cases}$$

and write $V_{n,m} = V_{+n} - V_{-m}$. We have

$$(F, e^{-t(K+V_{n,m})}G) = \int_{\mathbb{R}^d} \mathbb{E}^x[e^{-\int_0^t V_{n,m}(B_s)ds}(F(B_0), \mathscr{K}G(B_t))]dx. \tag{2.3.21}$$

Here $\mathscr{K} = \mathrm{I}_0^* e^{-\phi_{\mathrm{E}}(\int_0^t \delta_s \otimes \varphi(\cdot - B_s)ds)}\mathrm{I}_t$. Define the closed quadratic forms

$$\mathfrak{q}_{n,m}(F,F) = (K^{1/2}F, K^{1/2}F) + (V_{+n}^{1/2}F, V_{+n}^{1/2}F) - (V_{-m}^{1/2}F, V_{-m}^{1/2}F),$$
$$\mathfrak{q}_{n,\infty}(F,F) = (K^{1/2}F, K^{1/2}F) + (V_{+n}^{1/2}F, V_{+n}^{1/2}F) - (V_{-}^{1/2}F, V_{-}^{1/2}F),$$
$$\mathfrak{q}_{\infty,\infty}(F,F) = (K^{1/2}F, K^{1/2}F) + (V_{+}^{1/2}F, V_{+}^{1/2}F) - (V_{-}^{1/2}F, V_{-}^{1/2}F),$$

whose form domains respectively are $Q(\mathfrak{q}_{n,m}) = Q(K)$, $Q(\mathfrak{q}_{n,\infty}) = Q(K)$, and $Q(\mathfrak{q}_{\infty,\infty}) = Q(K) \cap Q(V_+)$. Clearly, $\mathfrak{q}_{n,m} \geq \mathfrak{q}_{n,m+1} \geq \mathfrak{q}_{n,m+2} \geq \cdots \geq \mathfrak{q}_{n,\infty}$ and $\mathfrak{q}_{n,m} \to \mathfrak{q}_{n,\infty}$ in the sense of quadratic forms on $\bigcup_{m=1}^{\infty} Q(\mathfrak{q}_{n,m}) = Q(K)$. Since $\mathfrak{q}_{n,\infty}$ is closed on $Q(K)$, by the monotone convergence theorem for forms, the associated positive self-adjoint operators satisfy $K + V_{+n} - V_{-m} \to K + V_{+n} \dot{-} V_-$ in strong resolvent sense, which implies that for all $t \geq 0$,

$$\exp\left(-t\left(K + V_{+n} - V_{-m}\right)\right) \to \exp\left(-t\left(K + V_{+n} \dot{-} V_-\right)\right) \tag{2.3.22}$$

strongly as $m \to \infty$. Similarly, we have $\mathfrak{q}_{n,\infty} \leq \mathfrak{q}_{n+1,\infty} \leq \mathfrak{q}_{n+2,\infty} \leq \cdots \leq \mathfrak{q}_\infty$ and $\mathfrak{q}_{n,\infty} \to \mathfrak{q}_{\infty,\infty}$ in the form sense on $\{F \in \bigcap_{n=1}^{\infty} Q(\mathfrak{q}_{n,\infty}) \mid \sup_{n\in\mathbb{N}} \mathfrak{q}_{n,\infty}(F,F) < \infty\} = Q(K) \cap Q(V_+)$. Hence by the monotone convergence theorem for forms again, we obtain

$$\exp(-t(K \dot{+} V_{+n} \dot{-} V_-)) \to \exp(-t(K \dot{+} V_+ \dot{-} V_-)) \tag{2.3.23}$$

strongly as $n \to \infty$. By taking first $m \to \infty$ and then $n \to \infty$, it can be proven that both sides of (2.3.21) converge, thus the left-hand is convergent by (2.3.22)–(2.3.23). On the other hand, let $F, G \in \mathcal{H}_\mathrm{N}$ and $F, G \geq 0$. Note that $(F(B_0), \mathscr{K} G(B_t)) \geq 0$. It follows that

$$\lim_{n\to\infty} \lim_{m\to\infty} \int_{\mathbb{R}^d} \mathbb{E}^x[e^{-\int_0^t V_{n,m}(B_s)ds}(F(B_0), \mathscr{K} G(B_t))]dx$$
$$= \int_{\mathbb{R}^d} \mathbb{E}^x[e^{-\int_0^t V(B_s)ds}(F(B_0), \mathscr{K} G(B_t))]dx$$

by the monotone convergence theorem for integrals for the limit of $m \to \infty$ and the dominated convergence theorem for $n \to \infty$. Hence the right-hand side of (2.3.21) converges to $\int_{\mathbb{R}^d} \mathbb{E}^x[e^{-\int_0^t V(B_s)ds}(F(B_0), \mathscr{K} G(B_t))]dx$ for every $F, G \in \mathcal{H}_\mathrm{N}$ by the decomposition $F = \mathfrak{R}F + \mathfrak{I}G$, $\mathfrak{R}F = \mathfrak{R}F_+ - \mathfrak{R}F_-$, $\mathfrak{I}F = \mathfrak{I}F_+ - \mathfrak{I}F_-$, $G = \mathfrak{R}G + \mathfrak{I}G$, $\mathfrak{R}G = \mathfrak{R}G_+ - \mathfrak{R}G_-$, and $\mathfrak{I}G = \mathfrak{I}G_+ - \mathfrak{I}G_-$. □

The functional integral representation (2.3.12) in Theorem 2.12 can further be extended to Kato-decomposable V, which will be done in the next section.

2.4 Nelson model with Kato-class potential

Theorem 2.12 gives a functional integral representation of $(F, e^{-tH_\mathrm{N}}G)$ in terms of Brownian motion $(B_t)_{t\geq 0}$ and the Euclidean field ϕ_E. In Theorem 2.18, we extended this formula to singular external potentials. In this section we extend it to the more singular class of Kato-decomposable potentials. When V is Kato-decomposable a functional integral representation of $(F, e^{-tL_N}G)$ can be actually constructed by using the $P(\phi)_1$-process $(X_t)_{t\geq 0}$ and the infinite dimensional Ornstein–Uhlenbeck process $(\xi_t)_{t\geq 0}$, but we need to assume that a ground state of H_p exists in order to define $(X_t)_{t\in\mathbb{R}}$. In Section 2.4, we do not assume existence of a ground state of H_p, and only use Assumption 2.19 below.

Assumption 2.19. The following conditions hold.
(1) (Dispersion relation) $\omega(k) = \omega_\nu(k) = \sqrt{|k|^2 + \nu^2}$, $\nu \geq 0$.
(2) (Charge distribution) $\varphi \in \mathscr{S}'(\mathbb{R}^d)$, $\overline{\hat{\varphi}(k)} = \hat{\varphi}(-k)$ and $\hat{\varphi}/\omega, \hat{\varphi}/\sqrt{\omega} \in L^2(\mathbb{R}^d)$.
(3) (External potential) V is Kato-decomposable.

First, we define H_N with a Kato-decomposable potential as a self-adjoint operator in $L^2(\mathbb{R}^d) \otimes L^2(\mathscr{Q})$, via the functional integral representation given by

$$(F, e^{-tH_N}G) = \int_{\mathbb{R}^d} \mathbb{E}^x[e^{-\int_0^t V(B_s)ds}(\mathrm{I}_0 F(x), e^{-\phi_E(K_t)}\mathrm{I}_t G(B_t))]dx,$$

where

$$K_t = \int_0^t \delta_s \otimes \varphi(\cdot - B_s)ds. \tag{2.4.1}$$

Thus we see that

$$(e^{-tH_N}G)(x) = \mathbb{E}^x[e^{-\int_0^t V(B_s)ds}\mathrm{I}_0^* e^{-\phi_E(K_t)}\mathrm{I}_t G(B_t)]. \tag{2.4.2}$$

Conversely, we show that a sufficient condition to define the right-hand side of (2.4.2) is that V is of Kato-decomposable. The idea used for Schrödinger operators with Kato-class potentials will be extended to the Nelson Hamiltonian.

Let V be a Kato-decomposable potential and define the family of operators

$$(T_t F)(x) = \mathbb{E}^x[e^{-\int_0^t V(B_r)dr}\mathrm{I}_0^* e^{-\phi_E(K_t)}\mathrm{I}_t F(B_t)].$$

Lemma 2.20. *Operator T_t is bounded on $\mathcal{H}_N$, for every $t > 0$.*

Proof. Let $F \in \mathcal{H}_N$. Since $\|\mathrm{I}_0 e^{-\phi_E(K_t)}\mathrm{I}_t F(B_t)\| \leq 2e^{E(t)}\|F(B_t)\|$, where

$$E(t) = \frac{t}{2}\|\hat{f}/\omega\|^2 + 2t(t \vee 1)(\|\hat{f}/\sqrt{\omega}\|^2 + \|\hat{f}/\omega\|^2), \tag{2.4.3}$$

is given by (1.5.21), we get from

$$\|T_t F\|^2_{\mathcal{H}_N} = \int_{\mathbb{R}^d} \|\mathbb{E}^x[e^{-\int_0^t V(B_r)dr}\mathrm{I}_0^* e^{-\phi_E(K_t)}\mathrm{I}_t F(B_t)]\|^2_{L^2(\mathscr{Q})}dx$$

that

$$\|T_tF\|^2_{\mathcal{H}_N} \le 2\int_{\mathbb{R}^d} \mathbb{E}^0[e^{-2\int_0^t V(B_r+x)dr}]\mathbb{E}^0[\|F(B_t+x)\|^2]e^{2E(t)}dx.$$

Since V is Kato-decomposable, we have $\sup_{x\in\mathbb{R}^d}\mathbb{E}^0[e^{-2\int_0^t V(B_r+x)dr}] = C < \infty$, and thus $\|T_tF\|^2_{\mathcal{H}_N} \le 4Ce^{2E(t)}\|F\|^2_{\mathcal{H}_N}$ results. □

In what follows, we show that $\{T_t : t \ge 0\}$ is a symmetric C_0-semigroup. To obtain this, we introduce the time shift operator

$$u_tf(x) = f(x_0 - t, \mathbf{x}), \quad x = (x_0, \mathbf{x}) \in \mathbb{R}\times\mathbb{R}^d$$

on $L^2(\mathbb{R}^d)$. It is straightforward that $u_t^* = u_{-t}$ and $u_t^*u_t = 1$. We denote the second quantization of u_t by $U_t = \Gamma_E(u_t)$, which acts on $L^2(\mathscr{Q}_E)$ and is a unitary map. We see that $U_t\mathrm{I}_s = \mathrm{I}_{s+t}$.

Lemma 2.21. *$T_sT_t = T_{s+t}$ holds for $s, t \ge 0$.*

Proof. By the definition of T_t, we have

$$T_sT_tF = \mathbb{E}^x\Big[e^{-\int_0^s V(B_r)dr}\mathrm{I}_0^*e^{-\phi_E(K_s)}\mathrm{I}_s\mathbb{E}^{B_s}[e^{-\int_0^t V(B_r)dr}\mathrm{I}_0^*e^{-\phi_E(K_t)}\mathrm{I}_tF(B_t)]\Big]. \tag{2.4.4}$$

By the formulae $\mathrm{I}_s\mathrm{I}_0^* = \mathrm{I}_s\mathrm{I}_s^*U_{-s}^* = \mathrm{E}_sU_{-s}^*$ and $\mathrm{I}_t = U_{-s}\mathrm{I}_{t+s}$, equality (2.4.4) can be written as

$$\mathbb{E}^x\Big[e^{-\int_0^s V(B_r)dr}\mathrm{I}_0^*e^{-\phi_E(K_s)}\mathrm{E}_s\mathbb{E}^{B_s}[e^{-\int_0^t V(B_r)dr}U_{-s}^*e^{-\phi_E(K_t)}U_{-s}\mathrm{I}_{t+s}F(B_t)]\Big]. \tag{2.4.5}$$

Since U_s is unitary, we have $U_{-s}^*e^{-\phi_E(K_t)}U_{-s} = e^{-\phi_E(u_{-s}^*K_t)}$ as an operator. The test function of the exponent $u_{-s}^*K_t$ is given by

$$u_{-s}^*K_t = \int_0^t \delta_{r+s}\otimes\varphi(\cdot - B_r)dr.$$

Moreover, by the Markov property of E_t, $t \in \mathbb{R}$, we may drop E_s in (2.4.5), and by the Markov property of $(B_t)_{t\ge0}$, we have

$$\begin{aligned}
T_sT_tF &= \mathbb{E}^x[e^{-\int_0^s V(B_r)dr}\mathrm{I}_0^*e^{-\phi_E(K_s)}\mathbb{E}^x[e^{-\int_s^{s+t} V(B_r)dr}e^{-\phi_E(K_s^{s+t})}\mathrm{I}_{s+t}F(B_{s+t})|\mathscr{F}_s^{\mathrm{BM}}]] \\
&= \mathbb{E}^x[e^{-\int_0^{s+t} V(B_r)dr}\mathrm{I}_0^*e^{-\phi_E(K_{s+t})}\mathrm{I}_{s+t}F(B_{s+t})] = T_{s+t}F,
\end{aligned}$$

where $K_s^{s+t} = \int_s^{s+t}\delta_r\otimes\varphi(\cdot - B_r)dr$ and $(\mathscr{F}_t^{\mathrm{BM}})_{t\ge0}$ denotes the natural filtration of $(B_t)_{t\ge0}$. □

Strong continuity of the map $t \mapsto T_t$ on $\mathcal{H}_{\mathrm{N}}$ can also be checked, while $T_0 = 1$ is trivial.

Theorem 2.22. *The semigroup $\{T_t : t \geq 0\}$ is a symmetric C_0-semigroup.*

Proof. Since we already know that $\{T_t : t \geq 0\}$ is a C_0-semigroup, it is enough to show that T_t is symmetric, i. e., $(F, T_t G) = (T_t F, G)$. Let $R = \Gamma_{\mathrm{E}}(r)$ be the second quantization of the reflection r, and $U_t = \Gamma_{\mathrm{E}}(u_t)$. We have

$$(F, T_t G) = \int_{\mathbb{R}^d} \mathbb{E}^x \left[e^{-\int_0^t V(B_s)ds} (U_t R \mathrm{I}_t F(B_0), U_t R e^{-\phi_{\mathrm{E}}(K_t)} \mathrm{I}_0 G(B_t)) \right] dx$$

$$= \int_{\mathbb{R}^d} \mathbb{E}^x \left[e^{-\int_0^t V(B_s)ds} (\mathrm{I}_t F(B_0), e^{-\phi_{\mathrm{E}}(u_t r K_t)} \mathrm{I}_0 G(B_t)) \right] dx,$$

where

$$u_t r K_t = \int_0^t \delta_{t-s} \otimes \varphi(\cdot - B_s) ds.$$

Noticing that $\dot{B}_s = B_{t-s} - B_t \stackrel{\mathrm{d}}{=} B_s$ for $0 \leq s \leq t$, we can replace B_s with $\dot{B}_s$ and obtain

$$(F, T_t G) = \int_{\mathbb{R}^d} \mathbb{E}^0 [e^{-\int_0^t V(\dot{B}_s + x)ds} (\mathrm{I}_t F(\dot{B}_0 + x), e^{-\phi_{\mathrm{E}}(u_t r \dot{K}_t)} \mathrm{I}_0 G(\dot{B}_t + x))] dx.$$

Here $\dot{K}_t = \int_0^t \delta_s \otimes \varphi(\cdot - \dot{B}_s) ds$. Swapping $\int_{\mathbb{R}^d} dx$ and $\int_{\mathscr{X}} d\mathcal{W}^0$, and changing the variable $-B_t + x$ to y, we have

$$(F, T_t G) = \int_{\mathbb{R}^d} \mathbb{E}^0 [e^{-\int_0^t V(B_{t-s} + y)ds} (\mathrm{I}_t F(B_t + y), e^{-\phi_{\mathrm{E}}(\tilde{K}_t)} \mathrm{I}_0 G(B_0 + y))] dy.$$

Here $\tilde{K}_t = \int_0^t \delta_{t-s} \otimes \varphi(\cdot - B_{t-s} - y) ds$. Thus we conclude that $(F, T_t G) = (T_t F, G)$ and the theorem follows. □

By Theorem 2.22, there exists a self-adjoint operator $H_{\mathrm{N}}^{\mathrm{Kato}}$ such that

$$T_t = e^{-tH_{\mathrm{N}}^{\mathrm{Kato}}}, \quad t \geq 0.$$

Definition 2.23 (Nelson Hamiltonian for Kato-class potential). We call the self-adjoint operator $H_{\mathrm{N}}^{\mathrm{Kato}}$ *Nelson Hamiltonian for Kato-class potential V.*

In what follows, we write again H_{N} for $H_{\mathrm{N}}^{\mathrm{Kato}}$, for notational simplicity.

The semigroup generated by $-\frac{1}{2}\Delta + V$ with Kato-decomposable potential V has a number of regularity properties. Let $K_t f(x) = \mathbb{E}^x[e^{-\int_0^t V(B_s)ds} f(B_t)]$ and suppose that $f \in L^p(\mathbb{R}^d)$ for some $1 \leq p \leq \infty$. Then $x \mapsto K_t f(x)$ is continuous for every $t > 0$ by

Proposition 4.109 in Volume 1. We can also have a counterpart of this fact for H_{N} with Kato-decomposable potential V.

Write

$$\mathrm{I}_{[0,t]} = \mathrm{I}_0^* e^{-\phi_{\mathrm{E}}(\int_0^t \delta_s \otimes \varphi(\cdot - B_s)ds)} \mathrm{I}_t.$$

The estimate

$$\|\mathrm{I}_{[0,t]}\| \leq 2e^{E(t)} \tag{2.4.6}$$

is established in (2.4.3). Employing the functional integral representation of the semigroup generated by H_{N}, we have

$$(F, e^{-tH_{\mathrm{N}}} G)_{\mathcal{H}_{\mathrm{N}}} = \int_{\mathbb{R}^d} \mathbb{E}^x \left[e^{-\int_0^t V(B_s)ds} (F(B_0), \mathrm{I}_{[0,t]} G(B_t))_{L^2(\mathcal{Q})} \right] dx. \tag{2.4.7}$$

From this, we obtain

$$\Phi_t(x) = (e^{-tH_{\mathrm{N}}} \Phi)(x) = \mathbb{E}^x [e^{-\int_0^t V(B_s)ds} \mathrm{I}_{[0,t]} \Phi(B_t)] \tag{2.4.8}$$

for almost every $x \in \mathbb{R}^d$. We remark that Φ_t is a vector in $L^2(\mathbb{R}^d \times \mathcal{Q})$. Hence it is meaningless to consider $\Phi_t(x)$ for each $x \in \mathbb{R}^d$. In the theorem below, we can show that the right-hand side of (2.4.8) gives the continuous version of $(\Psi, \Phi_t(x))_{L^2(\mathcal{Q})}$.

Theorem 2.24. *Let $\Psi \in L^2(\mathcal{Q})$ and $\Phi \in L^2(\mathbb{R}^d) \otimes L^2(\mathcal{Q})$. Then for every $t > 0$, the map $\mathbb{R}^d \ni x \mapsto \mathbb{E}^x[e^{-\int_0^t V(B_s)ds}(\Psi, \mathrm{I}_{[0,t]}\Phi(B_t))]$ is continuous, that is, $(\Psi, e^{-tH_{\mathrm{N}}}\Phi(x))_{L^2(\mathcal{Q})}$ has a continuous version.*

Proof. For arbitrary $t \geq 0$, it follows that

$$(\Psi, \Phi_t(x)) = \mathbb{E}^x [e^{-\int_0^t V(B_s)ds} (\Psi, \mathrm{I}_{[0,t]} \Phi(B_t))].$$

Write $\rho_t(x) = \mathbb{E}^x[e^{-\int_0^t V(B_s)ds}(\Psi, \mathrm{I}_{[0,t]}\Phi(B_t))]$. Since V is a Kato-decomposable potential, $\rho \in L^1(\mathbb{R}^d)$. We apply the smoothing effect of the heat semigroup discussed in Proposition 4.109 in Volume 1. Let $f \in L^1(\mathbb{R}^d)$ and define $(P_r f)(x) = \int_{\mathbb{R}^d} \Pi_r(x-y)f(y)dy$. The map $x \mapsto P_r f(x)$ is continuous, and furthermore

$$\mathbb{E}^x [e^{-\int_r^t V(B_s)ds} f(B_t)] = P_r e^{-(t-r)H_{\mathrm{P}}} f(x).$$

Using the Markov property of Brownian motion, we have

$$P_r \rho_{t-r}(x) = \mathbb{E}^x [e^{-\int_r^t V(B_s)ds} (\Psi, \mathrm{I}_0^* e^{-\phi(\int_r^t \delta_{s-r} \otimes \varphi(\cdot - B_s)ds)} \mathrm{I}_{t-r} \Phi(B_t))]$$

and $(P_r \rho_{t-r})(x)$ is continuous in x. We have

$$\rho_t(x) - \rho_t(y) = \rho_t(x) - (P_r\rho_{t-r})(x) + (P_r\rho_{t-r})(x) - (P_r\rho_{t-r})(y) + (P_r\rho_{t-r})(y) - \rho_t(y).$$

Since V is a Kato-decomposable potential and (2.4.6) holds, it is straightforward to see that $|\rho_t(x) - (P_r\rho_{t-r})(x)| \to 0$ as $r \to 0$ uniformly in x. Similarly, $|\rho_t(y) - (P_r\rho_{t-r})(y)| \to 0$ as $r \to 0$ uniformly in y. On the other hand, for each r and t, $(P_r\rho_{t-r})(x) - (P_r\rho_{t-r})(y) \to 0$ as $|x - y| \to 0$. From here, we conclude that $\rho(x)$ is continuous in x. □

2.5 Existence and uniqueness of the ground state

2.5.1 Uniqueness

First, we discuss the uniqueness of the ground state of H_N by means of the functional integral representation derived in the previous sections.

Proposition 2.25 (Positivity improving). *The semigroup $\{e^{-t\hat{H}_\mathrm{f}} : t > 0\}$ is positivity improving.*

Proof. The proposition follows from Corollary 1.54. □

Uniqueness of the ground state follows directly from the functional integral representation of the Nelson Hamiltonian and the Perron–Frobenius theorem.

Theorem 2.26 (Uniqueness of ground state). *Let either Assumption 2.2 or Assumption 2.19 hold. If H_N has a ground state, then it is unique.*

Proof. By Theorem 2.12, it suffices to show that $(F, e^{-tH_\mathrm{N}}G) \neq 0$ for every nonnegative F and G such that $F \not\equiv 0$ and $G \not\equiv 0$, since e^{-tH_N} is positivity preserving. Using (2.3.12), we have

$$(F, e^{-tH_\mathrm{N}}G) = \int_{\mathbb{R}^d} \mathbb{E}^x\big[e^{-\int_0^t V(B_s)ds}(\mathrm{I}_0F(B_0), e^{-\phi_\mathrm{E}(K)}\mathrm{I}_tG(B_t))_{L^2(\mathcal{Q}_\mathrm{E})}\big]dx,$$

where $K = \int_0^t \delta_s \otimes \varphi(\cdot - B_s)ds$. Suppose $(\mathrm{I}_0\Psi, e^{-\phi_\mathrm{E}(K)}\mathrm{I}_t\Phi)_{L^2(\mathcal{Q}_\mathrm{E})} = 0$ for some nonnegative Ψ and Φ. This means that

$$\mu_E(\mathrm{supp}[(e^{-\phi_\mathrm{E}(K)})\mathrm{I}_t\Phi] \cap \mathrm{supp}[\mathrm{I}_0\Phi]) = 0. \tag{2.5.1}$$

Since $\phi_\mathrm{E}(K) \in L^2(\mathcal{Q}_\mathrm{E})$, we see that $\mu_E(\{\phi_\mathrm{E} \in \mathcal{Q}_\mathrm{E} \mid e^{\phi_\mathrm{E}(K)} = 0\}) = 0$. Hence (2.5.1) implies that $\mu_E(\mathrm{supp}[\mathrm{I}_t\Psi] \cap \mathrm{supp}[\mathrm{I}_0\Phi]) = 0$, and thus $0 = (\mathrm{I}_t\Psi, \mathrm{I}_0\Phi) = (\Psi, e^{-t\hat{H}_\mathrm{f}}\Phi)$. Since $e^{-t\hat{H}_\mathrm{f}}$ is positivity improving, this is a contradiction. Hence $(\mathrm{I}_0\Psi, e^{-\phi_\mathrm{E}(K)}\mathrm{I}_t\Phi)_{L^2(\mathcal{Q}_\mathrm{E})} > 0$ and

$$(\mathrm{I}_0F(B_0), e^{-\phi_\mathrm{E}(K)}\mathrm{I}_tG(B_t))_{L^2(\mathcal{Q}_\mathrm{E})} > 0.$$

Thus $(F, e^{-tH_\mathrm{N}}G) > 0$ follows for any nonnegative F and G. □

If the ground state Ψ_g of H_N exists, from the proof of Theorem 2.26 we see that $\Psi_g = \Psi_g(x,\phi)$ is strictly positive for almost every $(x,\phi) \in \mathbb{R}^d \times \mathscr{Q}$. Thus we see that $(\Psi, \Psi_g(x))_{L^2(\mathscr{Q})} > 0$ for almost every $x \in \mathbb{R}^d$ for any $\Psi \geq 0$ but $\Psi \not\equiv 0$. We can also get to an improved statement.

Corollary 2.27. *Let either Assumption 2.2 with $V_+ \in L^1_{\mathrm{loc}}(\mathbb{R}^d)$ or Assumption 2.19 hold. Let $\Psi \in L^2(\mathscr{Q})$ be nonnegative and $\Psi \not\equiv 0$. Then $(\Psi, \Psi_g(x))_{L^2(\mathscr{Q})} > 0$ for all $x \in \mathbb{R}^d$.*

Proof. It has been established that $(\Psi, \Psi_g(x))_{L^2(\mathscr{Q})} \geq 0$ for any $x \in \mathbb{R}^d$. We show that $(\Psi, \Psi_g(x))_{L^2(\mathscr{Q})} \neq 0$ for all $x \in \mathbb{R}^d$. Suppose that $(\Psi, \Psi_g(x))_{L^2(\mathscr{Q})} = 0$ for some $x \in \mathbb{R}^d$. Since $\Psi_g = e^{-t(H_N - E)}\Psi_g$, for every fixed x we have the functional integral representation

$$(\Psi, \Psi_g(x))_{L^2(\mathscr{Q})} = e^{tE}\mathbb{E}_{\mathcal{W}}\left[e^{-\int_0^t V(B_s+x)ds}(\Psi, \mathrm{I}_{[0,t]}(x)\Psi_g(B_t + x))_{L^2(\mathscr{Q})}\right].$$

Here $\mathrm{I}_{[0,t]}(x) = \mathrm{I}_0 e^{-\phi_{\mathrm{E}}(\int_0^t \delta_s \otimes \varphi(\cdot - B_s - x)ds)}\mathrm{I}_t$. Since

$$e^{-\int_0^t V(B_s+x)ds}(\Psi, \mathrm{I}_{[0,t]}(x)\Psi_g(B_t + x))_{L^2(\mathscr{Q})} \geq 0 \quad \text{a. s.},$$

equality $(\Psi, \Psi_g(x))_{L^2(\mathscr{Q})} = 0$ implies that

$$e^{-\int_0^t V(B_s+x)ds}(\Psi, \mathrm{I}_{[0,t]}(x)\Psi_g(B_t + x))_{L^2(\mathscr{Q})} = 0 \quad \text{a. s.}$$

By Lemma 4.95 in Volume 1, $V_+ \in L^1_{\mathrm{loc}}(\mathbb{R}^d)$ implies that $e^{-\int_0^t V(B_s+x)ds} > 0$ a. s. This gives

$$(\Psi, \mathrm{I}_{[0,t]}(x)\Psi_g(B_t + x))_{L^2(\mathscr{Q})} = 0 \quad \text{a. s.}$$

Furthermore, $e^{-\phi_{\mathrm{E}}(\int_0^t \delta_s \otimes \varphi(\cdot - B_s)ds)} > 0$ a. s. on $\mathscr{X} \times \mathscr{Q}$, and we have

$$(\Psi, \mathrm{I}_0^* \mathrm{I}_t \Psi_g(B_t + x))_{L^2(\mathscr{Q})} = 0 \quad \text{a. s.}$$

Thus we conclude that

$$(\Psi, e^{-t\hat{H}_f}\Psi_g(B_t + x))_{L^2(\mathscr{Q})} = 0 \quad \text{a. s.}$$

Since $e^{-t\hat{H}_f}$ is positivity improving, we see that $\Psi_g(B_t + x) = 0$ a. s. on $\mathscr{X} \times \mathscr{Q}$. Hence

$$0 = \mathbb{E}_{\mathcal{W}}[(\Psi, \Psi_g(B_t + x))_{L^2(\mathscr{Q})}] = \int_{\mathbb{R}^d} \Pi_t(x - y)(\Psi, \Psi_g(y))_{L^2(\mathscr{Q})}dy.$$

Also, $\Pi_t(x - y) > 0$ implies that $(\Psi, \Psi_g(y))_{L^2(\mathscr{Q})} = 0$ for almost every $y \in \mathbb{R}^d$. Since Ψ is an arbitrary nonnegative vector, it follows that $\Psi_g(y) = 0$ for almost every $y \in \mathbb{R}^d$, which is a contradiction. Hence $(\Psi, \Psi_g(y))_{L^2(\mathscr{Q})} > 0$ for all $y \in \mathbb{R}^d$. □

2.5.2 Existence

We turn to the existence problem of a ground state of the Nelson Hamiltonian.

Theorem 2.28 (Existence of ground state). *In addition to Assumption 2.5, suppose that* $\int_{\mathbb{R}^d} |\hat{\varphi}(k)|^2/\omega(k)^3 dk < \infty$ *and*

$$\Sigma_{\mathrm{p}} - E_{\mathrm{p}} > \int_{\mathbb{R}^d} \frac{|\hat{\varphi}(k)|^2}{2\omega(k)^2} \frac{|k|^2}{2\omega(k) + |k|^2} dk \tag{2.5.2}$$

hold, where Σ_{p} *denotes the infimum of the essential spectrum of* H_{p}*. Then* H_{N} *has a ground state.*

Under $\int_{\mathbb{R}^d} |\hat{\varphi}(k)|^2/\omega(k)^3 dk < \infty$, we have the uniform bound

$$\int_{-\infty}^{0} ds \int_{0}^{\infty} |W(X_s - X_t, s - t)| dt \le \int_{\mathbb{R}^d} \frac{|\hat{\varphi}(k)|^2}{2\omega(k)^3} dk < \infty \tag{2.5.3}$$

on paths. In Section 2.7 we will also see that $\int_{\mathbb{R}^d} |\hat{\varphi}(k)|^2/\omega(k)^3 dk < \infty$ is crucial for a ground state to exist, and will show that in the physically important cases the existence of a ground state fails if the infrared regularity condition does not hold. Condition (2.5.2) is a restriction on the coupling constant, which in fact only comes to effect in the case where the particle Hamiltonian H_{p} has absolutely continuous spectrum. On physical grounds, (2.5.2) is not expected to be necessary. This is because the coupling to the quantum field should make the particle more heavy, enhancing the binding under V.

Lemma 2.29. *Let* $f \in L^2(\mathbb{R}^d)$ *and be a real-valued function. If* $\operatorname{ess\,inf}_{k \in K} f(k) > 0$ *for every compact set* $K \subset \mathbb{R}^d$*, then*

$$\lim_{T\to\infty} \frac{(f \otimes 1, e^{-(T+t)H_{\mathrm{N}}} f \otimes 1)}{(f \otimes 1, e^{-TH_{\mathrm{N}}} f \otimes 1)} = e^{-tE_{\mathrm{N}}}.$$

Proof. Recall that if μ is a measure on $\mathbb{R}$ with $\inf \operatorname{supp}(\mu) = E(\mu)$, then

$$E(\mu) = - \lim_{T\to\infty} \frac{1}{T} \ln\left(\int_{\mathbb{R}} e^{-Tx} \mu(dx) \right) \tag{2.5.4}$$

and

$$\lim_{T\to\infty} \frac{\int_{\mathbb{R}} e^{-(T+t)x} \mu(dx)}{\int_{\mathbb{R}} e^{-Tx} \mu(dx)} = e^{-tE(\mu)}. \tag{2.5.5}$$

Applying (2.5.5) to the spectral measure $\mu_{f\otimes 1}$ of H_{N} associated with vector $f \otimes 1$ yields the claim with E_{N} replaced by $\inf \operatorname{supp}(\mu_{f\otimes 1}) = E(\mu_{f\otimes 1})$. Thus it only remains to prove $E(\mu_{f\otimes 1}) = E_{\mathrm{N}}$. For this, define the class of functions

$$\mathscr{O} = \left\{ F \in L^2(\mathsf{P}_0) \,\middle|\, \operatorname{supp} F \subset \bigcup_{N,M>0} B_N(\mathbb{R}^d) \times B_M(\mathscr{M}_{-2}) \right\},$$

where $B_N(\mathbb{R}^d)$ and $B_M(\mathscr{M}_{-2})$ denote the balls centered in the origin in $\mathbb{R}^d$ and $\mathscr{M}_{-2}$ of radius N and M, respectively. This $\mathscr{O}$ is dense in $L^2(\mathsf{P}_0)$. Pick $g \in \mathscr{O}$. Since $e^{-tH_{\mathrm{N}}}$ is positivity preserving, we get

$$(g, e^{-TH_{\mathrm{N}}} g) \le (|g|, e^{-TH_{\mathrm{N}}} |g|) \le C^2 (f \otimes 1, e^{-TH_{\mathrm{N}}} f \otimes 1)$$

with

$$C = \frac{\operatorname{ess\,sup}_{(x,\xi)\in\mathbb{R}^d\times\mathscr{M}_{-2}} |g(x,\xi)|}{\operatorname{ess\,inf}_{(x,\xi)\in\operatorname{supp}|g|} f(x)}.$$

Equality (2.5.4) shows that $E(\mu_{f\otimes 1}) \le E(\mu_g)$, for all $g \in \mathscr{O}$. However, since $\mathscr{O}$ is a dense subset in $D(H_{\mathrm{N}})$, we have

$$E(\mu_{f\otimes 1}) \ge E_{\mathrm{N}} = \inf \left\{ E(\mu_g) \mid g \in \mathscr{O} \right\} \ge E(\mu_{f\otimes 1}),$$

and thus $E_{\mathrm{N}} = E(\mu_{f\otimes 1})$. □

Recall that Ψ_{p} is the normalized ground state of H_{p}, and for $T > 0$ define

$$\Psi_{\mathrm{g}}^T = \frac{e^{-TH_{\mathrm{N}}}(\Psi_{\mathrm{p}} \otimes 1)}{\|e^{-TH_{\mathrm{N}}}(\Psi_{\mathrm{p}} \otimes 1)\|}. \tag{2.5.6}$$

Since $\|\Psi_{\mathrm{g}}^T\| = 1$, Ψ_{g}^T has a subsequence $\Psi_{\mathrm{g}}^{T'}$ such that $\Psi_{\mathrm{g}}^{T'}$ is weakly convergent to a vector $\Psi_{\mathrm{g}}^{\infty}$. We reset T' to T, and defining

$$\gamma(T) = (\Psi_{\mathrm{p}} \otimes 1, \Psi_{\mathrm{g}}^T)^2$$

consider the limit of γ_T and more general cases for later use as $T \to \infty$ in what follows. Thus we choose $\rho \in L^2(\mathbb{R}^d)$ and define

$$\gamma_\rho(T) = \frac{(\rho \otimes 1, e^{-TH_{\mathrm{N}}}(\rho \otimes 1))^2}{\|e^{-TH_{\mathrm{N}}}(\rho \otimes 1)\|^2} \tag{2.5.7}$$

replacing Ψ_{p} by ρ in $\gamma(T)$. The next criterion is useful for showing the existence of a ground state of H_{N}.

Proposition 2.30 (Criterion for existence of ground state). *Suppose that* $\rho \in L^2(\mathbb{R}^d)$ *and* $\operatorname{ess\,inf}_{x\in K}\rho(x) > 0$ *for every compact set* $K \subset \mathbb{R}^d$. *Denote*

$$\lim_{T\to\infty} \gamma_\rho(T) = a.$$

If $a > 0$, *then* $H_{\rm N}$ *has a ground state.*

Proof. Since $a > 0$, for sufficiently large T it follows that $b < \sqrt{\gamma_\rho(T)}$ with some $b > 0$. Let $E(\cdot)$ be the spectral measure of $H_{\rm N}$ associated with $\rho \otimes 1$, and $\delta = \inf \operatorname{supp} E$. Then $\delta = E_N$ follows from the proof of Lemma 2.29. By the definition of δ, we see that $E([\delta, \delta+\varepsilon)) \neq 0$ for every $\varepsilon > 0$, and $E([\delta, \delta+\varepsilon))$ is strongly convergent to $E(\{\delta\})$ as $\varepsilon \downarrow 0$. We have

$$\sqrt{\gamma_\rho(T)} = \frac{\int_{[\delta,\delta+\varepsilon)} e^{-T(\lambda-\delta)} dE + \int_{[\delta+\varepsilon,\infty)} e^{-T(\lambda-\delta)} dE}{\left(\int_{[\delta,\infty)} e^{-2T(\lambda-\delta)} dE\right)^{1/2}}.$$

By the Schwarz inequality, we have

$$\frac{\int_{[\delta,\delta+\varepsilon)} e^{-T(\lambda-\delta)} dE}{\left(\int_{[\delta,\infty)} e^{-2T(\lambda-\delta)} dE\right)^{1/2}} \leq E([\delta, \delta+\varepsilon))^{1/2} \frac{\left(\int_{[\delta,\delta+\varepsilon)} e^{-2T(\lambda-\delta)} dE\right)^{1/2}}{\left(\int_{[\delta,\infty)} e^{-2T(\lambda-\delta)} dE\right)^{1/2}} \leq E([\delta, \delta+\varepsilon))^{1/2}.$$

Next consider $\frac{\int_{[\delta+\varepsilon,\infty)} e^{-T(\lambda-\delta)} dE}{\left(\int_{[\delta,\infty)} e^{-2T(\lambda-\delta)} dE\right)^{1/2}}$. Since the denominator and the numerator can be estimated as

$$\int\limits_{[\delta+\varepsilon,\infty)} e^{-T(\lambda-\delta)} dE \leq e^{-T\varepsilon},$$

$$\int\limits_{[\delta,\infty)} e^{-2T(\lambda-\delta)} dE \geq \int\limits_{[\delta,\delta+\varepsilon/2)} e^{-2T(\lambda-\delta)} dE \geq e^{-\varepsilon T} E([\delta, \delta+\varepsilon/2)),$$

we see that

$$\frac{\int_{[\delta+\varepsilon,\infty)} e^{-T(\lambda-\delta)} dE}{\left(\int_{[\delta,\infty)} e^{-2T(\lambda-\delta)} dE\right)^{1/2}} \leq \frac{e^{-T\varepsilon}}{e^{-T\varepsilon/2} E([\delta, \delta+\varepsilon/2))^{1/2}} \leq \frac{e^{-T\varepsilon/2}}{E([\delta, \delta+\varepsilon/2))^{1/2}}.$$

It follows that

$$\sqrt{\gamma_\rho(T)} \leq E([\delta, \delta+\varepsilon))^{1/2} + \frac{e^{-T\varepsilon/2}}{E([\delta, \delta+\varepsilon/2))^{1/2}}.$$

Take $T \to \infty$ on both sides above. We obtain $b \leq E([\delta, \delta+\varepsilon))$ for every $\varepsilon > 0$, and $b \leq E(\{\delta\})$ is obtained by taking the limit $\varepsilon \downarrow 0$. Hence $0 \neq E(\{\delta\}) = E(\{E_N\})$ and thus a ground state of $H_{\rm N}$ exists. □

Proposition 2.30 gives a sufficient condition for H_{N} to have a ground state. A similar argument can also be used to show absence of a ground states of H_{N}.

Proposition 2.31 (Criterion for absence of ground state). *Let $\rho \in L^2(\mathbb{R}^d)$ be a nonnegative function and not identically zero. If $a = \lim_{T\to\infty} \gamma_\rho(T) = 0$, then H_{N} has no ground state.*

Proof. We may suppose that $\inf \mathrm{Spec}(H_{\mathrm{N}}) = 0$, so that $\lim_{T\to\infty} e^{-TH_{\mathrm{N}}} = 1_{\{0\}}(H_{\mathrm{N}})$ in the strong sense. If 0 is an eigenvalue of H_{N}, i. e., H_{N} has a ground state Ψ_{g}, then $a = (\rho \otimes 1, \Psi_{\mathrm{g}}) > 0$, since Ψ_{g} is strictly positive by Theorem 2.26. Thus $a = 0$ implies that H_{N} has no ground state. □

Corollary 2.32. *Suppose that* $\operatorname{ess\,inf}_{x\in K} \Psi_{\mathrm{p}}(x) > 0$ *for every compact set $K \subset \mathbb{R}^d$. Then H_{N} has a ground state if and only if $\Psi_{\mathrm{g}}^\infty \neq 0$.*

Proof. Since Ψ_{g}^T is nonnegative, the weak limit Ψ_{g}^∞ is also nonnegative. Hence $\lim_{T\to\infty} \gamma(T) = (\Psi_{\mathrm{p}} \otimes 1, \Psi_{\mathrm{g}}^\infty)^2 > 0$ if $\Psi_{\mathrm{g}}^\infty \neq 0$ and $= 0$ if $\Psi_{\mathrm{g}}^\infty = 0$. The corollary follows then by Proposition 2.30. □

Remark 2.33. Note that since V is Kato-decomposable, by Lemma 4.179 in Volume 1, the ground state $\Psi_{\mathrm{p}}(x)$ is continuous and strictly positive. Hence $\operatorname{ess\,inf}_{x\in K} \Psi_{\mathrm{p}}(x) > 0$ for every compact set.

Proof of Theorem 2.28. Write Ψ_{p} instead of $\Psi_{\mathrm{p}} \otimes 1$. By Corollary 2.32, we need to prove that Ψ_{g}^T is not weakly convergent to zero. Let

$$S_{[a,b]} = \frac{1}{2} \int_a^b ds \int_a^b W(X_s - X_t, s - t) dt$$

and put

$$f(T, t) = (\Psi_{\mathrm{g}}^T, (e^{-tH_{\mathrm{p}}} \otimes P_0) \Psi_{\mathrm{g}}^T),$$

where P_0 denotes the projection onto the one-dimensional subspace generated by the constant function $1 \in L^2(\mathrm{G})$. We claim that

$$\liminf_{T\to\infty} f(T, t) \geq \exp\Big(-t\Big(E_{\mathrm{N}} + \int_{\mathbb{R}^d} \frac{|\hat{\varphi}(k)|^2}{2\omega(k)^2} dk \Big) - \int_{\mathbb{R}^d} \frac{|\hat{\varphi}(k)|^2 (1 + e^{-t\omega(k)})}{2\omega(k)^3} dk \Big). \tag{2.5.8}$$

To prove this, write

$$f(T, t) = \frac{(\Psi_{\mathrm{p}}, e^{-TH_{\mathrm{N}}} (e^{-tH_{\mathrm{p}}} \otimes P_0) e^{-TH_{\mathrm{N}}} \Psi_{\mathrm{p}})}{(\Psi_{\mathrm{p}}, e^{-(2T+t)H_{\mathrm{N}}} \Psi_{\mathrm{p}})} \frac{(\Psi_{\mathrm{p}}, e^{-(2T+t)H_{\mathrm{N}}} \Psi_{\mathrm{p}})}{(\Psi_{\mathrm{p}}, e^{-2TH_{\mathrm{N}}} \Psi_{\mathrm{p}})}.$$

The second ratio above converges to $e^{-E_N t}$ as $T \to \infty$ by Lemma 2.29 since $\Psi_p > 0$. The first ratio, denoted $g(t,T)$, can be written in terms of a functional integral. The denominator is

$$\mathbb{E}_{\mathcal{N}_0}[e^{S[-T,T+t]}]e^{-(2T+t)E_p} \tag{2.5.9}$$

due to (2.3.9) and shift invariance of $(X_t)_{t\geq 0}$. For the numerator of $g(t,T)$, notice that

$$(\Psi_p, e^{-TH_N}(e^{-tH_p} \otimes P_0)e^{-TH_N}\Psi_p) = (h_T, e^{-tH_p}h_T)_{L^2(\mathbb{R}^d)},$$

where $h_T(x) = (1, e^{-TH_N}\Psi_p)_{L^2(G)}(x)$. Also, notice that

$$\int_{\mathbb{R}^d} h_T(x)f(x)\Psi_p(x)dx = (f\Psi_p \otimes 1, e^{-TH_N}\Psi_p \otimes 1) = \mathbb{E}_{\mathcal{P}_0}[f(X_0)e^{-\int_0^t \xi_s(\varphi_{X_s})ds}]e^{-TE_p}$$
$$= \mathbb{E}_{\mathcal{N}_0}[f(X_0)\mathbb{E}_{\mathcal{G}}[e^{-\int_0^t \xi_s(\varphi_{X_s})ds}]]e^{-TE_p} = \int_{\mathbb{R}^d} \Psi_p(x)^2 f(x)\mathbb{E}^x_{\mathcal{N}_0}[e^{S[0,T]}]e^{-TE_p}dx.$$

Comparing the leftmost and right-hand side expressions, we have

$$h_T(x) = \Psi_p(x)F(x)e^{-TE_p}$$

with $F(x) = \mathbb{E}^x_{\mathcal{N}_0}[e^{S[0,T]}]$. Thus

$$(h_T, e^{-tH_p}h_T) = (F, e^{-tL_p}F)e^{-(2T+t)E_p} = \mathbb{E}_{\mathcal{N}_0}[F(X_0)F(X_t)]e^{-(2T+t)E_p}$$
$$= \int_{\mathbb{R}^d} \mathbb{E}^x_{\mathcal{N}_0}[\mathbb{E}^x_{\mathcal{N}_0}[e^{S[0,T]}]\mathbb{E}^{X_t}_{\mathcal{N}_0}[e^{S[0,T]}]]e^{-(2T+t)E_p}d\mathsf{N}_0.$$

By reflection symmetry,

$$(h_T, e^{-tH_p}h_T) = \int_{\mathbb{R}^d} \mathbb{E}^x_{\mathcal{N}_0}[\mathbb{E}^x_{\mathcal{N}_0}[e^{S[-T,0]}]\mathbb{E}^{X_t}_{\mathcal{N}_0}[e^{S[0,T]}]]e^{-(2T+t)E_p}d\mathsf{N}_0,$$

and by the Markov property,

$$(h_T, e^{-tH_p}h_T) = \int_{\mathbb{R}^d} \mathbb{E}^x_{\mathcal{N}_0}[\mathbb{E}^x_{\mathcal{N}_0}[e^{S[-T,0]}]\mathbb{E}^x_{\mathcal{N}_0}[e^{S[t,T+t]}|\sigma(X_t)]]e^{-(2T+t)E_p}d\mathsf{N}_0.$$

The independence of X_{-t}, $t \geq 0$ and X_s, $s \geq 0$, implies

$$(h_T, e^{-tH_p}h_T) = \int_{\mathbb{R}^d} \mathbb{E}^x_{\mathcal{N}_0}[\mathbb{E}^x_{\mathcal{N}_0}[e^{S[-T,0]+S[t,T+t]}|\sigma(X_t)]]e^{-(2T+t)E_p}d\mathsf{N}_0$$
$$= \int_{\mathbb{R}^d} \mathbb{E}^x_{\mathcal{N}_0}[e^{S[-T,0]+S[t,T+t]}]e^{-(2T+t)E_p}d\mathsf{N}_0$$

$$= \mathbb{E}_{\mathcal{N}_0}[e^{S[-T,0]+S[t,T+t]}]e^{-(2T+t)E_p}.$$

Finally, using (2.5.9) it follows that

$$g(T,t) = \frac{\mathbb{E}_{\mathcal{N}_0}[e^{S[-T,0]+S[t,T+t]}]}{\mathbb{E}_{\mathcal{N}_0}[e^{S[-T,T+t]}]} = \frac{\mathbb{E}_{\mathcal{N}_0}[e^{S_\Delta+S[-T,T+t]}]}{\mathbb{E}_{\mathcal{N}_0}[e^{S[-T,T+t]}]},$$

where $S_\Delta = S[-T,0] + S[t,T+t] - S[-T,T+t]$. Making use of the uniform pathwise estimate

$$|S_\Delta| \le \frac{1}{2}\int_{\mathbb{R}^d} \frac{|\hat\varphi(k)|^2}{2\omega(k)} dk \left(2\int_{-\infty}^{0}\int_{0}^{\infty} + 2\int_{0}^{t}\int_{t}^{\infty} + \int_0^t\int_0^t \right) e^{-|s-r|\omega(k)} ds dr$$
$$\le t\int_{\mathbb{R}^d} \frac{|\hat\varphi(k)|^2}{2\omega(k)^2} dk + \int_{\mathbb{R}^d} \frac{|\hat\varphi(k)|^2(1+2e^{-t\omega(k)})}{2\omega(k)^3} dk,$$

we can compare the numerator and denominator of $g(t,T)$ to find that

$$g(t,T) \ge \exp\left(-t\int_{\mathbb{R}^d} \frac{|\hat\varphi(k)|^2}{2\omega(k)^2} dk - \int_{\mathbb{R}^d} \frac{|\hat\varphi(k)|^2(1+2e^{-t\omega(k)})}{2\omega(k)^3} dk \right).$$

This proves (2.5.8), and we see that $\liminf_{T\to\infty} \|e^{-tH_p/2} \otimes P_0 \Psi_g^T\|$ is nonzero. In order to show that the family Ψ_g^T does not converge to zero, we replace $e^{-tH_p/2} \otimes P_0$ by a compact operator, which we will choose to be a spectral projection of $e^{-tH_p/2}$. Let $1_{[a,b]}(H_p)$ denote the projection onto the spectral subspace of H_p corresponding to $\mathrm{Spec}(H_p) \cap [a,b]$. By the definition of Σ_p and by standard properties of Schrödinger operators, H_p has only finitely many eigenvalues of finite multiplicity below $\Sigma_p - \delta$ for every $\delta > 0$, and thus $1_{[E_p,\Sigma_p-\delta]}(H_p) \otimes P_0$ is a finite rank operator. We have

$$(1_{[E_p,\Sigma_p-\delta]}(H_p) \otimes P_0)\Psi_g^T \to (1_{[E_p,\Sigma_p-\delta]}(H_p) \otimes P_0)\Psi_g^\infty$$

in strong sense as $T \to \infty$. On the other hand,

$$\left\| e^{-tH_p} 1_{(\Sigma_p-\delta,\infty)} \right\| \le e^{-t(\Sigma_p-\delta)}. \tag{2.5.10}$$

Hence

$$\left(\Psi_g^\infty, (e^{-tH_p} 1_{[E_p,\Sigma_p-\delta]}(H_p) \otimes P_0)\Psi_g^\infty\right)$$
$$= \lim_{T\to\infty} \left(\Psi_g^T, (e^{-tH_p} \otimes P_0)\Psi_g^T\right) - \left(\Psi_g^T, (e^{-tH_p} 1_{(\Sigma_p-\delta,\infty)}(H_p) \otimes P_0)\Psi_g^T\right)$$

and we estimate

$$\left(\Psi_g^\infty, (e^{-tH_p} 1_{[E_p,\Sigma_p-\delta]}(H_p) \otimes P_0)\Psi_g^\infty\right) \ge e^{-(E_N+C)t-C(t)} - e^{-t(\Sigma_p-\delta)t}$$

$$= e^{-(E_N+C)t}(e^{-C(t)} - e^{-(\Sigma_p-\delta-E_N-C)t}),$$

where we used (2.5.8) and (2.5.10), and the notations

$$C = \int_{\mathbb{R}^d} \frac{|\hat{\varphi}(k)|^2}{2\omega(k)^2} dk, \quad C(t) = \int_{\mathbb{R}^d} \frac{|\hat{\varphi}(k)|^2(1+e^{-t\omega(k)})}{2\omega(k)^3} dk.$$

By choosing δ small enough and t large enough, we find that Ψ_g^∞ is not zero provided

$$E_N < \Sigma_p - \int_{\mathbb{R}^d} \frac{|\hat{\varphi}(k)|^2}{2\omega(k)^2} dk. \tag{2.5.11}$$

The final step is to convert (2.5.11) to a condition involving the bottom E_p of the spectrum of H_p instead of that of H_N. This is achieved by the estimate

$$E_N \leq E_p - \int_{\mathbb{R}^d} \frac{|\hat{\varphi}(k)|^2}{2\omega(k)(\omega(k)+|k|^2/2)} dk \tag{2.5.12}$$

which will be proved in Lemma 2.34 below. Combining (2.5.12) and (2.5.11) completes the proof. □

A concluding lemma provides the missing piece in the proof above.

Lemma 2.34. *The estimate* (2.5.12) *holds.*

Proof. Instead of $L^2(\mathsf{P}_0)$, we show (2.5.12) on $\mathcal{H}_N = L^2(\mathbb{R}^d) \otimes \mathcal{F}_N$. Let $P_f = d\Gamma(k)$ be the momentum operator of $\mathcal{F}_N$, and define

$$\psi_f = e^{ix\otimes P_f}\Psi_p \otimes e^{-i\Pi(f)}\Omega_N,$$

where Ω_N is the Fock vacuum in $\mathcal{F}_N$, $\Pi(f) = i(a^*(f) - a(\bar{f}))$ is the conjugate momentum, and f will be determined below. By a direct computation,

$$\begin{aligned} E_N &\leq (\psi_f, H_N\psi_f) \\ &= E_p + \frac{1}{2}\Big(\int_{\mathbb{R}^d} k|f(k)|^2 dk\Big)^2 + \int_{\mathbb{R}^d} \Big(\Big(\omega(k) + \frac{|k|^2}{2}\Big)|f(k)|^2 + \frac{2\hat{\varphi}(k)\Re f(k)}{\sqrt{2\omega(k)}}\Big) dk. \end{aligned} \tag{2.5.13}$$

Let f be such that $f(-k) = f(k)$. The second term on the right-hand side of (2.5.13) is zero, and the last term is

$$\int_{\mathbb{R}^d} \Big(\omega(k) + \frac{1}{2}|k|^2\Big)\big(|f(k) + \Phi(k)|^2 - \Phi^2(k)\big) dk,$$

where

$$\Phi(k) = -\frac{\hat{\varphi}(k)}{\sqrt{2\omega(k)}(\omega(k)+|k|^2/2)}.$$

The minimizer $f(k) = \Phi(k)$ yields the bound (2.5.12). □

Corollary 2.35 (Existence of ground state for any coupling strength). *Let $\omega(k) = |k|$ and $\hat{\varphi}(k) = g 1_{\{\kappa \le |k| \le \Lambda\}}$ with coupling constant $g \in \mathbb{R}$. Suppose also that* $\mathrm{Spec}(H_\mathrm{p})$ *is purely discrete and* $\mathrm{ess\,inf}_{x\in K}\Psi_\mathrm{p}(x) > 0$ *for every compact set. Then for all $0 < \kappa < \Lambda$ and $g \in \mathbb{R}$, the operator H_N has a unique ground state.*

Proof. Since $\Sigma_\mathrm{p} - E_\mathrm{p} = \infty$ and $\int_{\mathbb{R}^d} |\hat{\varphi}(k)|^2/\omega(k)^3 dk < \infty$, the corollary follows by Theorem 2.28. □

For confining potentials V, the operator H has a purely discrete spectrum, which gives $\Sigma_\mathrm{p} - E_\mathrm{p} = \infty$.

2.6 Ground state expectations

2.6.1 General expressions

Throughout Section 2.6, we use Assumption 2.36 below.

Assumption 2.36. The following conditions hold:
(1) (Dispersion relation) $\omega(k) = \omega_\nu(k) = \sqrt{|k|^2+\nu^2}$, $\nu \ge 0$.
(2) (Charge distribution) $\varphi \in \mathscr{S}'(\mathbb{R}^d)$, $\overline{\hat{\varphi}(k)} = \hat{\varphi}(-k)$ and $\hat{\varphi}/\omega, \hat{\varphi}/\sqrt{\omega} \in L^2(\mathbb{R}^d)$.
(3) (Infrared regularity) $\int_{\mathbb{R}^d} |\hat{\varphi}(k)|^2/\omega(k)^3 dk < \infty$.
(4) (External potential) V is Kato-decomposable and H_p has a ground state Ψ_p with $H_\mathrm{p}\Psi_\mathrm{p} = E_\mathrm{p}\Psi_\mathrm{p}$.
(5) (Ground state) H_N has the unique ground state Ψ_g.

We write

$$d\mathcal{P}_T = \frac{1}{Z_T} e^{\int_{-T}^{T} \xi_s(\varphi_{X_s})ds} d\mathcal{P}_0$$

with normalizing constant Z_T. With these ingredients, the representation

$$(\Psi_\mathrm{g}^T, (f \otimes 1)\Psi_\mathrm{g}^T) = \mathbb{E}_{\mathcal{P}_T}[f(X_0)]$$

holds. Due to the linearity of the coupling, the field variables ξ can be integrated out in $\mathbb{E}_{\mathcal{P}_T}[f(X_0)]$ and this gives the functional integral representation

$$(\Psi_\mathrm{g}^T, (f \otimes 1)\Psi_\mathrm{g}^T) = \mathbb{E}_{\mathcal{N}_T}[f(X_0)],$$

where

$$d\mathcal{N}_T = \frac{1}{Z_T} e^{\frac{1}{2}\int_{-T}^{T} ds \int_{-T}^{T} W(X_s - X_t, s-t)dt} d\mathcal{N}_0. \tag{2.6.1}$$

Then formally,

$$(\Psi_g, (f \otimes 1)\Psi_g) = \lim_{T\to\infty} (\Psi_g^T, (f \otimes 1)\Psi_g^T) = \mathbb{E}_{\mathcal{N}}[f(X_0)]$$

holds with a probability measure $\mathcal{N}$. In the next theorem, we establish the existence of this measure.

Theorem 2.37 (Tightness). *The family of probability measures $\{\mathcal{N}_T\}_{T\geq 0}$ is tight. In particular, there exists a subsequence $(T_n')_{n\in\mathbb{N}}$ such that $\mathcal{N}_{T_n'}$ is weakly convergent to a probability measure $\mathcal{N}$ on $(\mathfrak{X}, \mathcal{B}(\mathfrak{X}))$.*

Proof. By the Prokhorov theorem, it suffices to show that
(1) $\lim_{\xi\to\infty} \sup_{T>0} \mathcal{N}_T(|X_0|^2 > \xi) = 0$,
(2) $\lim_{\delta\downarrow 0} \sup_{T>0} \mathcal{N}_T\left(\max_{|s-t|<\delta, -T\leq s,t\leq T} |X_s - X_t| > \varepsilon\right) = 0$ for any $\varepsilon > 0$.

We have

$$\mathcal{N}_T(|X_0|^2 > \xi) = (\Psi_g^T, 1_{\{|x|^2>\xi\}}\Psi_g^T).$$

Let $\varepsilon > 0$ be arbitrary. Since $\Psi_g^T \to \Psi_g$ strongly as $T \to \infty$, there exists $T_0 > 0$ such that for all $T > T_0$,

$$(\Psi_g^T, 1_{\{|x|^2>\xi\}}\Psi_g^T) \leq (\Psi_g, 1_{\{|x|^2>\xi\}}\Psi_g) + \varepsilon.$$

Hence

$$\mathcal{N}_T(|X_0|^2 > \xi) \leq \sup_{0\leq T\leq T_0} (\Psi_g^T, 1_{\{|x|^2>\xi\}}\Psi_g^T) + (\Psi_g, 1_{\{|x|^2>\xi\}}\Psi_g) + \varepsilon.$$

We estimate the first factor on the right-hand side above. Note that

$$(\Psi_g^T, 1_{\{|x|^2>\xi\}}\Psi_g^T) = \frac{\mathbb{E}_{\mathcal{N}_0}[1_{\{|X_0|^2>\xi\}} e^{\frac{1}{2}\int_{-T}^{T} ds \int_{-T}^{T} W(X_s-X_t,s-t)dt}]}{\mathbb{E}_{\mathcal{N}_0}[e^{\frac{1}{2}\int_{-T}^{T} ds \int_{-T}^{T} W(X_s-X_t,s-t)dt}]}. \tag{2.6.2}$$

Since

$$\frac{1}{2}\int_{-T}^{T} ds \int_{-T}^{T} W(X_s - X_t, s-t)dt \leq \int_{\mathbb{R}^d} \frac{|\hat{\varphi}(k)|^2}{2\omega(k)^3}\left(e^{-2T\omega(k)} - 1 + 2T\omega(k)\right) dk \leq aT + b$$

with some a, b, together with (2.6.2) it gives

$$(\Psi_g^T, 1_{\{|x|^2>\xi\}}\Psi_g^T) \leq \frac{\mathbb{E}_{\mathcal{N}_0}[1_{\{|X_0|^2>\xi\}}]e^{aT+b}}{e^{-(aT+b)}} = (\Psi_p, 1_{\{|x|^2>\xi\}}\Psi_p)e^{2(aT+b)}.$$

Thus

$$\sup_{0\le T\le T_0}(\Psi_g^T, 1_{\{|x|^2>\xi\}}\Psi_g^T) \to 0$$

as $\xi \to \infty$ and (1) follows.

To prove (2), it suffices to show that

$$\mathbb{E}_{\mathcal{N}_T}[|X_s - X_t|^{2n}] \le D|s-t|^n$$

for some $n \ge 1$ with a constant D. Let $t \ge s$. We have

$$\begin{aligned}\mathbb{E}_{\mathcal{N}_T}[|X_s - X_t|^{2n}] &= \sum_{\mu=1}^{d}\sum_{k=0}^{2n}\binom{2n}{k}(-1)^k \mathbb{E}_{\mathcal{N}_T}[(X_s^\mu)^{2n-k}(X_t^\mu)^k] \\ &= \sum_{\mu=1}^{d}\sum_{k=0}^{2n}\binom{2n}{k}(-1)^k (e^{-sH_{\mathrm{N}}}\Psi_g^T, x_\mu^{2n-k}e^{-(t-s)H_{\mathrm{N}}}x_\mu^k e^{+tH_{\mathrm{N}}}\Psi_g^T).\end{aligned}$$

Here we note that $H_{\mathrm{N}} = H_{\mathrm{N}}^{\mathrm{Kato}}$, and write $\Delta B_{t-s} = |B_0 - B_{t-s}|$. The last term can be represented by functional integration given by Theorem 2.12 in terms of Brownian motion $(B_t)_{t\ge0}$ and the Euclidean field ϕ_{E} as

$$\begin{aligned}&\mathbb{E}_{\mathcal{N}_T}[|X_s - X_t|^{2n}] \\ &= \int_{\mathbb{R}^d}\mathbb{E}^x\Big[\Delta B_{t-s}^{2n}e^{-\int_0^{t-s}V(B_r)dr}(\mathrm{I}_0 e^{-sH_{\mathrm{N}}}\Psi_g^T(B_0), e^{-\phi_{\mathrm{E}}(\int_0^{t-s}\delta_r\otimes\varphi(\cdot-B_r)dr)}\mathrm{I}_{t-s}e^{+tH_{\mathrm{N}}}\Psi_g^T(B_{t-s}))\Big]dx \\ &\le \int_{\mathbb{R}^d}\mathbb{E}^x\Big[\Delta B_{t-s}^{2n}e^{-\int_0^{t-s}V(B_r)dr}\|e^{-sH_{\mathrm{N}}}\Psi_g^T(B_0)\|\|\mathrm{I}_0^* e^{-\phi_{\mathrm{E}}(\int_0^{t-s}\delta_r\otimes\varphi(\cdot-B_r)dr)}\mathrm{I}_{t-s}\|\|e^{+tH_{\mathrm{N}}}\Psi_g^T(B_{t-s})\|\Big]dx,\end{aligned}$$

where the norms with no subscript are $L^2(\mathscr{Q})$-norms. We also note that

$$\|\mathrm{I}_0^* e^{-\phi_{\mathrm{E}}(\int_0^{t-s}\delta_r\otimes\varphi(\cdot-B_r)dr)}\mathrm{I}_{t-s}\| \le 2e^{(t-s)E(\hat\varphi)} = C. \tag{2.6.3}$$

Here $E(\hat\varphi)$ is given by (2.3.15). By the Schwarz inequality,

$$\begin{aligned}&\mathbb{E}_{\mathcal{N}_T}[|X_s - X_t|^{2n}] \\ &\le C\mathbb{E}\Big[\Delta B_{t-s}^{2n}\int_{\mathbb{R}^d}e^{-\int_0^{t-s}V(B_r+x)dr}\|e^{-sH_{\mathrm{N}}}\Psi_g^T(x)\|\|e^{tH_{\mathrm{N}}}\Psi_g^T(B_{t-s}+x)\|\Big]dx \\ &\le C\mathbb{E}\Big[\Delta B_{t-s}^{2n}\Big(\int_{\mathbb{R}^d}e^{-2\int_0^{t-s}V(B_r+x)dr}\|e^{-sH_{\mathrm{N}}}\Psi_g^T(x)\|^2dx\Big)^{1/2}\Big(\int_{\mathbb{R}^d}\|e^{tH_{\mathrm{N}}}\Psi_g^T(B_{t-s}+x)\|^2dx\Big)^{1/2}\Big] \\ &\le C(\mathbb{E}[\Delta B_{t-s}^{4n}])^{1/2}\Big(\mathbb{E}\Big[\int_{\mathbb{R}^d}e^{-2\int_0^{t-s}V(B_r+x)dr}\|e^{-sH_{\mathrm{N}}}\Psi_g^T(x)\|^2dx\int_{\mathbb{R}^d}\|e^{tH_{\mathrm{N}}}\Psi_g^T(B_{t-s}+x)\|^2dx\Big]\Big)^{1/2}\end{aligned}$$

$$\leq CD|s-t|^n\Big(\sup_{x\in\mathbb{R}^d}\mathbb{E}^x[e^{-2\int_0^{t-s}V(B_r)dr}]\Big)^{1/2}\|e^{-sH_N}\Psi_g^T\|_{\mathcal{H}_N}\|e^{tH_N}\Psi_g^T\|_{\mathcal{H}_N}.$$

Finally, noticing that $\|e^{-sH_N}\Psi_g^T\|\|e^{tH_N}\Psi_g^T\| \to \|\Psi_g\|^2$ as $T\to\infty$, we complete (2). □

We now establish an explicit formula for expectations $(\Psi_g, (g\otimes L)\Psi_g)$ of operators $g\otimes L$ as averages with respect to the probability measure $\mathcal{N}$. In order to state our main theorem, we introduce some special elements of $\mathcal{H}_M$. For every $T\in[0,\infty]$ and every path, define

$$\varphi_{T,X}^+(x) = -\int_0^T e^{-|s|\hat\omega}\varphi_{X_s}(x)ds,\ \text{i. e.,}\ \widehat{\varphi_{T,X}^+}(k) = -\int_0^T \hat\varphi(k)e^{-ik\cdot X_s}e^{-|s|\omega(k)}ds,$$

$$\varphi_{T,X}^-(x) = -\int_{-T}^0 e^{-|s|\hat\omega}\varphi_{X_s}(x)ds,\ \text{i. e.,}\ \widehat{\varphi_{T,X}^-}(k) = -\int_{-T}^0 \hat\varphi(k)e^{-ik\cdot X_s}e^{-|s|\omega(k)}ds,$$

where $\hat\omega=\omega(-i\nabla)$. We also define

$$\varphi_X^-(x) = -\int_{-\infty}^0 e^{-|s|\hat\omega}\varphi_{X_s}(x)ds \quad\text{and}\quad \varphi_X^+(x) = -\int_0^\infty e^{-|s|\hat\omega}\varphi_{X_s}(x)ds.$$

It follows that $\varphi_{T,X}^\pm\in\mathcal{H}_M$ and

$$(\varphi_{T,X}^-,\varphi_{T,X}^+)_{\mathcal{H}_M} = 2\int_{-T}^0 ds\int_0^T W(X_s-X_t,s-t)dt,$$

$$\|\varphi_{T,X}^\pm\|_{\mathcal{H}_M}^2 \leq \int_{\mathbb{R}^d}\frac{|\hat\varphi(k)|^2}{\omega(k)^3}dk<\infty.$$

Write

$$\mathcal{P}_0^{x,\xi} = \mathcal{N}_0^x\otimes\mathcal{G}^\xi,\quad (x,\xi)\in\mathbb{R}^d\times\mathscr{M}_{-2}.$$

Theorem 2.38 (Ground state expectations for bounded operators). *Let L be a bounded operator on $L^2(\mathrm{G})$, and $g\in L^\infty(\mathbb{R}^d)$ viewed as a multiplication operator. Then*

$$(\Psi_g,(g\otimes L)\Psi_g) = \mathbb{E}_{\mathcal{N}}\Big[\Big(:e^{\xi(\varphi_X^-)}:, L:e^{\xi(\varphi_X^+)}:\Big)_{L^2(\mathrm{G})}g(X_0)e^{-\int_{-\infty}^0 ds\int_0^\infty W(X_s-X_t,s-t)dt}\Big]. \tag{2.6.4}$$

Proof. By Theorem 2.7, we have

$$(F,\Psi_g^T) = \frac{1}{\|e^{-TL_N}1\|}(F,e^{-TL_N}1) = \frac{1}{\sqrt{Z_T}}\mathbb{E}_{\mathrm{G}\times\mathrm{N}_0}\Big[\overline{F(x,\xi)}\mathbb{E}_{\mathcal{P}_0}^{x,\xi}[e^{-\int_0^T\xi_s(\varphi_{X_s})ds}]\Big].$$

Hence we arrive at

$$\Psi_g^T(x,\xi) = \frac{1}{\sqrt{Z_T}} \mathbb{E}_{\mathcal{P}_0}^{x,\xi}\left[e^{-\int_0^T \xi_s(\varphi_{X_s})ds}\right] \tag{2.6.5}$$

for almost every $(x,\xi) \in \mathbb{R}^d \times \mathscr{M}_{-2}$. Notice that $\mathbb{E}_{\mathcal{P}_0}^{x,\xi} = \mathbb{E}_{\mathcal{N}_0}^{x}\mathbb{E}_{\mathcal{G}}^{\xi}$. We can compute $\mathbb{E}_{\mathcal{G}}^{\xi}[e^{-\int_0^T \xi_s(\varphi_{X_s})ds}]$ by using that (see Theorem 1.138)

$$\mathbb{E}_{\mathcal{G}}^{\xi}\Big[\int_0^T \xi_s(\varphi_{X_s})ds\Big] = \xi(\varphi_{T,X}^+),$$

$$\mathbb{E}_{\mathcal{G}}^{\xi}\Big[\Big(\int_0^T \xi_s(\varphi_{X_s})ds\Big)^2\Big] - \Big(\mathbb{E}_{\mathcal{G}}^{\xi}\Big[\int_0^T \xi_s(\varphi_{X_s})ds\Big]\Big)^2$$
$$= \frac{1}{2}\int_0^T ds \int_0^T dt \int_{\mathbb{R}^d} \frac{|\hat{\varphi}(k)|^2}{2\omega(k)} e^{-ik\cdot(X_s-X_t)}\big(e^{-|s-t|\omega(k)} - e^{-(|t|+|s|)\omega(k)}\big)dk.$$

Now the integration with respect to $\mathcal{G}^{\xi}$ in (2.6.5) can be carried out with the result

$$\Psi_g^T(x,\xi) = \frac{1}{\sqrt{Z_T}}\mathbb{E}_{\mathcal{N}_0}^{x}\left[e^{\xi(\varphi_{T,X}^+)} e^{\frac{1}{2}\int_0^T ds \int_0^T dt \int_{\mathbb{R}^d} \frac{|\hat{\varphi}(k)|^2}{2\omega(k)} e^{-ik\cdot(X_s-X_t)}\left(e^{-|s-t|\omega(k)} - e^{-(|t|+|s|)\omega(k)}\right)dk}\right]. \tag{2.6.6}$$

By the Wick ordering $:e^{\xi(f)}: = e^{-(1/4)\|f\|_{\mathcal{H}_M}^2} e^{\xi(f)}$ from (1.2.29), and then

$$e^{\xi(\varphi_{T,X}^+)} = :e^{\xi(\varphi_{T,X}^+)}: e^{\frac{1}{2}\int_0^T ds \int_0^T dt \int_{\mathbb{R}^d} \frac{|\hat{\varphi}(k)|^2}{2\omega(k)} e^{-ik\cdot(X_s-X_t)} e^{-(|t|+|s|)\omega(k)} dk}. \tag{2.6.7}$$

Hence, by combining (2.6.6) and (2.6.7),

$$\Psi_g^T(x,\xi) = \frac{1}{\sqrt{Z_T}}\mathbb{E}_{\mathcal{N}_0}^{x}[:e^{\xi(\varphi_{T,X}^+)}: e^{\frac{1}{2}\int_0^T ds \int_0^T W(X_s-X_t,s-t)dt}]. \tag{2.6.8}$$

By the reflection symmetry of X_t, we also have

$$\Psi_g^T(x,\xi) = \frac{1}{\sqrt{Z_T}}\mathbb{E}_{\mathcal{N}_0}^{x}[:e^{\xi(\varphi_{T,X}^-)}: e^{\frac{1}{2}\int_{-T}^0 ds \int_{-T}^0 W(X_s-X_t,s-t)dt}]. \tag{2.6.9}$$

Now we write (2.6.9) for the left and (2.6.8) for the right entry of the scalar product $(\Psi_g^T, (1\otimes L)\Psi_g^T)$, use independence of X_s, $s \le 0$ and X_t, $t \ge 0$, and we add and subtract the term $-\int_{-T}^0 ds \int_0^T W(X_s - X_t, s-t)dt$ in the exponent to obtain

$$(\Psi_g^T, (g\otimes L)\Psi_g^T)_{L^2(\mathrm{P}_0)} = \mathbb{E}_{\mathcal{N}_T}[(:e^{\xi(\varphi_{T,X}^-)}:, L:e^{\xi(\varphi_{T,X}^+)}:)g(X_0)e^{-\int_{-T}^0 ds \int_0^T W(X_s-X_t,s-t)dt}]. \tag{2.6.10}$$

This is the finite T version of (2.6.4). It remains to take the limit $T \to \infty$. On the left-hand side of (2.6.10), this is immediate since $\Psi_g^T \to \Psi_g$ in $L^2(\mathrm{P}_0)$ and L is continuous. On the right-hand side, we already know that $\mathcal{N}_{T'} \to \mathcal{N}$ weakly with some subsequence. Reset T' to T. Thus it only remains to show that the integrand of (2.6.10) is uniformly convergent in the paths. For the first factor, we find that $|\widehat{\varphi_{T,X}^{\pm}}(k)| \le |\hat{\varphi}(k)|/\omega(k)$ uniformly in T and the paths, and $\widehat{\varphi_{T,X}^{\pm}} \to \widehat{\varphi_X^{\pm}}$ strongly in $\mathcal{H}_M$ as $T \to \infty$ uniformly in paths. Thus $:e^{\xi(\varphi_{T,X}^+)}: \to :e^{\xi(\varphi_X^+)}:$ in $L^2(\mathrm{G})$ uniformly in paths. Since the same argument applies to $\varphi_{T,X}^-$ and L is continuous, $(:e^{\xi(\varphi_{T,X}^-)}:, L:e^{\xi(\varphi_{T,X}^+)}:)$ is uniformly convergent to $(:e^{\xi(\varphi_X^-)}:, L:e^{\xi(\varphi_X^+)}:)$ as $T \to \infty$. Moreover, the infrared regularity condition implies

$$\left|\int_{-T}^{0} ds \int_{0}^{T} W(X_s - X_t, s-t)dt\right| \le \int_{\mathbb{R}^d} \frac{|\hat{\varphi}(k)|^2}{2\omega(k)^3} dk < \infty$$

and hence

$$\int_{-T}^{0} ds \int_{0}^{T} W(X_s - X_t, s-t)dt \to \int_{-\infty}^{0} ds \int_{0}^{\infty} W(X_s - X_t, s-t)dt$$

as $T \to \infty$ uniformly in paths. Hence the integrand in (2.6.10) is uniformly convergent as $T \to \infty$. □

Most operators of physical interest are, however, not bounded. Therefore, we need to extend formula (2.6.4) to unbounded operators.

Theorem 2.39 (Ground state expectations for unbounded operators). *Let g be a measurable function on $\mathbb{R}^d$ and L a self-adjoint operator in $L^2(\mathrm{G})$ such that*

$$\mathbb{E}_{\mathcal{N}}\left[|g(X_0)|^2 \|L:e^{\xi(\varphi_X^{\pm})}:\|^2_{L^2(\mathrm{G})}\right] < \infty. \tag{2.6.11}$$

Then $\Psi_g \in D(g \otimes L)$ and (2.6.4) holds.

Proof. Without loss of generality, we can suppose that g is real-valued and

$$g_M(x) = \begin{cases} g(x), & |g(x)| < M, \\ M, & |g| \ge M. \end{cases}$$

Let $L_N = 1_{[-N,N]}(L)$. The operator $g_M \otimes L_N$ is bounded, hence (2.6.4) holds for $g_M \otimes L_N$. Let $I = \int_{\mathbb{R}^d} |\hat{\varphi}(k)|^2/\omega(k)^3 dk$. Using the Schwarz inequality, we have

$$\begin{aligned}\|(g_M \otimes L_N)\Psi_g\|^2 &= \mathbb{E}_{\mathcal{N}}\left[(:e^{\xi(\varphi_X^-)}:, L_N^2:e^{\xi(\varphi_X^+)}:)|g_M(X_0)|^2 e^{-\int_{-\infty}^0 ds \int_0^\infty W(X_s - X_t, s-t)dt}\right] \\ &\le e^I \mathbb{E}_{\mathcal{N}}\left[|g_M(X_0)|^2 \|L_N:e^{\xi(\varphi_X^-)}:\| \|L_N:e^{\xi(\varphi_X^+)}:\|\right] \\ &\le e^I \mathbb{E}_{\mathcal{N}}\left[|g(X_0)|^2 \|L:e^{\xi(\varphi_X^-)}:\| \|L:e^{\xi(\varphi_X^+)}:\|\right],\end{aligned}$$

which is bounded according to (2.6.11). By the monotone convergence theorem, this shows that $\Psi_g \in D(g\otimes L)$, and then we obtain that $(g_M\otimes L_N)\Psi_g \to (g\otimes L)\Psi_g$ as $M,N \to \infty$. Hence

$$(\Psi_g, (g_M \otimes L_N)\Psi_g) \to (\Psi_g, (g \otimes L)\Psi_g) \tag{2.6.12}$$

as $N,M \to \infty$. On the other hand, it follows from (2.6.11) that $:e^{\xi(\varphi_X^+)}: \in D(L)$ for $\mathcal{N}$-almost all paths. From this, we conclude

$$(:e^{\xi(\varphi_X^-)}:, L_N:e^{\xi(\varphi_X^+)}:) \to (:e^{\xi(\varphi_X^-)}:, L:e^{\xi(\varphi_X^+)}:)$$

for $\mathcal{N}$ almost all paths as $N \to \infty$. We also have

$$(:e^{\xi(\varphi_X^-)}:, L_N:e^{\xi(\varphi_X^+)}:) \le \|:e^{\xi(\varphi_X^-)}:\| \|L_N:e^{\xi(\varphi_X^+)}:\| \le e^{I/2}\|L:e^{\xi(\varphi_X^+)}:\|$$

for all paths, and the right-hand side is $\mathcal{N}$-integrable. Thus the Lebesgue dominated convergence theorem implies

$$\begin{aligned} &\mathbb{E}_{\mathcal{N}}[(:e^{\xi(\varphi_X^-)}:, L_N:e^{\xi(\varphi_X^+)}:)g_M(X_0)e^{-\int_{-\infty}^0 ds \int_0^\infty W(X_s-X_t,s-t)dt}] \\ &\quad \to \mathbb{E}_{\mathcal{N}}[(:e^{\xi(\varphi_X^-)}:, L:e^{\xi(\varphi_X^+)}:)g(X_0)e^{-\int_{-\infty}^0 ds \int_0^\infty W(X_s-X_t,s-t)dt}] \end{aligned} \tag{2.6.13}$$

as $M,N \to \infty$. By (2.6.12) and (2.6.13), the theorem follows. □

2.6.2 Ground state expectations for second quantized operators

Our next goal is to find similar expressions for the ground state expectation of second quantized operators. In Section 1.2.1, we introduced the second quantization $\Gamma(T)$ and the differential second quantization $d\Gamma(T)$ on a boson Fock space $\mathcal{F}_N$, and in Section 1.6.3, further on the function space $L^2(G)$. Here we use the notation $\Gamma(T)$, $d\Gamma(T)$ on $\mathcal{F}_N$, and $\tilde{\Gamma}(T)$, $d\tilde{\Gamma}(T)$ on $L^2(G)$. It is seen that

$$\Gamma(T) \cong \tilde{\Gamma}(F^{-1}TF), \quad d\Gamma(T) \cong d\tilde{\Gamma}(F^{-1}TF),$$

where $F : L^2(\mathbb{R}^d) \to L^2(\mathbb{R}^d)$ denotes Fourier transform. We have already seen that this unitary equivalence is derived from the Wiener–Itô–Segal isomorphism.

Theorem 2.40. *Suppose that L is a bounded self-adjoint operator on $\mathcal{H}_M$. Then $\Psi_g \in D(\tilde{\Gamma}(L)) \cap D(d\tilde{\Gamma}(L))$ and*

$$\begin{aligned} (\Psi_g, \tilde{\Gamma}(L)\Psi_g)_{L^2(P_0)} &= \mathbb{E}_{\mathcal{N}}[e^{\frac{1}{2}(\varphi_X^-, L\varphi_X^+)_{\mathcal{H}_M}} e^{-\int_{-\infty}^0 ds \int_0^\infty W(X_s-X_t,s-t)dt}], \\ (\Psi_g, d\tilde{\Gamma}(L)\Psi_g)_{L^2(P_0)} &= \frac{1}{2}\mathbb{E}_{\mathcal{N}}[(\varphi_X^-, L\varphi_X^+)_{\mathcal{H}_M}]. \end{aligned}$$

Proof. Note that

$$\|\tilde{\Gamma}(L){:}e^{\xi(\varphi_X^\pm)}{:}\|^2_{L^2(G)} = e^{\frac{1}{2}\|L\varphi_X^+\|^2_{\mathcal{H}_M}},$$
$$\|d\tilde{\Gamma}(L){:}e^{\xi(\varphi_X^\pm)}{:}\|^2_{L^2(G)} = \frac{1}{2}(\|L\varphi_X^\pm\|^2_{\mathcal{H}_M} + (\varphi_X^\pm, L\varphi_X^\pm)^2_{\mathcal{H}_M})e^{\frac{1}{2}\|\varphi_X^\pm\|^2_{\mathcal{H}_M}},$$

and $\|\varphi_X^\pm\|_{\mathcal{H}_M} \le \int_{\mathbb{R}^d} |\hat{\varphi}(k)|^2/\omega(k)^3 dk < \infty$. Since $\|Lf\|_{\mathcal{H}_M} \le \|L\|\|f\|_{\mathcal{H}_M}$ for all $f \in \mathcal{H}_M$, (2.6.11) is satisfied. By Theorem 2.39, the statement follows. □

Now we turn to specific cases of particular interest. Let $d\Gamma(1) = N$ and $d\tilde{\Gamma}(I) = \tilde{N}$ be the number operators.

Corollary 2.41 (Superexponential decay of boson number). *For every $\alpha \in \mathbb{C}$, we have that $\Psi_g \in D(e^{\alpha\tilde{N}})$ and*

$$(\Psi_g, e^{\alpha\tilde{N}}\Psi_g) = \mathbb{E}_{\mathcal{N}}[e^{-(1-e^\alpha)\int_{-\infty}^0 ds \int_0^\infty W(X_s - X_t, s-t)dt}].$$

Proof. Since $e^{\alpha\tilde{N}} = \tilde{\Gamma}(e^\alpha 1)$, Theorem 2.40 gives the result. □

By Corollary 2.41,

$$(\Psi_g, \tilde{N}^n\Psi_g) = \frac{d^n}{d\alpha^n}(\Psi_g, e^{\alpha\tilde{N}}\Psi_g)\lceil_{\alpha=0}.$$

In particular,

$$(\Psi_g, \tilde{N}\Psi_g) = \mathbb{E}_{\mathcal{N}}\left[\int_{-\infty}^0 ds \int_0^\infty W(X_s - X_t, s-t)dt\right]. \tag{2.6.14}$$

Corollary 2.42 (Pull-through formula). *We have*

$$(\Psi_g, \tilde{N}\Psi_g) = \int_{\mathbb{R}^d} \frac{|\hat{\varphi}(k)|^2}{2\omega(k)} \|(H_N - E_N + \omega(k))^{-1}e^{-ik\cdot x}\Psi_g\|^2 dk. \tag{2.6.15}$$

Proof. Computing (2.6.14), we obtain

$$(\Psi_g, \tilde{N}\Psi_g) = \int_{-\infty}^0 ds \int_0^\infty dt \int_{\mathbb{R}^d} \frac{|\hat{\varphi}(k)|^2}{2\omega(k)} e^{-|s-t|\omega(k)}\mathbb{E}_{\mathcal{N}}[e^{-ik\cdot(X_s - X_t)}]dk.$$

Since

$$\begin{aligned}\int_{-\infty}^0 ds \int_0^\infty e^{-|s-t|\omega(k)}\mathbb{E}_{\mathcal{N}}[e^{-ik\cdot(X_s-X_t)}]dt &= \int_{-\infty}^0 ds \int_0^\infty (e^{-ik\cdot x}\Psi_g, e^{-(t-s)(H_N - E_N + \omega(k))}e^{-ik\cdot x}\Psi_g)dt \\ &= \|(H_N - E_N + \omega(k))^{-1}e^{-ik\cdot x}\Psi_g\|^2,\end{aligned}$$

the corollary follows. □

Equality (2.6.15) is called the *pull-through formula*. It can also be obtained by the formal computation

$$(H_{\rm N} - E_{\rm N})a(k)\Psi_{\rm g} = [(H_{\rm N} - E_{\rm N}), a(k)]\Psi_{\rm g} = (-\omega(k) + [H_{\rm I}, a(k)])\Psi_{\rm g}$$

and

$$(\Psi_{\rm g}, \tilde{N}\Psi_{\rm g}) = \int_{\mathbb{R}^d} \|a(k)\Psi_{\rm g}\|^2 dk = \int_{\mathbb{R}^d} \|(H_{\rm N} - E_{\rm N} + \omega(k))^{-1}[H_{\rm I}, a(k)]\Psi_{\rm g}\|^2 dk.$$

Note that $[H_{\rm I}, a(k)] = \hat{\varphi}(k)e^{-ik\cdot x}/\sqrt{2\omega(k)}$. The formula is useful in studying the spectral properties of the Nelson model. This formal derivation above is justified, for instance, by Corollary 2.42.

The expectation for the number operator $\tilde{N} = {\rm d}\tilde{\Gamma}(1)$ can be extended to more general operators of the form ${\rm d}\tilde{\Gamma}(\hat{g})$.

Corollary 2.43. *Let g be a nonnegative measurable function such that*

$$\int_{\mathbb{R}^d} |g(k)| \frac{|\hat{\varphi}(k)|^2}{\omega(k)^3} dk < \infty.$$

Then $\Psi_{\rm g} \in D({\rm d}\tilde{\Gamma}(\hat{g}))$, where $\hat{g} = F^{-1}gF = g(-i\nabla)$, and

$$(\Psi_{\rm g}, {\rm d}\tilde{\Gamma}(\hat{g})\Psi_{\rm g}) = \int_{\mathbb{R}^d} \frac{|\hat{\varphi}(k)|^2}{2\omega(k)} g(k) dk \int_{-\infty}^{0} ds \int_{0}^{\infty} e^{-|s-t|\omega(k)} \mathbb{E}_{\mathcal{N}}[e^{-ik\cdot(X_s - X_t)}] dt. \tag{2.6.16}$$

In particular, the following hold:

(1) *(Pull-through formula)*

$$(\Psi_{\rm g}, {\rm d}\tilde{\Gamma}(\hat{g})\Psi_{\rm g}) = \int_{\mathbb{R}^d} |g(k)| \frac{|\hat{\varphi}(k)|^2}{2\omega(k)} \|(H_{\rm N} - E_{\rm N} + \omega(k))^{-1} e^{-ik\cdot x}\Psi_{\rm g}\|^2 dk.$$

(2) *(Upper bound)*

$$(\Psi_{\rm g}, {\rm d}\tilde{\Gamma}(\hat{g})\Psi_{\rm g}) \le \frac{1}{2} \int_{\mathbb{R}^d} |g(k)| \frac{|\hat{\varphi}(k)|^2}{\omega(k)^3} dk. \tag{2.6.17}$$

(3) *(Lower bound) If $\Psi_{\rm g} \in D(|x| \otimes 1)$, then there exists $C > 0$ such that*

$$\frac{1}{2} \int_{\mathbb{R}^d} |g(k)| \frac{|\hat{\varphi}(k)|^2}{\omega(k)^3} (1 - C|k|^2) dk \le (\Psi_{\rm g}, {\rm d}\tilde{\Gamma}(\hat{g})\Psi_{\rm g}).$$

Proof. It can be checked that

$$\|\mathrm{d}\tilde{\Gamma}(\hat{g}){:}e^{\xi(\varphi_X^{\pm})}{:}\|^2_{L^2(G)} \leq \frac{1}{2}\left(\|\hat{g}\varphi_X^{\pm}\|^2_{\mathcal{H}_M} + (\varphi_X^{\pm}, \hat{g}\varphi_X^{\pm})^2_{\mathcal{H}_M}\right) e^{\frac{1}{2}\|\varphi_X^{\pm}\|^2_{\mathcal{H}_M}} (1+I)\, e^I \int_{\mathbb{R}^d} |g(k)| \frac{|\hat{\varphi}(k)|^2}{\omega(k)^3} dk,$$

with $I = \int_{\mathbb{R}^d} |\hat{\varphi}(k)|^2/\omega(k)^3 dk$, and the right-hand side above is finite. Equality (2.6.16) follows from Theorem 2.40. The pull-through formula can be proven in a similar way to the case involving N. The upper bound (2.6.17) follows directly from the representation (2.6.16). From the estimate $1 - (|k|^2|x|^2)/2 \leq \cos k \cdot x$, we get

$$\mathbb{E}_{\mathcal{N}}[\cos(k \cdot (X_s - X_t))] \geq 1 - \frac{|k|^2}{2} \mathbb{E}_{\mathcal{N}}[X_t^2 + X_s^2 - 2X_t \cdot X_s] \geq 1 - |k|^2 \||x|\Psi_g\|^2.$$

The last bound above follows from

$$\mathbb{E}_{\mathcal{N}}[X_s \cdot X_t] = \|e^{-(H_N - E_N)(|s-t|/2)}|x|\Psi_g\|^2 \geq 0.$$

Writing $C = \||x|\Psi_g\|^2$, we have

$$(\Psi_g, \mathrm{d}\tilde{\Gamma}(\hat{g})\Psi_g) \geq \frac{1}{2} \int_{\mathbb{R}^d} |g(k)| \frac{|\hat{\varphi}(k)|^2}{\omega(k)^3} (1 - C|k|^2) dk.$$

This proves the lower bound. □

Corollary 2.44 (Infrared divergence). *Let $\tilde{M} \subset \mathbb{R}^d$ be a compact set and write $M = \mathbb{R}^d \setminus \tilde{M}$. If $d \leq 3$, $\Psi_g \in D(|x| \otimes 1)$, $0 < \kappa < \Lambda < \infty$, and*

$$\hat{\varphi}(k) = \begin{cases} 0, & |k| < \kappa, \\ 1, & \kappa \leq |k| \leq \Lambda, \\ 0, & |k| > \Lambda. \end{cases}$$

Then

$$\lim_{\kappa \to 0} (\Psi_g, \mathrm{d}\tilde{\Gamma}(\hat{1}_M)\Psi_g) \begin{cases} < \infty, & 0 \notin M, \\ = \infty, & 0 \in M. \end{cases}$$

In particular,

$$\lim_{\kappa \to 0} (\Psi_g, \tilde{N}\Psi_g) = \infty. \tag{2.6.18}$$

Proof. By Corollary 2.43, we have

$$(\Psi_g, \mathrm{d}\tilde{\Gamma}(\hat{1}_M)\Psi_g) \geq \int_M \frac{|\hat{\varphi}(k)|^2}{2\omega(k)^3} (1 - C|k|^2) dk.$$

The corollary follows from

$$\int_M \frac{|\hat{\varphi}(k)|^2}{2\omega(k)^3} dk \begin{cases} = \infty, & 0 \in M, \\ < \infty, & 0 \notin M. \end{cases}$$

□

Remark 2.45. Since the informal description of $d\Gamma(g)$ is

$$d\tilde{\Gamma}(g(-i\nabla)) \cong d\Gamma(g) = \int_{\mathbb{R}^d} g(k)a^*(k)a(k)dk,$$

the above results can be compactly written as

$$\frac{|\hat{\varphi}(k)|^2}{2\omega(k)^3}(1 - C|k|^2) \le \|a(k)\Psi_g\|^2 \le \frac{|\hat{\varphi}(k)|^2}{2\omega(k)^3}. \tag{2.6.19}$$

The quantity in the middle of (2.6.19) is the density at momentum k of the expected number of bosons.

We can also estimate the boson number distribution of the ground state Ψ_g from both sides.

Corollary 2.46 (Boson number distribution). *Let P_n be the projection onto the nth Fock space component. Then*

$$(\Psi_g, P_n\Psi_g)_{L^2(P_0)} = \frac{1}{n!}\mathbb{E}_{\mathcal{N}}[I_X^n e^{-I_X}],$$

where

$$I_X = \frac{1}{2}(\varphi_X^+, \varphi_X^-)_{\mathcal{H}_M} = \int_{-\infty}^0 ds \int_0^\infty W(X_s - X_t, s - t)dt.$$

Proof. By the definition of Wick exponential, we have

$$(:e^{\xi(\varphi_X^-)}:, P_n :e^{\xi(\varphi_X^+)}:)_{L^2(G)} = \frac{1}{2^n n!}(\varphi_X^+, \varphi_X^-)_{\mathcal{H}_M}^n.$$

The corollary follows from Theorem 2.38. □

Denote by $p_n = (\Psi_g, P_n\Psi_g)$ the probability of n bosons occurring in the ground state of H_N. Recall that $\omega(k) = \sqrt{|k|^2 + \nu^2}$.

Corollary 2.47. *Suppose that $\varphi \ge 0$ and $\nu > 0$. Then there exists $0 < D$ such that*

$$\frac{D^n}{n!}e^{-I/2} \le p_n \le \frac{(I/2)^n}{n!}e^{I/2}, \tag{2.6.20}$$

where $I = \int_{\mathbb{R}^d} |\hat{\varphi}(k)|^2/\omega(k)^3 dk$.

Proof. The right-hand side of (2.6.20) is obvious by Corollary 2.46. Note that

$$W(X-Y,t)=\frac{1}{2}\int_{|t|}^{\infty}(\varphi_X,e^{-r\hat{\omega}}\varphi_Y)dr$$

and

$$(\varphi_X,e^{-r\hat{\omega}}\varphi_Y)=\frac{1}{2\sqrt{\pi}}\int_0^{\infty}\frac{re^{-r^2/(4p)}}{p^{3/2}}(\varphi_X,e^{-p\hat{\omega}^2}\varphi_Y)dp.$$

Since $e^{-r\hat{\omega}^2}$ is positivity preserving and $\varphi\geq 0$, we see that $W(X_s-X_t,s-t)\geq 0$, and hence $I_X\geq 0$. The left-hand side follows from

$$p_n\geq\frac{1}{n!}e^{-I/2}\mathbb{E}_{\mathcal{N}}[I_X^n]\geq\frac{1}{n!}e^{-I/2}(\mathbb{E}_{\mathcal{N}}[I_X])^n.$$

The needed D is the expectation of the double integral above. □

Let g be the multiplication operator by g. Although we mentioned the unitary equivalence $d\Gamma(g)\cong d\tilde{\Gamma}(\hat{g})$, we are interested in investigating $d\Gamma(\hat{g})$, i. e., $d\Gamma(g(-i\nabla))$ in $\mathcal{H}_{\mathrm{N}}$ and $d\tilde{\Gamma}(g)$ in $L^2(\mathsf{P}_0)$.

Corollary 2.48. *Let $g\in L^\infty(\mathbb{R}^d)\cap L^1(\mathbb{R}^d)$ be real-valued, and $\hat{\varphi}/\omega^{3/2}\in L^1(\mathbb{R}^d)$. Then*

$$(\Psi_g,d\tilde{\Gamma}(g)\Psi_g)=\frac{1}{2(2\pi)^{d/2}}\int_{-\infty}^{0}ds\int_0^{\infty}dt\int_{\mathbb{R}^d\times\mathbb{R}^d}J(k,k')\mathbb{E}_{\mathcal{N}}[e^{i(k\cdot X_s-k'\cdot X_t)}]dkdk',$$

where

$$J(k,k')=\frac{\overline{\hat{\varphi}(k)}\hat{\varphi}(k')}{\sqrt{\omega(k)\omega(k')}}e^{-\omega(k)|s|-\omega(k')|t|}\check{g}(k-k').$$

In particular, we have the upper bound

$$(\Psi_g,d\tilde{\Gamma}(g)\Psi_g)\leq\frac{1}{2(2\pi)^d}\|\hat{\varphi}/\omega^{3/2}\|_{L^1}^2\|g\|_{L^1}.\tag{2.6.21}$$

Proof. In order to apply Theorem 2.40, we calculate $(\varphi_X^-,g\varphi_X^+)_{\mathcal{H}_M}$. The result is

$$(\varphi_X^-,g\varphi_X^+)_{\mathcal{H}_M}=\frac{1}{(2\pi)^{d/2}}\int_{-\infty}^{0}ds\int_0^{\infty}dt\int_{\mathbb{R}^d\times\mathbb{R}^d}J(k,k')dkdk'.$$ □

Remark 2.49. The informal description of $d\Gamma(g(-i\nabla))$ is

$$d\tilde{\Gamma}(g) \cong d\Gamma(g(-i\nabla)) = \int_{\mathbb{R}^d} g(x)a^*(x)a(x)dx.$$

Let $M \subset \mathbb{R}^d$ be a measurable set. The expression $(\Psi_g, d\Gamma(1_M(-i\nabla))\Psi_g)$ counts the expected number of bosons with position inside M. It is common to write

$$d\Gamma(1_M(-i\nabla)) = \int_M a^*(x)a(x)dx.$$

A special case of Corollary 2.48 is as follows.

Corollary 2.50. *Let $M \subset \mathbb{R}^d$ be a compact set and write* $\mathrm{Vol}\, M = \int_M dx$. *Suppose that $\hat{\varphi}/\omega^{3/2} \in L^1(\mathbb{R}^d)$. Then*

$$(\Psi_g, d\Gamma(1_M(-i\nabla))\Psi_g) \le \frac{1}{2(2\pi)^d}\|\hat{\varphi}/\omega^{3/2}\|_{L^1}^2 \,\mathrm{Vol}\, M.$$

In particular,

$$\lim_{\mathrm{Vol}\, M\to\infty} \frac{(\Psi_g, d\Gamma(1_M(-i\nabla))\Psi_g)}{\mathrm{Vol}\, M} \le \frac{1}{2(2\pi)^d}\|\hat{\varphi}/\omega^{3/2}\|_{L^1}^2 < \infty.$$

In what follows, for notational simplicity we use the notations $d\Gamma(g)$ and $\Gamma(g)$ in the function spaces $L^2(\mathscr{M}_{-2}, dG)$ for $d\tilde{\Gamma}(\hat{g})$ and $\tilde{\Gamma}(\hat{g})$, respectively.

2.6.3 Ground state expectation for fractional powers of the number operator

For more refined information, we also consider the functional $(\Psi_g, \tilde{N}^\alpha\Psi_g)$ for an arbitrary number $\alpha \in \mathbb{R}$ and derive an expression in terms of the measure $\mathcal{N}$. Let $(N_t)_{t\ge 0}$ be a Poisson process with intensity 1 on a probability space $(\Omega, \mathscr{F}, P)$, and recall that

$$\mathbb{E}_P[e^{-uN_t}] = e^{t(e^{-u}-1)}, \quad u \ge 0. \tag{2.6.22}$$

Let $f \in C^\infty((0,\infty))$ with $f \ge 0$, and recall from Section 3.6.2 in Volume 1, that f is called a Bernstein function whenever $(-1)^n \frac{d^n f}{dx^n}(x) \le 0$ for all $n \in \mathbb{N}$. We denote the set of Bernstein functions by $\mathcal{B}$, and by $\mathcal{B}_0$ the set of Bernstein functions with vanishing right limits at zero. For simplicity, in this section we denote

$$W = \int_{-\infty}^0 ds \int_0^\infty dt \int_{\mathbb{R}^d} \frac{|\hat{\varphi}(k)|^2}{2\omega(k)} e^{-|s-t|\omega(k)} e^{-ik\cdot(X_s-X_t)} dk \quad \text{and} \quad \mathrm{I} = \int_{\mathbb{R}^d} \frac{|\hat{\varphi}(k)|^2}{\omega(k)^3} dk.$$

Also, throughout this section we assume that

$$W \geq 0 \tag{2.6.23}$$

almost surely. Hence, together with (2.6.23), we have in fact $0 \leq W \leq I/2$ a. s.

Example 2.51. A case of $\varphi \in \mathscr{S}'(\mathbb{R}^d)$ such that $W \geq 0$ is the following. Suppose that $\varphi \in L^2(\mathbb{R}^d)$ and $\varphi(x) \geq 0$ for all $x \in \mathbb{R}^d$. By Fourier transform,

$$\begin{aligned}\int_{\mathbb{R}^d} \frac{|\hat{\varphi}(k)|^2}{\omega(k)} e^{-|t|\omega(k)} e^{-ik(x-y)} dk &= \left(\varphi(\cdot - x), \omega(-i\nabla)^{-1} e^{-|t|\omega(-i\nabla)} \varphi(\cdot - y)\right) \\ &= \int_{|t|}^{\infty} \left(\varphi(\cdot - x), e^{-s\omega(-i\nabla)} \varphi(\cdot - y)\right) ds\end{aligned}$$

for every $x, y \in \mathbb{R}^d$ and $t \in \mathbb{R}$. Since $e^{-s\omega(-i\nabla)}$ is positivity improving and $\varphi \geq 0$ by assumption, we have (2.6.23).

Corollary 2.52. *Let $m \in \mathbb{Z}$ and $G \in \mathcal{B}_0$. Then*

$$\begin{aligned} m \geq 1 &\Longrightarrow (\Psi_g, \tilde{N}^m G(\tilde{N}) \Psi_g) = \sum_{r=1}^{m} S(m,r) \mathbb{E}_{\mathcal{N}} \mathbb{E}_P \left[W^r G(N_W + r)\right], \\ m = 0 &\Longrightarrow (\Psi_g, G(\tilde{N}) \Psi_g) = \mathbb{E}_{\mathcal{N}} \mathbb{E}_P \left[G(N_W)\right], \\ m \leq -1 &\Longrightarrow (\Psi_g, (\tilde{N}+1)^m G(\tilde{N}+1) \Psi_g) = \mathbb{E}_{\mathcal{N}} \mathbb{E}_P \left[(N_W+1)^m G(N_W+1)\right], \end{aligned}$$

where

$$S(m,r) = \frac{(-1)^r}{r!} \sum_{s=1}^{r} (-1)^s \binom{r}{s} s^m$$

are the Stirling numbers *of the second kind, and N_W is N_t evaluated at the random time $t = W$.*

Proof. We first prove the case $m \geq 0$. Let

$$\rho(\beta) = \mathbb{E}_{\mathcal{N}}[e^{-(1-e^{-\beta})W}] = (\Psi_g, e^{-\beta\tilde{N}} \Psi_g)$$

for $\beta \in \mathbb{C}$. It suffices to consider only $m \geq 1$. There exists a Lévy measure λ such that

$$G(u) = \int_0^{\infty} (1 - e^{-uy}) \lambda(dy), \tag{2.6.24}$$

and

$$\tilde{N}^m = (-1)^m \frac{d^m}{d\beta^m} e^{-\beta\tilde{N}} \Big\lceil_{\beta=0}. \tag{2.6.25}$$

Let $\rho^{(m)} = d^m\rho/d\beta^m$. By Corollary 2.41 and a combination of (2.6.24)–(2.6.25), we have

$$(\Psi_g, \tilde{N}^m G(\tilde{N})\Psi_g) = (-1)^m \int_0^\infty (\rho^{(m)}(0) - \rho^{(m)}(\beta))\lambda(d\beta).$$

We note that in general that the mth derivative of $g(x) = f(e^x)$ is

$$g^{(m)}(x) = \sum_{r=1}^m e^{rx} S(m,r) f^{(r)}(e^x).$$

Since $\rho(\beta) = f(e^{-\beta})$ with $f(x) = \mathbb{E}_{\mathcal{N}}[e^{-(1-x)W}]$, we have

$$\partial^m \rho(\beta) = (-1)^m \sum_{r=1}^m S(m,r) e^{-r\beta} \mathbb{E}_{\mathcal{N}}\left[W^r e^{-(1-e^{-\beta})W}\right].$$

This yields

$$(\Psi_g, \tilde{N}^m G(\tilde{N})\Psi_g) = \mathbb{E}_{\mathcal{N}}\left[\int_0^\infty \sum_{r=1}^m S(m,r) W^r (1 - e^{-r\beta} e^{-(1-e^{-\beta})W})\lambda(d\beta)\right].$$

Using (2.6.22) gives

$$1 - e^{-r\beta} e^{-(1-e^{-\beta})W} = \mathbb{E}_P\left[1 - e^{-\beta(N_W + r)}\right].$$

Hence

$$(\Psi_g, \tilde{N}^m G(\tilde{N})\Psi_g) = \mathbb{E}_{\mathcal{N}}\left[\sum_{r=1}^m S(m,r) W^r \mathbb{E}_P\left[\int_0^\infty (1 - e^{-\beta(N_W + r)})\lambda(d\beta)\right]\right],$$

and the corollary follows for $m > 0$. Next we consider the case $m < 0$ by using a similar strategy. By a combination of the Laplace transform,

$$\int_{[0,\infty)^m} e^{-\sum_{j=1}^m \beta_j(\tilde{N}+1)} \prod_{j=1}^m d\beta_j = (\tilde{N}+1)^{-m}$$

and

$$G(\tilde{N}+1) = \int_0^\infty (1 - e^{-\beta(\tilde{N}+1)})\lambda(d\beta),$$

we obtain

$$(\Psi_g, (\tilde{N}+1)^{-m} G(\tilde{N}+1)\Psi_g)$$

$$= \int_{[0,\infty)^m} \prod_{j=1}^m d\beta_j \int_0^\infty (\Psi_g, (e^{-\sum_{j=1}^m \beta_j(\tilde{N}+1)} - e^{-(\sum_{j=1}^m \beta_j+\beta)(\tilde{N}+1)})\Psi_g)\lambda(d\beta)$$

$$= \int_{[0,\infty)^m} \prod_{j=1}^m d\beta_j \int_0^\infty e^{-\sum_{j=1}^m \beta_j} \mathbb{E}_{\mathcal{N}} \left[e^{-(1-e^{-\sum_{j=1}^m \beta_j})W} - e^{-(1-e^{-\sum_{j=1}^m \beta_j+\beta})W} e^{-\beta} \right] \lambda(d\beta).$$

The second equality follows by Corollary 2.41. In terms of the Poisson process $(N_t)_{t\geq 0}$, we can further rewrite as

$$= \int_{[0,\infty)^m} \prod_{j=1}^m d\beta_j \int_0^\infty \mathbb{E}_{\mathcal{N}} \mathbb{E}_P \left[e^{-\sum_{j=1}^m \beta_j(N_W+1)} - e^{-(\sum_{j=1}^m \beta_j+\beta)(N_W+1)} \right] \lambda(d\beta)$$

$$= \int_{[0,\infty)^m} \prod_{j=1}^m d\beta_j \int_0^\infty \mathbb{E}_{\mathcal{N}} \mathbb{E}_P \left[e^{-\sum_{j=1}^m \beta_j(N_W+1)} (1 - e^{-\beta(N_W+1)}) \right] \lambda(d\beta).$$

Integrating with respect to $\lambda(d\beta)$ and then with $\prod_{j=1}^m d\beta_j$, we finally obtain

$$= \int_{[0,\infty)^m} \prod_{j=1}^m \mathbb{E}_{\mathcal{N}} \mathbb{E}_P \left[e^{-\sum_{j=1}^m \beta_j(N_W+1)} G(N_W+1) \right] d\beta_j = \mathbb{E}_{\mathcal{N}} \mathbb{E}_P \left[(N_W+1)^{-m} G(N_W+1) \right]. \quad \square$$

In order to avoid singularities, in Corollary 2.52 we replaced $\tilde{N}$ with $\tilde{N}+1$ for the case $m < 0$, however, any positive multiple of 1 can be used. By the same corollary, we also have the result below.

Corollary 2.53 (Fractional powers of boson number). *Let $m \in \mathbb{Z}$ and $0 \leq \alpha < 2$. Then*

$$\begin{aligned} m \geq 1 &\implies (\Psi_g, \tilde{N}^{m+\frac{\alpha}{2}} \Psi_g) = \sum_{r=1}^m S(m,r) \mathbb{E}_{\mathcal{N}} \mathbb{E}_P \left[W^r (N_W + r)^{\alpha/2} \right], \\ m = 0 &\implies (\Psi_g, \tilde{N}^{\alpha/2} \Psi_g) = \mathbb{E}_{\mathcal{N}} \mathbb{E}_P \left[N_W^{\alpha/2} \right], \\ m \leq -1 &\implies (\Psi_g, (\tilde{N}+1)^{m+\frac{\alpha}{2}} \Psi_g) = \mathbb{E}_{\mathcal{N}} \mathbb{E}_P \left[(N_W+1)^{m+\alpha/2} \right]. \end{aligned}$$

The above corollary allows deriving the asymptotic behavior of the ground state expectation of noninteger powers of the boson number operator in the strong coupling limit. We introduce a coupling constant g in the Nelson Hamiltonian. It gives in $\mathcal{H}_N$,

$$H_N = H_p \otimes 1 + 1 \otimes H_f + gH_I$$

or in $L^2(\mathcal{P}_0)$,

$$L_N = L_p \otimes 1 + 1 \otimes \tilde{H}_f + g\tilde{H}_I.$$

Corollary 2.54. *If $G \in \mathcal{B}_0$ is a Bernstein function with $G(0) = 0$, then*

$$(\Psi_g, G(\tilde{N})\Psi_g) \leq G(g^2 \mathrm{I}/2).$$

Proof. Since any Bernstein function G is increasing and concave, by Jensen's inequality and assumption (2.6.23), we obtain more generally that

$$(\Psi_g, G(\tilde{N})\Psi_g) \leq G((\Psi_g, \tilde{N}\Psi_g)) \leq G(\mathrm{I}/2),$$

since $(\Psi_g, \tilde{N}\Psi_g) = \mathbb{E}_{\mathcal{N}}[W]$. □

By Corollary 2.54, we have an upper bound of $\lim_{g\to\infty}(\Psi_g, G(\tilde{N})\Psi_g)/G(g^2)$ in terms of $\lim_{g\to\infty} G(g^2\mathrm{I}/2)/G(g^2)$.

Corollary 2.55. *Let $0 < \alpha < 2$. Then*

$$\lim_{g\to\infty} \frac{(\Psi_g, \tilde{N}^{\alpha/2}\Psi_g)}{g^\alpha} \leq (\mathrm{I}/2)^{\alpha/2}$$

and

$$\lim_{g\to\infty} \frac{(\Psi_g, \log(1 + \tilde{N}^{\alpha/2})\Psi_g)}{\alpha \log g} \leq 1.$$

Proof. Set $G(u) = u^{\alpha/2}$ to obtain the first statement follows, and $G(u) = \log(1 + u^{\alpha/2})$ to get the second. □

For the special case $(\Psi_g, \tilde{N}^k\Psi_g)$, $k \geq 1$, we have both upper and lower bounds.

Corollary 2.56. *Let $k = m + \frac{\alpha}{2} \geq 1$, $m \in \mathbb{N}$ and $0 \leq \alpha < 2$. Then there exists A independent of g such that*

$$(\mathrm{I}/2 - A)^k \leq \lim_{g\to\infty} \frac{(\Psi_g, \tilde{N}^k\Psi_g)}{g^{2k}} \leq (\mathrm{I}/2)^k.$$

Proof. By Jensen's inequality, we have

$$\mathbb{E}_P\left[(N_W + r)^{\alpha/2}\right] \leq (\mathbb{E}_P[N_W + r])^{\alpha/2} = (\mathrm{I}/2 + r)^{\alpha/2}.$$

In particular, it follows that

$$(\Psi_g, \tilde{N}^{m+\frac{\alpha}{2}}\Psi_g) \leq \sum_{r=1}^{m} S(m,r)(\mathrm{I}/2 + r)^{\alpha/2}(\mathrm{I}/2)^r,$$

implying $\lim_{g\to\infty}(\Psi_g, \tilde{N}^{m+\frac{\alpha}{2}}\Psi_g)/g^{2m+\alpha} \leq S(m,m)(\mathrm{I}/2)^k = (\mathrm{I}/2)^k$. This gives the upper bound. Again, from Jensen's inequality it follows that

$$(\Psi_g, \tilde{N}^k \Psi_g) \geq (\Psi_g, \tilde{N}\Psi_g)^k.$$

Hence the lower bound can be derived from (3) of Corollary 2.43. □

2.6.4 Ground state expectations of field operators

For $\beta \in \mathbb{R}$ and $g \in \mathcal{H}_M$ next we consider $(\Psi_g, e^{\beta\xi(g)}\Psi_g)_{L^2(\mathsf{P}_0)}$, i. e., the *moment generating function* of the random variable $\xi \mapsto \xi_0(g)$.

Corollary 2.57 (Exponential moments). *Let $\beta \in \mathbb{C}$. Then $(\Psi_g, e^{\beta\xi(g)}\Psi_g)_{L^2(\mathsf{P}_0)}$ is finite and*

$$(\Psi_g, e^{\beta\xi(g)}\Psi_g)_{L^2(\mathsf{P}_0)} = \mathbb{E}_{\mathcal{N}}[e^{\frac{\beta^2}{2}I_g - \beta I_X}],$$

where

$$I_g = \int_{\mathbb{R}^d} \frac{|\hat{g}(k)|^2}{2\omega(k)} dk \quad \text{and} \quad I_X = \int_{\mathbb{R}^d} \frac{\hat{\varphi}(k)\overline{\hat{g}(k)}}{2\omega(k)} dk \int_{-\infty}^{\infty} e^{-\omega(k)|s|} e^{-ik\cdot X_s} ds.$$

Proof. Note that

$$(:e^{\xi(\varphi_X^-)}:, e^{\beta\xi_0(g)} :e^{\xi(\varphi_X^+)}:) = e^{\frac{\beta^2}{2}I_g - \beta I_X} e^{\int_{-\infty}^0 ds \int_0^\infty W(X_s - X_t, s-t)dt}.$$

The statement is then a direct consequence of Theorem 2.39. □

From this corollary, we immediately get

$$(\Psi_g, \xi(g)^n \Psi_g)_{L^2(\mathsf{P}_0)} = \frac{d^n}{d\beta^n}(\Psi_g, e^{\beta\xi(g)}\Psi_g)_{L^2(\mathsf{P}_0)}\lceil_{\beta=0} \quad \text{for all } n \in \mathbb{N}.$$

Let H_n be the *Hermite polynomials* of order n defined by

$$H_n(x) = (-1)^n e^{x^2} \frac{d^n}{dx^n} e^{-x^2}. \tag{2.6.26}$$

The first few examples are

$$H_0(x) = 1, \quad H_1(x) = 2x, \quad H_2(x) = 4x^2 - 2, \quad H_3(x) = 8x^3 - 12x.$$

We use the well-known generating function to write

$$e^{\frac{\beta^2}{2}I_g - \beta I_X} = \sum_{n=0}^{\infty} H_n\left(iI_X/\sqrt{I_g}\right) \frac{\left(i\beta\sqrt{I_g}\right)^n}{n!}.$$

Hence we obtain

$$\frac{d^n}{d\beta^n} e^{\frac{\beta^2}{2}I_g - \beta I_X} \lceil_{\beta=0} = i^n H_n\left(iI_X/\sqrt{I_g}\right) I_g^{n/2}.$$

In the next corollary, we look at the mean and variance of the random variable $\xi \mapsto \xi_0(g)$ for $g \in \mathcal{H}_M$ by making use of Corollary 2.57.

Corollary 2.58 (Average field strength and field fluctuations). *The following hold:*
(1) *(Average field strength) For* $n = 1$,

$$(\Psi_g, \xi(g)\Psi_g) = -\int_{\mathbb{R}^d} \frac{\hat{\varphi}(k)\overline{\hat{g}(k)}}{2\omega(k)} dk \int_{-\infty}^{\infty} e^{-\omega(k)|s|} \mathbb{E}_{\mathcal{N}}[e^{-ik\cdot X_s}] ds.$$

(2) *(Field fluctuations) For* $n = 2$,

$$(\Psi_g, \xi(g)^2\Psi_g) = \int_{\mathbb{R}^d} \frac{|\hat{g}(k)|^2}{2\omega(k)} dk + \mathbb{E}_{\mathcal{N}}\left[\left(\int_{\mathbb{R}^d} \frac{\hat{\varphi}(k)\overline{\hat{g}(k)}}{2\omega(k)} dk \int_{-\infty}^{\infty} e^{-\omega(k)|s|} e^{-ik\cdot X_s} ds\right)^2\right].$$

Remark 2.59.
(1) By using the previous result and the Schwarz inequality, we find that

$$(\Psi_g, \xi(g)^2\Psi_g) - (\Psi_g, \xi(g)\Psi_g)^2 \geq \int_{\mathbb{R}^d} \frac{|\hat{g}(k)|^2}{2\omega(k)} dk = (1, \xi(g)^2 1) - (1, \xi(g)1)^2.$$

The latter term represents the fluctuations of the free field. It is then seen that fluctuations increase by coupling the field to the particle.
(2) Note that

$$\mathbb{E}_{\mathcal{N}}[e^{-ik\cdot X_s}] = \int_{\mathbb{R}^d} \Psi_p(x)^2 \lambda(x)^2 e^{-ik\cdot x} dx,$$

where $\lambda(x)^2 = \int_{\mathscr{M}_{-2}} \Psi_g^2(\xi, x) d\mathsf{G}$ is the density of $\mathcal{N}$ with respect to N_0, and Ψ_p^2 is the density of N_0 with respect to Lebesgue measure and the square of the ground state of H_p.
(3) Writing $\chi = \Psi_p^2\lambda^2$ for the position density of the particle, and taking g to be a delta function in momentum space and in position space, respectively, we find

$$(\Psi_g, \xi(k)\Psi_g)_{L^2(\mathsf{P}_0)} = -\frac{\hat{\varphi}(k)\hat{\chi}(k)}{(2\pi)^{d/2}\omega^2(k)} \quad k \in \mathbb{R}^d,$$

and

$$(\Psi_g, \xi(x)\Psi_g)_{L^2(\mathsf{P}_0)} = (\chi * V_\omega * \varphi)(x) \quad x \in \mathbb{R}^d, \tag{2.6.27}$$

respectively. Here V_ω denotes the Fourier transform of $-1/\omega^2$ and is the Coulomb potential for massless bosons, i. e., for $\omega(k) = |k|$. Equality (2.6.27) gives the classical field generated by a particle with position distribution $\chi(x)dx$.

2.6.5 Gaussian domination

A further important application of Corollary 2.57 is a Gaussian domination property of the ground state Ψ_g. It is well known that the nth eigenfunction h_n of the harmonic oscillator $-\frac{1}{2}\Delta+\frac{1}{2}|x|^2$ is given by $H_n(x)e^{-|x|^2/2}$, where H_n denotes the Hermite polynomial of order n. Hence $h_n(x) \sim e^{-|x|^2/2}$ as $|x| \to \infty$, and

$$\lim_{\beta\uparrow 1/2} \|e^{\beta|x|^2} h_n\| = \infty.$$

The ground state of the Nelson model has a similar property, which we can show by using the probability measure $\mathcal{N}$.

Lemma 2.60. *Let $\beta > 0$ and $g \in \mathscr{S}'(\mathbb{R}^d)$ be such that $\hat{g}/\omega, \hat{g}/\sqrt{\omega} \in L^2(\mathbb{R}^d)$. Then*

$$(\Psi_g, e^{-\beta\xi(g)^2}\Psi_g) = \frac{1}{\sqrt{1+\beta\|\hat{g}/\sqrt{\omega}\|^2}}\mathbb{E}_{\mathcal{N}}\left[e^{-\frac{\beta K(g)^2}{1+\beta\|\hat{g}/\sqrt{\omega}\|}}\right],$$

where $K(g)$ denotes the random variable defined by

$$K(g) = \frac{1}{2}\int_{-\infty}^{\infty} (e^{-|r|\omega}e^{-ik\cdot B_r}\hat{\varphi}/\sqrt{\omega}, \hat{g}/\sqrt{\omega})dr.$$

Proof. For convenience, we use the same notations I_g and I_X as in Corollary 2.57. First, since $\hat{\varphi}/\omega \in L^2(\mathbb{R}^d)$ and $\hat{g}/\omega \in L^2(\mathbb{R}^d)$, it follows that $|K(g)| \le \|\hat{\varphi}/\omega\|\|\hat{g}/\omega\| < \infty$. By Corollary 2.57, we see that

$$(\Psi_g, e^{ik\beta\xi(g)}\Psi_g) = \mathbb{E}_{\mathcal{N}}\left[e^{-k^2\beta^2 I_g/2}e^{-ik\beta I_X}\right].$$

Thus

$$(\Psi_g, e^{-\beta^2\xi(g)^2/2}\Psi_g) = \frac{1}{\sqrt{2\pi}}\int_{\mathbb{R}} e^{-k^2/2}(\Psi_g, e^{ik\beta\xi(g)}\Psi_g)dk$$
$$= \frac{1}{\sqrt{2\pi}}\mathbb{E}_{\mathcal{N}}\left[\int_{\mathbb{R}} e^{-k^2/2}e^{-k^2\beta^2 I_g/2}e^{-ik\beta I_X}dk\right] = \frac{1}{\sqrt{1+\beta^2 I_g}}\mathbb{E}_{\mathcal{N}}\left[e^{-\frac{\beta^2 K(g)^2/2}{1+\beta^2 I_g}}\right].$$

Replacing $\beta^2/2$ by $\beta > 0$, the claim follows. □

Corollary 2.61 (Gaussian domination of the ground state). *Let $g \in \mathscr{S}'(\mathbb{R}^d)$ such that $\hat{g}/\omega, \hat{g}/\sqrt{\omega} \in L^2(\mathbb{R}^d)$, and $\beta < 1/\|\hat{g}/\sqrt{\omega}\|^2$. Then $\Psi_g \in D(e^{(\beta/2)\xi(g)^2})$ and*

$$\|e^{(\beta/2)\xi(g)^2}\Psi_g\|^2 = \frac{1}{\sqrt{1-\beta\|\hat{g}/\sqrt{\omega}\|^2}}\mathbb{E}_{\mathcal{N}}\left[e^{\frac{\beta K(g)^2}{1-\beta\|\hat{g}/\sqrt{\omega}\|^2}}\right].$$

In particular,

$$\lim_{\beta\uparrow(2\|\hat{g}/\sqrt{\omega}\|^2)^{-1}} \|e^{\beta\xi(g)^2}\Psi_g\| = \infty.$$

Proof. Write $\mathbb{C}_+ = \{z \in \mathbb{C} \mid \Re z > 0\}$ and $\mathbb{C}_- = \{z \in \mathbb{C} \mid \Re z < 0\}$, and take the open disc $B = \{z \in \mathbb{C} \mid |z| < 1/(2I_g)\}$. Define

$$\rho(z) = \frac{1}{\sqrt{1+2zI_g}}\mathbb{E}_{\mathcal{N}}\left[e^{-\frac{zK(g)^2}{1+2zI_g}}\right], \quad z > 0.$$

Since $|K(g)| \leq \|\hat{g}/\sqrt{\omega}\|\|\hat{\varphi}/\sqrt{\omega}\|$ uniformly in the paths, the function ρ can be analytically continued to a function $\tilde{\rho}$ on $\mathbb{C}_+ \cup B$. Let $B_\delta(w) \subset \mathbb{C}$ be the ball of radius δ centered at w. Suppose that $\delta < 2I_g$ and choose $w \in \mathbb{R} \cap B$ such that

$$(B_\delta(w) \cup \mathbb{C}_-) \cap B \neq \emptyset.$$

The analytic function $\tilde{\rho}(z)$ can be expanded as

$$\tilde{\rho}(z) = \sum_{n=0}^{\infty}(z-w)^n b_n(w), \quad z \in B_\delta(w) \cap B. \tag{2.6.28}$$

On the other hand, since $\Psi_g \in D(\xi^2(g))$, the function $\mathbb{C} \ni z \mapsto (\Psi_g, e^{-z\xi(g)^2}\Psi_g) \in \mathbb{C}$ is differentiable on $\mathbb{C}_+$. Hence

$$(\Psi_g, e^{-z\xi(g)^2}\Psi_g) = \sum_{n=0}^{\infty}(z-w)^n \frac{1}{n!}\int_0^\infty (-\lambda)^n e^{-w\lambda} dE_\lambda, \tag{2.6.29}$$

where E_λ denotes the spectral measure of the positive self-adjoint operator $\xi(g)^2$ associated with Ψ_g. Comparing (2.6.28) and (2.6.29), we obtain

$$b_n(w) = \frac{1}{n!}\int_0^\infty (-\lambda)^n e^{-w\lambda} dE_\lambda, \quad n = 0, 1, 2, \ldots \tag{2.6.30}$$

Inserting (2.6.30) into (2.6.28), we have

$$\tilde{\rho}(z) = \sum_{n=0}^{\infty}(z-w)^n \frac{1}{n!}\int_0^\infty (-\lambda)^n e^{-w\lambda} dE_\lambda, \quad z \in B_\delta(w) \cap B.$$

Moreover, the right-hand side is absolutely convergent for every $z \in B_\delta(w) \cap B$. For $z \in B_\delta(w) \cap B \cap \mathbb{R}$, we have

$$\int_0^M e^{-z\lambda} dE_\lambda \le \sum_{n=0}^{\infty} |z-w|^n \frac{1}{n!}\left|\int_0^M (-\lambda)^n e^{-w\lambda} dE_\lambda\right| = \sum_{n=0}^{\infty} |z-w|^n \frac{1}{n!}\left|\int_0^M \lambda^n e^{-w\lambda} dE_\lambda\right|$$

$$\le \sum_{n=0}^{\infty} |z-w|^n \frac{1}{n!}\left|\int_0^\infty \lambda^n e^{-w\lambda} dE_\lambda\right| = \sum_{n=0}^{\infty} |z-w|^n \frac{1}{n!}\left|\int_0^\infty (-\lambda)^n e^{-w\lambda} dE_\lambda\right| < \infty.$$

This implies that $\lim_{M\to\infty} \int_0^M e^{-z\lambda} dE_\lambda < \infty$ for $z \in B_\delta(w) \cap B \cap \mathbb{R}$. Thus we conclude that $\Psi_g \in D(e^{-(z/2)\xi(g)^2})$ and $\|e^{-(z/2)\xi(g)^2}\Psi_g\|^2 = \tilde{\rho}(z)$ follows for $z \in B_\delta \cap B \cap \mathbb{R}$. Here we notice that for every $\delta < 1/(2I_g)$ we can choose $w \in \mathbb{R} \cap B$ such that $\mathbb{C}_- \cap B \cap B_\delta(w) \neq \emptyset$. □

2.7 Infrared divergence

As seen above, in order to give a rigorous mathematical definition of the Nelson model, an infrared and an ultraviolet cut-off was necessary to make sure that divergences do not occur. To control the ultraviolet part, we introduced a charge distribution φ making the coupling to the field sufficiently regular, and to control the infrared part we required the integral $\int_{\mathbb{R}^d} \frac{|\hat{\varphi}(k)|^2}{\omega(k)^3} dk$ to be finite. Now we discuss the model in the conditions of infrared singularity, while keeping the ultraviolet cutoff in place. In Section 2.7, we use Assumption 2.62 below.

Assumption 2.62. The following conditions hold:
(1) (Dispersion relation) $\omega(k) = |k|$.
(2) (Charge distribution) $\varphi \in \mathscr{S}'(\mathbb{R}^d)$, $\overline{\hat{\varphi}(k)} = \hat{\varphi}(-k)$ and $\hat{\varphi}/\omega, \hat{\varphi}/\sqrt{\omega} \in L^2(\mathbb{R}^d)$.
(3) (External potential) V satisfies the bound

$$V(x) \ge C|x|^{2\beta}$$

with constant $C > 0$ and exponent $\beta > 0$.

In particular, we consider here the massless Nelson model. Under the conditions above, H_p has a unique, strictly positive ground state Ψ_p at eigenvalue E_p, and

$$\Psi_\mathrm{p}(x) \le e^{-C|x|^{\beta+1}}$$

holds with a constant C. Under the infrared regularity condition

$$\int_{\mathbb{R}^d} \frac{|\hat{\varphi}(k)|^2}{\omega(k)^3} dk < \infty,$$

we have seen in the previous sections that H_N has a unique, strictly positive ground state, and we have obtained detailed information on its properties. What we discuss here is whether there is any ground state in the same space when the infrared regularity condition is not guaranteed.

Theorem 2.63 (Absence of ground state). *Let $d = 3$, $\varphi \in L^1(\mathbb{R}^3)$, $\varphi \geq 0$ be such that $\varphi \not\equiv 0$. Then H_N has no ground state.*

Remark 2.64. Under the assumptions in Theorem 2.63, we are in the situation of infrared singularity. Indeed, $\int_{\mathbb{R}^3} \frac{|\hat{\varphi}(k)|^2}{\omega(k)^3} dk = \infty$, since $\hat{\varphi}(0) > 0$ and $\hat{\varphi}$ is continuous.

Recall that $\gamma(T) = (\Psi_p \otimes 1, \Psi_g^T)^2$. In order to prove Theorem 2.63, it suffices to show that

$$\lim_{T\to\infty} \gamma(T) = 0$$

by Proposition 2.31. We use the notation $S[a, b] = \int_a^b ds \int_a^b W(X_t - X_s, t - s)dt$ for the pair potential W of the Nelson model. Since

$$\gamma(T) = \frac{(\Psi_p \otimes 1, e^{-TL_N}\Psi_p \otimes 1)^2}{(\Psi_p \otimes 1, e^{-2TL_N}\Psi_p \otimes 1)},$$

we can express $\gamma(T)$ in terms of the $P(\phi)_1$-process $(X_t)_{t\in\mathbb{R}}$ associated with the self-adjoint operator $L_p = U_{\Psi_p}(H_p - E_p)U_{\Psi_p}^{-1}$ as

$$\gamma(T) = \frac{(\mathbb{E}_{\mathcal{N}_0}[e^{S[0,T]}])^2}{\mathbb{E}_{\mathcal{N}_0}[e^{S[-T,T]}]}. \tag{2.7.1}$$

Here we used the shift invariance of the process, i. e.,

$$\mathbb{E}_{\mathcal{N}_0}[e^{S[-T,T]}] = \mathbb{E}_{\mathcal{N}_0}[e^{S[0,2T]}].$$

Let $\mathcal{N}_T$ be the probability measure given in (2.6.1).

Lemma 2.65. *It follows that*

$$\gamma(T) \leq \mathbb{E}_{\mathcal{N}_T}[e^{-\int_{-T}^0 ds \int_0^T W(X_s - X_t, s-t)dt}]. \tag{2.7.2}$$

Proof. The numerator of (2.7.1) can be estimated by the Schwarz inequality with respect to $d\mathsf{N}_0$ and the reflection symmetry of the process as

$$(\mathbb{E}_{\mathcal{N}_0}[e^{S[0,T]}])^2 \leq \int_{\mathbb{R}^3} (\mathbb{E}_{\mathcal{N}_0}^x[e^{S[0,T]}])^2 d\mathsf{N}_0 = \int_{\mathbb{R}^3} (\mathbb{E}_{\mathcal{N}_0}^x[e^{S[0,T]}])(\mathbb{E}_{\mathcal{N}_0}^x[e^{S[-T,0]}])d\mathsf{N}_0.$$

Since X_{-s}, $s \geq 0$, and X_t, $t \geq 0$, are independent, we also see that

$$(\mathbb{E}_{\mathcal{N}_0}[e^{S[0,T]}])^2 \leq \int_{\mathbb{R}^3} \mathbb{E}_{\mathcal{N}_0}^x[e^{S[0,T]+S[-T,0]}]d\mathsf{N}_0 = \mathbb{E}_{\mathcal{N}_0}[e^{S[0,T]+S[-T,0]}].$$

Since, moreover, $S[0,T] + S[-T,0] = S[-T,T] - \int_{-T}^{0} ds \int_0^T W(X_s - X_t, s-t)dt$, we have

$$(\mathbb{E}_{\mathcal{N}_0}[e^{S[0,T]}])^2 \le \mathbb{E}_{\mathcal{N}_0}[e^{S[-T,T]-\int_{-T}^{0} ds \int_0^T W(X_s-X_t,s-t)dt}].$$

Thus

$$\gamma(T) \le \frac{\mathbb{E}_{\mathcal{N}_0}[e^{S[-T,T]-\int_{-T}^{0} ds \int_0^T W(X_s-X_t,s-t)dt}]}{\mathbb{E}_{\mathcal{N}_0}[e^{S[-T,T]}]} = \mathbb{E}_{\mathcal{N}_T}[e^{-\int_{-T}^{0} ds \int_0^T W(X_s-X_t,s-t)dt}],$$

and the lemma follows. ☐

The pair potential $W(x,t)$ can be computed explicitly. Using the integral kernel

$$e^{-|t|\sqrt{-\Delta}}(x,y) = \frac{1}{\pi^2}\frac{|t|}{(|x-y|^2+|t|^2)^2},$$

we obtain

$$\begin{aligned} W(x-y,t-s) &= \frac{1}{2}\int_{|s-t|}^{\infty} d|T| \int_{\mathbb{R}^3} e^{-ik\cdot(y-x)} e^{-|T|\omega(k)} \overline{\hat{\varphi}(k)}\hat{\varphi}(k)dk \\ &= \frac{1}{4\pi^2}\int_{\mathbb{R}^3} du \int_{\mathbb{R}^3} \frac{\varphi(u)\varphi(v)}{|x-y+u-v|^2+|s-t|^2} dv > 0. \end{aligned} \tag{2.7.3}$$

To show that $\gamma(T) \to 0$, we first restrict to the set

$$A_T = \left\{\omega \in \mathfrak{X} \,\middle|\, |X_t(\omega)| \le T^\lambda, |t| \le T\right\}$$

with some $\lambda < 1$. We split up the right-hand side of (2.7.2) by restricting to A_T and $\mathfrak{X} \setminus A_T$, and show that the corresponding expectations converge to zero separately.

Lemma 2.66. *It follows that*

$$\lim_{T\to\infty} \mathbb{E}_{\mathcal{N}_T}\left[1_{A_T} e^{-\int_{-T}^{0} ds \int_0^T W(X_s-X_t,s-t)dt}\right] = 0. \tag{2.7.4}$$

Proof. By using the estimate

$$|X_s - X_t + x - y|^2 + |s-t|^2 \le 8T^{2\lambda} + 2|x-y|^2 + |s-t|^2$$

on A_T, and

$$\int_{-T}^{0} ds \int_0^T \frac{1}{a^2+|s-t|^2} dt \ge \log\left(\frac{a^2+T^2/2}{a^2}\right),$$

we obtain by (2.7.3) that

$$\begin{aligned}
&\int_{-T}^{0} ds \int_{0}^{T} W(X_s - X_t, s-t)dt \\
&= \int_{-T}^{0} ds \int_{0}^{T} dt \int_{\mathbb{R}^3} dx \int_{\mathbb{R}^3} \frac{\varphi(x)\varphi(y)}{|X_s - X_t + x - y|^2 + |s-t|^2} dy \\
&\geq \int_{-T}^{0} ds \int_{0}^{T} dt \int_{\mathbb{R}^3} dx \int_{\mathbb{R}^3} \frac{\varphi(x)\varphi(y)}{8T^{2\lambda} + 2|x-y|^2 + |s-t|^2} dy \\
&= \int_{\mathbb{R}^3} dx \int_{\mathbb{R}^3} \varphi(x)\varphi(y) \log\left(\frac{8T^{2\lambda} + 2|x-y|^2 + T^2/2}{8T^{2\lambda} + 2|x-y|^2}\right) dy.
\end{aligned}$$

The right-hand side in the formula above diverges as $T \to \infty$, since $\lambda < 1$. This gives (2.7.4). □

Lemma 2.67. *It follows that*

$$\lim_{T\to\infty} \mathbb{E}_{\mathcal{N}_T}\left[1_{\mathfrak{X}\setminus A_T} e^{-\int_{-T}^{0} ds \int_0^T W(X_s - X_t, s-t)dt}\right] = 0.$$

Proof. Note that

$$\int_{-T}^{0} ds \int_{0}^{T} W(X_s - X_t, s-t)dt \leq \frac{T}{2}\|\hat{\varphi}/\omega\|^2.$$

We have

$$\begin{aligned}
\gamma(T) &\leq e^{(T/4)\|\hat{\varphi}/\omega\|^2} \frac{\mathbb{E}_{\mathcal{N}_0}[1_{\mathfrak{X}\setminus A_T} e^{S[-T,T]}]}{\mathbb{E}_{\mathcal{N}_0}[e^{S[-T,T]}]} \\
&\leq e^{(T/4)\|\hat{\varphi}/\omega\|^2} \frac{(\mathbb{E}_{\mathcal{N}_0}[e^{S[-T,T]}])^{1/2}}{\mathbb{E}_{\mathcal{N}_0}[e^{S[-T,T]}]} (\mathbb{E}_{\mathcal{N}_0}[1_{\mathfrak{X}\setminus A_T}])^{1/2}.
\end{aligned} \tag{2.7.5}$$

Moreover, there exists a constant $\delta > 0$ such that

$$-T\delta\|\hat{\varphi}/\omega\|^2 \leq |S[-T,T]| \leq T\delta\|\hat{\varphi}/\omega\|^2.$$

Thus

$$\frac{(\mathbb{E}_{\mathcal{N}_0}[e^{2S[-T,T]}])^{1/2}}{\mathbb{E}_{\mathcal{N}_0}[e^{S[-T,T]}]} \leq e^{2T\delta\|\hat{\varphi}/\omega\|^2}.$$

To complete the proof, we argue that this exponential growth is balanced by the second factor of (2.7.5). The estimate

$$\mathbb{E}_{\mathcal{N}_0}[1_{\mathfrak{X}\setminus A_T}] \le T^{-\lambda}\sqrt{a+Tb}\exp(-cT^{\lambda(\beta+1)}) \tag{2.7.6}$$

is obtained in Lemma 2.68. Here $a, b, c > 0$. By choosing $1/(\beta+1) < \lambda < 1$, the right-hand side of (2.7.6) goes to zero as $T \to \infty$. □

Now we are ready to prove the absence of a ground state.

Proof of Theorem 2.63. We have

$$\gamma(T) \le \mathbb{E}_{\mathcal{N}_T}[e^{-\int_{-T}^0 ds \int_0^T W(X_s - X_t, s-t)dt}] = \mathbb{E}_{\mathcal{N}_T}[1_{A_T}F] + \mathbb{E}_{\mathcal{N}_T}[1_{\mathfrak{X}\setminus A_T}F],$$

where $F = e^{-\int_{-T}^0 ds \int_0^T W(X_s - X_t, s-t)dt}$. Since $\lim_{T\to\infty}\mathbb{E}_{\mu_T}[1_{A_T}] = 0$ by Lemma 2.65 and $\lim_{T\to\infty}\mathbb{E}_{\mu_T}[1_{\mathfrak{X}\setminus A_T}] = 0$ by Lemma 2.66, the theorem follows. □

To prove (2.7.6), we make use of the path properties of the $P(\phi)_1$-process $(X_t)_{t\ge 0}$. Note that

$$\mathbb{E}_{\mathcal{N}_0}[1_{\mathfrak{X}\setminus A_T}] = \mathcal{N}_0\left(\sup_{t\in[-T,T]}|X_t| \ge T^\lambda\right).$$

Lemma 2.68. *Estimate* (2.7.6) *holds.*

Proof. Suppose that $f \in C^\infty(\mathbb{R}^d)$ is an even function, and

$$f(x)\begin{cases} = |x|, & |x| \ge T^\lambda, \\ \le |x|, & T^\lambda - 1 < |x| < T^\lambda, \\ = 0, & |x| \le T^\lambda - 1. \end{cases}$$

We have

$$\mathbb{E}_{\mathcal{N}_0}[1_{\mathfrak{X}\setminus A_T}] = \mathbb{E}_{\mathcal{N}_0}[1_{\{\sup_{|s|<T}|X_s|>T^\lambda\}}] = \mathbb{E}_{\mathcal{N}_0}[1_{\{\sup_{|s|<T}|f(X_s)|>T^\lambda\}}].$$

By the reflection symmetry of $(X_t)_{t\in\mathbb{R}}$,

$$\mathbb{E}_{\mathcal{N}_0}[1_{\{\sup_{|s|<T}|f(X_s)|>T^\lambda\}}] = 2\mathbb{E}_{\mathcal{N}_0}[1_{\{\sup_{0\le s\le T}|f(X_s)|>T^\lambda\}}]$$

and by the Dirichlet principle,

$$\mathbb{E}_{\mathcal{N}_0}\left[\sup_{|s|<T}|f(X_s)| > T^\lambda\right] \le \frac{2e}{T^\lambda}\left((f,f)_{L^2(\mathrm{N}_0)} + T(L_\mathrm{p}^{1/2}f, L_\mathrm{p}^{1/2}f)_{L^2(\mathrm{N}_0)}\right)^{1/2}. \tag{2.7.7}$$

We estimate the right-hand side of (2.7.7). First, note that $f\Psi_\mathrm{g} \in D(H_\mathrm{p})$ and

$$H_\mathrm{p} f\Psi_\mathrm{g} = -\Delta f \cdot \Psi_\mathrm{g} - \nabla f \cdot \nabla\Psi_\mathrm{g} + E_\mathrm{p} f\Psi_\mathrm{g}.$$

From this, we also see that

$$(L_\mathrm{p}^{1/2}f, L_\mathrm{p}^{1/2}f)_{L^2(\mathrm{N}_0)} = (f\Psi_\mathrm{p}, -\Delta f \cdot \Psi_\mathrm{g} - \nabla f \cdot \nabla\Psi_\mathrm{p})_{L^2(\mathrm{N}_0)}.$$

Using the superexponential decay $\Psi_{\rm p}(x) \le e^{-C|x|^{\beta+1}}$, we obtain

$$\|f\Psi_{\rm p}\|^2 = \int_{\mathbb{R}^3} f(x)^2\Psi_{\rm p}^2(x)dx \le e^{-2CT^{\lambda(\beta+1)}} \int_{\mathbb{R}^3} |x|^2 e^{-2C|x|^{\beta+1}} dx.$$

Note that $\nabla f, \Delta f \in L^\infty(\mathbb{R}^3)$. We have

$$(L_{\rm p}^{1/2} f, L_{\rm p}^{1/2} f)_{L^2(\mathrm{N}_0)} \le C'\|f\Psi_{\rm p}\|(\|\nabla\Psi_{\rm p}\| + \|\Psi_{\rm p}\|) \le C'' e^{-CT^{\lambda(\beta+1)}}(\|\nabla\Psi_{\rm p}\| + \|\Psi_{\rm p}\|).$$

Similarly,

$$(f, f)_{L^2(\mathrm{N}_0)} \le e^{-2CT^{\lambda(\beta+1)}} \int_{\mathbb{R}^3} |x|^2 e^{-2\gamma|x|^{\beta+1}} dx.$$

Hence

$$\mathbb{E}_{\mathcal{N}_T}[1_{\mathfrak{X}\setminus A_T}] \le T^{-\lambda}\sqrt{a + Tbe^{-cT^{\lambda(\beta+1)}}}$$

with constants a, b, c. □

2.8 Gibbs measure associated with the ground state

2.8.1 Local convergence and Gibbs measures

In Theorem 2.37, we have shown that the sequence of measures $(\mathcal{N}_T)_{T\ge 0}$ is tight. The weak limit $\mathcal{N}$ of $\mathcal{N}_{T_n'}$ for a subsequence $(T_n')_{n\in\mathbb{N}}$ exists, and by using the probability measure $\mathcal{N}$ we were able to express and estimate a variety of ground state expectations in Section 2.6. In this section we show that another family of probability measures μ_T defined below has a limit μ_∞ as $T \to \infty$ in the local sense. We note that in this case no time subsequences will be needed. By using μ_∞, we can again derive further ground state expectations. In Section 2.8 we use Assumption 2.69 below.

Assumption 2.69. The following conditions hold.
(1) (Dispersion relation) $\omega(k) = \omega_\nu(k) = \sqrt{|k|^2 + \nu^2}$, $\nu \ge 0$.
(2) (Charge distribution) $\varphi \in \mathscr{S}'(\mathbb{R}^d)$, $\overline{\hat\varphi(k)} = \hat\varphi(-k)$ and $\hat\varphi/\omega, \hat\varphi/\sqrt{\omega} \in L^2(\mathbb{R}^d)$.
(3) (External potential) $V \in \mathcal{R}_{\rm Kato}$ or V is Kato-decomposable.
(4) (Ground state) $H_{\rm N}$ has the unique ground state $\Psi_{\rm g}$.

Note that $\Psi_{\rm g} > 0$. We consider a d-dimensional two-sided Brownian motion $(B_t)_{t\in\mathbb{R}}$ on the probability space $(\mathfrak{X}, \mathcal{B}(\mathfrak{X}), \mathcal{W}^x)$. Fix a nonnegative function $f \in L^2(\mathbb{R}^d)$ and let

$$\mathscr{L}_T = f(B_{-T})f(B_T)e^{\frac{1}{2}\int_{-T}^T ds\int_{-T}^T W(B_s-B_t,s-t)dt} e^{-\int_{-T}^T V(B_s)ds}. \tag{2.8.1}$$

Define the probability measure μ_T on $(\mathscr{X}, \mathcal{B}(\mathscr{X}))$ by

$$\mu_T(A) = \frac{1}{Z_T} \int_{\mathbb{R}^d} \mathbb{E}^x[1_A \mathscr{L}_T] dx, \quad A \in \mathcal{B}(\mathscr{X}), \tag{2.8.2}$$

where Z_T is the normalizing constant such that

$$\frac{1}{Z_T} \int_{\mathbb{R}^d} \mathbb{E}^x[\mathscr{L}_T] dx = 1.$$

We call the measure μ_T the *finite volume Gibbs measure* associated with the Nelson model. Using the functional integral representation, we see that

$$(f \otimes 1, e^{-2TH_N} f \otimes 1) = \int_{\mathbb{R}^d} \mathbb{E}^x[\mathscr{L}_T] dx.$$

Our goal in this section is to show that μ_T converges in the local sense as $T \to \infty$ and (an infinite volume) *Gibbs measure* exists. This method will also be useful in constructing a limit measure on a path space with jumps to discuss the relativistic Pauli–Fierz model in Section 3.9 and the spin-boson model in Section 4.4 below.

Let $\mathscr{G}_T = \sigma(B_r, -T \leq r \leq T)$ be the natural filtration of $(B_t)_{t\in\mathbb{R}}$. The set

$$\mathscr{G} = \bigcup_{s\geq 0} \mathscr{G}_s$$

is a finitely additive family of sets. We notice that

$$\mathcal{B}(\mathscr{X}) = \sigma(\mathscr{G}).$$

We consider the family of probability spaces $(\mathscr{X}, \mathcal{B}(\mathscr{X}), \mu_T)$, $T > 0$, where the probability measure μ_T is given by (2.8.2).

Definition 2.70 (Local convergence). Let μ_∞ be a probability measure on $(\mathscr{X}, \mathcal{B}(\mathscr{X}))$. The family of probability measures $\{\mu_T\}_{T\geq 0}$ on $(\mathscr{X}, \mathcal{B}(\mathscr{X}))$ is said to converge to μ_∞ in the *local* sense if $\mu_T(A) \to \mu_\infty(A)$ as $T \to \infty$, for every $A \in \mathscr{G}$.

We begin with an outline of showing local convergence. Let

$$\Psi_g^T = \frac{e^{-T(H_N - E_N)} f \otimes 1}{\|e^{-T(H_N - E_N)} f \otimes 1\|} \tag{2.8.3}$$

and

$$f_T = e^{-T(H_N - E_N)} f \otimes 1.$$

Note that in (2.5.6) we used Ψ_g^T given by (2.8.3) with f replaced by Ψ_p; in this section, however, Ψ_g^T is defined by (2.8.3). First, by using Ψ_g^T we define a family of probability measure ρ_T on $(\mathscr{X}, \mathscr{G}_T)$, $T > 0$. Using functional integration, we will see that

$$\rho_T(A) = \mu_T(A), \quad A \in \mathscr{G}_t,\ t \le T. \tag{2.8.4}$$

Next, by using the ground state Ψ_g, we define a finitely additive set function μ on $(\mathscr{X}, \mathscr{G})$, and with its extension to a probability measure μ_∞ on $(\mathscr{X}, \sigma(\mathscr{G}))$, we get the probability space

$$(\mathscr{X}, \sigma(\mathscr{G}), \mu_\infty).$$

Making use of the strong convergence of Ψ_g^T to Ψ_g as $T \to \infty$, we prove that

$$\rho_T(A) \to \mu(A), \quad T \to \infty,$$

for $A \in \mathscr{G}$, which, together with (2.8.4), will imply that

$$\mu_T(A) \to \mu_\infty(A), \quad A \in \mathscr{G}.$$

Through the construction of μ_∞, we will also obtain an explicit form of this probability measure for $A \in \mathscr{G}$.

Define

$$\mathrm{I}_{[-T,T]} = e^{-\int_{-T}^{T} V(B_s)ds}\, \mathrm{I}_{-T}^* e^{-\phi_{\mathrm{E}}(\int_{-T}^{T} \delta_s \otimes \varphi(\cdot - B_s)ds)} \mathrm{I}_T.$$

Note that $\mathrm{I}_{[-T,T]} : L^2(\mathscr{Q}) \to L^2(\mathscr{Q})$ is an almost surely bounded linear operator. Define a finitely additive set function $\mu : \mathscr{G} \to [0, \infty)$ by

$$\mu(A) = e^{2E_{\mathrm{N}}T} \int_{\mathbb{R}^d} \mathbb{E}^x \left[1_A (\Psi_g(B_{-T}), \mathrm{I}_{[-T,T]} \Psi_g(B_T))\right] dx, \quad A \in \mathscr{G}_T.$$

Lemma 2.71. *The set function μ is well defined, i. e., for $A \in \mathscr{G}_T \subset \mathscr{G}_S$,*

$$\begin{aligned}\mu(A) &= e^{2E_{\mathrm{N}}T} \int_{\mathbb{R}^d} \mathbb{E}^x \left[1_A (\Psi_g(B_{-T}), \mathrm{I}_{[-T,T]} \Psi_g(B_T))\right] dx \\ &= e^{2E_{\mathrm{N}}S} \int_{\mathbb{R}^d} \mathbb{E}^x \left[1_A (\Psi_g(B_{-S}), \mathrm{I}_{[-S,S]} \Psi_g(B_S))\right] dx.\end{aligned}$$

Proof. Let $\mu_{(T)} = \mu\lceil_{\mathscr{G}_T}$, which is also a probability measure on the measurable space $(\mathscr{X}, \mathscr{G}_T)$. Let $-S < -T = t_0 < t_1 < \cdots < t_n = T < S$, $n \in \mathbb{N}$, be an arbitrary set of time-points. By Theorem 2.15, the family of finite dimensional distributions is given by

$$\mu_{(T)}^{t_0,\dots,t_n}(A_0 \times \cdots \times A_n) = \mu_{(T)}(B_{t_0} \in A_0, \dots, B_{t_n} \in A_n)$$

$$= e^{2E_N T} \int_{\mathbb{R}^d} \mathbb{E}^x\Big[\Big(\prod_{j=0}^{n} 1_{A_j}(B_{t_j})\Big) (\Psi_g(B_{-T}), I_{[-T,T]}\Psi_g(B_T))\Big] dx$$
$$= \Big(\Psi_g, 1_{A_0} e^{-(t_1-t_0)(H_N-E_N)} \cdots e^{-(t_n-t_{n-1})(H_N-E_N)} 1_{A_n} \Psi_g\Big),$$

where $A_j \subset \mathcal{B}(\mathbb{R}), j = 1, \ldots, A_n$. Using the identity $e^{-(t_0+S)(H_N-E_N)}\Psi_g = \Psi_g$, we have

$$= \Big(\Psi_g, e^{-(t_0+S)(H_N-E_N)} 1_{A_0} \cdots 1_{A_n} e^{-(S-t_n)(H_N-E_N)} \Psi_g\Big)$$
$$= e^{2E_N S} \int_{\mathbb{R}^d} \mathbb{E}^x\Big[\Big(\prod_{j=0}^{n} 1_{A_j}(B_{t_j})\Big) (\Psi_g(B_{-S}), I_{[-S,S]}\Psi_g(B_S))\, dx\Big]$$
$$= \mu_{(S)}^{t_0,\ldots,t_n}(A_0 \times \cdots \times A_n).$$

Furthermore, the family of finite-dimensional distributions $\{\mu_{(T)}^{\Lambda}, \Lambda \subset [-T,T], \#\Lambda < \infty\}$ satisfies the Kolmogorov's consistency condition

$$\mu_{(T)}^{t_0,\ldots,t_n}(A_0 \times \cdots \times A_n) = \mu_{(T)}^{t_0,\ldots,t_n,t_{n+1},\ldots,t_{n+l}}\Big(A_0 \times \cdots \times A_n \times \prod^{l} \mathbb{R}^d\Big).$$

Let $\pi_\Lambda : \mathbb{R}^{[-T,T]} \to \mathbb{R}^{\#\Lambda}$ be a projection, where $\Lambda \subset [-T,T]$ and $\#\Lambda < \infty$. Define

$$\mathcal{A} = \Big\{\pi_\Lambda^{-1}(E) \in \mathbb{R}^{[-T,T]} \mid E \in \mathcal{B}(\mathbb{R}^{\#\Lambda}), \Lambda \subset [-T,T], \#\Lambda < \infty\Big\}.$$

By the Kolmogorov's extension theorem, there exists a unique measure m on the measurable space $(\mathbb{R}^{[-T,T]}, \sigma(\mathcal{A}))$ such that

$$m\big(\{\omega \in \mathbb{R}^{[-T,T]} \mid \omega(t_0) \in A_0, \ldots, \omega(t_n) \in A_{t_n}\}\big)$$
$$= \mu_{(T)}^{\Lambda}(A_0 \times \cdots \times A_{t_n}) = \mu_{(S)}^{\Lambda}(A_0 \times \cdots \times A_{t_n}),$$

where $\Lambda = \{t_0, \ldots, t_n\} \subset [-T,T] \subset [-S,S]$. Note that $\mathcal{A} \ni A_{t_1,\ldots,t_n}^{E_1,\ldots,E_n}$, with the cylinder sets

$$A_{t_1,\ldots,t_n}^{E_1,\ldots,E_n} = \Big\{\omega \in \mathfrak{X} \mid (\omega(t_0), \ldots, \omega(t_n)) \in \prod_{j=1}^{n} E_j, E_j \in \mathcal{B}(\mathbb{R}), j = 1, \ldots, n, n \geq 1\Big\}$$

and the σ-field generated by the cylinder sets as above coincides with $\mathcal{G}_T$. Also, $\sigma(\mathcal{A}) \supset \mathcal{G}_T$, and it follows that $\mu_{(S)}(A) = m(A) = \mu_{(T)}(A)$ for all $A \in \mathcal{G}_T$, by the uniqueness of the extension. □

Let $\mathfrak{X}_T = C([-T,T])$ and $\|f\|_T = \sup_{s\in[-T,T]} |f(s)|$. Then $(\mathfrak{X}_T, \|\cdot\|_T)$ is a complete metric space, hence a Polish space. Moreover, $(\mathfrak{X}, \rho_{\mathfrak{X}})$ is also a Polish space, where

$$\rho_{\mathfrak{X}}(f,g) = \sum_{n=1}^{\infty} \frac{1}{2^n} \|f-g\|_n \wedge 1, \quad f, g \in \mathfrak{X}.$$

We define a probability measure ν_T on $(\mathfrak{X}_T, \mathcal{B}(\mathfrak{X}_T))$ by

$$\nu_T(A) = e^{2E_N T} \int_{\mathbb{R}^d} \mathbb{E}^x \left[1_A(\Psi_g(B_{-T}), I_{[-T,T]}\Psi_g(B_T))\right] dx, \quad A \in \mathcal{B}(\mathfrak{X}_T).$$

Let $\mathcal{G}'_T = \mathcal{B}(\mathfrak{X}_T)$. Let $\pi_T : \mathfrak{X} \to \mathfrak{X}_T$ be the projection defined by $\pi_T f = f|_{[-T,T]}$. For $0 < S < T$, let $\pi_{S,T} : \mathfrak{X}_T \to \mathfrak{X}_S$ be also the projection defined by $\pi_{S,T}(f) = f|_{[-S,S]}$. We have $\pi_T^{-1}(\mathcal{G}'_T) = \mathcal{G}_T$. For $A \in \mathcal{G}$, there exists T such that $A \in \mathcal{G}_T = \pi_T^{-1}(\mathcal{G}'_T)$. We also have

$$\mu(A) = \nu_T(\pi_T(A)), \quad A \in \mathcal{G}_T.$$

This is well defined by Lemma 2.71.

Lemma 2.72. *The finitely additive set function μ is a countably additive measure on $(\mathfrak{X}, \mathcal{G})$.*

Proof. Suppose that $E_n \in \mathcal{G}$ satisfy $E_1 \supset E_2 \supset \cdots$ and $\lim_{n\to\infty} \mu(E_n) = \alpha > 0$. To prove that μ is countably additive, it suffices to show that $\bigcap_{n=1}^\infty E_n \neq \emptyset$. Since $E_n \in \mathcal{G}_{T_n}$, we can write $E_n = \pi_{T_n}^{-1}(E'_n)$ with $E'_n \in \mathcal{G}'_{T_n}$. We may assume that $T_n < T_{n+1}$ and $T_n \to \infty$. Since $\mathfrak{X}_T$ is a Polish space and ν_T is a probability measure, ν_T is regular. Hence, there exists a compact set $K'_n \subset E'_n$ such that $\nu_{T_n}(E'_n \setminus K'_n) < \alpha/2^n$. Let $K_n = \pi_{T_n}^{-1}(K'_n)$, $D_n = \bigcap_{j=1}^n K_j$, and $D = \bigcap_{n=1}^\infty D_n$. Since $D \subset \bigcap_{n=1}^\infty E_n$, it suffices to show that $D \neq \emptyset$. We have

$$\begin{aligned} \alpha - \mu(D_n) &\le \mu(E_n) - \mu(D_n) \le \mu(E_n \setminus D_n) = \mu\Big(\bigcup_{j=1}^n (E_n \setminus K_j)\Big) = \mu\Big(\pi_{T_n}^{-1}\Big(\bigcup_{j=1}^n (E'_n \setminus K'_j)\Big)\Big) \\ &= \nu_{T_n}\Big(\bigcup_{j=1}^n (E'_n \setminus K'_j)\Big) \le \sum_{j=1}^n \nu_{T_n}(E'_n \setminus K'_j) \le \sum_{j=1}^n \nu_{T_j}(E'_j \setminus K'_j) = \sum_{j=1}^n \frac{\alpha}{2^j} < \alpha. \end{aligned}$$

Thus $0 < \mu(D_n)$, and hence $D_n \neq \emptyset$. Take $f_n \in D_n$, i. e., $f_n \in \bigcap_{j=1}^n K_j$. Then for any $\ell \le n$, we have $f_n \in K_\ell$. Fix $\ell = 1$. Then $\pi_{T_1}(f_n) \in K'_1$ for all $n \ge 1$. Since K'_1 is compact, there exists a subsequence $(f_{n'})$ such that $\pi_{T_1}(f_{n'}) \to h_1 \in K'_1$. Extracting subsequences inductively, we obtain a subsequence $(f_m)_m$ such that $\pi_{T_\ell}(f_m) \to h_\ell \in K'_\ell$ for every ℓ. Since $\pi_{T_\ell}(f_m) = \pi_{T_\ell,T_{\ell+1}}(\pi_{T_{\ell+1}}(f_m))$, passing to the limit yields $h_\ell = \pi_{T_\ell,T_{\ell+1}}(h_{\ell+1})$. Hence there exists $g \in \mathfrak{X}$ satisfying $\pi_{T_\ell}(g) = h_\ell$ for all ℓ. Since $h_\ell \in K'_\ell$, we have $g \in K_\ell = \pi_{T_\ell}^{-1}(K'_\ell)$ for every ℓ. Therefore $g \in \bigcap_{\ell=1}^\infty K_\ell \subset D$, and hence $D \neq \emptyset$. □

By Lemma 2.72 above, there exists a unique probability measure μ_∞ on $(\mathfrak{X}, \sigma(\mathcal{G}))$ such that $\mu_\infty(A) = \mu(A)$ for $A \in \mathcal{G}$, by the Hopf's extension theorem. We define the set function $\rho_T : \mathcal{G}_T \to [0, \infty)$ by

$$\rho_T(A) = e^{2E_N s} \int_{\mathbb{R}^d} \mathbb{E}^x \left[1_A\Big(\frac{f_{T-s}(B_{-s})}{\|f_T\|}, I_{[-s,s]}\frac{f_{T-s}(B_s)}{\|f_T\|}\Big)\right] dx$$

for $A \in \mathcal{G}_s$ with $s \le T$.

Lemma 2.73. *The set function ρ_T is well defined, and*

$$\begin{aligned}\rho_T(A) &= e^{2E_N r}\int_{\mathbb{R}^d}\mathbb{E}^x\Big[1_A\Big(\frac{f_{T-r}(B_{-r})}{\|f_T\|},I_{[-r,r]}\frac{f_{T-r}(B_r)}{\|f_T\|}\Big)\Big]dx\\ &= e^{2E_N s}\int_{\mathbb{R}^d}\mathbb{E}^x\Big[1_A\Big(\frac{f_{T-s}(B_{-s})}{\|f_T\|},I_{[-s,s]}\frac{f_{T-s}(B_s)}{\|f_T\|}\Big)\Big]dx, \end{aligned}\tag{2.8.5}$$

holds for all $r \le s \le T$ and $A \in \mathscr{G}_r$.

Proof. The proof of the second statement is similar to the proof of Lemma 2.71. Denote the left-hand side of (2.8.5) by $\rho_{(r)}(A)$, and the right-hand side by $\rho_{(s)}(A)$. The finite-dimensional distributions of $\rho_{(r)}$ are given by

$$\begin{aligned}\rho_{(r)}^{t_0,\dots,t_n}(A_0\times\cdots\times A_n) &= \rho_{(r)}(B_{t_0}\in A_0,\dots,B_{t_n}\in A_n)\\ &= \frac{e^{2E_N r}}{\|f_T\|^2}\int_{\mathbb{R}^d}\mathbb{E}^x\Big[\Big(\prod_{j=0}^n 1_{A_j}(B_{t_j})\Big)\big(f_{T-r}(B_{-r}),I_{[-r,r]}f_{T-r}(B_r)\big)\Big]dx.\end{aligned}$$

By Theorem 2.15, the right-hand side above can be represented as

$$\begin{aligned}&= \frac{\big(f_{T-r},e^{-(t_0+r)(H_N-E_N)}1_{A_0}e^{-(t_1-t_0)(H_N-E_N)}\cdots 1_{A_n}e^{-(r-t_n)(H_N-E_N)}f_{T-r}\big)}{\|f_T\|^2}\\ &= \frac{\big(f\otimes 1,e^{-(T+t_0)(H_N-E_N)}1_{A_0}e^{-(t_1-t_0)(H_N-E_N)}\cdots 1_{A_n}e^{-(T-t_n)(H_N-E_N)}f\otimes 1\big)}{\|f_T\|^2}\\ &= \frac{\big(f_{T-s},e^{-(t_0+s)(H_N-E_N)}1_{A_0}e^{-(t_1-t_0)(H_N-E_N)}\cdots 1_{A_n}e^{-(s-t_n)(H_N-E_N)}f_{T-s}\big)}{\|f_T\|^2}\\ &= \frac{e^{2E_N s}}{\|f_T\|^2}\int_{\mathbb{R}^d}\mathbb{E}^x\Big[\Big(\prod_{j=0}^n 1_{A_j}(B_{t_j})\Big)\big(f_{T-s}(B_{-s}),I_{[-s,s]}f_{T-s}(B_s)\big)\Big]dx\\ &= \rho_{(s)}^{t_0,\dots,t_n}(A_0\times\cdots\times A_n).\end{aligned}$$

Note that both $\rho_{(r)}^\Lambda$ and $\rho_{(s)}^\Lambda$, $\Lambda\subset[-T,T]$, $\#\Lambda<\infty$, satisfy the consistency condition. Note also that $\rho_{(r)}\lceil_{\mathscr{G}_r}$ and $\rho_{(s)}\lceil_{\mathscr{G}_r}$ are probability measures on $(\mathscr{X},\mathscr{G}_r)$. By the Kolmogorov's extension theorem, we see that $\rho_{(r)}(A)=\rho_{(s)}(A)$ for every $A\in\mathscr{G}_r\subset\mathscr{G}_s$, which proves the lemma. □

By Lemma 2.73, ρ_T is a probability measure on $(\mathscr{X},\mathscr{G}_T)$.

Lemma 2.74. *Let $s\le T$ and $A\in\mathscr{G}_s$. Then $\rho_T(A)=\mu_T(A)$.*

Proof. For $\Lambda=\{t_0,t_1,\dots,t_n\}\subset[-s,s]$ and $A_0\times\cdots\times A_n\in\times_{j=0}^n\mathcal{B}(\mathbb{R}^d)$, we define

$$\rho_T^\Lambda(A_0\times\cdots\times A_n)=\rho_T(B_{t_0}\in A_0,\dots,B_{t_n}\in A_n)$$

$$= \frac{e^{2E_N s}}{\|f_T\|^2} \int_{\mathbb{R}^d} \mathbb{E}^x \left[\left(\prod_{j=0}^n 1_{A_j}(B_{t_j}) \right) (f_{T-s}(B_{-s}), \mathrm{I}_{[-s,s]} f_{T-s}(B_s)) \right] dx$$

and

$$\mu_T^\Lambda(A_0 \times \cdots \times A_n) = \mu_T(B_{t_0} \in A_0, \ldots, B_{t_n} \in A_n) = \frac{1}{Z_T} \int_{\mathbb{R}^d} \mathbb{E}^x \left[\left(\prod_{j=0}^n 1_{A_j}(B_{t_j}) \right) \mathscr{L}_T \right] dx.$$

Both ρ_T^Λ and μ_T^Λ are probability measures on $((\mathbb{R}^d)^{\#\Lambda}, \mathcal{B}(\mathbb{R}^d)^{\#\Lambda})$. We have

$$\begin{aligned} \mu_T^\Lambda(A_0 \times \cdots \times A_n) &= \frac{(f \otimes 1, e^{-(t_0+T)H_N} 1_{A_0} e^{-(t_1-t_0)H_N} 1_{A_1} \cdots 1_{A_n} e^{-(T-t_n)H_N} f \otimes 1)}{\|f_T\|^2} \\ &= \frac{e^{2E_N s} (f_{T-s}, e^{-(t_0+s)H_N} 1_{A_0} e^{-(t_1-t_0)H_N} 1_{A_1} \cdots 1_{A_n} e^{-(s-t_n)H_N} f_{T-s})}{\|f_T\|^2} \end{aligned}$$

by the definition of f_{T-s}. The right-hand side above can be expressed as

$$= e^{2E_N s} \int_{\mathbb{R}^d} \mathbb{E}^x \left[\left(\prod_{j=0}^n 1_{A_j}(B_j) \right) \left(\frac{f_{T-s}(B_0)}{\|f_T\|}, \mathrm{I}_{[-s,s]} \frac{f_{T-s}(B_s)}{\|f_T\|} \right) \right] dx.$$

It follows that $\rho_T^\Lambda(A_0 \times \cdots \times A_n) = \mu_T^\Lambda(A_0 \times \cdots \times A_n)$. Note that

$$\begin{aligned} \mu_T \lceil_{\mathscr{G}_T} (B_{t_0} \in A_0 \times \cdots \times B_{t_n} \in A_n) &= \mu_T^{t_0, \ldots, t_n}(A_0 \times \cdots \times A_n) \\ &= \rho_T^{t_0, \ldots, t_n}(A_0 \times \cdots \times A_n) = \rho_T(B_{t_0} \in A_0 \times \cdots \times B_{t_n} \in A_n). \end{aligned}$$

The probability measures μ_T^Λ and ρ_T^Λ satisfy the Kolmogorov's consistency condition. By the Kolmogorov's extension theorem, $\rho_T = \mu_T \lceil_{\mathscr{G}_T}$ follows from the uniqueness of the extension. ◻

Lemma 2.75. *If $A \in \mathscr{G}$, then $\lim_{T\to\infty} \mu_T(A) = \mu_\infty(A)$.*

Proof. Suppose that $A \in \mathscr{G}_s$ for some s. By Lemma 2.74, we have

$$\lim_{T\to\infty} \mu_T(A) = \lim_{T\to\infty} \rho_T(A) = \lim_{T\to\infty} e^{2E_N s} \int_{\mathbb{R}^d} \mathbb{E}^x \left[1_A \left(\frac{f_{T-s}(B_{-s})}{\|f_T\|}, \mathrm{I}_{[-s,s]} \frac{f_{T-s}(B_s)}{\|f_T\|} \right) \right] dx.$$

Since $f_T / \|f_T\| \to \Psi_g$ strongly as $T \to \infty$, we have

$$\lim_{T\to\infty} \mu_T(A) = e^{2E_N s} \int_{\mathbb{R}^d} \mathbb{E}^x \left[1_A \left(\Psi_g(B_{-s}), \mathrm{I}_{[-s,s]} \Psi_g(B_s) \right) \right] dx = \mu_\infty(A). \tag{2.8.6}$$

◻

Remark 2.76. In a similar way to Lemma 2.71, it is seen that for $A \in \mathscr{F}^{\mathrm{BM}}_{[a,b]} \subset \mathscr{F}^{\mathrm{BM}}_{[-S,S]}$,

$$\mu(A) = e^{E_{\mathrm{N}}(b-a)} \int_{\mathbb{R}^d} \mathbb{E}^x \left[1_A(\Psi_{\mathrm{g}}(B_a), \mathrm{I}_{[a,b]} \Psi_{\mathrm{g}}(B_b)) \right] dx$$
$$= e^{2E_{\mathrm{N}}S} \int_{\mathbb{R}^d} \mathbb{E}^x \left[1_A(\Psi_{\mathrm{g}}(B_{-S}), \mathrm{I}_{[-S,S]} \Psi_{\mathrm{g}}(B_S)) \right] dx.$$

In addition to this, we can also show that

$$\rho_T(A) = e^{E_{\mathrm{N}}(b-a)} \int_{\mathbb{R}^d} \mathbb{E}^x \left[1_A \left(\frac{f_{T+a}(B_a)}{\|f_T\|}, \mathrm{I}_{[a,b]} \frac{f_{T-b}(B_b)}{\|f_T\|} \right) \right] dx$$

holds for all $-T < a < b < T$ and $A \in \mathscr{F}^{\mathrm{BM}}_{[a,b]}$. From these facts, for $A \in \mathscr{F}^{\mathrm{BM}}_{[a,b]}$ we get $\lim_{T\to\infty} \mu_T(A) = \mu_\infty(A)$ and

$$\mu_\infty(A) = e^{E_{\mathrm{N}}(b-a)} \int_{\mathbb{R}^d} \mathbb{E}^x \left[1_A \left(\Psi_{\mathrm{g}}(B_a), \mathrm{I}_{[a,b]} \Psi_{\mathrm{g}}(B_b) \right) \right] dx. \tag{2.8.7}$$

Finally, we obtain the local convergence of $\{\mu_T\}_{T>0}$.

Theorem 2.77 (Local convergence). *The family of probability measures $\{\mu_T\}_{T\geq 0}$ converges to μ_∞ on $(\mathscr{X}, \mathcal{B}(\mathscr{X}))$ in the local sense, i. e., $\mu_T(A) \to \mu_\infty(A)$ as $T \to \infty$ for all $A \in \mathscr{G}$, and the limit measure μ_∞ is independent of f.*

Proof. By Lemma 2.75, it follows that $\mu_T(A) \to \mu_\infty(A)$ for $A \in \mathscr{G}$. To show that μ_∞ is independent of the choice of f, suppose that μ'_∞ is a local limit of μ'_T defined by μ_T with f replaced by f' such that $0 \leq f'$ and $f' \in L^2(\mathbb{R}^d)$. By the construction of μ_∞, we have $\mu_\infty(A) = \mu'_\infty(A)$ for $A \in \mathscr{G}$ by (2.8.6). Uniqueness of the extension implies that $\mu_\infty = \mu'_\infty$, thus μ_∞ is independent of f. □

The probability measure μ_∞ is the *Gibbs measure* associated with the Nelson model.

Corollary 2.78. *Let f, g be bounded measurable functions on $\mathbb{R}^d$. Then it follows that for $a \leq b$,*

$$\mathbb{E}_{\mu_\infty}[f(B_a)g(B_b)] = (f\Psi, e^{-(b-a)(H_{\mathrm{N}}-E_{\mathrm{N}})} g\Psi). \tag{2.8.8}$$

Proof. Let $G_1, G_2 \in \mathcal{B}(\mathbb{R}^d)$. We have $1_{G_1}(B_a)1_{G_2}(B_b) = 1_{B_a^{-1}(G_1)\cap B_b^{-1}(G_2)}$ and by Remark 2.76 we find that

$$\mathbb{E}_{\mu_\infty}[1_{G_1}(B_a)1_{G_2}(B_b)] = \mu_\infty(B_a^{-1}(G_1) \cap B_b^{-1}(G_2))$$
$$= e^{E_{\mathrm{N}}(b-a)} \int_{\mathbb{R}^d} \mathbb{E}^x \left[1_{B_a^{-1}(G_1)\cap B_b^{-1}(G_2)} \left(\Psi_{\mathrm{g}}(B_a), \mathrm{I}_{[a,b]} \Psi_{\mathrm{g}}(B_b) \right) \right] dx$$

$$= e^{E_N(b-a)} \int_{\mathbb{R}^d} \mathbb{E}^x \left[(1_{G_1}(B_a)\Psi_g(B_a), I_{[a,b]} 1_{G_2}(B_b)\Psi_g(B_b)) \right] dx$$

$$= (1_{G_1}\Psi, e^{-(b-a)(H_N - E_N)} 1_{G_2}\Psi).$$

Hence a limiting argument yields the corollary. □

By a limiting argument, (2.8.8) can be extended to unbounded functions.

Corollary 2.79. *Suppose that $\Psi \in D(|x|)$. Then it follows that for $a \le b$,*

$$\mathbb{E}_{\mu_\infty}[|B_a||B_b|] = (|x|\Psi, e^{-(b-a)(H_N - E_N)}|x|\Psi).$$

Furthermore it follows that

$$\mathbb{E}_{\mu_\infty}[B_a \cdot B_b] = (x\Psi, e^{-(b-a)(H_N - E_N)} x\Psi).$$

Proof. Let $f_n(x) = |x|$ for $|x| \le n$ and $f_n(x) = n$ for $|x| \ge n$. Since f_n is bounded,

$$\mathbb{E}_{\mu_\infty}[f_n(B_a) f_n(B_b)] = (f_n\Psi, e^{-(b-a)(H_N - E_N)} f_n\Psi).$$

Since $e^{-(b-a)(H_N - E_N)}$ is positivity improving and

$$(f_n\Psi, e^{-(b-a)(H_N - E_N)} f_n\Psi) \le (|x|\Psi, e^{-(b-a)(H_N - E_N)}|x|\Psi),$$

the monotone convergence theorem yields that

$$\lim_{n\to\infty} (f_n\Psi, e^{-(b-a)(H_N - E_N)} f_n\Psi) = (|x|\Psi, e^{-(b-a)(H_N - E_N)}|x|\Psi).$$

Hence the monotone convergence theorem yields again that

$$\lim_{n\to\infty} \mathbb{E}_{\mu_\infty}[f_n(B_a) f_n(B_b)] = \mathbb{E}_{\mu_\infty}[|B_a||B_b|]$$

and the first statement follows. The second statement can be also show in a similar manner. □

2.8.2 $P(\phi)_1$-process associated with the Nelson Hamiltonian

In Section 2.8.1 we defined the probability measure μ_∞ on $(\mathscr{X}, \mathcal{B}(\mathscr{X}))$ by using the ground state Ψ_g of H_N. This can be extended to a probability measure $\hat{\mu}$ on $\mathscr{X} \times \mathfrak{Y}$. Recall that $\mathfrak{Y} = C(\mathbb{R}; \mathscr{M}_{-2})$ and $(\xi_t)_{t\in\mathbb{R}}$ is the infinite-dimensional Ornstein–Uhlenbeck process on $(\mathfrak{Y}, \mathcal{B}(\mathfrak{Y}))$. See (1.6.24). We write

$$M = \mathbb{R}^d \times \mathscr{M}_{-2}$$

and define the measure

$$dxdG = dM$$

on M. We regard $\mathfrak{X} \times \mathfrak{Y}$ as $C(\mathbb{R}; M)$, where the metric ρ is introduced by the sum of those of $\mathfrak{X}$ and $\mathfrak{Y}$, i. e., $\rho = \rho_{\mathfrak{X}} + \rho_{\mathfrak{Y}}$. See (1.6.25) for $\rho_{\mathfrak{Y}}$. Then $\mathfrak{X} \times \mathfrak{Y}$ is a Polish space. Also, we define an M-valued random process on $(\mathfrak{X} \times \mathfrak{Y}, \mathcal{B}(\mathfrak{X} \times \mathfrak{Y}), \mathcal{W}^0 \times \mathcal{G})$ by

$$X_t = (B_t, \xi_t).$$

Denote

$$\mathrm{I}_{[a,b]} = e^{-\int_a^b V(B_s)ds} e^{-\int_a^b \xi_s(\varphi(\cdot - B_s))ds}.$$

Using $\mathrm{I}_{[a,b]}$, we have the functional integral representation

$$(F, e^{-(b-a)H_{\mathrm{N}}} G)_{L^2(M)} = \int_M \mathbb{E}^{x,\xi}_{\mathcal{W}\times\mathcal{G}} \left[(F(X_a), \mathrm{I}_{[a,b]} G(X_b)) \right] dM$$

of the semigroup defined on $L^2(M)$. Let

$$\mathcal{F}_{[a,b]} = \sigma(X_r, r \in [a,b]).$$

We define the set function $\hat{\mu}$ by

$$\hat{\mu}(A) = e^{E_{\mathrm{N}}(b-a)} \int_M \mathbb{E}^{x,\xi}_{\mathcal{W}\times\mathcal{G}} \left[1_A \Psi_{\mathrm{g}}(X_a) \mathrm{I}_{[a,b]} \Psi_{\mathrm{g}}(X_b) \right] dM, \quad A \in \mathcal{F}_{[a,b]}.$$

It is seen that $\hat{\mu}$ is well defined in the sense that

$$\begin{aligned} & e^{E_{\mathrm{N}}(b-a)} \int_M \mathbb{E}^{x,\xi}_{\mathcal{W}\times\mathcal{G}} \left[1_A \Psi_{\mathrm{g}}(X_a), \mathrm{I}_{[a,b]} \Psi_{\mathrm{g}}(X_b) \right] dM \\ & \quad = e^{E_{\mathrm{N}}(b'-a')} \int_M \mathbb{E}^{x,\xi}_{\mathcal{W}\times\mathcal{G}} \left[1_A \Psi_{\mathrm{g}}(X_{a'}) \mathrm{I}_{[a',b']} \Psi_{\mathrm{g}}(X_{b'}) \right] dM \end{aligned}$$

for a', b' such that $[a,b] \subset [a',b']$. This can be proven in the same way as Lemma 2.73. Moreover, it is shown that $\hat{\mu}$ is countably additive. Consider

$$\hat{\mathcal{G}} = \bigcup_{T \geq 0} \mathcal{F}_{[-T,T]}$$

Note that $\sigma(\hat{\mathcal{G}}) = \mathcal{B}(\mathfrak{X} \times \mathfrak{Y})$. The set function $\hat{\mu}$ can be extended to a probability measure on $(\mathfrak{X} \times \mathfrak{Y}, \sigma(\hat{\mathcal{G}}))$, which we denote by $\hat{\mu}_\infty$. Let

$$\hat{\mu}_T(A) = e^{E_N(b-a)} \int_M \mathbb{E}^{x,\xi}_{\mathcal{W}\times\mathcal{G}}\left[1_A \frac{f_{T+a}(X_a)}{\|f_T\|}, I_{[a,b]} \frac{f_{T-b}(X_b)}{\|f_T\|}\right] dM,$$

where $f_T = e^{-T(H_N - E_N)} f \otimes 1$ with $0 \le f$ and $f \in L^2(\mathbb{R}^d)$. We note that $f_{T+a}/\|f_T\| \to e^{aE_N}\Psi_g$ as $T \to \infty$. We have the following result.

Theorem 2.80 (Local convergence). *The family of probability measures $\{\hat{\mu}_T\}_{T\ge 0}$ converges to $\hat{\mu}_\infty$ on $(\mathfrak{X}\times\mathfrak{Y}, \mathcal{B}(\mathfrak{X}\times\mathfrak{Y}))$ in the local sense, i. e., $\hat{\mu}_T(A) \to \hat{\mu}_\infty(A)$ as $T \to \infty$ for all $A \in \hat{\mathscr{G}}$, and the limit measure μ is independent of f.*

Proof. The proof is similar to that of Lemma 2.75 and is left to the reader. □

Note that

$$(f\Psi_g, e^{-|s-t|(H_N - E_N)} g\Psi_g)_{L^2(M)} = (f, e^{-|s-t|L} g)_{L^2(M, \Psi^2 dM)},$$

where

$$L = \frac{1}{\Psi_g}(H_N - E_N)\Psi_g.$$

Hence by the definition of $\hat{\mu}$ we see that for $0 \le t_0 \le t_1 \le \cdots \le t_n$ and bounded functions $f_0, \ldots, f_n$ and f,

$$\mathbb{E}_{\hat{\mu}}\left[\prod_{j=0}^n f_j(X_{t_j})\right] = (f_0, e^{-(t_1-t_0)L} f_1 \cdots f_{n-1} e^{-(t_n-t_{n-1})L} f_n)_{L^2(M, \Psi_g^2 dM)}, \tag{2.8.9}$$

$$\mathbb{E}_{\hat{\mu}}[f(X_s)] = (1, f)_{L^2(M, \Psi_g^2 dM)}, \quad s \ge 0. \tag{2.8.10}$$

Theorem 2.81 ($P(\phi)_1$-process). *The random process $(X_t)_{t\ge 0}$ on the probability space $(\mathfrak{X} \times \mathfrak{Y}, \mathcal{B}(\mathfrak{X} \times \mathfrak{Y}), \hat{\mu}_\infty)$ is a Markov process under the natural filtration, with invariant measure $\Psi_g^2 dM$.*

Proof. Define $p(t, X, A) = (e^{-tL} 1_A)(X)$ for all $t \ge 0$, $X \in M$ and $A \in \mathcal{B}(M)$. Clearly, $p(t, X, M) = 1$ and $p(t, \cdot, A)$ is Borel measurable. Writing $p(t, X, A) = \int_M 1_A(Y) p(t, X, dY)$ gives $\int_M f(Y) p(t, X, dY) = (e^{-tL} f)(X)$ for every bounded function f. With the semigroup property $e^{-sL} e^{-tL} 1_A(X) = e^{-(s+t)L} 1_A(X)$, we have

$$\begin{aligned} p(s+t, X, A) &= e^{-(s+t)L} 1_A(X) = e^{-sL} e^{-tL} 1_A(X) \\ &= \int_M e^{-tL} 1_A(Y) p(s, X, dY) = \int_M p(t, Y, A) p(s, X, dY), \end{aligned}$$

i. e., the Chapman–Kolmogorov identity is verified. Hence $p : [0, \infty) \times M \times \mathcal{B}(M) \to \mathbb{R}$ is a probability transition kernel. By (2.8.9), we also see that

$$\mathbb{E}_{\hat{\mu}_\infty}[1_{A_0}(X_0) \cdots 1_{A_n}(X_{t_n})] = (1_{A_0}, e^{-(t_1-t_0)L} 1_{A_1} \cdots 1_{A_{n-1}} e^{-(t_n-t_{n-1})L} 1_{A_n})_{L^2(M, \Psi_g^2 dM)}$$

$$= \int_{M^{(n+1)}} \Big(\prod_{j=0}^{n} 1_{A_j}(X_j)\Big)\Big(\prod_{j=1}^{n} p(t_j - t_{j-1}, X_{j-1}, dX_j)\Big) 1_{A_0}(X_0)\Psi_{\mathrm{g}}^2(X_0)dM(X_0).$$

Thus $(X_t)_{t\geq 0}$ is a Markov process with a probability transition kernel $p(t, X, A)$. Moreover, we have

$$\int_M p(t,X,A)\Psi_{\mathrm{g}}^2(X)dM = (1, e^{-tL}1_A)_{L^2(M,\Psi_{\mathrm{g}}^2 dM)} = (\Psi_{\mathrm{g}}, e^{-t(H_{\mathrm{N}}-E_{\mathrm{N}})}1_A\Psi_{\mathrm{g}})_{L^2(M)}$$
$$= (\Psi_{\mathrm{g}}, 1_A\Psi_{\mathrm{g}})_{L^2(M)} = \int_M 1_A(X)\Psi_{\mathrm{g}}^2(X)dM,$$

and thus $\Psi_{\mathrm{g}}^2 dM$ is an invariant measure for $(X_t)_{t\geq 0}$. □

The random process $(X_t)_{t\geq 0}$ is called the *$P(\phi)_1$-process* associated with the Nelson Hamiltonian.

2.8.3 Applications to ground state expectations

Likewise as with $\mathcal{N}$, the Gibbs measure μ_∞ associated with the Nelson model can be used to derive ground state expectations. As seen in (1.5.7) in Chapter 1, in order to factorize $e^{-t\hat{H}_{\mathrm{f}}}$, we needed an extra Hilbert space $\mathcal{H}_E$ in addition to $\mathcal{H}_M$, and the map $\tau_t : \mathcal{H}_M \to \mathcal{H}_E$ so that $\tau_s^*\tau_t = e^{-|t-s|\hat{\omega}}$. This proved to be a key formula allowing us to write $\mathrm{I}_t = \Gamma(\tau_t)$ and $\mathrm{I}_s^*\mathrm{I}_t = e^{-|t-s|\hat{H}_{\mathrm{f}}}$, which were useful in deriving a functional representation. Using the same idea, we construct a functional integral representation of $(e^{-TH_{\mathrm{N}}}F, e^{-\beta d\Gamma(\rho)}e^{-TH_{\mathrm{N}}}F)$ here, with a nonnegative measurable function ρ. To achieve this, we use the extra Hilbert space

$$\mathcal{H}_\rho = L^2(\mathbb{R}^{d+2})$$

and its Fourier image

$$\hat{\mathcal{H}}_\rho = FL^2(\mathbb{R}^{d+2}),$$

where F denotes Fourier transform on $\mathcal{H}_\rho$. We will denote the scalar products on $\hat{\mathcal{H}}_\rho$ and $\mathcal{H}_\rho$ by $(\cdot,\cdot)_{\hat{\rho}}$ and $(\cdot,\cdot)_\rho$, respectively. We define a family of Gaussian random variables $\{\phi_\rho(f), f \in L^2_{\mathrm{real}}(\mathbb{R}^{d+2})\}$ on a probability space $(\mathcal{Q}_\rho, \Sigma_\rho, \mu_\rho)$ indexed by $L^2_{\mathrm{real}}(\mathbb{R}^{d+2})$. For $\hat{f} \in \hat{\mathcal{H}}_\rho$, the variables of $\hat{f}$ will be denoted by $(k, k_0, k_1) \in \mathbb{R}^d \times \mathbb{R} \times \mathbb{R}$. Also, we define a family of isometries $q_s : \mathcal{H}_E \to \mathcal{H}_\rho$ by

$$\widehat{q_s f}(k, k_0, k_1) = \frac{\sqrt{2}}{\sqrt{\pi}} \frac{e^{-isk_1}}{\sqrt{\omega(k)^2 + |k_0|^2}} \sqrt{\frac{\rho(k)}{\rho(k)^2 + |k_1|^2}}\hat{f}(k, k_0).$$

Lemma 2.82. *It follows that*

$$q_s^* q_t = e^{-|s-t|\rho(-i\nabla)\otimes 1}, \quad s,t \in \mathbb{R}.$$

Here we used the identification $L^2(\mathbb{R}^{d+1}) \cong L^2(\mathbb{R}^d) \otimes L^2(\mathbb{R})$.

Proof. We have

$$(q_s f, q_t g)_{\mathcal{H}_\rho} = (\widehat{q_s f}, \widehat{q_t g})_{\hat{\mathcal{H}}_\rho} = \frac{2}{\pi} \int_{\mathbb{R}^{d+2}} \overline{\hat{f}(k,k_0)} \hat{g}(k,k_0) \frac{\rho(k)}{\omega(k)^2 + |k_0|^2} \frac{e^{i(s-t)k_1}}{\rho(k)^2 + |k_1|^2} dk dk_0 dk_1. \tag{2.8.11}$$

Using the formula

$$\frac{1}{\pi} \int_{\mathbb{R}} \frac{e^{i(s-t)k_1}}{\rho(k)^2 + |k_1|^2} dk_1 = \frac{e^{-|s-t|\rho(k)}}{\rho(k)}$$

and integrating (2.8.11) with respect to k_1, we have

$$(q_s f, q_t g)_{\mathcal{H}_\rho} = 2 \int_{\mathbb{R}^{d+1}} \overline{\hat{f}(k,k_0)} \hat{g}(k,k_0) \frac{e^{-|s-t|\rho(k)}}{\omega(k)^2 + |k_0|^2} dk dk_0 = (\hat{f}, e^{-|s-t|\rho\otimes 1} \hat{g})_{\hat{\mathcal{H}}_E}.$$ □

Next we define a family of second quantizations Ξ_s by

$$\Xi_t = \Gamma_{\mathrm{int}}(q_t) : L^2(\mathscr{Q}) \to L^2(Q_\rho), \quad t \in \mathbb{R}.$$

Let $\hat{\rho} = \rho(-i\nabla)$. We see that Ξ_s is an isometry for every t and, furthermore, it factorizes $e^{-t d\Gamma(\hat{\rho}\otimes 1)}$.

Lemma 2.83. *It follows that*

$$\Xi_s^* \Xi_t = e^{-|t-s| d\Gamma(\hat{\rho}\otimes 1)}, \quad s,t \in \mathbb{R}, \tag{2.8.12}$$

and the intertwining property

$$\mathrm{I}_t e^{-s d\Gamma(\hat{\rho})} = e^{-s d\Gamma(\hat{\rho}\otimes 1)} \mathrm{I}_t, \quad s \geq 0, t \in \mathbb{R} \tag{2.8.13}$$

holds.

Proof. The factorization formula (2.8.12) follows from the identity $q_s^* q_t = e^{-|s-t|\hat{\rho}\otimes 1}$. It is seen that

$$(f, \tau_t e^{-s\hat{\rho}} g)_{\mathcal{H}_E} = \frac{\sqrt{2}}{\sqrt{\pi}} \int_{\mathbb{R}^{d+1}} \frac{e^{-s\rho(k)} e^{-itk_0} \overline{\hat{f}(k,k_0)} \hat{g}(k,k_0)}{\omega(k)^2 + |k_0|^2} dk dk_0$$

$$= (\hat{f}, e^{-s\rho\otimes 1}\widehat{\tau_t g})_{\hat{\mathcal{H}}_E} = (f, e^{-s\hat{\rho}\otimes 1}\tau_t g)_{\mathcal{H}_E}.$$

Hence $\tau_t e^{-s\rho} = e^{-s\rho\otimes 1}\tau_t$, and from this (2.8.13) follows. □

Theorem 2.84. *Let ρ be a positive function on $\mathbb{R}^d$. For every $F, G \in L^2(\mathcal{Q})$ and $\beta > 0$, we have*

$$(e^{-TH_{\mathrm{N}}}F, e^{-\beta d\Gamma(\hat{\rho})}e^{-TH_{\mathrm{N}}}G) = \int_{\mathbb{R}^d} \mathbb{E}^x\Big[e^{-\int_{-T}^T V(B_s)ds}\big(e^{-\phi_{\mathrm{E}}(L_{-T})}\mathrm{I}_{-T}F(B_{-T}), e^{-\beta d\Gamma(\hat{\rho}\otimes 1)}e^{-\phi_{\mathrm{E}}(L_{+T})}\mathrm{I}_T G(B_T)\big)\Big]dx \tag{2.8.14}$$

or, equivalently,

$$(e^{-TH_{\mathrm{N}}}F, e^{-\beta d\Gamma(\hat{\rho})}e^{-TH_{\mathrm{N}}}G) = \int_{\mathbb{R}^d} \mathbb{E}^x\Big[e^{-\int_{-T}^T V(B_s)ds}(\Xi_0 \mathrm{I}_{-T}F(B_{-T}), e^{-\phi_\rho(K_T^\rho)}\Xi_\beta \mathrm{I}_T G(B_T))_{L^2(\mathcal{Q}_\rho)}\Big]dx, \tag{2.8.15}$$

where

$$L_{-T} = \int_{-T}^0 \tau_s\varphi(\cdot - B_s)ds, \quad L_{+T} = \int_0^T \tau_s\varphi(\cdot - B_s)ds,$$

and

$$K_T^\rho = \int_{-T}^0 q_0\tau_s\varphi(\cdot - B_s)ds + \int_0^T q_\beta\tau_s\varphi(\cdot - B_s)ds.$$

Proof. Let V be Kato-decomposable or $V \in C_0^\infty(\mathbb{R}^d)$. It follows that

$$e^{-TH_{\mathrm{N}}}F(x) = \mathbb{E}^x\Big[e^{-\int_0^T V(B_s)ds}\mathrm{I}_0^* e^{-\phi_{\mathrm{E}}(\int_0^T \tau_s\varphi(\cdot - B_s)ds)}\mathrm{I}_T F(B_T)\Big]$$

almost every $x \in \mathbb{R}^d$. Hence we have

$$(e^{-TH_{\mathrm{N}}}F, e^{-\beta d\Gamma(\hat{\rho})}e^{-TH_{\mathrm{N}}}G) = (F, e^{-TH_{\mathrm{N}}}e^{-\beta d\Gamma(\hat{\rho})}e^{-TH_{\mathrm{N}}}G)$$
$$= \int_{\mathbb{R}^d} \mathbb{E}_\mu \mathbb{E}^x\Big[\overline{F(B_0)}e^{-\int_0^T V(B_s)ds}\mathrm{I}_0^* e^{-\phi_{\mathrm{E}}(\int_0^T \tau_s\varphi(\cdot - B_s)ds)}\mathrm{I}_T$$
$$\times \mathbb{E}^{B_T}\Big[e^{-\beta d\Gamma(\hat{\rho})}e^{-\int_0^T V(B_s)ds}\mathrm{I}_0^* e^{-\phi_{\mathrm{E}}(\int_0^T \tau_s\varphi(\cdot - B_s)ds)}\mathrm{I}_T G(B_T)\Big]\Big]dx.$$

Since $\mathrm{I}_T\mathrm{I}_0^* = E_T U_{-T}^*$ and $\mathrm{I}_T = U_{-T}\mathrm{I}_{2T}$, by the Markov property of Brownian motion and the intertwining property $e^{-\beta d\Gamma(\hat{\rho}\otimes 1)}\mathrm{I}_T = \mathrm{I}_T e^{-\beta d\Gamma(\hat{\rho})}$, we have

$$
\begin{aligned}
&(e^{-TH_{\mathrm{N}}}F, e^{-\beta d\Gamma(\hat{\rho})}e^{-TH_{\mathrm{N}}}G) \\
&= \int_{\mathbb{R}^d} \mathbb{E}_\mu \mathbb{E}^x \Big[\overline{F(B_0)} e^{-\int_0^T V(B_s)ds} \mathrm{I}_0^* e^{-\phi_{\mathrm{E}}(\int_0^T \tau_s\varphi(\cdot-B_s)ds)} e^{-\beta d\Gamma(\hat{\rho}\otimes 1)} \\
&\quad \times \mathbb{E}^x\Big[e^{-\int_0^T V(B_{s+T})ds} E_T U_{-T}^* e^{-\phi_{\mathrm{E}}(\int_0^T \tau_s\varphi(\cdot-B_{s+T})ds)} U_{-T}\mathrm{I}_{2T}G(B_{2T}) | \mathscr{F}_T^{\mathrm{BM}}\Big]\Big] dx.
\end{aligned}
$$

Since $U_{-T}^* e^{-\phi_{\mathrm{E}}(\int_0^T \tau_s\varphi(\cdot-B_{s+T})ds)} U_{-T} = e^{-\phi_{\mathrm{E}}(\int_0^T \tau_{s+T}\varphi(\cdot-B_{s+T})ds)}$, and employing the Markov property of E_s, we have

$$
\begin{aligned}
&(e^{-TH_{\mathrm{N}}}F, e^{-\beta d\Gamma(\hat{\rho})}e^{-TH_{\mathrm{N}}}G) \\
&= \int_{\mathbb{R}^d} \mathbb{E}_\mu \mathbb{E}^x \Big[\overline{F(B_0)} a_V \mathrm{I}_0^* e^{-\phi_{\mathrm{E}}(\int_0^T \tau_s\varphi(\cdot-B_s)ds)} e^{-\beta d\Gamma(\hat{\rho}\otimes 1)} e^{-\phi_{\mathrm{E}}(\int_T^{2T} \tau_s\varphi(\cdot-B_s)ds)} \mathrm{I}_{2T}G(B_{2T})\Big] dx \\
&= \int_{\mathbb{R}^d} \mathbb{E}^x \Big[a_V \Big(e^{-\phi_{\mathrm{E}}(\int_0^T \tau_s\varphi(\cdot-B_s)ds)}\Phi_0, e^{-\beta d\Gamma(\hat{\rho}\otimes 1)} e^{-\phi_{\mathrm{E}}(\int_T^{2T} \tau_s\varphi(\cdot-B_s)ds)}\Phi_{2T}\Big)\Big] dx.
\end{aligned}
\tag{2.8.16}
$$

Here we set $a_V = e^{-\int_0^{2T} V(B_s)ds}$, $\Phi_0 = \mathrm{I}_0 F(B_0)$ and $\Phi_{2T} = \mathrm{I}_{2T}G(B_{2T})$ for a notational simplicity. We factorize $e^{-\beta d\Gamma(\hat{\rho}\otimes 1)}$ as $e^{-\beta d\Gamma(\hat{\rho}\otimes 1)} = \Xi_0^*\Xi_\beta$ and insert this into the above identity (2.8.16) to see that

$$
\begin{aligned}
&(e^{-TH_{\mathrm{N}}}F, e^{-\beta d\Gamma(\hat{\rho})}e^{-TH_{\mathrm{N}}}G) \\
&= \int_{\mathbb{R}^d} \mathbb{E}^x \Big[a_V \Big(\Xi_0 e^{-\phi_{\mathrm{E}}(\int_0^T \tau_s\varphi(\cdot-B_s)ds)}\Xi_0^*\Xi_0\Phi_0, \Xi_\beta e^{-\phi_{\mathrm{E}}(\int_T^{2T} \tau_s\varphi(\cdot-B_s)ds)}\Xi_\beta^*\Xi_\beta\Phi_{2T}\Big)\Big] dx.
\end{aligned}
$$

Again using the identity

$$
\Xi_t e^{-\phi_{\mathrm{E}}(\int_0^T \tau_s\varphi(\cdot-B_s)ds)}\Xi_t^* = E_t^q e^{-\phi_{\mathrm{E}}(\int_0^T q_t\tau_s\varphi(\cdot-B_s)ds)} E_t^q,
$$

where $E_t^q = \Xi_t\Xi_t^*$ is a projection, we have

$$
\begin{aligned}
&(e^{-TH_{\mathrm{N}}}F, e^{-\beta d\Gamma(\hat{\rho})}e^{-TH_{\mathrm{N}}}G) \\
&= \int_{\mathbb{R}^d} \mathbb{E}^x \Big[a_V \Big(E_0^q e^{-\phi_\rho(\int_0^T q_0\tau_s\varphi(\cdot-B_s)ds)} E_0^q\Xi_0\Phi_0, E_\beta^q e^{-\phi_\rho(\int_T^{2T} q_\beta\tau_s\varphi(\cdot-B_s)ds)} E_\beta^q\Xi_\beta\Phi_{2T}\Big)\Big] dx,
\end{aligned}
$$

and, furthermore, by the Markov property of E_t^q which can be obtained similarly as for E_t, we drop E_t^q in the above identity. Thus

$$
\begin{aligned}
&(e^{-TH_{\mathrm{N}}}F, e^{-\beta d\Gamma(\hat{\rho})}e^{-TH_{\mathrm{N}}}G) \\
&= \int_{\mathbb{R}^d} \mathbb{E}^x \Big[a_V \Big(e^{-\phi_\rho(\int_0^T q_0\tau_s\varphi(\cdot-B_s)ds)}\Xi_0\Phi_0, e^{-\phi_\rho(\int_T^{2T} q_\beta\tau_s\varphi(\cdot-B_s)ds)}\Xi_\beta\Phi_{2T}\Big)\Big] dx.
\end{aligned}
\tag{2.8.17}
$$

Using a time shift by $-T$ in (2.8.17), equality (2.8.15) follows, and (2.8.14) is obtained from (2.8.16). For general $V \in \mathcal{R}_{\text{Kato}}$, one can apply a limiting argument. □

Proposition 2.85. *Let ρ be a positive function on $\mathbb{R}^d$. Define $f_T = e^{-TH_N} f \otimes 1$. Then it follows that*

$$\frac{(f_T, e^{-\beta d\Gamma(\rho)} f_T)}{\|f_T\|^2} = \mathbb{E}_{\mu_T}\left[e^{-\int_{-T}^0 ds \int_0^T W_\beta(B_s - B_t, s-t)dt}\right],$$

where

$$W_\beta(x,t) = \int_{\mathbb{R}^d} \frac{|\hat{\varphi}(k)|^2}{2\omega(k)} e^{-|t|\omega(k)} e^{-ik\cdot x}(1 - e^{-\beta\rho(k)})dk.$$

Proof. By Theorem 2.84, we have

$$(f_T, e^{-\beta d\Gamma(\hat{\rho})} f_T) = \int_{\mathbb{R}^d} \mathbb{E}^x\left[e^{-\int_{-T}^T V(B_s)ds}(1, e^{-\phi_\rho(K_T^\rho)}1)_{L^2(\mathcal{Q}_\rho)}\right] dx. \tag{2.8.18}$$

A computation gives $(1, e^{-\phi_\rho(K_T^\rho)}1) = e^{\frac{1}{2}\|K_T^\rho\|_\rho^2}$ and

$$\begin{aligned}\|K_T^\rho\|_\rho^2 &= \left\|\int_{-T}^0 q_0 \tau_s \varphi(\cdot - B_s)ds\right\|^2 + \left\|\int_0^T q_\beta \tau_s \varphi(\cdot - B_s)ds\right\|^2 \\ &\quad + 2\Re\left(\int_{-T}^0 q_0 \tau_s \varphi(\cdot - B_s)ds, \int_0^T q_\beta \tau_s \varphi(\cdot - B_s)ds\right).\end{aligned}$$

Since $\widehat{q_0^* q_\beta f}(k, k_0) = e^{-\beta\rho(k)} \hat{f}(k, k_0)$, we have

$$\begin{aligned}\|K_T^\rho\|_\rho^2 &= \int_{-T}^0 ds \int_{-T}^0 dt \int_{\mathbb{R}^d} \frac{|\hat{\varphi}(k)|^2}{2\omega(k)} e^{-|s-t|\omega(k)} e^{-ik\cdot(B_s - B_t)} dk \\ &\quad + \int_0^T ds \int_0^T dt \int_{\mathbb{R}^d} \frac{|\hat{\varphi}(k)|^2}{2\omega(k)} e^{-|s-t|\omega(k)} e^{-ik\cdot(B_s - B_t)} dk \\ &\quad + 2\int_{-T}^0 ds \int_0^T dt \int_{\mathbb{R}^d} \frac{|\hat{\varphi}(k)|^2}{2\omega(k)} e^{-|s-t|\omega(k)} e^{-\beta\rho(k)} e^{-ik\cdot(B_s - B_t)} dk \\ &= \int_{-T}^T ds \int_{-T}^T W(B_s - B_t, s-t)dt - 2\int_{-T}^0 ds \int_0^T W_\beta(B_s - B_t, s-t)dt.\end{aligned}$$

On the other hand,

$$\|f_T\|^2 = \int_{\mathbb{R}^d} \mathbb{E}^x \left[e^{-\int_{-T}^{T} V(B_s)ds} e^{\frac{1}{2}\int_{-T}^{T} ds \int_{-T}^{T} W(B_s - B_t, s-t)dt} \right] dx.$$

It follows that

$$\frac{(f_T, e^{-\beta d\Gamma(\rho)} f_T)}{\|f_T\|^2} = \frac{\int_{\mathbb{R}^d} \mathbb{E}^x \left[e^{-\int_{-T}^{T} V(B_s)ds} e^{\frac{1}{2}\int_{-T}^{T} ds \int_{-T}^{T} W dt} e^{-\int_{-T}^{0} ds \int_0^T W_\beta dt} \right] dx}{\int_{\mathbb{R}^d} \mathbb{E}^x \left[e^{-\int_{-T}^{T} V(B_s)ds} e^{\frac{1}{2}\int_{-T}^{T} ds \int_{-T}^{T} W dt} \right] dx}$$
$$= \mathbb{E}_{\mu_T} \left[e^{-\int_{-T}^{0} ds \int_0^T W_\beta(B_s - B_t, s-t)dt} \right].$$

Hence the proposition follows. □

By this proposition, in particular, we have

$$\frac{(f_T, e^{-\beta N} f_T)}{\|f_T\|^2} = \mathbb{E}_{\mu_T} \left[e^{-(1-e^{-\beta})\int_{-T}^{0} ds \int_0^T W(B_s - B_t, s-t)dt} \right].$$

The expectation $(\Psi_g, e^{-\beta N}\Psi_g)$ expressed in terms of $\mathcal{N}$ in Corollary 2.41, can also be computed in terms of μ_∞.

Corollary 2.86 (Superexponential decay of boson number). *For every $\beta \in \mathbb{C}$, we have $\Psi_g \in D(e^{-\beta N})$ and*

$$(\Psi_g, e^{-\beta N}\Psi_g) = \mathbb{E}_{\mu_\infty} \left[\exp\left(-(1-e^{-\beta}) \int_{-\infty}^{0} ds \int_0^\infty W(B_s - B_t, s-t)dt \right) \right]. \tag{2.8.19}$$

Proof. Let $\beta > 0$. For ease of notation, we write

$$W_T = -(1-e^{-\beta}) \int_{-T}^{0} ds \int_0^T W(B_s - B_t, s-t)dt.$$

Note that for every $\delta > 0$ there is S_δ such that $|W_T - W_\infty| \le \delta$ for all $T > S_\delta$, uniformly in the paths. We have

$$\mathbb{E}_{\mu_T}[e^{W_T}] - \mathbb{E}_{\mu_\infty}[e^{W_\infty}] = \mathbb{E}_{\mu_T}[e^{W_T}] - \mathbb{E}_{\mu_T}[e^{W_\infty}] + \mathbb{E}_{\mu_T}[e^{W_\infty}] - \mathbb{E}_{\mu_\infty}[e^{W_\infty}]. \tag{2.8.20}$$

Also,

$$|\mathbb{E}_{\mu_T}[e^{W_T}] - \mathbb{E}_{\mu_T}[e^{W_\infty}]| \le C\delta \tag{2.8.21}$$

with a constant C. The second term on the right-hand side of (2.8.20) can be estimated as

$$|\mathbb{E}_{\mu_T}[e^{W_\infty}] - \mathbb{E}_{\mu_\infty}[e^{W_\infty}]| \le |\mathbb{E}_{\mu_T}[e^{W_\infty}] - \mathbb{E}_{\mu_T}[e^{W_{S_\delta}}]| \tag{2.8.22}$$

$$+ \left|\mathbb{E}_{\mu_T}[e^{W_{S_\delta}}] - \mathbb{E}_{\mu_\infty}[e^{W_{S_\delta}}]\right| \tag{2.8.23}$$

$$+ \left|\mathbb{E}_{\mu_\infty}[e^{W_{S_\delta}}] - \mathbb{E}_{\mu_\infty}[e^{W_\infty}]\right|. \tag{2.8.24}$$

For (2.8.22) and (2.8.24), we have the same upper bound as in (2.8.21). The term (2.8.23) converges to zero since $\mu_T \to \mu_\infty$ as $T \to \infty$ in the local sense. Thus the corollary follows for $\beta > 0$. We can extend this to all $\beta \in \mathbb{C}$ in the same way as in Corollary 2.61. Let $\mathbb{C}_+ = \{z \in \mathbb{C} \mid \Re z > 0\}$ and $\mathbb{C}_- = \{z \in \mathbb{C} \mid \Re z < 0\}$. Denote the right-hand side of (2.8.19) by $g(\beta)$, and the left-hand side by $f(\beta)$. Informally, g can be extended to the whole complex plane $\mathbb{C}$ with $g = f$ on $\mathbb{C}_+$ and then f can be also extended to $\mathbb{C}$, and thus $g = f$ holds on $\mathbb{C}$ by the identity theorem. Since $|W_\infty| \le \frac{1}{2}\int_{\mathbb{R}^d} |\hat{\varphi}(k)|^2/\omega(k)^3 dk < \infty$, the function g can be analytically continued in $\mathbb{C}$. We denote its analytic continuation by $\tilde{g}$. Let $\beta_0 > 0$, and consider the ball $B_r(z_0) = \{z \in \mathbb{C} \mid |z - z_0| < r\}$. Fix an arbitrary R such that $R > \beta_0$. We see that

$$\tilde{g}(\beta) = \sum_{n=0}^{\infty} (\beta - \beta_0)^n b_n(\beta_0) \tag{2.8.25}$$

for $\beta \in B_R(\beta_0)$, and (2.8.25) is absolutely convergent. Denote by $\nu(d\rho)$ the spectral measure of N associated with Ψ_g. We have

$$f(\beta) = \int_0^\infty e^{-\beta\rho}\nu(d\rho), \quad \beta \in \mathbb{C}_+.$$

Note that $f(\beta)$ is analytic on $\mathbb{C}_+$. We see that

$$f(\beta) = \sum_{n=0}^{\infty} \frac{(\beta - \beta_0)^n}{n!} \int_0^\infty (-\rho)^n e^{-\beta_0\rho}\nu(d\rho), \quad \beta \in B_{\beta_0}(\beta_0).$$

Since $f = \tilde{g}$ on $B_{\beta_0}(\beta_0)$, we also see that

$$b_n(\beta_0) = \frac{1}{n!} \int_0^\infty (-\rho)^n e^{-\beta_0\rho}\nu(d\rho).$$

It follows that

$$\tilde{g}(\beta) = \sum_{n=0}^{\infty} \frac{(\beta - \beta_0)^n}{n!} \int_0^\infty \rho^n e^{-\beta_0\rho}\nu(d\rho), \quad \beta \in B_R(\beta_0). \tag{2.8.26}$$

Notice that the right-hand side of (2.8.26) is absolutely convergent and $B_R(\beta_0) \cap \mathbb{C}_- \neq \emptyset$. In particular, for $\beta \in \mathbb{R} \cap B_R(\beta_0) \cap \mathbb{C}_-$, i. e., for

$$-(R - \beta_0) < \beta < 0, \tag{2.8.27}$$

by Fatou's lemma we have for every $M > 0$,

$$\int_0^M e^{-\beta\rho}\nu(d\rho) \le \sum_{n=0}^{\infty} \frac{(\beta-\beta_0)^n}{n!} \int_0^M \rho^n e^{-\beta_0\rho}\nu(d\rho) \le \sum_{n=0}^{\infty} \frac{(\beta-\beta_0)^n}{n!} \int_0^{\infty} \rho^n e^{-\beta_0\rho}\nu(d\rho) < \infty.$$

Thus $\int_0^\infty e^{-\beta\rho}\nu(d\rho) < \infty$ for β with (2.8.27), which implies that $\Psi_g \in D(e^{-(\beta/2)N})$ and (2.8.19) holds for β with (2.8.27). Since R is an arbitrary number, we get (2.8.19) for all $\beta \in \mathbb{R} \cup \mathbb{C}_+$. From this, it follows that $|(\Psi_g, e^{-\beta N}\Psi_g)| < \int_0^\infty e^{-\Re\beta\rho}\nu(d\rho) < \infty$ for all $\beta \in \mathbb{C}$. The function f can be extended in $\mathbb{C}$ to a function $\tilde{f}$. In particular, $\tilde{f} = \tilde{g}$ on $D = \mathbb{R} \cup \mathbb{C}_+$ and $\tilde{f}$ has a convergent series expansion in D. That is, $\int_0^\infty e^{-z\rho}\nu(d\rho) = \sum_{n=0}^\infty (-1)^n (z-\beta)^n \int_0^\infty e^{-\beta\rho}\rho^n\nu(d\rho)$ for $z \in B_K(\beta)$, for every K and $\beta < 0$. Hence $\tilde{f}$ is also analytic on $\mathbb{C}$, and $\tilde{g} = \tilde{f}$ on $\mathbb{C}$ follows. □

We can also estimate $(\Psi_g, N\Psi_g)$ by using the probability measure μ_∞ which is the same as in Section 2.6.2. For $m \ge 0$, we have

$$(\Psi_g, N^m\Psi_g) = (-1)^m \frac{d^m}{d\beta^m} \mathbb{E}_{\mu_\infty}\left[e^{-(1-e^{-\beta})\int_{-\infty}^0 ds \int_0^\infty W dt} \right]\Bigg|_{\beta=0}.$$

From this, we see that there exist positive constants a and b such that

$$-a + b\mathrm{I} \le (\Psi_g, N\Psi_g) \le \frac{g^2}{2}\mathrm{I}, \tag{2.8.28}$$

where $\mathrm{I} = \int_{\mathbb{R}^d} |\hat{\varphi}(k)|^2/\omega(k)^3 dk$, and it can be seen that $\lim_{\mathrm{I}\to\infty}(\Psi_g, N\Psi_g) = \infty$.

In Section 2.6.5 we showed a Gaussian domination of the ground state Ψ_g using the path measure $\mathcal{N}$. We can also see this by using the Gibbs measure μ_∞.

Corollary 2.87 (Gaussian domination of the ground state). *Take $g \in \mathscr{S}'(\mathbb{R}^d)$ such that $\hat{g}/\sqrt{\omega}, \hat{g}/\omega \in L^2(\mathbb{R}^d)$, and $\beta < 1/\|\hat{g}/\sqrt{\omega}\|^2$. Then $\Psi_g \in D(e^{(\beta/2)\phi(g)^2})$ and*

$$\|e^{(\beta/2)\phi(g)^2}\Psi_g\|^2 = \frac{1}{\sqrt{1-\beta\|\hat{g}/\sqrt{\omega}\|^2}} \mathbb{E}_{\mu_\infty}\left[\exp\left(\frac{\beta K(g)^2}{1-\beta\|\hat{g}/\sqrt{\omega}\|^2} \right) \right],$$

where $K(g)$ denotes the random variable defined by

$$K(g) = \frac{1}{2}\int_{-\infty}^{\infty} \left(e^{-|r|\omega} e^{-ik\cdot B_r}\hat{\varphi}/\sqrt{\omega}, \hat{g}/\sqrt{\omega} \right) dr.$$

In particular, it follows that

$$\lim_{\beta\uparrow(2\|\hat{g}/\sqrt{\omega}\|^2)^{-1}} \|e^{\beta\phi(g)^2}\Psi_g\| = \infty.$$

Proof. Let $f \in L^2(\mathbb{R}^d)$ and $f \geq 0$, and write $f_T = e^{-TH_N} f \otimes 1$. For every $\beta \in \mathbb{C}$, we have

$$\frac{(f_T, e^{ik\beta\phi(g)} f_T)}{\|f_T\|^2} = \frac{1}{\|f_T\|^2} \int_{\mathbb{R}^d} \mathbb{E}^x \left[f(B_{-T}) f(B_T) e^{-\int_{-T}^T V(B_s)ds} S_T \right] dx,$$

where

$$S_T = \left(1, e^{-\phi_E(\int_{-T}^0 \tau_s \varphi(\cdot - B_s)ds)} e^{ik\beta\phi_E(\tau_0 g)} e^{-\phi_E(\int_0^T \tau_s \varphi(\cdot - B_s)ds)} 1\right)_{L^2(\mathscr{Q}_E)}.$$

A computation gives

$$S_T = e^{-\frac{k^2\beta^2}{4}\|\hat{g}/\sqrt{\omega}\|^2} e^{-ik\beta \int_{\mathbb{R}^d} \left(\frac{\hat{\varphi}(k)\overline{\hat{g}(k)}}{2\omega(k)} \int_{-T}^T e^{-\omega(k)|s|} e^{-ik\cdot B_s} ds \right) dk} e^{\frac{1}{2}\int_{-T}^T ds \int_{-T}^T W(B_s - B_t, s-t)dt}.$$

Hence it follows that

$$\frac{(f_T, e^{ik\beta\phi(g)} f_T)}{\|f_T\|^2} = \mathbb{E}_{\mu_T} \left[e^{-\frac{k^2\beta^2}{4}\|\hat{g}/\sqrt{\omega}\|^2} e^{-ik\beta \int_{\mathbb{R}^d} \left(\frac{\hat{\varphi}(k)\overline{\hat{g}(k)}}{2\omega(k)} \int_{-T}^T e^{-\omega(k)|s|} e^{-ik\cdot B_s} ds \right) dk} \right].$$

Taking the limit $T \to \infty$, we obtain

$$(\Psi_g, e^{ik\beta\phi(g)} \Psi_g) = \mathbb{E}_{\mu_\infty} \left[e^{-k^2\beta^2 I_g/2} e^{-ik\beta I_B} \right], \tag{2.8.29}$$

where $I_g = \|\hat{g}/\sqrt{\omega}\|^2/2$ and $I_B = \int_{\mathbb{R}^d} \left(\frac{\hat{\varphi}(k)\overline{\hat{g}(k)}}{2\omega(k)} \int_{-\infty}^{\infty} e^{-\omega(k)|s|} e^{-ik\cdot B_s} ds \right) dk$. Thus using (2.8.29), in the same way as in Lemma 2.60, we can show that for $\beta < 0$,

$$(\Psi_g, e^{\beta\phi(g)^2} \Psi_g) = \frac{1}{\sqrt{1 - \beta\|\hat{g}/\sqrt{\omega}\|^2}} \mathbb{E}_{\mu_\infty} \left[\exp\left(\frac{\beta K(g)^2}{1 - \beta\|\hat{g}/\sqrt{\omega}\|} \right) \right].$$

Furthermore, we can extend this to $\beta < 1/\|\hat{g}/\sqrt{\omega}\|^2$ in the same way as in Corollary 2.61. □

2.9 Carmona-type estimates

2.9.1 Exponential decay of bound states: upper bound

In this section we briefly discuss the decay of bound states for potentials in the Carmona classes discussed in Section 4.3.8 in Volume 1, for Schrödinger operators.

We consider the following class of potentials. See also Definition 4.175 in Volume 1.

Definition 2.88. The potential classes $\mathbb{V}^{\text{upper}}$ and $\mathbb{V}^{\text{lower}}$ are defined by the following properties:

($\mathbb{V}^{\text{upper}}$) $V \in \mathbb{V}^{\text{upper}}$ if and only if $V = W - U$ such that

(1) $U \geq 0$ and $U \in L^p(\mathbb{R}^d)$ for $p > d/2$ and $1 \leq p < \infty$;
(2) $W \in L^1_{\mathrm{loc}}(\mathbb{R}^d)$ and $W_\infty = \inf_{x\in\mathbb{R}^d} W(x) > -\infty$.

($\mathbb{V}^{\mathrm{lower}}$) $V \in \mathbb{V}^{\mathrm{lower}}$ if and only if $V = W - U$ such that
(1) $U \geq 0$ and $U \in L^p(\mathbb{R}^d)$ for $p > d/2$ and $1 \leq p < \infty$;
(2) $W \geq 0$ and $W \in \mathcal{K}_{\mathrm{loc}}(\mathbb{R}^d)$.

Lemma 2.89 (Carmona-type estimate). *Suppose $\int_{\mathbb{R}^d} |\hat{\varphi}(k)|^2/\omega(k)^3 dk < \infty$ and let Assumption 2.16 hold. Let $V = U - W \in \mathbb{V}^{\mathrm{upper}}$ and $H_{\mathrm{N}}\Phi_{\mathrm{b}} = E\Phi_{\mathrm{b}}$. Then for every $t, a > 0$ and every $0 < \alpha < 1/2$, there exist constants $D_1, D_2, D_3 > 0$ such that*

$$\|\Phi_{\mathrm{b}}(x)\|_{L^2(\mathscr{Q})} \leq t^{-d/2} D_1 e^{D_2\|U\|_p t} e^{\tilde{E}t}(D_3 e^{-\frac{\alpha}{4}\frac{a^2}{t}} e^{-tW_\infty} + e^{-tW_a(x)})\|\Phi_{\mathrm{b}}\|_{\mathcal{H}_{\mathrm{N}}},$$

where $\tilde{E} = E + E(\hat{\varphi})$ and $W_a(x) = \inf\{W(y) \mid |x - y| < a\}$.

Proof. Writing $\Phi_{\mathrm{b}} = e^{tE} e^{-tH_{\mathrm{N}}}\Phi_{\mathrm{b}}$, the functional integral representation gives

$$\Phi_{\mathrm{b}}(x) = e^{tE}\mathbb{E}^x\left[\mathrm{I}_0^* e^{-\phi_{\mathrm{E}}(\int_0^t \delta_s\otimes\varphi(\cdot-B_s)ds)}\mathrm{I}_t e^{-\int_0^t V(B_s)ds}\Phi_{\mathrm{b}}(B_t)\right].$$

From this, it follows directly that

$$\|\Phi_{\mathrm{b}}(x)\|_{L^2(\mathscr{Q})} \leq e^{t(E+E(\hat{\varphi}))}\mathbb{E}^x\left[e^{-\int_0^t V(B_s)ds}\|\Phi_{\mathrm{b}}(B_t)\|_{L^2(\mathscr{Q})}\right].$$

The lemma follows in a similar way to Carmona's estimate for Schrödinger operators in Lemma 4.176 in Volume 1. □

From the above lemma, we can draw decay results similarly to the proof of eigenfunction decay for Schrödinger operators. For $V = W - U \in \mathbb{V}^{\mathrm{upper}}$, define $\Sigma = \liminf_{|x|\to\infty} V(x)$. Notice that $\Sigma = \liminf_{|x|\to\infty} W(x)$, and $\Sigma \geq W_\infty$ holds. Since the proof is similar to those of Corollaries 4.177 and 4.178 in Volume 1, we only state the results.

Corollary 2.90 (Upper bound of exponential decay). *Suppose $\int_{\mathbb{R}^d} |\hat{\varphi}(k)|^2/\omega(k)^3 dk < \infty$ and let Assumption 2.16 hold. Let $V = W - U \in \mathbb{V}^{\mathrm{upper}}$ and $H_{\mathrm{N}}\Phi_{\mathrm{b}} = E\Phi_{\mathrm{b}}$.*

(1) *Suppose that $W(x) \geq \gamma|x|^{2n}$ outside a compact set K, for some $n > 0$ and $\gamma > 0$. Take $0 < \alpha < 1/2$. Then there exists a constant $C_1 > 0$ such that*

$$\|\Phi_{\mathrm{b}}(x)\|_{L^2(\mathscr{Q})} \leq C_1 \exp\left(-\frac{\alpha c}{16}|x|^{n+1}\right)\|\Phi_{\mathrm{b}}\|_{\mathcal{H}_{\mathrm{N}}},$$

where $c = \inf_{x\in\mathbb{R}^d\setminus K} W_{|x|/2}(x)/|x|^{2n}$.

(2) *(Decaying potential) Suppose that $\Sigma - E > E(\hat{\varphi})$, $\Sigma > W_\infty$, and let $0 < \beta < 1$. Then there exists a constant $C_2 > 0$ such that*

$$\|\Phi_{\mathrm{b}}(x)\|_{L^2(\mathscr{Q})} \leq C_2 \exp\left(-\frac{\beta}{8\sqrt{2}}\frac{\Sigma - E - E(\hat{\varphi})}{\sqrt{\Sigma - W_\infty}}|x|\right)\|\Phi_{\mathrm{b}}\|_{\mathcal{H}_{\mathrm{N}}}.$$

(3) *(Confining potential) Suppose that* $\lim_{|x|\to\infty} W(x) = \infty$. *Then there exist constants* $C, \delta > 0$ *such that*

$$\|\Phi_{\mathrm{b}}(x)\|_{L^2(\mathscr{Q})} \leq C \exp\left(-\delta|x|\right) \|\Phi_{\mathrm{b}}\|_{\mathcal{H}_{\mathrm{N}}}.$$

2.9.2 Exponential decay of bound states: lower bound

By the Carmona-type estimate in the previous section, we can show not only upper but also lower bounds of positive eigenfunctions. In Section 2.9.2 we employ Assumption 2.16 and the existence of the unique ground Ψ_{g}. Here we derive an exponential lower bound of the ground state of H_{N}. First, we recall some estimates of probabilities of sets of paths discussed in (4.3.85) of Section 4.3.8 in Volume 1. Let

$$P(a, [c, d], t) = \left\{ \sup_{0 \leq s \leq t} |B_s| \leq a \right\} \cap \{B_t \in [c, d]\}.$$

The following statement has been established in Lemma 4.180 in Volume 1: Suppose that $a > 0$, $\alpha > 0$, and $t > 0$ satisfy $\alpha < a/2$ and $a^2/t > \beta$. Here β is the unique positive solution of

$$F(\xi) = 1 - (e^{-21\xi/8} + e^{-5\xi/8} + e^{-165\xi/8}) = 0.$$

Then for every $x \in [-(a - \alpha), a - \alpha]$, we have

$$\mathcal{W}(P(a, [x - \alpha, x + \alpha], t)) \geq \frac{\alpha}{\sqrt{2\pi t}} F\left(\frac{a^2}{t}\right) e^{-\frac{a^2}{2t}}. \tag{2.9.1}$$

We recall that $\mathrm{I}_{[0,t]}(x) = \mathrm{I}_0^* e^{-\Phi_{\mathrm{E}}(\int_0^t \delta_s \otimes \varphi(\cdot - B_s - x) ds)} \mathrm{I}_t$.

Lemma 2.91. *Suppose that* $V \in \mathbb{V}^{\mathrm{lower}}$ *and let* $K \subset \mathbb{R}^d$ *be compact. Then there exists* $\varepsilon_K > 0$ *such that*

$$\inf_{B_t + x \in K} \left(1, \mathrm{I}_{[0,t]}(x) \Psi_{\mathrm{g}}(B_t + x)\right)_{L^2(\mathscr{Q})} \geq \varepsilon_K \exp\left(-\frac{1}{\varepsilon_K} \left(\int_0^t ds \int_0^t W(s - r, B_s - B_r) dr \right)^{1/2} M_g \right).$$

Here $M_g = \sup_{x \in \mathbb{R}^d} \|\Psi_{\mathrm{g}}(x)\|_{L^2(\mathscr{Q})}$.

Proof. Fix x and let $P(A) = \frac{1}{N} \mathbb{E}_{\mu_{\mathrm{E}}}[1_A \mathrm{I}_t \Psi_{\mathrm{g}}(B_t + x)]$, $A \in \Sigma_{\mathrm{E}}$, be a probability measure on $(\mathscr{Q}_{\mathrm{E}}, \Sigma_{\mathrm{E}})$, where

$$N = \mathbb{E}_{\mu_{\mathrm{E}}}[\mathrm{I}_t \Psi_{\mathrm{g}}(B_t + x)]$$

is the normalizing constant. Note that $\mathrm{I}_t\Psi_{\mathrm{g}}(B_t + x) > 0$. Since V is Kato-decomposable, by Theorem 2.24 and Corollary 2.27, we get that $\mathbb{E}_{\mu_{\mathrm{E}}}[\mathrm{I}_t\Psi_{\mathrm{g}}(y)] = (1, \Psi_{\mathrm{g}}(y))$ is continuous in y and $(1, \Psi_{\mathrm{g}}(y)) > 0$. Hence there exists $\varepsilon_K > 0$ such that

$$\inf_{y\in K}\Big(1, \Psi_{\mathrm{g}}(y)\Big) \geq \varepsilon_K \tag{2.9.2}$$

and

$$N \geq \inf_{B_t+x\in K}\Big(1, \Psi_{\mathrm{g}}(B_t + x)\Big) \geq \varepsilon_K. \tag{2.9.3}$$

By Jensen's inequality, we have

$$\begin{aligned}\Big(1, \mathrm{I}_{[0,t]}(x)\Psi_{\mathrm{g}}(B_t + x)\Big) &= N\mathbb{E}_P[e^{-\phi_{\mathrm{E}}(\int_0^t \delta_s\otimes\varphi(\cdot-B_s-x)ds)}] \\ &\geq Ne^{-\mathbb{E}_P[\phi_{\mathrm{E}}(\int_0^t \delta_s\otimes\varphi(\cdot-B_s-x)ds)]} = Ne^{-\frac{1}{N}(\phi_{\mathrm{E}}(\int_0^t \delta_s\otimes\varphi(\cdot-B_s-x)ds),\mathrm{I}_t\Psi_{\mathrm{g}}(B_t))}.\end{aligned} \tag{2.9.4}$$

Since

$$\left|\left(\phi_{\mathrm{E}}\Big(\int_0^t \delta_s\otimes\varphi(\cdot - B_s - x)ds\Big), \mathrm{I}_t\Psi_{\mathrm{g}}(B_t)\right)\right| \leq \left(\int_0^t ds\int_0^t W(s-r, B_s - B_r)dr\right)^{1/2} M_g,$$

together with (2.9.3)–(2.9.4), the result follows. □

Lemma 2.92. *Let $V = W - U \in \mathcal{V}^{\mathrm{lower}}$. Suppose that $a_1\ldots, a_d$, $\alpha_1, \ldots, \alpha_d$, $b_1\ldots, b_d$, and t satisfy $a_j^2/t > \beta$, $a_j/2 > \alpha_j$, and $|[-a_j, a_j] \cap [-x_j - b_j, -x_j + b_j]| > 2\alpha_j$. Then*

$$(1, \Psi_{\mathrm{g}}(x)) \geq e^{-tW_a(x)}e^{tE}\varepsilon_K e^{-\sqrt{t}\|\hat{\varphi}/\omega\|M_g/\varepsilon_K}\prod_{j=1}^d \frac{\alpha_j}{\sqrt{2\pi t}}F\Big(\frac{a_j^2}{t}\Big)e^{-\frac{a_j^2}{2t}}, \tag{2.9.5}$$

where $W_a(x) = \sup\{W(y) \mid |y_j - x_j| < a_j, j = 1, \ldots, d\}$ and ε_K is given in Lemma 2.91 with the compact set $K = [-b_1, b_1] \times\cdots\times [-b_d, b_d]$.

Proof. Let $A = \bigcap_{j=1}^d P(a_j, [-b_j - x_j, b_j - x_j], t)$. By (2.9.1), we see that for $\alpha_j < a_j/2$ and $a_j^2/t > \beta$,

$$\mathcal{W}\Big(\bigcap_{j=1}^d P(a_j, [k_j, k_j + 2\alpha], t)\Big) \geq \prod_{j=1}^d \frac{\alpha_j}{\sqrt{2\pi t}}F\Big(\frac{a_j^2}{t}\Big)e^{-\frac{a_j^2}{2t}},$$

for any $k_j \in [-a_j, a_j]$. By the assumption, there exists $k_j \in [-a_j, a_j]$ such that $[k_j, k_j + 2\alpha] \subset [-b_j - x_j, b_j - x_j]$. We have

$$\mathcal{W}\Big(\bigcap_{j=1}^{d} P(a_j, [-b_j - x_j, b_j - x_j], t)\Big) \geq \mathcal{W}\Big(\bigcap_{j=1}^{d} P(a_j, [k_j, k_j + 2\alpha], t)\Big) \geq \prod_{j=1}^{d} \frac{a_j}{\sqrt{2\pi t}} F\Big(\frac{a_j^2}{t}\Big) e^{-\frac{a_j^2}{2t}} \tag{2.9.6}$$

and $\int_0^t ds \int_0^t W(s-r, B_s - B_r)dr \leq t\|\hat{\varphi}/\omega\|^2$ uniformly in paths. By Lemma 2.91, we see that

$$\begin{aligned}(1, \Psi_g(x)) &= e^{tE}\mathbb{E}\Big[e^{-\int_0^t V(x+B_s)ds}(1, \mathrm{I}_{[0,t]}(x)\Psi_g(x + B_t))\Big] \\ &\geq e^{tE}\mathbb{E}\Big[1_A e^{-\int_0^t W(x+B_s)ds}(1, \mathrm{I}_{[0,t]}(x)\Psi_g(x + B_t))\Big] \\ &\geq e^{tE}\mathbb{E}[1_A e^{-\int_0^t W_a(x)ds}\varepsilon_K e^{-\sqrt{t}\|\hat{\varphi}/\omega\|M_g/\varepsilon_K}] \\ &\geq e^{tE}e^{-tW_a(x)}\varepsilon_K e^{-\sqrt{t}\|\hat{\varphi}/\omega\|M_g/\varepsilon_K}\mathcal{W}\Big(\bigcap_{j=1}^{d} P(a_j, [k_j, k_j + 2\alpha], t)\Big).\end{aligned}$$

By (2.9.6), the lemma follows. □

Theorem 2.93 (Lower bound of exponential decay). *Let $V = W - U \in \mathbb{V}^{\text{lower}}$. Suppose that $W(x) \leq \gamma|x|^{2m}$ outside a compact set with $\gamma > 0$ and $m > 1$. Then there exist constants $\delta, D > 0$ such that $\Psi_g(x) \geq De^{-\delta|x|^{m+1}}$.*

Proof. The proof is a minor modification of Theorem 4.182 in Volume 1. We fix a sufficiently large compact set $K \subset \mathbb{R}^d$. Since $\sup_{x\in\mathbb{R}^d} \|\Psi_g(x)\|_{L^2(\mathscr{Q})} < \infty$, it suffices to show for large enough $|x|$. We also have

$$\|\Psi_g(x)\|^2 = \sum_{n=0}^{\infty} \|\Psi_g(x)^{(n)}\|^2 \geq \|\Psi_g(x)^{(0)}\|^2 = (1, \Psi_g(x))^2.$$

Thus it is sufficient to estimate $(1, \Psi_g(x))$ from below. We set $t = |x|^{-(m-1)}$, $a_j = 1 + |x_j|$, $\alpha_j = 1/2$, and $b_j = 1$ for $j = 1, \ldots, d$. Hence the assumptions in Lemma 2.92 are satisfied. We also see that $tW_a(x) \leq \gamma 2^{4m}|x|^{m+1}$. Inserting this into (2.9.5), we have

$$\begin{aligned}(1, \Psi_g(x)) &\geq \varepsilon_K e^{-\gamma 2^{4m}|x|^{m+1}} e^{E|x|^{-(m-1)}} e^{-|x|^{-(m-1)/2}\|\hat{\varphi}/\omega\|M_g/\varepsilon_K} \prod_{j=1}^{d}\Big(\frac{|x|^{(m-1)/2}}{4\sqrt{2\pi}}\Big) e^{-\frac{1}{2}(1+|x_j|)^2|x|^{m-1}} \\ &\geq \varepsilon_K e^{-(\gamma 2^{4m}+2)|x|^{m+1}} e^{E|x|^{-(m-1)}} e^{-|x|^{-(m-1)/2}\|\hat{\varphi}/\omega\|M_g/\varepsilon_K} \geq \frac{\varepsilon_K}{2} e^{-(\gamma 2^{4m}+2)|x|^{m+1}}.\end{aligned}$$

Here a large enough $|x|$ is taken so that both inequalities $F((1 + |x_j|)^2|x|^{m-1}) > \frac{1}{2}$ and $e^{E|x|^{-(m-1)}} e^{-|x|^{-(m-1)/2}\|\hat{\varphi}/\omega\|M_g/\varepsilon_K} > \frac{1}{2}$ hold. From this, we obtain $(1, \Psi_g(x)) \geq De^{-\delta|x|^{m+1}}$. □

Remark 2.94.

(1) In Theorem 2.93, we do not need to assume $\int_{\mathbb{R}^d} |\varphi(k)|^2/\omega(k)^3 dk < \infty$.

(2) Estimate (2.9.2) can be also shown for any strictly positive eigenfunction Φ of H_{N}, i. e., $\inf_{y\in K}(1,\Phi(y))_{L^2(\mathscr{Q})} \geq c_K$ with a strictly positive constant c_K. Hence, under the same assumptions as in Theorem 2.93, it can be shown that $\|\Phi(x)\|_{L^2(\mathscr{Q})} \geq De^{-\delta|x|^{m+1}}$.

2.10 Martingale properties and applications

2.10.1 Martingale properties

In Sections 4.6.3 and 4.8.2 in Volume 1, we have obtained a martingale related to the semigroup and process associated with the relativistic Schrödinger operators. In a similar way, we can construct a martingale related to the semigroup and process associated with the Schrödinger operator $H_{\mathrm{p}} = -\frac{1}{2}\Delta + V$. For an eigenfunction φ of H_{p}, we can define the random process

$$h_t(x) = e^{tE_{\mathrm{p}}}e^{-\int_0^t V(B_r+x)dr}\varphi(B_t + x), \quad t \geq 0.$$

As it was shown there, $\mathbb{E}[h_t(x)] = f(x)$ for all $t \geq 0$ and the random process $(h_t(x))_{t\geq 0}$ is a martingale with respect to the filtration $(\mathscr{F}_t^{BM})_{t\geq 0}$. This construction can be extended to the Nelson Hamiltonian, which we present now.

In Section 2.10.1 we work under Assumption 2.19, so that the potential V is Kato-decomposable. Consider an eigenfunction and eigenvalue satisfying

$$H_{\mathrm{N}}\Phi = E\Phi,$$

and define

$$H_t(x) = e^{tE}e^{-\int_0^t V(B_r+x)dr}e^{-\phi_{\mathrm{E}}(\int_0^t \delta_r\otimes\varphi(\cdot-x-B_r)dr)}\mathrm{I}_t\Phi(B_t + x). \tag{2.10.1}$$

Here $(H_t(x))_{t\geq 0}$ is a random process on $(\mathscr{X} \times \mathscr{Q}_{\mathrm{E}}, \mathcal{B}(\mathscr{X}) \times \Sigma_{\mathrm{E}}, \mathcal{W} \times \mu_{\mathrm{E}})$ parametrized by $x \in \mathbb{R}^d$. By the functional integral representation, we have

$$(\Psi, \Phi) = (\Psi, e^{-t(H_{\mathrm{N}}-E)}\Phi) = \int_{\mathbb{R}^d} \mathbb{E}_{\mu_{\mathrm{E}}}[\bar{\Psi}(x)\mathbb{E}_{\mathcal{W}}[\mathrm{I}_0^* H_t(x)]]dx,$$

implying $\Phi(x) = (e^{-t(H_{\mathrm{N}}-E)}\Phi)(x) = \mathbb{E}_{\mathcal{W}}[\mathrm{I}_0^* H_t(x)]$. Define the filtration $(\mathscr{M}_t)_{t\geq 0}$ by

$$\mathscr{M}_t = \mathscr{F}_t^{BM} \times \Sigma_{(-\infty,t]}, \quad t \geq 0.$$

Theorem 2.95 (Martingale property). *The random process $(H_t(x))_{t\geq 0}$ is a martingale with respect to $(\mathscr{M}_t)_{t\geq 0}$.*

Proof. We have

$$\mathbb{E}_{\mu_{\mathrm{E}}}\mathbb{E}_{\mathcal{W}}\left[H_t(x)|\mathscr{M}_s\right] = e^{tE}e^{-\int_0^s V(B_r+x)dr}e^{-\phi_{\mathrm{E}}(\int_0^s \delta_r\otimes\varphi(\cdot-x-B_r)dr)} \\ \times \mathbb{E}_{\mu_{\mathrm{E}}}\mathbb{E}_{\mathcal{W}}\left[e^{-\int_s^t V(B_r+x)dr}e^{-\phi_{\mathrm{E}}(\int_s^t \delta_r\otimes\varphi(\cdot-x-B_r)dr)}\mathrm{I}_t\Phi(B_t+x)|\mathscr{M}_s\right].$$

We compute the conditional expectation on the right-hand side. Using the Markov property of $(B_t)_{t\geq 0}$, we obtain

$$\mathbb{E}_{\mu_{\mathrm{E}}}\mathbb{E}_{\mathcal{W}}\left[e^{-\int_s^t V(B_r+x)dr}e^{-\phi_{\mathrm{E}}(\int_s^t \delta_r\otimes\varphi(\cdot-x-B_r)dr)}\mathrm{I}_t\Phi(B_t+x)|\mathscr{M}_s\right] \\ = \mathbb{E}_{\mu_{\mathrm{E}}}\left[\mathbb{E}_{\mathcal{W}}^{B_s}\left[e^{-\int_0^{t-s} V(B_r+x)dr}e^{-\phi_{\mathrm{E}}(\int_s^t \delta_r\otimes\varphi(\cdot-x-B_{r-s})dr)}\mathrm{I}_t\Phi(B_{t-s}+x)|\Sigma_{(-\infty,s]}\right]\right].$$

From the Markov property of the projection $\mathrm{E}_s = \mathrm{I}_s\mathrm{I}_s^*$, it furthermore follows that

$$= \mathbb{E}_{\mu_{\mathrm{E}}}\left[\mathbb{E}_{\mathcal{W}}^{B_s}\left[e^{-\int_0^{t-s} V(B_r+x)dr}e^{-\phi_{\mathrm{E}}(\int_s^t \delta_r\otimes\varphi(\cdot-x-B_{r-s})dr)}\mathrm{I}_t\Phi(B_{t-s}+x)|\Sigma_s\right]\right] \\ = \mathrm{E}_s\mathbb{E}_{\mathcal{W}}^{B_s}\left[e^{-\int_0^{t-s} V(B_r+x)dr}e^{-\phi_{\mathrm{E}}(\int_s^t \delta_r\otimes\varphi(\cdot-x-B_{r-s})dr)}\mathrm{I}_t\Phi(B_{t-s}+x)\right].$$

Denote $U_s = \Gamma_{\mathrm{E}}(u_s)$, where $u_s : \mathscr{E} \to \mathscr{E}$ denotes the time-shift operator defined by $u_s f(t,x) = f(t+s,x)$. Since $E_s = \mathrm{I}_s\mathrm{I}_0^* U_{-s}$, we have

$$= \mathrm{I}_s\mathrm{I}_0^* U_{-s}\mathbb{E}_{\mathcal{W}}^{B_s}\left[e^{-\int_0^{t-s} V(B_r+x)dr}e^{-\phi_{\mathrm{E}}(\int_s^t \delta_r\otimes\varphi(\cdot-x-B_{r-s})dr)}\mathrm{I}_t\Phi(B_{t-s}+x)\right] \\ = \mathrm{I}_s\mathrm{I}_0^*\mathbb{E}_{\mathcal{W}}^{B_s}\left[e^{-\int_0^{t-s} V(B_r+x)dr}e^{-\phi_{\mathrm{E}}(\int_s^t \delta_{r-s}\otimes\varphi(\cdot-x-B_{r-s})dr)}\mathrm{I}_{t-s}\Phi(B_{t-s}+x)\right] \\ = \mathrm{I}_s\mathrm{I}_0^*\mathbb{E}_{\mathcal{W}}^{B_s}\left[e^{-\int_0^{t-s} V(B_r+x)dr}e^{-\phi_{\mathrm{E}}(\int_0^{t-s} \delta_r\otimes\varphi(\cdot-x-B_r)dr)}\mathrm{I}_{t-s}\Phi(B_{t-s}+x)\right].$$

We conclude that

$$\mathbb{E}_{\mu_{\mathrm{E}}}\mathbb{E}_{\mathcal{W}}\left[H_t(x)|\mathscr{M}_s\right] = e^{sE}e^{-\int_0^s V(B_r+x)dr}e^{-\phi_{\mathrm{E}}(\int_0^s \delta_r\otimes\varphi(\cdot-x-B_r)dr)}e^{-(t-s)(H_{\mathrm{N}}-E)}\Phi(B_s+x) = H_s(x).$$

Hence $(H_t(x))_{t\geq 0}$ is a martingale. □

2.10.2 Exponential decay of bound states

Let τ be a stopping time with respect to $\mathscr{M}_t$. Then $H_{t\wedge\tau}(x)$ is also a martingale, in particular, $\mathbb{E}_{\mu_{\mathrm{E}}}\mathbb{E}_{\mathcal{W}}[H_t(x)] = \mathbb{E}_{\mu_{\mathrm{E}}}\mathbb{E}_{\mathcal{W}}[H_{t\wedge\tau}(x)]$.

Lemma 2.96. *Let $H_{\mathrm{N}}\Phi = E\Phi$. Then $\|\Phi(\cdot)\|_{L^2(\mathscr{Q})} \in L^\infty(\mathbb{R}^d)$.*

Proof. It follows that $\Phi(x) = \mathbb{E}_{\mu_{\mathrm{E}}}\mathbb{E}_{\mathcal{W}}[\mathrm{I}_0^* H_t(x)]$ for every $t > 0$. Since

$$\|\mathrm{I}_0^* e^{-\phi_{\mathrm{E}}(\int_0^t \delta_s\otimes\varphi(\cdot-B_s)ds)}\mathrm{I}_t\| \leq 2e^{E(t)}, \tag{2.10.2}$$

where $E(t) = \frac{t}{2}\|\hat{\varphi}/\sqrt{\omega}\|^2 + 2t(1 \vee t)(\|\hat{\varphi}/\sqrt{\omega}\|^2 + \|\hat{\varphi}/\omega\|^2)$, we obtain

$$\|\Phi(x)\| \leq 2e^{tE}e^{E(t)}\left(\mathbb{E}_{\mathcal{W}}^x[e^{-2\int_0^t V(B_r)dr}]\right)^{1/2}\left(\mathbb{E}_{\mathcal{W}}^x[\|\Phi(B_t)\|^2]\right)^{1/2}.$$

Since $\sup_{x\in\mathbb{R}^d}\mathbb{E}_{\mathcal{W}}^x[e^{-2\int_0^t V(B_r)dr}] < \infty$ and $\mathbb{E}_{\mathcal{W}}^x[\|\Phi(B_t)\|^2] \leq C\|\Phi\|$, the lemma follows. □

Remark 2.97. We note that if $\int_{\mathbb{R}^d}|\hat{\varphi}(k)|^2/\omega(k)^3dk < \infty$, then (2.10.2) can be replaced with

$$\|\mathrm{I}_0^* e^{-\phi_{\mathrm{E}}(\int_0^t \delta_s\otimes\varphi(\cdot-B_s)ds)}\mathrm{I}_t\| \leq 2e^{tE(\hat{\varphi})}, \tag{2.10.3}$$

where $E(\hat{\varphi})$ is given by (2.3.15).

Theorem 2.98 (Upper bound of exponential decay). *Assume* $\int_{\mathbb{R}^d}|\hat{\varphi}(k)|^2/\omega(k)^3dk < \infty$. *Let* $H_{\mathrm{N}}\Phi = E\Phi$. *If one of the conditions*

(1) $\lim_{|x|\to\infty} V(x) = \infty$;

(2) $\lim_{|x|\to\infty} V_-(x) + E + E(\hat{\varphi}) < 0$

holds, then there exist constants $c, C > 0$ *such that*

$$\|\Phi(x)\|_{L^2(\mathcal{Q})} \leq Ce^{-c|x|}.$$

Proof. Suppose (1) holds. Let $\tau_R = \inf\{t \mid |B_t| > R\}$, which is a stopping time with respect to the natural filtration of Brownian motion $(B_t)_{t\geq 0}$. Consider

$$W_R(x) = \inf\{V(y) \mid |x-y| < R\}.$$

Note that $W_R(x) \leq V(x+y)$ for $|y| < R$, and

$$W_R(x) - E - E(\hat{\varphi}) \to \infty \quad \text{as } |x| \to \infty.$$

Let $\Psi \in L^2(\mathcal{Q})$. The process $(\mathrm{I}_0\Psi \cdot H_t(x))_{t\geq 0}$ is also a martingale since

$$\mathbb{E}_{\mu_{\mathrm{E}}}\mathbb{E}_{\mathcal{W}}[\mathrm{I}_0\Psi\cdot H_t(x)|\mathcal{M}_s] = \mathrm{I}_0\Psi\cdot\mathbb{E}_{\mu_{\mathrm{E}}}\mathbb{E}_{\mathcal{W}}[H_t(x)|\mathcal{M}_s] = \mathrm{I}_0\Psi\cdot H_s(x), \quad 0 < s < t.$$

Hence it follows that

$$(\mathrm{I}_0\Psi, \Phi(x))_{L^2(\mathcal{Q})} = \mathbb{E}_{\mu_{\mathrm{E}}}\mathbb{E}_{\mathcal{W}}[\mathrm{I}_0\Psi\cdot H_0(x)] = \mathbb{E}_{\mu_{\mathrm{E}}}\mathbb{E}_{\mathcal{W}}[\mathrm{I}_0\Psi\cdot H_{t\wedge\tau_R}(x)].$$

Using estimate (2.10.3), we get

$$\begin{aligned}\|\Phi(x)\| &= \sup_{\Psi\in L^2(\mathcal{Q}),\|\Psi\|=1}|(\mathrm{I}_0\Psi,\Phi(x))| = \sup_{\Psi\in L^2(\mathcal{Q}),\|\Psi\|=1}\left|\mathbb{E}_{\mu_{\mathrm{E}}}\mathbb{E}_{\mathcal{W}}[\mathrm{I}_0\Psi\cdot H_{t\wedge\tau_R}(x)]\right| \\ &\leq \|\mathbb{E}_{\mathcal{W}}[\mathrm{I}_0^*H_{t\wedge\tau_R}(x)]\| \leq 2\mathbb{E}_{\mathcal{W}}\left[e^{-\int_0^{t\wedge\tau_R}(V(B_r+x)-E-E(\hat{\varphi}))dr}\right]\sup_{x\in\mathbb{R}^d}\|\Psi_{\mathrm{g}}(x)\|.\end{aligned}$$

We estimate the above expectation as

$$\mathbb{E}_{\mathcal{W}}[e^{-\int_0^{t\wedge\tau_R}(V(B_r+x)-E-E(\hat{\varphi}))dr}] \le \mathbb{E}_{\mathcal{W}}[e^{-(t\wedge\tau_R)(W_R(x)-E-E(\hat{\varphi}))}]$$
$$= \mathbb{E}_{\mathcal{W}}[1_{\{\tau_R<t\}}e^{-\int_0^{t\wedge\tau_R}(W_R(x)-E-E(\hat{\varphi}))dr}] + \mathbb{E}_{\mathcal{W}}[1_{\{\tau_R\ge t\}}e^{-\int_0^{t\wedge\tau_R}(W_R(x)-E-E(\hat{\varphi}))dr}].$$

The bound

$$\mathbb{E}_{\mathcal{W}}[1_{\{\tau_R\ge t\}}e^{-\int_0^{t\wedge\tau_R}(W_R(x)-E-E(\hat{\varphi}))dr}] \le e^{-t(W_R(x)-E-E(\hat{\varphi}))}$$

is obtained trivially. We also see that

$$\mathbb{E}_{\mathcal{W}}[1_{\{\tau_R<t\}}e^{-\int_0^{t\wedge\tau_R}(W_R(x)-E-E(\hat{\varphi}))dr}] \le \mathbb{E}_{\mathcal{W}}[1_{\{\tau_R<t\}}] = \mathbb{E}_{\mathcal{W}}[1_{\{|B_t|\ge R\}}]$$
$$= \int_{|y|\ge R}\Pi_t(y)dy = \frac{S_{d-1}}{(2\pi)^{d/2}}\int_{R/\sqrt{t}}^{\infty} e^{-r^2/2}r^{d-1}dr \le c_1 e^{-c_2R^2/t}.$$

Let $R = p|x|$ with $0 < p < 1$, and $t = \delta|x|$, where we assume that δ is a sufficiently small positive constant. Since $W_{p|x|}(x) - E - E(\hat{\varphi}) \to \infty$ as $|x| \to \infty$, the theorem follows.

Next suppose (2) holds. Let $\tau_R(x) = \inf\{t \ge 0 \mid |B_t + x| \le R\}$. Similarly to (1), it suffices to estimate $\mathbb{E}_{\mathcal{W}}[e^{\int_0^{t\wedge\tau_R(x)}(V_-(B_r+x)+E+E(\hat{\varphi}))dr}]$. Let R be sufficiently large. By the assumption, $|x| > R$ implies that

$$V_-(x) + E + E(\hat{\varphi}) < -\varepsilon < 0.$$

Hence

$$\mathbb{E}_{\mathcal{W}}[e^{\int_0^{t\wedge\tau_R(x)}(V_-(B_r+x)+E+E(\hat{\varphi}))dr}] \le e^{-t\varepsilon} + \mathbb{E}_{\mathcal{W}}^x[1_{\{t\ge\tau_R(0)\}}].$$

We have

$$\mathbb{E}_{\mathcal{W}}^x[1_{\{t\ge\tau_R(0)\}}] = \mathbb{E}_{\mathcal{W}}[1_{\{|B_t+x|\le R|\}}] = \frac{1}{(2\pi t)^{d/2}}\int_{|z|\le R} e^{-|z-x|^2/2t}dz$$
$$= \frac{1}{(2\pi)^{d/2}}\int_{|u|\le R/t} e^{-|u-\frac{x}{\sqrt{t}}|^2/2}du = \frac{1}{(2\pi)^{d/2}}\int_{|u|\le R/t} e^{-\frac{1}{2}(|u|^2-2\frac{u\cdot x}{\sqrt{|x|}}+\frac{|x|^2}{t})}du.$$

Setting $t = R = |x|$ gives

$$\mathbb{E}_{\mathcal{W}}^x[1_{\{t\ge\tau_R(0)\}}] \le \frac{1}{(2\pi)^{d/2}}e^{\sqrt{|x|}-\frac{1}{2}|x|}\int_{|u|\le 1} e^{-\frac{1}{2}|u|^2}du \le c_1e^{-c_2|x|}. \qquad \square$$

2.11 Ultraviolet divergence

2.11.1 Energy renormalization

In this section we consider the problem of removing the ultraviolet cutoff in the Nelson model by using functional integration techniques. After having dealt with the infrared divergence in Section 2.7, this is the second case of encountering singularities that we set to discuss at length.

The version of the Nelson model considered in Sections 2.11.1–2.11.5 is fixed by Assumption 2.99 below. Apart from this, we choose throughout this section space dimensionality

$$d = 3.$$

Assumption 2.99. The following conditions hold:
(1) (Dispersion relation) $\omega(k) = |k|$.
(2) (External potential) V is a continuous bounded function.

Note that in what follows, $V \equiv 0$ is a possible choice without changing the statements. We introduce hard infrared and ultraviolet cutoffs. The ultraviolet cutoff function which we use is

$$\hat{\varphi}_\Lambda(k) = 1_{\{|k|\le\Lambda\}}(k),$$

with ultraviolet cutoff parameter Λ. Furthermore, we introduce a coupling constant g and write the Hamiltonian as

$$H_{\mathrm{N}} = H_{\mathrm{p}} \otimes 1 + 1 \otimes H_{\mathrm{f}} + gH_{\mathrm{I}}.$$

Consider the operator

$$G_\Lambda = -i\frac{g}{\sqrt{2}}\Big(a_M(\hat{\varphi}_{\kappa,\Lambda}\beta e^{ik\cdot x}) - a_M^*(\hat{\varphi}_{\kappa,\Lambda}\beta e^{-ik\cdot x})\Big), \tag{2.11.1}$$

with

$$\beta(k) = \frac{1}{\omega(k) + |k|^2/2} \tag{2.11.2}$$

and

$$\hat{\varphi}_{\kappa,\Lambda}(k) = 1_{\{\kappa\le|k|\le\Lambda\}}(k).$$

Here $\kappa > 0$ is the infrared cutoff parameter. The operator (2.11.1) is formally written as

$$G_\Lambda = -i\frac{g}{\sqrt{2}}\int_{\mathbb{R}^3}\frac{\hat{\varphi}_{\kappa,\Lambda}(k)}{\sqrt{\omega(k)}}\beta(k)\left(e^{ik\cdot x}a(k) - e^{-ik\cdot x}a^*(k)\right)dk.$$

Note that $\hat{\varphi}_{\kappa,\Lambda}\beta \in \hat{\mathcal{H}}_M$, as a consequence of the infrared cutoff $\kappa > 0$, and that s-$\lim_{\Lambda\to\infty}\hat{\varphi}_{\kappa,\Lambda}\beta$ exists in $\hat{\mathcal{H}}_M$.

The conjugation map with respect to the unitary operator e^{iG_Λ} is called the *Gross transform*. A direct calculation shows that the field operators transform as

$$e^{-iG_\Lambda}a_M(\hat{f})e^{iG_\Lambda} = a_M(\hat{f}) - \frac{g}{\sqrt{2}}\int_{\mathbb{R}^3}\frac{\hat{f}(k)}{\sqrt{\omega(k)}}\frac{\hat{\varphi}_{\kappa,\Lambda}(k)}{\sqrt{\omega(k)}}\beta(k)e^{-ik\cdot x}dk,$$

$$e^{-iG_\Lambda}a_M^*(\hat{f})e^{iG_\Lambda} = a_M^*(\hat{f}) - \frac{g}{\sqrt{2}}\int_{\mathbb{R}^3}\frac{\hat{f}(k)}{\sqrt{\omega(k)}}\frac{\hat{\varphi}_{\kappa,\Lambda}(k)}{\sqrt{\omega(k)}}\beta(k)e^{+ik\cdot x}dk.$$

Hence it follows that

$$e^{-iG_\Lambda}H_\mathrm{I}e^{iG_\Lambda} = H_\mathrm{I} - g\left\|\frac{\hat{\varphi}_{\kappa,\Lambda}}{\sqrt{\omega}}\sqrt{\beta}\right\|^2.$$

For the momentum operator, we obtain

$$e^{-iG_\Lambda}pe^{iG_\Lambda} = p - gA_\Lambda - gA_\Lambda^*,$$

where

$$A_\Lambda = -\frac{1}{\sqrt{2}}\int_{\mathbb{R}^3}a(k)k\beta(k)e^{ik\cdot x}\frac{\hat{\varphi}_{\kappa,\Lambda}(k)}{\sqrt{\omega(k)}}dk, \quad \text{and} \quad A_\Lambda^* = -\frac{1}{\sqrt{2}}\int_{\mathbb{R}^3}a^*(k)k\beta(k)e^{-ik\cdot x}\frac{\hat{\varphi}_{\kappa,\Lambda}(k)}{\sqrt{\omega(k)}}dk,$$

and

$$\begin{aligned}&e^{-iG_\Lambda}p^2e^{iG_\Lambda}\\&\quad = p^2 - g(2pA_\Lambda + 2A_\Lambda^*p) + g^2(A_\Lambda^2 + 2A_\Lambda^*A_\Lambda + A_\Lambda^{*2}) - g[p, A_\Lambda^*] - g[A_\Lambda, p] + g^2[A_\Lambda, A_\Lambda^*].\end{aligned}$$

A calculation gives

$$[p, A_\Lambda^*] = C_\Lambda^* = \frac{1}{\sqrt{2}}\int_{\mathbb{R}^3}a^*(k)e^{-ik\cdot x}|k|^2\beta(k)\frac{\hat{\varphi}_{\kappa,\Lambda}(k)}{\sqrt{\omega(k)}}dk,$$

$$[p, A_\Lambda] = C_\Lambda = \frac{1}{\sqrt{2}}\int_{\mathbb{R}^3}a(k)e^{-ik\cdot x}|k|^2\beta(k)\frac{\hat{\varphi}_{\kappa,\Lambda}(k)}{\sqrt{\omega(k)}}dk,$$

$$[A_\Lambda, A_\Lambda^*] = \frac{1}{2}\left\||k|\beta\frac{\hat{\varphi}_{\kappa,\Lambda}}{\sqrt{\omega}}\right\|^2.$$

Furthermore, we have

$$e^{-iG_\Lambda}H_\mathrm{f}e^{iG_\Lambda} = H_\mathrm{f} + gB_\Lambda + gB_\Lambda^* + \frac{g^2}{2}\|\beta\hat{\varphi}_{\kappa,\Lambda}\|^2,$$

where

$$B_\Lambda = -\frac{1}{\sqrt{2}} \int_{\mathbb{R}^3} a(k)\omega(k)\beta(k)e^{ik\cdot x} \frac{\hat{\varphi}_{\kappa,\Lambda}(k)}{\sqrt{\omega(k)}} dk,$$

$$B_\Lambda^* = -\frac{1}{\sqrt{2}} \int_{\mathbb{R}^3} a^*(k)\omega(k)\beta(k)e^{-ik\cdot x} \frac{\hat{\varphi}_{\kappa,\Lambda}(k)}{\sqrt{\omega(k)}} dk.$$

Hence, with $H_0 = H_\mathrm{p} \otimes 1 + 1 \otimes H_\mathrm{f}$,

$$e^{-iG_\Lambda} H_\mathrm{N} e^{iG_\Lambda} = H_0 - g(pA_\Lambda + A_\Lambda^* p) + \frac{g^2}{2}(A_\Lambda^2 + A_\Lambda^{*2} + 2A_\Lambda A_\Lambda^*) + gH_\mathrm{I}^\kappa + E_\Lambda, \tag{2.11.3}$$

where $H_\mathrm{I}^\kappa = H_\mathrm{I} + B_\Lambda + B_\Lambda^* - \frac{1}{2}C_\Lambda - \frac{1}{2}C_\Lambda^*$. We have the expression

$$H_\mathrm{I}^\kappa = \frac{1}{\sqrt{2}} \left(a_M^*(\{1_{|k|<\kappa\}} e^{-ik\cdot x}) + a_M(\{1_{|k|<\kappa\}} e^{ik\cdot x}) \right),$$

i. e., H_I^κ is given by H_I with $\hat{\varphi}_\Lambda$ replaced by $1_{\{|k|<\kappa\}}$. We emphasize that H_I^κ is independent of the ultraviolet cutoff parameter Λ. The additive constant

$$E_\Lambda = -g^2 \left\| \sqrt{\beta} \frac{\hat{\varphi}_{\kappa,\Lambda}}{\sqrt{\omega}} \right\|^2 + \frac{g^2}{4} \left\| |k|\beta \frac{\hat{\varphi}_{\kappa,\Lambda}}{\sqrt{\omega}} \right\|^2 + \frac{g^2}{2} \|\beta \hat{\varphi}_{\kappa,\Lambda}\|^2 = -\frac{g^2}{2} \int_{\mathbb{R}^3} \frac{|\hat{\varphi}_{\kappa,\Lambda}(k)|^2}{\omega(k)} \beta(k) dk$$

diverges as $-\log\Lambda$ when $\Lambda \to \infty$. We denote the right-hand side of (2.11.3) by $\tilde{H}_\mathrm{N}$.

Theorem 2.100 (Existence of Nelson Hamiltonian without ultraviolet cutoff).
(1) *There exists a self-adjoint operator* $\tilde{H}_\mathrm{N}^\mathrm{ren}$ *such that*

$$\lim_{\Lambda\to\infty} e^{-t(\tilde{H}_\mathrm{N} - E_\Lambda)} = e^{-t\tilde{H}_\mathrm{N}^\mathrm{ren}}, \quad t > 0,$$

where the limit is in the uniform topology.
(2) *The unitary operator* e^{iG_Λ} *is strongly convergent to* e^{iG_∞} *as* $\Lambda \to \infty$, *where*

$$G_\infty = -i\frac{g}{\sqrt{2}} \int_{\mathbb{R}^3} \frac{1_{|k|\geq\kappa}(k)\beta(k)}{\sqrt{\omega(k)}} \left(e^{ik\cdot x} a(k) - e^{-ik\cdot x} a^*(k) \right) dk,$$

and then

$$\operatorname*{s-lim}_{\Lambda\to\infty} e^{-t(H_\mathrm{N} - E_\Lambda)} = e^{iG_\infty} e^{-t\tilde{H}_\mathrm{N}^\mathrm{ren}} e^{-iG_\infty}, \quad t > 0. \tag{2.11.4}$$

Operator $\tilde{H}_\mathrm{N}^\mathrm{ren}$ is formally given by

$$\tilde{H}_\mathrm{N}^\mathrm{ren} = H_0 - g(pA + A^* p) + \frac{g^2}{2} \left(A^2 + A^{*2} + 2A^* A \right) + gH_\mathrm{I}^\kappa, \tag{2.11.5}$$

and

$$A = A(x) = -\frac{1}{\sqrt{2}} \int_{\mathbb{R}^3} a(k) k\beta(k) e^{ik\cdot x} \frac{1}{\sqrt{\omega(k)}} dk,$$

$$A^* = A^*(x) = -\frac{1}{\sqrt{2}} \int_{\mathbb{R}^3} a^*(k) k\beta(k) e^{-ik\cdot x} \frac{1}{\sqrt{\omega(k)}} dk.$$

For large enough $|k|$, we have the behavior $\sqrt{|k|}\beta(k) \sim 1/|k|^{3/2}$. Note that the test function of A is not an L^2-function due to

$$\beta_\infty(k) = \frac{1}{\sqrt{\omega(k)}} k\beta(k) e^{ik\cdot x} \notin L^2(\mathbb{R}^3).$$

In particular, we see that

$$[A(x), A^*(y)] = \infty$$

and

$$\tilde{H}_{\mathrm{N}}^{\mathrm{ren}} \neq \frac{1}{2}(p - g(A + A^*))^2 + V + H_{\mathrm{f}}.$$

Therefore (2.11.5) is a formal description only, however, we know that

$$\|A(x)\Psi\| \leq \|\beta_\infty / \sqrt{\omega_\infty}\| \|H_\infty^{1/2}\Psi\|, \tag{2.11.6}$$

where $\omega_\infty(k) = |k|$ for $|k| \geq 1$ and $= 1$ for $|k| < 1$, and $H_\infty = d\Gamma(\omega_\infty)$. Note that

$$\|\beta_\infty / \sqrt{\omega_\infty}\| = \int_{|k|<1} \beta(k)^2 dk + \int_{|k|\geq 1} \frac{\beta(k)^2}{|k|^2} dk < \infty.$$

In spite of the fact that $\beta_\infty \notin L^2(\mathbb{R}^3)$, using (2.11.6) the operator A can be defined on $D(H_\infty^{1/2})$, however, A^* cannot. By a further computation, it can be seen that the quadratic form

$$\begin{aligned}(f,h) \mapsto \mathfrak{q}_N(f,h) &= (f, H_0 h) - g\left((pf, Ah) + (Af, ph)\right) \\ &\quad + \frac{g^2}{2}\left((A^2 f, h) + 2(Af, Ah) + (f, A^2 h)\right) + g(f, H_{\mathrm{I}}^{\kappa} h)\end{aligned}$$

on $D(\bar{H}_0^{1/2}) \times D(\bar{H}_0^{1/2})$ is well defined, and bounded from below uniformly with respect to Λ, where $\bar{H}_0 = H_{\mathrm{p}} + H_{\mathrm{f}} - E_{\mathrm{p}}$. This can be checked by using the estimates

$$\|p_\mu (\bar{H}_0 + 1)^{-1/2}\| < \infty, \tag{2.11.7}$$

$$\|(\bar{H}_0 + 1)^{-1/2} A^2 (\bar{H}_0 + 1)^{-1/2}\| < \infty. \tag{2.11.8}$$

Thus there exists a self-adjoint operator associated with the quadratic form q_N, which is the definition of $\tilde{H}_{\mathrm{N}}^{\mathrm{ren}}$. In the next section, we prove (2.11.4) by using functional integrals.

2.11.2 Regularized interaction

Instead of using the cutoff function $\hat{\varphi}_\Lambda$, we define $H_{\mathrm{N}}^{\varepsilon}$ by H_{N} with cutoff function

$$\hat{\varphi}(k) = e^{-\varepsilon|k|^2/2}, \quad \varepsilon \geq 0.$$

From now on, we consider the Nelson Hamiltonian with N-body Schrödinger operator of the form

$$H_{\mathrm{p}} = -\frac{1}{2}\sum_{j=1}^{N} \Delta_j + V.$$

Here Δ_j denotes the 3-dimensional Laplacian for the jth particle. Define the regularized Hamiltonian

$$H_{\mathrm{N}}^{\varepsilon} = H_{\mathrm{p}} \otimes 1 + 1 \otimes H_{\mathrm{f}} + gH_{\mathrm{I}}^{\varepsilon}, \quad \varepsilon > 0, \tag{2.11.9}$$

with $H_{\mathrm{I}}^{\varepsilon} = \int_{\mathbb{R}^{3N}}^{\oplus} \overline{H_{\mathrm{I}}^{\varepsilon}(x)} dx$ and

$$H_{\mathrm{I}}^{\varepsilon}(x) = \frac{1}{\sqrt{2}} \sum_{j=1}^{N} \left(a_M^*(\hat{\varphi} e^{ik\cdot x_j}) + a_M(\tilde{\hat{\varphi}} e^{-ik\cdot x_j}) \right).$$

Formally, it is written as

$$H_{\mathrm{I}}^{\varepsilon}(x) = \sum_{j=1}^{N} \int_{\mathbb{R}^3} \frac{1}{\sqrt{2\omega(k)}} \left(\hat{\varphi}(k) e^{ik\cdot x_j} a^*(k) + \hat{\varphi}(-k) e^{-ik\cdot x_j} a(k) \right) dk.$$

The Nelson Hamiltonian $H_{\mathrm{N}}^{\varepsilon}$ is a self-adjoint operator acting in the Hilbert space $L^2(\mathbb{R}^{3N}) \otimes \mathscr{F}_{\mathrm{N}}$. Here $\varepsilon > 0$ is regarded as the ultraviolet cutoff parameter. The Nelson Hamiltonian is also defined in $L^2(\mathbb{R}^{3N}) \otimes L^2(\mathscr{Q})$ by

$$\hat{H}_{\mathrm{N}}^{\varepsilon} = H_{\mathrm{p}} \otimes 1 + 1 \otimes \hat{H}_{\mathrm{f}} + g\hat{H}_{\mathrm{I}}^{\varepsilon},$$

where

$$\hat{H}_{\mathrm{I}}^{\varepsilon} = \sum_{j=1}^{N} \int_{\mathbb{R}^3}^{\oplus} \phi(\varphi(\cdot - x_j)) dx_j,$$

and $\phi(f)$ can be extended by $\phi(f) = \phi(\Re f) + i\phi(\Im f)$ for $f \in \mathcal{H}_M$. We also write $H_{\mathrm{N}}^{\varepsilon}$ for $\hat{H}_{\mathrm{N}}^{\varepsilon}$. The main purpose is to consider the limit $\varepsilon \downarrow 0$ of $H_{\mathrm{N}}^{\varepsilon}$. We show that this limit can be sensibly defined by an energy renormalization. Define

$$E_\varepsilon = -\frac{g^2}{2} N \int_{\mathbb{R}^3} \frac{e^{-\varepsilon|k|^2}}{\omega(k)} \beta(k) dk. \tag{2.11.10}$$

Notice that $E_\varepsilon \to -\infty$ as $\varepsilon \downarrow 0$. In the following, we will fix a time interval $[-T, T]$ and track the dependence on T of estimates.

A functional integral representation of $(F, e^{-2TH_{\mathrm{N}}^{\varepsilon}} G)$ is given in terms of Brownian motion and Euclidean field by Theorem 2.12. In particular, for $F = f \otimes 1$ and $G = h \otimes 1$, it can be described in terms of two-sided $3N$-dimensional Brownian motion $(B_t)_{t\in\mathbb{R}}$, $B_t = (B_t^1, \ldots, B_t^N)$, as

$$(f \otimes 1, e^{-2TH_{\mathrm{N}}^{\varepsilon}} h \otimes 1) = \int_{\mathbb{R}^{3N}} \mathbb{E}^x \left[\overline{f(B_{-T})} h(B_T) e^{-\int_{-T}^{T} V(B_s) ds} e^{\frac{g^2}{2} S_\varepsilon^T} \right] dx,$$

where

$$S_\varepsilon^T = \sum_{i,j=1}^{N} \int_{-T}^{T} ds \int_{-T}^{T} W_\varepsilon(B_s^i - B_t^j, s - t) dt \tag{2.11.11}$$

and

$$W_\varepsilon(x, t) = \int_{\mathbb{R}^3} \frac{1}{2\omega(k)} e^{-\varepsilon|k|^2} e^{-ik\cdot x} e^{-|t|\omega(k)} dk.$$

Definition 2.101 (Regularized pair potential and pair interaction). We call $W_\varepsilon(x, t)$ the *regularized pair potential*, and $g^2 S_\varepsilon^T$ in (2.11.11) the *regularized pair interaction.*

Consider the function

$$\varphi_\varepsilon(x, t) = \int_{\mathbb{R}^3} \frac{e^{-\varepsilon|k|^2} e^{-ik\cdot x - |t|\omega(k)}}{2\omega(k)} \beta(k) dk, \quad \varepsilon \geq 0, \tag{2.11.12}$$

where $\beta(k)$ is given by (2.11.2).

Lemma 2.102. *There exists a functional S_0^{ren} such that*

$$\lim_{\varepsilon\downarrow 0} \mathbb{E}^x \left[e^{-\int_{-T}^{T} V(B_s) ds} e^{\frac{g^2}{2}(S_\varepsilon^T - 4NT\varphi_\varepsilon(0,0))} \right] = \mathbb{E}^x \left[e^{-\int_{-T}^{T} V(B_s) ds} e^{\frac{g^2}{2} S_0^{\mathrm{ren}}} \right]. \tag{2.11.13}$$

Notice that $W_\varepsilon(x, t)$ is smooth and $W_\varepsilon(x, t) \to W_0(x, t)$ as $\varepsilon \downarrow 0$ for every $(x, t) \neq (0, 0)$, where

$$W_0(x,t) = \int_{\mathbb{R}^3} \frac{1}{2\omega(k)} e^{-ik\cdot x} e^{-|t|\omega(k)} dk.$$

It is seen, however, that $W_\varepsilon(0,0) \to \infty$ as $\varepsilon \downarrow 0$, i. e., $W_0(x,t)$ has a power-like singularity at $(0,0)$, thus (2.11.13) is nontrivial to obtain. We will prove Lemma 2.102 through a sequence of lemmas below. As $\varepsilon \downarrow 0$, only the diagonal part of the interaction has a singular term. Fix $0 < \tau \le T$ and denote

$$\langle t \rangle = \begin{cases} T, & t \ge T, \\ t, & |t| < T, \\ -T, & t \le -T. \end{cases}$$

We decompose the regularized interaction into diagonal and off-diagonal parts as

$$S_\varepsilon^T = S_\varepsilon^{\mathrm{D},T} + S_\varepsilon^{\mathrm{OD},T},$$

where

$$S_\varepsilon^{\mathrm{D},T} = 2\sum_{i,j=1}^N \int_{-T}^T ds \int_s^{\langle s+\tau\rangle} W_\varepsilon(B_s^i - B_t^j, s-t) dt$$

and

$$S_\varepsilon^{\mathrm{OD},T} = 2\sum_{i,j=1}^N \int_{-T}^T ds \int_{\langle s+\tau\rangle}^T W_\varepsilon(B_s^i - B_t^j, s-t) dt.$$

Here the quantity $S_\varepsilon^{\mathrm{D},T}$ is the integral over a neighborhood of the diagonal region $\{(t,t) \in \mathbb{R}^2 \mid |t| \le T\}$, and $S_\varepsilon^{\mathrm{OD}}$ on its complement. Notice that for $\tau = T$ we have $S_\varepsilon^{\mathrm{OD},T} = 0$. It is easy to show that the pathwise limit

$$\lim_{\varepsilon\downarrow 0} S_\varepsilon^{\mathrm{OD},T} = S_0^{\mathrm{OD},T} = 2\sum_{i,j=1}^N \int_{-T}^T ds \int_s^{\langle s+\tau\rangle} W_0(B_s^i - B_t^j, s-t) dt$$

exists. A representation in terms of a stochastic integral will help us deal with the more difficult term $S_\varepsilon^{\mathrm{D},T}$. We give an upper bound of $\varphi_\varepsilon(x,t)$ in the lemma below.

Lemma 2.103. *There exists a constant $c > 0$ such that the following bounds on φ_ε hold uniformly in ε, x, and t:*

$$|\varphi_\varepsilon(x,t)| \le c(1 + |\log(|t| \wedge 1)|),$$
$$|\nabla_x \varphi_\varepsilon(x,t)| \le c|t|^{-1}, \quad t \ne 0,$$
$$|\nabla_x \varphi_\varepsilon(x,t)| \le c|x|^{-1}, \quad |x| \ne 0,$$

where $\nabla\varphi_\varepsilon = (\partial_1\varphi_\varepsilon, \partial_2\varphi_\varepsilon, \partial_3\varphi_\varepsilon)$ *denotes the gradient of* φ_ε. *Furthermore, similar bounds hold for the function* $\varphi_0 - \varphi_\varepsilon$ *with a constant* $c'(\varepsilon) > 0$ *such that* $\lim_{\varepsilon\to 0} c'(\varepsilon) = 0$*:*

$$|\varphi_\varepsilon(x,t) - \varphi_0(x,t)| \leq c'(\varepsilon)(1 + |\log(|t| \wedge 1)|),$$
$$|\nabla_x\varphi_\varepsilon(x,t) - \nabla_x\varphi_0(x,t)| \leq c'(\varepsilon)|t|^{-1}, \quad t \neq 0,$$
$$|\nabla_x\varphi_\varepsilon(x,t) - \nabla_x\varphi_0(x,t)| \leq c'(\varepsilon)|x|^{-1}, \quad |x| \neq 0.$$

Proof. After integration over the angular variables, we obtain

$$\varphi_\varepsilon(x,t) = 2\pi \int_0^\infty \frac{e^{-\varepsilon r^2 - r|t|}}{2+r} \frac{\sin(r|x|)}{r|x|} dr. \tag{2.11.14}$$

Hence

$$|\varphi_\varepsilon(x,t)| \leq 2\pi C \int_0^\infty \frac{e^{-r|t|}}{2+r} dr = 2\pi C \int_0^\infty \frac{e^{-r}}{2|t| + r} dr,$$

where we used that $\sin a/a \leq C$ for some $C > 0$. Thus

$$|\varphi_\varepsilon(x,t)| \leq 2\pi C \int_0^1 \frac{1}{2|t| + r} dr + 2\pi C \int_1^\infty \frac{e^{-r}}{r} dr$$

which gives the bound

$$|\varphi_\varepsilon(x,t)| \leq C(1 + |\log(|t| \wedge 1)|).$$

The first bound on the gradient follows directly by

$$|\nabla_x\varphi_\varepsilon(x,t)| \leq \int_{\mathbb{R}^3} \frac{1}{2(|k| + |k|^2/2)} e^{-\varepsilon|k|^2} e^{-|k||t|} dk \leq 2\pi \int_0^\infty \frac{r^2}{r + r^2/2} e^{-rt} dr \leq c \int_0^\infty e^{-rt} dr.$$

Next consider the second bound on the gradient. Differentiation in (2.11.14) gives

$$\begin{aligned}\nabla_x\varphi_\varepsilon(x,t) &= 2\pi \int_0^\infty \frac{e^{-\varepsilon r^2 - r|t|}}{r(2+r)} \left(\frac{\cos(r|x|)rx}{|x|^2} - \frac{\sin(r|x|)x}{|x|^3} \right) dr \\ &= \frac{2\pi x}{|x|^2} \int_0^\infty \frac{e^{-\varepsilon r^2/|x|^2 - |t|r/|x|}}{r(2|x| + r)} (r\cos r - \sin r)\, dr,\end{aligned}$$

where we made the change of variable $r \mapsto r|x|$. The integral is uniformly bounded in ε and t; indeed

$$\Big|\int_0^\infty \frac{e^{-\varepsilon r^2/|x|^2-|t|r/|x|}}{r(2|x|+r)}(r\cos r-\sin r)\,dr\Big|$$
$$\le \int_0^1 \frac{|r\cos r-\sin r|}{r^2}dr+\Big|\int_1^\infty \frac{e^{-\varepsilon r^2/|x|^2-|t|r/|x|}}{(2|x|+r)}\cos r\,dr\Big|+\Big|\int_1^\infty \frac{e^{-\varepsilon r^2/|x|^2-|t|r/|x|}}{r(2|x|+r)}\sin r dr\Big|$$
$$\le \int_0^1 \frac{Cr^3}{r^2}dr+\Big|\int_1^\infty \frac{e^{-\varepsilon r^2/|x|^2-|t|r/|x|}}{(2|x|+r)}\cos rdr\Big|+\int_1^\infty \frac{1}{r^2}dr,$$

using $|r\cos r-\sin r|\le Cr^3$ for a constant C and all $r\in[0,1]$. It remains to show that the integral

$$J=\int_1^\infty \frac{e^{-\varepsilon r^2/|x|^2-|t|r/|x|}}{(2|x|+r)}\cos rdr$$

is bounded. We have

$$J=-\frac{e^{-\varepsilon/|x|^2-|t|/|x|}}{(2|x|+1)}\sin 1-\int_1^\infty \frac{d}{dr}\left(\frac{e^{-\varepsilon r^2/|x|^2-|t|r/|x|}}{(2|x|+r)}\right)\sin rdr.$$

Thus

$$|J|\le \sin 1+\int_1^\infty \Big|\frac{d}{dr}\left(\frac{e^{-\varepsilon r^2/|x|^2-|t|r/|x|}}{(2|x|+r)}\right)\Big|dr\le \sin 1+\int_1^\infty \frac{dr}{r^2}<\infty. \qquad \square$$

In order to subtract the singular part from $S_\varepsilon^{\mathrm{D},T}$, we define

$$X_\varepsilon^T=2\sum_{i\ne j}^N\int_{-T}^T \varphi_\varepsilon(B_s^i-B_s^j,0)ds,$$
$$Y_\varepsilon^T=2\sum_{i,j=1}^N\int_{-T}^T ds\int_s^{\langle s+\tau\rangle}\nabla\varphi_\varepsilon(B_s^i-B_t^j,s-t)dB_t^j,$$
$$Z_\varepsilon^T=-2\sum_{i,j=1}^N\int_{-T}^T \varphi_\varepsilon(B_s^i-B_{\langle s+\tau\rangle}^j,s-\langle s+\tau\rangle)ds.$$

A heuristic idea to subtract the singular part is to apply the Itô formula to the random variable $\varphi_\varepsilon(B_s^i-B_t^j,s-t)$. Fix s, i, and j with $i\ne j$. We formally have

$$d\varphi_\varepsilon=\nabla\varphi_\varepsilon dB_t^i+(\frac{1}{2}\Delta+\frac{d}{dt})\varphi_\varepsilon dt.$$

Since $(\frac{1}{2}\Delta + \frac{d}{dt})\varphi_\varepsilon = -W_\varepsilon = -W_\varepsilon(B_s^i - B_t^j, s - t)$, we see that

$$d\varphi_\varepsilon = \nabla\varphi_\varepsilon dB_t^j - W_\varepsilon dt,$$

and hence it follows that

$$\int_s^T W_\varepsilon dt = \int_s^T \nabla\varphi_\varepsilon dB_t^j + \varphi_\varepsilon(B_s^i - B_s^j, 0) - \varphi_\varepsilon(B_s^i - B_T^j, s - T).$$

By symmetry, we can also see that

$$\int_{-T}^T W_\varepsilon dt = 2\int_s^T \nabla\varphi_\varepsilon dB_t^j - 2\varphi_\varepsilon(B_s^i - B_T^j, s - T) + 2\varphi_\varepsilon(B_s^i - B_s^j, 0)$$

and finally,

$$\int_{-T}^T ds \int_{-T}^T W_\varepsilon dt = 2\int_{-T}^T ds \int_s^T \nabla\varphi_\varepsilon dB_t^j - 2\int_{-T}^T \varphi_\varepsilon(B_s^i - B_T^j, s - T)ds + 2\int_{-T}^T \varphi_\varepsilon(B_s^i - B_s^j, 0)ds.$$

Hence we have

$$\begin{aligned}\sum_{i,j=1}^N \int_{-T}^T ds \int_{-T}^T W_\varepsilon dt &= 2\sum_{i,j=1}^N \int_{-T}^T ds \int_s^T \nabla\varphi_\varepsilon dB_t^j - 2\sum_{i,j=1}^N \int_{-T}^T \varphi_\varepsilon(B_s^i - B_T^j, s - T)ds \\ &\quad + 2\sum_{i\neq j}^N \int_{-T}^T \varphi_\varepsilon(B_s^i - B_s^j, 0)ds + +4T\varphi_\varepsilon(0, 0).\end{aligned}$$

In the lemma below, we rigorously derive the formula above.

Lemma 2.104. *We have*

$$S_\varepsilon^{\mathrm{D},T} = 4TN\varphi_\varepsilon(0, 0) + X_\varepsilon^T + Y_\varepsilon^T + Z_\varepsilon^T$$

for all $\varepsilon > 0$.

Proof. Note that $\varphi_\varepsilon(x, t)$ solves the equation

$$\left(\partial_t + \frac{1}{2}\Delta_x\right)\varphi_\varepsilon(x, t) = -W_\varepsilon(x, t), \quad x \in \mathbb{R}^3,\ t \geq 0. \tag{2.11.15}$$

Fix i and j. We have $\varphi_\varepsilon(x, t) = \int_{\mathbb{R}^3} \frac{e^{-\varepsilon|k|^2}}{2\omega(k)}\beta(k)\rho(x, |t|)dk$, where

$$\rho(x, t) = \rho(x, t, k) = e^{-ik\cdot x}e^{-t\omega(k)}.$$

Notice that

$$\int_{\mathbb{R}^3} \frac{e^{-\varepsilon|k|^2}}{2\omega(k)}\beta(k)\rho(x,|t|,k)dk = \int_{\mathbb{R}^3} \frac{e^{-\varepsilon|k|^2}}{2\omega(k)}\beta(k)\rho(-x,|t|,k)dk. \tag{2.11.16}$$

Let $t \geq s$. The Itô formula yields

$$\begin{aligned}\rho(B_t^j,t) - \rho(B_s^j,s) &= \int_s^t \nabla\rho(B_r^j,r)dB_r^j + \int_s^t \left(\partial_t + \frac{1}{2}\Delta_x\right)\rho(B_r^j,r)dr \\ &= \int_s^t -ik\rho(B_r^j,r)dB_r^j + \int_s^t (-\omega(k) - |k|^2/2)\rho(B_r^j,r)dr.\end{aligned}$$

Multiplying by $e^{ik\cdot B_s^i}e^{s\omega(k)}$ on both sides above, we get

$$\begin{aligned}&\rho(B_t^j - B_s^i, t-s) - \rho(B_s^j - B_s^i, 0) \\ &= e^{ik\cdot B_s^i}\int_s^t -i\rho(B_r^j, r-s)k\cdot dB_r^j + \int_s^t (-\omega(k) - |k|^2/2)\rho(B_r^j - B_s^i, r-s)dr \\ &= \int_s^t -i\rho(B_r^j - B_s^i, r-s)k\cdot dB_r^j + \int_s^t (-\omega(k) - |k|^2/2)\rho(B_r^j - B_s^i, r-s)dr.\end{aligned}$$

Here we used that $e^{ik\cdot B_s^i}$ and $\int_s^t -i\rho(B_r^j, r-s)k\cdot dB_r^j$ are independent as $t \geq s$. Multiplying by $\frac{e^{-\varepsilon|k|^2}}{2\omega(k)}\beta(k)$ on both sides and integrating with respect to k gives

$$\begin{aligned}&\int_{\mathbb{R}^3} \frac{e^{-\varepsilon|k|^2}}{2\omega(k)}\beta(k)\left(\rho(B_t^j - B_s^i,|t-s|) - \rho(B_s^j - B_s^i,0)\right)dk \\ &= \int_s^t \left(\int_{\mathbb{R}^3} \frac{e^{-\varepsilon|k|^2}}{2\omega(k)}\beta(k)(-ik)\rho(B_r^j - B_s^i,|r-s|)dk\right)\cdot dB_r^j \\ &\quad + \int_s^t \left(\int_{\mathbb{R}^3} \frac{e^{-\varepsilon|k|^2}}{2\omega(k)}\beta(k)(-\omega(k) - |k|^2/2)\rho(B_r^j - B_s^i,|r-s|)dk\right)dr \\ &= \int_s^t \nabla\varphi_\varepsilon(B_r^j - B_s^i,|r-s|)dB_r^j + \int_s^t \left(\partial_t + \frac{1}{2}\Delta_x\right)\varphi_\varepsilon(B_r^j - B_s^i, r-s)dr.\end{aligned}$$

Hence by using reflection symmetry (2.11.16), we obtain

$$\varphi_\varepsilon(B^i_s - B^j_t, s-t) - \varphi_\varepsilon(B^i_s - B^j_s, 0)$$
$$= \int_s^t \nabla\varphi_\varepsilon(B^i_s - B^j_r, s-r)\cdot dB^j_r + \int_s^t \left(\partial_t + \frac{1}{2}\Delta\right)\varphi_\varepsilon(B^i_s - B^j_r, s-r)dr.$$

Denote $t = \langle s+\tau\rangle$. We have

$$\varphi_\varepsilon(B^i_s - B^j_{\langle s+\tau\rangle}, s - \langle s+\tau\rangle) - \varphi_\varepsilon(B^i_s - B^j_s, 0)$$
$$= \int_s^{\langle s+\tau\rangle} \nabla\varphi_\varepsilon(B^i_s - B^j_t, s-t)dB^j_t + \int_s^{\langle s+\tau\rangle} \left(\partial_t + \frac{1}{2}\Delta\right)\varphi_\varepsilon(B^i_s - B^j_t, s-t)dt.$$

By (2.11.15), for each i, j,

$$\int_s^{\langle s+\tau\rangle} W_\varepsilon(B^i_s - B^j_t, s-t)dt$$
$$= \varphi_\varepsilon(B^i_s - B^j_s, 0) - \varphi_\varepsilon(B^i_s - B^j_{\langle s+\tau\rangle}, s - \langle s+\tau\rangle) + \int_s^{\langle s+\tau\rangle} \nabla\varphi_\varepsilon(B^i_s - B^j_t, s-t)dB^j_t$$

follows. Inserting this expression into $S^{\mathrm{D},T}_\varepsilon$, proves the claim. □

Lemma 2.104 suggests the definition

$$S^{\mathrm{ren}}_\varepsilon = S^T_\varepsilon - 4NT\varphi_\varepsilon(0,0),$$

as a renormalized action, and this can be expressed as

$$S^{\mathrm{ren}}_\varepsilon = S^{\mathrm{OD},T}_\varepsilon + X^T_\varepsilon + Y^T_\varepsilon + Z^T_\varepsilon. \tag{2.11.17}$$

We will estimate each term of (2.11.17). The estimates of $S^{\mathrm{OD},T}_\varepsilon$ and Z^T_ε are straightforward.

Lemma 2.105. *Let $\varepsilon \geq 0$. There exist constants c_Z and c_S such that*

$$|S^{\mathrm{OD},T}_\varepsilon| \leq c_S(T+1),$$
$$|Z^T_\varepsilon| \leq c_Z T$$

uniformly in the paths and in $\varepsilon \geq 0$.

Proof. We see that

$$|Z^T_\varepsilon| \leq 4\pi N^2 \left(\int_{-T}^{T-\tau} ds \int_0^\infty \frac{e^{-r\tau}}{1+r/2}dr + \int_{T-\tau}^{T} ds \int_0^\infty \frac{e^{-r(T-s)}}{1+r/2}dr \right) \leq c_Z T$$

with some $c_z > 0$. It can be also seen that

$$|S_\varepsilon^{\mathrm{OD},T}| \le \int_{-T}^{T-\tau} ds \int_{\tau+s}^{T} dt \int_{\mathbb{R}^3} \frac{1}{2\omega(k)} e^{-(t-s)\omega(k)} dk.$$

Since

$$\int_{\mathbb{R}^3} \frac{1}{2\omega(k)} e^{-(t-s)\omega(k)} dk \le \frac{c}{(t-s)^2}$$

with a constant c. It follows that

$$|S_\varepsilon^{\mathrm{OD},T}| \le \int_{-T}^{T-\tau} ds \int_{\tau+s}^{T} dt \frac{c}{(t-s)^2} = c\left(\left(\frac{2T}{\tau} - 1\right) + \log\left(\frac{\tau}{2T}\right)\right)$$

and the bound $|S_\varepsilon^{\mathrm{OD},T}| \le c_s(T+1)$ follows. □

Next we estimate X_ε^T.

Lemma 2.106. *Let $\varepsilon \ge 0$. There exists a constant c_X independent of ε, such that for all $a > 0$ and $T > 0$,*

$$\sup_{x\in\mathbb{R}^{3N}} \mathbb{E}^x[e^{a|X_\varepsilon^T|}] \le e^{c_X(a^2+a)(T+\sqrt{T})}.$$

Proof. Notice that

$$X_\varepsilon^T = \sum_{i\neq j}^{N} \int_{-T}^{T} \frac{4\pi}{|B_s^i - B_s^j|} ds \int_0^\infty \frac{\sin(r|B_s^i - B_s^j|)}{r + r^2/2} e^{-\varepsilon r^2} dr, \quad \varepsilon \ge 0.$$

We have

$$X_\varepsilon^T \le \sum_{i\neq j}^{N} 4\pi \int_{-T}^{T} \left(\int_0^1 \frac{1}{1+r/2} dr + \int_1^\infty \frac{1}{(r+r^2/2)|B_s^i - B_s^j|} dr \right) ds.$$

Hence employing

$$a = 2\pi \int_1^\infty \frac{1}{r + r^2/2} dr, \quad b = 8N(N-1)\pi \log(3/2),$$

we have

$$|X_\varepsilon^T| \le a \sum_{i\neq j}^{N} \int_{-T}^{T} \frac{1}{|B_s^i - B_s^j|} ds + bT.$$

The lemma follows from Corollary 2.110 below. □

Finally, we estimate the crucial Y_ε^T term. By the stochastic Fubini theorem, we can interchange the stochastic and Lebesgue integrals when $\varepsilon > 0$. Thus Y_ε^T has the representation

$$Y_\varepsilon^T = \sum_{j=1}^{N} \int_{-T}^{T} \Phi_{\varepsilon,t}^j dB_t^j, \quad \varepsilon > 0,$$

where $\Phi_{\varepsilon,t} = (\Phi_{\varepsilon,t}^1, \ldots, \Phi_{\varepsilon,t}^N)$ is a process with values in $\mathbb{R}^{3N}$ given by

$$\Phi_{\varepsilon,t}^j = 2\sum_{i=1}^{N} \int_{\langle t-\tau\rangle}^{t} \nabla\varphi_\varepsilon(B_s^i - B_t^j, s-t)ds.$$

For $\varepsilon = 0$, define

$$Y_0^T = \sum_{j=1}^{N} \int_{-T}^{T} \Phi_{0,t}^j dB_t^j.$$

Lemma 2.107. *Let $\varepsilon \geq 0$ and $\frac{1}{2} < \theta < 1$. Then there exists a constant $c_Y = c_Y(\theta)$, independent of ε, such that*

$$\sup_{x\in\mathbb{R}^{3N}} \mathbb{E}^x[e^{aY_\varepsilon^T}] \leq e^{c_Y(a^2 + a^{\frac{2}{1-\theta}})(T+1)} \quad \textit{for all } a > 0,$$
$$\lim_{\varepsilon\downarrow 0} \mathbb{E}^x[|Y_\varepsilon^T - Y_0^T|^2] = 0 \quad \textit{for all } x \in \mathbb{R}^{3N}.$$

Proof. We have

$$\int_{-T}^{T} |\Phi_{\varepsilon,t}|^2 dt \leq 4\sum_{i=1}^{N} \int_{-T}^{T} \left[\sum_{j=1}^{N} \int_{\langle t-\tau\rangle}^{t} |\nabla\varphi_\varepsilon(B_s^i - B_t^j, s-t)|ds\right]^2 dt$$
$$\leq 4c^2 N \sum_{i,j=1}^{N} \int_{-T}^{T} \left[\int_{\langle t-\tau\rangle}^{t} |B_s^i - B_t^j|^{-\theta} |s-t|^{-(1-\theta)} ds\right]^2 dt,$$

where we used Jensen's inequality, Lemma 2.103, and an interpolation to obtain the bound

$$|\nabla\varphi_\varepsilon(x,t)| \leq c|x|^{-\theta}|t|^{-(1-\theta)}, \quad \theta \in [0,1],$$

which is uniform in $\varepsilon \in [0,1]$. With a suitable $\frac{1}{2} < \theta < 1$, the Schwarz inequality applied to the latter integral gives

$$\int_{-T}^{T} |\Phi_{\varepsilon,t}|^2 dt \leq 4c^2 N \sum_{i,j=1}^{N} \int_{-T}^{T} \left[\int_{\langle t-\tau\rangle}^{t} |B_s^i - B_t^j|^{-2\theta} ds\right] \left(\int_{\langle t-\tau\rangle}^{t} |s-t|^{-2(1-\theta)} ds\right) dt$$

$$\leq 4c^2\tau^{2\theta-1}N\sum_{i,j=1}^{N}\int_{-T}^{T}\left[\int_{\langle t-\tau\rangle}^{t}|B_s^i-B_t^j|^{-2\theta}ds\right]dt$$
$$\leq 4c^2\tau^{2\theta-1}NQ,$$

where c is the constant in Lemma 2.103, which is independent of ε, and

$$Q=\sum_{i,j=1}^{N}\int_{-T}^{T}ds\int_{s}^{\langle s+\tau\rangle}|B_s^i-B_t^j|^{-2\theta}dt.$$

By the Girsanov theorem,

$$\left(\mathbb{E}^x\left[e^{aY_\varepsilon^T}\right]\right)^2\leq\mathbb{E}^x\left[e^{2a\int_{-T}^{T}\Phi_{\varepsilon,t}dB_t-\frac{1}{2}(2a)^2\int_{-T}^{T}|\Phi_{\varepsilon,t}|^2dt}\right]\mathbb{E}^x\left[e^{2a^2\int_{-T}^{T}|\Phi_{\varepsilon,t}|^2dt}\right]$$
$$=\mathbb{E}^x\left[e^{2a^2\int_{-T}^{T}|\Phi_{\varepsilon,t}|^2dt}\right]\leq\mathbb{E}^x\left[e^{\gamma Q}\right],\tag{2.11.18}$$

where $\gamma=8c^2Na^2\tau^{2\theta-1}$, and we recall that we have chosen $\frac{1}{2}<\theta<1$. By Proposition 2.108 and Corollary 2.109 below, we obtain that

$$\sup_{x\in\mathbb{R}^{3N}}\mathbb{E}^x\left[e^{\gamma Q}\right]\leq e^{(\gamma+\gamma^{\frac{1}{1-\theta}})(a+bT)}\tag{2.11.19}$$

with nonnegative constants a and b, and then it follows that

$$\sup_{\varepsilon\in(0,1]}\sup_{x\in\mathbb{R}^{3N}}\mathbb{E}^x[e^{2aY_\varepsilon^T}]\leq e^{(a^2+a^{\frac{2}{1-\theta}})(c_1+c_2T)}$$

with nonnegative constants c_1 and c_2, for all $a\in\mathbb{R}$. For all $\varepsilon>0$, we see that

$$\int_{-T}^{T}|\Phi_{\varepsilon,t}-\Phi_{0,t}|^2dt\leq 4c_\varepsilon^2N\sum_{i,j=1}^{N}\int_{-T}^{T}\left[\int_{\langle t-\tau\rangle}^{t}|B_s^i-B_t^j|^{-2\theta}ds\right]\left(\int_{\langle t-\tau\rangle}^{t}|s-t|^{-2(1-\theta)}ds\right)dt$$
$$\leq 4c_\varepsilon^2\tau^{2\theta-1}N\sum_{i,j=1}^{N}\int_{-T}^{T}\left[\int_{\langle t-\tau\rangle}^{t}|B_s^i-B_t^j|^{-2\theta}ds\right]dt$$
$$\leq 4c_\varepsilon^2\tau^{2\theta-1}NQ,$$

where, as above, we used Lemma 2.103 and an interpolation to obtain the bound

$$|\nabla_x\varphi_\varepsilon(x,t)-\nabla_x\varphi_0(x,t)|\leq c_\varepsilon|x|^{-\theta}|t|^{-(1-\theta)},\quad\theta\in[0,1]$$

for all $\varepsilon>0$, and where the constant c_ε is such that $c_\varepsilon\to 0$ as $\varepsilon\downarrow 0$. The process Φ_ε converges to Φ_0 in $L^2([-T,T],\mathbb{R}^{3N})$ almost surely for all $x\in\mathbb{R}^{3N}$. The almost sure convergence is then a consequence of the fact that $Q<+\infty$ almost surely under $\mathcal{W}^x$

for all $x \in \mathbb{R}^{3N}$, which we have already shown. The convergence of Φ_ε implies also the convergence of Y_ε^T to Y_0^T, at least in $L^2(\mathfrak{X}, \mathcal{W}^x)$, □

We show a general proposition concerning an estimate of pair potentials.

Proposition 2.108. *Suppose that $(B_t^j)_{t\geq 0}$, $j = 1, \ldots, N$, are d-dimensional Brownian motions, and $(B_t^i)_{t\geq 0}$ and $(B_t^j)_{t\geq 0}$ are independent for $i \neq j$. Let f be a nonnegative function, write $f_*(t) = \operatorname{ess\,sup}_{t\leq s\leq T} f(s)$ for $t \leq T$, and choose $1 \leq \alpha < 2$. Then the following hold:*

(1) *For each j, there exist nonnegative constants a_α and b_α such that*

$$\mathbb{E}\Big[\exp\Big(\int_0^T ds \int_0^s \frac{f(s-t)}{|B_s^j - B_t^j|^\alpha} dt\Big)\Big] \leq \exp\Big(a_\alpha \int_0^T ds \Big(\int_0^s f_*(t)dt\Big)^{\frac{2}{2-\alpha}} + b_\alpha \int_0^T ds \int_0^s f_*(t) t^{-\alpha/2} dt\Big).$$

(2) *Let $i \neq j$. Then there exist nonnegative constants a'_α and b'_α such that*

$$\sup_{(x,y)\in\mathbb{R}^{2d}} \mathbb{E}\Big[\exp\Big(\int_0^T ds \int_0^s \frac{f(s-t)}{|B_s^i + x - B_t^j - y|^\alpha} dt\Big)\Big] \leq \exp\Big(a'_\alpha T\Big(\int_0^T f(t)dt\Big)^{\frac{2}{2-\alpha}} + b'_\alpha T^{1-\alpha/2} \int_0^T f(t)dt\Big).$$

(3) *For each j, there exist nonnegative constants a''_α and b''_α such that*

$$\sup_{x\in\mathbb{R}^d} \mathbb{E}^x\Big[\exp\Big(\int_0^T \frac{f(s)}{|B_s^j|^\alpha} ds\Big)\Big] \leq \exp\Big(a''_\alpha \int_0^T f_*(t)^{\frac{2}{2-\alpha}} dt + b''_\alpha \int_0^T f_*(t) t^{-\alpha/2} dt\Big).$$

We can apply Proposition 2.108 to estimate $\sup_{x\in\mathbb{R}^{3N}} \mathbb{E}^x[e^{\gamma Q}]$ and $\sup_{x\in\mathbb{R}^{3N}} \mathbb{E}^x[e^{\alpha|X_\varepsilon^T|}]$.

Corollary 2.109. *Let $1 \leq \alpha < 2$ and $\gamma \geq 0$. Then there exist nonnegative constants a and b such that*

$$\sup_{x\in\mathbb{R}^{dN}} \mathbb{E}^x\Big[\exp\Big(\gamma \sum_{i,j=1}^N \int_0^T ds \int_{s-\tau}^s \frac{1}{|B_s^i - B_t^j|^\alpha} dt\Big)\Big] \leq \exp\Big((\gamma^{\frac{2}{2-\alpha}} + \gamma)(a + bT)\Big). \qquad (2.11.20)$$

In particular, estimate (2.11.19) follows with nonnegative constants a and b dependent on τ, and $\lim_{\tau\to 0} a = \lim_{\tau\to 0} b = 0$.

Proof. We use notation as in Proposition 2.108. Let $f(t) = 1_{[0,\tau]}(t)$. We have

$$\int_0^T ds \int_0^s \frac{f(s-t)}{|B_s^i - B_t^j|^\alpha} dt = \int_0^T ds \int_{s-\tau}^s \frac{1}{|B_s^i - B_t^j|^\alpha} dt.$$

Since $f_*(t) = f(t)$, we see that

$$\begin{aligned}\int_0^T ds\Big(\int_0^s f_*(t)dt\Big)^{2/(2-\alpha)} &= \int_0^\tau s^{2/(2-\alpha)} ds + \int_\tau^T \tau^{2/(2-\alpha)} ds \\ &= \Big(\frac{2-\alpha}{4-\alpha}\tau^{(4-\alpha)/(2-\alpha)} - \tau^{(4-\alpha)/(2-\alpha)}\Big) + \tau^{2/(2-\alpha)} T = a_1(\tau) + a_2(\tau)T\end{aligned}$$

and

$$\begin{aligned}\int_0^T ds \int_0^s f_*(t) t^{-\alpha/2} dt &= \int_0^\tau \frac{1}{1-\alpha/2} s^{1-\alpha/2} ds + \int_\tau^T \frac{1}{1-\alpha/2}\tau^{1-\alpha/2} ds \\ &= \Big(\frac{\tau^{2-\alpha/2}}{(1-\alpha/2)(2-\alpha/2)} - \frac{\tau^{2-\alpha/2}}{1-\alpha/2}\Big) + \frac{\tau^{1-\alpha/2}}{1-\alpha/2} T = b_1(\tau) + b_2(\tau)T.\end{aligned}$$

By Proposition 2.108, for each j we have

$$\begin{aligned}\mathbb{E}\Big[\exp\gamma\Big(\int_0^T ds \int_{s-\tau}^s \frac{1}{|B_s^j - B_t^j|^\alpha} dt\Big)\Big] &\le \exp\Big(a_\alpha \gamma^{\frac{2}{2-\alpha}}(a_1(\tau) + a_2(\tau)T) + b_\alpha\gamma(b_1(\tau) + b_2(\tau)T)\Big) \\ &\le \exp\Big((\gamma^{\frac{2}{2-\alpha}} + \gamma)(d_1 + d_2 T)\Big),\end{aligned} \tag{2.11.21}$$

with nonnegative constants d_1 and d_2 independent of γ. We can also estimate the case when $i \neq j$. Since

$$a_\alpha' T\Big(\int_0^T f(t)dt\Big)^{\frac{2}{2-\alpha}} + b_\alpha' T^{1-\alpha/2}\int_0^T f(t)dt = a_\alpha' T\tau^{2/(2-\alpha)} + b_\alpha' T^{1-\alpha/2}\tau,$$

it follows by Proposition 2.108 again that

$$\begin{aligned}&\sup_{(x,y)\in\mathbb{R}^{2d}} \mathbb{E}\Big[\exp\gamma\Big(\int_0^T ds \int_{s-\tau}^s \frac{1}{|B_s^i + x - B_t^j - y|^\alpha} dt\Big)\Big] \\ &\le \exp\big(a_\alpha'\gamma^{2/(2-\alpha)} T\tau^{2/(2-\alpha)} + b_\alpha'\gamma T^{1-\alpha/2}\tau\big) \le \exp\Big((\gamma^{\frac{2}{2-\alpha}} + \gamma)(d_1' + d_2' T)\Big)\end{aligned} \tag{2.11.22}$$

with nonnegative constants d_1' and d_2' independent of γ. Here we used that $T^{1-\alpha/2} \le T+1$. Let $X_{i,j} = \int_0^T ds \int_{s-\tau}^S \frac{1}{|B_s^i - B_t^j|^\alpha} dt$ and write $Y_1 = X_{1,1}, \dots, Y_N = X_{1,N}, Y_{N+1} = X_{2,1}, \dots, Y_{2N} = X_{2,N}, \dots, Y_{N^2} = X_{N,N}$ for convenience. Applying the Schwarz inequality, we have

$$\sup_{x\in\mathbb{R}^{dN}} \mathbb{E}^x\left[\exp\left(\gamma \sum_{i,j=1}^{N} \int_0^T ds \int_{s-\tau}^{s} \frac{1}{|B_s^i - B_t^j|^\alpha} dt\right)\right]$$

$$= \sup_{x\in\mathbb{R}^{dN}} \mathbb{E}^x\left[\exp\left(\gamma \sum_{k=1}^{N^2} Y_k\right)\right] \le \sup_{x\in\mathbb{R}^{dN}} \prod_{k=1}^{N^2} \left(\mathbb{E}^x\left[\exp(2^k\gamma Y_k)\right]\right)^{1/2^k}. \quad (2.11.23)$$

By (2.11.21)–(2.11.22), each $\mathbb{E}^x[\exp(2^k\gamma Y_k)]$ can be estimated and we have the bound (2.11.20). By the definitions of d_1, d_1', d_2, and d_2', it can be seen that $\lim_{\tau\to 0} d_1 = \lim_{\tau\to 0} d_1' = \lim_{\tau\to 0} d_2 = \lim_{\tau\to 0} d_2' = 0$, and then $\lim_{\tau\to 0} a = \lim_{\tau\to 0} b = 0$ follows. Finally, we have

$$\mathbb{E}^x\left[e^{\gamma Q}\right] = \mathbb{E}^x\left[e^{\gamma \sum_{i,j=1}^N \int_{-T}^T ds \int_s^{(s+\tau)} |B_s^i - B_t^j|^{-2\theta} dt}\right]$$

$$= \mathbb{E}^x\left[e^{\sum_{i,j=1}^N \int_0^{2T} ds \int_s^{(s+\tau)\wedge 2T} |B_s^i - B_t^j|^{-2\theta} dt}\right] \le \mathbb{E}^x\left[e^{\sum_{i,j=1}^N \int_0^{2T} ds \int_s^{s+\tau} |B_s^i - B_t^j|^{-2\theta} dt}\right]$$

and the right-hand side above can be estimated as in (2.11.20) with T and α replaced by $2T$ and 2θ, respectively. □

Corollary 2.110. *There exist nonnegative constants a and b such that*

$$\sup_{x\in\mathbb{R}^{dN}} \mathbb{E}^x\left[\exp\left(\gamma \sum_{i\neq j}^{N} \int_0^T \frac{1}{|B_s^i - B_s^j|} ds\right)\right] \le \exp\left(a\gamma^2 T + b\gamma\sqrt{T}\right).$$

In particular, $\sup_{x\in\mathbb{R}^{3N}} \mathbb{E}^x[e^{\alpha|X_\varepsilon^T|}] \le \exp\left(c_X(\alpha^2 + \alpha)(T + \sqrt{T})\right)$.

Proof. Let $i \neq j$ and fix i and j. Since $(B_t^j)_{t\ge 0}$ and $(B_t^i)_{t\ge 0}$ are independent, $(W_t^{ij})_{t\in\mathbb{R}} = (B_t^i - B_t^j)_{t\in\mathbb{R}}$ is a Brownian motion. By (3) of Proposition 2.108, we have

$$\mathbb{E}^x\left[\exp\left(\gamma \int_0^T \frac{1}{|W_s^{ij}|^\alpha} ds\right)\right] \le \exp\left(a_\alpha \gamma^{2/(2-\alpha)} T + b_\alpha \gamma T^{1-\alpha/2}\right).$$

Putting $\alpha = 1$, in a similar way as in (2.11.23) we prove the corollary. □

Now we can estimate $S_\varepsilon^{\text{ren}}$ for $\varepsilon \ge 0$. Note that

$$S_0^{\text{ren}} = S_0^{\text{OD},T} + X_0^T + Y_0^T + Z_0^T.$$

Lemma 2.111. *Let $1/2 < \theta < 1$. Then there exists a constant c_{ren} independent of $\varepsilon \ge 0$ such that for all $\alpha \in \mathbb{R}$ and $f, h \in L^2(\mathbb{R}^{3N})$ we have for all $\varepsilon > 0$ that*

$$\int_{\mathbb{R}^{3N}} \mathbb{E}^x\left[f(B_{-T})h(B_T)e^{-\int_{-T}^T V(B_s)ds} e^{\alpha S_\varepsilon^{\text{ren}}}\right] dx \le \|f\|\|h\| e^{4T\|V\|_\infty} e^{c_{\text{ren}}\left(\alpha^{\frac{2}{1-\theta}} + \alpha^2 + \alpha\right)(T+1)}.$$

Proof. Recall the decomposition $S_\varepsilon^{\text{ren}} = S_\varepsilon^{\text{OD},T} + X_\varepsilon^T + Y_\varepsilon^T + Z_\varepsilon^T$. By Schwarz inequality, we have

$$\int_{\mathbb{R}^{3N}} \mathbb{E}^x\Big[|f(B_{-T})h(B_T)|e^{-\int_{-T}^T V(B_s)ds} e^{\alpha S_\varepsilon^{\text{ren}}}\Big]dx$$

$$\leq \|f\|\|h\| \sup_{x\in\mathbb{R}^{3N}} \left(\mathbb{E}^x\Big[e^{-2\int_{-T}^T V(B_s)ds} e^{2\alpha(S_\varepsilon^{\text{OD},T}+X_\varepsilon^T+Y_\varepsilon^T+Z_\varepsilon^T)}\Big]\right)^{1/2}.$$

By Lemmas 2.105–2.107, and the fact that V is bounded, we see that there exists a constant c_{ren} such that

$$\sup_{x\in\mathbb{R}^{3N}} \mathbb{E}^x\left[e^{-2\int_{-T}^T V(B_s)ds} e^{2\alpha(S_\varepsilon^{\text{OD},T}+X_\varepsilon^T+Y_\varepsilon^T+Z_\varepsilon^T)}\right] \leq e^{4T\|V\|_\infty} e^{2c_{\text{ren}}(\alpha^{\frac{2}{1-\theta}}+\alpha^2+\alpha)(T+1)}, \tag{2.11.24}$$

and the lemma follows. □

2.11.3 Removal of ultraviolet cutoff on Fock vacuum

In this section we show that $H_{\text{N}}^\varepsilon + g^2 N\varphi_\varepsilon(0,0)$ converges to a self-adjoint operator $H_{\text{N}}^{\text{ren}}$ as $\varepsilon \downarrow 0$.

Lemma 2.112. *If $\alpha \in \mathbb{R}$, then for every $x \in \mathbb{R}^{3N}$,*

$$\lim_{\varepsilon\downarrow 0} \mathbb{E}^x[|e^{\alpha U_\varepsilon^T} - e^{\alpha U_0^T}|] = 0, \quad U = S^{\text{OD}}, X, Y, Z. \tag{2.11.25}$$

Proof. Let $U = X$. We obtain that $|X_\varepsilon^T| \leq bT + \int_{-T}^T V_{\text{Coul}}(B_s)ds$ with $b > 0$, where

$$V_{\text{Coul}}(x) = C\sum_{i\neq j}^N \frac{1}{|x^i - x^j|}$$

with a constant C, and the fact that

$$\mathbb{E}^x\left[|e^{\alpha X_\varepsilon^T} - e^{\alpha X_0^T}|\right] \leq 2\sup_{x\in\mathbb{R}^d} \mathbb{E}^x\left[e^{\alpha(bT+\int_{-T}^T V_{\text{Coul}}(B_s)ds)}\right] < \infty.$$

Since $X_\varepsilon^T \to X_0^T$ a. s. with respect to $\mathcal{W}^x$ for every $x \in \mathbb{R}^{3N}$, the Lebesgue dominated convergence theorem implies (2.11.25).

Let $U = Y$. It suffices to show that $\sup_{x\in\mathbb{R}^{3N}} \mathbb{E}^x[|e^{\alpha(Y_\varepsilon^T - Y_0^T)} - 1|] \to 0$. We have

$$\mathbb{E}^x\left[\left(e^{\alpha(Y_\varepsilon^T-Y_0^T)} - 1\right)^2\right] = \mathbb{E}^x\left[e^{2\alpha(Y_\varepsilon^T-Y_0^T)}\right] + 1 - 2\mathbb{E}^x\left[e^{\alpha(Y_\varepsilon^T-Y_0^T)}\right].$$

We will show below that $\lim_{\varepsilon\downarrow 0}\mathbb{E}^x[e^{\alpha(Y_\varepsilon^T-Y_0^T)}]=1$. Define the random process $(\delta\Phi_t)_{t\in\mathbb{R}}=(\Phi_{\varepsilon,t}-\Phi_{0,t})_{t\in\mathbb{R}}$ so that

$$Y_\varepsilon^T-Y_0^T=\int_{-T}^{T}\delta\Phi_t dB_t.$$

By the Girsanov theorem,

$$\mathbb{E}^x\left[e^{\alpha\int_0^{2T}\delta\Phi_t dB_t-\frac{\alpha^2}{2}\int_0^{2T}|\delta\Phi_t|^2dt}\right]=1 \tag{2.11.26}$$

for every $\alpha\in\mathbb{R}$, hence it follows that

$$\left(\mathbb{E}^x\left[e^{\alpha(Y_\varepsilon^T-Y_0^T)}\right]-1\right)^2\le\mathbb{E}^x\left[e^{2\alpha\int_{-T}^{T}\delta\Phi_t dB_t}\right]\mathbb{E}^x\left[\left(1-e^{-\frac{\alpha^2}{2}\int_{-T}^{T}|\delta\Phi_t|^2dt}\right)^2\right]. \tag{2.11.27}$$

We see that by (2.11.26) again

$$\sup_{x\in\mathbb{R}^{3N}}\mathbb{E}^x\left[e^{2\alpha\int_{-T}^{T}\delta\Phi_t dB_t}\right]\le\sup_{x\in\mathbb{R}^{3N}}\left(\mathbb{E}^x\left[e^{4\alpha^2\int_{-T}^{T}|\delta\Phi_t|^2dt}\right]\right)^{1/2} \tag{2.11.28}$$

and, furthermore,

$$\mathbb{E}^x\left[\left(1-e^{-\frac{\alpha^2}{2}\int_{-T}^{T}|\delta\Phi_t|^2dt}\right)^2\right]\le\mathbb{E}^x\left[\left|\frac{\alpha^2}{2}\int_{-T}^{T}|\delta\Phi_t|^2dt\right|^2\right]\to 0 \tag{2.11.29}$$

as $\varepsilon\downarrow 0$. Here (2.11.29) can be shown by Lemma 2.107. The right-hand side of (2.11.28) is uniformly bounded in ε, which can be proven in the same way as in the proof of Lemma 2.111. Hence (2.11.27) converges to zero as $\varepsilon\downarrow 0$ and (2.11.25) for $U=Y$ follows.

Let $U=Z$. It suffices to show that $\sup_{x\in\mathbb{R}^{3N}}\mathbb{E}^x[|e^{\alpha(Z_\varepsilon^T-Z_0^T)}-1|]\to 0$. We have

$$Z_\varepsilon^T-Z_0^T=2\sum_{i,j=1}^{N}\int_{-T}^{T}ds\int_{\mathbb{R}^3}e^{-ik\cdot(B_s^i+x^i-B_{\langle s+\tau\rangle}^j-x^j)}e^{-(\langle s+\tau\rangle-s)\omega(k)}\frac{1-e^{-\varepsilon|k|^2}}{\omega(k)}\beta(k)dk.$$

Let $\eta_\varepsilon=\alpha(Z_\varepsilon^T-Z_0^T)$. It can be directly seen that $|\eta_\varepsilon|^n\le c^n\alpha^nT^n\varepsilon^n$ for a constant c. We have

$$|\mathbb{E}^x[e^{\eta_\varepsilon}]-1|\le\sum_{n=1}^{\infty}\frac{1}{n!}\mathbb{E}^x[|\eta_\varepsilon|^n]\le\sum_{n=1}^{\infty}\frac{1}{n!}c^nT^n\varepsilon^n\to 0$$

as $\varepsilon\downarrow 0$, uniformly in $x\in\mathbb{R}^{3N}$. Thus (2.11.25) for $U=Z$ follows. For $U=S^{\mathrm{OD}}$, we obtain (2.11.25) in a similar way. □

Lemma 2.113. *Let $a \in \mathbb{R}$ and $f, h \in L^2(\mathbb{R}^{3N})$. Then*

$$\lim_{\varepsilon\downarrow 0} \int_{\mathbb{R}^{3N}} \mathbb{E}^x[f(B_{-T})h(B_T)e^{-\int_{-T}^T V(B_s)ds} e^{aS_\varepsilon^{\mathrm{ren}}}]dx = \int_{\mathbb{R}^{3N}} \mathbb{E}^x[f(B_{-T})h(B_T)e^{-\int_{-T}^T V(B_s)ds} e^{aS_0^{\mathrm{ren}}}]dx.$$

Proof. Writing $\theta_\varepsilon(x) = \mathbb{E}^x[(e^{aS_\varepsilon^{\mathrm{ren}}} - e^{aS_0^{\mathrm{ren}}})^2]dx$ and telescoping, we obtain

$$\begin{aligned}&\left|\int_{\mathbb{R}^{3N}} \mathbb{E}^x\left[f(B_{-T})h(B_T)e^{-\int_{-T}^T V(B_s)ds}\left(e^{aS_\varepsilon^{\mathrm{ren}}} - e^{aS_0^{\mathrm{ren}}}\right)\right]dx\right| \\ &\quad \le e^{2T\|V\|_\infty} \int_{\mathbb{R}^{3N}} |f(x)| \left(\mathbb{E}^x\left[|h(B_T)|^2\right]\right)^{1/2} \theta_\varepsilon(x)^{1/2} dx.\end{aligned}$$

Note that from Lemma 2.111 it follows that $\sup_{x\in\mathbb{R}^{3N}} \theta_\varepsilon(x) < \infty$, and by Lemma 2.112 we see that $\lim_{\varepsilon\downarrow 0} \theta_\varepsilon(x) = 0$ for each $x \in \mathbb{R}^3$. Hence the Lebesgue dominated convergence theorem implies that the second term converges to zero, and the lemma follows. □

Lemma 2.114. *It follows that*

$$\lim_{\varepsilon\downarrow 0}\left(f\otimes 1, e^{-2T(H_N^\varepsilon + g^2 N\varphi_\varepsilon(0,0))} h\otimes 1\right) = \int_{\mathbb{R}^{3N}} \mathbb{E}^x\left[\overline{f(B_{-T})}h(B_T)e^{-\int_{-T}^T V(B_s)ds} e^{\frac{g^2}{2}S_0^{\mathrm{ren}}}\right]dx, \tag{2.11.30}$$

where

$$\begin{aligned} S_0^{\mathrm{ren}} &= 2\sum_{i\neq j}^N \int_{-T}^T \varphi_0(B_s^i - B_s^j, 0)ds - 2\sum_{i,j=1}^N \int_{-T}^T \varphi_0(B_s^i - B_T^j, s-T)ds \\ &\quad + 2\sum_{i,j=1}^N \int_{-T}^T \left(\int_{-T}^t \nabla\varphi_0(B_s^i - B_t^j, s-t)ds\right) dB_t, \end{aligned} \tag{2.11.31}$$

and the integrands are given by

$$\varphi_0(X,t) = \int_{\mathbb{R}^3} \frac{e^{-ik\cdot x}e^{-|t|\omega(k)}}{2\omega(k)}\beta(k)dk, \quad \nabla\varphi_0(X,t) = \int_{\mathbb{R}^3} \frac{-ie^{-ik\cdot x}e^{-|t|\omega(k)}}{2\omega(k)}\beta(k)kdk.$$

Proof. We have

$$(f\otimes 1, e^{-2T(H_N^\varepsilon + g^2 N\varphi_\varepsilon(0,0))} h\otimes 1) = \int_{\mathbb{R}^3} \mathbb{E}^x\left[\overline{f(B_{-T})}h(B_T)e^{-\int_{-T}^T V(B_s)ds} e^{\frac{g^2}{2}S_\varepsilon^{\mathrm{ren}}}\right]dx.$$

The right-hand side converges to $\int_{\mathbb{R}^3} \mathbb{E}^x[\overline{f(B_{-T})}h(B_T)e^{-\int_{-T}^T V(B_s)ds} e^{\frac{g^2}{2}S_0^{\mathrm{ren}}}]dx$ as $\varepsilon \downarrow 0$. Thus (2.11.30) follows. We also see that

$$S_0^{\rm ren} = 2\sum_{i\neq j}^{N}\int_{-T}^{T}\varphi_0(B_s^i - B_s^j, 0)ds + 2\sum_{i,j=1}^{N}\int_{-T}^{T}\left(\int_{\langle t-\tau\rangle}^{t}\nabla\varphi_0(B_s^i - B_t^j, s-t)ds\right)dB_t$$
$$-2\sum_{i,j=1}^{N}\int_{-T}^{T}\varphi_0(B_s^i - B_{\langle s+\tau\rangle}^j, s - \langle s+\tau\rangle)ds + S_0^{\rm OD,\mathit{T}}. \tag{2.11.32}$$

Taking $\tau = T$, we obtain $S_0^{\rm OD,\mathit{T}} = 0$ and (2.11.31) follows. □

Now we extend the result above from functions of the form $f\otimes 1$ to functions of the form $f\otimes F(\phi(f_1),\dots,\phi(f_n))$ with $F\in\mathscr{S}(\mathbb{R}^n)$.

Lemma 2.115. *Let $\rho_j\in\mathscr{M}$ for $j=1,2$, $f,h\in L^2(\mathbb{R}^{3N})$ and $\alpha,\beta\in\mathbb{C}$. Then*

$$\lim_{\varepsilon\downarrow 0}\left(f\otimes e^{\alpha\phi(\rho_1)}1, e^{-2T(H_{\rm N}^\varepsilon+g^2N\varphi_\varepsilon(0,0))}h\otimes e^{\beta\phi(\rho_2)}1\right)$$
$$=\int_{\mathbb{R}^{3N}}\mathbb{E}^x\left[\overline{f(B_{-T})}h(B_T)e^{-\int_{-T}^T V(B_s)ds}e^{\frac{g^2}{2}S_0^{\rm ren}+\frac14\xi}\right]dx,$$

where

$$\xi=\xi(g)=\bar\alpha^2\|\hat\rho_1/\sqrt{\omega}\|^2+\beta^2\|\hat\rho_2/\sqrt{\omega}\|^2+2\bar\alpha\beta(\hat\rho_1/\sqrt{\omega}, e^{-2T\omega}\hat\rho_2/\sqrt{\omega})$$
$$+2\bar\alpha g\sum_{j=1}^N\int_{-T}^T ds\int_{\mathbb{R}^3}\frac{\widehat{\rho_1}(k)}{\sqrt{\omega(k)}}e^{-|s-T|\omega(k)}e^{-ik\cdot B_s^j}dk$$
$$+2\beta g\sum_{j=1}^N\int_{-T}^T ds\int_{\mathbb{R}^3}\frac{\widehat{\rho_2}(k)}{\sqrt{\omega(k)}}e^{-|s+T|\omega(k)}e^{-ik\cdot B_s^j}dk.$$

Proof. By the functional integral representation in Theorem 2.12 we have

$$(f\otimes e^{\alpha\phi(\rho_1)}1, e^{-2T(H_{\rm N}^\varepsilon+g^2N\varphi_\varepsilon(0,0))}h\otimes e^{\beta\phi(\rho_2)}1)$$
$$=\int_{\mathbb{R}^{3N}}\mathbb{E}^x\Big[\overline{f(B_{-T})}h(B_T)e^{-\int_{-T}^T V(B_s)ds}$$
$$\times\mathbb{E}_{\mu_E}[e^{\bar\alpha\phi_{\rm E}(\delta_{-T}\otimes\rho_1)}e^{\beta\phi_{\rm E}(\delta_T\otimes\rho_2)}e^{-g\phi_{\rm E}(\sum_{j=1}^N\int_{-T}^T\delta_s\otimes\varphi_\varepsilon(\cdot-B_s^j)ds)}]\Big]e^{-2Tg^2N\varphi_\varepsilon(0,0)}dx.$$

It can be directly seen that

$$\mathbb{E}_{\mu_E}\left[e^{\bar\alpha\phi_{\rm E}(\delta_{-T}\otimes\rho_1)}e^{\beta\phi_{\rm E}(\delta_T\otimes\rho_2)}e^{-g\phi_{\rm E}(\sum_{j=1}^N\int_{-T}^T\delta_s\otimes\varphi_\varepsilon(\cdot-B_s^j)ds)}\right]e^{-2Tg^2N\varphi_\varepsilon(0,0)}=e^{\frac{g^2}{2}S_\varepsilon^{\rm ren}+\frac14\xi_\varepsilon},$$

where ξ_ε is defined by ξ with $e^{-ik\cdot B_s^j}/\sqrt{\omega(k)}$ replaced by $e^{-ik\cdot B_s^j}e^{-\varepsilon|k|^2/2}/\sqrt{\omega(k)}$. Thus

$$(f \otimes e^{\alpha\phi(\rho_1)}1, e^{-2T(H_N^\varepsilon + g^2 N\varphi_\varepsilon(0,0))} h \otimes e^{\beta\phi(\rho_2)}1) = \int_{\mathbb{R}^{3N}} \mathbb{E}^x \left[\overline{f(B_{-T})} h(B_T) e^{-\int_{-T}^T V(B_s)ds} e^{\frac{g^2}{2} S_\varepsilon^{\rm ren} + \frac{1}{4}\xi_\varepsilon} \right] dx.$$

Notice that with a constant C we have $\xi_\varepsilon \le C$ uniformly in the paths and $\varepsilon \ge 0$. Hence we can complete the proof of the lemma in a similar way to Lemma 2.114. □

Consider the dense subspace of $\mathcal{H}_N$ given by

$$\mathcal{D} = \text{L.H.}\Big\{ f \otimes F(\phi(f_1), \dots, \phi(f_n)), f \otimes 1 \mid f \in L^2(\mathbb{R}^{3N}),\ F \in \mathscr{S}(\mathbb{R}^n),\\ f_j \in C_0^\infty(\mathbb{R}^3),\ 1 \le j \le n,\ n \ge 1 \Big\}. \tag{2.11.33}$$

Lemma 2.116. *For $\Phi = f \otimes F(\phi(u_1), \dots, \phi(u_n))$, $\Psi = h \otimes G(\phi(v_1), \dots, \phi(v_m)) \in \mathcal{D}$,*

$$\lim_{\varepsilon \downarrow 0} (\Phi, e^{-2T(H_N^\varepsilon + g^2 N\varphi_\varepsilon(0,0))}\Psi) \\ = \frac{1}{(2\pi)^{(n+m)/2}} \int_{\mathbb{R}^{n+m}} \overline{\hat{F}(k_1)} \hat{G}(k_2) dk_1 dk_2 \int_{\mathbb{R}^{3N}} \mathbb{E}^x \left[\overline{f(B_{-T})} h(B_T) e^{-\int_{-T}^T V(B_s)ds} e^{\frac{g^2}{2} S_0^{\rm ren} + \frac{1}{4}\xi(k_1,k_2)} \right] dx \tag{2.11.34}$$

holds, where

$$\xi(k_1,k_2) = -\|k_1 \cdot u/\sqrt{\omega}\|^2 - \|k_2 \cdot v/\sqrt{\omega}\|^2 - 2(k_1 \cdot u/\sqrt{\omega}, e^{-2T\omega} k_2 \cdot v/\sqrt{\omega}) \\ - 2ig \sum_{j=1}^N \int_{-T}^T ds \int_{\mathbb{R}^3} \frac{k_1 \cdot \hat{u}(k)}{\sqrt{\omega(k)}} e^{-|s-T|\omega(k)} e^{-ik\cdot B_s^j} dk \\ + 2ig \sum_{j=1}^N \int_{-T}^T ds \int_{\mathbb{R}^3} \frac{k_2 \cdot \hat{v}(k)}{\sqrt{\omega(k)}} e^{-|s+T|\omega(k)} e^{-ik\cdot B_s^j} dk,$$

with $u = (u_1, \dots, u_n)$, $v = (v_1, \dots, v_m)$, and $k_1 \in \mathbb{R}^n$, $k_2 \in \mathbb{R}^m$.

Proof. We have

$$F(\phi(u_1), \dots, \phi(u_n)) = \frac{1}{(2\pi)^{n/2}} \int_{\mathbb{R}^n} \hat{F}(l) e^{i\phi(l\cdot u)} dl,$$

$$G(\phi(v_1), \dots, \phi(v_m)) = \frac{1}{(2\pi)^{m/2}} \int_{\mathbb{R}^m} \hat{G}(l) e^{i\phi(l\cdot v)} dl.$$

Hence

$$(\Phi, e^{-2T(H_N^\varepsilon + g^2 N\varphi_\varepsilon(0,0))}\Psi) \\ = \frac{1}{(2\pi)^{(n+m)/2}} \int_{\mathbb{R}^{m+n}} \overline{\hat{F}(k_1)} \hat{G}(k_2) (f \otimes e^{-i\phi(k_1\cdot u)}, e^{-2T(H_N^\varepsilon + g^2 N\varphi_\varepsilon(0,0))} h \otimes e^{-i\phi(k_2\cdot v)}) dk_1 dk_2.$$

Thus the statement follows by Lemma 2.115. □

The functional integral representation of $\lim_{\varepsilon\downarrow 0}(\Phi, e^{-2T(H_{\mathrm{N}}^{\varepsilon}+g^2N\varphi_\varepsilon(0,0))}\Psi)$ for arbitrary $\Phi, \Psi \in L^2(\mathbb{R}^d)\otimes L^2(\mathscr{Q})$ will be given in Section 2.11.5 in terms of exponentials of creation and annihilation operators.

2.11.4 Uniform lower bound and removal of ultraviolet cutoff

In this section we show a crucial lower bound on $H_{\mathrm{N}}^{\varepsilon} + g^2N\varphi_\varepsilon(0,0)$, uniform in $\varepsilon > 0$, and complete the removal of the ultraviolet cutoff.

Corollary 2.117. *There exists a constant $C \in \mathbb{R}$ such that $H_{\mathrm{N}}^{\varepsilon} + g^2N\varphi_\varepsilon(0,0) > C$ uniformly in $\varepsilon > 0$.*

Proof. Consider the quadratic function

$$v(x^1,\dots,x^N) = \sum_{j=1}^{N}|x^j|^2.$$

We denote $H_{\mathrm{N}}^{\varepsilon}$ with V replaced by $av + V$ by $H_{\mathrm{N}}^{\varepsilon}(a)$ for $a \geq 0$. Since V is bounded,

$$-\frac{1}{2}\sum_{j=1}^{N}\Delta_j + av + V, \quad a > 0,$$

has a compact resolvent, which implies that $H_{\mathrm{N}}^{\varepsilon}(a)$ for $a > 0$ has a unique ground state $\Psi_{\mathrm{g}}(a)$ by Theorem 2.28. Since $\Psi_{\mathrm{g}}(a) > 0$ almost surely due to the positivity improving property of $e^{-tH_{\mathrm{N}}^{\varepsilon}(a)}$, it can be shown that $(f\otimes 1, \Psi_{\mathrm{g}}(a)) \neq 0$ for every $0 \leq f \in L^2(\mathbb{R}^{3N})$, where $f \not\equiv 0$. Thus

$$\inf \operatorname{Spec}\left(H_{\mathrm{N}}^{\varepsilon}(a) + g^2N\varphi_\varepsilon(0,0)\right) = -\lim_{T\to\infty}\frac{1}{T}\log(f\otimes 1, e^{-T(H_{\mathrm{N}}^{\varepsilon}(a)+g^2N\varphi_\varepsilon(0,0))}f\otimes 1). \quad (2.11.35)$$

Using the functional integral representation, we have

$$(f\otimes 1, e^{-2T(H_{\mathrm{N}}^{\varepsilon}(a)+g^2N\varphi_\varepsilon(0,0))}f\otimes 1) = \int_{\mathbb{R}^{3N}} \mathbb{E}^x[f(B_{-T})f(B_T)e^{-\int_{-T}^{T}(av+V)(B_s)ds}e^{S_\varepsilon^{\mathrm{ren}}}]dx.$$

For $\frac{1}{2} < \theta < 1$, by Lemma 2.111 there exists a constant c_{ren} such that

$$\begin{aligned}(f\otimes 1, e^{-2T(H_{\mathrm{N}}^{\varepsilon}(a)+g^2N\varphi_\varepsilon(0,0))}f\otimes 1) &\leq \int_{\mathbb{R}^{3N}} \mathbb{E}^x\left[|f(B_{-T})||f(B_T)|e^{-\int_{-T}^{T}(av+V)(B_s)ds}e^{S_\varepsilon^{\mathrm{ren}}}\right]dx \\ &\leq \|f\|^2 e^{2T\|V\|_\infty}e^{c_{\mathrm{ren}}\left(g^{\frac{4}{1-\theta}}+g^4+g^2\right)(T+1)},\end{aligned}$$

which implies, together with (2.11.35), that

$$\inf \operatorname{Spec}\left(H_{\mathrm{N}}^{\varepsilon}(\alpha) + g^2 N \varphi_\varepsilon(0,0)\right) + b \geq 0, \quad \alpha > 0. \tag{2.11.36}$$

Here $b = c_{\mathrm{ren}}(g^{\frac{4}{1-\theta}} + g^4 + g^2) + \|V\|_\infty$ and b is independent of α. Thus

$$|(F, e^{-2T(H_{\mathrm{N}}^{\varepsilon}(\alpha)+g^2 N\varphi_\varepsilon(0,0))} G)| \leq \|F\|\|G\| e^{2Tb} \tag{2.11.37}$$

follows for every $\alpha > 0$. Let $F, G \in \mathcal{H}_{\mathrm{N}}$. By the functional integral representation again, we have

$$(F, e^{-2TH_{\mathrm{N}}^{\varepsilon}(\alpha)} G) = \int_{\mathbb{R}^3} \mathbb{E}^x \left[e^{-\int_{-T}^{T} (\alpha v + V)(B_s) ds} \left(F_{-T}(B_{-T}), e^{-\phi_{\mathrm{E}}(\int_{-T}^{T} \sum_{j=1}^{N} \delta_s \otimes \varphi(\cdot - B_s^j) ds)} G_T(B_T) \right) \right] dx,$$

where $F_{-T} = \mathrm{I}_{-T} F$ and $G_T = \mathrm{I}_T G$. The Lebesgue dominated convergence theorem furthermore implies that

$$\lim_{\alpha \downarrow 0} (F, e^{-2T(H_{\mathrm{N}}^{\varepsilon}(\alpha)+g^2 N\varphi_\varepsilon(0,0))} G) = (F, e^{-2T(H_{\mathrm{N}}^{\varepsilon}(0)+g^2 N\varphi_\varepsilon(0,0))} G).$$

Taking the limit as $\alpha \downarrow 0$ on both sides of (2.11.37), we have

$$|(F, e^{-2T(H_{\mathrm{N}}^{\varepsilon}(0)+g^2 N\varphi_\varepsilon(0,0))} G)| \leq \|F\|\|G\| e^{2Tb}.$$

This implies that (2.11.36) also holds for $\alpha = 0$. Since $H_{\mathrm{N}}^{\varepsilon} = H_{\mathrm{N}}^{\varepsilon}(0)$ and V is bounded, we obtain

$$\inf \operatorname{Spec}(H_{\mathrm{N}}^{\varepsilon} + g^2 N \varphi_\varepsilon(0,0)) + b \geq 0.$$

Setting $C = -b$ yields the corollary. □

We arrived at the main theorem in this section.

Theorem 2.118 (Removal of ultraviolet cutoff). *There exists a self-adjoint operator $H_{\mathrm{N}}^{\mathrm{ren}}$ such that*

$$\operatorname*{s-lim}_{\varepsilon \downarrow 0} e^{-t(H_{\mathrm{N}}^{\varepsilon} - E_\varepsilon)} = e^{-tH_{\mathrm{N}}^{\mathrm{ren}}}, \quad t \geq 0,$$

where the renormalization term E_ε is given by

$$E_\varepsilon = -g^2 N \hat{\varphi}_\varepsilon(0,0) = -g^2 N \int_{\mathbb{R}^3} \frac{e^{-\varepsilon|k|^2}}{2\omega(k)} \beta(k) dk.$$

Proof. Let $F, G \in \mathcal{H}_{\mathrm{N}}$ and $C_\varepsilon(F, G) = (F, e^{-t(H_{\mathrm{N}}^{\varepsilon}+g^2 N\varphi_\varepsilon(0,0))} G)$. By Lemma 2.115, we obtain that $C_\varepsilon(F, G)$ is convergent as $\varepsilon \downarrow 0$, for every $F, G \in \mathcal{D}$, where $\mathcal{D}$ is given by (2.11.33). Employing the uniform bound

$$\|e^{-t(H_N^\varepsilon+g^2N\varphi_\varepsilon(0,0))}\| < e^{-tC}$$

obtained from Corollary 2.117, and since $\mathcal{D}$ is dense in $\mathcal{H}_N$, we see that $(C_\varepsilon(F,G))_{\varepsilon>0}$ is a Cauchy sequence for $F,G \in \mathcal{H}_N$. Let $C_0(F,G) = \lim_{\varepsilon\downarrow 0} C_\varepsilon(F,G)$. Hence we get $|C_0(F,G)| \le e^{-tC}\|F\|\|G\|$. The Riesz representation theorem implies that there exists a bounded operator T_t such that

$$C_0(F,G) = (F, T_tG), \quad F,G \in \mathcal{H}_N.$$

Thus $\text{s-lim}_{\varepsilon\downarrow 0} e^{-t(H_N^\varepsilon+g^2N\varphi_\varepsilon(0,0))} = T_t$ follows. Furthermore, we also have that

$$\operatorname*{s-lim}_{\varepsilon\downarrow 0} e^{-t(H_N^\varepsilon+g^2N\varphi_\varepsilon(0,0))}e^{-s(H_N^\varepsilon+g^2N\varphi_\varepsilon(0,0))} = \operatorname*{s-lim}_{\varepsilon\downarrow 0} e^{-(t+s)(H_N^\varepsilon+g^2N\varphi_\varepsilon(0,0))} = T_{t+s}.$$

Since the left-hand side above is T_tT_s, the semigroup property of T_t follows. Since $e^{-t(H_N^\varepsilon+g^2N\varphi_\varepsilon(0,0))}$ is a symmetric semigroup, T_t is also symmetric. By the functional integral representation (2.11.34), the functional (F, T_tG) is continuous at $t = 0$ for every $F, G \in \mathcal{D}$. Since $\mathcal{D}$ is dense in $\mathcal{H}_N$ and $\|T_t\|$ is uniformly bounded in a neighborhood of $t = 0$, it also follows that T_t is strongly continuous at $t = 0$. The semigroup version of Stone's theorem implies that there exists a self-adjoint operator H_N^{ren}, bounded from below, such that $T_t = e^{-tH_N^{\mathrm{ren}}}$, $t \ge 0$. The proof is completed by setting $E_\varepsilon = -g^2N\varphi_\varepsilon(0,0)$. □

We can also determine the explicit form of the pair interaction associated with H_N^{ren}.

Corollary 2.119 (Pair interaction associated with renormalized Nelson Hamiltonian). *The pair interaction associated with H_N^{ren} is given by $g^2S_0^{\mathrm{ren}}$.*

Proof. By Lemma 2.114, we see that

$$(f\otimes 1, e^{-2TH_N^{\mathrm{ren}}} h\otimes 1) = \int_{\mathbb{R}^{3N}} \mathbb{E}^x\left[\overline{f(B_{-T})}h(B_T)e^{-\int_{-T}^T V(B_s)ds}e^{\frac{g^2}{2}S_0^{\mathrm{ren}}}\right]dx.$$ □

2.11.5 Functional integral representation of the ultraviolet renormalized Nelson model

We apply Proposition 1.100 to construct a functional integral representation of $e^{-tH_N^{\mathrm{ren}}}$ in terms of exponentials of creation and annihilation operators.

We recall that $(\phi(f), f \in \mathscr{M})$ and $(\phi_E(F), F \in \mathscr{E})$ are families of Gaussian random variables indexed by $\mathscr{M}$ and $\mathscr{E}$, respectively, and

$$\phi_E(F) \cong \frac{1}{\sqrt{2}}(a_E^*(\hat{F}) + a_E(\tilde{\hat{F}})), \quad \hat{F} \in \hat{\mathcal{H}}_E, \tag{2.11.38}$$

$$\phi(f) \cong \frac{1}{\sqrt{2}}(a_M^*(\hat{f}) + a_M(\tilde{\hat{f}})), \quad \hat{f} \in \hat{\mathcal{H}}_M, \tag{2.11.39}$$

hold. Here $a_E^*(\hat{F})$, $a_M^*(\hat{f})$ and $a_E(\hat{F})$, $a_M(\hat{f})$ are creation and annihilation operators on the boson Fock spaces $\mathcal{F}_b(\hat{\mathcal{H}}_E)$ and $\mathcal{F}_b(\hat{\mathcal{H}}_M)$, respectively, satisfying

$$[a_M^*(\hat{f}), a_M(\hat{g})] = (\bar{\hat{f}}, \hat{g})_{\hat{\mathcal{H}}_M}, \quad [a_E^*(\hat{F}), a_E(\hat{G})] = (\bar{\hat{F}}, \hat{G})_{\hat{\mathcal{H}}_E}.$$

We have already given a functional integral representation of $(F, e^{-TH_N^{\mathrm{ren}}} G)$ for $F, G \in \mathcal{D}$, where $\mathcal{D}$ is given in (2.11.33). This will now be extended to $\mathcal{H}_N$.

Theorem 2.120. *Let $F, G \in \mathcal{H}_N$ and*

$$U_T(k) = -\frac{g}{\sqrt{2}} \int_{-T}^{T} e^{-|s+T|\omega(k)} e^{-ik\cdot B_s} ds, \tag{2.11.40}$$

$$\tilde{U}_T(k) = -\bar{U}_{-T}(k) = -\frac{g}{\sqrt{2}} \int_{-T}^{T} e^{-|s-T|\omega(k)} e^{ik\cdot B_s} ds. \tag{2.11.41}$$

We have $U_T, \tilde{U}_T, U_T/\sqrt{\omega}, \tilde{U}_T/\sqrt{\omega} \in \hat{\mathcal{H}}_M$ a. s., i. e., $U_T/\sqrt{\omega}, \tilde{U}_T/\sqrt{\omega}, U_T/\omega, \tilde{U}_T/\omega \in L^2(\mathbb{R}^3)$ a. s., and it follows that

$$(F, e^{-2TH_N^{\mathrm{ren}}} G)_{\mathcal{H}_N} = \int_{\mathbb{R}^3} \mathbb{E}^x \left[e^{-\int_{-T}^{T} V(B_s) ds} e^{\frac{g^2}{2} S_0^{\mathrm{ren}}} (F(B_{-T}), A\tilde{A} G(B_T))_{L^2(\mathcal{Q})} \right] dx, \tag{2.11.42}$$

where A and $\tilde{A}$ are bounded operators on $L^2(\mathcal{Q})$ a. s. defined by

$$A = e^{a_M^*(U_T)} e^{-TH_f}, \quad \tilde{A} = e^{-TH_f} e^{a_M(\tilde{U}_T)}. \tag{2.11.43}$$

Proof. Let $\varrho_\varepsilon = (e^{-\varepsilon|k|^2/2})^\vee = e^{-|x|^2/2\varepsilon}/\varepsilon$ for $\varepsilon > 0$. We have

$$\begin{aligned}&(F, e^{-2T(H_N^\varepsilon - E_\varepsilon)} G)\\ &= \int_{\mathbb{R}^3} \mathbb{E}^x \left[e^{-\int_{-T}^{T} V(B_s) ds} e^{\frac{g^2}{2} S_\varepsilon^{\mathrm{ren}}} (F(B_{-T}), \mathrm{I}_{-T}^* e^{-g\phi_E(\int_{-T}^{T} \delta_s \otimes \varrho_\varepsilon(\cdot - B_s) ds)} \mathrm{I}_T G(B_T)) \right] dx.\end{aligned}$$

Let

$$u_{s,\varepsilon}(k_0, k) = \widehat{\delta_s \otimes \varrho_\varepsilon}(k_0, k) = \frac{e^{-isk_0} e^{-\varepsilon|k|^2/2}}{\sqrt{2\pi}}.$$

Using the identification

$$\phi_E\left(\int_{-T}^{T} \delta_s \otimes \varrho_\varepsilon(\cdot - B_s) ds \right) \cong \frac{1}{\sqrt{2}} \left[a_E^*\left(\int_{-T}^{T} u_{s,\varepsilon} e^{-ik\cdot B_s} ds \right) + a_E\left(\int_{-T}^{T} \tilde{u}_{s,\varepsilon} e^{ik\cdot B_s} ds \right) \right]$$

and the identity

$$\frac{1}{2}\left\| \int_{-T}^{T} u_{s,\varepsilon} e^{-ik\cdot B_s} ds \right\|^2_{\hat{\mathcal{H}}_E} = \left\| \frac{1}{k_0^2 + |k|^2} \int_{-T}^{T} u_{s,\varepsilon} e^{-ik\cdot B_s} ds \right\|^2_{L^2(\mathbb{R}^4)}$$

$$= \int_{-T}^{T} ds \int_{-T}^{T} dt \int_{\mathbb{R}^4} \frac{\overline{u_{t,\varepsilon}} u_{s,\varepsilon} e^{-ik\cdot(B_s - B_t)}}{k_0^2 + |k|^2} dk_0 dk = S_\varepsilon^T,$$

the Baker–Campbell–Hausdorff formula yields

$$e^{-g\phi_{\mathrm{E}}(\int_{-T}^{T} \delta_s \otimes \varrho_\varepsilon(\cdot - B_s) ds)} = e^{\frac{g^2}{2} S_\varepsilon^T} e^{-a_E^*(\frac{g}{\sqrt{2}} \int_{-T}^{T} u_{s,\varepsilon} e^{-ik\cdot B_s} ds)} e^{-a_E(\frac{g}{\sqrt{2}} \int_{-T}^{T} \bar{u}_{s,\varepsilon} e^{ik\cdot B_s} ds)}.$$

We can furthermore compute by using the intertwining property as

$$\mathrm{I}_{-T}^* e^{-g\phi_{\mathrm{E}}(\int_{-T}^{T} u_{s,\varepsilon}(\cdot - B_s) ds)} \mathrm{I}_T = e^{\frac{g^2}{2} S_\varepsilon^T} e^{a_M^*(U_T^\varepsilon)} e^{-2TH_{\mathrm{f}}} e^{a_M(\tilde{U}_T^\varepsilon)},$$

where

$$U_T^\varepsilon(k) = -\frac{g}{\sqrt{2}} \int_{-T}^{T} e^{-|s+T|\omega(k)} e^{-\varepsilon|k|^2/2} e^{-ik\cdot B_s} ds, \tag{2.11.44}$$

$$\tilde{U}_T^\varepsilon(k) = -\bar{U}_{-T}^\varepsilon(k) = -\frac{g}{\sqrt{2}} \int_{-T}^{T} e^{-|s-T|\omega(k)} e^{-\varepsilon|k|^2/2} e^{ik\cdot B_s} ds. \tag{2.11.45}$$

Thus we have a functional integral representation of $e^{-2T(H_{\mathrm{N}}^\varepsilon - E_\varepsilon)}$ in terms of exponentials of annihilation operators and creation operators, $e^{a_M^*(U_T^\varepsilon)} e^{-2TH_{\mathrm{f}}} e^{a_M(\tilde{U}_T^\varepsilon)}$ by

$$(F, e^{-2T(H_{\mathrm{N}}^\varepsilon - E_\varepsilon)} G) = \int_{\mathbb{R}^3} \mathbb{E}^x \left[e^{-\int_{-T}^{T} V(B_s) ds} e^{\frac{g^2}{2} S_\varepsilon^{\mathrm{ren}}} (F(B_{-T}), A_\varepsilon \tilde{A}_\varepsilon G(B_T)) \right] dx,$$

where A_ε is defined by A with U_T replaced by U_T^ε, and $\tilde{A}_\varepsilon$ by $\tilde{A}$ with $\tilde{U}_T$ replaced by $\tilde{U}_T^\varepsilon$. Write

$$\psi(B_s, s) = e^{-|s+T|\omega(k)} e^{-ik\cdot B_s}.$$

By the Itô formula, we have

$$\psi(B_T, T) - \psi(B_{-T}, -T) = \int_{-T}^{T} \nabla\psi(B_s, s) dB_s - \frac{1}{\beta(k)} \int_{-T}^{T} e^{-|s+T|\omega(k)} e^{-ik\cdot B_s} ds.$$

Here we recall that $\beta = \beta(k) = 1/(\omega(k) + |k|^2/2)$. We have

$$\frac{U_T}{\omega} = \frac{g}{\sqrt{2}}\frac{\beta\alpha(T)}{\omega} - \frac{g}{\sqrt{2}}\int_{-T}^{T}\frac{\beta\nabla\psi(B_s, s)}{\omega}dB_s. \tag{2.11.46}$$

Here

$$\alpha(T) = \psi(B_T, T) - \psi(B_{-T}, -T) = e^{-2|T|\omega(k)}e^{-ik\cdot B_T} - e^{-ik\cdot B_{-T}}.$$

Notice that

$$|\alpha(T)| \le \Omega(k), \tag{2.11.47}$$

where

$$\Omega(k) = 2 \wedge (2|T|\omega(k) + |k||B_T - B_{-T}|) \le \begin{cases} 2, \\ 2|k|\left(|T| + \frac{|B_T - B_{-T}|}{2}\right). \end{cases}$$

For every $k \in \mathbb{R}^3$, $\Omega(k)$ is a random variable. We show that the right-hand side of (2.11.46) is an L^2-function with respect to k almost surely. It is trivial to see that

$$\int_{\mathbb{R}^3}\left|\frac{\beta(k)\alpha(T)}{\omega(k)}\right|^2 dk \le \int_{\mathbb{R}^3}\frac{\Omega(k)^2}{\omega(k)^2(\omega(k) + |k|^2/2)^2}dk < \infty.$$

Next we prove that $\int_{-T}^{T}\frac{\beta(k)\nabla\psi(B_s,s)}{\omega(k)}dB_s \in L^2(\mathbb{R}^3)$. By the Itô isometry,

$$\begin{aligned}\mathbb{E}^x\left[\int_{\mathbb{R}^3}\left|\int_{-T}^{T}\frac{\beta(k)\nabla\psi(B_s, s)}{\omega(k)}dB_s\right|^2 dk\right] &= \int_{\mathbb{R}^3}\mathbb{E}^x\left[\left|\int_{-T}^{T}\frac{\beta(k)\nabla\psi(B_s, s)}{\omega(k)}ds\right|^2\right]dk \\ &= \int_{-T}^{T}ds\int_{\mathbb{R}^3}\frac{e^{-2|s+T|\omega(k)}|k|^2}{\omega(k)^2(\omega(k) + |k|^2/2)^2}dk = \int_{\mathbb{R}^3}\frac{(1 - e^{-4T\omega(k)})|k|^2}{2\omega(k)^3(\omega(k) + |k|^2/2)^2}dk < \infty.\end{aligned}$$

Thus it follows that $\mathbb{E}^x[\int_{\mathbb{R}^3}|\frac{U_T(k)}{\omega(k)}|^2 dk] < \infty$, which implies $U_T/\omega \in L^2(\mathbb{R}^3)$ and $\tilde{U}_T/\omega \in L^2(\mathbb{R}^3)$ a. s. Similarly, we can also show $U_T/\sqrt{\omega} \in L^2(\mathbb{R}^3)$ and $\tilde{U}_T/\sqrt{\omega} \in L^2(\mathbb{R}^3)$ a. s. Hence the first statement is proven.

Let $Q_\varepsilon = A_\varepsilon\tilde{A}_\varepsilon$ for $\varepsilon \ge 0$ with $Q_0 = A\tilde{A}$, and

$$P_\varepsilon(x) = \mathbb{E}^x\left[e^{-\int_{-T}^{T}V(B_s)ds}e^{\frac{g^2}{2}S_\varepsilon^{\mathrm{ren}}}(F(B_{-T}), Q_\varepsilon G(B_T))\right].$$

In Lemmas 2.125–2.126 below, we prove the crucial facts

$$(1)\quad P_\varepsilon \in L^1(\mathbb{R}^3_x) \quad \varepsilon \geq 0, \tag{2.11.48}$$

$$(2)\quad \operatorname*{s-lim}_{\varepsilon\downarrow 0} P_\varepsilon = P_0 \quad \text{in } L^1. \tag{2.11.49}$$

By (2.11.48)–(2.11.49), we see that

$$\begin{aligned}(F, e^{-2TH_{\mathrm{N}}^{\mathrm{ren}}} G) &= \lim_{\varepsilon\downarrow 0}(F, e^{-2T(H_{\mathrm{N}}^{\varepsilon}-E_\varepsilon)}G) = \lim_{\varepsilon\downarrow 0}\int_{\mathbb{R}^3} P_\varepsilon(x)dx = \int_{\mathbb{R}^3} P_0(x)\\ &= \int_{\mathbb{R}^3} \mathbb{E}^x\left[e^{-\int_{-T}^T V(B_s)ds} e^{\frac{g^2}{2}S_0^{\mathrm{ren}}}(F(B_{-T}), A\tilde{A}G(B_T))\right] dx.\end{aligned}$$ □

We give a remark on conditions $U_T, \tilde{U}_T, U_T/\sqrt{\omega}, \tilde{U}_T/\sqrt{\omega} \in \hat{\mathcal{H}}_M$ a. s. Condition $U_T, \tilde{U}_T \in \hat{\mathcal{H}}_M$ implies that both $a_M^*(U_T)$ and $a_M(\tilde{U}_T)$ are well defined, and condition $U_T/\sqrt{\omega}, \tilde{U}_T/\sqrt{\omega} \in \hat{\mathcal{H}}_M$ implies that both A and $\tilde{A}$ are bounded. We proceed through a sequence of lemmas to prove (2.11.48)–(2.11.49).

Lemma 2.121. *Let $n \in \mathbb{N}$ and $T > 0$. Then there exist $D_j = D_j(\xi)$, $j = 1, \ldots, 5$, independent of $(x, g, n) \in \mathbb{R}^3 \times \mathbb{R} \times \mathbb{N}$, such that $\lim_{\xi\to 0} D_j(\xi) = \infty$ for $j = 1, 2$ and $\lim_{\xi\to 0} D_j(\xi) = 0$ for $j = 3, 4, 5$, and*

$$\mathbb{E}^x\left[\|e^{a_M^*(U_T^\varepsilon)}e^{-2TH_{\mathrm{f}}}e^{a_M(\tilde{U}_T^\varepsilon)}\|^n\right] \leq 2^n D e^{g^2 nD_1 + g^4 n^2 D_2}, \quad \varepsilon \geq 0$$

with

$$D = \left(\frac{1}{(1-g^2nD_3)^{1/2}\cos^{1/16}(g^2nD_4)\cos^{1/16}(g^2nD_5)}\right)^3.$$

Here we chose ξ such that $(1-g^2nD_3)^{1/2}\cos^{1/16}(g^2nD_4)\cos^{1/16}(g^2nD_5) > 0$.

Proof. We prove the case $\varepsilon = 0$ and $T \geq 1$. Noticing that $e^{-\varepsilon|k|^2/2} < 1$, we can show the lemma for $\varepsilon > 0$ in the same way as for $\varepsilon = 0$. For $T < 1$, by using the estimate for $t < 1$ in Corollary 1.34, we can prove this in a similar way to the case $T \geq 1$. We have the estimate

$$\mathbb{E}^x\left[\|e^{a_M^*(U_T)}e^{-2TH_{\mathrm{f}}}e^{a_M(\tilde{U}_T)}\|^n\right] \leq 2^n\mathbb{E}^x\left[e^{4n\|U_T/\sqrt{\omega}\|_\omega^2}\right]$$

by Corollary 1.34. Here $\|\cdot\|_\omega = \|\cdot\| + \|\cdot/\sqrt{\omega}\|$. We estimate $\mathbb{E}^x[e^{\delta\|U_T/\sqrt{\omega}\|_\omega^2}]$ for $\delta \geq 0$. Denote $g^2/2 = a$. By the Itô formula (2.11.46), we have

$$\left\|\frac{U_T}{\sqrt{\omega}}\right\|_\omega^2 \leq 2a\left\|\frac{\beta a(T)}{\sqrt{\omega}}\right\|_\omega^2 + 2a\left\|\int_{-T}^T\frac{\beta\nabla\psi(B_s,s)}{\sqrt{\omega}}dB_s\right\|_\omega^2 = 2aA_T + 2aB_T.$$

Thus

$$\mathbb{E}^x[e^{\delta\|U_T/\sqrt{\omega}\|_\omega^2}] \leq \mathbb{E}^x[e^{2a\delta(A_T+B_T)}] \leq (\mathbb{E}^x[e^{4a\delta A_T}])^{1/2}(\mathbb{E}^x[e^{4a\delta B_T}])^{1/2}.$$

By Lemmas 2.123–2.124 below, we have the bounds

$$\mathbb{E}^x[e^{4a\delta A_T}] \le \mathbb{E}^x[e^{4a\delta C_A}], \tag{2.11.50}$$

$$\mathbb{E}^x[e^{4a\delta B_T}] \le \left(\mathbb{E}^x[Ce^{CC_1}]\mathbb{E}^x[e^{CC_2}]\right)^{1/4}, \tag{2.11.51}$$

with

$$C_A = \int_{\mathbb{R}^3} \frac{2\Omega(k)^2}{(\omega(k)+|k|^2/2)^2}\left(\frac{1}{\omega(k)}+\frac{1}{\omega(k)^2}\right)dk,$$

$$C_1 = 2\alpha\int_0^\infty \frac{2(\mathrm{Tr}\wedge 1)}{(1+r/2)^2}dr + 8\alpha^2\left(\int_0^\xi \frac{r}{(1+r/2)^2}dr\right)^2\left(\int_{-T}^T |B_s-B_{-T}|^2 ds + 8T^3\right)$$
$$+ 8T\times 8\alpha^2\left(\int_\xi^\infty \frac{1}{(1+r/2)^2}dr\right)^2 + 8\alpha^2\left(\int_0^\infty \frac{(2T\sqrt{r})\wedge\frac{1}{\sqrt{r}}}{(1+r/2)}dr\right)^2,$$

$$C_2 = 2\alpha\int_0^\infty \frac{2(\mathrm{Tr}\wedge 1)}{r(1+r/2)^2}dr + 8\alpha^2\left(\int_0^\xi \frac{1}{(1+r/2)^2}dr\right)^2\left(\int_{-T}^T |B_s-B_{-T}|^2 ds + 8T^3\right)$$
$$+ 8T\times 8\alpha^2\left(\int_\xi^\infty \frac{1}{r(1+r/2)^2}dr\right)^2 + 8\alpha^2\left(\int_0^\infty \frac{(2T\sqrt{r})\wedge\frac{1}{\sqrt{r}}}{r(1+r/2)}dr\right)^2,$$

where C is a positive constant independent of $(x,g,n) \in \mathbb{R}^3\times\mathbb{R}\times\mathbb{N}$, and $\xi > 0$. Here $\alpha = 16a\delta$. We notice that C_A, C_1 and C_2 depend on $w \in \mathfrak{X}$ due to $\Omega(\cdot)$. Hence

$$\mathbb{E}^x[e^{\delta\|U_T/\sqrt{\omega}\|_\omega^2}] \le \left(\mathbb{E}^x[e^{4a\delta C_A}]\right)^{1/2}\left(\mathbb{E}^x[e^{CC_1}]\mathbb{E}^x[e^{CC_2}]\right)^{1/8}.$$

We estimate $\mathbb{E}^x[e^{4a\delta C_A}]$ and $\mathbb{E}^x[e^{CC_1}]\mathbb{E}^x[e^{CC_2}]$ separately.

$$C_A \le \int_{|k|>\xi} \frac{8}{(\omega(k)+|k|^2/2)^2}\left(\frac{1}{\omega(k)}+\frac{1}{\omega(k)^2}\right)dk$$
$$+ \int_{|k|\le\xi} \frac{2(2|T|+|B_T-B_{-T}|)^2|k|^2}{(\omega(k)+|k|^2/2)^2}\left(\frac{1}{\omega(k)}+\frac{1}{\omega(k)^2}\right)dk$$

and

$$\int_{|k|\le\xi} \frac{2(2|T|+|B_T-B_{-T}|)^2|k|^2}{(\omega(k)+|k|^2/2)^2}\left(\frac{1}{\omega(k)}+\frac{1}{\omega(k)^2}\right)dk$$

$$\leq \int_{|k|\leq\xi} \frac{16|T|^2|k|^2}{(\omega(k)+|k|^2/2)^2}\left(\frac{1}{\omega(k)}+\frac{1}{\omega(k)^2}\right)dk$$
$$+ |B_T - B_{-T}|^2 \int_{|k|\leq\xi} \frac{4|k|^2}{(\omega(k)+|k|^2/2)^2}\left(\frac{1}{\omega(k)}+\frac{1}{\omega(k)^2}\right)dk.$$

Hence we conclude that

$$C_A \leq C_A^1 + |B_T - B_{-T}|^2 C_A^2, \tag{2.11.52}$$

where

$$C_A^1 = C_A^1(\xi) = \int_{|k|>\xi} \frac{8}{(\omega(k)+|k|^2/2)^2}\left(\frac{1}{\omega(k)}+\frac{1}{\omega(k)^2}\right)dk$$
$$+ \int_{|k|\leq\xi} \frac{16|T|^2|k|^2}{(\omega(k)+|k|^2/2)^2}\left(\frac{1}{\omega(k)}+\frac{1}{\omega(k)^2}\right)dk,$$
$$C_A^2 = C_A^2(\xi) = \int_{|k|\leq\xi} \frac{4|k|^2}{(\omega(k)+|k|^2/2)^2}\left(\frac{1}{\omega(k)}+\frac{1}{\omega(k)^2}\right)dk.$$

Note that $C_A^1 < \infty$ and $C_A^2 < \infty$, and $\lim_{\xi\to 0} C_A^2(\xi) = 0$ and $\lim_{\xi\to 0} C_A^1(\xi) = \infty$. In general, if $1/4T > c > 0$ then we have

$$\mathbb{E}^x[e^{c|B_T - B_{-T}|^2}] = \frac{1}{(2\pi T)^3} \int_{\mathbb{R}^3\times\mathbb{R}^3} e^{c|x-y|^2} e^{-|x|^2/2T} e^{-|y|^2/2T} dxdy$$
$$\leq \frac{1}{(2\pi T)^3} \int_{\mathbb{R}^3\times\mathbb{R}^3} e^{-(\frac{1}{2T}-2c)(|x|^2+|y|^2)} dxdy = \frac{1}{(1-4cT)^3}. \tag{2.11.53}$$

Since $\mathbb{E}^x[e^{4a\delta C_A}] \leq e^{4a\delta C_A^1}\mathbb{E}^x[e^{4a\delta|B_{-T}-B_T|^2 C_A^2}]$, we obtain that $\mathbb{E}^x[e^{4a\delta C_A}] < \infty$ if $4a\delta C_A^2 < 1/4T$. We can control C_A^2 by varying ξ, leading to

$$\mathbb{E}^x[e^{4a\delta C_A}] \leq e^{4a\delta C_A^1}\mathbb{E}^x[e^{4a\delta|B_{-T}-B_T|^2 C_A^2}] \leq \frac{e^{4a\delta C_A^1}}{(1-16a\delta C_A^2 T)^3},$$

for ξ such that $1 - 16a\delta C_A^2 T > 0$. Hence

$$\mathbb{E}^x[e^{8ng^2 C_A}] \leq \frac{e^{8ng^2 C_A^1}}{(1-32ng^2 T C_A^2)^3} \tag{2.11.54}$$

follows for ξ such that $1 - 32ng^2 TC_A^2 > 0$. Next we estimate C_1 and C_2 through similar computations as applied to C_A. We have

$$\mathbb{E}^X[e^{CC_1}] = e^{CD_1(\xi)}\mathbb{E}^0[e^{8a^2CQ_\xi \int_{-T}^T |B_s-B_{-T}|^2 ds}],$$

where $Q_\xi = \left(\int_0^\xi \frac{r}{(1+r/2)^2}dr\right)^2$ and

$$D_1(\xi) = 2\alpha \int_0^\infty \frac{2(\mathrm{Tr}\wedge 1)}{(1+r/2)^2}dr$$
$$+ 8\alpha^2 \left\{8T^3\left(\int_0^\xi \frac{r}{(1+r/2)^2}dr\right)^2 + 8T\left(\int_\xi^\infty \frac{1}{(1+r/2)^2}dr\right)^2 + \left(\int_0^\infty \frac{(2T\sqrt{r})\wedge \frac{1}{\sqrt{r}}}{(1+r/2)}dr\right)^2\right\}.$$

Note that $\lim_{\xi\to 0} Q_\xi 0$ and $\lim_{\xi\to 0} D_1(\xi) = \infty$. Similarly,

$$\mathbb{E}^X[e^{CC_2}] = e^{CD_2(\xi)}\mathbb{E}^0[e^{8a^2C\tilde{Q}_\xi \int_{-T}^T |B_s-B_{-T}|^2 ds}],$$

where $\tilde{Q}_\xi = \left(\int_0^\xi \frac{1}{(1+r/2)^2}dr\right)^2$ and

$$D_2(\xi) = 2\alpha \int_0^\infty \frac{2(\mathrm{Tr}\wedge 1)}{r(1+r/2)^2}dr$$
$$+ 8\alpha^2 \left\{8T^3\left(\int_0^\xi \frac{1}{(1+r/2)^2}dr\right)^2 + 8T\left(\int_\xi^\infty \frac{1}{r(1+r/2)^2}dr\right)^2 + \left(\int_0^\infty \frac{(2T\sqrt{r})\wedge \frac{1}{\sqrt{r}}}{r(1+r/2)}dr\right)^2\right\}.$$

In general, for a d-dimensional Brownian motion $(B_t)_{t\geq 0}$ and $u > 0$ satisfying $u < \frac{\pi^2}{8T^2}$, we have

$$\mathbb{E}^0\left[\exp\left(u\int_0^T |B_s|^2 ds\right)\right] = (\cos(\sqrt{2u}\,T))^{-d/2}.$$

Using this formula, we can compute

$$\mathbb{E}^0\left[\exp\left(8\alpha^2 CX\int_{-T}^T |B_s - B_{-T}|^2 ds\right)\right] = \frac{1}{\cos(8\alpha\sqrt{CX}\,T)^{3/2}}, \quad X = Q_\xi, \tilde{Q}_\xi.$$

Then we obtain

$$\mathbb{E}[e^{4a\delta B_T}] \leq \left(\frac{e^{C(D_1(\xi)+D_2(\xi))}}{\cos^{3/2}(8\alpha\sqrt{CQ_\xi}T)\cos^{3/2}(8\alpha\sqrt{C\tilde{Q}_\xi}T)}\right)^{1/4}. \tag{2.11.55}$$

Together with (2.11.54) and (2.11.55), we have

$$\mathbb{E}^x[\|e^{a_M^*(U_T)}e^{-2TH_f}e^{a_M(\tilde{U}_T)}\|^n] \leq 2^n \frac{e^{4ng^2C_A^1}}{(1-32ng^2TC_A^2)^{3/2}} \frac{e^{C(D_1(\xi)+D_2(\xi))}}{\cos^{3/16}(8\alpha\sqrt{CQ_\xi}T)\cos^{3/16}(8\alpha\sqrt{C\tilde{Q}_\xi}T)}.$$

Then the lemma is proven. □

The proof of Lemma 2.121 also gives the corollary below.

Corollary 2.122. *For every $T > 0$, there exist constants D_1 and D_2 independent of x such that for every $\varepsilon \geq 0$ it follows that $\mathbb{E}^x[e^{\|U_T^\varepsilon/\sqrt{\omega}\|_\omega^2}] \leq D_1 e^{D_2}$.*

Next we prove the bounds (2.11.50) and (2.11.51) used in the proof of Lemma 2.121.

Lemma 2.123. *It follows that $\mathbb{E}^x[e^{4a\delta A_T}] \leq \mathbb{E}^x[e^{4a\delta C_A}]$.*

Proof. By direct inspection, we have

$$A_T = \left\|\frac{\beta a(T)}{\sqrt{\omega}}\right\|_\omega^2 \leq 2\left\|\frac{\beta a(T)}{\omega}\right\|^2 + 2\left\|\frac{\beta a(T)}{\sqrt{\omega}}\right\|^2 \leq \int_{\mathbb{R}^3} \frac{2\Omega(k)^2}{(\omega(k)+|k|^2/2)^2}\left(\frac{1}{\omega(k)} + \frac{1}{\omega(k)^2}\right) dk = C_A.$$ □

Next we estimate the expectation of the exponent of the random process

$$B_T = \left\|\int_{-T}^{T} \frac{\beta\nabla\psi(B_s, s)}{\sqrt{\omega}} dB_s\right\|_\omega^2, \quad T \geq 0,$$

which is defined by the absolute value of Hilbert space-valued stochastic integrals. The estimate is not straightforward.

Lemma 2.124. *It follows that $\mathbb{E}^x[e^{4a\delta B_T}] \leq (\mathbb{E}^x[e^{CC_1}]\mathbb{E}^x[e^{CC_2}])^{1/4}$.*

Proof. We have $B_T \leq 2B_T^1 + 2B_T^2$, where

$$B_T^1 = \left\|\int_{-T}^{T} \frac{\beta\nabla\psi(B_s, s)}{\sqrt{\omega}} dB_s\right\|^2 \quad \text{and} \quad B_T^2 = \left\|\int_{-T}^{T} \frac{\beta\nabla\psi(B_s, s)}{\omega} dB_s\right\|^2.$$

We see that

$$\mathbb{E}^x[e^{4a\delta B_T}] \leq \left(\mathbb{E}^x[e^{16a\delta B_T^1}]\right)^{1/2}\left(\mathbb{E}^x[e^{16a\delta B_T^2}]\right)^{1/2}.$$

We only consider B_T^1, the case B_T^2 can be dealt with similarly. Let

$$X_t = \int_{-T}^{t} \frac{\beta\nabla\psi(B_r, r)}{\sqrt{\omega}} dB_r.$$

Itô formula yields

$$dX_t^2 = 2dX_t \cdot X_t + dX_t \cdot dX_t = 2X_t \frac{\beta\nabla\psi(B_t,t)}{\sqrt{\omega}} dB_t + \left|\frac{\beta\nabla\psi(B_t,t)}{\sqrt{\omega}}\right|^2 dt.$$

It implies

$$B_T^1 = \|X_T\|^2 = 2\int_{-T}^{T} \Phi_s dB_s + \int_{-T}^{T} \phi_s ds,$$

where $(\Phi_s)_{s\in\mathbb{R}}$ is an $\mathbb{R}^3$-valued random process defined by

$$\Phi_s = \left(\int_{-T}^{s} \frac{\beta\nabla\psi(B_r,r)}{\sqrt{\omega}} dB_r, \frac{\beta\nabla\psi(B_s,s)}{\sqrt{\omega}}\right)_{L^2(\mathbb{R}^3)}, \quad s \in \mathbb{R},$$

and ϕ_s is given by

$$\phi_s = \left\|\frac{\beta\nabla\psi(B_s,s)}{\sqrt{\omega}}\right\|^2.$$

Denote $\alpha = 16a\delta$. By the Girsanov theorem,

$$\begin{aligned}\mathbb{E}^x[e^{\alpha B_T^1}] &= \mathbb{E}^x\left[e^{2\alpha\int_{-T}^T \Phi_s dB_s + \alpha\int_{-T}^T \phi_s ds}\right] = \mathbb{E}^x\left[e^{2\alpha\int_{-T}^T \Phi_s dB_s - 4\alpha^2\int_{-T}^T |\Phi_s|^2 ds + 4\alpha^2\int_{-T}^T |\Phi_s|^2 ds + \alpha\int_{-T}^T \phi_s ds}\right] \\ &\le \left(\mathbb{E}^x\left[e^{4\alpha\int_{-T}^T \Phi_s dB_s - 8\alpha^2\int_{-T}^T |\Phi_s|^2 ds}\right]\right)^{1/2} \left(\mathbb{E}^x\left[e^{8\alpha^2\int_{-T}^T |\Phi_s|^2 ds + 2\alpha\int_{-T}^T \phi_s ds}\right]\right)^{1/2},\end{aligned}$$

thus it is enough to estimate the second factor since the first equals one.

First, it can be seen that

$$\int_{-T}^{T} \|\phi_s\|^2 ds \le C_1 \int_{-T}^{T} ds \int_0^\infty \frac{e^{-|s+T|r} r}{(1+r/2)^2} dr = c_1 \int_0^\infty \frac{1-e^{-2\mathrm{Tr}}}{(1+r/2)^2} dr \le c_2 \int_0^\infty \frac{2(\mathrm{Tr} \wedge 1)}{(1+r/2)^2} dr < \infty$$

for some constants $c_1, c_2 > 0$. Second, we estimate $\int_{-T}^T |\Phi_s|^2 ds$. Inserting formula (2.11.46) into $\int_{-T}^T \Phi_s dB_s$ gives

$$\begin{aligned}\int_{-T}^{T} |\Phi_s|^2 ds &= \int_{-T}^{T} \left|\left(\int_{-T}^{s} \frac{\beta\nabla\psi(B_r,r)}{\sqrt{\omega}} dB_r, \frac{\beta\nabla\psi(B_s,s)}{\sqrt{\omega}}\right)\right|^2 ds \\ &= \int_{-T}^{T} \left|\left(\frac{\sqrt{2}}{g}\frac{U_s}{\sqrt{\omega}} - \frac{\beta a(s)}{\sqrt{\omega}}, \frac{\beta\nabla\psi(B_s,s)}{\sqrt{\omega}}\right)\right|^2 ds\end{aligned}$$

$$\leq \frac{4}{g^2}\int_{-T}^{T}\left|\left(\frac{U_s}{\sqrt{\omega}}, \frac{\beta\nabla\psi(B_s,s)}{\sqrt{\omega}}\right)\right|^2 ds + 2\int_{-T}^{T}\left|\left(\frac{\beta\alpha(s)}{\sqrt{\omega}}, \frac{\beta\nabla\psi(B_s,s)}{\sqrt{\omega}}\right)\right|^2 ds. \quad (2.11.56)$$

Here U_s and $\alpha(s)$ are given by

$$U_s = -\frac{g}{\sqrt{2}}\int_{-T}^{s} e^{-|s-T|\omega}e^{-ikB_s}ds,$$

$$\alpha(s) = \psi(B_s,s) - \psi(B_{-T},-T) = e^{-|s+T|\omega}e^{-ikB_s} - e^{-ikB_{-T}}.$$

We have the bound $|\alpha(s)| \leq \Omega(k,s)$, where

$$\Omega(k,s) = 2 \wedge (|s+T|\omega(k) + |k||B_s - B_{-T}|) \leq \begin{cases} 2, \\ |k|(|T+s| + |B_s - B_{-T}|). \end{cases}$$

We estimate both terms at the right-hand side of (2.11.56) separately. In the calculations below, C will denote a positive constant which is independent of g and may differ from line to line. The integrand is estimated directly as

$$\left|\left(\frac{\beta\alpha(s)}{\sqrt{\omega}}, \frac{\beta\nabla\psi(B_s,s)}{\sqrt{\omega}}\right)\right| \leq \int_{\mathbb{R}^3}\frac{\Omega(k,s)e^{-|s+T|\omega(k)}|k|}{\omega(k)(\omega(k)+|k|^2/2)^2}dk \leq C\int_0^\infty \frac{e^{-|s+T|r}\Omega(r,s)}{(1+r/2)^2}dr$$

and then

$$\int_{-T}^{T}\left|\left(\frac{\beta\alpha(s)}{\sqrt{\omega}}, \frac{\beta\nabla\psi(B_s,s)}{\sqrt{\omega}}\right)\right|^2 ds \leq C\int_{-T}^{T}\left(\int_0^\infty \frac{e^{-|s+T|r}\Omega(r,s)}{(1+r/2)^2}dr\right)^2 ds.$$

In order to avoid the divergence of the integral at the origin $r = 0$, we introduce a regularization parameter $\xi > 0$. Near the origin, we employ the estimate

$$\Omega(r,s) \leq r(|T+s| + |B_s - B_{-T}|),$$

whereas away from the origin we use the uniform bound

$$\Omega(r,s) \leq 2.$$

These two estimates allow us to control the integrand uniformly over the entire domain. Using the bound $|\alpha(s)| \leq \Omega(k,s)$ we obtain

$$\int_{-T}^{T}\left(\int_0^\infty \frac{e^{-|s+T|r}\Omega(r,s)}{(1+r/2)^2}dr\right)^2 ds$$

$$\leq C\int_{-T}^{T}\left(\int_{0}^{\xi}\frac{e^{-|s+T|r}\Omega(r,s)}{(1+r/2)^2}dr\right)^2 ds + C\int_{-T}^{T}\left(\int_{\xi}^{\infty}\frac{e^{-|s+T|r}\Omega(r,s)}{(1+r/2)^2}dr\right)^2 ds$$

$$\leq C\int_{-T}^{T}\left(\int_{0}^{\xi}\frac{e^{-|s+T|r}r|B_s-B_{-T}|}{(1+r/2)^2}dr\right)^2 ds + C\int_{-T}^{T}\left(\int_{0}^{\xi}\frac{e^{-|s+T|r}2rT}{(1+r/2)^2}dr\right)^2 ds$$

$$+ C\int_{-T}^{T}\left(\int_{\xi}^{\infty}\frac{2e^{-|s+T|r}}{(1+r/2)^2}dr\right)^2 ds$$

$$= CQ_\xi\left(\int_{-T}^{T}|B_s-B_{-T}|^2 ds + 8T^3\right) + 8TCR_\xi. \tag{2.11.57}$$

Here $Q_\xi = \left(\int_0^\xi \frac{r}{(1+r/2)^2}dr\right)^2$ and $R_\xi = \left(\int_\xi^\infty \frac{1}{(1+r/2)^2}dr\right)^2$. Next we estimate

$$\int_{-T}^{T}\left|\left(\frac{U_s}{\sqrt{\omega}}, \frac{\beta\nabla\psi(B_s,s)}{\sqrt{\omega}}\right)\right|^2 ds = \frac{g^2}{2}\int_{-T}^{T}\left|\int_{-T}^{s}\left(\frac{\psi(B_t,t)}{\sqrt{\omega}}, \frac{\beta\nabla\psi(B_s,s)}{\sqrt{\omega}}\right)dt\right|^2 ds. \tag{2.11.58}$$

The integrand is estimated as

$$\left|\left(\frac{\psi(B_t,t)}{\sqrt{\omega}}, \frac{\beta\nabla\psi(B_s,s)}{\sqrt{\omega}}\right)\right| \leq \int_{\mathbb{R}^3}\frac{|k|e^{-|t+T|\omega(k)}e^{-|s+T|\omega(k)}}{\omega(k)(\omega(k)+|k|^2/2)^2}dk \leq C\int_0^\infty \frac{re^{-|t+T|r}e^{-|s+T|r}}{(1+r/2)^2}dr.$$

Hence it follows that

$$\int_{-T}^{T}\left|\int_{-T}^{s}\left(\frac{\psi(B_t,t)}{\sqrt{\omega}}, \frac{\beta\nabla\psi(B_s,s)}{\sqrt{\omega}}\right)dt\right|^2 ds$$

$$\leq C\int_{-T}^{T}ds\int_{-T}^{s}dt\int_{-T}^{s}dp\int_0^\infty dr\int_0^\infty \frac{re^{-|t+T|r}e^{-|s+T|r}}{(1+r/2)}\frac{r'e^{-|p+T|r'}e^{-|s+T|r'}}{(1+r'/2)}dr'$$

$$\leq C\int_{-T}^{T}ds\int_{-T}^{T}dt\int_{-T}^{T}dp\int_0^\infty dr\int_0^\infty \frac{re^{-|t+T|r}e^{-|s+T|r}}{(1+r/2)}\frac{r'e^{-|p+T|r'}e^{-|s+T|r'}}{(1+r'/2)}dr'$$

$$\leq C\int_{-T}^{T}ds\int_0^\infty dr\int_0^\infty \frac{e^{-|s+T|r}(1-e^{-2\mathrm{Tr}})}{(1+r/2)}\frac{e^{-|s+T|r'}(1-e^{-2\mathrm{Tr}'})}{(1+r'/2)}dr'$$

$$= C\int_0^\infty dr\int_0^\infty \frac{(1-e^{-2\mathrm{Tr}})(1-e^{-2\mathrm{Tr}'})}{(r+r')(1+r/2)(1+r'/2)}dr'$$

$$\leq C\left(\int_0^\infty \frac{1-e^{-2\mathrm{Tr}}}{\sqrt{r}(1+r/2)}dr\right)^2 \leq C\left(\int_0^\infty \frac{(2T\sqrt{r})\wedge\frac{1}{\sqrt{r}}}{(1+r/2)}dr\right)^2 < \infty. \tag{2.11.59}$$

By (2.11.57) and (2.11.59), we see that

$$\mathbb{E}^x\left[e^{aB_T^1}\right] \leq \left(\mathbb{E}^x[e^{CC_1}]\right)^{1/8},$$

where C is independent of g by (2.11.56) and (2.11.58). In the same way as above, we can also see that

$$\mathbb{E}^x\left[e^{aB_T^2}\right] \leq \left(\mathbb{E}^x[e^{CC_2}]\right)^{1/8}.$$

Hence $\mathbb{E}^x[e^{4a\delta B_T}] \leq \left(\mathbb{E}^x[e^{CC_1}]\mathbb{E}^x[e^{CC_2}]\right)^{1/8}$ holds true. □

The time shift defined by $B_s \to B_{s+T}$ allows to write

$$(F, e^{-2T(H_N^\varepsilon - E_\varepsilon)}G) = \int_{\mathbb{R}^3} P_\varepsilon(x)dx = \int_{\mathbb{R}^3} \mathbb{E}^x\left[e^{-\int_0^{2T} V(B_s)ds} e^{\frac{g^2}{2}\tilde{S}_\varepsilon^{\mathrm{ren}}}(F(B_0), \tilde{Q}_\varepsilon G(B_{2T}))\right]dx,$$

where $\tilde{S}_\varepsilon^{\mathrm{ren}}$ and $\tilde{Q}_\varepsilon$ are defined by $S_\varepsilon^{\mathrm{ren}}$ and Q_ε shifted by T, i. e.,

$$\begin{aligned}
\tilde{S}_\varepsilon^{\mathrm{ren}} &= -2\int_0^{2T} \varphi_\varepsilon(B_s - B_{2T}, s-2T)ds + 2\int_0^{2T} ds \int_s^{2T} \nabla\varphi_\varepsilon(B_s - B_t, s-t)dB_t,\\
\tilde{Q}_\varepsilon &= e^{a_M^*(S_T)}e^{-2TH_\mathrm{f}}e^{a_M(\tilde{S}_T)},\\
S_T^\varepsilon &= -\frac{g}{\sqrt{2}}\int_0^{2T} e^{-|s|\omega(k)}e^{-\varepsilon|k|^2/2}e^{-ik\cdot B_s}ds,\\
\tilde{S}_T^\varepsilon &= -\frac{g}{\sqrt{2}}\int_0^{2T} e^{-|s-2T|\omega(k)}e^{-\varepsilon|k|^2/2}e^{ik\cdot B_s}ds.
\end{aligned}$$

We also define S_T and $\tilde{S}_T$ by S_T^ε and $\tilde{S}_T^\varepsilon$ with $\varepsilon = 0$, respectively. Using this formula, we prove (2.11.48) and (2.11.49). The proof is in some sense straightforward, using the properties of exponentials of annihilation and creation operators discussed in Section 1.2.7. In particular, the uniform continuity of the map $f \mapsto e^{a^*(f)}e^{-TH_\mathrm{f}}$ for any $t > 0$ is a key ingredient. It results in the uniform bound of $\mathbb{E}^x[\|Q_\varepsilon - Q_0\|^n]$ for every $n \in \mathbb{N}$, and we can show that $\int_{\mathbb{R}^3}|P_\varepsilon(x) - P_0(x)|dx \to 0$ as $\varepsilon \downarrow 0$.

We show now the claim announced in (2.11.48).

Lemma 2.125. $P_\varepsilon \in L^1(\mathbb{R}_x^3)$, *for every* $\varepsilon \geq 0$.

Proof. By Lemma 2.121 and the bound (2.11.24), there exist C_1 and C_2, independent of $x \in \mathbb{R}$, such that

$$\mathbb{E}^x\left[e^{g^2\tilde{S}_\varepsilon^{\mathrm{ren}}}\|\tilde{Q}_\varepsilon\|^2\right] \le \left(\mathbb{E}^x\left[e^{2g^2\tilde{S}_\varepsilon^{\mathrm{ren}}}\right]\right)^{1/2}\left(\mathbb{E}^x\left[\|\tilde{Q}_\varepsilon\|^4\right]\right)^{1/2} \le C_1 e^{C_2}.$$

We find that

$$\begin{aligned}\int_{\mathbb{R}^3}|P_\varepsilon(x)|dx &\le e^{2T\|V\|_\infty}\int_{\mathbb{R}^3}\|F(x)\|\mathbb{E}^x\left[e^{g^2\tilde{S}_\varepsilon^{\mathrm{ren}}}\|\tilde{Q}_\varepsilon\|\|G(B_T)\|\right]dx\\ &\le e^{2T\|V\|_\infty}\int_{\mathbb{R}^3}\|F(x)\|\left(\mathbb{E}^x\left[e^{g^2\tilde{S}_\varepsilon^{\mathrm{ren}}}\|\tilde{Q}_\varepsilon\|^2\right]\right)^{1/2}\left(\mathbb{E}^x\left[\|G(B_{2T})\|^2\right]\right)^{1/2}dx\\ &\le C_1 e^{C_2}e^{2T\|V\|_\infty}\|F\|\|G\|.\end{aligned}$$

□

Next we also show (2.11.49).

Lemma 2.126. $\text{s-lim}_{\varepsilon\downarrow 0} P_\varepsilon = P_0$ *in* L^1.

Proof. We suppose that $T \ge 1$; for $T < 1$, the proof is similar by a scaling argument. We have

$$\begin{aligned}&\|P_\varepsilon - P_0\|_{L^1(\mathbb{R}^3)}\\ &= e^{2T\|V\|_\infty}\int_{\mathbb{R}^3}\|F(x)\|\left(\mathbb{E}^x\left[\|e^{\frac{g^2}{2}\tilde{S}_\varepsilon^{\mathrm{ren}}}\tilde{Q}_\varepsilon - e^{\frac{g^2}{2}\tilde{S}_0^{\mathrm{ren}}}\tilde{Q}_0\|^2\right]\right)^{1/2}\left(\mathbb{E}^x\left[\|G(B_T)\|^2\right]\right)^{1/2}dx.\end{aligned} \tag{2.11.60}$$

We show that the first expectation above converges to zero as $\varepsilon \downarrow 0$ uniformly with respect to $x \in \mathbb{R}^3$. We have the estimate

$$\|\tilde{Q}_\varepsilon - \tilde{Q}_0\| \le 2J\left\|\frac{S_T^\varepsilon}{\sqrt{\omega}} - \frac{S_T}{\sqrt{\omega}}\right\|_\omega,$$

where $J = 2e^{2(\|S_T^\varepsilon/\sqrt{\omega}\|_\omega + \|S_T/\sqrt{\omega}\|_\omega+1)^2}(e^{2\|S_T/\sqrt{\omega}\|_\omega^2} + e^{2\|S_T^\varepsilon/\sqrt{\omega}\|_\omega^2})$. This bound follows by Corollaries 1.34–1.35. Hence

$$\begin{aligned}\mathbb{E}^x\left[\|\tilde{Q}_\varepsilon - \tilde{Q}_0\|^4\right] &\le 2^4\left(\mathbb{E}^x\left[\left\|\frac{S_T^\varepsilon}{\sqrt{\omega}} - \frac{S_T}{\sqrt{\omega}}\right\|_\omega^8\right]\right)^{1/2}\left(\mathbb{E}^x J^8\right)^{1/2}\\ &\le 2^4 C_1 e^{C_2}\left(\mathbb{E}^x\left[\left\|\frac{S_T^\varepsilon}{\sqrt{\omega}} - \frac{S_T}{\sqrt{\omega}}\right\|_\omega^8\right]\right)^{1/2}\end{aligned}$$

with constants C_1 and C_2. The second estimate above follows by Lemma 2.121 and the bound (2.11.24). We show that $\lim_{\varepsilon\downarrow 0}\sup_{x\in\mathbb{R}^3}\mathbb{E}^x[\|\frac{S_T^\varepsilon}{\sqrt{\omega}} - \frac{S_T}{\sqrt{\omega}}\|_\omega^8] = 0$, i. e.,

$$\lim_{\varepsilon\downarrow 0}\sup_{x\in\mathbb{R}^3}\mathbb{E}^x\left[\left(\int_{\mathbb{R}^3}\frac{|S_T(k) - S_T^\varepsilon(k)|^2}{\omega(k)^n}dk\right)^4\right] = 0, \quad n = 1, 2.$$

Take the case $n = 1$, and denote

$$\tilde{\psi}_\varepsilon(B_s, s) = e^{-\varepsilon|k|^2/2} e^{-|s|\omega(k)} e^{-ik\cdot B_s}$$

for $\varepsilon \geq 0$. The Itô formula gives

$$\frac{S_T^\varepsilon}{\sqrt{\omega(k)}} = \frac{g}{\sqrt{2}} \frac{\beta(\tilde{\psi}_\varepsilon(B_{2T}, 2T) - \tilde{\psi}_\varepsilon(B_0, 0))}{\sqrt{\omega(k)}} - \frac{g}{\sqrt{2}} \int_0^{2T} \frac{\beta\nabla\tilde{\psi}_\varepsilon(B_s, s)}{\sqrt{\omega(k)}} \cdot dB_s. \tag{2.11.61}$$

Let $\Psi_\varepsilon(B_s, s) = \tilde{\psi}_\varepsilon(B_s, s) - \tilde{\psi}_\varepsilon(B_0, 0)$ and $\alpha_\varepsilon(T) = \Psi_\varepsilon(B_{2T}, 2T) - \Psi_0(B_{2T}, 2T)$. Notice that $|\alpha_\varepsilon(T)| \leq 2(1 - e^{-\varepsilon|k|^2/2})$. Hence we have

$$\begin{aligned}
\mathbb{E}^x\left[\left(\int_{\mathbb{R}^3} \frac{|S_T(k) - S_T^\varepsilon(k)|^2}{\omega(k)} dk\right)^4\right] &= \mathbb{E}^x\left[\left(\int_{\mathbb{R}^3} \left|\frac{g}{\sqrt{2}} \frac{\beta\alpha_\varepsilon(T)}{\sqrt{\omega(k)}} - \frac{g}{\sqrt{2}} \int_0^{2T} \frac{\beta\nabla\Psi_\varepsilon(B_s, s)}{\sqrt{\omega(k)}} dB_s\right|^2 dk\right)^4\right] \\
&\leq C\mathbb{E}^x\left[\left(\int_{\mathbb{R}^3} \left|\frac{\beta\alpha_\varepsilon(T)}{\sqrt{\omega(k)}}\right|^2 dk + \int_{\mathbb{R}^3} \left|\int_0^{2T} \frac{\beta\nabla\Psi_\varepsilon(B_s, s)}{\sqrt{\omega(k)}} dB_s\right|^2 dk\right)^4\right]
\end{aligned}$$

with a constant $C > 0$. Denote $X_T(k) = \int_0^{2T} \frac{\beta\nabla\Psi_\varepsilon(B_s,s)}{\sqrt{\omega(k)}} dB_s$. By the Schwarz inequality, we have

$$\begin{aligned}
\mathbb{E}^x\left[\left(\int_{\mathbb{R}^3} |X_T(k)|^2 dk\right)^4\right] &\leq \int_{\mathbb{R}^{3\times 4}} \prod_{n=1}^4 (\mathbb{E}^x[|X_T(k_n)|^8])^{1/4} dk_1 dk_2 dk_3 dk_4 \\
&= \left(\int_{\mathbb{R}^3} \left(\mathbb{E}^x[|X_T(k)|^8]\right)^{1/4} dk\right)^4
\end{aligned}$$

Hence by the Burkholder–Davis–Gundy inequality,

$$\begin{aligned}
\mathbb{E}^x\left[\left|X_T(k)\right|^8\right] &\leq 28^4 (2T)^3 \mathbb{E}^x\left[\int_0^{2T} \left|\frac{\beta\nabla\Psi_\varepsilon(B_s, s)}{\sqrt{\omega(k)}}\right|^8 ds\right] \\
&\leq 28^4 (2T)^3 \int_0^{2T} \frac{|k|^8 e^{-8|s|\omega(k)} (1 - e^{-\varepsilon|k|^2/2})^8}{\omega(k)^4 (\omega(k) + |k|^2/2)^8} ds \\
&\leq \frac{28^4 (2T)^3}{8} \frac{|k|^3 (1 - e^{-\varepsilon|k|^2/2})^8}{(\omega(k) + |k|^2/2)^8}.
\end{aligned}$$

Thus we obtain

$$\int_{\mathbb{R}^3} \left(\mathbb{E}^x \left[\left| X_T(k) \right|^8 \right] \right)^{1/4} dk \le C \int_{\mathbb{R}^3} \left(\frac{(1 - e^{-\varepsilon|k|^2/2})^8}{(1 + |k|/2)^3 (\omega(k) + |k|^2/2)^5} \right)^{1/4} dk$$

$$\le C \int_0^\infty \frac{(1 - e^{-\varepsilon r^2/2})^2 r^2}{(1 + r/2)^{3/4} (r + r^2/2)^{5/4}} dr \to 0$$

as $\varepsilon \downarrow 0$. This gives

$$\limsup_{\varepsilon \downarrow 0} \sup_{x \in \mathbb{R}^3} \mathbb{E}^x \left[\left(\int_{\mathbb{R}^3} \left| \int_0^{2T} \frac{\beta \nabla \Psi_\varepsilon(B_s, s)}{\sqrt{\omega(k)}} dB_s \right|^2 dk \right)^4 \right] = 0.$$

On the other hand, we also have

$$\mathbb{E}^x \left[\left(\int_{\mathbb{R}^3} \left| \frac{\beta a_\varepsilon(T)}{\sqrt{\omega(k)}} \right|^2 dk \right)^4 \right] \le \left(\int_{\mathbb{R}^3} \left(\mathbb{E}^x \left[\left| \frac{\beta a_\varepsilon(T)}{\sqrt{\omega(k)}} \right|^8 \right] \right)^{1/4} dk \right)^4$$

and

$$\mathbb{E}^x \left[\left| \frac{\beta a_\varepsilon(T)}{\sqrt{\omega(k)}} \right|^8 \right] \le \frac{(1 - e^{-\varepsilon|k|^2/2})^8}{\omega(k)^4 (\omega(k) + |k|^2/2)^8}.$$

It follows that

$$\mathbb{E}^x \left[\left(\int_{\mathbb{R}^3} \left| \frac{\beta a_\varepsilon(T)}{\sqrt{\omega(k)}} \right|^2 dk \right)^4 \right] \le \left(\int_0^\infty \frac{(1 - e^{-\varepsilon r^2})^2}{r(1 + r/2)^2} dr \right)^4 \to 0$$

as $\varepsilon \downarrow 0$. Hence for a given $\delta > 0$, there exists ε_0 such that for all $\varepsilon \le \varepsilon_0$,

$$\sup_{x \in \mathbb{R}^3} \mathbb{E}^x \left[e^{g^2 \tilde{S}_0^{\mathrm{ren}}} \| \tilde{Q}_\varepsilon - \tilde{Q}_0 \|^2 \right] \le \delta e^C. \tag{2.11.62}$$

Furthermore, for every $\delta > 0$ there exists ε_0' such that for all $\varepsilon \le \varepsilon_0'$,

$$\mathbb{E}^x \left[\left| e^{\frac{g^2}{2} \tilde{S}_\varepsilon^{\mathrm{ren}}} - e^{\frac{g^2}{2} \tilde{S}_0^{\mathrm{ren}}} \right|^2 \| \tilde{Q}_0 \|^2 \right] \le \left(\mathbb{E}^x \left[\left| e^{\frac{g^2}{2} \tilde{S}_\varepsilon^{\mathrm{ren}}} - e^{\frac{g^2}{2} \tilde{S}_0^{\mathrm{ren}}} \right|^4 \right] \right)^{1/2} \left(\mathbb{E}^x \left[\| \tilde{Q}_0 \|^4 \right] \right)^{1/2} \le \delta C_1 e^{C_2}.$$

These estimates follow by Corollary 2.122 and (2.11.24). Hence for a given $\delta > 0$, we obtain

$$\sup_{x \in \mathbb{R}^3} \mathbb{E}^x \left[\left| e^{\frac{g^2}{2} \tilde{S}_\varepsilon^{\mathrm{ren}}} - e^{\frac{g^2}{2} \tilde{S}_0^{\mathrm{ren}}} \right|^2 \| \tilde{Q}_0 \|^2 \right] \le \delta C_1 e^{C_2} \tag{2.11.63}$$

for a sufficiently small ε. Due to (2.11.62)–(2.11.63), we see that

$$\sup_{x\in\mathbb{R}^3} \mathbb{E}^x \left[\left| e^{\frac{g^2}{2}\tilde{S}_\varepsilon^{\mathrm{ren}}} \tilde{Q}_\varepsilon - e^{\frac{g^2}{2}\tilde{S}_0^{\mathrm{ren}}} \tilde{Q}_0 \right|^2 \right]$$

$$\leq \sup_{x\in\mathbb{R}^3} \mathbb{E}^x \left[e^{g^2 \tilde{S}_0^{\mathrm{ren}}} \|\tilde{Q}_\varepsilon - \tilde{Q}_0\|^2 \right] + \sup_{x\in\mathbb{R}^3} \mathbb{E}^x \left[\left| e^{\frac{g^2}{2}\tilde{S}_\varepsilon^{\mathrm{ren}}} - e^{\frac{g^2}{2}\tilde{S}_0^{\mathrm{ren}}} \right|^2 \|\tilde{Q}_0\|^2 \right] \leq \delta D \qquad (2.11.64)$$

with a constant $D > 0$ and sufficiently small $\varepsilon > 0$. Together with (2.11.64), we conclude from (2.11.60) that

$$\|P_\varepsilon - P_0\|_{L^1(\mathbb{R}^3)} \leq \delta D e^{2T\|V\|_\infty} \|F\| \|G\|. \qquad \square$$

We see that $e^{-tH_{\mathrm{N}}^{\mathrm{ren}}}$ is the strong limit of positivity preserving operators $e^{-t(H_{\mathrm{N}}^{\varepsilon}-E_\varepsilon)}$. Hence $e^{-tH_{\mathrm{N}}^{\mathrm{ren}}}$ is at least positivity preserving. By using formula (2.11.42), we can furthermore show that in fact the semigroup is positivity improving.

Corollary 2.127. *For every $t > 0$, the operator $e^{-tH_{\mathrm{N}}^{\mathrm{ren}}}$ is positivity improving. In particular, if a ground state of $H_{\mathrm{N}}^{\mathrm{ren}}$ exists, then it is unique.*

Proof. Let $t = 2T$ and $\Phi \in L^2(\mathscr{Q})$ be nonnegative. This Φ can be approximated by functions $(\Phi_n)_{n\in\mathbb{N}} \subset \mathscr{S}(\mathscr{Q})$ such that $\Phi_n = F_n(\phi(f_1^n), \ldots, \phi(f_{m_n}^n))$, where $F_n \in \mathscr{S}(\mathbb{R}^{m_n})$ is a nonnegative function and $f_i^j \in \mathscr{M}$. Suppose that $\Psi = F_n(\phi(f_1), \ldots, \phi(f_m)) \in \mathscr{S}(\mathscr{Q})$. For $g \in \mathscr{M}$, we have

$$e^{a_M(g)}\Psi = F_n\left(\phi(f_1) + (g, f_1)_{\mathcal{H}_M}, \ldots, \phi(f_m) + (g, f_m)_{\mathcal{H}_M}\right) \geq 0.$$

We conclude that $\overline{e^{-TH_{\mathrm{f}}} e^{a_M(g)}}$ is positivity improving for every $T > 0$. In particular, the bounded operator $A\tilde{A} = e^{a_M^*(U_T)} e^{-2TH_{\mathrm{f}}} e^{a_M(\tilde{U}_T)}$ is also positivity improving for all $T > 0$. Let $F, G \in L^2(\mathbb{R}^3 \times \mathscr{Q})$ be nonnegative functions. By formula (2.11.42), we have

$$(F, e^{-2TH_{\mathrm{N}}^{\mathrm{ren}}} G) = \int_{\mathbb{R}^3} \mathbb{E}^x \left[e^{-\int_{-T}^{T} V(B_s) ds} e^{\frac{g^2}{2} \tilde{S}_0^{\mathrm{ren}}} (F(B_{-T}), A\tilde{A} G(B_T)) \right] dx > 0. \qquad \square$$

2.11.6 Gibbs measures and applications

In this section we show that the family of probability measures $\{\mu_T^{\mathrm{ren}}\}_{T>0}$ defined below has a limit $\mu_\infty^{\mathrm{ren}}$ as $T \to \infty$ in the local sense. By using $\mu_\infty^{\mathrm{ren}}$, we can derive further ground state expectations. In Section 2.11.6 we use Assumption 2.128 below.

Assumption 2.128. The following conditions hold:
(1) (Dispersion relation) $\omega(k) = |k|$;
(2) (External potential) V is a continuous bounded function;
(3) (Number of particles) $N = 1$;
(4) (Infrared cutoff) Function $\xi_\lambda : \mathbb{R}^3 \to \mathbb{R}$ satisfies

(i) $\xi_{\lambda=0}(k) = 1$;

(ii) $\int_{\{|k|\le 1\}} |\xi_\lambda(k)|^2/\omega(k)^3 dk < \infty$ for $\lambda > 0$;

(iii) $H_{\mathrm{I}}^{\varepsilon}(x) = \frac{1}{\sqrt{2}}(a_M^*(\hat{\varphi}e^{ik\cdot x}\xi_\lambda) + a_M(\tilde{\hat{\varphi}}e^{-ik\cdot x}\xi_\lambda))$;

(5) (Ground state) For $\lambda > 0$, $H_{\mathrm{ren}} = H_{N=1}^{\mathrm{ren}}$ has the normalized ground state Ψ_{ren}.

The examples of infrared cutoff functions are (1) $\xi_\lambda(k) = 1_{\{|k|\ge\lambda\}}$, (2) $\xi_\lambda(k) = \sqrt{\omega(k)/\omega_\lambda(k)}$ with $\omega_\lambda(k) = \sqrt{|k|^2+\lambda^2}$. We regard $\lambda \ge 0$ as a parameter, i. e., in the case of $\lambda > 0$, infrared cutoff is introduced, and in the case of $\lambda = 0$, no infrared cutoff is introduced.

Remark 2.129 (Existence of ground state and infrared cutoff). The existence of the ground state is significantly dependent on infrared cutoffs. It is established that H_{ren} has the ground state for $\lambda > 0$ but no ground state for $\lambda = 0$.

Let

$$H_\varepsilon = H_{\mathrm{p}} \otimes 1 + 1 \otimes H_{\mathrm{f}} + gH_{\mathrm{I}}^{\varepsilon},$$

and the renormalization term with the infrared cutoff be given by

$$E_\varepsilon = -\frac{g^2}{2}\int_{\mathbb{R}^3} \frac{e^{-\varepsilon|k|^2}|\xi_\lambda(k)|^2}{\omega(k)}\beta(k)dk.$$

For $\lambda = 0$, it was shown that $H_\varepsilon - E_\varepsilon \to H_{\mathrm{ren}}$ as $\varepsilon \to 0$ in the sense of strong semigroup in Theorem 2.118. In the same way as for $\lambda > 0$, we can also show that $H_\varepsilon - E_\varepsilon$ converges as $\varepsilon \to 0$, and we denote the limit by the same notation H_{ren}. Note also that $\Psi_{\mathrm{ren}} > 0$ and the ground state is unique.

We consider a two-sided Brownian motion on the probability space $(\mathcal{X}, \mathcal{B}(\mathcal{X}), \mathcal{W}^x)$. Fix a nonnegative function $f \in L^2(\mathbb{R}^3)$ and let

$$\mathcal{L}_T^{\mathrm{ren}} = \mathcal{L}_T^{\mathrm{ren}}(\lambda) = f(B_{-T})f(B_T)e^{\frac{g^2}{2}S_0^{\mathrm{ren}}(T)}e^{-\int_{-T}^T V(B_s)ds}.$$

Here $S_0^{\mathrm{ren}}(T) = S_0^{\mathrm{ren}}(\lambda, T)$ is given by (2.11.32) with infrared cutoff $\lambda \ge 0$, i. e.,

$$S_0^{\mathrm{ren}}(T) = 2\int_{-T}^{T}\left(\int_{-T}^{t} \nabla\varphi_0(B_s - B_t, s-t)ds\right)dB_t - 2\int_{-T}^{T}\varphi_0(B_s - B_t, s-T)ds, \qquad (2.11.65)$$

with

$$\varphi_0(x,t) = \varphi_0(\lambda, x, t) = \int_{\mathbb{R}^3} \frac{e^{-ik\cdot x - |t|\omega(k)}}{2\omega(k)}|\xi_\lambda(k)|^2\beta(k)dk.$$

Define the probability measure $\mu_T^{\rm ren}$ on $(\mathscr{X}, \mathcal{B}(\mathscr{X}))$ by

$$\mu_T^{\rm ren}(A) = \frac{1}{Z_T} \int_{\mathbb{R}^3} \mathbb{E}^x[1_A \mathscr{L}_T^{\rm ren}]dx, \quad A \in \mathcal{B}(\mathscr{X}),$$

where Z_T is the normalizing constant such that $\frac{1}{Z_T}\int_{\mathbb{R}^3} \mathbb{E}^x[\mathscr{L}_T^{\rm ren}]dx = 1$. We call the measure $\mu_T^{\rm ren}$ the finite volume Gibbs measure associated with the renormalized Nelson model. Using the functional integral representation, we see that

$$(f \otimes 1, e^{-2TH_{\rm ren}} f \otimes 1) = \int_{\mathbb{R}^3} \mathbb{E}^x[\mathscr{L}_T^{\rm ren}]dx.$$

We shall show that $\mu_T^{\rm ren}$ converges to a probability measure $\mu_\infty^{\rm ren}$ in the local sense as $T \to \infty$ and Gibbs measure exists. The procedure of the proof of the convergence is, however, similar to the case of the Nelson Hamiltonian discussed in Section 2.8, and hence we show only an outline of the proof.

Let $\mathscr{G}_T = \mathscr{F}_{[-T,T]}^{\rm BM} = \sigma(B_r, -t \le r \le t)$ and, as usual, set

$$\mathscr{G} = \bigcup_{s \ge 0} \mathscr{G}_s.$$

We consider the family of probability spaces $(\mathscr{X}, \mathcal{B}(\mathscr{X}), \mu_T^{\rm ren})$, $T > 0$. As seen below, similarly as in Lemmas 2.71–2.74, we can show the convergence of $\mu_T^{\rm ren}$ as $T \to \infty$. Let

$$f_T = e^{-T(H_{\rm ren} - E_{\rm ren})} f \otimes 1,$$

where $E_{\rm ren} = \inf \operatorname{Spec}(H_{\rm ren})$. Define

$$\mathrm{I}_{[-T,T]}^{\rm ren} = e^{-\int_{-T}^T V(B_s)ds} e^{\frac{g^2}{2} S_0^{\rm ren}(T)} e^{a_M^*(U_T)} e^{-2TH_{\rm f}} e^{a_M(\tilde{U}_T)}.$$

Here

$$U_T(k) = -\frac{g}{\sqrt{2}} \int_{-T}^{T} \xi_\lambda(k) e^{-|s+T|\omega(k)} e^{-ik\cdot B_s} ds,$$

$$\tilde{U}_T(k) = -\frac{g}{\sqrt{2}} \int_{-T}^{T} \xi_\lambda(k) e^{-|s-T|\omega(k)} e^{ik\cdot B_s} ds.$$

Define a finitely additive set function $\mu^{\rm ren} : \mathscr{G} \to [0, \infty)$ by

$$\mu^{\rm ren}(A) = e^{2E_{\rm ren}T} \int_{\mathbb{R}^3} \mathbb{E}^x\Big[1_A\Big(\Psi_{\rm ren}(B_{-T}), \mathrm{I}_{[-T,T]}^{\rm ren} \Psi_{\rm ren}(B_T)\Big)\Big]dx, \quad A \in \mathscr{F}_{[-T,T]}^{\rm BM}.$$

The set function $\mu^{\rm ren}$ is well defined, i. e., for $A \in \mathscr{G}_T \subset \mathscr{G}_S$,

$$\mu^{\text{ren}}(A) = e^{2E_{\text{ren}}T} \int_{\mathbb{R}^3} \mathbb{E}^x \left[1_A\left(\Psi_{\text{ren}}(B_{-T}), \mathrm{I}^{\text{ren}}_{[-T,T]} \Psi_{\text{ren}}(B_T) \right) \right] dx$$
$$= e^{2E_{\text{ren}}S} \int_{\mathbb{R}^3} \mathbb{E}^x \left[1_A\left(\Psi_{\text{ren}}(B_{-S}), \mathrm{I}^{\text{ren}}_{[-S,S]} \Psi_{\text{ren}}(B_S) \right) \right] dx.$$

The set function μ^{ren} is also countably additive on $(\mathfrak{X}, \mathscr{G})$. Thus, there exists a unique probability measure μ^{ren}_∞ on $(\mathfrak{X}, \sigma(\mathscr{G}))$ such that

$$\mu^{\text{ren}} = \mu^{\text{ren}}_\infty \lceil_{\mathscr{G}}$$

by Hopf's extension theorem. We define the set function $\rho^{\text{ren}}_T : \mathscr{G}_T \to [0, \infty)$ by

$$\rho^{\text{ren}}_T(A) = e^{2E_{\text{ren}}s} \int_{\mathbb{R}^3} \mathbb{E}^x \left[1_A\left(\frac{f_{T-s}(B_{-s})}{\|f_T\|}, \mathrm{I}^{\text{ren}}_{[-s,s]} \frac{f_{T-s}(B_s)}{\|f_T\|} \right) \right] dx$$

for $A \in \mathscr{F}^{\text{BM}}_{[-s,s]}$ with $s \le T$. The set function ρ^{ren}_T is well defined, and

$$\rho^{\text{ren}}_T(A) = e^{2E_{\text{ren}}r} \int_{\mathbb{R}^3} \mathbb{E}^x \left[1_A\left(\frac{f_{T-r}(B_{-r})}{\|f_T\|}, \mathrm{I}^{\text{ren}}_{[-r,r]} \frac{f_{T-r}(B_r)}{\|f_T\|} \right) \right] dx$$
$$= e^{2E_{\text{ren}}s} \int_{\mathbb{R}^3} \mathbb{E}^x \left[1_A\left(\frac{f_{T-s}(B_{-s})}{\|f_T\|}, \mathrm{I}^{\text{ren}}_{[-s,s]} \frac{f_{T-s}(B_s)}{\|f_T\|} \right) \right] dx$$

holds for all $r \le s \le T$ and $A \in \mathscr{G}_r$. Then ρ^{ren}_T is a probability measure on $(\mathfrak{X}, \mathscr{G}_T)$. We can also see the relationship between ρ^{ren}_T and μ^{ren}_T. Let $s \le T$ and $A \in \mathscr{G}_s$. We have

$$\rho^{\text{ren}}_T(A) = \mu^{\text{ren}}_T(A).$$

Finally, we obtain the local convergence of $\{\mu^{\text{ren}}_T\}_{T>0}$ below.

Theorem 2.130 (Local convergence). *The family of probability measures $\{\mu^{\text{ren}}_T\}_{T\ge 0}$ converges to μ^{ren}_∞ on $(\mathfrak{X}, \mathcal{B}(\mathfrak{X}))$ in the local sense for arbitrary $\lambda > 0$, i. e., $\mu^{\text{ren}}_T(A) \to \mu^{\text{ren}}_\infty(A)$ as $T \to \infty$ for all $A \in \mathscr{G}$.*

Proof. Suppose that $A \in \mathscr{G}_s$ for some s. We have

$$\mu^{\text{ren}}_T(A) = \rho^{\text{ren}}_T(A) = e^{2E_{\text{ren}}s} \int_{\mathbb{R}^3} \mathbb{E}^x \left[1_A\left(\frac{f_{T-s}(B_{-s})}{\|f_T\|}, \mathrm{I}^{\text{ren}}_{[-s,s]} \frac{f_{T-s}(B_s)}{\|f_T\|} \right) \right] dx.$$

Since $\lambda > 0$, H_{ren} has the ground state. Hence $f_T/\|f_T\| \to \Psi_{\text{ren}}$ strongly as $T \to \infty$, and we have

$$\lim_{T\to\infty} \mu^{\text{ren}}_T(A) = e^{2E_{\text{ren}}s} \int_{\mathbb{R}^3} \mathbb{E}^x \left[1_A\left(\Psi_{\text{ren}}(B_{-s}), \mathrm{I}^{\text{ren}}_{[-s,s]} \Psi_{\text{ren}}(B_s) \right) \right] dx = \mu^{\text{ren}}_\infty(A). \qquad \square$$

The probability measure $\mu_\infty^{\rm ren}$ is the Gibbs measure associated with the renormalized Nelson model. Similar to Corollaries 2.78 and 2.79, we have the corollaries below.

Corollary 2.131. *Let f and g be bounded measurable functions on $\mathbb{R}^d$. Then it follows that for $a \le b$,*

$$\mathbb{E}_{\mu_\infty}[f(B_a)g(B_b)] = (f\Psi_{\rm ren}, e^{-(b-a)(H_{\rm ren}-E_{\rm ren})}g\Psi_{\rm ren}).$$

Proof. The proof is the same as that of Corollary 2.78. □

Corollary 2.132. *Suppose that $\Psi_{\rm ren} \in D(|x|)$. Then it follows that for $a \le b$,*

$$\mathbb{E}_{\mu_\infty}[|B_a||B_b|] = (|x|\Psi_{\rm ren}, e^{-(b-a)(H_{\rm ren}-E_{\rm ren})}|x|\Psi_{\rm ren}).$$

Furthermore, it follows that

$$\mathbb{E}_{\mu_\infty}[B_a \cdot B_b] = (x\Psi_{\rm ren}, e^{-(b-a)(H_{\rm ren}-E_{\rm ren})}x\Psi_{\rm ren}).$$

Proof. The proof is same as that of Corollary 2.79. □

Now we shall investigate expectation values of self-adjoint operator O with respect to $\Psi_{\rm ren}$ by the Gibbs measure we constructed. In order to investigate $(\Psi_{\rm ren}, O\Psi_{\rm ren})$, we discuss a functional integral representation of $(e^{-TH_{\rm ren}}F, e^{-d\Gamma(\rho)}e^{-TH_{\rm ren}}F)$ with $F = f \otimes 1$, since $e^{-TH_{\rm ren}}f \otimes 1/\|e^{-TH_{\rm ren}}f \otimes 1\| \to \Psi_{\rm ren}$ as $T \to \infty$ for $0 < f \in L^2(\mathbb{R}^3)$.

Lemma 2.133. *Suppose that ρ is a positive function on $\mathbb{R}^3$ such that*

$$\int_{\mathbb{R}^3} \frac{1-e^{-\rho(k)}}{\omega(k)^3}|\xi_\lambda(k)|^2 dk < \infty. \tag{2.11.66}$$

Then it follows that

$$\frac{(f_T, e^{-d\Gamma(\rho)}f_T)}{\|f_T\|^2} = \mathbb{E}_{\mu_T^{\rm ren}}\left[\exp\left(-g^2\int_{-T}^{0} ds \int_0^T W_\rho(B_s - B_t, s-t)dt\right)\right],$$

where

$$W_\rho(x,t) = \int_{\mathbb{R}^3} \frac{e^{-|t|\omega(k)}}{2\omega(k)} e^{-ik\cdot x}(1-e^{-\rho(k)})|\xi_\lambda(k)|^2 dk.$$

Proof. By Proposition 2.85, we have

$$\frac{(e^{-T(H_\varepsilon-E_\varepsilon)}f \otimes 1, e^{-d\Gamma(\rho)}e^{-T(H_\varepsilon-E_\varepsilon)}f \otimes 1)}{\|e^{-T(H_\varepsilon-E_\varepsilon)}f \otimes 1\|^2}$$
$$= \frac{\int_{\mathbb{R}^3} \mathbb{E}^x\left[e^{-\int_{-T}^T V(B_s)ds}e^{\frac{g^2}{2}(S_\varepsilon^T - 4\varphi_\varepsilon(0,0))}e^{-g^2\int_{-T}^0 ds\int_0^T W_\rho^\varepsilon(B_s-B_t,s-t)dt}\right]dx}{\int_{\mathbb{R}^3} \mathbb{E}^x\left[e^{-\int_{-T}^T V(B_s)ds}e^{\frac{g^2}{2}(S_\varepsilon^T-4\varphi_\varepsilon(0,0))}\right]dx},$$

where $\varphi_\varepsilon(x,t)$ is given by (2.11.12) with infrared cutoff $\lambda \ge 0$, namely

$$\varphi_\varepsilon(x,t) = \int_{\mathbb{R}^3} \frac{e^{-\varepsilon|k|^2} e^{-ik\cdot x - |t|\omega(k)}}{2\omega(k)} \beta(k)|\xi_\lambda(k)|^2 dk,$$

and $S_\varepsilon^T = S_\varepsilon^{\mathrm{D},T} + S_\varepsilon^{\mathrm{OD},T}$, where

$$S_\varepsilon^{\mathrm{D},T} = 2\int_{-T}^{T} ds \int_s^{\langle s+\tau\rangle} W_\rho^\varepsilon(B_s - B_t, s-t)dt,$$

$$S_\varepsilon^{\mathrm{OD},T} = 2\int_{-T}^{T} ds \int_{\langle s+\tau\rangle}^{T} W_\rho^\varepsilon(B_s - B_t, s-t)dt,$$

with

$$W_\rho^\varepsilon(X,t) = \int_{\mathbb{R}^3} \frac{e^{-\varepsilon|k|^2}}{2\omega(k)} e^{-ik\cdot X} e^{-|t|\omega(k)}(1-e^{-\rho(k)})|\xi_\lambda(k)|^2 dk.$$

Since

$$\left|\int_{-T}^{0} ds \int_0^T W_\rho^\varepsilon(B_s - B_t, s-t)dt\right| \le \int_{\mathbb{R}^3} \frac{1-e^{-\rho(k)}}{\omega(k)^3}|\xi_\lambda(k)|^2 dk < \infty$$

uniformly in paths, we have

$$\begin{aligned}
\frac{(f_T, e^{-\mathrm{d}\Gamma(\rho)} f_T)}{\|f_T\|^2} &= \lim_{\varepsilon\to 0} \frac{(e^{-T(H_\varepsilon - E_\varepsilon)} f\otimes 1, e^{-\mathrm{d}\Gamma(\rho)} e^{-T(H_\varepsilon - E_\varepsilon)} f \otimes 1)}{\|e^{-T(H_\varepsilon - E_\varepsilon)} f\otimes 1\|^2} \\
&= \frac{\int_{\mathbb{R}^3} \mathbb{E}^x \left[e^{-\int_{-T}^T V(B_s)ds} e^{\frac{g^2}{2} S_0^{\mathrm{ren}}(T)} e^{-g^2 \int_{-T}^0 ds \int_0^T W_\rho(B_s - B_t, s-t)dt}\right]}{\int_{\mathbb{R}^3} \mathbb{E}^x \left[e^{-\int_{-T}^T V(B_s)ds} e^{\frac{g^2}{2} S_0^{\mathrm{ren}}(T)}\right]} \\
&= \mathbb{E}_{\mu_T^{\mathrm{ren}}}[e^{-g^2 \int_{-T}^0 ds \int_0^T W_\rho(B_s - B_t, s-t)dt}].
\end{aligned}$$

Then the desired result is obtained. □

Theorem 2.134. *Suppose that ρ is a positive function on $\mathbb{R}^3$ and (2.11.66) holds. Let $\lambda > 0$. Then*

$$(\Psi_{\mathrm{ren}}, e^{-\mathrm{d}\Gamma(\rho)}\Psi_{\mathrm{ren}}) = \mathbb{E}_{\mu_\infty^{\mathrm{ren}}}\left[\exp\left(-g^2 \int_{-\infty}^0 ds \int_0^\infty W_\rho(B_s - B_t, s-t)dt\right)\right].$$

Proof. By Lemma 2.133, we see that

$$(\Psi_{\mathrm{ren}}, e^{-\mathrm{d}\Gamma(\rho)}\Psi_{\mathrm{ren}}) = \lim_{T\to\infty} \frac{(f_T, e^{-\mathrm{d}\Gamma(\rho)} f_T)}{\|f_T\|^2}$$
$$= \lim_{T\to\infty} \mathbb{E}_{\mu_T^{\mathrm{ren}}}\left[e^{-g^2\int_{-T}^0 ds\int_0^T W_\rho(B_s-B_t,s-t)dt}\right] = \mathbb{E}_{\mu_\infty^{\mathrm{ren}}}[e^{-g^2\int_{-\infty}^0 ds\int_0^\infty W_\rho(B_s-B_t,s-t)dt}].$$

Then we show the theorem. □

Using Theorem 2.134 and the functional integral representation of $(F, e^{-TH_{\mathrm{ren}}}G)$ due to Theorem 2.120, we can show the superexponential decay of the boson number in the ground state. The number operator N can be decomposed as

$$N = N_0 + N_\infty, \tag{2.11.67}$$

where $N_0 = \mathrm{d}\Gamma(1_{\{|k|\le 1\}})$ denotes the number operator for lower momentum and $N_\infty = \mathrm{d}\Gamma(1_{\{|k|>1\}})$ for higher momentum.

Lemma 2.135. *Let $\lambda \ge 0$. Then $e^{\beta N_\infty}e^{-2TH_{\mathrm{ren}}}$ is bounded for $T > \beta$.*

Proof. We have

$$(F, e^{\beta N_\infty}e^{-2TH_{\mathrm{ren}}}G)$$
$$= \int_{\mathbb{R}^3} \mathbb{E}^x\left[e^{-\int_{-T}^T V(B_s)ds}e^{\frac{g^2}{2}S_0^{\mathrm{ren}}}(F(B_{-T}), e^{\beta N_\infty}e^{a_M^*(U_T)}e^{-2TH_{\mathrm{f}}}e^{a_M(\tilde U_T)}G(B_T))\right].$$

It can be seen that

$$e^{\beta N_\infty}e^{a_M^*(U_T)}e^{-2TH_{\mathrm{f}}}e^{a_M(\tilde U_T)} = e^{a_M^*(\beta 1_{\{|k|>1\}}U_T)}e^{\beta N_\infty}e^{-2TH_{\mathrm{f}}}e^{a_M(\tilde U_T)}$$

and

$$e^{\beta N_\infty}e^{-2TH_{\mathrm{f}}} = e^{-\frac{T}{2}H_{\mathrm{f}}}e^{\beta N_\infty}e^{-TH_{\mathrm{f}}}e^{-\frac{T}{2}H_{\mathrm{f}}} = e^{-\frac{T}{2}H_{\mathrm{f}}}\Gamma(e^{-T|k|+\beta 1_{\{|k|>1\}}})e^{-\frac{T}{2}H_{\mathrm{f}}}$$
$$= e^{-\frac{T}{2}H_{\mathrm{f}}}\Gamma(e^{1_{\{|k|>1\}}(-T|k|+\beta)})\Gamma(e^{-T|k|1_{\{|k|<1\}}})e^{-\frac{T}{2}H_{\mathrm{f}}}.$$

From this, we can see that $e^{\beta N_\infty}e^{a_M^*(U_T)}e^{-2TH_{\mathrm{f}}}e^{a_M(\tilde U_T)}$ is bounded for $T > \beta$ a. e. We can also show that

$$\mathbb{E}^x[\|e^{a_M^*(\beta 1_{\{|k|>1\}}U_T)}e^{-\frac{T}{2}H_{\mathrm{f}}}e^{\beta N_\infty}e^{-TH_{\mathrm{f}}}e^{-\frac{T}{2}H_{\mathrm{f}}}e^{a_M(\tilde U_T)}\|^n] < \infty$$

for $T > \beta$ in a similar manner to Lemma 2.121, and hence

$$|(F, e^{\beta N_\infty}e^{-2TH_{\mathrm{ren}}}G)| \le C\|F\|\|G\|$$

is proven in the same way as Lemma 2.125, where C is a constant. This implies that $e^{\beta N_\infty}e^{-2TH_{\rm ren}}$ is bounded for $T > \beta$. □

In the case of $\lambda = 0$, we expect that $(\Psi_{\rm ren}, N_0\Psi_{\rm ren}) = \infty$ by an infrared divergence. We shall show this in Corollary 2.137 below. We can however show the superexponential decay of boson number of $\Psi_{\rm ren}$ for $\lambda > 0$.

Corollary 2.136 (Superexponential decay of boson number). *Let $\lambda > 0$. Then*

$$\Psi_{\rm ren} \in D(e^{\beta N})$$

for arbitrary $\beta \in \mathbb{C}$.

Proof. We have $\Psi_{\rm ren} = e^{2TE_{\rm ren}}e^{-2TH_{\rm ren}}\Psi_{\rm ren}$ for $T > 0$. Let $\beta > 0$. Hence $\|e^{\beta N_\infty}\Psi_{\rm ren}\| = e^{2TE_{\rm ren}}\|e^{\beta N_\infty}e^{-2TH_{\rm ren}}\Psi_{\rm ren}\| < \infty$ by Lemma 2.135. Hence

$$\Psi_{\rm ren} \in D(e^{\beta N_\infty}), \quad \forall \beta \in \mathbb{R}. \tag{2.11.68}$$

Next we estimate $e^{\beta N_0}\Psi_{\rm ren}$. By Theorem 2.134, we see that

$$(\Psi_{\rm ren}, e^{-\beta N_0}\Psi_{\rm ren}) = \mathbb{E}_{\mu_\infty^{\rm ren}}[e^{-(1-e^{-\beta})g^2\int_{-\infty}^0 ds\int_0^\infty W_0(B_s-B_t,s-t)dt}], \tag{2.11.69}$$

where

$$W_0(x,t) = W_{\beta 1_{\{|k|\le 1\}}}(x,t) = \int_{|k|\le 1}\frac{e^{-|t|\omega(k)}}{\omega(k)}e^{-ik\cdot x}|\xi_\lambda(k)|^2dk.$$

Since $\lambda > 0$,

$$\left|\int_{-\infty}^0 ds\int_0^\infty W_0(B_s-B_t,s-t)dt\right| \le \int_{|k|\le 1}\frac{|\xi_\lambda(k)|^2}{\omega(k)^3}dk < \infty.$$

By an analytic continuation with respect to β, we can see that

$$|(\Psi_{\rm ren}, e^{\beta N_0}\Psi_{\rm ren})| = \left|\mathbb{E}_{\mu_\infty^{\rm ren}}[e^{-(1-e^{\beta})g^2\int_{-\infty}^0 ds\int_0^\infty W_0(B_s-B_t,s-t)dt}]\right| < \infty$$

for $\beta \in \mathbb{C}$. In particular,

$$\Psi_{\rm ren} \in D(e^{\beta N_0}), \quad \forall \beta \in \mathbb{C}. \tag{2.11.70}$$

By (2.11.68) and (2.11.70), we see that

$$|(\Psi_{\rm ren}, e^{\beta N}\Psi_{\rm ren})| = |(\Psi_{\rm ren}, e^{\beta N_0}e^{\beta N_\infty}\Psi_{\rm ren})| \le \|e^{\beta N_0}\Psi_{\rm ren}\|\|e^{\beta N_\infty}\Psi_{\rm ren}\| < \infty.$$

Then the proof is complete. □

The expectation of the number operator with respect to the ground state for the Nelson Hamiltonian H_{N} with ultraviolet cutoff diverges as infrared cutoff is removed. This is shown in Corollary 2.44. In the corollary below, we can also show that Ψ_{ren} has the same property.

Corollary 2.137 (Infrared divergence). *Suppose that* $\Psi_{\mathrm{ren}} \in D(|x| \otimes 1)$. *Then*

$$\lim_{\lambda \downarrow 0} (\Psi_{\mathrm{ren}}, N\Psi_{\mathrm{ren}}) = \infty.$$

Proof. We have $(\Psi_{\mathrm{ren}}, N\Psi_{\mathrm{ren}}) \geq (\Psi_{\mathrm{ren}}, N_0\Psi_{\mathrm{ren}})$. Taking the derivative of (2.11.69) at $\beta = 0$ we have

$$\begin{aligned}(\Psi_{\mathrm{ren}}, N_0\Psi_{\mathrm{ren}}) &= \mathbb{E}_{\mu_\infty^{\mathrm{ren}}}\Bigg[\int_{-\infty}^{0} ds \int_{0}^{\infty} W_0(B_s - B_t, s-t)dt\Bigg] \\ &= \int_{-\infty}^{0} ds \int_{0}^{\infty} dt \int_{\{|k|\leq 1\}} \frac{|\xi_\lambda(k)|^2}{2\omega(k)} e^{-|s-t|\omega(k)} \mathbb{E}_{\mu_\infty^{\mathrm{ren}}}[e^{-ik\cdot(B_s - B_t)}]dk \\ &= \int_{-\infty}^{0} ds \int_{0}^{\infty} dt \int_{\{|k|\leq 1\}} \frac{|\xi_\lambda(k)|^2}{2\omega(k)} e^{-|s-t|\omega(k)} \mathbb{E}_{\mu_\infty^{\mathrm{ren}}}[\cos(k\cdot(B_s - B_t))]dk.\end{aligned}$$

Inequality

$$\mathbb{E}_{\mu_\infty^{\mathrm{ren}}}[\cos(k\cdot(B_s - B_t))] \geq 1 - \frac{|k|^2}{2}\mathbb{E}_{\mu_\infty^{\mathrm{ren}}}[B_t^2 + B_s^2 - 2B_tB_s] \geq 1 - |k|^2\||x|\Psi_{\mathrm{ren}}\|^2$$

follows from Corollary 2.132, namely

$$\mathbb{E}_{\mu_\infty^{\mathrm{ren}}}[B_s \cdot B_t] = (x\Psi_{\mathrm{ren}}, e^{-|s-t|(H_{\mathrm{ren}} - E_{\mathrm{ren}})} x\Psi_{\mathrm{ren}}) \geq 0.$$

Writing $C = \||x|\Psi_{\mathrm{ren}}\|^2$, we have

$$(\Psi_{\mathrm{ren}}, N_0\Psi_{\mathrm{ren}}) \geq \frac{1}{2}\int_{\{|k|\leq 1\}} \frac{|\xi_\lambda(k)|^2}{\omega(k)^3}(1 - C|k|^2)dk.$$

Since $\xi_\lambda(k) \to 1$ as $\lambda \to 0$, the right-hand side above diverges as $\lambda \to 0$. This proves the corollary. □

We can also see Gaussian domination of Ψ_{ren} by using the Gibbs measure $\mu_\infty^{\mathrm{ren}}$.

Corollary 2.138 (Gaussian domination of the ground state). *Let* $\lambda > 0$. *Take* $h \in \mathscr{S}'(\mathbb{R}^3)$ *such that* $\hat{h}/\sqrt{\omega} \in L^2(\mathbb{R}^3)$, $\xi_\lambda \hat{h}/\omega^2 \in L^1(\mathbb{R})$, *and* $\beta < 1/\|\hat{h}/\sqrt{\omega}\|^2$. *Then* $\Psi_{\mathrm{ren}} \in D(e^{(\beta/2)\phi(h)^2})$ *and*

$$\|e^{(\beta/2)\phi(h)^2}\Psi_{\mathrm{ren}}\|^2 = \frac{1}{\sqrt{1-\beta\|\hat{h}/\sqrt{\omega}\|^2}}\mathbb{E}_{\mu_\infty^{\mathrm{ren}}}\left[\exp\left(\frac{\beta g^2 K(h)^2}{1-\beta\|\hat{h}/\sqrt{\omega}\|^2}\right)\right],$$

where $K(h)$ denotes the random variable defined by

$$K(h) = \frac{1}{2}\int_{-\infty}^{\infty}\left(\xi_\lambda e^{-|r|\omega}e^{-ik\cdot B_r}/\sqrt{\omega}, \hat{h}/\sqrt{\omega}\right)dr.$$

In particular, it follows that

$$\lim_{\beta\uparrow(2\|\hat{h}/\sqrt{\omega}\|^2)^{-1}}\|e^{\beta\phi(h)^2}\Psi_{\mathrm{ren}}\| = \infty.$$

Proof. For every $\beta\in\mathbb{R}$, we have

$$\begin{aligned}\frac{(f_T, e^{ik\beta\phi(g)}f_T)}{\|f_T\|^2} &= \lim_{\varepsilon\to 0}\frac{(e^{-T(H_\varepsilon-E_\varepsilon)}f\otimes 1, e^{ik\beta\phi(g)}e^{-T(H_\varepsilon-E_\varepsilon)}f\otimes 1)}{\|e^{-T(H_\varepsilon-E_\varepsilon)}f\otimes 1\|^2}\\ &= \frac{1}{\|f_T\|^2}\int_{\mathbb{R}^3}\mathbb{E}^x\left[f(B_{-T})f(B_T)e^{-\int_{-T}^T V(B_s)ds}S_T\right]dx,\end{aligned}$$

where

$$S_T = e^{-\frac{k^2\beta^2}{4}\|\hat{h}/\sqrt{\omega}\|^2}e^{-ik\beta g\int_{\mathbb{R}^3}\frac{\overline{\hat{h}(k)}}{2\omega(k)}\xi_\lambda(k)\left(\int_{-T}^T e^{-\omega(k)|s|}e^{-ik\cdot B_s}ds\right)dk}e^{\frac{g^2}{2}S_0^{\mathrm{ren}}}.$$

Hence it follows that

$$\frac{(f_T, e^{ik\beta\phi(h)}f_T)}{\|f_T\|^2} = \mathbb{E}_{\mu_T^{\mathrm{ren}}}\left[e^{-\frac{k^2\beta^2}{4}\|\hat{h}/\sqrt{\omega}\|^2}e^{-ik\beta g\int_{\mathbb{R}^3}\frac{\overline{\hat{h}(k)}}{2\omega(k)}\xi_\lambda(k)\left(\int_{-T}^T e^{-\omega(k)|s|}e^{-ik\cdot B_s}ds\right)dk}\right].$$

Taking the limit as $T\to\infty$, we obtain

$$(\Psi_{\mathrm{ren}}, e^{ik\beta\phi(h)}\Psi_{\mathrm{ren}}) = \mathbb{E}_{\mu_\infty}\left[e^{-k^2\beta^2 I_h/2}e^{-ik\beta g I_B}\right], \tag{2.11.71}$$

where $I_h = \|\hat{h}/\sqrt{\omega}\|^2/2$ and $I_B = \int_{\mathbb{R}^3}\frac{\overline{\hat{h}(k)}}{2\omega(k)}\xi_\lambda(k)\left(\int_{-\infty}^{\infty}e^{-\omega(k)|s|}e^{-ik\cdot B_s}ds\right)dk$. Thus using (2.11.71), in the same way as in Lemma 2.60 we can show that for $\beta < 0$,

$$(\Psi_{\mathrm{ren}}, e^{\beta\phi(h)^2}\Psi_{\mathrm{ren}}) = \frac{1}{\sqrt{1-\beta\|\hat{h}/\sqrt{\omega}\|^2}}\mathbb{E}_{\mu_\infty}\left[\exp\left(\frac{\beta g^2 K(h)^2}{1-\beta\|\hat{h}/\sqrt{\omega}\|}\right)\right].$$

Furthermore, we can extend this to $\beta < 1/\|\hat{h}/\sqrt{\omega}\|^2$ in the same way as in Corollary 2.61. □

2.11.7 Weak coupling limit and removal of ultraviolet cutoff

We consider the weak coupling limit for the massless Nelson model with many particles. For simplicity, we assume $d = 3$ and no external potential. The *N-particle Nelson Hamiltonian* is defined by

$$H_{\mathrm{N}} = -\frac{1}{2}\sum_{j=1}^{N} \Delta_j \otimes 1 + 1 \otimes H_{\mathrm{f}} + gH_{\mathrm{I}},$$

where H_{I} is defined in (2.11.9) with $\hat{\varphi}$ replaced with (2.11.72) below. With given parameter $\Lambda > 0$, we define the ultraviolet cutoff function of H_{N} to be

$$\hat{\varphi}(k) = \frac{1}{(2\pi)^{3/2}} 1_{|k|\le\Lambda}(k). \tag{2.11.72}$$

Also, we write $H_{\mathrm{I}}(\Lambda)$ for H_{I} with cutoff function (2.11.72). We also introduce the scaling parameter $\kappa > 0$ such that $a_M(f) \to \kappa a_M(f)$ and $a_M^*(f) \to \kappa a_M^*(f)$. The *scaled Nelson Hamiltonian* is then defined by

$$H_{\mathrm{N}}(\kappa, \Lambda) = H_{\mathrm{p}} + g\kappa H_{\mathrm{I}}(\Lambda) + \kappa^2 H_{\mathrm{f}}. \tag{2.11.73}$$

We call the limit of $H_{\mathrm{N}}(\kappa, \Lambda)$ as $\Lambda \to \infty$ the *weak coupling limit*. Here we consider the asymptotic behavior of $e^{-TH_{\mathrm{N}}(\kappa,\Lambda)}$ as $\Lambda \to \infty$ and $\kappa \to \infty$ simultaneously.

Before going to a rigorous computation, we give a heuristic description of the asymptotic behavior. First, note that $e^{-T\kappa^2 H_{\mathrm{f}}} \to P_0$ as $\kappa \to \infty$, with P_0 being the projection onto the subspace spanned by the Fock vacuum. One can expect that $e^{-TH_{\mathrm{N}}(\kappa,\Lambda)} \to e^{-TH_\infty(\Lambda)} \otimes P_0$ as $\kappa \to \infty$ with a self-adjoint operator $H_\infty(\Lambda)$ in $L^2(\mathbb{R}^{3N})$. It can indeed be seen that

$$(f \otimes 1, e^{-TH_{\mathrm{N}}(\kappa,\Lambda)} g \otimes 1) = \int_{\mathbb{R}^3} \mathbb{E}^x \left[\overline{f(B_0)} g(B_T) e^{(g^2/2)\int_0^T ds \int_0^T W(B_s - B_t, s-t)dt} \right] dx,$$

where

$$W(B_s - B_t, s-t) = \sum_{i,j=1}^{N} \frac{1}{2(2\pi)^3} \int_{|k|\le\Lambda} \frac{1}{\omega(k)} e^{-ik\cdot(B_s^i - B_t^j)} \kappa^2 e^{-\kappa^2|s-t|\omega(k)} dk.$$

Since as $\kappa \to \infty$,

$$\kappa^2 e^{-\kappa^2|s-t|\omega} \to \frac{2}{\omega}\delta(t-s),$$

it follows that

$$\lim_{\Lambda\to\infty}\lim_{\kappa\to\infty}\left[\frac{g^2}{2}\int_0^T ds\int_0^T W(B_s-B_t,s-t)dt-g^2TNE(\Lambda)\right]=\frac{g^2}{4\pi}\sum_{i<j}^N\int_0^T\frac{1}{|B_t^i-B_t^j|}dt,$$

where the renormalized term $E(\Lambda)$ is defined by the diagonal term

$$E(\Lambda)=\frac{1}{2(2\pi)^3}\int_{|k|\le\Lambda}\frac{1}{\omega(k)^2}dk.$$

We then have

$$\lim_{\Lambda\to\infty}\lim_{\kappa\to\infty}(f\otimes 1,e^{-T(H_{\mathrm{N}}(\kappa,\Lambda)-g^2NE(\Lambda))}g\otimes 1)=(f,e^{-TH_\infty}g)_{L^2(\mathbb{R}^{3N})},$$

where

$$H_\infty=-\frac{1}{2}\sum_{j=1}^N\Delta_j-\frac{g^2}{4\pi}\sum_{i<j}^N\frac{1}{|x_i-x_j|}. \tag{2.11.74}$$

We obtain now (2.11.74) rigorously.

Theorem 2.139 (Weak coupling limit and removal of ultraviolet cutoff). *We have*

$$\operatorname*{s-lim}_{\Lambda\to\infty}\operatorname*{s-lim}_{\kappa\to\infty}e^{-T(H_{\mathrm{N}}(\kappa,\Lambda)-g^2NE(\Lambda))}=e^{-TH_\infty}\otimes P_0.$$

Proof. Let

$$\mathscr{D}=\mathrm{L.H.}\left\{f\otimes F(\phi(f_1),\dots,\phi(f_n)),f\otimes 1\mid f\in\mathscr{S}(\mathbb{R}^3),F\in\mathscr{S}(\mathbb{R}^n),n\ge 1\right\}.$$

Then $\mathscr{D}$ is dense in $\mathcal{H}_{\mathrm{N}}$. It suffices to show that

$$\lim_{\Lambda\to\infty}\lim_{\kappa\to\infty}(F,e^{-T(H_{\mathrm{N}}(\kappa,\Lambda)-g^2NE(\Lambda))}G)=(F,e^{-TH_\infty}\otimes P_0G)$$

for $F,G\in\mathscr{D}$. Since

$$f\otimes F(\phi(f_1),\dots,\phi(f_n))=\frac{1}{(2\pi)^{n/2}}\int_{\mathbb{R}^n}\check{F}(k)e^{-i\sum_{j=1}^n k_j\phi(f_j)}dk,$$

$$g\otimes G(\phi(g_1),\dots,\phi(g_m))=\frac{1}{(2\pi)^{m/2}}\int_{\mathbb{R}^m}\check{G}(k)e^{-i\sum_{i=1}^m l_i\phi(g_i)}dl,$$

we have

$$\begin{aligned}&(F,e^{-TH_{\mathrm{N}}(\kappa,\Lambda)}G)\\&=\frac{1}{(2\pi)^{\frac{n+m}{2}}}\int_{\mathbb{R}^{n+m}}\check{F}(k)\check{G}(l)\left(f\otimes e^{-i\sum_{j=1}^n k_j\cdot\phi(f_j)},e^{-TH_{\mathrm{N}}(\kappa,\Lambda)}g\otimes e^{-i\sum_{i=1}^m l_i\cdot\phi(g_i)}\right)dkdl.\end{aligned}$$

Thus it suffices to consider the limit

$$\lim_{\Lambda\to\infty}\lim_{\kappa\to\infty}(f_1\otimes e^{-i\phi(g_1)}, e^{-T(H_N(\kappa,\Lambda)-g^2NE(\Lambda))}f_2\otimes e^{-i\phi(g_2)}). \tag{2.11.75}$$

The functional integral representation of (2.11.75) is given by

$$\int_{\mathbb{R}^3}\mathbb{E}^x\left[\overline{f_1(B_0)}f_2(B_T)(e^{-i\phi_E(\delta_0\otimes g_1)}, e^{-i\phi_E(\delta_{\kappa^2T}\otimes g_2)}e^{-\kappa g\phi_E(\sum_{j=1}^N\int_0^T\delta_{\kappa^2s}\otimes\varphi(\cdot-B_s^j)ds)})\right]dx.$$

The field variable above can be integrated out and we have

$$(f_1\otimes e^{-i\phi(g_1)}, e^{-T(H_N(\kappa,\Lambda)-g^2NE(\Lambda))}f_2\otimes e^{-i\phi(g_2)}) = \int_{\mathbb{R}^3}\mathbb{E}^x[\overline{f_1(B_0)}f_2(B_T)e^{\frac{1}{4}S}]dx,$$

where $S = -S_1 + 2iS_2 + S_3$ with

$$S_1 = \|\delta_0\otimes g_1 - \delta_{\kappa^2T}\otimes g_2\|^2_{\mathcal{H}_E},$$

$$S_2 = \kappa g\sum_{j=1}^N\left(\delta_0\otimes g_1 - \delta_{\kappa^2T}\otimes g_2, \int_0^T\delta_{\kappa^2s}\otimes\varphi(\cdot - B_s^j)ds\right)_{\mathcal{H}_E},$$

$$S_3 = \kappa^2g^2\left\|\sum_{j=1}^N\int_0^T\delta_s\otimes\varphi(\cdot - B_s^j)ds\right\|^2_{\mathcal{H}_E}.$$

We see that

$$\lim_{\kappa\to\infty}S_1 = \|g_1\|^2_{\mathcal{H}_M} + \|g_2\|^2_{\mathcal{H}_M}.$$

Next consider S_2. Note that for a bounded continuous function $h(s)$,

$$\lim_{\kappa\to\infty}\int_0^T h(s)\kappa^2e^{-|s-t|\kappa^2}ds = 2h(t) \tag{2.11.76}$$

for $t < T$. For every j,

$$\lim_{\kappa\to\infty}\kappa\left(\delta_0\otimes g_1, \int_0^T\delta_{\kappa^2s}\otimes\varphi(\cdot - B_s^j)ds\right)$$

$$= \lim_{\kappa\to\infty}\frac{1}{(2\pi)^3}\int_0^T ds\int_{|k|\le\Lambda}\frac{\hat{g}_1(k)}{\omega(k)}e^{-ik\cdot B_s^j}\kappa e^{-\kappa^2|s|\omega(k)}dk = 0,$$

and similarly,

$$\lim_{\kappa\to\infty}\kappa\left(\delta_{\kappa^2T}\otimes g_2, \int_0^T\delta_{\kappa^2s}\otimes\varphi(\cdot - B_s^j)ds\right) = 0.$$

Hence $\lim_{\kappa\to\infty} S_2 = 0$. Finally, we estimate S_3. By (2.11.76), it follows that

$$\lim_{\kappa\to\infty} S_3 = \lim_{\kappa\to\infty} \sum_{i,j=1}^{N} \frac{1}{(2\pi)^3} \int_0^T ds \int_0^T dt \int_{|k|\le\Lambda} \frac{1}{\omega(k)} e^{-ik\cdot(B_s^i - B_t^j)} \kappa^2 e^{-|s-t|\omega(k)\kappa^2} dk$$

$$= \sum_{i,j=1}^{N} \frac{2}{(2\pi)^3} \int_0^T dt \int_{|k|\le\Lambda} \frac{1}{\omega(k)^2} e^{-ik\cdot(B_t^i - B_t^j)} dk.$$

Hence we have

$$\lim_{\kappa\to\infty} (f_1 \otimes e^{-i\phi(g_1)}, e^{-T(H_N(\kappa,\Lambda) - g^2 NE(\Lambda))} f_2 \otimes e^{-i\phi(g_2)})$$

$$= \int_{\mathbb{R}^3} \mathbb{E}^x \left[\overline{f_1(B_0)} f_2(B_T) e^{-\frac{1}{4}(\|g_1\|^2_{\mathcal{H}_M} + \|g_2\|^2_{\mathcal{H}_M})} e^{\int_0^T V_\Lambda(B_s^1,\ldots,B_s^N) ds} \right] dx,$$

where

$$V_\Lambda(x_1,\ldots,x_N) = \sum_{i<j}^{N} \frac{g^2}{(2\pi)^3} \int_{|k|\le\Lambda} \frac{1}{\omega(k)^2} e^{-ik(x_i - x_j)} dk.$$

Note that

$$V_\Lambda(x_1,\ldots,x_N) = \sum_{i<j}^{N} \frac{g^2}{2\pi^2} \frac{1}{|x_i - x_j|} \int_0^{\Lambda|x_i - x_j|} \frac{\sin r}{r} dr.$$

We prove that

$$|V_\Lambda(x_1,\ldots,x_N)| \le \frac{g^2}{4\pi} \sum_{i<j}^{N} \frac{2}{|x_i - x_j|} \tag{2.11.77}$$

in Lemma 2.140 below, and thus see that

$$\lim_{\Lambda\to\infty} V_\Lambda(x_1,\ldots,x_N) = \frac{g^2}{4\pi} \sum_{i<j}^{N} \frac{1}{|x_i - x_j|} \tag{2.11.78}$$

for every $(x_1,\ldots,x_N) \in \{(x_1,\ldots,x_N) \in \mathbb{R}^{3N} \mid x_i \ne x_j, i,j = 1,\ldots,N\}$. Denote the right-hand side of (2.11.78) by V. By the Lebesgue dominated convergence theorem,

$$\lim_{\Lambda\to\infty} \int_{\mathbb{R}^3} \mathbb{E}^x \left[\overline{f_1(B_0)} f_2(B_T) e^{-\frac{1}{4}(\|g_1\|^2_{\mathcal{H}_M} + \|g_2\|^2_{\mathcal{H}_M})} e^{+\int_0^T V_\Lambda(B_s^1,\ldots,B_s^N) ds} \right] dx$$

$$= (f_1 \otimes e^{-i\phi(g_1)}, (e^{-TH_\infty} \otimes P_0) f_2 \otimes e^{-i\phi(g_2)}).$$

□

Lemma 2.140. *Estimate* (2.11.77) *holds.*

Proof. Let $\mu > 0$ and define $V_{\Lambda,\mu}$ by V_Λ with $\omega(k)$ replaced by $\sqrt{|k|^2+\mu^2}$. It follows that

$$V_{\Lambda,\mu}(x_1,\dots,x_N) = \sum_{i<j}^{N} \frac{g^2}{2\pi^2}\frac{1}{|x_i - x_j|}\int_0^{\Lambda|x_i-x_j|} \frac{r}{r^2+\mu^2|x_i-x_j|^2}\sin r dr.$$

Define the closed curves γ_{ij}, $1 \le i,j \le N$, by

$$\gamma_{ij} = \{z \in \mathbb{R} \mid -\Lambda|x_i - x_j| \le z \le \Lambda|x_i - x_j|\} \cup C_{ij},$$

where $C_{ij} = \{\Lambda|x_i - x_j|e^{i\theta} \in \mathbb{C} \mid 0 \le \theta \le \pi\}$ denotes a semicircle in $\mathbb{C}$. Let

$$\rho(u) = \frac{u}{u^2 + |x_i - x_j|^2\mu^2}e^{iu}.$$

Since

$$\int_0^{\Lambda|x_i-x_j|} \frac{r \sin r dr}{r^2+\mu^2|x_i-x_j|^2} = \frac{1}{2}\Im\int_{-\Lambda|x_i-x_j|}^{\Lambda|x_i-x_j|}\rho(u)du = \frac{1}{2}\Im\Big(\int_{\gamma_{ij}}\rho(u)du - \int_{C_{ij}}\rho(u)du\Big),$$

we see that

$$\begin{aligned}
V_{\Lambda,\mu}(x_1,\dots,x_N) &= \frac{g^2}{4\pi^2}\Im\sum_{i<j}^{N}\frac{1}{|x_i-x_j|}\int_{-\Lambda|x_i-x_j|}^{\Lambda|x_i-x_j|}\frac{u}{u^2+|x_i-x_j|^2\mu^2}e^{iu}du \\
&= \frac{2\pi g^2}{4\pi^2}\sum_{i<j}^{N}\frac{1}{|x_i-x_j|}\operatorname{Res}\Big(\frac{ue^{iu}}{u^2+|x_i-x_j|^2\mu^2}; i\mu|x_i-x_j|\Big) \\
&\quad - \frac{g^2}{4\pi^2}\Im\sum_{i<j}^{N}\frac{1}{|x_i-x_j|}\int_0^{\pi}\frac{\Lambda^2|x_i-x_j|^2 ie^{2i\theta}e^{i\Lambda|x_i-x_j|e^{i\theta}}}{\Lambda^2|x_i-x_j|^2e^{2i\theta}+|x_i-x_j|^2\mu^2}d\theta,
\end{aligned}$$

where $\operatorname{Res}(f(u); z)$ denotes the residue of $f(u)$ at $u = z$. A computation gives

$$\operatorname{Res}\Big(\frac{ue^{iu}}{u^2+|x_i-x_j|^2\mu^2}; i\mu|x_i-x_j|\Big) = \frac{1}{2}e^{-\mu|x_i-x_j|}$$

and

$$\Big|\int_0^{\pi}\frac{\Lambda^2|x_i-x_j|^2 ie^{2i\theta}e^{i\Lambda|x_i-x_j|e^{i\theta}}}{\Lambda^2|x_i-x_j|^2e^{2i\theta}+|x_i-x_j|^2\mu^2}d\theta\Big| \le \pi\Big|\frac{\Lambda^2}{\Lambda^2-\mu^2}\Big|.$$

Hence

$$|V_{\Lambda,\mu}(x_1,\ldots,x_N)| \le \frac{g^2}{4\pi}\sum_{i<j}^N \frac{1}{|x_i-x_j|}\left(e^{-\mu|x_i-x_j|} + \left|\frac{\Lambda^2}{\Lambda^2-\mu^2}\right|\right). \tag{2.11.79}$$

Taking the limit as $\mu \to 0$ on both sides above, the lemma follows. □

Remark 2.141. Choosing $\omega(k) = \sqrt{|k|^2+\nu^2}$ leads to the *Yukawa potential* instead of the Coulomb potential, i. e.,

$$\operatorname*{s-lim}_{\Lambda\to\infty}\operatorname*{s-lim}_{\kappa\to\infty} e^{-T(H(\kappa,\Lambda)-g^2NE(\Lambda))} = e^{-TH_\infty}\otimes P_0,$$

with

$$H_\infty = -\frac{1}{2}\sum_{j=1}^N \Delta_j - \frac{g^2}{4\pi}\sum_{i<j}^N \frac{e^{-\nu|x_i-x_j|}}{|x_i-x_j|}.$$

This can be seen by a minor modification of the case $\mu = 0$.

2.12 Translation invariant Nelson model

2.12.1 Definition of translation invariant Nelson model

In this section we study the translation invariant Nelson model with a single quantum particle in d spatial dimensions. With coupling constant $g \in \mathbb{R}$, we write the Nelson Hamiltonian as

$$H_\mathrm{N} = H_\mathrm{p}\otimes 1 + 1\otimes H_\mathrm{f} + gH_\mathrm{I}.$$

In Sections 2.12.1–2.12.7 we use Assumption 2.16, i. e.,

(1) (Dispersion relation) $\omega(k) = \omega_\nu(k) = \sqrt{|k|^2+\nu^2}$, $\nu \ge 0$.

(2) (Charge distribution) $\varphi \in \mathscr{S}'(\mathbb{R}^d)$, $\overline{\hat\varphi(k)} = \hat\varphi(-k)$ and $\hat\varphi/\omega, \hat\varphi/\sqrt{\omega} \in L^2(\mathbb{R}^d)$.

Furthermore, we suppose throughout this section that $V = 0$. Thus

$$H_\mathrm{p} = -\frac{1}{2}\Delta.$$

Write

$$P_{\mathrm{f}\mu} = d\Gamma(k_\mu), \quad \mu = 1,\ldots,d,$$

for the *field momentum operator*. The *total momentum operator* $\mathrm{P}_\mu^{\mathrm{tot}}$, $\mu = 1,\ldots,d$, on $\mathcal{F}_\mathrm{N}$ is defined by the sum of particle and field momenta,

$$\mathrm{P}_\mu^{\mathrm{tot}} = -i\nabla_{x_\mu} \otimes 1 + 1 \otimes P_{\mathrm{f}\mu}.$$

It is straightforward to see that

$$[H_\mathrm{N}, \mathrm{P}_\mu^{\mathrm{tot}}] = 0, \quad \mu = 1, \ldots, d.$$

This has the implication that H_N is decomposable on the spectrum of the total momentum operator and

$$\mathrm{Spec}(\mathrm{P}_\mu^{\mathrm{tot}}) = \mathbb{R}$$

holds. We denote $P_\mathrm{f}^2 = \sum_{\mu=1}^d (P_{\mathrm{f}\mu})^2$.

Proposition 2.142 (Absence of ground state). *If $V = 0$, then H_N has no ground state.*

Proof. We use proof by contradiction by supposing, to the contrary, that H_N has a ground state Ψ_g satisfying $H_\mathrm{N}\Psi_\mathrm{g} = E\Psi_\mathrm{g}$. Since e^{-tH_N} is positivity improving, the ground state is unique up to multiplicative constants. Let E_T denote the spectral projection of $(\mathrm{P}^{\mathrm{tot}})^2$, which has a purely absolutely continuous spectrum, in particular, it has no point spectrum. We can find $A, B \subset [0, \infty)$ such that $A \cap B = \emptyset$, and $\Psi_\mathrm{g}^1 = E_T[A]\Psi_\mathrm{g} \neq 0$ and $\Psi_\mathrm{g}^2 = E_T[B]\Psi_\mathrm{g} \neq 0$. Hence we see that $H_\mathrm{N}\Psi_\mathrm{g}^1 = E\Psi_\mathrm{g}^1$, $H_\mathrm{N}\Psi_\mathrm{g}^2 = E\Psi_\mathrm{g}^2$, and $(\Psi_\mathrm{g}^1, \Psi_\mathrm{g}^2) = 0$. This contradicts the uniqueness of the ground state, thus H_N has no ground state. □

Definition 2.143 (Nelson Hamiltonian with a fixed total momentum in $\mathcal{F}_\mathrm{N}$). For every $p \in \mathbb{R}^d$, the *Nelson Hamiltonian with a fixed total momentum p* is defined by

$$H_\mathrm{N}(p) = \frac{1}{2}(p - P_\mathrm{f})^2 + gH_\mathrm{I}(0) + H_\mathrm{f}$$

with domain $D(H_\mathrm{N}(p)) = D(H_\mathrm{f}) \cap D(P_\mathrm{f}^2)$, where

$$H_\mathrm{I}(0) = \frac{1}{\sqrt{2}}(a_M^*(\hat{\varphi}) + a_M(\tilde{\hat{\varphi}})).$$

We also use the formal notation of $H_\mathrm{I}(0)$ as

$$H_\mathrm{I}(0) = \int_{\mathbb{R}^d} \frac{1}{\sqrt{2\omega(k)}}(\hat{\varphi}(k)a^*(k) + \hat{\varphi}(-k)a(k))dk$$

for convenience.

Proposition 2.144 (Self-adjointness). *For every $p \in \mathbb{R}^d$, $H_\mathrm{N}(p)$ is self-adjoint and essentially self-adjoint on any core of $\frac{1}{2}P_\mathrm{f}{}^2 + H_\mathrm{f}$.*

Proof. Since $H_\mathrm{N}(p) = \frac{1}{2}P_\mathrm{f}{}^2 + H_\mathrm{f} - pP_\mathrm{f} + gH_\mathrm{I}(0) + \frac{1}{2}p^2$ and $\hat{\varphi}/\sqrt{\omega}, \hat{\varphi}/\omega \in L^2(\mathbb{R}^d)$, we see that for every $\varepsilon > 0$ there exists $b_\varepsilon > 0$ such that

$$\|(-pP_\mathrm{f} + gH_\mathrm{I}(0))\Phi\| \le \varepsilon\left\|\left(\frac{1}{2}P_\mathrm{f}{}^2 + H_\mathrm{f}\right)\Phi\right\| + b_\varepsilon\|\Phi\|$$

for all $\Phi \in D(P_\mathrm{f}{}^2) \cap D(H_\mathrm{f})$. Hence the proposition follows by the Kato–Rellich theorem. □

Define the unitary operator $\mathscr{T} : \mathcal{H}_\mathrm{N} = L^2(\mathbb{R}^d_x) \otimes \mathcal{F}_\mathrm{N} \to L^2(\mathbb{R}^d_p) \otimes \mathcal{F}_\mathrm{N}$ by

$$\mathscr{T} = (F \otimes 1) \int_{\mathbb{R}^d}^{\oplus} e^{ix\cdot P_\mathrm{f}} dx, \tag{2.12.1}$$

with F denoting Fourier transform from $L^2(\mathbb{R}^d_x)$ to $L^2(\mathbb{R}^d_p)$. For $\Psi \in \mathcal{H}_\mathrm{N}$, we see that

$$(\mathscr{T}\Psi)(p) = \frac{1}{\sqrt{(2\pi)^d}} \int_{\mathbb{R}^d} e^{-ix\cdot p} e^{ix\cdot P_\mathrm{f}} \Psi(x) dx.$$

Theorem 2.145 (Fiber decomposition). *It follows that*

$$\mathscr{T}\left(\int_{\mathbb{R}^d}^{\oplus} \mathcal{F}_\mathrm{N} dp\right)\mathscr{T}^{-1} = \mathcal{H}_\mathrm{N} \quad \text{and} \quad \mathscr{T}\left(\int_{\mathbb{R}^d}^{\oplus} H_\mathrm{N}(p) dp\right)\mathscr{T}^{-1} = H_\mathrm{N}.$$

Proof. Define the self-adjoint operator $\bar{L}$ by $\bar{L} = \int_{\mathbb{R}^d}^{\oplus} H_\mathrm{N}(p)dp$. We show that

$$H_\mathrm{N} = \mathscr{T}^{-1}\bar{L}\mathscr{T}.$$

Let $\mathscr{D}_\infty = \mathrm{L.H.}\{f \otimes \Phi \mid f \in C_0^\infty(\mathbb{R}^d), \Phi \in \mathcal{F}_{\mathrm{N,fin}}\}$. For $F, G \in \mathscr{D}_\infty$, we have

$$\begin{aligned}\left(\mathscr{T}F, \frac{1}{2}(p - P_\mathrm{f})^2 \mathscr{T}G\right) &= \left((F \otimes 1)e^{ix\cdot P_\mathrm{f}}F, \frac{1}{2}(p - P_\mathrm{f})^2 (F \otimes 1)e^{ix\cdot P_\mathrm{f}} \mathscr{T}G\right) \\ &= \left(e^{ixP_\mathrm{f}}F, \frac{1}{2}(-i\nabla_x - P_\mathrm{f})^2 e^{ix\cdot P_\mathrm{f}} G\right) = \left(F, -\frac{1}{2}\Delta_x G\right)\end{aligned}$$

and

$$(\mathscr{T}F, H_\mathrm{I}(0)\mathscr{T}G) = (e^{ix\cdot P_\mathrm{f}}F, H_\mathrm{I}(0)e^{ix\cdot P_\mathrm{f}}G) = (F, H_\mathrm{I}(x)G).$$

Finally, it is trivial to see that

$$(\mathscr{T}F, H_\mathrm{f}\mathscr{T}G) = (F, H_\mathrm{f}G).$$

Combining with the above, we have $(F, H_\mathrm{N}G) = (\mathscr{T}F, \bar{L}\mathscr{T}G)$. Hence

$$H_\mathrm{N} = \mathscr{T}^{-1}\bar{L}\mathscr{T}$$

on $\mathscr{D}_\infty$. Since $\mathscr{D}_\infty$ is a core of H_N and $\bar{L}$ is self-adjoint, it follows that $\mathscr{T}$ maps $D(H_N)$ onto $D(\bar{L})$ with $\mathscr{T}H_N\mathscr{T}^{-1} = \bar{L}$. □

Our aim is to construct the functional integral representation of $e^{-tH_N(p)}$, which will be done in Theorem 2.147 below. First, we define $H_N(p)$ in $L^2(\mathscr{Q})$ for $\mathcal{F}_N$.

Definition 2.146 (Nelson Hamiltonian with a fixed total momentum in $L^2(\mathscr{Q})$). For every $p \in \mathbb{R}^d$, the Nelson Hamiltonian with a fixed total momentum p is defined by

$$\hat{H}_N(p) = \frac{1}{2}(p - \hat{P}_f)^2 + g\phi(\varphi) + H_f \tag{2.12.2}$$

with domain $D(\hat{H}_N(p)) = D(\hat{H}_f) \cap D(\hat{P}_f^{\,2})$, where $\hat{P}_f = d\Gamma(-i\nabla)$.

Hamiltonians $\hat{H}_N(p)$ and $H_N(p)$ are unitary equivalent, and in what follows we use the notation $H_N(p)$ for $\hat{H}_N(p)$.

2.12.2 Functional integral representation

Using the explicit form of $H_N(p)$ discussed in the previous section, now we turn to its functional integral representation. The main idea of the construction is to start from $e^{-tH_N}e^{-i\xi\cdot \mathrm{P}^{\mathrm{tot}}} \cong \int_{\mathbb{R}^d}^{\oplus} e^{-tH_N(p)}e^{-i\xi p}dp$ and write

$$(F, e^{-tH_N}e^{-i\xi\cdot \mathrm{P}^{\mathrm{tot}}}G) = \int_{\mathbb{R}^d} (\mathscr{T}F(p), e^{-tH_N(p)}\mathscr{T}G(p))e^{-i\xi\cdot p}dp.$$

Thus formally regarding $Z(p) = (\mathscr{T}F(p), e^{-tH_N(p)}\mathscr{T}G(p))$ as an $L^2(\mathbb{R}^d_p)$-function and $\hat{Z}(\xi) = (F, e^{-tH_N}e^{-i\xi\cdot \mathrm{P}^{\mathrm{tot}}}G)$ as an $L^2(\mathbb{R}^d_\xi)$-function, we arrive at

$$(\mathscr{T}F(p), e^{-tH_N(p)}\mathscr{T}G(p)) = \frac{1}{(2\pi)^d}\int_{\mathbb{R}^d} (F, e^{-tH_N}e^{-i\xi\cdot \mathrm{P}^{\mathrm{tot}}}G)e^{i\xi\cdot p}d\xi.$$

For given $\Psi, \Phi \in L^2(\mathscr{Q})$, taking suitable sequences F_s and G_r such that $\mathscr{T}F_s(p) \to (2\pi)^{-d/2}\Psi$ and $\mathscr{T}G_r(p) \to (2\pi)^{-d/2}\Phi$ as $s \to 0$ and $r \to 0$, we can then derive a functional integral representation of $(\Psi, e^{-tH_N(p)}\Phi)$ by taking the limit in

$$(F_s(p), e^{-tH_N(p)}G_r(p)) = \frac{1}{(2\pi)^d}\int_{\mathbb{R}^d} (\mathscr{T}F_s, e^{-tH_N}e^{-i\xi\cdot \mathrm{P}^{\mathrm{tot}}}\mathscr{T}G_r)e^{i\xi\cdot p}d\xi.$$

Theorem 2.147 (Functional integral representation for Nelson Hamiltonian with a fixed total momentum). *Let $p \in \mathbb{R}^d$ and define*

$$u_t(p) = e^{i(p-\hat{P}_f)\cdot B_t} \tag{2.12.3}$$

for $(p,t) \in \mathbb{R}^d \times \mathbb{R}$. *If* $\Psi, \Phi \in L^2(\mathscr{Q})$, *then*

$$(\Psi, e^{-2tH_{\mathrm{N}}(p)}\Phi)_{L^2(\mathscr{Q})} = \mathbb{E}^0\left[(\mathrm{I}_{-t}u_{-t}(p)\Psi, e^{-g\phi_{\mathrm{E}}(\int_{-t}^{t}\delta_s\otimes\varphi(\cdot-B_s)ds)}\mathrm{I}_t u_t(p)\Phi)_{L^2(\mathscr{Q}_{\mathrm{E}})}\right] \quad (2.12.4)$$

$$= \mathbb{E}^0\left[(\mathrm{I}_0\Psi, e^{-g\phi_{\mathrm{E}}(\int_0^{2t}\delta_s\otimes\varphi(\cdot-B_s)ds)}\mathrm{I}_{2t}u_{2t}(p)\Phi)_{L^2(\mathscr{Q}_{\mathrm{E}})}\right]. \quad (2.12.5)$$

Proof. In this proof, we replace $2t$ by t. Write $F_s = \Pi_s \otimes \Psi$ and $G_r = \Pi_r \otimes \Phi$, where $\Pi_s(x) = (2\pi s)^{-d/2}\exp(-|x|^2/2s)$ is the heat kernel, and $\Psi, \Phi \in L^2(\mathscr{Q})$. By the fact that $H_{\mathrm{N}} \cong \mathscr{T}^{-1}(\int_{\mathbb{R}^d}^{\oplus} H_{\mathrm{N}}(p)dp)\mathscr{T}$, we have

$$(F_s, e^{-tH_{\mathrm{N}}}e^{-i\xi\cdot\mathrm{p}^{\mathrm{tot}}}G_r) = \int_{\mathbb{R}^d} e^{-i\xi\cdot p}((\mathscr{T}F_s)(p), e^{-tH_{\mathrm{N}}(p)}(\mathscr{T}G_r)(p))dp$$

for every $\xi \in \mathbb{R}^d$. Note that $\lim_{s\to0}(\mathscr{T}F_s)(p) = (2\pi)^{-d/2}\Psi$ strongly in $L^2(\mathscr{Q})$ for every $p \in \mathbb{R}^d$. Hence

$$\lim_{s\to0}(F_s, e^{-tH_{\mathrm{N}}}e^{-i\xi\cdot\mathrm{p}^{\mathrm{tot}}}G_r) = \frac{1}{(2\pi)^{d/2}}\int_{\mathbb{R}^d} e^{-i\xi\cdot p}(\Psi, e^{-tH_{\mathrm{N}}(p)}(\mathscr{T}G_r)(p))dp. \quad (2.12.6)$$

Using that $e^{-i\xi\cdot\mathrm{p}^{\mathrm{tot}}}G_r(x) = \Pi_r(x-\xi)e^{-i\xi\cdot P_{\mathrm{f}}}\Phi$, we obtain

$$(F_s, e^{-tH_{\mathrm{N}}}e^{-i\xi\cdot\mathrm{p}^{\mathrm{tot}}}G_r) = \int_{\mathbb{R}^d}\mathbb{E}^x\left[\Pi_s(0)\Pi_r(B_t-\xi)(\mathrm{I}_0\Psi, e^{-\phi_{\mathrm{E}}(\int_0^t\delta_s\otimes\varphi(\cdot-B_s)ds)}\mathrm{I}_t e^{-i\xi\cdot\hat{P}_{\mathrm{f}}}\Phi)\right]dx.$$

It follows from $\Pi_s(x) \to \delta(x)$ as $s \to 0$ that

$$\lim_{s\to0}(F_s, e^{-tH_{\mathrm{N}}}e^{-i\xi\cdot\mathrm{p}^{\mathrm{tot}}}G_r) = \mathbb{E}^0\left[\Pi_r(B_t-\xi)(\mathrm{I}_0\Psi, e^{-\phi_{\mathrm{E}}(\int_0^t\delta_s\otimes\varphi(\cdot-B_s)ds)}\mathrm{I}_t e^{-i\xi\cdot P_{\mathrm{f}}}\Phi)\right]. \quad (2.12.7)$$

Combining (2.12.6) and (2.12.7) leads to

$$\frac{1}{(2\pi)^{d/2}}\int_{\mathbb{R}^3} e^{-i\xi\cdot p}(\Psi, e^{-tH_{\mathrm{N}}(p)}(\mathscr{T}G_r)(p))$$
$$= \mathbb{E}^0\left[\Pi_r(B_t-\xi)(\mathrm{I}_0\Psi, e^{-\phi_{\mathrm{E}}(\int_0^t\delta_s\otimes\varphi(\cdot-B_s)ds)}\mathrm{I}_t e^{-i\xi\cdot P_{\mathrm{f}}}\Phi)\right]. \quad (2.12.8)$$

Due to the bounds $|(\Psi, e^{-tH_{\mathrm{N}}(p)}(\mathscr{T}G_r)(p))| \le \|(\mathscr{T}G_r)(p)\|$ and $\|(\mathscr{T}G_r)(\cdot)\| \in L^2(\mathbb{R}^d)$, we have $(\Psi, e^{-tH_{\mathrm{N}}(p)}(\mathscr{T}G_r)(p)) \in L^2(\mathbb{R}_p^d)$. Taking the inverse Fourier transform on both sides of (2.12.8) with respect to p gives

$$(\Psi, e^{-tH_{\mathrm{N}}(p)}(\mathscr{T}G_r)(p))$$
$$= \frac{1}{(2\pi)^{d/2}}\mathbb{E}^0\left[\int_{\mathbb{R}^d} e^{i\xi\cdot p}\Pi_r(B_t-\xi)(\mathrm{I}_0\Psi, e^{-\phi_{\mathrm{E}}(\int_0^t\delta_s\otimes\varphi(\cdot-B_s)ds)}\mathrm{I}_t e^{-i\xi\cdot P_{\mathrm{f}}}\Phi)\right]d\xi \quad (2.12.9)$$

for almost every $p \in \mathbb{R}^d$. Since both sides of (2.12.9) are continuous in p, the equality stays valid for all $p \in \mathbb{R}^d$. After taking $r \to 0$ on both sides, we arrive at (2.12.5). Equality (2.12.4) is derived from the time shift of (2.12.5). □

The following result comes as an immediate corollary to the functional integral representation. Write $E_N(p) = \inf \mathrm{Spec}(H_N(p))$.

Corollary 2.148 (Properties of $E_N(p)$).
(1) *If $p = 0$, then $e^{-tH_N(0)}$ is positivity improving.*
(2) *The ground state of $H_N(0)$ is unique whenever it exists.*
(3) *The energy comparison inequality $E_N(0) \le E_N(p)$ holds.*
(4) *The map $\mathbb{R}^d \ni p \mapsto E_N(p) \in \mathbb{R}$ is continuous.*

Proof. We have

$$(\Psi, e^{-tH_N(0)}\Phi) = \mathbb{E}^0\left[\left(\mathrm{I}_0\Psi, e^{-g\phi_E(\int_0^t \delta_s\otimes\varphi(\cdot-B_s)ds)}\mathrm{I}_t e^{-iB_t\cdot\hat{P}_f}\Phi\right)\right].$$

Since $e^{-iB_t\cdot\hat{P}_f} = \Gamma(e^{-B_t\cdot\nabla})$ is the second quantization of the shift operator $e^{-B_t\cdot\nabla}$ by $B_t \in \mathbb{R}^d$, $e^{-iB_t\cdot\hat{P}_f}$ is positivity preserving, while $\mathrm{I}_0^* e^{-g\phi_E(\int_0^t \delta_s\otimes\varphi(\cdot-B_s)ds)}\mathrm{I}_t$ is positivity improving. Thus $e^{-tH_N(0)}$ is positivity improving, and the Perron–Frobenius theorem implies that the ground state is unique. We have

$$|(\Psi, e^{-tH_N(p)}\Phi)| \le \mathbb{E}^0\left[\left(|\Psi|, \mathrm{I}_0^* e^{-g\phi_E(\int_0^t \delta_s\otimes\varphi(\cdot-B_s)ds)}\mathrm{I}_t e^{-iB_t\cdot\hat{P}_f}|\Phi|\right)\right] = (|\Psi|, e^{-tH_N(0)}|\Phi|).$$

It can be seen that

$$e^{-tE_N(p)} = \sup_{\Psi\neq0,\Phi\neq0} \frac{(\Psi, e^{-tH_N(p)}\Phi)}{\|\Psi\|\|\Phi\|} \le \sup_{\Psi\neq0,\Phi\neq0} \frac{(|\Psi|, e^{-tH_N(0)}|\Phi|)}{\|\Psi\|\|\Phi\|} \le e^{-tE_N(0)},$$

and (3) follows. Finally, we prove (4). We have

$$(\Psi, (e^{-tH_N(p)} - e^{-tH_N(q)})\Phi) = \mathbb{E}^0\left[\left(\mathrm{I}_0\Psi, e^{-g\phi_E(\int_0^t \delta_s\otimes\varphi(\cdot-B_s)ds)}\mathrm{I}_t e^{-iB_t\cdot\hat{P}_f}\Phi\right)\left(e^{ip\cdot B_t} - e^{iq\cdot B_t}\right)\right].$$

Inserting the identity $e^{ip\cdot B_t} - e^{iq\cdot B_t} = i\int_{qB_t}^{pB_t} e^{i\theta} d\theta$, we see that

$$|(\Psi, (e^{-tH_N(p)} - e^{-tH_N(q)})\Phi)_{\mathcal{F}_N}| \le C\|\Psi\|\|\Phi\|\mathbb{E}^0\left[|p-q||B_t|\right]$$

Here the constant C is given by

$$C = 2e^{g^2(\frac{t}{2}\|\hat{\varphi}/\sqrt{\omega}\|^2 + 2t(t\vee1)(\|\hat{\varphi}/\sqrt{\omega}\|^2 + \|\hat{\varphi}/\omega\|^2))}.$$

This can be obtained from the estimate

$$\|\mathrm{I}_0^* e^{-g\phi_E(\int_0^t \delta_s\otimes\varphi(\cdot-B_s)ds)}\mathrm{I}_t\Phi\|_{L^2(\mathcal{Q}_E)} \le C\|\Phi\|_{L^2(\mathcal{Q}_E)}$$

which was shown in Corollary 1.88. Hence $e^{-tH_N(p)}$ is uniformly continuous in p, which implies (4). □

Corollary 2.149 (Ground state energy of the Nelson Hamiltonian with $V = 0$). *Let $V = 0$. Then the ground state energy of the Nelson Hamiltonian E_N coincides with the ground state energy of the Nelson Hamiltonian with zero total momentum, that is, $E_N(0) = E_N$.*

Proof. From the energy comparison inequality $E_N(0) \leq E_N(p)$, it follows that

$$(\Phi, H_N\Phi) = \int_{\mathbb{R}^d} (\Phi(p), H_N(p)\Phi(p))dp \geq E_N(0)\|\Phi\|^2$$

for every $\Phi \in D(H_N)$. This implies the bound $E_N(0) \leq E_N$. To show that $E_N \leq E_N(p)$ for all $p \neq 0$, suppose that $E_N > E_N(p)$ for some $p \neq 0$. Since $E_N(p)$ is continuous in p, $E_N > E_N(q)$ for every $q \in \mathbb{R}^d$ such that $|p - q| \leq \delta$ with some $\delta > 0$. Let $\Phi_\delta = \int_{\mathbb{R}^d}^{\oplus} \Phi(q)1_{\{|p-q|\leq\delta\}}dp$, where we assume that $\Phi(q)$ satisfies that

$$E_N(q) \leq \frac{(\Phi(q), H(q)\Phi(q))}{\|\Phi(q)\|^2} < E_N$$

for $|p - q| < \delta$. We have

$$(\Phi_\delta, H_N\Phi_\delta) = \int_{|q-p|\leq\delta} (\Phi(q), H(q)\Phi(q))dq < E_N\|\Phi_\delta\|^2.$$

This contracts that E_N is the bottom of the spectrum of H_N. Thus $E_N \leq E_N(p)$ follows for every $p \neq 0$. From $E_N(0) \leq E_N \leq E_N(p)$ and the continuity of $E_N(\cdot)$, it follows that $E_N = E_N(0)$. □

2.12.3 Existence of ground state

Consider

$$\Delta(p) = \inf \mathrm{Spec}_{\mathrm{ess}}(H_N(p)) - E_N(p).$$

It is known that

$$\Delta(p) = \inf_{k\in\mathbb{R}^d} (E_N(p - k) + \omega_\nu(k)) - E_N(p). \tag{2.12.10}$$

For the massive boson case, when $\nu > 0$, from here follows that $\Delta(p) > 0$ for sufficiently small $|p|$, thus $H_N(p)$ has a ground state at least for sufficiently small $|p|$. For massless bosons $\omega_0(0) = 0$, and the situation is less clear since now $\Delta(p) = 0$. We will investigate this problem next.

Choose $\omega_0 = \omega$, and let $\Psi_g(p) = \Psi_g(p, \nu)$ be the normalized ground state of $H_N(p)$ for $\nu > 0$. Taking a subsequence ν', we see that $\Psi_g(p, \nu')$ weakly converges to a function Ψ. A criterion for the existence of a ground state of $H_N(p)$ for $\nu = 0$ is that Ψ is nonzero. A sufficient condition is

$$\|N^{1/2}\Psi_g(p)\| < 1, \tag{2.12.11}$$

uniformly in $\nu > 0$, in which case $H_N(p)$ for $\nu = 0$ does have a ground state. To obtain the bound in (2.12.11), a standard strategy is to apply the pull-through formula

$$\|N^{1/2}\Psi_g(p)\|^2 = \frac{g^2}{2}\int_{\mathbb{R}^d} \left\|(H_N(p-k) + \omega_\nu(k) - E_N(p))^{-1}\frac{\hat{\varphi}(k)}{\sqrt{\omega_\nu(k)}}\Psi_g(p)\right\|^2 dk. \tag{2.12.12}$$

An estimate gives

$$\|N^{1/2}\Psi_g(p)\|^2 \le \frac{g^2}{2}\frac{\|\hat{\varphi}/\sqrt{\omega_\nu}\|^2}{|\Delta(p)|^2} \tag{2.12.13}$$

which, however, is not conclusive since the right-hand side of (2.12.13) diverges as $\nu \to 0$ makes $\Delta(p) \to 0$. We show that functional integration answers this question elegantly for $p = 0$.

Let $p = 0$, and define $\Psi_g^T = e^{-TH_N(0)}1/\|e^{-TH_N(0)}1\|$ and

$$\gamma(T) = (1, \Psi_g^T)^2 = \frac{(1, e^{-TH_N(0)}1)^2}{(1, e^{-2TH_N(0)}1)}.$$

Lemma 2.150. *There exists a ground state of $H_N(0)$ if and only if $\lim_{T\to\infty}\gamma(T) > 0$.*

Proof. This is proven in the same way as Propositions 2.30 and 2.31. □

Using Lemma 2.150, we can show the existence of a ground state of $H_N(0)$. By the functional integral representation, we have

$$(1, e^{-TH_N(0)}1) = \mathbb{E}\left[e^{\frac{g^2}{2}\int_0^T ds \int_0^T W dt}\right], \tag{2.12.14}$$

where

$$W = W(B_s - B_t, s-t) = \int_{\mathbb{R}^d} \frac{|\hat{\varphi}(k)|^2}{2\omega(k)^3} e^{-ik\cdot(B_s-B_t)} e^{-|s-t|\omega(k)} dk.$$

Theorem 2.151 (Existence of ground state). *If $\int_{\mathbb{R}^d} |\hat{\varphi}(k)|^2/\omega(k)^3 dk < \infty$, then $H_N(0)$ has a ground state and it is unique.*

Proof. Uniqueness follows from Corollary 2.148. By (2.12.14), we have

$$\gamma(T) = \frac{\left(\mathbb{E}\left[e^{\frac{g^2}{2}\int_0^T ds \int_0^T W dt}\right]\right)^2}{\mathbb{E}\left[e^{\frac{g^2}{2}\int_{-T}^T ds \int_{-T}^T W dt}\right]}.$$

By the reflection symmetry of Brownian motion,

$$\gamma(T) = \frac{\mathbb{E}\left[e^{\frac{g^2}{2}\int_0^T ds \int_0^T W dt}\right]\mathbb{E}\left[e^{\frac{g^2}{2}\int_{-T}^0 ds \int_{-T}^0 W dt}\right]}{\mathbb{E}\left[e^{\frac{g^2}{2}\int_{-T}^T ds \int_{-T}^T W dt}\right]},$$

and, by using the independence of B_s and B_t for $s < 0 < t$, the expression can be rearranged into

$$\gamma(T) = \frac{\mathbb{E}\left[e^{\frac{g^2}{2}\int_0^T ds \int_0^T W dt + \frac{g^2}{2}\int_{-T}^0 ds \int_{-T}^0 W dt}\right]}{\mathbb{E}\left[e^{\frac{g^2}{2}\int_{-T}^T ds \int_{-T}^T W dt}\right]} = \frac{\mathbb{E}\left[e^{\frac{g^2}{2}\int_{-T}^T ds \int_{-T}^T W dt - g^2\int_{-T}^0 ds \int_0^T W dt}\right]}{\mathbb{E}\left[e^{\frac{g^2}{2}\int_{-T}^T ds \int_{-T}^T W dt}\right]}.$$

Since $\int_{-T}^0 ds \int_0^T W dt \le \frac{1}{2}\int_{\mathbb{R}^d} |\hat{\varphi}(k)|^2/\omega(k)^3 dk$, we obtain

$$\gamma(T) \ge \exp\left(-\frac{g^2}{2}\int_{\mathbb{R}^d} \frac{|\hat{\varphi}(k)|^2}{\omega(k)^3} dk\right) > 0 \tag{2.12.15}$$

for all $T > 0$. Hence the theorem follows. □

In the proof of Theorem 2.151 above, $W(B_s - B_t, s - t)$ is *the pair potential of the Nelson Hamiltonian with a fixed total momentum.*

2.12.4 Gibbs measure associated with the ground state of Nelson model with zero total momentum

A Gibbs measure can be defined to the process describing the Nelson model with zero total momentum in a similar way to (2.8.2). Let $p = 0$, $E_{\mathrm{N}}(0) = E$, and denote the ground state of $H_{\mathrm{N}}(0)$ simply by Ψ_{g}. Define the probability measure ν_T on $(\mathscr{X}, \mathcal{B}(\mathscr{X}))$ by

$$\nu_T(A) = \mathbb{E}[1_A e^{\frac{g^2}{2}\int_{-T}^T ds \int_{-T}^T W(B_s - B_t, s-t)dt}], \quad A \in \mathcal{B}(\mathscr{X}),$$

which we call the *finite volume Gibbs measure* associated with the Nelson model with zero total momentum. Recall that we proved that the similar family of measures $\{\mu_T\}_{T\ge 0}$

in (2.8.2) associated with the Nelson model is tight. For $\{\nu_T\}_{T\geq 0}$, however, it is not straightforward to show tightness. The strategy to show the local convergence of $\{\nu_T\}_{T\geq 0}$ is, however, a minor modification of the arguments in Section 2.8. We show an the outline of this argument.

Let $\mathscr{G}_T = \mathscr{F}^{\mathrm{BM}}_{[-T,T]} = \sigma(B_r, -t \leq r \leq t)$ be the natural filtration of $(B_t)_{t\in\mathbb{R}}$, and define $\mathscr{G} = \bigcup_{0\leq s} \mathscr{G}_S$. Also, define the family of probability spaces $(\mathscr{X}, \mathcal{B}(\mathscr{X}), \nu_T)$, $T > 0$, and the finitely additive set function $\nu : \mathscr{G} \to [0,\infty)$,

$$\nu(A) = e^{2ET}\mathbb{E}\left[1_A(\Psi_g, \mathrm{I}_{[-T,T]}\Psi_g)\right], \quad A \in \mathscr{G}_T.$$

Here

$$\mathrm{I}_{[-T,T]} = \mathrm{I}^*_{-T} e^{-g\phi_\mathrm{E}(\int_{-T}^T \delta_s\otimes\varphi(\cdot-B_s)ds)}\mathrm{I}_T.$$

The set function ν is well defined in the sense that for $A \in \mathscr{G}_T \subset \mathscr{G}_S$, $T \leq S$,

$$\nu(A) = e^{2ET}\mathbb{E}\left[1_A(\Psi_g, \mathrm{I}_{[-T,T]}\Psi_g)\right] = e^{2ES}\mathbb{E}\left[1_A(\Psi_g, \mathrm{I}_{[-S,S]}\Psi_g)\right]$$

holds. This can be checked similarly to Lemma 2.71. It can be also shown that ν is countably additive. By Hopf's extension theorem, ν can be uniquely extended to a probability measure ν_∞ on $(\mathscr{X}, \mathcal{B}(\mathscr{X}))$. Note that $\sigma(\mathscr{G}) = \mathcal{B}(\mathscr{X})$. Furthermore, define the set function $\varrho_T : \mathscr{G}_T \to [0,\infty)$ by

$$\varrho_T(A) = e^{2Es}\mathbb{E}\left[1_A\left(\frac{1_{T-s}}{\|1_T\|}, \mathrm{I}_{[-s,s]}\frac{1_{T-s}}{\|1_T\|}\right)\right]$$

for $A \in \mathscr{G}_s$ with $s \leq T$. Here $1_t = e^{-t(H_\mathrm{N}(0)-E)}1$. Similarly to Lemma 2.73, it can be also seen that ϱ_T is well defined and hence ϱ_T is a probability measure on $(\mathscr{X}, \mathscr{G}_T)$. Moreover, it follows that

$$\varrho_T(A) = \nu_T(A) \tag{2.12.16}$$

for all $A \in \mathscr{G}_s$ with $s \leq T$.

Theorem 2.152 (Local convergence). *The family of probability measures $\{\nu_T\}_{T\geq 0}$ converges to ν_∞ on $(\mathscr{X}, \mathcal{B}(\mathscr{X}))$ in the local sense, i. e., $\nu_T(A) \to \nu_\infty(A)$ as $T \to \infty$ for all $A \in \mathscr{G}$.*

Proof. Suppose that $A \in \mathscr{G}_s$ with some s. We have

$$\lim_{T\to\infty} \nu_T(A) = \lim_{T\to\infty} \varrho_T(A) = \lim_{T\to\infty} e^{2Es}\mathbb{E}\left[1_A\left(\frac{1_{T-s}(B_{-s})}{\|1_T\|}, \mathrm{I}_{[-s,s]}\frac{1_{T-s}(B_s)}{\|1_T\|}\right)\right].$$

Since $1_T/\|1_T\| \to \Psi_g$ strongly as $T \to \infty$, we have

$$\lim_{T\to\infty} \nu_T(A) = e^{2Es}\mathbb{E}\left[1_A\left(\Psi_g, \mathrm{I}_{[-s,s]}\Psi_g\right)\right] = \nu_\infty(A).$$ □

We call ν_∞ the *Gibbs measure* associated with the Nelson model with zero total momentum. Using ν_∞, we can also prove the superexponential decay of the ground state Ψ_g of $H_N(0)$.

Corollary 2.153 (Superexponential decay of boson number). *Let $\beta \in \mathbb{C}$. Then $\Psi_g \in D(e^{-\beta N})$ and*

$$(\Psi_g, e^{-\beta N}\Psi_g) = \mathbb{E}_{\nu_\infty}\left[e^{-(1-e^{-\beta})\int_{-\infty}^0 ds \int_0^\infty W(B_s - B_t, s-t)dt}\right], \quad \beta \in \mathbb{C}.$$

In particular, $\Psi_g \in D(e^{\beta N})$ for all $\beta > 0$.

Proof. The proof is similar to Corollary 2.86. □

We can also prove Gaussian domination of the ground state Ψ_g of $H_N(0)$.

Corollary 2.154 (Gaussian domination of the ground state). *Take $g \in \mathscr{S}'(\mathbb{R}^d)$ such that $\hat{g}/\sqrt{\omega}, \hat{g}/\omega \in L^2(\mathbb{R}^d)$, and $\beta < 1/\|\hat{g}/\sqrt{\omega}\|^2$. Then $\Psi_g \in D(e^{(\beta/2)\phi(g)^2})$ and*

$$\|e^{(\beta/2)\phi(g)^2}\Psi_g\|^2 = \frac{1}{\sqrt{1-\beta\|\hat{g}/\sqrt{\omega}\|^2}}\mathbb{E}_{\nu_\infty}\left[\exp\left(\frac{\beta K(g)^2}{1-\beta\|\hat{g}/\sqrt{\omega}\|^2}\right)\right],$$

where $K(g)$ denotes the random variable defined by

$$K(g) = \frac{1}{2}\int_{-\infty}^{\infty}\left(e^{-|r|\omega}e^{-ik\cdot B_r}\hat{\varphi}/\sqrt{\omega}, \hat{g}/\sqrt{\omega}\right)dr.$$

In particular, it follows that

$$\lim_{\beta \uparrow (2\|\hat{g}/\sqrt{\omega}\|^2)^{-1}} \|e^{\beta\phi(g)^2}\Psi_g\| = \infty.$$

Proof. The proof is similar to that of Corollary 2.87. □

2.12.5 $P(\phi)_1$-process associated with Nelson Hamiltonian with zero total momentum

We consider the Nelson Hamiltonian with zero total momentum $H_N(0)$ in this section. We denote its ground state by

$$\Psi_g = \Psi_g(0).$$

In Section 2.12.4 we constructed a probability measure ν_∞ defined on $(\mathfrak{X}, \mathcal{B}(\mathfrak{X}))$. In this section we extend this measure to a probability measure $\hat{\nu}$ on $\mathfrak{X}\times\mathfrak{Y}$. This is a similar procedure to Section 2.8.2 for the Nelson Hamiltonian. In terms of an infinite-dimensional Ornstein–Uhlenbeck process $(\xi_t)_{t\in\mathbb{R}}$, we have

$$(\Psi, e^{-tH_{\mathrm{N}}(0)}\Phi)_{L^2(\mathrm{G})} = \int\limits_{\mathscr{M}_{-2}} \mathbb{E}^{0,\xi}_{\mathcal{W}\times\mathcal{G}}\left[\hat{\tau}_{B_0}\Psi(\xi_0)e^{-g\int_0^t \xi_s(\varphi(\cdot-B_s))ds}\hat{\tau}_{B_t}\Phi(\xi_t)\right]d\mathrm{G}, \tag{2.12.17}$$

where we write

$$\hat{\tau}_x = e^{-i\hat{P}_{\mathrm{f}}\cdot x}.$$

We note that

$$(\hat{\tau}_x F)(\xi(f_1),\dots,\xi(f_n)) = F(\xi(f_1(\cdot - x)),\dots,\xi(f_n(\cdot - x)))$$

for $F \in \mathscr{L}$, where

$$\mathscr{L} = \left\{F(\xi(f_1),\dots,\xi(f_n)) \mid F \in \mathscr{S}(\mathbb{R}^n), f_j \in C_0^\infty(\mathbb{R}^d), j = 1,\dots,n, n \geq 1\right\}.$$

Since $\hat{\tau}_x\xi(f) = \xi(f(\cdot - x))$, we can regard $\hat{\tau}_x$ as a map from $\mathscr{M}_{-2} \to \mathscr{M}_{-2}$. Thus $(\hat{\tau}_x\xi)(f) = \xi(f(\cdot - x))$, and for $F \in \mathscr{L}$ we see that $\hat{\tau}_x F(\xi) = F(\hat{\tau}_x\xi)$. For every $F \in L^2(\mathrm{G})$, there exists a sequence $(F_n)_{n\in\mathbb{N}} \subset \mathscr{L}$ such that $F_n \to F$ in $L^2(\mathrm{G})$ as $n \to \infty$. We have $(\hat{\tau}_x F_n)(\xi) = F_n(\hat{\tau}_x\xi)$. Taking a subsequence n_k, we have $(\hat{\tau}_x F_{n_k})(\xi) \to (\hat{\tau}_x F)(\xi)$ for almost every $\xi \in \mathscr{M}_{-2}$ and $F_{n_k}(\hat{\tau}_x\xi) \to F(\hat{\tau}_x\xi)$ a. s. $\xi \in \mathscr{M}_{-2}$, implying $\hat{\tau}_x F(\xi) = F(\hat{\tau}_x\xi)$ a. s. Hence (2.12.17) can be rewritten as

$$(\Psi, e^{-tH_{\mathrm{N}}(0)}\Phi)_{L^2(\mathrm{G})} = \int\limits_{\mathscr{M}_{-2}} \mathbb{E}^{0,\xi}_{\mathcal{W}\times\mathcal{G}}\left[\Psi(\hat{\tau}_{B_0}\xi_0)e^{-g\int_0^t \hat{\tau}_{B_s}\xi_s(\varphi)ds}\Phi(\hat{\tau}_{B_t}\xi_t)\right]d\mathrm{G}. \tag{2.12.18}$$

We now define an $\mathscr{M}_{-2}$-valued random process $(Y_t)_{t\in\mathbb{R}}$ on $(\mathfrak{X}\times\mathfrak{Y}, \mathcal{B}(\mathfrak{X}\times\mathfrak{Y}))$ by

$$Y_t = \hat{\tau}_{B_t}\xi_t. \tag{2.12.19}$$

Hence we also have

$$(\Psi, e^{-tH_{\mathrm{N}}(0)}\Phi)_{L^2(\mathrm{G})} = \int\limits_{\mathscr{M}_{-2}} \mathbb{E}^{0,\xi}_{\mathcal{W}\times\mathcal{G}}\left[\Psi(Y_0)e^{-g\int_0^t \hat{\tau}_{B_s}Y_s(\varphi)ds}\Phi(Y_t)\right]d\mathrm{G}. \tag{2.12.20}$$

Let $\mathscr{F}_{[-T,T]} = \sigma(Y_r, -t \leq r \leq t)$ and define $\hat{\mathscr{G}} = \bigcup_{0\leq s}\mathscr{F}_{[-s,s]}$. Also, define the finitely additive set function $\hat{\nu} : \hat{\mathscr{G}} \to [0,\infty)$,

$$\hat{\nu}(A) = e^{E(b-a)}\int\limits_{\mathscr{M}_{-2}} \mathbb{E}^{0,\xi}_{\mathcal{W}\times\mathcal{G}}\left[1_A\Psi_g(Y_0)\mathrm{I}_{[a,b]}\Psi_g(Y_t)\right]d\mathrm{G}, \quad A \in \mathscr{F}_{[a,b]}.$$

Here

$$\mathrm{I}_{[a,b]} = e^{-g\int_a^b Y_s(\varphi)ds}.$$

The set function $\hat{\nu}$ is well defined in the sense that for $A \in \mathscr{F}_{[a,b]} \subset \mathscr{F}_{[a',b']}$,

$$\hat{\nu}(A) = e^{E(b-a)} \int_{\mathscr{M}_{-2}} \mathbb{E}^{0,\xi}_{\mathcal{W}\times\mathscr{G}} [1_A \Psi_g(Y_0) \mathrm{I}_{[a,b]} \Psi_g(Y_t)]\, dG$$
$$= e^{E(b'-a')} \int_{\mathscr{M}_{-2}} \mathbb{E}^{0,\xi}_{\mathcal{W}\times\mathscr{G}} [1_A \Psi_g(Y_0) \mathrm{I}_{[a',b']} \Psi_g(Y_t)]\, dG$$

holds. This can be checked similarly to Lemma 2.71. Moreover, it can be seen that $\hat{\nu}$ is countably additive. By Hopf's extension theorem, $\hat{\nu}$ can be uniquely extended to a probability measure on $(\mathfrak{X}\times\mathfrak{Y}, \mathcal{B}(\mathfrak{X}\times\mathfrak{Y}))$. Note that $\sigma(\hat{\mathscr{G}}) = \mathcal{B}(\mathfrak{X}\times\mathfrak{Y})$. We denote the extension by $\hat{\nu}_\infty$. Moreover, define the finitely additive set function $\hat{\nu}_T : \mathscr{G}_T \to \mathbb{R}$,

$$\hat{\nu}_T(A) = e^{E(b-a)} \int_{\mathscr{M}_{-2}} \mathbb{E}^{0,\xi}_{\mathcal{W}\times\mathscr{G}} \left[1_A \frac{1_{T+a}(Y_a)}{\|1_T\|} \mathrm{I}_{[-T,T]} \frac{1_{T-b}(Y_b)}{\|1_T\|} \right] dG$$

for $A \in \mathscr{F}_{[a,b]}$ with $b \le T$. Here $1_t = e^{-t(H_N(0)-E_N(0))}1$. Since $1_{t+a}/\|1_t\| \to e^{aE_N(0)}\Psi$ as $t \to \infty$, we can prove the local convergence of $\hat{\nu}_T$ as $T \to \infty$ in a similar way to Theorem 2.80 for the Nelson model.

Theorem 2.155 (Local convergence). *The family of probability measures $\{\hat{\nu}_T\}_{T\ge 0}$ converges to the probability measure $\hat{\nu}_\infty$ as $T \to \infty$ on the probability space $(\mathfrak{X}\times\mathfrak{Y}, \mathcal{B}(\mathfrak{X}\times\mathfrak{Y}))$ in the local sense, i. e., $\hat{\nu}_T(A) \to \hat{\nu}_\infty(A)$ as $T \to \infty$ for all $A \in \hat{\mathscr{G}}$.*

Note that $(f\Psi_g, e^{-|s-t|(H_N(0)-E_N(0))} g\Psi_g)_{L^2(G)} = (f, e^{-|s-t|L(0)} g)_{L^2(\mathscr{M}_{-2}, \Psi_g^2 dG)}$, where

$$L(0) = \frac{1}{\Psi_g}(H_N(0) - E_N(0))\Psi_g.$$

Hence by the definition of $\hat{\mu}$ we see that for $0 \le t_0 \le t_1 \le \cdots \le t_n$,

$$\mathbb{E}_{\hat{\nu}}[\prod_{j=0}^n f_j(Y_{t_j})] = (f_0, e^{-(t_1-t_0)L(0)} f_1 \cdots f_{n-1} e^{-(t_n-t_{n-1})L(0)} f_n)_{L^2(\mathscr{M}_{-2}, \Psi_g^2 dG)}, \tag{2.12.21}$$

$$\mathbb{E}_{\hat{\nu}}[f(Y_s)] = (1, f)_{L^2(\mathscr{M}_{-2}, \Psi_g^2 dG)} \quad s \ge 0. \tag{2.12.22}$$

Theorem 2.156 ($P(\phi)_1$-process). *The random process $(Y_t)_{t\ge 0}$ on $(\mathfrak{X}\times\mathfrak{Y}, \mathcal{B}(\mathfrak{X}\times\mathfrak{Y}), \hat{\nu}_\infty)$ is a Markov process under the natural filtration, with invariant measure $\Psi_g^2 dG$.*

Proof. The proof is similar to that of Theorem 2.81, we only give an outline. Defining $p(t, Y, A) = (e^{-tL(0)} 1_A)(Y)$ for $t \ge 0$, $Y \in \mathscr{M}_{-2}$, and $A \in \mathcal{B}(\mathscr{M}_{-2})$, we have immediately that $p(t, Y, \mathscr{M}_{-2}) = 1$ and $p(t, \cdot, A)$ is Borel measurable. Since $\int_M f(Z) p(t, X, dZ) = (e^{-tL(0)} f)(Y)$ for a bounded function f, the semigroup property $e^{-sL} e^{-tL} 1_A(Y) = e^{-(s+t)L} 1_A(Y)$ yields

$$\int_M e^{-tL} 1_A(Y) p(s, Y, dZ) = \int_M P(t, Z, A) P(s, Y, dZ),$$

checking the Chapman–Kolmogorov identity. Hence $p : [0, \infty) \times \mathscr{M}_{-2} \times \mathcal{B}(\mathscr{M}_{-2}) \to \mathbb{R}$ is a probability transition kernel. We also see by (2.12.21) that

$$\begin{aligned}
&\mathbb{E}_{\hat{\nu}_\infty}[1_{A_0}(Y_0) \cdots 1_{A_n}(Y_{t_n})] \\
&= \int_{M^{(n+1)}} \Big(\prod_{j=0}^{n} 1_{A_j}(Y_j)\Big)\Big(\prod_{j=q}^{n} p(t_j - t_{j-1}, Y_{j-1} dY_j)\Big) 1_{A_0}(Y_0) \Psi_{\mathrm{g}}^2(Y_0) dG(Y_0).
\end{aligned}$$

Thus $(Y_t)_{t\geq 0}$ is a Markov process with a probability transition kernel $p(t, Y, A)$. Furthermore,

$$\int_{\mathscr{M}_{-2}} p(t, Y, A) \Psi_{\mathrm{g}}^2(Y) dG = (\Psi_{\mathrm{g}}, e^{-t(H_{\mathrm{N}}(0) - E_{\mathrm{N}}(0))} 1_A \Psi_{\mathrm{g}})_{L^2(\mathscr{M}_{-2})} = \int_{\mathscr{M}_{-2}} 1_A(Y) \Psi_{\mathrm{g}}^2(Y) dG,$$

and thus $\Psi_{\mathrm{g}}^2 dG$ is an invariant measure for $(Y_t)_{t\geq 0}$. □

The process $(Y_t)_{t\geq 0}$ is called the *$P(\phi)_1$-process* associated with the Nelson Hamiltonian with zero total momentum.

2.12.6 Removal of ultraviolet cutoff

In this section we set again $d = 3$, and similarly to the discussion in Section 2.11 of the Nelson model, we remove the ultraviolet cutoff of $H_{\mathrm{N}}(p)$ for every $p \in \mathbb{R}^3$.

We define the regularized $H_{\mathrm{N}}^{\varepsilon}(p)$ by $H_{\mathrm{N}}(p)$ with $\hat{\varphi}(k)$ replaced by $e^{-\varepsilon |k|^2/2}$ as in Section 2.11. By the functional integral representation, we have

$$(1, e^{-2TH_{\mathrm{N}}^{\varepsilon}(p)} 1) = \mathbb{E}\left[e^{ip\cdot(B_T - B_{-T})} e^{\frac{g^2}{2} S_\varepsilon^T}\right],$$

with

$$S_\varepsilon^T = \int_{-T}^{T} ds \int_{-T}^{T} W_\varepsilon(B_s - B_t, s - t) dt$$

given by the pair potential

$$W_\varepsilon(x, t) = \int_{\mathbb{R}^3} \frac{1}{2\omega(k)} e^{-\varepsilon|k|^2} e^{-ik\cdot x} e^{-|t|\omega(k)} dk, \quad W_\varepsilon : \mathbb{R}^3 \times \mathbb{R} \to \mathbb{R}.$$

Note that $\mathbb{E}[e^{ip\cdot(B_T-B_{-T})}e^{\frac{g^2}{2}S_\varepsilon^T}] \in \mathbb{R}$ for all $p \in \mathbb{R}^3$. From this, we can derive the lemma below in a similar way to Lemma 2.114. Denote

$$\varphi_\varepsilon(x,t) = \int_{\mathbb{R}^3} \frac{e^{-\varepsilon|k|^2} e^{-ik\cdot x-|t|\omega(k)}}{2\omega(k)} \beta(k)dk, \quad \varepsilon \geq 0.$$

Lemma 2.157. *It follows that*

$$\lim_{\varepsilon\downarrow 0}(1, e^{-2T(H_{\mathrm{N}}^\varepsilon(p)+g^2\varphi_\varepsilon(0,0))}1) = \mathbb{E}\left[e^{ip\cdot(B_T-B_{-T})}e^{\frac{g^2}{2}S_0^{\mathrm{ren}}}\right], \tag{2.12.23}$$

where

$$S_0^{\mathrm{ren}} = -2\int_{-T}^{T} \varphi_0(B_T-B_s, T-s)ds + 2\int_{-T}^{T}\left(\int_{-T}^{t} \nabla\varphi_0(B_t-B_s, t-s)ds\right)\cdot dB_t.$$

Furthermore, there exists a constant c independent of ε such that

$$\left|(1, e^{-2T(H_{\mathrm{N}}(p)+g^2\varphi_\varepsilon(0,0))}1)\right| \leq e^{c(T+1)}. \tag{2.12.24}$$

Proof. Equation (2.12.23) is obtained by Lemma 2.114, and (2.12.24) follows from Lemma 2.111. □

To remove the ultraviolet cutoff, we need to obtain a uniform lower bound of $H_{\mathrm{N}}^\varepsilon(p)$ with respect to ε. This can be done by using the existence of the ground state for $p = 0$ and the energy comparison inequality $E_{\mathrm{N}}(p) \leq E_{\mathrm{N}}(0)$.

Lemma 2.158. *There exists $C \in \mathbb{R}$ such that $H_{\mathrm{N}}^\varepsilon(p) + g^2\varphi_\varepsilon(0,0) > -C$, uniformly in $\varepsilon > 0$ and $p \in \mathbb{R}^3$.*

Proof. Let $\Psi_{\mathrm{g}}^\varepsilon$ be the ground state of $H_{\mathrm{N}}^\varepsilon(0)$. Since $e^{-2TH_{\mathrm{N}}^\varepsilon(0)}$ is positivity improving, we see that $(1, \Psi_{\mathrm{g}}^\varepsilon) \neq 0$ and then

$$E_{\mathrm{N}}^\varepsilon(0) + g^2\varphi_\varepsilon(0,0) = -\lim_{T\to\infty}\frac{1}{2T}\log(1, e^{-2T(H_{\mathrm{N}}^\varepsilon(0)+g^2\varphi_\varepsilon(0,0))}1).$$

By (2.12.24), we have $-\lim_{T\to\infty}\frac{1}{2T}\log(1, e^{-2T(H_{\mathrm{N}}^\varepsilon(0)-g^2\varphi_\varepsilon(0,0))}1) \geq -C$, where C is independent of $\varepsilon > 0$. Hence we obtain that $E_{\mathrm{N}}^\varepsilon(0) + g^2\varphi_\varepsilon(0,0) \geq -C$. This bound can be extended to $p \in \mathbb{R}^3$. By the energy comparison inequality $E_{\mathrm{N}}^\varepsilon(0) \leq E_{\mathrm{N}}^\varepsilon(p)$, we then get that

$$E_{\mathrm{N}}^\varepsilon(p) - g^2\varphi_\varepsilon(0,0) \geq -C.$$

□

Theorem 2.159 (Removal of ultraviolet cutoff). *Let $p \in \mathbb{R}^3$ and*

$$E_\varepsilon = -g^2\varphi_\varepsilon(0,0) = -g^2\int_{\mathbb{R}^3}\frac{e^{-\varepsilon|k|^2}}{2\omega(k)}\beta(k)dk.$$

Then there exists a self-adjoint operator $H_{\mathrm{N}}^{\mathrm{ren}}(p)$ *such that*

$$\operatorname*{s-lim}_{\varepsilon\downarrow 0} e^{-T(H_{\mathrm{N}}^{\varepsilon}(p)-E_\varepsilon)} = e^{-TH_{\mathrm{N}}^{\mathrm{ren}}(p)}, \quad T \ge 0.$$

Proof. For $\Phi, \Psi \in \mathscr{S}(\mathscr{Q})$, the limit $\lim_{\varepsilon\downarrow 0}(\Phi, e^{-2T(H_{\mathrm{N}}^{\varepsilon}(p)+g^2\varphi_\varepsilon(0,0))}\Psi)$ exists. This is proven in the same way as Lemma 2.116, and the theorem follows similarly as Theorem 2.118. □

Furthermore, a functional integral representation of $(F, e^{-2TH_{\mathrm{N}}^{\mathrm{ren}}(p)}G)$ can also be given in terms of exponentials of creation operators and annihilation operators.

Theorem 2.160 (Functional integral representation of Nelson Hamiltonian without ultraviolet cutoff with a fixed total momentum). *Suppose that* $\Phi, \Psi \in L^2(\mathscr{Q})$. *Let* A *and* $\tilde{A}$ *be given by* (2.11.43), *and* $u_T(p) = e^{i(p-\hat{P}_{\mathrm{f}})\cdot B_T}$, *i. e.,* $A = e^{a_M^*(U_T)}e^{-TH_{\mathrm{f}}}$ *and* $\tilde{A} = e^{-TH_{\mathrm{f}}}e^{a_M(\tilde{U}_T)}$ *with*

$$U_T(k) = -\frac{g}{\sqrt{2}}\int_{-T}^{T} e^{-|s+T|\omega(k)}e^{-ik\cdot B_s}ds,$$

$$\tilde{U}_T(k) = -\bar{U}_{-T}(k) = -\frac{g}{\sqrt{2}}\int_{-T}^{T} e^{-|s-T|\omega(k)}e^{ik\cdot B_s}ds.$$

Then

$$(\Phi, e^{-2TH_{\mathrm{N}}^{\mathrm{ren}}(p)}\Psi)_{L^2(\mathscr{Q})} = \mathbb{E}^0\left[e^{\frac{g^2}{2}S_0^{\mathrm{ren}}}(u_{-T}(p)\Phi, A\tilde{A}u_T(p)\Psi)\right]. \tag{2.12.25}$$

Proof. The proof is similar to that of Theorem 2.120. By the Baker–Campbell–Hausdorff formula, we obtain

$$\mathrm{I}_{-T}^* e^{-g\phi(\int_{-T}^{T}\delta_s\otimes\varrho_\varepsilon(\cdot-B_s)ds)}\mathrm{I}_T = e^{\frac{g^2}{2}S_\varepsilon^T}e^{a_M^*(U_T^\varepsilon)}e^{-2TH_{\mathrm{f}}}e^{a_M(\tilde{U}_T^\varepsilon)},$$

where $\varrho_\varepsilon = (e^{-\varepsilon|k|^2/2})^\vee$, and U_T^ε and $\tilde{U}_T^\varepsilon$ are given by (2.11.44) and (2.11.45), respectively, i. e.,

$$U_T^\varepsilon(k) = -\frac{g}{\sqrt{2}}\int_{-T}^{T} e^{-|s+T|\omega(k)}e^{-\varepsilon|k|^2/2}e^{-ik\cdot B_s}ds,$$

$$\tilde{U}_T^\varepsilon(k) = -\bar{U}_{-T}^\varepsilon(k) = -\frac{g}{\sqrt{2}}\int_{-T}^{T} e^{-|s-T|\omega(k)}e^{-\varepsilon|k|^2/2}e^{ik\cdot B_s}ds.$$

Thus we have the functional integral representation

$$(F, e^{-2T(H_{\mathrm{N}}^{\varepsilon}(p)-E_\varepsilon)}G) = \mathbb{E}^0\left[e^{\frac{g^2}{2}S_\varepsilon^{\mathrm{ren}}}(u_{-T}(p)\Phi, e^{a_M^*(U_T^\varepsilon)}e^{-2TH_{\mathrm{f}}}e^{a_M(\tilde{U}_T^\varepsilon)}u_T(p)\Psi)\right].$$

The limiting argument as $\varepsilon \downarrow 0$ is similar to the proof of Theorem 2.120, and we see that

$$(u_{-T}(p)\Phi, e^{a_M^*(U_T^\varepsilon)}e^{-2TH_{\mathrm{f}}}e^{a_M(\tilde{U}_T^\varepsilon)}u_T(p)\Psi) \to (u_{-T}(p)\Phi, e^{a_M^*(U_T)}e^{-2TH_{\mathrm{f}}}e^{a_M(\tilde{U}_T)}u_T(p)\Psi)$$

and

$$|(u_{-T}(p)\Phi, e^{a_M^*(U_T^\varepsilon)}e^{-2TH_{\mathrm{f}}}e^{a_M(\tilde{U}_T^\varepsilon)}u_T(p)\Psi)| < C$$

with a constant C independent of $\varepsilon \geq 0$. □

Write

$$E_{\mathrm{N}}^{\mathrm{ren}}(p) = \inf \operatorname{Spec}(H_{\mathrm{N}}^{\mathrm{ren}}(p)).$$

The following result is immediate by Theorem 2.159.

Corollary 2.161 (Properties of $E_{\mathrm{N}}^{\mathrm{ren}}(p)$).
(1) *If $p = 0$, then $e^{-TH_{\mathrm{N}}^{\mathrm{ren}}(0)}$ is positivity improving.*
(2) *The ground state of $H_{\mathrm{N}}^{\mathrm{ren}}(0)$ is unique whenever it exists.*
(3) *The energy comparison inequality $E_{\mathrm{N}}^{\mathrm{ren}}(0) \leq E_{\mathrm{N}}^{\mathrm{ren}}(p)$ holds.*
(4) *The map $\mathbb{R}^d \ni p \mapsto E_{\mathrm{N}}^{\mathrm{ren}}(p) \in \mathbb{R}$ is continuous.*

Proof. Let $\Phi, \Psi \in L^2(\mathscr{Q})$. We have

$$(\Phi, e^{-2TH_{\mathrm{N}}^{\mathrm{ren}}(0)}\Psi) = \mathbb{E}^0\left[e^{\frac{g^2}{2}S_0^{\mathrm{ren}}}\left(\Phi, e^{iB_{-T}\cdot\hat{P}_{\mathrm{f}}}A\tilde{A}e^{-iB_T\cdot\hat{P}_{\mathrm{f}}}\Psi\right)\right]. \tag{2.12.26}$$

Since $e^{-iB_T\cdot\hat{P}_{\mathrm{f}}}$ is positivity preserving and $A\tilde{A} = e^{a_M^*(U_T)}e^{-2TH_{\mathrm{f}}}e^{a_M(\tilde{U}_T)}$ is positivity improving, we see that $e^{-2TH_{\mathrm{N}}(0)}$ is positivity improving. This implies that the ground state is unique by the Perron–Frobenius theorem. By Theorem 2.160, we have the bound

$$\left|(\Psi, e^{-2TH_{\mathrm{N}}(p)}\Phi)\right| \leq \mathbb{E}^0\left[\left(|\Psi|, e^{iB_{-T}\cdot\hat{P}_{\mathrm{f}}}A\tilde{A}e^{-iB_T\cdot\hat{P}_{\mathrm{f}}}|\Phi|\right)\right] = (|\Psi|, e^{-2TH_{\mathrm{N}}(0)}|\Phi|),$$

and (3) follows. Finally, we prove (4). We have

$$(\Psi, (e^{-2TH_{\mathrm{N}}^{\mathrm{ren}}(p)} - e^{-2TH_{\mathrm{N}}^{\mathrm{ren}}(q)})\Phi) = \mathbb{E}^0\left[\left(\mathrm{I}_0\Psi, e^{iB_{-T}\cdot\hat{P}_{\mathrm{f}}}A\tilde{A}e^{-iB_T\cdot\hat{P}_{\mathrm{f}}}\Phi\right)\left(e^{ip\cdot(B_T-B_{-T})} - e^{iq\cdot(B_T-B_{-T})}\right)\right].$$

Using the identity $e^{ip\cdot(B_T-B_{-T})} - e^{iq\cdot(B_T-B_{-T})} = i\int_{q\cdot(B_T-B_{-T})}^{p\cdot(B_T-B_{-T})} e^{i\theta}d\theta$, we get

$$\begin{aligned}|(\Psi, (e^{-2TH_{\mathrm{N}}^{\mathrm{ren}}(p)} - e^{-2TH_{\mathrm{N}}^{\mathrm{ren}}(q)})\Phi)| &\leq \|\Psi\|\|\Phi\|\mathbb{E}^0\left[\|A\tilde{A}\||p-q||B_T - B_{-T}|\right] \\ &\leq |p-q|\|\Psi\|\|\Phi\|\left(\mathbb{E}^0\left[\|A\tilde{A}\|^2\right]\right)^{1/2}\mathbb{E}^0\left[|B_T - B_{-T}|^2\right]^{1/2}.\end{aligned}$$

Hence $e^{-2TH_{\mathrm{N}}^{\mathrm{ren}}(p)}$ is uniformly continuous in p, which implies statement (4). □

Lemma 2.162. *If $V = 0$, then*

$$H_{\mathrm{N}}^{\mathrm{ren}} = \hat{\mathscr{F}}^{-1}\Big(\int_{\mathbb{R}^3}^{\oplus} H_{\mathrm{N}}^{\mathrm{ren}}(p)dp\Big)\hat{\mathscr{F}}.$$

Here $\hat{\mathscr{F}} : \mathcal{H}_{\mathrm{N}} = L^2(\mathbb{R}_x^d) \otimes L^2(\mathscr{Q}) \to L^2(\mathbb{R}_p^d) \otimes L^2(\mathscr{Q})$ is the unitary map given by

$$\hat{\mathscr{F}} = (F \otimes 1)\int_{\mathbb{R}^d}^{\oplus} e^{ix\cdot\hat{P}_{\mathrm{f}}}dx.$$

Proof. It is enough to show that

$$(F, e^{-TH_{\mathrm{N}}^{\mathrm{ren}}}G)_{\mathcal{H}_{\mathrm{N}}} = \int_{\mathbb{R}^3} (\hat{\mathscr{F}}F(p), e^{-TH_{\mathrm{N}}^{\mathrm{ren}}(p)}\hat{\mathscr{F}}G(p))_{L^2(\mathscr{Q})}dp$$

for every $F, G \in \mathcal{H}_{\mathrm{N}}$. By Theorem 2.145, we show that

$$(F, e^{-T(H_{\mathrm{N}}^{\varepsilon}-E_{\varepsilon})}G) = \int_{\mathbb{R}^3} (\hat{\mathscr{F}}F(p), e^{-T(H_{\mathrm{N}}^{\varepsilon}(p)-E_{\varepsilon})}\hat{\mathscr{F}}G(p))dp.$$

We take the limit as $\varepsilon \downarrow 0$ on both sides and obtain

$$(F, e^{-TH_{\mathrm{N}}^{\mathrm{ren}}}G) = \lim_{\varepsilon\downarrow 0}\int_{\mathbb{R}^3} (\hat{\mathscr{F}}F(p), e^{-T(H_{\mathrm{N}}^{\varepsilon}(p)-E_{\varepsilon})}\hat{\mathscr{F}}G(p))dp.$$

Notice that $H_{\mathrm{N}}^{\varepsilon}(p) - E_{\varepsilon} \geq -C$ with a constant C independent of $\varepsilon > 0$ and $p \in \mathbb{R}^3$. We have

$$\big|(\hat{\mathscr{F}}F(p), e^{-T(H_{\mathrm{N}}^{\varepsilon}(p)-E_{\varepsilon})}\hat{\mathscr{F}}G(p))\big| \leq \|F(p)\|\|G(p)\|e^{tC}$$

and $\|F(p)\|\|G(p)\|e^{Ct} \in L^2(\mathbb{R}^3{}_p)$. The Lebesgue dominated convergence theorem yields

$$\lim_{\varepsilon\downarrow 0}\int_{\mathbb{R}^3} (\hat{\mathscr{F}}F(p), e^{-T(H_{\mathrm{N}}^{\varepsilon}(p)-E_{\varepsilon})}\hat{\mathscr{F}}G(p))dp = \int_{\mathbb{R}^3} (\hat{\mathscr{F}}F(p), e^{-TH_{\mathrm{N}}^{\mathrm{ren}}(p)}\hat{\mathscr{F}}G(p))dp,$$

and the lemma follows. □

Corollary 2.163 (Ground state energy of the renormalized Nelson Hamiltonian with $V = 0$). *Let $V = 0$. Then the ground state energy of the Nelson Hamiltonian without ultraviolet cutoff $E_{\mathrm{N}}^{\mathrm{ren}}$ coincides with the ground state energy of the Nelson Hamiltonian without ultraviolet cutoff with zero total momentum $E_{\mathrm{N}}^{\mathrm{ren}}(0)$, i. e., $E_{\mathrm{N}}^{\mathrm{ren}}(0) = E_{\mathrm{N}}^{\mathrm{ren}}$.*

Proof. The proof is the same as Corollary 2.149, and we skip it. □

In the $d = 2$ case under an infrared cutoff, we can make the following interesting observation. In principle, we could to do the same renormalization procedure as for $d = 3$, however, this is unnecessary since now $\varphi_\varepsilon(0,0)$ converges to the finite number $\varphi_0(0,0)$ as $\varepsilon \downarrow 0$. An important consequence of Lemma 2.150 is then the existence of a ground state of $H_{\mathrm{N}}^{\mathrm{ren}}(0)$.

Let $\kappa > 0$ be the infrared cutoff parameter, and define $H_{\mathrm{N}}^{\varepsilon,\kappa}(p)$ by $H_{\mathrm{N}}^{\varepsilon}(p)$ with cutoff function $e^{-\varepsilon|k|^2/2}$ replaced by $e^{-\varepsilon|k|^2/2}1_{\{|k|\geq\kappa\}}$. For $d = 2$, the renormalization term is given by

$$E_{\varepsilon,\kappa} = -g^2 \int_{\mathbb{R}^2} \frac{e^{-\varepsilon|k|^2}}{2\omega(k)} \beta(k) 1_{\{|k|\geq\kappa\}} dk.$$

However, using the infrared cutoff, we have $|E_{\varepsilon,\kappa}| < \infty$ for all $0 \leq \varepsilon$. It can be also shown that there exists $H_{\mathrm{N}}^{\mathrm{ren},\kappa}(p)$ such that

$$\operatorname*{s-lim}_{\varepsilon\downarrow 0} e^{-TH_{\mathrm{N}}^{\varepsilon,\kappa}(p)} = e^{-TH_{\mathrm{N}}^{\mathrm{ren},\kappa}(p)}.$$

Theorem 2.164 (Existence of ground state). *Let $d = 2$ and assume $\kappa > 0$. Then for every value of g the operator $H_{\mathrm{N}}^{\mathrm{ren},\kappa}(0)$ has a ground state $\Psi_g(0)$, such that $(1, \Psi_g(0)) \neq 0$.*

Proof. By (2.12.15), we have

$$\frac{(1, e^{-TH_{\mathrm{N}}^{\varepsilon,\kappa}(0)}1)^2}{(1, e^{-2TH_{\mathrm{N}}^{\varepsilon,\kappa}(0)}1)} \geq \exp\left(-\frac{g^2}{2} \int_{\{|k|\geq\kappa\}} \frac{e^{-\varepsilon|k|^2}}{\omega(k)^3} dk\right).$$

Taking the limit as $\varepsilon \downarrow 0$ on both sides gives

$$\frac{(1, e^{-TH_{\mathrm{N}}^{\mathrm{ren}}(0)}1)^2}{(1, e^{-2TH_{\mathrm{N}}^{\mathrm{ren}}(0)}1)} > \exp\left(-\frac{g^2}{2} \int_{\{|k|\geq\kappa\}} \frac{1}{\omega(k)^3} dk\right), \tag{2.12.27}$$

and hence

$$\lim_{T\to\infty} \frac{(1, e^{-TH_{\mathrm{N}}^{\mathrm{ren}}(0)}1)^2}{(1, e^{-2TH_{\mathrm{N}}^{\mathrm{ren}}(0)}1)} > \exp\left(-\frac{g^2}{2} \int_{\{|k|\geq\kappa\}} \frac{1}{\omega(k)^3} dk\right) > 0. \tag{2.12.28}$$

On the other hand, we see that

$$\lim_{T\to\infty} \frac{(1, e^{-TH_{\mathrm{N}}^{\mathrm{ren}}(0)}1)^2}{(1, e^{-2TH_{\mathrm{N}}^{\mathrm{ren}}(0)}1)} = \|P_g 1\|^2,$$

where P_g denotes the projection to the subspace Ker $(H_{\mathrm{N}}^{\mathrm{ren}}(0) - E^{\mathrm{ren}}(0))$. From (2.12.28), we readily obtain $\|P_g 1\|^2 > 0$, which implies that $H_{\mathrm{N}}^{\mathrm{ren}}(0)$ has a ground state $\Psi_g(0)$ such that $(1, \Psi_g(0)) \neq 0$. □

2.12.7 Ground state energy and ultraviolet renormalization term

Let

$$E_\varepsilon(g^2) = \inf \operatorname{Spec}(H_N^\varepsilon(0))$$

be the bottom of the spectrum of the Nelson Hamiltonian with zero total momentum. Suppose that formally $E_\varepsilon(g^2)$ can be expanded in g^2 as

$$E_\varepsilon(g^2) = E_\varepsilon(0) + a_2 g^2 + a_4 g^4 + \cdots,$$

and the ground state as

$$\Psi_g = 1 + g\phi_1 + g^2\phi_2 + \cdots.$$

Note that $E_\varepsilon(0) = 0$. From the eigenvalue equation $H_N^\varepsilon(0)\Psi_g = E_\varepsilon(g^2)\Psi_g$, we get $\phi_1 = -(\frac{1}{2}P_f^2 + H_f)^{-1}H_I^\varepsilon(0)1$ and

$$a_2 = (1, H_I^\varepsilon(0)\phi_1) = -\left(H_I^\varepsilon(0)1, \left(\frac{1}{2}P_f^2 + H_f\right)^{-1} H_I^\varepsilon(0)1\right).$$

Recall that

$$\varphi_\varepsilon(x,t) = \int_{\mathbb{R}^3} \frac{e^{-\varepsilon|k|^2} e^{-ik\cdot x - |t|\omega(k)}}{2\omega(k)} \beta(k)dk, \quad \varepsilon \geq 0.$$

Hence $a_2 = E_\varepsilon = -\varphi_\varepsilon(0,0)$ is obtained, giving

$$\lim_{g\to 0} \frac{E_\varepsilon(g^2) - E_\varepsilon(0)}{g^2} = E_\varepsilon.$$

The ultraviolet renormalization term E_ε coincides with the coefficient of g^2 in the expansion of the ground state energy $E_\varepsilon(g^2)$ of the Nelson Hamiltonian with zero total momentum. Furthermore, we expect that

$$\lim_{\varepsilon\downarrow 0} |E_\varepsilon(g^2) - g^2 E_\varepsilon| < \infty.$$

For the remainder of this section, we discuss rigorously the above informal calculations. Putting $p = 0$, we have

$$(1, e^{-2TH_N^\varepsilon(0)}1) = \mathbb{E}\left[e^{\frac{g^2}{2}S_\varepsilon^T}\right].$$

Lemma 2.165. *It follows that*

$$E_\varepsilon(g^2) = -\lim_{T\to\infty} \frac{1}{2T} \log(1, e^{-2TH_N^\varepsilon(0)}1).$$

In particular,

$$E_\varepsilon(g^2) = -\lim_{T\to\infty}\frac{1}{2T}\log\mathbb{E}\left[e^{\frac{g^2}{2}S_\varepsilon^T}\right].$$

Proof. This follows from Lemma 1.56. □

We derive a precise estimate of the expectation $\mathbb{E}[e^{\frac{g^2}{2}S_\varepsilon^{\mathrm{ren}}}]$ in the lemma below in order to estimate $(E_\varepsilon(g^2) - E_\varepsilon(0))/g^2$.

Lemma 2.166. *Let $\frac{1}{2} < \theta < 1$. Then there exist constants $b, c > 0$, independent of g, such that for every $\varepsilon > 0$,*

$$\mathbb{E}\left[e^{\frac{g^2}{2}S_\varepsilon^{\mathrm{ren}}}\right] \le \exp\left[b\left(c + (g^4 + g^{\frac{4}{1-\theta}})(T+1) + g^2\log T\right) + c(\tau)(g^2/2)T\right], \tag{2.12.29}$$

where $c(\tau) = 8\pi\int_0^\infty e^{-\varepsilon r^2}e^{-\tau r}dr$. In particular, the ground state energy of $H_{\mathrm{N}}^{\mathrm{ren}}$ and $H_{\mathrm{N}}^{\mathrm{ren}}(0)$ are estimated as

$$E_{\mathrm{N}}^{\mathrm{ren}}(0) = E_{\mathrm{N}}^{\mathrm{ren}} \ge -\frac{b}{2}(g^4 + g^{\frac{4}{1-\theta}}) - \frac{c(\tau)}{4}g^2. \tag{2.12.30}$$

Proof. Let $S_\varepsilon^{\mathrm{ren}} = S_\varepsilon^{\mathrm{OD}} + Y_\varepsilon^T + Z_\varepsilon^T$. It has been seen before that for every $\frac{1}{2} < \theta < 1$,

$$\mathbb{E}[e^{aY_\varepsilon^T}] \le e^{(a^2 + a^{\frac{2}{1-\theta}})(T+1)c_Y} \tag{2.12.31}$$

with a constant c_Y. We estimate $\mathbb{E}[e^{aZ_\varepsilon^T}]$. There exists a constant $M > 0$ such that

$$|\varphi_\varepsilon(B_s - B_T, s - T)| \le |\varphi_\varepsilon(0, s-T)| < M$$

for all T, and

$$\varphi_\varepsilon(0,u) \le \frac{C}{u}$$

with a constant C. We have

$$|Z_\varepsilon^T| \le 2\int_0^1 \varphi_\varepsilon(0,u)du + 2\int_1^{2T}\varphi_\varepsilon(0,u)du \le 2M + \int_1^{2T}\frac{1}{u}du = 2M + C\log(2T). \tag{2.12.32}$$

Finally, we compute $S_\varepsilon^{\mathrm{OD}}$. We have

$$\begin{aligned}|S_\varepsilon^{\mathrm{OD}}| &\le 2\int_{-T}^{T-\tau}ds\int_{s+\tau}^{T}dt\int_{\mathbb{R}^3}\frac{1}{2\omega(k)}e^{-\varepsilon|k|^2}e^{-|s-t|\omega(k)}dk\\ &= 4\pi\int_0^\infty e^{-\varepsilon r^2}\frac{e^{-\tau r}}{r}\left(e^{-(2T-\tau)r} - 1 + (2T-\tau)r\right)dr \le c(\tau)T.\end{aligned} \tag{2.12.33}$$

The bound (2.12.29) follows by the Schwarz inequality

$$\mathbb{E}[e^{(g^2/2)(S_\varepsilon^{\mathrm{OD}}+Y_\varepsilon^T+Z_\varepsilon^T)}] \le \left(\mathbb{E}[e^{g^2 Y_\varepsilon^T}]\right)^{1/2}\left(\mathbb{E}[e^{g^2(S_\varepsilon^{\mathrm{OD}}+Z_\varepsilon^T)}]\right)^{1/2},$$

and (2.12.31)–(2.12.33). From

$$E_{\mathrm{N}}^{\mathrm{ren}}(0) = -\lim_{T\to\infty}\frac{1}{2T}\log\mathbb{E}\Big[e^{\frac{g^2}{2}S_0^{\mathrm{ren}}}\Big],$$

(2.12.30) also follows. □

The constants b and c given in Lemma 2.166 depend on τ. Now we show a key lemma.

Lemma 2.167. *Let $\frac{1}{2} < \theta < 1$. Also, let $b > 0$ and $c(\tau)$ be as in Lemma 2.166. Then*

$$\left|\frac{E_\varepsilon(g^2)}{g^2} + \varphi_\varepsilon(0,0)\right| \le \frac{1}{2}\left(b(g^2 + g^{\frac{4}{1-\theta}}) + \frac{1}{2}c(\tau)\right). \tag{2.12.34}$$

Proof. We have

$$E_\varepsilon(g^2) = -\lim_{T\to\infty}\frac{1}{2T}\log\mathbb{E}\left[e^{\frac{g^2}{2}(S_\varepsilon^{\mathrm{ren}}+4T\varphi_\varepsilon(0,0))}\right]$$

and

$$E_\varepsilon(g^2) = -g^2\varphi_\varepsilon(0,0) - \lim_{T\to\infty}\frac{1}{2T}\log\mathbb{E}\left[e^{\frac{g^2}{2}S_\varepsilon^{\mathrm{ren}}}\right].$$

Hence

$$\frac{1}{g^2}|E_\varepsilon(g^2) + g^2\varphi_\varepsilon(0,0)| \le \frac{1}{g^2}\lim_{T\to\infty}\frac{1}{2T}\log\mathbb{E}\left[e^{\frac{g^2}{2}S_\varepsilon^{\mathrm{ren}}}\right].$$

Using Lemma 2.166, we obtain (2.12.34). □

Theorem 2.168 (Ultraviolet renormalization term). *It follows that*

$$\lim_{g\to 0}\frac{E_\varepsilon(g^2)}{g^2} = E_\varepsilon \tag{2.12.35}$$

and

$$\lim_{\varepsilon\downarrow 0}\left|E_\varepsilon(g^2) - g^2 E_\varepsilon\right| < \infty. \tag{2.12.36}$$

Proof. By (2.12.34), we see that

$$\lim_{g\to 0}\left|\frac{E_\varepsilon(g^2)}{g^2} - E_\varepsilon\right| \le \frac{1}{4}c(\tau)$$

holds for every $\tau > 0$. Also, $\lim_{\tau\to\infty} c(\tau) = 0$ implies (2.12.35). Furthermore, (2.12.36) can be derived from (2.12.34) and the fact that $\lim_{\varepsilon\downarrow 0} c(\tau) < \infty$. □

Corollary 2.169. *Let $H_{\mathrm{N}}^{\varepsilon}$ be the Nelson Hamiltonian with $V \equiv 0$. Then*

$$E_{\varepsilon} = \lim_{g\to 0} \frac{\inf \mathrm{Spec}(H_{\mathrm{N}}^{\varepsilon})}{g^2}.$$

Proof. This follows from $\inf \mathrm{Spec}(H_{\mathrm{N}}^{\varepsilon}) = E_{\varepsilon}(g^2)$ and Theorem 2.168. □

2.12.8 Gibbs measures and applications

In Section 2.11.2 we have constructed the functional integral representation of the inner product $(\Phi, e^{-TH_{\mathrm{N}}^{\mathrm{ren}}(p)}\Psi)$. In particular, $e^{-TH_{\mathrm{N}}^{\mathrm{ren}}(0)}$ is positivity improving. Hence in the same way as for $H_{\mathrm{N}}^{\mathrm{ren}}$, we can construct the Gibbs measure associated with $H_{\mathrm{N}}^{\mathrm{ren}}(0)$ and we shall show that the family of probability measures $\{\nu_T^{\mathrm{ren}}\}_{T>0}$ defined below has a limit $\nu_{\infty}^{\mathrm{ren}}$ as $T \to \infty$ in the local sense. Let

$$H_{\mathrm{N}}^{\varepsilon}(0) = \frac{1}{2}P_{\mathrm{f}}^2 + H_{\mathrm{f}} + gH_{\mathrm{I}}^{\varepsilon}(0),$$

and the renormalization term with the infrared cutoff ξ_λ be given by

$$E_{\varepsilon} = -\frac{g^2}{2}\int_{\mathbb{R}^3} \frac{e^{-\varepsilon|k|^2}|\xi_\lambda(k)|^2}{\omega(k)}\beta(k)dk.$$

In Section 2.12.8 we suppose the assumptions below.

Assumption 2.170. The following conditions hold:
(1) (Dispersion relation) $\omega(k) = |k|$;
(2) (Infrared cutoff) Function $\xi_\lambda : \mathbb{R}^3 \to \mathbb{R}$ satisfies
  (i) $\xi_{\lambda=0}(k) = 1$;
  (ii) $\int_{\{|k|\le 1\}} |\xi_\lambda(k)|^2/\omega(k)^3 dk < \infty$ for $\lambda > 0$;
  (iii) $H_{\mathrm{I}}^{\varepsilon}(0) = \frac{1}{\sqrt{2}}(a_M^*(\hat{\varphi}\xi_\lambda) + a_M(\tilde{\hat{\varphi}}\xi_\lambda))$;
(3) (Ground state) For $\lambda > 0$, $H_{\mathrm{N}}^{\mathrm{ren}}(0)$ has the normalized ground state $\Psi_{\mathrm{ren}}(0)$.

For $\lambda = 0$, it was shown that $H_{\mathrm{N}}^{\varepsilon}(0) - E_{\varepsilon} \to H_{\mathrm{N}}^{\mathrm{ren}}(0)$ as $\varepsilon \to 0$ in the sense of the strong semigroup in Theorem 2.159. In the same way as for $\lambda > 0$, we can also show that $H_{\mathrm{N}}^{\varepsilon}(0) - E_{\varepsilon}$ converges as $\varepsilon \to 0$, and we denote the limit by the same notation $H_{\mathrm{N}}^{\mathrm{ren}}(0)$. Let

$$\mathscr{M}_T^{\mathrm{ren}} = \mathscr{M}_T^{\mathrm{ren}}(\lambda) = e^{\frac{g^2}{2}S_0^{\mathrm{ren}}(T)},$$

where $S_0^{\mathrm{ren}}(T) = S_0^{\mathrm{ren}}(\lambda, T)$ is given by (2.11.65). Let $(B_t)_{t\in\mathbb{R}}$ be a two-sided Brownian motion on the probability space $(\mathscr{X}, \mathcal{B}(\mathscr{X}), \mathcal{W}^x)$. Define the probability measure ν_T^{ren} on $(\mathscr{X}, \mathcal{B}(\mathscr{X}))$ by

$$\nu_T^{\rm ren}(A) = \frac{1}{Z_T}\mathbb{E}^0[1_A \mathscr{M}_T^{\rm ren}],$$

where Z_T is the normalizing constant such that $\frac{1}{Z_T}\mathbb{E}^0[\mathscr{M}_T^{\rm ren}] = 1$. We call the measure $\nu_T^{\rm ren}$ the finite volume Gibbs measure associated with the renormalized Nelson model with zero total momentum. Using the functional integral representation, we see that

$$(1, e^{-2TH_{\rm N}^{\rm ren}(0)}1) = \mathbb{E}^0[\mathscr{M}_T^{\rm ren}].$$

We shall show that $\nu_T^{\rm ren}$ converges to $\nu_\infty^{\rm ren}$ in the local sense as $T \to \infty$ and Gibbs measure exists. We can prove this in a similar way as in the case of the renormalized Nelson Hamiltonian discussed in Section 2.11.6, by replacing $H_{\rm N}^{\rm ren}$ and $f \otimes 1$ with $H_{\rm N}^{\rm ren}(0)$ and 1, respectively. Let $\mathscr{G}_T = \sigma(B_r, -t \le r \le t)$ and set $\mathscr{G} = \bigcup_{s\ge 0}\mathscr{G}_s$. We consider the family of probability spaces $(\mathscr{X}, \mathcal{B}(\mathscr{X}), \nu_T^{\rm ren})$, $T > 0$. Let

$$1_T = e^{-T(H_{\rm N}^{\rm ren}(0) - E_{\rm N}^{\rm ren}(0))}1,$$

where $E_{\rm N}^{\rm ren}(0) = \inf {\rm Spec}(H_{\rm N}^{\rm ren}(0))$. For $\lambda > 0$, it follows that $\lim_{T\to\infty} 1_T/\|1_T\| = \Psi_{\rm ren}$. Using this fact, we can obtain the local convergence of $\{\nu_T^{\rm ren}\}_{T>0}$ below.

Theorem 2.171 (Local convergence). *The family of probability measures $\{\nu_T^{\rm ren}\}_{T\ge 0}$ converges to $\nu_\infty^{\rm ren}$ on $(\mathscr{X}, \mathcal{B}(\mathscr{X}))$ in the local sense for arbitrary $\lambda > 0$.*

Notice that

$$\nu_\infty^{\rm ren}(A) = e^{2sE_{\rm N}^{\rm ren}(0)}\mathbb{E}^0\left[1_A\left(\Psi_{\rm ren}(0), {\rm I}_{[-s,s]}^{\rm ren}\Psi_{\rm ren}(0)\right)\right]dx$$

for $A \in \mathscr{G}_s$. The probability measure $\nu_\infty^{\rm ren}$ is the Gibbs measure associated with the renormalized Nelson model with zero total momentum.

Now we shall investigate expectation values of self-adjoint operator O with respect to $\Psi_{\rm ren}$ by the Gibbs measure.

Lemma 2.172. *Suppose that ρ is a positive function on $\mathbb{R}^3$ such that*

$$\int_{\mathbb{R}^3} \frac{1-e^{-\rho(k)}}{\omega(k)^3}|\xi_\lambda(k)|^2 dk < \infty. \tag{2.12.37}$$

Then it follows that

$$\frac{(1_T, e^{-{\rm d}\Gamma(\rho)}1_T)}{\|1_T\|^2} = \mathbb{E}_{\nu_T^{\rm ren}}\left[\exp\left(-g^2\int_{-T}^{0} ds \int_0^T W_\rho(B_s - B_t, s-t)dt\right)\right],$$

where

$$W_\rho(x,t) = \int_{\mathbb{R}^3} \frac{e^{-|t|\omega(k)}}{2\omega(k)} e^{-ik\cdot x}(1-e^{-\rho(k)})|\xi_\lambda(k)|^2 dk.$$

By Lemma 2.172 and a limiting argument as in Theorem 2.134, we have the theorem below.

Theorem 2.173. *Suppose that ρ is a positive function on $\mathbb{R}^3$ and (2.12.37) holds. Let $\lambda > 0$. Then*

$$(\Psi_{\mathrm{ren}}(0), e^{-\mathrm{d}\Gamma(\rho)}\Psi_{\mathrm{ren}}(0)) = \mathbb{E}_{\nu_\infty^{\mathrm{ren}}}\left[\exp\left(-g^2\int_{-\infty}^{0} ds\int_{0}^{\infty} W_\rho(B_s - B_t, s-t)dt\right)\right].$$

Using Theorem 2.173 and the functional integral representation of $(F, e^{-TH_{\mathrm{N}}^{\mathrm{ren}}(0)}G)$ due to Theorem 2.160, we can show the superexponential decay of boson number in the ground state. Recall that the number operator N can be decomposed as $N = N_0 + N_\infty$. See (2.11.67).

Corollary 2.174 (Superexponential decay of boson number). *Let $\lambda > 0$. Then*

$$\Psi_{\mathrm{ren}}(0) \in D(e^{\beta N})$$

for arbitrary $\beta \in \mathbb{C}$.

Proof. In a similar way as in Lemma 2.135, we can show that $e^{\beta N_\infty}e^{-2TH_{\mathrm{N}}^{\mathrm{ren}}(0)}$ is bounded for $T > \beta$. Furthermore, $(\Psi_{\mathrm{ren}}, e^{\beta N_0}\Psi_{\mathrm{ren}}) < \infty$ for any $\beta > 0$ can be proven by using the Gibbs measure $\nu_\infty^{\mathrm{ren}}$ in a similar way as in Corollary 2.136. □

We can also see Gaussian domination of $\Psi_{\mathrm{ren}}(0)$ by using the Gibbs measure $\nu_\infty^{\mathrm{ren}}$.

Corollary 2.175 (Gaussian domination of the ground state). *Let $\lambda > 0$. Take $h \in \mathscr{S}'(\mathbb{R}^3)$ such that $\hat{h}/\sqrt{\omega} \in L^2(\mathbb{R}^3)$, $\xi_\lambda\hat{h}/\omega^2 \in L^1(\mathbb{R})$, and $\beta < 1/\|\hat{h}/\sqrt{\omega}\|^2$. Then $\Psi_{\mathrm{ren}}(0) \in D(e^{(\beta/2)\phi(h)^2})$ and*

$$\|e^{(\beta/2)\phi(h)^2}\Psi_{\mathrm{ren}}(0)\|^2 = \frac{1}{\sqrt{1-\beta\|\hat{h}/\sqrt{\omega}\|^2}}\mathbb{E}_{\nu_\infty^{\mathrm{ren}}}\left[\exp\left(\frac{\beta g^2 K(h)^2}{1-\beta\|\hat{h}/\sqrt{\omega}\|^2}\right)\right],$$

where $K(h)$ denotes the random variable defined by

$$K(h) = \frac{1}{2}\int_{-\infty}^{\infty}\left(\xi_\lambda e^{-|r|\omega}e^{-ik\cdot B_r}/\sqrt{\omega}, \hat{h}/\sqrt{\omega}\right)dr.$$

In particular, it follows that

$$\lim_{\beta\uparrow(2\|\hat{h}/\sqrt{\omega}\|^2)^{-1}} \|e^{\beta\phi(h)^2}\Psi_{\mathrm{ren}}(0)\| = \infty.$$

2.13 Polaron model

2.13.1 Definition of the polaron model

The *polaron model* is related to the Nelson model, and describes the interaction of electrons with the quantized vibrations of a crystal lattice. A conduction electron in an ionic crystal or a polar semiconductor are prototype cases of a polaron. In this section we discuss the three-dimensional polaron model and study its ultraviolet renormalization. The main steps of the discussion are similar to the Nelson Hamiltonian.

While the state space of bosons of the Nelson Hamiltonian is given by $\mathcal{F}_{\mathrm{b}}(\hat{\mathcal{H}}_M)$, for the polaron we use $\mathcal{F}_{\mathrm{b}}(L^2(\mathbb{R}^3))$. The *polaron Hamiltonian*

$$H_{\mathrm{pol}} = H_{\mathrm{p}} \otimes 1 + 1 \otimes N + g \int_{\mathbb{R}^3}^{\oplus} \Phi(x) dx$$

is a self-adjoint operator on the Hilbert space

$$\mathcal{H}_{\mathrm{pol}} = L^2(\mathbb{R}^3) \otimes \mathcal{F}_{\mathrm{b}}(L^2(\mathbb{R}^3)).$$

Here $g \in \mathbb{R}$ is a coupling constant, N is the phonon number operator, and

$$\Phi(x) = \frac{1}{\sqrt{2}}\left(a^*(\hat{\varphi}e^{-ik\cdot x}/\omega) + a(\tilde{\hat{\varphi}}e^{ik\cdot x}/\omega)\right),$$

is the interaction term, with a and a^* denoting the annihilation and creation operators, respectively, acting in $\mathcal{F}_{\mathrm{b}}(L^2(\mathbb{R}^3))$. They satisfy

$$[a(\hat{f}), a^*(\hat{g})] = (\bar{\hat{f}}, \hat{g})_{L^2(\mathbb{R}^3)}, \quad [a(\hat{f}), a(\hat{g})] = 0 = [a^*(\hat{f}), a^*(\hat{g})].$$

We use Assumption 2.176 below in Section 2.13 until otherwise stated.

Assumption 2.176. The following conditions hold:
(1) (Dispersion relation) $\omega(k) = |k|$;
(2) (Charge distribution) $\varphi \in \mathscr{S}'(\mathbb{R}^3)$, $\overline{\hat{\varphi}(k)} = \hat{\varphi}(-k)$ and $\hat{\varphi}/\omega \in L^2(\mathbb{R}^3)$;
(3) (External potential) $V \in \mathcal{R}_{\mathrm{Kato}}$.

Note that the charge distribution of the polaron model is $\hat{\varphi}/\omega$ and differs from the expression $\hat{\varphi}/\sqrt{\omega}$ in the Nelson model. It can shown that H_{pol} is self-adjoint on $D(\Delta \otimes 1) \cap D(1 \otimes N)$ and bounded from below by Kato–Rellich theorem; moreover, it is essentially self-adjoint on any core of $-\frac{1}{2}\Delta \otimes 1 + 1 \otimes N$.

2.13.2 Functional integral representation

To define the polaron Hamiltonian on function space, let $(\phi(f); f \in L^2_{\mathrm{real}}(\mathbb{R}^3))$ be a family of Gaussian random variables on $(\mathscr{Q}, \Sigma, \mu)$ indexed by $L^2_{\mathrm{real}}(\mathbb{R}^3)$, and similarly $(\phi_{\mathrm{E}}(F); F \in L^2_{\mathrm{real}}(\mathbb{R}^4))$ on $(\mathscr{Q}_{\mathrm{E}}, \Sigma_E, \mu_E)$ indexed by $L^2_{\mathrm{real}}(\mathbb{R}^4)$. The polaron Hamiltonian H_{pol} is then defined on $L^2(\mathbb{R}^3) \otimes L^2(\mathscr{Q})$ by

$$\hat{H}_{\mathrm{pol}} = H_{\mathrm{p}} \otimes 1 + 1 \otimes \hat{N} + \int_{\mathbb{R}^3}^{\oplus} \phi(\tilde{\varphi}(\cdot - x))dx,$$

where

$$\tilde{\varphi} = (\hat{\varphi}/\omega)^{\vee}.$$

We will keep using the same notations H_{pol} for $\hat{H}_{\mathrm{pol}}$, and $\mathcal{H}_{\mathrm{pol}}$ for $L^2(\mathbb{R}^3) \otimes L^2(\mathscr{Q})$. As for the Nelson model, the $V = 0$ case gives a translation invariant model. Thus similarly the polaron Hamiltonian can be decomposed through the fiber Hamiltonians $H_{\mathrm{pol}}(p)$ with total momentum $p \in \mathbb{R}^3$, giving

$$H_{\mathrm{pol}} \cong \int_{\mathbb{R}^3}^{\oplus} H_{\mathrm{pol}}(p)dp.$$

Here

$$H_{\mathrm{pol}}(p) = \frac{1}{2}(p - P_{\mathrm{f}})^2 + N + g\Phi, \quad p \in \mathbb{R}^3,$$

and $\Phi = \Phi(0)$. The state space of phonons is $\mathcal{F}_{\mathrm{b}}(L^2(\mathbb{R}^3))$. Similarly to the Nelson model, using the polaron state space, there exists a family of isometries

$$\bar{\tau}_s : L^2_{\mathrm{real}}(\mathbb{R}^3) \to L^2_{\mathrm{real}}(\mathbb{R}^4), \quad s \in \mathbb{R},$$

such that

$$\bar{\tau}_s^* \bar{\tau}_t = e^{-|t-s|} \quad s, t \in \mathbb{R}.$$

They are given by

$$\widehat{\bar{\tau}_t f}(k_0, k) = \frac{e^{-itk_0}}{\sqrt{\pi}} \frac{1}{\sqrt{1 + |k_0|^2}} \hat{f}(k), \quad (k_0, k) \in \mathbb{R} \times \mathbb{R}^3.$$

A straightforward calculation shows that

$$(\bar{\tau}_s f, \bar{\tau}_t g) = (\widehat{\bar{\tau}_s f}, \widehat{\bar{\tau}_t g}) = \frac{1}{\pi} \int_{\mathbb{R}} \frac{e^{-ik_0(t-s)}}{1 + |k_0|^2} dk_0 \int_{\mathbb{R}^3} \bar{\hat{f}}(k)\hat{g}(k)dk = e^{-|t-s|}(f, g).$$

We denote the second quantization of $\bar{\tau}_t$ by

$$\bar{\mathrm{I}}_t = \Gamma_{\mathrm{Int}}(\bar{\tau}_t).$$

Thus $\bar{\mathrm{I}}_t : L^2(\mathscr{Q}) \to L^2(\mathscr{Q}_{\mathrm{E}})$ is an isometry for every $t \in \mathbb{R}$, and the factorization formula

$$\bar{\mathrm{I}}_s^* \bar{\mathrm{I}}_t = e^{-|t-s|\hat{N}}, \quad s, t \in \mathbb{R},$$

holds. In particular, we have $(F, e^{-t\hat{N}} G)_{L^2(\mathscr{Q})} = (\bar{\mathrm{I}}_0 F, \bar{\mathrm{I}}_t G)_{L^2(\mathscr{Q}_{\mathrm{E}})}$.

Theorem 2.177 (Functional integral representation of polaron Hamiltonian).

(1) *If $F, G \in \mathcal{H}_{\mathrm{pol}}$, then*

$$(F, e^{-2TH_{\mathrm{pol}}} G)_{\mathcal{H}_{\mathrm{pol}}} = \int_{\mathbb{R}^3} \mathbb{E}^x \left[e^{-\int_{-T}^T V(B_s) ds} (\bar{\mathrm{I}}_{-T} F(B_{-T}), e^{-\phi_{\mathrm{E}}\left(\int_{-T}^T \bar{\tau}_s \tilde{\varphi}(\cdot - B s) ds\right)} \bar{\mathrm{I}}_T G(B_T)) \right] dx.$$

(2) *Let $\Psi, \Phi \in L^2(\mathscr{Q})$, and $p \in \mathbb{R}^3$. Then*

$$(\Psi, e^{-2TH_{\mathrm{pol}}(p)} \Phi)_{L^2(\mathscr{Q})} = \mathbb{E}^0 \left[(\bar{\mathrm{I}}_{-T} u_{-T}(p) \Psi(B_{-T}), e^{-\phi_{\mathrm{E}}\left(\int_{-T}^T \bar{\tau}_s \tilde{\varphi}(\cdot - B s) ds\right)} \bar{\mathrm{I}}_T u_T(p) \Phi(B_T)) \right].$$

Here $u_T(p) = e^{i(p - \hat{P}_{\mathrm{f}}) \cdot B_T}$.

Proof. The proof of (1) is the same as that of $(F, e^{-2TH_{\mathrm{N}}} G)$ by replacing I_t and τ_t with $\bar{\mathrm{I}}_t$ and $\bar{\tau}_t$, respectively. The proof of (2) is also the same as that of $(\Psi, e^{-2TH_{\mathrm{N}}(p)} \Phi)$, and we skip the details. □

2.13.3 Removal of ultraviolet cutoff

In this section we discuss the ultraviolet problem for the polaron Hamiltonian. Due to the more regular function $\hat{\varphi}/\omega$ appearing in the polaron Hamiltonian, the situation simplifies from the Nelson Hamiltonian's case where we had $\hat{\varphi}/\sqrt{\omega}$. Also, since the free field term of the polaron Hamiltonian is given by the number operator N, there is no infrared divergence in this case.

Denote the polaron Hamiltonian with $\hat{\varphi}(k) = e^{-\varepsilon|k|^2/2}$ by $H^\varepsilon_{\mathrm{pol}}$. The vacuum expectation of $e^{-TH^\varepsilon_{\mathrm{pol}}}$ is given by

$$(f \otimes 1, e^{-TH^\varepsilon_{\mathrm{pol}}} h \otimes 1) = \int_{\mathbb{R}^3} \mathbb{E}^x [\overline{f(B_0)} h(B_T) e^{\frac{g^2}{2} S^\varepsilon_{\mathrm{pol}}} e^{-\int_0^t V(B_s) ds}] dx,$$

where

$$S^{\varepsilon}_{\rm pol}(B_s - B_t, s - t) = \int_0^T ds \int_0^T W^{\varepsilon}_{\rm pol}(B_s - B_t, s - t)dt$$

is the pair interaction associated with the polaron Hamiltonian with the pair potential

$$W^{\varepsilon}_{\rm pol}(x, t) = e^{-|t|} \int_{\mathbb{R}^3} \frac{1}{2\omega(k)^2} e^{-\varepsilon|k|^2} e^{-ik\cdot x} dk.$$

Also, we have

$$W^0_{\rm pol}(x, t) = e^{-|t|} \int_{\mathbb{R}^3} \frac{1}{2\omega(k)^2} e^{-ik\cdot x} dk = 2\pi \frac{e^{-|t|}}{|x|} \int_0^\infty \frac{\sin u}{u} du = \pi^2 \frac{e^{-|t|}}{|x|},$$

and clearly $W^{\varepsilon}_{\rm pol}(x, t) \to W^0_{\rm pol}(x, t)$ for every (x, t) as $\varepsilon \downarrow 0$. Let

$$S^0_{\rm pol}(B_t - B_s, t - s) = \pi^2 \int_0^T ds \int_0^T \frac{e^{-|t-s|}}{|B_t - B_s|} dt.$$

Then

$$\lim_{\varepsilon\downarrow 0} \mathbb{E}\left[\left|e^{\frac{g^2}{2} S^{\varepsilon}_{\rm pol}} - e^{\frac{g^2}{2} S^0_{\rm pol}}\right|\right] = 0.$$

For $f, h \in L^2(\mathbb{R}^3)$, it is then immediate to get

$$\lim_{\varepsilon\downarrow 0}(f \otimes 1, e^{-TH^{\varepsilon}_{\rm pol}} h \otimes 1) = \int_{\mathbb{R}^3} \mathbb{E}^x[\overline{f(B_0)} h(B_T) e^{\frac{g^2}{2} S^0_{\rm pol}} e^{-\int_0^t V(B_s)ds}] dx. \tag{2.13.1}$$

It is seen that no ultraviolet renormalization is needed in (2.13.1). Thus the theorem below can be proven in the same way as for the Nelson Hamiltonian.

Theorem 2.178 (Removal of ultraviolet cutoff).
(1) *There exists a self-adjoint operator* $H^{\rm ren}_{\rm pol}$ *such that*

$$\text{s-}\lim_{\varepsilon\downarrow 0} e^{-TH^{\varepsilon}_{\rm pol}} = e^{-TH^{\rm ren}_{\rm pol}}, \quad T \geq 0.$$

(2) *If* $p \in \mathbb{R}^3$, *then there exists a self-adjoint operator* $H^{\rm ren}_{\rm pol}(p)$ *such that*

$$\text{s-}\lim_{\varepsilon\downarrow 0} e^{-TH^{\varepsilon}_{\rm pol}(p)} = e^{-TH^{\rm ren}_{\rm pol}(p)}, \quad T \geq 0.$$

By using the Baker–Campbell–Hausdorff formula, we can also derive a functional integral representation of $(\Phi, e^{-TH^{\rm ren}_{\rm pol}} \Psi)$ in terms of exponentials of creation and annihilation operators. Let

$$V_T(k) = -\frac{g}{\sqrt{2}} \int_{-T}^{T} e^{-|s+T|} \frac{e^{-ik\cdot B_s}}{\omega(k)} ds \quad \text{and} \quad \tilde{V}_T(k) = -\frac{g}{\sqrt{2}} \int_{-T}^{T} e^{-|s-T|} \frac{e^{ik\cdot B_s}}{\omega(k)} ds,$$

and define

$$B = e^{a^*(V_T)} e^{-TN} \quad \text{and} \quad \tilde{B} = e^{-TN} e^{a(\tilde{V}_T)}.$$

We present the next result without proof since one can argue similarly to the case of the Nelson Hamiltonian with no cutoff.

Theorem 2.179 (Functional integral representation of polaron Hamiltonians without ultraviolet cutoff). *Let*

$$S^0_{\text{pol}}(B_s - B_t, s - t) = \pi^2 \int_{-T}^{T} ds \int_{-T}^{T} \frac{e^{-|s-t|}}{|B_s - B_t|} dt.$$

(1) *We have $V_T, \tilde{V}_T \in L^2(\mathbb{R}^3_k)$ a. s.*
(2) *If $F, G \in \mathcal{H}_{\text{pol}}$, then*

$$(F, e^{-2TH^{\text{ren}}_{\text{pol}}} G)_{\mathcal{H}_{\text{pol}}} = \int_{\mathbb{R}^3} \mathbb{E}^x \left[e^{-\int_{-T}^{T} V(B_s) ds} e^{\frac{g^2}{2} S^0_{\text{pol}}} (F(B_{-T}), B\tilde{B}G(B_T)) \right] dx.$$

(3) *If $\Phi, \Psi \in L^2(\mathcal{Q})$, then*

$$(\Phi, e^{-2TH^{\text{ren}}_{\text{pol}}(p)} \Psi)_{L^2(\mathcal{Q})} = \mathbb{E}^0 \left[e^{\frac{g^2}{2} S^0_{\text{pol}}} (u_{-T}(p)\Phi, B\tilde{B} u_T(p)\Psi) \right].$$

Here $u_T(p) = e^{i(p - \hat{P}_f)\cdot B_T}$.

With $E_{\text{pol}}(p) = \inf \text{Spec}(H_{\text{pol}}(p))$ and $E^{\text{ren}}_{\text{pol}}(p) = \inf \text{Spec}(H^{\text{ren}}_{\text{pol}}(p))$, the functional integral representation gives similar properties as seen before in the Nelson model's case.

Corollary 2.180 (Properties of $E_{\text{pol}}(p)$ and $E^{\text{ren}}_{\text{pol}}(p)$).
(1) *If $p = 0$, then $e^{-TH_{\text{pol}}(0)}$ and $e^{-TH^{\text{ren}}_{\text{pol}}(0)}$ are positivity improving.*
(2) *The ground states of $H_{\text{pol}}(0)$ and $H^{\text{ren}}_{\text{pol}}(0)$ are in each case unique whenever they exist.*
(3) *The energy comparison inequalities $E_{\text{pol}}(0) \le E_{\text{pol}}$ and $E^{\text{ren}}_{\text{pol}}(0) \le E^{\text{ren}}_{\text{pol}}(p)$ hold.*
(4) *The maps $\mathbb{R}^3 \ni p \mapsto E_{\text{pol}}(p) \in \mathbb{R}$ and $\mathbb{R}^3 \ni p \mapsto E^{\text{ren}}_{\text{pol}}(p) \in \mathbb{R}$ are continuous.*

Proof. The proof is the same as the proofs of Corollaries 2.148 and 2.161. □

2.14 Functional central limit theorem

2.14.1 Gibbs measures with no external potential

In this section, contrary to Section 2.12.2, beginning with a functional integral representation of the semigroup generated by the Nelson Hamiltonian, we define a Hamiltonian equivalent to a single particle translation invariant Nelson Hamiltonian with total momentum zero, i. e., $p = 0$, and consider a Gibbs measure associated with this Hamiltonian. This construction of the Gibbs measure can be regraded as a construction of a Gibbs measure with external potential $V = 0$ in Section 5.2 in Volume 1. We shall assume a general conditions on a pair potential considered in Section 2.14. We assume that

$$W(x,t) = \frac{1}{2}\int_{\mathbb{R}^d} \frac{|\hat{\varphi}(k)|^2}{\omega(k)} e^{-ik\cdot x} e^{-\omega(k)|t|} dk \tag{2.14.1}$$

with general ω and $\hat{\varphi}$ satisfying Assumption 2.181 below.

Assumption 2.181. The following conditions hold:
(1) (Dispersion relation) $\omega(k) \geq 0$, $\omega(k) = \omega(-k)$.
(2) (Charge distribution) $\varphi \in \mathscr{S}'(\mathbb{R}^3)$, $\hat{\varphi}(k) = \overline{\hat{\varphi}(-k)}$, and $\hat{\varphi}/\sqrt{\omega}, \hat{\varphi}/\omega \in L^2(\mathbb{R}^d)$.
(3) (Infrared regularity) $\int_{\mathbb{R}^d} |\hat{\varphi}(k)|^2/\omega^3(k)dk < \infty$.

In this section we use a d-dimensional Brownian motion over the whole time. We recall that $W(B_s - B_t, s-t)$ is uniformly bounded with respect to paths as a function of x and t, and

$$\int_{-\infty}^{\infty} |W(B_s - B_t, s-t)| ds \leq \int_{\mathbb{R}^d} \frac{|\hat{\varphi}(k)|^2}{\omega(k)^2} dk, \tag{2.14.2}$$

$$\int_{-\infty}^{0} ds \int_{0}^{\infty} |W(B_s - B_t, s-t)| dt \leq \frac{1}{2}\int_{\mathbb{R}^d} \frac{|\hat{\varphi}(k)|^2}{\omega(k)^3} dk. \tag{2.14.3}$$

We define a measure on $(\mathfrak{X}, \mathcal{B}(\mathfrak{X}))$ by

$$d\mathcal{P}_T^0 = \frac{1}{Z_T} \exp\left(\int_{-T}^{T} ds \int_{-T}^{T} W(B_s - B_t, s-t) dt \right) d\mathcal{W}^0, \tag{2.14.4}$$

where

$$Z_T = \mathbb{E}_{\mathcal{W}^0}\left[\exp\left(\int_{-T}^{T} ds \int_{-T}^{T} W(B_s - B_t, s-t) dt \right) \right].$$

For $x \in \mathbb{R}^d$, let $\hat{\tau}_x$ be the shift by x on $\mathcal{M}_{-2}$ such that

$$(\hat{\tau}_x \xi)(f) = \xi(\theta_x f)$$

for $\xi \in \mathcal{M}_{-2}$, where $(\theta_x f)(\cdot) = f(\cdot - x)$.

Lemma 2.182. *For every $x \in \mathbb{R}^d$, the map $\hat{\tau}_x$ can be extended to a unitary on $L^2(\mathrm{G})$, and $x \mapsto \hat{\tau}_x$ is a strongly continuous group on $L^2(\mathrm{G})$.*

Proof. First, note that for $f, g \in \mathcal{M}$,

$$\begin{aligned}(f, \theta_x g)_{\mathcal{M}} &= \int_{\mathbb{R}^d} \overline{\hat{f}(k)} \widehat{\theta_x g}(k) \frac{1}{\omega(k)} dk = \int_{\mathbb{R}^d} \overline{\hat{f}(k)} e^{ikx} \hat{g}(k) \frac{1}{\omega(k)} dk \\ &= \int_{\mathbb{R}^d} \overline{\widehat{\theta_{-x} f}(k)} \hat{g}(k) \frac{1}{\omega(k)} dk = (\theta_{-x} f, g)_{\mathcal{M}}.\end{aligned}$$

Similarly, $(\theta_x f, \theta_x g)_{\mathcal{M}} = (f, g)_{\mathcal{M}}$. We have

$$\begin{aligned}(e^{i\xi(f)}, \hat{\tau}_x e^{i\xi(g)})_{L^2(\mathrm{G})} &= \exp\left\{-\frac{1}{4}((f,f)_{\mathcal{M}} + (f, \theta_x g)_{\mathcal{M}} + (\theta_x g, f)_{\mathcal{M}} + (\theta_x g, \theta_x g)_{\mathcal{M}})\right\} \\ &= \exp\left\{-\frac{1}{4}((\theta_{-x} f, \theta_{-x} f)_{\mathcal{M}} + (\theta_{-x} f, g)_{\mathcal{M}} + (g, \theta_{-x} f)_{\mathcal{M}} + (g,g)_{\mathcal{M}})\right\} \\ &= (\theta_{-x} e^{i\xi(f)}, e^{i\xi(g)})_{L^2(\mathrm{G})}.\end{aligned}$$

Since the functions of the form $e^{i\xi_s(f)}$ span $L^2(\mathrm{G})$, the equality of the first two lines above shows strong continuity of the map $x \mapsto \hat{\tau}_x$, and the semigroup property $\hat{\tau}_{x+y} = \hat{\tau}_x \hat{\tau}_y$. Moreover, the equality of the first and last lines shows that the adjoint of $\hat{\tau}_x$ is given by $\hat{\tau}_{-x}$. Since, clearly, $\hat{\tau}_x \hat{\tau}_{-x} = 1$, $\hat{\tau}_x$ is a unitary map. □

Let $\mathcal{L} = \{F(\xi(f_1), \dots, \xi(f_n)) \mid F \in \mathcal{S}(\mathbb{R}^n), f_j \in C_0^\infty(\mathbb{R}^d), j = 1, \dots, n, n \geq 1\}$. Note that $\mathcal{L}$ is dense in $L^2(\mathrm{G})$. We also have

$$\hat{\tau}_x F(\xi(f_1), \dots, \xi(f_n)) = F(\xi(\theta_x f_1), \dots, \xi(\theta_x f_n)).$$

For $T > 0$, define measures on $\mathfrak{X} \otimes \mathfrak{Y}$ by

$$\mathcal{P}_0^W = \mathcal{W}^0 \otimes \mathcal{G}, \tag{2.14.5}$$

$$d\mathcal{P}_T^W = \frac{1}{Z_T} \exp\left(-\int_{-T}^{T} \hat{\tau}_{B_s} \xi_s(\varphi) ds\right) d\mathcal{P}_0^W, \tag{2.14.6}$$

where Z_T is the normalizing constant

$$Z_T = \mathbb{E}_{\mathcal{P}_0^W}\left[\exp\left(-\int_{-T}^{T} \hat{\tau}_{B_s} \xi_s(\varphi) ds\right)\right].$$

Next lemma can be proven in a similar way showing the fact $\varrho_T = \nu_T$ in (2.12.16) in Section 2.12.4. We recall that $\mathscr{G}_T = \bigcup_{0\le s\le T} \mathscr{F}^{\mathrm{BM}}_{[-s,s]}$.

Lemma 2.183. *Let $A \in \mathscr{G}_t$. Then $\mathcal{P}^W_T(A) = \mathcal{P}^0_T(A)$ for $t \le T$.*

Proof. Since $\int_{-T}^T \hat{\tau}_{B_s}\xi_s(\varphi)ds$ is a Gaussian random variable on $\mathfrak{Y}$ with mean zero and covariance

$$\mathbb{E}_{\mathcal{G}}\left[\left|\int_{-T}^{T} \hat{\tau}_{B_s}\xi_s(\varphi)ds\right|^2\right] = \frac{1}{2}\int_{-T}^{T} ds \int_{-T}^{T} W(B_s - B_t, s-t)dt,$$

the lemma follows. □

We are interested in the limit of $\mathcal{P}^0_T$ as $T \to \infty$. In light of Lemma 2.183, we will study the family of measures $(\mathcal{P}^W_T)_{T\ge 0}$ instead of $(\mathcal{P}^0_T)_{T\ge 0}$. At first sight, this seems more difficult as the state space $\mathbb{R}^d \times \mathscr{M}_{-2}$ is now infinite-dimensional. However, the great advantage is that $\mathcal{P}^W_T$ is tightly connected to a Markov process, as we will now see.

Definition 2.184. We define the operator P_t on $L^2(\mathbb{R}^d) \otimes L^2(\mathrm{G})$ by

$$(F, P_t G)_{L^2(\mathbb{R}^d)\otimes L^2(\mathrm{G})} = \int_{\mathbb{R}^d} \mathbb{E}_{\mathcal{W}^x\otimes\mathcal{G}}[F(B_0,\xi_0)e^{-\int_0^t \hat{\tau}_{B_s}\xi_s(\varphi)ds}G(B_t,\xi_t)]dx \tag{2.14.7}$$

for $F, G \in L^2(\mathbb{R}^d) \otimes L^2(\mathrm{G})$.

In a similar way to Corollary 1.88, we have

$$\left|\int_{\mathbb{R}^d} \mathbb{E}_{\mathcal{W}^x\otimes\mathcal{G}}[F(B_0,\xi_0)e^{-\int_0^t \hat{\tau}_{B_s}\xi_s(\varphi)ds}G(B_t,\xi_t)]dx\right| \le \|F\|\|G\|e^{t\mathrm{I}}, \tag{2.14.8}$$

where $\mathrm{I} = \int_{\mathbb{R}^d} |\hat{\varphi}(k)|^2/\omega(k)^3 dk$.

Remark 2.185. We have already shown in Theorem 2.12 that the right-hand side of (2.14.7) is a functional integral representation of the semigroup generated by the Nelson Hamiltonian defined on $L^2(\mathbb{R}^d)\otimes L^2(\mathrm{G})$. One of the main purpose of this section is, however, to derive the translation invariant Nelson Hamiltonian with total momentum zero from the right-hand side of (2.14.7).

Proposition 2.186. *We have*

$$P_t = e^{-tH}, \quad t \ge 0, \tag{2.14.9}$$

where H is given by

$$H = -\frac{1}{2}\Delta_x + \hat{H}_{\mathrm{f}} + H_{\mathrm{I}}, \tag{2.14.10}$$

where $\hat{H}_{\mathrm{f}} = d\Gamma(\omega(-i\nabla))$ *is the free field Hamiltonian and*

$$(H_{\mathrm{I}}f)(x,\xi) = (\hat{\tau}_x\xi(\varphi))f(x,\xi) = \xi(\varphi(\cdot - x))f(x,\xi).$$

Moreover,

$$\|P_tf\|^2_{L^2(\mathrm{G})}(x) = \mathbb{E}_{\mathcal{W}^x\otimes\mathcal{G}}[f(B_{-t},\xi_{-t})e^{-\int_{-t}^t \hat{\tau}_{B_s}\xi_s(\varphi)ds}f(B_t,\xi_t)] \tag{2.14.11}$$

for $f \in L^2(\mathbb{R}^d)\otimes L^2(\mathrm{G})$.

Proof. Equation (2.14.9) is due to Theorem 2.12. For (2.14.11), note that for $g \in C_0^\infty(\mathbb{R}^d)$, the definition of P_t and the semigroup property give

$$\begin{aligned}
\int_{\mathbb{R}^d} g(x)\|P_tf\|^2_{L^2(\mathrm{G})}(x)dx &= (P_tf, gP_tf) = (f, P_tgP_tf)\\
&= \int_{\mathbb{R}^d} \mathbb{E}_{\mathcal{W}^x\otimes\mathcal{G}}[f(B_0,\xi_0)e^{-\int_0^t \hat{\tau}_{B_s}\xi_s(\varphi)ds}g(B_t)e^{-\int_t^{2t} \hat{\tau}_{B_s}\xi_s(\varphi)ds}f(B_{2t},\xi_{2t})]dx\\
&= \int_{\mathbb{R}^d} \mathbb{E}_{\mathcal{W}^x\otimes\mathcal{G}}[f(B_0,\xi_0)e^{-\int_0^{2t} \hat{\tau}_{B_s}\xi_s(\varphi)ds}g(B_t)f(B_{2t},\xi_{2t})]dx.
\end{aligned}$$

Since both $(B_t)_{t\in\mathbb{R}}$ and $(\xi_t)_{t\in\mathbb{R}}$ are invariant under the time shift, we can replace B_s by B_{s-t} and ξ_s with ξ_{s-t} above, and we find

$$\int_{\mathbb{R}^d} g(x)\|P_tf\|^2_{L^2(\mathrm{G})}(x)dx = \int_{\mathbb{R}^d} g(x)\mathbb{E}_{\mathcal{W}^x\otimes\mathcal{G}}[f(B_{-t},\xi_{-t})e^{-\int_{-t}^t \hat{\tau}_{B_s}\xi_s(\varphi)ds}f(B_t,\xi_t)]dx.$$

Since $g \in C_0^\infty(\mathbb{R}^d)$ is arbitrary, we obtain the claim. □

We have already seen in Proposition 2.4 that H in (2.14.10) is the Nelson Hamiltonian and is self-adjoint on $D(H) = D(-\Delta)\cap D(\hat{H}_{\mathrm{f}})$, and for each $x \in \mathbb{R}^d$, $\hat{H}_{\mathrm{f}} + H_{\mathrm{I}}(x)$ is a self-adjoint operator in $L^2(\mathrm{G})$ with the domain $D(\hat{H}_{\mathrm{f}})$ and bounded from below uniformly in x.

We shall extend P_t on some spaces which are not necessarily Hilbert spaces. Hence we need to work with the following additional function spaces:

$$\begin{aligned}
L^\infty(\mathbb{R}^d; L^2(\mathrm{G})) &= \{f : \mathbb{R}^d\times\mathscr{M}_{-2}\to\mathbb{C} \mid \operatorname*{ess\,sup}_{x\in\mathbb{R}^d}\|f(x,\cdot)\|_{L^2(\mathrm{G})} < \infty\},\\
L^\infty(\mathbb{R}^d\times\mathscr{M}_{-2}) &= \{f : \mathbb{R}^d\times\mathscr{M}_{-2}\to\mathbb{C} \mid \operatorname*{ess\,sup}_{(x,\xi)\in\mathbb{R}^d\times\mathscr{M}_{-2}}|f(x,\xi)| < \infty\}.
\end{aligned}$$

Clearly, $L^\infty(\mathbb{R}^d\times\mathscr{M}_{-2}) \subset L^\infty(\mathbb{R}^d; L^2(\mathrm{G}))$, and unitary equivalences

$$L^2(\mathbb{R}^d)\otimes L^2(\mathrm{G}) \cong L^2(\mathbb{R}^d; L^2(\mathrm{G})) \cong L^2(\mathbb{R}^d\times\mathrm{G})$$

are used without further notice. Moreover, we write

$$C(\mathbb{R}^d; L^2(\mathrm{G})) = \{f : \mathbb{R}^d \times \mathscr{M}_{-2} \to \mathbb{C} \mid \mathbb{R}^d \ni x \mapsto f(x,\cdot) \in L^2(\mathrm{G}) \text{ is continuous}\},$$
$$C_{\mathrm{b}}(\mathbb{R}^d; L^2(\mathrm{G})) = C(\mathbb{R}^d; L^2(\mathrm{G})) \cap L^\infty(\mathbb{R}^d; L^2(\mathrm{G})).$$

Here $C(\mathbb{R}^d; L^2(\mathrm{G}))$ is a Banach space with the norm $\|f\|_{L^\infty(\mathbb{R}^d;L^2(\mathrm{G}))}$. We need to study P_t on two closed subspaces of $C(\mathbb{R}^d; L^2(\mathrm{G}))$. The first is

$$C_0(\mathbb{R}^d; L^2(\mathrm{G})) = \left\{ f \in C_{\mathrm{b}}(\mathbb{R}^d; L^2(\mathrm{G})) \,\Big|\, \lim_{|x|\to\infty} \|f(x)\|_{L^2(\mathrm{G})} = 0 \right\}.$$

The second subspace is $\mathcal{T}$. To define $\mathcal{T}$, we need to define U by

$$U : L^2(\mathrm{G}) \to C(\mathbb{R}^d; L^2(\mathrm{G})), \quad (Uf)(x) = (\hat{\tau}_x f)(\cdot). \tag{2.14.12}$$

Definition 2.187. Subspace $\mathcal{T}$ is the Hilbert space defined by

$$\mathcal{T} = UL^2(\mathrm{G}) = \{Uf \mid f \in L^2(\mathrm{G})\}$$

equipped with the scalar product $(f,g)_{\mathcal{T}} = (U^{-1}f, U^{-1}g)_{L^2(\mathrm{G})}$.

Note that for $\mathcal{T} \ni f = f(x,\xi) = (Ug)(x,\xi) = (\hat{\tau}_x g)(\xi)$,

$$\|f\|_{\mathcal{T}} = \|U^{-1}f\|_{L^2(\mathrm{G})} = \|g\|_{L^2(\mathrm{G})} = \|\hat{\tau}_x g\|_{L^2(\mathrm{G})} = \sup_{y\in\mathbb{R}^d} \|\hat{\tau}_y g\|_{L^2(\mathrm{G})} = \|f\|_{L^\infty(\mathbb{R}^d;L^2(\mathrm{G}))},$$

and U is a unitary operator from $L^2(\mathrm{G})$ onto $\mathcal{T}$. We notice that

$$(U^{-1}f)(x,\xi) = f(\hat{\tau}_x \xi). \tag{2.14.13}$$

Note that $\mathcal{T} \not\subset L^2(\mathbb{R}^d) \otimes L^2(\mathrm{G})$ and $C(\mathbb{R}^d; L^2(\mathrm{G})) \not\subset L^2(\mathbb{R}^d) \otimes L^2(\mathrm{G})$. Replacing F and G in Definition 2.188 with Uf and Ug, respectively, we define P_t on $\mathcal{T}$.

Definition 2.188 (P_t on $\mathcal{T}$). Let $g \in L^2(\mathrm{G})$. We define P_t on $\mathcal{T}$ by

$$P_t(Ug)(x,\xi) = \mathbb{E}_{\mathcal{W}^x \otimes \mathcal{G}^\xi}[e^{-\int_0^t \hat{\tau}_{B_s}\xi_s(\varphi)ds} g(\hat{\tau}_{B_t}\xi_t)]. \tag{2.14.14}$$

We denote it by the same notation P_t. In what follows, we define $P_t f$ for a map f on $\mathbb{R}^d \times \mathscr{M}_{-2}$ by (2.14.14) with Ug replaced by f and use the same notation P_t unless any confusion may arise. The examples are functions in $L^\infty(\mathbb{R}^d; L^2(\mathrm{G}))$ and $C_0(\mathbb{R}^d; L^2(\mathrm{G}))$.

We need a further auxiliary result.

Lemma 2.189.

(1) *$\{P_t : t \ge 0\}$ is a C_0-semigroup on $L^\infty(\mathbb{R}^d; L^2(\mathrm{G}))$.*

(2) *If $f \in L^\infty(\mathbb{R}^d \times \mathscr{M}_{-2}) \cap C(\mathbb{R}^d; L^2(\mathrm{G}))$, then $P_t f \in C_{\mathrm{b}}(\mathbb{R}^d; L^2(\mathrm{G}))$.*

(3) *If* $f \in L^\infty(\mathbb{R}^d \times \mathscr{M}_{-2}) \cap C(\mathbb{R}^d; L^2(\mathrm{G}))$, *then* $\lim_{t\to 0} \sup_{x\in K} \|P_t f(x) - f(x)\|_{L^2(\mathrm{G})} = 0$ *for any compact set* $K \subset \mathbb{R}^d$.

Proof. (1) Once we have shown that P_t is bounded on $L^\infty(\mathbb{R}^d; L^2(\mathrm{G}))$, the semigroup property follows from the Markov property of $(\xi_t)_{t\in\mathbb{R}}$ and $(B_t)_{t\ge 0}$. To show boundedness, note that $|P_t f| \le P_t|f|$ pointwise. Therefore it suffices to consider nonnegative functions. First, let $f \in L^\infty(\mathbb{R}^d \times \mathscr{M}_{-2})$. We fix a not necessarily continuous, but bounded, path $X : [-t,t] \to \mathbb{R}^d$ and define the bilinear form

$$(f,g)_X = \mathbb{E}_{\mathcal{G}}[f(X_{-t},\xi_{-t})e^{-\int_{-t}^t \hat{\tau}_{X_s}\xi_s(\varphi)ds} g(X_t,\xi_t)].$$

For two paths X and $\tilde{X}$,

$$|(f,g)_X - (f,g)_{\tilde{X}}| \le \|f\|_\infty \|g\|_\infty \mathbb{E}_{\mathcal{G}}[(e^{-\int_{-t}^t \hat{\tau}_{X_s}\xi_s(\varphi)ds} - e^{-\int_{-t}^t \hat{\tau}_{\tilde{X}_s}\xi_s(\varphi)ds})^2]. \tag{2.14.15}$$

Gaussian integration with respect to $\mathcal{G}$ then shows that $X \mapsto (f,g)_X$ is continuous from $L^\infty([-t,t],\mathbb{R})$ to $\mathbb{R}$. Now let X be a continuous path, and define $X_s^{(n)} = \sum_{j=-n}^{n-1} X_{tj/n} 1_{[tj/n, t(j+1)/n)}(s)$. Then

$$|(f,g)_{X^{(n)}}| \le e^{\mathrm{I}} \|f(X_{-t})\|_{L^2(\mathrm{G})} \|g(X_t)\|_{L^2(\mathrm{G})}$$

for bounded f and g; $X^{(n)} \to X$ uniformly on $[-t,t]$ as $n \to \infty$, where $\mathrm{I} = (1/4)\int_{\mathbb{R}^d} |\hat{\varphi}(k)|^2/\omega(k)^3 dk$, and thus the above inequality remains valid when we replace $X^{(n)}$ by X. For $f,g \in L^2(\mathbb{R}^d; L^2(\mathrm{G}))$, the monotone convergence theorem implies

$$|(f,g)_X| \le e^{\mathrm{I}} \|f(X_{-t})\|_{L^2(\mathrm{G})} \|g(X_t)\|_{L^2(\mathrm{G})} \tag{2.14.16}$$

for all $X \in \mathfrak{X}$. Using this in (2.14.11) shows (1).

(2) Choose $x,y \in \mathbb{R}^d$, $f,g \in L^\infty(\mathbb{R}^d \times \mathscr{M}_{-2})$ and write $f_y(x,\xi) = f(x+y,\xi)$. Then

$$\begin{aligned}
&\|P_t f(x) - P_t f(x+y)\|_{L^2(\mathrm{G})}^2 \\
&\quad = \mathbb{E}_{\mathrm{G}}[(\mathbb{E}_{\mathcal{W}^x \otimes \mathcal{G}^\xi}[e^{-\int_0^t \hat{\tau}_{B_s}\xi_s(\varphi)ds} f(B_t,\xi_t) - e^{-\int_0^t \hat{\tau}_{(B_s+y)}\xi_s(\varphi)ds} f_y(B_t,\xi_t)])^2] \\
&\quad = \mathbb{E}_{\mathcal{W}^x}[(f,f)_B + (f_y,f_y)_{B+y} - 2(f,f_y)_{B+1_{\{t\ge 0\}}y}].
\end{aligned}$$

Using (2.14.15), (2.14.16), and the continuity of $B \mapsto f(B)$, we see that each term in the last line above converges to $(f,f)_B$ as $r \to 0$. It follows that the integrand in the first line converges to zero pathwise, and the second claim follows by the dominated convergence theorem.

(3) Define $(Q_t f)(x,\xi) = \mathbb{E}_{\mathcal{W}^x \otimes \mathcal{G}^\xi}[f(B_t,\xi_t)]$. If f is bounded, a calculation similar to that in (2.14.15) shows $\|Q_t f - P_t f\|_{L^\infty(\mathbb{R}^d; L^2(\mathrm{G}))} \to 0$ as $t \to 0$. Therefore, we only need to show that $\|Q_t f - f\|_{L^2(\mathrm{G})}$ vanishes uniformly on compact sets as $t \to 0$. Write $f_t(x,\xi) = f(x,\xi_t)$.

Note that f_t is a function on $\mathbb{R}^d \times \mathfrak{Y}$ while f is a function on $\mathbb{R}^d \times \mathscr{M}_{-2}$. By the Schwarz inequality, we get $\|Q_t f(x) - f(x)\|^2_{L^2(G)} \le \mathbb{E}_{\mathcal{W}^x}[\|f_t(B_t) - f_0(B_0)\|^2_{L^2(\mathcal{G})}]$. Moreover,

$$\begin{aligned}\|f_t(B_t) - f_0(B_0)\|_{L^2(\mathcal{G})} &\le \|f_t(B_t) - f_t(B_0)\|_{L^2(\mathcal{G})} + \|f_t(B_0) - f_0(B_0)\|_{L^2(\mathcal{G})}\\ &= \|f(B_t) - f(B_0)\|_{L^2(G)} + \|f_t(B_0) - f_0(B_0)\|_{L^2(\mathcal{G})}.\end{aligned}$$

It thus suffices to show that

$$\lim_{t\to 0} \|f_t(x) - f_0(x)\|_{L^2(\mathcal{G})} = 0 \tag{2.14.17}$$

uniformly in x on compact sets and

$$\lim_{t\to 0} \mathbb{E}_{\mathcal{W}^x}[\|f(B_t) - f(B_0)\|^m_{L^2(G)}] = 0 \tag{2.14.18}$$

uniformly on compact sets for $m = 1, 2$. To prove (2.14.17), assume the contrary. Then there exist bounded sequences $(x_n)_{n\in\mathbb{N}} \subset \mathbb{R}^d$, $(t_n)_{n\in\mathbb{N}} \subset [0,\infty)$ with $t_n \to 0$, and $\|f_{t_n}(x_n) - f_0(x_n)\|_{L^2(\mathcal{G})} > \delta$ for all n. By compactness, we may assume that x_n converges to $x \in \mathbb{R}^d$. Then

$$\begin{aligned}\delta &< \|f_{t_n}(x_n) - f_0(x_n)\|_{L^2(\mathcal{G})}\\ &\le \|f_{t_n}(x_n) - f_{t_n}(x)\|_{L^2(\mathcal{G})} + \|f_{t_n}(x) - f_0(x)\|_{L^2(\mathcal{G})} + \|f_0(x) - f_0(x_n)\|_{L^2(\mathcal{G})}\\ &= \|f_{t_n}(x) - f_0(x)\|_{L^2(\mathcal{G})} + 2\|f(x_n) - f(x)\|_{L^2(G)}.\end{aligned}$$

We assume the continuity of f. Thus we can choose n_0 large enough such that $\|f(x_n) - f(x)\|_{L^2(G)} < \delta/3$, for all $n > n_0$, and the above calculation shows that $\|f_{t_n}(x) - f_0(x)\|_{L^2(\mathcal{G})} > \delta/3$ for all such n. Since $\|f_{t_n}(x) - f_0(x)\|_{L^2(\mathcal{G})} = \|e^{-t_n H_f} f(x) - f(x)\|_{L^2(\mathcal{G})}$ must converge to zero by the strong continuity of the semigroup $e^{-t\hat{H}_f}$, this is a contradiction and thus (2.14.17) holds. For (2.14.18), note that $x \mapsto f(x)$ is uniformly continuous on compact sets. Thus for $\varepsilon > 0$, we can choose $\delta > 0$ such that $\|f(x) - f(\tilde{x})\|_{L^2(G)} < \varepsilon/2$ whenever $|x - \tilde{x}| < \delta$. By the properties of Brownian motion, there exists $t_0 > 0$ such that for $0 \le t < t_0$,

$$\mathcal{W}^x(|B_t - x| > \delta) < \frac{\varepsilon}{2^{m+1}\|f\|^m_{L^\infty(\mathbb{R}^d;L^2(G))}}.$$

Then $\mathbb{E}_{\mathcal{W}^x}[\|f(x_t) - f(x_0)\|^m_{L^2(G)}] < \varepsilon$ for these t, uniformly in x. This proves (2.14.18). □

By Lemma 2.189, we can see an interesting property of P_t. It can be shown in the proposition below that P_t is also strongly continuous one-parameter semigroup on the Banach space $C_0(\mathbb{R}^d; L^2(G))$.

Proposition 2.190.

(1) *For each $t \ge 0$, P_t leaves $C_0(\mathbb{R}^d; L^2(G))$ invariant.*

(2) *$\{P_t : t \ge 0\}$ is a C_0-semigroup on $C_0(\mathbb{R}^d; L^2(G))$.*

Proof. We shall first show that $\{P_t : y \geq 0\}$ is a C_0-semigroup on $C_0 = C_0(\mathbb{R}^d; L^2(\mathrm{G}))$. We use (2) and (3) of Lemma 2.189 for an approximation and therefore must first show that $L^\infty(\mathbb{R}^d \times \mathscr{M}_{-2}) \cap C_0$ is dense in C_0. Let $f \in C_0$ and take $f_R = (f \wedge -R) \vee R$ for $R \geq 0$; $L^\infty(\mathscr{M}_{-2})$ is dense in $L^2(\mathrm{G})$, and thus for each $x \in \mathbb{R}^d$ and all $\varepsilon > 0$,

$$R_x(\varepsilon) = \inf\{R \geq 0 \mid \|f(x) - f_R(x)\|_{L^2(\mathrm{G})} \leq \varepsilon\} < \infty, \tag{2.14.19}$$

where $R_x(\varepsilon)$ is bounded on compact subsets of $\mathbb{R}^d$. To see this, assume the contrary. Then there exists $(x_n)_{n\in\mathbb{N}} \subset \mathbb{R}^d$ with $x_n \to x$, and $R_n = R_{x_n}(\varepsilon) > n$ for all n. Choosing n_0 large enough such that $\|f(x_n) - f(x)\|_{L^2(\mathrm{G})} < \varepsilon/3$ for all $n > n_0$ gives

$$\begin{aligned}
\varepsilon &= \|f(x_n) - f_{R_n}(x_n)\|_{L^2(\mathrm{G})} \\
&\leq \|f(x_n) - f(x)\|_{L^2(\mathrm{G})} + \|f(x) - f_{R_n}(x)\|_{L^2(\mathrm{G})} + \|f_{R_n}(x) - f_{R_n}(x_n)\|_{L^2(\mathrm{G})} \\
&\leq \|f(x) - f_{R_n}(x)\|_{L^2(\mathrm{G})} + \frac{2\varepsilon}{3} \leq \|f(x) - f_n(x)\|_{L^2(\mathrm{G})} + \frac{2\varepsilon}{3}
\end{aligned}$$

for each $n > n_0$. This implies $R_x(\varepsilon/3) = \infty$, in contradiction to (2.14.19). Thus $R_x(\varepsilon)$ is bounded on compact sets. Since $f \in C_0$, $R_x(\varepsilon) = 0$ for $|x|$ large enough, and thus $R_x(\varepsilon)$ is bounded on all $\mathbb{R}^d$. Thus bounded functions are dense in C_0. Next we show that P_t leaves C_0 invariant. Let $f \in C_0$. From (2.14.11) and (2.14.16), we have

$$\begin{aligned}
&\|P_t f(x)\|_{L^2(\mathrm{G})} \leq e^{\mathrm{I}} \mathbb{E}_{\mathcal{W}^x}[\|f(B_{-t})\|_{L^2(\mathrm{G})} \|f(B_t)\|_{L^2(\mathrm{G})}] \\
&\quad = e^{\mathrm{I}} \mathbb{E}_{\mathcal{W}^x}[1_{\{|B_t| \leq |x|/2\}} \|f(B_{-t})\|_{L^2(\mathrm{G})} \|f(B_t)\|_{L^2(\mathrm{G})}] + e^{\mathrm{I}} \mathbb{E}_{\mathcal{W}^x}[1_{\{|B_t| > |x|/2\}} \|f(B_{-t})\|_{L^2(\mathrm{G})} \|f(B_t)\|_{L^2(\mathrm{G})}];
\end{aligned}$$

$\mathcal{W}^0(|B_t| \geq R)$ decays exponentially in R for all t, and since $f \in C_0 \subset L^\infty(\mathbb{R}^d, L^2(\mathrm{G}))$, the term in the second line above also decays exponentially. The last line is bounded by a constant times $\sup_{|y| > |x|/2} \|f(x)\|_{L^2(\mathrm{G})}$, and this term decays to zero as $|x| \to \infty$, by the assumption $f \in C_0$. In a similar way, we obtain the strong continuity from (3) of Lemma 2.189. Thus P_t is a strongly continuous semigroup of bounded operators on $C_0 = C_0(\mathbb{R}; L^2(\mathrm{G}))$. □

Next we investigate $\{P_t : t \geq 0\}$ on the Hilbert space $\mathcal{T}$.

Theorem 2.191.
(1) *For each $t \geq 0$, P_t leaves $\mathcal{T}$ invariant.*
(2) *$\{P_t : t \geq 0\}$ is a C_0-semigroup on $\mathcal{T}$.*

Proof. The semigroup property of $\{P_t : t \geq 0\}$ follows from Markov property of $(B_t)_{t\geq 0}$ and $(\xi_t)_{t\in\mathbb{R}}$. We shall show that P_t is strongly continuous. Choose $Uf \in \mathcal{T}$. We find

$$(P_t Uf)(x, \xi) = \mathbb{E}_{\mathcal{W}^x \otimes \mathcal{G}^\xi}[e^{-\int_0^t \hat{\tau}_{B_s} \xi_s(\varphi) ds} \hat{\tau}_{B_t} f(\xi_t)] = \mathbb{E}_{\mathcal{W}^0 \otimes \mathcal{G}^\xi}[e^{-\int_0^t \hat{\tau}_{B_s + x} \xi_s(\varphi) ds} \hat{\tau}_{B_t + x} f(\xi_t)].$$

The semigroup property $\hat{\tau}_{B_s + x} = \hat{\tau}_{B_s} \hat{\tau}_x$ yields that

$$(P_tUf)(x,\xi) = \mathbb{E}_{\mathcal{W}^0\otimes\mathcal{G}^{\hat{\tau}_x\xi}}[e^{-\int_0^t \hat{\tau}_{B_s}\xi_s(\varphi)ds}\hat{\tau}_{B_t}f(\xi_t)] = U\Big(\mathbb{E}_{\mathcal{W}^0\otimes\mathcal{G}^{\xi}}[e^{-\int_0^t \hat{\tau}_{B_s}\xi_s(\varphi)ds}\hat{\tau}_{B_t}f(\xi_t)]\Big). \tag{2.14.20}$$

We conclude that P_t leaves $\mathcal{T}$ invariant. It is seen that $L^\infty(\mathbb{R}^d; L^2(\mathrm{G})) \cap \mathcal{T}$ is dense in $\mathcal{T}$. Strong continuity then follows directly from (3) of Lemma 2.189. □

We shall derive the generator of $P_t : \mathcal{T} \to \mathcal{T}$. The generator of P_t on $\mathcal{T}$ is denoted by $-H_\mathcal{T}$. Hence $P_t = e^{-tH_\mathcal{T}}$ for $t \geq 0$.

Corollary 2.192. *Let $H(0)$ be a single particle translation invariant Nelson Hamiltonian* (2.12.2) *on $L^2(\mathrm{G})$ with total momentum $p = 0$; $H(0) = \frac{1}{2}d\Gamma(-i\nabla)^2 + \hat{H}_\mathrm{f} + H_\mathrm{I}$. Then*

$$H_\mathcal{T} = UH(0)U^{-1}.$$

i. e., $\mathcal{T} \cong L^2(\mathrm{G})$ and $H_\mathcal{T} \cong H(0)$. In particular, $D(H_\mathcal{T}) = UD(H(0))$ and

$$H_\mathcal{T}F = (-\frac{1}{2}\Delta_x + \hat{H}_\mathrm{f} + H_\mathrm{I})F$$

for $F \in U\mathcal{L}$.

Proof. Let $f, g \in \mathcal{L}$. Since by (2.14.20) we have

$$U^{-1}P_tUf(x,\xi) = \mathbb{E}_{\mathcal{W}^0\otimes\mathcal{G}^{\xi}}[e^{-\int_0^t \hat{\tau}_{B_s}\xi_s(\varphi)ds}(Uf)(B_t,\xi_t)], \tag{2.14.21}$$

we can see that

$$\begin{aligned}(Uf, P_tUg)_\mathcal{T} &= (f, U^{-1}P_tUg)_{L^2(\mathrm{G})} = (f, \mathbb{E}_{\mathcal{W}^0\otimes\mathcal{G}^{\xi}}[e^{-\int_0^t \hat{\tau}_{B_s}\xi_s(\varphi)ds}(Ug)(B_t,\xi_t)])_{L^2(\mathrm{G})}\\ &= \mathbb{E}_{\mathcal{W}^0\otimes\mathcal{G}}[(Uf)(B_0,\xi_0)e^{-\int_0^t \hat{\tau}_{B_s}\xi_s(\varphi)ds}(Ug)(B_t,\xi_t)].\end{aligned}$$

In the same way as in Remark 2.13, we can see that

$$\lim_{t\to 0}\frac{1}{t}(Uf, P_tUg - Ug)_\mathcal{T} = (f, -U^{-1}(-\frac{1}{2}\Delta_x + \hat{H}_\mathrm{f} + H_\mathrm{I})Ug)_{L^2(\mathrm{G})}.$$

Then

$$-U^{-1}(-\frac{1}{2}\Delta_x + \hat{H}_\mathrm{f} + H_\mathrm{I})U = -U^{-1}H_\mathcal{T}U$$

on $\mathcal{L}$. Taking $f \in C_0^\infty(\mathbb{R}^d)$, we have

$$Ue^{i\xi(f)}(x,\xi) = \hat{\tau}_x e^{i\xi(f)} = e^{i\xi(f(\cdot-x))} = g(x,\xi).$$

We set $f_j = \partial_{x_j} f$ and $f_{jj} = \partial_{x_j}^2 f$. Since $\partial_{x_j} g(x,\xi) = -i\xi(f_j(\cdot - x))g(x,\xi)$, we find that

$$-\frac{1}{2}\Delta_x g(x,\xi) = -\frac{1}{2}\sum_{j=1}^{d}(i\xi(f_{jj}(\cdot - x)) - \xi(f_j(\cdot - x))^2)g(x,\xi)$$

and

$$U^{-1}(-\frac{1}{2}\Delta_x)Ue^{i\xi(f)} = -\frac{1}{2}\sum_{j=1}^{d}(i\xi(f_{jj}) - \xi(f_j)^2)e^{i\xi(f)}.$$

On the other hand, since $d\Gamma(-i\nabla_j)e^{i\xi(f)} = \xi(f_j)e^{i\xi(f)}$, we have

$$(d\Gamma(-i\nabla_j))^2 g(x,\xi) = (-i\xi(f_{jj}) + \xi(f_j)^2)e^{i\xi(f)}.$$

Hence

$$U^{-1}(-\frac{1}{2}\Delta_x)Ue^{i\xi(f)} = \frac{1}{2}(d\Gamma(-i\nabla))^2 e^{i\xi(f)},$$

and we conclude that

$$U^{-1}\Big(-\frac{1}{2}\Delta_x + \hat{H}_{\mathrm{f}} + H_{\mathrm{I}}\Big)Ue^{i\xi(f)} = \Big(\frac{1}{2}d\Gamma(-i\nabla)^2 + \hat{H}_{\mathrm{f}} + H_{\mathrm{I}}\Big)e^{i\xi(f)} = H(0)e^{i\xi(f)}.$$

Since $\mathscr{L} \ni f$ is given by

$$f(\xi(f_1),\dots,\xi(f_n)) = (2\pi)^{-n/2}\int_{\mathbb{R}^n}\hat{f}(k)e^{i\xi(\sum_{j=1}^n k_j f_j)}dk,$$

we have

$$U^{-1}\Big(-\frac{1}{2}\Delta_x + \hat{H}_{\mathrm{f}} + H_{\mathrm{I}}\Big)Uf = H(0)f = U^{-1}H_{\mathcal{T}}Uf.$$

Since $\mathscr{L}$ is a core for $H(0)$, $H_{\mathcal{T}} = UH(0)U^{-1}$ follows. □

By Corollary 2.192, we can now get some spectral information on $H_{\mathcal{T}}$.

Theorem 2.193 (Ground state of $H_{\mathcal{T}}$). *Operator $H_{\mathcal{T}}$ has the ground state $\Psi_{\mathrm{g}} \in \mathcal{T}$ and Ψ_{g} can be chosen strictly positive.*

Proof. Since $H_{\mathcal{T}} = UH(0)U^{-1}$ and $H(0)$ has the unique and strictly positive ground state $\Psi_{\mathrm{g}}(0)$ by Theorem 2.151, $H_{\mathcal{T}}$ also has the unique and strictly positive ground state Ψ_{g}, i. e., $\Psi_{\mathrm{g}} = U\Psi_{\mathrm{g}}(0)$. □

Next we consider the limit of the families $(\mathcal{P}_T^W)_{T>0}$ on $\mathfrak{X} \times \mathfrak{Y}$ and $(\mathcal{P}_T^0)_{T>0}$ on $\mathfrak{X}$. For a bounded interval $I \subset \mathbb{R}$, let $\mathscr{F}_I = \sigma((B_t,\xi_t), t \in I)$. Let

$$\hat{\mathscr{G}} = \bigcup_{s>0}\mathscr{F}_{[-s,s]}. \tag{2.14.22}$$

We write

$$E = \inf \mathrm{Spec}(H_T).$$

We define $\mathcal{P}_\infty^W$ to be the unique probability measure on $(\mathfrak{X} \times \mathfrak{Y}, \mathcal{B}(\mathfrak{X} \times \mathfrak{Y}))$ such that for every $T > 0$ and $A \in \mathscr{F}_{[-T,T]}$,

$$\mathcal{P}_\infty^W(A) = e^{2TE} \mathbb{E}_{\mathcal{P}_0^W} [\Psi_g(B_{-T}, \xi_{-T}) e^{-\int_{-T}^T \hat{\tau}_{B_s} \xi_s(\varphi) ds} \Psi_g(B_T, \xi_T) 1_A]. \tag{2.14.23}$$

We are now ready to state and prove the main result of this section.

Theorem 2.194 (Local convergence).
(1) *The family of probability measures $\{\mathcal{P}_T^W\}_{T\geq 0}$ converges to $\mathcal{P}_\infty^W$ as $T \to \infty$ on the probability space $(\mathfrak{X} \times \mathfrak{Y}, \mathcal{B}(\mathfrak{X} \times \mathfrak{Y}))$ in the local sense, i. e., $\mathcal{P}_T^W(A) \to \mathcal{P}_\infty^W(A)$ for all $A \in \hat{\mathscr{G}}$.*
(2) *The family of probability measures $\{\mathcal{P}_T^0\}_{T>0}$ converges to $\mathcal{P}_\infty^0$ as $T \to \infty$ on the probability space $(\mathfrak{X}, \mathcal{B}(\mathfrak{X}))$ in the local sense, i. e., $\mathcal{P}_T^0(A) \to \mathcal{P}_\infty^0(A)$ for all $A \in \bigcup_{t\geq 0} \mathscr{F}_{[-t,t]}^{\mathrm{BM}}$.*

Proof. The proof of (1) is similar to that of Theorem 2.155, and (2) to Theorem 2.152. In the proofs of Theorems 2.152 and 2.155, replacing ϱ_T and ν_T by $\mathcal{P}_T^0$ and $\mathcal{P}_T^W$, respectively, we can prove this theorem. □

2.14.2 Diffusive behavior

By Theorem 2.194, we know that limit measures $\mathcal{P}_\infty^0$ and $\mathcal{P}_\infty^W$ exist. In this section we discuss the peculiar property of these measures that on a large scale under this measure paths behave like Brownian motion, however, with a reduced diffusion constant.

The classical *central limit theorem* yields that the partial sum

$$S_n = \frac{1}{\sqrt{n}} \sum_{i=1}^n X_i$$

of independently and identically distributed random variables X_i with mean zero and covariance $\mathbb{E}[X_i^2] = \sigma^2$ weakly converges to the Gaussian random variable with mean zero and covariance σ^2. We see that for a continuous bounded function g,

$$\lim_{n\to\infty} \mathbb{E}[g(S_n)] \to \mathbb{E}[g(X)],$$

where X is a Gaussian random variable with covariance σ. We are interested in the asymptotic behavior of

$$\frac{1}{\sqrt{n}} \gamma \cdot B_{tn}$$

as $t \to \infty$ for $\gamma \in \mathbb{R}^d$ under the probability measure P_∞^W.

We define two random processes $(X_t)_{t\in\mathbb{R}}$ and $(Y_t)_{t\in\mathbb{R}}$ in this section. Let $\mathscr{F}_\infty$ be defined by (2.14.22). First, $(X_t)_{t\in\mathbb{R}}$ is an $\mathbb{R}^d \times \mathscr{M}_{-2}$-valued random process on the probability space $(\mathfrak{X} \times \mathfrak{Y}, \mathscr{F}_\infty, P_\infty^W)$ defined by

$$X_t = (B_t, \xi_t) \quad t \in \mathbb{R}. \tag{2.14.24}$$

Second, $(Y_t)_{t\in\mathbb{R}}$ is an $\mathscr{M}_{-2}$-valued random process on $(\mathfrak{X} \times \mathfrak{Y}, \mathscr{F}_\infty, P_\infty^W)$ defined by

$$Y_t = \hat{\tau}_{B_t}\xi_t, \quad t \in \mathbb{R}. \tag{2.14.25}$$

From (2.14.13), it follows that

$$f(Y_t) = (Uf)(X_t), \quad (U^{-1}g)(Y_t) = g(X_t) \tag{2.14.26}$$

for $f \in L^2(\mathrm{G})$ and $g \in \mathcal{T}$. We often use this formulae in what follows. Using $(Y_t)_{t\in\mathbb{R}}$, we can rewrite $(Uf, e^{-tH_\mathcal{T}}Ug)_\mathcal{T}$ for $f, g \in L^2(\mathrm{G})$ in (2.14.14) by

$$(Uf, e^{-tH_\mathcal{T}}Ug)_\mathcal{T} = \mathbb{E}_{\mathcal{W}^0\otimes\mathcal{G}}[f(Y_0)e^{-\int_0^t Y_s(\varphi)ds}g(Y_t)], \tag{2.14.27}$$

where $Y_s(\varphi) = (\hat{\tau}_{B_t}\xi_t)(\varphi)$. We define

$$L = \frac{1}{\Psi_\mathrm{g}}(H_\mathcal{T} - E)\Psi_\mathrm{g}.$$

The domain of L is

$$D(L) = \{f \in \mathcal{T} \mid U^{-1}f\Psi_\mathrm{g} \in D(\hat{H}_\mathrm{f}) \cap D(d\Gamma(-i\nabla))\}.$$

We set $L^2(\mathscr{M}_{-2}, (U^{-1}\Psi_\mathrm{g})^2\mathrm{G}) = L^2((U^{-1}\Psi_\mathrm{g})^2\mathrm{G})$ and $L^2(\mathscr{M}_{-2}, \Psi_\mathrm{g}^2\mathrm{G}) = L^2(\Psi_\mathrm{g}^2\mathrm{G})$.

Lemma 2.195. *Take $f, g \in L^2(\mathrm{G})$. The following claims* (1) *and* (2) *below hold:*

(1)

$$\mathbb{E}_{\mathcal{P}_\infty^W}[f(Y_s)g(Y_t)] = (\bar{f}, e^{-|s-t|U^{-1}LU}g)_{L^2((U^{-1}\Psi_\mathrm{g})^2\mathrm{G})}.$$

In particular,

$$\mathbb{E}_{\mathcal{P}_\infty^W}[f(Y_t)] = \int_{\mathscr{M}_{-2}} f(U^{-1}\Psi_\mathrm{g})^2 d\mathrm{G}, \tag{2.14.28}$$

$$\mathbb{E}_{\mathcal{P}_\infty^W}[f(Y_t)g(Y_t)] = \int_{\mathscr{M}_{-2}} fg(U^{-1}\Psi_\mathrm{g})^2 d\mathrm{G}. \tag{2.14.29}$$

(2)

$$\mathbb{E}_{\mathcal{P}_\infty^W}[(Uf)(X_s)(Ug)(X_t)] = (\bar{Uf}, e^{-|s-t|L}Ug)_{L^2(\Psi_\mathrm{g}^2\mathrm{G})}.$$

In particular,

$$\mathbb{E}_{\mathcal{P}_\infty^W}[(Uf)(X_t)] = \int_{\mathcal{M}_{-2}} (Uf)\Psi_g^2 dG, \tag{2.14.30}$$

$$\mathbb{E}_{\mathcal{P}_\infty^W}[(Uf)(X_t)(Ug)(X_t)] = \int_{\mathcal{M}_{-2}} (Uf)(Ug)\Psi_g^2 dG. \tag{2.14.31}$$

Proof. Since we have

$$\mathbb{E}_{\mathcal{P}_\infty^W}[f(Y_s)g(Y_t)] = e^{|s-t|E}(\Psi_g \overline{Uf}, e^{-|s-t|H_{\mathcal{T}}}(\Psi_g Ug))_{\mathcal{T}}$$

in the same way as (2.12.21), claim (1) follows. Since $(Uf)(X_t) = f(Y_t)$ and

$$\begin{aligned}(Uf, e^{-|s-t|L}Ug)_{L^2(\Psi_g^2 G)} &= \int_{\mathcal{M}_{-2}} \overline{(Uf)(x,\xi)}(e^{-|s-t|L}Ug)(x,\xi)\Psi_g^2(x,\xi)dG(\xi)\\ &= \int_{\mathcal{M}_{-2}} \overline{f(\hat{\tau}_x\xi)}(U^{-1}e^{-|s-t|L}Ug)(\hat{\tau}_x\xi)(U^{-1}\Psi_g(\hat{\tau}_x\xi))^2 dG(\xi)\\ &= \int_{\mathcal{M}_{-2}} \overline{f(\hat{\tau}_0\xi)}(U^{-1}e^{-|s-t|L}Ug)(\hat{\tau}_0\xi)(U^{-1}\Psi_g(\hat{\tau}_0\xi))^2 dG(\xi)\\ &= \int_{\mathcal{M}_{-2}} \overline{f(\xi)}(U^{-1}e^{-|s-t|L}Ug)(\xi)(U^{-1}\Psi_g(\xi))^2 dG(\xi)\\ &= (f, e^{-|s-t|U^{-1}LU}g)_{L^2((U^{-1}\Psi_g)^2 G)},\end{aligned}$$

claim (2) follows. □

Lemma 2.196 (Markov properties). (1) $Y = (Y_t)_{t\geq 0}$ *is a stationary Markov process with respect to the natural filtration. The generator of* Y *is* $-U^{-1}LU$ *and the invariant measure is* $(U^{-1}\Psi_g)G$. *(2)* $X = (X_t)_{t\geq 0}$ *is a stationary Markov process with respect to the natural filtration. The generator of* X *is* $-L$.

Proof. By Lemma 2.195, it holds that

$$\mathbb{E}_{\mathcal{P}_\infty^W}[f(Y_s)g(Y_t)] = (\bar{f}, e^{-|s-t|U^{-1}LU}g)_{L^2((U^{-1}\Psi_g)^2 G)},$$

and we can also show that

$$\mathbb{E}_{\mathcal{P}_\infty^W}\left[\prod_{j=0}^n f_j(Y_{t_j})\right] = (\bar{f}_0, e^{-|t_1-t_0|K}f_1 \cdots f_{n-1}e^{-|t_n-t_{n-1}|K}f_n)_{L^2((U^{-1}\Psi_g)^2 G)},$$

where we set $K = U^{-1}LU$. Hence it can be proven that Y is a stationary Markov process with respect to the natural filtration in the same way as in Theorem 2.156. Similar to

(1), we see that $\mathbb{E}_{\mathcal{P}_\infty^W}[(Uf)(X_s)(Ug)(X_t)] = (Uf, e^{-|s-t|L}Ug)_{L^2(\Psi_g^2 G)}$ by Lemma 2.195. Then (2) follows in the same way as Theorem 2.156. □

Definition 2.197 (Differential operator in $\mathcal{T}$). We define the differential operator in $\mathcal{T}$ by

$$-i\hat{\nabla}_x = Ud\Gamma(-i\nabla)U^{-1}.$$

Since $H_{\mathcal{T}} = UH(0)U^{-1}$, we see that $\Psi_g \in D(\hat{\nabla}_{x_j})$. Define

$$\tilde{L} = \frac{1}{\Psi_g}\Big(-\frac{1}{2}\Delta_x + \hat{H}_f + H_I\Big)\Psi_g$$

on $C^2(\mathbb{R}^d; L^2(G))$ by

$$\tilde{L}f(x,\xi) = \frac{1}{\Psi_g}\Big(-\frac{1}{2}\Delta_x\Psi_g(x,\xi)f(x,\xi) + \hat{H}_f\Psi_g(x,\xi)f(x,\xi) + H_I(x,\xi)\Psi_g(x,\xi)f(x,\xi)\Big)$$

It can be seen that $\tilde{L}\lceil_{\mathcal{T}} = L$. We define the map $h_\gamma : \mathbb{R}^d \times \mathscr{M}_{-2} \to \mathbb{R}$ by

$$h_\gamma(x,\xi) = \gamma\cdot x, \quad \gamma \in \mathbb{R}^d. \tag{2.14.32}$$

Note that h_γ is independent of ξ. We have

$$\tilde{L}h_\gamma = \gamma\cdot\frac{\hat{\nabla}_x\Psi_g}{\Psi_g}.$$

With

$$\rho = U^{-1}(\gamma\cdot\frac{\hat{\nabla}_x\Psi_g}{\Psi_g}) \in L^2(G), \tag{2.14.33}$$

we have

$$\tilde{L}h_\gamma(X_t) = (U\rho)(X_t) = \rho(Y_t).$$

We are interested in proving a central limit theorem for the process $(\gamma\cdot B_t)_{t\geq 0}$ under $\mathcal{P}_\infty^W$ or, equivalently, for the process $(\gamma\cdot B_t)_{t\geq 0}$ under $\mathcal{P}_\infty^0$. The fundamental tool will be the martingale central limit theorem. Let

$$M_t = h_\gamma(X_t) - h_\gamma(X_0) - \int_0^t \tilde{L}h_\gamma(X_s)ds. \tag{2.14.34}$$

Since $h_\gamma(X_t) = \gamma\cdot B_t$, $h_\gamma(X_0) = 0$ and $\tilde{L}h_\gamma(X_s) = \rho(Y_s)$, (2.14.34) can be also written as

$$\gamma\cdot B_t = M_t + \int_0^t \rho(Y_s)ds. \tag{2.14.35}$$

Lemma 2.198. *The process $(M_t)_{t\geq 0}$ is a martingale with respect to the natural filtration of $(X_t)_{t\geq 0}$.*

Proof. The operator $\tilde{L}$ is the generator of the Markov process $(X_t)_{t\geq 0}$. The lemma follows in a similar way as Theorem 2.93 in Volume 1. Let $\mathcal{B}_t = \sigma(X_r, 0 \leq r \leq t)$. We define $\tilde{P}_t$ by

$$(\tilde{P}_t F)(x, \xi) = \mathbb{E}_{P^W_\infty}[f(X_t)|X_0 = (x, \xi)]. \tag{2.14.36}$$

Let $f = h_\gamma$ for simplicity. By the Markov property of $(X_t)_{t\geq 0}$, we have $\mathbb{E}_{P^W_\infty}[f(X_t)|\mathcal{B}_s] = \tilde{P}_{t-s}f(X_s)$. It is trivial to see that

$$\mathbb{E}_{P^W_\infty}[M_t|\mathcal{B}_s] = M_s + \mathbb{E}_{P^W_\infty}\left[f(X_t) - f(X_s) - \int_s^t \tilde{L}f(X_r)dr \,\middle|\, \mathcal{B}_s \right]. \tag{2.14.37}$$

Since $f \in D(\tilde{L})$, we have $\frac{d}{dt}\tilde{P}_t f = \tilde{L}\tilde{P}_t f = \tilde{P}_t \tilde{L} f$ and $\tilde{P}_t f - f = \int_0^t \tilde{L}\tilde{P}_s f ds$. It follows that

$$\begin{aligned}\mathbb{E}_{P^W_\infty}\left[f(X_t) - f(X_s) - \int_s^t \tilde{L}f(X_r)dr \,\middle|\, \mathcal{B}_s \right] &= \tilde{P}_{t-s}f(X_s) - f(X_s) - \int_s^t \tilde{L}\tilde{P}_{r-s}f(X_s)dr \\ &= \tilde{P}_{t-s}f(X_s) - f(X_s) - \int_0^{t-s} \tilde{L}\tilde{P}_r f(X_s)dr = 0.\end{aligned}$$

Hence the second term on the right-hand side of (2.14.37) is zero and the martingale property is proven. □

Lemma 2.199. *We have*

$$\mathbb{E}_{\mathcal{P}^W_\infty}[\rho(Y_t)^2] = \|\gamma \cdot \hat{\nabla}_x \Psi_g\|_{\mathcal{T}}, \tag{2.14.38}$$

$$\mathbb{E}_{\mathcal{P}^W_\infty}[M_t^2] = |\gamma|^2 t. \tag{2.14.39}$$

Proof. By (2.14.33) and (2.14.28), we can show that

$$\mathbb{E}_{\mathcal{P}^W_\infty}[\rho(Y_t)^2] = \int_{\mathcal{M}_{-2}} (U^{-1}\gamma \cdot \frac{\hat{\nabla}_x \Psi_g}{\Psi_g})^2 (U^{-1}\Psi_g)^2 dG = \|\gamma \cdot \hat{\nabla}_x \Psi_g\|_{\mathcal{T}}.$$

We have

$$(Lh_\gamma^2)(x, \xi) = |\gamma|^2 + 2h_\gamma(x, \xi)(\gamma \cdot \frac{\hat{\nabla}_x \Psi_g(x, \xi)}{\Psi_g(x, \xi)}) = |\gamma|^2 + 2h_\gamma(x, \xi)(Lh_\gamma)(x, \xi). \tag{2.14.40}$$

For simplicity, let $h_\gamma = f$, and $\mathbb{E}[\ldots] = \mathbb{E}_{\mathcal{P}^W_\infty}[\ldots]$. Write $g(X_s) = (\tilde{L}f)(X_s)$. We have

$$\mathbb{E}[M_t^2] = \mathbb{E}[f^2(X_t)] + \mathbb{E}[f^2(X_0)] + 2\int_0^t ds \int_s^t \mathbb{E}[g(X_s)g(X_r)]dr$$
$$- 2\mathbb{E}[f(X_0)f(X_t)] + 2\int_0^t \mathbb{E}[f(X_0)g(X_r)]dr - 2\int_0^t \mathbb{E}[f(X_t)g(X_r)]dr. \tag{2.14.41}$$

Since

$$\mathbb{E}[f(X_0)\int_0^t g(X_r)dr] = \mathbb{E}[f(X_0)(f(X_t) - f(X_0))],$$

the second, fourth, and fifth terms of (2.14.41) combined give $-\mathbb{E}[f^2(X_0)]$. Also, since

$$\int_0^t \mathbb{E}[g(X_s)g(X_r)]dr = \mathbb{E}[g(X_s)(f(X_t) - f(X_s))],$$

the third term of (2.14.41) decomposes into two expressions, one of which cancels the last term of (2.14.41), while the other gives the last term in the formula. Then it follows that

$$\mathbb{E}[M_t^2] = \mathbb{E}[f^2(X_t)] - \mathbb{E}[f^2(X_0)] - 2\int_0^t \mathbb{E}[f(X_s)(Lf)(X_s)]ds \tag{2.14.42}$$

and, by (2.14.40), we have

$$\mathbb{E}[M_t^2] = \int_0^t \mathbb{E}[(Lh_\gamma^2)(X_s) - 2h_\gamma(X_s)(Lh_\gamma)(X_s)]ds = |\gamma|^2 t. \qquad \square$$

Consider the decomposition (2.14.35). We have $\mathbb{E}_{\mathcal{P}_\infty^W}[(\gamma \cdot B_t)^2] < \infty$ for all t by Proposition 2.194.

Lemma 2.200. *We have*

$$\lim_{t\to\infty} \frac{1}{t}\mathbb{E}_{\mathcal{P}_\infty^W}[(\gamma \cdot B_t)^2] = |\gamma|^2 - 2(\gamma \cdot \hat{\nabla}_X \Psi_g, (H_\mathcal{T} - E)^{-1}\gamma \cdot \hat{\nabla}_X \Psi_g)_\mathcal{T}. \tag{2.14.43}$$

Proof. By (2.14.35),

$$\mathbb{E}_{\mathcal{P}_\infty^W}[(\gamma \cdot B_t)^2] = \mathbb{E}_{\mathcal{P}_\infty^W}[M_t^2] - \mathbb{E}_{\mathcal{P}_\infty^W}\left[\left(\int_0^t \rho(Y_s)ds\right)^2\right] + 2\mathbb{E}_{\mathcal{P}_\infty^W}\left[\gamma \cdot B_t \int_0^t \rho(Y_s)ds\right]. \tag{2.14.44}$$

The third term in (2.14.44) is zero. This can be seen as follows. We have

$$\mathbb{E}_{\mathcal{P}_\infty^W}\left[\gamma \cdot B_t \int_0^t \rho(Y_s)ds\right] = \mathbb{E}_{\mathcal{W}^0}[\gamma \cdot B_t I(B)], \tag{2.14.45}$$

where

$$I(B) = \mathbb{E}_{\mathcal{G}}\left[\Psi_g(Y_0)e^{-\int_0^t Y_s(\varphi)ds}\left(\int_0^t \rho(Y_s)ds\right)\Psi_g(Y_t)\right]. \tag{2.14.46}$$

Put $\tilde{B}_s = B_{t-s} - B_t$. Then by the reversibility of $\mathcal{G}$ and the fact that $\mathcal{G}$ is invariant under the constant shift by $\hat{\tau}_{B_t}$, we have $I(\tilde{X}) = I(X)$. Moreover, $\tilde{X}_t = -B_t$, $\mathcal{W}^0$-a. s. and $\mathcal{W}^0$ is invariant under the transformation $B \mapsto \tilde{B}$. Thus

$$\mathbb{E}_{\mathcal{W}^0}[\gamma \cdot B_t I(X)] = -\mathbb{E}_{\mathcal{W}^0}[\gamma \cdot B_t I(X)] = 0. \tag{2.14.47}$$

From (2.14.39), we know that $\mathbb{E}_{\mathcal{P}^W_\infty}[M_t^2] = |\gamma|^2 t$. We finally estimate $\mathbb{E}_{\mathcal{P}^W_\infty}\left[\left(\int_0^t \rho(Y_s)ds\right)^2\right]$. In general, we can see that $\lim_{t\to\infty}(f, e^{-tK}g)/t = 2(f, K^{-1}g)$. Hence we have

$$\begin{aligned}
\frac{1}{t}\mathbb{E}_{\mathcal{P}^W_\infty}\left[\left(\int_0^t \rho(Y_s)ds\right)^2\right] &= \frac{1}{t}\int_0^t ds \int_0^t \mathbb{E}_{\mathcal{P}^W_\infty}[\rho(Y_s)\rho(Y_r)]dr \\
&= \frac{1}{t}\int_0^t ds \int_0^t (\rho, e^{-|s-r|U^{-1}LU}\rho)_{L^2((U^{-1}\Psi_g)^2\mathrm{G})}dr \\
&\overset{t\to\infty}{\to} 2(\rho, U^{-1}L^{-1}U\rho)_{L^2((U^{-1}\Psi_g)^2\mathrm{G})} = 2(\gamma \cdot \hat{\nabla}_x \Psi_g, (H_{\mathcal{T}} - E)^{-1}\gamma \cdot \hat{\nabla}_x \Psi_g)_{\mathcal{T}}.
\end{aligned}$$

Note that the last quantity is automatically finite; this follows from (2.14.44). So we have shown (2.14.43). □

We will need the following abstract result, where a random process of the form $\int_0^t F(Z_s)ds$ can be asymptotically approximated by some martingale.

Proposition 2.201 (Functional central limit theorem). *Let $(S, \mathcal{S}, \pi)$ be a probability space. Suppose that $Z = (Z_t)_{t\in\mathbb{R}}$ is an S-valued Markov process on a filtered probability space $(\Omega, \mathcal{F}, (\mathcal{F}_t)_{t\geq 0}, P)$, and K is the generator of Z. Assume that*

(1) *$(Z_t)_{t\in\mathbb{R}}$ is reversible and stationary, and e^{-tK} is positivity improving for some $t > 0$;*
(2) *π is an invariant measure for Z, i. e., $\mathbb{E}_P[f(Z_t)] = \int_S \mathbb{E}_{P^s}[f(Z_t)]d\pi(s) = \int_S f(s)d\pi(s)$ and $F \in L^2(S)$ and $\mathbb{E}_\pi[F] = 0$;*
(3) *F is in the domain of $K^{-1/2}$, where K is the generator of the process $(Y_t)_{t\in\mathbb{R}}$.*

Let

$$F_t = \int_0^t F(Z_s)ds. \tag{2.14.48}$$

Then there exists a square integrable martingale $(N_t)_{t\geq 0}$ with respect to $(\mathscr{F}_t)_{t\geq 0}$, with stationary increments, such that

$$\lim_{t\to\infty}\frac{1}{\sqrt{t}}\sup_{0\leq s\leq t}|F_s - N_s| = 0 \quad \textit{in probability with respect to } P,$$

$$\lim_{t\to\infty}\frac{1}{t}\mathbb{E}_P[|F_t - N_t|^2] = 0.$$

Applying Proposition 2.201, we can approximate $\int_0^t \rho(Y_s)ds$ by a martingale.

Lemma 2.202. *There exists a martingale $(N_t)_{t\geq 0}$ with respect to the natural filtration of $(Y_t)_{t\geq 0}$ such that*

$$\lim_{t\to\infty}\frac{1}{\sqrt{t}}\sup_{0\leq s\leq t}\left|\int_0^s \rho(Y_r)dr - N_s\right| = 0, \quad \lim_{t\to\infty}\frac{1}{t}\mathbb{E}_{P_\infty^W}\left[\left|\int_0^t \rho(Y_s)ds - N_t\right|^2\right] = 0. \tag{2.14.49}$$

Proof. We apply Proposition 2.201. By Lemma 2.195, $Y = (Y_t)_{t\geq 0}$ is a stationary and reversible Markov process with respect to the natural filtration and the semigroup associated with Y is $e^{-tU^{-1}LU}$, and it is positivity improving. Hence assumption (1) is satisfied. By Lemma 2.195 again, $(U^{-1}\Psi_g)^2\mathrm{G}$ is the invariant measure for Y and

$$\int_{\mathscr{M}_{-2}} \rho(U^{-1}\Psi_g)^2 d\mathrm{G} = (\Psi_g, (H_{\mathcal{T}} - E)\Psi_g)_{\mathcal{T}} = 0.$$

Then assumption (2) is satisfied. Finally, $\rho \in D(\tilde{L}^{-1/2})$ and assumption (3) is satisfied. Then the proof is completed by Proposition 2.201 under the correspondences: Z and Y, $(\Omega, \mathscr{F}, (\mathscr{F}_t)_{t\geq 0}, P)$ and $(\mathfrak{X}\times\mathfrak{Y}, \mathscr{F}_\infty, (\mathcal{B}_t)_{t\geq 0}, P_\infty^W)$ with $\mathcal{B}_t = \sigma(Y_r, 0\leq r\leq t)$, $(S, \mathcal{S}, \pi)$ and $(\mathscr{M}_{-2}, \mathcal{B}(\mathscr{M}_{-2}), (U^{-1}\Psi_g)^2\mathrm{G})$, F and ρ, and finally, K and $\tilde{L}$. □

Furthermore, we need a version of the martingale central limit theorem.

Proposition 2.203 (Central limit theorem for martingale). *Let $(N_n)_{n\in\mathbb{N}}$ be a martingale on a filtered probability space $(\Omega, \mathscr{F}, (\mathscr{F}_n)_{n\in\mathbb{N}}, P)$ with $\sigma^2 = \lim_{n\to\infty}\frac{1}{n}\mathbb{E}[N_n^2]$, and assume that $(N_n)_{n\in\mathbb{N}}$ has stationary increments. Then*

$$\lim_{n\to\infty}\frac{1}{\sqrt{n}}N_{[tn]} = \sigma^2 B_t, \tag{2.14.50}$$

in the sense of weak convergence, i. e., for any continuous bounded function g on $\mathbb{R}$,

$$\lim_{n\to\infty}\mathbb{E}_P\left[g\left(\frac{1}{\sqrt{n}}N_{[tn]}\right)\right] = \mathbb{E}_{\mathcal{W}^0}[g(\sigma^2 B_t)].$$

Theorem 2.204 (Central limit theorem). *It follows that*

$$\lim_{n\to\infty}\frac{1}{\sqrt{n}}\gamma\cdot B_{tn} = (\gamma, D\gamma)B_t^1$$

in the sense of weak convergence, i. e., for any continuous bounded function g on $\mathbb{R}$,

$$\lim_{n\to\infty} \mathbb{E}_{P^W_\infty}\Big[g\Big(\frac{1}{\sqrt{n}}\gamma\cdot B_{tn}\Big)\Big] = \mathbb{E}_{\mathcal{W}^0}[g((\gamma, D\gamma)B^1_t)],$$

where $(B^1_t)_{t\geq 0}$ is a 1-dimensional Brownian motion and $D = (D_{ij})_{1\leq i,j\leq d}$ a diffusion matrix with elements

$$D_{ij} = \delta_{ij} - 2(\hat{\nabla}_i\Psi_g, (H_{\mathcal{T}} - E)^{-1}\hat{\nabla}_j\Psi_g)_{\mathcal{T}}, \quad i,j = 1,\ldots,d. \tag{2.14.51}$$

Proof. We decompose $\gamma\cdot B_t$ as $\gamma\cdot B_t = M_t + \int_0^t \rho(Y_s)ds$. By Lemma 2.202, there exists a martingale $(N_t)_{t\geq 0}$ with respect to the natural filtration $(\mathcal{B}_t)_{t\geq 0}$ of $(Y_t)_{t\geq 0}$ such that (2.14.49) holds. We have

$$\gamma\cdot B_t = M_t + N_t + \epsilon_t,$$

where $\epsilon_t = \int_0^t \rho(Y_s)ds - N_t$. Since $\sigma(X_t) = \sigma(Y_t)$, $(M_t)_{t\geq 0}$ is a martingale with respect to $(\mathcal{B}_t)_{t\geq 0}$. Then $(M_t + N_t)_{t\geq 0}$ is a martingale with respect to $(\mathcal{B}_t)_{t\geq 0}$, since both $(M_t)_{t\geq 0}$ and $(N_t)_{t\geq 0}$ are martingales with respect to $(\mathcal{B}_t)_{t\geq 0}$. We have

$$\frac{1}{t}\int_0^t \mathbb{E}_{P^W_\infty}[|M_s + N_s|^2]ds = \frac{1}{t}\int_0^t \mathbb{E}_{P^W_\infty}\Big[|\gamma\cdot B_s|^2 - 2\epsilon_s\gamma\cdot B_s + |\epsilon_s|^2\Big]ds \to (\gamma, D\gamma)$$

as $t\to\infty$ by (2.14.49) and (2.14.43). By Proposition 2.203, we have

$$\lim_{n\to\infty}\frac{1}{\sqrt{n}}(M_{[tn]} + N_{[tn]}) = (\gamma, D\gamma)B^1_t$$

in the sense of weak convergence. Let $g\in C^2(\mathbb{R})$. By the Taylor expansion, we have

$$\begin{aligned}\mathbb{E}_{P^W_\infty}\Big[g(\frac{1}{\sqrt{n}}\gamma\cdot B_{[nt]})\Big] &= \mathbb{E}_{P^W_\infty}\Big[g(\frac{1}{\sqrt{n}}(M_{[nt]} + N_{[nt]} + \epsilon_{[nt]}))\Big]\\ &= \mathbb{E}_{P^W_\infty}\Big[g(\frac{1}{\sqrt{n}}(M_{[nt]} + N_{[nt]})) + \frac{1}{\sqrt{n}}\epsilon_{[nt]}g'(X)\Big] \to \mathbb{E}_{\mathcal{W}}[g((\gamma, D\gamma)B^1_t)].\end{aligned} \tag{2.14.52}$$

Here $X = M_{[nt]} + N_{[nt]} + \theta\frac{1}{\sqrt{n}}\epsilon_{[nt]}$ with some $-1<\theta<1$. Next g be a bounded continuous function on $\mathbb{R}$. In this case we can prove (2.14.52) by a simple approximation procedure. There exists a sequence $g_n\in C^2(\mathbb{R})$ such that $\lim_{m\to\infty}\|g - g_m\|_\infty = 0$. For any $\epsilon > 0$, there exists m_0 such that for any $m > m_0$,

$$\begin{aligned}&\Big|\mathbb{E}_{P^W_\infty}\Big[g(\frac{1}{\sqrt{n}}\gamma\cdot B_{[nt]})\Big] - \mathbb{E}_{\mathcal{W}}[g((\gamma, D\gamma)B^1_t)]\Big|\\ &= \Big|\mathbb{E}_{P^W_\infty}\Big[g(\frac{1}{\sqrt{n}}\gamma\cdot B_{[nt]}) - g_m(\frac{1}{\sqrt{n}}\gamma\cdot B_{[nt]}) + g_m(\frac{1}{\sqrt{n}}\gamma\cdot B_{[nt]})\Big]\end{aligned}$$

$$- \mathbb{E}_{\mathcal{W}}\Big[g((\gamma, D\gamma)B_t^1) - g_m((\gamma, D\gamma)B_t^1) + g_m((\gamma, D\gamma)B_t^1)\Big]\Big|$$
$$\leq 2\epsilon + \Big|\mathbb{E}_{P_\infty^W}\Big[g_m(\frac{1}{\sqrt{n}}\gamma \cdot B_{[nt]})\Big] - \mathbb{E}_{\mathcal{W}}[g_m((\gamma, D\gamma)B_t^1)]\Big|.$$

Taking the limit as $n \to \infty$ on both sides above, we conclude that

$$\lim_{n\to\infty} \mathbb{E}_{P_\infty^W}\Big[g\Big(\frac{1}{\sqrt{n}}\gamma \cdot B_{[nt]}\Big)\Big] = \mathbb{E}_{\mathcal{W}}[g((\gamma, D\gamma)B_t^1)].$$ □

2.14.3 Diffusion matrix and effective mass

The diffusion matrix D derived from a central limit theorem in Section 2.14.2 coincides with the so-called *effective mass*. We shall show this in this section. In terms of translation invariant Nelson Hamiltonian with total momentum zero, $H(0)$, the matrix can be represented as

$$D_{ij} = \delta_{ij} - 2(P_{\mathrm{f}}^i \Psi_{\mathrm{g}}(0), (H(0) - E(0))^{-1} P_{\mathrm{f}}^j \Psi_{\mathrm{g}}(0))_{L^2(G)},$$

where $P_{\mathrm{f}}^j = d\Gamma(-i\nabla_j)$ is the field momentum operator. Let $\Psi_{\mathrm{g}}(p)$ be the ground state of $H(p)$ with ground state energy $E(p)$, namely

$$H(p)\Psi_{\mathrm{g}}(p) = E(p)\Psi_{\mathrm{g}}(p) \tag{2.14.53}$$

for $p \in \mathbb{R}^d$. Formally, we expect that

$$E(p) = E(0) + \frac{1}{2m_{\mathrm{eff}}}|p|^2 + O(|p|^3).$$

Here $O(|p|^3)/|p|^\alpha \to 0$ as $|p| \to 0$ for $\alpha < 2$. Also, m_{eff} is called the effective mass. Suppose that $E(p)$ and $\Psi_{\mathrm{g}}(p)$ are differentiable with respect to p. Then m_{eff} is the Hessian of $E(p)$, i. e.,

$$\frac{1}{m_{\mathrm{eff}}} = \Delta_p E(p)\Big|_{p=0}.$$

Let

$$\frac{1}{m_{ij}} = \frac{\partial^2 E(p)}{\partial p_i \partial p_j}\Big|_{p=0}$$

in a heuristic way in what follows, and we shall give the definition of the effective mass. We notice that

$$H(p) = \frac{1}{2}(p - P_{\mathrm{f}})^2 + H_{\mathrm{f}} + H_{\mathrm{I}}(0)$$

has the reflection symmetry, $H(p) \cong H(-p)$. For simplicity, we write $\Psi_g(p) = \Psi(p)$ and $P_f^i = P_i$. Moreover, we write $\partial_{p_i}\Psi(p) = \Psi_i(p)$, $\partial_{p_i}\partial_{p_j}\Psi(p) = \Psi_{ij}(p)$, $\partial_{p_i}E(p) = E_i(p)$, $\partial_{p_i}\partial_{p_j}E(p) = E_{ij}(p)$. Take the derivative ∂_{p_i} on both sides of (2.14.53). We have

$$(p_i - P_i)\Psi(p) + H(p)\Psi_i(p) = E_i(p)\Psi(p) + E(p)\Psi_i(p). \tag{2.14.54}$$

Substituting $p = 0$, we obtain

$$-P_i\Psi(0) + H(0)\Psi_i(0) = E_i(0)\Psi(0) + E(0)\Psi_i(0).$$

From the reflection symmetry, it follows that $E_i(0) = 0$ and

$$-P_i\Psi(0) + H(0)\Psi_i(0) = E(0)\Psi_i(0).$$

We can derive that

$$\Psi_i(0) = (H(0) - E(0))^{-1}P_i\Psi(0).$$

Take the derivative ∂_{p_j} on both sides of (2.14.54). We have

$$\begin{aligned}&\delta_{ij}\Psi(p) + (p_i - P_i)\Psi_j(p) + (p_j - P_j)\Psi_i(p) + H(p)\Psi_{ij}(p)\\&\quad = E_{ij}(p)\Psi(p) + E_i(p)\Psi_j(p) + E_j(p)\Psi_i(p) + E(p)\Psi_{ij}(p).\end{aligned}$$

Substituting $p = 0$, we get

$$\delta_{ij}\Psi(0) - P_i\Psi_j(p) - P_j\Psi_i(p) + H(0)\Psi_{ij}(0) = E_{ij}(0)\Psi(0) + E(0)\Psi_{ij}(p).$$

Take the scalar product with $\Psi(0)$ on both sides above. We have

$$\delta_{ij} - (\Psi(0), P_j\Psi_i(0) + P_i\Psi_j(0)) = E_{ij}(0).$$

Inserting $\Psi_i(0) = (H(0) - E(0))^{-1}P_i\Psi(0)$ yields

$$E_{ij}(0) = \delta_{ij} - 2(P_i\Psi(0), (H(0) - E(0))^{-1}P_j\Psi(0))_{L^2(G)}.$$

We conclude that

$$\frac{1}{m_{ij}} = \delta_{ij} - 2(P_f^i\Psi_g(0), (H(0) - E(0))^{-1}P_f^j\Psi_g(0))_{L^2(G)}.$$

We now define the effective mass by the inverse of $\sum_{i=1}^d \frac{1}{m_{ii}}$.

Definition 2.205 (Effective mass). The effective mass m_{eff} of $H(p)$ is defined by

$$\frac{1}{m_{\mathrm{eff}}} = d - 2\sum_{i=1}^d (P_f^i\Psi_g(0), (H(0) - E(0))^{-1}P_f^i\Psi_g(0))_{L^2(G)}.$$

Using the diffusion matrix $D = (D_{ij})_{1\leq i,j\leq d}$, we have the theorem below.

Theorem 2.206 (Effective mass and diffusion matrix). *Let D be the diffusion matrix given by* (2.14.51). *Then*

$$m_{\text{eff}} = \frac{1}{\operatorname{Tr} D}.$$

Proof. By Corollary 2.192 and Definition 2.197, we see that

$$(P_{\mathrm{f}}^{i}\Psi_{\mathrm{g}}(0), (H(0) - E(0))^{-1} P_{\mathrm{f}}^{i}\Psi_{\mathrm{g}}(0))_{L^2(G)} = (\hat{\nabla}_i \Psi_{\mathrm{g}}, (H_{\mathcal{T}} - E)^{-1} \hat{\nabla}_j \Psi_{\mathrm{g}})_{\mathcal{T}}.$$

Then the theorem is proven. □

3 The Pauli–Fierz model by path measures

3.1 Preliminaries

3.1.1 Introduction

It has been seen in the previous section that functional integration in the framework of Euclidean quantum field theory is a powerful tool in the spectral analysis of the Nelson scalar quantum field model. In this section, we consider a model of nonrelativistic quantum electrodynamics (QED) by using functional integration. This involves a stochastic integral with a Euclidean quantum field version of a quantized radiation field. A significant point is that the theory is nonrelativistic in the motion of the particle only, while the quantized radiation field is necessarily relativistic. In addition, it also has the feature that describes the particle "dressed" in a cloud of bosons, while its bare charge is conserved.

The Hamiltonian of nonrelativistic QED is defined as a self-adjoint operator on a Hilbert space. Its spectrum is of basic interest since it involves perturbations of embedded eigenvalues in the continuous spectrum. Recently, progress in the spectral analysis of nonrelativistic QED has produced an understanding of aspects of binding, spectral scattering theory, resonances, and effective mass to a fine degree. The functional integration introduced in this section offers a new point of view in the spectral analysis of nonrelativistic QED. Importantly, this approach is free of perturbative elements, and it leads to a proof of existence and uniqueness of the ground state of the Hamiltonian in a surprisingly simple manner. In addition to its applications in spectral analysis, we believe that functional integration is an interesting mathematical method in its own right. As we will see, the course of developing this theory gives rise to a new class of objects in probability theory and new problems.

In this section, we consider the *Pauli–Fierz Hamiltonian* H_{PF} in nonrelativistic QED, in which a quantum system of a low energy electron interacting through a minimally coupling with a massless quantized radiation field is described. In this model, the radiation field is quantized in the Coulomb gauge instead of the Lorentz gauge, derivative coupling is involved in its interaction by the minimal coupling, and an ultraviolet cutoff is imposed to define the model as a self-adjoint operator on a Hilbert space.

The existence of a ground state Ψ_{g} of the Pauli–Fierz Hamiltonian has been established by Bach–Fröhlich–Sigal [32] and Griesemer–Lieb–Loss [130]. For the details of the assumptions on the external potential, we refer to the original paper. In particular, no infrared cutoff is assumed and no restriction on the values of the coupling constant is imposed.

Theorem 3.1. *Under given conditions on V, there exists a ground state Ψ_{g} of H_{PF}, and* $(\Psi_{\mathrm{g}}, N\Psi_{\mathrm{g}}) < \infty$.

https://doi.org/10.1515/9783112231494-003

We emphasize that in this theorem no infrared regularity condition such as (2.2.6) in the case of the Nelson model is assumed.

Conventional quantum electrodynamics uses the Lagrangian density $\mathscr{L}_{\mathrm{QED}}(x)$. In the language of Feynman diagrams, the leading term of the effective mass is computed from the self-energy of the electron and the g-factor shift by vertex diagrams. Both diagrams include virtual photons, and give corrections to a bare mass and the g-factor. On the other hand, the effective charge e_{eff} is computed from the self-energy of photons. The photon self-energy diagram can be interpreted as the emission of the pairs of virtual electrons and positrons. The assumption that the energy of the electron is low implies that no emission of pairs of virtual electrons and positrons takes place. In the Pauli–Fierz model, thus no diagram of self-energy of photons is included, and the effective charge equals the bare charge and the number of electrons stays constant.

Here is an outline of cases discussed in terms of functional integration techniques in this chapter:

(1) spinless Pauli–Fierz model,
(2) relativistic Pauli–Fierz model,
(3) Pauli–Fierz model with spin 1/2,
(4) translation invariant Pauli–Fierz model.

In Chapter 2, the path measure was constructed on a Hilbert space-valued path space, and the functional integrals of semigroups are expressed over an infinite-dimensional Ornstein–Uhlenbeck process. Intuitively, this is an analogue of the finite-dimensional Ornstein–Uhlenbeck process. For the more complicated Pauli–Fierz model, however, we adopt the Euclidean quantum field method described in Chapter 1. In this case, the functional integrals are constructed on the product measure space of a Lévy measure and a Gaussian measure associated with the Euclidean quantum field version of the quantized radiation field. Moreover, in this case we will have to deal with a Hilbert space-valued stochastic integral with a time dependent integrand.

3.1.2 Lagrangian QED

As in the case of Nelson's model discussed in the previous chapter, also here we start by the formulation of the model in Fock space and then proceed to an equivalent formulation in terms of path integrals. In this section, we review the Lagrangian formalism of quantum electrodynamics.

We begin by some notation of the relativistically covariant theory used only in this section. Let

$$g_{\mu\nu} = \begin{cases} +1, & \mu = \nu = 0, \\ -1, & \mu = \nu = 1,2,3, \\ 0, & \text{otherwise} \end{cases}$$

and write $x = (t, \mathbf{x}) = (x^0, x^1, x^2, x^3) \in \mathbb{R}^4$, $x_\mu = g_{\mu\nu}x^\nu$. Here and in what follows, in this section we understand that summation is performed over repeated indices. The index μ runs from 0 to 3 and j from 1 to 3. Also, we put $\partial x^\mu = \partial_\mu$ and $\partial x_\mu = \partial^\mu$, involving $\partial^0 = \partial_0$ and $\partial^j = -\partial_j$. The 4×4 *gamma matrices* are defined by

$$\gamma^\mu = \begin{pmatrix} 0 & \sigma^\mu \\ \sigma^\mu & 0 \end{pmatrix},$$

where $\sigma^0 = 1_{2\times 2}$ and σ^μ, $\mu = 1, 2, 3$ are the 2×2 Pauli matrices.

Let $\mathbf{A}^\mu = \mathbf{A}^\mu(x)$ stand for the radiation field and write $\mathbf{A}_\mu = \mathbf{A}_\mu(x) = g_{\mu\nu}\mathbf{A}^\nu(x)$, i. e., $\mathbf{A}^0 = \mathbf{A}_0$ and $\mathbf{A}^j = -\mathbf{A}_j$. The spinor field will be denoted by $\Psi = \Psi(x)$, and we write $\overline{\Psi} = \overline{\Psi(x)} = \Psi^* \gamma^0$. The Lagrangian density in QED terms is given by

$$\mathscr{L}_{\mathrm{QED}} = \overline{\Psi}(i\gamma^\mu \partial_\mu - m)\Psi - \frac{1}{4}F_{\mu\nu}F^{\mu\nu} - e\overline{\Psi}\gamma^\mu \Psi \mathbf{A}_\mu.$$

Here, $e \in \mathbb{R}$ and $m > 0$ denote the charge and mass of the electron, respectively, $F^{\mu\nu}$ is the antisymmetric second-rank field tensor (field strength tensor)

$$F^{\mu\nu} = \partial^\mu \mathbf{A}^\nu - \partial^\nu \mathbf{A}^\mu$$

and $F_{\mu\nu} = g_{\mu\alpha}g_{\nu\beta}F^{\alpha\beta} = \partial_\mu \mathbf{A}_\nu - \partial_\nu \mathbf{A}_\mu$. The *longitudinal component* of the radiation field will be denoted by $\Phi = \mathbf{A}^0$ and the transversal by $\mathbf{A} = (\mathbf{A}^1, \mathbf{A}^2, \mathbf{A}^3)$. Let $\dot{X} = \partial_0 X$. Then the electrostatic field $\mathbb{E} = (\mathbb{E}^1, \mathbb{E}^2, \mathbb{E}^3)$ is defined by

$$\mathbb{E}^j = F^{j0} = -\partial_j \Phi - \dot{\mathbf{A}}^j,$$

while the magnetic field is the curl of the transversal component,

$$\mathbb{B} = (\mathbb{B}^1, \mathbb{B}^2, \mathbb{B}^3) = \nabla \times \mathbf{A},$$

where $\nabla = (\partial_1, \partial_2, \partial_3)$. In particular, the field tensor is obtained as

$$F^{\mu\nu} = \begin{pmatrix} 0 & -\mathbb{E}^1 & -\mathbb{E}^2 & -\mathbb{E}^3 \\ \mathbb{E}^1 & 0 & -\mathbb{B}^3 & \mathbb{B}^2 \\ \mathbb{E}^2 & \mathbb{B}^3 & 0 & -\mathbb{B}^1 \\ \mathbb{E}^3 & -\mathbb{B}^2 & \mathbb{B}^1 & 0 \end{pmatrix}, \quad F_{\mu\nu} = \begin{pmatrix} 0 & \mathbb{E}^1 & \mathbb{E}^2 & \mathbb{E}^3 \\ -\mathbb{E}^1 & 0 & -\mathbb{B}^3 & \mathbb{B}^2 \\ -\mathbb{E}^2 & \mathbb{B}^3 & 0 & -\mathbb{B}^1 \\ -\mathbb{E}^3 & -\mathbb{B}^2 & \mathbb{B}^1 & 0 \end{pmatrix}.$$

By straightforward computation, using the explicit forms of $F_{\mu\nu}$ and $F^{\mu\nu}$ it follows that

$$-\frac{1}{4}F_{\mu\nu}F^{\mu\nu} = \frac{1}{2}(\mathbb{E}^2 - \mathbb{B}^2).$$

From this, the Lagrangian density is obtained as

$$\mathscr{L}_{\mathrm{QED}} = \overline{\Psi}(i\gamma^\mu\partial_\mu - m)\Psi + \frac{1}{2}(\mathbb{E}^2 - \mathbb{B}^2) + e\overline{\Psi}\gamma^j\Psi\mathbf{A}^j - e\Psi^*\Psi\Phi$$

using the fact that $\overline{\Psi}\gamma^0\Psi = \Psi^*\Psi$. From the Lagrangian, we can derive the Euler–Lagrange equations through the variational principle

$$\text{(M)} \quad \partial_\mu F^{\mu\nu} = e\bar{\Psi}\gamma^\nu\Psi, \tag{3.1.1}$$

$$\text{(D)} \quad (i\gamma^\mu\partial_\mu - m)\Psi = e\gamma^\mu\mathbf{A}_\mu\Psi. \tag{3.1.2}$$

Equation (3.1.2) is the *Dirac equation* and (3.1.1) is written as

$$\square\mathbf{A}^\nu - \partial^\nu\partial_\mu\mathbf{A}^\mu = e\bar{\Psi}\gamma^\nu\Psi, \tag{3.1.3}$$

where $\square = \partial_0\partial_0 - \partial_j\partial_j$, and is equivalent to

$$(\nu = 0) \quad \nabla\cdot\mathbb{E} = e\Psi^*\Psi, \tag{3.1.4}$$

$$(\nu = j) \quad \dot{\mathbb{E}}^j = (\nabla\times\mathbb{B})^j - e\bar{\Psi}\gamma^j\Psi. \tag{3.1.5}$$

Moreover, the definitions of $\mathbb{E}$ and $\mathbb{B}$ give

$$\dot{\mathbb{B}} = -\nabla\times\mathbb{E}, \tag{3.1.6}$$

$$\nabla\cdot\mathbb{B} = 0. \tag{3.1.7}$$

The system of (3.1.4)–(3.1.7) are the celebrated *Maxwell equations.*

We derive now the QED Hamiltonian by using Legendre transforms. To see this from $\mathscr{L}_{\mathrm{QED}}$, we identify the conjugate momenta

$$\Pi^0 = \frac{\partial\mathscr{L}_{\mathrm{QED}}}{\partial\dot{\mathbf{A}}_0} = 0,$$

$$\Pi^j = \frac{\partial\mathscr{L}_{\mathrm{QED}}}{\partial\dot{\mathbf{A}}_j} = \mathbb{E}^j$$

and

$$\frac{\partial\mathscr{L}_{\mathrm{QED}}}{\partial\dot{\Psi}} = i\Psi^*.$$

The canonical momentum Π^0 vanishes. Thus the Legendre transform gives the Hamiltonian density $\mathscr{H}_{\mathrm{QED}} = \mathscr{H}_{\mathrm{QED}}(x)$ by

$$\begin{aligned}\mathscr{H}_{\mathrm{QED}} &= \Pi^j\dot{\mathbf{A}}_j + i\Psi^*\dot{\Psi} - \mathscr{L}_{\mathrm{QED}}\\ &= \Psi^*\alpha^j(-i\partial_j + e\mathbf{A}_j)\Psi + m\beta\Psi^*\Psi + \frac{1}{2}(\mathbb{E}^2 + \mathbb{B}^2) + e\Psi^*\Psi\Phi + \nabla\Phi\cdot\mathbb{E}.\end{aligned}$$

Here, $\alpha^j = \gamma^0\gamma^j$ and $\beta = \gamma^0$.

Next, in order to derive the Pauli–Fierz Hamiltonian as an analogue of full QED, instead of the Lorentz gauge the *Coulomb gauge* condition

$$\nabla \cdot \mathbf{A} = 0 \tag{3.1.8}$$

is used. This gauge is not covariant and (3.1.3) reduces to

$$\Box \mathbf{A}^\nu - \partial^\nu \dot{\Phi} = e\bar{\Psi}\gamma^\nu \Psi.$$

Notice that under the *Lorentz gauge* condition $\partial_\mu \mathbf{A}^\mu = 0$, (3.1.3) reduces to the wave equation

$$\Box \mathbf{A}^\nu = e\bar{\Psi}\gamma^\nu \Psi.$$

The Coulomb gauge condition (3.1.8) yields

$$\nabla \cdot \mathbb{E} = -\Delta \Phi = e\Psi^* \Psi,$$

where $\Delta = \partial_j \partial_j$ denotes the 3-dimensional Laplacian and the second equality is derived from the Euler–Lagrange equations (3.1.4)–(3.1.5). This is the *Poisson equation* and the longitudinal component Φ in the Coulomb gauge becomes

$$\Phi(t, \mathbf{x}) = e \int_{\mathbb{R}^3} \frac{\Psi^*(t, \mathbf{y})\Psi(t, \mathbf{y})}{4\pi|\mathbf{x} - \mathbf{y}|} d\mathbf{y}.$$

Decomposing the electric field $\mathbb{E} = -\nabla\Phi - \dot{\mathbf{A}}$ into a longitudinal and a transversal component, we furthermore have

$$\int_{\mathbb{R}^3} \mathbb{E}^2 d\mathbf{x} = -\int_{\mathbb{R}^3} \Phi \cdot \Delta\Phi d\mathbf{x} + \int_{\mathbb{R}^3} \dot{\mathbf{A}}^2 d\mathbf{x}$$

by the Coulomb gauge condition. Moreover,

$$-\int_{\mathbb{R}^3} \Phi \cdot \Delta\Phi d\mathbf{x} = +e^2 \int_{\mathbb{R}^3} \frac{\Psi^*(t, \mathbf{y})\Psi(t, \mathbf{y})\Psi^*(t, \mathbf{x})\Psi(t, \mathbf{x})}{4\pi|\mathbf{x} - \mathbf{y}|} d\mathbf{x}d\mathbf{y}.$$

The term $e\Psi^*\Psi\Phi + \nabla\Phi \cdot \mathbb{E}$ in $\mathscr{H}_{\mathrm{QED}}$ vanishes since by integration,

$$\int_{\mathbb{R}^3} \nabla\Phi \cdot \mathbb{E} d\mathbf{x} = -\int_{\mathbb{R}^3} \Phi\nabla \cdot \mathbb{E} d\mathbf{x} = -e \int_{\mathbb{R}^3} \Psi^*\Psi\Phi d\mathbf{x}.$$

Finally, the Hamiltonian

$$H_{\mathrm{QED}} = \int_{\mathbb{R}^3} \mathscr{H}_{\mathrm{QED}} d\mathbf{x}$$

of full QED with the Coulomb gauge is given by

$$H_{\mathrm{QED}} = \int_{\mathbb{R}^3} \Psi^* \alpha^j(-i\partial_j + e\mathbf{A}_j)\Psi d\mathbf{x} + m\beta \int_{\mathbb{R}^3} \Psi^*\Psi d\mathbf{x}$$
$$+ \frac{1}{2}\int_{\mathbb{R}^3} (\dot{\mathbf{A}}^2 + \mathbb{B}^2) d\mathbf{x} + e^2 \int_{\mathbb{R}^3\times\mathbb{R}^3} \frac{\Psi^*(t,\mathbf{y})\Psi(t,\mathbf{y})\Psi^*(t,\mathbf{x})\Psi(t,\mathbf{x})}{4\pi|\mathbf{x}-\mathbf{y}|} d\mathbf{x}d\mathbf{y}.$$

3.1.3 Classical variant of nonrelativistic QED

In this section, we derive a classical version of the nonrelativistic QED Hamiltonian using the Lagrange formalism, which is an analogue of the discussion in the preceding section. Here, $\overline{\Psi}(i\gamma^\mu\partial_\mu - m)\Psi$ is replaced by the N-particle classical kinetic energy $(1/2)\sum_{j=1}^N m_j\dot{q}_j^2$, $e\overline{\Psi}\gamma^j\Psi$ by the current density j, and $e\Psi^*\Psi$ by the density ρ. The Pauli–Fierz Hamiltonian in nonrelativistic QED will be given by quantization of the classical version and the spinor field is changed to electrons governed by the Schrödinger operator.

Let N electrons interact with a classical electromagnetic field described by the *Maxwell equations*. Denote by $B(t,x) \in \mathbb{R}^3$ the classical magnetic field and by $E(t,x) \in \mathbb{R}^3$ the classical electric field. Consider the Maxwell equations with form factor λ,

$$\nabla \cdot E = \sum_{j=1}^N e_j\lambda(\cdot - x_j), \tag{3.1.9}$$

$$\dot{E} = \nabla \times B - \sum_{j=1}^N e_j\lambda(\cdot - x_j)\dot{q}_j, \tag{3.1.10}$$

$$\dot{B} = -\nabla \times E, \tag{3.1.11}$$

$$\nabla \cdot B = 0. \tag{3.1.12}$$

Here, $x_j = x_j(t) \in \mathbb{R}^3$ denotes the position and e_j the charge of the electron labeled by j. The form factor describes the charge distribution satisfying $\lambda \geq 0$ and $\int_{\mathbb{R}^3} \lambda(x)dx = 1$. We define the current by

$$(j,\rho) = \left(\sum_{j=1}^N e_j\lambda(\cdot - x_j)\dot{q}_j, \sum_{j=1}^N e_j\lambda(\cdot - x_j)\right);$$

which satisfies the *continuity equation* $\dot{\rho} + \nabla\cdot j = 0$. The *equation of motion* of the electron is given by

$$m_j\ddot{q}_j = e_j(E + \dot{q}_j \times B),$$

where m_j is the mass of the jth electron.

We next rewrite the Maxwell equations by introducing a vector potential A and a scalar potential ϕ. By (3.1.12), $A = (A_1(t,x), A_2(t,x), A_3(t,x))$ is required to satisfy

$$B = \nabla \times A.$$

Furthermore, from (3.1.11) it follows that $\nabla \times (E + \dot{A}) = 0$. The scalar potential $\phi = \phi(t,x)$ is given by

$$E = -\dot{A} - \nabla\phi. \tag{3.1.13}$$

With this, the Lagrangian becomes

$$\mathscr{L}_{\mathrm{PF}} = \frac{1}{2}\sum_{j=1}^{N} m_j \dot{q}_j^2 + \int_{\mathbb{R}^3} \frac{1}{2}(E^2 - B^2)dx + \int_{\mathbb{R}^3} j \cdot A dx + \int_{\mathbb{R}^3} (-\rho\phi) dx,$$

supplemented with the Coulomb gauge condition

$$\nabla \cdot A = 0.$$

The system (3.1.9)–(3.1.12) thus turns into

$$\begin{aligned} \Box A &= \nabla\dot{\phi} - j, \\ \Delta\phi &= -\rho. \end{aligned}$$

Inserting the definition of E (3.1.13) and taking into account the Coulomb gauge, we furthermore have

$$\int_{\mathbb{R}^3} \frac{1}{2}(E^2 - B^2)dx = \int_{\mathbb{R}^3} \left\{\frac{1}{2}\left(\dot{A}^2 - B^2\right) + \frac{1}{2}\rho\phi\right\} dx.$$

By the Poisson equation $\Delta\phi = -\rho$, we have the Coulomb potential regularized by ρ,

$$\frac{1}{2}\int_{\mathbb{R}^3} \rho\phi dx = \frac{1}{2}\sum_{i,j=1}^{N} \int_{\mathbb{R}^3} \frac{e_i e_j \lambda(x_i - y)\lambda(x_j - y')}{4\pi|y - y'|} dy dy'.$$

Now we define the Hamiltonian associated with the Lagrangian $\mathscr{L}_{\mathrm{PF}}$ through Legendre transform. We introduce the canonical momenta p_j and Π_i by

$$p_j = \frac{\partial \mathscr{L}_{\mathrm{PF}}}{\partial \dot{q}_j} = m_j \dot{q}_j + e_j \int_{\mathbb{R}^3} A(x)\lambda(x - x_j)dx$$

and

$$\Pi_i = \frac{\delta \mathscr{L}_{\mathrm{PF}}}{\delta \dot{A}_i} = \dot{A}_i, \quad i = 1, 2, 3.$$

In terms of these variables, the Hamiltonian associated with $\mathscr{L}_{\mathrm{PF}}$ becomes

$$H_{\mathrm{classical}} = \sum_{j=1}^{N} p_j \dot{q}_j + \int_{\mathbb{R}^3} \Pi \cdot \dot{A} dx - \mathscr{L}_{\mathrm{PF}}.$$

A straightforward computation yields

$$\begin{aligned} H_{\mathrm{classical}} &= \sum_{j=1}^{N} \frac{1}{2m_j} \left(p_j - e_j \int_{\mathbb{R}^3} A(x)\lambda(x - x_j) dx \right)^2 + \frac{1}{2} \int_{\mathbb{R}^3} (\dot{A}^2 + B^2) dx \\ &\quad + \frac{1}{2} \sum_{i,j=1}^{N} \int_{\mathbb{R}^3 \times \mathbb{R}^3} \frac{e_i e_j \lambda(x_i - y)\lambda(x_j - y')}{4\pi |y - y'|} dy dy'. \end{aligned}$$

This Hamiltonian contains the convolution $\int_{\mathbb{R}^3} A(x)\lambda(x - x_j)dx$, the field energy $\frac{1}{2}\int_{\mathbb{R}^3}(\dot{A}^2 + B^2)dx$, the regularized Coulomb potential, and the minimal interaction. In the next section, we leave the realm of classical electrodynamics and quantize $H_{\mathrm{classical}}$. We will define a self-adjoint operator H_{PF} on a suitable Hilbert space $\mathcal{H}_{\mathrm{PF}}$. In the quantum description, H_{PF} will be defined by $H_{\mathrm{classical}}$ replaced by $p_j \to -i\partial_j, j = 1, \ldots, N$, the smeared Coulomb potential goes into a multiplication operator given by a real function V,

$$\int_{\mathbb{R}^3} A_\mu(x)\lambda(x - x_j)dx \to \hat{A}_\mu(\tilde{\varphi}(\cdot - x)),$$

and

$$\frac{1}{2} \int_{\mathbb{R}^3} (\dot{A}^2 + B^2) dx \to H_{\mathrm{rad}}.$$

Here, $\hat{A}_\mu(\tilde{\varphi}(\cdot - x))$ and H_{rad} are the quantized radiation field and the free field Hamiltonian on the boson Fock space, respectively, which are defined and discussed in detail in the next section.

3.2 The Pauli–Fierz model in nonrelativistic QED

3.2.1 The Pauli–Fierz model in Fock space

We now turn to defining the Pauli–Fierz Hamiltonian rigorously as a self-adjoint operator bounded from below on a Hilbert space over $\mathbb{C}$. For simplicity, we assume only one electron interacting with the field.

Let

$$\mathcal{H}_{\mathrm{PF}} = L^2(\mathbb{R}^3) \otimes \mathcal{F}_{\mathrm{rad}}$$

be the Hilbert space describing the joint electron–photon state vectors. Here,

$$\mathcal{F}_{\rm rad} = \mathcal{F}_{\rm rad}(L^2(\mathbb{R}^3 \times \{\pm\}))$$

is the boson Fock space over $L^2(\mathbb{R}^3 \times \{\pm\})$. The elements of the set $\{\pm\}$ account for the fact that a photon is a *transversal wave* perpendicular to the direction of its propagation, thus it has two components. The Fock vacuum in $\mathcal{F}_{\rm rad}$ is denoted by $\Omega_{\rm ph}$. We identify $\mathcal{H}_{\rm PF}$ as the set of $\mathcal{F}_{\rm rad}$-valued L^2-functions on $\mathbb{R}^3$,

$$\mathcal{H}_{\rm PF} \cong \int_{\mathbb{R}^3}^{\oplus} \mathcal{F}_{\rm rad} dx;$$

this will be used without further notice in what follows. Let $a(f)$ and $a^*(f)$ be the annihilation and creation operators, respectively, on $\mathcal{F}_{\rm rad}$ smeared by $f \in L^2(\mathbb{R}^3 \times \{\pm\})$. Then

$$[a(f), a^*(g)] = \sum_{j=\pm} (\bar{f}_j, g_j), \quad [a(f), a(g)] = 0, \quad [a^*(f), a^*(g)] = 0.$$

We use the identification $L^2(\mathbb{R}^3 \times \{\pm\}) \cong L^2(\mathbb{R}^3) \oplus L^2(\mathbb{R}^3)$ and set

$$a^\sharp(f, +) = a^\sharp(f \oplus 0), \quad a^\sharp(f, -) = a^\sharp(0 \oplus f),$$

where $a^\sharp$ stands for either operator. Then $a^\sharp(f) = \sum_{j=\pm} a^\sharp(f_j, j)$ for $f = f_1 \oplus f_2$. In terms of (formal) kernels $a^\sharp(k, j)$, $a^\sharp(f, j)$ will be written more conveniently as

$$a^\sharp(f, j) = \int_{\mathbb{R}^3} a^\sharp(k, j) f(k) dk.$$

The finite particle subspace of $\mathcal{F}_{\rm rad}$ is given by

$$\mathcal{F}_{\rm rad,fin} = \{\Psi = \{\Psi^{(n)}\}_{n=0}^{\infty} \mid \Psi^{(m)} = 0 \text{ for all } m > N \text{ with some } N\}.$$

Next, we define the quantized radiation field with a given form factor $\hat{\varphi}$. Put

$$\wp_\mu^j(x) = \frac{\hat{\varphi}(k)}{\sqrt{\omega(k)}} e_\mu^j(k) e^{-ikx} \in L^2(\mathbb{R}_k^3),$$
$$\tilde{\wp}_\mu^j(x) = \frac{\hat{\varphi}(-k)}{\sqrt{\omega(k)}} e_\mu^j(k) e^{ik\cdot x} \in L^2(\mathbb{R}_k^3),$$

for $x \in \mathbb{R}^3$, $j = \pm$ and $\mu = 1, 2, 3$, where ω is the dispersion relation for massless photons defined by

$$\omega(k) = \omega_{\nu=0}(k) = |k|.$$

Here, $\hat{\varphi}$ denotes the Fourier transform of the charge distribution φ. The vectors $e^+(k)$ and $e^-(k)$ are called *polarization vectors*, that is, $e^+(k)$, $e^-(k)$, and $k/|k|$ form a right-hand system at $k \in \mathbb{R}^3$:

$$e^i(k) \cdot e^j(k) = \delta_{ij}1, \quad e^j(k) \cdot k = 0, \quad e^+(k) \times e^-(k) = k/|k|.$$

In Proposition 3.6, we will see that the spectral analysis is independent of the choice of polarization vectors, i. e., the Hamiltonians defined through different polarizations are unitary equivalent. Thus we may fix the polarization vectors as it is most convenient.

Definition 3.2 (Quantized radiation field). The *quantized radiation field* with form factor $\hat{\varphi}$ is defined by

$$A_\mu(x) = \frac{1}{\sqrt{2}} \sum_{j=\pm} (a^*(\wp^j_\mu(x), j) + a(\tilde{\wp}^j_\mu(x), j))$$

and the *conjugate momentum* by

$$\Pi_\mu(x) = \frac{i}{\sqrt{2}} \sum_{j=\pm} (a^*(\wp^j_\mu(x), j) - a(\tilde{\wp}^j_\mu(x), j)).$$

In the case of $\hat{\varphi}/\sqrt{\omega} \in L^2(\mathbb{R}^d)$ and $\overline{\hat{\varphi}(k)} = \hat{\varphi}(-k)$, $A_\mu(x)$ is symmetric and, moreover, essentially self-adjoint on the finite particle subspace of $\mathcal{F}_{\mathrm{rad}}$, $\mathcal{F}_{\mathrm{rad,fin}}$, by the Nelson analytic vector theorem (Proposition 1.16), (1.2.14) and (1.2.15). The condition $\overline{\hat{\varphi}(k)} = \hat{\varphi}(-k)$ is necessary and sufficient for φ to be real. A physically reasonable choice of φ requires a positive function. We denote the closure of $A_\mu(x)\lceil_{\mathcal{F}_{\mathrm{rad,fin}}}$ by the same symbol. Write

$$A_\mu = \int_{\mathbb{R}^3}^{\oplus} A_\mu(x)dx, \quad A = (A_1, A_2, A_3).$$

Operator A_μ is self-adjoint on

$$D(A_\mu) = \left\{ F \in \mathcal{H}_{\mathrm{PF}} \,\middle|\, F(x) \in D(A_\mu(x)) \text{ and } \int_{\mathbb{R}^3} \|A_\mu(x)F(x)\|^2_{\mathcal{F}_{\mathrm{rad}}} dx < \infty \right\}$$

and acts by

$$(A_\mu F)(x) = A_\mu(x)F(x), \quad F \in D(A_\mu),$$

for almost every $x \in \mathbb{R}^3$. In terms of the kernel $a^\sharp(k, j)$, the quantized radiation field $A_\mu(x)$ can be written as

$$A_\mu(x) = \sum_{j=\pm} \frac{1}{\sqrt{2}} \int_{\mathbb{R}^3} e^j_\mu(k) \left(\frac{\hat{\varphi}(k)}{\sqrt{\omega(k)}} e^{-ik\cdot x} a^*(k, j) + \frac{\hat{\varphi}(-k)}{\sqrt{\omega(k)}} e^{ik\cdot x} a(k, j) \right) dk$$

and the conjugate momentum by

$$\Pi_\mu(x) = \sum_{j=\pm} \frac{i}{\sqrt{2}} \int_{\mathbb{R}^3} e^j_\mu(k) \left(\frac{\hat{\varphi}(k)}{\sqrt{\omega(k)}} e^{-ik\cdot x} a^*(k,j) - \frac{\hat{\varphi}(-k)}{\sqrt{\omega(k)}} e^{ik\cdot x} a(k,j) \right) dk.$$

Since $k \cdot e^j(k) = 0$, the polarization vectors introduced above are chosen in such a way that $\sum_{\mu=1}^3 \nabla_\mu \wp_j^\mu(x) = 0$, implying the Coulomb gauge condition

$$\sum_{\mu=1}^3 \nabla_\mu \cdot A_\mu = 0.$$

This in turn yields $\sum_{\mu=1}^3 [\nabla_\mu, A_\mu] = 0$.

Next, we introduce the free field Hamiltonian on $\mathcal{F}_{\text{rad}}$. The free field Hamiltonian H_{rad} on $\mathcal{F}_{\text{rad}}$ is given in terms of the second quantization

$$H_{\text{rad}} = \mathrm{d}\Gamma(\omega \oplus \omega),$$

formally corresponding to

$$\sum_{j=\pm} \int_{\mathbb{R}^3} \omega(k) a^*(k,j) a(k,j) dk.$$

It leaves the n-particle subspace $\mathcal{F}_{\text{rad}}^{(n)} = L^2_{\text{sym}}((\mathbb{R}^3 \oplus \mathbb{R}^3)^n)$ invariant and acts as

$$\left(H_{\text{rad}} \begin{pmatrix} f_1 \\ f_2 \end{pmatrix} \right) (k_1, \ldots, k_n) = \left(\sum_{j=1}^n \omega(k_j) \right) \begin{pmatrix} f_1(k_1, \ldots, k_n) \\ f_2(k_1, \ldots, k_n) \end{pmatrix}.$$

The Pauli–Fierz model describes the minimal interaction between an electron and the quantized radiation field, in which the electron is assumed to carry low energy. The electron, as in the case of the Nelson model, is described by the Schrödinger operator

$$H_{\text{p}} = -\frac{1}{2}\Delta + V$$

in $L^2(\mathbb{R}^3)$. The assumptions on V will be specified later on below. The Hamiltonian for the noninteracting quantized radiation field and electron is thus given by

$$H_{\text{PF},0} = -\frac{1}{2}\Delta \otimes 1 + 1 \otimes H_{\text{rad}}$$

with domain

$$D_{\text{PF}} = D(H_{\text{PF},0}) = D\left(-\frac{1}{2}\Delta \otimes 1\right) \cap D(1 \otimes H_{\text{rad}}).$$

The interaction is obtained by minimal coupling $-i\nabla_\mu \otimes 1 \mapsto -i\nabla_\mu \otimes 1 - eA_\mu$.

Definition 3.3 (Pauli–Fierz Hamiltonian in Fock space). The Pauli–Fierz Hamiltonian in nonrelativistic quantum electrodynamics is defined by the operator

$$H_{\mathrm{PF}} = \frac{1}{2}(-i\nabla \otimes 1 - eA)^2 + V \otimes 1 + 1 \otimes H_{\mathrm{rad}}$$

in $\mathcal{H}_{\mathrm{PF}}$.

We define $H_{\mathrm{PF}}(A)$ and $H_{\mathrm{PF,I}}$ by

$$H_{\mathrm{PF}}(A) = \frac{1}{2}(-i\nabla \otimes 1 - eA)^2,$$

$$H_{\mathrm{PF,I}} = -e(-i\nabla \otimes 1) \cdot A + \frac{e^2}{2}A^2 + V \otimes 1.$$

Clearly,

$$H_{\mathrm{PF}} = H_{\mathrm{PF,0}} + H_{\mathrm{PF,I}}$$

by the commutation relation $\sum_{\mu=1}^{3}[\nabla_\mu \otimes 1, A_\mu] = 0$. In what follows, we omit the tensor notation $\otimes$ to ease the notation. In Sections 3.2–3.3, unless otherwise stated, we suppose the following assumptions.

Assumption 3.4. The following conditions hold:
(1) (Charge distribution) $\varphi \in \mathscr{S}'(\mathbb{R}^3)$ satisfies that

$$\text{(i)}\quad \hat{\varphi}(-k) = \overline{\hat{\varphi}(k)}, \qquad \text{(ii)}\quad \sqrt{\omega}\hat{\varphi}, \hat{\varphi}/\sqrt{\omega}, \hat{\varphi}/\omega \in L^2(\mathbb{R}^3), \tag{3.2.1}$$

(2) (External potential) $V \in \mathcal{R}_{\mathrm{Kato}}$, i. e., there exist $0 \le a < 1$ and $0 \le b$ such that

$$\|Vf\| \le a\| - (1/2)\Delta f\| + b\|f\|, \quad f \in D(-(1/2)\Delta). \tag{3.2.2}$$

As was stated before, condition (i) of (1) means that $\varphi(x) = \bar{\varphi}(x)$, and hence φ is real and it ensures that H_{PF} is symmetric. By condition (ii) of (1), $(-i\nabla \otimes 1) \cdot A$ and A^2 are relatively bounded with respect to $H_{\mathrm{PF,0}}$. Finally, by (2) and the Kato–Rellich theorem, H_{p} is self-adjoint on $D(-(1/2)\Delta)$. As a consequence, $H_{\mathrm{PF,I}}$ is relatively bounded with respect to $H_{\mathrm{PF,0}}$. As in Section 3.5.1 below, we can prove the self-adjointness of H_{PF}.

Theorem 3.5 (Self-adjointness). *The Pauli–Fierz Hamiltonian H_{PF} is self-adjoint on D_{PF} and essentially self-adjoint on any core of $H_{\mathrm{PF,0}}$.*

In fact, for small values of $|e|$ it is not hard to see self-adjointness of H_{PF} on D_{PF} as we have

$$\|H_{\mathrm{PF,I}}\Psi\| \le (a + a_1|e| + a_2|e|^2)\|H_{\mathrm{PF,0}}\Psi\| + b(e)\|\Psi\|$$

with a given in (3.2.2) and constants a_1, a_2 and $b(e)$ whose existence is made sure by (ii) of (3.2.1). Thus for e such that $(a + a_1|e| + a_2|e|^2) < 1$, H_{PF} is self-adjoint on D_{PF} by the

Kato–Rellich theorem. A more challenging problem is to show that H_{PF} is self-adjoint for all $e \in \mathbb{R}$. For the moment, we assume that $|e|$ is sufficiently small.

Remarkably, Pauli–Fierz Hamiltonians with different polarization vectors are equivalent to each other. We show this next. Let $e^{\pm}$ and $\eta^{\pm}$ be polarization vectors, and $H_{\mathrm{PF}}(e^{\pm})$ and $H_{\mathrm{PF}}(\eta^{\pm})$ the corresponding Pauli–Fierz Hamiltonians, respectively.

Proposition 3.6. *Hamiltonians $H_{\mathrm{PF}}(e^{\pm})$ and $H_{\mathrm{PF}}(\eta^{\pm})$ are unitary equivalent.*

Proof. Since for each $k \in \mathbb{R}^3$ both polarization vectors form orthogonal bases on the plane perpendicular to the vector k, there exists ϑ_k such that

$$\begin{pmatrix} e^+(k) \\ e^-(k) \end{pmatrix} = \begin{pmatrix} \cos\vartheta_k 1_3 & -\sin\vartheta_k 1_3 \\ \sin\vartheta_k 1_3 & \cos\vartheta_k 1_3 \end{pmatrix} \begin{pmatrix} \eta^+(k) \\ \eta^-(k) \end{pmatrix}, \quad \text{i. e.,} \quad \begin{pmatrix} e^+_\mu(k) \\ e^-_\mu(k) \end{pmatrix} = \mathscr{R}_k \begin{pmatrix} \eta^+_\mu(k) \\ \eta^-_\mu(k) \end{pmatrix},$$

where $\mathscr{R}_k = \left(\begin{smallmatrix} \cos\vartheta_k & -\sin\vartheta_k \\ \sin\vartheta_k & \cos\vartheta_k \end{smallmatrix}\right)$. Define $R : L^2(\mathbb{R}^3; \mathbb{C}^2) \to L^2(\mathbb{R}^3; \mathbb{C}^2)$, for almost every k, by

$$R\begin{pmatrix} f \\ g \end{pmatrix}(k) = \mathscr{R}_k \begin{pmatrix} f(k) \\ g(k) \end{pmatrix}$$

and $\Gamma(R) : \mathcal{F}_{\mathrm{rad}} \to \mathcal{F}_{\mathrm{rad}}$ by the second quantization of R. Then $\Gamma(R)$ is a unitary map on $\mathcal{F}_{\mathrm{rad}}$. Note that

$$R\begin{pmatrix} \eta^+_\mu f \\ \eta^-_\mu f \end{pmatrix} = \begin{pmatrix} e^+_\mu f \\ e^-_\mu f \end{pmatrix},$$

which implies that $\Gamma(R) H_{\mathrm{PF}}(\eta^{\pm}) \Gamma(R)^{-1} = H_{\mathrm{PF}}(e^{\pm})$. □

3.2.2 The Pauli–Fierz model in function space

We introduce a $\mathscr{Q}$-space associated with the quantized radiation field and reformulate the Pauli–Fierz Hamiltonian on $L^2(\mathbb{R}^3) \otimes L^2(\mathscr{Q})$ instead of $\mathcal{H}_{\mathrm{PF}}$. Furthermore, we introduce the Euclidean quantum field associated with the Pauli–Fierz Hamiltonian to derive the functional integral representation of $e^{-tH_{\mathrm{PF}}}$. A slight modification of the setup discussed in Chapters 1 and 2 is needed as the scalar field $(\phi(f), f \in \mathscr{M})$ is indexed by $f \in \mathscr{M}$ while the quantized radiation field is indexed by $L^2_{\mathrm{real}}(\mathbb{R}^3)$. In particular, the family of isometries $\tau_t : \mathscr{M} \to \mathscr{E}$ will be modified to

$$\mathrm{j}_t : L^2_{\mathrm{real}}(\mathbb{R}^3) \to L^2_{\mathrm{real}}(\mathbb{R}^4).$$

For a real-valued $f \in L^2(\mathbb{R}^3)$,

$$A_\mu(f) = \frac{1}{\sqrt{2}} \sum_{j=\pm} \int_{\mathbb{R}^3} e^j_\mu(k)(\hat{f}(k) a^*(k,j) + \hat{f}(-k) a(k,j)) dk.$$

With this notation, we write $A_\mu(x) = A_\mu(\tilde{\varphi}(\cdot - x))$, where $\tilde{\varphi} = (\hat{\varphi}/\sqrt{\omega})\check{}$. The relations

$$\text{(mean)} \quad (\Omega_{\mathrm{PF}}, A_\mu(f)\Omega_{\mathrm{PF}}) = 0, \tag{3.2.3}$$

$$\text{(covariance)} \quad (\Omega_{\mathrm{PF}}, A_\mu(f)A_\nu(g)\Omega_{\mathrm{PF}}) = \frac{1}{2}\int_{\mathbb{R}^3} \delta^\perp_{\mu\nu}(k)\overline{\hat{f}(k)}\hat{g}(k)dk \tag{3.2.4}$$

are immediate. Here, $\delta^\perp_{\mu\nu}(k)$ is the *transversal delta function* given by

$$\delta^\perp_{\mu\nu}(k) = \delta_{\mu\nu} - \frac{k_\mu k_\nu}{|k|^2}.$$

The identity

$$\delta^\perp_{\mu\nu}(k) = \sum_{j=\pm} e^j_\mu(k)e^j_\nu(k) \tag{3.2.5}$$

follows from the fact that the 3×3 matrix

$$\begin{pmatrix} e_1^+(k) & e_1^-(k) & k_1/|k| \\ e_2^+(k) & e_2^-(k) & k_2/|k| \\ e_3^+(k) & e_3^-(k) & k_3/|k| \end{pmatrix}$$

is orthogonal. Note that relation (3.2.5) is independent of the choice of polarization vectors $e^\pm(k)$. We define the 3×3 matrix $\delta^\perp(k)$ by

$$\delta^\perp(k) = (\delta^\perp_{\mu\nu}(k))_{1\le\mu,\nu\le 3}.$$

In order to have a functional integral representation of $(F, e^{-tH_{\mathrm{PF}}}G)$, we construct probability spaces $(\mathcal{Q}_\beta, \Sigma_\beta, \mu_\beta)$, $\beta = 0, 1$, and Gaussian random variables $\hat{A}_\beta(\mathbf{f})$ indexed by $\mathbf{f} \in \oplus^3 L^2_{\mathrm{real}}(\mathbb{R}^{3+\beta})$ of mean zero

$$\mathbb{E}_{\mu_\beta}[\hat{A}_\beta(\mathbf{f})] = 0 \tag{3.2.6}$$

and covariance

$$\mathbb{E}_{\mu_\beta}[\hat{A}_\beta(\mathbf{f})\hat{A}_\beta(\mathbf{g})] = \mathfrak{q}_\beta(\mathbf{f}, \mathbf{g}). \tag{3.2.7}$$

Here, the bilinear forms $\mathfrak{q}_\beta$ on $(\oplus^3 L^2_{\mathrm{real}}(\mathbb{R}^{3+\beta})) \times (\oplus^3 L^2_{\mathrm{real}}(\mathbb{R}^{3+\beta}))$ are defined by

$$\mathfrak{q}_0(\mathbf{f}, \mathbf{g}) = \frac{1}{2}\int_{\mathbb{R}^3} \overline{\hat{\mathbf{f}}(k)} \cdot \delta^\perp(k)\hat{\mathbf{g}}(k)dk, \tag{3.2.8}$$

$$\mathfrak{q}_1(\mathbf{f}, \mathbf{g}) = \frac{1}{2}\int_{\mathbb{R}^4} \overline{\hat{\mathbf{f}}(\mathbf{k})} \cdot \delta^\perp(k)\hat{\mathbf{g}}(\mathbf{k})d\mathbf{k}, \tag{3.2.9}$$

where in what follows we set $\mathbf{k} = (k, k_0) \in \mathbb{R}^3 \times \mathbb{R}$. Note that $\delta^\perp(k)$ in (3.2.9) is independent of $k_0 \in \mathbb{R}$. The definitions of (3.2.8) and (3.2.9) are motivated by (3.2.3) and (3.2.4). It is, however, not straightforward to construct Gaussian random variables such as (3.2.6) and (3.2.7) since, as is seen in the definition, q_β is degenerate; q_β has nonzero null space. In order to construct a $\mathscr{Q}$-space for the Pauli–Fierz Hamiltonian, we have to take a quotient space with respect to the null space of q_β.

Proposition 3.7 (Gaussian process and quantized radiation field). *There exist probability spaces* $(\mathscr{Q}_\beta, \Sigma_\beta, \mu_\beta)$, $\beta = 0, 1$, *and Gaussian random variables* $(\hat{A}_\beta(\mathbf{f}), \mathbf{f} \in \oplus^3 L^2_{\mathrm{real}}(\mathbb{R}^{3+\beta}))$ *satisfying* (3.2.6) *and* (3.2.7).

Proof. Let $\mathscr{N}_\beta = \{\mathbf{f} \in \oplus^3 L^2_{\mathrm{real}}(\mathbb{R}^{3+\beta}) \mid \mathrm{q}_\beta(\mathbf{f}, \mathbf{f}) = 0\}$. Define the Hilbert space

$$\mathscr{K}_\beta = (\oplus^3 L^2_{\mathrm{real}}(\mathbb{R}^{3+\beta}))/\mathscr{N}_\beta$$

as a quotient space, and take the canonical map $\pi_\beta : \oplus^3 L^2_{\mathrm{real}}(\mathbb{R}^{3+\beta}) \to \mathscr{K}_\beta$. Then

$$(\pi_\beta(\mathbf{f}), \pi_\beta(\mathbf{g}))_{\mathscr{K}_\beta} = \mathrm{q}_\beta(\mathbf{f}, \mathbf{g}).$$

By Theorem 1.68, there exists a probability space $(\mathscr{Q}_\beta, \Sigma_\beta, \mu_\beta)$ and a family of Gaussian random variables $\{\phi_\beta(\pi_\beta(\mathbf{f})) \mid \pi_\beta(\mathbf{f}) \in \mathscr{K}_\beta\}$ such that Σ_β is full and

$$\mathbb{E}_{\mu_\beta}[\phi_\beta(\pi_\beta(\mathbf{f}))] = 0,$$
$$\mathbb{E}_{\mu_\beta}[\phi_\beta(\pi_\beta(\mathbf{f}))\phi_\beta(\pi_\beta(\mathbf{g}))] = \mathrm{q}_\beta(\mathbf{f}, \mathbf{g}).$$

Let, furthermore,

$$\hat{A}_\beta(\mathbf{f}) = \phi_\beta(\pi_\beta(\mathbf{f})), \quad \mathbf{f} \in \oplus^3 L^2_{\mathrm{real}}(\mathbb{R}^{3+\beta}).$$

Then $\hat{A}_\beta(\mathbf{f})$ satisfies (3.2.6) and (3.2.7). □

Define the μth component of $\hat{A}_\beta$ by

$$\hat{A}_{\beta,\mu}(f) = \hat{A}_\beta\Big(\bigoplus_{\nu=1}^{3} \delta_{\mu\nu} f\Big), \quad f \in L^2(\mathbb{R}^{3+\beta}).$$

In what follows, we denote

$$\begin{array}{llll} \text{(Minkowskian)} & \hat{A} = \hat{A}_0, & \mathrm{q} = \mathrm{q}_0, & \mathscr{Q} = \mathscr{Q}_0, \\ \text{(Euclidean)} & \hat{A}_{\mathrm{E}} = \hat{A}_1, & \mathrm{q}_{\mathrm{E}} = \mathrm{q}_1, & \mathscr{Q}_{\mathrm{E}} = \mathscr{Q}_1 \end{array}$$

using the subscript E for Euclidean objects.

In the same way as for the scalar field, we define the second quantization $\Gamma_{\beta\beta'}(T)$ for $T \in \mathscr{C}(L^2(\mathbb{R}^{3+\beta}) \to L^2(\mathbb{R}^{3+\beta'}))$ such that

$$T\mathcal{N}_\beta \subset \mathcal{N}_{\beta'}. \tag{3.2.10}$$

Here and in what follows, for the operator $S : L^2(\mathbb{R}^{3+\beta}) \to L^2(\mathbb{R}^{3+\beta'})$ we use the same notation S as for the operator

$$\oplus^3 S : \oplus^3 L^2(\mathbb{R}^{3+\beta}) \to \oplus^3 L^2(\mathbb{R}^{3+\beta'}), \quad (f_1, f_2, f_3) \mapsto (Sf_1, Sf_2, Sf_3)$$

for notational convenience, and we write simply $\hat{A}_\beta(T\mathbf{f})$ for $\hat{A}_\beta((\oplus^3 T)\mathbf{f})$, etc. By (3.2.10), $\pi_\beta(\mathbf{f}) \mapsto \pi_{\beta'}(T\mathbf{f})$ defines a map from $\mathscr{K}_\beta$ to $\mathscr{K}_{\beta'}$. Let thus

$$\Gamma_{\beta\beta'} : \mathscr{C}(L^2(\mathbb{R}^{3+\beta}) \to L^2(\mathbb{R}^{3+\beta'})) \to \mathscr{C}(L^2(\mathscr{Q}_\beta) \to L^2(\mathscr{Q}_{\beta'}))$$

be the functor defined by $\Gamma_{\beta\beta'}(T)1 = 1$ and

$$\Gamma_{\beta\beta'}(T){:}\prod_{i=1}^n \hat{A}_\beta(\mathbf{f}_i){:} = {:}\prod_{i=1}^n \hat{A}_{\beta'}(T\mathbf{f}_i){:}.$$

We write $\Gamma_{00} = \Gamma$, $\Gamma_{11} = \Gamma_{\mathrm{E}}$, and $\Gamma_{01} = \Gamma_{\mathrm{Int}}$.

Our goal here is to discuss the Euclidean quantum field associated with the Pauli–Fierz model. We introduce a family of isometries connecting the Minkowski field and the Euclidean quantum field, as well as the scalar field discussed in Section 1.1. In contrast with the $\mathscr{Q}$-space representation of the scalar field indexed by vectors in $\mathscr{M}$, the $\mathscr{Q}$-space representation $\hat{A}(\mathbf{f})$ for the Pauli–Fierz model is indexed by $\mathbf{f} \in \oplus^3 L^2(\mathbb{R}^3)$. So instead of the isometry $\tau_t : \mathscr{M} \to \mathscr{E}$ with the property $\tau_s^* \tau_t = e^{-|s-t|\omega(-i\nabla)}$, we define the isometry

$$\mathrm{j}_t : L^2_{\mathrm{real}}(\mathbb{R}^3) \to L^2_{\mathrm{real}}(\mathbb{R}^4)$$

through τ_t by pull-back from the operator $\mathcal{H}_M \to \mathcal{H}_E$ to the operator $L^2(\mathbb{R}^3) \to L^2(\mathbb{R}^4)$. Let $i_{-1/2} : L^2(\mathbb{R}^3) \to \mathcal{H}_M$ and $i_{-1} : L^2(\mathbb{R}^4) \to \mathcal{H}_E$ be given by

$$\widehat{i_{-1/2}f}(k) = \sqrt{\omega(k)}\hat{f}(k),$$
$$\widehat{i_{-1}f}(\mathbf{k}) = \frac{1}{\sqrt{2}}\sqrt{\omega(k)^2 + |k_0|^2}\hat{f}(\mathbf{k}).$$

Both $i_{-1/2}$ and i_{-1} are unitary operators.

Definition 3.8. Define the family of isometries $\mathrm{j}_t : L^2_{\mathrm{real}}(\mathbb{R}^3) \to L^2_{\mathrm{real}}(\mathbb{R}^4)$ by

$$\mathrm{j}_t = (i_{-1})^{-1} \circ \tau_t \circ i_{-1/2}, \quad t \in \mathbb{R}.$$

By this definition, j_t can be represented as

$$\widehat{\mathrm{j}_t f}(\mathbf{k}) = \frac{e^{-itk_0}}{\sqrt{\pi}} \frac{\sqrt{\omega(k)}}{\sqrt{\omega(k)^2 + |k_0|^2}}\hat{f}(k)$$

or in the position representation,

$$\mathrm{j}_t f(x, x_0) = \frac{1}{\sqrt{\pi}(2\pi)^2} \int_{\mathbb{R}^4} e^{-ik_0(t-x_0)+ik\cdot x} \frac{\sqrt{\omega(k)}}{\sqrt{\omega(k)^2 + |k_0|^2}} \hat{f}(k) d\mathbf{k}.$$

Note that for $f \in L^2_{\mathrm{real}}(\mathbb{R}^3)$, $\overline{\mathrm{j}_t f} = \mathrm{j}_t f$, meaning that j_t preserves realness, i. e., $\mathrm{j}_t : L^2_{\mathrm{real}}(\mathbb{R}^3) \to L^2_{\mathrm{real}}(\mathbb{R}^4)$, and $\mathrm{j}_t \mathscr{N}_0 \subset \mathscr{N}_1$. In fact, it follows that $\mathrm{q}_1(\mathrm{j}_t \mathbf{f}, \mathrm{j}_t \mathbf{f}) = \mathrm{q}_0(\mathbf{f}, \mathbf{f})$. By using the definition, it can also be seen that

$$\begin{aligned}(\mathrm{j}_s f, \mathrm{j}_t g) &= \frac{1}{\pi} \int_{\mathbb{R}^3} \omega(k) \bar{\hat{f}}(k) \hat{g}(k) \left(\int_{\mathbb{R}} \frac{e^{-i(t-s)k_0}}{\omega(k)^2 + k_0^2} dk_0 \right) dk \\ &= \frac{1}{\pi} \int_{\mathbb{R}^3} \omega(k) \bar{\hat{f}}(k) \hat{g}(k) \frac{\pi e^{-|s-t|\omega(k)}}{\omega(k)} dk = (\hat{f}, e^{-|s-t|\omega} \hat{g}),\end{aligned}$$

where we used

$$\int_{\mathbb{R}} \frac{e^{-itx}}{a^2 + x^2} dx = \frac{\pi e^{-|t|a}}{a}, \quad a > 0.$$

Hence

$$\mathrm{j}_s^* \mathrm{j}_t = e^{-|s-t|\hat{\omega}}, \quad s, t \in \mathbb{R}, \tag{3.2.11}$$

where $\hat{\omega} = \omega(-i\nabla)$.

Definition 3.9. Define the family of isometries $\mathrm{J}_t : L^2(\mathscr{Q}) \to L^2(\mathscr{Q}_{\mathrm{E}})$ by the second quantization of j_t, i. e.,

$$\mathrm{J}_t = \Gamma_{\mathrm{Int}}(\mathrm{j}_t).$$

By (3.2.11),

$$\mathrm{J}_t^* \mathrm{J}_s = \Gamma(e^{-|s-t|\hat{\omega}}) = e^{-|s-t|\hat{H}_{\mathrm{rad}}}, \quad s, t \in \mathbb{R},$$

holds with a self-adjoint operator $\hat{H}_{\mathrm{rad}}$. As in the case of the scalar field, J_t plays an important role in the functional integral representation, which connects Minkowskian quantum fields with Euclidean quantum fields.

Note the intertwining property of J_t. Let T be a bounded multiplication operator $(Tf)(k) = T(k)f(k)$ on $L^2(\mathbb{R}^3)$, and $\hat{T}$ the bounded pseudo-differential operator $T(-i\nabla)$. It follows that

$$\mathrm{j}_t \hat{T} f = (\hat{T} \otimes 1) \mathrm{j}_t f,$$

where $\hat{T} \otimes 1$ is a bounded operator on $L^2(\mathbb{R}^4)$ under the identification

$$L^2(\mathbb{R}^4) \cong L^2(\mathbb{R}^3) \otimes L^2(\mathbb{R}).$$

Thus

$$\mathrm{J}_t\Gamma(\hat{T}) = \Gamma_{\mathrm{E}}(\hat{T}\otimes 1)\mathrm{J}_t. \tag{3.2.12}$$

Let h be a real-valued multiplication operator on $L^2(\mathbb{R}^3)$ and define $\hat{h} = h(-i\nabla)$. Hence

$$\mathrm{J}_t\Gamma(e^{-i\hat{h}}) = \Gamma_{\mathrm{E}}(e^{-i\hat{h}\otimes 1})\mathrm{J}_t$$

follows from (3.2.12). Then

$$\mathrm{J}_t d\Gamma(\hat{h}) = d\Gamma_{\mathrm{E}}(\hat{h}\otimes 1)\mathrm{J}_t$$

also holds. We summarize these results below.

Proposition 3.10 (Intertwining property). *Let $\hat{h} = h(-i\nabla)$ be the self-adjoint operator with real-valued symbol h. Then $\mathrm{J}_t\Gamma(e^{-i\hat{h}}) = \Gamma_{\mathrm{E}}(e^{-i\hat{h}\otimes 1})\mathrm{J}_t$ and $\mathrm{J}_t d\Gamma(\hat{h}) = d\Gamma_{\mathrm{E}}(\hat{h}\otimes 1)\mathrm{J}_t$ hold.*

The Wiener–Itô–Segal isomorphism can be extended to the isomorphism θ_{PF} between $L^2(\mathscr{Q})$ and $\mathcal{F}_{\mathrm{rad}}$. Let $\theta_{\mathrm{PF}} : \mathcal{F}_{\mathrm{rad}} \to L^2(\mathscr{Q})$ be defined by

$$\begin{aligned}
&\theta_{\mathrm{PF}}\Omega_{\mathrm{PF}} = 1,\\
&\theta_{\mathrm{PF}}:\prod_{i=1}^{n} A(\mathbf{f}_i):\Omega_{\mathrm{PF}} = :\prod_{i=1}^{n}\hat{A}(\mathbf{f}_i):.
\end{aligned}$$

By the commutation relations, we see that

$$\left(:\prod_{i=1}^{n} A(\mathbf{f}_i):\Omega_{\mathrm{PF}}, :\prod_{j=1}^{n} A(\mathbf{g}_j):\Omega_{\mathrm{PF}}\right) = \left(:\prod_{i=1}^{n}\hat{A}(\mathbf{f}_i):, :\prod_{j=1}^{n}\hat{A}(\mathbf{g}_j):\right).$$

Thus θ_{PF} can be extended to the unitary operator from $\mathcal{F}_{\mathrm{rad}}$ to $L^2(\mathscr{Q})$. We denote $1\otimes\theta_{\mathrm{PF}} : \mathcal{H}_{\mathrm{PF}} \to L^2(\mathbb{R}^3)\otimes L^2(\mathscr{Q})$ by θ_{PF} for simplicity. Note that the inverse Fourier transform of $g(k,x) = e^{-ik\cdot x}\hat{\varphi}(k)/\sqrt{\omega(k)}$ equals $\check{g}(y,x) = \tilde{\varphi}(y-x)$, where

$$\tilde{\varphi} = (\hat{\varphi}/\sqrt{\omega})\check{} \in L^2_{\mathrm{real}}(\mathbb{R}^3).$$

Recall that $A_\mu(x) = A_\mu(\tilde{\varphi}(\cdot - x))$ for each x by the definition of $A_\mu(f)$. Notice that the test function of $A_\mu(f)$ is $\hat{f}$ instead of f. Hence the relations

$$\begin{aligned}
&\theta_{\mathrm{PF}}A_\mu(x)\theta_{\mathrm{PF}}^{-1} = \hat{A}_\mu(\tilde{\varphi}(\cdot - x)),\\
&\theta_{\mathrm{PF}}H_{\mathrm{rad}}\theta_{\mathrm{PF}}^{-1} = \hat{H}_{\mathrm{rad}},
\end{aligned}$$

follow directly. As a result,

$$\theta_{\mathrm{PF}}H_{\mathrm{PF}}\theta_{\mathrm{PF}}^{-1} = \frac{1}{2}(-i\nabla - e\hat{A})^2 + V + \hat{H}_{\mathrm{rad}},$$

with $\hat{A} = (\hat{A}_1, \hat{A}_2, \hat{A}_3)$ and

$$\hat{A}_\mu = \int_{\mathbb{R}^3}^{\oplus} \hat{A}_\mu(\tilde{\varphi}(\cdot - x))dx, \quad \mu = 1, 2, 3.$$

In what follows, we use the same notation H_{PF} for $\theta_{\mathrm{PF}} H_{\mathrm{PF}} \theta_{\mathrm{PF}}^{-1}$ and H_{rad} for $\hat{H}_{\mathrm{rad}}$.

Definition 3.11 (Pauli–Fierz Hamiltonian in function space). The *Pauli–Fierz Hamiltonian in function space* is defined by

$$H_{\mathrm{PF}} = \frac{1}{2}(-i\nabla - e\hat{A})^2 + V + H_{\mathrm{rad}}$$

in $L^2(\mathbb{R}^3) \otimes L^2(\mathcal{Q})$.

We use the shorthand $\mathcal{H}_{\mathrm{PF}}$ for $L^2(\mathbb{R}^3) \otimes L^2(\mathcal{Q})$ and $\mathcal{H}_{\mathrm{E}} = L^2(\mathbb{R}^3) \otimes L^2(\mathcal{Q}_{\mathrm{E}})$ standing for the Euclidean space.

Remark 3.12. Note that in the $\mathcal{Q}$-space representation the test function f of $\hat{A}_\mu(f)$ is in the position representation but the test function f of $A_\mu(f)$ is in the momentum representation. For instance, the dispersion relation in Fock representation is $\omega(k) = |k|$, while it becomes $\omega(-i\nabla) = |-i\nabla|$ in $\mathcal{Q}$-space representation.

3.2.3 Markov property

In the scalar quantum field theory discussed in Chapter 1, the family of projections $\mathrm{E}_t = \mathrm{I}_t \mathrm{I}_t^*$, $t \in \mathbb{R}$, was defined and it was seen that the Markov property plays an important role in the functional integral representation. Here, we address a similar problem for the Pauli–Fierz model.

In the Pauli–Fierz model, we will keep to the same notation e_t, E_t, etc., similar to the objects for the scalar field. Let

$$e_t = \mathrm{j}_t \mathrm{j}_t^*, \quad t \in \mathbb{R}.$$

Then $\{e_t\}_{t \in \mathbb{R}}$ is a family of projections from $L^2_{\mathrm{real}}(\mathbb{R}^4)$ to $\mathrm{Ran}\, \mathrm{j}_t$. The latter is a closed subspace of $L^2_{\mathrm{real}}(\mathbb{R}^4)$. Define

$$U_{[a,b]} = \overline{\mathrm{L.H.}\, \{f \in L^2_{\mathrm{real}}(\mathbb{R}^4) \mid f \in \mathrm{Ran}\, \mathrm{j}_t \text{ for some } t \in [a, b]\}}.$$

Let $e_{[a,b]} : L^2_{\mathrm{real}}(\mathbb{R}^4) \to U_{[a,b]}$ denote the orthogonal projection. In the same way as for the scalar quantum field theory, it follows that for $a \le b \le t \le c \le d$,

$$(1) \quad e_a e_b e_c = e_a e_c, \qquad (2) \quad e_{[a,b]} e_t e_{[c,d]} = e_{[a,b]} e_{[c,d]}.$$

The projections on $L^2(\mathscr{Q}_\mathrm{E})$ similarly are

$$\mathrm{E}_t = \mathrm{J}_t\mathrm{J}_t^* = \Gamma_\mathrm{E}(e_t), \quad \mathrm{E}_{[a,b]} = \Gamma_\mathrm{E}(e_{[a,b]}).$$

Let $\Sigma_{[a,b]}$ be the minimal σ-field generated by

$$\{\hat{A}_{\mathrm{E}_\mu}(f) \in L^2(\mathscr{Q}_\mathrm{E}) \mid f \in U_{[a,b]}, \mu = 1,2,3\}.$$

The set of $\Sigma_{[a,b]}$-measurable functions in $L^2(\mathscr{Q}_\mathrm{E})$ will be denoted by $\mathscr{E}_{[a,b]}$. The projection $\mathrm{E}_{[a,b]}$ and the set of $\Sigma_{[a,b]}$-measurable functions $\mathscr{E}_{[a,b]}$ satisfy

$$(1) \quad \mathrm{Ran}\,\mathrm{E}_{[a,b]} = \mathscr{E}_{[a,b]}, \qquad (2) \quad \mathrm{E}_{[a,b]}\mathrm{E}_t\mathrm{E}_{[c,d]} = \mathrm{E}_{[a,b]}\mathrm{E}_{[c,d]}$$

for $a \le b \le t \le c \le d$. Equation (2) above is the Markov property for the present model.

Next, we turn to discussing the family of projections E_t, $t \in \mathbb{R}$. Let

$$\|F\|_p = \left(\int_{\mathscr{Q}_\beta} |F(\phi)|^p d\mu_\beta\right)^{1/p}$$

be L^p-norm on $(\mathscr{Q}_\beta, \mu_\beta)$ and $(\cdot,\cdot)_2$ the $L^2(\mathscr{Q}_\beta)$-scalar product. We drop the subscript β unless confusion may arise. As it was seen, $\Gamma_\beta(T)$ for $\|T\| \le 1$ is a contraction on $L^2(\mathscr{Q}_\beta)$. It also has the *hypercontractivity* property. The following sharper result is available. See Theorem 1.67.

Proposition 3.13 (L^p–L^q boundedness). *Let $T : L^2(\mathbb{R}^{3+\beta}) \to L^2(\mathbb{R}^{3+\beta})$ be a contraction such that*

$$\| \oplus^3 T\|^2 \le (p-1)(q-1)^{-1} \le 1 \tag{3.2.13}$$

for some $1 \le p \le q$. Then $\Gamma_\beta(T)$ is a contraction from $L^p(\mathscr{Q}_\beta)$ to $L^q(\mathscr{Q}_\beta)$, i. e., for $\Phi \in L^p(\mathscr{Q}_\beta)$, $\Gamma_\beta(T)\Phi \in L^q(\mathscr{Q}_\beta)$, and

$$\|\Gamma_\beta(T)\Phi\|_q \le \|\Phi\|_p.$$

Note that $\Gamma(e^{-t\tilde{\omega}_\nu})$, $\nu > 0$, is a contraction from L^p to L^q if t is sufficiently large but it fails to be so for $\nu = 0$.

Lemma 3.14. *Suppose that $\nu > 0$ and let $a \le b < t < c \le d$, $F \in \mathscr{E}_{[a,b]}$, and $G \in \mathscr{E}_{[c,d]}$. Take $1 \le r < \infty$, $r < p$, and $r < q$. Suppose, moreover, that $e^{-2\nu(c-b)} \le (p/r-1)(q/r-1) \le 1$, $F \in L^p(\mathscr{Q}_\mathrm{E})$, and $G \in L^q(\mathscr{Q}_\mathrm{E})$. Then $FG \in L^r(\mathscr{Q}_\mathrm{E})$ and $\|FG\|_r \le \|F\|_p\|G\|_q$. In particular, for r such that*

$$r \in \left[1, \frac{2}{1+e^{-\nu(c-b)}}\right],$$

we have $\|FG\|_r \le \|F\|_2\|G\|_2$.

Proof. Let

$$F_N = \begin{cases} F, & |F| < N, \\ 0, & |F| \geq N, \end{cases} \quad \text{and} \quad G_N = \begin{cases} G, & |G| < N, \\ 0, & |G| \geq N. \end{cases}$$

Then $|F_N|^r \in \mathscr{E}_{[a,b]}$, $|G_N|^r \in \mathscr{E}_{[c,d]}$, and

$$\int_{\mathscr{Q}_{\mathrm{E}}} |F_N|^r |G_N|^r d\mu_{\mathrm{E}} = (E_{[a,b]}|F_N|^r, E_{[c,d]}|G_N|^r)_2 = (|F_N|^r, \Gamma_{\mathrm{E}}(e_{[a,b]}e_{[c,d]})|G_N|^r)_2.$$

Note that $e_{[a,b]}e_{[c,d]}$ satisfies

$$\|e_{[a,b]}e_{[c,d]}\|^2 = \|e_{[a,b]}e_b e_c e_{[c,d]}\|^2 \leq \|\mathrm{j}_b^* \mathrm{j}_c\|^2 = \|e^{-|c-b|\omega_\nu}\|^2 \leq e^{-2\nu(c-b)} \leq (p/r-1)(q/r-1).$$

Thus by the Hölder inequality,

$$\|F_N G_N\|_r^r \leq \| |F_N|^r \|_{q/r} \|\Gamma_{\mathrm{E}}(e_{[a,b]}e_{[c,d]})|G_N|^r\|_s, \tag{3.2.14}$$

where $1 = 1/s + r/q$. Since

$$\|e_{[a,b]}e_{[c,d]}\|^2 \leq (p/r-1)(q/r-1) = (p/r-1)(s-1)^{-1} \leq 1,$$

by Proposition 3.13 it is seen that $\|\Gamma_{\mathrm{E}}(e_{[a,b]}e_{[c,d]})|G_N|^r\|_s \leq \| |G_N|^r \|_{p/r}$. Together with (3.2.14), this yields

$$\|F_N G_N\|_r \leq \|F_N\|_q \|G_N\|_p \leq \|F\|_q \|G\|_p. \tag{3.2.15}$$

Taking the limit as $N \to \infty$ on both sides of (3.2.15), by monotone convergence the lemma follows. □

An immediate consequence is as follows.

Corollary 3.15. *Let $\nu > 0$ and $F, G \in L^2(\mathscr{Q})$. Suppose that $\Phi \in \bigcap_{p \geq 1} L^p(\mathscr{Q}_{\mathrm{E}})$. Then $(\mathrm{J}_a F)(\mathrm{J}_b G)\Phi \in L^1(\mathscr{Q}_{\mathrm{E}})$ and*

$$\int_{\mathscr{Q}_{\mathrm{E}}} |(\mathrm{J}_a F)(\mathrm{J}_b G)\Phi| d\mu_{\mathrm{E}} \leq \|\Phi\|_s \|F\|_2 \|G\|_2 \tag{3.2.16}$$

for arbitrary $a \neq b$, where

$$s = s(a,b) = \frac{2}{1 - e^{-\nu(b-a)}}. \tag{3.2.17}$$

If $a \neq b$, then $\mathrm{J}_a^* \Phi \mathrm{J}_b$ is a bounded operator and $\|\mathrm{J}_a^* \Phi \mathrm{J}_b\| \leq \|\Phi\|_s$.

Proof. Let $a < b$, and $r = \frac{2}{1+e^{-\nu(b-a)}} > 1$ and $s > 1$ be such that $1/r + 1/s = 1$, i. e., $s = r/(r-1)$. Without loss of generality, we can assume that Φ is real-valued. Then by Lemma 3.14,

$$|(\mathrm{J}_a F, \Phi \mathrm{J}_b G)_2| \le \|\Phi\|_s \|(\mathrm{J}_a F)(\mathrm{J}_b G)\|_r \le \|\Phi\|_s \|\mathrm{J}_a F\|_2 \|\mathrm{J}_b G\|_2 = \|\Phi\|_s \|F\|_2 \|G\|_2.$$

Then the corollary follows. □

Example 3.16. Since $e^{sA_{\mathrm{E}}(f)} \in L^1(\mathscr{Q}_{\mathrm{E}})$ for any $s \in \mathbb{R}$, we have

$$e^{A_{\mathrm{E}}(f)} \in \bigcap_{p\ge1} L^p(\mathscr{Q}_{\mathrm{E}}).$$

Let $\nu > 0$, and $a \ne b$. By Corollary 3.15, we can see that $\mathrm{J}_a^* e^{A_{\mathrm{E}}(f)} \mathrm{J}_b$ is a bounded operator for $f \in \oplus^2 L^2(\mathbb{R}^3)$ such that

$$\|\mathrm{J}_a^* e^{A_{\mathrm{E}}(f)} \mathrm{J}_b\| \le \|e^{sA_{\mathrm{E}}(f)}\|_{L^1(\mathscr{Q}_{\mathrm{E}})}^{1/s},$$

where s is given by (3.2.17).

In the case where $\nu = 0$, we can also show the boundedness of $\mathrm{J}_a^* e^{A_{\mathrm{E}}(f)} \mathrm{J}_b$, but we need an additional condition on f.

Theorem 3.17. *Suppose that $f, f/\sqrt{\omega} \in \oplus^2 L^2(\mathbb{R}^3)$, and $a \ne b$. Then $\mathrm{J}_a^* e^{A_{\mathrm{E}}(f)} \mathrm{J}_b$ is a bounded operator.*

Proof. This is a minor modification of Theorem 1.87. Then we omit the proof. □

3.3 Functional integral representation for the Pauli–Fierz Hamiltonian

3.3.1 Hilbert space-valued stochastic integrals

In this section, we define a Hilbert space-valued stochastic integral. It will be first explained in some generality and then applied to the Pauli–Fierz model. Using the Trotter product formula and the factorization formula $e^{-|s-t|H_{\mathrm{rad}}} = \mathrm{J}_t^* \mathrm{J}_s$, we derive the functional integral representation of $e^{-tH_{\mathrm{PF}}}$.

Let $\mathscr{K}$ be a Hilbert space and define

$$C^n(\mathbb{R}^3; \mathscr{K}) = \{f : \mathbb{R}^3 \to \mathscr{K} \mid f \text{ is } n \text{ times continuously strongly differentiable}\}$$

and

$$C_{\mathrm{b}}^n(\mathbb{R}^3; \mathscr{K}) = \Big\{f \in C^n(\mathbb{R}^3; \mathscr{K}) \,\Big|\, \sup_{|z|\le n, x\in\mathbb{R}^3} \|\partial^z f(x)\|_{\mathscr{K}} < \infty\Big\},$$

where $|z| = z_1 + z_2 + z_3$ for $z = (z_1, z_2, z_3)$ and $\partial^z = \partial_{x_1}^{z_1}\partial_{x_2}^{z_2}\partial_{x_3}^{z_3}$ denotes strong derivative. We set

$$L^2(\mathscr{X}) = L^2(\mathscr{X}, d\mathcal{W}^x), \quad L^2(\mathfrak{X}) = L^2(\mathfrak{X}, d\mathcal{W}^x),$$

where we recall that $\mathscr{X} = C([0,\infty);\mathbb{R}^3)$ and $\mathfrak{X} = C(\mathbb{R};\mathbb{R}^3)$. The proof of the following lemma is straightforward and similar to the case of real-valued processes.

Lemma 3.18. *Let $f \in C_{\mathrm{b}}^1(\mathbb{R}\times\mathbb{R}^3;\mathscr{K})$. The sequence defined by*

$$\mathrm{J}_n^\mu(f) = \sum_{j=1}^{2^n} f\left(\frac{j-1}{2^n}t, B_{\frac{j-1}{2^n}t}\right)\left(B^\mu_{\frac{j}{2^n}t} - B^\mu_{\frac{j-1}{2^n}t}\right)$$

is a Cauchy sequence in $L^2(\mathscr{X})\otimes\mathscr{K}$.

Definition 3.19 (Hilbert space-valued stochastic integral). For $f \in C_{\mathrm{b}}^1(\mathbb{R}\times\mathbb{R}^3;\mathscr{K})$, the limit

$$\int_0^t f(s, B_s)dB_s^\mu = \operatorname*{s-lim}_{n\to\infty} \mathrm{J}_n^\mu(f)$$

defines a *$\mathscr{K}$-valued stochastic integral.*

By the above definition,

$$\mathbb{E}\left[\left(\int_0^t f(s,B_s)dB_s^\mu, \int_0^t g(s,B_s)dB_s^\nu\right)_{\mathscr{K}}\right] = \delta_{\mu\nu}\mathbb{E}\left[\int_0^t (f(s,B_s), g(s,B_s))_{\mathscr{K}} ds\right].$$

In particular, the Itô isometry

$$\mathbb{E}\left[\left\|\int_0^t f(s,B_s)dB_s^\mu\right\|_{\mathscr{K}}^2\right] = \mathbb{E}\left[\int_0^t \|f(s,B_s)\|_{\mathscr{K}}^2 ds\right]$$

holds. Similar to the case of real-valued processes, the result below follows.

Corollary 3.20. *Let $f \in C_{\mathrm{b}}^2(\mathbb{R}^3;\mathscr{K})$ and*

$$S_n^\mu(f) = \sum_{j=1}^{2^n} (f(B_{\frac{j}{2^n}t}) + f(B_{\frac{j-1}{2^n}t}))(B^\mu_{\frac{j}{2^n}t} - B^\mu_{\frac{j-1}{2^n}t}).$$

Then

$$\operatorname*{s-lim}_{n\to\infty} S_n^\mu(f) = \int_0^t f(B_s)dB_s^\mu + \frac{1}{2}\int_0^t \partial_\mu f(B_s)ds$$

holds in $L^2(\mathscr{X})\otimes\mathscr{K}$.

Example 3.21. We will make the following specific choices for the functional integral representation of the Pauli–Fierz model below. Let $\hat{\lambda}, \omega\hat{\lambda} \in L^2(\mathbb{R}^3)$ and define the map

$$\xi : \mathbb{R}^3 \ni x \mapsto \lambda(\cdot - x) \in L^2(\mathbb{R}^3).$$

Thus $\xi \in C_{\mathrm{b}}^1(\mathbb{R}^3; L^2(\mathbb{R}^3))$ by the assumption $\omega\hat{\lambda} \in L^2(\mathbb{R}^3)$, and we can define an $L^2(\mathbb{R}^3)$-valued stochastic integral

$$\int_0^t \xi(B_s)dB_s^\mu = \int_0^t \lambda(\cdot - B_s)dB_s^\mu.$$

Later on, we will construct the functional integral through the Euclidean quantum field, and will use an $L^2(\mathbb{R}^4)$-valued stochastic integral of the form

$$\int_0^t \mathrm{j}_s\tilde{\varphi}(\cdot - B_s)dB_s^\mu. \tag{3.3.1}$$

However, since

$$\frac{\|\mathrm{j}_t f - \mathrm{j}_s f\|^2}{|s-t|^2} = 2\left(\hat{f}, \frac{(1-e^{-|s-t|\omega})}{|s-t|}\hat{f}\right)\frac{1}{|s-t|}$$

diverges as $t \to s$ for $\hat{f} \in D(\omega)$, $\mathbb{R} \times \mathbb{R}^3 \ni (s,x) \mapsto \mathrm{j}_s\tilde{\varphi}(\cdot - x) \in L^2(\mathbb{R}^4)$ is not strongly differentiable in $s \in \mathbb{R}$. Then $\mathrm{j}_s\tilde{\varphi}(\cdot - x) \notin C_{\mathrm{b}}^1(\mathbb{R}_s \times \mathbb{R}_x^3; L^2(\mathbb{R}^4))$. Therefore, we need to give a proper definition of (3.3.1).

Lemma 3.22. *If $\hat{\lambda}, \omega\hat{\lambda} \in L^2(\mathbb{R}^3)$, then for each $\mu = 1, 2, 3$,*

$$S_n^\mu(\lambda) = \sum_{j=1}^{2^n} \mathrm{j}_{(j-1)t/2^n}\lambda(\cdot - B_{(j-1)t/2^n})(B_{jt/2^n}^\mu - B_{(j-1)t/2^n}^\mu), \quad n = 1, 2, 3, \ldots,$$

is a Cauchy sequence in $L^2(\mathscr{X}) \otimes L^2(\mathbb{R}^4)$.

Proof. Fix an μ. Write $S_n = S_n^\mu(\lambda)$ and $\eta_* = \mathrm{j}_*\lambda(\cdot - B_*)$. Then $S_{n+1} - S_n = \sum_{m=1}^{2^n} a_m$, where

$$a_m = (\eta_{(2m-1)t/2^{n+1}} - \eta_{(2m-2)t/2^{n+1}})(B_{2mt/2^{n+1}}^\mu - B_{(2m-1)t/2^{n+1}}^\mu).$$

Thus $\mathbb{E}^x[(a_i, a_j)_{L^2(\mathbb{R}^4)}] = 0$ for $i \neq j$, since $B_{2jt/2^{n+1}}^\mu - B_{(2j-1)t/2^{n+1}}^\mu$ is independent of the rest of a_i and a_j for $j > i$, and $\mathbb{E}^x[B_{2jt/2^{n+1}}^\mu - B_{(2j-1)t/2^{n+1}}^\mu] = 0$. Hence

$$\mathbb{E}^x[\|S_{n+1} - S_n\|^2] = \sum_{m=1}^{2^n} \mathbb{E}^x[\|\eta_{(2m-1)t/2^{n+1}} - \eta_{(2m-2)t/2^{n+1}}\|^2]\frac{t}{2^{n+1}}.$$

Since $\|\mathrm{j}_t f - \mathrm{j}_s g\|^2 = \|f-g\|^2 + 2(\hat{f}, (1-e^{-|s-t|\omega})\hat{g})$ and $\|\lambda(\cdot - X) - \lambda(\cdot - Y)\| \le |X-Y|\|\omega\hat{\lambda}\|$, we have

$$\|\eta_{(2m-1)t/2^{n+1}} - \eta_{(2m-2)t/2^{n+1}}\|^2 \le \|\omega\hat{\lambda}\|^2 |B^\mu_{(2m-1)t/2^{n+1}} - B^\mu_{(2m-2)t/2^{n+1}}|^2 + 2\frac{t}{2^{n+1}}\|\lambda\|\|\omega\hat{\lambda}\|.$$

Finally, we conclude that

$$\left(\mathbb{E}^X[\|S_m - S_n\|^2]\right)^{1/2} \le \left(\frac{2\|\lambda\|\|\omega\hat{\lambda}\| + \|\omega\hat{\lambda}\|^2}{2}\right)^{1/2} \sum_{j=n+1}^{m} \frac{t}{2^{(j+1)/2}},$$

thus $(S_n)_{n\in\mathbb{N}}$ is a Cauchy sequence. □

Definition 3.23 ($L^2(\mathbb{R}^4)$-valued stochastic integral). Let $\hat{\lambda}, \omega\hat{\lambda} \in L^2(\mathbb{R}^3)$. Then

$$\int_0^t \mathrm{j}_s\lambda(\cdot - B_s)dB^\mu_s = \operatorname*{s-lim}_{n\to\infty} S^\mu_n, \quad \mu = 1,2,3,$$

defines an $L^2(\mathbb{R}^4)$-valued stochastic integral, where the strong limit is in the topology of $L^2(\mathscr{X}) \otimes L^2(\mathbb{R}^4)$.

By the definition, it is seen that

$$\mathbb{E}^X\left[\left(\int_0^t \mathrm{j}_s\lambda(\cdot - B_s)dB^\mu_s, \int_0^t \mathrm{j}_s\rho(\cdot - B_s)dB^\nu_s\right)\right] = \delta_{\mu\nu}\mathbb{E}^X\left[\int_0^t (\lambda(\cdot - B_s), \rho(\cdot - B_s))ds\right].$$

The integral $\int_0^t \mathrm{j}_s\lambda(\cdot - B_s)dB^\mu_s$ can also be defined as a limit of a sequence of $L^2(\mathbb{R}^4)$-valued stochastic integrals, which will be used in the construction of the functional integral representation for the Pauli–Fierz Hamiltonian.

Corollary 3.24. *Suppose that $\hat{\lambda}, \omega\hat{\lambda} \in L^2(\mathbb{R}^3)$ and let*

$$\tilde{S}^\mu_n = \sum_{k=1}^{2^n} \int_{(k-1)t/2^n}^{kt/2^n} \mathrm{j}_{(k-1)t/2^n}\lambda(\cdot - B_s)dB^\mu_s.$$

Then

$$\operatorname*{s-lim}_{n\to\infty} \tilde{S}^\mu_n = \int_0^t \mathrm{j}_s\lambda(\cdot - B_s)dB^\mu_s$$

in $L^2(\mathscr{X}) \otimes L^2(\mathbb{R}^4)$.

Proof. Using that j_s is an isometry, we obtain

$$\mathbb{E}^x\left[\left\|\tilde{S}_n^\mu - \int_0^t \mathrm{j}_s\lambda(\cdot - B_s)dB_s^\mu\right\|^2\right] = \sum_{k=1}^{2^n}\int_{(k-1)t/2^n}^{kt/2^n} \mathbb{E}^x[\|\lambda(\cdot - B_s) - \lambda(\cdot - B_{(k-1)t/2^n})\|^2]ds \le C\frac{t}{2^n}.$$

Then the corollary follows. □

3.3.2 Functional integral representation

The functional integral representation for the decoupled Hamiltonian comes about surprisingly directly. A combination of the functional integral representation for e^{-tH_p} and the equality $e^{-tH_\mathrm{rad}} = \mathrm{J}_0^*\mathrm{J}_t$ gives

$$(F, e^{-tH_{\mathrm{PF},0}}G)_{\mathcal{H}_\mathrm{PF}} = (F, e^{-tH_\mathrm{p}}e^{-tH_\mathrm{rad}}G)_{\mathcal{H}_\mathrm{PF}} = (\mathrm{J}_0F, e^{-tH_\mathrm{p}}\mathrm{J}_tG)_{\mathcal{H}_\mathrm{E}}.$$

Proposition 3.25. *Let $F, G \in \mathcal{H}_\mathrm{PF}$. Then*

$$(F, e^{-tH_{\mathrm{PF},0}}G)_{\mathcal{H}_\mathrm{PF}} = \int_{\mathbb{R}^3} \mathbb{E}^x[e^{-\int_0^t V(B_s)ds}(\mathrm{J}_0F(B_0), \mathrm{J}_tG(B_t))_{L^2(\mathcal{Q}_\mathrm{E})}]dx.$$

Our next goal is to derive a functional integral representation for e^{-tH_PF}. Since

$$H_\mathrm{PF}(\hat{A}) = \frac{1}{2}(-i\nabla - e\hat{A})^2$$

has a similar form as $H_\mathrm{p}(a) = \frac{1}{2}(-i\nabla - a)^2$ in the Schrödinger Hamiltonian with a vector potential a, the procedure of obtaining this functional integral will narrow down to a combination of the construction for $e^{-tH_\mathrm{p}(a)}$. We have seen that

$$(f, e^{-tH_\mathrm{p}(a)}g) = \int_{\mathbb{R}^3} \mathbb{E}^x[\overline{f(B_0)}e^{-i\int_0^t a(B_s)\circ dB_s}g(B_t)]dx. \tag{3.3.2}$$

The proof of formula (3.3.2) is proceeded as follows. Define the family of integral operators $\varrho_s : L^2(\mathbb{R}^d) \to L^2(\mathbb{R}^d)$ by

$$\varrho_s f(x) = \int_{\mathbb{R}^d} \Pi_s(x-y)e^{ih(x,y)}f(y)dy, \quad s \ge 0,$$

where $\varrho_0 f(x) = f(x)$ and $h(x,y) = (1/2)(a(x) + a(y)) \cdot (x-y)$. Notice that ϱ_s is symmetric and $\|\varrho_s\| \le 1$. For $f, g \in L^2(\mathbb{R}^d)$,

$$(g, \varrho_{t/2^n}^{2^n} f) = \int_{\mathbb{R}^d} \overline{g(x)}e^{i\sum_{j=1}^{2^n} h(x_{j-1},x_j)}f(x_n)\Big(\prod_{j=1}^{2^n}\Pi_{t/2^n}(x_j - x_{j-1})\Big)\prod_{j=0}^{2^n} dx_j$$

with $x_0 = x$. Therefore,

$$\lim_{n\to\infty}(g, \varrho_{t/2^n}^{2^n} f) = \int_{\mathbb{R}^d} \mathbb{E}^x\left[\overline{g(B_0)} f(B_t) \exp\left(-i\int_0^t a(B_s)\circ dB_s\right)\right] dx.$$

By the Riesz representation theorem, there exists a C_0-semigroup $\{S_t : t \geq 0\}$ such that

$$\lim_{n\to\infty}(g, \varrho_{t/2^n}^{2^n} f) = (g, S_t f).$$

It is directly seen that $\lim_{s\to 0}(d/ds)(g, \varrho_s f) = (g, -H(a)f)$, and hence

$$\lim_{t\to 0}(g, t^{-1}(1-\varrho_t)f) = (g, H_{\mathrm{p}}(a)f).$$

Thus we conclude that

$$\left(\frac{1}{t}(e^{-tA}-1)g, f\right) = \int_0^1 (-H_{\mathrm{p}}(a)g, e^{-tsA}f)ds. \tag{3.3.3}$$

As $t \to 0$ on both sides above, we get $(g, Af) = (H_{\mathrm{p}}(a)g, f)$, implying that $Ag = H_{\mathrm{p}}(a)g$. Hence $A = H_{\mathrm{p}}(a)$ and (3.3.2) follows for $V = 0$. Let now V be continuous and bounded. From the Trotter product formula, we have

$$\begin{aligned}(f, e^{-tH_{\mathrm{p}}(a)}g) &= \lim_{n\to\infty}(f, (e^{-(t/n)H_{\mathrm{p}}(a)}e^{-(t/n)V})^n g) \\ &= \int_{\mathbb{R}^d} \mathbb{E}^x[\overline{f(B_0)} e^{-i\int_0^t a(B_s)\circ dB_s} e^{-\int_0^t V(B_s)ds}]dx.\end{aligned}$$

At first glance, it can be expected that the functional integral representation for $(F, e^{-tH_{\mathrm{PF}}(\hat{A})}G)$ goes similarly with a replaced by $\hat{A}$ which is $L^2(\mathscr{Q})$-valued multiplication operator. There is, however, a difference. Although in the classical case $\int_0^t a(B_s)\circ dB_s$ in (3.3.2), integrand a does not depend on time s explicitly, in the Pauli–Fierz model there is dependence on s and the integral $\int_0^t \mathrm{j}_s\tilde{\varphi}(\cdot - B_s)dB_s^\mu$ appears instead of the Stratonovich integral. This time dependence originates from the semigroup of the free field Hamiltonian $e^{-tH_{\mathrm{rad}}}$. An extra difficulty is that $\mathrm{j}_s\hat{\varphi}(\cdot - x)$ is not differentiable in s.

Here is an outline of our construction of the functional integral representation of $(F, e^{-tH_{\mathrm{PF}}}G)$. Replicating the second proof of functional integral representation for Schrödinger operator with vector potentials we notice that

$$e^{-tH_{\mathrm{PF}}(\hat{A})} = \operatorname*{s-lim}_{n\to\infty}(P_{t/n})^n$$

with a family of suitable integral operators $P_s : L^2(\mathbb{R}^3; L^2(\mathscr{Q})) \to L^2(\mathbb{R}^3; L^2(\mathscr{Q}))$, $s \geq 0$. Thus by virtue of $\mathrm{J}_s^*\mathrm{J}_t = e^{-|s-t|H_{\mathrm{rad}}}$, we have

$$e^{-tH_{\mathrm{PF}}} = \operatorname*{s\text{-}lim}_{n\to\infty} \lim_{m_1\to\infty} \cdots \lim_{m_n\to\infty} \mathrm{J}_0^* \left(\prod_{j=1}^{n} \mathrm{J}_{tj/n} (P_{t/nm_j})^{m_j} \mathrm{J}_{tj/n}^* \right) \mathrm{J}_t,$$

and, by using the Markov property of $\mathrm{E}_s = \mathrm{J}_s \mathrm{J}_s^*$, we deduce

$$\begin{aligned} e^{-tH_{\mathrm{PF}}} &= \operatorname*{s\text{-}lim}_{n\to\infty} \lim_{m_1\to\infty} \cdots \lim_{m_n\to\infty} \mathrm{J}_0^* \left(\prod_{j=1}^{n} \mathrm{E}_{tj/n} (P_{t/nm_j,j})^{m_j} \mathrm{E}_{tj/n} \right) \mathrm{J}_t \\ &= \operatorname*{s\text{-}lim}_{n\to\infty} \lim_{m_1\to\infty} \cdots \lim_{m_n\to\infty} \mathrm{J}_0^* \left(\prod_{j=1}^{n} (P_{t/nm_j,j})^{m_j} \right) \mathrm{J}_t, \end{aligned}$$

where $P_{S,T}$ denotes an operator acting on the Euclidean space $\mathcal{H}_{\mathrm{E}}$. Finally, we compute $\prod_{j=1}^{n} (P_{t/nm_j,j})^{m_j}$ which gives the functional integral formula of $(F, e^{-tH_{\mathrm{PF}}} G)$. In what follows, we will turn this argument rigorous. We first show the functional integral representation for the Pauli–Fierz Hamiltonian with sufficiently smooth and bounded V. After establishing it, we extend it to more general potentials V.

Theorem 3.26 (Functional integral representation for Pauli–Fierz Hamiltonian). *We suppose that $V \in C_0^\infty(\mathbb{R}^3)$. Then*

$$(F, e^{-tH_{\mathrm{PF}}} G) = \int_{\mathbb{R}^3} \mathbb{E}^x [e^{-\int_0^t V(B_s) ds} (\mathrm{J}_0 F(B_0), e^{-ie\hat{A}_{\mathrm{E}}(K_t)} \mathrm{J}_t G(B_t))_{L^2(\mathcal{Q}_{\mathrm{E}})}] dx. \tag{3.3.4}$$

Here, K_t denotes the $\oplus^3 L^2(\mathbb{R}^3)$-valued stochastic integral given by

$$K_t = \bigoplus_{\mu=1}^{3} \int_0^t \mathrm{j}_s \tilde{\varphi}(\cdot - B_s) dB_s^\mu.$$

Proof. For the main part of the proof, we may and do assume $V = 0$, and in this proof we write $\hat{A}_{\mathrm{E}_s}(x) = \hat{A}_{\mathrm{E}}(\oplus^3 \mathrm{j}_s \tilde{\varphi}(\cdot - x))$. Define the family of symmetric contraction operators $P_s : \mathcal{H}_{\mathrm{PF}} \to \mathcal{H}_{\mathrm{PF}}$ by

$$P_s F(x) = \int_{\mathbb{R}^3} \Pi_s(x-y) e^{i\mathfrak{h}(x,y)} F(y) dy, \quad s > 0,$$

with $P_0 F = F$, where $\Pi_s(x) = (2\pi s)^{-3/2} \exp(-|x|^2/2s)$ is the heat kernel and

$$\mathfrak{h}(x,y) = \frac{e}{2} \left(\hat{A}(x) + \hat{A}(y) \right) (x-y).$$

Here, P_s is the counterpart of the quantum field for ϱ_s defined above for a Schrödinger operator with a vector potential a. By a direct computation,

$$(F,(P_{t/2^n})^{2^n}G)=\int_{\mathbb{R}^3}dx\int_{(\mathbb{R}^3)^{2^n}}\overline{F(x)}\exp\left(i\sum_{j=1}^{2^n}\mathfrak{h}_j\right)G(x_{2^n})\left(\prod_{j=1}^{2^n}\Pi_{t/2^n}(x_{j-1}-x_j)\right)\prod_{j=1}^{2^n}dx_j$$

with $x=x_0$, $\mathfrak{h}_j=\mathfrak{h}_j(x_{j-1},x_j)=(e/2)(\hat{A}(x_{j-1})+\hat{A}(x_j))\cdot(x_{j-1}-x_j)$. This can be expressed by using Brownian motion as

$$(F,(P_{t/2^n})^{2^n}G)=\int_{\mathbb{R}^3}\mathbb{E}^x[(F(B_0),e^{-ie\hat{A}(L_n)}G(B_t))]dx,$$

where

$$L_n=\bigoplus_{\mu=1}^{3}\sum_{m=0}^{2^n}(\tilde{\varphi}(\cdot-B_{t\frac{m}{2^n}})+\tilde{\varphi}(\cdot-B_{t\frac{m-1}{2^n}}))(B^\mu_{t\frac{m}{2^n}}-B^\mu_{t\frac{m-1}{2^n}}).$$

It is seen that

$$L_n\to L_t=\bigoplus_{\mu=1}^{3}\int_0^t\tilde{\varphi}(\cdot-B_s)dB^\mu_s$$

as $n\to\infty$ strongly in $\oplus^3\left(L^2(\mathscr{X})\otimes L^2(\mathbb{R}^3)\right)$, implying that for $F,G\in\mathcal{H}_{\mathrm{PF}}$,

$$\lim_{n\to\infty}(F,(P_{t/2^n})^{2^n}G)=\int_{\mathbb{R}^3}\mathbb{E}^x[(F(B_0),e^{-ie\hat{A}(L_t)}G(B_t))]dx. \tag{3.3.5}$$

From (3.3.5), it follows that $|\lim_{n\to\infty}(F,(P_{t/2^n})^{2^n}G)|\le\|F\|\|G\|$. Hence for each $t\ge0$, there exists a symmetric bounded operator S_t such that

$$\lim_{n\to\infty}(F,(P_{t/2^n})^{2^n}G)=(F,S_tG).$$

Since $(P_{t/2^n})^{2^n}$ is uniformly bounded as $\|(P_{t/2^n})^{2^n}\|\le1$, the above weak convergence improves to

$$\operatorname*{s-lim}_{n\to\infty}(P_{t/2^n})^{2^n}=S_t,\quad t\ge0. \tag{3.3.6}$$

Furthermore, by (3.3.5),

$$(F,S_tG)=\int_{\mathbb{R}^3}\mathbb{E}^x[(F(B_0),e^{-ie\hat{A}(L_t)}G(B_t))]dx. \tag{3.3.7}$$

Thus

$$\begin{aligned}(F,S_sS_tG)&=\lim_{n\to\infty}\lim_{m\to\infty}(F,(P_{s/2^n})^{2^n}(P_{t/2^m})^{2^m}G)\\&=\int_{\mathbb{R}^3}\mathbb{E}^x[(F(B_0),e^{-ie\hat{A}(L_{(s+t)})}G(B_t))]dx=(F,S_{s+t}G),\end{aligned}$$

giving the semigroup property $S_s S_t = S_{s+t}$, $s,t \geq 0$. It can also be seen that $\text{w-lim}_{t\to 0} S_t = 1$ by (3.3.7), which further implies that $\text{s-lim}_{t\to 0} S_t = 1$, since $\lim_{t\to 0}\|S_t F\|^2 = \lim_{t\to 0}(F, S_{2t}F) = \|F\|^2$. Finally, it is trivial to see that $S_0 = 1$. Putting these together, we conclude that $\{S_t : t \geq 0\}$ is a symmetric C_0-semigroup, thus there exists a unique self-adjoint operator $\hat{H}_{\mathrm{PF}}(\hat{A})$ such that $S_t = e^{-t\hat{H}_{\mathrm{PF}}(\hat{A})}$, $t \geq 0$, i. e.,

$$(F, e^{-t\hat{H}_{\mathrm{PF}}(\hat{A})}G) = \int_{\mathbb{R}^3} \mathbb{E}^x[(F(B_0), e^{-ie\hat{A}(L_t)}G(B_t))]dx. \tag{3.3.8}$$

Furthermore, we have

$$\lim_{t\to 0}(F, t^{-1}(1 - P_t)G) = (F, H_{\mathrm{PF}}(\hat{A})G) \tag{3.3.9}$$

for $F, G \in C_0^\infty(\mathbb{R}^3)\hat{\otimes}\mathcal{F}_{\mathrm{rad,fin}}$ in Lemma 3.27 below. This leads to

$$(t^{-1}(e^{-t\hat{H}_{\mathrm{PF}}(\hat{A})} - 1)F, G) = \lim_{n\to\infty}(t^{-1}((P_{t/2^n})^{2^n} - 1)F, G) = -\int_0^1 (H_{\mathrm{PF}}(\hat{A})F, e^{-ts\hat{H}_{\mathrm{PF}}(\hat{A})}G)ds. \tag{3.3.10}$$

In the second equality above, we used (3.3.9). Since

$$\|H_{\mathrm{PF}}(\hat{A})F\| \leq C(\|-\Delta F\| + \|H_{\mathrm{rad}}F\| + \|F\|) \tag{3.3.11}$$

for $F \in D((-1/2)\Delta) \cap D(H_{\mathrm{f}})$, equality (3.3.10) can be immediately extended to vectors $F \in D((-1/2)\Delta) \cap D(H_{\mathrm{f}})$ and $G \in D(\hat{H}_{\mathrm{PF}}(\hat{A}))$. We made use of the assumptions $\sqrt{\omega}\hat{\varphi} \in L^2(\mathbb{R}^3)$ and $\hat{\varphi}/\omega \in L^2(\mathbb{R}^3)$ in (3.3.11). Take now $t \downarrow 0$ on both sides of equality (3.3.10). Then it holds that $(F, \hat{H}_{\mathrm{PF}}(\hat{A})G) = (H_{\mathrm{PF}}(\hat{A})F, G)$ for vectors $F \in D((-1/2)\Delta) \cap D(H_{\mathrm{f}})$ and $G \in D(\hat{H}_{\mathrm{PF}}(\hat{A}))$, which implies that $\hat{H}_{\mathrm{PF}}(\hat{A}) \supset H_{\mathrm{PF}}(\hat{A})\lceil_{D_{\mathrm{PF}}}$ as $\hat{H}_{\mathrm{PF}}(\hat{A})$ is self-adjoint. Define

$$\hat{H}_{\mathrm{PF}} = \hat{H}_{\mathrm{PF}}(\hat{A}) \dotplus H_{\mathrm{rad}}.$$

The Trotter product formula and the factorization $e^{-(t/n)H_{\mathrm{rad}}} = \mathrm{J}^*_{kt/n}\mathrm{J}_{(k+1)t/n}$ yield that

$$(F, e^{-t\hat{H}_{\mathrm{PF}}}G) = \lim_{n\to\infty}(F, (e^{-(t/n)\hat{H}_{\mathrm{PF}}(\hat{A})}e^{-(t/n)H_{\mathrm{rad}}})^n G) = \lim_{n\to\infty}\left(\mathrm{J}_0 F, \left(\prod_{i=0}^{n-1} R_i\right)\mathrm{J}_t G\right),$$

where $R_j = \mathrm{J}_{jt/n}e^{-(t/n)\hat{H}_{\mathrm{PF}}(\hat{A})}\mathrm{J}^*_{jt/n}$. Using the definition of $e^{-t\hat{H}_{\mathrm{PF}}(\hat{A})}$, we get

$$\begin{aligned}
\mathrm{J}_s e^{-t\hat{H}_{\mathrm{PF}}(\hat{A})}\mathrm{J}_s^* G(x) &= \underset{n\to\infty}{\text{s-lim}}\, \mathrm{J}_s (P_{t/2^n})^{2^n}\mathrm{J}_s^* G(x) \\
&= \underset{n\to\infty}{\text{s-lim}} \int_{(\mathbb{R}^3)^{2^n}} \mathrm{J}_s \exp\left(i\sum_{j=1}^{2^n}\mathfrak{h}_j\right)\mathrm{J}_s^* G(x_{2^n})\left(\prod_{j=1}^{2^n}\Pi_{t/2^n}(x_{j-1} - x_j)\right)\prod_{j=1}^{2^n}dx_j
\end{aligned}$$

with $x = x_0$. We are halfway through the trying argument. Write $\delta_j = \delta_j(n,t,n_j) = t/(n2^{n_j})$ for $j = 0,1,\dots,n-1$ and define $P_{s,j} : \mathcal{H}_{\mathrm{E}} \to \mathcal{H}_{\mathrm{E}}$ by P_s with $\mathfrak{h}(x,y)$ replaced by the Euclidean version $\mathfrak{h}^{tj/n}(x,y)$ given by

$$\mathfrak{h}^s(x,y) = \frac{e}{2}\left(\hat{A}_{\mathrm{E}_s}(x) + \hat{A}_{\mathrm{E}_s}(y)\right)\cdot(x-y)$$

as

$$P_{s,j}F(x) = \int_{\mathbb{R}^3} \Pi_s(x-y)e^{i\mathfrak{h}^{tj/n}(x,y)}F(y)dy.$$

We have

$$\begin{aligned}\left(\mathrm{J}_0F, \left(\prod_{i=0}^{n-1} R_i\right)\mathrm{J}_tG\right) &= \lim_{n_0\to\infty}\cdots\lim_{n_{n-1}\to\infty}\left(\mathrm{J}_0F, \prod_{i=0}^{n-1}(\mathrm{E}_i(P_{\delta_i,i})^{2^{n_i}}\mathrm{E}_i)\mathrm{J}_tG\right)\\ &= \lim_{n_0\to\infty}\cdots\lim_{n_{n-1}\to\infty}\left(\mathrm{J}_0F, \left(\prod_{i=0}^{n-1}(P_{\delta_i,i})^{2^{n_i}}\right)\mathrm{J}_tG\right).\end{aligned}$$

Here, in the first equality, we used the fact that

$$\mathrm{J}_s \exp\left(i\sum_{j=1}^{2^n}\mathfrak{h}_j\right)\mathrm{J}_s^* = \mathrm{E}_s \exp\left(i\sum_{j=1}^{2^n}\mathfrak{h}_j^s\right)\mathrm{E}_s,$$

where $\mathfrak{h}_j^s$ is defined by $\mathfrak{h}_j^s = \mathfrak{h}^s(x_{j-1},x_j)$, and in the second equality we used the Markov property of E_s. As a result, we have

$$\left(\mathrm{J}_0F, \left(\prod_{i=0}^{n-1} R_i\right)\mathrm{J}_tG\right) = \lim_{n_0\to\infty}\cdots\lim_{n_{n-1}\to\infty}\int_{\mathbb{R}^3}\mathbb{E}^x[(\mathrm{J}_0F(B_0), e^{-ie\hat{A}_{\mathrm{E}}(K)}\mathrm{J}_tG(B_t))]dx$$

with $K = K(n_0,n_1,\dots,n_{n-1},n)$ given by

$$K = \bigoplus_{\mu=1}^{3}\sum_{j=0}^{n-1}\sum_{m=1}^{2^{n_j}}\mathrm{j}_{\frac{tj}{n}}\left(\tilde{\varphi}(\cdot - B_{\frac{t}{n}(\frac{m}{2^{n_j}}+j)}) + \tilde{\varphi}(\cdot - B_{\frac{t}{n}(\frac{m-1}{2^{n_j}}+j)})\right)\left(B^\mu_{\frac{t}{n}(\frac{m}{2^{n_j}}+j)} - B^\mu_{\frac{t}{n}(\frac{m-1}{2^{n_j}}+j)}\right)$$

and

$$K(n_0,n_1,\dots,n_{n-1},n) \to \bigoplus_{\mu=1}^{3}\sum_{j=0}^{n-1}\int_{tj/n}^{t(j+1)/n}\mathrm{j}_{tj/n}\tilde{\varphi}(\cdot - B_s)\circ dB_s^\mu$$

as $n_0,n_1,\dots,n_{n-1}\to\infty$ in $\oplus^3(L^2(\mathscr{X})\otimes L^2(\mathbb{R}^3))$. Finally, as $n\to\infty$, we have

$$(F, e^{-t\hat{H}_{\mathrm{PF}}}G) = \int_{\mathbb{R}^3}\mathbb{E}^x[(\mathrm{J}_0F(B_0), e^{-ie\hat{A}_{\mathrm{E}}(K_t)}\mathrm{J}_tG(B_t))]dx. \tag{3.3.12}$$

By the construction of $\hat{H}_{\mathrm{PF}}$,

$$\hat{H}_{\mathrm{PF}} \supset \frac{1}{2}(-i\nabla - e\hat{A})^2 + H_{\mathrm{rad}}\lceil_{D_{\mathrm{PF}}}. \tag{3.3.13}$$

In the next section, we will show that

$$\hat{H}_{\mathrm{PF}} = \frac{1}{2}(-i\nabla - e\hat{A})^2 + H_{\mathrm{rad}}$$

and $\hat{H}_{\mathrm{PF}}$ is self-adjoint on D_{PF}. The functional integral representation of $e^{-tH_{\mathrm{PF}}}$ including nonzero V can be obtained by the Trotter product formula

$$(F, e^{-tH_{\mathrm{PF}}}G) = \lim_{n\to\infty}(F, (e^{-(t/n)\hat{H}_{\mathrm{PF}}(\hat{A})}e^{-(t/n)V}e^{-(t/n)H_{\mathrm{rad}}})^n G).$$

This completes the proof. □

Lemma 3.27. *Equation* (3.3.9) *holds true for* $F, G \in C_0^\infty(\mathbb{R}^3)\hat{\otimes}\mathcal{F}_{\mathrm{rad,fin}}$.

Proof. It is directly seen that

$$\frac{d}{ds}(F, P_s G) = \int_{\mathbb{R}^3} dx \int_{\mathbb{R}^3} \Pi_s(x-y)\left(F(x), \frac{1}{2}\Delta_y e^{i\mathfrak{h}(x,y)}G(y)\right) dy.$$

Here, we used that $(d\Pi_s/ds)(x-y) = (1/2)\Delta_y \Pi_s(x-y)$. Since

$$\begin{aligned}
\partial_{y_\mu} e^{i\mathfrak{h}} G &= (i\partial_{y_\mu}\mathfrak{h}\cdot G + \partial_{y_\mu} G)e^{i\mathfrak{h}},\\
\partial_{y_\mu}^2 e^{i\mathfrak{h}} G &= \{\partial_{y_\mu}^2 G + 2i\partial_{y_\mu}\mathfrak{h}\cdot\partial_{y_\mu} G + (i\partial_{y_\mu}^2\mathfrak{h} + (i\partial_{y_\mu}\mathfrak{h})^2)G\}e^{i\mathfrak{h}},\\
\partial_{y_\mu}\mathfrak{h}(x,y) &= \frac{e}{2}\{\partial_{y_\mu}\hat{A}(y)\cdot(x-y) - (\hat{A}_\mu(x) + \hat{A}_\mu(y))\},\\
\partial_{y_\mu}^2\mathfrak{h}(x,y) &= e\left\{\frac{1}{2}\partial_{y_\mu}^2\hat{A}(y)\cdot(x-y) - \partial_{y_\mu}\hat{A}_\mu(y)\right\},
\end{aligned}$$

we have $(F(x), \Delta_y e^{i\mathfrak{h}(x,y)}G(y)) = I_\mu^+(x,y) + I_\mu^-(x,y)$, where

$$\begin{aligned}
I_\mu^+(x,y) &= \left(F(x), e^{i\mathfrak{h}(x,y)}\left(e\hat{A}^\mu(y)i\partial_{y_\mu} + i\frac{e}{2}\hat{A}^{\mu\mu}(y) + \frac{e^2}{2}\hat{A}^\mu(y)(\hat{A}_\mu(x) + \hat{A}_\mu(y))\right)G(y)\right)\cdot(x-y)\\
&\quad - \frac{e^2}{4}(F(x), e^{i\mathfrak{h}(x,y)}(\hat{A}^\mu(y)\cdot(x-y))^2 G(y)),\\
I_\mu^-(x,y) &= \left(F(x), e^{i\mathfrak{h}(x,y)}\left(\Delta_{y_\mu} - e(\hat{A}_\mu(x) + \hat{A}_\mu(y))i\partial_{y_\mu} - \frac{e^2}{4}(\hat{A}_\mu(x) + \hat{A}_\mu(y))^2 - ie\hat{A}_\mu^\mu(y)\right)G(y)\right).
\end{aligned}$$

Here, $\hat{A}^\mu(y) = \partial_{y_\mu}\hat{A}(y) = (\hat{A}_1^\mu(y), \hat{A}_2^\mu(y), \hat{A}_3^\mu(y))$, and in the Fock representation,

$$\hat{A}_\nu^\mu(y) \cong \frac{1}{\sqrt{2}}\sum_{j=\pm}\int_{\mathbb{R}^3} e_\nu^j(k)\left(\frac{-ik_\mu\hat{\varphi}(k)}{\sqrt{\omega(k)}}e^{-iky}a^*(k,j) + \frac{ik_\mu\hat{\varphi}(-k)}{\sqrt{\omega(k)}}e^{iky}a(k,j)\right)dk.$$

We have the bound

$$|I_\mu^+(x,y)| \le \|F(x)\|(\|N^{1/2}\partial_{y_\mu} G(y)\| + \|NG(y)\|)|x-y|^2,$$

which implies that

$$\lim_{t\to 0} \int_{\mathbb{R}^3} dx \int_{\mathbb{R}^3} \Pi_t(x-y) I_\mu^+(x,y)dy = 0,$$

since $\lim_{t\to 0}\int_{\mathbb{R}^3}\Pi_t(x)|x|^n dx = 0$ for $n > 0$. It can be also seen that the map $y \mapsto I_\mu^-(x,y)$ is continuous for each $x \in \mathbb{R}^3$ and it follows that

$$\lim_{y\to x} I_\mu^-(x,y) = -(F(x), (-i\nabla_{x_\mu} - e\hat{A}_\mu(x))^2 G(x)).$$

Hence we have

$$\lim_{t\to 0} \int_{\mathbb{R}^3} dx \int_{\mathbb{R}^3} \Pi_t(x-y) I_\mu^-(x,y)dy = \int_{\mathbb{R}^3} I_\mu^-(x,x)dx = -\int_{\mathbb{R}^3} (F(x), (-i\nabla_{x_\mu} - e\hat{A}_\mu)^2 G(x))dx,$$

since $\lim_{t\to 0}\int_{\mathbb{R}^3}\Pi_t(x)f(x)dx = f(0)$ for any measurable bounded function f being continuous at $x = 0$. Together with them, we have

$$\begin{aligned}\lim_{t\to 0}(F, t^{-1}(1-P_t)G) &= -\lim_{t\to 0+}\frac{d}{dt}(F, P_t G)\\ &= -\lim_{t\to 0}\sum_{\mu=1}^{3}\frac{1}{2}\int_{\mathbb{R}^3} dx \int_{\mathbb{R}^3} \Pi_t(x-y)(I_\mu^-(x,y) + I_\mu^+(x,y))dy = (F, H_{\mathrm{PF}}(\hat{A})G).\end{aligned}$$

Thus the lemma follows. □

A rewarding application of the functional integral representation of $(F, e^{-tH_{\mathrm{PF}}}G)$ is a diamagnetic inequality as the exponent $-ie\hat{A}_{\mathrm{E}}(K_t)$ of $(F, e^{-tH_{\mathrm{PF}}}G)$ is purely imaginary.

Corollary 3.28 (Diamagnetic inequality). *Under the conditions of Theorem* 3.26, *it follows that*

$$|(F, e^{-tH_{\mathrm{PF}}}G)| \le (|F|, e^{-tH_{\mathrm{PF},0}}|G|).$$

Proof. By the functional integral representation in Theorem 3.26, we plainly have

$$|(F, e^{-tH_{\mathrm{PF}}}G)| \le \int_{\mathbb{R}^3} \mathbb{E}^x[e^{-\int_0^t V(B_s)ds}(\mathrm{J}_0|F(B_0)|, \mathrm{J}_t|G(B_t)|)]dx = (|F|, e^{-tH_{\mathrm{PF},0}}|G|).$$

Here, we used that $|\mathrm{J}_s F| \le \mathrm{J}_s|F|$ since J_s is positivity preserving. □

The diamagnetic inequality shows that coupling the particle to the quantized radiation field by minimal interaction increases the ground state energy of the noninteracting system.

Corollary 3.29 (Enhanced ground state energy). *We have*

$$\inf \mathrm{Spec}(H_\mathrm{p}) \leq \inf \mathrm{Spec}(H_\mathrm{PF}).$$

Proof. Corollary 3.28 implies that $\inf \mathrm{Spec}(H_{\mathrm{PF},0}) \leq \inf \mathrm{Spec}(H_\mathrm{PF})$. Since

$$\inf \mathrm{Spec}(H_{\mathrm{PF},0}) = \inf \mathrm{Spec}(H_\mathrm{p}),$$

the corollary follows. □

3.3.3 Extension to general external potential

In Theorem 3.26, we assumed smoothness of the external potential. In this section, we offer an extension of the functional integral representation to a wider potential class. For convenience, in this section we denote H_PF with $V \equiv 0$ by

$$H_0 = \frac{1}{2}(-i\nabla - e\hat{A})^2 + H_\mathrm{rad}.$$

First, consider a form boundedness property of H_0. Let

$$\mathrm{sgn}\, F(x) = \begin{cases} F(x)/\|F(x)\|_{L^2(\mathscr{Q})} & \text{if } \|F(x)\|_{L^2(\mathscr{Q})} \neq 0, \\ 0 & \text{if } \|F(x)\|_{L^2(\mathscr{Q})} = 0. \end{cases}$$

Lemma 3.30. *Suppose that V is $-(1/2)\Delta$-form bounded with a relative bound a. Then $|V|$ is H_0-form bounded with a relative bound smaller than a.*

Proof. The proof is similar to that of Lemma 2.17. Let $\psi \in C_0^\infty(\mathbb{R}^3)$ and $\psi > 0$. Substituting $F = \mathrm{sgn}((H_0+z)^{-1/2}G)\cdot\psi$ in the diamagnetic inequality $|(F, e^{-tH_0}G)| \leq (|F|, e^{-t(-(1/2)\Delta)}|G|)$, we obtain

$$(\psi, \|(H_0 + z)^{-1/2}G\|_{L^2(\mathscr{Q})})_{L^2(\mathbb{R}^3)} \leq \left(\psi, \left(-\frac{1}{2}\Delta + z\right)^{-1/2} \|G\|_{L^2(\mathscr{Q})}\right)_{L^2(\mathbb{R}^3)}.$$

Hence we have

$$\frac{\||V|^{1/2}(H_0 + z)^{-1/2}G\|_{\mathcal{H}_\mathrm{PF}}}{\|G\|_{\mathcal{H}_\mathrm{PF}}} \leq \frac{\left\||V|^{1/2}(-\frac{1}{2}\Delta + z)^{-1/2}\|G(\cdot)\|_{L^2(\mathscr{Q})}\right\|_{L^2(\mathbb{R}^3)}}{\|G\|_{\mathcal{H}_\mathrm{PF}}}.$$

Taking the supremum of both sides above with respect to $\|G\|_{\mathcal{H}_\mathrm{PF}} \neq 0$, we further obtain

$$\| |V|^{1/2}(H_0+z)^{-1/2}\|_{\mathcal{H}_{\rm PF}} \le \left\| |V|^{1/2}\left(-\frac{1}{2}\Delta+z\right)^{-1/2}\right\|_{L^2(\mathbb{R}^3)}.$$

Then the lemma follows. □

Remark 3.31. In a similar way as in Lemma 3.30, we can show that if V is relatively bounded with respect to $-(1/2)\Delta$ with a relative bound a, then V is relatively bounded with respect to H_0 with a relative bound a.

Theorem 3.32 (Functional integral representation for Pauli–Fierz Hamiltonian with singular external potential). *Take $H_0 \dot{+} V_+ \dot{-} V_-$ with $V_+ \in L^1_{\rm loc}(\mathbb{R}^3)$ and V_- is $-(1/2)\Delta$-relatively form bounded with a relative bound strictly less than* 1. *Equality* (3.3.4) *holds for $H_0 \dot{+} V_+ \dot{-} V_-$.*

Proof. The proof is a minor modification of that of the Feynman–Kac formula for Schrödinger operator with singular external potentials and singular vector potentials in Theorem 2.18. Let

$$V_{+n}(x)=\begin{cases} V_+(x), & V_+(x)<n,\\ n, & V_+(x)\ge n,\end{cases} \qquad V_{-m}(x)=\begin{cases} V_-(x), & V_-(x)<m,\\ m, & V_-(x)\ge m.\end{cases}$$

Set $V_{n,m}=V_{+n}-V_{-m}$ and $h=\frac{1}{2}(-i\nabla-e\hat{A})^2+H_{\rm rad}$. Then

$$(F, e^{-t(h+V_{n,m})}G)=\int_{\mathbb{R}^3}\mathbb{E}^x\left[e^{-\int_0^t V_{n,m}(B_s)\,ds}(F(B_0),\mathscr{L}G(B_t))\right]dx, \tag{3.3.14}$$

where $\mathscr{L}=\mathrm{J}_0^* e^{-ie\hat{A}_{\rm E}(K_t)}\mathrm{J}_t$. Define the closed quadratic forms

$$\begin{aligned} \mathfrak{q}_{n,m}(F,F)&=(h^{1/2}F,h^{1/2}F)+(V_{+n}^{1/2}F,V_{+n}^{1/2}F)-(V_{-m}^{1/2}F,V_{-m}^{1/2}F),\\ \mathfrak{q}_{n,\infty}(F,F)&=(h^{1/2}F,h^{1/2}F)+(V_{+n}^{1/2}F,V_{+n}^{1/2}F)-(V_{-}^{1/2}F,V_{-}^{1/2}F),\\ \mathfrak{q}_{\infty,\infty}(F,F)&=(h^{1/2}F,h^{1/2}F)+(V_{+}^{1/2}F,V_{+}^{1/2}F)-(V_{-}^{1/2}F,V_{-}^{1/2}F), \end{aligned}$$

whose form domains are respectively $Q(\mathfrak{q}_{n,m})=Q(h)$, $Q(\mathfrak{q}_{n,\infty})=Q(h)$, and $Q(\mathfrak{q}_{\infty,\infty})=Q(h)\cap Q(V_+)$. By the same arguments involving monotone convergence of quadratic forms as the proof of Theorem 2.18, we can conclude that for all $t\ge 0$,

$$\exp\left(-t\left(h+V_{+n}-V_{-m}\right)\right)\to\exp\left(-t\left(h+V_{+n}\dot{-}V_-\right)\right) \tag{3.3.15}$$

as $m\to\infty$, and

$$\exp(-t(h\dot{+}V_{+n}\dot{-}V_-))\to\exp(-t(h\dot{+}V_+\dot{-}V_-)) \tag{3.3.16}$$

as $n\to\infty$ strongly. By taking first $n\to\infty$ and then $m\to\infty$, it can be proven that both sides of (3.3.14) converge. That is, the left-hand side of (3.3.14) converges by (3.3.15) and (3.3.16). We have the inequality

$$\int_{\mathbb{R}^3} \mathbb{E}^x[e^{-\int_0^t V_{n,m}(B_s)\,ds} |(F(B_0), \mathscr{L}G(B_t))|]dx$$
$$\leq \int_{\mathbb{R}^3} \mathbb{E}^x[e^{-\int_0^t V_{n,\infty}(B_s)\,ds}(|F(B_0)|, e^{-tH_{\mathrm{rad}}}|G(B_t)|)]dx < \infty.$$

Here, the finiteness of the second term can be derived from the proof of Theorem 2.18. Since

$$|e^{-\int_0^t V_{n,m}(B_s)\,ds}(F(B_0), \mathscr{L}G(B_t))| \leq e^{-\int_0^t V_{n,\infty}(B_s)\,ds}(|F(B_0)|, e^{-tH_{\mathrm{rad}}}|G(B_t)|),$$

the dominated convergence theorem yields that

$$\lim_{m\to\infty} \int_{\mathbb{R}^3} \mathbb{E}^x[e^{-\int_0^t V_{n,m}(B_s)\,ds} (F(B_0), \mathscr{L}G(B_t))]dx = \int_{\mathbb{R}^3} \mathbb{E}^x[e^{-\int_0^t V_{n,\infty}(B_s)\,ds}(F(B_0), \mathscr{L}G(B_t))]dx.$$

Furthermore, since

$$\int_{\mathbb{R}^3} \mathbb{E}^x[e^{-\int_0^t V_{n,\infty}(B_s)\,ds} |(F(B_0), \mathscr{L}G(B_t))|]dx$$
$$\leq \int_{\mathbb{R}^3} \mathbb{E}^x[e^{-\int_0^t V_{n,\infty}(B_s)\,ds}(|F(B_0)|, e^{-tH_{\mathrm{rad}}}|G(B_t)|)]dx < \infty$$

and

$$\left|e^{-\int_0^t V_{n',\infty}(B_s)\,ds} (F(B_0), \mathscr{L}G(B_t))\right| \leq e^{-\int_0^t V_{n,\infty}(B_s)\,ds}(|F(B_0)|, e^{-tH_{\mathrm{rad}}}|G(B_t)|)$$

for $n \leq n'$, the dominated convergence theorem again yields that

$$\lim_{n\to\infty} \int_{\mathbb{R}^3} \mathbb{E}^x[e^{-\int_0^t V_{n,\infty}(B_s)\,ds} (F(B_0), \mathscr{L}G(B_t))]dx = \int_{\mathbb{R}^3} \mathbb{E}^x[e^{-\int_0^t V(B_s)\,ds}(F(B_0), \mathscr{L}G(B_t))]dx.$$

Thus the right-hand side of (3.3.14) converges to

$$\int_{\mathbb{R}^3} \mathbb{E}^x[e^{-\int_0^t V(B_s)\,ds}(F(B_0), \mathscr{L}G(B_t))]dx$$

for any $F, G \in \mathcal{H}_{\mathrm{PF}}$ as first $m \to \infty$ and then $n \to \infty$. Then the proof is complete. □

3.4 The Pauli–Fierz model with Kato-class potential

We consider the Pauli–Fierz model with Kato-class potential V. In Section 3.4 we use Assumption 3.33 below.

Assumption 3.33. The following conditions hold:
(1) (Charge distribution) $\varphi \in \mathscr{S}'(\mathbb{R}^3)$, $\hat{\varphi}(-k) = \overline{\hat{\varphi}(k)}$ and $\sqrt{\omega}\hat{\varphi}, \hat{\varphi}/\sqrt{\omega}, \hat{\varphi}/\omega \in L^2(\mathbb{R}^3)$;
(2) (External potential) V is Kato-decomposable.

First, we are interested in defining H_{PF} with a Kato-class potential as a self-adjoint operator. This will be done through the functional integral representation established in the previous section. The expression

$$(F, e^{-tH_{\mathrm{PF}}}G) = \int_{\mathbb{R}^3} \mathbb{E}^x[e^{-\int_0^t V(B_s)ds}(\mathrm{J}_0F(x), e^{-ie\hat{A}_{\mathrm{E}}(K_t)}\mathrm{J}_tG(B_t))]dx$$

implies

$$(e^{-tH_{\mathrm{PF}}}G)(x) = \mathbb{E}^x[e^{-\int_0^t V(B_s)ds}\mathrm{J}_0^*e^{-ie\hat{A}_{\mathrm{E}}(K_t)}\mathrm{J}_tG(B_t)]. \tag{3.4.1}$$

Conversely, we shall show that a sufficient condition to define the right-hand side of (3.4.1) is that V is of Kato-class. This idea used for Schrödinger operators with Kato-class potentials will be extended to the Pauli–Fierz Hamiltonian in this section.

Let V be a Kato-decomposable potential and define the family of operators

$$(T_tF)(x) = \mathbb{E}^x[e^{-\int_0^t V(B_r)dr}\mathrm{J}_0^*e^{-ie\hat{A}_{\mathrm{E}}(K_t)}\mathrm{J}_tF(B_t)].$$

Lemma 3.34. *Operator T_t is bounded on $\mathcal{H}_{\mathrm{PF}}$ for each $t \geq 0$.*

Proof. Let $F \in \mathcal{H}_{\mathrm{PF}}$. We show that

$$\|T_tF\|^2_{\mathcal{H}_{\mathrm{PF}}} = \int_{\mathbb{R}^3} \|\mathbb{E}^x[e^{-\int_0^t V(B_r)dr}\mathrm{J}_0^*e^{-ie\hat{A}_{\mathrm{E}}(K_t)}\mathrm{J}_tF(B_t)]\|^2_{L^2(\mathscr{Q})}dx < \infty. \tag{3.4.2}$$

By the Schwarz inequality, we have

$$\|T_tF\|^2_{\mathcal{H}_{\mathrm{PF}}} \leq \int_{\mathbb{R}^3} \mathbb{E}^0[e^{-2\int_0^t V(B_r+x)dr}]\mathbb{E}^0[\|F(B_t+x)\|^2]dx.$$

Since V is of Kato-class, we have $\sup_{x\in\mathbb{R}^3} \mathbb{E}^0[e^{-2\int_0^t V(B_r+x)dr}] = C < \infty$, and thus

$$\|T_tF\|^2_{\mathcal{H}_{\mathrm{PF}}} \leq C\|F\|^2_{\mathcal{H}_{\mathrm{PF}}}. \qquad \square$$

From now on, we shall show that $\{T_t : t \geq 0\}$ is a symmetric C_0-semigroup. To do that, we introduce the time shift operator u_t on $L^2(\mathbb{R}^4)$ by

$$u_tf(x) = f(x_0 - t, \mathbf{x}), \quad x = (x_0, \mathbf{x}) \in \mathbb{R} \times \mathbb{R}^3.$$

It is straightforward that $u_t^* = u_{-t}$ and $u_t^* u_t = 1$. We denote the second quantization of u_t denoted by $U_t = \Gamma_{\mathrm{E}}(u_t)$ which acts on $L^2(\mathscr{Q}_{\mathrm{E}})$ and is unitary. The relationship between j_s and u_t is given next.

Lemma 3.35. *It follows that $u_t \mathrm{j}_s = \mathrm{j}_{s+t}$ for every $t, s \in \mathbb{R}$.*

Proof. In the position representation, j_s is given by

$$\mathrm{j}_s f(x) = \frac{1}{\sqrt{\pi}(2\pi)^2} \int_{\mathbb{R}^4} e^{+i(k_0(x_0-s)+k\cdot x)} \frac{\sqrt{\omega(k)}}{\sqrt{\omega(k)^2 + |k_0|^2}} \hat{f}(k) d\mathbf{k}.$$

Hence $u_t \mathrm{j}_s f(x) = \mathrm{j}_{s+t} f(x)$. □

Lemma 3.35 implies the formula $U_t \mathrm{J}_s = \mathrm{J}_{s+t}$.

Lemma 3.36. *It follows that $T_s T_t = T_{s+t}$ for $s, t \geq 0$.*

Proof. By the definition of T_t, we have

$$T_s T_t F = \mathbb{E}^x \left[e^{-\int_0^s V(B_r)dr} \mathrm{J}_0^* e^{-ie\hat{A}_{\mathrm{E}}(K_s)} \mathrm{J}_s \mathbb{E}^{B_s} [e^{-\int_0^t V(B_r)dr} \mathrm{J}_0^* e^{-ie\hat{A}_{\mathrm{E}}(K_t)} \mathrm{J}_t F(B_t)] \right]. \tag{3.4.3}$$

By the formulae, $\mathrm{J}_s \mathrm{J}_0^* = \mathrm{J}_s (U_{-s} \mathrm{J}_s)^* = \mathrm{J}_s \mathrm{J}_s^* U_{-s}^* = \mathrm{E}_s U_{-s}^*$ and $\mathrm{J}_t = U_{-s} \mathrm{J}_{t+s}$, we see that (3.4.3) is equal to

$$\mathbb{E}^x \left[e^{-\int_0^s V(B_r)dr} \mathrm{J}_0^* e^{-ie\hat{A}_{\mathrm{E}}(K_s)} \mathrm{E}_s \mathbb{E}^{B_s} [e^{-\int_0^t V(B_r)dr} U_{-s}^* e^{-ie\hat{A}_{\mathrm{E}}(K_t)} U_{-s} \mathrm{J}_{t+s} F(B_t)] \right]. \tag{3.4.4}$$

Now we compute $U_{-s}^* e^{-ie\hat{A}_{\mathrm{E}}(K_t)} U_{-s}$. Since U_s is unitary, we have

$$U_{-s}^* e^{-ie\hat{A}_{\mathrm{E}}(K_t)} U_{-s} = e^{-ie\hat{A}_{\mathrm{E}}(u_{-s}^* K_t)}$$

as an operator. The test function of the exponent $u_{-s}^* K_t$ is given by

$$u_{-s}^* K_t = \bigoplus_{\mu=1}^3 \int_0^t \mathrm{j}_{r+s} \tilde{\varphi}(\cdot - B_r) dB_r^\mu.$$

Moreover, by the Markov property of E_t, $t \in \mathbb{R}$, we may neglect E_s in (3.4.4), and by the Markov property of $(B_t)_{t \geq 0}$ we have

$$\begin{aligned}(T_s T_t F)(x) &= \mathbb{E}^x [e^{-\int_0^s V(B_r)dr} \mathrm{J}_0^* e^{-ie\hat{A}_{\mathrm{E}}(K_s)} \mathbb{E}^x [e^{-\int_s^{s+t} V(B_r)dr} e^{-ie\hat{A}_{\mathrm{E}}(K_s^{s+t})} \mathrm{J}_{s+t} F(B_{s+t}) | \mathscr{F}_s^{\mathrm{BM}}]] \\ &= \mathbb{E}^x [e^{-\int_0^{s+t} V(B_r)dr} \mathrm{J}_0^* e^{-ie\hat{A}_{\mathrm{E}}(K_{s+t})} \mathrm{J}_{s+t} F(B_{s+t})] = T_{s+t} F,\end{aligned}$$

where $(\mathscr{F}_t^{\mathrm{BM}})_{t \geq 0}$ denotes the natural filtration of $(B_t)_{t \geq 0}$. □

Strong continuity of T_t on $\mathcal{H}_{\mathrm{PF}}$ can be checked similarly as previously done, while $T_0 = 1$ is trivial.

Theorem 3.37. *Semigroup $\{T_t : t \geq 0\}$ is a symmetric C_0-semigroup.*

Proof. It was shown that $\{T_t : t \geq 0\}$ is a C_0-semigroup. Hence it is sufficient to show that T_t is symmetric for each $t \geq 0$. Recall that $\mathrm{R} = \Gamma(r)$ is the second quantization of the reflection r and the second quantization of time-shift u_t is denoted by $U_t = \Gamma(u_t)$. We have

$$(F, T_t G) = (U_t \mathrm{R} F, U_t \mathrm{R} T_t G) = \int_{\mathbb{R}^3} \mathbb{E}^x \left[e^{-\int_0^t V(B_s)ds} \left(\mathrm{J}_t F(B_0), e^{-ie\hat{A}_{\mathrm{E}}(u_t r K_t)} \mathrm{J}_0 G(B_t) \right) \right] dx.$$

Notice that $u_t r K_t = \int_0^t \mathrm{j}_{t-s} \tilde{\varphi}(\cdot - B_s) dB_s^\mu$. Since $\dot{B}_s = B_{t-s} - B_t \stackrel{\mathrm{d}}{=} B_s$ for any s such that $0 \leq s \leq t$ with fixed t, we have

$$(F, T_t G) = \int_{\mathbb{R}^3} \mathbb{E}^x \left[e^{-\int_0^t V(\dot{B}_s)ds} \left(\mathrm{J}_t F(\dot{B}_0), e^{-ie\hat{A}_{\mathrm{E}}(u_t r \dot{K}_t)} \mathrm{J}_0 G(\dot{B}_t) \right) \right] dx.$$

Here, $\dot{K}_t = \bigoplus_{\mu=1}^3 \dot{K}_t^\mu = \bigoplus_{\mu=1}^3 \int_0^t \mathrm{j}_{t-s} \tilde{\varphi}(\cdot - \dot{B}_s) d\dot{B}_s^\mu$. Thus

$$\begin{aligned} \dot{K}_t^\mu &= \lim_{m\to\infty} \sum_{j=1}^{2^m} \mathrm{j}_{t-t_{j-1}} \tilde{\varphi}(\cdot - B_{t-t_{j-1}} + B_t - x) \left(B_{t-t_{j-1}}^\mu - B_{t-t_{j-1}}^\mu \right) \\ &= -\lim_{m\to\infty} \sum_{j=1}^{2^m} \mathrm{j}_{t_j} \tilde{\varphi}(\cdot - B_{t_j} + B_t - x) \left(B_{t_j}^\mu - B_{t_{j-1}}^\mu \right) = \lim_{m\to\infty} K_t^{\mu,m}. \end{aligned}$$

Here, $t_j = tj/2^m$. Note that

$$\lim_{m\to\infty} \sum_{j=1}^{2^m} \mathrm{j}_{t_{j-1}} \tilde{\varphi}(\cdot - B_{t_{j-1}} + y) \left(B_{t_j}^\mu - B_{t_{j-1}}^\mu \right) = \int_0^t \mathrm{j}_s \tilde{\varphi}(\cdot - B_s) dB_s^\mu.$$

It is, however, shown by the Coulomb gauge condition (3.1.8) and the fact that j_t is an isometry that $K_t^{\mu,m}$ also converges to $\int_0^t \mathrm{j}_s \tilde{\varphi}(\cdot - B_s) dB_s^\mu$ as $m \to \infty$. Exchanging integrals $\int_{\mathscr{X}} d\mathcal{W}^0$ and $\int_{\mathbb{R}^3} dx$ and changing the variable x to $y + B_t$, we have

$$\begin{aligned} (F, T_t G) &= \lim_{m\to\infty} \int_{\mathbb{R}^3} \mathbb{E}^x \left[e^{-\int_0^t V(\dot{B}_s)ds} \left(\mathrm{J}_t F(\dot{B}_0), e^{-ie\hat{A}_{\mathrm{E}}(K_t^m)} \mathrm{J}_0 G(\dot{B}_t) \right) \right] dx \\ &= \mathbb{E}^0 \left[\int_{\mathbb{R}^3} e^{-\int_0^t V(B_{t-s}+y)ds} \left(\mathrm{J}_t F(B_t + y), e^{+ie\hat{A}_{\mathrm{E}}(K_t)} \mathrm{J}_0 G(y) \right) dy \right]. \end{aligned}$$

Then

$$(F, T_t G) = \int_{\mathbb{R}^3} \mathbb{E}^y \left[e^{-\int_0^t V(B_s)ds} \left(\mathrm{J}_0^* e^{-ie\hat{A}_\mathrm{E}(K_t)} \mathrm{J}_t F(y + B_t), G(y) \right) \right] dy = (T_t F, G)$$

and T_t is symmetric. Then the proof is complete. □

By Theorem 3.37 and Stone's theorem for semigroups, there exists a self-adjoint operator $H_{\mathrm{PF}}^{\mathrm{Kato}}$ such that

$$T_t = e^{-tH_{\mathrm{PF}}^{\mathrm{Kato}}}, \quad t \geq 0.$$

Definition 3.38 (Pauli–Fierz Hamiltonian for Kato-class potential). We call the self-adjoint operator $H_{\mathrm{PF}}^{\mathrm{Kato}}$ the *Pauli–Fierz Hamiltonian for Kato-class potential* V.

Similarly as for H_{PF}, with $V \in \mathcal{R}_{\mathrm{Kato}}$, the Pauli–Fierz Hamiltonian for Kato-class potential V, $H_{\mathrm{PF}}^{\mathrm{Kato}}$, also satisfies the diamagnetic inequality.

Corollary 3.39 (Diamagnetic inequality). *It follows that*

$$|(F, e^{-tH_{\mathrm{PF}}^{\mathrm{Kato}}} G)| \leq (|F|, e^{-t(H_{\mathrm{p}}^{\mathrm{Kato}} + H_{\mathrm{rad}})} |G|),$$

where $H_{\mathrm{p}}^{\mathrm{Kato}}$ is the Schrödinger operator with a Kato-class potential.

Proof. Since J_0^* and J_t are positivity preserving, we have $|\mathrm{J}_0^* F| \leq \mathrm{J}_0^* |F|$, $|\mathrm{J}_t F| \leq \mathrm{J}_t |F|$, and

$$|e^{-tH_{\mathrm{PF}}^{\mathrm{Kato}}} F| \leq \mathbb{E}^x [e^{-\int_0^t V(B_r)dr} \mathrm{J}_0^* \mathrm{J}_t |F(B_t)|].$$

The right-hand side above is just $e^{-t(H_{\mathrm{p}}^{\mathrm{Kato}} + H_{\mathrm{rad}})} |F|$. □

For the notational convenience, we use notation H_{PF} for $H_{\mathrm{PF}}^{\mathrm{Kato}}$ unless confusions arise in what follows.

3.5 Applications of functional integral representations

3.5.1 Self-adjointness of the Pauli–Fierz Hamiltonian

One of the basic problems in the spectral analysis of Hamiltonians under consideration is to show their essential self-adjointness or self-adjointness. In Section 3.5.1 we assume Assumption 3.4. As we have pointed out in the previous section, for a sufficiently small e the Pauli–Fierz operator H_{PF} is self-adjoint on D_{PF}. We will show in this section that H_{PF} is self-adjoint on D_{PF} for all $e \in \mathbb{R}$. Recall that $D_{\mathrm{PF}} = D(-(1/2)\Delta) \cap D(H_{\mathrm{rad}})$.

We have already constructed a functional integral representation for $\hat{H}_{\mathrm{PF}}$, however, it remains to be shown that $\hat{H}_{\mathrm{PF}} = H_{\mathrm{PF}}$. In the proof of Theorem 3.26, we defined the self-adjoint operator $\hat{H}_{\mathrm{PF}} = \hat{H}_{\mathrm{PF}}(\hat{A}) \dotplus H_{\mathrm{rad}}$ and have shown that $\hat{H}_{\mathrm{PF}} \supset H_{\mathrm{PF}}\lceil_{D_{\mathrm{PF}}}$ in (3.3.13), and

$$(F, e^{-t\hat{H}_{\mathrm{PF}}}G) = \int_{\mathbb{R}^3} \mathbb{E}^x[(\mathrm{J}_0 F(B_0), e^{-ie\hat{A}_{\mathrm{E}}(K_t)}\mathrm{J}_t G(B_t))]dx \tag{3.5.1}$$

in (3.3.12). Let $R_t(F, G)$ be the quadratic form defined by the right-hand side of (3.5.1) in which there is no restriction on the value of e. The right-hand side of (3.5.1) is well defined for any $e \in \mathbb{R}$. There exists a semigroup S_t such that $R_t(F, G) = (F, S_t G)$ and its generator $\hat{H}_{\mathrm{PF}}$ satisfies $\hat{H}_{\mathrm{PF}} \supset H_{\mathrm{PF}}\lceil_{D_{\mathrm{PF}}}$. Clearly, for a weak enough coupling, $\hat{H}_{\mathrm{PF}} = H_{\mathrm{PF}}\lceil_{D_{\mathrm{PF}}}$. We will show that S_t leaves D_{PF} invariant and from that conclude that $\hat{H}_{\mathrm{PF}}$ is essentially self-adjoint on D_{PF}. This will imply that H_{PF} is essentially self-adjoint on this joint domain. Finally, we will show that H_{PF} is closed on D_{PF} and, therefore, H_{PF} is self-adjoint on D_{PF}.

We start by a general result.

Lemma 3.40. *Let K be a nonnegative self-adjoint operator. Suppose that there exists a dense domain D such that $D \subset D(K)$ and $e^{-tK}D \subset D$ for all $t \geq 0$. Then $K\lceil_D$ is essentially self-adjoint.*

Proof. It suffices to show that for some $\lambda > 0$, $\mathrm{Ran}((\lambda + K)\lceil_D)$ is dense. Suppose the opposite. Then there exists nonzero f such that $(f, (\lambda + K)\psi) = 0$ for all $\psi \in D$. We have

$$\frac{d}{dt}(f, e^{-tK}\psi) = (f, -Ke^{-tK}\psi) = \lambda(f, e^{-tK}\psi).$$

Thus $(f, e^{-tK}\psi) = e^{\lambda t}(f, \psi)$. If $(f, \psi) \neq 0$, then $\lim_{t\to\infty} |(f, e^{-tK}\psi)| = \infty$, contradicting the fact that e^{-tK} is a contraction. Hence $(f, \psi) = 0$ for all $\psi \in D(K)$, but (f, ψ) cannot equal zero for all $\psi \in D(K)$, since $D(K)$ is dense. Hence we conclude that $\mathrm{Ran}((\lambda + K)\lceil_D)$ is dense. □

The next result is an important application of Lemma 3.40 to the Pauli–Fierz Hamiltonian.

Lemma 3.41. *Suppose that*

$$|(H_{\mathrm{PF},0}F, e^{-t\hat{H}_{\mathrm{PF}}}G)| \leq C_G\|F\| \tag{3.5.2}$$

for all $F, G \in D(H_{\mathrm{PF},0}) = D_{\mathrm{PF}}$ with some constant C_G which may depend on G. Then H_{PF} is essentially self-adjoint on D_{PF}.

Proof. Inequality (3.5.2) yields that $e^{-t\hat{H}_{\mathrm{PF}}}G \in D(H_{\mathrm{PF},0})$ by the Riesz representation theorem, i. e., $e^{-t\hat{H}_{\mathrm{PF}}}D_{\mathrm{PF}} \subset D_{\mathrm{PF}}$. Thus Lemma 3.40 implies that $\hat{H}_{\mathrm{PF}}\lceil_{D_{\mathrm{PF}}}$ is essentially self-adjoint. The lemma follows from $H_{\mathrm{PF}} = \hat{H}_{\mathrm{PF}}$ on D_{PF}. □

In order to show (3.5.2), we need an inequality for $\int_0^t \mathrm{j}_s\tilde{\varphi}(\cdot - B_s)dB_s^\mu$ similar to the BDG inequality. With an $\mathbb{R}^d$-valued function $a = (a_1, \dots, a_d)$, random process $Y_t^\mu = \int_0^t a_\mu(s, B_s)dB_s^\mu$ satisfies

$$d(Y_t^\mu)^2 = 2Y_t^\mu a_\mu(t, B_t)dB_t^\mu + a_\mu(t, B_t)^2 dt$$

and the BDG inequality

$$\mathbb{E}^x\left[\left|\int_0^t a(s, B_s)dB_s\right|^{2m}\right] \le (m(2m-1))^m t^{m-1}\mathbb{E}^x\left[\int_0^t |a(s, B_s)|^{2m} ds\right]$$

holds. We show its infinite-dimensional version below. Let

$$\hat{\omega}_{\mathrm{E}} = \omega(-i\nabla) \otimes 1,$$

under the identification $L^2(\mathbb{R}^4) \cong L^2(\mathbb{R}^3) \otimes L^2(\mathbb{R})$.

Lemma 3.42 (BDG-type inequality). *If $\omega^{(k-1)/2}\hat{\varphi} \in L^2(\mathbb{R}^3)$, then*

$$\mathbb{E}^x\left[\left\|\hat{\omega}_{\mathrm{E}}^{k/2}\int_0^t \mathrm{j}_s\tilde{\varphi}(\cdot - B_s)dB_s^\mu\right\|_{L^2(\mathbb{R}^3)}^{2m}\right] \le \frac{(2m)!}{2^m} t^m \|\omega^{(k-1)/2}\hat{\varphi}\|_{L^2(\mathbb{R}^3)}^{2m}.$$

In particular,

$$\sup_{x\in\mathbb{R}^3} \mathbb{E}^x[\|\hat{\omega}_{\mathrm{E}}^{k/2} K_t\|_{\oplus^3 L^2(\mathbb{R}^3)}^{2m}] \le \frac{3(2m)!}{2^m} t^m \|\omega^{(k-1)/2}\hat{\varphi}\|_{L^2(\mathbb{R}^3)}^{2m},$$

where $K_t = \bigoplus_{\mu=1}^3 \int_0^t \mathrm{j}_s\tilde{\varphi}(\cdot - B_s)dB_s^\mu$.

Proof. Let $(X_t)_{t\ge0} = (X_t^1, X_t^2, X_t^3)_{t\ge0}$ be a random process defined by

$$X_t^\mu = \left\|\int_0^t \mathrm{j}_s\lambda(\cdot - B_s)dB_s^\mu\right\|_{L^2(\mathbb{R}^3)}^2$$

with $\lambda \in L^2(\mathbb{R}^3)$. By the definition of j_s, we have

$$X_t^\mu = \lim_{n\to\infty} \int_{\mathbb{R}^4} \frac{1}{\pi}\mu(\mathbf{k})^2|\lambda(k)|^2\left|\sum_{j=1}^n \int_{t(j-1)/n}^{tj/n} e^{-ik\cdot B_s}e^{-ik_0\frac{t(j-1)}{n}} dB_s^\mu\right|^2 d\mathbf{k},$$

where $\mu(\mathbf{k}) = (\frac{\omega(k)}{\omega(k)^2+|k_0|^2})^{1/2}$. Since

$$\lim_{n\to\infty}\mathbb{E}\left|\sum_{j=1}^{n}\int_{t(j-1)/n}^{tj/n} e^{-ikB_s}e^{-ik_0\frac{t(j-1)}{n}}dB_s^\mu-\int_0^t e^{-ikB_s}e^{-ik_0 s}dB_s^\mu\right|^2=0,$$

there exists a subsequence n' such that

$$\sum_{j=1}^{n'}\int_{t(j-1)/n'}^{tj/n'} e^{-ik\cdot B_s}e^{-ik_0\frac{t(j-1)}{n'}}dB_s^\mu\to\int_0^t e^{-ik\cdot B_s}e^{-ik_0 s}dB_s^\mu$$

almost surely. Hence we obtain that

$$X_t^\mu=\frac{1}{\pi}\int_{\mathbb{R}^3}\mu(\mathbf{k})^2|\lambda(k)|^2|Y_t^\mu|^2d\mathbf{k},\tag{3.5.3}$$

where Y_t^μ is the stochastic integral defined by $Y_t^\mu=\int_0^t e^{-isk_0}e^{-ik\cdot B_s}dB_s^\mu$. The Itô formula then gives

$$d|Y_t^\mu|^2=2\operatorname{Re}(Y_t^\mu e^{ik_0t}e^{ik\cdot B_t})dB_t^\mu+dt.\tag{3.5.4}$$

Inserting (3.5.4) into (3.5.3), it is seen that

$$X_t^\mu=2\operatorname{Re}\frac{1}{\pi}\int_{\mathbb{R}^3}\mu(\mathbf{k})^2|\lambda(k)|^2\left(\int_0^t Y_s^\mu e^{ik_0s}e^{ik\cdot B_s}dB_s^\mu\right)d\mathbf{k}+t\|\lambda\|^2,$$

where we used that $\int_{\mathbb{R}^4}\mu(\mathbf{k})^2d\mathbf{k}=\pi$. Hence the stochastic differential equation

$$dX_t^\mu=Z_t^\mu dB_t^\mu+\|\lambda\|^2dt$$

is obtained, where

$$Z_t^\mu=2\operatorname{Re}\left(\mathrm{j}_t\lambda(\cdot-B_t),\int_0^t\mathrm{j}_s\lambda(\cdot-B_s)dB_s^\mu\right)_{L^2(\mathbb{R}^3)}.$$

An application of the Itô formula gives

$$d(X_t^\mu)^m=m(X_t^\mu)^{m-1}Z_t^\mu dB_t^\mu+\left(m(X_t^\mu)^{m-1}\|\lambda\|^2+\frac{1}{2}m(m-1)(X_t^\mu)^{m-2}(Z_t^\mu)^2\right)dt.$$

Taking expectation on both sides, we have

$$\mathbb{E}^x[(X_t^\mu)^m]=m\|\lambda\|^2\int_0^t\mathbb{E}^x[(X_s^\mu)^{m-1}]ds+\frac{m(m-1)}{2}\int_0^t\mathbb{E}^x[(X_s^\mu)^{m-2}(Z_s^\mu)^2]ds.$$

By the bound $|Z_t^\mu| \le 2\|\lambda\|\sqrt{X_t^\mu}$, it follows that

$$\mathbb{E}^x[(X_t^\mu)^m] \le \frac{2m(2m-1)}{2}\|\lambda\|^2 \int_0^t \mathbb{E}^x[(X_s^\mu)^{m-1}]ds,$$

and by induction, furthermore,

$$\mathbb{E}^x[(X_t^\mu)^m] \le \frac{(2m)!}{2^m}\|\lambda\|^{2m}t^m$$

is obtained. Notice that the intertwining property $(\omega(-i\nabla)\otimes 1)\mathrm{j}_s = \mathrm{j}_s\omega(-i\nabla)$ holds. Write $\lambda = \omega(-i\nabla)^{k/2}\tilde{\varphi}$. Then $\|\lambda\| = \|\omega^{(k-1)/2}\hat{\varphi}\|$ and the proposition follows. □

Before checking (3.5.2), note that J_t intertwines the free field Hamiltonian H_{rad} and $\mathrm{d}\Gamma_{\mathrm{E}}(\hat{\omega}_{\mathrm{E}})$, i. e.,

$$\mathrm{J}_t H_{\mathrm{rad}} = \mathrm{d}\Gamma_{\mathrm{E}}(\hat{\omega}_{\mathrm{E}})\mathrm{J}_t. \tag{3.5.5}$$

Define

$$\hat{\Pi}_{\mathrm{E}}(\mathbf{f}) = i[N, \hat{A}_{\mathrm{E}}(\mathbf{f})]. \tag{3.5.6}$$

Then $\hat{\Pi}_{\mathrm{E}}(\mathbf{f})$ is the conjugate momentum of $\hat{A}_{\mathrm{E}}(\mathbf{f})$, i. e., the canonical commutation relation

$$[\hat{A}_{\mathrm{E}}(\mathbf{f}), \hat{\Pi}_{\mathrm{E}}(\mathbf{g})] = 2i\mathrm{q}_{\mathrm{E}}(\mathbf{f}, \mathbf{g})$$

is satisfied for all $\mathbf{f}, \mathbf{g} \in \oplus^3 L^2(\mathbb{R}^4)$. Let G be an analytic vector for $\hat{A}_{\mathrm{E}}(\mathbf{f})$. Then

$$e^{-i\hat{A}_{\mathrm{E}}(\mathbf{f})}G = \sum_{n=0}^{\infty} \frac{(-i\hat{A}_{\mathrm{E}}(\mathbf{f}))^n}{n!}G \tag{3.5.7}$$

holds. By the commutation relations,

$$\begin{aligned}[\mathrm{d}\Gamma_{\mathrm{E}}(\hat{\omega}_{\mathrm{E}}), \hat{A}_{\mathrm{E}}(\mathbf{f})] &= -i\hat{\Pi}_{\mathrm{E}}(\hat{\omega}_{\mathrm{E}}\mathbf{f}),\\ [[\mathrm{d}\Gamma_{\mathrm{E}}(\hat{\omega}_{\mathrm{E}}), \hat{A}_{\mathrm{E}}(\mathbf{f})], \hat{A}_{\mathrm{E}}(\mathbf{f})] &= -2\mathrm{q}_{\mathrm{E}}(\hat{\omega}_{\mathrm{E}}\mathbf{f}, \mathbf{f}),\end{aligned}$$

together with (3.5.7), we, moreover, have the identity

$$\mathrm{d}\Gamma_{\mathrm{E}}(\hat{\omega}_{\mathrm{E}})e^{-i\hat{A}_{\mathrm{E}}(\mathbf{f})}G = e^{-i\hat{A}_{\mathrm{E}}(\mathbf{f})}(\mathrm{d}\Gamma_{\mathrm{E}}(\hat{\omega}_{\mathrm{E}}) - \hat{\Pi}_{\mathrm{E}}(\hat{\omega}_{\mathrm{E}}\mathbf{f}) + \mathrm{q}_{\mathrm{E}}(\hat{\omega}_{\mathrm{E}}\mathbf{f}, \mathbf{f}))G \tag{3.5.8}$$

for $\mathbf{f} \in D(\hat{\omega}_{\mathrm{E}})$. Furthermore, by a limiting argument, it can be seen that $e^{-i\hat{A}_{\mathrm{E}}(\mathbf{f})}$ with $\mathbf{f} \in D(\hat{\omega}_{\mathrm{E}})$ leaves $D(\mathrm{d}\Gamma_{\mathrm{E}}(\hat{\omega}_{\mathrm{E}}))$ invariant and the identity (3.5.8) holds for $G \in D(\mathrm{d}\Gamma_{\mathrm{E}}(\hat{\omega}_{\mathrm{E}}))$. Put

$$\mathfrak{J}_t = \mathrm{J}_0^* e^{-ie\hat{A}_{\mathrm{E}}(K_t)}\mathrm{J}_t.$$

Lemma 3.43. *Let $F, G \in D(H_{\mathrm{rad}})$. Then we have*

$$|(H_{\mathrm{rad}}F, e^{-t\hat{H}_{\mathrm{PF}}}G)| \le C\Big((\sqrt{t}+t)\|(H_{\mathrm{rad}}+1)^{1/2}G\| + \|H_{\mathrm{rad}}G\|\Big)\|F\|.$$

In particular, $e^{-tH_{\mathrm{PF}}}D_{\mathrm{PF}} \subset D(H_{\mathrm{rad}})$.

Proof. We see that J_t leaves $D(H_{\mathrm{rad}})$ invariant, and for $G \in D(H_{\mathrm{rad}})$

$$[H_{\mathrm{rad}}, \mathrm{J}_t]G = \mathrm{J}_0^* e^{-ie\hat{A}_{\mathrm{E}}(K_t)}\mathrm{C}_t\mathrm{J}_tG$$

by (3.5.5) and (3.5.8), where

$$\mathrm{C}_t = -e\hat{\Pi}_{\mathrm{E}}(\hat{\omega}_{\mathrm{E}}K_t) + e^2\mathrm{q}_{\mathrm{E}}(\hat{\omega}_{\mathrm{E}}K_t, K_t). \tag{3.5.9}$$

Let $F, G \in D(H_{\mathrm{rad}})$. Then the functional integral representation yields

$$(H_{\mathrm{rad}}F, e^{-t\hat{H}_{\mathrm{PF}}}G) = \int_{\mathbb{R}^3} \mathbb{E}^x\left[(F(B_0), [H_{\mathrm{rad}}, \mathrm{J}_t]G(B_t))\right]dx + (F, e^{-t\hat{H}_{\mathrm{PF}}}H_{\mathrm{rad}}G).$$

Furthermore, the right-hand side above can be estimated as

$$|(H_{\mathrm{rad}}F, e^{-t\hat{H}_{\mathrm{PF}}}G)| \le \|F\|\left(\int_{\mathbb{R}^3}(\mathbb{E}^x[\|[H_{\mathrm{rad}}, \mathrm{J}_t]G_t\|])^2\,dx\right)^{1/2} + \|F\|\|H_{\mathrm{rad}}G\|.$$

Note the bound $\|[H_{\mathrm{rad}}, \mathrm{J}_t]G\| \le \|\mathrm{C}_t\mathrm{J}_tG\|$. We estimate each term in (3.5.9) separately. We have

$$\|\hat{\Pi}_{\mathrm{E}}(\hat{\omega}_{\mathrm{E}}K_t)\mathrm{J}_t\Psi\| \le C(\|\hat{\omega}_{\mathrm{E}}^{1/2}K_t\| + \|\hat{\omega}_{\mathrm{E}}K_t\|)\|(H_{\mathrm{rad}}+1)^{1/2}\Psi\|,$$
$$|\mathrm{q}_{\mathrm{E}}(\hat{\omega}_{\mathrm{E}}K_t, K_t)| \le \|\hat{\omega}_{\mathrm{E}}^{1/2}K_t\|^2$$

with a constant C. Applying our BDG-type inequality to $\|\hat{\omega}_{\mathrm{E}}K_t\|^{2n}$, we have

$$\int_{\mathbb{R}^3}(\mathbb{E}^x[\|[H_{\mathrm{rad}}, \mathrm{J}_t]G_t\|])^2dx \le C'(t+t^2)\|(H_{\mathrm{rad}}+1)^{1/2}G\|^2$$

with a constant C'. Hence the lemma follows. □

We next show that $e^{-t\hat{H}_{\mathrm{PF}}}D_{\mathrm{PF}} \subset D((-1/2)\Delta)$. Let

$$\mathscr{S}(\mathscr{Q}) = \left\{F(\hat{A}(\mathbf{f}_1)\dots\hat{A}(\mathbf{f}_n))\,\middle|\,F \in \mathscr{S}(\mathbb{R}^n), \mathbf{f}_1,\dots,\mathbf{f}_n \in \oplus^3L^2_{\mathrm{real}}(\mathbb{R}^3), n \ge 1\right\}. \tag{3.5.10}$$

Lemma 3.44. *Let $\omega^{3/2}\hat{\varphi} \in L^2(\mathbb{R}^3)$. Then for $F \in D(\Delta)$ and $G \in D_{\mathrm{PF}}$,*

$$|(\Delta F, e^{-t\hat{H}_{\mathrm{PF}}}G)| \le C\Big(\| -\Delta G\| + (\sqrt{t}+t)\|(-\Delta + H_{\mathrm{rad}} + 1)G\|\Big)\|F\| \tag{3.5.11}$$

with a constant $C > 0$. In particular, $e^{-t\hat{H}_{\mathrm{PF}}}D_{\mathrm{PF}} \subset D((-1/2)\Delta)$.

Proof. Suppose that $F = f \otimes \Phi$ and $G = g \otimes \Psi$, where $f, g \in C_0^\infty(\mathbb{R}^3)$ and $\Phi, \Psi \in \mathscr{S}(\mathscr{Q})$. By a functional integration, we have

$$(-\Delta F, e^{-t\hat{H}_{\mathrm{PF}}}G) = -\int_{\mathbb{R}^3} \mathbb{E}[(\mathrm{J}_0 F(x), e^{-ie\hat{A}_{\mathrm{E}}(K_t)} U \mathrm{J}_t G(B_t + x))]dx,$$

where

$$U = \Delta + \sum_{\mu=1}^{3}(-ie\hat{A}_{\mathrm{E}}(K_{\mu\mu}) - e^2\hat{A}_{\mathrm{E}}(K_\mu)^2 - 2ie\hat{A}_{\mathrm{E}}(K_\mu)\cdot\nabla_\mu)$$

and

$$K_\mu = \bigoplus_{\nu=1}^{3}\int_0^t \mathrm{j}_t\tilde{\varphi}_\mu(\cdot - B_s)dB_s^\nu, \quad \tilde{\varphi}_\mu = (-ik_\mu\hat{\varphi}/\sqrt{\omega})\check{},$$

$$K_{\mu\mu} = \bigoplus_{\nu=1}^{3}\int_0^t \mathrm{j}_t\tilde{\varphi}_{\mu\mu}(\cdot - B_s)dB_s^\nu, \quad \tilde{\varphi}_{\mu\mu} = (-k_\mu^2\hat{\varphi}/\sqrt{\omega})\check{}.$$

Since we assumed that $\omega^{3/2}\hat{\varphi} \in L^2(\mathbb{R}^3)$, it follows that $k_\mu^2\hat{\varphi}/\sqrt{\omega} \in L^2(\mathbb{R}^3)$, and thus $\tilde{\varphi}_{\mu\mu}$ is well defined. Hence

$$|(-\Delta F, e^{-t\hat{H}_{\mathrm{PF}}}G)| \le \|F\|\left(\int_{\mathbb{R}^3} |\mathbb{E}\,[\|U\mathrm{J}_t G(B_t + x)\|]|^2\,dx\right)^{1/2}. \tag{3.5.12}$$

We estimate $\mathbb{E}\,[\|U\mathrm{J}_t G(B_t + x)\|]$. By the BDG-type inequality again, we have

$$\sup_{x\in\mathbb{R}^3}\mathbb{E}^x[\|\hat{\omega}_{\mathrm{E}}^{n/2}K_{\mu\mu}\|^{2m}] \le C' t^m \|\omega^{(n+3)/2}\hat{\varphi}\|^{2m} \tag{3.5.13}$$

and

$$\sup_{x\in\mathbb{R}^3}\mathbb{E}^x[\|\hat{\omega}_{\mathrm{E}}^{n/2}K_{\mu}\|^{2m}] \le C'' t^m \|\omega^{(n+1)/2}\hat{\varphi}\|^{2m} \tag{3.5.14}$$

with constants C' and C''. Let $\|f\|_a = \|\omega^{a/2}f\|$ and

$$A = \|f\|_{-1} + \|f\|_0, \quad B = (\|f\|_{-1} + \|f\|_0)(\|g\|_1 + \|g\|_2).$$

Combining the inequalities,

$$\|a^\sharp(f)\Psi\| \le A\|(H_{\rm rad}+1)^{1/2}\Psi\|,$$
$$\|a^\sharp(f)a^\sharp(g)\Psi\| \le C_1B\|(H_{\rm rad}+1)^{1/2}\Psi\| + C_2A\|(H_{\rm rad}+1)\Psi\|$$

with (3.5.13) and (3.5.14), we can estimate the right-hand side of (3.5.12) and show (3.5.11). By a limiting argument, the lemma follows for $F \in D(\Delta)$ and $G \in D_{\rm PF}$. □

Lemma 3.45. *Let $\omega^{3/2}\hat{\varphi} \in L^2(\mathbb{R}^3)$. Then $H_{\rm PF}$ is essentially self-adjoint on $D_{\rm PF}$.*

Proof. By Lemmas 3.43 and 3.44, we obtain that $|(H_{\rm PF,0}F, e^{-t\hat{H}_{\rm PF}}G)| \le C_G\|F\|$ for $F, G \in D(H_{\rm PF,0}) = D_{\rm PF}$ with a constant C_G depending on G. Then $e^{-t\hat{H}_{\rm PF}}$ leaves $D_{\rm PF}$ invariant. This implies that $\hat{H}_{\rm PF}$ is essentially self-adjoint on $D_{\rm PF}$ and then such is also $H_{\rm PF}$ since $H_{\rm PF} = \hat{H}_{\rm PF}$ on $D_{\rm PF}$. □

Next, we show self-adjointness of $H_{\rm PF}$ on $D_{\rm PF}$ without assuming $\omega^{3/2}\hat{\varphi} \in L^2(\mathbb{R}^3)$.

Lemma 3.46. *Suppose that $\omega^{3/2}\hat{\varphi} \in L^2(\mathbb{R}^3)$. Then $H_{\rm PF,0}(\hat{H}_{\rm PF}+z)^{-1}$ is bounded for every $z \in \mathbb{C}$ with $\operatorname{Im} z \neq 0$.*

Proof. We prove that both of $\Delta(\hat{H}_{\rm PF}+z)^{-1}$ and $H_{\rm rad}(\hat{H}_{\rm PF}+z)^{-1}$ are bounded. Note that $H_{\rm PF} = \hat{H}_{\rm PF}$ on $D_{\rm PF}$. Let $p_\mu = -i\nabla_\mu$. In this proof, all numbered constants are nonnegative. We have

$$\|p^2\Psi\|^2 \le C_1(\|(p-e\hat{A})^2\Psi\|^2 + \|\hat{A}\cdot(p-e\hat{A})\Psi\|^2 + \|\hat{A}^2\Psi\|^2).$$

We estimate $\|\hat{A}\cdot(p-e\hat{A})\Psi\|^2$ as

$$\begin{aligned}\|\hat{A}\cdot(p-e\hat{A})\Psi\|^2 &\le C_2\sum_{\mu=1}^{3}((p-e\hat{A})_\mu\Psi, (H_{\rm rad}+1)(p-e\hat{A})_\mu\Psi)\\ &= C_2((p-e\hat{A})^2\Psi, (H_{\rm rad}+1)\Psi) + C_2\sum_{\mu=1}^{3}((p-e\hat{A})_\mu\Psi, -e[H_{\rm rad}, \hat{A}_\mu]\Psi).\end{aligned}$$

Since $\|[H_{\rm rad}, \hat{A}_\mu]\Psi\| \le C_2'\|(H_{\rm rad}+1)^{1/2}\Psi\|$, we have

$$\|p^2\Psi\|^2 \le C_3(\|(p-e\hat{A})^2\Psi\|^2 + \|H_{\rm rad}\Psi\|^2 + \|\Psi\|^2). \tag{3.5.15}$$

We note that

$$\|\hat{H}_{\rm PF}\Psi\|^2 = \left\|\frac{1}{2}(p-e\hat{A})^2\Psi\right\|^2 + \|H_{\rm rad}\Psi\|^2 + \operatorname{Re}((p-e\hat{A})^2\Psi, H_{\rm rad}\Psi).$$

Since

$$\operatorname{Re}((p-e\hat{A})^2\Psi, H_{\rm rad}\Psi)$$

$$
\begin{aligned}
&= \sum_{\mu=1}^{3} \{((p - e\hat{A})_\mu \Psi, H_{\mathrm{rad}}(p - e\hat{A})_\mu \Psi) + \mathrm{Re}((p - e\hat{A})_\mu \Psi, -e[H_{\mathrm{rad}}, \hat{A}_\mu]\Psi)\} \\
&\geq \sum_{\mu=1}^{3} \mathrm{Re}((p - e\hat{A})_\mu \Psi, -e[H_{\mathrm{rad}}, \hat{A}_\mu]\Psi) \\
&\geq -\varepsilon \|(p - e\hat{A})^2 \Psi\|^2 - C_4 \frac{1}{4\varepsilon}(\|H_{\mathrm{rad}}^{1/2}\Psi\|^2 + \|\Psi\|^2)
\end{aligned}
$$

for any $\varepsilon > 0$, by the trivial bound $\|H_{\mathrm{rad}}^{1/2}\Psi\|^2 \leq \eta \|H_{\mathrm{rad}}\Psi\|^2 + \frac{1}{4\eta}\|\Psi\|^2$ for any $\eta > 0$, we further have

$$
\|\hat{H}_{\mathrm{PF}}\Psi\|^2 \geq C_5 \|(p - e\hat{A})^2 \Psi\|^2 + C_6 \|H_{\mathrm{rad}}\Psi\|^2 - C_7 \|\Psi\|^2.
$$

Together with (3.5.15), this gives

$$
\|p^2 \Psi\|^2 + \|H_{\mathrm{rad}}\Psi\|^2 \leq C_8(\|\hat{H}_{\mathrm{PF}}\Psi\|^2 + \|\Psi\|^2). \tag{3.5.16}
$$

Thus $\Delta(\hat{H}_{\mathrm{PF}} + z)^{-1}$ and $H_{\mathrm{rad}}(\hat{H}_{\mathrm{PF}} + z)^{-1}$ are bounded. □

Theorem 3.47 (Self-adjointness). *We have the following:*
(1) *H_{PF} is self-adjoint on D_{PF} and essentially self-adjoint on any core of $H_{\mathrm{PF},0}$;*
(2) *the functional integral representation*

$$
(F, e^{-tH_{\mathrm{PF}}} G) = \int_{\mathbb{R}^3} \mathbb{E}^x[e^{-\int_0^t V(B_s)ds} (\mathrm{J}_0 F(B_0), e^{-ie\hat{A}_{\mathrm{E}}(K_t)} \mathrm{J}_t G(B_t))]dx \tag{3.5.17}
$$

holds.

Proof. First, we assume that $\omega^{3/2}\hat{\varphi} \in L^2(\mathbb{R}^3)$. Take $V = 0$. By Lemma 3.46,

$$
\|H_{\mathrm{PF},0}\Psi\| \leq C' \|(\hat{H}_{\mathrm{PF}} + z)\Psi\| \tag{3.5.18}
$$

holds with some C' for all $\Psi \in D(\hat{H}_{\mathrm{PF}})$. Let $(\Psi_n)_{n=1}^{\infty} \subset D(H_{\mathrm{PF},0})$ be such that $\Psi_n \to \Psi$ and $\hat{H}_{\mathrm{PF}}\Psi_n \to \Phi$ for some Φ in strong sense as $n \to \infty$. Note that $D(\hat{H}_{\mathrm{PF}}) \supset D(H_{\mathrm{PF},0}) = D_{\mathrm{PF}}$. Due to (3.5.18), $(H_{\mathrm{PF},0}\Psi_n)_{n\in\mathbb{N}}$ is a Cauchy sequence and then $\Psi \in D(H_{\mathrm{PF},0})$ since $H_{\mathrm{PF},0}$ is closed. In particular, $\hat{H}_{\mathrm{PF}}$ is thus closed on D_{PF}. This implies that $H_{\mathrm{PF}}\lceil_{D_{\mathrm{PF}}}$ is closed and H_{PF} is self-adjoint since we already proven in Lemma 3.45 that $H_{\mathrm{PF}}\lceil_{D_{\mathrm{PF}}}$ is essentially self-adjoint and $\hat{H}_{\mathrm{PF}} = H_{\mathrm{PF}}$ on D_{PF}.

Next, let $V \neq 0$. As mentioned in Remark 3.31, V is relatively bounded with respect to $\hat{H}_{\mathrm{PF}}$ with a relative bound strictly smaller than 1. Self-adjointness of H_{PF} on D_{PF} follows then by the Kato–Rellich theorem.

Finally, we remove the assumption $\omega^{3/2}\hat{\varphi} \in L^2(\mathbb{R}^3)$. We divide $\hat{\varphi}$ as

$$
\hat{\varphi} = \hat{\varphi}_\Lambda + \hat{\varphi}_{\Lambda^\perp},
$$

where $\hat{\varphi}_\Lambda = \hat{\varphi}1_\Lambda$ and $\hat{\varphi}_{\Lambda^\perp} = \hat{\varphi} - \hat{\varphi}_\Lambda$ with the indicator function 1_Λ on $|k| \le \Lambda$. Write $\hat{A} = \hat{A}_\Lambda + \hat{A}_{\Lambda^\perp}$, where $\hat{A}_\Lambda$ has the form factor $\hat{\varphi}_\Lambda$ instead of $\hat{\varphi}$, and $\hat{A}_{\Lambda^\perp}$ does $\hat{\varphi}_{\Lambda^\perp}$. Then $H_{\rm PF}$ can be split off like $H_{\rm PF} = H_{\rm PF}(\Lambda) + H_{\Lambda^\perp}$, where

$$H_{\rm PF}(\Lambda) = \frac{1}{2}(p - e\hat{A}_\Lambda)^2 + V + H_{\rm rad}, \quad H_{\Lambda^\perp} = -e\hat{A}_{\Lambda^\perp} \cdot (p - e\hat{A}_\Lambda) + \frac{e^2}{2}\hat{A}^2_{\Lambda^\perp}.$$

Since $\omega^{3/2}\hat{\varphi}_\Lambda \in L^2(\mathbb{R}^3)$, $H_{\rm PF}(\Lambda)$ is self-adjoint on $D_{\rm PF}$. We shall show below that $H_{\Lambda^\perp}$ is relatively bounded with respect to $H_{\rm PF}(\Lambda)$ with a relative bound strictly smaller than one for the large Λ. The estimate is similar to the proof of Lemma 3.46. Since for any $\delta > 0$,

$$\|e\hat{A}_{\Lambda^\perp} \cdot (p - e\hat{A}_\Lambda)\Psi\| \le \delta\|(p - e\hat{A}_\Lambda)^2\Psi\| + \frac{e^2}{2\delta}\|\hat{A}^2_{\Lambda^\perp}\Psi\|,$$

we have

$$\|H_{\Lambda^\perp}\Psi\| \le \delta\|(p - e\hat{A}_\Lambda)^2\Psi\| + e^2\left(\frac{1}{2\delta} + 1\right)\|\hat{A}^2_{\Lambda^\perp}\Psi\|.$$

Similar to the estimate in the proof of Lemma 3.46, we have for any $\varepsilon > 0$ and $\eta > 0$,

$$\left\|\frac{1}{2}(p - e\hat{A}_\Lambda)^2\Psi\right\|^2 + \|H_{\rm rad}\Psi\|^2 - \varepsilon\|(p - e\hat{A}_\Lambda)^2\Psi\|^2 - \frac{C_1}{4\varepsilon}\left(\eta\|H_{\rm rad}\Psi\|^2 + \frac{1}{4\eta}\|\Psi\|^2\right)$$
$$\le \|H_{\rm PF}(\Lambda)\Psi\|^2,$$

where C_1 depends on $\|\sqrt{\omega}\hat{\varphi}_\Lambda\|$ and $\|\hat{\varphi}_\Lambda/\sqrt{\omega}\|$, but is independent of $\|\omega^{3/2}\hat{\varphi}_\Lambda\|$. In particular, $\lim_{\Lambda\to\infty} C_1 < \infty$. From this, we have

$$\frac{1 - 4\varepsilon}{4}\|(p - e\hat{A}_\Lambda)^2\Psi\|^2 + \left(1 - C_1\frac{\eta}{4\varepsilon}\right)\|H_{\rm rad}\Psi\|^2 \le \|H_{\rm PF}(\Lambda)\Psi\|^2 + \frac{C_1}{16\varepsilon\eta}\|\Psi\|^2. \quad (3.5.19)$$

Let ε be sufficiently small and choose η such that $1 - \frac{C_1\eta}{4\varepsilon} > 0$. Thus we obtain that

$$\|(p - e\hat{A}_\Lambda)^2\Psi\|^2 \le \frac{4}{1 - 4\varepsilon}\|H_{\rm PF}(\Lambda)\Psi\|^2 + \frac{4C_1}{16\eta\varepsilon(1 - 4\varepsilon)}\|\Psi\|^2.$$

Next, we estimate $\|\hat{A}^2_{\Lambda^\perp}\Psi\|$. We have

$$\|\hat{A}^2_{\Lambda^\perp}\Psi\|^2 \le C_2(\|H_{\rm rad}\Psi\|^2 + \|\Psi\|^2),$$

where C_2 depends on $\|\omega^{a/2}\hat{\varphi}_{\Lambda^\perp}\|$, $a = -2, -1, 0, 1$. Thus $\lim_{\Lambda\to\infty} C_2 = 0$ and by (3.5.19),

$$\|\hat{A}^2_{\Lambda^\perp}\Psi\|^2 \le \left(\frac{C_2}{1 - C_1\frac{\eta}{4\varepsilon}}\right)\|H_{\rm PF}(\Lambda)\Psi\|^2 + C_3\|\Psi\|^2.$$

Hence we are led to the bounds

$$\|(p - e\hat{A}_\Lambda)^2\Psi\|^2 \le C_4(\|H_{\mathrm{PF}}(\Lambda)\Psi\|^2 + \|\Psi\|^2),$$
$$\|\hat{A}^2_{\Lambda^\perp}\Psi\|^2 \le C_5(\|H_{\mathrm{PF}}(\Lambda)\Psi\|^2 + \|\Psi\|^2),$$

where $\lim_{\Lambda\to\infty} C_4 < \infty$ and $\lim_{\Lambda\to\infty} C_5 = 0$. From the bound,

$$\|H_{\Lambda^\perp}\Psi\| \le \left(\delta C_4 + e^2\left(\frac{1}{2\delta} + 1\right)C_5\right)\|H_{\mathrm{PF}}(\Lambda)\Psi\| + C_6\|\Psi\|$$

it follows that for sufficiently large Λ and small δ such that

$$\delta C_4 + e^2\left(\frac{1}{2\delta} + 1\right)C_5 < 1,$$

$H_{\Lambda^\perp}$ is relatively bounded with respect to $H_{\mathrm{PF}}(\Lambda)$ with a relative bound strictly smaller than 1. Thus H_{PF} is self-adjoint on D_{PF}. The inequality

$$\|H_{\mathrm{PF}}\Psi\| + \|\Psi\| \le C''(\|H_{\mathrm{PF},0}\Psi\| + \|\Psi\|) \tag{3.5.20}$$

can be derived directly without effort with some constant C''. Let D be a core of $H_{\mathrm{PF},0}$. For any $\Psi \in D(H_{\mathrm{PF}}) = D(H_{\mathrm{PF},0})$, there exists a sequence $(\Psi_n)_{n\in\mathbb{N}} \subset D$ such that $\Psi_n \to \Psi$ and $H_{\mathrm{PF},0}\Psi \to H_{\mathrm{PF},0}\Psi$. This implies that $(H_{\mathrm{PF}}\Psi_n)_{n=1}^\infty$ is a Cauchy sequence by (3.5.20). Thus $H_{\mathrm{PF}}\Psi_n$ converges to some Φ as $n \to \infty$. Since H_{PF} is closed, $\Phi = H_{\mathrm{PF}}\Psi$. Thus D is a core of H_{PF}. Since (3.5.17) with $e^{-tH_{\mathrm{PF}}}$ replaced by $e^{-t\hat{H}_{\mathrm{PF}}}$ is verified, (3.5.17) holds since $\hat{H}_{\mathrm{PF}} = H_{\mathrm{PF}}$. □

3.5.2 Positivity improving and uniqueness of the ground state

The main question addressed in this section is whether the ground state of the Pauli–Fierz Hamiltonian is unique. In Section 3.5.2 we assume either Assumption 3.4 or 3.33. As seen earlier on, a way of proving this is through the Perron–Frobenius theorem, relying on the positivity improving property of the semigroup. However, in the functional integral representation of $e^{-tH_{\mathrm{PF}}}$ the phase factor $e^{-ie\hat{A}_{\mathrm{E}}(K_t)}$ appears, therefore, intuition would say that the semigroup does not even preserve realness. Nevertheless, we can show that this semigroup is unitary equivalent with a positivity improving operator, and that will allow us to conclude that the ground state of H_{PF} is unique whenever it exists.

Consider the multiplication operator $T_t = e^{itx}$, $t \in \mathbb{R}$, in $L^2(\mathbb{R})$ with respect to x. Although T_t is not a positivity preserving operator, the operator $\hat{F}T_t\hat{F}^{-1}$, obtained under Fourier transform $\hat{F}$ on $L^2(\mathbb{R})$, is a shift operator, i. e.,

$$(f, \hat{F}T_t\hat{F}^{-1}g) = (f, g(\cdot + t)) \ge 0$$

for nonnegative functions f and g. This implies that $\hat{F}T_t\hat{F}^{-1}$ is a positivity preserving, but not positivity improving operator. We extend this argument to the Pauli–Fierz model by introducing a transformation on $L^2(\mathscr{Q})$ corresponding to Fourier transform on $L^2(\mathbb{R})$.

We return to the Fock space $\mathcal{F}_b$ in Chapter 1, in which we defined the Segal field $\Phi(f)$ and its conjugate $\Pi(f)$. Since for $\alpha \in \mathbb{R}$,

$$e^{i\alpha N}a^*(f)e^{-i\alpha N} = a^*(e^{i\alpha}f),$$
$$e^{i\alpha N}a(f)e^{-i\alpha N} = a(e^{-i\alpha}f);$$

in particular, for $\alpha = \pi/2$ we have

$$e^{i(\pi/2)N}a^*(f)e^{-i(\pi/2)N} = ia^*(f),$$
$$e^{i(\pi/2)N}a(f)e^{-i(\pi/2)N} = -ia(f).$$

Thus Φ and Π are related by the unitary operator $e^{i(\pi/2)N}$ as

$$e^{i(\pi/2)N}\Phi(f)e^{-i(\pi/2)N} = \Pi(f).$$

This may suggest that $e^{i(\pi/2)N}$ can be regarded as a Fourier transform on $\mathcal{F}_b$. By the argument above, we define the unitary operator $\mathfrak{S}_\beta$ on $L^2(\mathscr{Q}_\beta)$ by

$$\mathfrak{S}_\beta = \exp\left(-i\frac{\pi}{2}N_\beta\right), \quad \beta = 0, 1,$$

where N_β denotes the number operator in $L^2(\mathscr{Q}_\beta)$. We simply write $\mathfrak{S}_\beta$ for the unitary operator $1 \otimes \mathfrak{S}_\beta$ on $L^2(\mathbb{R}^3) \otimes L^2(\mathscr{Q}_\beta)$ as long as there is no risk of confusion. We have

$$\hat{\Pi}_\beta(\mathbf{f}) = \mathfrak{S}_\beta^{-1}\hat{A}_\beta(\mathbf{f})\mathfrak{S}_\beta$$

and write $\hat{\Pi}_0 = \hat{\Pi}$, $\hat{\Pi}_1 = \hat{\Pi}_{\mathrm{E}}$, $\mathfrak{S}_0 = \mathfrak{S}$, and $\mathfrak{S}_1 = \mathfrak{S}_{\mathrm{E}}$. Clearly, $\hat{\Pi}_{\mathrm{E}}$ coincides with the expression under (3.5.6). It is clear that

$$[\hat{A}_\beta(\mathbf{f}), \hat{\Pi}_\beta(\mathbf{g})] = 2i\mathfrak{q}_\beta(\mathbf{f}, \mathbf{g})$$

and, importantly,

$$e^{i\hat{\Pi}_\beta(\mathbf{g})}1 = e^{-\mathfrak{q}_\beta(\mathbf{g},\mathbf{g})}e^{-\hat{A}_\beta(\mathbf{g})}1.$$

Moreover, the Weyl relation

$$e^{i\hat{\Pi}_\beta(\mathbf{g})}e^{i\hat{A}_\beta(\mathbf{f})} = e^{i2\mathfrak{q}_\beta(\mathbf{f},\mathbf{g})}e^{i\hat{A}_\beta(\mathbf{f})}e^{i\hat{\Pi}_\beta(\mathbf{g})}$$

holds.

Lemma 3.48. *Let* $\mathbf{f} \in \oplus^3 L^2_{\mathrm{real}}(\mathbb{R}^3)$. *Then the unitary operator* $e^{i\hat{\Pi}(\mathbf{f})}$ *is positivity preserving.*

Proof. Let $F, G \in \mathscr{S}(\mathscr{Q})$ (see (3.5.10)) be such that

$$F = (2\pi)^{-n/2} \int_{\mathbb{R}^n} \hat{F}(k) e^{i\sum_{j=1}^n k_j \hat{A}(\mathbf{f}_j)} dk,$$

$$G = (2\pi)^{-m/2} \int_{\mathbb{R}^m} \hat{G}(k) e^{i\sum_{j=1}^m k_j \hat{A}(\mathbf{g}_j)} dk,$$

with positive $F \in \mathscr{S}(\mathbb{R}^n)$ and $G \in \mathscr{S}(\mathbb{R}^m)$. Using the fact that

$$(e^{i\hat{A}(\mathbf{g}_1)}, e^{i\hat{\Pi}(\mathbf{f})} e^{i\hat{A}(\mathbf{g}_2)}) = (e^{i\hat{A}(\mathbf{g}_1-\mathbf{g}_2)}, e^{i\hat{\Pi}(\mathbf{f})}) e^{2i\mathrm{q}(\mathbf{g}_2,\mathbf{f})},$$

it is directly seen that

$$\begin{aligned}(F, e^{i\hat{\Pi}(\mathbf{f})} G) = \frac{1}{(2\pi)^{n+m}} \int_{\mathbb{R}^{n+m}} & \overline{\hat{F}(k)} \hat{G}(k') \exp\left(i \sum_{j=1}^n \mathrm{q}(\mathbf{f}_j, \mathbf{f}) k_j\right) \\ & \times \exp\left(i \sum_{j=1}^m \mathrm{q}(\mathbf{g}_j, \mathbf{f}) k'_j\right) \exp\left(-\frac{1}{2}\mathrm{q}(\mathbf{f}, \mathbf{f})\right) \mathcal{G}(k, k') dk dk',\end{aligned}$$

where

$$\mathcal{G}(k, k') = \exp\left(-\frac{1}{2} \sum_{i,j=1}^n \sum_{i',j'=1}^m \mathrm{q}(k_i \mathbf{f}_i - k'_{i'} \mathbf{g}_{i'}, k_j \mathbf{f}_j - k'_{j'} \mathbf{g}_{j'})\right).$$

Hence $(F, e^{i\hat{\Pi}(\mathbf{f})} G) \geq 0$. Since any nonnegative function can be approximated by functions in $\mathscr{S}(\mathscr{Q})$, the lemma follows. □

Note the intertwining property $\mathrm{J}_t \mathfrak{S} = \mathfrak{S}_{\mathrm{E}} \mathrm{J}_t$ and $\mathfrak{S}_{\mathrm{E}}^{-1} e^{-ie\hat{A}_{\mathrm{E}}(K_t)} \mathfrak{S}_{\mathrm{E}} = e^{-ie\hat{\Pi}_{\mathrm{E}}(K_t)}$. Using the functional integral representation (3.3.4), we have

$$(F, \mathfrak{S}^{-1} e^{-tH_{\mathrm{PF}}} \mathfrak{S} G) = \int_{\mathbb{R}^3} \mathbb{E}^x [e^{-\int_0^t V(B_s) ds} (\mathrm{J}_0 F(B_0), e^{-ie\hat{\Pi}_{\mathrm{E}}(K_t)} \mathrm{J}_t G(B_s))] dx.$$

Although Proposition 1.52 shows that $e^{-tH_{\mathrm{rad}}}$ is positivity preserving, we can show a stronger statement below.

Proposition 3.49 (Positivity improving for $e^{-tH_{\mathrm{rad}}}$). *The semigroup* $\{e^{-tH_{\mathrm{rad}}} : t > 0\}$ *is positivity improving.*

Proof. The semigroup $\{e^{-tH_{\mathrm{rad}}} : t > 0\}$ is positivity preserving, and 1 is a simple eigenvector of H_{rad} with $H_{\mathrm{rad}} 1 = 0$, and strictly positive, i. e., $1 > 0$. Then the proposition follows from Proposition 1.53. □

This proposition implies that $\mathrm{J}_t^*\mathrm{J}_0$ is positivity improving. We show furthermore that $\mathrm{J}_t^* e^{ie\hat{\Pi}_{\mathrm{E}}(K_t)}\mathrm{J}_0$ is positivity improving. The key factorization identity to achieving that is

$$\mathrm{J}_t^* e^{i\hat{\Pi}_{\mathrm{E}}(\mathbf{f})}\mathrm{J}_0\Psi = cAB\Psi, \tag{3.5.21}$$

for $\Psi \in \mathscr{S}(\mathscr{Q})$, where $c = e^{-(\mathrm{q}_{\mathrm{E}}(\mathbf{f},\mathbf{f})+\mathrm{q}(\mathrm{j}_0^*\mathbf{f},\mathrm{j}_0^*\mathbf{f}))}$ is a constant, $A = \mathrm{J}_t^* e^{-\hat{A}_{\mathrm{E}}(\mathbf{f})}\mathrm{J}_0$, and $B = e^{i\hat{\Pi}(\mathrm{j}_0^*\mathbf{f})}e^{\hat{A}(\mathrm{j}_0^*\mathbf{f})}$. We will see below that there exists A_M and B_M such that $A \geq A_M$ and $B \geq B_M$, moreover, A_M is positivity improving and B_M positivity preserving, which shows that the closure of cAB is also positivity improving, since

$$\overline{cAB\lceil_{\mathscr{S}(\mathscr{Q})}} \geq cA_M B_M$$

and $B_M G \neq 0$ if $G \neq 0$. First, we show (3.5.21). Note that

$$\mathrm{J}_0 G = G(\hat{A}_{\mathrm{E}}(\mathrm{j}_0\mathbf{g}_1), \ldots, \hat{A}_{\mathrm{E}}(\mathrm{j}_0\mathbf{g}_m))1.$$

By the commutation relations,

$$e^{i\hat{\Pi}_{\mathrm{E}}(\mathbf{f})}\mathrm{J}_0 G1 = G(\hat{A}_{\mathrm{E}}(\mathrm{j}_0\mathbf{g}_1) + 2\mathrm{q}_{\mathrm{E}}(\mathrm{j}_0\mathbf{g}_1, \mathbf{f}), \ldots, \hat{A}_{\mathrm{E}}(\mathrm{j}_0\mathbf{g}_m) + 2\mathrm{q}_{\mathrm{E}}(\mathrm{j}_0\mathbf{g}_m, \mathbf{f}))e^{i\hat{\Pi}_{\mathrm{E}}(\mathbf{f})}1.$$

From $e^{i\hat{\Pi}_{\mathrm{E}}(\mathbf{f})}1 = e^{-\mathrm{q}_{\mathrm{E}}(\mathbf{f},\mathbf{f})}e^{-\hat{A}_{\mathrm{E}}(\mathbf{f})}1$, it then follows that

$$e^{i\hat{\Pi}_{\mathrm{E}}(\mathbf{f})}\mathrm{J}_0 G1 = e^{-\mathrm{q}_{\mathrm{E}}(\mathbf{f},\mathbf{f})}e^{-\hat{A}_{\mathrm{E}}(\mathbf{f})}G(\hat{A}_{\mathrm{E}}(\mathrm{j}_0\mathbf{g}_1) + 2\mathrm{q}_{\mathrm{E}}(\mathrm{j}_0\mathbf{g}_1, \mathbf{f}), \ldots, \hat{A}_{\mathrm{E}}(\mathrm{j}_0\mathbf{g}_m) + 2\mathrm{q}_{\mathrm{E}}(\mathrm{j}_0\mathbf{g}_m, \mathbf{f}))1$$

and from $\mathrm{q}_{\mathrm{E}}(\mathrm{j}_0\mathbf{g}, \mathbf{f}) = \mathrm{q}(\mathbf{g}, \mathrm{j}_0^*\mathbf{f})$, this is further

$$= e^{-\mathrm{q}_{\mathrm{E}}(\mathbf{f},\mathbf{f})}e^{-\hat{A}_{\mathrm{E}}(\mathbf{f})}G(\hat{A}_{\mathrm{E}}(\mathrm{j}_0\mathbf{g}_1) + 2\mathrm{q}(\mathbf{g}_1, \mathrm{j}_0^*\mathbf{f}), \ldots, \hat{A}_{\mathrm{E}}(\mathrm{j}_0\mathbf{g}_m) + 2\mathrm{q}(\mathbf{g}_m, \mathrm{j}_0^*\mathbf{f}))1. \tag{3.5.22}$$

Here,

$$\widehat{\mathrm{j}_t^* h}(k) = \frac{2}{\sqrt{2\pi}} \int_{\mathbb{R}} \frac{|k|}{|k|^2 + |k_0|^2} e^{itk_0}\hat{h}(\mathbf{k})dk_0, \quad t \in \mathbb{R}.$$

The right-hand side of (3.5.22) can be written as

$$e^{-\mathrm{q}_{\mathrm{E}}(\mathbf{f},\mathbf{f})}e^{-\hat{A}_{\mathrm{E}}(\mathbf{f})}\mathrm{J}_0 e^{i\hat{\Pi}(\mathrm{j}_0^*\mathbf{f})}Ge^{-i\hat{\Pi}(\mathrm{j}_0^*\mathbf{f})}1.$$

Using again $e^{-i\hat{\Pi}(\mathbf{f})}1 = e^{-\mathrm{q}(\mathbf{f},\mathbf{f})}e^{\hat{A}(\mathbf{f})}1$ and noting that G commutes with $e^{\hat{A}(\mathrm{j}_0^*\mathbf{f})}$, we finally have

$$e^{i\hat{\Pi}_{\mathrm{E}}(\mathbf{f})}\mathrm{J}_0 G1 = e^{-\mathrm{q}_{\mathrm{E}}(\mathbf{f},\mathbf{f})-\mathrm{q}(\mathrm{j}_0^*\mathbf{f},\mathrm{j}_0^*\mathbf{f})}e^{-\hat{A}_{\mathrm{E}}(\mathbf{f})}\mathrm{J}_0 e^{i\hat{\Pi}(\mathrm{j}_0^*\mathbf{f})}e^{\hat{A}(\mathrm{j}_0^*\mathbf{f})}G1.$$

This gives (3.5.21). From this, we have the operator identity

$$\mathrm{J}_t^* e^{i\hat{\Pi}_{\mathrm{E}}(\mathbf{f})}\mathrm{J}_0 = e^{-(q_{\mathrm{E}}(\mathbf{f},\mathbf{f})+q(\mathrm{j}_0^*\mathbf{f},\mathrm{j}_0^*\mathbf{f}))}\overline{(\mathrm{J}_t^* e^{-\hat{A}_{\mathrm{E}}(\mathbf{f})}\mathrm{J}_0 e^{i\hat{\Pi}(\mathrm{j}_0^*\mathbf{f})} e^{\hat{A}(\mathrm{j}_0^*\mathbf{f})})\lceil \mathscr{S}(\mathscr{Q})}.$$

Note that $e^{-\hat{A}_{\mathrm{E}}(\mathbf{f})}$ and $e^{\hat{A}(\mathrm{j}_0^*\mathbf{f})}$ are unbounded.

Define the bounded functions obtained by the truncations

$$(e^{-\hat{A}_{\mathrm{E}}(\mathbf{f})})_M = \begin{cases} e^{-\hat{A}_{\mathrm{E}}(\mathbf{f})}, & e^{-\hat{A}_{\mathrm{E}}(\mathbf{f})} < M, \\ M, & e^{-\hat{A}_{\mathrm{E}}(\mathbf{f})} \geq M, \end{cases}$$

$$(e^{-\hat{A}(\mathrm{j}_0^*\mathbf{f})})_M = \begin{cases} e^{-\hat{A}(\mathrm{j}_0^*\mathbf{f})}, & e^{-\hat{A}(\mathrm{j}_0^*\mathbf{f})} < M, \\ M, & e^{-\hat{A}(\mathrm{j}_0^*\mathbf{f})} \geq M. \end{cases}$$

Lemma 3.50. *For $t > 0$, $\mathrm{J}_t^*(\mathrm{e}^{-\hat{A}_{\mathrm{E}}(\mathbf{f})})_M\mathrm{J}_0$ is positivity improving.*

Proof. It suffices to show that $(F, \mathrm{J}_t^*(\mathrm{e}^{-\hat{A}_{\mathrm{E}}(\mathbf{f})})_M\mathrm{J}_0 G) \neq 0$ for any nonnegative F and G, but $F \not\equiv 0$ and $G \not\equiv 0$ since $\mathrm{J}_t^*(\mathrm{e}^{-\hat{A}_{\mathrm{E}}(\mathbf{f})})_M\mathrm{J}_0$ is positivity preserving. Suppose

$$(F, \mathrm{J}_t^*(\mathrm{e}^{-\hat{A}_{\mathrm{E}}(\mathbf{f})})_M\mathrm{J}_0 G) = ((e^{-\hat{A}_{\mathrm{E}}(\mathbf{f})})_M \mathrm{J}_t F, \mathrm{J}_0 G) = 0 \tag{3.5.23}$$

for some nonnegative F and G. (3.5.23) implies that

$$\mu_E(\mathrm{supp}[(e^{-\hat{A}_{\mathrm{E}}(\mathbf{f})})_M \mathrm{J}_t F] \cap \mathrm{supp}[\mathrm{J}_0 G]) = 0. \tag{3.5.24}$$

As $\hat{A}_{\mathrm{E}}(\mathbf{f}) \in L^2(\mathscr{Q}_{\mathrm{E}})$,

$$\mu_E(\{\hat{A}_{\mathrm{E}} \in \mathscr{Q}_{\mathrm{E}} \mid (e^{-\hat{A}_{\mathrm{E}}(\mathbf{f})})_M = 0\}) = 0,$$

(3.5.24) implies that

$$\mu_E(\mathrm{supp}[\mathrm{J}_t F] \cap \mathrm{supp}[\mathrm{J}_0 G]) = 0,$$

and thus

$$0 = (\mathrm{J}_t F, \mathrm{J}_0 G) = (F, e^{-tH_{\mathrm{rad}}} G).$$

Since $e^{-tH_{\mathrm{rad}}}$ is positivity improving by Proposition 3.49, this is in contradiction with the assumption. □

Theorem 3.51 (Positivity improving for $e^{-tH_{\mathrm{PF}}}$). *Operator $\mathfrak{S}^{-1}e^{-tH_{\mathrm{PF}}}\mathfrak{S}$ is positivity improving for every $t > 0$.*

Proof. Let f be real valued. Notice that if $f \geq g$, then $Tf \geq Tg$ for any positivity preserving operator T. For a nonnegative $G \in \mathscr{S}(\mathscr{Q})$,

$$\mathrm{J}_t^* e^{i\hat{\Pi}_{\mathrm{E}}(\mathbf{f})}\mathrm{J}_0 G \geq e^{-(q_{\mathrm{E}}(\mathbf{f},\mathbf{f})+q(\mathrm{j}_0^*\mathbf{f},\mathrm{j}_0^*\mathbf{f}))}\mathrm{J}_t^*(e^{-\hat{A}_{\mathrm{E}}(\mathbf{f})})_M \mathrm{J}_0 e^{i\hat{\Pi}(\mathrm{j}_0^*\mathbf{f})}(e^{\hat{A}(\mathrm{j}_0^*\mathbf{f})})_M G = S_M G.$$

Since S_M is bounded and any nonnegative function can be approximated by a function in $\mathscr{S}(\mathscr{Q})$, we have for an arbitrary nonnegative G,

$$\mathrm{J}_t^* e^{i\hat{\Pi}_{\mathrm{E}}(\mathbf{f})}\mathrm{J}_0 G \geq S_M G.$$

Since $\mathrm{J}_t^*(e^{-\hat{A}_{\mathrm{E}}(\mathbf{f})})_M \mathrm{J}_0$ is positivity improving by Lemma 3.51 and $e^{i\hat{\Pi}(\mathrm{j}_0^*\mathbf{f})}(e^{\hat{A}(\mathrm{j}_0^*\mathbf{f})})_M$ is positivity preserving with $e^{i\hat{\Pi}(\mathrm{j}_0^*\mathbf{f})}(e^{\hat{A}(\mathrm{j}_0^*\mathbf{f})})_M G \not\equiv 0$ for $G \geq 0$ but $G \not\equiv 0$, S_M is positivity improving and so is $\mathrm{J}_t^* e^{i\hat{\Pi}_{\mathrm{E}}(\mathbf{f})}\mathrm{J}_0$. In particular, $\mathfrak{S}^{-1}\mathrm{J}_t^*\mathfrak{S}$ is positivity improving as well. Let $F, G \geq 0$, not identically vanishing. Note that

$$(F, \mathfrak{S}^{-1}e^{-tH_{\mathrm{PF}}}\mathfrak{S}G) = \int_{\mathbb{R}^3} \mathbb{E}^x[e^{-\int_0^t V(B_s)ds}(\mathfrak{S}^{-1}\mathrm{J}_t^*\mathfrak{S}F(B_0), G(B_t))]dx. \tag{3.5.25}$$

We show that (3.5.25) is strictly positive. Let

$$D_G = \{x \in \mathbb{R}^3 \mid G(x,\cdot) \not\equiv 0\}, \quad D_F = \{x \in \mathbb{R}^3 \mid F(x,\cdot) \not\equiv 0\}$$

and

$$D_{FG} = \{\omega \in \mathscr{X} \mid B_0(\omega) \in D_F, B_t(\omega) \in D_G\}.$$

In order to complete the proof of the theorem, it suffices to see that the measure of $D_{FG} \subset \mathscr{X}$ is positive as $\mathfrak{S}^{-1}\mathrm{J}_t\mathfrak{S}$ is positivity improving. We have

$$\int_{\mathbb{R}^3} \mathbb{E}^x[1_{D_{FG}}]dx = \int_{D_F} \mathbb{E}^x[1_{D_G}(B_t)]dx = (2\pi t)^{-3/2}\int_{D_F} dx \int_{D_G} e^{-|x-y|^2/2t}dy > 0.$$

Thus it is obtained that the right-hand side of (3.5.25) is strictly positive and the theorem follows. □

We proved that $\mathfrak{S}e^{-tH_{\mathrm{PF}}}\mathfrak{S}^{-1}$ is positivity improving. A direct consequence of this property is uniqueness of the ground state of H_{PF}.

Corollary 3.52 (Uniqueness of ground state). *The ground state Ψ_{g} of H_{PF} satisfies $\mathfrak{S}^{-1}\Psi_{\mathrm{g}} > 0$, and it is unique up to multiple constants.*

Proof. It follows from the Perron–Frobenius theorem and Theorem 3.51. □

By Theorem 3.51 again, we have $(f \otimes 1, \mathfrak{S}\Psi_{\mathrm{g}}) \neq 0$. This allows deriving an expression of the ground state energy $\inf \mathrm{Spec}(H_{\mathrm{PF}})$ from which some of its properties follow. Write $E(e^2) = \inf \mathrm{Spec}(H_{\mathrm{PF}})$.

Theorem 3.53 (Concavity of ground state energy). *The function $e^2 \mapsto E(e^2)$ is monotone increasing, continuous, and concave.*

Proof. For a strictly positive $f > 0$, we have

$$E(e^2) = \lim_{t\to\infty}\left(-\frac{1}{t}\log(f\otimes 1, e^{-tH_{\mathrm{PF}}}f\otimes 1)\right)$$
$$= \lim_{t\to\infty}\left(-\frac{1}{t}\log\int_{\mathbb{R}^3}\mathbb{E}^x[f(B_0)f(B_t)e^{-\int_0^t V(B_s)ds}e^{-(e^2/2)\mathfrak{q}_{\mathrm{E}}(K_t,K_t)}]dx\right). \tag{3.5.26}$$

As $e^{-(e^2/2)\mathfrak{q}_{\mathrm{E}}(K_t,K_t)}$ is log-convex in e^2, $E(\cdot)$ is concave. Thus $E(e^2)$ is continuous on $(0,\infty)$. Since $E(e^2)$ is continuous at $e = 0$, $E \in C(\mathbb{R}^+)$ and $E(e^2)$ can be expressed as $E(e^2) = \int_0^{e^2}\rho(t)dt$ with a suitable positive function $\rho(t)$. This implies that $E(e^2)$ is monotone increasing in e^2. □

Notice that we know that H_{PF} has a unique ground state Ψ_{g} for any value of the coupling constant $e \in \mathbb{R}$. Since $0 \neq (f\otimes 1, \mathfrak{S}^{-1}\Psi_{\mathrm{g}}) = (f\otimes 1, \Psi_{\mathrm{g}})$ for any $e \in \mathbb{R}$ and not identically zero $f \geq 0$, (3.5.26) follows.

3.5.3 Spatial decay of bound states

Similar to the Nelson model in the previous chapter, we can derive the spatial exponential decay of bound states Φ_{b} of the Pauli–Fierz Hamiltonian. Recall the class $\mathbb{V}^{\mathrm{upper}}$ defined in Definition 2.88, and let $H_{\mathrm{PF}}\Phi_{\mathrm{b}} = E\Phi_{\mathrm{b}}$. We have the following Carmona-type estimate for the Pauli–Fierz Hamiltonian.

Lemma 3.54. *Suppose* (1) *of Assumption* 3.4 *holds. If* $V \in \mathbb{V}^{\mathrm{upper}}$, *then for any* $t, a > 0$ *and every* $0 < \alpha < 1/2$, *there exist constants* $D_1, D_2, D_3 > 0$ *such that*

$$\|\Phi_{\mathrm{b}}(x)\|_{L^2(\mathscr{Q})} \leq t^{-d/2}D_1 e^{D_2\|U\|_p t}e^{Et}(D_3 e^{-\frac{\alpha}{4}\frac{a^2}{t}}e^{-tW_\infty} + e^{-tW_a(x)})\|\Phi_{\mathrm{b}}\|_{\mathcal{H}_{\mathrm{PF}}},$$

where $W_a(x) = \inf\{W(y) \mid |x-y| < a\}$.

Proof. Since $\Phi_{\mathrm{b}} = e^{tE}e^{-tH_{\mathrm{PF}}}\Phi_{\mathrm{b}}$, the functional integral representation yields

$$\Phi_{\mathrm{b}}(x) = \mathbb{E}^x[J_0^* e^{-ie\hat{A}_{\mathrm{E}}(K_t)}J_t e^{-\int_0^t V(B_s)ds}\Phi_{\mathrm{b}}(B_t)].$$

From this, we directly obtain that

$$\|\Phi_{\mathrm{b}}(x)\|_{L^2(\mathscr{Q})} \leq \mathbb{E}^x[e^{-\int_0^t V(B_s)ds}\|\Phi_{\mathrm{b}}(B_t)\|_{L^2(\mathscr{Q})}].$$

Then the lemma follows similarly as Carmona's estimate of Schrödinger operators in Lemma 4.176 in Volume 1. □

This lemma implies a similar result of decay of the ground state as for Schrödinger operators. For $V = W - U \in \mathbb{V}^{\mathrm{upper}}$, recall that $\Sigma = \liminf_{|x|\to\infty} V(x)$. We only state the results.

Corollary 3.55 (Exponential decay of bound states). *Suppose* (1) *of Assumption* 3.4 *holds. Let* $V = W - U \in \mathbb{V}^{\mathrm{upper}}$ *and* $E = \inf \mathrm{Spec}(H_{\mathrm{PF}})$.

(1) *Suppose that* $W(x) \geq \gamma|x|^{2n}$ *outside a compact set* K, *for some* $n > 0$ *and* $\gamma > 0$. *Take* $0 < a < 1/2$. *Then there exists a constant* $C_1 > 0$ *such that*

$$\|\Phi_{\mathrm{b}}(x)\|_{L^2(\mathcal{Q})} \leq C_1 \exp\left(-\frac{ac}{16}|x|^{n+1}\right) \|\Phi_{\mathrm{b}}\|_{\mathcal{H}_{\mathrm{PF}}},$$

where $c = \inf_{x\in\mathbb{R}^3\setminus K} W_{|x|/2}(x)/|x|^{2n}$.

(2) *(Decaying potential) Suppose that* $\Sigma > E$, $\Sigma > W_\infty$, *and* $0 < \beta < 1$. *Then there exists a constant* $C_2 > 0$ *such that*

$$\|\Phi_{\mathrm{b}}(x)\|_{L^2(\mathcal{Q})} \leq C_2 \exp\left(-\frac{\beta}{8\sqrt{2}}\frac{(\Sigma - E)}{\sqrt{\Sigma - W_\infty}}|x|\right) \|\Phi_{\mathrm{b}}\|_{\mathcal{H}_{\mathrm{PF}}}.$$

(3) *(Confining potential) Suppose that* $\lim_{|x|\to\infty} W(x) = \infty$. *Then there exist constants* $C, \delta > 0$ *such that*

$$\|\Phi_{\mathrm{b}}(x)\|_{L^2(\mathcal{Q})} \leq C \exp(-\delta|x|) \|\Phi_{\mathrm{b}}\|_{\mathcal{H}_{\mathrm{PF}}}.$$

3.6 Path measure associated with the ground state of Pauli–Fierz Hamiltonian

3.6.1 Path measure with double stochastic integrals

In Sections 3.6.1 and 3.6.2, we assume either Assumption 3.4 or 3.33 unless otherwise stated. That is, external potential V is $V \in \mathcal{R}_{\mathrm{Kato}}$ or Kato-decomposable. The functional integral method for nonrelativistic QED provides a new class of objects for probability theory. An infinite volume path measure is derived from the Nelson model discussed in Chapter 2. The Pauli–Fierz model also yields a path measure with densities dependent on a double stochastic integral. Its limit can be applied to studying the ground state of the Pauli–Fierz model. As the first step, we show the existence of path measures with densities dependent on a double stochastic integral.

Let Ψ_{g} be the ground state of H_{PF}, which is unique and $\mathfrak{S}^{-1}\Psi_{\mathrm{g}} > 0$ by Corollary 3.52. Although the existence of the ground state has been proven, the proof is not constructive and, therefore, a direct study of its properties is impossible. As an alternative, the functional integral representation can be used, and ground state expectations $(\Psi_{\mathrm{g}}, \mathcal{O}\Psi_{\mathrm{g}})$ for various operators $\mathcal{O}$ can be expressed in terms of averages of the path measure.

Let M be a positive Borel measurable function on $\mathbb{R}^3$. Write

$$\mathcal{M} = d\Gamma(M(-i\nabla)) \tag{3.6.1}$$

and consider $\mathcal{O} = e^{-\beta\mathcal{M}}$ in the discussion above. The strategy is to construct a sequence $\{\Psi_g^t\}_{t>0}$ such that $\Psi_g^t \to \Psi_g$ as $t \to \infty$. One candidate is

$$\Psi_g^t = \|e^{-tH_{\mathrm{PF}}}(f \otimes 1)\|^{-1} e^{-tH_{\mathrm{PF}}}(f \otimes 1)$$

with $f \geq 0$, since $f \otimes 1$ overlaps with Ψ_g by Corollary 3.52, i. e.,

$$(\Psi_g, f \otimes 1) = (\mathfrak{S}^{-1}\Psi_g, f \otimes 1) > 0.$$

This gives

$$\lim_{t\to 0}(\Psi_g^t, e^{-\beta\mathcal{M}}\Psi_g^t) = (\Psi_g, e^{-\beta\mathcal{M}}\Psi_g),$$

i. e.,

$$(\Psi_g, e^{-\beta\mathcal{M}}\Psi_g) = \lim_{t\to 0} \frac{(f \otimes 1, e^{-tH_{\mathrm{PF}}} e^{-\beta\mathcal{M}} e^{-tH_{\mathrm{PF}}}(f \otimes 1))}{(e^{-tH_{\mathrm{PF}}}(f \otimes 1), e^{-tH_{\mathrm{PF}}}(f \otimes 1))}. \tag{3.6.2}$$

To have a functional integral representation of the right-hand side of (3.6.2), we need to extend the functional integration $(F, e^{-tH_{\mathrm{PF}}} e^{-\beta\mathcal{M}} e^{-sH_{\mathrm{PF}}} G)$. It was a key step before that $e^{-tH_{\mathrm{rad}}}$ could be decomposed as $e^{-|s-t|H_{\mathrm{rad}}} = \mathrm{J}_s^*\mathrm{J}_t$, leading to the functional integral representation of $(F, e^{-tH_{\mathrm{PF}}} G)$. Our goal is to extend this idea to $\mathcal{M}$.

Let $\iota_t : L^2(\mathbb{R}^4) \to L^2(\mathbb{R}^5)$ be defined by

$$\widehat{\iota_t f}(\mathbf{k}, k_1) = \frac{e^{-ik_1 t}}{\sqrt{\pi}} \frac{\sqrt{M(k)}}{\sqrt{M(k)^2 + |k_1|^2}} \hat{f}(\mathbf{k}), \quad (\mathbf{k}, k_1) = (k, k_0, k_1) \in \mathbb{R}^3 \times \mathbb{R} \times \mathbb{R}. \tag{3.6.3}$$

Then the factorization formula

$$\iota_s^* \iota_t = e^{-|s-t|(M(-i\nabla)\otimes 1)}, \quad s, t \in \mathbb{R},$$

is satisfied. Here, operator $M(-i\nabla) \otimes 1$ is defined under the identification $L^2(\mathbb{R}^4) \cong L^2(\mathbb{R}^3) \otimes L^2(\mathbb{R})$. Let $(\mathcal{Q}_2, \Sigma_2, \mu_2)$ be a probability space and $(\hat{A}_2(\mathbf{f}), \mathbf{f} \in \oplus^3 L^2(\mathbb{R}^5))$ denotes a family of Gaussian random variables indexed by $\mathbf{f} \in \oplus^3 L^2(\mathbb{R}^5)$ with mean zero and covariance given by

$$\mathbb{E}_{\mu_2}[\hat{A}_2(\mathbf{f})\hat{A}_2(\mathbf{g})] = \frac{1}{2}\int_{\mathbb{R}^5} \overline{\hat{\mathbf{f}}(\mathbf{k}, k_1)} \cdot \delta^\perp(k)\hat{\mathbf{g}}(\mathbf{k}, k_1) d\mathbf{k} dk_1 = \mathrm{q}_2(\mathbf{f}, \mathbf{g}).$$

Define the contraction operator $\Xi_t : L^2(\mathscr{Q}_{\rm E}) \to L^2(\mathscr{Q}_2)$ by the second quantization of ι_t, i. e.,

$$\Xi_t = \Gamma_{\rm int}(\iota_t).$$

Then the factorization formula

$$e^{-|s-t|d\Gamma_{\rm E}(M(-i\nabla)\otimes 1)} = \Xi_t^* \Xi_s, \quad s, t \in \mathbb{R}, \tag{3.6.4}$$

follows. In Section 3.2.3, we discuss the Markov property of projections e_t and E_t. Similarly, projections $\iota_t \iota_t^*$ and $\Xi_t \Xi_t^*$ also have Markov property.

Theorem 3.56 (Euclidean Green functions). *Let* $0 = t_0 < t_1 < \cdots < t_n = t$ *and* $0 = s_0 < s_1 < \cdots < s_n = s$. *Then*

$$\left(F, \left(\prod_{j=1}^n e^{-(t_j - t_{j-1})H_{\rm PF}} e^{-(s_j - s_{j-1})\mathcal{M}}\right) G\right)_{\mathcal{H}_{\rm PF}}$$
$$= \int_{\mathbb{R}^3} \mathbb{E}^x \left[e^{-\int_0^t V(B_s)ds} (\Xi_0 \mathrm{J}_0 F(B_0), e^{-ie\hat{A}_{\rm E}(K_t(s_0,s_1,\ldots,s_n))} \Xi_s \mathrm{J}_t G(B_t))_{L^2(\mathscr{Q}_2)} \right] dx, \tag{3.6.5}$$

where

$$K_t(s_0, s_1, \ldots, s_n) = \bigoplus_{\mu=1}^3 \sum_{j=1}^n \int_{t_{j-1}}^{t_j} \iota_{s_{j-1}} \mathrm{j}_s \tilde{\varphi}(\cdot - B_s) dB_s^\mu.$$

Proof. We give an outline of the proof. Let $T_j = (t_j - t_{j-1})/M_j$. Substitute

$$e^{-(t_j - t_{j-1})H_{\rm PF}} = \operatorname*{s-lim}_{M_j\to\infty} \lim_{m_1\to\infty} \cdots \lim_{m_n\to\infty} \mathrm{J}_{t_{j-1}}^* \left(\prod_{i=1}^{M_j} (\mathrm{J}_{iT_j} (P_{T_j/2^{m_i}})^{2^{m_i}} \mathrm{J}_{iT_j}^*) \right) \mathrm{J}_{t_j}$$

into the left-hand side of (3.6.5). The left-hand side of (3.6.5) can be expressed as the limit of

$$T(M_1, m_1^1, \ldots, m_{M_1}^1, \ldots, M_n, m_1^n, \ldots, m_{M_n}^n) = \left(\mathrm{J}_{t_0} F, \left(\prod_{i=1}^n \eta_i \zeta_i \right) \mathrm{J}_{t_n} G \right), \tag{3.6.6}$$

where

$$\eta_j = \eta_j(t_{j-1}, t_j, m_1^j, \ldots, m_{M_j}^j, M_j) = \prod_{i=1}^{M_j} \left(\mathrm{J}_{iT_j} (P_{T_j/2^{m_i^j}})^{2^{m_i^j}} \mathrm{J}_{iT_j} \right),$$
$$\zeta_j = \mathrm{J}_{t_j} e^{-(s_j - s_{j-1})\mathcal{M}} \mathrm{J}_{t_j}^*.$$

In other words, T converges to the left-hand side of (3.6.5) as $m_i^j \to \infty$ for $i = 1, \ldots, n$, $j = 1, \ldots, M_j$, $M_j \to \infty$ for $j = 1, \ldots, n$ and $n \to \infty$. Note that

$$\mathrm{J}_{t_j} e^{-(s_j - s_{j-1})\mathcal{M}} \mathrm{J}_{t_j}^* = \Xi_{s_{j-1}}^* \Xi_{s_j} \mathrm{E}_{t_j}.$$

Inserting this identity into (3.6.6), and using the Markov property of both $E_t = \mathrm{J}_t \mathrm{J}_t^*$ and $\Xi_t \Xi_t^*$, we obtain the theorem. □

Next, we construct a functional integral representation for other types of Green functions. Define

$$L^2_\infty(\mathcal{Q}) = \{\Phi(\hat{A}(\mathbf{f}_1), \dots, \hat{A}(\mathbf{f}_n)) \mid \Phi \in L^\infty(\mathbb{R}^n), \mathbf{f}_j \in \oplus^3 L^2(\mathbb{R}^3), j = 1, \dots, n, n \geq 0\}.$$

Theorem 3.57 (Euclidean Green functions). *Let $0 = t_0 < t_1 < \dots < t_n = t$. Suppose that $F_j = f_j \otimes \Phi_j(\hat{A}(\mathbf{f}_1^j), \dots, \hat{A}(\mathbf{f}_{n_j}^j)) \in L^\infty(\mathbb{R}^3) \otimes L^2_\infty(\mathcal{Q})$, $1 \leq j \leq n-1$ and $F_0, F_n \in \mathcal{H}_{\mathrm{PF}}$. Then*

$$\begin{aligned}
&\left(F_0, \left(\prod_{j=1}^n e^{-(t_j - t_{j-1})H_{\mathrm{PF}}} F_j\right)\right) \\
&= \int_{\mathbb{R}^3} \mathbb{E}^x \left[e^{-\int_0^t V(B_s)ds} \left(\mathrm{J}_0 F_0(B_0), e^{-ie\hat{A}_{\mathrm{E}}(K_t)} \left(\prod_{j=1}^{n-1} \tilde{F}_j(B_{t_j}) \right) \mathrm{J}_t F_n(B_t) \right)_{L^2(\mathcal{Q}_2)} \right] dx,
\end{aligned} \tag{3.6.7}$$

where $\tilde{F}_j(x) = f_j(x) \Phi_j(\hat{A}_{\mathrm{E}}(\mathrm{j}_{t_j} \mathbf{f}_1^j), \dots, \hat{A}_{\mathrm{E}}(\mathrm{j}_{t_j} \mathbf{f}_{n_j}^j))$.

Proof. First, assume that f_j and Φ_j, $j = 1, \dots, n$, are sufficiently smooth functions with a compact support. Note that

$$\mathrm{J}_s F_j \mathrm{J}_s^* = f_j \Phi_j(\hat{A}_{\mathrm{E}}(\mathrm{j}_s \mathbf{f}_1^j), \dots, \hat{A}_{\mathrm{E}}(\mathrm{j}_s \mathbf{f}_{n_j}^j)) \tag{3.6.8}$$

as a bounded operator. Thus substituting (3.6.8) into the left-hand side of (3.6.7), the theorem follows. For a general F_j, a limiting argument can be used. □

As an application, we show the functional integral presentation of $(\Psi_{\mathrm{g}}, e^{-\beta N} \Psi_{\mathrm{g}})$, where N is the boson number operator and $\beta > 0$. Let

$$(1, T1)_{L^2(\mathcal{Q}_{\mathrm{E}})} = \langle T \rangle_{\mathrm{vac}}.$$

Set $M = 1$ in (3.6.1). Then $\mathcal{M} = N$ and let $\iota_t : L^2(\mathbb{R}^4) \to L^2(\mathbb{R}^5)$ be the family of isometries defined in (3.6.3). Write

$$Y = Y(t, \beta) = \bigoplus_{\mu=1}^3 \left(\int_{-t}^0 \iota_0 \mathrm{j}_s \tilde{\varphi}(\cdot - B_s) dB_s^\mu + \int_0^t \iota_\beta \mathrm{j}_s \tilde{\varphi}(\cdot - B_s) dB_s^\mu \right),$$

$$Z = Z(t) = \bigoplus_{\mu=1}^3 \int_{-t}^t \mathrm{j}_s \tilde{\varphi}(\cdot - B_s) dB_s^\mu.$$

We note that B^{μ}_{-s} and B^{μ}_{t} are independent, for all $s, t > 0$. Let $f \geq 0$ be fixed. Using Theorem 3.57, one can compute

$$(\Psi^t_{\mathrm{g}}, e^{-\beta \mathcal{M}} \Psi^t_{\mathrm{g}}) = \frac{\int_{\mathbb{R}^3} \mathbb{E}^x[f(B_{-t}) f(B_t) e^{-\int_{-t}^t V(B_s) ds} \langle e^{-ie\hat{A}_{\mathrm{E}}(Y)} \rangle_{\mathrm{vac}}] dx}{\int_{\mathbb{R}^3} \mathbb{E}^x[f(B_{-t}) f(B_t) e^{-\int_{-t}^t V(B_s) ds} \langle e^{-ie\hat{A}_{\mathrm{E}}(Z)} \rangle_{\mathrm{vac}}] dx}. \tag{3.6.9}$$

Notice that

$$\langle \exp\left(-ie\hat{A}_{\mathrm{E}}(Y)\right) \rangle_{\mathrm{vac}} = \exp\left(-\frac{e^2}{2} \mathrm{q}_2\,(Y, Y)\right).$$

As q_{E} is a bilinear form, the identity

$$\mathrm{q}_{\mathrm{E}}\,(Y, Y) = \mathrm{q}_{\mathrm{E}}(Z, Z) + 2(e^{-\beta} - 1)\mathrm{q}_{\mathrm{E}}(Z_{(-t,0)}, Z_{(0,t)}) \tag{3.6.10}$$

holds, where

$$Z_{(S,T)} = \bigoplus_{\mu=1}^{3} \int_S^T \mathrm{j}_s \tilde{\varphi}(\cdot - B_s) dB^{\mu}_s.$$

Define the family of finite volume Gibbs measures,

$$d\mu_t = \frac{1}{N_t} \left(\int_{\mathbb{R}^3} f(x + B_{-t}) f(x + B_t) e^{-\int_{-t}^t V(x + B_s) ds} dx \right) e^{-(e^2/2)\mathrm{q}_{\mathrm{E}}(Z,Z)} d\mathcal{W}^0, \tag{3.6.11}$$

as probability measures on $(\mathcal{X}, \mathcal{B}(\mathcal{X}))$. Note that we use the notation $\mathcal{W}^0$ for a two-sided Wiener measure unless confusion may arise. In (3.6.11), N_t denotes the denominator of the right-hand side of (3.6.9), regarded as the normalizing constant such that $\int_{\mathcal{X}} d\mu_t = 1$, and $\mathrm{q}_{\mathrm{E}}(Z, Z)$ is independent of x.

Corollary 3.58. *It follows that*

$$(\Psi_{\mathrm{g}}, e^{-\beta N} \Psi_{\mathrm{g}}) = \lim_{t \to \infty} \mathbb{E}_{\mu_t} [e^{+e^2(1-e^{-\beta})\mathrm{q}_{\mathrm{E}}(Z_{(-t,0)}, Z_{(0,t)})}].$$

Proof. By (3.6.10) and (3.6.9), we have

$$(\Psi^t_{\mathrm{g}}, e^{-\beta N} \Psi^t_{\mathrm{g}}) = \mathbb{E}_{\mu_t} [e^{e^2(1-e^{-\beta})\mathrm{q}_{\mathrm{E}}(Z_{(-t,0)}, Z_{(0,t)})}].$$

Since $\Psi^t_{\mathrm{g}} \to \Psi_{\mathrm{g}}$ as $t \to \infty$, the corollary follows. □

The measure $d\mu_t$ includes pair interaction $e^2 \mathrm{q}_{\mathrm{E}}(Z, Z)$ associated with the Pauli–Fierz Hamiltonian which has the formal expression

$$q_E(Z,Z) \overset{\text{formal}}{=} \sum_{\mu,\nu=1}^{3} \int_{-t}^{t} dB_s^\mu \int_{-t}^{t} W_{\mu\nu}(B_s - B_r, s-r)dB_r^\nu, \tag{3.6.12}$$

where $W_{\mu\nu}$ is given by

$$W_{\mu\nu}(X,T) = \int_{\mathbb{R}^3} \delta_{\mu\nu}^\perp(k) \frac{|\hat{\varphi}(k)|^2}{2\omega(k)} e^{-|T|\omega(k)} e^{-ik\cdot X} dk. \tag{3.6.13}$$

Here, $W_{\mu\nu}(X,T)$ is called the pair potential of the Pauli–Fierz Hamiltonian. Moreover,

$$q_E(Z_{(-t,0)}, Z_{(0,t)}) \overset{\text{formal}}{=} \sum_{\mu,\nu=1}^{3} \int_0^t dB_s^\mu \int_{-t}^{0} W_{\mu\nu}(B_s - B_r, s-r)dB_r^\nu.$$

Equation (3.6.13) expresses the effect of the quantum field. Recall that in the Nelson model, instead of the double stochastic integral a double Riemann integral appears. The formal expression (3.6.12), however, is not well defined, since the integrand is not admissible. In the next section, we find a suitable expression of the integral.

3.6.2 Expression in terms of iterated stochastic integrals

We define the iterated stochastic integral S_T by

$$S_T = \frac{1}{2} \int_{\mathbb{R}^3} \frac{|\hat{\varphi}(k)|^2}{\omega(k)} \left(2T + \int_{-T}^{T} e^{ik\cdot B_s} dB_s \cdot \delta^\perp(k) \int_{-T}^{s} e^{-\omega(k)(s-r)} e^{-ik\cdot B_r} dB_r \right) dk. \tag{3.6.14}$$

Here S_T is a well-defined expression and we use it to replace (3.6.12). We will refer to (3.6.14) as the double stochastic integral with the diagonal removed.

Theorem 3.59 (Iterated stochastic integral). *Suppose that $\hat{\varphi}$ is rotation invariant. Then we have*

$$\langle e^{-ie\hat{A}(Z)} \rangle_{\text{vac}} = e^{-e^2 S_T}.$$

Proof. We replace the time interval $[-T,T]$ with $[0,2T]$ by shift invariance, and reset $2T$ by T for notational convenience. Using the definition of Z and the Lebesgue dominated convergence theorem gives

$$\langle \exp(-ie\hat{A}(Z)) \rangle_{\text{vac}} = \lim_{n\to\infty} \left\langle \exp\left(-ie \sum_{j=1}^{n} \hat{A}_E(\mathrm{j}_{\Delta_j} \tilde{\varphi}(\cdot - B_{\Delta_j}))\Delta B_j \right) \right\rangle_{\text{vac}},$$

where we set $\Delta B_j = B_{jT/n} - B_{(j-1)T/n}$ and $\Delta_j = (j-1)T/n, j = 1, \ldots, n$. Since $\hat{A}_E(\mathbf{f})$ is Gaussian, it can be computed as

$$\langle \exp(-ie\hat{A}(Z))\rangle_{\text{vac}} = \lim_{n\to\infty} \exp\left(-\frac{e^2}{2}\int_{\mathbb{R}^3} \frac{|\hat{\varphi}(k)|^2}{2\omega(k)} \sum_{j=1}^n\sum_{l=1}^n e^{-|\Delta_j-\Delta_l|\omega(k)} e^{ik\cdot(B_{\Delta_j}-B_{\Delta_l})}\Delta B_j \delta^\perp(k)\Delta B_l dk\right).$$

Now the integral $\int_{\mathbb{R}^3}\dots dk$ above can be computed as

$$\int_{\mathbb{R}^3} \frac{|\hat{\varphi}(k)|^2}{2\omega(k)} \sum_{j=1}^n\sum_{l=1}^n e^{-|\Delta_j-\Delta_l|\omega(k)} e^{ik\cdot(B_{\Delta_j}-B_{\Delta_l})}\Delta B_j \delta^\perp(k)\Delta B_l dk$$

$$= \sum_{j=1}^n \Delta B_j \left(\int_{\mathbb{R}^3} \frac{|\hat{\varphi}(k)|^2}{2\omega(k)}\delta^\perp(k)dk\right)\Delta B_j \tag{3.6.15}$$

$$+ \sum_{j=1}^n \int_{\mathbb{R}^3} \frac{|\hat{\varphi}(k)|^2}{\omega(k)} e^{-\Delta_j\omega(k)}e^{ik\cdot B_{\Delta_j}}\Delta B_j\delta^\perp(k)\left(\sum_{l=1}^{j-1} e^{\Delta_l\omega(k)}e^{-ik\cdot B_{\Delta_l}}\Delta B_l\right)dk. \tag{3.6.16}$$

For the diagonal term (3.6.15), we note that

$$\int_{\mathbb{R}^3} \frac{|\hat{\varphi}(k)|^2}{2\omega(k)}\delta^\perp_{\mu\nu}(k)dk = \delta_{\mu\nu}\frac{2}{3}\int_{\mathbb{R}^3}\frac{|\hat{\varphi}(k)|^2}{2\omega(k)}dk$$

by rotation invariance of $\hat{\varphi}$. As $n\to\infty$, $\sum_{\mu=1}^3\sum_{j=1}^n |\Delta B_j^\mu|^2 \to 3T$ almost surely. Thus we find

$$\lim_{n\to\infty}\sum_{j=1}^n \Delta B_j\left(\int_{\mathbb{R}^3}\frac{|\hat{\varphi}(k)|^2}{2\omega(k)}\delta^\perp(k)dk\right)\Delta B_j = T\int_{\mathbb{R}^3}\frac{|\hat{\varphi}(k)|^2}{\omega(k)}dk$$

almost surely. For the off-diagonal term (3.6.16), we start by noting that by the definition of the Itô integral for locally bounded functions $f, g : \mathbb{R}\times\mathbb{R}^3 \to \mathbb{R}$,

$$\mathbb{E}\left[\int_0^t \left|f(s,B_s)\int_0^s g(r,B_r)dB_r^\mu\right|^2 ds\right] = \int_0^t ds\mathbb{E}[|f(s,B_s)|^2]\int_0^s \mathbb{E}[|g(r,B_r)|^2]dr < \infty.$$

Hence the stochastic integral of $\rho(s) = f(s,B_s)\int_0^s g(r,B_r)dB_r^\mu$ exists and for every $k\in\mathbb{R}^3$,

$$\lim_{n\to\infty}\sum_{j=1}^n f(\Delta_j, B_{\Delta_j})\Delta B_j\delta^\perp(k)\int_0^{\Delta_j} g(r,B_r)dB_r = \int_0^T f(s,B_s)dB_s\delta^\perp(k)\int_0^s g(r,B_r)\,dB_r \tag{3.6.17}$$

strongly in $L^2(\mathfrak{X})$. By the independence of increments of Brownian motion and the fact that $\sum_{\mu=1}^3 \mathbb{E}[(\Delta B_j^\mu)^2] = 3/n$ and $\mathbb{E}[\Delta B_j^\mu] = 0$, we can estimate the difference of (3.6.17) and the off-diagonal term (3.6.16) to get

$$\mathbb{E}\Big[\sum_{j=1}^{n} f(\Delta_j, B_{\Delta_j})\Delta B_j \delta^\perp(k)\Big(\int_0^{\Delta_j} g(r,B_r)\,dB_r - \sum_{l=1}^{j} g(\Delta_l, B_{\Delta_l})\Delta B_l\Big)\Big]^2$$

$$\leq \|f^2\|_\infty \frac{3}{n}\sum_{\nu=1}^{3}\sum_{j=1}^{n}\mathbb{E}\Big[\Big|\int_0^{\Delta_j} g(r,B_r)\,dB_r^\nu - \sum_{l=1}^{j} g(\Delta_l, B_{\Delta_l})\Delta B_l^\nu\Big|^2\Big]. \tag{3.6.18}$$

The right-hand side above converges to zero as $n \to \infty$. Combining (3.6.17) and (3.6.18), we find that for every $k \in \mathbb{R}^3$,

$$\lim_{n\to\infty}\sum_{j=1}^{n} f(\Delta_j, B_{\Delta_j})\Delta B_j \delta^\perp(k)\Big(\sum_{l=1}^{j} g(\Delta_l, B_{\Delta_l})\Delta B_l\Big) = \int_0^t f(s,B_s)\delta^\perp(k)\Big(\int_0^s g(r,B_r)\,dB_r\Big)dB_s \tag{3.6.19}$$

strongly in $L^2(\mathscr{X}, d\mathcal{W}^x)$. By putting $f(t,x) = e^{ik\cdot x}e^{-\omega(k)t}$, $g(t,x) = e^{-ik\cdot x}e^{\omega(k)t}$ in (3.6.19), it follows that

$$\lim_{n\to\infty}\sum_{j=1}^{n} e^{-\Delta_j\omega(k)}e^{ik\cdot B_{\Delta_j}}\Delta B_j\delta^\perp(k)\left(\sum_{l=1}^{j-1} e^{\Delta_l\omega(k)}e^{-ik\cdot B_{\Delta_l}}\Delta B_l\right)$$

$$= \int_0^T e^{ik\cdot B_s}\delta^\perp(k)\Big(\int_0^s e^{-\omega(k)(s-r)}e^{-ik\cdot B_r}\,dB_r\Big)dB_s.$$

Hence

$$\lim_{n\to\infty}(3.6.16) = \int_{\mathbb{R}^3} dk\frac{|\hat{\varphi}(k)|^2}{\omega(k)}\int_0^T e^{ik\cdot B_s}dB_s\cdot\delta^\perp(k)\int_0^s e^{-\omega(k)(s-r)}e^{-ik\cdot B_r}\,dB_r. \qquad \square$$

We give a comment on Theorem 3.59. Let

$$X_t^\mu = \int_{-T}^{t} \mathrm{j}_s\tilde{\varphi}(\cdot - B_s)\,dB_s^\mu.$$

Then, by applying Itô formula, we obtain

$$d(X_t^\mu)^2 = 2X_t^\mu\,dX_t^\mu + dX_t^\mu\,dX_t^\mu.$$

Since X_t^μ is an Itô integral, its quadratic variation is given by

$$dX_t^\mu\,dX_t^\mu = |\mathrm{j}_t\tilde{\varphi}(\cdot - B_t)|^2\,dt.$$

Therefore,

$$d(X_t^\mu)^2 = 2X_t^\mu\,\mathrm{j}_t\tilde{\varphi}(\cdot - B_t)\,dB_t^\mu + |\mathrm{j}_t\tilde{\varphi}(\cdot - B_t)|^2\,dt.$$

Integrating over $[-T, T]$, we obtain

$$(X_t^\mu)^2 = 2\int_{-T}^{T} X_s^\mu\, \mathrm{j}_s\tilde{\varphi}(\cdot - B_s)\, dB_s^\mu + \int_{-T}^{T} |\mathrm{j}_s\tilde{\varphi}(\cdot - B_s)|^2\, ds.$$

This representation provides an alternative derivation of Theorem 3.59. Here is a concluding summary.

Proposition 3.60. *Let μ_T be the family of probability measures on $\mathfrak{X}$ defined in* (3.6.11). *Then the following identity holds:*

$$d\mu_T = \frac{1}{\hat{Z}_T}\left(\int_{\mathbb{R}^3} f(B_{-T} + x) f(B_T + x) e^{-\int_{-T}^{T} V(B_s+x)\, ds} dx\right) e^{-e^2 \hat{S}_T} d\mathcal{W}^0,$$

where $\hat{S}_T$ is defined by S_T with the diagonal part removed

$$\hat{S}_T = \frac{1}{2}\int_{\mathbb{R}^3} dk \frac{|\hat{\varphi}(k)|^2}{\omega(k)} \int_{-T}^{T} e^{ik\cdot B_s} dB_s \delta^\perp(k) \int_{-T}^{s} e^{-\omega(k)(s-r)} e^{-ik\cdot B_r} dB_r$$

and $\hat{Z}_T$ denotes the normalizing constant such that $\int_{\mathfrak{X}} d\mu_T = 1$.

Here $\hat{S}_T$ may be symbolically written as

$$\hat{S}_T = \sum_{\mu,\nu=1}^{3} \int_{[-T,T]^2 \setminus \{s=r\}} \left(\int_{\mathbb{R}^3} W_{\mu\nu}(B_s - B_r, s - r) ds dr\right) dB_s^\mu dB_r^\nu.$$

3.6.3 Weak convergence and Gibbs measures

In this section, we show tightness of the family of probability measures $\{\mu_t\}_{t\geq 0}$ on $(\mathfrak{X}, \mathcal{B}(\mathfrak{X}))$. In Section 3.6.3 we suppose Assumption 3.33.

Lemma 3.61. *Let $f_1, \dots, f_{n-1} \in L^\infty(\mathbb{R}^3)$, $0 \leq f \in L^2(\mathbb{R}^3)$, and $-T = t_0 \leq t_1 \leq \cdots \leq t_n = T$. Then the Euclidean Green function is expressed as*

$$\frac{(f\otimes 1, e^{-(t_1-t_0)H_{\mathrm{PF}}}(f_1\otimes 1)\cdots(f_{n-1}\otimes 1)e^{-(t_n-t_{n-1})H_{\mathrm{PF}}} f\otimes 1)}{(f\otimes 1, e^{-2TH} f\otimes 1)} = \mathbb{E}_{\mu_T}\left[\prod_{j=1}^{n-1} f_j(B_{t_j})\right].$$

Proof. By Theorem 3.57, we can directly see that

$$(f\otimes 1, e^{-(t_1-t_0)H_{\mathrm{PF}}}(f_1\otimes 1)\cdots(f_{n-1}\otimes 1)e^{-(t_n-t_{n-1})H_{\mathrm{PF}}} f\otimes 1)$$

$$= \int_{\mathbb{R}^3} \mathbb{E}^x \left[f(B_{-T}) f(B_T) \left(\prod_{j=1}^{n-1} f_j(B_{t_j}) \right) e^{-\int_{-T}^{T} V(B_s) ds} \langle e^{-ie\hat{A}_E(Z)} \rangle_{\mathrm{vac}} \right] dx.$$

By the definition of the measure μ_T, the lemma follows. □

An immediate result of s-$\lim_{t\to\infty} \Psi_g^t = \Psi_g$ and Lemma 3.61 is as follows. Let $\rho, \rho_1, \rho_2 \in L^\infty(\mathbb{R}^3)$. Then for $t > s$,

$$\lim_{T\to\infty} \mathbb{E}_{\mu_T}[\rho(B_0)] = (\Psi_g, (\rho \otimes 1)\Psi_g),$$

$$\lim_{T\to\infty} \mathbb{E}_{\mu_T}[\rho_1(B_s)\rho_2(B_t)] = (\Psi_g, (\rho_1 \otimes 1)e^{-(t-s)H_{\mathrm{PF}}}(\rho_2 \otimes 1)\Psi_g)e^{(t-s)E(H_{\mathrm{PF}})},$$

where $E(H_{\mathrm{PF}}) = \inf \mathrm{Spec}(H_{\mathrm{PF}})$ denotes the ground state energy of H_{PF}. In the next theorem, we can show the existence of Gibbs measure μ_∞ which is the weak limit of a subsequence of $\{\mu_T\}_{t\geq 0}$.

Theorem 3.62 (Tightness). *Assume that there exists the ground state of H_{PF}. Then there exists a subsequence T' such that $\mu_{T'}$ has a weak limit μ_∞ on $(\mathcal{X}, \mathcal{B}(\mathcal{X}))$ as $T' \to \infty$.*

Proof. The proof is a modification of that of Theorem 2.37. By the Prokhorov theorem, it suffices to show (1) and (2) below:
(1) $\lim_{\Lambda\to\infty} \sup_{T\geq 0} \mu_T(|B_0|^2 > \Lambda) = 0$;
(2) for any $\varepsilon > 0$, $\lim_{\delta\downarrow 0} \sup_{T\geq 0} \mu_T(\max_{|s-t|<\delta, -T\leq s,t\leq T} |B_t - B_s| > \varepsilon) = 0$.

We have

$$\mu_T(|B_0|^2 > \Lambda) = (\Psi_g^T, 1_{\{|x|^2>\Lambda\}}\Psi_g^T).$$

Note that $\Psi_g^T \to \Psi_g$ strongly as $T \to \infty$. Set $\alpha(T) = (\Psi_g^T, 1_{\{|x|^2>\Lambda\}}\Psi_g^T)$ and, without loss of generality, assume that $T \geq 1$. Let $\varepsilon > 0$ be given. Since as a bounded operator $\|1_{\{|x|^2>\Lambda\}} \otimes 1\| \leq 1$, there exists $T^* > 0$ independent of Λ such that

$$\alpha(T) \leq (\Psi_g, 1_{\{|x|^2>\Lambda\}}\Psi_g) + \varepsilon \quad \forall T > T^*.$$

Thus

$$\sup_{1\leq T} \alpha(T) \leq \sup_{1\leq T\leq T^*} \alpha(T) + (\Psi_g, 1_{\{|x|^2>\Lambda\}}\Psi_g) + \varepsilon.$$

We shall estimate $\sup_{1\leq T\leq T^*} \alpha(T)$. Let $\mathfrak{S} = e^{-i(\pi/2)N}$. Since $\mathfrak{S}^{-1}e^{-TH_{\mathrm{PF}}}\mathfrak{S}$ is positivity improving and $\mathfrak{S}^{-1}(f \otimes 1) = f \otimes 1$, we see that

$$\|e^{-TH_{\mathrm{PF}}}(f \otimes 1)\| = \|\mathfrak{S}^{-1}e^{-TH_{\mathrm{PF}}}\mathfrak{S}(f \otimes 1)\| > 0$$

for all $T \geq 0$. Then $\inf_{1\leq T\leq T^*} \|e^{-TH_{\mathrm{PF}}}(f \otimes 1)\| > 0$. Denote the left-hand side by c. Since T^* is independent of Λ, c is also independent of Λ. Thus

$$\alpha(T) \leq c^{-2}(e^{-TH_{\mathrm{PF}}}(f \otimes 1), 1_{\{|x|^2>\Lambda\}} e^{-TH_{\mathrm{PF}}}(f \otimes 1)).$$

By the functional integral representation and Schwarz inequality, we have

$$\begin{aligned}(e^{-TH_{\mathrm{PF}}}(f \otimes 1), 1_{\{|x|^2>\Lambda\}} e^{-TH_{\mathrm{PF}}}(f \otimes 1)) &= \int_{\mathbb{R}^3} \mathbb{E}^x[f(B_0)f(B_{2T})1_{\{|x|^2>\Lambda\}}(B_T)e^{-\int_0^{2T} V(B_s)ds} e^{-S_{2T}}]dx \\ &\leq V_M \int_{\mathbb{R}^3} (\mathbb{E}^x[f(B_{2T})^2])^{1/2} f(x)(\mathbb{E}^x[1_{\{|x|^2>\Lambda\}}(B_T)])^{1/4} dx,\end{aligned}$$

where S_{2T} is given by (3.6.14) and $V_M = \sup_{x\in\mathbb{R}^3}(\mathbb{E}^x[e^{-4\int_0^{2T} V(B_s)ds}])^{1/4}$. We see by the Schwarz inequality again that

$$(e^{-TH_{\mathrm{PF}}}(f \otimes 1), 1_{\{|x|^2>\Lambda\}} e^{-TH_{\mathrm{PF}}}(f \otimes 1)) \leq V_M \|f\| \left(\int_{\mathbb{R}^3} (\mathbb{E}^x[1_{\{|x|^2>\Lambda\}}(B_T)])^{1/2} |f(x)|^2 dx \right)^{1/2}.$$

Since $1 \leq T \leq T^*$, the bound

$$\int_{\mathbb{R}^3} (\mathbb{E}^x[1_{\{|x|^2>\Lambda\}}(B_T)])^{1/2} f(x)^2 dx \leq (2\pi)^{-3/4} \int_{\mathbb{R}^3} f(x)^2 \left(\int_{\mathbb{R}^3} dy e^{-|x-y|^2/(2T^*)} 1_{\{|x|^2>\Lambda\}}(y) \right)^{1/2} dx$$

is obtained. Denote the right-hand side above by $C_f(\Lambda)$. Thus

$$\sup_{1\leq T\leq T^*} \alpha(T) \leq V_M C_f(\Lambda)^{1/2} \|f\|/c^2.$$

Since c and V_M are independent of Λ and $C_f(\Lambda) \to 0$ as $\Lambda \to \infty$, we obtain

$$\lim_{\Lambda\to\infty} \sup_{T\geq 1} \alpha(T) \leq \lim_{\Lambda\to\infty} \left(\frac{1}{c^2} \|f\| V_M C_f(\Lambda)^{1/2} + (\Psi_{\mathrm{g}}, 1_{\{|x|^2>\Lambda\}} \Psi_{\mathrm{g}}) \right) + \varepsilon = \varepsilon.$$

Since ε is arbitrary, (1) follows. Claim (2) can be proven in the same way as Theorem 2.37. □

Using the functional integration representation of $e^{-tH_{\mathrm{PF}}}$, in Corollary 3.55 we show that if $V(x) = |x|^{2n}$, then $\|\Psi_{\mathrm{g}}(x)\|_{\mathcal{F}_{\mathrm{rad}}} \leq C_1 e^{-C_2|x|^{n+1}}$, while if $V(x) = -1/|x|$, then $\|\Psi_{\mathrm{g}}(x)\|_{\mathcal{F}_{\mathrm{rad}}} \leq C_3 e^{-C_4|x|}$ for appropriate constants C_j.

Corollary 3.63. *Assume that $\|\Psi_{\mathrm{g}}(x)\|_{\mathcal{F}_{\mathrm{rad}}} \leq Ce^{-c|x|^\gamma}$ for some positive constants C, c, and γ. Then*

$$\mathbb{E}_{\mu_\infty}[e^{c|B_0|^\gamma}] < \infty. \tag{3.6.20}$$

Proof. Let

$$\rho_m(x) = \begin{cases} e^{c|x|^\gamma}, & e^{c|x|^\gamma} \le m, \\ m, & e^{c|x|^\gamma} > m \end{cases}$$

be the truncated function $e^{c|x|^\gamma}$. Then $(\Psi_g, (\rho_m \otimes 1)\Psi_g) = \mathbb{E}_{\mu_\infty}[\rho_m(B_0)]$ follows. By a limiting argument, as $m \to \infty$, (3.6.20) follows. □

Using this measure μ_∞, the expectation $\langle e^{-\beta N}\rangle_{\mathrm{vac}}$ can be formally represented as

$$\langle e^{-\beta N}\rangle_{\mathrm{vac}} = \mathbb{E}_{\mu_\infty}[e^{e^2(1-e^{-\beta})\int_{-\infty}^0 dB_s^\mu \int_0^\infty W_{\mu\nu}(B_s - B_r, s-r) dB_r^\nu}].$$

It is, however, not easy to control $\int_{-t}^0 dB_s^\mu \int_0^t W_{\mu\nu}(B_s - B_r, s-r)dB_r^\nu$ as $t \to \infty$. For the Nelson Hamiltonian, the role of the double stochastic integral is taken by a double Riemann integral and then

$$\left|\int_{-t}^0 ds \int_0^t W(B_s - B_r, s-r)dr\right| \le \frac{1}{2}\int_{\mathbb{R}^3} \frac{\hat{\varphi}(k)^2}{\omega(k)^3} dk.$$

Thus if the integral of the right-hand side is bounded, then the left-hand side is uniformly bounded with respect to the path and t.

3.6.4 Gaussian domination of ground states

We can show the Gaussian domination of the ground state of the Pauli–Fierz Hamiltonian in the same way as the Nelson Hamiltonian. In Section 3.6.4 we suppose either Assumption 3.4 or 3.33, and that H has the ground state Ψ_g, i. e.,

$$H\Psi_g = E\Psi_g, \quad E = \inf \mathrm{Spec}(H).$$

Hence $\Psi_g^t = e^{-tH} f \otimes 1 / \|e^{-tH} f \otimes 1\|$ converges to Ψ_g as $t \to \infty$. We shall show the Gaussian domination of Ψ_g^t for $t < \infty$. We define the self-adjoint operator $\hat{A}_\xi$ in $\mathcal{H}_{\mathrm{PF}}$ by $\hat{A}_\xi = \int_{\mathbb{R}^3}^{\oplus} \hat{A}(\xi(\cdot - x))dx$, where $\xi \in \bigoplus_{\mu=1}^3 L^2_{\mathrm{real}}(\mathbb{R}^3)$. Then we have

$$(\Psi_g, e^{-i\beta\hat{A}_\xi}\Psi_g) = \lim_{t\to 0} \frac{(e^{-tH} f \otimes 1, e^{-i\beta\hat{A}_\xi} e^{-tH} f \otimes 1)}{(e^{-tH} f \otimes 1, e^{-tH} f \otimes 1)}, \quad \beta \in \mathbb{R}.$$

By using the measure μ_t, we can see that

$$\frac{(e^{-tH} f \otimes 1, e^{-i\beta\hat{A}_\xi} e^{-tH} f \otimes 1)}{(e^{-tH} f \otimes 1, e^{-tH} f \otimes 1)} = \mathbb{E}_{\mu_t}[e^{-\frac{1}{2}(2ie\beta \mathrm{q}_{\mathrm{E}}(Z, \mathrm{j}_0\xi) + \beta^2 \mathrm{q}_{\mathrm{E}}(\mathrm{j}_0\xi, \mathrm{j}_0\xi))}].$$

Here, $Z = Z(t) = \bigoplus_{\mu=1}^3 \int_0^t \mathrm{j}_s \tilde{\varphi}(\cdot - B_s) dB_s^\mu$ and $\mathrm{q}_{\mathrm{E}}(Z, \mathrm{j}_0\xi)$ is real.

Lemma 3.64. *Suppose that* $\beta < (2\mathrm{q_E}(\mathrm{j}_0\xi, \mathrm{j}_0\xi))^{-1}$. *Then* $\Psi_\mathrm{g}^t \in D(e^{(\beta/2)\hat{A}_\xi^2})$ *and*

$$\|e^{(\beta/2)\hat{A}_\xi^2}\Psi_\mathrm{g}^t\|^2 = (1 - 2\beta\mathrm{q_E}(\mathrm{j}_0\xi, \mathrm{j}_0\xi))^{-1/2}\mathbb{E}_{\mu_t}\left[\exp\left(\frac{\beta e^2\mathrm{q_E}(Z, \mathrm{j}_0\xi)^2}{(1 - 2\beta\mathrm{q_E}(\mathrm{j}_0\xi, \mathrm{j}_0\xi))}\right)\right]. \tag{3.6.21}$$

Proof. The proof is a minor modification of Lemma 2.60 and Corollary 2.61. We have

$$(\Psi_\mathrm{g}^t, e^{-ik\hat{A}_\xi}\Psi_\mathrm{g}^t) = \mathbb{E}_{\mu_t}\left[e^{-iek\mathrm{q_E}(Z,\mathrm{j}_0\xi)}\right]e^{-\frac{1}{2}k^2\mathrm{q_E}(\mathrm{j}_0\xi,\mathrm{j}_0\xi)}, \quad k \in \mathbb{R}.$$

By the Gaussian transformation with respect to k, we see that

$$(\Psi_\mathrm{g}^t, e^{-\hat{A}_\xi^2/2}\Psi_\mathrm{g}^t) = \frac{1}{\sqrt{1 + \mathrm{q_E}(\mathrm{j}_0\xi, \mathrm{j}_0\xi)}}\mathbb{E}_{\mu_t}\left[\exp\left(-\frac{e^2\mathrm{q_E}(Z, \mathrm{j}_0\xi)^2}{2(1 + \mathrm{q_E}(\mathrm{j}_0\xi, \mathrm{j}_0\xi))}\right)\right].$$

Replacing ξ with $\sqrt{-2\beta}\xi$ for $\beta < 0$, we have (3.6.21) with $\beta < 0$. We can extend this to $\beta < (2\mathrm{q_E}(\mathrm{j}_0\xi, \mathrm{j}_0\xi))^{-1}$ by an analytic continuation in the same way as in the proof of Corollary 2.61. □

For the Nelson Hamiltonian H_N, we can construct the equality

$$\|e^{(\beta/2)\xi(g)^2}\Psi_\mathrm{g}\|^2 = \frac{1}{\sqrt{1 - \beta\|\hat{g}/\sqrt{\omega}\|^2}}\mathbb{E}_\mathcal{N}\left[\exp\left(\frac{\beta K(g)^2}{1 - \beta\|\hat{g}/\sqrt{\omega}\|^2}\right)\right]$$

in terms of the path measure $\mathcal{N}$, by which we can show the Gaussian domination of the ground state and, moreover, see the behavior of $\|e^{(\beta/2)\xi(g)^2}\Psi_\mathrm{g}\|$ at the critical point $\beta = \frac{1}{\|\hat{g}/\sqrt{\omega}\|^2}$, i. e., $\lim_{\beta\uparrow\frac{1}{2\|\hat{g}/\sqrt{\omega}\|^2}}\|e^{\beta\xi(g)^2}\Psi_\mathrm{g}\| = \infty$. Unfortunately, it is not, however, straightforward to represent the expectation of $e^{\beta\hat{A}_\xi^2}$ with respect to the ground state Ψ_g in terms of the measure μ_∞. The reason is why we do not have a uniform bound of the pair potential $W_{\mu\nu}$ with respect to paths. The Gaussian domination of the ground state, however, can be shown by using Ψ_g^t, but only for sufficiently small $|\beta|$, and unfortunately we do not mention anything on the critical behavior of $\|e^{(\beta/2)\hat{A}_\xi^2}\Psi_\mathrm{g}\|$ like the ground state of the Nelson Hamiltonian.

Theorem 3.65 (Gaussian domination of the ground state). *Let* $\beta < (2\mathrm{q_E}(\mathrm{j}_0\xi, \mathrm{j}_0\xi))^{-1}$. *Then* $\Psi_\mathrm{g} \in D(e^{(\beta/4)\hat{A}_\xi^2})$ *follows.*

Proof. By Lemma 3.64, we have the uniform bound

$$\|e^{(\beta/2)\hat{A}_\xi^2}\Psi_\mathrm{g}^t\|^2 \le \frac{1}{\sqrt{1 - 2\beta\mathrm{q_E}(\mathrm{j}_0\xi, \mathrm{j}_0\xi)^2}}$$

in t. Thus there exists a subsequence t' such that $\|e^{(\beta/4)\hat{A}_\xi^2}\Psi_g^{t'}\|^2$ converges to some c as $t' \to \infty$. We reset t' as t. We claim that $\{e^{(\beta/4)\hat{A}_\xi^2}\Psi_g^t\}_{t\geq 0}$ is a Cauchy sequence. Directly, we have

$$\|e^{(\beta/4)\hat{A}_\xi^2}\Psi_g^t - e^{(\beta/4)\hat{A}_\xi^2}\Psi_g^s\|^2 = \|e^{(\beta/4)\hat{A}_\xi^2}\Psi_g^t\|^2 + \|e^{(\beta/4)\hat{A}_\xi^2}\Psi_g^s\|^2 - 2(\Psi_g^s, e^{(\beta/2)\hat{A}_\xi^2}\Psi_g^t).$$

Note that Ψ_g^t strongly converges to Ψ_g as $t \to \infty$. Since the uniform bound of $\|e^{\beta\hat{A}_\xi^2}\Psi_g^t\|^2$ implies that

$$(\Psi_g^s, e^{(\beta/2)\hat{A}_\xi^2}\Psi_g^t) = (\Psi_g^s - \Psi_g^t, e^{(\beta/2)\hat{A}_\xi^2}\Psi_g^t) + \|e^{(\beta/4)\hat{A}_\xi^2}\Psi_g^t\|^2 \to c$$

as $t, s \to \infty$, we obtain that

$$\lim_{t,s\to\infty} \|e^{(\beta/4)\hat{A}_\xi^2}\Psi_g^t - e^{(\beta/4)\hat{A}_\xi^2}\Psi_g^s\| = 0$$

and $e^{(\beta/4)\hat{A}_\xi^2}\Psi_g^t$, $t > 0$, is a convergent sequence. Hence the closedness of $e^{(\beta/4)\hat{A}_\xi^2}$ yields the desired results. □

3.7 Translation invariant Pauli–Fierz model

In this section, we consider the translation invariant Pauli–Fierz Hamiltonian. This is obtained by setting the external potential V identically zero, resulting in the fact that H_{PF} commutes with the total momentum operator. A general argument allows then decomposing H_{PF} as $H_{\mathrm{PF}} = \int_{\mathbb{R}^3}^{\oplus} H_{\mathrm{PF}}(p)dp$. Moreover, for each $p \in \mathbb{R}^3$, the fiber Hamiltonian $H_{\mathrm{PF}}(p)$ is self-adjoint on $\mathcal{F}_{\mathrm{rad}}$. We will specify the exact form of $H_{\mathrm{PF}}(p)$ and its domain. Although H_{PF} with $V \equiv 0$ has no ground state, the fiber Hamiltonian $H_{\mathrm{PF}}(p)$ may have ground states and a functional integral representation of $e^{-tH_{\mathrm{PF}}(p)}$ can be established to study the spectrum of $H_{\mathrm{PF}}(p)$.

We start by defining the fiber Hamiltonian. As said, the standing assumptions throughout Section 3.7 are as follows.

Assumption 3.66. The following conditions hold:
(1) (Charge distribution) $\varphi \in \mathscr{S}'(\mathbb{R}^3)$ and $\hat{\varphi}(-k) = \overline{\hat{\varphi}(k)}$ $\sqrt{\omega}\hat{\varphi}, \hat{\varphi}/\sqrt{\omega}, \hat{\varphi}/\omega \in L^2(\mathbb{R}^3)$;
(2) (External potential) $V = 0$.

Put

$$P_{\mathrm{f}\mu} = d\Gamma(k_\mu \oplus k_\mu) = \sum_{j=\pm}\int_{\mathbb{R}^3} k_\mu a^*(k,j)a(k,j)dk, \quad \mu = 1, 2, 3,$$

which describes the field momentum. The *total momentum operator* $\mathrm{P}^{\mathrm{tot}}$ on $\mathcal{H}_{\mathrm{PF}}$ is defined by the sum of the momentum operator for the particle and that of field,

$$\mathrm{P}_\mu^{\rm tot} = -i\nabla_{x_\mu} \otimes 1 + 1 \otimes P_{\mathrm{f}\mu}, \quad \mu = 1,2,3.$$

Recall that $\wp_\mu^j(x) = \hat{\varphi}(k)e_\mu^j(k)e^{-ik\cdot x}/\sqrt{\omega(k)}$. Since

$$[-i\nabla_{x_\nu}, a^*(\wp_\mu^j, j)] = a^*(-k_\nu \wp_\mu^j, j),$$

$$[P_{\mathrm{f}\nu}, a^*(\wp_\mu^j, j)] = a^*(k_\nu \wp_\mu^j, j),$$

it follows that

$$[H_{\rm PF}, \mathrm{P}_\mu^{\rm tot}] = 0, \quad \mu = 1,2,3.$$

This leads to a decomposition $H_{\rm PF}$ of the spectrum of the total momentum operator $\mathrm{Spec}(\mathrm{P}_\mu^{\rm tot}) = \mathbb{R}$, $\mu = 1,2,3$. We denote $\sum_{\mu=1}^3 P_{\mathrm{f}\mu}^2$ by P_{f}^2. In the same way as in Proposition 2.142, we can also show that $H_{\rm PF}$ with $V = 0$ has no ground state. We state this as a proposition.

Proposition 3.67. *Hamiltonian $H_{\rm PF}$ has no ground state.*

Proof. If $H_{\rm PF}$ has a ground state, then it must be unique. Then the proposition follows in the same way as that of Proposition 2.142. □

Definition 3.68 (Pauli–Fierz Hamiltonian with a fixed total momentum).
The *Pauli–Fierz Hamiltonian with a fixed total momentum* is

$$H_{\rm PF}(p) = \frac{1}{2}(p - P_{\mathrm{f}} - eA(0))^2 + H_{\rm rad}, \quad p \in \mathbb{R}^3,$$

with domain

$$D(H_{\rm PF}(p)) = D(H_{\rm rad}) \cap D(P_{\mathrm{f}}^2),$$

where $A_\mu(0) = A_\mu(x = 0)$ is given by

$$A_\mu(0) = \frac{1}{\sqrt{2}} \sum_{j=\pm} \int_{\mathbb{R}^3} e_\mu^j(k)\left(a^*(k,j)\frac{\hat{\varphi}(k)}{\sqrt{\omega(k)}} + a(k,j)\frac{\hat{\varphi}(-k)}{\sqrt{\omega(k)}}\right)dk.$$

Here, $p \in \mathbb{R}^3$ is called the *total momentum.*

It is easy to see that $H_{\rm PF}(p)$ is self-adjoint for sufficiently small coupling constants e as a consequence of the Kato–Rellich theorem. In fact, by splitting off $H_{\rm PF}(p)$ as

$$H_{\rm PF}(p) = H_{\rm PF,0}(p) + H_{\rm PF,I}(p),$$

with

$$H_{\rm PF,0}(p) = H_{\rm rad} + \frac{1}{2}(p - P_{\mathrm{f}})^2,$$

$$H_{\rm PF,I}(p) = -e(p - P_{\mathrm{f}})A(0) + \frac{e^2}{2}A(0)^2,$$

we readily see that $H_{\mathrm{PF,I}}(p)$ is relatively bounded with respect to $H_{\mathrm{PF,0}}(p)$. We give the relationship between H_{PF} and $H_{\mathrm{PF}}(p)$. Define the unitary operator

$$\mathscr{T} : \mathcal{H}_{\mathrm{PF}} = L^2(\mathbb{R}^3_x) \otimes \mathcal{F}_{\mathrm{rad}} \to L^2(\mathbb{R}^3_p) \otimes \mathcal{F}_{\mathrm{rad}}$$

by

$$\mathscr{T} = (\hat{F} \otimes 1) \int_{\mathbb{R}^3}^{\oplus} \exp(ixP_{\mathrm{f}})dx, \tag{3.7.1}$$

denoting by $\hat{F}$ the Fourier transformation from $L^2(\mathbb{R}^3_x)$ to $L^2(\mathbb{R}^3_p)$. For $\Psi \in \mathcal{H}_{\mathrm{PF}}$,

$$(\mathscr{T}\Psi)(p) = \frac{1}{\sqrt{(2\pi)^3}} \int_{\mathbb{R}^3} e^{-ixp} e^{ixP_{\mathrm{f}}} \Psi(x)dx.$$

Let $M_\mu f(x) = x_\mu f(x)$, $\mu = 1, 2, 3$, be the multiplication by x_μ in $L^2(\mathbb{R}^3)$. Write $M = (M_1, M_2, M_3)$ and

$$\mathcal{F}_{\mathrm{rad}}^{\infty} = \mathrm{L.H.} \left\{ \prod_{i=1}^{n} a^*(f_i)\Omega_{\mathrm{PF}}, \Omega_{\mathrm{PF}} \,\middle|\, f_i \in \oplus^2 C_0^{\infty}(\mathbb{R}^3), i = 1, \ldots, n, n \geq 1 \right\} \tag{3.7.2}$$

and

$$D_\infty = C_0^\infty(\mathbb{R}^3) \hat{\otimes} \mathcal{F}_{\mathrm{rad}}^{\infty}. \tag{3.7.3}$$

We define $H_{\mathrm{PF}}(p)$ for arbitrary values of $e \in \mathbb{R}$ and $p \in \mathbb{R}^3$ as a self-adjoint operator. As was mentioned above, for sufficiently small e, $H_{\mathrm{PF}}(p)$ is defined as a self-adjoint operator, however, for arbitrary values of e, this is not clear. Define

$$L = \overline{\frac{1}{2}(M \otimes 1 - 1 \otimes P_{\mathrm{f}} - e1 \otimes A(0))^2 + 1 \otimes H_{\mathrm{rad}} \lceil_{D_\infty}}.$$

We have $\mathscr{T} H_{\mathrm{PF}} \Phi = L \mathscr{T} \Phi$ for $\Phi \in D_\infty$. Since by Theorem 3.47, D_∞ is a core of H_{PF}, and L is closed, we see that $\mathscr{T}$ maps $D(H_{\mathrm{PF}})$ onto $D(L)$ with $\mathscr{T} H_{\mathrm{PF}} \mathscr{T}^{-1} = L$. Note also that

$$D(L) = \mathscr{T} D(H_{\mathrm{PF}}) = D((\overline{M \otimes 1 - 1 \otimes P_{\mathrm{f}}})^2) \cap D(1 \otimes H_{\mathrm{rad}}).$$

Theorem 3.69 (Fiber decomposition and self-adjointness). *For each $p \in \mathbb{R}^3$, $H_{\mathrm{PF}}(p)$ is a nonnegative self-adjoint operator, and*

$$\mathscr{T} \left(\int_{\mathbb{R}^3}^{\oplus} \mathcal{F}_{\mathrm{rad}} dp \right) \mathscr{T}^{-1} = \mathcal{H}_{\mathrm{PF}}, \quad \mathscr{T} \left(\int_{\mathbb{R}^3}^{\oplus} H_{\mathrm{PF}}(p) dp \right) \mathscr{T}^{-1} = H_{\mathrm{PF}}.$$

Proof. Define the quadratic form

$$Q_p(\Psi,\Phi) = \frac{1}{2}\sum_{\mu=1}^{3}((p - P_\mathrm{f} - eA(0))_\mu\Psi, (p - P_\mathrm{f} - eA(0))_\mu\Phi) + (H_\mathrm{rad}^{1/2}\Psi, H_\mathrm{rad}^{1/2}\Phi)$$

on $\mathcal{F}_\mathrm{rad} \times \mathcal{F}_\mathrm{rad}$ with form domain $\bigcap_{\mu=1}^{3}(D(P_{\mathrm{f}\mu}) \cap D(A_\mu(0))) \cap D(H_\mathrm{rad}^{1/2})$. Since Q_p is densely defined and nonnegative, there exists a positive self-adjoint operator $L(p)$ such that $Q_p(\Psi,\Phi) = (L(p)^{1/2}\Psi, L(p)^{1/2}\Phi)$. Note that

$$L(p) = H_\mathrm{PF}(p) \tag{3.7.4}$$

on $D(H_\mathrm{rad}) \cap D(P_\mathrm{f}^2)$. Define $\tilde{L} = \int_{\mathbb{R}^3}^{\oplus} L(p)dp$. For $\phi, \psi \in D_\infty$, we have

$$(\mathscr{T}\psi, L\mathscr{T}\phi) = (\mathscr{T}\psi, \tilde{L}\mathscr{T}\phi).$$

Hence $\mathscr{T}^{-1}L\mathscr{T} = \mathscr{T}^{-1}\tilde{L}\mathscr{T}$ on D_∞ and then $H_\mathrm{PF} = \mathscr{T}^{-1}\tilde{L}\mathscr{T}$ on D_∞. Since D_∞ is a core of H_PF and $\tilde{L}$ is self-adjoint, it follows that $\mathscr{T}$ maps $D(H_\mathrm{PF})$ onto $D(\tilde{L})$ with

$$\mathscr{T}H_\mathrm{PF}\mathscr{T}^{-1} = \tilde{L}. \tag{3.7.5}$$

Next, we prove the self-adjointness of $H_\mathrm{PF}(p)$. The operator $H_{\mathrm{PF},0}(p)$ is self-adjoint on $D(H_{\mathrm{PF},0}(p)) = D(P_\mathrm{f}^2) \cap D(H_\mathrm{rad})$ and

$$\mathscr{T}\left(\int_{\mathbb{R}^3}^{\oplus} H_{\mathrm{PF},0}(p)dp\right)\mathscr{T}^{-1} = H_{\mathrm{PF},0}.$$

Thus we have for $F \in \mathcal{H}_\mathrm{PF}$ such that $(\mathscr{T}F)(p) = f(p)\Phi$ with $f \in C_0^\infty(\mathbb{R}^3)$ and $\Phi \in \mathcal{F}_\mathrm{rad}^\infty$, by the inequality $\|H_{\mathrm{PF},0}F\| \le C\|(H_\mathrm{PF}+1)F\|$ derived from Theorem 3.47 and the closed graph theorem, and (3.7.5),

$$\int_{\mathbb{R}^3} |f(p)|^2\|H_{\mathrm{PF},0}(p)\Phi\|^2 dp \le C^2 \int_{\mathbb{R}^3} |f(p)|^2\|(L(p)+1)\Phi\|^2 dp.$$

Hence

$$\|H_{\mathrm{PF},0}(p)\Phi\| \le C\|(L(p)+1)\Phi\| \tag{3.7.6}$$

follows for almost every $p \in \mathbb{R}^3$, since $f \in C_0^\infty(\mathbb{R}^d)$ is arbitrary. Since both sides of (3.7.6) are continuous in p, this inequality holds for all $p \in \mathbb{R}^3$. Thus $H_{\mathrm{PF},0}(p)(L(p)+1)^{-1}$ is bounded, and so is also $H_{\mathrm{PF},0}(p)e^{-tL(p)}$, which implies that $e^{-tL(p)}$ leaves $D(H_\mathrm{rad})\cap D(P_\mathrm{f}^2)$ invariant. This means that $L(p)$ is essentially self-adjoint on $D(H_\mathrm{rad})\cap D(P_\mathrm{f}^2)$ by Lemma 3.40. Moreover, (3.7.6) implies that $L(p)$ is closed on $D(H_\mathrm{rad}) \cap D(P_\mathrm{f}^2)$, and hence $L(p)$ is self-adjoint on $D(H_\mathrm{rad}) \cap D(P_\mathrm{f}^2)$. By the basic inequality,

$$\|L(p)\Phi\| \le C'(\|H_{\mathrm{PF},0}(p)\Phi\| + \|\Phi\|), \quad \Phi \in D(H_{\mathrm{PF},0}(p)),$$

it is clear that $L(p)$ is essentially self-adjoint on any core of $H_{\mathrm{PF},0}(p)$. By (3.7.4), the theorem follows. □

Self-adjointness of $H_{\mathrm{PF}}(p)$ ensures that $\{e^{-tH_{\mathrm{PF}}(p)} : t \ge 0\}$ is a well-defined C_0-semigroup. As in the previous section, we move to from Fock representation to Schrödinger representation in order to construct a functional integral representation. Then $H_{\mathrm{PF}}(p)$ becomes

$$H_{\mathrm{PF}}(p) = \frac{1}{2}(p - \hat{P}_{\mathrm{f}} - e\hat{A}(0))^2 + H_{\mathrm{rad}}$$

on $L^2(\mathscr{Q})$, where $\hat{A}(0) = \hat{A}(x = 0) = (\hat{A}_1(\tilde{\varphi}), \hat{A}_2(\tilde{\varphi}), \hat{A}_3(\tilde{\varphi}))$, and recall that $\hat{P}_{\mathrm{f}} = d\Gamma(-i\nabla)$ and $H_{\mathrm{rad}} = d\Gamma(\omega(-i\nabla))$. We remark that we also use notation H_{rad} for $d\Gamma(\omega(-i\nabla))$. The functional integral representation of $e^{-tH_{\mathrm{PF}}(p)}$ can be also constructed as an application of that of $e^{-tH_{\mathrm{PF}}}$.

Theorem 3.70 (Functional integral representation for Pauli–Fierz Hamiltonian with a fixed total momentum). *Let* $\Psi, \Phi \in L^2(\mathscr{Q})$. *Then*

$$(\Psi, e^{-tH_{\mathrm{PF}}(p)}\Phi) = \mathbb{E}^0[e^{ip\cdot B_t}(\mathrm{J}_0\Psi, e^{-ie\hat{A}_{\mathrm{E}}(K_t)}\mathrm{J}_t e^{-i\hat{P}_{\mathrm{f}}\cdot B_t}\Phi)_{L^2(\mathscr{Q}_{\mathrm{E}})}]. \tag{3.7.7}$$

Proof. Write $F_s = \Pi_s \otimes \Psi$ and $G_r = \Pi_r \otimes \Phi$, where Π_s is the three-dimensional heat kernel, $\Pi_s(x) = (2\pi s)^{-3/2}\exp(-|x|^2/2s)$, and $\Psi, \Phi \in L^2(\mathscr{Q})$. Due to the fact that $H_{\mathrm{PF}} = \mathscr{T}(\int_{\mathbb{R}^3}^{\oplus} H_{\mathrm{PF}}(p)dp)\mathscr{T}^{-1}$, we have

$$(F_s, e^{-tH_{\mathrm{PF}}}e^{-i\xi \mathrm{p}^{\mathrm{tot}}}G_r) = \int_{\mathbb{R}^3} e^{-i\xi p}((\mathscr{T}F_s)(p), e^{-tH_{\mathrm{PF}}(p)}(\mathscr{T}G_r)(p))dp$$

for any $\xi \in \mathbb{R}^3$. Note that $\lim_{s\to 0}(\mathscr{T}F_s)(p) = (2\pi)^{-3/2}\Psi$ strongly in $L^2(\mathscr{Q})$ for each $p \in \mathbb{R}^3$. Hence

$$\lim_{s\to 0}(F_s, e^{-tH_{\mathrm{PF}}}e^{-i\xi \mathrm{p}^{\mathrm{tot}}}G_r) = \frac{1}{\sqrt{(2\pi)^3}}\int_{\mathbb{R}^3}(\Psi, e^{-tH_{\mathrm{PF}}(p)}e^{-i\xi p}(\mathscr{T}G_r)(p))dp. \tag{3.7.8}$$

Using the fact $e^{i\xi \mathrm{p}^{\mathrm{tot}}}G_r(x) = \Pi_r(x - \xi)e^{-i\xi\hat{P}_{\mathrm{f}}}\Phi$, we obtain by the functional integral representation in Theorem 3.26 that

$$(F_s, e^{-tH_{\mathrm{PF}}}e^{-i\xi \mathrm{p}^{\mathrm{tot}}}G_r) = \int_{\mathbb{R}^3}\mathbb{E}^x[\Pi_s(x)\Pi_r(B_t - \xi)(\mathrm{J}_0\Psi, e^{-ie\hat{A}_{\mathrm{E}}(K_t)}\mathrm{J}_t e^{-i\xi\hat{P}_{\mathrm{f}}}\Phi)]dx.$$

Then it follows that, due to $\Pi_s(x) \to \delta(x)$ as $s \to 0$,

$$\lim_{s\to 0}(F_s, e^{-tH_{\mathrm{PF}}}e^{-i\xi \mathrm{p}^{\mathrm{tot}}}G_r) = \mathbb{E}^0[\Pi_r(B_t - \xi)(\mathrm{J}_0\Psi, e^{-ie\hat{A}_{\mathrm{E}}(K_t)}\mathrm{J}_t e^{-i\xi\hat{P}_{\mathrm{f}}}\Phi)]. \tag{3.7.9}$$

Combining (3.7.8) and (3.7.9) leads to

$$\frac{1}{\sqrt{(2\pi)^3}}\int_{\mathbb{R}^3} e^{-i\xi p}(\Psi, e^{-tH_{\mathrm{PF}}(p)}(\mathscr{T}G_r)(p))dp = \mathbb{E}^0[\Pi_r(B_t-\xi)(\mathrm{J}_0\Psi, e^{-ie\hat{A}_{\mathrm{E}}(K_t)}\mathrm{J}_t e^{-i\xi\hat{P}_{\mathrm{f}}}\Phi)]. \tag{3.7.10}$$

Since $|(\Psi, e^{-tH_{\mathrm{PF}}(p)}(\mathscr{T}G_r)(p))| \le \|(\mathscr{T}G_r)(p)\|$ and $\|(\mathscr{T}G_r)(\cdot)\| \in L^2(\mathbb{R}^3)$, we have $(\Psi, e^{-tH_{\mathrm{PF}}(p)}(\mathscr{T}G_r)(p)) \in L^2(\mathbb{R}^3_p)$. Taking the inverse Fourier transform on both sides of (3.7.10) with respect to p gives

$$(\Psi, e^{-tH_{\mathrm{PF}}(p)}(\mathscr{T}G_r)(p)) = \frac{1}{\sqrt{(2\pi)^3}}\mathbb{E}^0\left[\int_{\mathbb{R}^3} e^{i\xi p}\Pi_r(B_t-\xi)(\mathrm{J}_0\Psi, e^{-ie\hat{A}_{\mathrm{E}}(K_t)}\mathrm{J}_t e^{-i\xi\hat{P}_{\mathrm{f}}}\Phi)d\xi\right] \tag{3.7.11}$$

for almost every $p \in \mathbb{R}^3$. Since both sides of (3.7.11) are continuous in p, the equality stays valid for all $p \in \mathbb{R}^3$. After taking $r \to 0$ on both sides, we arrive at (3.7.7). □

By Theorem 3.70, the functional integral representation of $e^{-tH_{\mathrm{PF}}(p)}$ can be formally written as

$$(\Psi, e^{-tH_{\mathrm{PF}}(p)}\Phi) \overset{\text{formal}}{=} \mathbb{E}^0\left[\int_{\mathscr{Q}_{\mathrm{E}}} \overline{\mathrm{J}_0\Psi} e^{i\int_0^t (p-\mathrm{d}\Gamma_{\mathrm{E}}(-i\nabla\otimes 1)-e\hat{A}_{\mathrm{E}}(\tilde{\varphi}(\cdot-B_s)))dB_s}\mathrm{J}_t\Phi d\mu_{\mathrm{E}}\right].$$

In the exponent, the expression $p-\mathrm{d}\Gamma_{\mathrm{E}}(-i\nabla\otimes 1)-e\hat{A}_{\mathrm{E}}(\tilde{\varphi}(\cdot-B_s))$ appears suggestive of the Euclidean version of $p-\hat{P}_{\mathrm{f}}-e\hat{A}(\tilde{\varphi})$.

Since intuitively $e^{-i\hat{P}_{\mathrm{f}}\cdot B_t}$ is a shift operator on $L^2(\mathscr{Q})$, it can be expected to be a positivity preserving.

Lemma 3.71. *Let $\xi \in \mathbb{R}^3$. Then $e^{-i\hat{P}_{\mathrm{f}}\xi}$ is positivity preserving. In particular, it follows that $|e^{-i\hat{P}_{\mathrm{f}}\xi}\Psi| \le e^{-i\hat{P}_{\mathrm{f}}\xi}|\Psi|$.*

Proof. Let

$$F = F(\hat{A}(f_1),\dots,\hat{A}(f_n)) = (2\pi)^{-n/2}\int_{\mathbb{R}^n}\hat{F}(k)e^{i\sum_{j=1}^n k_j\hat{A}(f_j)}dk,$$

$$G = G(\hat{A}(g_1),\dots,\hat{A}(g_m)) = (2\pi)^{-m/2}\int_{\mathbb{R}^m}\hat{G}(k)e^{i\sum_{j=1}^m k_j\hat{A}(g_j)}dk,$$

with $0 \le F \in \mathscr{S}(\mathbb{R}^n)$ and $0 \le G \in \mathscr{S}(\mathbb{R}^m)$. Employing

$$(e^{i\hat{A}(g_1)}, e^{i\hat{P}_{\mathrm{f}}\xi}e^{i\hat{A}(g_2)}) = (e^{i\hat{A}(g_1)}, e^{i\hat{A}(g_2(\cdot+\xi))}),$$

we see that

$$(F, e^{i\hat{P}_{\mathrm{f}}\xi}G) = \big(F, G(\hat{A}(g_1(\cdot+\xi)),\dots,\hat{A}(g_m(\cdot+\xi)))\big) \ge 0.$$

Hence by a limiting argument, $(F, e^{i\hat{P}_{\mathrm{f}}\xi}G) \ge 0$ for arbitrary nonnegative $F, G \in L^2(\mathscr{Q})$. □

An analogue of the diamagnetic inequality for $e^{-tH_{\rm PF}}$ can be derived from the functional integral representation for $e^{-tH_{\rm PF}(p)}$.

Corollary 3.72 (Diamagnetic inequality). *It follows that*

$$|(\Psi, e^{-tH_{\rm PF}(p)}\Phi)| \le (|\Psi|, e^{-t(\hat{P}_{\rm f}^2+H_{\rm rad})}|\Phi|).$$

Proof. This follows from Lemma 3.71 and

$$|(\Psi, e^{-tH_{\rm PF}(p)}\Phi)| \le \mathbb{E}[(\mathrm{J}_0|\Psi|, \mathrm{J}_t e^{-iB_t\hat{P}_{\rm f}}|\Phi|)] = (|\Psi|, e^{-t(\hat{P}_{\rm f}^2+H_{\rm rad})}|\Phi|). \qquad \square$$

From this diamagnetic inequality, however, we can only deduce the trivial energy comparison inequality $\inf \mathrm{Spec}(\hat{P}_{\rm f}^2 + H_{\rm rad}) \le \inf \mathrm{Spec}(H_{\rm PF}(p)) = 0$. However, combining the unitary transformation $e^{-i(\pi/2)N}$ with the functional integral representation (3.7.7), we obtain an interesting result. Denote $E_{\rm PF}(p) = \inf \mathrm{Spec}(H_{\rm PF}(p))$.

Corollary 3.73 (Properties of $E_{\rm PF}(p)$). *It follows that:*
(1) *Let $p = 0$ and $\mathfrak{S} = e^{-i(\pi/2)N}$. Then $\mathfrak{S}^{-1}e^{-tH_{\rm PF}(0)}\mathfrak{S}$ is positivity improving.*
(2) *The ground state of $H_{\rm PF}(0)$ is unique whenever it exists.*
(3) *$E_{\rm PF}(0) \le E_{\rm PF}(p)$ holds.*
(4) *The map $p \mapsto E_{\rm PF}(p)$ is continuous and $E_{\rm PF}(0) = \inf \mathrm{Spec}(H_{{\rm PF},V=0})$, where $H_{{\rm PF},V=0}$ is $H_{\rm PF}$ with $V = 0$.*

Proof. In the case of $p = 0$, we remark that $e^{ip\cdot B_t} = 1$. Then we have

$$(\Psi, \mathfrak{S}^{-1}e^{tH_{\rm PF}(0)}\mathfrak{S}\Phi) = \mathbb{E}^0[(\mathrm{J}_0\Psi, e^{-ie\hat{\Pi}_{\rm E}(K_t)}\mathrm{J}_t e^{-i\hat{P}_{\rm f}\cdot B_t}\Phi)].$$

Since $\mathrm{J}_0^* e^{-ie\hat{\Pi}_{\rm E}(K_t)}\mathrm{J}_t$ is positivity improving and $e^{-i\hat{P}_{\rm f}B_t}$ is positivity preserving by Lemma 3.71, (1) follows. Claim (2) is implied by (1) and the Perron–Frobenius theorem. We have

$$|(\Psi, \mathfrak{S}^{-1}e^{-tH_{\rm PF}(p)}\mathfrak{S}\Phi)| \le \mathbb{E}^0[(\mathrm{J}_0|\Psi|, e^{-ie\hat{\Pi}_{\rm E}(K_t)}\mathrm{J}_t e^{-i\hat{P}_{\rm f}B_t}|\Phi|)] = (|\Psi|, \mathfrak{S}^{-1}e^{-tH_{\rm PF}(0)}\mathfrak{S}|\Phi|).$$

This yields (3). Finally, (4) is derived in the same way as Corollary 2.148. $\square$

Finally, we show the vacuum expectation $(1, e^{-tH_{\rm PF}(p)}1)$ as a corollary of Theorem 3.70.

Corollary 3.74. *It follows that*

$$(1, e^{-tH_{\rm PF}(p)}1) = \mathbb{E}^0[e^{ip\cdot B_t}e^{-(e^2/2)\mathrm{q}_{\rm E}(Z,Z)}],$$

where $\mathrm{q}_{\rm E}(Z,Z)$ is given in (3.6.11).

As was mentioned in the previous section, $q_E(Z,Z)$ has the formal expression $q_E(Z,Z) = \sum_{\mu,\nu=1}^{3} \int_{-t}^{t} dB_s^\mu \int_{-t}^{t} dB_r^\nu W_{\mu\nu}(B_s - B_r, s-r)$, where $W_{\mu\nu}$ is given by

$$W_{\mu\nu}(X,T) = \int_{\mathbb{R}^3} \delta_{\mu\nu}^\perp(k) \frac{|\hat{\varphi}(k)|^2}{2\omega(k)} e^{-|T|\omega(k)} e^{-ik\cdot X} dk$$

and it is called *the pair potential of the Pauli–Fierz Hamiltonian with a fixed total momentum.*

3.8 The Pauli–Fierz model with spin

We assume Assumption 3.4 in Sections 3.8.1–3.8.8, and Assumption 3.33 in Section 3.8.9, unless otherwise stated.

3.8.1 Counting measures

In the previous sections, we considered the spinless version of the Pauli–Fierz Hamiltonian ignoring the spin of the electron. In this section, we define the Hamiltonian including spin 1/2 and construct its functional integral representation with a scalar integrand dependent on a jump process which is a so-called spin process. To derive a functional integral representation of the Pauli–Fierz Hamiltonian with spin, we review classical cases in which the Feynman–Kac formula for Schrödinger operators with spin is investigated. As is established in classical cases in addition to Brownian motion, we need a Lévy process for presenting spin. We give this 3-dimensional Lévy process $(Z_t)_{t\geq 0}$ on a probability space $(\mathcal{S}, \mathcal{B}_\mathcal{S}, P)$ with characteristics (b, A, ν) such that $\nu(\mathbb{R}^3 \setminus \{0\}) = 1$. For $I \in \mathcal{B}(\mathbb{R}^3)$, let

$$N(t,I) = |\{0 \leq s \leq t \mid \Delta Z_s \in I\}|$$

be the counting measure associated with $(Z_t)_{t\geq 0}$. Define the random process $(N_t)_{t\geq 0}$ on $(\mathcal{S}, \mathcal{B}_\mathcal{S}, P)$ by $N_t = N(t, \mathbb{R}^3 \setminus \{0\})$. Note that $(N_t)_{t\geq 0}$ is a Poisson process with intensity 1. Consider the measure

$$dN_t = \int_{\mathbb{R}^3\setminus\{0\}} N(dtdz)$$

on $[0,\infty) = \mathbb{R}^+$. The compensator of N_t is given by t and the expectation of exponent e^{-aN_t} is given by $\mathbb{E}_P[e^{-aN_t}] = e^{t(e^{-a}-1)}$. We have for $f = f(s,\tau)$, $(s,\tau) \in \mathbb{R}^+ \times \mathcal{S}$, independently of $z \in \mathbb{R}^3 \setminus \{0\}$, that

$$\mathbb{E}_P\left[\int_0^{t+} f(s,\tau) dN_s\right] = \mathbb{E}_P\left[\int_0^{t} f(s,\tau) ds\right]$$

and, furthermore,

$$\mathbb{E}_P\left[\int_0^{t+} f(s,N_s)dN_s\right] = \mathbb{E}_P\left[\int_0^{t} f(s,N_s)ds\right] = \int_0^t \sum_{n=0}^{\infty} f(s,n)\frac{s^n}{n!}e^{-s}ds.$$

Here, $\int_0^{t+} = \int_{[0,t]}$. Let $N+\alpha = (N_t+\alpha)_{t\geq 0}$ and $\mathbb{E}_P^{\alpha}[f(N)] = \mathbb{E}_P[f(N+\alpha)]$.

3.8.2 Review of classical cases

Let us review Schrödinger operators with spin 1/2. The spin will be described in terms of a $\mathbb{Z}_2$-valued process and the Schrödinger operator is defined on $L^2(\mathbb{R}^3\times\mathbb{Z}_2;\mathbb{C})$ instead of $L^2(\mathbb{R}^3;\mathbb{C}^2)$. Let $\sigma_1, \sigma_2, \sigma_3$ be the 2×2 *Pauli matrices* given by

$$\sigma_1 = \begin{pmatrix} 0 & 1 \\ 1 & 0 \end{pmatrix}, \quad \sigma_2 = \begin{pmatrix} 0 & -i \\ i & 0 \end{pmatrix}, \quad \sigma_3 = \begin{pmatrix} 1 & 0 \\ 0 & -1 \end{pmatrix}. \tag{3.8.1}$$

Let $\mathbb{Z}_2$ be the set of the square roots of identity, i. e.,

$$\mathbb{Z}_2 = \{\theta_1, \theta_2\} = \{-1, 1\},$$

where $\theta_\alpha = (-1)^\alpha$. The particle and spin yield an $(\mathbb{R}^3\times\mathbb{Z}_2)$-valued random process $(q_t)_{t\geq 0}$ on the product space $(\mathscr{X}\times S, \mathcal{B}(\mathscr{X})\times\mathcal{B}_S, \mathcal{W}^x\times P)$ by

$$q_t = (B_t, \theta_{N_t}), \quad t\geq 0,$$

where $\theta_{N_t} = (-1)^{N_t}$ is called the *spin process*. For simplicity, we write $\mathbb{E}^{x,\alpha}_{\mathcal{W}\times P} = \mathbb{E}^{x,\alpha}$ for $(x,\alpha)\in\mathbb{R}^3\times\{1,2\}$. The generator of random process $(q_t)_{t\geq 0}$ is

$$-\left(-\frac{1}{2}\Delta\otimes 1 + 1\otimes\sigma_{\mathrm{F}}\right)$$

under the identification $L^2(\mathbb{R}^3\times\mathbb{Z}_2)\cong L^2(\mathbb{R}^3)\otimes\mathbb{C}^2$ given by

$$L^2(\mathbb{R}^3;\mathbb{C}^2)\ni\begin{pmatrix} f(x,+1) \\ f(x,-1)\end{pmatrix}\mapsto f(x,\theta)\in L^2(\mathbb{R}^3\times\mathbb{Z}_2;\mathbb{C}), \tag{3.8.2}$$

where

$$\sigma_{\mathrm{F}} = \frac{1}{2}(\sigma_3 + i\sigma_2)(\sigma_3 - i\sigma_2).$$

Actually, $\sigma_{\mathrm{F}} = -\sigma_1 + 1$, and hence we have

$$(f, e^{-t(-\frac{1}{2}\Delta-\sigma_1)}g) = e^t\sum_{\alpha=1,2}\int_{\mathbb{R}^3}\mathbb{E}^{x,\alpha}[\bar{f}(q_0)g(q_t)]dx.$$

Let $a \in (C_b^2(\mathbb{R}^3))^3$ and $V \in L^\infty(\mathbb{R}^3)$. We call the operator

$$H_S(a) = \frac{1}{2}(\sigma \cdot (-i\nabla - a))^2 + V \tag{3.8.3}$$

on $L^2(\mathbb{R}^3;\mathbb{C}^2)$ a Schrödinger operator with a vector potential a and spin 1/2. Set $D = (D_1, D_2, D_3)$ with $D_\mu = (-i\nabla - a)_\mu$ and

$$\sigma = (\sigma_1, \sigma_2, \sigma_3).$$

Since $\sigma_1\sigma_2 = i\sigma_3$, $\sigma_2\sigma_3 = i\sigma_1$ and $\sigma_3\sigma_1 = i\sigma_2$, we have

$$(\sigma \cdot D)^2 = D^2 + i\sigma_1(D_2D_3 - D_3D_2) + i\sigma_2(D_3D_1 - D_1D_3) + i\sigma_3(D_1D_2 - D_2D_1).$$

Using relations $D_2D_3 - D_3D_2 = i(\nabla\times a)_1$, $D_3D_1 - D_1D_3 = i(\nabla\times a)_2$, and $D_1D_2 - D_2D_1 = i(\nabla\times a)_3$, we expand (3.8.3) to obtain

$$H_S(a) = H_p(a) - \frac{1}{2}\sigma \cdot b, \tag{3.8.4}$$

where $b = (b_1, b_2, b_3) = \nabla \times a$ is a magnetic field. Let

$$a \in (C_b^2(\mathbb{R}^3))^3, \quad b \in (L^\infty(\mathbb{R}^3))^3, \quad V \in L^\infty(\mathbb{R}^3). \tag{3.8.5}$$

Then $H_S(a)$ is self-adjoint on $D(-\Delta)$ and bounded below. Now we transform $H_S(a)$ into an operator on the set of $\mathbb{C}$-valued functions on a suitable Hilbert space. By the identification (3.8.2), $H_S(a)$ can be reduced to the self-adjoint operator $H_{\mathbb{Z}_2}(a)$ defined in $L^2(\mathbb{R}^3\times\mathbb{Z}_2)$ to obtain

$$(H_{\mathbb{Z}_2}(a)f)(x,\theta) = \left(H_p(a) - \frac{1}{2}\theta b_3(x)\right)f(x,\theta) - \frac{1}{2}(b_1(x) - i\theta b_2(x))f(x,-\theta),$$

where $x \in \mathbb{R}^3$ and $\theta \in \mathbb{Z}_2$. Thus $H_S(a)$ can be regarded as $H_{\mathbb{Z}_2}(a)$ on the space of $\mathbb{C}$-valued functions with the configuration space $\mathbb{R}^3 \times \mathbb{Z}_2$. Suppose (3.8.5) and that $\int_0^t ds \int_{\mathbb{R}^3} |\log \frac{1}{2}\sqrt{b_1(y)^2 + b_2(y)^2}|\Pi_s(y-x)dy < \infty$ for all $(x,t) \in \mathbb{R}^3 \times \mathbb{R}^+$. Then

$$(f, e^{-tH_{\mathbb{Z}_2}(a)}g) = e^t \sum_{\alpha=1,2} \int_{\mathbb{R}^3} \mathbb{E}^{x,\alpha}\left[\overline{f(q_0)}g(q_t)e^{Z_t}\right]dx \tag{3.8.6}$$

holds. Here, exponent Z_t consists of 4 parts: vector potential a, external potential V, the diagonal part of spin interaction b_3, and the off-diagonal part of spin interaction $b_1 - ib_2$, namely

$$Z_t = -i\int_0^t a(B_s)\circ dB_s - \int_0^t V(B_s)ds + \int_0^t \frac{1}{2}\theta_{N_s} b_3(B_s)ds$$

$$+\int_0^{t+} \log\left(\frac{1}{2}(b_1(B_s) - i\theta_{N_s} b_2(B_s))\right) dN_s.$$

Here, $\int_0^{t+} \dots dN_s$ describes the sum of jump points of integrand,

$$\int_0^{t+} \log\left(\frac{1}{2}(b_1(B_s) - i\theta_{N_s} b_2(B_s))\right) dN_s = \sum_{\substack{r\in[0,t]\\ N_{r+}\neq N_{r-}}} \log\left(\frac{1}{2}(b_1(B_r) - i\theta_{N_{r-}} b_2(B_r))\right).$$

We apply classical cases to construct a functional integral representation of the Pauli–Fierz Hamiltonian with spin in the next section.

3.8.3 Definition of the Pauli–Fierz Hamiltonian with spin

In this section, we choose a specific $\mathscr{Q}$ space instead of the abstract setup. This will serve the purpose of drawing sufficient regularity of $\hat{A}$ with respect to $\phi \in \mathscr{Q}$. Let $\mathscr{S}_{\mathrm{real}}(\mathbb{R}^{3+\beta})$, $\beta = 0, 1$, be the set of real-valued Schwarz test functions on $\mathbb{R}^{3+\beta}$ and put $\mathscr{S}_\beta = \oplus^3 \mathscr{S}_{\mathrm{real}}(\mathbb{R}^{3+\beta})$. Put

$$\mathscr{Q}_\beta = \mathscr{S}_\beta',$$

with $\mathscr{S}_\beta'$ denoting the dual space of $\mathscr{S}_\beta$. Also, denote the pairing between elements of $\mathscr{Q}_\beta$ and $\mathscr{S}_\beta$ by $\langle\langle \phi, \mathbf{f}\rangle\rangle \in \mathbb{R}$, where $\phi \in \mathscr{Q}_\beta$ and $\mathbf{f} \in \mathscr{S}_\beta$. By the Bochner–Minlos theorem, there exists a complete probability space $(\mathscr{Q}_\beta, \Sigma_\beta, \mu_\beta)$ such that Σ_β is the smallest σ-field generated by $\{\langle\langle \phi, \mathbf{f}\rangle\rangle \mid f \in \mathscr{S}_\beta\}$ and $\langle\langle \phi, \mathbf{f}\rangle\rangle$ is a Gaussian random variable with mean zero and covariance

$$\mathbb{E}_{\mu_\beta}[\langle\langle \phi, \mathbf{f}\rangle\rangle \langle\langle \phi, \mathbf{g}\rangle\rangle] = \mathrm{q}_\beta(\mathbf{f}, \mathbf{g}). \tag{3.8.7}$$

We can extend $\langle\langle \phi, \mathbf{f}\rangle\rangle$ to more general $\mathbf{f}$. For any $\mathbf{f} = \Re\mathbf{f} + i\Im\mathbf{f} \in \oplus^3\mathscr{S}(\mathbb{R}^{3+\beta})$, put $\langle\langle \phi, \mathbf{f}\rangle\rangle = \langle\langle \phi, \Re\mathbf{f}\rangle\rangle + i\langle\langle \phi, \Im\mathbf{f}\rangle\rangle$. Since $\mathscr{S}(\mathbb{R}^{3+\beta})$ is dense in $L^2(\mathbb{R}^{3+\beta})$ and the inequality

$$\mathbb{E}_{\mu_\beta}[|\langle\langle \phi, \mathbf{f}\rangle\rangle|^2] \le \|\mathbf{f}\|^2_{\oplus^3 L^2(\mathbb{R}^{3+\beta})}$$

holds by (3.8.7), we define $\langle\langle \phi, \mathbf{f}\rangle\rangle$ for $\mathbf{f} \in \oplus^3 L^2(\mathbb{R}^{3+\beta})$ by $\langle\langle \phi, \mathbf{f}\rangle\rangle = \lim_{n\to\infty}\langle\langle \phi, \mathbf{f}_n\rangle\rangle$ in $L^2(\mathscr{Q}_\beta)$, where $(\mathbf{f}_n)_{n\in\mathbb{N}} \subset \oplus^3\mathscr{S}(\mathbb{R}^{3+\beta})$ is any sequence such that s-$\lim_{n\to\infty} \mathbf{f}_n = \mathbf{f}$. Thus the quantized radiation field $\hat{A}_\beta(\mathbf{f})$ with $\mathbf{f} \in \oplus^3 L^2(\mathbb{R}^{3+\beta})$ is realized as

$$\left(\hat{A}_\beta(\mathbf{f})F\right)(\phi) = \langle\langle \phi, \mathbf{f}\rangle\rangle F(\phi), \quad \phi \in \mathscr{Q}_\beta,$$

in $L^2(\mathscr{Q}_\beta)$, with domain

$$D(\hat{A}_\beta(\mathbf{f})) = \left\{F \in L^2(\mathcal{Q}_\beta) \,\middle|\, \mathbb{E}_{\mu_\beta}[|\langle\langle\phi, \mathbf{f}\rangle\rangle F(\phi)|^2] < \infty\right\}.$$

We define $\hat{A}_\beta(\mathbf{f}, \phi)$ for each $\phi \in \mathcal{Q}_\beta$ by

$$\hat{A}_\beta(\mathbf{f}, \phi) = \langle\langle\phi, \mathbf{f}\rangle\rangle, \quad \mathbf{f} \in \oplus^3 L^2(\mathbb{R}^{3+\beta}).$$

In the same way as in the previous section, we put

$$\begin{array}{llll} \text{(Minkowskian)} & \hat{A} = \hat{A}_0, & \mu = \mu_0, & \mathcal{Q} = \mathcal{Q}_0, \\ \text{(Euclidean)} & \hat{A}_{\mathrm{E}} = \hat{A}_1, & \mu_{\mathrm{E}} = \mu_1, & \mathcal{Q}_{\mathrm{E}} = \mathcal{Q}_1. \end{array}$$

The Hilbert space consisting of state vectors of the Pauli–Fierz Hamiltonian with spin 1/2 is

$$\mathcal{H}_{\mathrm{S}} = L^2(\mathbb{R}^3; \mathbb{C}^2) \otimes L^2(\mathcal{Q}).$$

The quantized radiation field $\hat{A} = (\hat{A}_1, \hat{A}_2, \hat{A}_3)$ in $\mathcal{H}_{\mathrm{S}}$ with cutoff function $\tilde{\varphi}$ is defined by

$$\hat{A}_\mu = \int_{\mathbb{R}^3}^{\oplus} \hat{A}_\mu(\tilde{\varphi}(\cdot - x))dx, \quad \mu = 1, 2, 3,$$

under the identification $\mathcal{H}_{\mathrm{S}} \cong \int_{\mathbb{R}^3}^{\oplus} \mathbb{C}^2 \otimes L^2(\mathcal{Q})dx$. The *quantized magnetic field* is defined by the curl of $\hat{A}$ as usual:

$$\hat{B} = (\hat{B}_1, \hat{B}_2, \hat{B}_3) = \mathrm{rot}_x \hat{A}.$$

It is straightforward to see that

$$\hat{B}_\mu(x) = \sum_{\lambda,\alpha,\nu=1}^{3} \hat{A}_\lambda(\delta_{\lambda\nu}\varepsilon^{\mu\alpha\nu}\partial_{x_\alpha}\tilde{\varphi}(\cdot - x)), \tag{3.8.8}$$

where $\varepsilon^{\alpha\beta\gamma}$ denotes the antisymmetric tensor defined by

$$\varepsilon^{\alpha\beta\gamma} = \begin{cases} 1, & (\alpha\beta\gamma) \text{ is an even permutation of (123)}, \\ -1, & (\alpha\beta\gamma) \text{ is an odd permutation of (123)}, \\ 0, & \text{otherwise}. \end{cases}$$

For example, $\varepsilon^{132} = -1$, $\varepsilon^{231} = 1$, and $\varepsilon^{112} = 0$. We define $\hat{B}_\mu(f)$ with a test function $f \in L^2(\mathbb{R}^3)$ by

$$\hat{B}_\mu(f) = \sum_{\lambda,\alpha,\nu=1}^{3} \hat{A}_\lambda(\delta_{\lambda\nu}\varepsilon^{\mu\alpha\nu}\partial_{x_\alpha} f).$$

Here, $\partial_{x_\alpha} f = \partial_{x_\alpha} f(\cdot - x)\lceil_{x=0}$, hence $\widehat{\partial_{x_\alpha} f} = -ik_\alpha \hat{f}(k)$ follows. The Euclidean version of $\hat{B}_\mu(g)$ with test function $g \in L^2(\mathbb{R}^4)$ is defined by

$$\hat{B}_{\mathrm{E}_\mu}(g) = \sum_{\lambda,\alpha,\nu=1}^{3} \hat{A}_{\mathrm{E}_\lambda}(\delta_{\lambda\nu}\varepsilon^{\mu\alpha\nu}\partial_{x_\alpha} g). \tag{3.8.9}$$

We will use this terminology later.

Definition 3.75 (Pauli–Fierz Hamiltonian with spin 1/2). The *Pauli–Fierz Hamiltonian with spin* 1/2 is defined by

$$H_{\mathrm{PF}}^{\mathrm{S}} = \frac{1}{2}(-i\nabla - e\hat{A})^2 + V + H_{\mathrm{rad}} - \frac{e}{2}\sigma \cdot \hat{B}. \tag{3.8.10}$$

The usual definition by minimal coupling,

$$H_{\mathrm{PF}}^{\mathrm{S}} = \frac{1}{2}\left(\sigma \cdot (-i\nabla - e\hat{A})\right)^2 + V + H_{\mathrm{rad}},$$

indeed coincides with (3.8.10), which can be seen by expanding the square above and making use of the identity $(\sigma \cdot a)(\sigma \cdot b) = a \cdot b + i\sigma \cdot (a \times b)$.

Remark 3.76 (Fock representations). Indeed, $\hat{B}_\mu(f)$ corresponds to

$$B_\mu(f) = \frac{1}{\sqrt{2}}\sum_{j=\pm}\left(a^*(\eta_\mu^j \hat{f}, j) - a(\eta_\mu^j \tilde{\hat{f}}, j)\right)$$

in Fock representation on $\mathcal{F}_{\mathrm{rad}}$, where

$$\eta^j(k) = -ik \times e^j(k), \quad j = \pm.$$

Then in Fock representation, the Pauli–Fierz Hamiltonian with spin 1/2 is realized as

$$\frac{1}{2}(-i\nabla - eA)^2 + V + H_{\mathrm{rad}} - \frac{e}{2}\sigma \cdot B \tag{3.8.11}$$

on $L^2(\mathbb{R}^3; \mathbb{C}^2) \otimes \mathcal{F}_{\mathrm{rad}}$, where the quantized magnetic field B is given by

$$B_\mu = \int_{\mathbb{R}^3}^{\oplus} B_\mu(x)dx,$$

$$B_\mu(x) = \sum_{j=\pm}\frac{1}{\sqrt{2}}\int_{\mathbb{R}^3} \eta_\mu^j(k)\left(\frac{\hat{\varphi}(k)}{\sqrt{\omega(k)}}e^{-ik\cdot x}a^*(k,j) - \frac{\hat{\varphi}(-k)}{\sqrt{\omega(k)}}e^{ik\cdot x}a(k,j)\right)dk.$$

Proposition 3.77 (Self-adjointness). *Hamiltonian $H_{\mathrm{PF}}^{\mathrm{S}}$ is self-adjoint on D_{PF} and bounded from below. Moreover, it is essentially self-adjoint on any core of $H_{\mathrm{PF},0} = H_{\mathrm{p}} \otimes 1 + 1 \otimes H_{\mathrm{rad}}$.*

Proof. Let $H_{\mathrm{PF}}^{\mathrm{S}} = H_{\mathrm{PF}} - \frac{e}{2}\sigma \cdot \hat{B}$. It is shown in Theorem 3.47 that H_{PF} is self-adjoint on D_{PF}. From the bound $\|H_{\mathrm{PF}}\Psi\|^2 + C\|\Psi\|^2 \geq C'\|H_{\mathrm{rad}}\Psi\|^2$ given in (3.5.16) and $\| - \frac{e}{2}\sigma \cdot \hat{B}\Psi\| \leq \varepsilon\|H_{\mathrm{rad}}\Psi\| + C_\varepsilon\|\Psi\|$ for arbitrary $\varepsilon > 0$ and a constant C_ε, we know that the spin interaction part $-\frac{e}{2}\sigma\cdot\hat{B}$ is infinitesimally small with respect to H_{PF}. Then the proposition follows. □

3.8.4 Symmetry and polarization

In this section, we take the Fock representation (3.8.11). When the form factor $\hat{\varphi}$ and the external potential V are rotation-invariant, i. e., $\hat{\varphi}(\mathscr{R}k) = \hat{\varphi}(k)$ and $V(\mathscr{R}x) = V(x)$ for any $\mathscr{R} \in \mathrm{O}(3)$, then $H_{\mathrm{PF}}^{\mathrm{S}}$ has the symmetry

$$\mathrm{SU}(2) \times \mathrm{SO}_{\mathrm{part}}(3) \times \mathrm{SO}_{\mathrm{field}}(3) \times \text{helicity},$$

where SU(2) and $\mathrm{SO}_{\mathrm{part}}(3)$ come from the spin and the *angular momentum* of the particle, respectively, while the $\mathrm{SO}_{\mathrm{field}}(3)$ and helicity part respectively follow from the angular momentum and the helicity of photons.

Let $\mathscr{R} \in \mathrm{SO}(3)$ and $\hat{k} = k/|k|$. Note that $\mathscr{R}\hat{k} = \widehat{\mathscr{R}k}$. Thus the orthogonal bases $\{e^+(\mathscr{R}k), e^-(\mathscr{R}k), \widehat{\mathscr{R}k}\}$ and $\{\mathscr{R}e^+(k), \mathscr{R}e^-(k), \mathscr{R}\hat{k}\}$ in $\mathbb{R}^3$ at $\mathscr{R}k$ satisfy

$$\begin{pmatrix} e^+(\mathscr{R}k) \\ e^-(\mathscr{R}k) \\ \widehat{\mathscr{R}k} \end{pmatrix} = \begin{pmatrix} \cos\vartheta 1_3 & -\sin\vartheta 1_3 & 0 \\ \sin\vartheta 1_3 & \cos\vartheta 1_3 & 0 \\ 0 & 0 & 1_3 \end{pmatrix} \begin{pmatrix} \mathscr{R}e^+(k) \\ \mathscr{R}e^-(k) \\ \mathscr{R}\hat{k} \end{pmatrix}, \tag{3.8.12}$$

where 1_3 is the 3×3 unit matrix and $\vartheta = \vartheta(\mathscr{R}, k)$ satisfies $\cos\vartheta = \mathscr{R}e^+(k) \cdot e^+(\mathscr{R}k)$. Let $\mathscr{R} = \mathscr{R}(n, \phi) \in \mathrm{SO}(3)$ be the rotation around $n \in \mathrm{S}^2 = \{k \in \mathbb{R}^3 \mid |k| = 1\}$ by the angle $\phi \in \mathbb{R}$, and $\det\mathscr{R} = 1$. Also, let

$$\ell_k = k \times (-i\nabla_k) = (\ell_{k1}, \ell_{k2}, \ell_{k3})$$

be the triplet of angular momentum operators in $L^2(\mathbb{R}_k^3)$. Then (3.8.12) can be rewritten as

$$e^{i\phi n\cdot\ell_k}\begin{pmatrix} e^+(k) \\ e^-(k) \end{pmatrix} = e^{-i\vartheta\Sigma_2}\begin{pmatrix} \mathscr{R} & 0 \\ 0 & \mathscr{R} \end{pmatrix}\begin{pmatrix} e^+(k) \\ e^-(k) \end{pmatrix},$$

where

$$\Sigma_2 = i\begin{pmatrix} 0 & -1_3 \\ 1_3 & 0 \end{pmatrix}.$$

In order to discuss the symmetry of $H_{\mathrm{PF}}^{\mathrm{S}}$, we introduce *coherent polarization vectors* in given directions. We make the following.

Definition 3.78 (Coherent polarization vectors). The polarization vectors $e^{\pm}$ are coherent polarization vectors in direction $n \in S^2$ whenever there exists $z \in \mathbb{Z}$ such that for any $\phi \in [0, 2\pi)$ and any k with $\hat{k} \neq n$,

$$\begin{pmatrix} e^+(\mathscr{R}k) \\ e^-(\mathscr{R}k) \end{pmatrix} = \begin{pmatrix} \cos(z\phi)1_3 & -\sin(z\phi)1_3 \\ \sin(z\phi)1_3 & \cos(z\phi)1_3 \end{pmatrix} \begin{pmatrix} \mathscr{R} & 0 \\ 0 & \mathscr{R} \end{pmatrix} \begin{pmatrix} e^+(k) \\ e^-(k) \end{pmatrix}, \tag{3.8.13}$$

or componentwise

$$\begin{pmatrix} e_\mu^+(\mathscr{R}k) \\ e_\mu^-(\mathscr{R}k) \end{pmatrix} = \begin{pmatrix} \cos(z\phi) & -\sin(z\phi) \\ \sin(z\phi) & \cos(z\phi) \end{pmatrix} \begin{pmatrix} (\mathscr{R}e^+(k))_\mu \\ (\mathscr{R}e^-(k))_\mu \end{pmatrix},$$

where $\mathscr{R} = \mathscr{R}(n, \phi)$.

If polarization vectors $e^{\pm}$ are coherent polarization vectors in direction $n \in S^2$, then we have

$$\exp\{i\phi\,(z\Sigma_2 + n \cdot \ell_k)\} \begin{pmatrix} e_\mu^+(k) \\ e_\mu^-(k) \end{pmatrix} = \begin{pmatrix} (\mathscr{R}e^+(k))_\mu \\ (\mathscr{R}e^-(k))_\mu \end{pmatrix}. \tag{3.8.14}$$

Here is an example of coherent polarization vectors in direction $n \in S^2$.

Example 3.79. Let $n_3 = (0, 0, 1)$ and $S_\parallel^2 = \{(\sqrt{1 - z^2}, 0, z) \in S^2 \mid -1 \leq z \leq 1\}$. Take any polarization vectors on $S_\parallel^2$: $e^{\pm}(k)$ for $k \in S_\parallel^2$. For $k \in S^2 \setminus S_\parallel^2$, there exists a unique $h_k \in S_\parallel^2$ and $0 < \phi < 2\pi$ such that $\mathscr{R}(n_3, \phi)h_k = k$. Then we define $e^{\pm}(k)$ for $k \in S^2 \setminus S_\parallel^2$ by

$$\begin{pmatrix} e^+(k) \\ e^-(k) \end{pmatrix} = \begin{pmatrix} \cos(z\phi)1_3 & -\sin(z\phi)1_3 \\ \sin(z\phi)1_3 & \cos(z\phi)1_3 \end{pmatrix} \begin{pmatrix} \mathscr{R}(n_3, \phi)e^+(h_k) \\ \mathscr{R}(n_3, \phi)e^-(h_k) \end{pmatrix}. \tag{3.8.15}$$

It is readily checked that $e^{\pm}$ satisfies (3.8.13) with $(n_3, z) \in S^2 \times \mathbb{Z}$.

Example 3.80. Let $n \in S^2$, and

$$e^+(k) = \hat{k} \times n / \sin\vartheta, \quad e^-(k) = \hat{k} \times e^+(k),$$

where $\vartheta = \cos^{-1}(\hat{k}n)$. Then, since $\mathscr{R} = \mathscr{R}(n, \phi)$ satisfies $\mathscr{R}n = n$ and $\mathscr{R}u \times \mathscr{R}v = \mathscr{R}(u \times v)$, $e^{\pm}(k)$ obeys (3.8.13) with $(n, 0) \in S^2 \times \mathbb{Z}$, i. e., $\mathscr{R}e^{\pm}(k) = e^{\pm}(\mathscr{R}k)$. In particular, when $(n, z) = (n_3, 0) \in S^2 \times \mathbb{Z}$, $\sin\vartheta = \sqrt{k_1^2 + k_2^2}$ and

$$e^+(k) = \frac{(-k_2, k_1, 0)}{\sqrt{k_1^2 + k_2^2}}, \quad e^-(k) = \frac{(k_3k_1, -k_2k_3, k_1^2 + k_2^2)}{|k|\sqrt{k_1^2 + k_2^2}}.$$

Let $\mathfrak{H}_2 : L^2(\mathbb{R}^3) \oplus L^2(\mathbb{R}^3) \to L^2(\mathbb{R}^3) \oplus L^2(\mathbb{R}^3)$ be given by

$$\mathfrak{H}_2 = i \begin{pmatrix} 0 & -1_{L^2(\mathbb{R}^3)} \\ 1_{L^2(\mathbb{R}^3)} & 0 \end{pmatrix}. \tag{3.8.16}$$

Suppose that polarization vectors are coherent polarization vectors in direction $n \in S^2$ and $z \in \mathbb{Z}$. We define

$$S_f = d\Gamma(z\mathfrak{H}_2)$$

and

$$L_f = (L_{f,1}, L_{f,2}, L_{f,3}) = d\Gamma(\ell_k). \tag{3.8.17}$$

Here S_f is called the *helicity* and L_f the *angular momentum of the field*. The helicity S_f can be formally written as

$$S_f = i \int_{\mathbb{R}^3} z(a^*(k,-)a(k,+) - a^*(k,+)a(k,-))dk. \tag{3.8.18}$$

For $(n, z) \in S^2 \times \mathbb{Z}$, define $J_f = J_f(n, z)$ by

$$J_f = n \cdot L_f + S_f$$

and $J_p = J_p(n)$ by

$$J_p = n \cdot \ell_x + \frac{1}{2} n \cdot \sigma. \tag{3.8.19}$$

This J_p is the angular momentum plus spin for the particle. Write

$$\mathrm{J} = J_p \otimes 1 + 1 \otimes J_f.$$

Clearly, $\mathrm{J} = \mathrm{J}(n, z)$ is defined for each $(n, z) \in S^2 \times \mathbb{Z}$.

Proposition 3.81 (J-invariance of H_{PF}^S). *If the polarization vectors are coherent in direction n, and $\hat{\varphi}$ and V are rotation-invariant, then*

$$e^{i\phi \mathrm{J}} H_{PF}^S e^{-i\phi \mathrm{J}} = H_{PF}^S$$

for every $\phi \in \mathbb{R}$.

Proof. Write $a^\sharp(\binom{f}{g})$ for $a^\sharp(f \oplus g)$. Notice that for a rotation-invariant f,

$$e^{i\phi J_f} a^* \left(f e^{-ik\cdot x} \begin{pmatrix} e_\mu^+ \\ e_\mu^- \end{pmatrix} \right) e^{-i\phi J_f} = a^* \left(f e^{i\phi(z\Sigma_2 + n\ell_k)} e^{-ik\cdot x} \begin{pmatrix} e_\mu^+ \\ e_\mu^- \end{pmatrix} \right).$$

Since the polarization vectors are coherent, we have by (3.8.14),

$$= a^* \left(f e^{-i\mathscr{R}k\cdot x} \begin{pmatrix} (\mathscr{R}e^+)_\mu \\ (\mathscr{R}e^-)_\mu \end{pmatrix} \right) = \sum_{\nu=1}^{3} \mathscr{R}_{\mu\nu} a^* \left(f e^{-ik\mathscr{R}^{-1}x} \begin{pmatrix} e_\nu^+ \\ e_\nu^- \end{pmatrix} \right), \tag{3.8.20}$$

where $\mathscr{R} = \mathscr{R}(n, \phi) = (\mathscr{R}_{\mu\nu})_{1 \le \mu,\nu \le 3}$. By (3.8.20), we see that the field part transforms as

$$e^{i\phi J_f} H_{\text{rad}} e^{-i\phi J_f} = H_{\text{rad}},$$
$$e^{i\phi J_f} A_\mu(x) e^{-i\phi J_f} = (\mathscr{R}A)_\mu(\mathscr{R}^{-1}x).$$

For $x \in \mathbb{R}^3$, define $x \cdot \sigma = \sum_{\mu=1}^3 x_\mu \sigma_\mu$. Since $\sigma_\mu \sigma_\nu = \delta_{\mu\nu} I + i \sum_{\lambda=1}^3 \varepsilon_{\mu\nu\lambda}\sigma_\lambda$, we obtain the identity

$$(x \cdot \sigma)(y \cdot \sigma) = (x \cdot y)I + i(x \times y) \cdot \sigma.$$

In particular, $(n \cdot \sigma)^2 = I$ for $|n| = 1$. Using $(n \cdot \sigma)^2 = I$, we obtain

$$U = \cos\frac{\theta}{2} I + i \sin\frac{\theta}{2} (n \cdot \sigma).$$

Introduce $c = \cos\frac{\theta}{2}$, $s = \sin\frac{\theta}{2}$, $N = n\cdot\sigma$, and $A = x\cdot\sigma$. Then $U = cI + isN$ and $U^{-1} = cI - isN$. A direct computation gives

$$UAU^{-1} = c^2 A + s^2 NAN + isc(NA - AN).$$

We also see that $NA - AN = 2i(n \times x) \cdot \sigma$ and $NAN = (-x + 2n(n \cdot x)) \cdot \sigma$. We obtain

$$U(x \cdot \sigma)U^{-1} = \Big(x \cos\theta - (n \times x)\sin\theta + n(n \cdot x)(1 - \cos\theta)\Big) \cdot \sigma.$$

The vector in parentheses is exactly the Rodrigues formula for a rotation by angle $-\theta$ about axis $n \in \mathbb{R}^3$. For any $x \in \mathbb{R}^3$,

$$U(x \cdot \sigma)U^{-1} = (R(n, -\theta)x) \cdot \sigma,$$

where $R(n, -\theta)$ is the rotation in $\mathbb{R}^3$ about axis n by angle $-\theta$. Hence we have

$$U\sigma_\mu U^{-1} = \sum_{\nu=1}^3 R_{\mu\nu}\sigma_\nu,$$

where $R = R(n, -\theta) \in SO(3)$. Equivalently, if $\sigma = (\sigma_1, \sigma_2, \sigma_3)$ is regarded as a row vector, $U\sigma U^{-1} = \sigma R^{-1}$. Then the particle part transforms as

$$e^{i\phi n\cdot \ell_x} x_\mu e^{-i\phi n \cdot \ell_x} = \sum_{\nu=1}^3 \mathscr{R}_{\mu\nu} x_\nu,$$
$$e^{i\phi n\cdot \ell_x} (-i\nabla_x)_\mu e^{-i\phi n \cdot \ell_x} = \sum_{\nu=1}^3 \mathscr{R}_{\mu\nu} (-i\nabla_x)_\nu,$$
$$e^{i\phi n\cdot (1/2)\sigma} \sigma_\mu e^{-i\phi n \cdot (1/2)\sigma} = \sum_{\nu=1}^3 \mathscr{R}_{\mu\nu} \sigma_\nu.$$

Together with all the identities above, we have

$$e^{i\phi J} H_{\text{PF}}^{\text{S}} e^{-i\phi J} = \frac{1}{2}\left(\mathscr{R}\sigma \cdot \mathscr{R}\left(-i\nabla - e\hat{A}(x)\right)\right)^2 + H_{\text{rad}} + V(\mathscr{R}x) = H_{\text{PF}}^{\text{S}}. \qquad \square$$

From this proposition, it is clear that $H_{\mathrm{PF}}^{\mathrm{S}}$ with coherent polarization vectors has the symmetry

$$\mathrm{SU}(2) \times \mathrm{SO}_{\mathrm{part}}(3) \times \mathrm{SO}_{\mathrm{field}}(3) \times \text{helicity}. \tag{3.8.21}$$

Denote the set of half integers by $\mathbb{Z}_{1/2} = \{w/2 \mid w \in \mathbb{Z}\}$. For each $(n, z) \in \mathrm{S}^2 \times \mathbb{Z}$, notice that

$$\mathrm{Spec}(n \cdot (\ell_x + (1/2)\sigma)) = \mathbb{Z}_{1/2}, \tag{3.8.22}$$

$$\mathrm{Spec}(n \cdot L_{\mathrm{f}}) = \mathbb{Z}, \tag{3.8.23}$$

$$\mathrm{Spec}(S_{\mathrm{f}}) = \begin{cases} \mathbb{Z}, & z \neq 0, \\ 0, & z = 0. \end{cases} \tag{3.8.24}$$

Thus for each $(n, z) \in \mathrm{S}^2 \times \mathbb{Z}$,

$$\mathrm{Spec}(\mathrm{J}) = \mathbb{Z}_{1/2} \tag{3.8.25}$$

and we have the theorem below.

Theorem 3.82. *If the polarization vectors are coherent in direction n, and $\hat{\varphi}$ and V are rotation-invariant, then $\mathcal{H}_{\mathrm{S}}$ and $H_{\mathrm{PF}}^{\mathrm{S}}$ can be decomposed as*

$$\mathcal{H}_{\mathrm{S}} = \bigoplus_{w \in \mathbb{Z}_{1/2}} \mathcal{H}_{\mathrm{S}}(w), \quad H_{\mathrm{PF}}^{\mathrm{S}} = \bigoplus_{w \in \mathbb{Z}_{1/2}} H_{\mathrm{PF}}^{\mathrm{S}}(w). \tag{3.8.26}$$

Here $\mathcal{H}_{\mathrm{S}}(w)$ is the subspace spanned by eigenvectors of J associated with eigenvalue $w \in \mathbb{Z}_{1/2}$ and $H_{\mathrm{PF}}^{\mathrm{S}}(w) = H_{\mathrm{PF}}^{\mathrm{S}} \lceil_{\mathcal{H}_{\mathrm{S}}(w)}$.

Proof. This follows from Proposition 3.81 and the fact that $\mathrm{Spec}(\mathrm{J}) = \mathbb{Z}_{1/2}$. □

Next, we consider general polarization vectors, i. e., not necessarily coherent sets. The Pauli–Fierz Hamiltonians with different polarization vectors, however, are unitary equivalent. Denote the Pauli–Fierz Hamiltonian with polarization vectors $e^{\pm}$ by $H_{\mathrm{PF}}^{\mathrm{S}}(e^{\pm})$. Combining Proposition 3.6 and Theorem 3.82, we have the corollary below.

Corollary 3.83 (Fiber decomposition of Pauli–Fierz Hamiltonian with spin; general polarization). *Suppose that $\hat{\varphi}$ and V are rotation-invariant, and $e^{\pm}$ are coherent polarization vectors. Let $\eta^{\pm}$ be arbitrary polarization vectors. Then $H_{\mathrm{PF}}^{\mathrm{S}}(\eta^{\pm})$ is unitary equivalent to $\oplus_{w \in \mathbb{Z}_{1/2}} H_{\mathrm{PF}}^{\mathrm{S}}(e^{\pm}, w)$.*

By using the symmetries of the Pauli–Fierz Hamiltonian, we can show the degeneracy of ground states of $H_{\mathrm{PF}}^{\mathrm{S}}$. Assume that V is rotation-invariant and the polarization vectors $e^{\pm}$ are given by

$$e^{+}(k) = \frac{(-k_2, k_1, 0)}{\sqrt{k_1^2 + k_2^2}}, \quad e^{-}(k) = \hat{k} \times e^{+}(k). \tag{3.8.27}$$

These are coherent in direction n_3 and their helicity is zero. Let $\Lambda : \mathbb{R}^3 \to \mathbb{R}^3$ be the flip defined by

$$\Lambda\begin{pmatrix} k_1 \\ k_2 \\ k_3 \end{pmatrix} = \begin{pmatrix} k_1 \\ -k_2 \\ k_3 \end{pmatrix}.$$

Consider the unitary operators $L^2(\mathbb{R}^3; \mathbb{C}^2) \to L^2(\mathbb{R}^3; \mathbb{C}^2)$ given by

$$\tilde{u} : \begin{pmatrix} f \\ g \end{pmatrix} \mapsto \begin{pmatrix} f \circ \Lambda \\ g \circ \Lambda \end{pmatrix}, \quad u : \begin{pmatrix} f \\ g \end{pmatrix} \mapsto \begin{pmatrix} -f \circ \Lambda \\ g \circ \Lambda \end{pmatrix}. \tag{3.8.28}$$

A computation gives

$$u^{\sharp^{-1}} k_\mu u^\sharp = \begin{cases} k_\mu, & \mu = 1,3, \\ -k_\mu, & \mu = 2, \end{cases} \qquad u^{\sharp^{-1}} \nabla_\mu u^\sharp = \begin{cases} \nabla_\mu, & \mu = 1,3, \\ -\nabla_\mu, & \mu = 2, \end{cases}$$

where $u^\sharp = u$ or $\tilde{u}$. From this and

$$e^-(k) = \frac{(-k_3k_1, -k_2k_3, k_1^2 + k_2^2)}{|k|\sqrt{k_1^2 + k_2^2}},$$

we have for rotation-invariant f and g,

$$u^{-1}\begin{pmatrix} e_\mu^+ f \\ e_\mu^- g \end{pmatrix} = \begin{cases} \begin{pmatrix} e_\mu^+ f \\ e_\mu^- g \end{pmatrix}, & \mu = 1,3, \\ -\begin{pmatrix} e_\mu^+ f \\ e_\mu^- g \end{pmatrix}, & \mu = 2. \end{cases}$$

Then the second quantization $\Gamma(u)$ of u induces the unitary operator on $\mathcal{F}_{\mathrm{rad}}$ and, for rotation-invariant $\hat{\varphi}$, we obtain

$$\Gamma(u)^{-1} A_\mu(x) \Gamma(u) = \begin{cases} A_\mu(\Lambda x), & \mu = 1,3, \\ -A_\mu(\Lambda x), & \mu = 2, \end{cases} \tag{3.8.29}$$

and

$$\Gamma(u)^{-1} H_{\mathrm{rad}} \Gamma(u) = H_{\mathrm{rad}}. \tag{3.8.30}$$

Next, we consider the transformation on $\sigma = (\sigma_1, \sigma_2, \sigma_3)$ given by

$$\sigma_\mu \mapsto \sigma_2 \sigma_\mu \sigma_2 = \begin{cases} -\sigma_\mu, & \mu = 1,3, \\ \sigma_\mu, & \mu = 2. \end{cases} \tag{3.8.31}$$

Under the identification $L^2(\mathbb{R}^3; \mathbb{C}^2) \cong \mathbb{C}^2 \otimes L^2(\mathbb{R}^3)$, we define $\tau = \sigma_2 \otimes \tilde{u} : L^2(\mathbb{R}^3; \mathbb{C}^2) \to L^2(\mathbb{R}^3; \mathbb{C}^2)$. This satisfies

$$\tau^{-1}(\sigma_\mu \otimes f)\tau = \begin{cases} -\sigma_\mu \otimes f \circ \Lambda^{-1}, & \mu = 1, 3, \\ \sigma_\mu \otimes f \circ \Lambda^{-1}, & \mu = 2, \end{cases} \tag{3.8.32}$$

and

$$\tau^{-1}(\sigma_\mu \otimes \nabla_\mu)\tau = -\sigma_\mu \otimes \nabla_\mu, \quad \mu = 1, 2, 3. \tag{3.8.33}$$

We finally define the unitary operator $\mathfrak{J} : \mathcal{H}_S \to \mathcal{H}_S$ by

$$\mathfrak{J} = \tau \otimes \Gamma(u).$$

Combining (3.8.29)–(3.8.33), for rotation-invariant $\hat{\varphi}$ and V

$$\begin{aligned} &\mathfrak{J}^{-1}\sigma_\mu(-i\nabla_\mu - eA_\mu)\mathfrak{J} = -\sigma_\mu(-i\nabla_\mu - eA_\mu), \\ &\mathfrak{J}^{-1}H_{\mathrm{rad}}\mathfrak{J} = H_{\mathrm{rad}}, \\ &\mathfrak{J}^{-1}V\mathfrak{J} = V \end{aligned}$$

are obtained. From these relations, we can show the theorem below.

Theorem 3.84 (Reflection symmetry of $H_{\mathrm{PF}}^{\mathrm{S}}$). *If $\hat{\varphi}$ and V are rotation-invariant, and the polarization vectors $e^\pm$ are given by (3.8.27), then $H_{\mathrm{PF}}^{\mathrm{S}}(z)$ and $H_{\mathrm{PF}}^{\mathrm{S}}(-z)$ are unitary equivalent.*

Proof. Since $e^\pm$ is coherent in direction $(0, 0, 1)$ and its helicity is zero, J is of the form $\mathrm{J} = (\ell_{x,3} + \frac{1}{2}\sigma_3) \otimes 1 + 1 \otimes L_{\mathrm{f},3}$. It follows that

$$\mathfrak{J}^{-1}\mathrm{J}\mathfrak{J} = -\mathrm{J}.$$

This implies that $\mathfrak{J}$ maps $\mathcal{H}_S(w)$ onto $\mathcal{H}_S(-w)$. Furthermore,

$$\mathfrak{J}^{-1}H_{\mathrm{PF}}^{\mathrm{S}}\mathfrak{J} = \frac{1}{2}(-\sigma \cdot (-i\nabla - eA))^2 + V + H_{\mathrm{rad}} = H_{\mathrm{PF}}^{\mathrm{S}}.$$

Thus $\mathfrak{J}^{-1}H_{\mathrm{PF}}^{\mathrm{S}}(w)\mathfrak{J} = H_{\mathrm{PF}}^{\mathrm{S}}(-w)$ follows. □

An interesting application of Theorem 3.84 is to estimate the multiplicity of bound states of $H_{\mathrm{PF}}^{\mathrm{S}}$.

Corollary 3.85 (Degeneracy of bound states). *Suppose that V and $\hat{\varphi}$ are rotation-invariant. Let M be the multiplicity of eigenvalues of $H_{\mathrm{PF}}^{\mathrm{S}}$. Then M is an even number. In particular, whenever a ground states exists, it is degenerate.*

Proof. By the unitary equivalence, we may suppose that the polarization vectors of $H_{\mathrm{PF}}^{\mathrm{S}}$ are given by (3.8.27). Thus $H_{\mathrm{PF}}^{\mathrm{S}} = \bigoplus_{w \in \mathbb{Z}_{1/2}} H_{\mathrm{PF}}^{\mathrm{S}}(w)$. Let Ψ be a bound state of $H_{\mathrm{PF}}^{\mathrm{S}}$ with eigenvalue a. Thus Ψ is a bound state of $H_{\mathrm{PF}}^{\mathrm{S}}(w)$ with some $w \in \mathbb{Z}_{1/2}$, $H_{\mathrm{PF}}^{\mathrm{S}}(w)\Psi = a\Psi$. Theorem 3.84 implies $H_{\mathrm{PF}}^{\mathrm{S}}(w) \cong H_{\mathrm{PF}}^{\mathrm{S}}(-w)$, and thus there exists $\Phi \in \mathcal{H}_S(-w)$ such that $H_{\mathrm{PF}}^{\mathrm{S}}(-w)\Phi = a\Phi$. Thus the multiplicity of a is even. □

3.8.5 Scalar representations

As in the classical case (3.8.4), in order to construct the functional integral representation of $(F, e^{-tH_{\mathrm{PF}}^{\mathrm{S}}}G)$ with a scalar integrand, we introduce a two-valued spin variable $\theta \in \mathbb{Z}_2 = \{-1, +1\} = \{\theta_1, \theta_2\}$ and redefine $H_{\mathrm{PF}}^{\mathrm{S}}$ in order to reduce it to a scalar operator. Since

$$H_{\mathrm{PF}}^{\mathrm{S}} = \frac{1}{2}(-i\nabla - e\hat{A})^2 + V + H_{\mathrm{rad}} - \frac{e}{2}\begin{pmatrix} \hat{B}_3 & \hat{B}_1 - i\hat{B}_2 \\ \hat{B}_1 + i\hat{B}_2 & -\hat{B}_3 \end{pmatrix},$$

our Hamiltonian can be regarded as an operator acting on $L^2(\mathbb{R}^3 \times \mathbb{Z}_2) \otimes L^2(\mathscr{Q})$ and is defined by

$$(H_{\mathrm{PF}}^{\mathbb{Z}_2} F)(\theta) = \left(\frac{1}{2}(-i\nabla - e\hat{A})^2 + V + H_{\mathrm{rad}} + \hat{H}_{\mathrm{d}}(\theta)\right) F(\theta) + \hat{H}_{\mathrm{od}}(-\theta) F(-\theta)$$

with $\theta \in \mathbb{Z}_2$ for $F = \binom{F(+1)}{F(-1)} \in L^2(\mathbb{R}^3 \times \mathbb{Z}_2) \otimes L^2(\mathscr{Q})$, where each component is $F(\theta) \in L^2(\mathbb{R}^3) \times L^2(\mathscr{Q})$. Here, $\hat{H}_{\mathrm{d}}$ and $\hat{H}_{\mathrm{od}}$ denote the diagonal, respectively, off-diagonal parts of the spin interaction explicitly given by

$$\hat{H}_{\mathrm{d}}(\theta) = \int_{\mathbb{R}^3}^{\oplus} \hat{H}_{\mathrm{d}}(x, \theta) dx, \quad \hat{H}_{\mathrm{od}}(-\theta) = \int_{\mathbb{R}^3}^{\oplus} \hat{H}_{\mathrm{od}}(x, -\theta) dx,$$

where

$$\hat{H}_{\mathrm{d}}(x, \theta) = -\frac{e}{2}\theta \hat{B}_3(x), \quad \hat{H}_{\mathrm{od}}(x, -\theta) = -\frac{e}{2}\left(\hat{B}_1(x) - i\theta \hat{B}_2(x)\right).$$

Here, $\hat{B}_\mu(x)$ is defined in (3.8.8).

Now we define the Pauli–Fierz Hamiltonian with spin 1/2 on $L^2(\mathbb{R}^3 \times \mathbb{Z}_2) \otimes L^2(\mathscr{Q})$ instead of $\mathcal{H}_{\mathrm{S}}$. Furthermore, we treat $\hat{A}$ and $\hat{B}$ or $\hat{H}_{\mathrm{d}}$ and $\hat{H}_{\mathrm{od}}$ as not necessarily dependent vectors. This is the same idea as the one applied to Schrödinger operator $H_{\mathbb{Z}_2}(a, b)$. Write

$$\mathscr{K} = L^2(\mathbb{R}^3 \times \mathbb{Z}_2) \otimes L^2(\mathscr{Q})$$

and its Euclidean version

$$\mathscr{K}_{\mathrm{E}} = L^2(\mathbb{R}^3 \times \mathbb{Z}_2) \otimes L^2(\mathscr{Q}_{\mathrm{E}}).$$

Definition 3.86 (Pauli–Fierz Hamiltonian with spin 1/2 on $\mathscr{K}$). We define the *Pauli–Fierz Hamiltonian with spin* 1/2 on $\mathscr{K}$ by $H_{\mathrm{PF}}^{\mathbb{Z}_2}$.

The key idea of constructing a functional integral representation of $e^{-tH_{\mathrm{PF}}^{\mathbb{Z}_2}}$ is to use the identity

$$\mathscr{K} \cong \int_{\mathscr{Q}}^{\oplus} L^2(\mathbb{R}^3 \times \mathbb{Z}_2) d\mu. \tag{3.8.34}$$

In other words, we regard $\mathscr{K}$ as the set of $L^2(\mathbb{R}^3 \times \mathbb{Z}_2)$-valued L^2-functions on $\mathscr{Q}$. We make the decomposition

$$H_{\mathrm{PF}}^{\mathbb{Z}_2} = \int_{\mathscr{Q}}^{\oplus} K(\phi) d\mu(\phi) + H_{\mathrm{rad}},$$

where $K(\phi)$ is a self-adjoint operator on $L^2(\mathbb{R}^3 \times \mathbb{Z}_2)$ for each $\phi \in \mathscr{Q}$. We construct a functional integral representation of $e^{-tH_{\mathrm{PF}}^{\mathbb{Z}_2}}$ through the functional integrals of $e^{-tK(\phi)}$ and $e^{-tH_{\mathrm{rad}}}$, and the Trotter product formula. In order to do that, we will use the identity

$$(F, e^{-t\int_{\mathscr{Q}}^{\oplus} K(\phi)d\mu} G) = \int_{\mathscr{Q}} \big(F(\phi), e^{-tK(\phi)} G(\phi)\big)\, d\mu(\phi)$$

while we have already done this of $(F(\phi), e^{-tK(\phi)} G(\phi))$, $\phi \in \mathscr{Q}$, in the classical case $(f, e^{-tH_{\mathbb{Z}_2}(a)} g)$ in (3.8.6).

First, we make the fiber decomposition of the quantized radiation field $\hat{A}$ and the quantized magnetic field $\hat{B}$ on $\mathscr{Q}$. For each $\phi \in \mathscr{Q}$, define

$$\hat{A}_\mu(\phi) = \int_{\mathbb{R}^3}^{\oplus} \hat{A}_\mu(x,\phi) dx, \quad \hat{B}_\mu(\phi) = \int_{\mathbb{R}^3}^{\oplus} \hat{B}_\mu(x,\phi) dx,$$

where

$$\hat{A}_\mu(x,\phi) = \hat{A}_\mu(\tilde{\varphi}(\cdot - x), \phi), \quad \hat{B}_\mu(x,\phi) = (\nabla_x \times \hat{A}(\tilde{\varphi}(\cdot - x), \phi))_\mu.$$

Recall that $\hat{A}_\mu(\tilde{\varphi}(\cdot - x), \phi) = \langle\!\langle \phi, \bigoplus_{\nu=1}^3 \delta_{\mu\nu} \tilde{\varphi}(\cdot - x)\rangle\!\rangle$. For each fiber $\phi \in \mathscr{Q}$, define the Hamiltonian $K(\phi)$ on $L^2(\mathbb{R}^3 \times \mathbb{Z}_2)$ by

$$(K(\phi)F)(x,\theta) = \Big(\frac{1}{2}(-i\nabla - e\hat{A}(\phi))^2 + V + \hat{H}_{\mathrm{d}}(\phi)\Big) F(x,\theta) + \hat{H}_{\mathrm{od}}(\phi) F(x,-\theta). \tag{3.8.35}$$

Here,

$$\hat{H}_{\mathrm{d}}(\phi) = \int_{\mathbb{R}^3}^{\oplus} \hat{H}_{\mathrm{d}}(x,\theta,\phi) dx, \quad \hat{H}_{\mathrm{od}}(\phi) = \int_{\mathbb{R}^3}^{\oplus} \hat{H}_{\mathrm{od}}(x,-\theta,\phi) dx,$$

where

$$\hat{H}_{\mathrm{d}}(x,\theta,\phi) = -\frac{e}{2}\theta \hat{B}_3(x,\phi), \quad \hat{H}_{\mathrm{od}}(x,-\theta,\phi) = -\frac{e}{2}(\hat{B}_1(x,\phi) - i\theta \hat{B}_2(x,\phi)).$$

To prevent the off-diagonal part $\hat{H}_{\mathrm{od}}$ vanish, we introduce a regularization $H_{\mathrm{PF},\varepsilon}^{\mathbb{Z}_2}$ of $H_{\mathrm{PF}}^{\mathbb{Z}_2}$ by

$$H^{\mathbb{Z}_2}_{\mathrm{PF},\varepsilon}F(\theta) = \left(\frac{1}{2}(-i\nabla - e\hat{A})^2 + V + H_{\mathrm{rad}} + \hat{H}_{\mathrm{d}}(\theta)\right)F(\theta) + \Psi_\varepsilon(\hat{H}_{\mathrm{od}}(-\theta))F(-\theta),$$

where $\Psi_\varepsilon(X) = X + \varepsilon\psi_\varepsilon(|X|)$ and $\psi_\varepsilon \in C_0^\infty(\mathbb{R})$ is given by

$$\psi_\varepsilon(z) = \psi_\varepsilon(|z|) = \begin{cases} 1, & |z| < \varepsilon/2, \\ \leq 1, & \varepsilon/2 \leq |z| \leq \varepsilon, \\ 0, & |z| > \varepsilon. \end{cases}$$

Also, let $K_\varepsilon(\phi)$ be the counterpart of $K(\phi)$ with $\hat{H}_{\mathrm{od}}(\phi)$ replaced by $\Psi_\varepsilon(\hat{H}_{\mathrm{od}}(\phi))$, i. e.,

$$(K_\varepsilon(\phi)F)(x,\theta) = \left(\frac{1}{2}(-i\nabla - e\hat{A}(\phi))^2 + V + \hat{H}_{\mathrm{d}}(\phi)\right)F(x,\theta) + \Psi_\varepsilon(\hat{H}_{\mathrm{od}}(\phi))F(x,-\theta).$$

Recall that $\theta_z = (-1)^z$ for $z \in \mathbb{Z}$ and $(q_t)_{t\geq 0} = (B_t, \theta_{N_t})_{t\geq 0}$ is an $(\mathbb{R}^3 \times \mathbb{Z}_2)$-valued random process on $\mathscr{X} \times \mathcal{S}$.

Lemma 3.87. *Suppose that $V \in \mathcal{R}_{\mathrm{Kato}}$. If $\tilde{\varphi} \in C_0^\infty(\mathbb{R}^3)$, then for every $\phi \in \mathscr{Q}$, $K_\varepsilon(\phi)$ is self-adjoint on $D(-\Delta) \otimes \mathbb{Z}_2$ and for $g \in L^2(\mathbb{R}^3 \times \mathbb{Z}_2)$,*

$$(e^{-tK_\varepsilon(\phi)}g)(x,\theta_\alpha) = e^t\mathbb{E}^{x,\alpha}[e^{-\int_0^t V(B_s)ds}e^{Z_t(\phi,\varepsilon)}g(q_t)],$$

where

$$Z_t(\phi,\varepsilon) = -ie\int_0^t \hat{A}(\tilde{\varphi}(\cdot - B_s),\phi)dB_s - \int_0^t \hat{H}_{\mathrm{d}}(B_s,\theta_{N_s},\phi)ds \int_0^{t+} \log\left(-\Psi_\varepsilon(\hat{H}_{\mathrm{od}}(B_s,-\theta_{N_{s-}},\phi))\right)dN_s.$$

Proof. As $\tilde{\varphi} \in C_0^\infty(\mathbb{R}^3)$, we have $\hat{A}_\mu(\phi) = \langle\langle \phi, \bigoplus_{\nu=1}^3 \delta_{\mu\nu}\tilde{\varphi}(\cdot - x)\rangle\rangle \in C_{\mathrm{b}}^\infty(\mathbb{R}^3_x)$, for every $\phi \in \mathscr{Q}$. Thus $K_\varepsilon(\phi)$ is the Pauli operator with spin 1/2 with a smooth bounded vector potential $\hat{A}(\phi)$, and the off-diagonal part is perturbed by the bounded operator $\varepsilon\psi_\varepsilon(\hat{H}_{\mathrm{od}}(\phi))$. Hence it is self-adjoint on $D(-\Delta) \otimes \mathbb{Z}_2$ and its functional integral representation follows from the Feynman–Kac formula for the Schrödinger operator with vector potentials and spin, namely $(f, e^{-tH_{\mathbb{Z}_2}(a)}g) = e^t\sum_{\alpha=1,2}\int_{\mathbb{R}^3}\mathbb{E}^{x,\alpha}[\overline{f(q_0)}g(q_t)e^{Z_t}]dx$ in (3.8.6). □

Next, we define the operator $K_\varepsilon(\hat{A})$ on $\mathscr{H}$ through $K_\varepsilon(\phi)$ and the constant fiber direct integral representation (3.8.34) of $\mathscr{H}$. Take $\tilde{\varphi} \in C_0^\infty(\mathbb{R}^3)$ and define the self-adjoint operator $K_\varepsilon(\hat{A})$ on $\mathscr{H}$ by

$$K_\varepsilon(\hat{A}) = \int_{\mathscr{Q}}^{\oplus} K_\varepsilon(\phi)d\mu(\phi),$$

that is, $(K_\varepsilon(\hat{A})F)(\phi) = K_\varepsilon(\phi)F(\phi)$ with domain

$$D(K_\varepsilon(\hat{A})) = \left\{ F \in \mathscr{K} \;\middle|\; \int_{\mathscr{Q}} \|K_\varepsilon(\phi)F(\phi)\|^2_{L^2(\mathbb{R}^3\times\mathbb{Z}_2)} d\mu(\phi) < \infty \right\}.$$

Then we are able to define the self-adjoint operator K_ε by

$$K_\varepsilon = K_\varepsilon(\hat{A}) \;\dot{+}\; H_{\mathrm{rad}}.$$

In what follows, we construct the functional integral representation of e^{-tK_ε} and show that $e^{-tK_\varepsilon} = e^{-tH^{\mathbb{Z}_2}_{\mathrm{PF},\varepsilon}}$. Let us define the finite particle subspace of $L^2(\mathscr{Q}_\beta)$ by

$$L^2_{\mathrm{fin}}(\mathscr{Q}_\beta) = \left\{ \bigoplus_{n=0}^{\infty} L^2_n(\mathscr{Q}_\beta) \ni \{\Phi^{(n)}\}_{n=0}^{\infty} \mid \exists m \text{ such that } \Phi^{(k)} = 0 \text{ for any } k \geq m \right\},$$

where $\mathscr{Q}_\beta = \mathscr{Q}$ or $\mathscr{Q}_{\mathrm{E}}$, and define the dense subspace by

$$\mathscr{K}^\infty = C_0^\infty(\mathbb{R}^3 \times \mathbb{Z}_2) \,\hat{\otimes}\, L^2_{\mathrm{fin}}(\mathscr{Q})$$

and its Euclidean version by

$$\mathscr{K}^\infty_{\mathrm{E}} = C_0^\infty(\mathbb{R}^3 \times \mathbb{Z}_2) \,\hat{\otimes}\, L^2_{\mathrm{fin}}(\mathscr{Q}_{\mathrm{E}}).$$

Lemma 3.88. *It follows that*

$$\operatorname*{s-lim}_{\varepsilon\to 0} e^{-tH^{\mathbb{Z}_2}_{\mathrm{PF},\varepsilon}} = e^{-tH^{\mathbb{Z}_2}_{\mathrm{PF}}}. \tag{3.8.36}$$

Suppose that $\tilde{\varphi} \in C_0^\infty(\mathbb{R}^3)$. Then $H^{\mathbb{Z}_2}_{\mathrm{PF},\varepsilon} = K_\varepsilon$ and in particular it follows that

$$(F, e^{-tH^{\mathbb{Z}_2}_{\mathrm{PF}}} G) = \lim_{\varepsilon\downarrow 0} (F, e^{-tK_\varepsilon} G). \tag{3.8.37}$$

Proof. It is seen that $K_\varepsilon = H^{\mathbb{Z}_2}_{\mathrm{PF},\varepsilon}$ on $\mathscr{K}^\infty$, implying that $K_\varepsilon = H^{\mathbb{Z}_2}_{\mathrm{PF},\varepsilon}$ as a self-adjoint operator since $\mathscr{K}^\infty$ is a core of $H^{\mathbb{Z}_2}_{\mathrm{PF},\varepsilon}$. Moreover, $H^{\mathbb{Z}_2}_{\mathrm{PF},\varepsilon} \to H^{\mathbb{Z}_2}_{\mathrm{PF}}$ on $\mathscr{K}^\infty$ as $\varepsilon \to 0$ and $\mathscr{K}^\infty$ is a common core of the sequence $(H^{\mathbb{Z}_2}_{\mathrm{PF},\varepsilon})_{\varepsilon\geq 0}$. Thus (3.8.36) follows. □

By this lemma, to have a functional integral representation of $(F, e^{-tH^{\mathbb{Z}_2}_{\mathrm{PF}}} G)$ for any $\tilde{\varphi}$ defined in Assumption 3.4, it suffices to construct a functional integral representation of the right-hand side of (3.8.37) and to take approximation arguments on $\hat{\varphi}$. To obtain a functional integral representation of e^{-tK_ε}, we apply the Trotter product formula, i. e.,

$$e^{-tK_\varepsilon} = \operatorname*{s-lim}_{n\to\infty} (e^{-(t/n)K_\varepsilon(\hat{A})} e^{-(t/n)H_{\mathrm{rad}}})^n.$$

By the property of $J_s J_t^* = e^{-|s-t|H_{\mathrm{rad}}}$ for $s, t \in \mathbb{R}$, it is reduced to

$$e^{-tK_\varepsilon} = \operatorname*{s-lim}_{n\to\infty} \mathrm{J}_0^* \Big(\prod_{i=0}^{n-1} \mathrm{J}_{ti/n} e^{-(t/n)K_\varepsilon(\hat{A})} \mathrm{J}_{ti/n}^* \Big) \mathrm{J}_t.$$

From this, we can construct a functional integral representation of e^{-tK_ε}. Define the Euclidean version of $\hat{H}_\mathrm{d}(x,\theta)$ and $\hat{H}_\mathrm{od}(x,-\theta)$ by

$$\hat{H}_{\mathrm{E,d}}(x,\theta,s) = -\frac{e}{2}\theta \hat{B}_{\mathrm{E}_3}(\mathrm{j}_s\tilde{\varphi}(\cdot - x)), \tag{3.8.38}$$

$$\hat{H}_{\mathrm{E,od}}(x,-\theta,s) = -\frac{e}{2}(\hat{B}_{\mathrm{E}_1}(\mathrm{j}_s\tilde{\varphi}(\cdot - x)) - i\theta \hat{B}_{\mathrm{E}_2}(\mathrm{j}_s\tilde{\varphi}(\cdot - x))). \tag{3.8.39}$$

3.8.6 Fock representations

We will use Fock representations to construct functional integral representations. Here, we review them for reader's convenience. In Fock representation, $\hat{H}_\mathrm{d}$ and $\hat{H}_\mathrm{od}$ can be represented in terms of annihilation operators and creation operators in $\mathcal{F}_\mathrm{rad}$ as

$$\hat{H}_\mathrm{d}(x,\theta) \cong \frac{1}{\sqrt{2}} \sum_{j=\pm} \{a^*(S_j\hat{\varphi}e^{-ik\cdot x}/\sqrt{\omega}, j) - a(S_j\tilde{\hat{\varphi}}e^{ik\cdot x}/\sqrt{\omega}, j)\},$$

$$\hat{H}_\mathrm{od}(x,-\theta) \cong \frac{1}{\sqrt{2}} \sum_{j=\pm} \{a^*(T_j\hat{\varphi}e^{-ik\cdot x}/\sqrt{\omega}, j) - a(T_j\tilde{\hat{\varphi}}e^{ik\cdot x}/\sqrt{\omega}, j)\},$$

where

$$S_j = S_j(\theta) = -\frac{e}{2}\theta\eta_3^j, \quad T_j = T_j(\theta) = -\frac{e}{2}(\eta_1^j - i\theta\eta_2^j). \tag{3.8.40}$$

Since $\eta^j(k) = -ik \times e^j(k)$ and

$$|\eta_\mu^j(k)| \le |k|, \quad \mu = 1,2,3,$$

we have bounds which are used frequently in this section:

$$\Big\| \sum_{j=\pm} a^*\Big(S_j \frac{\hat{\varphi}}{\sqrt{\omega}} e^{-ik\cdot x}, j\Big)\Phi \Big\| \le |e|a\|(N+1)^{1/2}\Phi\|,$$

$$\Big\| \sum_{j=\pm} a\Big(S_j \frac{\hat{\varphi}}{\sqrt{\omega}} e^{-ik\cdot x}, j\Big)\Phi \Big\| \le |e|a\|N^{1/2}\Phi\|,$$

$$\Big\| \sum_{j=\pm} a^*\Big(T_j \frac{\hat{\varphi}}{\sqrt{\omega}} e^{-ik\cdot x}, j\Big)\Phi \Big\| \le 2|e|a\|(N+1)^{1/2}\Phi\|,$$

$$\Big\| \sum_{j=\pm} a\Big(T_j \frac{\hat{\varphi}}{\sqrt{\omega}} e^{-ik\cdot x}, j\Big)\Phi \Big\| \le 2|e|a\|N^{1/2}\Phi\|.$$

Here, N denotes the number operator in $L^2(\mathscr{Q})$, and $a = \||k|\hat{\varphi}/\sqrt{\omega}\|/2$. Thus we have

$$\|\hat{H}_{\rm d}(x,\theta)\Phi\| \le \sqrt{2}|e|a\|(N+1)^{1/2}\Phi\|,$$
$$\|\hat{H}_{\rm od}(x,-\theta)\Phi\| \le 2\sqrt{2}|e|a\|(N+1)^{1/2}\Phi\|.$$

Moreover, we give a Fock representation of Euclidean versions. Let a_E and a_E^* be the annihilation and creation operators, respectively, in $\mathscr{F}_{\rm b}(L^2(\mathbb{R}^4\times\{\pm\}))$. Then

$$\hat{H}_{\rm E,d}(x,\theta,s) \cong \frac{1}{\sqrt{2}}\sum_{j=\pm}\left\{a_E^*\left(S_j\Big(\hat{\rm j}_s\frac{\hat{\varphi}}{\sqrt{\omega}}\Big)e^{-ik\cdot x},j\right) - a_E\left(S_j\Big(\hat{\rm j}_s\frac{\tilde{\hat{\varphi}}}{\sqrt{\omega}}\Big)e^{ik\cdot x},j\right)\right\},$$
$$\hat{H}_{\rm E,od}(x,-\theta,s) \cong \frac{1}{\sqrt{2}}\sum_{j=\pm}\left\{a_E^*\left(T_j\Big(\hat{\rm j}_s\frac{\hat{\varphi}}{\sqrt{\omega}}\Big)e^{-ik\cdot x},j\right) - a_E\left(T_j\Big(\hat{\rm j}_s\frac{\tilde{\hat{\varphi}}}{\sqrt{\omega}}\Big)e^{ik\cdot x},j\right)\right\}.$$

Here, $\hat{\rm j}_s = F_4{\rm j}_sF_3^{-1} : L^2(\mathbb{R}^3) \to L^2(\mathbb{R}^4)$, $s \in \mathbb{R}$, is the family of isometries such that $\hat{\rm j}_s\hat{\rm j}_t = e^{-|s-t|\omega}$, where F_3 and F_4 are the Fourier transform on $L^2(\mathbb{R}^3)$ and on $L^2(\mathbb{R}^4)$, respectively. Hence it follows that

$$\hat{\rm j}_s\hat{f} = \widehat{{\rm j}_sf}$$

for $f \in L^2(\mathbb{R}^3)$. Then test functions

$$S_j(\hat{\rm j}_s\frac{\hat{\varphi}}{\sqrt{\omega}})e^{-ik\cdot x},\quad T_j(\hat{\rm j}_s\frac{\hat{\varphi}}{\sqrt{\omega}})e^{-ik\cdot x} \in L^2(\mathbb{R}^4_{(k_0,k)}),\quad j=\pm,$$

are explicitly given by

$$S_j\Big(\hat{\rm j}_s\frac{\hat{\varphi}}{\sqrt{\omega}}\Big)e^{-ik\cdot x} = -\frac{e}{2}\theta\frac{e^{-isk_0-ik\cdot x}\eta_3^j(k)\hat{\varphi}(k)}{\sqrt{\pi}\sqrt{\omega(k)^2+|k_0|^2}},$$
$$T_j\Big(\hat{\rm j}_s\frac{\hat{\varphi}}{\sqrt{\omega}}\Big)e^{-ik\cdot x} = -\frac{e}{2}\theta\frac{e^{-isk_0-ik\cdot x}(\eta_1^j(k)-i\theta\eta_2^j(k))\hat{\varphi}(k)}{\sqrt{\pi}\sqrt{\omega(k)^2+|k_0|^2}}.$$

From these expressions, we see that

$$\left|S_j\Big(\hat{\rm j}_s\frac{\hat{\varphi}}{\sqrt{\omega}}\Big)e^{-ik\cdot x}\right| \le \frac{|e|}{2\sqrt{\pi}}\frac{|k||\hat{\varphi}(k)|}{\sqrt{\omega(k)^2+|k_0|^2}} \le \frac{|e|}{2\sqrt{\pi}}|\hat{\varphi}(k)|,$$
$$\left|T_j\Big(\hat{\rm j}_s\frac{\hat{\varphi}}{\sqrt{\omega}}\Big)e^{-ik\cdot x}\right| \le \frac{|e|}{\sqrt{\pi}}\frac{|k||\hat{\varphi}(k)|}{\sqrt{\omega(k)^2+|k_0|^2}} \le \frac{|e|}{\sqrt{\pi}}|\hat{\varphi}(k)|.$$

We also have bounds:

$$\left\|\sum_{j=\pm}a_E^*\left(S_j\Big(\hat{\rm j}_s\frac{\hat{\varphi}}{\sqrt{\omega}}\Big)e^{-ik\cdot x},j\right)\Phi\right\| \le |e|a\|(N+1)^{1/2}\Phi\|,$$

$$\left\| \sum_{j=\pm} a_E\left(S_j\left(\hat{\jmath}_s \frac{\hat{\varphi}}{\sqrt{\omega}}\right)e^{-ik\cdot x}, j\right)\Phi\right\| \le |e|a\|N^{1/2}\Phi\|,$$

$$\left\| \sum_{j=\pm} a_E^*\left(T_j\left(\hat{\jmath}_s \frac{\hat{\varphi}}{\sqrt{\omega}}\right)e^{-ik\cdot x}, j\right)\Phi\right\| \le 2|e|a\|(N+1)^{1/2}\Phi\|,$$

$$\left\| \sum_{j=\pm} a_E\left(T_j\left(\hat{\jmath}_s \frac{\hat{\varphi}}{\sqrt{\omega}}\right)e^{-ik\cdot x}, j\right)\Phi\right\| \le 2|e|a\|N^{1/2}\Phi\|.$$

Hence we have

$$\|\hat{H}_{\mathrm{E,d}}(x,\theta,s)\Phi\| \le \sqrt{2}|e|a\|(N+1)^{1/2}\Phi\|,$$
$$\|\hat{H}_{\mathrm{E,od}}(x,-\theta,s)\Phi\| \le 2\sqrt{2}|e|a\|(N+1)^{1/2}\Phi\|.$$

3.8.7 Preparation of functional integral representations

To construct a functional integral representation of $(F, e^{-tH_{\mathrm{PF}}^{\mathbb{Z}_2}}G)$, we prepare some technical lemmas. We can predict the form of a functional integral representation $(F, e^{-tH_{\mathrm{PF}}^{\mathbb{Z}_2}}G)$ from the classical case (3.8.6), which is unfortunately rather complicated than the spinless case $(F, e^{-tH_{\mathrm{PF}}}G)$. To avoid complicated computations, we introduce self-adjoint operators L and L_R for $R > 0$ which satisfy

$$(|F|, e^{-tL_R}|G|)| \uparrow (|F|, e^{-tL}|G|) \quad \text{as } R \uparrow \infty$$

and diamagnetic type inequality

$$|(F, e^{-tH_{\mathrm{PF}}^{\mathbb{Z}_2}}G)| \le (|F|, e^{-tL}|G|).$$

Applying this inequality, we can avoid technical difficulties to construct a functional integral representation of $(F, e^{-tH_{\mathrm{PF}}^{\mathbb{Z}_2}}G)$.

We introduce two cutoff functions for $R > 0$. Let $\chi_R^- \in C^\infty(\mathbb{R})$ be defined by:

(1) $\chi_R^-(x) = x$ for $x > -R + 1$;
(2) $-R \le \chi_R^-(x) \le -R + 1$ for $-R \le x \le -R + 1$;
(3) $\chi_R^-(x) = -R$ for $x < -R$.

We also define $\chi_R^+ \in C^\infty(\mathbb{R})$ by:

(1) $\chi_R^+(x) = R$ for $x > R$;
(2) $R - 1 \le \chi_R^+(x) \le R$ for $R - 1 \le x \le R$;
(3) $\chi_R^+(x) = x$ for $x < R - 1$.

Furthermore, we suppose a uniform Lipschitz continuity of both χ_R^- and χ_R^+ such that $|\chi_R^\pm(x) - \chi_R^\pm(y)| \le c|x-y|$ with some c for any $x, y \in \mathbb{R}$. We define $\Psi_\varepsilon(|\hat{H}_{\rm od}(-\theta)|)_R$ and $\hat{H}_{\rm d}(\theta)_R$ by

$$\Psi_\varepsilon(|\hat{H}_{\rm od}(-\theta)|)_R = \chi_R^+\left(\Psi_\varepsilon(|\hat{H}_{\rm od}(-\theta)|)\right), \tag{3.8.41}$$

$$\hat{H}_{\rm d}(\theta)_R = \chi_R^-\left(\hat{H}_{\rm d}(\theta)\right). \tag{3.8.42}$$

Hence $\hat{H}_{\rm d}(\theta)_R$ is bounded below and $\Psi_\varepsilon(|\hat{H}_{\rm od}(-\theta)|)_R$ is bounded. Through the fiber direct integral, K_ε is defined. We also define another useful self-adjoint operator in a similar manner to K_ε. Let $\tilde{\varphi} \in C_0^\infty(\mathbb{R}^3)$. We set for each $\phi \in \mathcal{Q}$,

$$(L_R'(\phi)F)(\theta) = (H_{\rm p} + \hat{H}_{\rm d}(\phi,\theta)_R)F(\theta) - \Psi_\varepsilon(|\hat{H}_{\rm od}(\phi,-\theta)|)_R F(-\theta)$$

for each $R > 0$, and

$$(L'(\phi)F)(\theta) = (H_{\rm p} + \hat{H}_{\rm d}(\phi,\theta))F(\theta) - \Psi_\varepsilon(|\hat{H}_{\rm od}(\phi,-\theta)|)F(-\theta).$$

Then $L_R'(\phi)$ and $L'(\phi)$ are self-adjoint and we define

$$L' = \int_{\mathcal{Q}}^{\oplus} L'(\phi)d\mu(\phi), \quad L_R' = \int_{\mathcal{Q}}^{\oplus} L_R'(\phi)d\mu(\phi).$$

We also define

$$L_R = L_R' \dotplus H_{\rm rad}$$

for each $R > 0$ and

$$L = L' \dotplus H_{\rm rad}.$$

Actually, we can show that both L_R and L are self-adjoint on $D_{\rm PF}$,

$$(L_R'F)(\theta) = (H_{\rm p} + \hat{H}_{\rm d}(\theta)_R)F(\theta) - \Psi_\varepsilon(|\hat{H}_{\rm od}(-\theta)|)_R F(-\theta)$$

for each $R > 0$, and

$$(L'F)(\theta) = (H_{\rm p} + \hat{H}_{\rm d}(\theta))F(\theta) - \Psi_\varepsilon(|\hat{H}_{\rm od}(-\theta)|)F(-\theta).$$

Furthermore, in the Fock representation L_R and L are given by

$$L_R = H_{\rm p} + H_{\rm rad} - \begin{pmatrix} (\frac{e}{2}\hat{B}_3)_R & \Psi_\varepsilon\left(\frac{|e|}{2}\sqrt{\hat{B}_1^2 + \hat{B}_2^2}\right)_R \\ \Psi_\varepsilon\left(\frac{|e|}{2}\sqrt{\hat{B}_1^2 + \hat{B}_2^2}\right)_R & -(\frac{e}{2}\hat{B}_3)_R \end{pmatrix},$$

$$L = H_{\mathrm{p}} + H_{\mathrm{rad}} - \begin{pmatrix} \frac{e}{2}\hat{B}_3 & \Psi_\varepsilon\left(\frac{|e|}{2}\sqrt{\hat{B}_1^2+\hat{B}_2^2}\right) \\ \Psi_\varepsilon\left(\frac{|e|}{2}\sqrt{\hat{B}_1^2+\hat{B}_2^2}\right) & -\frac{e}{2}\hat{B}_3 \end{pmatrix}.$$

Remark 3.89. We give a remark on $|\hat{H}_{\mathrm{od}}(-\theta)|$. In the Fock representation, we have $|\hat{H}_{\mathrm{od}}(-\theta)| = \frac{|e|}{2}\sqrt{\hat{B}_1^2+\hat{B}_2^2}$, which leads that $|\hat{H}_{\mathrm{od}}(-\theta)|$ is independent of θ.

The family of self-adjoint operators L_R, $R > 0$, have also a common core $\mathscr{K}^\infty$ and $\lim_{R\to\infty} L_R F = KF$ for $F \in \mathscr{K}^\infty$. Hence

$$e^{-tL_R} \to e^{-tL}$$

strongly as $R \to \infty$. The functional integral representation of $e^{-tL_R'(\phi)}$ can be constructed by (3.8.6) and that of e^{-tL_R} can be done by the Trotter product formula.

Lemma 3.90. *Assume that $V \in L^\infty(\mathbb{R}^3)$ and $\tilde{\varphi} \in C_0^\infty(\mathbb{R}^3)$. Let $F, G \in \mathscr{K}_E$. Then*

$$\sum_{\alpha=1,2} \int_{\mathbb{R}^3} \mathbb{E}^{x,\alpha}\left[e^{-\int_0^t V(B_r)dr}(F(q_0), \mathrm{E}_s e^{L_t(\varepsilon,s)}\mathrm{E}_s G(q_t))\right] dx$$

is finite and

$$(F, \mathrm{J}_s e^{-tL_R'}\mathrm{J}_s^* G) = e^t \sum_{\alpha=1,2} \int_{\mathbb{R}^3} \mathbb{E}^{x,\alpha}\left[e^{-\int_0^t V(B_r)dr}(F(q_0), \mathrm{E}_s e^{L_t(\varepsilon,s)}\mathrm{E}_s G(q_t))\right] dx. \tag{3.8.43}$$

Here,

$$L_t(\varepsilon, s) = -\int_0^t \hat{H}_{\mathrm{E,d}}(B_r, \theta_{N_r}, s)_R dr + \int_0^{t+} \log \Psi_\varepsilon(|\hat{H}_{\mathrm{E,od}}(B_r, -\theta_{N_r}, s)|)_R dN_r,$$

with $\hat{H}_{\mathrm{E,d}}(x,\theta,s)$ and $\hat{H}_{\mathrm{E,od}}(x,-\theta,s)$ given by (3.8.38) *and* (3.8.39)*, respectively. The truncated functions $\hat{H}_{\mathrm{E,d}}(B_r, \theta_{N_r}, s)_R$ and $\Psi_\varepsilon(|\hat{H}_{\mathrm{E,od}}(x,-\theta,s)|)_R$ are defined in a similar way to* (3.8.41) *and* (3.8.42)*, respectively.*

Proof. Let us set

$$V_M = \sup_{x\in\mathbb{R}^3} \mathbb{E}^x[e^{-2\int_0^t V(B_s)ds}] < \infty.$$

Notice that the right-hand side of (3.8.43) is finite, since $\hat{H}_{\mathrm{d}}(x,\theta,s)_R$ is bounded below and $\Psi_\varepsilon(|\hat{H}_{\mathrm{od}}(x,-\theta,s)|)_R$ is bounded; $e^{L_t(\varepsilon,s)}$ is a bounded function. For each $(x,\alpha,w,\tau) \in \mathbb{R}^3 \times \{1,2\} \times \mathscr{X} \times S$, we have

$$|(F(q_0), \mathrm{E}_s e^{L_t(\varepsilon,s)}\mathrm{E}_s G(q_t))| \le \|F(q_0)\|\|G(q_t)\|\|e^{L_t(\varepsilon,s)}\|.$$

Here, $\|e^{L_t(\varepsilon,s)}\|$ is the operator norm of bounded operator $e^{L_t(\varepsilon,s)}$ on $L^2(\mathscr{Q}_{\mathrm{E}})$. Then it follows that

$$|\mathrm{RHS}\ (3.8.43)| \le e^t \sum_{\alpha=1,2} \int_{\mathbb{R}^3} \mathbb{E}^{x,\alpha}\left[e^{-\int_0^t V(B_r)dr} \|F(q_0)\|\|G(q_t)\|\|e^{L_t(\varepsilon,s)}\|\right] dx. \tag{3.8.44}$$

We will also prove in Lemma 3.91 below that there exists a random variable A_t on $(\mathcal{S}, \mathcal{B}_{\mathcal{S}}, P)$ such that:

(1) $\|e^{L_t(\varepsilon,s)}\| \le A_t$;
(2) A_t is independent of $(x, \alpha, w) \in \mathbb{R}^3 \times \{1,2\} \times \mathscr{X}$;
(3) $\mathbb{E}_P[A_t^2] < \infty$.

By (1)–(3) above and (3.8.44),

$$|\mathrm{RHS}\ (3.8.43)| \le \|G\|\|F\|V_M^{1/2}(\mathbb{E}_P[A_t^2])^{1/2} < \infty. \tag{3.8.45}$$

Next, we prove the equality (3.8.43). Note that L_R' is defined by a direct fiber integral representation and by Lemma 3.87, we have

$$(\mathrm{J}_s^* F, e^{-tL_R'}\mathrm{J}_s^* G) = e^t \sum_{\alpha=1,2} \int_{\mathbb{R}^3} \mathbb{E}^{x,\alpha}\left[e^{-\int_0^t V(B_r)dr} \mathbb{E}_\mu\left[\overline{(\mathrm{J}_s^* F)(\phi, q_0)} e^{L_t(\phi,\varepsilon)} (\mathrm{J}_s^* G)(\phi, q_t)\right]\right] dx.$$

Here, we used Fubini's lemma and

$$L_t(\phi, \varepsilon) = -\int_0^t \hat{H}_{\mathrm{d}}(\phi, B_s, \theta_{N_s})_R ds + \int_0^{t+} \log \Psi_\varepsilon(|\hat{H}_{\mathrm{od}}(\phi, B_s, -\theta_{N_s})|)_R dN_s.$$

Let

$$L_t(\varepsilon) = -\int_0^t \hat{H}_{\mathrm{d}}(B_s, \theta_{N_s})_R ds + \int_0^{t+} \log \Psi_\varepsilon(|\hat{H}_{\mathrm{od}}(B_s, -\theta_{N_s})|)_R dN_s.$$

Given that $\mathrm{J}_s^* F, e^{L_t(\varepsilon)}\mathrm{J}_s^* G(q_t) \in L^2(\mathscr{Q})$, we rewrite as

$$(\mathrm{J}_s^* F, e^{-tL_R'}\mathrm{J}_s^* G) = e^t \sum_{\alpha=1,2} \int_{\mathbb{R}^3} \mathbb{E}^{x,\alpha}\left[e^{-\int_0^t V(B_r)dr} (F(q_0), \mathrm{J}_s e^{L_t(\varepsilon)} \mathrm{J}_s^* G(q_t))\right] dx.$$

We can see that $\mathrm{J}_s e^{L_t(\varepsilon)}\mathrm{J}_s^* = \mathrm{E}_s e^{L_t(\varepsilon,s)}\mathrm{E}_s$ by a limiting argument, leading to (3.8.43). Then the proof can be completed. □

Now we have to prove (1)–(3) used in the proof of Lemma 3.90.

Lemma 3.91. *Operator $e^{L_t(\varepsilon,s)}$ is bounded, and there exists a random variable $A_t = A_t(\tau)$ on $(\mathcal{S}, \mathcal{B}_{\mathcal{S}}, P)$ satisfying* (1)–(3) *in the proof of Lemma* 3.90.

Proof. Since $R < \infty$, we have $|e^{L_t(\varepsilon,s)}| \le e^{tR}e^{\int_0^{t+}\log R dN_s}$. For each $\tau \in S$, the number of jumps of map $s \mapsto N_s(\tau)$ for $0 \le s \le t$ is denoted by $N = N(\tau)$. Hence $\int_0^{t+}\log R dN_s = \log R^N$ and then $|e^{L_t(\varepsilon,s)}| \le e^{tR}R^N$. Set $A_t = A_t(\tau) = e^{tR}R^{N(\tau)}$. Then $\mathbb{E}_P[A_t^2] = e^{2tR}\sum_{N=0}^{\infty}\frac{t^N R^{2N}}{N!}e^{-t} = e^{t(R^2+2R-1)} < \infty$. Then A_t satisfies (1)–(3). □

In the next lemma, we construct a functional integral representation of e^{-tL_R} for $R > 0$. Let

$$Y_R = -\int_0^t \hat{H}_{\mathrm{E,d}}(B_s, \theta_{N_s}, s)_R ds + \int_0^{t+} \log \Psi_\varepsilon(|\hat{H}_{\mathrm{E,od}}(B_s, -\theta_{N_{s-}}, s)|)_R dN_s.$$

Lemma 3.92. *Suppose that $V \in L^\infty(\mathbb{R}^3)$ and let $\tilde{\varphi} \in C_0^\infty(\mathbb{R}^3)$. Then for every $t \ge 0$ and all $F, G \in \mathscr{K}$, it follows that $\mathrm{J}_0^* e^{Y_R}\mathrm{J}_t$ is a bounded operator on $L^2(\mathscr{Q})$ for every $(x, \alpha, w, \tau) \in \mathbb{R}^3 \times \{1,2\} \times \mathscr{X} \times S$ and*

$$(F, e^{-tL_R}G) = e^t \sum_{\alpha=1,2}\int_{\mathbb{R}^3} \mathbb{E}^{x,\alpha}\left[e^{-\int_0^t V(B_s)ds}(F(q_0), \mathrm{J}_0^* e^{Y_R}\mathrm{J}_t G(q_t))\right] dx. \tag{3.8.46}$$

Proof. Similarly as in the proof of Lemma 3.91, for each $\tau \in S$, the number of jumps of map $s \mapsto N_s(\tau)$ for $0 \le s \le t$ is denoted by $N = N(\tau)$. Fix $\tau \in S$. Hence by virtue of cutoff parameter $R > 0$, it can be seen that $\|\mathrm{J}_0^* e^{Y_R}\mathrm{J}_t\Phi\| \le e^{tR}R^{N(\tau)}\|\Phi\|$, which implies that $\mathrm{J}_0^* e^{Y_R}\mathrm{J}_t$ is bounded for every $(x, \alpha, w, \tau) \in \mathbb{R}^3 \times \{1,2\} \times \mathscr{X} \times S$.

By the Trotter product formula, we have

$$(F, e^{-tL_R}G) = \lim_{n\to\infty}\left(F, \left(e^{-\frac{t}{n}L_R'}e^{-\frac{t}{n}H_{\mathrm{rad}}}\right)^n G\right) = \lim_{n\to\infty}\left(F, \mathrm{J}_0^*\left(\prod_{i=0}^{n-1}\mathrm{J}_{it/n}e^{-\frac{t}{n}L_R'}\mathrm{J}_{it/n}^*\right)\mathrm{J}_t G\right).$$

By the Markov property of $\mathrm{J}_s\mathrm{J}_s^* = \mathrm{E}_s$ and Lemma 3.90, we can see that

$$(F, e^{-tL_R}G) = \lim_{n\to\infty}\sum_{\alpha=1,2}\int_{\mathbb{R}^3}\mathbb{E}^{x,\alpha}\left[e^{-\int_0^t V(B_s)ds}(F(q_0), \mathrm{J}_0^* e^{Y_R(n)}\mathrm{J}_t G(q_t))\right] dx, \tag{3.8.47}$$

where

$$Y_R(n) = -\sum_{i=1}^n \int_{(i-1)t/n}^{it/n} \hat{H}_{\mathrm{E,d}}(B_s, \theta_{N_s}, it/n)_R ds + \sum_{i=1}^n \int_{(i-1)t/n}^{it/n+} \log \Psi_\varepsilon(|\hat{H}_{\mathrm{E,od}}(B_s, -\theta_{N_{s-}}, it/n)|)_R dN_s.$$

Then

$$|(F(q_0), \mathrm{J}_0^* e^{Y_R(n)}\mathrm{J}_t G(q_t))| \le \|F(q_0)\|\|G(q_t)\|e^{tR}R^{N(\tau)} \tag{3.8.48}$$

and

$$\sum_{\alpha=1,2}\int_{\mathbb{R}^3}\mathbb{E}^{x,\alpha}\left[e^{-\int_0^t V(B_s)ds}\|F(q_0)\|\|G(q_t)\|e^{tR}R^{N(\tau)}\right]dx$$
$$\leq \|F\|\|G\|V_M^{1/2}(\mathbb{E}_P[e^{2tR}R^{2N}])^{1/2} \leq \|F\|\|G\|V_M^{1/2}e^{\frac{t}{2}(R^2+2R-1)} < \infty. \tag{3.8.49}$$

Since the right-hand side above is independent of n, it is sufficient to prove (3.8.46) for $F, G \in \mathscr{K}^\infty$, hence we suppose that $F, G \in \mathscr{K}^\infty$ for the rest of this proof. We can see that

$$\left\|\sum_{i=1}^n \int_{(i-1)t/n}^{it/n} \hat{H}_{\mathrm{E,d}}(B_s, \theta_{N_s}, it/n)_R ds F\right\| \leq \sum_{i=1}^n \int_{(i-1)t/n}^{it/n} \left\|\hat{H}_{\mathrm{E,d}}(B_s, \theta_{N_s}, it/n)F\right\| ds$$
$$\leq tC\||k|\hat{\varphi}/\sqrt{\omega}\|\|(N+1)^{1/2}F\|$$

and, similarly,

$$\left\|\int_0^t \hat{H}_{\mathrm{E,d}}(B_s, \theta_{N_s}, s)_R ds F\right\| \leq tC\||k|\hat{\varphi}/\sqrt{\omega}\|\|(N+1)^{1/2}F\|$$

with some constant C. By the uniform Lipschitz continuity of function $\chi_{\bar{R}}$, we see that $|\chi_{\bar{R}}(x) - \chi_{\bar{R}}(y)| \leq c|x-y|$ with some constant c, we can also see that

$$\left\|\left(\sum_{i=1}^n \int_{(i-1)t/n}^{it/n} \hat{H}_{\mathrm{E,d}}(B_s, \theta_{N_s}, it/n)_R ds - \int_0^t \hat{H}_{\mathrm{E,d}}(B_s, \theta_{N_s}, s)_R ds\right)F\right\|$$
$$\leq c\sum_{i=1}^n \int_{(i-1)t/n}^{it/n} \left\||\hat{H}_{\mathrm{E,d}}(B_s, \theta_{N_s}, it/n) - \hat{H}_{\mathrm{E,d}}(B_s, \theta_{N_s}, s)|F\right\| ds$$
$$\leq c\sum_{i=1}^n \left(\int_{(i-1)t/n}^{it/n} \left\||\hat{H}_{\mathrm{E,d}}(B_s, \theta_{N_s}, it/n) - \hat{H}_{\mathrm{E,d}}(B_s, \theta_{N_s}, s)|F\right\|^2 ds\right)^{1/2} \frac{\sqrt{t}}{\sqrt{n}}.$$

It can be seen that

$$\left\||\hat{H}_{\mathrm{E,d}}(B_s, \theta_{N_s}, it/n) - \hat{H}_{\mathrm{E,d}}(B_s, \theta_{N_s}, s)|F\right\|^2 \leq c'\frac{t}{n}\||k|\hat{\varphi}\|^2\|(N+1)^{1/2}F\|^2$$

with some constant c'. Then

$$\left\|\left(\sum_{i=1}^n \int_{(i-1)t/n}^{it/n} \hat{H}_{\mathrm{E,d}}(B_s, \theta_{N_s}, it/n)_R ds - \int_0^t \hat{H}_{\mathrm{E,d}}(B_s, \theta_{N_s}, s)_R ds\right)F\right\|$$
$$\leq cc'\sum_{i=1}^n \||k|\hat{\varphi}\|\frac{t\sqrt{t}}{n\sqrt{n}}\|(N+1)^{1/2}F\| = cc'\||k|\hat{\varphi}\|\frac{t\sqrt{t}}{\sqrt{n}}\|(N+1)^{1/2}F\| \to 0$$

as $n \to \infty$. Thus it can be straightforwardly seen that

$$\exp\left(-\sum_{i=1}^{n}\int_{(i-1)t/n}^{it/n} \hat{H}_{\mathrm{E,d}}(B_s, \theta_{N_s}, it/n)_R ds\right)F \to \exp\left(-\int_0^t \hat{H}_{\mathrm{E,d}}(B_s, \theta_{N_s}, s)_R ds\right)F \quad (3.8.50)$$

by the expansion $e^X = \sum_{j=0}^{\infty} \frac{X^j}{j!}$ on $\mathscr{K}^\infty$ for

$$X = -\int_0^t \hat{H}_{\mathrm{E,d}}(B_s, \theta_{N_s}, s)_R ds \quad \text{or} \quad X = -\sum_{i=1}^{n}\int_{(i-1)t/n}^{it/n} \hat{H}_{\mathrm{E,d}}(B_s, \theta_{N_s}, s)_R ds.$$

Next, we consider $\exp(\sum_{i=1}^n \int_{(i-1)t/n}^{it/n+} \log \Psi_\varepsilon(|\hat{H}_{\mathrm{E,od}}(B_s, -\theta_{N_{s-}}, it/n)|)_R dN_s)$. Points of discontinuity of the map $r \mapsto N_r(\tau)$ are denoted by $s_1 = s_1(\tau), \ldots, s_N = s_N(\tau) \in (0, \infty)$. For a sufficiently large n, the number of discontinuous points located in the interval $(t(i-1)/n, ti/n]$ is at most one. Then by taking n large enough and denoting $(n(s_i), n(s_i) + t/n]$ for the interval containing s_i, we get

$$\exp\left(\sum_{i=1}^{n}\int_{(i-1)t/n}^{it/n+} \log \Psi_\varepsilon(|\hat{H}_{\mathrm{E,od}}(B_s, -\theta_{N_{s-}}, it/n)|)_R dN_s\right) = \prod_{i=1}^{N} \chi_R^+(|\phi_i| + \varepsilon\psi_\varepsilon(|\phi_i|)),$$

where $\phi_i = \hat{H}_{\mathrm{E,od}}(B_{s_i}, -\theta_{N_{s_i-}}, n(s_i))$. By the Lipschitz continuity of χ_R^+ and ψ_ε, we have

$$\begin{aligned}|\Psi_\varepsilon(x)_R - \Psi_\varepsilon(y)_R| &\le c|\Psi_\varepsilon(x) - \Psi_\varepsilon(y)| \le c|x - y + \varepsilon\psi_\varepsilon(x) - \varepsilon\psi(y)| \\ &\le c|x-y| + c\varepsilon|\psi_\varepsilon(x) - \psi(y)| \le c|x-y| + cc'\varepsilon|x-y|\end{aligned}$$

with some constants c and c'. Hence

$$\begin{aligned}&\left\|\left(\Psi_\varepsilon(|\hat{H}_{\mathrm{E,od}}(B_s, -\theta_{N_{s-}}, n(s_i))|)_R - \Psi_\varepsilon(|\hat{H}_{\mathrm{E,od}}(B_s, -\theta_{N_{s-}}, s_i)|)_R\right)F\right\| \\ &\quad \le C\left\|\left(|\hat{H}_{\mathrm{E,od}}(B_s, -\theta_{N_{s-}}, n(s_i))| - |\hat{H}_{\mathrm{E,od}}(B_s, -\theta_{N_{s-}}, s_i)|\right)F\right\|.\end{aligned}$$

Note that

$$\begin{aligned}&\left\|\left(|\hat{H}_{\mathrm{E,od}}(B_s, -\theta_{N_{s-}}, n(s_i))| - |\hat{H}_{\mathrm{E,od}}(B_s, -\theta_{N_{s-}}, s_i)|\right)F\right\|^2 \\ &\quad \le \mathbb{E}_\mu[|\hat{H}_{\mathrm{E,od}}(B_s, -\theta_{N_{s-}}, n(s_i)) - \hat{H}_{\mathrm{E,od}}(B_s, -\theta_{N_{s-}}, s_i)|^2|F|^2] \\ &\quad \le C\||k|\hat{\varphi}\|^2|n(s_i) - s_i|\|(N+1)^{1/2}F\|^2\end{aligned}$$

with some $C > 0$. Clearly, $n(s_i) \to s_i$ as $n \to \infty$. Then

$$\lim_{n\to\infty} \hat{H}_{\mathrm{E,od}}(B_{s_i}, -\theta_{N_{s_i-}}, n(s_i)) = \hat{H}_{\mathrm{E,od}}(B_{s_i}, -\theta_{N_{s_i-}}, s_i)$$

on $\mathscr{K}^\infty$, and we have

$$\lim_{n\to\infty} \exp\left(\sum_{i=1}^{n} \int_{(i-1)t/n}^{it/n+} \log \Psi_\varepsilon(|\hat{H}_{\mathrm{E,od}}(B_s, -\theta_{N_{s-}}, it/n)|)_R dN_s \right)$$
$$= \prod_{i=1}^{N} \Psi_\varepsilon(|\hat{H}_{\mathrm{E,od}}(B_{s_i}, -\theta_{N_{s_i-}}, s_i)|)_R. \tag{3.8.51}$$

Then by (3.8.50) and (3.8.51), we can see that for each $(x, \alpha, w, \tau) \in \mathbb{R}^3 \times \{1, 2\} \times \mathscr{X} \times \mathcal{S}$,

$$\left(F(q_0), \mathrm{J}_0^* e^{Y_R(n)} \mathrm{J}_t G(q_t)\right) \to \left(F(q_0), \mathrm{J}_0^* e^{Y_R} \mathrm{J}_t G(q_t)\right)$$

as $n \to \infty$. Together with (3.8.48) and (3.8.49), the Lebesgue dominated convergence theorem yields that

$$\sum_{\alpha=1,2} \int_{\mathbb{R}^3} \mathbb{E}^{x,\alpha}\left[e^{-\int_0^t V(B_r)dr} \left(F(q_0), \mathrm{J}_0^* e^{Y_R(n)} \mathrm{J}_t G(q_t)\right)\right] dx$$
$$\to \sum_{\alpha=1,2} \int_{\mathbb{R}^3} \mathbb{E}^{x,\alpha}\left[e^{-\int_0^t V(B_r)dr} \left(F(q_0), \mathrm{J}_0^* e^{Y_R} \mathrm{J}_t G(q_t)\right)\right] dx$$

for $F, G \in \mathscr{K}^\infty$. Then the proof is complete. □

Let us define

$$Y = -\int_0^t \hat{H}_{\mathrm{E,d}}(B_s, \theta_{N_s}, s) ds + \int_0^{t+} \log \Psi_\varepsilon(|\hat{H}_{\mathrm{E,od}}(B_s, -\theta_{N_{s-}}, s)|) dN_s. \tag{3.8.52}$$

We already see that $\mathrm{J}_0^* e^{Y_R} \mathrm{J}_t$ is bounded, but it is not trivial to see that $\mathrm{J}_0^* e^{Y} \mathrm{J}_t$ is bounded.

Lemma 3.93. *For each $(x, \alpha, w, \tau) \in \mathbb{R}^3 \times \{1, 2\} \times \mathscr{X} \times \mathcal{S}$, $\mathrm{J}_0^* e^Y \mathrm{J}_t$ is bounded on $L^2(\mathscr{Q})$. In particular, $\mathrm{J}_0^* e^Y \mathrm{J}_t \Psi \in L^2(\mathscr{Q})$ for $\Psi \in L^2(\mathscr{Q})$.*

Proof. We introduce proofs for massless and massive cases. For massive cases, we can apply the hypercontractivity of $e^{-tH_{\mathrm{rad}}}$, but for the massless case we use the Baker–Campbell–Hausdorff formula.

(Massive case) Suppose that $\omega(k) = \sqrt{|k|^2 + \nu^2}$ with $\nu > 0$. To see the boundedness of $\mathrm{J}_0^* e^Y \mathrm{J}_t$, by the hypercontractivity of $e^{-tH_{\mathrm{rad}}}$ it is sufficient to show that

$$e^Y \in \bigcap_{p>0} L^p(\mathscr{Q}_{\mathrm{E}}). \tag{3.8.53}$$

It is known that

$$\|\mathrm{J}_0^* e^Y \mathrm{J}_t\| \le \|e^Y\|_{L^q(\mathscr{Q}_{\mathrm{E}})}, \quad q = \frac{2}{1 - e^{-\nu t}}.$$

Condition (3.8.53) is equivalent to

$$e^{pY} \in L^1(\mathscr{Q}_{\mathrm{E}}) \quad \text{for any } p \in \mathbb{N}. \tag{3.8.54}$$

We shall prove (3.8.54). Note that

$$\|e^{\frac{p}{2}Y}\|^2_{L^1(\mathscr{Q}_{\rm E})} \le \left\|e^{-p\int_0^t \hat{H}_{\rm E,d}(B_r,\theta_{N_r},r)dr}\right\|^2 \left\|e^{p\int_0^t |\log \Psi_\varepsilon(|\hat{H}_{\rm E,od}(B_r,-\theta_{N_{r-}},r)|)|dN_r}\right\|^2.$$

We estimate the right-hand side of this expression. It follows that

$$\int_0^t \hat{H}_{\rm E,d}(B_r,\theta_{N_r},r)dr = \hat{B}_{\rm E_3}\left(-\frac{e}{2}\int_0^t \theta_{N_r}{\rm j}_r\tilde{\varphi}(\cdot - B_r)dr\right).$$

Since $\hat{B}_{\rm E_\mu}(f)$ is a Gaussian random variable with mean zero and covariance,

$$\mathbb{E}_{\mu_{\rm E}}[\hat{B}_{\rm E_\mu}(f)\hat{B}_{\rm E_\nu}(g)] = \frac{1}{2}\int_{\mathbb{R}^4} \overline{\hat{f}(\mathbf{k})}\hat{g}(\mathbf{k})|k|^2\delta_{\mu\nu}(k)d\mathbf{k},$$

we have

$$\|e^{-p\int_0^t \hat{H}_{\rm E,d}(B_r,\theta_{N_r},r)dr}\|^2 \le c_1,$$

where

$$c_1 = \exp\left(t^2\frac{p^2e^2}{4}\||k|\hat{\varphi}/\sqrt{\omega}\|^2\right)$$

and c_1 is thus independent of $(x,\theta) \in \mathbb{R}^3 \times \mathbb{Z}_2$. Next, consider the off-diagonal spin interaction $\|e^{p\int_0^t |\log\Psi_\varepsilon(|\hat{H}_{\rm E,od}(B_r,-\theta_{N_{r-}},r)|)|dN_r}\|$. Write $\hat{B}_{\rm E_\mu}(t) = \hat{B}_{\rm E_\mu}({\rm j}_t\tilde{\varphi}(\cdot - B_t))$ for notational convenience. For each $\tau \in \mathcal{S}$, there exists $N = N(\tau) \in \mathbb{N}$ and points of discontinuity of $r \mapsto N_r(\tau)$, $s_1 = s_1(\tau), \dots, s_N = s_N(\tau) \in (0,\infty)$, such that

$$\int_0^t |\log\Psi_\varepsilon(|\hat{H}_{\rm E,od}(B_r,-\theta_{N_{r-}},r)|)|dN_r = \sum_{j=1}^N |\log\Psi_\varepsilon(|\hat{H}_{\rm E,od}(B_{s_j},-\theta_{N_{s_j-}},s_j)|)|.$$

Since $|\Psi_\varepsilon(w)| \le |w| + \varepsilon$, we have

$$\left\|e^{p\int_0^t |\log\Psi_\varepsilon(|\hat{H}_{\rm E,od}(B_r,-\theta_{N_{r-}},r)|)|dN_r}\right\|^2 \le \left(1, e^{2p\sum_{i=1}^N \log\left(\frac{|e|}{\sqrt{2}}\sqrt{\hat{B}_{\rm E_1}(s_i)^2+\hat{B}_{\rm E_2}(s_i)^2+\varepsilon^2}\right)}1\right),$$

where we used that $|a+ib|+\varepsilon \le \sqrt{2}\sqrt{a^2+b^2+\varepsilon^2}$, $a,b,\varepsilon \in \mathbb{R}$. Then the right-hand side is computed as

$$\left\|e^{p\int_0^t |\log\Psi_\varepsilon(|\hat{H}_{\rm E,od}(B_r,-\theta_{N_{r-}},r)|)|dN_r}\right\|^2 \le \left(\frac{|e|}{\sqrt{2}}\right)^{2pN}\sum_{m=0}^{pN}\varepsilon^{2(pN-m)}\sum_{{\rm comb}_m}\|\underbrace{\hat{B}_{\rm E_\#}\cdots\hat{B}_{\rm E_\#}}_{m\text{-fold}}1\|^2,$$

where $\sum_{\mathrm{comb}_m}$ denotes summation over the 2^m terms in the expansion of the product $\prod_{i=1}^m (\hat{B}_{\mathrm{E}_1}(s_i)^2 + \hat{B}_{\mathrm{E}_2}(s_i)^2)$, $\hat{B}_{\mathrm{E}_\#}$ denotes one of $\hat{B}_{\mathrm{E}_\mu}(s_i)$, $\mu = 1, 2$, $i = 1, \ldots, pN$. Thus we obtain

$$\left\| e^{p \int_0^t |\log(\Psi_\varepsilon(|\hat{H}_{\mathrm{E,od}}(B_r, -\theta_{N_{r-}}, r)|))| dN_r} \right\|^2 \le c_2(\tau), \tag{3.8.55}$$

where

$$c_2(\tau) = \left(\frac{|e|}{\sqrt{2}}\right)^{2pN} \sum_{m=0}^{pN} \varepsilon^{2(pN-m)} 2^m (\sqrt{2})^{2m} m! \, \||k|\hat{\varphi}/\sqrt{\omega}\|^{2m}.$$

Note that $c_2(\tau)$ is independent of $(x, \alpha, w) \in \mathbb{R}^3 \times \{1, 2\} \times \mathscr{X}$. Write $C_t(\tau) = c_1 c_2(\tau)$. Then $\|\mathrm{J}_0^* e^Y \mathrm{J}_t\| \le C_t(\tau)$. This completes the proof.

(Massless case) Suppose that $\omega(k) = |k|$, i. e., $\nu = 0$. In what follows, we show the lemma for massless cases, but it also covers the massive case. The idea is taken from Section 1.2.8, where we define an operator $W(t)$ and show its boundedness. As will be seen below, the estimate of $\mathrm{J}_0^* e^Y \mathrm{J}_t$ can be reduced to the estimate of $W(t)$. Let $Y = Y_1 + Y_2$, where

$$Y_1 = -\int_0^t \hat{H}_{\mathrm{E,d}}(B_r, \theta_{N_r}, r) dr,$$

$$Y_2 = \int_0^{t+} \log(\Psi_\varepsilon(|\hat{H}_{\mathrm{E,od}}(B_r, -\theta_{N_{r-}}, r)|)) dN_r.$$

Using the Baker–Campbell–Hausdorff formula, we expand $\mathrm{J}_0^* e^Y \mathrm{J}_t$. Thus, formally, we have to estimate

$$\mathrm{J}_0^* e^Y \mathrm{J}_t = \mathrm{J}_0^* e^{\frac{B^\dagger}{\sqrt{2}} + \frac{B}{\sqrt{2}}} \prod_{j=1}^N \left(\left| \frac{C_j^\dagger}{\sqrt{2}} + \frac{C_j}{\sqrt{2}} \right| + \varepsilon \psi_\varepsilon(j) \right) \mathrm{J}_t,$$

where in the Fock representation $B^\sharp$ and $C_i^\sharp$ are defined by

$$B^\dagger = -\sum_{j=\pm} a_E^* \left(\int_0^t S_j(\theta_{N_r}) \hat{\mathrm{j}}_r \left(\frac{\hat{\varphi}}{\sqrt{\omega}}\right) e^{-ik \cdot B_r} dr, j \right),$$

$$B = -\sum_{j=\pm} a_E \left(\int_0^t S_j(\theta_{N_r}) \hat{\mathrm{j}}_r \left(\frac{\hat{\varphi}}{\sqrt{\omega}}\right) e^{ik \cdot B_r} dr, j \right),$$

$$C_i^\dagger = \sum_{j=\pm} a_E^* \left(T_j(\theta_{N_{s_i}}) \hat{\mathrm{j}}_{s_i} \left(\frac{\hat{\varphi}}{\sqrt{\omega}}\right) e^{-ik \cdot B_{s_i}}, j \right),$$

$$C_i = \sum_{j=\pm} a_E \left(T_j(\theta_{N_{s_i}}) \hat{\mathrm{j}}_{s_i} \left(\frac{\hat{\varphi}}{\sqrt{\omega}}\right) e^{ik \cdot B_{s_i}}, j \right),$$

$$\psi_\varepsilon(j) = \psi_\varepsilon(|C_j^\dagger + C_j|/\sqrt{2}).$$

Note that $\psi_\varepsilon(j)$ is bounded with bound $\|\psi_\varepsilon(j)\| \le 1$ for any j.

Now fix $(w,\tau) \in \mathscr{X} \times S$. Similarly as in massive cases, there exist $N = N(\tau) \in \mathbb{N}$ and points $0 \le s_1 < \cdots < s_N \le t$, $s_j = s_j(\tau)$, $j = 1,\dots,N$, depending on τ such that $s \mapsto N_s(\tau)$ is not continuous. Then by taking n large enough and denoting $(n(s_i), n(s_i) + t/n]$ for the interval containing s_i, we get

$$e^{Y_2} = \prod_{i=1}^N (|\phi_i| + \varepsilon\psi_\varepsilon(|\phi_i|)),$$

where $\phi_i = \hat{H}_{\mathrm{E,od}}(B_{s_i}, -\theta_{N_{s_i-}}, n(s_i))$. We have

$$\prod_{i=1}^N (|\phi_i| + \varepsilon\psi_\varepsilon(|\phi_i|)) \le 1 + \prod_{i=1}^N (\phi_i + \varepsilon\psi_\varepsilon(\phi_i))\overline{(\phi_i + \varepsilon\psi_\varepsilon(\phi_i))}$$

and $(\phi_i + \varepsilon\psi_\varepsilon(\phi_i))\overline{(\phi_i + \varepsilon\psi_\varepsilon(\phi_i))} \le (1+\varepsilon)\phi_i\overline{\phi_i} + \varepsilon^2$. Then

$$|e^{Y_2}| \le 1 + \prod_{i=1}^N ((1+\varepsilon)\phi_i\overline{\phi_i} + \varepsilon^2).$$

Hence it follows that

$$|(F, \mathrm{J}_0^* e^Y \mathrm{J}_t G)| \le \left(|F|, \mathrm{J}_0^* e^{Y_1} \left(1 + \prod_{i=1}^N ((1+\varepsilon)\phi_i\overline{\phi_i} + \varepsilon^2) \right) \mathrm{J}_t |G| \right). \tag{3.8.56}$$

We shall show that $\mathrm{J}_0^* e^{Y_1} \prod_{i=1}^k \phi_i \mathrm{J}_t$ is bounded. Commutation relations are given by

$$[B, B^\dagger] = \frac{e^2}{4} \int_0^t dr \int_0^t dr' \int_{\mathbb{R}^3} \frac{|\hat{\varphi}(k)|^2}{|k|} d(k) e^{-|r-r'|\omega(k)} e^{-ik\cdot(B_r - B_{r'})} dk,$$

$$[C_i, C_j^\dagger] = \frac{e^2}{4} \int_{\mathbb{R}^3} \frac{|\hat{\varphi}(k)|^2}{|k|} c(k) e^{-|s_i - s_j|\omega(k)} e^{-ik\cdot(B_{s_j} - B_{s_i})} dk,$$

$$[B, C_i^\dagger] = -\frac{e^2}{4} \int_0^t dr \int_{\mathbb{R}^3} \frac{|\hat{\varphi}(k)|^2}{|k|} b(k) e^{-|r-s_i|\omega(k)} e^{-ik\cdot(B_{s_i} - B_r)} dk,$$

$$[C_i, B^\dagger] = -\frac{e^2}{4} \int_0^t dr \int_{\mathbb{R}^3} \frac{|\hat{\varphi}(k)|^2}{|k|} b(k) e^{-|r-s_i|\omega(k)} e^{-ik\cdot(B_r - B_{s_i})} dk.$$

Here,

$$c(k) = \sum_{j=\pm} (|\eta_1^j(k)|^2 + |\eta_2^j(k)|^2), \tag{3.8.57}$$

$$b(k) = \sum_{j=\pm} \theta_{N_r} \eta_3^j(k)(\eta_1^j(k) - i\theta_{N_{s_i}} \eta_2^j(k)), \tag{3.8.58}$$

$$d(k) = \sum_{j=\pm} |\eta_3^j(k)|^2. \tag{3.8.59}$$

Since the Baker–Campbell–Hausdorff formula yields that

$$e^{Y_1} = e^{\beta} e^{\frac{B^\dagger}{\sqrt{2}}} e^{\frac{B}{\sqrt{2}}},$$

where $\beta = [B, B^\dagger]/4$. We have

$$\mathrm{J}_0^* e^{Y_1} \prod_{i=1}^k \phi_i \mathrm{J}_t = \mathrm{J}_0^* e^{\beta} e^{\frac{B^\dagger}{\sqrt{2}}} e^{\frac{B}{\sqrt{2}}} \prod_{i=1}^k \left(\frac{C_{j_i}^\dagger}{\sqrt{2}} + \frac{C_{j_i}}{\sqrt{2}} \right) \mathrm{J}_t. \tag{3.8.60}$$

We apply Wick's theorem to compute the commutator between $\prod_{i=1}^k (\frac{C_{j_i}^\dagger}{\sqrt{2}} + \frac{C_{j_i}}{\sqrt{2}})$ and $e^{\frac{B^\dagger}{\sqrt{2}}} e^{\frac{B}{\sqrt{2}}}$. Set $j_i = i$ for simplicity. Hence

$$\prod_{i=1}^k \left(\frac{C_i^\dagger}{\sqrt{2}} + \frac{C_i}{\sqrt{2}} \right) = \sum_{p=0}^{[k/2]} \frac{1}{2^p} \sum_{i_1 \dots, i_{2p}}^{\mathrm{pair}} \left(\prod_{l=1}^p C_{i_{2l-1}, i_{2l}} \right) c_{i_1,\dots,i_{2k}}^k.$$

Here, $C_{i,j} = [C_i, C_j^\dagger]$, we recall that $\sum_{i_1 \dots, i_{2p}}^{\mathrm{pair}}$ denotes summation over all p-pairs chosen from $\{1, \dots, k\}$, and

$$c_{i_1,\dots,i_{2p}}^k = 2^{-(k-2p)/2} \sum_{l=0}^{k-2p} \sum_{\{n_1,\dots,n_l\} \subset \{i_1,\dots,i_{2p}\}^c} \prod_{i \in \{n_1,\dots,n_l\}} C_i^\dagger \prod_{i \in \{n_1,\dots,n_l\}^c \cap \{i_1,\dots,i_{2p}\}^c} C_i.$$

Hence

$$e^{\frac{B^\dagger}{\sqrt{2}}} e^{\frac{B}{\sqrt{2}}} \prod_{i=1}^k \left(\frac{C_i^\dagger}{\sqrt{2}} + \frac{C_i}{\sqrt{2}} \right) \sum_{p=0}^{[k/2]} \frac{1}{2^p} \sum_{i_1 \dots, i_{2p}}^{\mathrm{pair}} \left(\prod_{l=1}^p C_{i_{2l-1}, i_{2l}} \right) \sum_{l=0}^{k-2p} \sum_{\{n_1,\dots,n_l\} \subset \{i_1,\dots,i_{2p}\}^c} X e^{\frac{B^\dagger}{\sqrt{2}}} e^{\frac{B}{\sqrt{2}}} Z, \tag{3.8.61}$$

where $X = \prod_{i \in \{n_1,\dots,n_l\}} (C_i^\dagger + y_i)$ and $Z = \prod_{i \in \{n_1,\dots,n_l\}^c \cap \{i_1,\dots,i_{2p}\}^c} C_i$ with $y_j = [B, C_j^\dagger]$. Let us define operators in $\mathcal{F}_{\mathrm{rad}}$ by

$$B^\dagger(l) = -\sum_{j=\pm} a^* \left(\int_0^t e^{-|r-l|\omega} S_j(\theta_{N_r}) \frac{\hat{\varphi}}{\sqrt{\omega}} e^{-ik\cdot B_r} dr, j \right),$$

$$B(l) = -\sum_{j=\pm} a \left(\int_0^t e^{-|r-l|\omega} S_j(\theta_{N_r}) \frac{\hat{\varphi}}{\sqrt{\omega}} e^{ik\cdot B_r} dr, j \right),$$

$$C_i^\dagger(l) = \sum_{j=\pm} a^*\left(e^{-|s_i-l|\omega}T_j(\theta_{N_{s_i}})\frac{\hat{\varphi}}{\sqrt{\omega}}e^{-ik\cdot B_{s_i}}, j\right),$$

$$C_i(l) = \sum_{j=\pm} a\left(e^{-|s_i-l|\omega}T_j(\theta_{N_{s_i}})\frac{\hat{\varphi}}{\sqrt{\omega}}e^{ik\cdot B_{s_i}}, j\right).$$

Then commutation relations are

$$[B(l), B^\dagger(l')] = \frac{e^2}{4}\int_0^t dr\int_0^t dr'\int_{\mathbb{R}^3} e^{-(|r-l|+|r-l'|)\omega(k)}\frac{|\hat{\varphi}(k)|^2}{|k|}d(k)e^{-ik\cdot(B_r-B_{r'})}dk,$$

$$[C_i(l), C_j^\dagger(l')] = \frac{e^2}{4}\int_{\mathbb{R}^3} e^{-(|s_i-l|+|s_j-l'|)\omega(k)}\frac{|\hat{\varphi}(k)|^2}{|k|}c(k)e^{-ik\cdot(B_{s_j}-B_{s_i})}dk,$$

$$[B(l), C_i^\dagger(l')] = \frac{e^2}{4}\int_0^t dr\int_{\mathbb{R}^3} e^{-(|r-l|+|s_i-l'|)\omega(k)}\frac{|\hat{\varphi}(k)|^2}{|k|}b(k)e^{-ik\cdot(B_{s_i}-B_r)}dk,$$

$$[C_i(l), B^\dagger(l')] = \frac{e^2}{4}\int_0^t dr\int_{\mathbb{R}^3} e^{-(|s_i-l|+|r-l'|)\omega(k)}\frac{|\hat{\varphi}(k)|^2}{|k|}b(k)e^{-ik\cdot(B_r-B_{s_i})}dk.$$

Here, $c(k)$, $b(k)$, and $d(k)$ are given by (3.8.57)–(3.8.59), respectively. Employing the intertwining properties

$$J_0^*B^\dagger = B^\dagger(0)J_0^*,\quad J_0B(0) = BJ_0,\quad J_0^*C_i^\dagger = C_i^\dagger(0)J_0^*,\quad J_0C_i(0) = C_iJ_0,$$

and factorization formula $J_0^*J_t = e^{-tH_{\mathrm{rad}}}$, we see by (3.8.61) that

$$\begin{aligned}W(k) &= J_0^* e^{\frac{B^\dagger}{\sqrt{2}}}e^{\frac{B}{\sqrt{2}}}\prod_{i=1}^k\left(\frac{C_i^\dagger}{\sqrt{2}}+\frac{C_i}{\sqrt{2}}\right)J_t\\ &= \sum_{p=0}^{[k/2]}\frac{1}{2^p}\sum_{i_1,\dots,i_{2p}}^{\mathrm{pair}}\left(\prod_{l=1}^p C_{i_{2l-1},i_{2l}}\right)\sum_{l=0}^{k-2p}\sum_{\{n_1,\dots,n_l\}\subset\{i_1,\dots,i_{2p}\}^c} X(0)e^{\frac{B^\dagger(0)}{\sqrt{2}}}e^{-tH_{\mathrm{rad}}}\times e^{\frac{B(t)}{\sqrt{2}}}Z(t),\end{aligned}$$

where $X(0)$ and $Z(t)$ are defined by X and Z with C_j replaced by $C_j(0)$ and $C_j(t)$, respectively. Let

$$\alpha = \frac{|e|}{\sqrt{2}}(\||k|\hat{\varphi}/\sqrt{\omega}\| + \||k|\hat{\varphi}/\omega\|),\quad \beta = \frac{|e|}{\sqrt{2}}\||k|\hat{\varphi}/\sqrt{\omega}\|.$$

Then $\|[C_i(l), C_j^\dagger(l)]\| \le \beta^2$ for any i, j. Hence we have the bound by Theorem 1.43. Let $\xi(l,m)$ and $\xi_r(l,m)$ be in Lemma 1.41, i. e.,

$$\xi(l,m) = \sum_{j,j'=0}^\infty \frac{(|e|\gamma)^{j+j''}\sqrt{(j+l)!}\sqrt{(j'+m)!}}{j!j'!}$$

and

$$\xi_r(l,m) = \sum_{j,j'=0}^{\infty} \frac{(|e|\gamma)^{j+j''}\sqrt{(j+l)!}\sqrt{(j'+m)!}}{r^{(j+l)/2}r^{(j'+m)/2}j!j'!}.$$

Here, $\gamma = \||k|\hat{\varphi}/\sqrt{\omega}\| + \||k|\hat{\varphi}/\omega\|$.

Case $t \geq 1$. We have

$$\|W(k)\Psi\| \leq w(k)\left\|e^{-\frac{1}{2}(t-1)H_{\mathrm{rad}}}\right\|^2 \|\Psi\|, \tag{3.8.62}$$

where

$$w(k) = k!\sum_{p=0}^{[k/2]}\sum_{l+m+2p=k}\frac{\left(\frac{\beta^2}{4}\right)^p (a+y)^l a^m}{p!l!m!}\xi(l,m)$$

and $y = t\beta^2$.

Case $t < 1$. For any $0 < r < t$, we have

$$\|W(k)\Psi\| \leq w(r,k)\left\|e^{-\frac{1}{2}(t-r)H_{\mathrm{rad}}}\right\|^2 \|\Psi\|, \tag{3.8.63}$$

where

$$w(r,k) = k!\sum_{p=0}^{[k/2]}\sum_{l+m+2p=k}\frac{\left(\frac{\beta^2}{4}\right)^p (a+y)^l a^m}{p!l!m!}\xi_r(l,m).$$

Hence $W(k)$ is bounded.

Since

$$\begin{aligned}
&\mathrm{J}_0^* e^{Y_1}\left(1+\prod_{i=1}^N((1+\varepsilon)\phi_i\overline{\phi_i}+\varepsilon^2)\right)\mathrm{J}_t \\
&\quad = \mathrm{J}_0^* e^{Y_1}\mathrm{J}_t + \sum_{k=0}^{N}\sum_{\{i_1,\dots,i_k\}\subset[N]}\varepsilon^{2(N-k)}(1+\varepsilon)^k \mathrm{J}_0^* e^{Y_1}\left(\prod_{j=1}^k \phi_{i_j}\bar{\phi}_{i_j}\right)\mathrm{J}_t,
\end{aligned}$$

where $[N] = \{1,\dots,N\}$, in a similar manner to (3.8.62) and (3.8.63) we see the bound

$$\left\|\mathrm{J}_0^* e^{Y_1}\left(\prod_{j=1}^k \phi_{i_j}\bar{\phi}_{i_j}\right)\mathrm{J}_t\right\| \leq w(r,2k)\left\|e^{-\frac{1}{2}(t-r)H_{\mathrm{rad}}}\right\|^2,$$

with $0 < r < t$ for $t < 1$ and $r = 1$ for $t \geq 1$, and have

$$\left\|\mathrm{J}_0^* e^{Y_1}\left(1+\prod_{i=1}^N((1+\varepsilon)\phi_i\overline{\phi_i}+\varepsilon^2)\right)\mathrm{J}_t\right\|$$

$$\leq \left(w(r,0) + \sum_{k=0}^{N} \frac{N!}{(N-k)!k!} \varepsilon^{2(N-k)} (1+\varepsilon)^k w(r,2k) \right) \left\| e^{-\frac{1}{2}(t-r)H_{\mathrm{rad}}} \right\|^2 .$$

Then the proof is complete. □

Lemma 3.94. *Let $V \in L^\infty(\mathbb{R}^3)$ and $\hat{\varphi} \in C_0^\infty(\mathbb{R}^3)$. Then for every $t \geq 0$ and all $F, G \in \mathscr{K}$, it follows that*

$$(F, e^{-tL} G) = e^t \sum_{\alpha=1,2} \int_{\mathbb{R}^3} \mathbb{E}^{x,\alpha} [e^{-\int_0^t V(B_s) ds} (F(q_0), \mathrm{J}_0^* e^Y \mathrm{J}_t G(q_t))] dx. \tag{3.8.64}$$

Proof. Let $F, G \in \mathscr{K}$ be such that $F \geq 0$ and $G \geq 0$. By Lemma 3.92, we have

$$(F, e^{-tL_R} G) = e^t \sum_{\alpha=1,2} \int_{\mathbb{R}^3} \mathbb{E}^{x,\alpha} [e^{-\int_0^t V(B_s) ds} (F(q_0), \mathrm{J}_0^* e^{Y_R} \mathrm{J}_t G(q_t))] dx. \tag{3.8.65}$$

Here, $0 \leq g(R) = (F(q_0), \mathrm{J}_0^* e^{Y_R} \mathrm{J}_t G(q_t))$ and $g(R)$ is a monotone increasing function on R. Hence $(F, e^{-tL_R} G)$ is also increasing on R and $(F, e^{-tL_R} G) \uparrow (F, e^{-tL} G)$ as $R \uparrow \infty$. For each $\phi \in \mathscr{Q}$, it follows that

$$\mathrm{J}_0 F(q_0) \cdot e^{Y_R} \mathrm{J}_t G(q_t) \to \mathrm{J}_0 F(q_0) \cdot e^Y \mathrm{J}_t G(q_t)$$

as $R \to \infty$. Then the monotone convergence theorem yields that the function on the right-hand side above $\mathrm{J}_0 F(q_0) \cdot e^Y \mathrm{J}_t G(q_t)$ is finite for a. e. $(\phi, x, \alpha, w, \tau) \in \mathscr{Q} \times \mathbb{R}^3 \times \{1,2\} \times \mathscr{X} \times S$, and

$$\begin{aligned}
(F, e^{-tL} G) &= \lim_{R\to\infty} (F, e^{-tL_R} G) = e^t \lim_{R\to\infty} \sum_{\alpha=1,2} \int_{\mathbb{R}^3} \mathbb{E}^{x,\alpha} [e^{-\int_0^t V(B_s) ds} (F(q_0), \mathrm{J}_0^* e^{Y_R} \mathrm{J}_t G(q_t))] dx \\
&= e^t \sum_{\alpha=1,2} \int_{\mathbb{R}^3} \mathbb{E}^{x,\alpha} [\lim_{R\to\infty} e^{-\int_0^t V(B_s) ds} (F(q_0), \mathrm{J}_0^* e^{Y_R} \mathrm{J}_t G(q_t))] dx \\
&= e^t \sum_{\alpha=1,2} \int_{\mathbb{R}^3} \mathbb{E}^{x,\alpha} \left[e^{-\int_0^t V(B_s) ds} \mathbb{E}_\mu [\mathrm{J}_0 F(q_0) \cdot e^Y \mathrm{J}_t G(q_t)] \right] dx.
\end{aligned}$$

For general $F, G \in \mathscr{K}$, the lemma is proven by decomposing F and G as a linear sum of positive functions, $F = \Re F_+ - \Re F_- + i(\Im F_+ - \Im F_-)$ and $G = \Re G_+ - \Re G_- + i(\Im G_+ - \Im G_-)$. Finally, we show that $\mathrm{J}_0^* e^Y \mathrm{J}_t G(q_t) \in L^2(\mathscr{Q})$ for each $(x, \alpha, w, \tau) \in \mathbb{R}^3 \times \{1,2\} \times \mathscr{X} \times S$ in Lemma 3.93. Then $\mathbb{E}_\mu [\mathrm{J}_0 F(q_0) \cdot e^Y \mathrm{J}_t G(q_t)] = (F(q_0), \mathrm{J}_0^* e^Y \mathrm{J}_t G(q_t))$ follows, and the lemma is proven. □

3.8.8 Functional integral representations

In this section, we state the main results on functional integral representation of semigroup $(F, e^{-tH_{\rm PF}^{\mathbb{Z}_2}} G)$. Let $X_t = Y_t(1) + Y_t(2) + Y_t(3)$, where

$$Y_t(1) = -ie\hat{A}_{\rm E}\left(\bigoplus_{\mu=1}^{3} \int_0^t \mathrm{j}_s \tilde{\varphi}(\cdot - B_s) dB_s^\mu\right),$$

$$Y_t(2) = -\int_0^t \hat{H}_{\rm E,d}(B_s, \theta_{N_s}, s) ds,$$

$$Y_t(3) = \int_0^{t+} W_E(B_s, -\theta_{N_{s-}}, s) dN_s.$$

Here, $W_E(B_s, -\theta_{N_{s-}}, s) = \log(-\hat{H}_{\rm E,od}(B_s, -\theta_{N_{s-}}, s))$. We shall see that X_t turns to be the exponent of integral kernel of $e^{-tH_{\rm PF}^{\mathbb{Z}_2}}$. Here, $Y_t(1)$ describes the part of quantized radiation field, $Y_t(2)$ the diagonal part of spin interaction, and $Y_t(3)$ the off-diagonal part of spin interaction. We, furthermore, define $X_t(\varepsilon)$ by X_t with $Y_t(3)$ replaced by $Y_t(3,\varepsilon)$, i. e., $X_t(\varepsilon) = Y_t(1) + Y_t(2) + Y_t(3,\varepsilon)$, where

$$Y_t(3, \varepsilon) = \int_0^{t+} W_{\rm E,\varepsilon}(B_s, -\theta_{N_{s-}}, s) dN_s$$

and $W_{\rm E,\varepsilon}(B_s, -\theta_{N_{s-}}, s) = \log(-\Psi_\varepsilon(\hat{H}_{\rm E,od}(B_s, -\theta_{N_{s-}}, s)))$.

Lemma 3.95. *For each $(x, \alpha, w, \tau) \in \mathbb{R}^3 \times \{1,2\} \times \mathscr{X} \times S$, $\mathrm{J}_0^* e^{X_t(\varepsilon)} \mathrm{J}_t$ is a bounded operator.*

Proof. Let $\Phi, \Psi \in L^2(\mathscr{Q})$. Using $Y = Y_t(2) + \int_0^{t+} \log \Psi_\varepsilon(|\hat{H}_{\rm E,od}(B_s, -\theta_{N_{s-}}, s)|) dN_s$ defined in (3.8.52), we see that

$$|(\Phi, \mathrm{J}_0^* e^{X_t(\varepsilon)} \mathrm{J}_t \Psi)| \le (|\Phi|, \mathrm{J}_0^* e^{Y} \mathrm{J}_t |\Psi|) \le \|\Phi\| \|\Psi\| \|\mathrm{J}_0^* e^{Y} \mathrm{J}_t\|$$

follows from Lemma 3.93. Then the lemma is proven. □

Theorem 3.96 (Functional integral representation for Pauli–Fierz Hamiltonian with spin 1/2). *For every $t \ge 0$ and all $F, G \in \mathscr{H}$, it follows that*

$$(F, e^{-tH_{\rm PF}^{\mathbb{Z}_2}} G) = \lim_{\varepsilon \downarrow 0} (F, e^{-tH_{\rm PF,\varepsilon}^{\mathbb{Z}_2}} G) \tag{3.8.66}$$

and

$$(F, e^{-tH_{\rm PF,\varepsilon}^{\mathbb{Z}_2}} G) = e^t \sum_{\alpha=1,2} \int_{\mathbb{R}^3} \mathbb{E}^{x,\alpha}[e^{-\int_0^t V(B_s) ds} (F(q_0), \mathrm{J}_0^* e^{X_t(\varepsilon)} \mathrm{J}_t G(q_t))] dx. \tag{3.8.67}$$

Proof. Since

$$e^{-tH_{\mathrm{PF},\varepsilon}^{\mathbb{Z}_2}} \to e^{-tH_{\mathrm{PF}}^{\mathbb{Z}_2}}$$

strongly as $\varepsilon \to 0$, (3.8.66) follows. Now we turn to proving (3.8.67). Suppose that $\tilde{\varphi} \in C_0^\infty(\mathbb{R}^3)$ and $V \in L^\infty(\mathbb{R}^3)$. Write

$$X_{S,T}(\varepsilon,s) = -ie\int_S^T \hat{A}_{\mathrm{E}}(\mathrm{j}_s\tilde{\varphi}(\cdot - B_r))dB_r - \int_S^T \hat{H}_{\mathrm{E,d}}(B_r,\theta_{N_r},s)dr + \int_S^{T+} W_{\mathrm{E},\varepsilon}(B_r,-\theta_{N_{r-}},s)dN_r.$$

Define $S_{t,s}^\varepsilon : \mathscr{K}_{\mathrm{E}}^\infty \to \mathscr{K}_{\mathrm{E}}^\infty$ by

$$(S_{t,s}^\varepsilon G)(x,\theta_\alpha) = e^t\mathbb{E}^{x,\alpha}[e^{-\int_0^t V(B_r)dr}e^{X_{0,t}(\varepsilon,s)}G(q_t)].$$

It can be seen that $S_{t,s}^\varepsilon$ has the property:

$$S_{t,s}^\varepsilon S_{t',s'}^\varepsilon G(x,\theta_\alpha) = e^{t+t'}\mathbb{E}^{x,\alpha}[e^{-\int_0^{t+t'} V(B_r)dr}e^{X_{0,t}(\varepsilon,s)+X_{t,t+t'}(\varepsilon,s')}G(q_{t+t'})]. \tag{3.8.68}$$

Note that for $s_1 \le \cdots \le s_n$,

$$e^{X_{0,t_1}(\varepsilon,s_1)+X_{t_1,t_1+t_2}(\varepsilon,s_2)+\cdots+X_{t_1+\cdots+t_{n-1},t_1+\cdots+t_n}(\varepsilon,s_n)} \in \mathscr{E}_{[s_1,s_n]}. \tag{3.8.69}$$

Since

$$(F,e^{-tH_{\mathrm{PF},\varepsilon}^{\mathbb{Z}_2}}G) = (F,e^{-tK_\varepsilon}G),$$

similar to the proof of Theorem 1.98, by the Trotter product formula, (3.8.68), (3.8.69) and the Markov property of E_s, $s \in \mathbb{R}$, we obtain

$$\begin{aligned}(F,e^{-tH_{\mathrm{PF},\varepsilon}^{\mathbb{Z}_2}}G) &= \lim_{n\to\infty}\left(F,\mathrm{J}_0^*\Big(\prod_{i=0}^{n-1} S_{t/n,it/n}^\varepsilon\Big)\mathrm{J}_t G\right)\\ &= e^t\lim_{n\to\infty}\sum_{\alpha=1,2}\int_{\mathbb{R}^3}\mathbb{E}^{x,\alpha}\left[e^{-\int_0^t V(B_r)dr}\left(F(q_0),\mathrm{J}_0^* e^{X_t^n(\varepsilon)}\mathrm{J}_t G(q_t)\right)\right]dx,\end{aligned} \tag{3.8.70}$$

where $X_t^n(\varepsilon) = Y_t^n(1) + Y_t^n(2) + Y_t^n(3,\varepsilon)$ with

$$Y_t^n(1) = -ie\hat{A}_{\mathrm{E}}\left(\bigoplus_{\mu=1}^3\sum_{j=1}^n\int_{t_{j-1}}^{t_j}\mathrm{j}_{t_{j-1}}\tilde{\varphi}(\cdot - B_s)dB_s^\mu\right),$$

$$Y_t^n(2) = -\sum_{j=1}^n\int_{t_{j-1}}^{t_j}\hat{H}_{\mathrm{E,d}}(B_s,\theta_{N_s},t_{j-1})ds,$$

$$Y_t^n(3,\varepsilon) = \sum_{j=1}^{n} \int_{t_{j-1}}^{t_j} W_{E,\varepsilon}(B_s, -\theta_{N_{s-}}, t_{j-1}) dN_s,$$

and $t_j = jt/n$. Put

$$\langle F, T_n G\rangle = e^t \sum_{\alpha=1,2} \int_{\mathbb{R}^3} \mathbb{E}^{x,\alpha} \left[e^{-\int_0^t V(B_r)dr} \left(F(q_0), J_0^* e^{X_t^n(\varepsilon)} J_t G(q_t) \right) \right] dx,$$

$$\langle F, TG\rangle = e^t \sum_{\alpha=1,2} \int_{\mathbb{R}^3} \mathbb{E}^{x,\alpha} \left[e^{-\int_0^t V(B_r)dr} \left(F(q_0), J_0^* e^{X_t(\varepsilon)} J_t G(q_t) \right) \right] dx.$$

Notice that, by the definition of $\langle F, T_n G\rangle$, we see that

$$|\langle F, T_n G\rangle| = \left| \left(F, J_0^* \left(\prod_{i=0}^{n-1} J_{it/n} e^{-\frac{t}{n} K_\varepsilon(\hat{A})} J_{it/n}^* \right) J_t G \right) \right| \le \|F\| \|G\|.$$

Suppose that $F_m \to F$ and $G_m \to G$ as $m \to \infty$. Then by telescoping, we can see that

$$\begin{aligned} |\langle F, T_n G\rangle - \langle F_m, T_n G_m\rangle| &\le |\langle F, T_n G\rangle - \langle F_m, T_n G\rangle| + |\langle F_m, T_n G\rangle - \langle F_m, T_n G_m\rangle| \\ &\le \|F - F_m\| \|G\| + \|G - G_m\| \|F_m\|. \end{aligned}$$

Using the facts that $|J_t F| \le J_t |F|$, $|e^{\pm Y_t(1)} F| = |F|$ and $|e^{Y_t(2)+Y_t(3,\varepsilon)} F| = e^Y |F|$, we see that

$$\begin{aligned} |\langle F, TG\rangle| &\le e^t \sum_{\alpha=1,2} \int_{\mathbb{R}^3} \mathbb{E}^{x,\alpha} \left[e^{-\int_0^t V(B_r)dr} \left(|e^{-Y_t(1)} J_0 F(q_0)|, e^{Y_t(2)} |e^{Y_t(3)}| J_t |G(q_t)| \right) \right] dx \\ &\le e^t \sum_{\alpha=1,2} \int_{\mathbb{R}^3} \mathbb{E}^{x,\alpha} \left[e^{-\int_0^t V(B_r)dr} \left(J_0 |F(q_0)|, e^Y J_t |G(q_t)| \right) \right] dx \\ &= (|F|, e^{-tL} |G|) \le a \|F\| \|G\|, \end{aligned}$$

where $a = e^{-t \inf \mathrm{Spec}(L)}$. Hence we have

$$\begin{aligned} |\langle F_m, TG_m\rangle - \langle F, TG\rangle| &\le |\langle F_m, TG_m\rangle - \langle F, TG_m\rangle| + |\langle F, TG_m\rangle - \langle F, TG\rangle| \\ &\le a\|F_m - F\| \|G_m\| + a\|F\| \|G - G_m\|. \end{aligned}$$

Suppose that $\|F - F_m\| < \varepsilon$ and $\|G - G_m\| < \varepsilon$. Together with them we have

$$\begin{aligned} &|\langle F, T_n G\rangle - \langle F, TG\rangle| \\ &\le |\langle F, T_n G\rangle - \langle F_m, T_n G_m\rangle| + |\langle F_m, T_n G_m\rangle - \langle F_m, TG_m\rangle| + |\langle F_m, TG_m\rangle - \langle F, TG\rangle| \\ &\le \varepsilon(\|G\| + a\|G_m\| + \|F_m\| + a\|F\|) + |\langle F_m, T_n G_m\rangle - \langle F_m, TG_m\rangle|. \end{aligned} \tag{3.8.71}$$

It can be concluded from (3.8.71) that it is sufficient to show the lemma for arbitrary F, G included in some dense domain. We shall show the lemma for $F, G \in \mathscr{K}^\infty$. We claim that:

(1) For each $(x, \alpha, w, \tau) \in \mathbb{R}^3 \times \{1, 2\} \times \mathscr{X} \times \mathcal{S}$, there exists $C_t(\tau)$ such that

$$|(F, \mathrm{J}_0^* e^{X_t^n(\varepsilon)} \mathrm{J}_t G)_{L^2(\mathscr{Q})}| \le C_t(\tau),$$

where $C_t(\tau)$ is independent of $(x, \alpha, w) \in \mathbb{R}^3 \times \{1, 2\} \times \mathscr{X}$, $\varepsilon > 0$ and n, and $\mathbb{E}_P[C_t^2] < \infty$.

(2) For each $(x, \alpha, \tau) \in \mathbb{R}^3 \times \{1, 2\} \times \mathcal{S}$,

$$\lim_{n\to\infty} \mathbb{E}_{\mathcal{W}}^x[(F(q_0), \mathrm{J}_0^* e^{X_t^n(\varepsilon)} \mathrm{J}_t G(q_t))] = \mathbb{E}_{\mathcal{W}}^x[(F(q_0), \mathrm{J}_0^* e^{X_t(\varepsilon)} \mathrm{J}_t G(q_t))].$$

The proofs of (1) and (2) will be given in Lemmas 3.97 and 3.98 below, respectively. We set

$$\text{RHS } (3.8.70) = e^t \sum_{\alpha=1,2} \int_{\mathbb{R}^3} \mathbb{E}^{x,\alpha}[\xi_n] dx, \tag{3.8.72}$$

where $\xi_n = e^{-\int_0^t V(B_s)ds} (F(q_0), \mathrm{J}_0^* e^{X_t^n(\varepsilon)} \mathrm{J}_t G(q_t))$. Thus we have

$$\mathbb{E}_{\mathcal{W}}^x[|\xi_n|] \le \mathbb{E}_{\mathcal{W}}^x[e^{-\int_0^t V(B_s)ds} C_t(\tau) \|F(q_0)\| \|G(q_t)\|]$$

and

$$\sum_{\alpha=1,2} \int_{\mathbb{R}^3} \mathbb{E}^{x,\alpha}[e^{-\int_0^t V(B_s)ds} C_t(\tau) \|F(q_0)\| \|G(q_t)\|] dx \le V_M^{1/2} (\mathbb{E}_P[C_t^2])^{1/2} \|F\| \|G\| < \infty.$$

Since $\mathbb{E}_{\mathcal{W}}^x[\xi_n] \to \mathbb{E}_{\mathcal{W}}^x[\xi_\infty]$ as $n \to \infty$ for each $(x, \alpha, \tau) \in \mathbb{R}^3 \times \{1, 2\} \times \mathcal{S}$, by the Lebesgue dominated convergence theorem,

$$\lim_{n\to\infty} \sum_{\alpha=1,2} \int_{\mathbb{R}^3} \mathbb{E}_P^\alpha[\mathbb{E}_{\mathcal{W}}^x[\xi_n]] dx = \sum_{\alpha=1,2} \int_{\mathbb{R}^3} \mathbb{E}_P^\alpha[\mathbb{E}_{\mathcal{W}}^x[\xi_\infty]] dx$$

follows. For $V \in \mathcal{R}_{\text{Kato}}$, by a limiting argument, we can show the theorem. Finally, for $\hat{\varphi}$ in Assumption 3.4, we can see (3.8.67) by an approximation, which is also shown in Lemma 3.99 below. □

Lemma 3.97. *Statement* (1) *of the proof of Theorem* 3.96 *is true.*

Proof. For each $(x, \alpha, w, \tau) \in \mathbb{R}^3 \times \{1, 2\} \times \mathscr{X} \times \mathcal{S}$, it can be directly seen that $F, G \in D(e^{X_t(\varepsilon)} \mathrm{J}_0) \cap D(e^{X_t^n(\varepsilon)} \mathrm{J}_0)$ for any $n \in \mathbb{N}$. We get the inequality

$$\|Y_t(2)\Phi\| \le \gamma \sqrt{m+1} \|\Phi\|$$

for $\Phi \in L^2_m(\mathscr{Q})$, where $\gamma = \sqrt{2}t|e|\||k|\hat{\varphi}/\sqrt{\omega}\|$. Since $L^2_{\rm fin}(\mathscr{Q})$ is the set of analytic vectors for $e^{Y^n_t(2)}$, for $F \in L^2_{\rm fin}(\mathscr{Q})$ we have $e^{Y^n_t(2)}F = \sum_{j=0}^\infty \frac{Y^n_t(2)^j}{j!}F$. We assume that $F \in L^2_{m'}(\mathscr{Q})$ and $G \in L^2_m(\mathscr{Q})$, hence $NF = m'F$ and $NG = mG$ for the number operator N. Then

$$\left\| \frac{Y^n_t(2)^j}{j!}F \right\| \leq \frac{\sqrt{(m+j-1)\cdots(m+1)m}}{j!}\gamma^j \|F\|.$$

Thus

$$\left\| e^{Y^n_t(2)}F \right\| \leq \sum_{j=0}^\infty \frac{\sqrt{(m+j-1)\cdots(m+1)m}}{j!}\gamma^j \|F\|.$$

On the other hand, by the definition of $Y^n_t(3,\varepsilon)$, we have

$$e^{Y^n_t(3,\varepsilon)} = \prod_{i=1}^n \exp\left(\int_{t(i-1)/n}^{ti/n+} W_{{\rm E},\varepsilon}(B_s, -\theta_{N_{s-}}, t(i-1)/n) dN_s \right). \tag{3.8.73}$$

For every $\tau \in \mathcal{S}$, there exists $N = N(\tau) \in \mathbb{N}$ such that the map $t \mapsto N_t(\tau)$ is not continuous at points $s_1 = s_1(\tau), \ldots, s_N = s_N(\tau)$. For a sufficiently large n, the number of discontinuous points located in the interval $(t(i-1)/n, ti/n]$ is at most one. Then by taking n large enough and putting $(n(s_i), n(s_i) + t/n]$ for the interval containing s_i, we get

$$e^{Y^n_t(3,\varepsilon)} = \prod_{i=1}^N (-\phi_i - \varepsilon\psi_\varepsilon(\phi_i)),$$

where $\phi_i = \hat{H}_{\rm E,od}(B_{s_i}, -\theta_{N_{s_i-}}, n(s_i))$. Fixing $\tau \in \mathcal{S}$ above, we shall estimate $|(F, {\rm J}_0^* e^{Y^n_t(1)} e^{Y^n_t(2)} \prod_{i=1}^N (-\phi_i - \varepsilon\psi_\varepsilon(\phi_i))G)|$. We have

$$\prod_{i=1}^N (-\phi_i - \varepsilon\psi_\varepsilon(\phi_i)) = (-1)^N \sum_{p=0}^N \varepsilon^{N-p} \sum_{\{j_1,\ldots,j_p\}\subset[N]} \left(\prod_{i\notin\{j_1,\ldots,j_p\}} \psi_i \right) \left(\prod_{i=1}^p \phi_{j_i} \right),$$

where $\psi_i = \psi_\varepsilon(\phi_{j_i})$ and $[N] = \{1, \ldots, N\}$, and

$$(F, {\rm J}_0^* e^{Y^n_t(1)} e^{Y^n_t(2)} e^{Y^n_t(3)} {\rm J}_t G) = \sum_{p=0}^N \varepsilon^{N-p} \sum_{\{j_1,\ldots,j_p\}\subset[N]} \left(e^{-Y^n_t(1)} {\rm J}_0 F, e^{Y^n_t(2)} \prod_{i\notin\{j_1,\ldots,j_p\}} \psi_i \prod_{i=1}^p \phi_{j_i} {\rm J}_t G \right).$$

Notice again that

$$\|\phi_{j_i}\Phi\| \leq 2\gamma\sqrt{m+1}\|\Phi\|$$

for $\Phi \in L^2_m(\mathscr{Q})$ and $i = 1, \ldots, p$. Since $\psi_\varepsilon \leq 1$, each term on the right-hand side above can be estimated as

$$\begin{aligned}&\left|\left(e^{-Y_t^n(1)}\mathrm{J}_0F, e^{Y_t^n(2)}\prod_{i\notin\{j_1,\dots,j_p\}}\psi_i\prod_{i=1}^p\phi_{j_i}\mathrm{J}_tG\right)\right|\\&\quad\le\sum_{j=0}^\infty\varepsilon^{N-p}\frac{\sqrt{(j+p+m-1)\cdots(m+1)m}}{j!}(2\gamma)^{j+p-m+1}\|F\|\|G\|.\end{aligned}$$

Thus

$$\begin{aligned}&|(F,\mathrm{J}_0^*e^{Y_t^n(1)}e^{Y_t^n(2)}e^{Y_t^n(3,\varepsilon)}\mathrm{J}_tG)|\\&\quad\le\sum_{p=0}^N\frac{\varepsilon^{N-p}N!}{p!(N-p)!}\sum_{j=0}^\infty\frac{\sqrt{(j+p+m-1)\cdots(m+1)m}}{j!}(2\gamma)^{j+p-m+1}\|F\|\|G\|\\&\quad\le A_t(N)\|F\|\|G\|,\end{aligned}$$

where

$$A_t(N)=\sum_{j=0}^\infty(1+\varepsilon)^N\frac{\sqrt{(j+N+m-1)\cdots(m+1)m}}{j!}(2\gamma)^{j+N-m+1}.$$

The number N depends on $\tau\in S$, then $A_t(N(\tau))$ turns to be a random process on $(\mathcal{S},\mathcal{B}_{\mathcal{S}},P)$. We set $C_t(\tau)=A_t(N(\tau))$. For each $(x,\alpha,w,\tau)\in\mathbb{R}^3\times\{1,2\}\times\mathscr{X}\times\mathcal{S}$, $|(F,\mathrm{J}_0^*e^{X_t^n(\varepsilon)}\mathrm{J}_tG)|$ is finite with the bound $|(F,\mathrm{J}_0^*e^{X_t^n(\varepsilon)}\mathrm{J}_tG)|\le C_t(\tau)$ and

$$\begin{aligned}\mathbb{E}_P[C_t^2]&=\sum_{N=0}^\infty\frac{t^NA_t(N)^2}{N!}e^{-t}\\&\le e^{-t}\sum_{N=0}^\infty\left(\sum_{j=0}^\infty(1+\varepsilon)^N(2\gamma)^{j+N-m+1}\frac{\sqrt{(j+N+m-1)\cdots(m+1)m}}{N!j!}\right)^2<\infty.\end{aligned}$$

Thus (1) follows. □

Lemma 3.98. *Statement* (2) *of the proof of Theorem* 3.96 *is true.*

Proof. We show the convergence of $\mathrm{J}_0^*e^{X_t^n(\varepsilon)}\mathrm{J}_t$ as $n\to\infty$. Let $F\in L^2_{\mathrm{fin}}(\mathscr{Q})$. We show the convergence of $e^{Y_t^n(1)}$, $e^{Y_t^n(2)}$ and $e^{Y_t^n(3,\varepsilon)}$, separately. Thus

$$\begin{aligned}\|e^{-Y_t(1)}\mathrm{J}_0F-e^{-Y_t^n(1)}\mathrm{J}_0F\|&\le|e|\left\|\hat{A}_{\mathrm{E}}\left(\bigoplus_{\mu=1}^3\sum_{j=1}^n\int_{t_{j-1}}^{t_j}(\mathrm{j}_{t_{j-1}}-\mathrm{j}_s)\tilde{\varphi}(\cdot-B_s)dB_s^\mu\right)F\right\|\\&\le|e|\left\|\bigoplus_{\mu=1}^3\sum_{j=1}^n\int_{t_{j-1}}^{t_j}(\mathrm{j}_{t_{j-1}}-\mathrm{j}_s)\tilde{\varphi}(\cdot-B_s)dB_s^\mu\right\|\|(N+1)^{1/2}F\|.\end{aligned}$$

Then it follows that

$$\mathbb{E}_{\mathcal{W}}^{x}[\|e^{-Y_t(1)}\mathrm{J}_0F - e^{-Y_t^n(1)}\mathrm{J}_0F\|^2] \le 3|e|\sum_{j=1}^{n}\int_{t_{j-1}}^{t_j}\|(\mathrm{j}_{t_{j-1}} - \mathrm{j}_s)\tilde{\varphi}(\cdot - B_s)\|^2 ds\|(N+1)^{1/2}F\|^2$$

$$= 3|e|\sum_{j=1}^{n}\int_{t_{j-1}}^{t_j}\left(\frac{\hat{\varphi}}{\sqrt{\omega}}, (1 - e^{-|s-t_{j-1}|\omega})\frac{\hat{\varphi}}{\sqrt{\omega}}\right) ds\|(N+1)^{1/2}F\|^2 \le 3|e|\frac{t}{n}\|\sqrt{\omega}\hat{\varphi}\|^2\|(N+1)^{1/2}F\|^2.$$

Hence we have

$$\lim_{n\to\infty}\mathbb{E}_{\mathcal{W}}^{x}[\|e^{-Y_t(1)}\mathrm{J}_0F - e^{-Y_t^n(1)}\mathrm{J}_0F\|] = 0. \tag{3.8.74}$$

On the other hand in the Fock representation, $Y_t(2)$ and $Y_t^n(2)$ are represented as

$$Y_t(2) = \frac{-1}{\sqrt{2}}\sum_{j=\pm}\left\{a_E^*\left(\int_0^t b_j^-(s)ds, j\right) - a_E\left(\int_0^t b_j^+(s)ds, j\right)\right\},$$

$$Y_t^n(2) = \frac{-1}{\sqrt{2}}\sum_{j=\pm}\sum_{i=1}^{n}\left\{a_E^*\left(\int_{t_{i-1}}^{t_i} b_j^-(t_{i-1})ds, j\right) - a_E\left(\int_{t_{i-1}}^{t_i} b_j^+(t_{i-1})ds, j\right)\right\}.$$

Here, $b_j^{\pm}(s) = S_j(\theta_{N_s})e^{\pm ik\cdot B_s}\hat{\mathrm{j}}_s\frac{\hat{\varphi}}{\sqrt{\omega}}$ and $b_j^{\pm}(t_{i-1}) = S_j(\theta_{N_s})e^{\pm ik\cdot B_s}\hat{\mathrm{j}}_{t_{i-1}}\frac{\hat{\varphi}}{\sqrt{\omega}}$. Then the distance between test functions of $Y_t(2)$ and $Y_t^n(2)$ can be estimated as

$$\left\|\int_0^t b_j^-(s)ds - \sum_{i=1}^{n}\int_{t_{i-1}}^{t_i} b_j^-(t_{i-1})ds\right\|^2 \le t\sum_{i=1}^{n}\int_{t_{i-1}}^{t_i}\left\|S_j(\theta_{N_s})e^{-ik\cdot B_s}\left(\hat{\mathrm{j}}_s\frac{\hat{\varphi}}{\sqrt{\omega}} - \hat{\mathrm{j}}_{t_{i-1}}\frac{\hat{\varphi}}{\sqrt{\omega}}\right)\right\|^2 ds.$$

Since

$$\left\|S_j(\theta_{N_s})e^{-ik\cdot B_s}\left(\hat{\mathrm{j}}_s\frac{\hat{\varphi}}{\sqrt{\omega}} - \hat{\mathrm{j}}_{t_{i-1}}\frac{\hat{\varphi}}{\sqrt{\omega}}\right)\right\|^2 \le 2|e|^2|s - t_{i-1}|\||k|\hat{\varphi}\|^2,$$

we see that

$$\left\|\int_0^t b_j^-(s)ds - \sum_{i=1}^{n}\int_{t_{i-1}}^{t_i} b_j^-(t_{i-1})ds\right\|^2 \le \frac{2|e|^2}{n}t^2\||k|\hat{\varphi}\|^2.$$

It is concluded that

$$\lim_{n\to\infty}\left\|\int_0^t b_j^-(s)ds - \sum_{i=1}^{n}\int_{t_{i-1}}^{t_i} b_j^-(t_{i-1})ds\right\| = 0. \tag{3.8.75}$$

Furthermore, by (3.8.75) it can be straightforwardly seen that

$$e^{Y_t^n(2)}F \to e^{Y_t(2)}F \tag{3.8.76}$$

due to the expansion $e^{Y_t^n(2)}F = \sum_{j=0}^{\infty} \frac{Y_t^n(2)^j}{j!}F$.

By the same argument used in the proof of Lemma 3.97, for every $\tau \in \mathcal{S}$, there exists $N = N(\tau) \in \mathbb{N}$ such that the map $t \mapsto N_t(\tau)$ is not continuous at points $s_1 = s_1(\tau), \ldots, s_N = s_N(\tau)$. We suppose that s_i is located in $(n(s_i), n(s_i) + t/n]$ for a sufficiently large n. We get

$$e^{Y_t^n(3,\varepsilon)} = \prod_{i=1}^{N}(-\phi_i - \varepsilon\psi_\varepsilon(\phi_i)),$$

where $\phi_i = \hat{H}_{\mathrm{E,od}}(B_{s_i}, -\theta_{N_{s_i-}}, n(s_i))$. In the Fock representation, we can also have

$$\hat{H}_{\mathrm{E,od}}(B_{s_i}, -\theta_{N_{s_i-}}, s_i) \cong \frac{1}{\sqrt{2}} \sum_{j=\pm} \left\{a_E^*\left(c_j^-(s_i), j\right) - a_E\left(c_j^+(s_i), j\right)\right\},$$

$$\hat{H}_{\mathrm{E,od}}(B_{s_i}, -\theta_{N_{s_i-}}, n(s_i)) \cong \frac{1}{\sqrt{2}} \sum_{j=\pm} \left\{a_E^*\left(c_j^-(n(s_i)), j\right) - a_E\left(c_j^+(n(s_i)), j\right)\right\},$$

where $c_j^\pm(t) = T_j(-\theta_{N_{s_i-}})(\mathrm{j}_t \frac{\hat{\varphi}}{\sqrt{\omega}})e^{\pm ik\cdot B_{s_i}}$. We then estimate the distance between their test functions as

$$\left\|c_j^-(n(s_i)) - c_j^-(s_i)\right\|^2 \le 2|e|^2 \left(\frac{|k|\hat{\varphi}}{\sqrt{\omega}}, (1 - e^{-|n(s_i)-s_i|\omega})\frac{|k|\hat{\varphi}}{\sqrt{\omega}}\right).$$

Clearly, $n(s_i) \to s_i$ as $n \to \infty$. Then

$$\lim_{n\to\infty} \hat{H}_{\mathrm{E,od}}(B_{s_i}, -\theta_{N_{s_i-}}, n(s_i)) = \hat{H}_{\mathrm{E,od}}(B_{s_i}, -\theta_{N_{s_i-}}, s_i),$$

$$\lim_{n\to\infty} \psi_\varepsilon(\hat{H}_{\mathrm{E,od}}(B_{s_i}, -\theta_{N_{s_i-}}, n(s_i))) = \psi_\varepsilon(\hat{H}_{\mathrm{E,od}}(B_{s_i}, -\theta_{N_{s_i-}}, s_i))$$

on $\mathscr{K}^\infty$, and we have

$$\lim_{n\to\infty} e^{Y_t^n(3,\varepsilon)} = \prod_{i=1}^{N}(-\Psi_\varepsilon(\hat{H}_{\mathrm{E,od}}(B_{s_i}, -\theta_{N_{s_i-}}, s_i))). \tag{3.8.77}$$

Then by (3.8.74), (3.8.76), and (3.8.77), we can see that for each $(x, \alpha, \tau) \in \mathbb{R}^3 \times \{1, 2\} \times \mathcal{S}$,

$$\mathbb{E}_{\mathcal{W}}^x\left[\left(F(q_0), \mathrm{J}_0^* e^{X_t^n(\varepsilon)} \mathrm{J}_t G(q_t)\right)\right] \to \mathbb{E}_{\mathcal{W}}^x\left[\left(F(q_0), \mathrm{J}_0^* e^{X_t(\varepsilon)} \mathrm{J}_t G(q_t)\right)\right]$$

as $n \to \infty$, $|\mathbb{E}_{\mathcal{W}}^x[(F(q_0), \mathrm{J}_0^* e^{X_t^n(\varepsilon)} \mathrm{J}_t G(q_t))]| \le C_t(\tau)\mathbb{E}_W^x[\|F(q_0)\|\|G(q_t)\|]$, and the dominated function $C_t(\tau)\|F(q_0)\|\|G(q_t)\|$ is integrable. The Lebesgue dominated convergence theorem yields

$$\sum_{\alpha=1,2} \int_{\mathbb{R}^3} \mathbb{E}^{x,\alpha}\left[e^{-\int_0^t V(B_r)dr}\left(F(q_0), \mathrm{J}_0^* e^{X_t^n(\varepsilon)} \mathrm{J}_t G(q_t)\right)\right] dx$$

$$\to \sum_{\alpha=1,2} \int_{\mathbb{R}^3} \mathbb{E}^{x,\alpha}\left[e^{-\int_0^t V(B_r)dr}\left(F(q_0), \mathrm{J}_0^* e^{X_t(\varepsilon)} \mathrm{J}_t G(q_t)\right)\right] dx$$

for $F, G \in \mathscr{K}^\infty$ as $n \to \infty$. Hence the proof of (2) is complete. □

Lemma 3.99. *Equation* (3.8.67) *is valid for* $\hat{\varphi}$ *in Assumption* 3.4.

Proof. It is sufficient to show (3.8.67) for $F, G \in \mathscr{K}^\infty$ by an approximation argument. Take a sequence $(\hat{\varphi}_n)_{n\in\mathbb{N}} \subset C_0^\infty(\mathbb{R}^3)$ such that $|k|\hat{\varphi}_n/\sqrt{\omega} \to |k|\hat{\varphi}/\sqrt{\omega}$ as $n \to \infty$. Then (3.8.67) is valid for each $\hat{\varphi}_n$. Indeed,

$$(F, e^{-tH_{\mathrm{PF}}^{\mathbb{Z}_2}(n)} G) = \sum_{\alpha=1,2} \int_{\mathbb{R}^3} \mathbb{E}^{x,\alpha}[e^{-\int_0^t V(B_s)ds}(F(q_0), \mathrm{J}_0^* e^{X_t(\varepsilon)_n} \mathrm{J}_t G(q_t))] dx.$$

Here, $X_t(\varepsilon)_n$ is defined by $X_t(\varepsilon)$ with $\hat{\varphi}$ replaced by $\hat{\varphi}_n$, and $H_{\mathrm{PF}}^{\mathbb{Z}_2}(n)$ by $H_{\mathrm{PF}}^{\mathbb{Z}_2}$ with $\hat{\varphi}$ replaced by $\hat{\varphi}_n$. It can be seen that $H_{\mathrm{PF}}^{\mathbb{Z}_2}(n) \to H_{\mathrm{PF}}^{\mathbb{Z}_2}$ as $n \to \infty$ on a common core $\mathscr{K}^\infty$. Then $e^{-tH_{\mathrm{PF}}^{\mathbb{Z}_2}(n)} \to e^{-tH_{\mathrm{PF}}^{\mathbb{Z}_2}}$ strongly as $n \to \infty$. Let $Y_t^n(1)$, $Y_t^n(2)$, and $Y_t^n(3,\varepsilon)$ be $Y_t(1)$, $Y_t(2)$, and $Y_t(3,\varepsilon)$ with $\hat{\varphi}$ replaced by $\hat{\varphi}_n$, respectively. By a similar approximation argument used in the proof of Lemma 3.98 above, we can show that:
(1) $\mathbb{E}_{\mathcal{W}}^X[\|(e^{Y_t^n(1)} - e^{Y_t(1)})F] \to 0$ as $\|(\hat{\varphi}_n - \hat{\varphi})/\sqrt{\omega}\| \to 0$;
(2) $\|(e^{Y_t^n(2)} - e^{Y_t(2)})F\| \to 0$ as $\||k|(\hat{\varphi}_n - \hat{\varphi})/\sqrt{\omega}\| \to 0$;
(3) $\|(e^{Y_t^n(3,\varepsilon)} - e^{Y_t(3,\varepsilon)})F\| \to$ as $\||k|(\hat{\varphi}_n - \hat{\varphi})/\sqrt{\omega}\| \to 0$

for $F \in L_{\mathrm{fin}}^2(\mathscr{Q})$. Hence

$$\begin{aligned}(F, e^{-tH_{\mathrm{PF}}^{\mathbb{Z}_2}} G) &= \lim_{n\to\infty} \sum_{\alpha=1,2} \int_{\mathbb{R}^3} \mathbb{E}^{x,\alpha}[e^{-\int_0^t V(B_s)ds}(F(q_0), \mathrm{J}_0^* e^{X_t(\varepsilon)_n} \mathrm{J}_t G(q_t))] dx \\ &= \sum_{\alpha=1,2} \int_{\mathbb{R}^3} \mathbb{E}^{x,\alpha}[e^{-\int_0^t V(B_s)ds}(F(q_0), \mathrm{J}_0^* e^{X_t(\varepsilon)} \mathrm{J}_t G(q_t))] dx\end{aligned}$$

for $F, G \in \mathscr{K}^\infty$. Then the proof is complete. □

As was mentioned above, we need the regularization $\Psi_\varepsilon(\hat{H}_{\mathrm{od}})$ of $\hat{H}_{\mathrm{od}}$ to prevent zeros of $\hat{H}_{\mathrm{od}}$. The zeros of $\hat{H}_{\mathrm{od}}$ produce the zeros of $e^{\int_0^{t+} \log(\hat{H}_{\mathrm{E,od}}(B_s, -\theta_{N_{s-}}, s))dN_s}$. In order to avoid $\lim_{\varepsilon\to 0}$ in the functional integral representation, instead of introducing regularization Ψ_ε, we introduce a subset W of $\mathbb{R}^3 \times \mathscr{X} \times S \times \mathscr{Q}_{\mathrm{E}}$ by

$$W = \left\{ \int_0^{t+} \log\left(\frac{|e|}{2}\sqrt{|b_1(s,x,B_s)|^2 + |b_2(s,x,B_s)|^2}\right) dN_s > -\infty \right\}, \tag{3.8.78}$$

where $b_\alpha(s, x, B_s) = \hat{B}_{\mathrm{E}_\alpha}(\mathrm{j}_s \tilde{\varphi}(\cdot - B_s - x))$.

Theorem 3.100 (Functional integral representation for Pauli–Fierz Hamiltonian with spin 1/2). *For every $t \geq 0$ and all $F, G \in \mathscr{K}$, it follows that*

$$(F, e^{-tH_{\mathrm{PF}}^{\mathbb{Z}_2}} G) = e^t \sum_{\alpha=1,2} \int_{\mathbb{R}^3} \mathbb{E}^{x,\alpha}[e^{-\int_0^t V(B_s)ds} (\mathrm{J}_0 F(q_0), e^{X_t} 1_W \mathrm{J}_t G(q_t))] dx. \tag{3.8.79}$$

Here, W is given by (3.8.78), *the exponent X_t is defined by*

$$X_t = -ie \int_0^t \hat{A}_{\mathrm{E}}(\mathrm{j}_s \tilde{\varphi}(\cdot - B_s)) dB_s - \int_0^t \hat{H}_{\mathrm{E,d}}(B_s, \theta_{N_s}, s) ds + \int_0^{t+} \log(-\hat{H}_{\mathrm{E,od}}(B_s, -\theta_{N_{s-}}, s)) dN_s,$$

and

$$\int_0^{t+} \log(-\hat{H}_{\mathrm{E,od}}(B_s, -\theta_{N_{s-}}, s)) dN_s = \sum_{\substack{r \in [0,t] \\ N_{r+} \neq N_{r-}}} \log(-\hat{H}_{\mathrm{E,od}}(B_r, -\theta_{N_{r-}}, r))$$

is possibly infinite.

Proof. By the Lebesgue dominated convergence theorem, we have

$$\begin{aligned} &\lim_{\varepsilon \to 0} \int_{\mathbb{R}^3} \mathbb{E}^{x,\alpha}[e^{-\int_0^t V(B_s)ds} (\mathrm{J}_0 F(q_0), e^{X_t(\varepsilon)} \mathrm{J}_t G(q_t))] dx \\ &\quad = \int_{\mathbb{R}^3} \mathbb{E}^{x,\alpha}[e^{-\int_0^t V(B_s)ds} \lim_{\varepsilon \to 0} (\mathrm{J}_0 F(q_0), e^{X_t(\varepsilon)} \mathrm{J}_t G(q_t))] dx. \end{aligned}$$

We can also show that

$$\lim_{\varepsilon \to 0} (\mathrm{J}_0 F(q_0), 1_{W^c} e^{X_t(\varepsilon)} \mathrm{J}_t G(q_t)) = 0$$

and

$$\lim_{\varepsilon \to 0} (\mathrm{J}_0 F(q_0), 1_W e^{X_t(\varepsilon)} \mathrm{J}_t G(q_t)) = (\mathrm{J}_0 F(q_0), 1_W e^{X_t} \mathrm{J}_t G(q_t)).$$

Then the theorem follows. □

By using the functional integral representation, we can estimate the ground state energy of $H_{\mathrm{PF}}^{\mathbb{Z}_2}$. Write

$$E(\hat{A}, \hat{B}_1, \hat{B}_2, \hat{B}_3) = \inf \mathrm{Spec}(H_{\mathrm{PF}}^{\mathbb{Z}_2}).$$

For the spinless Pauli–Fierz Hamiltonian H_{PF}, we have $\inf \mathrm{Spec}(H_{\mathrm{PF}}) = E(\hat{A}, 0, 0, 0)$ and the diamagnetic inequality $E(0, 0, 0, 0) \leq E(\hat{A}, 0, 0, 0)$ was already seen. We extend this inequality to $H_{\mathrm{PF}}^{\mathbb{Z}_2}$. Define

$$(H_{\mathrm{PF}}^{\mathbb{Z}_2}(0) F)(\theta) = (H_{\mathrm{p}} + H_{\mathrm{rad}} + \hat{H}_{\mathrm{d}}(\theta)) F(\theta) - |\hat{H}_{\mathrm{od}}(-\theta)| F(-\theta),$$

where

$$|\hat{H}_{\mathrm{od}}(-\theta)| = \frac{|e|}{2} \int_{\mathbb{R}^3}^{\oplus} \sqrt{\hat{B}_1^2(x) + \hat{B}_2^2(x)} dx$$

is independent of $\theta \in \mathbb{Z}_2$. Hamiltonian $H_{\mathrm{PF}}^{\mathbb{Z}_2}(0)$ corresponds to

$$H_{\mathrm{p}} + H_{\mathrm{rad}} - \begin{pmatrix} \frac{e}{2}\hat{B}_3 & \frac{|e|}{2}\sqrt{\hat{B}_1^2 + \hat{B}_2^2} \\ \frac{|e|}{2}\sqrt{\hat{B}_1^2 + \hat{B}_2^2} & -\frac{e}{2}\hat{B}_3 \end{pmatrix}$$

acting in $L^2(\mathbb{R}^3; \mathbb{C}^2) \otimes L^2(\mathscr{Q})$. Furthermore, to avoid zeroes of the off-diagonal part, we also define

$$(H_{\mathrm{PF},\varepsilon}^{\mathbb{Z}_2}(0)F)(\theta) = (H_{\mathrm{p}} + H_{\mathrm{rad}} + \hat{H}_{\mathrm{d}}(\theta))F(\theta) - \Psi_\varepsilon(|\hat{H}_{\mathrm{od}}(-\theta)|)F(-\theta).$$

Since the spin interaction is infinitesimally small with respect to the free Hamiltonian $H_{\mathrm{p}} + H_{\mathrm{rad}}$, $H_{\mathrm{PF}}^{\mathbb{Z}_2}(0)$ and $H_{\mathrm{PF},\varepsilon}^{\mathbb{Z}_2}(0)$ are self-adjoint on D_{PF} and bounded from below. The functional integral representation of $e^{-tH_{\mathrm{PF}}^{\mathbb{Z}_2}(0)}$ is given by

$$(F, e^{-tH_{\mathrm{PF}}^{\mathbb{Z}_2}(0)} G) = \lim_{\varepsilon \downarrow 0} e^t \sum_{\alpha=1,2} \int_{\mathbb{R}^3} \mathbb{E}^{x,\alpha}[e^{-\int_0^t V(B_s)ds} (\mathrm{J}_0 F(q_0), e^{X_t^\perp(\varepsilon)} \mathrm{J}_t G(q_t))] dx,$$

where

$$X_t^\perp(\varepsilon) = -\int_0^t \hat{H}_{\mathrm{E,d}}(B_s, \theta_{N_s}, s) ds + \int_0^{t+} \log(\Psi_\varepsilon(|\hat{H}_{\mathrm{E,od}}(B_s, s)|)) dN_s. \tag{3.8.80}$$

Corollary 3.101 (Diamagnetic inequality). *It follows that*

$$|(F, e^{-tH_{\mathrm{PF}}^{\mathbb{Z}_2}} G)| \leq (|F|, e^{-tH_{\mathrm{PF}}^{\mathbb{Z}_2}(0)} |G|) \tag{3.8.81}$$

and

$$\max_{(\alpha\beta\gamma)=(123),(231),(312)} E(0, \sqrt{\hat{B}_\alpha^2 + \hat{B}_\beta^2}, 0, \hat{B}_\gamma) \leq E(\hat{A}, \hat{B}_1, \hat{B}_2, \hat{B}_3). \tag{3.8.82}$$

Proof. Since $|\Psi_\varepsilon(\hat{H}_{\mathrm{E,od}})| \leq \Psi_\varepsilon(|\hat{H}_{\mathrm{E,od}}|)$, $|e^{X_t(\varepsilon)}| \leq e^{X_t^\perp(\varepsilon)}$, and $|\mathrm{J}_t G| \leq \mathrm{J}_t |G|$ as J_t is positivity preserving, by the functional integral representation of $e^{-tH_{\mathrm{PF}}^{\mathbb{Z}_2}}$, we have

$$|(F, e^{-tH_{\mathrm{PF}}^{\mathbb{Z}_2}} G)| \leq \lim_{\varepsilon \downarrow 0} (|F|, e^{-tH_{\mathrm{PF},\varepsilon}^{\mathbb{Z}_2}(0)} |G|) = (|F|, e^{-tH_{\mathrm{PF}}^{\mathbb{Z}_2}(0)} |G|).$$

Thus (3.8.81) follows. From this, $E(0, \sqrt{\hat{B}_1^2 + \hat{B}_2^2}, 0, \hat{B}_3) \le E(\hat{A}, \hat{B}_1, \hat{B}_2, \hat{B}_3)$ is obtained. We will show that

$$E(\hat{A}, \hat{B}_1, \hat{B}_2, \hat{B}_3) = E(\hat{A}, \hat{B}_2, \hat{B}_3, \hat{B}_1) = E(\hat{A}, \hat{B}_3, \hat{B}_1, \hat{B}_2) \tag{3.8.83}$$

by SU(2)-symmetry. Let $\mathscr{R} \in O(3)$ be such that

$$\mathscr{R}\begin{pmatrix} x_1 \\ x_2 \\ x_3 \end{pmatrix} = \begin{pmatrix} x_2 \\ x_3 \\ x_1 \end{pmatrix}.$$

Then there exists $(n, \phi) \in S^2 \times \mathbb{R}$ such that $\mathscr{R} = \mathscr{R}(n, \phi)$. Recall that $\mathscr{R}(n, \phi)$ describes the rotation around n by angle ϕ. Hence we see that

$$e^{i\phi n\cdot(1/2)\sigma}\sigma_\mu e^{-i\phi n\cdot(1/2)\sigma} = (\mathscr{R}\sigma)_\mu.$$

Now we write $H_{\mathrm{PF}}^{\mathbb{Z}_2}$ by $H_{\mathrm{PF}}^{\mathbb{Z}_2}(\hat{A}, \hat{B}_1, \hat{B}_2, \hat{B}_3)$. Thus we obtain

$$e^{i\phi n\cdot(1/2)\sigma} H_{\mathrm{PF}}^{\mathbb{Z}_2}(\hat{A}, \hat{B}_1, \hat{B}_2, \hat{B}_3) e^{-i\phi n\cdot(1/2)\sigma} = H_{\mathrm{PF}}^{\mathbb{Z}_2}(\hat{A}, \hat{B}_2, \hat{B}_3, \hat{B}_1),$$

which implies the first equality in (3.8.83). The second is proven in the same way. □

3.8.9 Translation invariant Pauli–Fierz Hamiltonian with spin

Finally, we study the translation invariant Pauli–Fierz Hamiltonian with spin 1/2, that is, $H_{\mathrm{PF}}^{\mathrm{S}}$ with $V \equiv 0$. We suppose Assumption 3.66 in Section 3.8.9.

Definition 3.102 (Pauli–Fierz Hamiltonian with spin 1/2 and a fixed total momentum). The *Pauli–Fierz Hamiltonian with spin and a fixed total momentum p* is defined by

$$H_{\mathrm{PF}}^{\mathrm{S}}(p) = \frac{1}{2}(p - P_{\mathrm{f}} - eA(0))^2 + H_{\mathrm{rad}} - \frac{e}{2}\sigma \cdot B(0), \quad p \in \mathbb{R}^3,$$

with domain $D(H_{\mathrm{PF}}^{\mathrm{S}}(p)) = D(H_{\mathrm{rad}}) \cap D(P_{\mathrm{f}}^2)$.

Theorem 3.103 (Self-adjointness). *Hamiltonian $H_{\mathrm{PF}}^{\mathrm{S}}(p)$ is self-adjoint and essentially self-adjoint on any core of $H_{\mathrm{rad}} + P_{\mathrm{f}}^2$.*

Proof. The proof is similar to that of Proposition 3.77. By (3.7.6), we have

$$\|H_{\mathrm{PF},0}(p)\Phi\| \le C\|(H_{\mathrm{PF}}(p) + 1)\Phi\|$$

for $\Phi \in D(H_{\mathrm{rad}}) \cap D(P_{\mathrm{f}}^2)$, with some constant C. From this inequality, we see that for any $\varepsilon > 0$ there exists $C_\epsilon > 0$ such that $\|H_{\mathrm{rad}}^{1/2}\Phi\| \le \varepsilon\|H_{\mathrm{PF}}(p)\Phi\| + C_\varepsilon\|\Phi\|$ for $\Phi \in D(H_{\mathrm{rad}}) \cap D(P_{\mathrm{f}}^2)$.

Since $\|\sigma \cdot \hat{B}(0)\Phi\| \le \|H_{\rm rad}^{1/2}\Phi\| + b_\varepsilon\|\Phi\|$, it follows that $\sigma \cdot \hat{B}(0)$ is infinitesimally small with respect to $H_{\rm PF}(p)$. Thus $H_{\rm PF}^{\rm S}(p)$ is self-adjoint on $D(H_{\rm rad}) \cap D(P_{\rm f}^2)$. Essential self-adjointness follows from the trivial inequality $\|H_{\rm PF}^{\rm S}(p)\Phi\| \le C\|(H_{\rm rad} + P_{\rm f}^2)\Phi\| + \|\Phi\|$. □

As in the case of $H_{\rm PF}(p)$, we have the theorem below.

Theorem 3.104 (Fiber decomposition). *It follows that*

$$L^2(\mathbb{R}^3; \mathbb{C}^2) \otimes \mathcal{F}_{\rm rad} = \mathscr{T}\left(\int_{\mathbb{R}^3}^{\oplus} \mathbb{C}^2 \otimes \mathcal{F}_{\rm rad} dp\right)\mathscr{T}^{-1}$$

and

$$H_{\rm PF}^{\rm S} = \mathscr{T}\left(\int_{\mathbb{R}^3}^{\oplus} H_{\rm PF}^{\rm S}(p) dp\right)\mathscr{T}^{-1},$$

where $\mathscr{T}$ is defined in (3.7.1).

In $\mathscr{Q}$-representation, $H_{\rm PF}^{\rm S}(p)$ is rewritten as

$$H_{\rm PF}^{\rm S}(p) = \frac{1}{2}(p - \hat{P}_{\rm f} - e\hat{A}(0))^2 + H_{\rm rad} - \frac{e}{2}\sigma \cdot \hat{B}(0),$$

where $\hat{B}(0) = \hat{B}(x = 0)$ defined in (3.8.8). Before going to construct the functional integral representation of $e^{-tH_{\rm PF}^{\rm S}(p)}$, we look at the symmetry properties of $H_{\rm PF}^{\rm S}(p)$.

Lemma 3.105. *Let $\hat{\varphi}$ be rotation-invariant. Then for any choice of the polarization vectors $H_{\rm PF}^{\rm S}(p)$ is unitary equivalent with $H_{\rm PF}^{\rm S}(\mathscr{R}^{-1}p)$, for all $\mathscr{R} \in$ SO(3).*

Proof. It is sufficient to show the lemma for an arbitrary rotation $\mathscr{R} = \mathscr{R}(m, \phi)$ with $(m, \phi) \in S^2 \times [0, 2\pi)$. For any polarization vectors $e^\pm$, define

$$h_{\rm f} = d\Gamma(\theta(\mathscr{R}, \cdot)\mathfrak{H}_2),$$

where $\theta(\mathscr{R}, k) = \cos^{-1}(e^+(\mathscr{R}k) \cdot \mathscr{R}e^+(k))$ and $\mathfrak{H}_2$ in (3.8.16). Formally,

$$h_{\rm f} = \sum_{j=\pm} \int_{\mathbb{R}^3} \theta(\mathscr{R}, k)(a^*(k, -)a(k, +) - a^*(k, +)a(k, -))dk.$$

Thus we obtain that

$$\begin{aligned}
e^{ih_{\rm f}} e^{i\phi m \cdot L_{\rm f}} H_{\rm rad} e^{-i\phi m \cdot L_{\rm f}} e^{-ih_{\rm f}} &= H_{\rm rad}, \\
e^{ih_{\rm f}} e^{i\phi m \cdot L_{\rm f}} P_{{\rm f},\mu} e^{-i\phi m \cdot L_{\rm f}} e^{-ih_{\rm f}} &= (\mathscr{R} P_{\rm f})_\mu, \\
e^{ih_{\rm f}} e^{i\phi m \cdot L_{\rm f}} A_\mu(0) e^{-i\phi m \cdot L_{\rm f}} e^{-ih_{\rm f}} &= (\mathscr{R} A(0))_\mu, \\
e^{ih_{\rm f}} e^{i\phi m \cdot L_{\rm f}} B_\mu(0) e^{-i\phi m \cdot L_{\rm f}} e^{-ih_{\rm f}} &= (\mathscr{R} B(0))_\mu,
\end{aligned}$$

$$e^{i\phi m\cdot(1/2)\sigma}\sigma_\mu e^{-i\phi m\cdot(1/2)\sigma} = (\mathscr{R}\sigma)_\mu.$$

From these identities, it follows that

$$\begin{aligned}
&e^{ih_{\mathrm f}}e^{i\phi m\cdot((1/2)\sigma+L_{\mathrm f})}H^{\mathrm S}_{\mathrm{PF}}(p)e^{-i\phi m\cdot((1/2)\sigma+L_{\mathrm f})}e^{-ih_{\mathrm f}}\\
&= \frac{1}{2}(p-\mathscr{R}P_{\mathrm f}-e\mathscr{R}A(0))^2 + H_{\mathrm{rad}} - \frac{1}{2}(\mathscr{R}\sigma)\cdot(\mathscr{R}B(0))\\
&= H^{\mathrm S}_{\mathrm{PF}}(\mathscr{R}^{-1}p),
\end{aligned} \tag{3.8.84}$$

proving the lemma. □

Let $E^{\mathrm S}(p,e^2) = \inf\mathrm{Spec}(H^{\mathrm S}_{\mathrm{PF}}(p))$. An immediate consequence of Lemma 3.105 is as follows.

Corollary 3.106 (Rotation invariance). *Let $\hat\varphi$ be rotation-invariant. Then for every $\mathscr{R}\in$ SO(3), it follows that $E^{\mathrm S}(\mathscr{R}p,e^2) = E^{\mathrm S}(p,e^2)$.*

Theorem 3.107 (Fiber decomposition of the Pauli–Fierz Hamiltonian with spin and a fixed total momentum). *Suppose that polarization vectors are coherent in direction $n\in S^2$ and $\hat\varphi$ is rotation invariant. Let $p/|p| = n$. Then the Hilbert space $\mathbb{C}^2\otimes\mathcal{F}_{\mathrm{rad}}$ and the self-adjoint operator $H^{\mathrm S}_{\mathrm{PF}}(p) = H^{\mathrm S}_{\mathrm{PF}}(|p|n)$ are decomposed as*

$$\mathbb{C}^2\otimes\mathcal{F}_{\mathrm{rad}} = \bigoplus_{w\in\mathbb{Z}_{1/2}}\mathcal{F}_{\mathrm{rad}}(w),\quad H^{\mathrm S}_{\mathrm{PF}}(p) = \bigoplus_{w\in\mathbb{Z}_{1/2}}H^{\mathrm S}_{\mathrm{PF}}(p,w). \tag{3.8.85}$$

Here, $\mathcal{F}_{\mathrm{rad}}(w)$ is the subspace spanned by eigenvectors of

$$\mathrm{J}_0 = n\cdot\frac{1}{2}\sigma + n\cdot L_{\mathrm f} + S_{\mathrm f}$$

associated with the eigenvalue $w\in\mathbb{Z}_{1/2}$ and $H^{\mathrm S}_{\mathrm{PF}}(|p|n,w) = H^{\mathrm S}_{\mathrm{PF}}(|p|n)\lceil_{\mathcal{F}_{\mathrm{rad}}(w)}$.

Proof. Let $\mathscr{R} = \mathscr{R}(n,\phi)$. Since the polarization vectors are coherent in direction n, it follows that $\theta(\mathscr{R},k) = \phi z$ with some $z\in\mathbb{Z}$, and then $h_{\mathrm f} = \phi S_{\mathrm f}$. Since $\mathscr{R}p = |p|\mathscr{R}n = |p|n = p$, we have from (3.8.84),

$$e^{i\phi\mathrm{J}_0}H^{\mathrm S}_{\mathrm{PF}}(p)e^{-i\phi\mathrm{J}_0} = H^{\mathrm S}_{\mathrm{PF}}(\mathscr{R}^{-1}p) = H^{\mathrm S}_{\mathrm{PF}}(p),\quad \phi\in\mathbb{R}. \tag{3.8.86}$$

Hence $H^{\mathrm S}_{\mathrm{PF}}(p)$ is reduced by J_0 and (3.8.85) follows. □

From this theorem, $H^{\mathrm S}_{\mathrm{PF}}(p)$ with coherent polarization vectors in direction $p/|p|$ and $p\neq 0$ has the symmetry

$$\mathrm{SU}(2)\times\mathrm{SO}_{\mathrm{field}}(3)\times\text{helicity}.$$

Let $\hat\varphi$ be rotation-invariant and $H^{\mathrm S}_{\mathrm{PF}}(p,e^\pm)$ be the Hamiltonian with coherent polarization vectors given in Example 3.80 with $n = n_3 = (0,0,1)$, i. e.,

$$e^{+}(k) = \frac{(-k_2, k_1, 0)}{\sqrt{k_1^2 + k_2^2}}, \quad e^{-}(k) = \hat{k} \times e^{+}(k). \tag{3.8.87}$$

Hamiltonian $H_{\mathrm{PF}}^{\mathrm{S}}(p)$ with arbitrary polarization vectors is unitary equivalent to $H_{\mathrm{PF}}^{\mathrm{S}}(p, e^{\pm})$,

$$H_{\mathrm{PF}}^{\mathrm{S}}(p) \cong H_{\mathrm{PF}}^{\mathrm{S}}(p, e^{\pm}).$$

Since there exists $\mathscr{R}$ such that $\mathscr{R}p = |p|n_3$, we have by Lemma 3.105,

$$H_{\mathrm{PF}}^{\mathrm{S}}(p, e^{\pm}) \cong H_{\mathrm{PF}}^{\mathrm{S}}(|p|n_3, e^{\pm}).$$

By (3.8.85), we obtain

$$H_{\mathrm{PF}}^{\mathrm{S}}(|p|n_3, e^{\pm}) = \bigoplus_{w \in \mathbb{Z}_{1/2}} H_{\mathrm{PF}}^{\mathrm{S}}(|p|n_3, e^{\pm}, w).$$

Then for rotation-invariant $\hat{\varphi}$, we have the unitary equivalence

$$H_{\mathrm{PF}}^{\mathrm{S}}(p) \cong \bigoplus_{w \in \mathbb{Z}_{1/2}} H_{\mathrm{PF}}^{\mathrm{S}}(|p|n_3, e^{\pm}, w).$$

We can show the theorem below in a similar way to Theorem 3.84.

Theorem 3.108 (Reflection symmetry). *Let $\hat{\varphi}$ be rotation-invariant and the coherent polarization vectors $e^{\pm}$ be given by* (3.8.87). *Then $H_{\mathrm{PF}}^{\mathrm{S}}(p, w) \cong H_{\mathrm{PF}}^{\mathrm{S}}(p, -w)$ for $w \in \mathbb{Z}_{1/2}$ and p with* $(0, 0, 1) = p/|p|$.

Proof. The proof is similar to that of Theorem 3.84. Let $\mathfrak{J}_0 = \sigma_2 \otimes \Gamma(u)$, where u is defined in (3.8.28). Thus we have

$$\mathfrak{J}_0^{-1} L_{\mathrm{f}\mu} \mathfrak{J}_0 = \begin{cases} L_{\mathrm{f}\mu}, & \mu = 2, \\ -L_{\mathrm{f}\mu}, & \mu = 1, 3, \end{cases}$$

$$\mathfrak{J}_0^{-1} P_{\mathrm{f}\mu} \mathfrak{J}_0 = \begin{cases} P_{\mathrm{f}\mu}, & \mu = 1, 3, \\ -P_{\mathrm{f}\mu}, & \mu = 2, \end{cases}$$

$$\mathfrak{J}_0^{-1} A_\mu(0) \mathfrak{J}_0 = \begin{cases} A_\mu(0), & \mu = 1, 3, \\ -A_\mu(0), & \mu = 2, \end{cases}$$

$$\mathfrak{J}_0^{-1} \sigma_\mu \mathfrak{J}_0 = \begin{cases} -\sigma_\mu, & \mu = 1, 3, \\ \sigma_\mu, & \mu = 2. \end{cases}$$

Since $n = n_3 = (0, 0, 1)$ and $p = |p|n_3$, and the helicity of $e^{\pm}$ is zero, we have $\mathrm{J}_0 = \frac{1}{2}\sigma_3 + L_{\mathrm{f},3}$ and

$$\mathfrak{J}_0^{-1} \mathrm{J}_0 \mathfrak{J}_0 = -\mathrm{J}_0. \tag{3.8.88}$$

We also see that $\mathfrak{J}_0^{-1}\sigma_\mu(p - P_\mathrm{f} - eA)_\mu\mathfrak{J}_0 = -\sigma_\mu(p - P_\mathrm{f} - eA)_\mu$ for $\mu = 1, 2, 3$. From here, it follows that

$$\mathfrak{J}_0^{-1} H_{\mathrm{PF}}^{\mathrm{S}}(p)\mathfrak{J}_0 = H_{\mathrm{PF}}^{\mathrm{S}}(p). \tag{3.8.89}$$

Note also that

$$H_{\mathrm{PF}}^{\mathrm{S}}(p)\phi = H_{\mathrm{PF}}^{\mathrm{S}}(p, w)\phi, \quad \phi \in \mathcal{F}_{\mathrm{rad}}(w). \tag{3.8.90}$$

Let $\phi \in \mathcal{F}_{\mathrm{rad}}(w)$. We have $\mathfrak{J}_0\phi \in \mathcal{F}_{\mathrm{rad}}(-w)$ by (3.8.88), and $\mathfrak{J}_0^{-1} H_{\mathrm{PF}}^{\mathrm{S}}(p, -w)\mathfrak{J}_0\phi = \mathfrak{J}_0^{-1} H_{\mathrm{PF}}^{\mathrm{S}}(p)\mathfrak{J}_0\phi$ by (3.8.90). Equation (3.8.89) implies $\mathfrak{J}_0^{-1} H_{\mathrm{PF}}^{\mathrm{S}}(p)\mathfrak{J}_0\phi = H_{\mathrm{PF}}^{\mathrm{S}}(p)\phi = H_{\mathrm{PF}}^{\mathrm{S}}(p, w)\phi$. Thus

$$\mathfrak{J}_0^{-1} H_{\mathrm{PF}}^{\mathrm{S}}(p, -w)\mathfrak{J}_0 = H_{\mathrm{PF}}^{\mathrm{S}}(p, w)$$

follows. □

Since the operators $H_{\mathrm{PF}}^{\mathrm{S}}(p)$ with $p \neq 0$ and different polarization vectors from (3.8.87) are unitary equivalent with $H_{\mathrm{PF}}^{\mathrm{S}}(n_3|p|)$ with $e^{\pm}$ in (3.8.87), we have

Corollary 3.109 (Degeneracy of bound states). *Let $\hat{\varphi}$ be rotation-invariant and M be the multiplicity of bound states of $H_{\mathrm{PF}}^{\mathrm{S}}(p)$. Then M is an even number. In particular, whenever a ground states exists, it is degenerate.*

The discussion mentioned above can be extended to the Hamiltonian without spin. With a rotation-invariant $\hat{\varphi}$, similar to (3.8.86) also $H_{\mathrm{PF}}(p)$ with $p/|p| = n$ and coherent polarization vectors in direction n satisfies

$$e^{i\phi J_\mathrm{f}} H_{\mathrm{PF}}(p) e^{-i\phi J_\mathrm{f}} = H_{\mathrm{PF}}(p), \quad \phi \in \mathbb{R}.$$

Thus $H_{\mathrm{PF}}(p)$ with coherent polarization vectors in direction $n = p/|p|$ has the symmetry

$$\mathrm{SO}_{\mathrm{field}}(3) \times \text{helicity}.$$

Since $\sigma(J_\mathrm{f}) = \mathbb{Z}$, this gives rise to the decomposition

$$\mathcal{F}_{\mathrm{rad}} = \bigoplus_{w \in \mathbb{Z}} \mathcal{F}_{\mathrm{rad}}^{0}(w), \quad H_{\mathrm{PF}}(p) = \bigoplus_{w \in \mathbb{Z}} H_{\mathrm{PF}}(p, w)$$

where $\mathcal{F}_{\mathrm{rad}}^{0}(w)$ denotes the subspace spanned by eigenvectors at eigenvalue $w \in \mathbb{Z}$ of J_f. By Corollary 3.73, as soon as $p = 0$, the ground state of $H_{\mathrm{PF}}(p = 0)$ is unique for any $e \in \mathbb{R}$. It is also known that $H_{\mathrm{PF}}(p)$ has the unique ground state for sufficiently small $|p|$ and $|e|$. We may suppose that the polarization vectors of $H_{\mathrm{PF}}(p)$ is coherent in direction $p/|p|$, since $H_{\mathrm{PF}}(p)$ with different polarization vectors are unitary equivalent. Suppose that the ground state $\Psi_\mathrm{g}(p)$ of $H_{\mathrm{PF}}(p)$ is unique. Then $\Psi_\mathrm{g}(p)$ must belong to some $\mathcal{F}_{\mathrm{rad}}^{0}(w)$.

The only possibility is $z = 0$, since $H_{\mathrm{PF}}(|p|n, z) \cong H_{\mathrm{PF}}(|p|n, -z)$, $z \in \mathbb{Z}$, by Lemma 3.108. This implies $J_{\mathrm{f}}\Psi_{\mathrm{g}}(p) = 0$.

We now construct the functional integral representation for the translation invariant Pauli–Fierz Hamiltonian with spin 1/2. As before, we transform $H_{\mathrm{PF}}^{\mathrm{S}}$ on $\mathbb{C}^2 \otimes \mathcal{F}_{\mathrm{rad}}$ to $H_{\mathrm{PF}}^{\mathbb{Z}_2}$ on $\ell^2(\mathbb{Z}_2) \otimes L^2(\mathcal{Q})$ which is defined by

$$(H_{\mathrm{PF}}^{\mathbb{Z}_2}(p)\Psi)(\theta) = (H_{\mathrm{PF}}(p) + \hat{H}_{\mathrm{d}}(0,\theta))\Psi(\theta) + \hat{H}_{\mathrm{od}}(0,-\theta)\Psi(-\theta),$$

where

$$\hat{H}_{\mathrm{d}}(0,\theta) = -\frac{e}{2}\theta\hat{B}_3(0),$$
$$\hat{H}_{\mathrm{od}}(-\theta,0) = -\frac{e}{2}(\hat{B}_1(0) - i\theta\hat{B}_2(0))$$

with $\theta \in \mathbb{Z}_2$. Moreover, $H_{\mathrm{PF},\varepsilon}^{\mathbb{Z}_2}(p)$ is defined by $H_{\mathrm{PF}}^{\mathbb{Z}_2}(p)$ with off-diagonal $\hat{H}_{\mathrm{od}}$ replaced by $\Psi_\varepsilon(\hat{H}_{\mathrm{od}})$, which is introduced to avoid the off-diagonal part of $H_{\mathrm{PF}}^{\mathbb{Z}_2}(p)$ becoming singular. The strategy of constructing the functional integral representation of $e^{-tH_{\mathrm{PF}}^{\mathrm{S}}(p)}$ is similar to that of the spinless case.

Theorem 3.110 (Functional integral representation for Pauli–Fierz Hamiltonian with spin 1/2 and a fixed total momentum). *Let $\Phi, \Psi \in \ell^2(\mathbb{Z}_2) \otimes L^2(\mathcal{Q})$.*

(1) *It follows that*

$$\lim_{\varepsilon\downarrow 0}(\Phi, e^{-tH_{\mathrm{PF},\varepsilon}^{\mathbb{Z}_2}(p)}\Psi) = (\Phi, e^{-tH_{\mathrm{PF}}^{\mathbb{Z}_2}(p)}\Psi)$$

and

$$(\Phi, e^{-tH_{\mathrm{PF},\varepsilon}^{\mathbb{Z}_2}(p)}\Psi) = e^t \sum_{\alpha=1,2} \mathbb{E}^{0,\alpha}[e^{ip\cdot B_t}(\mathrm{J}_0\Phi(\theta_\alpha), e^{X_t(\varepsilon)}\mathrm{J}_t e^{-i\hat{P}_{\mathrm{f}}\cdot B_t}\Psi(\theta_{N_t}))],$$

where the exponent $X_t(\varepsilon)$ is given by

$$X_t(\varepsilon) = -ie\int_0^t \hat{A}_{\mathrm{E}}(\mathrm{j}_s\tilde{\varphi}(\cdot - B_s))dB_s - \int_0^t \hat{H}_{\mathrm{E,d}}(B_s, \theta_{N_s}, s)ds$$
$$+ \int_0^{t+} \log(-\Psi_\varepsilon(\hat{H}_{\mathrm{E,od}}(B_s, -\theta_{N_{s-}}, s)))dN_s.$$

(2) *Let $W_0 \subset \mathscr{X} \times \mathcal{S} \times \mathcal{Q}_{\mathrm{E}}$ be defined by*

$$W_0 = \left\{\int_0^{t+} \log\left(\frac{|e|}{2}\sqrt{|b_1(s,B_s)|^2 + |b_2(s,B_s)|^2}\right)dN_s > -\infty\right\},$$

where $b_a(s,B_s)=\hat{B}_{\mathrm{E}_a}(\mathrm{j}_s\tilde{\varphi}(\cdot-B_s))$. *Then*

$$(\Phi, e^{-tH^{\mathrm{S}}_{\mathrm{PF}}(p)}\Psi)=e^t\sum_{\alpha=1,2}\mathbb{E}^{0,\alpha}[e^{ip\cdot B_t}(\mathrm{J}_0\Phi(\theta_\alpha),e^{X_t}1_{W_0}\mathrm{J}_t e^{-i\hat{P}_{\mathrm{f}}\cdot B_t}\Psi(\theta_{N_t}))],$$

where the exponent X_t *is given by*

$$X_t=-ie\int_0^t\hat{A}_{\mathrm{E}}(\mathrm{j}_s\tilde{\varphi}(\cdot-B_s))dB_s-\int_0^t\hat{H}_{\mathrm{E,d}}(B_s,\theta_{N_s},s)ds+\int_0^{t+}\log(-\hat{H}_{\mathrm{E,od}}(B_s,-\theta_{N_{s-}},s))dN_s.$$

Proof. The proof of (1) goes along the same lines as in Theorem 3.70, and for (2), in addition to the proof of Theorem 3.70, it can be shown by mimicking the proof of Theorem 3.100. □

From Theorem 3.110, we can derive energy inequalities in a similar manner to Corollary 3.72 for the spinless case. Write

$$E(p,\hat{A},\hat{B}_1,\hat{B}_2,\hat{B}_3)=\inf\mathrm{Spec}(H^{\mathbb{Z}_2}_{\mathrm{PF}}(p)),$$

and define $H^{\mathbb{Z}_2}_{\mathrm{PF}}(p,0)$ by

$$(H^{\mathbb{Z}_2}_{\mathrm{PF}}(p,0)\Psi)(\theta)=(H_{\mathrm{PF}}(p)+\hat{H}_{\mathrm{d}}(0,\theta))\Psi(\theta)-|\hat{H}_{\mathrm{od}}(0,-\theta)|\Psi(-\theta),$$

where $|\hat{H}_{\mathrm{od}}(0,-\theta)|=\frac{|e|}{2}\sqrt{\hat{B}_1(0)^2+\hat{B}_2(0)^2}$. This corresponds to

$$\frac{1}{2}(p-P_{\mathrm{f}})^2+H_{\mathrm{rad}}-\begin{pmatrix}\frac{e}{2}\hat{B}_3(0) & \frac{|e|}{2}\sqrt{\hat{B}_1(0)^2+\hat{B}_2(0)^2}\\ \frac{|e|}{2}\sqrt{\hat{B}_1(0)^2+\hat{B}_2(0)^2} & -\frac{e}{2}\hat{B}_3(0)\end{pmatrix}$$

in $\mathbb{C}^2\otimes L^2(\mathscr{Q})$. Note that $|\hat{H}_{\mathrm{od}}(0,-\theta)|$ is independent of θ. We have the energy comparison inequality

Corollary 3.111 (Diamagnetic inequality). *It follows that*

$$|(\Phi,e^{-tH^{\mathbb{Z}_2}_{\mathrm{PF}}(p)}\Psi)|\le(|\Phi|,e^{-tH^{\mathbb{Z}_2}_{\mathrm{PF}}(p,0)}|\Psi|)\tag{3.8.91}$$

and

$$\max_{(\alpha\beta\gamma)=(123),(231),(312)}E(0,0,\sqrt{\hat{B}_\alpha^2+\hat{B}_\beta^2},0,\hat{B}_\gamma)\le E(p,\hat{A},\hat{B}_1,\hat{B}_2,\hat{B}_3).\tag{3.8.92}$$

Proof. We have

$$\begin{aligned}|(\Phi,e^{-tH^{\mathrm{S}}_{\mathrm{PF}}(p)}\Psi)|&\le e^t\lim_{\varepsilon\downarrow0}\sum_{\theta\in\mathbb{Z}_2}\mathbb{E}^{0,\alpha}[(\mathrm{J}_0|\Phi(\theta_\alpha)|,e^{X_t^\perp(\varepsilon)}\mathrm{J}_t e^{-iP_{\mathrm{f}}\cdot B_t}|\Phi(\theta_{N_t})|)]\\&=\mathrm{RHS}\ (3.8.91),\end{aligned}$$

where $X_t^{\perp}(\varepsilon)$ is given by (3.8.80). Equation (3.8.92) is immediate from (3.8.91) and a similar argument to (3.8.82). □

3.9 Relativistic Pauli–Fierz model

3.9.1 Definition of relativistic Pauli–Fierz Hamiltonian

In quantum mechanics, the relativistic Schrödinger operator with a vector potential a is defined by

$$H_{\mathrm{R}}(a) = \sqrt{(-i\nabla - a)^2 + m^2} - m + V,$$

and the functional integral representation of $e^{-tH_{\mathrm{R}}(a)}$ is known. A key element in the construction of the Feynman–Kac formula of $e^{-tH_{\mathrm{R}}(a)}$ is to use the Lévy subordinator $(T_t)_{t\geq 0}$ on a suitable probability space

$$(\mathcal{T}, \mathcal{B}_{\mathcal{T}}, \nu)$$

such that

$$\mathbb{E}_\nu^0[\exp(-uT_t)] = \exp(-t(\sqrt{2u + m^2} - m)).$$

For simplicity, we write $\mathbb{E}^{x,0}$ for $\mathbb{E}^{x,0}_{\mathcal{W}\times\nu}$. Suppose that $V \in L^\infty(\mathbb{R}^3)$, $a \in (L^2_{\mathrm{loc}}(\mathbb{R}^3))^3$, and $\nabla \cdot a \in L^1_{\mathrm{loc}}(\mathbb{R}^3)$. Then we see that

$$(f, e^{-tH_{\mathrm{R}}(a)} g) = \int_{\mathbb{R}^3} \mathbb{E}^{x,0}[\overline{f(B_0)} g(B_{T_t}) e^{-i\int_0^{T_t} a(B_s)\circ dB_s} e^{-\int_0^t V(B_{T_s})ds}]dx.$$

In this section, the analogue version of the Pauli–Fierz model is defined and its functional integral representation is given. The relativistic Pauli–Fierz Hamiltonian is defined by

$$H_{\mathrm{RPF}} = \sqrt{(-i\nabla - eA)^2 + m^2} - m + V + H_{\mathrm{rad}}$$

on $\mathcal{H}_{\mathrm{PF}}$ as a self-adjoint operator. A rigorous definition of H_{RPF} has, however, a difficulty caused by the term containing the square root. In Sections 3.9.1–3.9.4, we use Assumption 3.112 below unless otherwise stated.

Assumption 3.112. The following conditions hold:
(1) (Charge distribution) $\varphi \in \mathscr{S}'(\mathbb{R}^3)$, $\hat{\varphi}(-k) = \overline{\hat{\varphi}(k)}$ and $\omega^{3/2}\hat{\varphi}$, $\sqrt{\omega}\hat{\varphi}$, $\hat{\varphi}/\omega \in L^2(\mathbb{R}^3)$;
(2) (External potential) V is such that there exist $0 \leq a < 1$ and $0 \leq b$ such that

$$\|Vf\| \leq a\|\sqrt{-\Delta}f\| + b\|f\|. \tag{3.9.1}$$

In Assumptions 3.112, we add the extra condition $\omega^{3/2}\hat{\varphi} \in L^2(\mathbb{R}^3)$ to Assumption 3.4, and instead of $V \in \mathcal{R}_{\text{Kato}}$, we suppose that V is relatively bounded with respect to $\sqrt{-\Delta}$. Although a standard way to define $(-i\nabla - eA)^2 + m^2$ as a self-adjoint operator is to take the self-adjoint operator associated with the quadratic form

$$F, G \mapsto \sum_{\mu=1}^{3} ((-i\nabla - eA)_\mu F, (-i\nabla - eA)_\mu G) + m^2(F, G),$$

instead of this, we proceed by finding a core of $(-i\nabla - eA)^2 + m^2$ by using functional integration. In Section 3.3.2, we defined the self-adjoint operator $\hat{H}_{\text{PF}}(\hat{A})$ by the generator of a C_0-semigroup, and saw that

$$\hat{H}_{\text{PF}}(\hat{A}) \supset \frac{1}{2}(-i\nabla - e\hat{A})^2 \lceil_{D_{\text{PF}}}.$$

Let $\mathscr{D} = D(\Delta) \cap C^\infty(N)$.

Lemma 3.113. *Suppose $\hat{\varphi}(-k) = \overline{\hat{\varphi}(k)}$ and $\omega^{3/2}\hat{\varphi}, \hat{\varphi}/\sqrt{\omega} \in L^2(\mathbb{R}^3)$. Then $\frac{1}{2}(-i\nabla - e\hat{A})^2$ is essentially self-adjoint on $\mathscr{D}$.*

Proof. By (3.3.8), we have

$$(F, e^{-t\hat{H}_{\text{PF}}(\hat{A})}G) = \int_{\mathbb{R}^3} \mathbb{E}^x \left[(F(B_0), e^{-ie\hat{A}(L_t)}G(B_t))\right] dx, \tag{3.9.2}$$

where we recall that $L_t = \bigoplus_{\mu=1}^3 \int_0^t \tilde{\varphi}(\cdot - B_s)dB_s^\mu$. By using (3.9.2), we shall show that $e^{-t\hat{H}_{\text{PF}}(\hat{A})}$ leaves $\mathscr{D}$ invariant. Then the lemma follows.

First, it can be proven that $e^{-t\hat{H}_{\text{PF}}(\hat{A})}\mathscr{D} \subset D(\Delta)$ in a similar manner to (3.5.11) of Lemma 3.44 with H_{rad} replaced by N. To see that $e^{-t\hat{H}_{\text{PF}}(\hat{A})}\mathscr{D} \subset C^\infty(N)$, take $n \in \mathbb{N}$ and $F, G \in D(N^n)$. We have

$$(N^n F, e^{-t\hat{H}_{\text{PF}}(\hat{A})}G) = \int_{\mathbb{R}^3} \mathbb{E}^x \left[(N^n F(B_0), e^{-ie\hat{A}(L_t)}G(B_t))\right] dx.$$

Note that

$$e^{ie\hat{A}(L_t)}Ne^{-ie\hat{A}(L_t)} = N - e\hat{\Pi}(L_t) + \frac{e^2}{2}\|L_t\|^2,$$

where $\hat{\Pi}(f) = i[N, \hat{A}(f)]$ is the conjugate momentum, and thus

$$(N^n F, e^{-t\hat{H}_{\text{PF}}(\hat{A})}G) = \int_{\mathbb{R}^3} \mathbb{E}^x \left[\left(F(B_0), e^{-ie\hat{A}(L_t)}\left(N - e\hat{\Pi}(L_t) + \frac{e^2}{2}\|L_t\|^2\right)^n G(B_t)\right)\right] dx. \tag{3.9.3}$$

By the BDG-type inequality,

$$\mathbb{E}^x\left[\left\|\int_0^t \tilde{\varphi}(\cdot - B_s)dB_s^\mu\right\|_{L^2(\mathbb{R}^3)}^{2n}\right] \leq \frac{(2n)!}{2^n} t^n \|\hat{\varphi}\|_{L^2(\mathbb{R}^3)}^{2n},$$

and we obtain

$$\int_{\mathbb{R}^3} \mathbb{E}^x\left[\left\|\left(N - e\hat{\Pi}(L_t) + \frac{e^2}{2}\|L_t\|^2\right)^n F(B_t)\right\|^2\right] dx \leq C_n^2\|(N+1)^n F\|^2 \tag{3.9.4}$$

with a constant C_n. A combination of (3.9.3) and (3.9.4) gives

$$|(N^n F, e^{-t\hat{H}_{\mathrm{PF}}(\hat{A})}G)| \leq C_n\|F\|\|(N+1)^n G\|.$$

This implies $e^{-t\hat{H}_{\mathrm{PF}}(\hat{A})}C^\infty(N) \subset C^\infty(N)$ and $e^{-t\hat{H}_{\mathrm{PF}}(\hat{A})}\mathscr{D} \subset \mathscr{D}$ follows. Hence $\hat{H}_{\mathrm{PF}}(\hat{A})$ is essentially self-adjoint on $\mathscr{D}$. □

We keep denoting the self-adjoint extension of $\hat{H}_{\mathrm{PF}}(\hat{A})\lceil_{\mathscr{D}}$ by the same symbol $\hat{H}_{\mathrm{PF}}(\hat{A})$ for simplicity, and we define $\sqrt{2\hat{H}_{\mathrm{PF}}(\hat{A}) + m^2}$ by the spectral resolution of $\hat{H}_{\mathrm{PF}}(\hat{A})$.

Definition 3.114 (Relativistic Pauli–Fierz Hamiltonian). The *relativistic Pauli–Fierz Hamiltonian* is defined by

$$H_{\mathrm{RPF}} = \sqrt{2\hat{H}_{\mathrm{PF}}(\hat{A}) + m^2} - m + H_{\mathrm{rad}} + V.$$

As in the classical case in addition to Brownian motion, we need a subordinator to construct a functional integral representation for $e^{-tH_{\mathrm{RPF}}}$. Recall that $(T_t)_{t\geq 0}$ is the subordinator on a probability space $(\mathcal{T}, \mathcal{B}_{\mathcal{T}}, \nu)$. Since

$$(F, e^{-t(\sqrt{2\hat{H}_{\mathrm{PF}}(\hat{A})+m^2}-m)}G) = \mathbb{E}_\nu^0[(F, e^{-T_t\hat{H}_{\mathrm{PF}}(\hat{A})}G)], \tag{3.9.5}$$

combining the functional integral representation of $e^{-t\hat{H}_{\mathrm{PF}}(\hat{A})}$ with (3.9.5), we immediately have that

$$(F, e^{-t(\sqrt{2\hat{H}_{\mathrm{PF}}(\hat{A})+m^2}-m)}G) = \int_{\mathbb{R}^3} \mathbb{E}^{x,0}\left[(F(B_0), e^{-ie\hat{A}(L_{T_t})}G(B_{T_t}))\right] dx. \tag{3.9.6}$$

3.9.2 Functional integral representations

Now we will construct the functional integral representation of $e^{-tH_{\mathrm{RPF}}}$ through the Trotter product formula. We set

$$T_{\mathrm{kin}} = \sqrt{2\hat{H}_{\mathrm{PF}}(\hat{A}) + m^2} - m. \tag{3.9.7}$$

Lemma 3.115. *Suppose $V \in C_0^\infty(\mathbb{R}^3)$. Then*

$$\left(F, \left(e^{-\frac{t}{2^n}T_{\rm kin}}e^{-\frac{t}{2^n}H_{\rm rad}}e^{-\frac{t}{2^n}V}\right)^{2^n} G\right) = \int_{\mathbb{R}^3} \mathbb{E}^{x,0}\left[\left(\mathrm{J}_0 F(B_{T_0}), e^{-ie\hat{A}_{\rm E}(K_t^{\rm rel}(n))}\mathrm{J}_t G(B_{T_t})\right) e^{-\sum_{j=0}^{2^n}\frac{t}{2^n}V(B_{T_{t_j}})}\right] dx,$$

where

$$K_t^{\rm rel}(n) = \bigoplus_{\mu=1}^3 \sum_{j=1}^{2^n} \int_{T_{t_{j-1}}}^{T_{t_j}} \mathrm{j}_{t_{j-1}}\tilde{\varphi}(\cdot - B_s)dB_s^\mu,$$

with $t_j = tj/2^n$, and $\int_{T_{t_{j-1}}}^{T_{t_j}} \mathrm{j}_{t_{j-1}}\tilde{\varphi}(\cdot - B_s)dB_s^\mu$ denotes an $L^2(\mathbb{R}^4)$-valued stochastic integral $\int_T^S \mathrm{j}_{t_{j-1}}\tilde{\varphi}(\cdot - B_s)dB_s^\mu$ evaluated at $T = T_{t_{j-1}}$ and $S = T_{t_j}$.

Proof. By the formula $\mathrm{J}_t^*\mathrm{J}_s = e^{-|s-t|H_{\rm rad}}$ and (3.9.6), we have

$$\left(F, \left(e^{-\frac{t}{2^n}T_{\rm kin}}e^{-\frac{t}{2^n}H_{\rm rad}}e^{-\frac{t}{2^n}V}\right)^{2^n} G\right) = \int_{\mathbb{R}^3} \mathbb{E}^{x,0}\left[U_n e^{-\sum_{j=0}^{2^n}\frac{t}{2^n}V(B_{T_{t_j}})}\right],$$

where

$$U_n = \left(\mathrm{J}_0 F(B_{T_0}), \prod_{j=1}^{2^n}\left(\mathrm{J}_{t_{j-1}} e^{-ie\hat{A}\left(\bigoplus_{\mu=1}^3 \int_{T_{t_{j-1}}}^{T_{t_j}} \tilde{\varphi}(\cdot - B_r)dB_r^\mu\right)}\mathrm{J}_{t_{j-1}}^*\right)\mathrm{J}_t G(B_{T_t})\right),$$

and we see that $\mathrm{J}_t e^{-ie\hat{A}(\bigoplus_{\mu=1}^3 \int_S^T \tilde{\varphi}(\cdot - B_r)dB_r^\mu)}\mathrm{J}_t^* = E_t e^{-ie\hat{A}_{\rm E}(\bigoplus_{\mu=1}^3 \int_S^T \mathrm{j}_t\tilde{\varphi}(\cdot - B_r)dB_r^\mu)}E_t$ by the definition of J_t and E_t. Then by the Markov property of projection $E_{\mathcal{O}}$, projection E_t's can be removed, and thus the lemma follows. □

The sequence $(K_t^{\rm rel}(n))_{t\geq 0}$ can be regarded as an $\oplus^3 L^2(\mathbb{R}^4)$-valued random process on the product probability space $(\mathscr{X} \times \mathcal{T}, \mathcal{B}(\mathscr{X}) \times \mathcal{B}_{\mathcal{T}}, \mathcal{W}^x \otimes \nu)$. By the Itô isometry, we have

$$\mathbb{E}^x\left[\|K_t^{\rm rel}(n)\|^2_{\oplus^3 L^2(\mathbb{R}^4)}\right] = 3T_t\|\hat{\varphi}/\sqrt{\omega}\|^2_{L^2(\mathbb{R}^3)} \tag{3.9.8}$$

for each T_t. We will show that $K_t^{\rm rel}(n)$ has a limit as $n \to \infty$ in some sense. Let $\mathscr{N}_\nu \in \mathcal{B}_{\mathcal{T}}$ be a null set, i. e., $\nu(\mathscr{N}_\nu) = 0$, such that for arbitrary $w \in \mathcal{T} \setminus \mathscr{N}_\nu$, the path $t \mapsto T_t(w)$ is nondecreasing and right-continuous, and has the left-limit.

Lemma 3.116. *For each $w \in \mathcal{T} \setminus \mathscr{N}_\nu$ and each $t \geq 0$, the sequence $(K_t^{\rm rel}(n))_{n\in\mathbb{N}}$ strongly converges in $L^2(\mathscr{X}) \otimes (\oplus^3 L^2(\mathbb{R}^4))$ as $n \to \infty$, i. e., there exists*

$$K_t^{\rm rel} \in L^2(\mathscr{X}) \otimes (\oplus^3 L^2(\mathbb{R}^4))$$

such that $\lim_{n\to\infty} \mathbb{E}^x[\|K_t^{\rm rel}(n) - K_t^{\rm rel}\|^2] = 0$.

Proof. Fix $w \in \mathcal{T} \setminus \mathcal{N}_\nu$. Set $\mathrm{I}_n = K_t^{\rm rel}(n)$. It is sufficient to show that $(\mathrm{I}_n)_{n\in\mathbb{N}}$ is a Cauchy sequence in $L^2(\mathscr{X}) \otimes (\oplus^3 L^2(\mathbb{R}^4))$. We have

$$\mathrm{I}_{n+1} - \mathrm{I}_n = \bigoplus_{\mu=1}^{3} \sum_{m=1}^{2^n} \int_{T_{t_{2m-1}}}^{T_{t_{2m}}} (\mathrm{j}_{t_{2m-1}} - \mathrm{j}_{t_{2m-2}})\tilde{\varphi}(\cdot - B_s) dB_s^\mu,$$

where $t_j = tj/2^{n+1}$. Thus

$$\mathbb{E}^x[\|\mathrm{I}_{n+1} - \mathrm{I}_n\|^2] = 3\sum_{m=1}^{2^n} \mathbb{E}^x\left[\int_{T_{t_{2m-1}}}^{T_{t_{2m}}} \|(\mathrm{j}_{t_{2m-1}} - \mathrm{j}_{t_{2m-2}})\tilde{\varphi}(\cdot - B_s)\|^2 ds\right]$$

by the Itô isometry (3.9.8). Notice that $\|(\mathrm{j}_t - \mathrm{j}_s)f\|^2 = 2(\hat{f}, (1 - e^{-|s-t|\omega})\hat{f})$. Thus

$$\mathbb{E}^x[\|\mathrm{I}_{n+1} - \mathrm{I}_n\|^2] \le 3\sum_{m=1}^{2^n} 2(\hat{\varphi}/\sqrt{\omega}, (1 - e^{-\frac{t}{2^{n+1}}\omega})\hat{\varphi}/\sqrt{\omega})(T_{t_{2m}} - T_{t_{2m-1}}).$$

Since $T_t = T_t(w)$ is not decreasing in t, $\sum_{m=1}^{2^n}(T_{t_{2m}} - T_{t_{2m-1}}) \le T_t$ follows. Thus

$$\mathbb{E}^x[\|\mathrm{I}_{n+1} - \mathrm{I}_n\|^2] \le 3T_t \frac{t}{2^n}\|\hat{\varphi}\|^2.$$

Hence we have

$$\mathbb{E}^x[\|\mathrm{I}_m - \mathrm{I}_n\|^2] \le \left(\sqrt{3tT_t}\|\hat{\varphi}\| \sum_{j=n+1}^{m} \left(\frac{1}{\sqrt{2}}\right)^j\right)^2$$

for $m > n$. The right-hand side above converges to zero as $n, m \to \infty$. Then the sequence $(\mathrm{I}_n)_{n\in\mathbb{N}}$ is a Cauchy sequence for almost surely $\mathcal{W}^x \otimes \nu$. Then the lemma follows. □

For each $\tau \in \mathcal{T}$, there exist jump points $0 < s_1 < \cdots < s_n \le t$. Then for each $\tau \in \mathcal{T}$, $\Delta_j = T_{s_j+} - T_{s_j} > 0$ for $j = 1, \ldots, n$ and integral $K_t^{\rm rel}$ is informally written as

$$K_t^{\rm rel} = \bigoplus_{\mu=1}^{3} \sum_{j=1}^{n} \int_{\Delta_j} \mathrm{j}_{s_j}\tilde{\varphi}(\cdot - B_s) dB_s^\mu = \bigoplus_{\mu=1}^{3} \int_0^{T_t} \mathrm{j}_{T_s^*}\tilde{\varphi}(\cdot - B_s) dB_s^\mu.$$

Here $T_s^* = \inf\{t \mid T_t = s\}$.

Lemma 3.117. *Let* $V = 0$. *Then*

$$(F, e^{-t(T_{\rm kin} \dotplus H_{\rm rad})} G) = \int_{\mathbb{R}^3} \mathbb{E}^{x,0}\left[\left(\mathrm{J}_0 F(B_{T_0}), e^{-ie\hat{A}_{\rm E}(K_t^{\rm rel})} \mathrm{J}_t G(B_{T_t})\right)\right] dx.$$

Proof. By the Trotter product formula and Lemma 3.115, we have

$$\begin{aligned}(F, e^{-t(T_{\rm kin} \dotplus H_{\rm rad})} G) &= \lim_{n\to\infty}\left(F, \left(e^{-t/2^n \hat{H}_{\rm PF}(\hat{A})} e^{-t/2^n H_{\rm rad}}\right)^{2^n} G\right) \\ &= \lim_{n\to\infty} \int_{\mathbb{R}^3} \mathbb{E}^{x,0}\left[(\overline{\mathrm{J}_0 F(B_0)}, e^{-ie\hat{A}_{\rm E}(K_t^{\rm rel}(n))} \mathrm{J}_t G(B_{T_t}))\right] dx.\end{aligned}$$

By Lemma 3.116 and a limiting argument, the lemma follows. □

The immediate consequence of Lemma 3.117 is the diamagnetic inequality. Let $F, G \in \mathcal{H}_{\rm PF}$. Then it follows that

$$\begin{aligned}|(F, e^{-t(T_{\rm kin} \dotplus H_{\rm rad})} G)| &\le (|F|, e^{-t(\sqrt{-\Delta+m^2}-m+H_{\rm rad})} |G|), \\ |(F, e^{-t(T_{\rm kin} \dotplus H_{\rm rad})} G)| &\le (\|F\|_{L^2(\mathscr{Q})}, e^{-t(\sqrt{-\Delta+m^2}-m)} \|G\|_{L^2(\mathscr{Q})})_{L^2(\mathbb{R}^3)}.\end{aligned}$$

In a similar way to Lemma 3.30, we also have the lemma below.

Lemma 3.118. *The following hold:*
(1) *If* V *is* $(\sqrt{-\Delta+m^2}-m)$*-form bounded with a relative bound* a*, then* $|V|$ *is also* $(T_{\rm kin} \dotplus H_{\rm rad})$*-form bounded with a relative bound smaller than* a.
(2) *If* V *is relatively bounded with respect to* $\sqrt{-\Delta+m^2}-m$ *with a relative bound* a*, then* V *is also relatively bounded with respect to* $T_{\rm kin} \dotplus H_{\rm rad}$ *with a relative bound* a.

Theorem 3.119 (Functional integral representation for relativistic Pauli–Fierz Hamiltonian). *It follows that*

$$(F, e^{-tH_{\rm RPF}} G) = \int_{\mathbb{R}^3} \mathbb{E}^{x,0}\left[e^{-\int_0^t V(B_{T_s})ds} (\mathrm{J}_0 F(B_0), e^{-ie\hat{A}_{\rm E}(K_t^{\rm rel})} \mathrm{J}_t G(B_{T_t}))\right] dx. \tag{3.9.9}$$

Proof. When V is bounded and continuous, the theorem can be proven by the Trotter formula

$$(F, e^{-tH_{\rm RPF}} G) = \lim_{n\to\infty}\left(F, \left(e^{-\frac{t}{2^n} T_{\rm kin}} e^{-\frac{t}{2^n} H_{\rm rad}} e^{-\frac{t}{2^n} V}\right)^{2^n} G\right).$$

Furthermore, it can be extended to a general V given in this theorem in the same way as Theorem 3.32. We omit the details. □

We can shift the time in the functional integral representation of (3.9.9). We see it in the corollary below. Now let $(B_t)_{t\in\mathbb{R}}$ be a 3-dimensional Brownian motion on the whole real line. We define $\int_{-T_t}^{0} \mathrm{j}_{T_s^*}\tilde{\varphi}(\cdot - B_s)dB_s^\mu$ by

$$\int_{-T_t}^{0} \mathrm{j}_{T_s^*}\tilde{\varphi}(\cdot - B_s)dB_s^\mu = \lim_{n\to\infty}\sum_{j=1}^{2^n}\int_{-T_{-(t_{j-1}-t)}}^{-T_{-(t_j-t)}} \mathrm{j}_{-(t_{j-1}-t)}\tilde{\varphi}(\cdot - B_s)dB_s^\mu.$$

Here, $t_j = tj/2^n$. In the same way as $\int_0^{T_t} \mathrm{j}_{T_s^*}\tilde{\varphi}(\cdot - B_s)dB_s^\mu$, the right-hand side above strongly converges in $L^2(\mathfrak{X})\otimes L^2(\mathbb{R}^4)$.

Corollary 3.120. *The functional integral representation of $e^{-2tH_{\mathrm{RPF}}}$ is given by*

$$(F, e^{-2tH_{\mathrm{RPF}}}G) = \int_{\mathbb{R}^3} \mathbb{E}^{x,0}\left[e^{-\int_0^t V(B_{T_s})ds - \int_{-t}^{0} V(B_{-T_{-s}})ds}\left(\mathrm{J}_{-t}F(B_{-T_t}), e^{-ie\hat{A}_{\mathrm{E}}(\widehat{K}_t^{\mathrm{rel}})}\mathrm{J}_t G(B_{T_t})\right)\right]dx.$$

Here,

$$\widehat{K}_t^{\mathrm{rel}} = \bigoplus_{\mu=1}^{3}\left(\int_0^{T_t} \mathrm{j}_{T_s^*}\tilde{\varphi}(\cdot - B_s)dB_s^\mu + \int_{-T_t}^{0} \mathrm{j}_{T_s^*}\tilde{\varphi}(\cdot - B_s)dB_s^\mu\right).$$

Proof. This is proven by means of the shift U_t in the field and the fact that $T_s - T_t = T_{s-t}$ in law. That is, for $t \le t_1 \le \cdots \le t_n$,

$$\begin{aligned}
&\mathbb{E}^0[f(T_{t_1} - T_t, \ldots, T_{t_n} - T_t)] \\
&= \mathbb{E}^0[f(T_{t_1} - T_t, T_{t_2} - T_{t_1} + T_{t_1} - T_t, \ldots, T_{t_n} - T_{t_{n-1}} + \cdots + T_{t_1} - T_t)] \\
&= \int_{\mathbb{R}^n} f(x_1, x_2 + x_1, \ldots, x_n + \cdots + x_1)\mu_{t_1,t}(dx_1)\cdots\mu_{t_n,t_{n-1}}(dx_n),
\end{aligned} \tag{3.9.10}$$

where μ_{t_i,t_j} is the distribution of $T_{t_i} - T_{t_j}$, and $\mu_{t_i,t_j} = \mu_{t_i-t_j}$ is satisfied. Here, μ_r is the distribution of T_r. On the other hand,

$$\begin{aligned}
&\mathbb{E}^0[f(T_{t_1-t}, \ldots, T_{t_n-t})] \\
&= \mathbb{E}^0[f(T_{t_1-t}, T_{t_2-t} - T_{t_1-t} + T_{t_1-t}, \ldots, T_{t_n-t} - T_{t_{n_1}-t} + \cdots + T_{t_1-t})] \\
&= \int_{\mathbb{R}^n} f(x_1, x_2 + x_1, \ldots, x_n + \cdots + x_1)\mu_{t_1,t}(dx_1)\cdots\mu_{t_n,t_{n-1}}(dx_n).
\end{aligned} \tag{3.9.11}$$

Comparing (3.9.10) with (3.9.11) above, we can see that

$$\mathbb{E}^0[f(T_{t_1} - T_t, \ldots, T_{t_n} - T_t)] = \mathbb{E}^0[f(T_{t_1-t}, \ldots, T_{t_n-t})] \tag{3.9.12}$$

for each t. By Theorem 3.119, we have

$$(F, e^{-2tH_{\mathrm{RPF}}}G) = \int_{\mathbb{R}^3} \mathbb{E}^{x,0}\left[\left(\mathrm{J}_0 F(B_{T_0}), e^{-ie\hat{A}_{\mathrm{E}}(K_{2t}^{\mathrm{rel}})}\mathrm{J}_{2t}G(B_{T_{2t}})\right) e^{-\int_0^{2t} V(B_{T_s})ds}\right] dx$$

and

$$= \int_{\mathbb{R}^3} \mathbb{E}^{x,0}\left[\left(\mathrm{J}_{-t}F(B_{T_0}), \mathrm{U}_t e^{-ie\hat{A}_{\mathrm{E}}(K_{2t}^{\mathrm{rel}})}\mathrm{U}_{-t}\mathrm{J}_t G(B_{T_{2t}})\right) e^{-\int_0^{2t} V(B_{T_s})ds}\right] dx.$$

By the shift of Brownian motion, $B_t \to B_{t-T_t}$, we have

$$= \int_{\mathbb{R}^3} \mathbb{E}^{x,0}\left[\left(\mathrm{J}_{-t}F(B_{-T_t}), e^{-i\alpha\hat{A}_{\mathrm{E}}(S)}\mathrm{J}_t G(B_{T_{2t}-T_t})\right) e^{-\int_0^{2t} V(B_{T_s-T_t})ds}\right] dx,$$

where $S = \lim_{n\to\infty} \bigoplus_{\mu=1}^3 \sum_{j=1}^{2\cdot 2^n} \int_{T_{t_{j-1}}-T_t}^{T_{t_j}-T_t} \mathrm{j}_{t_{j-1}-t}\tilde{\varphi}(\cdot - B_s)dB_s^\mu$ and, by (3.9.12), we can check that the following identities in law follow:

$$\int_0^{2t} V(B_{T_s-T_t})ds = \int_0^t V(B_{-(T_{t-s})})ds + \int_t^{2t} V(B_{T_{s-t}})ds = \int_{-t}^0 V(B_{-T_{-s}})ds + \int_0^t V(B_{T_s})ds.$$

Furthermore, with $F(s) = \tilde{\varphi}(\cdot - B_s)$, we have

$$\sum_{j=1}^{2\cdot 2^n} \int_{T_{t_{j-1}}-T_t}^{T_{t_j}-T_t} \mathrm{j}_{t_{j-1}-t}F(s)dB_s^\mu = \sum_{j=1}^{2^n} \int_{-T_{-(t_{j-1}-t)}}^{-T_{-(t_j-t)}} \mathrm{j}_{-(t_{j-1}-t)}F(s)dB_s^\mu + \sum_{j=2^n+1}^{2\cdot 2^n} \int_{T_{t_{j-1}-t}}^{T_{t_j-t}} \mathrm{j}_{t_{j-1}-t}F(s)dB_s^\mu.$$

Then the corollary follows. □

In what follows, we set

$$\int_0^t V(B_{T_s})ds + \int_{-t}^0 V(B_{-T_{-s}})ds = \int_{-t}^t V(B_{T_s})ds.$$

By Lemma 3.118, we can also extend the functional integral representation to a wider class of external potentials. Let $V = V_+ - V_-$ be such that V_- is relatively form bounded with respect to $\sqrt{-\Delta + m^2} - m$ with a relative bound strictly smaller than 1 and $V_+ \in L^1_{\mathrm{loc}}(\mathbb{R}^3)$. Then we define the *relativistic Pauli–Fierz Hamiltonian with singular external potential* by

$$H_{\mathrm{RPF}} = T_{\mathrm{kin}} \dotplus V_+ \dot{-} V_- \dotplus H_{\mathrm{rad}}.$$

Note that $D(T_{\mathrm{kin}}^{1/2}) \cap D(H_{\mathrm{rad}}^{1/2}) \cap D(V_+^{1/2}) \cap D(V_-^{1/2})$ is dense. Hence we can also have the following functional integral representation for such singular external potentials. The proof is similar to that of Theorem 3.32.

Theorem 3.121 (Functional integral representation for relativistic Pauli–Fierz Hamiltonian with singular external potential). *Suppose Assumption* 3.112 *holds, but instead of relative boundedness of* $V = V_+ - V_-$ *with respect to* $\sqrt{-\Delta}$, *it is supposed that* V_- *is relatively form bounded with respect to* $\sqrt{-\Delta + m^2} - m$ *with a relative bound strictly smaller than* 1 *and* $V_+ \in L^1_{\mathrm{loc}}(\mathbb{R}^3)$. *Then*

$$(F, e^{-tH_{\mathrm{RPF}}} G) = \int_{\mathbb{R}^3} \mathbb{E}^{x,0}\left[e^{-\int_0^t V(B_{T_s})ds}(\mathrm{J}_0 F(B_0), e^{-ie\hat{A}_{\mathrm{E}}(K_t^{\mathrm{rel}})}\mathrm{J}_t G(B_{T_t}))\right] dx. \tag{3.9.13}$$

By using this functional integral representation, we can obtain similar results to those of H_{PF}.

Corollary 3.122 (Diamagnetic inequality). *Let* $E(e) = \inf \mathrm{Spec}(H_{\mathrm{RPF}})$. *Under Assumption* 3.112 *or the assumptions of Theorem* 3.121, *it follows that*

$$|(F, e^{-tH_{\mathrm{RPF}}} G)| \le \left(|F|, e^{-t(\sqrt{-\Delta+m^2}-m \dotplus V_+ \dot{-} V_- + H_{\mathrm{rad}})}|G|\right)$$

and $E(0) \le E(e)$.

Proof. Since $|e^{-ie\hat{A}_{\mathrm{E}}(\widehat{K}_t^{\mathrm{rel}})}| = 1$, the corollary follows from the functional integral representation of $(F, e^{-tH_{\mathrm{RPF}}} G)$. □

Corollary 3.123 (Positivity improving). *Under Assumption* 3.112 *or the assumptions of Theorem* 3.121, $e^{-i(\pi/2)N} e^{-tH_{\mathrm{RPF}}} e^{i(\pi/2)N}$ *is positivity improving. In particular, the ground state of* H_{RPF} *is unique, if it exists.*

Proof. Since $\mathrm{J}_0^* e^{-ie\hat{A}_{\mathrm{E}}(\widehat{K}_t^{\mathrm{rel}})}\mathrm{J}_t$ is positivity improving, the corollary follows from the functional integral representation of $(F, e^{-tH_{\mathrm{RPF}}} G)$. □

3.9.3 Self-adjointness

In this section by using the functional integral representation derived in Theorem 3.119, we show the self-adjointness of H_{RPF} for arbitrary values of coupling constants. To prove this, we find an invariant domain D so that $D \subset D(H_{\mathrm{RPF}})$ and $e^{-tH_{\mathrm{RPF}}} D \subset D$ in a similar way to that of H_{PF}.

Let $\hat{\omega}_{\mathrm{E}} = \omega(-i\nabla) \otimes 1$ under $L^2(\mathbb{R}^4) \cong L^2(\mathbb{R}^3) \otimes L^2(\mathbb{R})$. By the Itô isometry, we have

$$\mathbb{E}^{x,0}\left[\|\hat{\omega}_{\mathrm{E}}^{\alpha/2} K_t^{\mathrm{rel}}\|^2_{\oplus^3 L^2(\mathbb{R}^4)}\right] = 3\mathbb{E}^{x,0}\left[\int_0^{T_t} \|\hat{\omega}^{\alpha/2}\tilde{\varphi}(\cdot - B_r)\|^2_{L^2(\mathbb{R}^3)} dr\right].$$

In particular,

$$\mathbb{E}^{x,0}\left[\|\hat{\omega}_{\mathrm{E}}^{\alpha/2} K_t^{\mathrm{rel}}\|^2_{\oplus^3 L^2(\mathbb{R}^4)}\right] \le 3\mathbb{E}_\nu^0[T_t]\|\omega^{(\alpha-1)/2}\hat{\varphi}\|^2_{L^2(\mathbb{R}^3)}$$

and the right-hand side above is finite in the case of $m > 0$, since $\mathbb{E}_\nu^0[T_t] < \infty$. We can also estimate forth moment $\mathbb{E}^{x,0}[\|\hat{\omega}_{\mathrm{E}}^{\alpha/2} K_t^{\mathrm{rel}}\|_{\oplus^3 L^2(\mathbb{R}^4)}^4]$ in the next lemma.

Lemma 3.124 (BDG-type inequality). *Let $\alpha > 0$ and $m > 0$, i. e., $\mathbb{E}_\nu^0[e^{-uT_t}] = \exp(-t(\sqrt{2u+m^2} - m))$. Suppose that $\hat{\varphi}/\sqrt{\omega}, \omega^{(\alpha-1)/2}\hat{\varphi} \in L^2(\mathbb{R}^3)$. Then the BDG-type inequality holds, namely*

$$\mathbb{E}^{x,0}\left[\|\hat{\omega}_{\mathrm{E}}^{\alpha/2} K_t^{\mathrm{rel}}\|_{\oplus^3 L^2(\mathbb{R}^4)}^4\right] \le C\|\omega^{(\alpha-1)/2}\hat{\varphi}\|_{L^2(\mathbb{R}^3)}^4,$$

where C is a constant.

Proof. Recall the BDG-type inequality

$$\mathbb{E}^x\left[\left\|\hat{\omega}_{\mathrm{E}}^{\alpha/2}\int_0^t \mathrm{j}_s\tilde{\varphi}(\cdot - B_s)dB_s^\mu\right\|_{L^2(\mathbb{R}^4)}^{2n}\right] \le \frac{(2n)!}{2^n} t^n \|\omega^{(\alpha-1)/2}\hat{\varphi}\|_{L^2(\mathbb{R}^3)}^{2n}. \tag{3.9.14}$$

Notice that $\hat{\omega}_{\mathrm{E}}^{\alpha/2} K_t^{\mathrm{rel}} = \text{s-lim}_{n\to\infty} \bigoplus_{\mu=1}^3 \sum_{j=1}^{2^n} a_j^\mu$ with

$$a_j^\mu = \int_{T_{t_{j-1}}}^{T_{t_j}} \mathrm{j}_{t_{j-1}}\lambda_\alpha(\cdot - B_s)dB_s^\mu,$$

and $\lambda_\alpha = \hat{\omega}^{\alpha/2}\tilde{\varphi}$ and $\hat{\lambda}_\alpha = \omega^{(\alpha-1)/2}\hat{\varphi}$. We fix a μ and set $a_j^\mu = a_j$ for simplicity. Here a_j and a_i are independent for $i \ne j$ and then we have

$$\mathbb{E}^{x,0}\left[\left\|\sum_{j=1}^{2^n} a_j\right\|_{L^2(\mathbb{R}^4)}^2\right] = \sum_{j,j'=1}^{2^n}\sum_{i,i'=1}^{2^n} \mathbb{E}^{x,0}\left[\left(\int_{\mathbb{R}^4} a_j(z)a_{j'}(z)dz\right)\left(\int_{\mathbb{R}^4} a_i(y)a_{i'}(y)dy\right)\right].$$

We divide the sum above into the diagonal part and the off diagonal part. Then we have

$$\begin{aligned}
&= \sum_{j=1}^{2^n} \mathbb{E}^{x,0}\left[\left(\int_{\mathbb{R}^4} a_j(z)^2 dz\right)^2\right] + \sum_{j=1}^{2^n} \mathbb{E}^{x,0}\left[\int_{\mathbb{R}^4} a_j(z)^2 dz\right] \sum_{i\ne j}^{2^n} \mathbb{E}^{x,0}\left[\int_{\mathbb{R}^4} a_i(z)^2 dz\right] \\
&\quad + \sum_{j=1}^{2^n}\sum_{i\ne j}^{2^n} \mathbb{E}^{x,0}\left[\int_{\mathbb{R}^4} a_j(z)a_i(z)dz \int_{\mathbb{R}^4} a_j(y)a_i(y)dy\right].
\end{aligned}$$

We estimate the first term of the right-hand side above. We have by (3.9.14)

$$\sum_{j=1}^{2^n} \mathbb{E}^{x,0}\left[\left(\int_{\mathbb{R}^4} a_j(z)^2 dz\right)^2\right] = \sum_{j=1}^{2^n} \mathbb{E}^{x,0}\left[\|a_j\|^4\right] \le 6\|\omega^{(\alpha-1)/2}\hat{\varphi}\|^4 \sum_{j=1}^{2^n} \mathbb{E}_\nu^0\left[\left|T_{\frac{t}{2^{n+1}}}\right|^2\right].$$

By using

$$\mathbb{E}_v^0[T_t^n] = \frac{te^{tm}}{\sqrt{2\pi}} \int_0^\infty \frac{s^n}{s^{3/2}} \exp\left(-\frac{1}{2}\left(\frac{t^2}{s} + m^2 s\right)\right) ds < \infty$$

for $n \geq 0$, and the assumption $m > 0$, we have

$$\sum_{j=1}^{2^n} \mathbb{E}^{x,0}\left[\left(\int_{\mathbb{R}^4} a_j(z)^2 dz\right)^2\right] \leq \frac{6t\|\omega^{(\alpha-1)/2}\hat{\varphi}\|^4}{2\sqrt{2\pi}} e^{\frac{mt}{2^{n+1}}} \int_0^\infty \sqrt{s} \exp\left(-\frac{1}{2}\left(\frac{(\frac{t}{2^{n+1}})^2}{s} + m^2 s\right)\right) ds.$$

The right-hand side converges to

$$\frac{3t}{\sqrt{2\pi}} \|\omega^{(\alpha-1)/2}\hat{\varphi}\|^4 \int_0^\infty \sqrt{s} \exp\left(-\frac{1}{2} m^2 s\right) ds$$

as $n \to \infty$. The second term is estimated as

$$\sum_{j=1}^{2^n} \mathbb{E}^{x,0}\left[\int_{\mathbb{R}^4} a_j(z)^2 dz\right] \sum_{i\neq j}^{2^n} \mathbb{E}^{x,0}\left[\int_{\mathbb{R}^4} a_j(z)^2 dz\right] \leq \left(\sum_{j=1}^{2^n} \mathbb{E}^{x,0}\left[\int_{\mathbb{R}^4} a_j(z)^2 dz\right]\right)^2.$$

By the Itô isometry, we have

$$\sum_{j=1}^{2^n} \mathbb{E}^{x,0}\left[\int_{\mathbb{R}^4} a_j(z)^2 dz\right] = \mathbb{E}^{x,0}\left[\int_0^{T_t} \|\mathrm{j}_s \lambda_\alpha(\cdot - B_s)\|^2 ds\right] \leq \mathbb{E}_v^0[T_t] \|\omega^{(\alpha-1)/2}\hat{\varphi}\|^2.$$

Hence

$$\sum_{j=1}^{2^n} \mathbb{E}^{x,0}\left[\int_{\mathbb{R}^4} a_j(z)^2 dz\right] \sum_{i\neq j}^{2^n} \mathbb{E}^{x,0}\left[\int_{\mathbb{R}^4} a_j(z)^2 dz\right] \leq (\mathbb{E}_v^0[T_t])^2 \|\omega^{(\alpha-1)/2}\hat{\varphi}\|^4.$$

Finally, we estimate the third term. We see that

$$\sum_{j=1}^{2^n} \sum_{i\neq j}^{2^n} \mathbb{E}^{x,0}\left[\int_{\mathbb{R}^4} a_j(z) a_i(z) dz \int_{\mathbb{R}^4} a_j(y) a_i(y) dy\right] \leq \int_{\mathbb{R}^4} dz \int_{\mathbb{R}^4} \left|\sum_{j=1}^{2^n} \mathbb{E}^{x,0}\left[a_j(z) a_j(y)\right]\right|^2 dy.$$

Note that $\mathbb{E}^{x,0}[a_j(z)a_j(y)] = \mathbb{E}^{x,0}[\int_{T_{t_{j-1}}}^{T_{t_j}} A_s(z,j) A_s(y,j) ds]$, where we set

$$A_s(z,j) = (\mathrm{j}_{t_{j-1}} \lambda_\alpha(\cdot - B_s))(z).$$

By the Schwarz inequality, we have

$$\leq \int_{\mathbb{R}^4} dz \int_{\mathbb{R}^4} \mathbb{E}^{X,0}\left[\left(\sum_{j=1}^{2^n} \int_{T_{t_{j-1}}}^{T_{t_j}} A_s(z,j)A_s(y,j)ds\right)^2\right] dy$$

$$\leq \int_{\mathbb{R}^4} dz \int_{\mathbb{R}^4} \mathbb{E}^{X,0}\left[\left(\sum_{j=1}^{2^n} \int_{T_{t_{j-1}}}^{T_{t_j}} A_s(z,j)^2 ds\right)\left(\sum_{j=1}^{2^n} \int_{T_{t_{j-1}}}^{T_{t_j}} A_s(y,j)^2 ds\right)\right] dy,$$

and Fubini's theorem yields that

$$= \mathbb{E}^{X,0}\left[\int_{\mathbb{R}^4}\left(\sum_{j=1}^{2^n} \int_{T_{t_{j-1}}}^{T_{t_j}} A_s(z,j)^2 ds\right) dz \int_{\mathbb{R}^4}\left(\sum_{j=1}^{2^n} \int_{T_{t_{j-1}}}^{T_{t_j}} A_s(y,j)^2 ds\right) dy\right]$$

$$= \mathbb{E}^{X,0}\left[T_t^2 \|\omega^{(\alpha-1)/2}\hat{\varphi}\|^4\right] = \mathbb{E}_\nu^0[T_t^2]\|\omega^{(\alpha-1)/2}\hat{\varphi}\|^4.$$

Then the lemma follows. □

Let $\mathrm{P}_\mu^{\mathrm{tot}} = -i\nabla_\mu \otimes 1 + 1 \otimes P_{\mathrm{f}\mu}$ be the total momentum operator in $\mathcal{H}_{\mathrm{PF}}$.

Lemma 3.125. *Let $V = 0$. Then $e^{-it\mathrm{P}_\mu^{\mathrm{tot}}} e^{-sH_{\mathrm{RPF}}} e^{it\mathrm{P}_\mu^{\mathrm{tot}}} = e^{-sH_{\mathrm{RPF}}}$.*

Proof. We set $\overline{P_{\mathrm{f}\mu}} = \mathrm{d}\Gamma_{\mathrm{E}}(-i\nabla_\mu \otimes 1)$. We have

$$\begin{aligned}(F, e^{-it\mathrm{P}_\mu^{\mathrm{tot}}} e^{-sH_{\mathrm{RPF}}} e^{it\mathrm{P}_\mu^{\mathrm{tot}}} G) \\ = \int_{\mathbb{R}^3} \mathbb{E}^{X,0}\left[(\mathrm{J}_0 F(B_{T_0}), e^{-it(-i\nabla_\mu)} e^{-it\overline{P_{\mathrm{f}\mu}}} e^{-i\hat{A}_{\mathrm{E}}(K_t^{\mathrm{rel}})} e^{it\overline{P_{\mathrm{f}\mu}}} e^{it(-i\nabla_\mu)} \mathrm{J}_t G(B_{T_t}))\right] dx.\end{aligned}$$

Since

$$e^{-it(-i\nabla_\mu)} e^{-it\overline{P_{\mathrm{f}\mu}}} e^{-i\hat{A}_{\mathrm{E}}(K_t^{\mathrm{rel}})} e^{it\overline{P_{\mathrm{f}\mu}}} e^{it(-i\nabla_\mu)} = e^{-i\hat{A}_{\mathrm{E}}(K_t^{\mathrm{rel}})},$$

the lemma follows. □

Lemma 3.126. *Let $V = 0$. For $F \in D(-i\nabla_\mu)$ and $G \in D(-i\nabla_\mu) \cap D(H_{\mathrm{rad}}^{1/2})$, it follows that*

$$(-i\nabla_\mu F, e^{-tH_{\mathrm{RPF}}} G) \leq C\left((\|\sqrt{\omega}\hat{\varphi}\| + \|\hat{\varphi}\|)\|(H_{\mathrm{rad}} + 1)^{1/2} G\| + \|-i\nabla_\mu G\|\right)\|F\|.$$

In particular, $e^{-tH_{\mathrm{RPF}}} G \in D(\sqrt{-\Delta})$ for $t \geq 0$.

Proof. Notice that $(e^{is(-i\nabla_\mu)} F, e^{-tH_{\mathrm{RPF}}} G) = (e^{-is\hat{P}_{\mathrm{f},\mu}} F, e^{-tH_{\mathrm{RPF}}} e^{-is\mathrm{P}_\mu^{\mathrm{tot}}} G)$. We have

$$(e^{is(-i\nabla_\mu)} F, e^{-tH_{\mathrm{RPF}}} G) = \int_{\mathbb{R}^3} \mathbb{E}^{X,0}\left[\left(\mathrm{J}_0 F(B_{T_0}), e^{+is\overline{P_{\mathrm{f}\mu}}} e^{-ie\hat{A}_{\mathrm{E}}(\mathrm{K})} e^{-is\overline{P_{\mathrm{f}\mu}}} \mathrm{J}_t e^{-is(-i\nabla_\mu)} G(B_{T_t})\right)\right] dx. \tag{3.9.15}$$

Here and in what follows in this proof, we set $\mathrm{K} = K_t^{\mathrm{rel}}$. We see that

$$e^{+is\overline{P_{\mathrm{f}\mu}}}e^{-ie\hat{A}_{\mathrm{E}}(\mathrm{K})}e^{-is\overline{P_{\mathrm{f}\mu}}} = e^{-ie\hat{A}_{\mathrm{E}}(e^{is(-i\nabla_\mu\otimes 1)}\mathrm{K})}.$$

Take the derivative at $s = 0$ on both sides of (3.9.15). We have

$$\begin{aligned}(\nabla_\mu F, e^{-tH_{\mathrm{RPF}}}G) &= \int_{\mathbb{R}^3} \mathbb{E}^{x,0}\left[\left(\mathrm{J}_0 F(B_{T_0}), -ie\hat{A}_{\mathrm{E}_\mu}(\nabla_\mu \mathrm{K})e^{-ie\hat{A}_{\mathrm{E}}(\mathrm{K})}\mathrm{J}_t G(B_{T_t})\right)\right] dx \\ &\quad + \int_{\mathbb{R}^3} \mathbb{E}^{x,0}\left[\left(\mathrm{J}_0 F(B_{T_0}), e^{-ie\hat{A}_{\mathrm{E}}(\mathrm{K})}\mathrm{J}_t(-\nabla_\mu G)(B_{T_t})\right)\right] dx. \end{aligned} \tag{3.9.16}$$

It is trivial to see that

$$\left|\int_{\mathbb{R}^3} \mathbb{E}^{x,0}\left[\left(\mathrm{J}_0 F(B_{T_0}), e^{-ie\hat{A}_{\mathrm{E}}(\mathrm{K})}\mathrm{J}_t(-\nabla_\mu G)(B_{T_t})\right)\right] dx\right| \le \|F\|\|\nabla_\mu G\|.$$

We can estimate the first term on the right-hand side of (3.9.16) as

$$\begin{aligned}&\left|\int_{\mathbb{R}^3} \mathbb{E}^{x,0}\left[\left(\mathrm{J}_0 F(B_{T_0}), \hat{A}_{\mathrm{E}_\mu}(\nabla_\mu \mathrm{K})e^{-ie\hat{A}_{\mathrm{E}_\mu}(\mathrm{K})}\mathrm{J}_t G(B_{T_t})\right)\right] dx\right| \\ &\le \int_{\mathbb{R}^3} \mathbb{E}^{x,0}\left[\|\hat{A}_{\mathrm{E}_\mu}(\nabla_\mu \mathrm{K})\mathrm{J}_0 F(B_{T_0})\|\|\mathrm{J}_t G(B_{T_t})\|\right] dx\end{aligned}$$

By the inequality $\|\hat{A}_{\mathrm{E}_\mu}(f)\Phi\| \le C(\|f\| + \|\hat{\omega}_{\mathrm{E}}^{-1/2}f\|)\|(H_{\mathrm{rad}}+1)^{1/2}\Phi\|$ with some constant $C > 0$, we have

$$\begin{aligned}&\le C\int_{\mathbb{R}^3} \mathbb{E}^{x,0}\left[(\|\nabla_\mu \mathrm{K}\| + \|\hat{\omega}_{\mathrm{E}}^{-1/2}\nabla_\mu \mathrm{K}\|)\|(H_{\mathrm{rad}}+1)^{1/2}F(B_{T_0})\|\|G(B_{T_t})\|\right] dx \\ &\le C(\|\omega^{1/2}\hat{\varphi}\| + \|\hat{\varphi}\|)\|(H_{\mathrm{rad}}+1)^{1/2}F\|\|G\|.\end{aligned}$$

Then the lemma follows. □

Lemma 3.127. *Let $V = 0$. Then for $F, G \in D(H_{\mathrm{rad}})$ it follows that*

$$(H_{\mathrm{rad}}F, e^{-tH_{\mathrm{RPF}}}G) \le \left(\|H_{\mathrm{rad}}G\| + |e|(\|\sqrt{\omega}\hat{\varphi}\| + \|\hat{\varphi}\|)\|(H_{\mathrm{rad}}+1)^{1/2}G\| + |e|^2\|\hat{\varphi}/\sqrt{\omega}\|^2\|G\|\right)\|F\|.$$

In particular, $e^{-tH_{\mathrm{RPF}}}G \in D(H_{\mathrm{rad}})$ for $t \ge 0$.

Proof. Let $\overline{H_{\mathrm{rad}}} = d\Gamma_{\mathrm{E}}(\hat{\omega}_{\mathrm{E}})$. Note that $\Pi_{\mathrm{E}}(\hat{\omega}f) = i[\overline{H_{\mathrm{rad}}}, \hat{A}_{\mathrm{E}}(f)]$ in the function space. We have

$$(H_{\mathrm{rad}}F, e^{-tH_{\mathrm{RPF}}}G) = \int_{\mathbb{R}^3} \mathbb{E}^{x,0}\left[\left(\mathrm{J}_0 F(B_{T_0}), e^{-ie\hat{A}_{\mathrm{E}}(K_t^{\mathrm{rel}})}S\mathrm{J}_t G(B_{T_t})\right)\right] dx,$$

where

$$S = e^{ie\hat{A}_E(K_t^{\mathrm{rel}})}\overline{H_{\mathrm{rad}}}e^{-ie\hat{A}_E(K_t^{\mathrm{rel}})} = \overline{H_{\mathrm{rad}}} - e\Pi_E(\hat{\omega}K_t^{\mathrm{rel}}) + e^2 g$$

with $g = \mathrm{q}_E(K_t^{\mathrm{rel}}, K_t^{\mathrm{rel}})$. It is trivial to see that

$$\left|\int_{\mathbb{R}^3} \mathbb{E}^{x,0}\left[\left(\mathrm{J}_0 F(B_{T_0}), e^{-ie\hat{A}_E(K_t^{\mathrm{rel}})}\overline{H_{\mathrm{rad}}}\mathrm{J}_t G(B_{T_t})\right)\right] dx\right| \le \|F\|\|H_{\mathrm{rad}}G\|.$$

In the same way as when obtaining the estimate of the first term of the right-hand side of (3.9.16), we can see that

$$\begin{aligned}&\left|\int_{\mathbb{R}^3} \mathbb{E}^{x,0}\left[\left(\mathrm{J}_0 F(B_{T_0}), e^{-ie\hat{A}_E(K_t^{\mathrm{rel}})}\Pi_E(\hat{\omega}K_t^{\mathrm{rel}})\mathrm{J}_t G(B_{T_t})\right)\right] dx\right| \\ &\quad \le C(\|\sqrt{\omega}\hat{\varphi}\| + \|\hat{\varphi}\|)\|F\|\|(H_{\mathrm{rad}} + 1)^{1/2}G\|\end{aligned}$$

with some constant $C > 0$. Here, we used Lemma 3.124. Finally, we see that $g \le C\|K_t^{\mathrm{rel}}\|^2$ and, by Lemma 3.124 again,

$$\begin{aligned}&\left|\int_{\mathbb{R}^3} \mathbb{E}^{x,0}\left[\left(\mathrm{J}_0 F(B_{T_0}), e^{-ie\hat{A}_E(K_t^{\mathrm{rel}})} g \mathrm{J}_t G(B_{T_t})\right)\right] dx\right| \\ &\quad \le C\left(\int_{\mathbb{R}^3} \mathbb{E}^{x,0}\left[\|K_t^{\mathrm{rel}}\|^4\right]\|F(x)\|^2 dx\right)^{1/2} \|G\| \le C\|\hat{\varphi}/\sqrt{\omega}\|^2\|F\|\|G\|.\end{aligned}$$

Then the lemma follows. □

We recall $D_\infty = C_0^\infty(\mathbb{R}^3) \otimes \mathcal{F}_{\mathrm{rad}}^\infty$. See (3.7.2) and (3.7.3).

Lemma 3.128. *Let $m > 0$. Then H_{RPF} is essentially self-adjoint on $D(\sqrt{-\Delta}) \cap D(H_{\mathrm{rad}})$.*

Proof. Suppose $V = 0$. Let $F \in D_\infty$. Since $\hat{\varphi}/\omega \in L^2(\mathbb{R}^3)$, we see that

$$\|(T_{\mathrm{kin}} \dotplus H_{\mathrm{rad}})F\|^2 \le C_1\|\sqrt{-\Delta}F\|^2 + C_2\|H_{\mathrm{rad}}F\|^2 + C_3\|F\|^2 \tag{3.9.17}$$

with some constants C_1, C_2, and C_3. Since D_∞ is a core of $\sqrt{-\Delta} + H_{\mathrm{rad}}$,

$$D(\sqrt{-\Delta}) \cap D(H_{\mathrm{rad}}) \supset D(T_{\mathrm{kin}} \dotplus H_{\mathrm{rad}}) \tag{3.9.18}$$

follows from a limiting argument. By Lemmas 3.126 and 3.127, we also see that

$$e^{-t(T_{\mathrm{kin}} \dotplus H_{\mathrm{rad}})}\left(D(\sqrt{-\Delta}) \cap D(H_{\mathrm{rad}})\right) \subset \left(D(\sqrt{-\Delta}) \cap D(H_{\mathrm{rad}})\right). \tag{3.9.19}$$

Equations (3.9.18) and (3.9.19) imply that $T_{\mathrm{kin}} \dotplus H_{\mathrm{rad}}$ is essentially self-adjoint on the domain $D(\sqrt{-\Delta}) \cap D(H_{\mathrm{rad}})$ by Lemma 3.40. Next, we suppose that V satisfies assumptions in the lemma. Then V is also relatively bounded with respect to $T_{\rm kin} \dot{+} H_{\rm rad}$ with a relative bound strictly smaller than one. Then the theorem follows by the Kato–Rellich theorem. □

We will be able to show that $H_{\rm RPF}$ is essentially self-adjoint on D_∞ by limiting arguments. We define additional two dense subspaces $\mathscr{E}$ and $\mathscr{D}_0$ such that

$$D_\infty \subset \mathscr{E} \subset \mathscr{D}_0 \subset D(\sqrt{-\Delta}) \cap D(H_{\rm rad}).$$

We shall show below that $H_{\rm RPF}\lceil_{\mathscr{D}_0}$ and $H_{\rm RPF}\lceil_{\mathscr{E}}$ are essentially self-adjoint, and finally we can conclude that D_∞ is a core of $H_{\rm RPF}$. Let

$$\mathscr{D} = \left\{\{\Psi^{(n)}\}_{n=0}^\infty \in \mathscr{H} \mid \Psi^{(n)} = 0 \text{ for all } n \geq n_0 \text{ with some } n_0 \geq 1\right\}$$

and we set $\mathscr{D}_0 = D(\sqrt{-\Delta}) \cap D(H_{\rm rad}) \cap \mathscr{D}$. We define

$$H_0 = T_{\rm kin} \dot{+} H_{\rm rad}.$$

Lemma 3.129. *Let $m > 0$. Then $\mathscr{D}$ is a core of H_0.*

Proof. Let $P_n = 1_{[0,n]}(N)$ for $n \in \mathbb{N}$. Take an arbitrary $\Psi \in D(\sqrt{-\Delta}) \cap D(H_{\rm rad})$. Hence $P_n\Psi \in \mathscr{D}$. We see that $P_n\Psi \to \Psi$ as $n \to \infty$. Since

$$\|H_0(P_n - P_m)\Psi\| \leq C(\|(P_n - P_m)\sqrt{-\Delta}\Psi\| + \|(P_n - P_m)H_{\rm rad}\Psi\|),$$

we also see that $(H_0P_n\Psi)_{n\in\mathbb{N}}$ is a Cauchy sequence in $\mathcal{H}_{\rm PF}$. From the closedness of H_0, it follows that $\Psi \in D(H_0)$ and $H_0P_n\Psi \to H_0\Psi$. Thus $\mathscr{D}$ is a core of H_0 by Lemma 3.128. □

Let

$$\mathcal{E} = \{\{\Psi^{(n)}\}_{n=0}^\infty \in \mathscr{D} \mid \Psi^{(n)}(\cdot, \mathbf{k}) \in C_0^\infty(\mathbb{R}^3) \text{ a. e. } \mathbf{k} \in \mathbb{R}^{3n}, n \geq 1\}.$$

Lemma 3.130. *Let $m > 0$. Then $\mathcal{E}$ is a core of H_0.*

Proof. Take an arbitrary $\Phi \in \mathscr{D}$. Let $0 \leq j \in C_0^\infty(\mathbb{R}^3)$ and $g \in C_0^\infty(\mathbb{R}^3)$ such that $\int_{\mathbb{R}^3} j(x)dx = 1$, $0 \leq g(x) \leq 1$ and $g(x) = 1$ for $|x| \leq 1$. For each $\varepsilon > 0$, we set $j_\varepsilon(x) = \varepsilon^{-3}j(x/\varepsilon)$ and $\Phi_{\varepsilon,L}^{(n)}(x, \mathbf{k}) = g(x/L)\int_{\mathbb{R}^3} j_\varepsilon(x - y)\Phi^{(n)}(y, \mathbf{k})dy$. Define $\Phi_{\varepsilon,L} = \{\Phi_{\varepsilon,L}^{(n)}\}_{n=0}^\infty$. We see that $\Phi_{\varepsilon,L} \to \Phi$, $-i\nabla_\mu\Phi_{\varepsilon,L} \to -i\nabla_\mu\Phi$ and $H_{\rm rad}\Phi_{\varepsilon,L} \to H_{\rm rad}\Phi$ strongly as $\varepsilon \downarrow 0$ and $L \to \infty$. Then by (3.9.17) and the closedness of H_0, we see that $\Phi \in D(H_0)$ and $\lim_{L\to\infty}\lim_{\varepsilon\downarrow 0} H_0\Phi_{\varepsilon,L} = H_0\Phi$ in $\mathcal{H}_{\rm PF}$. Thus the lemma follows from Lemma 3.129. □

Theorem 3.131. *Let $m > 0$. Then H_0 is essentially self-adjoint on D_∞.*

Proof. Suppose that $m > 0$. Let $F \in \mathcal{E}$. Then it follows that

$$\| - \Delta F\| + \|H_{\rm rad}F\| \leq C\|(-\Delta + H_{\rm rad} + 1)F\| \tag{3.9.20}$$

with some constant $C > 0$. Note that $-\Delta + H_{\mathrm{rad}}$ is essentially self-adjoint on D_∞. There exists a sequence $(F_n)_{n\in\mathbb{N}} \subset D_\infty$ such that $F_n \to F$, and $(-\Delta + H_{\mathrm{rad}})F_n \to (-\Delta + H_{\mathrm{rad}})F$ as $n \to \infty$. From (3.9.20), it follows that $-\Delta F_n \to -\Delta F$ and $H_{\mathrm{rad}}F_n \to H_{\mathrm{rad}}F$ as $n \to \infty$. Then we can also see that $(H_0F_n)_{n\in\mathbb{N}}$ is a Cauchy sequence by (3.9.17) and $H_0F_n \to H_0F$ as $n \to \infty$ follows. Thus D_∞ is a core of H_0 by Lemma 3.130. □

Furthermore, we can establish the self-adjointness of H_{RPF}. The key inequality to show the self-adjointness of H_{RPF} on $D(\sqrt{-\Delta}) \cap D(H_{\mathrm{rad}})$ is the following inequality.

Lemma 3.132. *Let $m > 0$ and $V = 0$. Then there exists a constant C such that*

$$\|\sqrt{-\Delta}F\|^2 + \|H_{\mathrm{rad}}F\|^2 \le C\|(T_{\mathrm{kin}} \dot{+} H_{\mathrm{rad}} + 1)F\|^2 \tag{3.9.21}$$

for all $F \in D(\sqrt{-\Delta}) \cap D(H_{\mathrm{rad}})$.

Proof. Suppose that $m = 0$. In the case of $m > 0$, the proof is parallel with that of $m = 0$, but rather easier. Let $F \in D_\infty$. We have

$$\|H_0F\|^2 = \||-i\nabla - eA|F\|^2 + \|H_{\mathrm{rad}}F\|^2 + 2\Re(|-i\nabla - eA|F, H_{\mathrm{rad}}F). \tag{3.9.22}$$

We estimate $\||-i\nabla - eA|F\|^2$ and $\Re(|-i\nabla - eA|F, H_{\mathrm{rad}}F)$ on the right-hand side of (3.9.22) from below. Since the operator $|-i\nabla - eA|$ is singular, we introduce an artificial positive mass $M > 0$ and

$$T_M = \sqrt{(-i\nabla - eA)^2 + M^2}.$$

Note that

$$|-i\nabla - eA| = T_M + (|-i\nabla - eA| - T_M).$$

Let E_λ be the spectral resolution of nonnegative self-adjoint operator $(-i\nabla - eA)^2$. Then

$$\begin{aligned}\|(|-i\nabla - eA| - T_M)F\|^2 &= \int_0^\infty \left|\frac{M^2}{\sqrt{\lambda + M^2} + \sqrt{\lambda}}\right| d\|E_\lambda F\|^2 + M^2\|F\|^2 \\ &\le (1 + M^2)\|F\|^2.\end{aligned}$$

Then $B = |-i\nabla - eA| - T_M$ is bounded, and $|-i\nabla - eA| = T_M + B$. We have

$$\Re(|-i\nabla - eA|F, H_{\mathrm{rad}}F) = \Re(T_MF, H_{\mathrm{rad}}F) + \Re(BF, H_{\mathrm{rad}}F).$$

Since $F \in D_\infty$, $H_{\mathrm{rad}}F \in D(-\Delta) \cap D(H_{\mathrm{rad}})$. In particular, $H_{\mathrm{rad}}F \in D(T_M)$ and then $H_{\mathrm{rad}}F \in D(T_M^{1/2})$. Furthermore, we show that

$$T_M^{1/2}F \in D(H_{\mathrm{rad}}) \tag{3.9.23}$$

in Lemma 3.133 below. So we can see that

$$\begin{aligned}\Re(|-i\nabla - eA|F, H_{\rm rad}F) &= (T_M^{1/2}F, H_{\rm rad}T_M^{1/2}F) + \Re(T_M^{1/2}F, [T_M^{1/2}, H_{\rm rad}]F) + \Re(BF, H_{\rm rad}F)\\ &\geq \Re(T_M^{1/2}F, [T_M^{1/2}, H_{\rm rad}]F) + \Re(BF, H_{\rm rad}F).\end{aligned}$$

We estimate $(BF, H_{\rm rad}F)$. Since $\|BF\| \leq M\|F\|$, we see that for each $\varepsilon > 0$ there exists $C_1 > 0$ such that

$$\Re(BF, H_{\rm rad}F) \geq -\varepsilon\|H_{\rm rad}F\|^2 - C_1\|F\|^2.$$

On the other hand, we estimate $\Re(T_M^{1/2}F, [T_M^{1/2}, H_{\rm rad}]F)$. Let $\varepsilon > 0$ be given. Then there exists $C_2 > 0$ such that

$$\begin{aligned}\Re(T_M^{1/2}F, [T_M^{1/2}, H_{\rm rad}]F) &\geq -c\|T_M^{1/2}F\|\,\|(H_{\rm rad}+1)F\|\\ &\geq -\varepsilon\||-i\nabla - eA|F\|^2 - \varepsilon\|H_{\rm rad}F\|^2 - C_2\|F\|^2.\end{aligned} \tag{3.9.24}$$

The first inequality of (3.9.24) is derived from

$$\|[T_M^{1/2}, H_{\rm rad}]F\| \leq c\|(H_{\rm rad}+1)^{1/2}F\| \tag{3.9.25}$$

with some constant $c > 0$. This is shown in Lemma 3.133 below. Hence we have

$$\Re(|-i\nabla - eA|F, H_{\rm rad}F) \geq -\varepsilon\||-i\nabla - eA|F\|^2 - 2\varepsilon\|H_{\rm rad}F\|^2 - (C_1 + C_2)\|F\|. \tag{3.9.26}$$

We next estimate $\||-i\nabla - eA|F\|$. Note that

$$\|-i\nabla_\mu F\|^2 = \|(-i\nabla_\mu - eA_\mu)F\|^2 + 2\Re(eA_\mu F, (-i\nabla_\mu - eA_\mu)F) + \|eA_\mu F\|^2.$$

For each $\varepsilon > 0$, there exist $C_5 > 0$ and $C_6 > 0$ such that

$$\begin{aligned}|\Re(AF, (-i\nabla - eA)F)| &\leq \varepsilon(\||-i\nabla - eA|F\|^2 + \|H_{\rm rad}F\|^2) + C_5\|F\|^2,\\ \|\sqrt{-\Delta}F\|^2 &\leq (1+\varepsilon)\||-i\nabla - eA|F\|^2 + \varepsilon\|H_{\rm rad}F\|^2 + C_6\|F\|^2.\end{aligned}$$

Hence we have

$$\||-i\nabla - eA|F\|^2 \geq \frac{1}{1+\varepsilon}\|\sqrt{-\Delta}F\|^2 - \frac{\varepsilon}{1+\varepsilon}\|H_{\rm rad}F\|^2 - \frac{C_6}{1+\varepsilon}\|F\|^2. \tag{3.9.27}$$

By (3.9.26), (3.9.27), and (3.9.22), we can see (3.9.21) for $F \in D_\infty$. Let $F \in D(H_{\rm RPF})$. Since $H_{\rm RPF}$ is essentially self-adjoint on D_∞, by a limiting argument we can see (3.9.21) for $F \in D(H_{\rm RPF})$. □

Lemma 3.133. *Equations* (3.9.23) *and* (3.9.25) *hold true.*

Proof. Note that $T_M^{1/2} = (2K + M^2)^{1/4}$, where $K = \frac{1}{2}(-i\nabla - eA)^2$. We have

$$(2K + M^2)^{\alpha/2} = C_\alpha \int_0^\infty (1 - e^{-\lambda(2K+M^2)}) \frac{d\lambda}{\lambda^{1+\alpha/2}}$$

for $0 \le \alpha < 2$ with some constant C_α. From this formula, we have

$$(F, T_M^{1/2} G) = C_{1/2} \int_0^\infty \left\{ (F, G) - e^{-\lambda M^2} \int_{\mathbb{R}^3} \mathbb{E}^x[(F(B_0), e^{-i\hat{A}(L_{2\lambda})} G(B_{2\lambda}))] dx \right\} \frac{d\lambda}{\lambda^{5/4}}.$$

Here, $L_{2\lambda} = \bigoplus_{\mu=1}^3 \int_0^{2\lambda} \tilde{\varphi}(\cdot - B_s) dB_s^\mu$. Let $F \in D(H_{\mathrm{rad}})$ and $G \in D_\infty$. Thus $H_{\mathrm{rad}} G \in D_\infty$. We have

$$(H_{\mathrm{rad}} F, T_M^{1/2} G) = C_{1/2} \int_0^\infty \left\{ (H_{\mathrm{rad}} F, G) - e^{-\lambda M^2} \int_{\mathbb{R}^3} \mathbb{E}^x[(H_{\mathrm{rad}} F(B_0), e^{-i\hat{A}(L_{2\lambda})} G(B_{2\lambda}))] dx \right\} \frac{d\lambda}{\lambda^{5/4}}.$$

Then

$$\begin{aligned} &(H_{\mathrm{rad}} F, T_M^{1/2} G) - (F, T_M^{1/2} H_{\mathrm{rad}} G) \\ &= -C_{1/2} \int_0^\infty \frac{e^{-\lambda M^2}}{\lambda^{5/4}} d\lambda \int_{\mathbb{R}^3} \mathbb{E}^x[(F(B_0), [H_{\mathrm{rad}}, e^{-i\hat{A}(L_{2\lambda})}] G(B_{2\lambda}))] dx. \end{aligned} \tag{3.9.28}$$

Note that $[H_{\mathrm{rad}}, e^{-i\hat{A}(L_{2\lambda})}] = e^{-i\hat{A}(L_{2\lambda})}(\hat{\Pi}(L_{2\lambda}) + \xi)$, where $\xi = \mathrm{q}_0(L_{2\lambda}, L_{2\lambda})$ and $\hat{\Pi}(L_{2\lambda}) = [H_{\mathrm{rad}}, \hat{A}(L_{2\lambda})]$. Thus we see that

$$\begin{aligned} &\left| \int_0^\infty \left\{ \int_{\mathbb{R}^3} \mathbb{E}^x[(F(B_0), [H_{\mathrm{rad}}, e^{-i\hat{A}(L_{2\lambda})}] G(B_{2\lambda}))] dx \right\} \frac{e^{-\lambda M^2} d\lambda}{\lambda^{5/4}} \right| \\ &\le C \int_0^\infty \frac{\sqrt{\lambda} + \lambda}{\lambda^{5/4}} e^{-\lambda M^2} d\lambda \|F\| \|(H_{\mathrm{rad}} + 1)^{1/2} G\| \end{aligned}$$

with some constant C. Then $|(H_{\mathrm{rad}} F, T_M^{1/2} G)| \le C\|F\|\|(H_{\mathrm{rad}} + 1) G\|$ with some constant $C > 0$. Hence $T_M^{1/2} G \in D(H_{\mathrm{rad}})$ follows.

Next, we shall prove (3.9.25). The proof of (3.9.25) is similar to that of (3.9.23). Let $G \in D_\infty$. From (3.9.28), it follows that $|(F, [H_{\mathrm{rad}}, T_M^{1/2}] G)| \le C\|F\|\|(1 + H_{\mathrm{rad}})^{1/2} G\|$. This implies (3.9.25). □

Theorem 3.134 (Self-adjointness). *Let $m \ge 0$. Then H_{RPF} is self-adjoint on $D(\sqrt{-\Delta}) \cap D(H_{\mathrm{rad}})$, and essentially self-adjoint on D_∞.*

Proof. Suppose that $V = 0$. We write $H^{(m)}$ for H_{RPF} to emphasize m-dependence. Let $m > 0$. Then $H^{(m)}$ is essentially self-adjoint on $D(\sqrt{-\Delta}) \cap D(H_{\mathrm{rad}})$. On the other hand, by (3.9.21), $H^{(m)}\lceil_{D(\sqrt{-\Delta})\cap D(H_{\mathrm{rad}})}$ is closed. Then $H^{(m)}$ is self-adjoint on $D(\sqrt{-\Delta}) \cap D(H_{\mathrm{rad}})$. Note that $H^{(0)} = H^{(m)} + (H^{(0)} - H^{(m)})$ and $H^{(0)} - H^{(m)}$ is bounded. Then $H^{(0)}$ is also self-adjoint on $D(\sqrt{-\Delta}) \cap D(H_{\mathrm{rad}})$ for $V = 0$. Finally, let V be a potential satisfying the assumptions given in this theorem. Then V is also relatively bounded with respect to $H^{(m)}$ with a relative bound strictly smaller than one. Then the theorem follows from Kato–Rellich theorem and Theorem 3.131. □

Example 3.135 (Hydrogen-like atom). Let $d = 3$. A spinless hydrogen-like atom is defined by introducing the Coulomb potential $V_{\mathrm{Coulomb}}(x) = -g/|x|$, $g > 0$, which is relatively form bounded with respect to $\sqrt{-\Delta + m^2}$ with a relative bound strictly smaller than 1 if $g \le 2/\pi$. Furthermore, if $g < 1/2$, V_{Coulomb} is relatively bounded with respect to $\sqrt{-\Delta + m^2}$ with a relative bound strictly smaller than 1. Let $\hat{A}_\Lambda$ be the quantized radiation field with the cutoff function $\hat{\varphi}(k) = 1_{|k|\le\Lambda}(k)/\sqrt{(2\pi)^3}$, where $\Lambda > 0$ describes a ultraviolet cutoff parameter. By Lemma 3.118, when $g < 2/\pi$, V is relatively form bounded with respect to $T_{\mathrm{kin}} \dotplus H_{\mathrm{rad}}$ and H_{RPF} is well defined as a self-adjoint operator. Furthermore, by Theorem 3.134 when $g < 1/2$, H_{RPF} is self-adjoint on $D(\sqrt{-\Delta}) \cap D(H_{\mathrm{rad}})$. All the statements mentioned above are true for arbitrary values of coupling constant $e \in \mathbb{R}$ and $\Lambda > 0$.

3.9.4 Nonrelativistic limit of relativistic Pauli–Fierz Hamiltonian

In the quantum mechanics, we can see that

$$\sqrt{-c^2\Delta + m^2c^4} - mc^2 + V \to -\frac{1}{2m}\Delta + V$$

as $c \to \infty$ strongly in the sense of semigroup. We introduce the speed of light c in H_{RPF}, and it turns to be

$$H_{\mathrm{RPF}} = \sqrt{c^2(-i\nabla - eA)^2 + m^2c^4} - mc^2 + V + H_{\mathrm{rad}}.$$

It is interesting to investigating the *nonrelativistic limit* of the relativistic Pauli–Fierz Hamiltonian. Here the nonrelativistic limit describes the nonrelativistic limit of the charged particle, hence we set $c = 1$ in A and H_{rad}. The subordinator $(T_t(c))_{t\ge0}$ with parameter c in $(\mathcal{T}, \mathcal{B}_{\mathcal{T}}, \nu)$ satisfies

$$\mathbb{E}^0_\nu[\exp(-uT_t(c))] = \exp(-t(\sqrt{2c^2u + m^2c^4} - mc^2)). \tag{3.9.29}$$

In Lemma 4.228 in Volume 1, it is shown that

$$\lim_{c\to\infty} \mathbb{E}^0_\nu[f(T_t(c))] = f(t/m)$$

for every bounded continuous function f. From this fact, we can show that

$$(f, e^{-t(\sqrt{-c^2\Delta+m^2c^4}-mc^2+V)}g) = \int_{\mathbb{R}^3} \mathbb{E}^{x,0}[\bar{f}(x)g(B_{T_t(c)})e^{-\int_0^t V(B_{T_s(c)})ds}]dx$$

$$\to \int_{\mathbb{R}^3} \mathbb{E}^x[\bar{f}(x)g(B_{t/m})e^{-\int_0^t V(B_{s/m})ds}]dx = (f, e^{-t(-\frac{1}{2m}\Delta+V)}g).$$

In a similar way to this, we shall show the nonrelativistic limit of the relativistic Pauli–Fierz Hamiltonian. The Pauli–Fierz Hamiltonian with particle mass $m > 0$ is defined by

$$H_{\mathrm{PF}} = \frac{1}{2m}(-i\nabla - eA)^2 + V + H_{\mathrm{rad}}.$$

Using $(T_t(c))_{t\geq 0}$, we can see that

$$(F, e^{-tH_{\mathrm{RPF}}}G) = \int_{\mathbb{R}^3} \mathbb{E}^{x,0}[e^{-\int_0^t V(B_{T_s(c)})ds}(\mathrm{J}_0F(x), e^{-ie\hat{A}_{\mathrm{E}}(K_t^{\mathrm{rel}}(c))}\mathrm{J}_tG(B_{T_t(c)}))]dx, \tag{3.9.30}$$

where $K_t^{\mathrm{rel}}(c)$ is defined by K_t^{rel} with T_t replaced by $T_t(c)$, and the functional integral representation of $e^{-tH_{\mathrm{PF}}}$ with mass m is given by

$$(F, e^{-tH_{\mathrm{PF}}}G) = \int_{\mathbb{R}^3} \mathbb{E}^x\Big[e^{-\int_0^t V(B_{s/m})ds}(\mathrm{J}_0F(x), e^{-ie\hat{A}_{\mathrm{E}}(K_t)}\mathrm{J}_tG(B_{t/m}))\Big]dx, \tag{3.9.31}$$

where $K_t = \bigoplus_{\mu=1}^3 \int_0^{t/m} \mathrm{j}_s\tilde{\varphi}(\cdot - B_s)\circ dB_s^\mu$. The purpose of this section is to show that $e^{-tH_{\mathrm{RPF}}} \to e^{-tH_{\mathrm{PF}}}$ as $c \to \infty$ strongly by using (3.9.30) and (3.9.31).

Lemma 3.136. *It follows that* $\lim_{c\to\infty} K_t^{\mathrm{rel}}(c) = K_t$ *strongly in* $L^2(\mathscr{X} \times \mathcal{T}) \otimes (\oplus^3 L^2(\mathbb{R}^4))$.

Proof. Let

$$\mathrm{I}_n(c) = \bigoplus_{\mu=1}^3 \sum_{j=1}^{2^n} \int_{T_{t_{j-1}}(c)}^{T_{t_j}(c)} \mathrm{j}_{t_{j-1}}\tilde{\varphi}(\cdot - B_s)\circ dB_s^\mu, \quad \mathrm{I}_n = \bigoplus_{\mu=1}^3 \sum_{j=1}^{2^n} \int_{t_{j-1}/m}^{t_j/m} \mathrm{j}_{t_{j-1}}\tilde{\varphi}(\cdot - B_s)\circ dB_s^\mu.$$

We see that $\mathrm{I}_n(c) \to K_t^{\mathrm{rel}}(c)$ and $\mathrm{I}_n \to K_t$ as $n \to \infty$ strongly in $L^2(\mathscr{X} \times \mathcal{T}) \otimes (\oplus^3 L^2(\mathbb{R}^4))$ and $L^2(\mathscr{X}) \otimes (\oplus^3 L^2(\mathbb{R}^4))$, respectively. We have

$$\|K_t^{\mathrm{rel}}(c) - K_t\| \leq \|K_t^{\mathrm{rel}}(c) - \mathrm{I}_n(c)\| + \|\mathrm{I}_n(c) - \mathrm{I}_n\| + \|\mathrm{I}_n - K_t\|.$$

Here $\|\cdot\|$ denotes the norm on $\oplus^3 L^2(\mathbb{R}^4)$. We have

$$\mathbb{E}^x[\|\mathrm{I}_n(c) - \mathrm{I}_k(c)\|^2] \le 3T_t(c)\|\hat{\varphi}/\sqrt{\omega}\|^2 \Big(\sum_{j=n+1}^{k} 2^{-j/2} \Big)^2.$$

From this, we have

$$\mathbb{E}^{x,0}[\|\mathrm{I}_n(c) - K_t^{\mathrm{rel}}(c)\|^2] \le 3\mathbb{E}^0[T_t(c)]\|\hat{\varphi}/\sqrt{\omega}\|^2 \Big(\sum_{j=n+1}^{\infty} 2^{-j/2} \Big)^2.$$

Since $\mathbb{E}^0[T_t(c)] = t/m$, which is independent of $c > 0$, we obtain

$$\mathbb{E}^{x,0}[\|\mathrm{I}_n(c) - K^{\mathrm{rel}}(c)\|^2] \le 3\frac{t}{m}\|\hat{\varphi}/\sqrt{\omega}\|^2 \Big(\sum_{j=n+1}^{\infty} 2^{-j/2} \Big)^2$$

and conclude that

$$\lim_{n\to\infty} \sup_{c>0} \mathbb{E}^{x,0}[\|\mathrm{I}_n(c) - K^{\mathrm{rel}}(c)\|^2] = 0. \tag{3.9.32}$$

Now we estimate $\|\mathrm{I}_n(c) - \mathrm{I}_n\|$. We have

$$\mathrm{I}_n(c) - \mathrm{I}_n = \bigoplus_{\mu=1}^{3} \sum_{j=1}^{2^n} \Big(\int_{T_{t_{j-1}}(c)}^{T_{t_j}(c)} \mathrm{j}_{t_{j-1}}\tilde{\varphi}(\cdot - B_s)dB_s^\mu - \int_{t_{j-1}/m}^{t_j/m} \mathrm{j}_{t_{j-1}}\tilde{\varphi}(\cdot - B_s)dB_s^\mu \Big).$$

We note that $s \to \int_a^s \mathrm{j}_{t_{j-1}}\tilde{\varphi}(\cdot - B_s)dB_s^\mu$ and $s \to \int_s^b \mathrm{j}_{t_{j-1}}\tilde{\varphi}(\cdot - B_s)dB_s^\mu$ are almost surely continuous. Hence

$$(S,T) \to \mathbb{E}^x\Big[\Big(\int_S^T \mathrm{j}_{t_{j-1}}\tilde{\varphi}(\cdot - B_s)dB_s^\mu, \int_{t_{j-1}/m}^{t_j/m} \mathrm{j}_{t_{j-1}}\tilde{\varphi}(\cdot - B_s)dB_s^\mu \Big)\Big]$$

is continuous, where $(\cdot,\cdot)$ denotes the scalar product on $L^2(\mathbb{R}^3)$. This implies that for every j,

$$\begin{aligned} &\mathbb{E}^{x,0}\Big[\Big(\int_{T_{t_{j-1}}(c)}^{T_{t_j}(c)} \mathrm{j}_{t_{j-1}}\tilde{\varphi}(\cdot - B_s)dB_s^\mu, \int_{t_{j-1}/m}^{t_j/m} \mathrm{j}_{t_{j-1}}\tilde{\varphi}(\cdot - B_s)dB_s^\mu \Big)\Big] \\ &\to \mathbb{E}^x\Big[\Big(\int_{t_{j-1}/m}^{t_j/m} \mathrm{j}_{t_{j-1}}\tilde{\varphi}(\cdot - B_s)dB_s^\mu, \int_{t_{j-1}/m}^{t_j/m} \mathrm{j}_{t_{j-1}}\tilde{\varphi}(\cdot - B_s)dB_s^\mu \Big)\Big] = \frac{(t_j - t_{j-1})}{m}\|\hat{\varphi}/\sqrt{\omega}\|^2 \end{aligned} \tag{3.9.33}$$

as $c \to \infty$, by Lemma 4.228 in Volume 1. We see that

$$\mathbb{E}^{x,0}[\|\mathrm{I}_n(c)-\mathrm{I}_n\|^2] = 3\sum_{j=1}^{2^n}\mathbb{E}^{x,0}\left[\left\|\int_{T_{t_{j-1}}(c)}^{T_{t_j}(c)} \mathrm{j}_{t_{j-1}}\tilde{\varphi}(\cdot - B_s)dB_s^\mu - \int_{t_{j-1}/m}^{t_j/m} \mathrm{j}_{t_{j-1}}\tilde{\varphi}(\cdot - B_s)dB_s^\mu\right\|^2\right]$$

and

$$\begin{aligned}
&\mathbb{E}^{x,0}\left[\left\|\int_{T_{t_{j-1}}(c)}^{T_{t_j}(c)} \mathrm{j}_{t_{j-1}}\tilde{\varphi}(\cdot - B_s)dB_s^\mu - \int_{t_{j-1}/m}^{t_j/m} \mathrm{j}_{t_{j-1}}\tilde{\varphi}(\cdot - B_s)dB_s^\mu\right\|^2\right]\\
&= \mathbb{E}^{x,0}\left[\left\|\int_{T_{t_{j-1}}(c)}^{T_{t_j}(c)} \mathrm{j}_{t_{j-1}}\tilde{\varphi}(\cdot - B_s)dB_s^\mu\right\|^2\right] + \mathbb{E}^{x,0}\left[\left\|\int_{t_{j-1}/m}^{t_j/m} \mathrm{j}_{t_{j-1}}\tilde{\varphi}(\cdot - B_s)dB_s^\mu\right\|^2\right]\\
&\quad - 2\mathbb{E}^{x,0}\left[\left(\int_{T_{t_{j-1}}(c)}^{T_{t_j}(c)} \mathrm{j}_{t_{j-1}}\tilde{\varphi}(\cdot - B_s)dB_s^\mu, \int_{t_{j-1}/m}^{t_j/m} \mathrm{j}_{t_{j-1}}\tilde{\varphi}(\cdot - B_s)dB_s^\mu\right)\right]\\
&= \frac{1}{m}\|\hat{\varphi}/\sqrt{\omega}\|^2(\mathbb{E}^0[T_{t_j}(c) - T_{t_{j-1}}(c)] + t_j - t_{j-1})\\
&\quad - 2\mathbb{E}^{x,0}\left[\left(\int_{T_{t_{j-1}}(c)}^{T_{t_j}(c)} \mathrm{j}_{t_{j-1}}\tilde{\varphi}(\cdot - B_s)dB_s^\mu, \int_{t_{j-1}/m}^{t_j/m} \mathrm{j}_{t_{j-1}}\tilde{\varphi}(\cdot - B_s)dB_s^\mu\right)\right].
\end{aligned}$$

Note that $\mathbb{E}^0[T_{t_j}(c) - T_{t_{j-1}}(c)] = t_j - t_{j-1}$ and (3.9.33). We can see that

$$\mathbb{E}^{x,0}[\|\mathrm{I}_n(c)-\mathrm{I}_n\|^2] \to 0$$

as $c \to \infty$ for each n. Let $\varepsilon > 0$ be arbitrary. There exists $n_0 \in \mathbb{N}$ such that for all $n > n_0$, $\mathbb{E}^{x,0}[\|K_t^{\mathrm{rel}}(c) - \mathrm{I}_n(c)\|^2] < \varepsilon^2$ and $\mathbb{E}^{x,0}[\|\mathrm{I}_n - K_t\|^2] < \varepsilon^2$ uniformly in c. We have for $n > n_0$,

$$\lim_{c\to\infty}(\mathbb{E}^{x,0}[\|K_t^{\mathrm{rel}}(c) - K_t\|]^2)^{1/2} \le 2\varepsilon + \lim_{c\to\infty}(\mathbb{E}^{x,0}[\|\mathrm{I}_n(c) - \mathrm{I}_n\|])^{1/2} = 2\varepsilon.$$

Thus the lemma is proven. □

Theorem 3.137 (Nonrelativistic limit). *Suppose that V is bounded and continuous. Then for every $t \ge 0$, it follows that*

$$\operatorname*{s-lim}_{c\to\infty} e^{-tH_{\mathrm{RPF}}} = e^{-tH_{\mathrm{PF}}}.$$

Proof. Suppose that $F, G \in C_0^\infty(\mathbb{R}^3) \otimes \mathcal{F}_{\mathrm{rad}}$. From Lemma 3.136 and

$$(F, e^{-tH_{\mathrm{RPF}}}G) = \int_{\mathbb{R}^3} \mathbb{E}^{x,0}\left[e^{-\int_0^t V(B_{T_s(c)})ds}(\mathrm{J}_0F(x), e^{-ie\hat{A}_{\mathrm{E}}(K_t^{\mathrm{rel}}(c))}\mathrm{J}_tG(B_{T_t(c)}))\right]dx,$$

it follows that

$$\lim_{c\to\infty}(F, e^{-tH_{\rm RPF}}G) = \int_{\mathbb{R}^3} \mathbb{E}^x\left[e^{-\int_0^t V(B_{s/m})ds}(\mathrm{J}_0F(x), e^{-ie\hat{A}_{\rm E}(K_t)}\mathrm{J}_tG(B_{t/m}))\right]dx = (F, e^{-tH_{\rm PF}}G).$$

Since $H_{\rm RPF} \geq \inf_{x\in\mathbb{R}^3} V(x) = g \geq -\infty$, $e^{-H_{\rm RPF}} \leq e^{-tg}$. Let $F, G \in \mathcal{H}_{\rm PF}$. There exist $F_n, G_n \in C_0^\infty(\mathbb{R}^3)\otimes\mathcal{F}_{\rm rad}$ such that $F_n \to F$ and $G_n \to G$ strongly as $n\to\infty$. Using the uniform bound $e^{-tH_{\rm RPF}} \leq e^{-tg}$, we can show $\lim_{c\to\infty}(F, e^{-tH_{\rm RPF}}G) = (F, e^{-tH_{\rm PF}}G)$. Finally, since the weak convergence of $e^{-tH_{\rm RPF}}$ implies the strong convergence, the theorem follows. □

3.9.5 Relativistic Pauli–Fierz model with relativistic Kato-class potential

Similar to Section 3.4, we can consider the relativistic Pauli–Fierz model with Kato-decomposable potential V. Arguments are parallel to those for the Pauli–Fierz model $H_{\rm PF}^{\rm Kato}$. Let us recall the d-dimensional relativistic Kato-class. Let

$$p_t^R(x) = \frac{1}{(2\pi)^d}\int_{\mathbb{R}^d} e^{-ix\cdot\xi}e^{-t(\sqrt{|\xi|^2+m^2}-m)}d\xi.$$

We define

$$g_R(x) = \int_0^\infty e^{-t}p_t^R(x)dt.$$

Then $g_R(x)$ gives the integral kernel of $(\sqrt{-\Delta+m^2}-m+1)^{-1}$. Potential V is of relativistic Kato-class whenever it satisfies

$$\lim_{\delta\downarrow 0}\sup_{x\in\mathbb{R}^d}\int_{|x-y|<\delta} g_R(x-y)V(y)dy = 0. \tag{3.9.34}$$

We denote this linear space by $\mathcal{K}_R(\mathbb{R}^d)$. Moreover, $V = V_+ - V_-$ is *relativistic Kato-decomposable* whenever $V_- \in \mathcal{K}_R(\mathbb{R}^d)$ and $V_+1_K \in \mathcal{K}_R(\mathbb{R}^d)$ for every compact set $K\subset\mathbb{R}^d$. In Section 4.6.2 of Volume 1, we introduced three classes of external potentials given by

$$\mathcal{V}_1 = \left\{V:\mathbb{R}^d\to\mathbb{R}\ \middle|\ \lim_{t\downarrow 0}\sup_{x\in\mathbb{R}^d}\int_0^t \mathbb{E}^{x,0}_{\mathcal{W}\times P}[V(B_{T_s})]ds = 0\right\},$$

$$\mathcal{V}_2 = \left\{V:\mathbb{R}^d\to\mathbb{R}\ \middle|\ \lim_{\lambda\to\infty}\sup_{x\in\mathbb{R}^d}\left((\sqrt{-\Delta+m^2}-m+\lambda)^{-1}V\right)(x) = 0\right\},$$

$$\mathcal{V}_3 = \left\{V:\mathbb{R}^d\to\mathbb{R}\ \middle|\ \lim_{\delta\downarrow 0}\sup_{x\in\mathbb{R}^d}\int_{|x-y|<\delta} g_R(x-y)V(y)dy = 0\right\}.$$

Also, we say that V is uniformly locally integrable whenever

$$\sup_{x\in\mathbb{R}^d}\int_{x+C_0}|V(y)|dy<\infty, \tag{3.9.35}$$

where C_0 denotes the d-dimensional cube centered at 0 with sides of unit length and $C_0+x=\{x+y\mid y\in C_0\}$. Let V be a nonnegative function. Due to Proposition 4.214 in Volume 1, we have:

(1) $\mathcal{V}_2\subset\mathcal{V}_1$ and $\mathcal{V}_2\subset\mathcal{V}_3$;
(2) if V is uniformly locally integrable, then $\mathcal{V}_3\subset\mathcal{V}_2$ and $\mathcal{V}_1\subset\mathcal{V}_2$.

Hence when V is uniformly locally integrable, we can conclude that the following are equivalent:

(1) $\lim_{t\downarrow 0}\sup_{x\in\mathbb{R}^d}\int_0^t\mathbb{E}^{x,0}_{\mathcal{W}\times P}[V(B_{T_s})]ds=0$;
(2) $\lim_{\lambda\to\infty}\sup_{x\in\mathbb{R}^d}((\sqrt{-\Delta+m^2}-m+\lambda)^{-1}V)(x)=0$;
(3) V is of relativistic Kato-class.

We can also show an exponential bound of relativistic Kato-decomposable potentials by using Khasminskii's lemma. Suppose that V is uniformly locally integrable, and relativistic Kato-decomposable. Then for any $t\geq 0$,

$$\sup_{x\in\mathbb{R}^d}\mathbb{E}^{x,0}\left[e^{\int_0^t V(B_{T_s})\,ds}\right]<\infty. \tag{3.9.36}$$

We use Assumption 3.138 below in Sections 3.9.5–3.9.7.

Assumption 3.138. The following conditions hold:
(1) (Charge distribution) $\varphi\in\mathscr{S}'(\mathbb{R}^3)$, $\hat{\varphi}(-k)=\overline{\hat{\varphi}(k)}$ and $\omega^{3/2}\hat{\varphi}$, $\sqrt{\omega}\hat{\varphi}$, $\hat{\varphi}/\omega\in L^2(\mathbb{R}^3)$;
(2) (External potential) V is a uniformly locally integrable and relativistic Kato-decomposable potential.

When Assumption 3.138 is satisfied, we can see that

$$r_t(F,G)=\int_{\mathbb{R}^3}\mathbb{E}^{x,0}\left[\left(\mathrm{J}_0F(B_{T_0}),e^{-ie\hat{A}_{\mathrm{E}}(K_t^{\mathrm{rel}})}e^{-\int_0^t V(B_{T_r})dr}\mathrm{J}_tG(B_{T_t})\right)\right]dx$$

is well defined for all $F,G\in\mathcal{H}_{\mathrm{PF}}$, and $|r_t(F,G)|\leq c_t\|F\|\|G\|$ follows with some constant c_t. Then the Riesz representation theorem yields that there exists a bounded operator S_t such that

$$r_t(F,G)=(F,S_tG)$$

for $F, G \in \mathcal{H}_{\mathrm{PF}}$ and $\|S_t\| \le c_t$. By the functional integral representation (3.9.9), we indeed see that $(S_t G)(x) = \mathbb{E}^{x,0}[\mathrm{J}_{[0,t]} G(B_{T_t})]$, where $\mathrm{J}_{[0,t]}$ is the operator given by

$$\mathrm{J}_{[0,t]} = \mathrm{J}_0^* e^{-\int_0^t V(B_{T_r})dr} e^{-ie\hat{A}_{\mathrm{E}}(K_t^{\mathrm{rel}})} \mathrm{J}_t.$$

Lemma 3.139. *The family of operators $\{S_t\}_{t\ge0}$ satisfies the semigroup property, i. e., $S_s S_t = S_{s+t}$ for all $s, t \ge 0$.*

Proof. We have $(F, S_s S_t G) = (F, \mathbb{E}^{x,0}[\mathrm{J}_{[0,s]} \mathbb{E}^{B_{T_s},0}[\mathrm{J}_{[0,t]} G(B_{T_t})]])$. We can show that

$$\mathbb{E}^{x,0}\left[\mathrm{J}_{[0,s]} \mathbb{E}^{B_{T_s},0}\left[\mathrm{J}_{[0,t]} G(B_{T_t})\right]\right] = \mathbb{E}^{x,0}\left[\mathrm{J}_{[0,s]} \mathrm{J}_0^* e^{-\int_s^{s+t} V(B_{T_r})dr} e^{-ie\hat{A}_{\mathrm{E}}(\mathrm{K}_0[s,s+t])} \mathrm{J}_t G(B_{T_{s+t}})\right],$$

where

$$\mathrm{K}_0[s, s+t] = \operatorname*{s-lim}_{n\to\infty} \bigoplus_{\mu=1}^{3} \sum_{j=1}^{2^n} \int_{T_{\frac{t}{2^n}(j-1)+s}}^{T_{\frac{t}{2^n}j+s}} \mathrm{j}_{\frac{t}{2^n}(j-1)} \tilde{\varphi}(\cdot - B_r) dB_r^{\mu}.$$

Since it is obtained that

$$\begin{aligned}
&\mathrm{J}_{[0,s]} \mathrm{J}_0^* e^{-\int_s^{s+t} V(B_{T_r})dr} e^{-ie\hat{A}_{\mathrm{E}}(\mathrm{K}_0[s,s+t])} \mathrm{J}_t G(B_{T_{s+t}}) \\
&= \mathrm{J}_0^* e^{-\int_0^{s+t} V(B_{T_r})dr} e^{-ie\hat{A}_{\mathrm{E}}(K_s^{\mathrm{rel}})} \mathrm{J}_s \mathrm{J}_0^* e^{-ie\hat{A}_{\mathrm{E}}(\mathrm{K}_0[s,s+t])} \mathrm{J}_t G(B_{T_{s+t}})
\end{aligned}$$

and $\mathrm{J}_s \mathrm{J}_0^* = \mathrm{U}_s \mathrm{J}_0 \mathrm{J}_0^* = \mathrm{U}_s E_s$, where we recall that $\mathrm{U}_t = \Gamma(u_t)$, we have

$$(F, S_s S_t G) = \left(F, \mathbb{E}^{x,0}\left[\mathrm{J}_0^* e^{-ie\hat{A}_{\mathrm{E}}(K_s^{\mathrm{rel}})} \mathrm{U}_s E_s e^{-ie\hat{A}_{\mathrm{E}}(\mathrm{K}_0[s,s+t])} e^{-\int_0^{s+t} V(B_{T_r})dr} \mathrm{J}_t G(B_{T_{s+t}})\right]\right).$$

By the Markov property of projection E_s, E_s can be deleted, and U_s satisfies

$$\mathrm{U}_s e^{-ie\hat{A}_{\mathrm{E}}(\mathrm{K}_0[s,s+t])} \mathrm{J}_t G(B_{T_{s+t}}) = e^{-ie\hat{A}_{\mathrm{E}}(\mathrm{K}[s,s+t])} \mathrm{J}_{s+t} G(B_{T_{s+t}}).$$

Here,

$$\mathrm{K}[s, s+t] = \operatorname*{s-lim}_{n\to\infty} \bigoplus_{\mu=1}^{3} \sum_{j=1}^{2^n} \int_{T_{\frac{t}{2^n}(j-1)+s}}^{T_{\frac{t}{2^n}j+s}} \mathrm{j}_{\frac{t}{2^n}(j-1)+s} \tilde{\varphi}(\cdot - B_r) dB_r^{\mu}.$$

Then we have

$$(F, S_s S_t G) = \left(F, \mathbb{E}^{x,0}\left[\mathrm{J}_0^* e^{-\int_0^{s+t} V(B_{T_r})dr} e^{-ie\hat{A}_{\mathrm{E}}(K_{s+t}^{\mathrm{rel}})} \mathrm{J}_{s+t} G(B_{T_{s+t}})\right]\right) = (F, S_{s+t} G).$$

Then the semigroup property, $S_s S_t = S_{s+t}$, follows. □

The strong continuity of $t \mapsto S_t$ on $\mathcal{H}_{\mathrm{PF}}$ can be checked easily, while $S_0 = 1$ is trivial.

Lemma 3.140. *Operator S_t is symmetric for every $t \ge 0$, i. e., $S_t^* = S_t$ for all $t \ge 0$.*

Proof. Recall that $\mathrm{R} = \Gamma(r)$ is the second quantization of the reflection r. Then

$$(F, S_t G) = (U_t \mathrm{R} F, U_t \mathrm{R} S_t G) = \int_{\mathbb{R}^3} \mathbb{E}^{x,0}\left[e^{-\int_0^t V(B_{T_s})ds}\left(\mathrm{J}_t F(B_{T_0}), e^{-ie\hat{A}_{\mathrm{E}}(u_t r K_t^{\mathrm{rel}})}\mathrm{J}_0 G(B_{T_t})\right)\right] dx.$$

Notice that $u_t r K_t^{\mathrm{rel}} = \lim_{n\to\infty}\sum_{j=1}^{2^n}\int_{T_{t_{j-1}}}^{T_{t_j}} \mathrm{j}_{t-t_{j-1}}\tilde{\varphi}(\cdot - B_s)dB_s^{\mu}$. Exchanging integrals $\int_{\mathscr{X}} d\mathcal{W}^0$ and $\int_{\mathbb{R}^3} dx$, and changing the variable x to $y - B_{T_t}$, we have

$$= \mathbb{E}^{0,0}\left[\int_{\mathbb{R}^3} dy e^{-\int_0^t V(B_{T_s} - B_{T_t} + y)ds}\left(\mathrm{J}_t F(y - B_{T_t}), e^{-ie\hat{A}_{\mathrm{E}}(u_t r \tilde{K}_t^{\mathrm{rel}})}\mathrm{J}_0 G(y)\right)\right],$$

where

$$u_t r \tilde{K}_t^{\mathrm{rel}} = \lim_{n\to\infty}\sum_{j=1}^{2^n}\int_{T_{t_{j-1}}}^{T_{t_j}} \mathrm{j}_{t-t_{j-1}}\tilde{\varphi}(\cdot - (B_s - B_{T_t} + y))dB_s^{\mu}.$$

Note that $B_{T_r} - B_{T_t} = B_{T_t - T_r}$ and $y - B_{T_t} = y + B_{T_t}$ in law. We fix μ. Thus

$$\sum_{j=1}^{2^n}\int_{T_{t_{j-1}}}^{T_{t_j}} \mathrm{j}_{t-t_{j-1}}\tilde{\varphi}(\cdot - (B_s - B_{T_t} + y))dB_s^{\mu}$$
$$= \lim_{m\to\infty}\sum_{i=1}^{2^m}\sum_{j=1}^{2^n} \mathrm{j}_{t-t_{j-1}}\tilde{\varphi}\left(\cdot - B_{T_t - T_{t_{j-1}} - (i-1)\Delta_{j-1}} - y\right)\left(B^{\mu}_{T_t - T_{t_{j-1}} - i\Delta_{j-1}} - B^{\mu}_{T_t - T_{t_{j-1}} - (i-1)\Delta_{j-1}}\right),$$

where $\Delta_{j-1} = \frac{1}{2^n}(T_{t_j} - T_{t_{j-1}})$. Since $T_t - T_s$ also has the same law as T_{t-s}, we can replace the right-hand side above with

$$-\lim_{m\to\infty}\sum_{i=1}^{2^m}\sum_{j=1}^{2^n} \mathrm{j}_{t-t_{j-1}}\tilde{\varphi}\left(\cdot - B_{T_{t-t_{j-1}} - \frac{i-1}{2^n}(T_{t-t_{j-1}} - T_{t-t_j})} - y\right)$$
$$\times\left(B^{\mu}_{T_{t-t_{j-1}} - \frac{i-1}{2^n}(T_{t-t_{j-1}} - T_{t-t_j})} - B^{\mu}_{T_{t-t_{j-1}} - \frac{i}{2^n}(T_{t-t_{j-1}} - T_{t-t_j})}\right).$$

By the definition of $\int_S^T \mathrm{j}_s\tilde{\varphi}(\cdot - B_s)dB_s^{\mu}$ and the Coulomb gauge condition (3.1.8), it follows that

$$= -\sum_{j=1}^{2^n}\int_{T_{t-t_j}}^{T_{t-t_{j-1}}} \mathrm{j}_{t-t_{j-1}}\tilde{\varphi}(\cdot - B_s - y)dB_s^{\mu} = -\sum_{j=1}^{2^n}\int_{T_{t_{j-1}}}^{T_{t_j}} \mathrm{j}_{t_j}\tilde{\varphi}(\cdot - B_s - y)dB_s^{\mu}.$$

Finally, we can see that

$$-\sum_{j=1}^{2^n}\int_{T_{t_{j-1}}}^{T_{t_j}} \mathrm{j}_{t_j}\tilde{\varphi}(\cdot - B_s - y)dB_s^\mu = -\sum_{j=1}^{2^n}\int_{T_{t_{j-1}}}^{T_{t_j}} \mathrm{j}_{t_{j-1}}\tilde{\varphi}(\cdot - B_s - y)dB_s^\mu.$$

Then

$$(F, S_t G) = \int_{\mathbb{R}^3} \mathbb{E}^{y,0}\left[e^{-\int_0^t V(B_{T_s})ds}\left(\mathrm{J}_0^* e^{-ie\hat{A}_{\mathrm{E}}(K_t^{\mathrm{rel}})}\mathrm{J}_t F(y + B_{T_t}), G(y)\right)\right] dy = (S_t F, G),$$

and the lemma follows. □

By Lemmas 3.139 and 3.140, it can be seen that $\{S_t : t \geq 0\}$ is a symmetric C_0-semigroup. We summarize it as the theorem below.

Theorem 3.141. *Semigroup $\{S_t : t \geq 0\}$ is a symmetric C_0-semigroup.*

By Theorem 3.141, there exists a self-adjoint operator $H_{\mathrm{RPF}}^{\mathrm{Kato}}$ such that

$$S_t = e^{-tH_{\mathrm{RPF}}^{\mathrm{Kato}}}, \quad t \geq 0.$$

Definition 3.142 (Relativistic Pauli–Fierz Hamiltonian for relativistic Kato-class potential). We call the self-adjoint operator $H_{\mathrm{RPF}}^{\mathrm{Kato}}$ the *relativistic Pauli–Fierz Hamiltonian for relativistic Kato-class potential.*

In the same way as in Corollaries 3.122 and 3.123, we can show the diamagnetic inequality and that $e^{-tH_{\mathrm{RPF}}^{\mathrm{Kato}}}$ is equivalent to a positivity improving operator.

Corollary 3.143 (Diamagnetic inequality). *Let $E(e) = \inf \mathrm{Spec}(H_{\mathrm{RPF}}^{\mathrm{Kato}})$. It follows that*

$$|(F, e^{-tH_{\mathrm{RPF}}^{\mathrm{Kato}}} G)| \leq \left(|F|, e^{-t(\sqrt{-\Delta+m^2}-m+V+H_{\mathrm{rad}})}|G|\right)$$

and $E(0) \leq E(e)$.

Corollary 3.144 (Positivity improving). *Operator $e^{-\mathrm{i}(\pi/2)N}e^{-tH_{\mathrm{RPF}}}e^{\mathrm{i}(\pi/2)N}$ is positivity improving for every $t \geq 0$. In particular, the ground state of $H_{\mathrm{RPF}}^{\mathrm{Kato}}$ is unique, if it exists.*

We write H_{RPF} for $H_{\mathrm{RPF}}^{\mathrm{Kato}}$ unless confusions arise.

3.9.6 Martingale properties

Let Φ_{b} be a bound state of H_{RPF} and $E \in \mathbb{R}$ the eigenvalue associated with Φ_{b}, namely

$$H_{\mathrm{RPF}}\Phi_{\mathrm{b}} = E\Phi_{\mathrm{b}}.$$

Recall that in Section 3.9.6 we suppose Assumption 3.138 holds. Hence V is relativistic Kato-decomposable. We shall study the spatial decay of $\|\Phi_{\mathrm{b}}(x)\|_{L^2(\mathscr{Q})}$ as $|x| \to \infty$ by a martingale property of some random process associated with H_{RPF}, which is similar to that of Nelson model discussed in Section 2.9. It is, however, rather more complicated than that of the Nelson model. For H_{PF}, the functional integral representation is represented in terms of Brownian motion and a Carmona-type estimate is available in Section 3.5.3. However, the functional integral representation for H_{RPF} is represented by the combination of Brownian motion and subordinator. Then it is not straightforward to apply the Carmona-type estimate to the estimate of the spatial decay of a bound state of H_{RPF}. Instead of this, to achieve our goal, we show the martingale property of the random process $(H_t(x))_{t\geq 0}$,

$$H_t(x) = e^{tE} e^{-\int_0^t V(B_{T_s}+x)ds} e^{-ie\hat{A}_{\mathrm{E}}(K_t^{\mathrm{rel}}(x))} \mathrm{J}_t \Phi_{\mathrm{b}}(B_{T_t}+x), \quad t \geq 0,$$

on $\mathscr{X} \times \mathcal{T} \times \mathscr{Q}_{\mathrm{E}}$. Here, $K_t^{\mathrm{rel}}(x)$ is defined by K_t^{rel} with B_s replaced by $B_s + x$, i. e., $K_t^{\mathrm{rel}}(x) = \bigoplus_{\mu=1}^3 \int_0^{T_t} \mathrm{j}_{T_s^*} \tilde{\varphi}(\cdot - B_s - x) dB_s^\mu$. Using the random process $(H_t(x))_{t\geq 0}$, bound state Φ_{b} can be represented as

$$\Phi_{\mathrm{b}}(x) = \mathbb{E}^{0,0}\left[\mathrm{J}_0^* H_t(x)\right]$$

for arbitrary $t \geq 0$. We can also obtain that

$$(u \otimes \Phi, \Phi_{\mathrm{b}}) = (u \otimes \Phi, e^{-t(H_{\mathrm{RPF}}-E)}\Phi_{\mathrm{b}}) = \int_{\mathbb{R}^3} \overline{u(x)} \mathbb{E}^{0,0}\mathbb{E}_{\mu_{\mathrm{E}}}\left[\mathrm{J}_0\Phi \cdot H_t(x)\right] dx.$$

Then we have $(\Phi, \Phi_{\mathrm{b}}(x))_{L^2(\mathscr{Q})} = \mathbb{E}^{0,0}\mathbb{E}_{\mu_{\mathrm{E}}}[\mathrm{J}_0\Phi \cdot H_t(x)]$.

Lemma 3.145. *It follows that* $\|\Phi_{\mathrm{b}}(\cdot)\|_{L^2(\mathscr{Q})} \in L^\infty(\mathbb{R}^3)$.

Proof. As $\Phi_{\mathrm{b}}(x) = \mathbb{E}^{0,0}\mathbb{E}_{\mu_{\mathrm{E}}}[\mathrm{J}_0^* H_t(x)]$ for arbitrary $t > 0$, we have

$$\|\Phi_{\mathrm{b}}(x)\|_{L^2(\mathscr{Q})} \leq e^{tE}\left(\mathbb{E}^{0,0}\left[e^{-2\int_0^t V(B_{T_s}+x)ds}\right]\right)^{1/2}\left(\mathbb{E}^{0,0}\left[\|\Phi_{\mathrm{b}}(B_{T_t}+x)\|^2\right]\right)^{1/2}.$$

We have $\sup_{x\in\mathbb{R}^3} \mathbb{E}^{0,0}[e^{-2\int_0^t V(B_{T_s}+x)ds}] < \infty$, since V is of relativistic Kato-class, and

$$\mathbb{E}^{0,0}\left[\|\Phi_{\mathrm{b}}(B_{T_t}+x)\|^2\right] = \int_{\mathbb{R}^3} \|\Phi_{\mathrm{b}}(x+y)\|^2 dy \int_0^\infty \frac{\rho_t(s)e^{-|y|^2/(2s)}}{(2\pi s)^{d/2}} ds \leq C\|\Phi_{\mathrm{b}}\|^2.$$

Here, ρ_t denotes the distribution of T_t, which is given by

$$\rho_t(s) = \frac{te^{tm}}{\sqrt{2\pi}} \frac{1}{s^{3/2}} \exp\left(-\frac{1}{2}\left(\frac{t^2}{s} + m^2 s\right)\right) 1_{\{s\geq 0\}}.$$

Then $\sup_{x\in\mathbb{R}^3} \|\Phi_{\mathrm{b}}(x)\|^2 \leq C\|\Phi_{\mathrm{b}}\|^2$ follows. ∎

We define a filtration under which $(H_t(x))_{t\geq 0}$ is a martingale. Let

$$\mathscr{F}^{(1)}_{[0,t]} = \{E \in \mathcal{B}(\mathscr{X}) \times \mathcal{B}_{\mathcal{T}} \mid E(\cdot, w_1) \in \sigma(B_r, 0 \leq r \leq T_t(w_1))\ \forall w_1 \in \mathcal{T}\}$$

and

$$\mathscr{F}^{(2)}_{[0,t]} = \{E \in \mathcal{B}(\mathscr{X}) \times \mathcal{B}_{\mathcal{T}} \mid E(w_2, \cdot) \in \sigma(T_r, 0 \leq r \leq t)\ \forall w_2 \in \mathscr{X}\}.$$

Here, $E(w_2, \cdot)$ (resp., $E(\cdot, w_1)$) is the section at $w_2 \in \mathscr{X}$ (resp., $w_1 \in \mathcal{T}$). Then $\mathscr{F}^{(1)}_{[0,t]}$ and $\mathscr{F}^{(2)}_{[0,t]}$ are sub-σ-fields, and we set $\mathscr{F}_{[0,t]} = \mathscr{F}^{(1)}_{[0,t]} \cap \mathscr{F}^{(2)}_{[0,t]}$ for $t \geq 0$. We define a filtration in $\mathcal{B}(\mathscr{X}) \times \mathcal{B}_{\mathcal{T}} \times \Sigma_{\mathrm{E}}$ by

$$(\mathscr{M}_t)_{t\geq 0} = \left(\mathscr{F}_{[0,t]} \times \Sigma_{(-\infty,t]}\right)_{t\geq 0}.$$

Theorem 3.146 (Martingale property). *The random process $(H_t(x))_{t\geq 0}$ is a martingale with respect to $(\mathscr{M}_t)_{t\geq 0}$, i. e.,*

$$\mathbb{E}^{0,0}\mathbb{E}_{\mu_{\mathrm{E}}}[H_t(x)|\mathscr{M}_s] = H_s(x), \quad t \geq s.$$

Proof. The idea of the proof is the same as that of Nelson model in Theorem 2.95. It is, however, more complicated than for the Nelson model. Let

$$\mathrm{K}^x[s,t] = \bigoplus_{\mu=1}^{3} \int_{T_s}^{T_t} \mathrm{j}_{T_s^*}\tilde{\varphi}(\cdot - B_r - x)dB_r^\mu,$$

$$\mathrm{K}^x_n[s,t] = \bigoplus_{\mu=1}^{3} \sum_{j=1}^{2^n} \int_{T_{\frac{(t-s)}{2^n}(j-1)+s}}^{T_{\frac{(t-s)}{2^n}j+s}} \mathrm{j}_{\frac{(t-s)}{2^n}(j-1)+s}\tilde{\varphi}(\cdot - B_r - x)dB_r^\mu.$$

Since $e^{-ie\hat{A}_{\mathrm{E}}(\mathrm{K}_s^{\mathrm{rel}}(x))}e^{-\int_0^s V(B_{T_r}+x)dr}$ is $\mathscr{M}_s$-measurable, we have

$$\begin{aligned}
&\mathbb{E}^{0,0}\mathbb{E}_{\mu_{\mathrm{E}}}[H_t(x)|\mathscr{M}_s] \\
&= e^{tE}e^{-ie\hat{A}_{\mathrm{E}}(\mathrm{K}_s^{\mathrm{rel}}(x))}e^{-\int_0^s V(B_{T_r}+x)dr}\mathbb{E}^{0,0}\mathbb{E}_{\mu_{\mathrm{E}}}\left[e^{-ie\hat{A}_{\mathrm{E}}(\mathrm{K}^x[s,t])}e^{-\int_s^t V(B_{T_r}+x)dr}\mathrm{J}_t\Phi_{\mathrm{b}}(B_{T_t}+x)|\mathscr{M}_s\right].
\end{aligned}$$

By the definition of $\mathrm{K}^x[s,t]$ and $\mathrm{K}^x_n[s,t]$, it is seen that

$$\begin{aligned}
&\mathbb{E}^{0,0}\mathbb{E}_{\mu_{\mathrm{E}}}\left[e^{-ie\hat{A}_{\mathrm{E}}(\mathrm{K}^x[s,t])}e^{-\int_s^t V(B_{T_r}+x)dr}\mathrm{J}_t\Phi_{\mathrm{b}}(B_{T_t}+x)|\mathscr{M}_s\right] \\
&= \lim_{n\to\infty}\mathbb{E}^{0,0}\mathbb{E}_{\mu_{\mathrm{E}}}\left[e^{-ie\hat{A}_{\mathrm{E}}(\mathrm{K}^x_n[s,t])}e^{-\int_s^t V(B_{T_r}+x)dr}\mathrm{J}_t\Phi_{\mathrm{b}}(B_{T_t}+x)|\mathscr{M}_s\right],
\end{aligned}$$

and then

$$\begin{aligned}
&\mathbb{E}^{0,0}\mathbb{E}_{\mu_{\mathrm{E}}}\left[e^{-ie\hat{A}_{\mathrm{E}}(\mathrm{K}^x_n[s,t])}e^{-\int_s^t V(B_{T_r}+x)dr}\mathrm{J}_t\Phi_{\mathrm{b}}(B_{T_t}+x)|\mathscr{M}_s\right] \\
&= \mathbb{E}_{\mu_{\mathrm{E}}}\left[\mathbb{E}_\nu^0\left[\mathbb{E}_{\mathcal{W}}^0\left[e^{-ie\hat{A}_{\mathrm{E}}(\mathrm{K}^x_n[s,t])}e^{-\int_s^t V(B_{T_r}+x)dr}\mathrm{J}_t\Phi_{\mathrm{b}}(B_{T_t}+x)|\mathscr{F}^{(1)}_{[0,s]}\right]|\mathscr{F}^{(2)}_{[0,s]}\right]|\Sigma_{[-\infty,s]}\right].
\end{aligned}$$

By the Markov property of Brownian motion, we see that

$$\begin{aligned}
&\mathbb{E}^0_{\mathcal{W}}\left[e^{-ie\hat{A}_{\mathrm{E}}(\mathrm{K}^x_n[s,t])}e^{-\int_s^t V(B_{T_r}+x)dr}\mathrm{J}_t\Phi_{\mathrm{b}}(B_{T_t}+x)|\mathscr{F}^{(1)}_{[0,s]}\right]\\
&=\lim_{n\to\infty}\mathbb{E}^{B_{T_s}}_{\mathcal{W}}\left[e^{-ie\hat{A}_{\mathrm{E}}(\mathrm{K}^{(1),x}_n[s,t])}e^{-\int_s^t V(B_{T_r-T_s}+x)dr}\mathrm{J}_t\Phi_{\mathrm{b}}(B_{T_t-T_s}+x)\right],
\end{aligned}$$

where $\mathbb{E}^{B_{T_s}}_{\mathcal{W}}$ means $\mathbb{E}^y_{\mathcal{W}}$ evaluated at $y=B_{T_s}$ and

$$\mathrm{K}^{(1),x}_n[s,t]=\bigoplus_{\mu=1}^3\sum_{j=1}^{2^n}\int_{T_{\frac{(t-s)}{2^n}(j-1)+s}-T_s}^{T_{\frac{(t-s)}{2^n}j+s}-T_s}\mathrm{j}_{\frac{(t-s)}{2^n}(j-1)+s}\tilde{\varphi}(\cdot-B_r-x)dB^\mu_r.$$

Since subordinator $(T_t)_{t\geq 0}$ is also a Markov process, we have

$$\begin{aligned}
&\mathbb{E}^0_{\nu}\left[\mathbb{E}^{B_{T_s}}_{\mathcal{W}}\left[e^{-ie\hat{A}_{\mathrm{E}}(\mathrm{K}^{(1),x}_n[s,t])}e^{-\int_s^t V(B_{T_r-T_s}+x)dr}\mathrm{J}_t\Phi_{\mathrm{b}}(B_{T_t-T_s}+x)\right]|\mathscr{F}^{(2)}_{[0,s]}\right]\\
&=\mathbb{E}^{T_s}_{\nu}\mathbb{E}^{B_{T_0}}_{\mathcal{W}}\left[e^{-ie\hat{A}_{\mathrm{E}}(\mathrm{K}^{(2),x}_n[s,t])}e^{-\int_s^t V(B_{T_{r-s}-T_0}+x)dr}\mathrm{J}_t\Phi_{\mathrm{b}}(B_{T_{t-s}-T_0}+x)\right],
\end{aligned}$$

where $\mathbb{E}^{T_s}_{\nu}$ also means $\mathbb{E}^y_{\nu}$ evaluated at $y=T_s$ and

$$\mathrm{K}^{(2),x}_n[s,t]=\bigoplus_{\mu=1}^3\sum_{j=1}^{2^n}\int_{T_{\frac{(t-s)}{2^n}(j-1)}-T_0}^{T_{\frac{(t-s)}{2^n}j}-T_0}\mathrm{j}_{\frac{(t-s)}{2^n}(j-1)+s}\tilde{\varphi}(\cdot-B_r-x)dB^\mu_r.$$

Again the Markov property of the Euclidean field yields that

$$\begin{aligned}
&\mathbb{E}_{\mu_{\mathrm{E}}}\left[\mathbb{E}^{T_s}_{\nu}\mathbb{E}^{B_{T_0}}_{\mathcal{W}}\left[e^{-ie\hat{A}_{\mathrm{E}}(\mathrm{K}^{(2),x}_n[s,t])}e^{-\int_s^t V(B_{T_{r-s}-T_0}+x)dr}\mathrm{J}_t\Phi_{\mathrm{b}}(B_{T_{t-s}-T_0}+x)\right]|\Sigma_{[-\infty,s]}\right]\\
&=\mathbb{E}_{\mu_{\mathrm{E}}}\left[\mathbb{E}^{T_s}_{\nu}\mathbb{E}^{B_{T_0}}_{\mathcal{W}}\left[e^{-ie\hat{A}_{\mathrm{E}}(\mathrm{K}^{(2),x}_n[s,t])}e^{-\int_s^t V(B_{T_{r-s}-T_0}+x)dr}\mathrm{J}_t\Phi_{\mathrm{b}}(B_{T_{t-s}-T_0}+x)\right]|\Sigma_s\right].
\end{aligned}$$

The right-hand side above equals to

$$E_s\mathbb{E}^{T_s}_{\nu}\mathbb{E}^{B_{T_0}}_{\mathcal{W}}\left[e^{-ie\hat{A}_{\mathrm{E}}(\mathrm{K}^{(2),x}_n[s,t])}e^{-\int_s^t V(B_{T_{r-s}-T_0}+x)dr}\mathrm{J}_t\Phi_{\mathrm{b}}(B_{T_{t-s}-T_0}+x)\right]$$

and, due to $E_s=\mathrm{J}_s\mathrm{J}_0^*\mathrm{U}_{-s}$, we have

$$=\mathrm{J}_s\mathrm{J}_0^*\mathrm{U}_{-s}\mathbb{E}^{T_s}_{\nu}\mathbb{E}^{B_{T_0}}_{\mathcal{W}}\left[e^{-ie\hat{A}_{\mathrm{E}}(\mathrm{K}^{(2),x}_n[s,t])}e^{-\int_s^t V(B_{T_{r-s}-T_0}+x)dr}\mathrm{J}_t\Phi_{\mathrm{b}}(B_{T_{t-s}-T_0}+x)\right].$$

Since U_{-s} is the shift by $-s$, we have

$$=\mathrm{J}_s\mathrm{J}_0^*\mathbb{E}^{T_s}_{\nu}\mathbb{E}^{B_{T_0}}_{\mathcal{W}}\left[e^{-ie\hat{A}_{\mathrm{E}}(\mathrm{K}^{(3),x}_n[s,t])}e^{-\int_s^t V(B_{T_{r-s}-T_0}+x)dr}\mathrm{J}_{t-s}\Phi_{\mathrm{b}}(B_{T_{t-s}-T_0}+x)\right],$$

where

$$\mathrm{K}_n^{(3),x}[s,t] = \bigoplus_{\mu=1}^{3} \sum_{j=1}^{2^n} \int_{T_{\frac{(t-s)}{2^n}(j-1)}-T_0}^{T_{\frac{(t-s)}{2^n}j}-T_0} \mathrm{j}_{\frac{(t-s)}{2^n}(j-1)} \tilde{\varphi}(\cdot - B_r - x) dB_r^{\mu}.$$

We notice that the random variable $T_t + y$ under ν has the same law as T_t under ν^y, i. e., $\mathbb{E}_\nu^y[f(T_t)] = \mathbb{E}_\nu^0[f(T_t + y)]$, we can see that

$$= \mathrm{J}_s \mathrm{J}_0^* \mathbb{E}_\nu^0 \mathbb{E}_{\mathcal{W}}^{B_{u+T_0}} \left[e^{-ie\hat{A}_{\mathrm{E}}(\mathrm{K}_n^{(3),x}[s,t])} e^{-\int_s^t V(B_{T_{r-s}-T_0}+x)dr} \mathrm{J}_{t-s} \Phi_{\mathrm{b}}(B_{T_{t-s}-T_0} + x) \right] \Big|_{u=T_s}$$
$$= \mathrm{J}_s \mathrm{J}_0^* \mathbb{E}_\nu^0 \mathbb{E}_{\mathcal{W}}^{B_{T_s}} \left[e^{-ie\hat{A}_{\mathrm{E}}(\mathrm{K}_n^{(4),x}[s,t])} e^{-\int_s^t V(B_{T_{r-s}}+x)dr} \mathrm{J}_{t-s} \Phi_{\mathrm{b}}(B_{T_{t-s}} + x) \right],$$

where

$$\mathrm{K}_n^{(4),x}[s,t] = \bigoplus_{\mu=1}^{3} \sum_{j=1}^{2^n} \int_{T_{\frac{(t-s)}{2^n}(j-1)}}^{T_{\frac{(t-s)}{2^n}j}} \mathrm{j}_{\frac{(t-s)}{2^n}(j-1)} \tilde{\varphi}(\cdot - B_r - x) dB_r^{\mu}.$$

Taking the limit $n \to \infty$, we finally obtain that

$$\begin{aligned} \mathbb{E}^{0,0} \mathbb{E}_{\mu_{\mathrm{E}}} [H_t(x) | \mathscr{M}_s] &= e^{sE} e^{-ie\hat{A}_{\mathrm{E}}(\mathrm{K}^x[0,s])} e^{-\int_0^s V(B_{T_r}+x)dr} \mathrm{J}_s \\ &\quad \times e^{(t-s)E} \mathbb{E}^{B_{T_s},0} \left[\mathrm{J}_0^* e^{-ie\hat{A}_{\mathrm{E}}(\mathrm{K}^x[0,t-s])} e^{-\int_0^{t-s} V(B_{T_r}+x)dr} \mathrm{J}_{t-s} \Phi_{\mathrm{b}}(B_{T_{t-s}} + x) \right]. \end{aligned}$$

Notice that

$$e^{(t-s)E} \mathbb{E}^{B_{T_s},0} \left[\mathrm{J}_0^* e^{-ie\hat{A}_{\mathrm{E}}(\mathrm{K}^x[0,t-s])} e^{-\int_0^{t-s} V(B_{T_r}+x)dr} \mathrm{J}_{t-s} \Phi_{\mathrm{b}}(B_{T_{t-s}} + x) \right] = \Phi_{\mathrm{b}}(B_{T_s} + x)$$

and hence

$$\mathbb{E}^{0,0} \mathbb{E}_{\mu_{\mathrm{E}}} [H_t(x) | \mathscr{M}_s] = e^{sE} e^{-ie\hat{A}_{\mathrm{E}}(K_s^{\mathrm{rel}}(x))} e^{-\int_0^s V(B_{T_r}+x)dr} \mathrm{J}_s \Phi_{\mathrm{b}}(B_{T_s} + x) = H_s(x).$$

Thus the proof is complete. □

Since we have showed that $(H_t(x))_{t\geq 0}$ is a martingale, for an arbitrary stopping time τ with respect to $(\mathscr{M}_t)_{t\geq 0}$, $(H_{t\wedge\tau}(x))_{t\geq 0}$ is also a martingale. By using this fact, we can show a spatial decay of bound state Φ_{b} of H_{RPF}. It will be seen it in the next section.

3.9.7 Spatial decay of bound states

In this section we show spacial decay of bound states under Assumption 3.138. Let $(z_t)_{t\geq 0}$ be the a 3-dimensional Lévy process on a probability space $(\mathscr{Z}, \mathcal{B}_{\mathscr{Z}}, \mathrm{Z}^x)$ such that

$\mathbb{E}_Z^x[e^{-iuz_t}] = e^{-t(\sqrt{|u|^2+m^2}-m)}e^{-iux}$ for $u \in \mathbb{R}^3$ and $x \in \mathbb{R}^3$. Hence the generator of $(z_t)_{t\geq 0}$ is given by $\sqrt{-\Delta+m^2}-m$. Spatial decay of bound state ϕ of the relativistic Schrödinger operator, $\sqrt{-\Delta+m^2}-m+V$, can be estimated by using the martingale technique. Define random process

$$(X_t)_{t\geq 0} = (e^{tE}e^{-\int_0^t V(z_s+x)ds}\phi(z_t+x))_{t\geq 0},$$

which is a martingale with respect to the filtration $\mathscr{G}_t = \sigma(z_s, 0 \leq s \leq t)$, and

$$\phi(x) = \mathbb{E}^0[X_t] = \mathbb{E}^0[X_{t\wedge\tau}] \tag{3.9.37}$$

for any stopping time τ. Using the fact that $\phi \in L^\infty(\mathbb{R}^3)$, we can directly see that

$$|\phi(x)| \leq e^{(t\wedge\tau)E}\|\phi\|_\infty \mathbb{E}^0[e^{-\int_0^{t\wedge\tau} V(z_s+x)ds}].$$

Choosing a suitable τ, we estimate $\mathbb{E}^0[e^{-\int_0^{t\wedge\tau} V(z_s+x)ds}]$.

On the other hand, for a bound state Φ_b of the relativistic Pauli–Fierz Hamiltonian, we see that $(H_t(x))_{t\geq 0}$ is a martingale and

$$\Phi_b(x) = \mathbb{E}^{0,0}[J_0^* H_t(x)]. \tag{3.9.38}$$

To express $\Phi_b(x)$, we need to add J_0^* to $H_t(x)$, which differs from the quantum-mechanical case (3.9.37). Such arguments were already presented for the Nelson Hamiltonian in Section 2.10.1.

Lemma 3.147. *Let τ be a stopping time with respect to the filtration $(\mathscr{M}_t)_{t\geq 0}$. Then*

$$\|\Phi_b(x)\|_{L^2(\mathscr{Q})} \leq C\mathbb{E}_Z^0\left[e^{-\int_0^{t\wedge\tau}(V(z_r+x)-E)dr}\right], \tag{3.9.39}$$

where $C = e^{(t\wedge\tau)E}\sup_{x\in\mathbb{R}^3}\|\Phi_b(x)\|_{L^2(\mathscr{Q})}$.

Proof. Let $\Phi \in L^2(\mathscr{Q})$. Since $(J_0\Phi \cdot H_t(x))_{t\geq 0}$ is an $L^2(\mathscr{Q}_E)$-valued martingale with respect to the filtration $(\mathscr{M}_t)_{t\geq 0}$, such is also $(J_0\Phi \cdot H_{t\wedge\tau}(x))_{t\geq 0}$. Then

$$\mathbb{E}^{0,0}\mathbb{E}_{\mu_E}[J_0\Phi \cdot H_t(x)] = \mathbb{E}^{0,0}\mathbb{E}_{\mu_E}[J_0\Phi \cdot H_{t\wedge\tau}(x)]$$

follows. It is immediate to see that

$$\left|\mathbb{E}^{0,0}[\mathbb{E}_{\mu_E}[J_0\Phi \cdot H_{t\wedge\tau}(x)]]\right| \leq C\|\Phi\|\mathbb{E}^{0,0}\left[e^{-\int_0^{t\wedge\tau}(V(B_{T_r}+x)-E)dr}\right],$$

where $C = e^{(t\wedge\tau)E}\sup_{x\in\mathbb{R}^3}\|\Phi_b(x)\|_{L^2(\mathscr{Q})}$. Since $B_{T_t} = z_t$ in law, we then have

$$\left|\mathbb{E}^{0,0}\mathbb{E}_{\mu_E}[J_0\Phi \cdot H_t(x)]\right| \leq C\|\Phi\|\mathbb{E}_Z^0\left[e^{-\int_0^{t\wedge\tau}(V(z_r+x)-E)dr}\right]. \tag{3.9.40}$$

From (3.9.40) and the fact

$$\|\Phi_b(x)\| = \sup_{\|\Phi\|=1} |(\Phi, \Phi_b(x))| = \sup_{\|\Phi\|=1} |(\Phi, \mathbb{E}^{0,0}[J_0^* H_t(x)])| = \sup_{\|\Phi\|=1} \left|\mathbb{E}^{0,0}\mathbb{E}_{\mu_E}[J_0\Phi \cdot H_t(x)]\right|,$$

the lemma follows. □

Let $\tau_R = \tau_R(x) = \inf\{s \mid |z_s + x| < R\}$ and $\tau_R' = \inf\{s \mid |z_s| > R\}$.

Lemma 3.148. *Both τ_R and τ_R' are stopping times with respect to $(\mathscr{M}_t)_{t\geq 0}$.*

Proof. It is sufficient to show that $E_t = \{\tau_R \leq t\} \in \mathscr{F}_{[0,t]}$ for arbitrary $t \geq 0$. The sector of E_t at $w' \in \mathcal{T}$ is given by

$$\{w \in \mathscr{X} \mid \sup_{0\leq s\leq t} |B_{T_s(w')}(w) + x| > R\} \in \sigma(B_r, 0 \leq r \leq T_t(w')).$$

Then $\{\tau_R \leq t\} \in \mathscr{F}^{(1)}_{[0,t]}$. The sector of E_t at $w \in \mathscr{X}$ is also given by

$$\left\{w' \in \mathcal{T} \mid \sup_{0\leq s\leq t} |B_{T_s(w')}(w) + x| > R\right\} \in \sigma(T_r, 0 \leq r \leq t).$$

Then $\{\tau_R \leq t\} \in \mathscr{F}^{(2)}_{[0,t]}$. Hence $\{\tau_R \leq t\} \in \mathscr{F}_{[0,t]}$. In a similar manner to τ_R, we can show that τ_R' is also a stopping time. □

Theorem 3.149 (Spatial decay of bound states). *The following hold:*

(1) *Suppose that* $\lim_{|x|\to\infty} V_-(x) + E = a < 0$. *Then:*

($m = 0$) *there exists a constant $C > 0$ such that*

$$\|\Phi_b(x)\|_{L^2(\mathscr{Q})} \leq \frac{C}{1 + |x|^4};$$

($m > 0$) *there exist constants $C > 0$ and $c > 0$ such that*

$$\|\Phi_b(x)\|_{L^2(\mathscr{Q})} \leq Ce^{-c|x|}.$$

(2) *Suppose that* $\lim_{|x|\to\infty} V(x) = \infty$. *Then there exist constants $C > 0$ and $c > 0$ such that*

$$\|\Phi_b(x)\|_{L^2(\mathscr{Q})} \leq Ce^{-c|x|}.$$

Proof. Let $\tau_R = \tau_R(x) = \inf\{s \mid |z_s + x| < R\}$ and $\tau_R' = \inf\{s \mid |z_s| > R\}$.

(1) Suppose that $V_-(x) + E < a + \epsilon < 0$ for all x such that $|x| > R$. By (3.9.39), we have $\|\Phi_b(x)\| \leq C\|\Phi_b\|_{\mathcal{H}_{PF}} \mathbb{E}_Z^0[e^{+(\epsilon+a)(t\wedge\tau_R(x))}]$ for $|x| > R$ with some constant $C > 0$. It can be seen that

$$\mathbb{E}_Z^0\left[e^{+(\epsilon+a)(t\wedge\tau_R(x))}\right] \le \frac{C_1}{1+|x|^4}, \qquad m=0,$$

$$\mathbb{E}_Z^0\left[e^{+(\epsilon+a)(t\wedge\tau_R(x))}\right] \le C_2 e^{-C_3|x|}, \qquad m>0.$$

Here, C_1, C_2, and C_3 are positive constants. Thus (1) follows.

(2) Let $W_R(x) = \inf\{V(y) \mid |x-y| < R\}$. Then it can be shown that

$$\mathbb{E}_Z^0\left[e^{(t\wedge\tau_R')E} e^{-\int_0^{t\wedge\tau_R'} V(z_r+x)dr}\right] \le e^{-t(W_R(x)-E)} + Ce^{-aR}e^{ct}$$

with some constants a, c, and C. Inserting $R = p|x|$ with any $0 < p < 1$, we see that $W_R(x) \to \infty$ as $|x| \to \infty$. Thus substituting $t = \delta|x|$ for sufficiently small $\delta > 0$ and $R = p|x|$ with some $0 < p < 1$, we can derive (2). □

3.9.8 Gaussian domination of ground states

By Corollaries 3.123 and 3.144, it is shown that $e^{i\frac{\pi}{2}\mathrm{N}}e^{-tH_{\mathrm{RPF}}}e^{-i\frac{\pi}{2}\mathrm{N}}$ is positivity improving for all $t > 0$ under Assumptions 3.112 and 3.138. In particular, $e^{i\frac{\pi}{2}\mathrm{N}}\Psi_{\mathrm{g}}$ is strictly positive and then the ground state Ψ_{g} of H_{RPF} is unique up to multiplication constants if it exists. In Section 3.9.8 we suppose either Assumption 3.112 or 3.138. For an arbitrary fixed $0 \le f \in L^2(\mathbb{R}^3)$, but $f \not\equiv 0$, we define

$$f_t = e^{-t(H_{\mathrm{RPF}}-E)}(f\otimes 1), \quad \Psi_{\mathrm{g}}^t = f_t/\|f_t\|.$$

Then it follows that $\Psi_{\mathrm{g}}^t \to \Psi_{\mathrm{g}}$ strongly as $t \to \infty$, since $(f\otimes 1, \Psi_{\mathrm{g}}) \neq 0$.

Let

$$\mathscr{L}_t = f(B_{-T_t})f(B_{T_t})e^{-\frac{e^2}{2}\mathrm{q}_{\mathrm{E}}(K_t^{\mathrm{rel}},K_t^{\mathrm{rel}})}e^{-\int_{-t}^t V(B_{T_s})ds}, \qquad t \ge 0.$$

We can formally write the pair interaction $W^{\mathrm{RPF}} = e^2\mathrm{q}_{\mathrm{E}}(K_t^{\mathrm{rel}}, K_t^{\mathrm{rel}})$ by

$$\mathrm{q}_{\mathrm{E}}(K_t^{\mathrm{rel}}, K_t^{\mathrm{rel}}) = \sum_{\mu,\nu=1}^3 \int_{-T_t}^{T_t} dB_s^\mu \int_{-T_t}^{T_t} W_{\mu\nu}(T_s^* - T_r^*, B_s - B_r)dB_r^\nu,$$

where the pair potential, $W_{\mu\nu}(t,X)$, is given by

$$W_{\mu\nu}(t,X) = \frac{1}{2}\int_{\mathbb{R}^3} \frac{|\hat\varphi(k)|^2}{\omega(k)}\left(\delta_{\mu\nu} - \frac{k_\mu k_\nu}{|k|^2}\right)e^{-ik\cdot X}e^{-\omega(k)|t|}dk.$$

For each $t \ge 0$, we define a family of probability measures μ_t, $t > 0$, on the measurable space $(\mathcal{X}\times\mathcal{T}, \mathcal{B}(\mathcal{X})\times\mathcal{B}_{\mathcal{T}})$ by

$$\mu_t(A) = \frac{1}{Z_t} \int_{\mathbb{R}^3} \mathbb{E}^{x,0} [1_A \mathscr{L}_t] \, dx, \quad t \geq 0.$$

Here, Z_t is the normalizing constant such that $\mu_t(\mathscr{X} \times \mathcal{T}) = 1$. This is the finite volume Gibbs measure associated with the relativistic Pauli–Fierz model. We shall discuss it in the next section. We define the self-adjoint operator $\hat{A}_\xi$ in $\mathcal{H}_{\mathrm{PF}}$ by $\hat{A}_\xi = \int_{\mathbb{R}^3}^{\oplus} \hat{A}(\xi(\cdot - x))dx$, where $\xi \in \oplus^3 L^2_{\mathrm{real}}(\mathbb{R}^3)$. Then we have

$$(\Psi_{\mathrm{g}}, e^{-i\beta\hat{A}_\xi}\Psi_{\mathrm{g}}) = \lim_{t\to 0} \frac{(e^{-tH_{\mathrm{RPF}}} f \otimes 1, e^{-i\beta\hat{A}_\xi} e^{-tH_{\mathrm{RPF}}} f \otimes 1)}{(e^{-tH_{\mathrm{RPF}}} f \otimes 1, e^{-tH_{\mathrm{RPF}}} f \otimes 1)}, \quad \beta \in \mathbb{R}.$$

Let $\beta \in \mathbb{R}$. Then it follows that

$$\frac{(e^{-tH_{\mathrm{RPF}}} f \otimes 1, e^{-i\beta\hat{A}_\xi} e^{-tH_{\mathrm{RPF}}} f \otimes 1)}{(e^{-tH_{\mathrm{RPF}}} f \otimes 1, e^{-tH_{\mathrm{RPF}}} f \otimes 1)} = \mathbb{E}_{\mu_t} [e^{-\frac{1}{2}(2e\beta i \Re \mathrm{q}_{\mathrm{E}}(K_t^{\mathrm{rel}}, \mathrm{j}_0\xi) + \beta^2 \mathrm{q}_{\mathrm{E}}(\mathrm{j}_0\xi, \mathrm{j}_0\xi))}].$$

Note that both $\mathrm{q}_{\mathrm{E}}(K_t^{\mathrm{rel}}, \mathrm{j}_0\xi)$ and $\mathrm{q}_{\mathrm{E}}(\mathrm{j}_0\xi, \mathrm{j}_0\xi)$ do not depend on x.

Lemma 3.150. *Suppose that $\beta < (2\mathrm{q}_{\mathrm{E}}(\mathrm{j}_0\xi, \mathrm{j}_0\xi))^{-1}$. Then $\Psi_{\mathrm{g}}^t \in D(e^{(\beta/2)\hat{A}_\xi^2})$ and*

$$\|e^{(\beta/2)\hat{A}_\xi^2}\Psi_{\mathrm{g}}^t\|^2 = (1 - 2\beta\mathrm{q}_{\mathrm{E}}(\mathrm{j}_0\xi, \mathrm{j}_0\xi))^{-1/2}\mathbb{E}_{\mu_t}\left[\exp\left(\frac{-\beta e^2 \mathrm{q}_{\mathrm{E}}(K_t^{\mathrm{rel}}, \mathrm{j}_0\xi)^2}{(1 - 2\beta\mathrm{q}_{\mathrm{E}}(\mathrm{j}_0\xi, \mathrm{j}_0\xi))}\right)\right].$$

Proof. The proof is a minor modification of Lemma 3.64. □

Thus in the same way as in Theorem 3.65, we have the theorem below.

Theorem 3.151 (Gaussian domination of the ground state). *Suppose that the ground state Ψ_{g} of H_{RPF} exists. Let $\beta < (2\mathrm{q}_{\mathrm{E}}(\mathrm{j}_0\xi, \mathrm{j}_0\xi))^{-1}$. Then $\Psi_{\mathrm{g}} \in D(e^{(\beta/4)\hat{A}_\xi^2})$ follows.*

3.9.9 Gibbs measure associated with the ground state of relativistic Pauli–Fierz Hamiltonian

In Section 3.9.9, we assume either Assumption 3.112 or 3.138, and that H_{RPF} has a ground state Ψ_{g}. In this section, we consider the two-sided subordinator $(T_t)_{t\in\mathbb{R}}$ defined by

$$T_t = \begin{cases} S_t, & t \geq 0, \\ -\bar{S}_{(-t)-}, & t < 0, \end{cases}$$

where $(S_t)_{t\geq 0}$ and $(\bar{S}_t)_{t\geq 0}$ are independent subordinators. This process can be realized as a coordinate process on $(\mathscr{X}_D, \mathcal{B}(\mathscr{X}_D), \tilde{\nu})$ for some probability measure $\tilde{\nu}$, where $\mathscr{X}_D = D(\mathbb{R})$ denotes the space of càdlàg paths on $\mathbb{R}$. We denote the coordinate process again by $(T_t)_{t\in\mathbb{R}}$. We set

$$\mathscr{X} = \mathfrak{X} \times \mathfrak{X}_D, \quad \mathrm{W}^x = \mathcal{W}^x \otimes \tilde{\nu}.$$

Let $(B_t)_{t\in\mathbb{R}}$ be a three-dimensional two-sided Brownian motion, and define

$$z_t = B_{T_t}, \quad t \in \mathbb{R}.$$

Then

$$t \mapsto z_t(\omega_1, \omega_2) = B_{T_t(\omega_2)}(\omega_1)$$

is a càdlàg path. Let $Z_t = (B_t, T_t) \in \mathbb{R}^{3+1}$ for $t \in \mathbb{R}$ and $\mathscr{G}_s = \sigma(Z_r\,;\, r \in [-s, s])$, and set $\mathscr{G} = \bigcup_{s\geq 0} \mathscr{G}_s$, which is a finitely additive family of sets. Then we may regard $\mathcal{Z}$ as a subset of $D(\mathbb{R}; \mathbb{R}^{3+1})$, and $\mathscr{G}_s$ as a sub-σ-algebra of $\mathcal{B}(D(\mathbb{R}; \mathbb{R}^{3+1}))$. We define a family of probability spaces $(\mathscr{X}, \sigma(\mathscr{G}), \mu_T)$ for $T > 0$, where μ_T is given by

$$\mu_T(A) = \frac{1}{Z_T} \int_{\mathbb{R}^3} \mathbb{E}^{x,0}_{\mathrm{W}} \left[1_A \mathscr{L}_T \right] dx.$$

Here

$$\mathscr{L}_T = f(z_{-T}) f(z_T) \exp\left(-\frac{e^2}{2} \mathrm{q}_{\mathrm{E}}(K^{\mathrm{rel}}_T, K^{\mathrm{rel}}_T) \right) \exp\left(-\int_{-T}^{T} V(z_s)\, ds \right).$$

This defines the finite volume Gibbs measure for the relativistic Pauli–Fierz model. To emphasize the relativistic nature, we rename μ_T as μ^{rel}_T. In a manner similar as for the Nelson model, one can show that there exists a probability measure μ_∞ on $(\mathscr{X}, \sigma(\mathscr{G}))$ such that $\mu^{\mathrm{rel}}_T \to \mu^{\mathrm{rel}}_\infty$ as $T \to \infty$ in the local sense.

The idea to show the convergence is the same as for the Nelson model. We show only the outline. First, we define the family of set functions ρ_T on $(\mathscr{X}, \mathscr{G}_T)$, $T > 0$, by

$$\rho_T(A) = e^{2Es} \int_{\mathbb{R}^3} \mathbb{E}^{x,0} \left[1_A \left(\frac{f_{T-s}(z_{-s})}{\|f_T\|}, \mathrm{J}_{[-s,s]} \frac{f_{T-s}(z_s)}{\|f_T\|} \right) \right] dx,$$

where $A \in \mathscr{G}_s$ and $f_T = e^{-T(H-E)}(f \otimes 1)$ and

$$\mathrm{J}_{[-s,s]} = \mathrm{J}^*_{-s} e^{-\int_{-s}^{s} V(z_r)dr} e^{-ie\hat{A}_{\mathrm{E}}(K^{\mathrm{rel}}_s)} \mathrm{J}_s.$$

We can show that ρ_T is well defined, and then it is a probability measure on $(\mathscr{X}, \mathscr{G}_T)$. Thus we define the probability space $(\mathscr{X}, \mathscr{G}_T, \rho_T)$. By using the functional integral representation for H_{RPF}, we can see that $\rho_T(A) = \mu^{\mathrm{rel}}_T(A)$ for $A \in \mathscr{G}_s$ for all $s \leq T$. Next, in terms of the ground state Ψ_{g}, we define a finitely additive set function μ^{rel} on $(\mathscr{X}, \mathscr{G})$ by

$$\mu^{\mathrm{rel}}(A) = e^{2ET} \int_{\mathbb{R}^3} \mathbb{E}^{x,0}_W \left[1_A (\Psi_{\mathrm{g}}(z_{-T}), \mathrm{J}_{[-T,T]} \Psi_{\mathrm{g}}(z_T)) \right] dx$$

for $A \in \mathscr{G}_T$. We can verify that μ^{rel} is countably additive, and the extension to the probability measure on $(\mathscr{X}, \sigma(\mathscr{G}))$ is denoted by $\mu^{\mathrm{rel}}_\infty$. Thus we define the probability space $(\mathscr{X}, \sigma(\mathscr{G}), \mu^{\mathrm{rel}}_\infty)$. By applying the fact that $f_T/\|f_T\|$ strongly converges to Ψ_{g} as $T \to \infty$, we prove that $\rho^{\mathrm{rel}}_T(A) \to \mu^{\mathrm{rel}}(A)$ as $T \to \infty$ for $A \in \mathscr{G}$, which implies that $\mu^{\mathrm{rel}}_T(A) \to \mu^{\mathrm{rel}}_\infty(A)$ for $A \in \mathscr{G}$, and μ^{rel}_T converges to the measure $\mu^{\mathrm{rel}}_\infty$ in the local sense. We state it as a theorem.

Theorem 3.152 (Local convergence). *Suppose that there exists a ground state of H_{RPF}. Then the probability measures μ^{rel}_T converge to $\mu^{\mathrm{rel}}_\infty$ on $(\mathscr{X}, \sigma(\mathscr{G}))$ in the local sense, i. e., $\mu^{\mathrm{rel}}_T(A) \to \mu^{\mathrm{rel}}_\infty(A)$ as $T \to \infty$ for each $A \in \mathscr{G}$.*

The probability measure $\mu^{\mathrm{rel}}_\infty$ is called the Gibbs measure for the relativistic Pauli–Fierz model.

3.9.10 Translation invariant relativistic Pauli–Fierz model

In the case of the relativistic Pauli–Fierz Hamiltonian with $V = 0$, it can be seen that $[H_{\mathrm{RPF}}, \mathrm{P}^{\mathrm{tot}}_\mu] = 0$, where $\mathrm{P}^{\mathrm{tot}}_\mu = -i\nabla_{x_\mu} \otimes 1 + 1 \otimes P_{\mathrm{f}\mu}$ is the total momentum operator. This means that H_{RPF} with $V = 0$ has no ground state, which is shown in the same way as the absence of a ground state of H_{PF} with $V = 0$, and there exists a self-adjoint operator $H_{\mathrm{RPF}}(p)$ in $\mathcal{F}_{\mathrm{rad}}$ such that

$$H_{\mathrm{RPF}} = \int_{\mathbb{R}^3}^{\oplus} H_{\mathrm{RPF}}(p)dp.$$

The self-adjoint operator $H_{\mathrm{RPF}}(p)$, $p \in \mathbb{R}^3$, is called the *relativistic Pauli–Fierz Hamiltonian with a total momentum p*. We can show similar results as for $H_{\mathrm{PF}}(p)$ by using the functional integral representation of $e^{-tH_{\mathrm{RPF}}}$. We use Assumption 3.153 below in Section 3.9.10.

Assumption 3.153. The following conditions hold:
(1) (Charge distribution) $\varphi \in \mathscr{S}'(\mathbb{R}^3)$, $\hat{\varphi}(-k) = \overline{\hat{\varphi}(k)}$ and $\omega^{3/2}\hat{\varphi}$, $\sqrt{\omega}\hat{\varphi}$, $\hat{\varphi}/\omega \in L^2(\mathbb{R}^3)$;
(2) (External potential) $V = 0$.

Theorem 3.154 (Functional integral representation for relativistic Pauli–Fierz model with a fixed total momentum). *Suppose Assumption* 3.153. *Let* $\Psi, \Phi \in L^2(\mathscr{Q})$ *and* $E_{\mathrm{RPF}}(p) = \inf \mathrm{Spec}(H_{\mathrm{RPF}}(p))$. *Then*

$$(\Psi, e^{-tH_{\mathrm{RPF}}(p)}\Phi) = \mathbb{E}^{0,0}\left[e^{ip\cdot B_{T_t}}\left(\mathrm{J}_0\Psi, e^{-ie\hat{A}_{\mathrm{E}}(K^{\mathrm{rel}}_t)}\mathrm{J}_t e^{-i\hat{P}_{\mathrm{f}}B_{T_t}}\Phi\right)\right]$$

and (1)–(4) *below are satisfied:*
(1) *Let* $p = 0$. *Then* $e^{i(\pi/2)N}e^{-tH_{\mathrm{RPF}}(0)}e^{-i(\pi/2)N}$ *is positivity improving.*
(2) *The ground state of* $H_{\mathrm{RPF}}(0)$ *is unique whenever it exists.*

(3) *Energy comparison inequality* $E_{\mathrm{RPF}}(0) \le E_{\mathrm{RPF}}(p)$ *holds.*
(4) *The map* $p \mapsto E_{\mathrm{RPF}}(p)$ *is continuous and* $E(0) = \inf \mathrm{Spec}(H_{\mathrm{RPF}\,V=0})$, *where* $H_{\mathrm{RPF}\,V=0}$ *is* H_{RPF} *with* $V = 0$.

Proof. The proof is similar to those of Theorem 3.70 and Corollary 3.73. □

We can see the explicit form of the fiber Hamiltonian $H_{\mathrm{RPF}}(p)$. Let

$$K(p) = (p - \hat{P}_{\mathrm{f}} - \hat{A}(0))^2 + m^2.$$

Then we have

$$(\Psi, e^{-tK(p)}\Phi) = e^{-tm^2}\mathbb{E}^{0,0}\left[e^{ip\cdot B_t}\left(\Psi, e^{-ie\hat{A}(L_t)}e^{-i\hat{P}_{\mathrm{f}}\cdot B_t}\Phi\right)\right].$$

Let $D = D(\hat{P}_{\mathrm{f}}^2) \cap D(H_{\mathrm{rad}})$ and $D_\infty = C_0^\infty(\mathbb{R}^3) \otimes \mathcal{F}_{\mathrm{rad}}^\infty$. See (3.7.2) for the definition of $\mathcal{F}_{\mathrm{rad}}^\infty$. Applying the functional integral representation, we show that $K(p)$ is essentially self-adjoint on D, i. e., it can be shown that $e^{-tK(p)}D \subset D$. Set $\bar{K}(p) = \overline{K(p)\lceil_D}$. We define $L_{\mathrm{RPF}}(p)$ by

$$L_{\mathrm{RPF}}(p) = \sqrt{\bar{K}(p)} \dotplus H_{\mathrm{rad}} - m$$

for each $p \in \mathbb{R}^3$.

Theorem 3.155 (Fiber decomposition). *Suppose Assumption* 3.153 *holds. Then*

$$H_{\mathrm{RPF}} \cong \int_{\mathbb{R}^3}^{\oplus} L_{\mathrm{RPF}}(p)dp.$$

In particular, $H_{\mathrm{RPF}}(p) = L_{\mathrm{RPF}}(p)$.

Proof. We define the unitary operator U on $\mathcal{H}_{\mathrm{PF}}$ by

$$(UF)(p) = (2\pi)^{-3/2}\int_{\mathbb{R}^3} e^{ipx}e^{-i\hat{P}_{\mathrm{f}}x}F(x)dx.$$

Then $U^{-1}(\int_{\mathbb{R}^3}^{\oplus}\bar{K}(p)dp)U = T_{\mathrm{kin}}^2$, where $T_{\mathrm{kin}} = \sqrt{2\hat{H}_{\mathrm{PF}}(\hat{A}) + m^2}$. Actually, it is shown that

$$(F, T_{\mathrm{kin}}^2 G) = \int_{\mathbb{R}^3} ((UF)(p), \bar{K}(p)(UG)(p))\, dp$$

for $F, G \in D_\infty$. We see that $U^{-1}(\int_{\mathbb{R}^3}^{\oplus} e^{-t\bar{K}(p)}dp)U = e^{-tT_{\mathrm{kin}}^2}$ for all $t \ge 0$. Employing the formula

$$T_{\rm kin}^{\alpha} = C_\alpha \int_0^\infty (1 - e^{-\lambda T_{\rm kin}^2}) \frac{d\lambda}{\lambda^{1+\alpha/2}},$$

we can see that for $F \in D_\infty$,

$$(F, T_{\rm kin} F) = C_1 \int_0^\infty \frac{d\lambda}{\lambda^{3/2}} \int_{\mathbb{R}^3} ((UF)(p), (1 - e^{-\lambda \bar{K}(p)})(UF)(p)) dp.$$

By Fubini's theorem, we have

$$(F, T_{\rm kin} F) = C_1 \int_{\mathbb{R}^3} dp \int ((UF)(p), (1 - e^{-\lambda \bar{K}(p)})(UF)(p)) \frac{d\lambda}{\lambda^{3/2}}.$$

Note that $(UF)(p) \in \mathcal{F}_{\rm rad}^{\infty}$ for each $p \in \mathbb{R}^3$. Hence $(UF)(p) \in D(\bar{K}(p)) \subset D(\sqrt{\bar{K}(p)})$, which implies

$$(F, T_{\rm kin} F) = \int_{\mathbb{R}^3} \left((UF)(p), \sqrt{\bar{K}(p)} (UF)(p) \right) dp. \tag{3.9.41}$$

By the polarization identity and (3.9.41), we have

$$(F, T_{\rm kin} G) = \int_{\mathbb{R}^3} \left((UF)(p), \sqrt{\bar{K}(p)} (UG)(p) \right) dp.$$

Furthermore, we see that

$$(F, (T_{\rm kin} + H_{\rm rad})\, G) = \int_{\mathbb{R}^3} ((UF)(p), L_{\rm RPF}(p)(UG)(p))\, dp,$$

which implies

$$T_{\rm kin} \dotplus H_{\rm rad} = U^{-1} \left(\int_{\mathbb{R}^3}^{\oplus} L_{\rm RPF}(p) dp \right) U \tag{3.9.42}$$

on D_∞. Since D_∞ is a core of the left-hand side of (3.9.42), the claim holds true for self-adjoint operators. Note that $T_{\rm kin} \dotplus H_{\rm rad} = T_{\rm kin} + H_{\rm rad}$ on $D(\sqrt{-\Delta}) \cap D(H_{\rm rad})$ and $T_{\rm kin} + H_{\rm rad}$ is self-adjoint on $D(\sqrt{-\Delta}) \cap D(H_{\rm rad})$, finishing the proof. □

Note that $D(|p - \hat{P}_{\rm f}|) = D(|\hat{P}_{\rm f}|)$ for all $p \in \mathbb{R}^3$. The essential self-adjointness of $H_{\rm RPF}(p)$ is established.

Lemma 3.156. *Let $m > 0$. Then $H_{\rm RPF}(p)$ is closable on $D(|\hat{P}_{\rm f}|) \cap D(H_{\rm rad})$.*

Proof. The proof is similar to show the essential self-adjointness of H_{RPF}. By using the functional integral representation,

$$e^{-tH_{\mathrm{RPF}}(p)}D(|\hat{P}_{\mathrm{f}}|) \cap D(H_{\mathrm{rad}}) \subset D(|\hat{P}_{\mathrm{f}}|) \cap D(H_{\mathrm{rad}})$$

can be shown. Then the lemma follows. □

We can furthermore specify the domain of $H_{\mathrm{RPF}}(p)$.

Theorem 3.157 (Self-adjointness). *Let* $m \geq 0$. *Then* $H_{\mathrm{RPF}}(p)$ *is self-adjoint on* $D(|\hat{P}_{\mathrm{f}}|) \cap D(H_{\mathrm{rad}})$ *and essentially self-adjoint on* $\mathcal{F}_{\mathrm{rad}}^{\infty}$.

Proof. We show the outline of the proof. Note that $H_{\mathrm{RPF}}(p) = L_{\mathrm{RPF}}(p)$. It can be seen that there exists a constant $C > 0$ such that for arbitrary $\Psi \in \mathcal{F}_{\mathrm{rad}}^{\infty}$,

$$\|\sqrt{\bar{K}(p)}\Psi\| \leq C(\||p - \hat{P}_{\mathrm{f}}|\Psi\| + \|H_{\mathrm{rad}}^{1/2}\Psi\| + \|\Psi\|).$$

Then we can derive that

$$\|L_{\mathrm{RPF}}(p)\Psi\| \leq C(\||p - \hat{P}_{\mathrm{f}}|\Psi\| + \|H_{\mathrm{rad}}\Psi\| + \|\Psi\|) \tag{3.9.43}$$

for $\Psi \in \mathcal{F}_{\mathrm{rad}}^{\infty}$. From (3.9.43), we can see that $\mathcal{F}_{\mathrm{rad}}^{\infty}$ is a core of $L_{\mathrm{RPF}}(p)$ for $m > 0$ by a limiting argument. In the case of $m = 0$,

$$|p - \hat{P}_{\mathrm{f}} - \hat{A}(0)| = \sqrt{\bar{K}(p)} + (|p - \hat{P}_{\mathrm{f}} - \hat{A}(0)| - \sqrt{\bar{K}(p)})$$

and $|p - \hat{P}_{\mathrm{f}} - \hat{A}(0)| - \sqrt{\bar{K}(p)}$ is bounded. Then $|p - \hat{P}_{\mathrm{f}} - \hat{A}(0)| + H_{\mathrm{rad}}$ is essentially self-adjoint on $\mathcal{F}_{\mathrm{rad}}^{\infty}$. The key inequality to show the self-adjointness of $L_{\mathrm{RPF}}(p)$ is

$$\||p - \hat{P}_{\mathrm{f}}|\Psi\|^2 + \|H_{\mathrm{rad}}\Psi\|^2 \leq C\|(L_{\mathrm{RPF}}(p) + 1)\Psi\|^2 \tag{3.9.44}$$

with some $C > 0$ for $\Psi \in \mathcal{F}_{\mathrm{rad}}^{\infty}$. This is shown by using the inequality

$$\|[\bar{K}(p)^{1/4}, H_{\mathrm{rad}}]\Psi\| \leq c\|(H_{\mathrm{rad}} + 1)^{1/2}\Psi\|. \tag{3.9.45}$$

Inequality (3.9.45) will be proven in Lemma 3.158 below. Thus we can see that $L_{\mathrm{RPF}}(p)\lceil_{D(|\hat{P}_{\mathrm{f}}|)\cap D(H_{\mathrm{rad}})}$ is closed by (3.9.44). Then $L_{\mathrm{RPF}}(p)$ is self-adjoint on the domain $D(|\hat{P}_{\mathrm{f}}|) \cap D(H_{\mathrm{rad}})$ for $m \geq 0$. □

Lemma 3.158. *Equation* (3.9.45) *holds true.*

Proof. The idea of the proof of (3.9.45) is similar to those of (3.9.23) and (3.9.25). We have

$$(\Phi, [H_{\mathrm{rad}}, \bar{K}(p)^{1/4}]\Psi) = (H_{\mathrm{rad}}\Phi, \bar{K}(p)^{1/4}\Psi) - (\Phi, \bar{K}(p)^{1/4}H_{\mathrm{rad}}\Psi)$$

The functional integral representation yields

$$(\Phi, [H_{\text{rad}}, \bar{K}(p)^{1/4}]\Psi) = \int_0^\infty \frac{e^{-m^2\lambda}}{\lambda^{5/4}} \mathbb{E}^0[e^{ip\cdot B_{2\lambda}}(\Phi(B_0), [H_{\text{rad}}, e^{-i\hat{A}(L_{2\lambda})}]e^{-iP_{\text{f}}\cdot B_{2\lambda}}\Psi(B_{2\lambda}))]d\lambda.$$

Since $[H_{\text{rad}}, e^{-i\hat{A}(L_{2\lambda})}] = e^{-i\hat{A}(L_{2\lambda})}(\hat{\Pi}(L_{2\lambda}) + \xi)$, we can derive the desired results. □

3.9.11 Nonrelativistic limit of translation invariant relativistic Pauli–Fierz Hamiltonian

Similarly as for the relativistic Pauli–Fierz Hamiltonian, we can establish the nonrelativistic limit of a translation invariant relativistic Pauli–Fierz Hamiltonian. We introduce the speed of light c in $H_{\text{RPF}}(p)$ by

$$H_{\text{RPF}}(p) = \sqrt{c^2(p - \hat{P}_{\text{f}} - \hat{A}(0))^2 + m^2c^4} - mc^2 + H_{\text{rad}}.$$

The subordinator $(T_t(c))_{t\geq 0}$ with parameter c in $(\mathcal{T}, \mathcal{B}_{\mathcal{T}}, \nu)$ satisfies (3.9.29). The translation invariant Pauli–Fierz Hamiltonian with particle mass $m > 0$ is defined by

$$H_{\text{PF}}(p) = \frac{1}{2m}(p - \hat{P}_{\text{f}} - \hat{A}(0))^2 + H_{\text{rad}}.$$

Using $(T_t(c))_{t\geq 0}$, we can see that

$$(\Psi, e^{-tH_{\text{RPF}}(p)}\Phi) = \mathbb{E}^{0,0}\left[e^{ip\cdot B_{T_t(c)}}\left(\text{J}_0\Psi, e^{-ie\hat{A}_{\text{E}}(K_t^{\text{rel}}(c))}\text{J}_t e^{-i\hat{P}_{\text{f}}\cdot B_{T_t(c)}}\Phi\right)\right],$$

where $K_t^{\text{rel}}(c)$ is defined by K_t^{rel} with T_t replaced by $T_t(c)$, and the functional integral representation of $e^{-tH_{\text{PF}}(p)}$ with mass m is given by

$$(\Psi, e^{-tH_{\text{PF}}(p)}\Phi) = \mathbb{E}^0\left[e^{ip\cdot B_{t/m}}(\text{J}_0\Psi, e^{-ie\hat{A}_{\text{E}}(K_t)}\text{J}_t e^{-i\hat{P}_{\text{f}}\cdot B_{t/m}}\Phi)\right]dx,$$

where $K_t = \bigoplus_{\mu=1}^3 \int_0^{t/m} \text{j}_s\tilde{\varphi}(\cdot - B_s) \circ dB_s^\mu$.

Theorem 3.159 (Nonrelativistic limit). *For every $t \geq 0$ and $p \in \mathbb{R}^3$, it follows that*

$$\operatorname*{s-lim}_{c\to\infty} e^{-tH_{\text{RPF}}(p)} = e^{-tH_{\text{PF}}(p)}.$$

Proof. Let $\Psi, \Phi \in \mathcal{F}_{\text{rad}}$. By Lemma 3.136, it is shown that $\lim_{c\to\infty} K_t^{\text{rel}}(c) = K_t$ strongly in $L^2(\mathscr{X} \times \mathcal{T}) \otimes (\oplus^3 L^2(\mathbb{R}^4))$. It follows that

$$\begin{aligned}\lim_{c\to\infty}(\Psi, e^{-tH_{\text{RPF}}(p)}\Phi) &= \lim_{c\to\infty}\mathbb{E}^{0,0}\left[e^{ip\cdot B_{T_t(c)}}\left(\text{J}_0\Psi, e^{-ie\hat{A}_{\text{E}}(K_t^{\text{rel}}(c))}\text{J}_t e^{-i\hat{P}_{\text{f}}B_{T_t(c)}}\Phi\right)\right] \\ &= \mathbb{E}^0\left[e^{ip\cdot B_{t/m}}(\text{J}_0\Psi, e^{-ie\hat{A}_{\text{E}}(K_t)}\text{J}_t e^{-i\hat{P}_{\text{f}}\cdot B_{t/m}}\Phi)\right]dx = (\Psi, e^{-tH_{\text{PF}}(p)}\Phi).\end{aligned}$$

Then the proof is complete. □

3.10 Weak coupling limit

3.10.1 Weak coupling limit of the dipole approximated Pauli–Fierz Hamiltonian

The weak coupling limit problem for the Pauli–Fierz Hamiltonian is particularly well suited to the use of functional integral techniques. We now introduce this method. To study the dimensional dependence, we extended the spatial dimension to d. The boson Fock space is given by

$$\mathcal{F}_{\mathrm{rad}} = \mathcal{F}_{\mathrm{rad}}(L^2(\mathbb{R}^d \times \{1, \dots, d-1\})).$$

We generalize the creation and annihilation operators to the d-dimensional setting, denoted by $a^\sharp(f,j)$ for $f \in L^2(\mathbb{R}^d)$ and $j = 1, \dots, d-1$, satisfying

$$[a(f,i), a^*(g,j)] = \delta_{ij}(\bar{f}, g), \quad [a(f,i), a(g,j)] = 0, \quad [a^*(f,i), a^*(g,j)] = 0.$$

As a consequence, the polarization vectors are given by $(e^1, \dots, e^{d-1})$, where

$$e^i(k) \cdot e^j(k) = \delta_{ij}, \quad e^i(k) \cdot k = 0, \quad 1 \le i,j \le d-1,$$

and the transversal delta function is given by

$$\delta^\perp_{\mu\nu}(k) = \sum_{j=1}^{d-1} e^j_\mu(k)\, e^j_\nu(k) = \delta_{\mu\nu} - \frac{k_\mu k_\nu}{|k|^2}.$$

Throughout Section 3.10, we assume that the coupling constant $e = 0$ for simplicity.

Let $\kappa > 0$ be a scaling parameter. The *scaled Pauli–Fierz Hamiltonian* is defined by

$$H_\kappa = \frac{1}{2}(-i\nabla - \kappa A)^2 + V + \kappa^2 H_{\mathrm{rad}}$$

on $\mathcal{H}_{\mathrm{PF}}$. We also define

$$H_\kappa(p) = \frac{1}{2}(p - P_{\mathrm{f}} - \kappa A(0))^2 + V + \kappa^2 H_{\mathrm{rad}}$$

on $\mathcal{F}_{\mathrm{rad}}$ for $p \in \mathbb{R}^d$. We refer to the limits of H_κ and $H_\kappa(p)$ as $\kappa \to \infty$ as the *weak coupling limit*. We introduce assumptions on charge distributions below.

Assumption 3.160. Distribution $\varphi \in \mathscr{S}'(\mathbb{R}^3)$ is such that $\sqrt{\omega}\hat{\varphi}, \hat{\varphi}/\sqrt{\omega}, \hat{\varphi}/\omega \in L^2(\mathbb{R}^3)$ and $\hat{\varphi}$ is rotation invariant.

Rotation invariance of $\hat{\varphi}$ is assumed only to simplify the computations. The results on the weak coupling limit can be obtained even without imposing this condition.

The dipole approximation physically corresponds to the neglect of recoil effects associated with both photons and the electron. Within this approximation, the Pauli–Fierz

Hamiltonian becomes exactly solvable, leading to a significant simplification of the weak coupling limit. Under the dipole approximation, the Pauli–Fierz Hamiltonian is obtained by replacing the vector potential $A(x)$ with $A(0)$, yielding

$$H_{\mathrm{dip}} = \frac{1}{2}(-i\nabla - A(0))^2 + V + H_{\mathrm{rad}}.$$

Similar to H_{PH}, it can be seen that H_{dip} is self-adjoint on $D(-\Delta) \cap D(H_{\mathrm{rad}})$ under Assumptions 3.4. We also define $H_{\mathrm{dip}}(p)$ for each $p \in \mathbb{R}^d$ by

$$H_{\mathrm{dip}}(p) = \frac{1}{2}(p - A(0))^2 + H_{\mathrm{rad}}. \tag{3.10.1}$$

Under (1) of Assumption 3.4, $H_{\mathrm{dip}}(p)$ is self-adjoint on $D(H_{\mathrm{rad}})$, and it follows that

$$H_{\mathrm{dip}} \cong \int_{\mathbb{R}^d}^{\oplus} H_{\mathrm{dip}}(p)dp. \tag{3.10.2}$$

Define

$$H_{\mathrm{dip},\kappa} = \frac{1}{2}(-i\nabla - \kappa A(0))^2 + V + \kappa^2 H_{\mathrm{rad}},$$
$$H_{\mathrm{dip},\kappa}(p) = \frac{1}{2}(p - \kappa A(0))^2 + \kappa^2 H_{\mathrm{rad}}.$$

We shall show the weak coupling limit of both $H_{\mathrm{dip},\kappa}$ and $H_{\mathrm{dip},\kappa}(p)$ in Section 3.10.1.

Suppose $\int_{\mathbb{R}^d} |\hat{\varphi}(k)|^2/\omega(k)^3 dk < \infty$. Then we define the unitary operator

$$\mathcal{U}_p = \exp\left(\frac{-i}{\kappa m_{\mathrm{eff}}} p \cdot \tilde{\Pi}\right), \quad p \in \mathbb{R}^d,$$

where

$$\tilde{\Pi}_\nu = \frac{i}{\sqrt{2}} \sum_{j=1}^{d-1} \left\{ a^*\left(e_\nu^r \frac{\hat{\varphi}}{\omega\sqrt{\omega}}, j\right) - a\left(e_\nu^r \frac{\tilde{\hat{\varphi}}}{\omega\sqrt{\omega}}, j\right)\right\}$$

and m_{eff} is the effective mass given by

$$m_{\mathrm{eff}} = 1 + \delta_m, \quad \delta_m = \frac{d-1}{d}\|\hat{\varphi}/\omega\|^2.$$

Lemma 3.161. *Suppose Assumption* 3.160 *holds and* $\int_{\mathbb{R}^d} |\hat{\varphi}(k)|^2/\omega(k)^3 dk < \infty$. *Then* $\mathcal{U}_p$ *maps* $D(H_{\mathrm{rad}})$ *onto itself and the following equality holds on* $D(H_{\mathrm{rad}})$:

$$\mathcal{U}_p^{-1} H_{\mathrm{dip},\kappa} \mathcal{U}_p = \frac{p^2}{2m_{\mathrm{eff}}} + \kappa^2\left(\frac{1}{2}A(0)^2 + H_{\mathrm{rad}}\right). \tag{3.10.3}$$

Proof. We see that

$$\mathcal{U}_p^{-1}\kappa A_\nu(0)\mathcal{U}_p = \kappa A_\nu(0) + \frac{\delta_m}{m_{\text{eff}}}p_\nu,$$

$$\mathcal{U}_p^{-1}\kappa^2 H_{\text{rad}}\mathcal{U}_p = \kappa^2 H_{\text{rad}} + \kappa\frac{1}{m_{\text{eff}}}p \cdot A(0) + \frac{1}{2}\frac{\delta_m}{m_{\text{eff}}^2}p^2.$$

By a direct computation, we can see (3.10.3) on $\mathcal{F}_{\text{rad,fin}}$. Since $\mathcal{F}_{\text{rad,fin}}$ is a core of the self-adjoint operator $\frac{p^2}{2m_{\text{eff}}}+\kappa^2\left(\frac{1}{2}A(0)^2 + H_{\text{rad}}\right)$, we can show that (3.10.3) holds true on $D(H_{\text{rad}})$ by a limiting argument. □

The lower bound of the spectrum of $H_{\text{dip},\kappa}(p)$ has been fully characterized as

$$\inf \operatorname{Spec}(H_{\text{dip},\kappa}(p)) = \frac{p^2}{2m_{\text{eff}}} + \kappa^2\mathcal{E}, \tag{3.10.4}$$

where $\mathcal{E}$ is explicitly given by Lemma 3.169 below. In particular,

$$\inf \operatorname{Spec}\left(\frac{1}{2}A(0)^2 + H_{\text{rad}}\right) = \mathcal{E}. \tag{3.10.5}$$

The existence and absence of a ground state for $H_{\text{dip}}(0)$ plays a fundamental role in the analysis of the weak coupling limit. The interplay between the infrared regular condition $\int_{\mathbb{R}^d} |\hat{\varphi}(k)|^2/\omega(k)^3 dk < \infty$ and the existence or absence of ground states for $H_{\text{dip}}(p)$ as a function of p has been thoroughly elucidated.

Proposition 3.162. *Let Assumption* 3.160 *hold.*
(1) *Suppose* $\int_{\mathbb{R}^d} |\hat{\varphi}(k)|^2/\omega(k)^3 dk < \infty$. *Then* $H_{\text{dip},\kappa}(p)$ *has a ground state and it is unique for all* $p \in \mathbb{R}^d$ *and* $\kappa > 0$.
(2) *Suppose* $\int_{\mathbb{R}^d} |\hat{\varphi}(k)|^2/\omega(k)^3 dk = \infty$. *Then* $H_{\text{dip},\kappa}(p)$ *has no ground state for* $p \neq 0$ *and* $\kappa > 0$, *but it has a ground state for* $p = 0$ *and* $\kappa > 0$, *and it is unique.*

A noteworthy implication of Proposition 3.162 is given below.

Corollary 3.163. *Let Assumption* 3.160 *hold.*
(1) *Suppose* $\int_{\mathbb{R}^d} |\hat{\varphi}(k)|^2/\omega(k)^3 dk < \infty$. *Then for all* $p \in \mathbb{R}^d$ *and* $\kappa > 0$,

$$H_{\text{dip},\kappa}(p) \cong \frac{p^2}{2m_{\text{eff}}} + \kappa^2 H_{\text{dip}}(0).$$

(2) *Suppose* $\int_{\mathbb{R}^d} |\hat{\varphi}(k)|^2/\omega(k)^3 dk = \infty$. *Then for* $p \neq 0$ *and* $\kappa > 0$,

$$H_{\text{dip},\kappa}(p) \ncong \frac{p^2}{2m_{\text{eff}}} + \kappa^2 H_{\text{dip}}(0).$$

Proof. Suppose $\int_{\mathbb{R}^d} |\hat{\varphi}(k)|^2/\omega(k)^3 dk < \infty$. Claim (1) follows from Lemma 3.161. Suppose $\int_{\mathbb{R}^d} |\hat{\varphi}(k)|^2/\omega(k)^3 dk = \infty$. By (2) of Proposition 3.162, $H_{\mathrm{dip},\kappa}(0)$ has a ground state, but $H_{\mathrm{dip},\kappa}(p)$ has no ground state. Then (2) follows. □

The ground state of $H_{\mathrm{dip}}(0)$ (if it exists) is denoted by Φ_{g}. Let P_{g} be the projection onto L.H.$\{\Phi_{\mathrm{g}}\}$.

Lemma 3.164 (Weak coupling limit of $H_{\mathrm{dip},\kappa}(p)$). *Suppose that Assumption* 3.160 *holds and* $\int_{\mathbb{R}^d} \frac{|\hat{\varphi}(k)|^2}{\omega(k)^3} dk < \infty$. *Then*

$$\lim_{\kappa\to\infty} e^{-T(H_{\mathrm{dip},\kappa}(p)-\kappa^2\mathcal{E})} = e^{-T\frac{p^2}{2m_{\mathrm{eff}}}} P_{\mathrm{g}}.$$

Proof. Lemma 3.161 leads to

$$e^{-T(H_{\mathrm{dip},\kappa}(p)-\kappa^2\mathcal{E})} = e^{-T\frac{p^2}{2m_{\mathrm{eff}}}} \mathcal{U}_p e^{-\kappa^2 T(H_{\mathrm{dip}}(0)-\mathcal{E})} \mathcal{U}_p^{-1}.$$

Letting $\kappa \to \infty$ yields $\mathcal{U}_p \to 1$ and $e^{-\kappa^2 T(H_{\mathrm{dip}}(0)-\mathcal{E})} \to P_{\mathrm{g}}$ strongly. Then the proof is complete. □

Theorem 3.165 (Weak coupling limit of $H_{\mathrm{dip},\kappa}$). *Suppose that Assumption* 3.160 *holds and* $\int_{\mathbb{R}^d} \frac{|\hat{\varphi}(k)|^2}{\omega(k)^3} dk < \infty$. *Let V be bounded. Then*

$$\lim_{\kappa\to\infty} e^{-T(H_{\mathrm{dip},\kappa}-\kappa^2\mathcal{E})} = e^{-TH_{\mathrm{eff}}} \otimes P_{\mathrm{g}},$$

where

$$H_{\mathrm{eff}} = -\frac{1}{2m_{\mathrm{eff}}}\Delta + V.$$

Proof. Let $V = 0$. Then the theorem follows from

$$(F, e^{-TH_{\mathrm{dip},\kappa}} G) = \int_{\mathbb{R}^d} (F(p), e^{-TH_{\mathrm{dip},\kappa}(p)} G(p))_{\mathcal{F}} dp.$$

Let V be bounded. By the Neumann expansion of the resolvent $(H_{\mathrm{dip},\kappa} - \kappa^2\mathcal{E} + z)^{-1}$, for sufficiently large $|z|$ we can show that

$$\lim_{\kappa\to\infty} (H_{\mathrm{dip},\kappa} - \kappa^2\mathcal{E} + z)^{-1} = (H_{\mathrm{eff}} + z)^{-1} \otimes P_{\mathrm{g}}$$

in the strong sense. It leads to the theorem via the formula

$$e^{-TK} f = \frac{(k-1)!}{T^{k-1}} \frac{1}{2\pi i} \lim_{n\to\infty} \int_{a-in}^{a+in} e^{sT}(K+s)^{-k} f ds,$$

which holds for a self-adjoint operator $K > 0$, for all $f \in D(K^2)$, $a > 0$, and $k \in \mathbb{N}$. □

3.10.2 Ground state energy of the Pauli–Fierz Hamiltonian

In Section 3.10.2, we derive the ground state energy of the dipole-approximated Pauli–Fierz Hamiltonian by means of the Feynman–Kac formula. We also show that it coincides with the ground state energy of the full Pauli–Fierz Hamiltonian at $p = 0$. These results provide a good example illustrating the usefulness of the Feynman–Kac formula. Set

$$K_{\mathrm{dip},t} = \bigoplus_{\mu=1}^{d} \int_0^t \mathrm{j}_{\kappa^2 s} \tilde{\varphi} dB_s^{\mu}$$

and

$$I_\kappa = \mathrm{J}_0^* e^{-i\kappa \hat{A}_{\mathrm{E}}(K_{\mathrm{dip},T})} \mathrm{J}_{\kappa^2 T}.$$

Note that the integrand $\mathrm{j}_{\kappa^2 s}\tilde{\varphi}$ is nonrandom. The operator I_κ is bounded and uniformly bounded with respect to κ as $\|I_\kappa\| \le 1$. We derive the infimum of the spectrum $\sigma(H_{\mathrm{dip},\kappa}(p))$ by means of the Feynman–Kac formula. To this end, we introduce the Hilbert space $\mathfrak{h}$ by

$$\mathfrak{h} = L^2([0,T]).$$

Let $\tilde{\mathfrak{h}} = \bigoplus^d \mathfrak{h}$ and the scalar product on $\tilde{\mathfrak{h}}$ be denoted by $\langle \cdot, \cdot \rangle$. Let $Z(s) = (Z_1(s), \ldots, Z_d(s))$ for $0 \le s \le T$ be the Gaussian random process on $(\mathcal{Q}_E, \mu_E)$ such that

$$\int_{\mathcal{Q}_E} Z_\mu(s) d\mu_E = 0,$$

$$\int_{\mathcal{Q}_E} Z_\mu(s) Z_\nu(t) d\mu_E = \delta_{\mu\nu} \frac{d-1}{d} \int_{\mathbb{R}^d} \frac{|\hat{\varphi}(k)|^2}{2\omega(k)} e^{-\kappa^2 |s-t| \omega(k)} dk.$$

In fact, we have the identity $Z_\mu(s) = \hat{A}_{\mathrm{E}}\left(\bigoplus_{\nu=1}^d \delta_{\mu\nu} \mathrm{j}_{\kappa^2 s}\tilde{\varphi}\right)$. Let Ω be the Fock vacuum in $\mathcal{F}_{\mathrm{rad}}$.

Lemma 3.166. *Suppose Assumption* 3.160 *holds. Then*

$$(\Omega, e^{-TH_{\mathrm{dip},\kappa}(p)} \Omega) = (1, e^{-\frac{1}{2}\langle p - \kappa Z, p - \kappa Z \rangle} 1).$$

Proof. We have

$$(\Omega, e^{-TH_{\mathrm{dip},\kappa}(p)} \Omega) = \mathbb{E}\left[(1, e^{-i\int_0^T (\kappa \hat{A}_{\mathrm{E}(K_{\mathrm{dip},T})} - p) dB_s} 1)\right] = (1, e^{-\frac{1}{2}\int_0^T (p - \kappa Z(s))^2 ds} 1) = (1, e^{-\frac{1}{2}\langle p - \kappa Z, p - \kappa Z \rangle} 1).$$

Thus the lemma is proven. □

We define the covariance operator $\tilde{C}_T : \tilde{\mathfrak{h}} \to \tilde{\mathfrak{h}}$ by

$$(1, \langle g, Z\rangle\langle Z, f\rangle 1)_{L^2(\mathscr{Q}_E)} = \langle g, \tilde{C}_T f\rangle, \quad f, g \in \tilde{\mathfrak{h}}.$$

Then

$$(\tilde{C}_T f)(t) = \int_0^T \tilde{\rho}(s-t) f(s) ds,$$

where $\tilde{\rho}(t)$ is the diagonal matrix $\tilde{\rho}(t) = \mathrm{diag}(\rho(t), \ldots, \rho(t))$ and

$$\rho(t) = \frac{d-1}{d} \int_{\mathbb{R}^d} \frac{|\hat{\varphi}(k)|^2}{2\omega(k)} e^{-|t|\kappa^2\omega(k)} dk.$$

We also define $C_T : \mathfrak{h} \to \mathfrak{h}$ by

$$(C_T u)(t) = \int_0^T \rho(s-t) u(s) ds.$$

The operator C_T is referred to as the truncated Wiener–Hopf operator on $\mathfrak{h}$. Then it follows that $\tilde{C}_T = \bigoplus^d C_T$. The lemma below can be verified immediately.

Lemma 3.167. *Suppose Assumption* 3.160 *holds. Then it follows that* ρ *is a nonnegative and even function and:*
(1) $\int_{\mathbb{R}} (1 + |t|) \rho(t) dt < \infty$;
(2) *the Fourier transform of* ρ *is given by*

$$\hat{\rho}(t) = \frac{d-1}{d} \int_{\mathbb{R}^d} \frac{\kappa^2 |\hat{\varphi}(k)|^2}{\kappa^4 \omega(k)^2 + t^2} dk$$

and $\int_{\mathbb{R}} \hat{\rho}(t) dt < \infty$.
(3) $\tilde{C}_T$ *is a nonnegative self-adjoint operator and belongs to the trace class on* $\mathfrak{h}$.

Proof. Claim (1) follows from the assumption $\int_{\mathbb{R}^d} \frac{|\hat{\varphi}(k)|^2}{\omega(k)^2} dk < \infty$, while (2) is straightforward. Since $\int_0^T \int_0^T \bar{f}(s) f(t) e^{-|s-t|\omega(k)} dsdt > 0$ for any $f \in \mathfrak{h}$, the operator C_T is a nonnegative self-adjoint operator. Since $\int_0^T \int_0^T |\rho(t-s)|^2 dsdt < \infty$, the operator C_T is compact. Moreover, since C_T is positive, it is of trace class. Then (3) is proven. □

Since $\hat{\rho}$ is a nonnegative function, $\log(1 + \kappa^2 \hat{\rho})$ is a real-valued function.

Proposition 3.168 (Ahiezer–Kac formula). *Suppose Assumption* 3.160 *holds. Then*

$$\lim_{T\to\infty}\frac{1}{T}\log\det(1+\kappa^2 C_T)=\frac{1}{2\pi}\int_{-\infty}^{\infty}\log(1+\kappa^2\hat{\rho}(t))dt=\frac{\kappa^2}{\pi}\int_{-\infty}^{\infty}G(t)dt, \tag{3.10.6}$$

where

$$G(t)=\frac{\frac{d-1}{d}\left\|\frac{t\hat{\varphi}}{t^2+\omega^2}\right\|^2}{1+\frac{d-1}{d}\left\|\frac{\hat{\varphi}}{\sqrt{t^2+\omega^2}}\right\|^2}.$$

Proof. Since $\log(1+\kappa^2\hat{\rho}(t))\le\kappa^2\hat{\rho}(t)$ and $\hat{\rho}\in L^1(\mathbb{R})$, we can see $\int_{-\infty}^{\infty}\log(1+\kappa^2\hat{\rho}(t))dt<\infty$. From Lemma 3.167 (3), it follows that $\log\det(1+\kappa^2C_T)=\mathrm{Tr}\log(1+\kappa^2C_T)\le\kappa^2\mathrm{Tr}C_T<\infty$. Claims (1) and (2) of Lemma 3.167 imply the Ahiezer–Kac formula

$$\lim_{T\to\infty}\frac{\det(1+\kappa^2C_T)}{\exp\left(\frac{T}{2\pi}\int_{-\infty}^{\infty}\log(1+\kappa^2\hat{\rho}(t))dt\right)}=\exp\left(\int_0^{\infty}s\left|\frac{1}{2\pi}\int_{-\infty}^{\infty}\log(1+\kappa^2\hat{\rho}(t))e^{its}dt\right|^2ds\right). \tag{3.10.7}$$

Then the first equality of (3.10.6) holds true. Note that $\left\|\frac{t\hat{\varphi}}{t^2+\omega^2}\right\|\le\frac{1}{2}\left\|\frac{\hat{\varphi}}{\omega}\right\|$ and $\left\|\frac{\hat{\varphi}}{\sqrt{t^2+\omega^2}}\right\|\le\left\|\frac{\hat{\varphi}}{\omega}\right\|$ for any $t\in\mathbb{R}$. Since G is continuous, $G(0)=0$, and

$$G(t)\le\frac{\frac{d-1}{d}\|\hat{\varphi}\|^2}{t^2+\frac{d-1}{d}\|\hat{\varphi}/\sqrt{1+\omega^2}\|^2}$$

for $t\ge1$, we conclude that the most right-hand side of (3.10.6) is finite. Note that $\hat{\rho}(t)=\frac{d-1}{d}\kappa^2\left\|\frac{\hat{\varphi}}{\sqrt{\kappa^4\omega^2+t^2}}\right\|^2$. Since $\left\|\frac{\hat{\varphi}}{\omega}\right\|<\infty$, $\hat{\rho}(t)$ is differentiable at $t\ne0$ and $\hat{\rho}(t)'=-2\frac{d-1}{d}t\kappa^2\left\|\frac{\hat{\varphi}}{\kappa^4\omega^2+t^2}\right\|^2$. Changing the variable, we have

$$\int_{-\infty}^{\infty}t\frac{d}{dt}\log(1+\kappa^2\hat{\rho}(t))dt=\int_{-\infty}^{\infty}\frac{\kappa^2t\hat{\rho}(t)'}{1+\kappa^2\hat{\rho}(t)}dt=-2\kappa^2\int_{-\infty}^{\infty}G(t)dt<\infty, \tag{3.10.8}$$

which implies

$$\int_{-\infty}^{\infty}\log(1+\kappa^2\hat{\rho}(t))dt=[t\log(1+\kappa^2\hat{\rho}(t))]_{-\infty}^{\infty}+2\kappa^2\int_{-\infty}^{\infty}G(t)dt.$$

We observe that $[t\log(1+\kappa^2\hat{\rho}(t))]_{-\infty}^{\infty}=0$ and the second equality of (3.10.6) holds true. □

We can compute $\inf\mathrm{Spec}(H_{\mathrm{dip},\kappa}(p))$ explicitly by using Proposition 3.168.

Theorem 3.169 (Ground state energy of $H_{\mathrm{dip},\kappa}(p)$). *Suppose Assumption* 3.160 *holds. Then*

$$\inf \mathrm{Spec}(H_{\mathrm{dip},\kappa}(p)) = \frac{p^2}{2m_{\mathrm{eff}}} + \frac{d}{2\pi}\int_{-\infty}^{\infty} \log(1+\kappa^2\hat{\rho}(t))dt$$

$$= \frac{p^2}{2m_{\mathrm{eff}}} + \kappa^2\frac{d}{2\pi}\int_{-\infty}^{\infty} \frac{\frac{d-1}{d}\left\|\frac{t\hat{\varphi}}{t^2+\omega^2}\right\|^2}{1+\frac{d-1}{d}\left\|\frac{\hat{\varphi}}{\sqrt{t^2+\omega^2}}\right\|^2}dt.$$

In particular,

$$\mathcal{E} = \frac{d}{2\pi}\int_{-\infty}^{\infty} \frac{\frac{d-1}{d}\left\|\frac{t\hat{\varphi}}{t^2+\omega^2}\right\|^2}{1+\frac{d-1}{d}\left\|\frac{\hat{\varphi}}{\sqrt{t^2+\omega^2}}\right\|^2}dt.$$

Proof. Since $e^{-TH_{\mathrm{dip},\kappa}(p)}$ is positivity improving, we have

$$\inf \mathrm{Spec}(H_{\mathrm{dip},\kappa}(p)) = -\lim_{T\to\infty}\frac{1}{T}\log(\Omega, e^{-TH_{\mathrm{dip},\kappa}(p)}\Omega).$$

By Lemma 3.166, we have

$$(\Omega, e^{-TH_{\mathrm{dip},\kappa}(p)}\Omega) = \det(1+\kappa^2\tilde{C}_T)^{-1/2}e^{-\frac{1}{2}\langle p,(1+\kappa^2\tilde{C}_T)^{-1}p\rangle}.$$

Hence,

$$\inf \mathrm{Spec}(H_{\mathrm{dip},\kappa}(p)) = \lim_{T\to\infty}\frac{1}{2T}\langle p,(1+\kappa^2\tilde{C}_T)^{-1}p\rangle - \lim_{T\to\infty}\frac{1}{T}\log\det(1+\kappa^2\tilde{C}_T)^{-1/2}$$

$$= \frac{p^2}{2}\lim_{T\to\infty}\frac{1}{T}\langle 1,(1+\kappa^2 C_T)^{-1}1\rangle_{\mathfrak{h}} + \frac{d}{2}\lim_{T\to\infty}\frac{1}{T}\log\det(1+\kappa^2 C_T).$$

We compute the above two limits separately. First, we can compute

$$\lim_{T\to\infty}\frac{d}{2T}\log\det(1+\kappa^2 C_T) = \kappa^2\frac{d}{2\pi}\int_{-\infty}^{\infty} G(t)dt$$

by Proposition 3.168. We have

$$\langle 1,(1+\kappa^2 C_T)^{-1}1\rangle_{\mathfrak{h}} = \int_0^T u_T(t)dt,$$

where $u_T(t) = (1+\kappa^2 C_T)^{-1}1(t)$. This u_T satisfies

$$u_T(t) + \kappa^2\int_0^T \rho(t-s)u_T(s)ds = 1$$

for all $t \in [0,T]$. Integrating both sides above with respect to dt over $[0,T]$, we have

$$\int_0^T u_T(t)dt + \kappa^2 \int_0^T dt \int_0^T \rho(t-s)u_T(s)ds = T.$$

It can be seen that

$$\kappa^2 \int_0^T dt \int_0^T \rho(t-s)u_T(s)ds = \frac{d-1}{d}\int_0^T u_T(s)ds \cdot \int_{\mathbb{R}^d} \frac{|\hat{\varphi}(k)|^2}{\omega(k)^2}dk$$
$$- \frac{d-1}{d}\int_{\mathbb{R}^d} \frac{|\hat{\varphi}(k)|^2}{2\omega(k)^2}\int_0^T u_T(s)\left(e^{-s\kappa^2\omega(k)} + e^{-(T-s)\kappa^2\omega(k)}\right)dsdk$$

and then

$$\frac{1}{T}\int_0^T u_T(t)dt\left(1 + \frac{d-1}{d}\left\|\frac{\hat{\varphi}}{\omega}\right\|^2\right)$$
$$= 1 + \frac{1}{T}\frac{d-1}{d}\int_{\mathbb{R}^d} \frac{|\hat{\varphi}(k)|^2}{2\omega(k)^2}\int_0^T u_T(s)\left(e^{-s\kappa^2\omega(k)} + e^{-(T-s)\kappa^2\omega(k)}\right)dsdk.$$

Let $\epsilon > 0$ be arbitrary. We have

$$\frac{1}{T}\frac{d-1}{d}\int_{\mathbb{R}^d} \frac{|\hat{\varphi}(k)|^2}{2\omega(k)^2}\int_0^T u_T(s)\left(e^{-s\kappa^2\omega(k)} + e^{-(T-s)\kappa^2\omega(k)}\right)dsdk = I_0 + I_1,$$

where

$$I_0 = \frac{1}{T}\frac{d-1}{d}\int_{|k|<\epsilon} \frac{|\hat{\varphi}(k)|^2}{2\omega(k)^2}\int_0^T u_T(s)\left(e^{-s\kappa^2\omega(k)} + e^{-(T-s)\kappa^2\omega(k)}\right)dsdk,$$
$$I_1 = \frac{1}{T}\frac{d-1}{d}\int_{|k|\geq\epsilon} \frac{|\hat{\varphi}(k)|^2}{2\omega(k)^2}\int_0^T u_T(s)\left(e^{-s\kappa^2\omega(k)} + e^{-(T-s)\kappa^2\omega(k)}\right)dsdk.$$

Let $\delta_\epsilon = \frac{d-1}{d}\int_{|k|<\epsilon}\frac{|\hat{\varphi}(k)|^2}{2\omega(k)^2}$. We have the bound

$$-2\delta_\epsilon \frac{1}{T}\int_0^T u_T(s)ds \leq I_0 \leq 2\delta_\epsilon \frac{1}{T}\int_0^T u_T(s)ds.$$

By the Schwarz inequality, we obtain

$$I_1 \le \frac{1}{T}\frac{d-1}{d}\int\limits_{|k|\ge\epsilon} \frac{|\hat{\varphi}(k)|^2}{2\omega(k)^2}\|u_T\|_{\mathfrak{h}}\|e^{-s\kappa^2\omega(k)} + e^{-(T-s)\kappa^2\omega(k)}\|_{\mathfrak{h}} dk$$
$$\le \frac{\sqrt{2(1+1/e)}}{\kappa^2\sqrt{\epsilon}\sqrt{T}}\frac{d-1}{d}\int\limits_{|k|\ge\epsilon}\frac{|\hat{\varphi}(k)|^2}{2\omega(k)^2}dk \to 0$$

as $T \to \infty$. Here we used that $\|u_T\|_{\mathfrak{h}} \le \|1\|_{\mathfrak{h}} = \sqrt{T}$ and

$$\|e^{-s\kappa^2\omega(k)} + e^{-(T-s)\kappa^2\omega(k)}\|_{\mathfrak{h}} \le \sqrt{\frac{1-e^{-2T\kappa^2\omega(k)} + 2T\kappa^2\omega(k)e^{-T\kappa^2\omega(k)}}{\kappa^2\omega(k)}} \le \frac{\sqrt{2(1+1/e)}}{\kappa\sqrt{\epsilon}}$$

for $|k| \ge \epsilon$. Hence

$$\frac{1}{1+\frac{d-1}{d}\left\|\frac{\hat{\varphi}}{\omega}\right\|^2 + 2\delta_\epsilon} \le \lim_{T\to\infty}\frac{1}{T}\langle 1, (1+\kappa^2 C_T)^{-1}1\rangle_{\mathfrak{h}} \le \frac{1}{1+\frac{d-1}{d}\left\|\frac{\hat{\varphi}}{\omega}\right\|^2 - 2\delta_\epsilon}.$$

Since δ_ϵ is arbitrary, we conclude that

$$\lim_{T\to\infty}\frac{1}{T}\langle 1, (1+\kappa^2 C_T)^{-1}1\rangle_{\mathfrak{h}} = \frac{1}{1+\frac{d-1}{d}\left\|\frac{\hat{\varphi}}{\omega}\right\|^2}.$$

Then the proof is complete. □

We determine the lower bound of the spectrum

$$E_\kappa = \inf \operatorname{Spec}(H_{\mathrm{dip},\kappa})$$

in the case $V = 0$.

Corollary 3.170 (Ground state energy of $H_{\mathrm{dip},\kappa}$). *Suppose Assumption* 3.160 *holds. Then we obtain that* $E_\kappa = \kappa^2\mathcal{E}$.

Proof. This follows from Corollary 3.73. □

4 Spin-boson model by path measures

4.1 Definitions

As discussed in Section 3.8, for the Pauli–Fierz Hamiltonian with spin $\frac{1}{2}$, a functional integral representation can be derived by using suitable Lévy processes. In this chapter we derive further functional integral representations for semigroups generated by the so-called spin-boson and Rabi models, which will be less technically demanding than for the Pauli–Fierz case.

The spin-boson model describes a two-state quantum system linearly coupled to a scalar quantum field. To study this model in a stochastic representation, we describe the spin states by the set $\mathbb{Z}_2 = \{-1, +1\}$ as in the Pauli–Fierz model with spin, and derive a Poisson-driven random process with càdlàg path space $\mathfrak{X}_D = D(\mathbb{R})$ indexed by $\mathbb{R}$. This will give the spin-process. On the path space $\mathfrak{X}_D$, we are then able to construct a probability measure μ_∞ associated with the unique ground state Ψ_g of the spin-boson Hamiltonian. Using this probability measure, we can represent ground state expectations for operators of interest. The strategy is similar as for the Nelson and Pauli–Fierz models discussed in the previous sections.

Consider the boson Fock space $\mathcal{F}_{SB} = \mathcal{F}_b(L^2(\mathbb{R}^d))$ over $L^2(\mathbb{R}^d)$, $d \geq 1$, and the Hilbert space

$$\mathcal{H}_{SB} = \mathbb{C}^2 \otimes \mathcal{F}_{SB}.$$

We begin by defining the spin-boson Hamiltonian as a self-adjoint operator acting in $\mathcal{H}_{SB}$. The free field Hamiltonian for the spin-boson model is given by

$$H_f = d\Gamma(\omega),$$

where $\omega(k) = |k|$. The operator

$$\phi_\varphi = \frac{1}{\sqrt{2}}\left(a^*(\hat{\varphi}/\sqrt{\omega}) + a(\tilde{\hat{\varphi}}/\sqrt{\omega})\right)$$

acting on the boson Fock space is the scalar field operator, where the form factor φ is required to satisfy the following conditions.

Assumption 4.1. The form factor $\varphi \in \mathscr{S}'(\mathbb{R}^d)$ satisfies $\overline{\hat{\varphi}(k)} = \hat{\varphi}(-k)$ and $\hat{\varphi}/\omega, \hat{\varphi}/\sqrt{\omega} \in L^2(\mathbb{R}^d)$.

Let $\sigma_x, \sigma_y, \sigma_z$ be the 2×2 Pauli matrices given by

$$\sigma_x = \begin{pmatrix} 0 & 1 \\ 1 & 0 \end{pmatrix}, \quad \sigma_y = \begin{pmatrix} 0 & -i \\ i & 0 \end{pmatrix}, \quad \sigma_z = \begin{pmatrix} 1 & 0 \\ 0 & -1 \end{pmatrix}.$$

https://doi.org/10.1515/9783112231494-004

Definition 4.2 (Spin-boson Hamiltonian). The *spin-boson Hamiltonian* is defined by the linear operator

$$H_{\mathrm{SB}} = \kappa\sigma_z \otimes 1 + 1 \otimes H_{\mathrm{f}} + \sigma_x \otimes \phi_\varphi$$

on $\mathcal{H}_{\mathrm{SB}}$, where $\kappa \geq 0$ is a parameter.

We note that the physical meaning of the parameter is that 2κ is the gap between the ground state energy and the energy of the first excited state of the two-level atom.

Proposition 4.3 (Self-adjointness). *Let Assumption 4.1 hold. Then H_{SB} is a self-adjoint operator on $D(H_{\mathrm{f}})$ and bounded from below. Furthermore, H_{SB} is essentially self-adjoint on any core of H_{f}.*

Proof. By the assumption, $\kappa\sigma_z \otimes 1 + \sigma_x \otimes \phi_\varphi$ is symmetric and infinitesimally small with respect to $1 \otimes H_{\mathrm{f}}$, hence the proposition follows by the Kato–Rellich theorem. □

Recall that the rotation group in $\mathbb{R}^3$ has an adjoint representation on SU(2). Let $n \in \mathbb{R}^3$ be a unit vector and $\theta \in [0, 2\pi)$. Thus $e^{(i/2)\theta n\cdot\sigma}$ satisfies

$$e^{(i/2)\theta n\cdot\sigma}\sigma_\mu e^{-(i/2)\theta n\cdot\sigma} = (\mathscr{R}_\theta\sigma)_\mu, \tag{4.1.1}$$

where $\mathscr{R}_\theta$ denotes the 3×3 matrix representing the rotation around n with angle θ, and $\sigma = (\sigma_x, \sigma_y, \sigma_z)$. Note that $n\cdot\sigma = n_1\sigma_x + n_2\sigma_y + \sigma_3\sigma_z$ for $n = (n_1, n_2, \sigma_3) \in \mathbb{R}^3$. In particular, for $n = (0, 1, 0)$ and $\theta = \pi/2$, we have

$$e^{(i/2)\theta n\cdot\sigma}\sigma_x e^{-(i/2)\theta n\cdot\sigma} = \sigma_z \quad \text{and} \quad e^{(i/2)\theta n\cdot\sigma}\sigma_z e^{-(i/2)\theta n\cdot\sigma} = -\sigma_x. \tag{4.1.2}$$

Let

$$U = \exp\left(i\frac{\pi}{4}\sigma_y\right) \otimes 1 = \frac{1}{\sqrt{2}}\begin{pmatrix} 1 & 1 \\ -1 & 1 \end{pmatrix} \otimes 1 \tag{4.1.3}$$

be a unitary operator on $\mathcal{H}_{\mathrm{SB}}$. By (4.1.2), the operator H_{SB} transforms as

$$\tilde{H}_{\mathrm{SB}} = UH_{\mathrm{SB}}U^* = -\kappa\sigma_x \otimes 1 + 1 \otimes H_{\mathrm{f}} + \sigma_z \otimes \phi_\varphi.$$

Then $\tilde{H}_{\mathrm{SB}}$ is realized as

$$\tilde{H}_{\mathrm{SB}} = \begin{pmatrix} H_{\mathrm{f}} + \phi_\varphi & -\kappa \\ -\kappa & H_{\mathrm{f}} - \phi_\varphi \end{pmatrix}.$$

Each diagonal part is the so-called van Hove Hamiltonian which will be discussed later. In particular, $\kappa = 0$ makes H_{SB} diagonal, obtained as a direct sum of van Hove Hamiltonians.

It is also a known fact that H_{SB} has a parity symmetry. Let

$$P = \sigma_z \otimes (-1)^N,$$

where N denotes the number operator in $\mathcal{F}_{\mathrm{SB}}$. From $\mathrm{Spec}(\sigma_z) = \{-1, 1\}$ and $\mathrm{Spec}(N) = \mathbb{N} \cup \{0\}$, it follows that $\mathrm{Spec}(P) = \{-1, 1\}$. In fact, P is a projection on $\mathcal{H}_{\mathrm{SB}}$. We identify $\mathcal{H}_{\mathrm{SB}}$ as

$$\mathcal{H}_{\mathrm{SB}} = \mathcal{F}_{\mathrm{SB}}^{+} \oplus \mathcal{F}_{\mathrm{SB}}^{-},$$

where $\mathcal{F}_{\mathrm{SB}}^{+}$ and $\mathcal{F}_{\mathrm{SB}}^{-}$ are identical copies of $\mathcal{F}_{\mathrm{SB}}$. Then each Pauli matrix $\theta_X = \left(\begin{smallmatrix} a & b \\ c & d \end{smallmatrix}\right)$ acts as

$$\theta_X \begin{pmatrix} \Psi(+) \\ \Psi(-) \end{pmatrix} = \begin{pmatrix} a\Psi(+) + b\Psi(-) \\ c\Psi(+) + d\Psi(-) \end{pmatrix}$$

for $\left(\begin{smallmatrix} \Psi(+) \\ \Psi(-) \end{smallmatrix}\right) \in \mathcal{F}_{\mathrm{SB}}^{+} \oplus \mathcal{F}_{\mathrm{SB}}^{-}$. Furthermore, $\mathcal{F}_{\mathrm{SB}}$ can be decomposed as

$$\mathcal{F}_{\mathrm{SB}} = \mathcal{F}_{\mathrm{SB}}^{\mathrm{e}} \oplus \mathcal{F}_{\mathrm{SB}}^{\mathrm{o}},$$

where $\mathcal{F}_{\mathrm{SB}}^{\mathrm{e}}$ and $\mathcal{F}_{\mathrm{SB}}^{\mathrm{o}}$ denote respectively the subspaces of $\mathcal{F}_{\mathrm{SB}}$ consisting of even and odd numbers of bosons, i. e., $\mathcal{F}_{\mathrm{SB}}^{\mathrm{e}} = \bigoplus_{m=0}^{\infty} \mathcal{F}_{\mathrm{SB}}^{(2m)}$ and $\mathcal{F}_{\mathrm{SB}}^{\mathrm{o}} = \bigoplus_{m=0}^{\infty} \mathcal{F}_{\mathrm{SB}}^{(2m+1)}$. The projections from $\mathcal{F}_{\mathrm{SB}}$ to $\mathcal{F}_{\mathrm{SB}}^{\mathrm{e}}$ and $\mathcal{F}_{\mathrm{SB}}^{\mathrm{o}}$ will be denoted by P_{e} and P_{o}, respectively. We will make use of the subspaces

$$\mathcal{H}_{\mathrm{SB}}^{+} = P_{\mathrm{e}}\mathcal{F}_{\mathrm{SB}}^{+} \oplus P_{\mathrm{o}}\mathcal{F}_{\mathrm{SB}}^{-} \quad \text{and} \quad \mathcal{H}_{\mathrm{SB}}^{-} = P_{\mathrm{o}}\mathcal{F}_{\mathrm{SB}}^{+} \oplus P_{\mathrm{e}}\mathcal{F}_{\mathrm{SB}}^{-}$$

of $\mathcal{F}_{\mathrm{SB}}^{+} \oplus \mathcal{F}_{\mathrm{SB}}^{-}$.

Lemma 4.4. *The following properties hold:*

(1) *$\mathcal{H}_{\mathrm{SB}}$ can be identified with $\mathcal{H}_{\mathrm{SB}}^{+} \oplus \mathcal{H}_{\mathrm{SB}}^{-}$, $\mathcal{H}_{\mathrm{SB}} \cong \mathcal{H}_{\mathrm{SB}}^{+} \oplus \mathcal{H}_{\mathrm{SB}}^{-}$, by the correspondence*

$$\mathcal{F}_{\mathrm{SB}}^{+} \oplus \mathcal{F}_{\mathrm{SB}}^{-} \ni \begin{pmatrix} \Psi(+) \\ \Psi(-) \end{pmatrix} \mapsto \begin{pmatrix} \Psi_{\mathrm{e}}(+) \\ \Psi_{\mathrm{o}}(-) \end{pmatrix} \oplus \begin{pmatrix} \Psi_{\mathrm{o}}(+) \\ \Psi_{\mathrm{e}}(-) \end{pmatrix} \in \mathcal{H}_{\mathrm{SB}}^{+} \oplus \mathcal{H}_{\mathrm{SB}}^{-},$$

where $\Psi_{\mathrm{e}}(\pm) = P_{\mathrm{e}}\Psi(\pm)$ and $\Psi_{\mathrm{o}}(\pm) = P_{\mathrm{o}}\Psi(\pm)$.

(2) *The commutator $[H_{\mathrm{SB}}, P] = 0$.*

(3) *$\mathcal{H}_{\mathrm{SB}}^{\pm}$ is the eigenspace associated with the eigenvalue ± 1 of P, i. e., $P\mathcal{H}_{\mathrm{SB}} = \mathcal{H}_{\mathrm{SB}}^{+}$ and $(1-P)\mathcal{H}_{\mathrm{SB}} = \mathcal{H}_{\mathrm{SB}}^{-}$.*

(4) *H_{SB} can be decomposed as $H_{\mathrm{SB}} = H_{\mathrm{SB}}\lceil_{\mathcal{H}_{\mathrm{SB}}^{+}} \oplus H_{\mathrm{SB}}\lceil_{\mathcal{H}_{\mathrm{SB}}^{-}}$.*

Proof. Establishing (1) is straightforward. We see that

$$\begin{aligned} [H_{\mathrm{SB}}, P] = [\sigma_x \otimes \phi_\varphi, P] &= \sigma_x\sigma_z \otimes \phi_\varphi(-1)^N - \sigma_z\sigma_x \otimes (-1)^N\phi_\varphi \\ &= -i\sigma_y \otimes (\phi_\varphi(-1)^N + (-1)^N\phi_\varphi). \end{aligned}$$

Let $\Psi \in \mathcal{F}_{\mathrm{SB}}^{+}$. Since $\phi_\varphi\Psi \in \mathcal{F}_{\mathrm{SB}}^{-}$, we have

$$(\phi_\varphi(-1)^N + (-1)^N\phi_\varphi)\Psi = (\phi_\varphi - \phi_\varphi)\Psi = 0.$$

Similarly, let $\Psi \in \mathcal{F}_{\mathrm{SB}}^{-}$. Since $\phi_\varphi\Psi \in \mathcal{F}_{\mathrm{SB}}^{+}$, we have

$$(\phi_\varphi(-1)^N + (-1)^N\phi_\varphi)\Psi = (-\phi_\varphi + \phi_\varphi)\Psi = 0.$$

Then $[H_{\mathrm{SB}}, P] = 0$ follows. Let $\Psi = \begin{pmatrix} \Psi_{\mathrm{e}}(+) \\ \Psi_{\mathrm{o}}(-) \end{pmatrix} \in \mathcal{H}_{\mathrm{SB}}^{+}$. Then $P\Psi = \Psi$ follows by a direct calculation, thus Ψ is an eigenvector of P with eigenvalue +1. Similarly, it follows that $\mathcal{H}_{\mathrm{SB}}^{-}$ is the eigenspace associated with eigenvalue −1 of P, and (3) also follows; (4) is obtained by a combination of (1), (2), and (3). □

Statement (2) of Lemma 4.4 is called the *parity symmetry*. We write

$$H_{\mathrm{SB}}^{\pm} = H_{\mathrm{SB}}\lceil_{\mathcal{H}_{\mathrm{SB}}^{\pm}}.$$

Thus H_{SB}^{+} is of parity plus and H_{SB}^{-} of parity minus. We also can derive the explicit forms of reduced Hamiltonian $H_{\mathrm{SB}}^{\pm}$.

Corollary 4.5. *It follows that*

$$H_{\mathrm{SB}}^{+} = \kappa \otimes (-1)^N + 1 \otimes H_{\mathrm{f}} + \sigma_x \otimes \phi_\varphi,$$
$$H_{\mathrm{SB}}^{-} = -\kappa \otimes (-1)^N + 1 \otimes H_{\mathrm{f}} + \sigma_x \otimes \phi_\varphi.$$

Proof. Since $H_{\mathrm{SB}}^{+} = PH_{\mathrm{SB}}P$ and $H_{\mathrm{SB}}^{-} = (1-P)H_{\mathrm{SB}}(1-P)$, the corollary is immediate. □

In Theorem 4.12 and Corollary 4.13 below, we will see that for $\kappa \neq 0$ the boson Hamiltonian H_{SB} has a unique ground state of parity minus. This results in an interesting fact. If $\inf \mathrm{Spec}(H_{\mathrm{SB}}^{+}) = \inf \mathrm{Spec}(H_{\mathrm{SB}}^{-})$, then H_{SB}^{+} has no ground state, which can be seen directly in a simple case. Let $\varphi = 0$ and $\kappa \neq 0$. Then $H_{\mathrm{SB}}^{\pm}$ can be regarded as operators in $\mathcal{F}_{\mathrm{SB}}$ with the expressions

$$H_{\mathrm{SB}}^{+} = \kappa(-1)^N + H_{\mathrm{f}} \quad \text{and} \quad H_{\mathrm{SB}}^{-} = -\kappa(-1)^N + H_{\mathrm{f}}.$$

It can be seen that $\inf \mathrm{Spec}(H_{\mathrm{SB}}^{-}) = -\kappa$ and $H_{\mathrm{SB}}^{-}\Omega_{\mathrm{b}} = -\kappa\Omega_{\mathrm{b}}$. On the other hand, H_{SB}^{+} can be reduced to the n-particle subspace $\mathcal{F}_{\mathrm{SB}}^{(n)}$ for each $n \geq 0$, and for $\Phi = \{\Phi^{(n)}\}_{n=0}^{\infty}$ we have

$$H_{\mathrm{SB}}^{+}\Phi = \kappa\Phi^{(0)} + \bigoplus_{n=1}^{\infty}\left((-1)^n\kappa + \sum_{j=1}^{n}\omega(k_j)\right)\Phi^{(n)}.$$

Since $\omega(k) = \sqrt{|k|^2 + \nu^2}$, we have

$$\inf \mathrm{Spec}(H_{\mathrm{SB}}^{+}) = \inf\left\{\kappa, (-1)^n\kappa + n\nu, n \in \mathbb{N}\right\} = -\kappa + \nu.$$

In particular, when $\nu = 0$, we have

$$\inf \operatorname{Spec}(H_{\mathrm{SB}}^{-}) = \inf \operatorname{Spec}(H_{\mathrm{SB}}^{+}) = -\kappa.$$

Thus H_{SB}^{+} has no ground state, since the multiplication operator

$$H_{\mathrm{SB}}^{+} \lceil_{\mathcal{F}_{\mathrm{SB}}^{(1)}} = -\kappa + |k|$$

in $\mathcal{F}_{\mathrm{SB}}^{(1)} = L^2(\mathbb{R}^d)$ has no ground state. These statements are also valid for $\varphi \neq 0$ under some further conditions.

Finally, for later use we show some related *spin-flip* properties.

Lemma 4.6. *For every* $\Psi, \Phi \in \mathcal{H}_{\mathrm{SB}}^{\pm}$, *we have*
(1) $(\Psi, \sigma_x \Phi) = 0$ *and* $(U^* \Psi, \sigma_z U^* \Phi) = 0$;
(2) $(\Psi, \Phi(\hat{f})\Phi) = 0$ *and* $(U^*\Psi, \Phi(\hat{f}) U^* \Phi) = 0$, *where* $\Phi(\hat{f}) = \frac{1}{\sqrt{2}}(a^*(\hat{f}) + a(\tilde{\hat{f}}))$.

Proof. (1) We see that σ_x is a spin-flip transform, i. e., $\sigma_x \mathcal{H}_{\mathrm{SB}}^{\pm} \subset \mathcal{H}_{\mathrm{SB}}^{\mp}$, which gives the first statement. The second statement follows by observing that $U \sigma_z U^* = \sigma_x$. To obtain (2), we show again that $\Phi(\hat{f}) \left(\mathcal{H}_{\mathrm{SB}}^{\pm} \cap D(\Phi(\hat{f}))\right) \subset \mathcal{H}_{\mathrm{SB}}^{\mp}$. The second part follows from $U \Phi(\hat{f}) U^* = \Phi(\hat{f})$. □

4.2 Functional integral representation for the spin-boson Hamiltonian

4.2.1 Preliminaries

As we have seen in Section 3.8, for the functional integral representation of $e^{-tH_{\mathrm{PF}}^{\mathbb{Z}_2}}$ generated by the Pauli–Fierz Hamiltonian with spin it was convenient to use a spin variable $\sigma \in \mathbb{Z}_2$ in a related space. This will prove to be useful also for $e^{-tH_{\mathrm{SB}}}$.

Let $(\phi(f), f \in L^2_{\mathrm{real}}(\mathbb{R}^d))$ be a family of Gaussian random variables indexed by $L^2_{\mathrm{real}}(\mathbb{R}^d)$ on a probability space (Q, Σ, μ). As before, we identify $\mathcal{F}_{\mathrm{SB}}$ with $L^2(\mathcal{Q})$. Under this equivalence, ϕ_φ can be identified with a Gaussian random variable $\phi(\tilde{\varphi})$ indexed by $\tilde{\varphi} = (\hat{\varphi}/\sqrt{\omega})\check{}$. Thus the spin-boson Hamiltonian $\tilde{H}_{\mathrm{SB}}$ can be equivalently defined on the Hilbert space $\mathbb{C}^2 \otimes L^2(\mathcal{Q})$ by

$$-\kappa\sigma_x \otimes 1 + 1 \otimes \hat{H}_{\mathrm{f}} + \sigma_z \otimes \phi(\tilde{\varphi}), \tag{4.2.1}$$

which we will also denote by $\tilde{H}_{\mathrm{SB}}$, and $\mathbb{C}^2 \otimes L^2(\mathcal{Q}) \cong L^2(\mathcal{Q}) \oplus L^2(\mathcal{Q})$ also by $\mathcal{H}_{\mathrm{SB}}$. Recall that $\hat{H}_{\mathrm{f}} = d\Gamma(\hat{\omega})$. For $\Psi = \begin{pmatrix} \Psi(+) \\ \Psi(-) \end{pmatrix} \in \mathcal{H}_{\mathrm{SB}}$, we have

$$\tilde{H}_{\mathrm{SB}}\Psi = \begin{pmatrix} (\hat{H}_{\mathrm{f}} + \phi(\tilde{\varphi}))\Psi(+) - \kappa\Psi(-) \\ (\hat{H}_{\mathrm{f}} - \phi(\tilde{\varphi}))\Psi(-) - \kappa\Psi(+) \end{pmatrix}.$$

Thus we can transform $\tilde{H}_{\mathrm{SB}}$ on $\mathcal{H}_{\mathrm{SB}}$ to the operator $H_{\mathrm{SB}}^{\mathbb{Z}_2}$ on

$$L^2(\mathbb{Z}_2; L^2(\mathcal{Q})) = \left\{ f : \mathbb{Z}_2 \to L^2(\mathcal{Q}) \,\middle|\, \sum_{\sigma \in \mathbb{Z}_2} \|f(\sigma)\|^2_{L^2(\mathcal{Q})} < \infty \right\}$$

by

$$(H_{\mathrm{SB}}^{\mathbb{Z}_2}\Psi)(\sigma) = \left(\hat{H}_{\mathrm{f}} + \sigma\phi(\tilde{\varphi})\right)\Psi(\sigma) - \kappa\Psi(-\sigma), \quad \sigma \in \mathbb{Z}_2. \tag{4.2.2}$$

In what follows, we write H_{f} for $\hat{H}_{\mathrm{f}}$ and also identify $\mathcal{H}_{\mathrm{SB}}$ with $L^2(\mathbb{Z}_2; L^2(\mathcal{Q}))$ through

$$\mathcal{H}_{\mathrm{SB}} \ni \begin{pmatrix} \Psi(+) \\ \Psi(-) \end{pmatrix} \mapsto \Psi(\sigma) = \begin{cases} \Psi(+), & \sigma = +1 \\ \Psi(-), & \sigma = -1 \end{cases} \in L^2(\mathbb{Z}_2; L^2(\mathcal{Q})),$$

and consider $H_{\mathrm{SB}}^{\mathbb{Z}_2}$ instead of $\tilde{H}_{\mathrm{SB}}$.

4.2.2 Spin process

Let $(N_t)_{t\in\mathbb{R}}$ be a two-sided Poisson process with unit intensity on a probability space $(\mathcal{S}, \mathcal{B}_{\mathcal{S}}, P)$, defined by

$$N_t = \begin{cases} \bar{N}_t, & t \geq 0, \\ \bar{\bar{N}}_{(-t)-}, & t < 0, \end{cases}$$

where $(\bar{N}_t)_{t\geq 0}$ is a càdlàg version of the Poisson process, and $(\bar{\bar{N}}_t)_{t\geq 0}$ a càglàd version such that $P(\bar{N}_0 = 0) = 1 = P(\bar{\bar{N}}_0 = 0)$. Here $\mathbb{R} \ni t \mapsto N_t \in \mathbb{Z}$ is a càdlàg function. We define the *spin process* by

$$(\sigma_t)_{t\in\mathbb{R}} = (\sigma(-1)^{N_t})_{t\in\mathbb{R}}, \quad \sigma \in \mathbb{Z}_2.$$

We write

$$\mathbb{E}^\sigma[f(\sigma_t)] = \mathbb{E}[f(\sigma(-1)^{N_t})], \quad \sigma \in \mathbb{Z}_2.$$

Since the spin process is indexed by the real line, we summarize some basic properties of $(N_t)_{t\in\mathbb{R}}$ below:

(1) (Independence) The random variables N_t and N_s are independent for all $s \leq 0 \leq t$, $s \neq t$.

(2) (Markov property) The random processes $(N_t)_{t\geq 0}$ and $(N_t)_{t\leq 0}$ have the Markov property with respect to the natural filtrations $\mathcal{N}_t^+ = \sigma(N_s, 0 \leq s \leq t)$ and $\mathcal{N}_t^- = \sigma(N_s, t \leq s \leq 0)$, respectively, i. e., for $s, t \geq 0$ and for a bounded function f,

$$\mathbb{E}_P^0\left[f(N_{t+s})|\mathcal{N}_s^+\right] = \mathbb{E}_P^{N_s}\left[N_t\right] \quad \text{and} \quad \mathbb{E}_P^0\left[f(N_{-t-s})|\mathcal{N}_{-s}^-\right] = \mathbb{E}_P^{N_{-s}}\left[N_{-t}\right].$$

(3) (Reflection symmetry) The random variables N_t and N_{-t} are identically distributed for all $t \in \mathbb{R}$, i. e.,

$$\mathbb{E}_P^0[f(N_{-t})] = \mathbb{E}_P^0[f(N_t)] = \sum_{n=0}^{\infty} f(n)\frac{t^n}{n!}e^{-t}, \quad t \geq 0.$$

Properties (1)–(3) can be shown directly by the definition of $(N_t)_{t\in\mathbb{R}}$.

Now we discuss some technical facts on the spin process. Consider $L^2(\mathbb{Z}_2)$, which can be identified with $\mathbb{C}^2$ under the map $L^2(\mathbb{Z}_2) \ni f \mapsto \binom{f(+1)}{f(-1)} \in \mathbb{C}^2$. Define the C_0-semigroup $P_t : \mathbb{C}^2 \to \mathbb{C}^2, t \geq 0$, under this identification by

$$(P_t f)(\sigma) = \mathbb{E}_P^\sigma[f(\sigma_t)], \quad \sigma \in \mathbb{Z}_2.$$

The generator of P_t is $\sigma_x - 1$ and thus

$$P_t = e^{t(\sigma_x - 1)}, \quad t \geq 0.$$

Write $S = -\sigma_x + 1$. We have for $0 \leq t_1 \leq \cdots \leq t_n$ and $f_j \in \mathbb{C}^2, j = 1, \ldots, n$,

$$\sum_{\sigma\in\mathbb{Z}_2} \mathbb{E}_P^\sigma\Big[\prod_{j=1}^n f_j(\sigma_{t_j})\Big] = (1, e^{-t_1 S} f_1 e^{-(t_2-t_1)S} f_2 \cdots e^{-(t_n-t_{n-1})S} f_n)_{L^2(\mathbb{Z}_2)}.$$

Proposition 4.7 (Shift invariance). *Let $f_j \in L^2(\mathbb{Z}_2), j = 1, \ldots, n+m$, $m, n \in \mathbb{N}$, and consider the arbitrary time-points $-\infty < t_1 \leq \cdots \leq t_n \leq 0 \leq t_{n+1} \leq \cdots \leq t_{n+m} < \infty$. Then*

$$\sum_{\sigma\in\mathbb{Z}_2} \mathbb{E}_P^\sigma\Big[\prod_{j=1}^{n+m} f_j(\sigma_{t_j})\Big] = \sum_{\sigma\in\mathbb{Z}_2} \mathbb{E}_P^\sigma\Big[\prod_{j=1}^{n+m} f_j(\sigma_{t_j+s})\Big];$$

for $-t_{n+m} \leq s \leq -t_1$.

Proof. Let $-t_1 \leq \cdots \leq -t_n \leq 0 \leq t_{n+1} \leq \cdots \leq t_{n+m}$. We show that

$$\sum_{\sigma\in\mathbb{Z}_2} \mathbb{E}_P^\sigma\Big[\prod_{i=1}^n f_i(\sigma_{-t_i}) \prod_{j=n+1}^{n+m} f_j(\sigma_{t_j})\Big] = \sum_{\sigma\in\mathbb{Z}_2} \mathbb{E}_P^\sigma\Big[\prod_{i=1}^n f_i(\sigma_{t_1-t_i}) \prod_{j=n+1}^{n+m} f_j(\sigma_{t_1+t_j})\Big], \tag{4.2.3}$$

from which the proposition follows. First, we compute the left-hand side of (4.2.3). By the independence of N_{-t} and N_s for $-t \leq 0 \leq s$ and the reflection symmetry $N_{-t} \stackrel{\mathrm{d}}{=} N_t$, we have

$$\begin{aligned}\sum_{\sigma\in\mathbb{Z}_2} \mathbb{E}_P^\sigma\Big[\prod_{i=1}^n f_i(\sigma_{-t_i}) \prod_{j=n+1}^{n+m} f_j(\sigma_{t_j})\Big] &= \sum_{\sigma\in\mathbb{Z}_2} \mathbb{E}_P^\sigma\Big[\prod_{i=1}^n f_i(\sigma_{-t_i})\Big]\mathbb{E}_P^\sigma\Big[\prod_{j=n+1}^{n+m} f_j(\sigma_{t_j})\Big] \\ &= \sum_{\sigma\in\mathbb{Z}_2} \mathbb{E}_P^\sigma\Big[\prod_{i=1}^n f_i(\sigma_{t_i})\Big]\mathbb{E}_P^\sigma\Big[\prod_{j=n+1}^{n+m} f_j(\sigma_{t_j})\Big]\end{aligned}$$

$$= (e^{-t_n S} f_n e^{-(t_{n-1}-t_n)S} \cdots e^{-(t_1-t_2)S} f_1, e^{-t_{n+1}S} f_{n+1} e^{-(t_{n+2}-t_{n+1})S} \cdots e^{-(t_{n+m}-t_{n+m-1})S} f_{n+m}). \quad (4.2.4)$$

Hence we can compute

$$\text{RHS } (4.2.4) = \left(1, f_1 e^{-(t_1-t_2)S} \cdots f_n e^{-(t_n+t_{n+1})S} f_{n+1} e^{-(t_{n+2}-t_{n+1})S} \cdots e^{-(t_{n+m}-t_{n+m-1})S} f_{n+m}\right). \quad (4.2.5)$$

On the other hand, we have

$$\sum_{\sigma\in\mathbb{Z}_2} \mathbb{E}_P^\sigma\Big[\prod_{i=1}^n f_i(\sigma_{t_1-t_i}) \prod_{i=n+1}^{n+m} f_i(\sigma_{t_1+t_j})\Big] = (4.2.5).$$

□

Let $\kappa > 0$ and consider the random process $(\sigma_{\kappa t})_{t\in\mathbb{R}}$. Then $(P_t f)(\sigma) = \mathbb{E}_P^\sigma[f(\sigma_{\kappa t})]$ also defines the C_0-semigroup on $\mathbb{C}^2$ with generator $\kappa\sigma_x$. Hence

$$P_t = e^{t\kappa(\sigma_x - 1)}.$$

We use this random process to construct a functional integral representation of the spin-boson Hamiltonian, and will also refer to $(\sigma_{\kappa t})_{t\in\mathbb{R}}$ as the spin process.

4.2.3 Functional integral representation

We can construct a functional integral representation of $e^{-tH_{\text{SB}}^{\mathbb{Z}_2}}$ by a modification of the similar construction for $e^{-tH_{\text{PF}}^{\mathbb{Z}_2}}$, using that simplifications occur for the spin-boson Hamiltonian. We identify $L^2(\mathbb{Z}_2; L^2(\mathscr{Q}))$ with $L^2(\mathbb{Z}_2 \times \mathscr{Q})$ and denote $\mathcal{H}_{\text{SB}} = L^2(\mathbb{Z}_2 \times \mathscr{Q})$.

Theorem 4.8 (Functional integral representation for spin-boson Hamiltonian). *For every $\Phi, \Psi \in \mathcal{H}_{\text{SB}}$, the expression*

$$(\Phi, e^{-TH_{\text{SB}}^{\mathbb{Z}_2}}\Psi) = e^{\kappa T} \sum_{\sigma\in\mathbb{Z}_2} \mathbb{E}_P^\sigma\left[(\mathrm{J}_0\Phi(\sigma_0), e^{-\phi_{\mathrm{E}}(\int_0^T \sigma_{\kappa s}\mathrm{j}_s\tilde{\varphi}ds)}\mathrm{J}_T\Psi(\sigma_{\kappa T}))\right] \quad (4.2.6)$$

holds.

Proof. We see by (4.2.2) that

$$(H_{\text{SB}}^{\mathbb{Z}_2}\Psi)(\sigma) = (H_{\text{f}} + \sigma\phi(\tilde{\varphi}))\Psi(\sigma) - \kappa\Psi(-\sigma), \quad \sigma \in \mathbb{Z}_2.$$

Then by Lemma 3.96 with σ_t replaced by $\sigma_{\kappa t}$, we have

$$(\Phi, e^{-TH_{\text{SB}}^{\mathbb{Z}_2}}\Psi) = e^{\kappa T} \sum_{\sigma\in\mathbb{Z}_2} \mathbb{E}_P^\sigma\Big[(\mathrm{J}_0\Phi(\sigma_0), e^{-\int_0^T \sigma_{\kappa s}\phi_{\mathrm{E}}(\mathrm{j}_s\tilde{\varphi})ds + \int_0^{T+}\log 1 dN_s}\mathrm{J}_T\Psi(\sigma_{\kappa T}))\Big].$$

Since $\int_0^{T+}\log 1 dN_s = 0$, the theorem follows. □

Remark 4.9. By the shift invariance of $(\sigma_t)_{t\in\mathbb{R}}$, we get that

$$(\Phi, e^{-TH_{\mathrm{SB}}^{\mathbb{Z}_2}}\Psi) = e^{\kappa T}\sum_{\sigma\in\mathbb{Z}_2}\mathbb{E}_P^\sigma\Big[(\mathrm{J}_{-r}\Phi(\sigma_{-\kappa r}), e^{-\phi_{\mathrm{E}}\left(\int_{-r}^{T-r}\sigma_{\kappa s}\mathrm{j}_s\tilde{\varphi}ds\right)}\mathrm{J}_{T-r}\Psi(\sigma_{\kappa(T-r)}))\Big]$$

for $0 \le r \le T$. In particular,

$$(\Phi, e^{-2TH_{\mathrm{SB}}^{\mathbb{Z}_2}}\Psi) = e^{2\kappa T}\sum_{\sigma\in\mathbb{Z}_2}\mathbb{E}_P^\sigma\left[(\mathrm{J}_{-T}\Phi(\sigma_{-\kappa T}), e^{-\phi_{\mathrm{E}}\left(\int_{-T}^{T}\sigma_{\kappa s}\mathrm{j}_s\tilde{\varphi}ds\right)}\mathrm{J}_{T}\Psi(\sigma_{\kappa T}))\right].$$

This will be used to construct a Gibbs measure for the spin-boson model.

Remark 4.10. We can give an alternative functional integral representation of $e^{-TH_{\mathrm{SB}}^{\mathbb{Z}_2}}$ in terms of the Poisson process $(N_t)_{t\ge 0}$. This is

$$(\Phi, e^{-TH_{\mathrm{SB}}^{\mathbb{Z}_2}}\Psi) = \begin{cases} e^T\sum_{\sigma\in\mathbb{Z}_2}\mathbb{E}_P^\sigma[(\mathrm{J}_0\Phi(\sigma_0), e^{-\phi_{\mathrm{E}}(\int_0^T\sigma_s\mathrm{j}_s\tilde{\varphi}ds)}\kappa^{N_T}\mathrm{J}_T\Psi(\sigma_T))], & \kappa\neq 0,\\ e^T\sum_{\sigma\in\mathbb{Z}_2}(\mathrm{J}_0\Phi(\sigma), e^{-\phi_{\mathrm{E}}(\sigma\int_0^T\mathrm{j}_s\tilde{\varphi}ds)}\mathrm{J}_T\Psi(\sigma)), & \kappa = 0.\end{cases} \tag{4.2.7}$$

The case $\kappa = 0$ follows by Theorem 4.8. If $\kappa \neq 0$, we have

$$(\Phi, e^{-TH_{\mathrm{SB}}^{\mathbb{Z}_2}}\Psi) = e^T\sum_{\sigma\in\mathbb{Z}_2}\mathbb{E}_P^\sigma\Big[(\mathrm{J}_0\Phi(\sigma_0), e^{-\int_0^T\sigma_s\phi_{\mathrm{E}}(\mathrm{j}_s\tilde{\varphi})ds+\int_0^{T+}\log\kappa dN_s}\mathrm{J}_T\Psi(\sigma_T))\Big]$$

by Lemma 3.87. Since $\int_0^{T+}\log\kappa dN_s = N_T\log\kappa$, the result follows. We emphasize that (4.2.6) is shift invariant, but the representation (4.2.7) is not.

By the functional integral representation, we see that

$$e^{-TH_{\mathrm{SB}}^{\mathbb{Z}_2}}\Phi(\sigma) = e^{\kappa T}\mathbb{E}_P^\sigma\left[\mathrm{J}_0^* e^{-\phi_{\mathrm{E}}(\int_0^T\sigma_{\kappa r}\mathrm{j}_r\tilde{\varphi}dr)}\mathrm{J}_T\Phi(\sigma_{\kappa T})\right]. \tag{4.2.8}$$

Denote

$$\mathbb{1}_{\mathrm{SB}} = \mathbb{1}_{L^2(\mathbb{Z}_2)}\otimes\mathbb{1}_{L^2(\mathscr{Q})}$$

for simplicity. Using Theorem 4.8, we can compute the vacuum expectation of the semigroup $e^{-TH_{\mathrm{SB}}^{\mathbb{Z}_2}}$.

Corollary 4.11 (Pair interaction and pair potential). *For every $T \ge 0$, we have*

$$(\mathbb{1}_{\mathrm{SB}}, e^{-TH_{\mathrm{SB}}^{\mathbb{Z}_2}}\mathbb{1}_{\mathrm{SB}}) = e^{\kappa T}\sum_{\sigma\in\mathbb{Z}_2}\mathbb{E}_P^\sigma\left[e^{\frac{1}{2}\int_0^T ds\int_0^T W(\sigma_{\kappa s}\sigma_{\kappa t}, s-t)dt}\right],$$

where $\int_0^T ds\int_0^T W(\sigma_{\kappa s}\sigma_{\kappa t}, s-t)dt$ is the pair interaction associated with spin-boson Hamiltonian with the pair potential

$$W(\eta, t) = \frac{\eta}{2}\int_{\mathbb{R}^d} e^{-|t|\omega(k)}\frac{|\hat{\varphi}(k)|^2}{\omega(k)}dk.$$

Proof. By Theorem 4.8, it is direct to see that

$$(1_{\mathrm{SB}}, e^{-TH_{\mathrm{SB}}^{\mathbb{Z}_2}} 1_{\mathrm{SB}}) = e^{\kappa T} \sum_{\sigma \in \mathbb{Z}_2} \mathbb{E}_P^\sigma \left[(1, e^{-\phi_{\mathrm{E}}\left(\int_0^T \sigma_{\kappa s} \mathrm{j}_s \tilde{\varphi} ds\right)} 1)_{L^2(\mathscr{Q}_{\mathrm{E}})} \right]$$

$$= e^{\kappa T} \sum_{\sigma \in \mathbb{Z}_2} \mathbb{E}_P^\sigma \left[e^{\frac{1}{2} \int_0^T ds \int_0^T W(\sigma_{\kappa s} \sigma_{\kappa t}, s-t) dt} \right]. \qquad \square$$

4.3 Existence and uniqueness of ground state

Now we make use of the functional integral representation established in the previous section to show existence and uniqueness of the ground state of the spin-boson Hamiltonian. Denote

$$\mathrm{I}_{\mathrm{SB}} = \int_{\mathbb{R}^d} \frac{|\hat{\varphi}(k)|^2}{\omega(k)^3} dk.$$

Whenever $\kappa = 0$, the Hamiltonian $H_{\mathrm{SB}}^{\mathbb{Z}_2}$ is diagonal, and we have

$$H_{\mathrm{SB}}^{\mathbb{Z}_2} = \begin{pmatrix} H_{\mathrm{f}} + \phi_\varphi & 0 \\ 0 & H_{\mathrm{f}} - \phi_\varphi \end{pmatrix}.$$

The operator $H_{\mathrm{f}} + \phi_\varphi$ is a van Hove Hamiltonian and has a unique ground state if and only if $\mathrm{I}_{\mathrm{SB}} < \infty$, which implies that $H_{\mathrm{SB}}^{\mathbb{Z}_2}$ with $\kappa = 0$ has a two-fold degenerate ground state if and only if $\mathrm{I}_{\mathrm{SB}} < \infty$.

Next we consider the case $\kappa \neq 0$. Denote $E = \inf \sigma(H_{\mathrm{SB}}^{\mathbb{Z}_2})$ and write

$$\Psi_{\mathrm{g}}^T = e^{-T(H_{\mathrm{SB}}^{\mathbb{Z}_2} - E)} 1_{\mathrm{SB}}, \quad T \geq 0,$$

and

$$\gamma(T) = \frac{(1_{\mathrm{SB}}, \Psi_{\mathrm{g}}^T)^2}{\|\Psi_{\mathrm{g}}^T\|^2} = \frac{(1_{\mathrm{SB}}, e^{-TH_{\mathrm{SB}}^{\mathbb{Z}_2}} 1_{\mathrm{SB}})^2}{(1_{\mathrm{SB}}, e^{-2TH_{\mathrm{SB}}^{\mathbb{Z}_2}} 1_{\mathrm{SB}})}.$$

By Propositions 2.30–2.31, a ground state of $H_{\mathrm{SB}}^{\mathbb{Z}_2}$ exists if and only if $\lim_{T \to \infty} \gamma(T) > 0$. By Corollary 4.11, we have

$$\|\Psi_{\mathrm{g}}^T\|^2 = e^{2TE} e^{2\kappa T} \sum_{\sigma \in \mathbb{Z}_2} \mathbb{E}_P^\sigma \left[e^{\frac{1}{2} \int_{-T}^T ds \int_{-T}^T W(\sigma_{\kappa s} \sigma_{\kappa t}, s-t) dt} \right],$$

$$(1_{\mathrm{SB}}, \Psi_{\mathrm{g}}^T) = e^{TE} e^{\kappa T} \sum_{\sigma \in \mathbb{Z}_2} \mathbb{E}_P^\sigma \left[e^{\frac{1}{2} \int_0^T ds \int_0^T W(\sigma_{\kappa s} \sigma_{\kappa t}, s-t) dt} \right].$$

Note that

$$\left|\int_{-T}^{0} ds \int_{0}^{T} W(\sigma_{\kappa s}\sigma_{\kappa t}, s-t)dt\right| \le \frac{1}{2}\int_{\mathbb{R}^d} \frac{|\hat{\varphi}(k)|^2}{\omega(k)^3} dk, \tag{4.3.1}$$

uniformly in T and in the paths.

Theorem 4.12 (Existence of ground state). *If $I_{SB} < \infty$, then $H_{SB}^{\mathbb{Z}_2}$ has a ground state and it is unique.*

Proof. For short, we write $W = W(\kappa_s\kappa_t, s-t)$, and

$$\int_{-T}^{T} ds \int_{-T}^{T} Wdt = \int_{-T}^{0} ds \int_{-T}^{0} Wdt + \int_{0}^{T} ds \int_{0}^{T} Wdt + 2\int_{-T}^{0} ds \int_{0}^{T} Wdt.$$

Using (4.3.1), we obtain

$$\|\Psi_g^T\|^2 \le e^{2TE} e^{2\kappa T} \sum_{\sigma\in\mathbb{Z}_2} \mathbb{E}_P^\sigma \left[e^{\frac{1}{2}\left(\int_{-T}^0 ds \int_{-T}^0 Wdt + \int_0^T ds \int_0^T Wdt + I_{SB}\right)}\right].$$

By the independence of N_t and N_{-s} for $-s < 0 < t$ and reflection symmetry of the paths, we furthermore obtain that

$$\begin{aligned} \|\Psi_g^T\|^2 &\le e^{2TE} e^{2\kappa T} \sum_{\sigma\in\mathbb{Z}_2} \left(\mathbb{E}_P^\sigma \left[e^{\frac{1}{2}\int_0^T ds \int_0^T W(\sigma_{\kappa s}\sigma_{\kappa t}, s-t)dt}\right]\right)^2 e^{\frac{1}{2}I_{SB}} \\ &\le \left(e^{TE} e^{\kappa T} \sum_{\sigma\in\mathbb{Z}_2} \mathbb{E}_P^\sigma \left[e^{\frac{1}{2}\int_0^T ds \int_0^T W(\sigma_{\kappa s}\sigma_{\kappa t}, s-t)dt}\right]\right)^2 e^{\frac{1}{2}I_{SB}} = (1, \Psi_g^T)^2 e^{\frac{1}{2}I_{SB}}. \end{aligned}$$

Hence $\gamma(T) \ge e^{-\frac{1}{2}I_{SB}}$ and a ground state Ψ_g of $H_{SB}^{\mathbb{Z}_2}$ exists. □

Corollary 4.13 (Positivity improving). *Let $\kappa \ne 0$. Then $e^{-tH_{SB}^{\mathbb{Z}_2}}$, $t > 0$, is positivity improving on $L^2(\mathbb{Z}_2 \times Q)$, i. e., $(\Psi, e^{-tH_{SB}^{\mathbb{Z}_2}}\Phi) > 0$ for $\Psi, \Phi \ge 0$, with $\Psi \not\equiv 0 \not\equiv \Phi$. In particular, if $H_{SB}^{\mathbb{Z}_2}$ has a ground state for $\kappa \ne 0$, then it is unique.*

Proof. Positivity improving follows by the functional integral representation in Theorem 4.8. The uniqueness follows by the Perron–Frobenius theorem. □

As was seen above, $\mathcal{H}_{SB}$ can be decomposed as $\mathcal{H}_{SB} = \mathcal{H}_{SB}^+ \oplus \mathcal{H}_{SB}^-$, and $H_{SB}^{\mathbb{Z}_2}$ can be reduced by $\mathcal{H}_{SB}^\pm$.

Corollary 4.14 (Parity of ground state). *Let Ψ_g be the ground state of $H_{SB}^{\mathbb{Z}_2}$. Then $\Psi_g \in \mathcal{H}_{SB}^-$, i. e., the parity of the ground state of the spin-boson Hamiltonian is -1.*

Proof. Let U be the unitary operator as in (4.1.3). Notice that $\Psi_g = U^*\Psi_g$, and thus

$$\Psi_g = \operatorname*{s-lim}_{T\to\infty} \frac{U^* e^{-TH_{SB}^{\mathbb{Z}_2}} 1_{SB}}{\|U^* e^{-TH_{SB}^{\mathbb{Z}_2}} 1_{SB}\|} = \operatorname*{s-lim}_{T\to\infty} \frac{e^{-TH_{SB}^{\mathbb{Z}_2}} U^* 1_{SB}}{\|e^{-TH_{SB}^{\mathbb{Z}_2}} U^* 1_{SB}\|}.$$

The function $1_{\mathrm{SB}} \in L^2(\mathbb{Z}_2; L^2(\mathscr{Q}))$ corresponds to $\left(\begin{smallmatrix}\Omega_b\\ \Omega_b\end{smallmatrix}\right) \in \mathcal{F}_{\mathrm{SB}}^+ \oplus \mathcal{F}_{\mathrm{SB}}^-$ and

$$U^* 1_{\mathrm{SB}} = \frac{1}{2}\begin{pmatrix}1 & -1\\ 1 & 1\end{pmatrix}\begin{pmatrix}\Omega_b\\ \Omega_b\end{pmatrix} = \begin{pmatrix}0\\ \Omega_b\end{pmatrix} \in \mathcal{H}_{\mathrm{SB}}^-.$$

Hence by the parity symmetry of $H_{\mathrm{SB}}^{\mathbb{Z}_2}$, (2) of Lemma 4.4, we have

$$Pe^{-TH_{\mathrm{SB}}^{\mathbb{Z}_2}} U^* 1_{\mathrm{SB}} = e^{-TH_{\mathrm{SB}}^{\mathbb{Z}_2}} PU^* 1_{\mathrm{SB}} = -e^{-TH_{\mathrm{SB}}^{\mathbb{Z}_2}} U^* 1_{\mathrm{SB}},$$

and thus $e^{-TH_{\mathrm{SB}}^{\mathbb{Z}_2}} U^* 1_{\mathrm{SB}} \in \mathcal{H}_{\mathrm{SB}}^-$. This implies $\Psi_g \in \mathcal{H}_{\mathrm{SB}}^-$. □

4.4 Gibbs measure associated with the ground state

4.4.1 Local convergence and Gibbs measures

In this section, we use the coordinate process of the two-sided Poisson process $(N_t)_{t\in\mathbb{R}}$. Let $\mathfrak{X}_D = D(\mathbb{R})$ be the space of càdlàg paths on $\mathbb{R}$. Let ρ_D denote the Skorokhod topology on $\mathfrak{X}_D$. Then $(\mathfrak{X}_D, \rho_D)$ is a Polish space. Let $\mathcal{B}(\mathfrak{X}_D)$ denote the Borel σ-field, which coincides with the σ-field generated by cylinder sets. Thus

$$N_\cdot : (\mathcal{S}, \mathcal{B}_{\mathcal{S}}, P) \to (\mathfrak{X}_D, \mathcal{B}(\mathfrak{X}_D))$$

is an $\mathfrak{X}_D$-valued random variable. We denote its image measure on $(\mathfrak{X}_D, \mathcal{B}(\mathfrak{X}_D))$ by $\mathcal{P}$, i. e.,

$$\mathcal{P}(A) = P(N_\cdot^{-1}(A)), \quad A \in \mathcal{B}(\mathfrak{X}_D).$$

We denote the coordinate process by the same symbol $(N_t)_{t\in\mathbb{R}}$, i. e.,

$$N_t(\omega) = \omega(t), \quad \omega \in \mathfrak{X}_D.$$

Then $\mathcal{P}(N_0 = 0) = 1$. We set

$$\sigma_{\kappa t} = \sigma(-1)^{N_{\kappa t}}.$$

Hence Theorem 4.8 can be reformulated in terms of $(\sigma_{\kappa t})_{t\geq 0}$ as

$$(\Phi, e^{-tH_{\mathrm{SB}}^{\mathbb{Z}_2}}\Psi) = e^{\kappa t} \sum_{\sigma\in\mathbb{Z}_2} \mathbb{E}_{\mathcal{P}}^{\sigma}\left[(\mathrm{J}_0\Phi(\sigma_0), e^{-\phi_{\mathrm{E}}(\int_0^t \sigma_{\kappa s}\mathrm{j}_s\tilde{\varphi}ds)}\mathrm{J}_t\Psi(\sigma_{\kappa t}))\right]. \tag{4.4.1}$$

Here we write $\mathbb{E}_{\mathcal{P}}^{\sigma}[f(\sigma_t)] = \mathbb{E}_{\mathcal{P}}[f(\sigma(-1)^{N_t})]$. Then (4.2.8) can be cast in the form

$$e^{-tH_{\mathrm{SB}}^{\mathbb{Z}_2}}\Psi(\sigma) = e^{\kappa t}\mathbb{E}_{\mathcal{P}}^{\sigma}\left[\mathrm{J}_0^* e^{-\phi_{\mathrm{E}}(\int_0^t \sigma_{\kappa r}\mathrm{j}_r\tilde{\varphi}dr)}\mathrm{J}_t\Psi(\sigma_{\kappa t})\right]. \tag{4.4.2}$$

Let $\mathscr{G}_T = \sigma(N_t; t \in [-T,T])$ be the family of sub-σ-fields of $\mathcal{B}(\mathfrak{X}_D)$, and define $\mathscr{G} = \bigcup_{s\geq 0}\mathscr{G}_{[-s,s]}$. Also, define the probability measure

$$\mu_T(A) = \frac{e^{2\kappa T}}{Z_T}\sum_{\sigma\in\mathbb{Z}_2}\mathbb{E}_{\mathcal{P}}^{\sigma}\left[1_A e^{\frac{1}{2}\int_{-T}^{T}ds\int_{-T}^{T}W(\sigma_{\kappa s}\sigma_{\kappa t},s-t)dt}\right], \quad A \in \sigma(\mathscr{G}),$$

on the measurable space $(\mathfrak{X}_D, \sigma(\mathscr{G}))$, where Z_T is the normalizing constant such that $\mu_T(\mathfrak{X}_D) = 1$. The probability measure μ_T is a *finite volume Gibbs measure* for the spin-boson model.

We can show the convergence of μ_T to a probability measure μ_∞ in the local sense as $T \to \infty$. Let μ be the additive set function on $\mathscr{G}$ defined by

$$\mu(A) = e^{2\kappa T}e^{2ET}\sum_{\sigma\in\mathbb{Z}_2}\mathbb{E}_{\mathcal{P}}^{\sigma}\left[(\Psi_g(\sigma_{-\kappa T}), \mathrm{J}_{[-T,T]}\Psi_g(\sigma_{\kappa T}))\,1_A\right], \quad A \in \mathscr{G}_{[-T,T]}.$$

Here $E = \inf\mathrm{Spec}(H_{\mathrm{SB}})$ and

$$\mathrm{J}_{[-T,T]} = \mathrm{J}_{-T}^{*}e^{-\phi_{\mathrm{E}}(\int_{-T}^{T}\sigma_{\kappa r}\mathrm{j}_r\tilde{\varphi}dr)}\mathrm{J}_T.$$

In the same way as in Lemma 2.72, we can prove that the set function μ is countably additive. Thus by Hopf's extension theorem, there exists a unique probability measure μ_∞ on $(\mathfrak{X}_D, \sigma(\mathscr{G}))$ such that $\mu_\infty(A) = \mu(A)$ for $A \in \mathscr{G}$. Note that $\sigma(\mathscr{G}) = \mathcal{B}(\mathfrak{X}_D)$. This is the *Gibbs measure* associated with the spin-boson Hamiltonian.

Theorem 4.15 (Local convergence). *The family of probability measures $\{\mu_T\}_{T>0}$ converges to μ_∞ on $(\mathfrak{X}_D, \mathcal{B}(\mathfrak{X}_D))$ in the local sense, i. e., $\mu_T(A) \to \mu_\infty(A)$ as $T \to \infty$ for all $A \in \mathscr{G}$.*

Proof. We show an outline of the proof. Define the set function $\rho_T : \mathscr{G}_T \to \mathbb{R}$ by

$$\rho_T(A) = e^{2Es}e^{2\kappa s}\sum_{\sigma\in\mathbb{Z}_2}\mathbb{E}_{\mathcal{P}}^{\sigma}\left[1_A\left(\frac{1_{T-s}(\sigma_{-\kappa s})}{\|1_T\|}, \mathrm{J}_{[-s,s]}\frac{1_{T-s}(\sigma_{\kappa s})}{\|1_T\|}\right)\right]$$

for $A \in \mathscr{G}_{[-s,s]}$ with $s \leq T$. Here

$$1_T = e^{-T(H_{\mathrm{SB}}^{\mathbb{Z}_2}-E)}1_{\mathrm{SB}}.$$

The set function ρ_T is well defined, i. e.,

$$\begin{aligned}\rho_T(A) &= e^{2Er}e^{2\kappa r}\sum_{\sigma\in\mathbb{Z}_2}\mathbb{E}_{\mathcal{P}}^{\sigma}\left[1_A\left(\frac{1_{T-r}(\sigma_{-\kappa r})}{\|1_T\|}, \mathrm{J}_{[-r,r]}\frac{1_{T-r}(\sigma_{\kappa r})}{\|1_T\|}\right)\right]\\ &= e^{2Es}e^{2\kappa s}\sum_{\sigma\in\mathbb{Z}_2}\mathbb{E}_{\mathcal{P}}^{\sigma}\left[1_A\left(\frac{1_{T-s}(\sigma_{-\kappa s})}{\|1_T\|}, \mathrm{J}_{[-s,s]}\frac{1_{T-s}(\sigma_{\kappa s})}{\|1_T\|}\right)\right]\end{aligned}$$

for all $r \leq s \leq T$ and $A \in \mathscr{G}_{[-r,r]}$. Then ρ_T is a probability measure on $(\mathfrak{X}_D, \sigma(\mathscr{G}_T))$. Let $s \leq T$ and $A \in \mathscr{G}_s$. Then $\rho_T(A) = \mu_T(A)$ and

$$\lim_{T\to\infty}\mu_T(A)=\lim_{T\to\infty}e^{2Es}e^{2\kappa s}\sum_{\sigma\in\mathbb{Z}_2}\mathbb{E}_{\mathcal{P}}^{\sigma}\left[1_A\left(\frac{1_{T-s}(\sigma_{-\kappa s})}{\|1_T\|},\mathrm{J}_{[-s,s]}\frac{1_{T-s}(\sigma_{\kappa s})}{\|1_T\|}\right)\right].$$

Since $1_T/\|1_T\|\to\Psi_{\mathrm{g}}$ strongly as $T\to\infty$, we have

$$\lim_{T\to\infty}\mu_T(A)=e^{2Es}e^{2\kappa s}\sum_{\sigma\in\mathbb{Z}_2}\mathbb{E}_{\mathcal{P}}^{\sigma}\left[1_A\left(\Psi_{\mathrm{g}}(\sigma_{-\kappa s}),\mathrm{J}_{[-s,s]}\Psi_{\mathrm{g}}(\sigma_{\kappa s})\right)\right]=\mu_\infty(A).\qquad\square$$

4.4.2 Ground state properties

Like in the sections above, the Gibbs measure μ_∞ can be used to derive the ground state properties of the Hamiltonian. We start by considering ground state expectations of the form $(\Psi_{\mathrm{g}},hF(\phi(f))\Psi_{\mathrm{g}})$ with suitable functions F and h. By the parity symmetry,

$$(\Psi_{\mathrm{g}},\sigma\Psi_{\mathrm{g}})_{L^2(\mathbb{Z}_2;L^2(\mathscr{Q}))}=(\Psi_{\mathrm{g}},\sigma_x\Psi_{\mathrm{g}})_{\mathbb{C}^2\otimes\mathcal{F}_{\mathrm{SB}}}=0.$$

Theorem 4.16. *Let f be a bounded $\mathscr{G}_t$-measurable function on $\mathfrak{X}_D$. Then*

$$\mathbb{E}_{\mu_\infty}[f]=e^{2\kappa t}e^{2Et}\sum_{\sigma\in\mathbb{Z}_2}\mathbb{E}_{\mathcal{P}}^{\sigma}\left[\left(\Psi_{\mathrm{g}}(\sigma_{-\kappa t}),\mathrm{J}_{[-t,t]}\Psi_{\mathrm{g}}(\sigma_{\kappa t})\right)f\right].\tag{4.4.3}$$

Proof. For $A\in\mathscr{G}_{[-t,t]}$, we have

$$\mathbb{E}_{\mu_\infty}[1_A]=\mu_\infty(A)=e^{2\kappa t}e^{2Et}\sum_{\sigma\in\mathbb{Z}_2}\mathbb{E}_{\mathcal{P}}^{\sigma}\left[\left(\Psi_{\mathrm{g}}(\sigma_{-\kappa t}),\mathrm{J}_{[-t,t]}\Psi_{\mathrm{g}}(\sigma_{\kappa t})\right)1_A\right]$$

and

$$\mathbb{E}_{\mu_\infty}\Big[\sum_{j=1}^n a_j 1_{A_j}\Big]=e^{2\kappa t}e^{2Et}\sum_{\sigma\in\mathbb{Z}_2}\mathbb{E}_{\mathcal{P}}^{\sigma}\Big[\left(\Psi_{\mathrm{g}}(\sigma_{-\kappa t}),\mathrm{J}_{[-t,t]}\Psi_{\mathrm{g}}(\sigma_{\kappa t})\right)\sum_{j=1}^n a_j 1_{A_j}\Big]$$

for every $A_j\in\mathscr{G}_{[-t,t]}$ and $a_j\geq 0$, $j=1,\dots,n$. Then (4.4.3) follows by a limiting argument. $\square$

An immediate consequence of Theorem 4.16 is the following.

Corollary 4.17 (Euclidean Green functions). *Let $f_j:\mathbb{Z}_2\to\mathbb{C}$, $j=0,\dots,n$, $n\in\mathbb{N}$, be bounded functions. Then for $t_0\leq t_1\leq\cdots\leq t_n$,*

$$\mathbb{E}_{\mu_\infty}\Big[\prod_{j=0}^n f_j(\sigma_{\kappa t_j})\Big]=\left(\Psi_{\mathrm{g}},f_0e^{-(t_1-t_0)(H_{\mathrm{SB}}^{\mathbb{Z}_2}-E)}f_1\cdots f_{n-1}e^{-(t_n-t_{n-1})(H_{\mathrm{SB}}^{\mathbb{Z}_2}-E)}f_n\Psi_{\mathrm{g}}\right).$$

In particular, for all bounded functions h,f and $g:\mathbb{Z}_2\to\mathbb{C}$ we have

$$\mathbb{E}_{\mu_\infty}\left[h(\sigma_0)\right]=(\Psi_{\mathrm{g}},h\Psi_{\mathrm{g}}),$$

$$\mathbb{E}_{\mu_\infty}[f(\sigma_{\kappa t})g(\sigma_{\kappa s})]=(f\Psi_{\mathrm{g}},e^{-|t-s|(H_{\mathrm{SB}}^{\mathbb{Z}_2}-E)}g\Psi_{\mathrm{g}}).$$

Proof. For $A_j \in \sigma(\mathscr{G}), j = 0, 1, \ldots, n$, it follows that

$$\mathbb{E}_{\mu_\infty}\Big[\prod_{j=0}^{n} 1_{A_j}(\sigma_{\kappa t_j})\Big] = e^{2\kappa t} e^{2Et} \sum_{\sigma\in\mathbb{Z}_2} \mathbb{E}_{\mathcal{P}}^{\sigma}\Big[(\Psi_g(\sigma_{\kappa t_0}), \mathrm{J}_{[-t,t]}\Psi_g(\sigma_{\kappa t_n})) \prod_{j=0}^{n} 1_{A_j}(\sigma_{\kappa t_j})\Big]$$

$$= \Big(\Psi_g, 1_{A_0} e^{-(t_1-t_0)(H_{\mathrm{SB}}^{\mathbb{Z}_2}-E)} 1_{A_1} \cdots e^{-(t_n-t_{n-1})(H_{\mathrm{SB}}^{\mathbb{Z}_2}-E)} 1_{A_n} \Psi_g\Big).$$

The corollary is obtained by a limiting argument. □

Lemma 4.18. *Let $F : \mathbb{R} \to \mathbb{R}$ be a bounded function, $f \in L^2_{\mathrm{real}}(\mathbb{R}^d)$, and $h : \mathbb{Z}_2 \to \mathbb{C}$ be another bounded function. Then*

$$(e^{-TH_{\mathrm{SB}}^{\mathbb{Z}_2}} 1_{\mathrm{SB}}, hF(\phi(f)) e^{-TH_{\mathrm{SB}}^{\mathbb{Z}_2}} 1_{\mathrm{SB}}) = e^{2\kappa T} \sum_{\sigma\in\mathbb{Z}_2} \mathbb{E}_{\mathcal{P}}^{\sigma} \mathbb{E}_{\mu_{\mathrm{E}}} \Big[h(\sigma_0) e^{-\phi_{\mathrm{E}}(\int_{-T}^{T} \sigma_{\kappa s} \mathrm{j}_s \tilde{\varphi} ds)} F(\phi_{\mathrm{E}}(\mathrm{j}_0 f))\Big].$$

Proof. Using $e^{-tH_{\mathrm{SB}}^{\mathbb{Z}_2}} \Psi(\sigma) = e^{\kappa t} \mathbb{E}_{\mathcal{P}}^{\sigma} [\mathrm{J}_0^* e^{-\phi_{\mathrm{E}}(\int_0^t \sigma_{\kappa r} \mathrm{j}_r \tilde{\varphi} dr)} \mathrm{J}_t \Psi(\sigma_{\kappa t})]$, we have

$$(e^{-TH_{\mathrm{SB}}^{\mathbb{Z}_2}} hF(\phi(f)) e^{-TH_{\mathrm{SB}}^{\mathbb{Z}_2}} 1_{\mathrm{SB}})(\sigma) = e^{2\kappa T} \mathbb{E}_{\mathcal{P}}^{\sigma} \big[\mathrm{J}_{[0,T]} h(\sigma_{\kappa T}) F(\phi(f)) \mathbb{E}_{\mathcal{P}}^{\sigma_{\kappa T}} [\mathrm{J}_{[0,T]} 1(\sigma_{\kappa T})]\big].$$

Note that $1_{\mathrm{SB}}(\sigma) = 1_{L^2(\mathscr{Q})}$ for every $\sigma \in \mathbb{Z}_2$. Thus we see that

$$\sum_{\sigma\in\mathbb{Z}_2} (1, (e^{-TH_{\mathrm{SB}}^{\mathbb{Z}_2}} hF(\phi(f)) e^{-TH_{\mathrm{SB}}^{\mathbb{Z}_2}} 1_{\mathrm{SB}})(\sigma))_{L^2(\mathscr{Q})}$$

$$= e^{2\kappa T} \sum_{\sigma\in\mathbb{Z}_2} \mathbb{E}_{\mathcal{P}}^{\sigma} \big[(1, \mathrm{J}_{[0,T]} h(\sigma_{\kappa T}) F(\phi(f)) \mathbb{E}_{\mathcal{P}}^{\sigma_{\kappa T}} [\mathrm{J}_{[0,T]} 1_{\mathrm{SB}}])\big]$$

$$= e^{2\kappa T} \sum_{\sigma\in\mathbb{Z}_2} \mathbb{E}_{\mathcal{P}}^{\sigma} \Big[\Big(1, X \mathbb{E}_{\mathcal{P}}^{\sigma_{\kappa T}} [\mathrm{J}_T F(\phi(f)) \mathrm{J}_0^* e^{-\phi_{\mathrm{E}}(\int_0^T \sigma_{\kappa s} \mathrm{j}_s \tilde{\varphi} ds)} 1_{\mathrm{SB}}]\Big)\Big].$$

Here $X = e^{-\phi_{\mathrm{E}}(\int_0^T \sigma_{\kappa s} \mathrm{j}_s \tilde{\varphi} ds)} h(\sigma_{\kappa T})$. By the Markov property of $\mathrm{E}_t = \mathrm{J}_t \mathrm{J}_t^*$, we also have

$$= e^{2\kappa T} \sum_{\sigma\in\mathbb{Z}_2} \mathbb{E}_{\mathcal{P}}^{\sigma} \Big[\Big(1, X F(\phi_{\mathrm{E}}(\mathrm{j}_T f)) \mathbb{E}_{\mathcal{P}}^{\sigma_{\kappa T}} [e^{-\phi_{\mathrm{E}}(\int_0^T \sigma_{\kappa s} \mathrm{j}_{s+T} \tilde{\varphi} ds)} 1_{\mathrm{SB}}]\Big)\Big],$$

and by the Markov property of $(N_t)_{t\in\mathbb{R}}$ we furthermore have

$$= e^{2\kappa T} \sum_{\sigma\in\mathbb{Z}_2} \mathbb{E}_{\mathcal{P}}^{\sigma} \Big[(1, X F(\phi_{\mathrm{E}}(\mathrm{j}_t f)) e^{-\phi_{\mathrm{E}}(\int_0^T \sigma_{\kappa(s+T)} \mathrm{j}_{s+T} \tilde{\varphi} ds)} 1)\Big]$$

$$= e^{2\kappa T} \sum_{\sigma\in\mathbb{Z}_2} \mathbb{E}_{\mathcal{P}}^{\sigma} \Big[(1, e^{-\phi_{\mathrm{E}}(\int_0^{2T} \sigma_{\kappa s} \mathrm{j}_s \tilde{\varphi} ds)} h(\sigma_{\kappa T}) F(\phi_{\mathrm{E}}(\mathrm{j}_T f)) 1)\Big].$$

Using shift invariance, the lemma follows. □

Theorem 4.19. *Let $\mathrm{I}_{\mathrm{SB}} < \infty$, $f \in L^2_{\mathrm{real}}(\mathbb{R}^d)$, $h : \mathbb{Z}_2 \to \mathbb{C}$ be a bounded function, and $\beta \in \mathbb{R}$. Then*

$$(\Psi_g, he^{i\beta\phi(f)}\Psi_g) = e^{-\frac{\beta^2}{4}\|f\|^2}\mathbb{E}_{\mu_\infty}\left[h(\sigma_0)e^{i\beta K(f)}\right], \tag{4.4.4}$$

where $K(f)$ is a random variable on $(\mathscr{X}_D, \sigma(\mathscr{G}))$ given by

$$K(f) = -\frac{1}{2}\int_{-\infty}^{\infty}(e^{-|r|\omega}\tilde{\varphi}/\sqrt{\omega}, \hat{f})\sigma_{\kappa r}dr.$$

Proof. Note that

$$(\Psi_g, he^{i\beta\phi(f)}\Psi_g) = \lim_{T\to\infty}(\Psi_g^T, he^{i\beta\phi(f)}\Psi_g^T),$$

where $\Psi_g^T = 1_T/\|1_T\|$, and we see that by Lemma 4.18,

$$\left(\Psi_g^T, he^{i\beta\phi(f)}\Psi_g^T\right) = \frac{e^{2\kappa T}}{Z_T}\sum_{\sigma\in\mathbb{Z}_2}\mathbb{E}_{\mathcal{P}}^{\sigma}\mathbb{E}_{\mu_{\mathrm{E}}}\left[h(\sigma_0)e^{-\phi_{\mathrm{E}}(\int_{-T}^{T}\sigma_{\kappa s}\mathrm{j}_s\tilde{\varphi}ds)}e^{i\beta\phi_{\mathrm{E}}(\mathrm{j}_0 f)}\right].$$

The expectation with respect to μ_{E} can be computed explicitly, and thus

$$(\Psi_g, he^{i\beta\phi(f)}\Psi_g) = \lim_{T\to\infty}e^{-\frac{\beta^2}{4}\|f\|^2}\mathbb{E}_{\mu_T}\left[h(\sigma_0)e^{-i\frac{\beta}{2}\int_{-T}^{T}(e^{-|s|\omega}\tilde{\varphi}/\sqrt{\omega},\hat{f})\sigma_{\kappa s}ds}\right].$$

Notice that

$$\left|\int_{-\infty}^{\infty}(e^{-|s|\omega}\tilde{\varphi}/\sqrt{\omega}, \hat{f})\sigma_{\kappa s}ds\right| \le 2\mathrm{I}_{\mathrm{SB}}^{1/2}\|\hat{f}\| < \infty.$$

By the local convergence of μ_T as $T \to \infty$ and employing a similar telescoping as in (2.8.22)–(2.8.24) in the proof of Corollary 2.86, we obtain the desired result. □

Using Theorem 4.19, the functionals $(\Psi_g, hF(\phi(f))\Psi_g)$ can be represented in terms of averages with respect to the Gibbs measure μ_∞. We consider the cases when F is a polynomial or a Schwartz test function. We will show in Corollary 4.27 below that $\Psi_g \in D(e^{+\beta N})$ for all $\beta > 0$, thus $\Psi_g \in D(\phi(f)^n)$ for every $n \in \mathbb{N}$.

Corollary 4.20 (Moments of field operators). *Assume $\mathrm{I}_{\mathrm{SB}} < \infty$, let $f \in L^2_{\mathrm{real}}(\mathbb{R}^d)$, and $h : \mathbb{Z}_2 \to \mathbb{C}$ be a bounded function. Also, let $H_n(x) = (-1)^n e^{x^2/2}\frac{d^n}{dx^n}e^{-x^2/2}$ be the Hermite polynomial of degree n. Then*

$$(\Psi_g, h\phi(f)^n\Psi_g) = i^n\mathbb{E}_{\mu_\infty}\left[h(\sigma_0)H_n\left(\frac{-iK(f)}{\|f\|/\sqrt{2}}\right)\right](\|f\|/\sqrt{2})^n, \quad n \in \mathbb{N}. \tag{4.4.5}$$

Proof. We have

$$\frac{1}{i^n}\frac{d^n}{d\beta^n}e^{-\beta^2\|f\|^2/4}e^{i\beta K(f)}\Big|_{\beta=0} = i^n H_n\left(\frac{-iK(f)}{\|f\|/\sqrt{2}}\right)(\|f\|/\sqrt{2})^n.$$

Using this and the equality

$$(\Psi_g, h\phi(f)^n\Psi_g) = \frac{1}{i^n}\frac{d^n}{d\beta^n}e^{-\frac{\beta^2}{4}\|f\|^2}\mathbb{E}_{\mu_\infty}[h(\sigma_0)e^{i\beta K(f)}]\Big\lceil_{\beta=0},$$

we obtain (4.4.5). ☐

In the next corollary, we express $(\Psi_g, hF(\phi(f))\Psi_g)$, $F \in \mathscr{S}(\mathbb{R})$, in terms of the Gibbs measure.

Corollary 4.21. *Let* $\mathrm{I}_{\mathrm{SB}} < \infty$, $f \in L^2_{\mathrm{real}}(\mathbb{R}^d)$, $F \in \mathscr{S}(\mathbb{R})$, *and* $h : \mathbb{Z}_2 \to \mathbb{C}$ *be a bounded function. Then*

$$(\Psi_g, hF(\phi(f))\Psi_g) = \mathbb{E}_{\mu_\infty}[h(\sigma_0)G(K(f))], \tag{4.4.6}$$

where $G = \check{F} * \check{g}$ *and* $g(\beta) = e^{-\beta^2\|f\|^2/4}$.

Proof. Since $F(\phi(f)) = \frac{1}{\sqrt{2\pi}}\int_{-\infty}^{\infty}\check{F}(\beta)e^{i\beta\phi(f)}d\beta$, we have

$$(\Psi_g, hF(\phi(f))\Psi_g) = \frac{1}{\sqrt{2\pi}}\int_{-\infty}^{\infty}\check{F}(\beta)e^{-\frac{\beta^2}{4}\|f\|^2}\mathbb{E}_{\mu_\infty}\left[h(\sigma_0)e^{i\beta K(f)}\right]d\beta.$$ ☐

The field fluctuations in the ground state can be defined for every real-valued function $f \in L^2_{\mathrm{real}}(\mathbb{R}^d)$ by

$$F(f) = (\Psi_g, \phi(f)^2\Psi_g) - (\Psi_g, \phi(f)\Psi_g)^2. \tag{4.4.7}$$

More generally, we also consider fluctuations of the form

$$G(f) = (\Psi_g, (\sigma\phi(f))^2\Psi_g) - (\Psi_g, \sigma\phi(f)\Psi_g)^2.$$

Corollary 4.22 (Average field strength and field fluctuations). *Assume* $\mathrm{I}_{\mathrm{SB}} < \infty$ *and let* $f \in L^2_{\mathrm{real}}(\mathbb{R}^d)$. *The following hold:*

(1) *(Average field strength)* $(\Psi_g, \sigma\phi(f)\Psi_g) = \mathbb{E}_{\mu_\infty}[\sigma_0 K(f)]$.
(2) *(Field fluctuations)* $(\Psi_g, (\sigma\phi(f))^2\Psi_g) = \mathbb{E}_{\mu_\infty}[(\sigma_0 K(f))^2] + \frac{1}{2}\|f\|^2$.

In particular, we have

(3) $G(f) = \mathbb{E}_{\mu_\infty}[(\sigma_0 K(f))^2] - (\mathbb{E}_{\mu_\infty}[\sigma_0 K(f)])^2 + \frac{1}{2}\|f\|^2$.
(4) $(\Psi_g, \phi(f)\Psi_g) = 0$ *and* $F(f) = (\Psi_g, \phi(f)^2\Psi_g) = \mathbb{E}_{\mu_\infty}[K(f)^2] + \frac{1}{2}\|f\|^2$.
(5) *Let* $f \not\equiv 0$, F_0 *and* G_0 *be given by* F *and* G *with* ϕ_φ *replaced by* 0, *then* (i) $F(f) > 0$ *and* $F(f) \geq F_0(f)$, (ii) $G(f) > 0$, *and* $G(f) \geq G_0(f)$.

Proof. Statements (1)–(3) easily follow from Corollary 4.20, which imply (4) for $\sigma = 1$. Using Schwarz inequality, we obtain (5) (ii), while (5) (i) is clear by (4). ☐

Note that to prove (1)–(2) in Corollary 4.22, we can proceed, alternatively, to first derive the equality $(\Psi_g, \phi(f)\Psi_g) = \mathbb{E}_{\mu_\infty}[K(f)]$ by using Corollary 4.20, and then from $\mathbb{E}_{\mu_\infty}[\sigma_0] = (\Psi_g, \sigma\Psi_g)$ to further obtain

$$(\Psi_g, \phi(f)\Psi_g) = -\frac{1}{2}(\hat{\varphi}/\omega, \hat{f})(\Psi_g, \sigma\Psi_g) = 0.$$

(Notice that, obviously, $\sigma_0^2 = 1$.) Thus in Corollary 4.22, we have equivalently

$$\mathbb{E}_{\mu_\infty}[(\sigma_0 K(f))^2] = \mathbb{E}_{\mu_\infty}[K(f)^2].$$

Next we show that $(\Psi_g, e^{\beta\phi(f)^2}\Psi_g) < \infty$ for some $\beta > 0$.

Theorem 4.23 (Gaussian domination of the ground state). *Assume* $I_{SB} < \infty$ *and let* $f \in L^2_{\mathrm{real}}(\mathbb{R}^d)$. *If* $\beta \in (-\infty, 1/\|f\|^2)$, *then* $\Psi_g \in D(e^{(\beta/2)\phi(f)^2})$ *and*

$$\|e^{(\beta/2)\phi(f)^2}\Psi_g\|^2 = \frac{1}{\sqrt{1-\beta\|f\|^2}}\mathbb{E}_{\mu_\infty}\left[e^{\frac{\beta K(f)^2}{1-\beta\|f\|^2}}\right].$$

In particular,

$$\lim_{\beta\uparrow 1/2\|f\|^2} \|e^{\beta\phi(f)^2}\Psi_g\| = \infty.$$

Proof. Let $B = \{z \in \mathbb{C} \mid |z| < 1/\|f\|^2\}$ and consider $\mathbb{C}_+ = \{z \in \mathbb{C} \mid \Re z > 0\}$. Define

$$\rho(z) = \frac{1}{\sqrt{1+z\|f\|^2}}\mathbb{E}_{\mu_\infty}\left[e^{-\frac{zK(f)^2}{1+z\|f\|^2}}\right], \quad z > 0.$$

Since $|K(f)| \le \|f\|\|\hat{\varphi}/\sqrt{\omega}\|$ uniformly in the paths, the function ρ can be analytically continued to a function $\tilde{\rho}$ on $\mathbb{C}_+ \cup B$. In the same way as in the proof of Corollary 2.61, we can show that $\Psi_g \in D(e^{-(z/2)\phi(f)^2})$ and $\|e^{-(z/2)\phi(f)^2}\Psi_g\|^2 = \tilde{\rho}(z)$ follows for $z \in \mathbb{C}_+ \cup B$. □

We can also derive the moments of fractional order. Define $|\phi(f)|^s = (\phi(f)^2)^{s/2}$ for $0 \le s \le 2$, and let λ_s be a Lévy measure on $(\mathbb{R}\setminus\{0\}, \mathcal{B}(\mathbb{R}\setminus\{0\}))$ such that

$$\int_0^\infty (1-e^{-yu})\lambda_s(dy) = u^{s/2}, \quad u > 0,$$

i. e.,

$$\lambda_s(dy) = \frac{s}{2\Gamma(1-s/2)}y^{-1-\frac{s}{2}}1_{(0,\infty)}(y)dy,$$

corresponding to the $s/2$-stable process. Let $\Lambda_{\hat{\varphi}} = \||\phi(f)|^{s/2}\Psi_g\|^2$.

Corollary 4.24 (Fractional moments of field operators). *Assume* $I_{SB} < \infty$, *and take* $f \in L^2_{real}(\mathbb{R}^d)$. *Then for* $0 < s < 2$,

$$\Lambda_{\hat\varphi} = \mathbb{E}_{\mu_\infty}\Big[\int_0^\infty \Big(1 - \frac{e^{-\frac{\beta K(f)^2}{1+\beta\|f\|^2}}}{\sqrt{1+\beta\|f\|^2}}\Big)\lambda_s(d\beta)\Big]. \tag{4.4.8}$$

In particular, $\Lambda_0 \le \Lambda_{\hat\varphi}$ *follows.*

Proof. Notice that

$$\Big(1 - \frac{e^{-\frac{\beta K(f)^2}{1+\beta\|f\|^2}}}{\sqrt{1+\beta\|f\|^2}}\Big)\beta^{-1-\frac{s}{2}} \le \Big(1 - \frac{e^{-\frac{\beta\|f\|^2 I_{SB}/4}{1+\beta\|f\|^2}}}{\sqrt{1+\beta\|f\|^2}}\Big)\beta^{-1-\frac{s}{2}} = \eta(\beta).$$

In a neighborhood of $\beta = 0$, we have that $\eta(\beta) = \beta^{-s/2} + o(\beta)$ locally uniformly. Then $\eta(\beta)$ is integrable in this region, and since $\eta(\beta) \le \text{const}\,\beta^{1+s/2}$, it follows that $\eta \in L^1([0,\infty))$. Then (4.4.8) is immediate by using Fubini's theorem. Since

$$\Lambda_0 = \int_0^\infty \Big(1 - \frac{1}{\sqrt{1+\beta\|f\|^2}}\Big)\lambda_s(d\beta),$$

$\Lambda_0 \le \Lambda_{\hat\varphi}$ also follows. □

Theorem 4.23 says that $\|e^{(\beta/2)\phi(f)^2}\Psi_g\| < \infty$ for $\beta < 1/\|f\|^2$. Using this fact, we can obtain explicit formulae for the exponential moments $(\Psi_g, e^{\beta\phi(f)}\Psi_g)$ of the field.

Corollary 4.25 (Exponential moments). *Assume* $I_{SB} < \infty$ *and let* $f \in L^2_{real}(\mathbb{R}^d)$. *Then* $\Psi_g \in D(e^{\beta\phi(f)})$ *and*

$$(\Psi_g, e^{\beta\phi(f)}\Psi_g) = (\Psi_g, \cosh(\beta\phi(f))\Psi_g) = e^{\frac{\beta^2}{4}\|f\|^2}\mathbb{E}_{\mu_\infty}\big[e^{\beta K(f)}\big], \tag{4.4.9}$$

$$(\Psi_g, \sigma e^{\beta\phi(f)}\Psi_g) = (\Psi_g, \sigma\sinh(\beta\phi(f))\Psi_g) = e^{\frac{\beta^2}{4}\|f\|^2}\mathbb{E}_{\mu_\infty}\big[\sigma_0 e^{\beta K(f)}\big]. \tag{4.4.10}$$

Proof. For simplicity, we reset βf to f. By using the generating function

$$e^{xy-\frac{1}{2}y^2} = \sum_{n=0}^\infty H_n(x)\frac{y^n}{n!}$$

of Hermite polynomials, summation in (4.4.5) gives

$$\lim_{M\to\infty}\Big(\Psi_g, \sum_{n=0}^M \frac{1}{n!}\phi(f)^n\Psi_g\Big) = e^{\frac{1}{4}\|f\|^2}\mathbb{E}_{\mu_\infty}\big[e^{K(f)}\big]. \tag{4.4.11}$$

We need to check that the left-hand side converges to $(\Psi_g, e^{\phi(f)}\Psi_g)$. Since by the spin flip property (2) in Lemma 4.6 for odd n we get $(\Psi_g, \phi(f)^n\Psi_g) = 0$, it suffices to show convergence of the terms with even exponents. By Theorem 4.23, we have that $\|e^{\phi(f)^2/(4\|f\|^2)}\Psi_g\| < \infty$. Let E_λ be the spectral measure of $\phi(f)$ with respect to Ψ_g. Then

$$\sum_{n=0}^{M} \frac{1}{(2n)!}(\Psi_g, \phi(f)^{2n}\Psi_g) = \int_{\mathbb{R}} \sum_{n=0}^{M} \frac{1}{(2n)!}\lambda^{2n} e^{-\lambda^2/(4\|f\|^2)} e^{\lambda^2/(4\|f\|^2)} dE_\lambda.$$

Since by Theorem 4.23 the function $e^{\lambda^2/(4\|f\|^2)}$ is integrable with respect to dE_λ, the sums $\sum_{n=0}^{M} \frac{1}{(2n)!}\lambda^{2n} e^{-\lambda^2/(4\|f\|^2)}$ are monotone increasing to $\cosh(\lambda)e^{-\lambda^2/(4\|f\|^2)}$ as $M \uparrow \infty$, which is a bounded function, hence the monotone convergence theorem yields

$$\lim_{M\to\infty} \int_{\mathbb{R}} \sum_{n=0}^{M} \frac{1}{n!}\lambda^n dE_\lambda = \int_{\mathbb{R}} e^\lambda dE_\lambda < \infty,$$

implying $\Psi_g \in D(e^{\phi(f)})$ and (4.4.9). Equality (4.4.10) is obtained similarly. □

Now we restrict attention to expectations of the form $(\Psi_g, e^{-\beta d\Gamma(\hat{\rho})}\Psi_g)$, where ρ is a real-valued function and $\hat{\rho} = \rho(-i\nabla)$. A fundamental example is $\rho = 1$, giving the boson number operator N. As for the Nelson model in Theorem 3.56, we obtain the expression

$$\frac{(\Psi_g^T, he^{-\beta d\Gamma(\hat{\rho})}\Psi_g^T)}{\|\Psi_g^T\|^2} = \mathbb{E}_{\mu_T}\left[h(\sigma_0)e^{-\int_{-T}^{0} ds \int_0^T W^{\rho,\beta}(\sigma_{\kappa s}\sigma_{\kappa t}, s-t)dt}\right],$$

with

$$W^{\rho,\beta}(\eta, T) = \frac{\eta}{2}\int_{\mathbb{R}^d} |\hat{\varphi}(k)|^2 e^{-|T|\omega(k)}(1 - e^{-\beta\rho(k)})dk.$$

Denote

$$W_\infty^{\rho,\beta} = \int_{-\infty}^{0} ds \int_0^\infty W^{\rho,\beta}(\sigma_{\kappa s}\sigma_{\kappa t}, s-t)dt,$$

and notice that $|W_\infty^{\rho,\beta}| \le \mathrm{I}_{\mathrm{SB}}/2 < \infty$ uniformly in the paths in $\mathfrak{X}_D$. Thus we have the theorem below.

Theorem 4.26. *Assume* $\mathrm{I}_{\mathrm{SB}} < \infty$ *and let* $h : \mathbb{Z}_2 \to \mathbb{C}$ *be a bounded function. Then*

$$(\Psi_g, he^{-\beta d\Gamma(\hat{\rho})}\Psi_g) = \mathbb{E}_{\mu_\infty}\left[h(\sigma_0)e^{-W_\infty^{\rho,\beta}}\right], \quad \beta > 0.$$

In particular,

$$(\Psi_g, he^{-\beta N}\Psi_g) = \mathbb{E}_{\mu_\infty}\left[h(\sigma_0)e^{-(1-e^{-\beta})W_\infty}\right], \tag{4.4.12}$$

where

$$W_\infty = \int_{-\infty}^{0} ds \int_{0}^{\infty} W(\sigma_{\kappa s}\sigma_{\kappa t}, s-t)dt.$$

The following result says that the distribution of the number of bosons in the ground state has a superexponentially short tail.

Corollary 4.27 (Super-exponential decay of boson number). *If* $\mathrm{I}_{\mathrm{SB}} < \infty$, *then* $\Psi_{\mathrm{g}} \in D(e^{\beta N})$ *for every* $\beta \in \mathbb{C}$, *and*

$$(\Psi_{\mathrm{g}}, e^{\beta N}\Psi_{\mathrm{g}}) = \mathbb{E}_{\mu_\infty}\left[e^{-(1-e^{\beta})W_\infty}\right]$$

holds. In particular, $\Psi_{\mathrm{g}} \in D(e^{+\beta N})$, *for all* $\beta > 0$.

Proof. The proof is similar to that of Corollary 2.86, and is left to the reader. □

By using Corollary 4.27, we can also derive the expectations of N^m, $m \in \mathbb{N}$, with respect to the ground state Ψ_{g}.

Corollary 4.28. *If* $\mathrm{I}_{\mathrm{SB}} < \infty$, *then*

$$(\Psi_{\mathrm{g}}, N^m\Psi_{\mathrm{g}}) = \sum_{r=1}^{m} S(m,r)\mathbb{E}_{\mu_\infty}\left[W_\infty^r\right], \quad m \in \mathbb{N},$$

where $S(m,r)$ *are the Stirling numbers of the second kind.*

Proof. It can be checked directly that

$$\frac{d^m}{d\beta^m}e^{-C(1-e^{-\beta})} = (-1)^m \sum_{r=1}^{m} S(r,m)e^{-r\beta}(-C)^r e^{-a(1-e^{-\beta})}.$$

Then the corollary follows from $(\Psi_{\mathrm{g}}, N^m\Psi_{\mathrm{g}}) = (-1)^m \frac{d^m}{d\beta^m}(\Psi_{\mathrm{g}}, e^{-\beta N}\Psi_{\mathrm{g}})\lceil_{\beta=0}$ and Corollary 4.27. □

Corollary 4.29. *If* $\mathrm{I}_{\mathrm{SB}} < \infty$, *then*

$$(\Psi_{\mathrm{g}}, (-1)^N\Psi_{\mathrm{g}}) = \mathbb{E}_{\mu_\infty}\left[e^{-2W_\infty}\right], \tag{4.4.13}$$

$$(\Psi_{\mathrm{g}}, h(-1)^N\Psi_{\mathrm{g}}) = \mathbb{E}_{\mu_\infty}\left[h(\sigma_0)e^{-2W_\infty}\right]. \tag{4.4.14}$$

In particular,

$$(\Psi_{\mathrm{g}}, (-1)^N\Psi_{\mathrm{g}}) \geq e^{-\mathrm{I}_{\mathrm{SB}}} > 0, \tag{4.4.15}$$

$$(\Psi_{\mathrm{g}}, \sigma(-1)^N\Psi_{\mathrm{g}}) = \mathbb{E}_{\mu_\infty}\left[\sigma_0 e^{-2W_\infty}\right] = -1 < 0.$$

Proof. Equality (4.4.13) is derived from (4.4.12) with $h(\sigma) = 1$ and $\beta = -i\pi$, and (4.4.14) with $h(\sigma) = \sigma$. Inequality (4.4.15) follows from an estimate of the right-hand side of (4.4.14). Noticing that $\Psi_g \in \mathcal{H}_{SB}^-$, we obtain $P\Psi_g = -\Psi_g$. In particular, this gives

$$\mathbb{E}_{\mu_\infty}\left[\sigma_0 e^{-2W_\infty}\right] = (\Psi_g, \sigma(-1)^N \Psi_g) = (\Psi_g, P\Psi_g) = -1. \qquad \square$$

4.4.3 van Hove representation

The spin-boson model with $\kappa = 0$ can be realized as the direct sum of the so-called van Hove Hamiltonian. Then all the spectral properties of the spin-boson Hamiltonian go over from the van Hove Hamiltonian. The *van Hove Hamiltonian* is defined by the self-adjoint operator

$$H_{\mathrm{vH}}(\hat{g}) = H_{\mathrm{f}} + \Phi(\hat{g}),$$

in Fock space $\mathcal{F}_{\mathrm{SB}}$. Here

$$\Phi(\hat{g}) = \frac{1}{\sqrt{2}}(a^*(\hat{g}) + a(\tilde{\hat{g}}))$$

is the scalar field with cutoff $\hat{g}$. Suppose that $\hat{g}/\omega \in L^2(\mathbb{R}^d)$ and define the conjugate momentum by

$$\Pi(\hat{g}) = \frac{i}{\sqrt{2}}(a^*(\hat{g}/\omega) - a(\tilde{\hat{g}}/\omega)).$$

Then we have

$$e^{i\Pi(\hat{g})} H_{\mathrm{vH}}(\hat{g}) e^{-i\Pi(\hat{g})} = H_{\mathrm{f}} - \frac{1}{2}\|\hat{g}/\sqrt{\omega}\|^2.$$

Hence we conclude that $H_{\mathrm{vH}}(\hat{g})$ has the unique ground state

$$\Psi_{\mathrm{g}}^{\mathrm{vH}}(\hat{g}) = e^{-i\Pi(\hat{g})}\Omega_{\mathrm{b}}$$

whenever $\hat{g}/\omega \in L^2(\mathbb{R}^d)$. It is also known that $H_{\mathrm{vH}}(\hat{g})$ has no ground state when $\hat{g}/\omega \notin L^2(\mathbb{R}^d)$. We summarize this in the proposition below.

Proposition 4.30. *Hamiltonian $H_{\mathrm{vH}}(\hat{g})$ has a ground state if and only if $\hat{g}/\omega \in L^2(\mathbb{R}^d)$.*

On the other hand, clearly, the spin-boson Hamiltonian H with $\kappa = 0$ is the direct sum of van Hove Hamiltonians since

$$H = \begin{pmatrix} H_{\mathrm{f}} + \Phi(\hat{\varphi}/\sqrt{\omega}) & 0 \\ 0 & H_{\mathrm{f}} - \Phi(\hat{\varphi}/\sqrt{\omega}) \end{pmatrix} \tag{4.4.16}$$

and $H_f \pm \Phi(\hat{\varphi}/\sqrt{\omega})$ are equivalent. Therefore, whenever $I_{SB} < \infty$, the ground state of H with $\kappa = 0$ can be obtained as

$$\Psi_g = \begin{pmatrix} \Psi_g^{vH}(\hat{\varphi}/\sqrt{\omega}) \\ \Psi_g^{vH}(-\hat{\varphi}/\sqrt{\omega}) \end{pmatrix}.$$

The operator (4.4.16) has, however, no ground state when $I_{SB} = \infty$. Thus in the case that $I_{SB} < \infty$ and $\kappa = 0$, we have

$$(\Psi_g, e^{i\beta\phi(f)}\Psi_g)_{\mathcal{H}_{SB}} = \frac{1}{2}\sum_{\sigma\in\mathbb{Z}_2}(\Psi_g^{vH}(\sigma\hat{\varphi}/\sqrt{\omega}), e^{i\beta\phi(\hat{f})}\Psi_g^{vH}(\sigma\hat{\varphi}/\sqrt{\omega}))_{\mathcal{F}_{SB}},$$

and the right-hand side above equals

$$(\Omega_b, e^{i\beta(\phi(\hat{f})+(\hat{\varphi}/\omega,\hat{f}))}\Omega_b)_{\mathcal{F}_{SB}} = e^{-\beta^2\|\hat{f}\|^2/4+i\beta(\hat{\varphi}/\omega,\hat{f})}.$$

When $\kappa \neq 0$, we can derive similar but nontrivial representations. Define the random boson field operator

$$\Psi(\hat{f}) = \phi(\hat{f}) + K(f)$$

on $\mathcal{F}_{SB}$, where $K(f) = -\frac{1}{2}\int_{-\infty}^{\infty}(e^{-|r|\omega}\hat{\varphi}/\sqrt{\omega},\hat{f})\sigma_{\kappa r}dr$ is the random variable on $(\mathfrak{X}_D, \mathcal{B}(\mathfrak{X}_D))$. Then we see that:

(1) $(\Omega_b, \Psi(\hat{f})\Omega_b) = K(f)$;

(2) $(\Omega_b, \Psi(\hat{f})^2\Omega_b) - (\Omega_b, \Psi(\hat{f})\Omega_b)^2 = \|f\|^2/2$;

(3) $(\Omega_b, e^{i\beta\Psi(\hat{f})}\Omega_b) = e^{-\beta^2\|f\|^2/4+i\beta K(f)}$.

Let

$$\chi = -\frac{1}{2}\hat{\varphi}(k)\sqrt{\omega(k)}\int_{-\infty}^{\infty} e^{-|s|\omega(k)}\sigma_{\kappa s}ds.$$

Note that $\chi \in L^2(\mathbb{R}^d)$ and

$$K(f) = (\chi, \hat{f});$$

moreover, $\chi/\omega \in L^2(\mathbb{R}^d)$ whenever $I_{SB} < \infty$, and $\chi = \sigma\hat{\varphi}/\sqrt{\omega}$ when $\kappa = 0$. We define the *random van Hove Hamiltonian* by $H_{vH}(\chi)$.

Theorem 4.31 (van Hove representation). *If $I_{SB} < \infty$, then*

$$(\Psi_g, e^{i\beta\phi(f)}\Psi_g) = \mathbb{E}_{\mu_\infty}\left[(\Omega_b, e^{i\beta\Psi(\hat{f})}\Omega_b)\right] = \mathbb{E}_{\mu_\infty}\left[(\Psi_g^{vH}(\chi), e^{i\beta\phi(\hat{f})}\Psi_g^{vH}(\chi))\right].$$

Proof. The first equality can be directly obtained by Theorem 4.19. The second follows from $e^{i\Pi(\chi)}\Psi(\hat{f})e^{-i\Pi(\chi)} = \phi(\hat{f})$. ◻

Corollary 4.32. *Suppose that $F \in \mathscr{S}(\mathbb{R})$. Then we have*

$$(\Psi_g, F(\phi(f))\Psi_g) = \mathbb{E}_{\mu_\infty}\left[(\Omega_b, F(\Psi(\hat{f}))\Omega_b)\right] = \mathbb{E}_{\mu_\infty}\left[(\Psi_g^{vH}(\chi), F(\phi(\hat{f}))\Psi_g^{vH}(\chi))\right],$$
$$\|e^{\beta\phi(f)^2/2}\Psi_g\|^2 = \mathbb{E}_{\mu_\infty}\left[\|e^{\beta\Psi(\hat{f})^2/2}\Omega_b\|^2\right] = \mathbb{E}_{\mu_\infty}\left[\|e^{\beta\phi(\hat{f})^2/2}\Psi_g^{vH}(\chi)\|^2\right].$$

Proof. This follows by Corollary 4.21 and Theorem 4.23. □

4.5 Rabi Hamiltonian

4.5.1 Definition of Rabi Hamiltonian

Cavity quantum electrodynamics has supplied evidence of stronger interactions than encountered in standard quantum electrodynamics. Experimental physicists study this interaction by coupling a two-level atom with a one-mode light in a mirror cavity. This can be realized as a one-photon mode spin-boson model. The Hamiltonian of the Rabi model is defined as a self-adjoint operator in $\mathbb{C}^2 \otimes L^2(\mathbb{R})$. Let a and a^* denote the single mode annihilation and creation operators given by

$$a = \frac{1}{\sqrt{2}}\left(\frac{1}{\sqrt{\omega}}\frac{d}{dx} + \sqrt{\omega}x\right) \quad \text{and} \quad a^* = \frac{1}{\sqrt{2}}\left(-\frac{1}{\sqrt{\omega}}\frac{d}{dx} + \sqrt{\omega}x\right).$$

Definition 4.33 (Rabi Hamiltonian). The *Rabi Hamiltonian* is defined by

$$H_{\text{Rabi}} = \kappa\sigma_z \otimes 1 + 1 \otimes \omega a^* a + g\sigma_x \otimes (a + a^*)$$

in the Hilbert space $\mathbb{C}^2 \otimes L^2(\mathbb{R})$. Here $\kappa > 0$ and $\omega > 0$ are respectively the atom transition and cavity resonance frequencies, and $g \in \mathbb{R}$ is a coupling constant.

The Hamiltonian H_{Rabi} has the purely discrete spectrum $\sigma(H_{\text{Rabi}}) = \{E_j(g)\}_{j=0}^{\infty}$, and the map

$$g \mapsto E_j(g)$$

gives rise to the *eigenvalue curves*. We are interested in studying spectral properties of H_{Rabi}, in particular, the crossing of eigenvalue curves. The absence of crossing of ground state energy $g \mapsto E_0(g)$ can be derived from the nondegeneracy of the ground state energy of H_{Rabi}. We will construct a functional integral representation of $e^{-tH_{\text{Rabi}}}$ to show that the ground state energy has multiplicity 1. This requires a minor modification of the spin-boson model discussed so far.

Denote $U = e^{(i\pi/4)\sigma_y}$, as for spin-boson model. Then

$$UH_{\text{Rabi}}U^{-1} = -\kappa\sigma_x \otimes 1 + 1 \otimes \omega a^* a + g\sigma_z \otimes (a + a^*).$$

Recall the ground state $\Psi_{\rm osc}(x) = (\omega/\pi)^{1/4}e^{-\omega x^2/2}$ of the classical harmonic oscillator $H_{\rm osc}$. Since $\Psi_{\rm osc}$ is strictly positive, we can define the ground state transform $U_{\rm g} : L^2(\mathbb{R}) \to L^2(\mathbb{R}, \Psi_{\rm osc}^2 dx)$ by $U_{\rm g}f = f/\Psi_{\rm osc}$, and we will denote the probability measure $\Psi_{\rm osc}^2 dx$ on $\mathbb{R}$ by $d\mu$. Thus $UH_{\rm Rabi}U^{-1}$ is mapped into the operator

$$U_{\rm g}UH_{\rm Rabi}U^{-1}U_{\rm g}^{-1} = -\kappa\sigma_x \otimes 1 + 1 \otimes \frac{1}{2}\left(-\frac{d^2}{dx^2} + 2\omega x\frac{d}{dx}\right) + g\sigma_z \otimes \sqrt{2\omega}x \tag{4.5.1}$$

in $\mathbb{C}^2 \otimes L^2(\mathbb{R}, d\mu)$. Formally, $U_{\rm g}UH_{\rm Rabi}U^{-1}U_{\rm g}^{-1}$ is the one-dimensional version of $\tilde{H}_{\rm SB}$ under the correspondence $H_{\rm f} \sim \frac{1}{2}(-\frac{d^2}{dx^2} + 2\omega x\frac{d}{dx})$ and $\phi(\tilde{\varphi}) \sim \sqrt{2\omega}x$. We make the identification

$$\mathbb{C}^2 \otimes L^2(\mathbb{R}) \cong L^2(\mathbb{R} \times \mathbb{Z}_2, d\mu) = \left\{f = f(x,\sigma) \middle| \sum_{\sigma\in\mathbb{Z}_2}\int_{\mathbb{R}} |f(x,\sigma)|^2 d\mu(x) < \infty\right\}$$

through

$$\mathbb{C}^2 \otimes L^2(\mathbb{R}) \ni \begin{pmatrix} f_+(x) \\ f_-(x) \end{pmatrix} \mapsto f(x,\sigma) \in L^2(\mathbb{R} \times \mathbb{Z}_2, d\mu).$$

Under this equivalence, (4.5.1) turns into the operator

$$(H_{\rm Rabi}^{\mathbb{Z}_2}f)(x,\sigma) = \left[\frac{1}{2}\left(-\frac{d^2}{dx^2} + 2\omega x\frac{d}{dx}\right) + g\sqrt{2\omega}\sigma x\right]f(x,\sigma) - \kappa f(x,-\sigma).$$

In what follows, we use $H_{\rm Rabi}^{\mathbb{Z}_2}$ instead of $H_{\rm Rabi}$. Let

$$L_{\rm osc} = \frac{1}{2}\left(-\frac{d^2}{dx^2} + 2\omega x\frac{d}{dx}\right)$$

and $(X_t)_{t\geq 0}$ be an Ornstein–Uhlenbeck process on Wiener space $(\mathscr{X}, \mathcal{B}(\mathscr{X}), \mathcal{W}^x)$. We have

$$\int_{\mathbb{R}} \mathbb{E}^x[X_t]d\mu(x) = 0 \quad \text{and} \quad \int_{\mathbb{R}} \mathbb{E}^x[X_sX_t]d\mu(x) = \frac{e^{-|t-s|\omega}}{2\omega}.$$

The generator of X_t is $-L_{\rm osc}$ and

$$(f, e^{-tL_{\rm osc}}g)_{L^2(\mathbb{R},d\mu)} = \int_{\mathbb{R}} \mathbb{E}^x\left[\overline{f(X_0)}g(X_t)\right]d\mu(x)$$

holds. The distribution $\rho_t(x,y)$ of X_t is given by

$$\rho_t(x,y) = \frac{1}{\Psi_{\rm osc}(x)}K_t(x,y)\Psi_{\rm osc}(y),$$

where $K_t(x,y)$ denotes the Mehler kernel

$$K_t(x,y) = \frac{1}{\sqrt{\pi(1-e^{-2t})}} \exp\left(\frac{4xye^{-t} - (x^2+y^2)(1+e^{-2t})}{2(1-e^{-2t})}\right).$$

We denote $\mathbb{E}^{x,\sigma}$ for $\mathbb{E}^{x,\sigma}_{\mathcal{W}\times\mathcal{P}}$.

Theorem 4.34 (Functional integral representation for Rabi Hamiltonian). *We have*

$$(f, e^{-tH^{\mathbb{Z}_2}_{\mathrm{Rabi}}} g) = e^{\kappa t} \sum_{\sigma\in\mathbb{Z}_2} \int_{\mathbb{R}} \mathbb{E}^{x,\sigma}\left[\overline{f(X_0,\sigma_0)} g(X_t, \sigma_{\kappa t}) e^{-g\sqrt{2\omega}\int_0^t \sigma_{\kappa s} X_s ds}\right] d\mu(x). \tag{4.5.2}$$

Proof. The Rabi Hamiltonian $H^{\mathbb{Z}_2}_{\mathrm{Rabi}}$ is the one-dimensional version of the spin-boson Hamiltonian $H^{\mathbb{Z}_2}_{\mathrm{SB}}$, obtaining the functional integral representation is similar to Theorem 4.8. □

Corollary 4.35. *The ground state of H_{Rabi} is unique. In particular, the ground state energy of H_{Rabi} has no crossing for all values of g and κ.*

Proof. Let $f, g \geq 0$, not identically zero. Then for sufficiently small $\varepsilon > 0$, the sets

$$C_f = \left\{(x,\sigma) \in \mathbb{R}\times\mathbb{Z}_2 \mid f(x,\sigma) > \varepsilon\right\},$$
$$C_g = \left\{(x,\sigma) \in \mathbb{R}\times\mathbb{Z}_2 \mid g(x,\sigma) > \varepsilon\right\}$$

have positive measure. We have by (4.5.2),

$$(f, e^{-tH^{\mathbb{Z}_2}_{\mathrm{Rabi}}} g) \geq \varepsilon^2 e^{\kappa t} \sum_{\sigma\in\mathbb{Z}_2} \int_{\mathbb{R}} \mathbb{E}^{x,\sigma}\left[1_{C_f}(X_0,\sigma_0) 1_{C_g}(X_t, \theta_{N_t}) e^{-g\sqrt{2\omega}\int_0^t \sigma_{\kappa s} X_s ds}\right] d\mu(x).$$

Since C_f is a subset of $\mathbb{R}\times\mathbb{Z}_2$, we have $C_f = \bigcup_{\sigma\in\mathbb{Z}_2}(C_f^\sigma, \sigma)$. Thus either C_f^{+1} or C_f^{-1} have at least a positive measure. Suppose that C_f^{+1} has a positive measure. Similarly, we see that $C_g = \bigcup_{\sigma\in\mathbb{Z}_2}(C_g^\sigma, \sigma)$ and suppose that C_g^{+1} has a positive measure. Let C be the set of paths starting from the inside of $(C_f^{+1}, +1)$ and arriving at the inside of $(C_g^{+1}, +1)$. We see that

$$\sum_{\sigma\in\mathbb{Z}_2} \int_{\mathbb{R}} \mathbb{E}^{x,\sigma}\left[1_C\right] d\mu(x) = \sum_{\sigma\in\mathbb{Z}_2} \int_{\mathbb{R}} \mathbb{E}^{x,\sigma}\left[1_{C_f^{+1}}(X_0) 1_{C_g^{+1}}(X_t) 1_{N_t=\mathrm{even}}\right] d\mu(x).$$

By using the distribution ρ_t of X_t, we get

$$\sum_{\sigma\in\mathbb{Z}_2} \int_{\mathbb{R}} \mathbb{E}^{x,\sigma}\left[1_C\right] d\mu(x) = \sum_{n=0}^{\infty} \frac{t^{2n} e^{-t}}{(2n)!} \int_{C_f^{+1}} dx \int_{C_g^{+1}} \Psi_{\mathrm{osc}}(x) K_t(x,y) \Psi_{\mathrm{osc}}(y) dy > 0.$$

Hence we conclude that C has a positive measure and

$$(f, e^{-tH^{\mathbb{Z}_2}_{\mathrm{Rabi}}} g) \geq \varepsilon^2 e^{\kappa t} \sum_{\sigma\in\mathbb{Z}_2} \int_{\mathbb{R}} \mathbb{E}^{x,\sigma}\left[1_C e^{-g\sqrt{2\omega}\int_0^t \sigma_{\kappa s} X_s ds}\right] d\mu(x) > 0.$$

Thus $e^{-tH^{\mathbb{Z}_2}_{\text{Rabi}}}$ is a positivity improving operator and uniqueness of the ground state follows by the Perron–Frobenius theorem. □

4.5.2 Local convergence and Gibbs measures

We set $\mathcal{H} = L^2(\mathbb{R} \times \mathbb{Z}_2, d\mu)$, $q_s = (X_s, \sigma_{\kappa s})$ for $s \geq 0$, and $L = H^{\mathbb{Z}_2}_{\text{Rabi}}$. We also assume that $\omega = 1$ for simplicity. Throughout, we assume that $\kappa > 0$. Then

$$(\phi, e^{-tL}\psi) = e^{\kappa t} \sum_{\sigma\in\mathbb{Z}_2} \int_{\mathbb{R}} \mathbb{E}^{x,\sigma} \left[\overline{\phi(q_0)}\psi(q_t) e^{-g\int_0^t W(q_s)ds} \right] d\mu(x). \tag{4.5.3}$$

Here $W(q_s) = \sqrt{2}\sigma_{\kappa t} X_s$. For later use, we record the following technical lemma.

Lemma 4.36. *We have*

$$\mathbb{E}^x \left[e^{-g\int_0^t W(q_s)ds} \right] = e^{-\sqrt{2}g(\int_0^t e^{-s}(-1)^{N_{\kappa s}}ds)x} e^{\frac{g^2}{4}\int_0^{(1-e^{-2t})/2} |\int_y^t (-1)^{N_{\kappa s}}ds|^2 dy}.$$

In particular,

$$\mathbb{E}^x \left[e^{-g\int_0^t W(q_s)ds} \right] \leq e^{\sqrt{2}|g|(1-e^{-t})x} e^{\frac{g^2}{4}\int_0^{(1-e^{-2t})/2} |t-y|^2 dy}.$$

Proof. Since $X_s \doteq e^{-s}(x + \frac{1}{\sqrt{2}}B_{e^{2s}-1})$, we see that

$$\begin{aligned}
\mathbb{E}^x \left[e^{-g\int_0^t W(q_s)ds} \right] &= \mathbb{E}^0 \left[e^{-\sqrt{2}g\int_0^t e^{-s}(x+\frac{1}{\sqrt{2}}B_{e^{2s}-1})(-1)^{N_{\kappa s}}ds} \right] \\
&= e^{-\sqrt{2}g(\int_0^t e^{-s}(-1)^{N_{\kappa s}}ds)x} \mathbb{E}^0 \left[e^{-g\int_0^t B_{(1-e^{-2s})/2}(-1)^{N_{\kappa s}}ds} \right] \\
&= e^{-\sqrt{2}g(\int_0^t e^{-s}(-1)^{N_{\kappa s}}ds)x} e^{\frac{g^2}{4}\|\int_0^t 1_{(1-e^{-2s})/2}(\cdot)(-1)^{N_{\kappa s}}ds\|^2_{L^2(\mathbb{R})}}.
\end{aligned}$$

This proves the claim. □

We next extend $(\sigma_t)_{t\geq 0}$ and $(X_t)_{t\geq 0}$ to processes indexed by the whole real line. We use the same notations as $(\sigma_t)_{t\in\mathbb{R}}$ and $(X_t)_{t\in\mathbb{R}}$. Accordingly, $(\sigma_t)_{t\in\mathbb{R}}$ can be regarded as the coordinate process on $(\mathcal{X}_D, \mathcal{B}(\mathcal{X}_D), \mathcal{P})$ and

$$(\phi, e^{-tL}\psi) = e^{\kappa t} \sum_{\sigma\in\mathbb{Z}_2} \int_{\mathbb{R}} \mathbb{E}^{x,\sigma} \left[\overline{\phi(q_0)} e^{-g\int_0^t W(q_s)ds} \psi(q_t) \right] d\mu(x). \tag{4.5.4}$$

Here $q_s = (X_s, \sigma_s)$ for $s \in \mathbb{R}$. Then

$$\begin{aligned}
&\sum_{\sigma\in\mathbb{Z}_2} \int_{\mathbb{R}} \mathbb{E}^{x,\sigma} \left[\overline{\phi(q_0)} e^{-g\int_0^t W(q_s)ds} \psi(q_t) \right] d\mu(x) \\
&= \sum_{\sigma\in\mathbb{Z}_2} \int_{\mathbb{R}} \mathbb{E}^{x,\sigma} \left[\overline{\phi(q_{-r})} e^{-g\int_0^t W(q_{s-r})ds} \psi(q_{t-r}) \right] d\mu(x)
\end{aligned}$$

for every $0 \le r \le t$. Let

$$W_\kappa(t,s) = \sigma_t \sigma_s e^{-|t-s|}. \tag{4.5.5}$$

Lemma 4.37. *We have*

$$(1, e^{-tL}1) = 2e^{\kappa t} \mathbb{E}_{\mathcal{P}}^{\sigma}\left[\exp\left(\frac{g^2}{2}\int_0^t ds \int_0^t dr W_\kappa(s,r)\right)\right].$$

Proof. By the Feynman–Kac formula (4.5.3), we obtain

$$(1, e^{-tL}1) = e^{\kappa t} \sum_{\sigma \in \mathbb{Z}_2} \mathbb{E}_{\mathcal{P}}^{\sigma}\left[\mathbb{E}^x\left[e^{-g\int_0^t \sigma_s e^{-s} B_{e^{2s}-1} ds}\right] \int_{\mathbb{R}} \frac{1}{\sqrt{\pi}} e^{-x^2 - (\sqrt{2}g \int_0^t \sigma_s e^{-s} ds)x} dx\right].$$

Since

$$\mathbb{E}^x\left[e^{-g\int_0^t \sigma_s e^{-s} B_{e^{2s}-1} ds}\right] = \exp\left(\frac{g^2}{2}\int_0^t ds \int_0^t dr \sigma_s \sigma_r e^{-(s+r)}(e^{2(s\wedge r)} - 1)\right),$$

$$\int_{\mathbb{R}} e^{-(\sqrt{2}g\int_0^t \sigma_s e^{-s} ds)x} d\mu(x) = \exp\left(\frac{g^2}{2}\left(\int_0^t \sigma_s e^{-s} ds\right)^2\right),$$

it follows that

$$(1, e^{-tL}1) = e^{\kappa t} \sum_{\sigma \in \mathbb{Z}_2} \mathbb{E}_{\mathcal{P}}^{\sigma}\left[\exp\left(\frac{g^2}{2}\int_0^t ds \int_0^t dr \sigma_s \sigma_r e^{-|s-r|}\right)\right].$$

This proves the lemma. □

Remark 4.38. (1) As $W_\kappa(s,r)$ is independent of σ, $\mathbb{E}_{\mathcal{P}}^{\sigma}\left[\exp\left(\frac{g^2}{2}\int_0^t ds \int_0^t dr W_\kappa(s,r)\right)\right]$ is likewise independent of σ.

(2) By the shift invariance of σ_s, we further have

$$\mathbb{E}_{\mathcal{P}}^{\sigma}\left[\exp\left(\frac{g^2}{2}\int_0^t ds \int_0^t dr W_\kappa(s,r)\right)\right] = \mathbb{E}_{\mathcal{P}}^{\sigma}\left[\exp\left(\frac{g^2}{2}\int_{-u}^{t-u} ds \int_{-u}^{t-u} dr W_\kappa(s,r)\right)\right]$$

for every $0 \le u \le t$. Consequently,

$$\begin{aligned}(e^{-tL}1, e^{-tL}1) &= 2e^{2\kappa t} \mathbb{E}_{\mathcal{P}}^{\sigma}\left[\exp\left(\frac{g^2}{2}\int_0^{2t} ds \int_0^{2t} dr W_\kappa(s,r)\right)\right] \\ &= 2e^{2\kappa t} \mathbb{E}_{\mathcal{P}}^{\sigma}\left[\exp\left(\frac{g^2}{2}\int_{-t}^{t} ds \int_{-t}^{t} dr W_\kappa(s,r)\right)\right].\end{aligned} \tag{4.5.6}$$

Let $b = a + \frac{x}{\sqrt{2}}$ and $b^\dagger - \frac{x}{\sqrt{2}}$. Then $b^\dagger b = -\frac{1}{2}\Delta + x\frac{d}{dx}$. We can also compute $(e^{-tL}1, e^{-\beta b^\dagger b}e^{-tL}1)$ for $\beta > 0$.

Lemma 4.39. *Let $\beta > 0$. Then*

$$(e^{-tL}1, e^{-\beta b^\dagger b}e^{-tL}1)$$
$$= 2e^{2\kappa t}\mathbb{E}_{\mathcal{P}}^{\sigma}\left[\exp\left(\frac{g^2}{2}\int_{-t}^{t}\int_{-t}^{t} W_\kappa(s,r)dsdr - g^2(1-e^{-\beta})\int_{-t}^{0}\int_{0}^{t} W_\kappa(s,r)dsdr\right)\right].$$

Proof. Since

$$(\phi, e^{-\beta b^\dagger b}\psi) = \sum_{\sigma\in\mathbb{Z}_2}\int_{\mathbb{R}} \bar{\phi}(X_0,\sigma)\mathbb{E}^x[\psi(X_\beta,\sigma)]d\mu(x),$$

it follows that

$$(e^{-tL}1, e^{-\beta b^\dagger b}e^{-tL}1) = \sum_{\sigma\in\mathbb{Z}_2}\int_{\mathbb{R}} (e^{-tL}1)(X_0,\sigma)\mathbb{E}^x[(e^{-tL}1)(X_\beta,\sigma)]d\mu(x).$$

It is straightforward to compute $(e^{-tL}1)(X_0,\sigma)$ and $(e^{-tL}1)(X_\beta,\sigma)$. Indeed,

$$\begin{aligned}(e^{-tL}1)(X_0,\sigma) &= e^{\kappa t}\mathbb{E}_{\mathcal{P}}^{\sigma}\mathbb{E}^x\left[e^{-\sqrt{2}g\int_0^t \sigma_s X_s^x ds}\right]\\ &= e^{\kappa t}\mathbb{E}_{\mathcal{P}}^{\sigma}\left[e^{-\sqrt{2}g\int_0^t \sigma_s e^{-s}dsx}\mathbb{E}_{\mathcal{W}}^{0}\left[e^{-g\int_0^t \sigma_s e^{-s}B_{e^{2s}-1}ds}\right]\right]\\ &= e^{\kappa t}\mathbb{E}_{\mathcal{P}}^{\sigma}\left[e^{-\sqrt{2}g\int_0^t \sigma_s e^{-s}dsx}e^{\frac{g^2}{2}\int_0^t ds\int_0^t dr\sigma_s\sigma_r e^{-(s+r)}(e^{2(s\wedge r)}-1)}\right].\end{aligned}$$

The evaluation of $\mathbb{E}^x[(e^{-tL}1)(X_\beta,\sigma)]$ is somewhat more involved than that of $(e^{-tL}1)(X_0,\sigma)$. We have

$$\mathbb{E}^x\left[(e^{-tL}1)(X_\beta,\sigma)\right] = e^{\kappa t}\mathbb{E}^x\left[\mathbb{E}_{\mathcal{P}}^{\sigma}\left[e^{-\sqrt{2}g\int_0^t \sigma_s e^{-s}dsX_\beta}e^{\frac{g^2}{2}\int_0^t ds\int_0^t dr\sigma_s\sigma_r e^{-(s+r)}(e^{2(s\wedge r)}-1)}\right]\right].$$

Substituting $X_s \doteq e^{-s}(x + \frac{1}{\sqrt{2}}B_{e^{2s}-1})$ into X_β once again, we find

$$\begin{aligned}&= e^{\kappa t}\mathbb{E}^0\left[\mathbb{E}_{\mathcal{P}}^{\sigma}\left[e^{-\sqrt{2}g\int_0^t \sigma_s e^{-s}dse^{-\beta}\left(x+\frac{1}{\sqrt{2}}B_{e^{2\beta}-1}\right)}e^{\frac{g^2}{2}\int_0^t ds\int_0^t dr\sigma_s\sigma_r e^{-(s+r)}(e^{2(s\wedge r)}-1)}\right]\right]\\ &= e^{\kappa t}\mathbb{E}_{\mathcal{P}}^{\sigma}\left[e^{-\left(\sqrt{2}g\int_0^t \sigma_s e^{-s}ds\right)e^{-\beta}x}e^{\frac{g^2}{2}\int_0^t ds\int_0^t dr\sigma_s\sigma_r e^{-(s+r)}(1-e^{-2\beta})}e^{\frac{g^2}{2}\int_0^t ds\int_0^t dr\sigma_s\sigma_r e^{-(s+r)}(e^{2(s\wedge r)}-1)}\right]\\ &= e^{\kappa t}\mathbb{E}_{\mathcal{P}}^{\sigma}\left[e^{-\left(\sqrt{2}g\int_0^t \sigma_{s-t} e^{-s}ds\right)e^{-\beta}x}e^{\frac{g^2}{2}\int_0^t ds\int_0^t dr\sigma_{s-t}\sigma_{r-t} e^{-(s+r)}(1-e^{-2\beta})}e^{\frac{g^2}{2}\int_0^t ds\int_0^t dr\sigma_{s-t}\sigma_r e^{-(s+r)}(e^{2(s\wedge r)}-1)}\right].\end{aligned}$$

In the last two lines we have shifted σ_s by t. Since σ_u for $0 \le u \le t$ and σ_{s-t} for $0 \le s \le t$ are independent, combining the above computations yields

$$
\begin{aligned}
&(e^{-tL}1, e^{-\beta b^\dagger b}e^{-tL}1) \\
&= \sum_{\sigma\in\mathbb{Z}_2} e^{2\kappa t}\int_{\mathbb{R}} \frac{e^{-x^2}}{\sqrt{\pi}}\mathbb{E}_{\mathcal{P}}^{\sigma}\Big[e^{-(\sqrt{2}g\int_0^t \sigma_s e^{-s}ds)x}e^{-(\sqrt{2}g\int_0^t \sigma_{s-t}e^{-s}ds)e^{-\beta}x}\, e^{\frac{g^2}{2}\int_0^t ds\int_0^t dr\sigma_s\sigma_r e^{-(s+r)}(e^{2(s\wedge r)}-1)} \\
&\quad\times e^{\frac{g^2}{2}\int_0^t ds\int_0^t dr\sigma_{s-t}\sigma_{r-t}e^{-(s+r)}(1-e^{-2\beta})}\, e^{\frac{g^2}{2}\int_0^t ds\int_0^t dr\sigma_{s-t}\sigma_{r-t}e^{-(s+r)}(e^{2(s\wedge r)}-1)}\Big]dx. \qquad (4.5.7)
\end{aligned}
$$

The terms involving x in the exponent can be rewritten as

$$
-\left(x+\frac{g}{\sqrt{2}}\int_0^t \sigma_s e^{-s}ds+\frac{g}{\sqrt{2}}e^{-\beta}\int_0^t \sigma_{s-t}e^{-s}ds\right)^2+\frac{g^2}{2}\left(\int_0^t \sigma_s e^{-s}ds+e^{-\beta}\int_0^t \sigma_{s-t}e^{-s}ds\right)^2.
$$

The first term on the right-hand side integrates to

$$
\frac{1}{\sqrt{\pi}}\int_{\mathbb{R}} e^{-(x+\frac{g}{\sqrt{2}}\int_0^t \sigma_s e^{-s}ds+\frac{g}{\sqrt{2}}e^{-\beta}\int_0^t \sigma_{s-t}e^{-s}ds)^2}dx = 1.
$$

The second term on the right-hand side is

$$
\begin{aligned}
&\left(\int_0^t \sigma_s e^{-s}ds+e^{-\beta}\int_0^t \sigma_{s-t}e^{-s}ds\right)^2 \\
&= \int_0^t\int_0^t \sigma_s\sigma_r e^{-(s+r)}dsdr + 2e^{-\beta}\int_0^t\int_0^t \sigma_{s-t}\sigma_r e^{-(s+r)}dsdr + e^{-2\beta}\int_0^t\int_0^t \sigma_{s-t}\sigma_{r-t}e^{-(s+r)}dsdr. \qquad (4.5.8)
\end{aligned}
$$

The terms independent of x in (4.5.7) are

$$
\begin{aligned}
&\int_0^t ds\int_0^t \sigma_s\sigma_r e^{-(s+r)}(e^{2(s\wedge r)}-1)dr + \int_0^t ds\int_0^t \sigma_{s-t}\sigma_{r-t}e^{-(s+r)}(1-e^{-2\beta})dr \\
&\quad + \int_0^t ds\int_0^t \sigma_{s-t}\sigma_{r-t}e^{-(s+r)}(e^{2(s\wedge r)}-1)dr. \qquad (4.5.9)
\end{aligned}
$$

Therefore,

$$
\begin{aligned}
&(4.5.8)+(4.5.9) \\
&= \int_0^t ds\int_0^t \sigma_{s-t}\sigma_{r-t}e^{-(s+r)}e^{2(s\wedge r)}dr + \int_0^t ds\int_0^t \sigma_s\sigma_r e^{-(s+r)}e^{2(s\wedge r)}dr \\
&\quad + 2e^{-\beta}\int_0^t\int_0^t \sigma_{s-t}\sigma_r e^{-(s+r)}dsdr
\end{aligned}
$$

$$= \int_0^t ds \int_0^t \sigma_{s-t}\sigma_{r-t}e^{-|s-r|}dr + \int_0^t ds \int_0^t \sigma_s\sigma_r e^{-|s-r|}dr + 2e^{-\beta}\int_0^t\int_0^t \sigma_{s-t}\sigma_r e^{-(s+r)}dsdr$$

$$= \int_{-t}^0 ds \int_{-t}^0 \sigma_s\sigma_r e^{-|s-r|}dr + \int_0^t ds \int_0^t \sigma_s\sigma_r e^{-|s-r|}dr + 2e^{-\beta}\int_{-t}^0\int_0^t \sigma_s\sigma_r e^{-|s-r|}dsdr.$$

Using the elementary decomposition $\int_{-t}^t\int_{-t}^t = \int_{-t}^0\int_{-t}^0 + \int_0^t\int_0^t + 2\int_{-t}^0\int_0^t$, we conclude that

$$(4.5.8) + (4.5.9) = \int_{-t}^t ds \int_{-t}^t \sigma_s\sigma_r e^{-|s-r|}dr - 2(1-e^{-\beta})\int_{-t}^0\int_0^t \sigma_s\sigma_r e^{-|s-r|}dsdr.$$

This proves the lemma. □

Let Ψ_g be the ground state of the Rabi Hamiltonian L.

Corollary 4.40. *Let P be the parity operator. Then $P\Psi_g = -\Psi_g$, and $g \mapsto E(g)$ is concave and differentiable almost everywhere in g.*

Proof. Since

$$E(g) = \lim_{t\to\infty} -\frac{1}{t}\log(1, e^{-tL}1),$$

it follows that $g \mapsto E(g)$ is concave and differentiable almost everywhere in g. Moreover, from $P1 = -1$ and $\Psi_g = \lim_{t\to\infty} e^{-tL}1/\|e^{-tL}1\|$, the second assertion follows. □

Define the probability measure μ_T^{Rabi} on $(\mathfrak{X}_D, \mathcal{B}(\mathfrak{X}_D))$ by

$$\mu_T^{\mathrm{Rabi}}(A) = \frac{1}{Z_T}\mathbb{E}_{\mathcal{P}}^{\sigma}\left[1_A e^{\frac{g^2}{2}\int_{-T}^T dt \int_{-T}^T ds W_\kappa(t,s)}\right], \quad A \in \mathcal{B}(\mathfrak{X}_D), \tag{4.5.10}$$

where Z_T is the normalizing constant such that $\mu_T(\mathfrak{X}_D) = 1$.

Recall that $\mathcal{G}_T = \sigma(N_t; t \in [-T, T])$ is the family of sub-σ-fields of $\mathcal{B}(\mathfrak{X}_D)$, and we define $\mathcal{G}$ by $\mathcal{G} = \bigcup_{s\geq 0}\mathcal{G}_{[-s,s]}$.

Theorem 4.41 (Local convergence). *The family of probability measures $\{\mu_T^{\mathrm{Rabi}}\}_{T>0}$ converges to $\mu_\infty^{\mathrm{Rabi}}$ on $(\mathfrak{X}_D, \mathcal{B}(\mathfrak{X}_D))$ in the local sense, i. e., $\mu_T(A) \to \mu_\infty(A)$ as $T \to \infty$ for all $A \in \mathcal{G}$.*

Proof. The proof is a minor modification of that of Theorem 4.15. □

Corollary 4.42. *Let f be a bounded $\mathcal{G}_t$-measurable function. Then*

$$\lim_{T\to\infty}\mathbb{E}_{\mu_T^{\mathrm{Rabi}}}[f] = \mathbb{E}_{\mu_\infty^{\mathrm{Rabi}}}[f].$$

Proof. It suffices to prove the assertion for nonnegative functions f. Since f is bounded and $\mathcal{G}_t$-measurable, there exists a sequence $\{f_n\}$ such that

$$\lim_{n\to\infty} \sup_{x\in\mathcal{D}} |f_n(x) - f(x)| = 0.$$

Here f_n is of the form $f_n = \sum_{j=1}^{m_n} a_j 1_{A_j}$ with $A_j \in \mathcal{G}_t$ and $a_j > 0$. Let $\epsilon > 0$ be arbitrary, and choose n so that $\sup_{x\in\mathcal{D}} |f_n(x) - f(x)| \le \epsilon$. Then

$$\begin{aligned}|\mathbb{E}_{\mu_T^{\mathrm{Rabi}}}[f] - \mathbb{E}_{\mu_\infty^{\mathrm{Rabi}}}[f]| &\le \mathbb{E}_{\mu_T^{\mathrm{Rabi}}}[|f - f_n|] + |\mathbb{E}_{\mu_T^{\mathrm{Rabi}}}[f_n] - \mathbb{E}_{\mu_\infty^{\mathrm{Rabi}}}[f_n]| + \mathbb{E}_{\mu_\infty^{\mathrm{Rabi}}}[|f_n - f|] \\ &\le 2\epsilon + |\mathbb{E}_{\mu_T^{\mathrm{Rabi}}}[f_n] - \mathbb{E}_{\mu_\infty^{\mathrm{Rabi}}}[f_n]|.\end{aligned}$$

By Theorem 4.41, it follows that $\lim_{T\to\infty} |\mathbb{E}_{\mu_T^{\mathrm{Rabi}}}[f] - \mathbb{E}_{\mu_\infty^{\mathrm{Rabi}}}[f]| \le 2\epsilon$. Since $\epsilon > 0$ is arbitrary, the conclusion follows. □

4.5.3 Ground state properties

In this section we present several applications of $\mu_\infty^{\mathrm{Rabi}}$. These may be viewed as one-mode analogues of corresponding results for the spin-boson model. For this reason, we give only outlines of the proofs. The expectation with respect to the ground Ψ_g is given by $\langle \mathcal{O} \rangle = (\Psi_g, \mathcal{O}\Psi_g)$.

Theorem 4.43 (Superexponential decay of boson number). *Let $\beta \in \mathbb{C}$. Then*

$$\langle e^{\beta b^\dagger b} \rangle = \mathbb{E}_{\mu_\infty^{\mathrm{Rabi}}}\left[e^{-g^2(1-e^{\beta}) \int_{-\infty}^0 \int_0^\infty W_\kappa(s,r)dsdr} \right], \tag{4.5.11}$$

$$\langle (b^\dagger b)^m \rangle = \sum_{l=1}^m a_l(m) g^{2l} \mathbb{E}_{\mu_\infty^{\mathrm{Rabi}}}\left[\left(\int_{-\infty}^0 \int_0^\infty W_\kappa(s,r)dsdr \right)^l \right]. \tag{4.5.12}$$

Here $a_l(m) = \frac{(-1)^l}{l!} \sum_{s=1}^l (-1)^s \binom{l}{s}$ are the Stirling numbers. In particular,

$$\langle (b^\dagger b)^m \rangle \le e^{2g^2} - 1$$

for every $m \ge 0$.

Proof. Since $(1, \Psi_g) > 0$, we have

$$\langle e^{-\beta b^\dagger b} \rangle = \lim_{t\to\infty} \frac{(e^{-tL}1, e^{-\beta b^\dagger b} e^{-tL}1)}{(e^{-tL}1, e^{-tL}1)}.$$

By Lemmas 4.37 and 4.39,

$$\begin{aligned}\langle e^{-\beta b^\dagger b} \rangle &= \lim_{t\to\infty} \frac{\mathbb{E}_{\mathcal{P}}^{\sigma}\left[e^{\frac{g^2}{2} \int_{-t}^t \int_{-t}^t W_\kappa(s,r)dsdr - g^2(1-e^{-\beta}) \int_{-t}^0 \int_0^t W_\kappa(s,r)dsdr} \right]}{\mathbb{E}_{\mathcal{P}}^{\sigma}\left[e^{\frac{g^2}{2} \int_{-t}^t \int_{-t}^t W_\kappa(s,r)dsdr} \right]} \\ &= \lim_{t\to\infty} \mathbb{E}_{\mu_t^{\mathrm{Rabi}}}\left[e^{-g^2(1-e^{-\beta}) \int_{-t}^0 \int_0^t W_\kappa(s,r)dsdr} \right].\end{aligned}$$

Observe that $\left|\int_{-\infty}^{0}\int_{0}^{\infty} W_\kappa(s,r)dsdr\right| \le 1$, and hence

$$\left|e^{-g^2(1-e^{-\beta})\int_{-\infty}^{0}\int_{0}^{\infty} W_\kappa(s,r)dsdr} - e^{-g^2(1-e^{-\beta})\int_{-S}^{0}\int_{0}^{S} W_\kappa(s,r)dsdr}\right| \le \epsilon$$

uniformly in the paths for all sufficiently large $S > 0$. Therefore, by Corollary 4.42,

$$\lim_{t\to\infty}\left|\mathbb{E}_{\mu_t^{\mathrm{Rabi}}}\left[e^{-g^2(1-e^{-\beta})\int_{-t}^{0}\int_{0}^{t} W_\kappa(s,r)dsdr}\right]\mathbb{E}_{\mu_\infty^{\mathrm{Rabi}}}\left[e^{-g^2(1-e^{-\beta})\int_{-\infty}^{0}\int_{0}^{\infty} W_\kappa(s,r)dsdr}\right]\right| = 0.$$

Hence

$$\langle e^{-\beta b^\dagger b}\rangle = \mathbb{E}_{\mu_\infty^{\mathrm{Rabi}}}\left[e^{-g^2(1-e^{-\beta})\int_{-\infty}^{0}\int_{0}^{\infty} W_\kappa(s,r)dsdr}\right] \quad \beta > 0. \tag{4.5.13}$$

Formula (4.5.11) now follows from analytic continuation of (4.5.13) in β. Formula (4.5.12) is obtained by differentiating (4.5.11) with respect to β at $\beta = 0$. □

A simple, but nontrivial, application is the following. Since the parity of Ψ_g is -1, we know that $\langle \sigma_x \otimes (-1)^{b^\dagger b}\rangle < 0$. As a consequence of Theorem 4.43, we may show that $\langle (-1)^{b^\dagger b}\rangle > 0$.

Corollary 4.44. *We have*

$$\langle (-1)^{b^\dagger b}\rangle = \mathbb{E}_{\mu_\infty^{\mathrm{Rabi}}}\left[e^{-2g^2\int_{-\infty}^{0}\int_{0}^{\infty} W_\kappa(s,r)dsdr}\right] > 0.$$

Proof. Set $\beta = i\pi$ in Theorem 4.43. Then the claim follows immediately. □

We now construct a functional integral representation of $\langle e^{i\beta x}\rangle$.

Theorem 4.45. *We have*

$$\langle e^{i\beta x}\rangle = e^{-\beta^2/4}\mathbb{E}_{\mu_\infty^{\mathrm{Rabi}}}\left[e^{i\beta K}\right],$$

where

$$K = -\frac{g}{\sqrt{2}}\int_{-\infty}^{\infty} \sigma_s e^{-|s|}ds.$$

Proof. The proof closely parallels that of Lemma 4.37. Since

$$(\phi, e^{i\beta x}\psi) = \sum_{\sigma\in\mathbb{Z}_2}\int_{\mathbb{R}} \bar{\phi}(x,\sigma)e^{i\beta x}\psi(x,\sigma)d\mu(x),$$

we have

$$(e^{-tL}1, e^{i\beta x}e^{-tL}1) = \sum_{\sigma\in\mathbb{Z}_2} e^{2\kappa t}\int_{\mathbb{R}} \mathbb{E}_{\mathcal{P}}^{\sigma}\Big[e^{-\sqrt{2}g\int_0^t \sigma_s e^{-s}dsx}e^{-\sqrt{2}g\int_0^t \sigma_{s-t}e^{-s}dsx}e^{i\beta x}$$
$$\times e^{\frac{g^2}{2}\int_0^t ds\int_0^t dr\sigma_s\sigma_r e^{-(s+r)}(e^{2(s\wedge r)}-1)}e^{\frac{g^2}{2}\int_0^t ds\int_0^t dr\sigma_{s-t}\sigma_{r-t}e^{-(s+r)}(e^{2(s\wedge r)}-1)}\Big]\frac{e^{-x^2}}{\sqrt{\pi}}dx.$$

The terms involving x in the exponent can be rewritten as

$$i\beta x - \left(x + \frac{g}{\sqrt{2}}\int_0^t \sigma_s e^{-s}ds + \frac{g}{\sqrt{2}}\int_0^t \sigma_{s-t}e^{-s}ds\right)^2 + \frac{g^2}{2}\left(\int_0^t \sigma_s e^{-s}ds + \int_0^t \sigma_{s-t}e^{-s}ds\right)^2.$$

The first term on the right-hand side integrates as

$$\frac{1}{\sqrt{\pi}}\int_{\mathbb{R}} e^{i\beta x-(x+\frac{g}{\sqrt{2}}\int_0^t \sigma_s e^{-s}ds+\frac{g}{\sqrt{2}}\int_0^t \sigma_{s-t}e^{-s}ds)^2}dx = e^{-\beta^2/4}e^{-i\beta(\frac{g}{\sqrt{2}}\int_0^t \sigma_s e^{-s}ds+\frac{g}{\sqrt{2}}\int_0^t \sigma_{s-t}e^{-s}ds)}.$$

The remaining steps are identical to those in the proof of Lemma 4.44. Hence we obtain

$$\langle e^{i\beta x}\rangle = \lim_{t\to\infty} e^{-\beta^2/4}\mathbb{E}_{\mu_t^{\rm Rabi}}\left[e^{i\beta K_t}\right],$$

where

$$K_t = -\frac{g}{\sqrt{2}}\int_{-t}^{t} \sigma_s e^{-|s|}ds.$$

The theorem now follows from Corollary 4.42. □

Theorem 4.46 (Gaussian domination of the ground state). *Let $\beta \in \mathbb{C}$ satisfy $|\beta| < 1$. Then*

$$\langle e^{\beta x^2}\rangle = \frac{1}{\sqrt{1-\beta}}\mathbb{E}_{\mu_\infty^{\rm Rabi}}\left[e^{\frac{\beta K^2}{1-\beta}}\right]. \tag{4.5.14}$$

In particular, $\lim_{\beta\uparrow 1}\|e^{\beta x^2/2}\Psi_{\rm g}\|^2 = \infty$.

Proof. By Theorem 4.45, we have

$$\langle e^{-\beta^2x^2/2}\rangle = \frac{1}{\sqrt{2\pi}}\int_{\mathbb{R}}(\Psi_{\rm g}, e^{ik\beta x}\Psi_{\rm g})e^{-k^2/2}dk = \frac{1}{\sqrt{2\pi}}\int_{\mathbb{R}} e^{-k^2\beta^2/4}\mathbb{E}_{\mu_\infty^{\rm Rabi}}\left[e^{ik\beta K}\right]e^{-k^2/2}dk$$
$$= \mathbb{E}_{\mu_\infty^{\rm Rabi}}\left[\frac{1}{\sqrt{2\pi}}\int_{\mathbb{R}} e^{-k^2\beta^2/4}e^{ik\beta K}e^{-k^2/2}dk\right] = \frac{1}{\sqrt{1+\beta^2/2}}\mathbb{E}_{\mu_\infty^{\rm Rabi}}\left[e^{-\frac{\beta^2K^2}{\beta^2+2}}\right].$$

By analytic continuation, we obtain (4.5.14) for $\beta \in \mathbb{C}$ with $|\beta| < 1$. The final assertion follows immediately. □

Let $\bar{L} = L - \inf\mathrm{Spec}(L)$.

Lemma 4.47. *(1) Suppose that $\phi, \psi \in \mathcal{H}$ and $f_j = f_j(x,\sigma) \in L^\infty(\mathbb{R} \times \mathbb{Z}_2)$ for $j = 0,1,\ldots,n$, and $0 < t_0 < t_1 < \cdots < t_n < t$. Then*

$$
\begin{aligned}
&(\phi, e^{-t_0\bar{L}} f_0 e^{-(t_1-t_0)\bar{L}} f_1 e^{-(t_2-t_1)\bar{L}} \cdots e^{-(t_n-t_{n-1})\bar{L}} f_n e^{-(t-t_n)\bar{L}} \psi) \\
&= e^{\kappa t} e^{Et} \sum_{\sigma\in\mathbb{Z}_2} \int_{\mathbb{R}} \mathbb{E}^{\sigma,x} \left[\bar{\phi}(q_0)\psi(q_t) \left(\prod_{j=0}^{n} f_j(q_{t_j}) \right) e^{-g\int_0^t W(q_s)ds} \right].
\end{aligned}
\tag{4.5.15}
$$

(2) Suppose that $g_j = g_j(\sigma) \in L^\infty(\mathbb{Z}_2)$ for $j = 0,1,\ldots,n$ and $0 < t_0 < t_1 < \cdots < t_n < t$. Then

$$
\begin{aligned}
&(1, e^{-t_0\bar{L}} g_0(\sigma_z) e^{-(t_1-t_0)\bar{L}} g_1(\sigma_z) e^{-(t_2-t_1)\bar{L}} \cdots e^{-(t_n-t_{n-1})\bar{L}} g_n(\sigma_z) e^{-(t-t_n)\bar{L}} 1) \\
&= e^{\kappa t} e^{Et} \sum_{\sigma\in\mathbb{Z}_2} \mathbb{E}^{\sigma}_{\mathcal{P}} \left[\left(\prod_{j=0}^{n} g_j(\sigma_{t_j}) \right) \int_{\mathbb{R}} \mathbb{E}^{x} \left[e^{-g\int_0^t W(q_s)ds} \right] d\mu(x) \right].
\end{aligned}
\tag{4.5.16}
$$

Proof. Statement (1) is a straightforward, and (2) is merely a special case of (1). □

By Lemma 4.36, the integrand in (4.5.16) is given by

$$
\mathbb{E}^{x} \left[e^{-g\int_0^t W(q_s)ds} \right] = e^{-g\left(\int_0^t e^{-s}(-1)^{N_{\kappa s}} ds\right)x} e^{\frac{g^2}{4}\int_0^{(1-e^{-2t})/2} \left|\int_y^t (-1)^{N_{\kappa s}} ds\right|^2 dy}.
$$

Theorem 4.48. *We have $\langle \sigma_z e^{-|t-s|\bar{L}} \sigma_z \rangle = \mathbb{E}_{\mu_\infty^{\mathrm{Rabi}}}[\sigma_t \sigma_s]$ for any $t, s \in \mathbb{R}$.*

Proof. By Lemma 4.47 and a limiting argument, we obtain

$$
\begin{aligned}
(\sigma_z \Psi_g, e^{-t\bar{L}} \sigma_z \Psi_g) &= \lim_{T\to\infty} \frac{1}{\|1_{T-t/2}\|^2} (\sigma_z 1_{T-t/2}, e^{-t\bar{L}} \sigma_z 1_{T-t/2}) \\
&= \lim_{T\to\infty} \frac{e^{2ET} e^{2T\kappa}}{\|1_{T-t/2}\|^2} \sum_{\sigma\in\mathbb{Z}_2} \mathbb{E}^{\sigma}_{\mathcal{P}} \left[\sigma_{-t/2}\sigma_{t/2} e^{\frac{g^2}{2}\int_{-T}^T dt \int_{-T}^T ds W_\kappa(t,s)} \right].
\end{aligned}
$$

Hence

$$
\begin{aligned}
(\sigma_z \Psi_g, e^{-t\bar{L}} \sigma_z \Psi_g) &= \lim_{T\to\infty} \frac{\|1_T\|^2}{\|1_{T-t/2}\|^2} \frac{e^{2ET} e^{2T\kappa}}{\|1_T\|^2} \sum_{\sigma\in\mathbb{Z}_2} \mathbb{E}^{\sigma}_{\mathcal{P}} \left[\sigma_{-t/2}\sigma_{t/2} e^{\frac{g^2}{2}\int_{-T}^T dt \int_{-T}^T ds W_\kappa(t,s)} \right] \\
&= \lim_{T\to\infty} \frac{\|1_T\|^2}{\|1_{T-t/2}\|^2} \frac{\mathbb{E}^{\sigma}_{\mathcal{P}} \left[\sigma_{-t/2}\sigma_{t/2} e^{\frac{g^2}{2}\int_{-T}^T dt \int_{-T}^T ds W_\kappa(t,s)} \right]}{\mathbb{E}^{\sigma}_{\mathcal{P}} \left[e^{\frac{g^2}{2}\int_{-T}^T dt \int_{-T}^T ds W_\kappa(t,s)} \right]} = \mathbb{E}_{\mu_\infty^{\mathrm{Rabi}}}[\sigma_{-t/2}\sigma_{t/2}].
\end{aligned}
$$

Therefore, for $t > s$,

$$
(\sigma_z \Psi_g, e^{-(t-s)\bar{L}} \sigma_z \Psi_g) = \mathbb{E}_{\mu_\infty^{\mathrm{Rabi}}}[\sigma_{-(t-s)/2}\sigma_{(t-s)/2}] = \mathbb{E}_{\mu_\infty^{\mathrm{Rabi}}}[\sigma_t \sigma_s]
$$

by shift invariance. □

In light of the foregoing analysis, it becomes apparent that the structural properties established for the spin-boson model admit a natural and essentially parallel realization in the Rabi model once it is reformulated through the Feynman–Kac framework.

From a broader perspective, this correspondence is far from accidental. Rather, it reflects a deeper unifying mechanism underlying both models: namely, that the probabilistic representation provided by the Feynman–Kac formula serves as a robust bridge between operator-theoretic formulations and path-integral descriptions, even in the presence of spin degrees of freedom. Within this framework, quantities that were originally derived in the spin-boson setting such as the construction of Gibbs measures, the characterization of ground states, and the analysis of their spatial or probabilistic properties can be transferred, with suitable modifications, to the Rabi model in a conceptually transparent manner. Thus, when viewed through the lens of the Feynman–Kac formalism, the Rabi model does not merely resemble the spin-boson model. It fits into the same analytical paradigm. This parallelism not only clarifies previously opaque aspects of the Rabi model but also reinforces the versatility of the Feynman–Kac approach as a unifying tool in quantum field theoretical models with internal degrees of freedom.

5 Notes and references

Notes to Chapter 1

1.1

A mathematical introduction to quantum field theory is given in [18, 26, 128, 144, 288, 303]. References [206, 270, 310–312] treat quantum field theory from the point of view of particle physics. Useful texts on quantum field theory include [21, 48, 55, 56, 99, 154, 255, 283, 278]. Reference [97] is a monumental book on quantum field theory using path integrals. We also refer to [4, 64, 75, 146, 276, 281, 304] for texts in this spirit. Reference [93] offers a concise summary of fermion functional integrals. Geometric and topological discussions on quantum field theory are [94, 110, 282, 302, 308]. Standard texts on QED are [152, 208, 211, 245, 280, 319]. Quantum field theory in curved space-time is discussed in [34, 47, 108]. References [39, 241] cover spectral methods in infinite-dimensional analysis.

1.2

The Fock space was introduced in [98]. The second quantization $d\Gamma(h)$ was first treated by Cook [71], where self-adjointness of the field operator $\Phi(f)$ and its conjugate momentum $\Pi(f)$ was shown. Nelson's analytic vector theorem (Proposition 1.16) appeared in [256]. The Segal–Bargmann space was introduced in [37, 284]. Wick product was introduced in [314]. The number operator bounds (1.2.14)–(1.2.15) can be generalized to

$$\|(N+1)^{-m/2}T(N+1)^{-n/2}\| \leq \|W\|_{L^2(\mathbb{R}^{d(m+n)})},$$

where $T = \int_{\mathbb{R}^{d(m+n)}} W(k_1,\ldots,k_m,p_1,\ldots,p_n)\prod_{j=1}^m a^*(k_j)\prod_{i=1}^n a(p_i)dkdp$; see [121] for details. Operators of the form

$$e^{a^\sharp(f)} = \sum_{n=0}^{\infty}\frac{1}{n!}a^\sharp(f)^n, \quad a^\sharp = a, a^*,$$

are useful to study functional integral representations as discussed in this chapter, but they are unbounded. There is much literature in which $e^{a^\sharp(f)}$ is treated, however, with few rigorous discussions including domain issues of $e^{a^\sharp(f)}$. In [143] several useful and interesting properties of $e^{a^\sharp(f)}$ are discussed, where Lemma 1.32 is proven and it is rigorously shown that $e^{a^*(f)}e^{-tH_{\mathrm{f}}}$ is bounded for $t > 0$ (Proposition 1.33).

1.3

Segal [285, 286] introduced $\mathscr{Q}$ space, which linked the theory of free quantum fields with probability. Using this expression, the spectrum of the $P(\phi)_2$-Hamiltonian has been studied, for instance, in the series of papers [122–127] by Glimm and Jaffe.

https://doi.org/10.1515/9783112231494-005

Let A be a bounded operator on $L^2(M, d\mu)$, where (M, μ) is a σ-finite measure space. Operator A is called *ergodic* if and only if it is positivity preserving and for all nonnegative $f, g \in L^2(M, d\mu)$ there is $n > 0$ for which $(f, A^n g) > 0$. A positivity improving operator A is clearly ergodic. Also, $\|A\|$ is a simple eigenvalue and the associated eigenvector is strictly positive if and only if A is ergodic. Proposition 1.53 is a special version of this fact; we refer to [275, XIII.12]. For abstract discussions of positivity improving operators, see [89, 249]. The double Markovianity and the hypercontractivity discussed in Section 1.3.3 are taken from [288, Sections I.4 and I.5]. The hypercontractivity was introduced by Nelson in [260, 261], where also Proposition 3.13 was proved in [261, Theorem 2]. The example is furthermore given in [261, Theorem 4] where $c > \sqrt{(p-1)/(q-1)}$ with $1 \le p \le q \le \infty$ and $\Gamma(c)$ is not bounded from $L^p(\mathbb{R})$ to $L^q(\mathbb{R})$. Hence it implies that the condition $\|T\| \le \sqrt{(p-1)/(q-1)} \le 1$ in Theorem 1.67 is best possible. We also refer the reader to [207, 289].

1.4

Theorem 1.68 is taken from [288, Section I.2]. Minlos theorem, Theorem 1.72, is an infinite-dimensional extension of Bochner theorem. Minlos theorem first appeared in [246], and the proof of Theorem 1.72 is taken from [155]. By virtue of the Minlos theorem, quantum field theory can be investigated from a measure-theoretical point of view. References [128, 288] point in this direction.

1.5

We refer to [33, Section 1.3] for homogeneous Sobolev spaces, where dualities, critical dimension $s = d/2$, and necessary and sufficient conditions for $\dot{H}^s(\mathbb{R}^d)$ to be a Hilbert space are discussed. Nelson emphasized the functional integral approach to quantum field theory in [260], and the fundamental results concerning the Markov property in quantum field theory were established in [261, 259]. We also refer to [288] for further aspects of Euclidean quantum field theory. Nelson's Euclidean quantum field and its lattice approximation due to [142] changed the main object studied in quantum field theory before.

1.6

As explained in this section, there exists no Gaussian measure on a real Hilbert space $\mathcal{H}$ with a given covariance. Instead of this, we consider the triplet $\mathscr{M}_{+2} \subset \mathscr{M} \subset \mathscr{M}_{-2}$ under the identification $\mathscr{M}^* = \mathscr{M}$, and construct the process $t \mapsto \xi_t \in \mathscr{M}_{-2}$ such that $\xi_t(f) = \langle\langle \xi_t, f \rangle\rangle$ is a Gaussian process for $f \in \mathscr{M}_{+2}$. Similar results appear in [61], where path continuity is considered. Gaussian and Borel measures on an infinite product probability space are studied in [73, 156, 318]. Reference [271] treats the Ornstein–Uhlenbeck semigroup in an infinite-dimensional L^2 setting, while [207] is a comprehensive study of Gaussian measures on a Hilbert space. We also refer to [82] for non-Gaussian measures. References [73, 212, 272] treat infinite-dimensional stochastic equations. For the Minlos–Sazonov theorem (Theorem 1.117), we refer to [212, Theorem 2.3.4].

Due to Bochner theorem, it is an established fact that positive definite functions are characteristic functions of finite measures. Note that the product and sum of positive definite functions are also positive definite. A large class of positive definite functions that was discovered by Pólya [239]. Given a function $\phi : [0,\infty) \to [0,1]$, satisfying (1) ϕ is continuous with $\phi(0) = 1$, (2) $\lim_{x\to\infty} \phi(x) = 0$, (3) ϕ is convex, an even extension of ϕ to the real line is positive definite. In other words, the function $f : \mathbb{R} \to [0,1]$ given by $f(x) = \phi(|x|)$ is positive definite. Using this, we see that the functions $f(x) = \max(0, 1-|x|)$ and $g(x) = e^{-|x|}$ are positive definite.

1.7

In our presentations we have followed [240, 262, 263, 81], we refer the reader to these sources for further detail and proofs, see also [10, 11]. For white noise theory, we refer to [155, 156, 264]. For quantum stochastic calculus, see the seminal [204, 268], and for a quantum Malliavin calculus, refer to [100].

Notes to Chapter 2

2.1

In 1964, Nelson demonstrated the mathematical existence of a meson theory with nonrelativistic nucleons in [257]. He introduced a system of Schrödinger particles coupled to a quantized relativistic field, which is nowadays called the Nelson model. He also introduced a quantum field theory model defined by

$$H = \frac{1}{(2m)} \int_{\mathbb{R}^d} \psi^*(x)(-\Delta)\psi(x)dx + H_{\mathrm{f}} + g \int_{\mathbb{R}^d} \psi(x)H_{\mathrm{I}}(x)\psi(x)dx,$$

where ψ is a nonrelativistic complex scalar field on $\mathcal{H} = \bigoplus_{n=0}^{\infty} \mathcal{H}_n$. The restriction $H\lceil_{\mathcal{H}_n}$ is the n-nucleon Nelson model. Since this model can be reduced to each $\mathcal{H}_n$, this is essentially the same as the Nelson model considered in this chapter.

2.2

As explained in this section, infrared and ultraviolet divergences are a ubiquitous difficulty not only in the Nelson model but in the general theory of quantum fields. In 1952, Miyatake [252] and van Hove [307] found that the ground state of a Hamiltonian in Fock space weakly converges to zero when the ultraviolet cutoff is removed. Shale [287] developed mathematical methods to study both infrared and ultraviolet divergences.

In [257], Nelson defined a self-adjoint operator with ultraviolet cutoff on the tensor product $L^2(\mathbb{R}^d)\otimes\mathcal{F}_b$ of Hilbert spaces. The solution of the Schrödinger equation diverges as the cutoff tends to infinity, but the divergence amounts merely to a complex infinite phase shift due to the self-energy of the particles. A canonical transformation, which is implemented by a unitary operator when the infrared cutoff is introduced, separates the divergent self-energy term. It is then shown that, after removing the ultraviolet cutoff and subtracting an infinite constant, a limit Hamiltonian exists. The infrared problem in the Nelson model was first studied by Fröhlich [104, 105]. We also refer to, e. g., [59, 134, 149, 203, 215, 291–293] for earlier results on the Nelson model.

2.3

The Schrödinger operator $H_p = -\frac{1}{2}\Delta + V$ with Kato-decomposable potential V can be defined through Feynman–Kac formula of e^{-tH_p}. Let Ψ_p be the positive ground state of H_p and define $L_p = \frac{1}{\Psi_p}(H_p - E_p)\Psi_p$ with $E_p = \inf \mathrm{Spec}(H_p)$ as a self-adjoint operator in $L^2(\mathbb{R}^d, \Psi_p^2 dx)$. Then there exists a diffusion process $(X_t)_{t\geq 0}$ such that

$$(f, e^{-tL_p}g)_{L^2(\mathbb{R}^d,\Psi_p^2 dx)} = \int_{\mathbb{R}^d} \mathbb{E}^x[f(\bar{X}_0)g(X_t)]\Psi_p(x)^2 dx.$$

The random process $(X_t)_{t\geq 0}$ is called the $P(\phi)_1$-process associated with H_p. This is discussed in detail in [231]. The functional integral representation of the semigroup generated by the Nelson Hamiltonian in terms of a combination of $(X_t)_{t\geq 0}$ and an $\mathscr{M}_{-2}$-valued random process $(\xi_t)_{t\geq 0}$ is due to [43]. Here the ground state transformation of the Nelson Hamiltonian is applied. A similar functional integral representation, with no quantized field, is discussed in [316].

To construct $(X_t)_{t\geq 0}$, the existence of a positive ground state of H_p is assumed. By using Brownian motion $(B_t)_{t\geq 0}$ instead of $(X_t)_{t\geq 0}$, we can construct a functional integral representation of e^{-tH_N} without any assumption on the existence of ground state of H_p. Then the integral kernel of $(F, e^{-tH_N}G)$ is given by

$$e^{-\int_0^t V(B_s)ds} \mathrm{I}_0^* e^{-\phi_E(\int_0^t \delta_s\otimes\varphi(\cdot-B_s)ds)}\mathrm{I}_t.$$

In the massive case, the bound

$$\|\mathrm{I}_0^* e^{-\phi_E(\int_0^t \delta_s\otimes\varphi(\cdot-B_s)ds)}\mathrm{I}_t\| \leq \|e^{-\phi_E(\int_0^t \delta_s\otimes\varphi(\cdot-B_s)ds)}\|_{L^p}$$

can be derived from hypercontractivity with $p = \frac{2}{1-e^{-t\nu}}$, where $\nu > 0$ is the mass of a single boson. On the other hand, in the massless cases the bound (2.3.18), namely

$$\|\mathrm{I}_0^* e^{-\phi_E(\int_0^t \delta_s\otimes\varphi(\cdot-B_s)ds)}\mathrm{I}_t\| \leq \frac{q}{q-1}e^{tE_q(\hat{\varphi})},$$

is derived from the Baker–Campbell–Hausdorff formula and estimates of exponentials of creation and annihilation operators. This is discussed in [243]. A diamagnetic type inequality in Corollary 2.14 is also obtained in [173].

In [115, 117] the functional integral representation of the semigroup generated by the Nelson Hamiltonian defined on a *static Lorentz manifold* is constructed. Here the static Lorentz manifold is a 4-dimensional pseudo-Riemannian manifold with line element

$$ds^2 = g_{00}(x)dt \otimes dt - \sum_{i,j} \gamma_{ij}(x)dx^i \otimes dx^j,$$

where $g_{00} > 0$ is time-independent and $\gamma = (\gamma_{ij})$ denotes a 3-dimensional Riemannian metric. There the diffusion process $(X_t)_{t\geq 0}$ associated with $H_{\mathrm{p}} = -\frac{1}{2}\Delta + V$ is replaced by a random process associated with the divergence form

$$-\sum_{\mu,\nu=1}^{3} \frac{1}{c(x)} \nabla_\mu A_{\mu\nu}(x) \nabla_\nu \frac{1}{c(x)} + V,$$

and the dispersion relation $\omega = \sqrt{-\Delta + m^2}$ expressed in the position representation is replaced by the pseudodifferential operator

$$\omega_{a,m} = \Big(- \sum_{\mu,\nu=1}^{3} \nabla_\mu a_{\mu\nu}(x) \nabla_\nu + m^2(x)\Big)^{1/2}.$$

Note that, importantly, the mass becomes a position-dependent function $m(x)$.

2.4

Kato-decomposable potentials provide a class of semigroups generated by H_{p} with singular potentials. In order to construct a path measure associated with the ground state of the Nelson Hamiltonian, the choice of Kato-decomposable potentials is also suitable. This is discussed in [43]. The absence of the ground state of the Nelson Hamiltonian is discussed under taking Kato-class potential; we refer to [236].

2.5

For the massive case, the existence of the ground state of the translation invariant Nelson Hamiltonian is studied in [105, 133]. In the massless case, Bach, Fröhlich, and Sigal [30, 31] proved the existence and uniqueness of the ground state for sufficiently weak couplings under the infrared regular condition; they treated more general models, including the Nelson model, and Arai and Hirokawa [22] for the spin-boson model. Spohn [299] and Gérard [112, 114] proved the existence of a ground state for arbitrary values of the coupling constant. Functional integral methods are used in [299], where Theorem 2.28 is proven, and operator theory is employed in [112]. External potentials V discussed in [112, 299] are confining. Let $1_{|x|\geq R}\Psi = \Psi_R$. We define the ionization energy by

$\Sigma = \lim_{R\to\infty} \Sigma_R$, where $\Sigma_R = \inf_{\Psi \in D(H_N), \Psi \neq 0} \frac{(\Psi_R, H_N \Psi_R)}{(\Psi_R, \Psi_R)}$. Let $\inf \mathrm{Spec}(H_N) = E_N$. Then the *binding condition* is given by

$$\Sigma - E_N > 0.$$

Under the binding condition, for general potentials including Coulomb potential, Griesemer, Lieb, and Loss [130], as well as Lieb and Loss [227], proved the existence of a ground state by a compactness argument. In [130] the Pauli–Fierz model in nonrelativistic quantum electrodynamics is treated, but the method can be applied to other models including the Nelson model; see, e. g., [279] for the Nelson model. A comprehensive reference for the existence of ground states in quantum field theory is Hiroshima [184].

In the infrared singular case, the Nelson Hamiltonian has no ground state as is proven in this chapter. However, it has a ground state when a non-Fock representation is used; see [20, 235]. In a series of papers [115–117], the Nelson model on a static Lorenz manifold is considered. In particular, the infrared problem is studied, and criteria for the existence and absence of ground states are shown. We also refer to [78–80, 109, 111, 113, 118, 157].

2.6

This material is taken from [43]. Physically, it is expected that the number of soft (i. e., low-energy) bosons in the ground state diverges when the infrared cutoff is removed, and thus a ground state cannot exist in Fock space. Equation (2.6.18) mathematically validates the first half of this intuition. The pull-through formula is quite useful to estimate $(\Psi_g, N\Psi_g)$, which is used in, e. g., [5, 30, 32, 104, 112, 172]. Superexponential decay of the number of bosons in the ground state of models in quantum field theory is also studied in [134, 296]. Ground state expectation for the fractional power of the number operator in Section 2.6.3 is studied in [188], and the Gaussian domination stated in Corollary 2.87 is investigated in [171, 181].

2.7

In the infrared singular case, i. e., when $\int_{\mathbb{R}^d} |\hat{\varphi}(k)|^2/\omega(k)^3 dk = \infty$, the absence of the ground state of the Nelson Hamiltonian is proven in [24, 79, 167, 235, 236]. Theorem 2.63 is due to [235, 236], where $|V(x) \geq |x|^{2\beta}$ for some $\beta > 0$ is needed. In [24, 79, 167], it is shown that $\|N^{1/2}\Psi_g\| = \infty$ follows from the pull-through formula, if H_N has a ground state under $\int_{\mathbb{R}^d} |\hat{\varphi}(k)|^2/\omega(k)^3 dk = \infty$. This contradicts $\Psi_g \in D(H_f)$ in the massive case and a ground state does not exist. In the massless case, further investigation is needed. It is not straightforward to show the absence of the ground state of the Nelson model on a static Lorentz manifold by operator theory. Since the dispersion relation is the pseudodifferential operator $\omega_{a,m} = (-\sum_{\mu,\nu=1}^3 \nabla_\mu a_{\mu\nu}(x)\nabla_\nu + m^2(x))^{1/2}$, it is then not clear how to estimate the decay rate of $\widehat{\omega_{a,m}^{\alpha}\varphi}(k)$ as $|k| \to \infty$. Hence the pull-through formula is not applicable. Instead of operator theory, one can apply the functional integral method introduced in this chapter to show the absence of a ground state.

2.8

Lemma 2.72 can be seen as a special case of the following general proposition.

Let $(X_n)_{n\in\mathbb{N}}$ be a projective system of Polish spaces with continuous maps $\pi_{m,n} : X_n \to X_m$ $(m \le n)$ satisfying $\pi_{k,m} \circ \pi_{m,n} = \pi_{k,n}$ $(k \le m \le n)$. Let

$$X = \varprojlim X_n = \left\{(x_n)_{n\in\mathbb{N}} \in \prod_{n=1}^{\infty} X_n \,\middle|\, \pi_{m,n}(x_n) = x_m \text{ for all } m \le n\right\},$$

and denote by $\pi_n : X \to X_n$ the canonical projections. Let $(\nu_n)_{n\in\mathbb{N}}$ be a projective family of probability measures, i. e., $\nu_m = \nu_n \circ \pi_{m,n}^{-1}$ $(m \le n)$. Let $\mathscr{G}$ be the algebra generated by cylinder sets $\pi_n^{-1}(A)$ with $A \in \mathcal{B}(X_n)$. Assume that μ is a finitely additive set function on $\mathscr{G}$ such that $\mu(\pi_n^{-1}(A)) = \nu_n(A)$. Let $E_n \in \mathscr{G}$ satisfy $E_1 \supset E_2 \supset \cdots$, $E_n = \pi_n^{-1}(E_n')$ with $E_n' \in \mathcal{B}(X_n)$, and $\lim_{n\to\infty} \mu(E_n) = a > 0$. Then $\bigcap_{n=1}^{\infty} E_n \neq \emptyset$ and hence μ is countably additive.

For the Nelson Hamiltonian, we define the family of probability measures μ_T, $T \ge 0$, by (2.8.1) with the pair potential W, and a limit measure μ_∞ exists (Theorem 2.77). Due to the infinite range pair potential W, μ_T does not define a Markov process. The pair potential W of the Nelson model is given by an iterated Lebesgue integral and the limit measure μ_∞ can be applied to derive ground state expectations of observables. This discussion was initiated by [44, 294, 296] and extended to other models by [171, 181]. The key ingredient to prove the existence of the limit measure is the existence of the ground state Ψ_g of the Nelson Hamiltonian. In [265], however, the existence of a limit measure is shown without assuming the existence of the ground state but W is modified. For the proper pair potential of the Nelson model, the existence, uniqueness, and properties of the Gibbs measure have been obtained under the assumption of a small coupling constant in [234] by using cluster expansion. This has been further developed to cover Gibbs measures on more singular processes involving double Itô-integrals related to the ultraviolet renormalized Nelson model in [140, 230], by using techniques of rough paths. Table 5.1 below summarizes the Gibbs measures associated with H_{N}, $H_{\mathrm{N}}(0)$, $H_{\mathrm{N}}^{\mathrm{ren}}$, and $H_{\mathrm{N}}^{\mathrm{ren}}(0)$.

An alternative proof of the existence of the limit measure is to show the tightness of the family of measures $\{\mu_T\}_{T\ge 0}$. This is carried out in [42] for the Pauli–Fierz Hamiltonian. It is, however, not straightforward to show the tightness of the family of measures with a pair potential including discontinuous paths. Sufficient conditions for the tightness of the family of measures on a discontinuous path space are comprehensively discussed in [46, 86].

Table 5.1: Nelson-type model.

	Gibbs measure	probability space	σ-field
H_{N}	$\mu_\infty, \mathcal{N}$	$\mathfrak{X}$	$\mathcal{B}(\mathfrak{X})$
$H_{\mathrm{N}}, p(\phi)_1$	$\hat{\mu}_\infty$	$\mathfrak{X} \times \mathfrak{Y}$	$\mathcal{B}(\mathfrak{X} \times \mathfrak{Y})$
$H_{\mathrm{N}}(0)$	$\nu_\infty, \mathcal{P}_\infty^0$	$\mathfrak{X}$	$\mathcal{B}(\mathfrak{X})$
$H_{\mathrm{N}}(0), p(\phi)_1$	$\hat{\nu}_\infty, \mathcal{P}_\infty^W$	$\mathfrak{X} \times \mathfrak{Y}$	$\mathcal{B}(\mathfrak{X} \times \mathfrak{Y})$
$H_{\mathrm{N}}^{\mathrm{ren}}$	$\mu_\infty^{\mathrm{ren}}$	$\mathfrak{X}$	$\mathcal{B}(\mathfrak{X})$
$H_{\mathrm{N}}^{\mathrm{ren}}(0)$	$\nu_\infty^{\mathrm{ren}}$	$\mathfrak{X}$	$\mathcal{B}(\mathfrak{X})$

2.9

Carmona-type estimate plays an important role in investigating decay properties of eigenvectors of Schrödinger operators. This is shown in Carmona [62]. It can be applied to estimate decay properties of bound states of the Nelson Hamiltonian. Such applications can be done not only for the Nelson Hamiltonian but also for the Pauli–Fierz Hamiltonian. We refer the reader to [190] for further details.

2.10

An eigenfunction φ of the Schrödinger operator $H_{\mathrm{p}} = -\frac{1}{2}\Delta + V$ defines a martingale with respect to the σ-field $(\mathscr{F}_t^{\mathrm{BM}})_{t\geq 0}$ by

$$h_t(x) = e^{tE} e^{-\int_0^t V(B_s+x)ds} \varphi(B_t + x), \quad t \geq 0,$$

where $(-\frac{1}{2}\Delta + V)\varphi = E\varphi$. It is seen that $\mathbb{E}[h_t(x)|\mathscr{F}_s^{\mathrm{BM}}] = \mathbb{E}[h_s(x)]$, thus $\varphi(x) = \mathbb{E}[h_0(x)] = \mathbb{E}[h_{t\wedge\tau}(x)]$ follows for every stopping time τ and $|\varphi(x)|$ can be estimated by choosing a suitable stopping time according to V. This is applied to the estimate of the spatial decay of eigenfunctions of the semirelativistic Schrödinger operators in [63, 185]. We also refer to [53, 54] for discussions on capacitary measures. This method is also applicable to models in quantum field theory with nonlocal kinetic terms; see [181].

2.11

Removal of the ultraviolet cutoff of the Nelson Hamiltonian (Theorem 2.100) has been proven first by Nelson [257] by using the so-called Gross transformation [131, 225]. Nelson [258] also tried to prove this by functional integration, and finally Gubinelli, Hiroshima, and Lőrinczi [139] proved this by path integral methods, where a stochastic Fubini theorem [309] was used and a uniform lower bound of $H_{\mathrm{N}}^{\varepsilon}$ with respect to ε was proven. Furthermore, Matte and Møller [243] gave an alternative proof by using path measures. Regarding functional integral representations of semigroups generated by renormalized Nelson Hamiltonians, Lemma 2.114 is due to [139, 141] and Theorem 2.120 is [243]. In Theorem 2.120, a functional integral representation of $(F, e^{-TH_{\mathrm{N}}^{\mathrm{ren}}} G)$ for arbitrary F, G

is given, where the uniqueness of the ground state can be also derived by the Perron–Frobenius theorem.

Proposition 2.108 is due to Bley and Thomas [51, 50] where the case of $0 < \alpha < 2$ is estimated. Using this formula, the ground state energies of related models including polaron model can be estimated. In order to derive Proposition 2.108, the Clark–Ocone formula is applied; see, e. g., [213, 214, 262]. Let $f : \mathbb{R} \to \mathbb{R}$ be absolutely continuous. Then the fundamental theorem of calculus is given by $f(x) = f(0) + \int_0^x f'(t)dt$ a. e. A counterpart of this in stochastic analysis is the Clark–Ocone formula (see Theorem 1.166). Let F be a Brownian functional which is $\mathcal{F}_T^{\mathrm{BM}}$-measurable. Then under some conditions, there exists an $(\mathcal{F}_t^{\mathrm{BM}})$-adapted process $(\rho_t)_{t\geq 0}$ such that

$$F = \mathbb{E}[F] + \int_0^T \rho_t dB_t.$$

Thus

$$\mathbb{E}[e^F] = e^{\mathbb{E}[F]}\mathbb{E}[e^{\int_0^T \rho_t dB_t}] = e^{\mathbb{E}[F]}\mathbb{E}[e^{\int_0^T \rho_t dB_t + \frac{p}{2}\int_0^T |\rho_t|^2 dt - \frac{p}{2}\int_0^T |\rho_t|^2 dt}].$$

Then by Schwarz inequality and the same trick used in (2.11.18) with the Girsanov theorem, it follows that

$$\begin{aligned} e^{\mathbb{E}[F]} &\leq e^{\mathbb{E}[F]}\mathbb{E}[e^{p\int_0^T \rho_t dB_t + \frac{p^2}{2}\int_0^T |\rho_t|^2 dt}]^{1/p}\mathbb{E}[e^{-\frac{qp}{2}\int_0^T |\rho_t|^2 dt}]^{1/q} \\ &= e^{\mathbb{E}[F]}\mathbb{E}[e^{-\frac{qp}{2}\int_0^T |\rho_t|^2 dt}]^{1/q} \end{aligned}$$

for $1/p + 1/q = 1$. Using this bound and some estimates on the heat kernel, Proposition 2.108 can be obtained. The bounds (2.11.7)–(2.11.8) are presented in [5, 172].

The existence of a ground state of the renormalized Nelson Hamiltonian $H_{\mathrm{N}}^{\mathrm{ren}}$ is proven in [189] for arbitrary values of coupling constants, in [172] for sufficiently small coupling constants, and in [5] for the massive case. A path measure associated with the ground state of the renormalized Nelson Hamiltonian discussed in Section 2.11.6 is taken from [189] where Gaussian domination and superexponential decay of the number of boson of the ground state are also shown. If the infrared singular condition

$$\int_{\mathbb{R}^3} \frac{|\xi_\lambda(k)|^2}{\omega(k)^3} dk = \infty$$

is satisfied, then H_{ren} has no ground state. This is also proven in [189].

In [76, 77], a weak coupling limit of the form $H_{\mathrm{p}} + \Lambda\phi(f) + \Lambda^2 H_{\mathrm{f}}$ is discussed. In [175] the weak coupling limit with a simultaneous removal of ultraviolet cutoff is studied by using functional integrals. Related works on the scaling limit in quantum field theory are [17, 19, 84, 267, 306]. In particular, in [17] a general criterion of convergence in the scaling limit is found.

The spatial decay of the ground state of the renormalized Nelson model has been rigorously established in [190]. More precisely, the renormalized Nelson model with a scalar potential $V(x) \sim |x|^{2n}$ it is shown that the ground state admits sharp two-sided bounds of the form

$$C_1 e^{-(1+\varepsilon)|x|^{n+1}} \leq \|\Psi_g(x)\|_{\mathcal{F}} \leq C_2 e^{-(1-\varepsilon)|x|^{n+1}}$$

for sufficiently large $|x|$ and any $\varepsilon > 0$, with suitable constants $C_1, C_2 > 0$. This result provides a precise characterization of the superexponential decay of the ground state and, in particular, yields a satisfactory and essentially optimal description of its spatial behavior. We define the space of Lipschitz continuous paths

$$C_T(x,y) = \{\gamma \in \mathrm{Lip}([0,T];\mathbb{R}^3) \mid \gamma(0) = y,\ \gamma(T) = x\}, \quad x,y \in \mathbb{R}^3,\ T > 0,$$

as well as the length functional, defined on the union of all such path spaces,

$$L_V(\gamma) = \int_0^T \sqrt{2V(\gamma(s))}\, |\dot{\gamma}(s)|\, ds, \quad \gamma \in C_T(x,y).$$

The value of $L_V(\gamma)$ is invariant under Lipschitz reparametrizations of γ. It is therefore natural to define the geodesic distances $d(x,y) = \inf_{\gamma\in C_T(x,y)} L_V(\gamma)$ and $\varrho(x) = d(x,0)$ for $x,y \in \mathbb{R}^3$. In the study of eigenfunctions of Schrödinger operators, $d(x,y)$ is commonly referred to as the *Agmon distance*. The proof relies on a comprehensive and delicate exploitation of the Agmon metric in combination with the Feynman–Kac representation. Notably, the derivation of the lower bound is achieved here for the first time, marking a substantial advance in the study of spatial decay properties.

2.12

The translation invariant Nelson model and related models have been studied in, e. g., [9, 59, 61, 104, 105, 133, 134, 253]. The functional integral approach to this model has been developed, and Theorem 2.147 is due to [180, 295]. The identity (2.12.10) is due to [105, 295], and the pull-through formula (2.12.12) is studied in [105]. The removal of the ultraviolet cutoff of translation invariant Nelson Hamiltonian is parallel with that of the Nelson Hamiltonian; see, e. g., [183]. The renormalization term

$$-\frac{g^2}{2}\int_{\mathbb{R}^3} \frac{|\hat{\varphi}(k)|^2}{\omega(k)} \frac{1}{\omega(k) + |k|^2/2} dk$$

introduced by Nelson [257] coincides with the coefficient of g^2 in the Taylor expansion of the ground state energy $E_{\mathrm{N}}(g^2)$ of the translation invariant Nelson Hamiltonian with zero total momentum $H_{\mathrm{N}}(0) = \frac{1}{2}P_{\mathrm{f}}^2 + H_{\mathrm{f}} + gH_{\mathrm{I}}(0)$ by a formal perturbation theory. On the other hand, the ground state energy is represented as $E_{\mathrm{N}}(g^2) = -\lim_{t\to 0} \frac{1}{t} \log(f \otimes$

$1, e^{-tH_N(0)}f \otimes 1)$, and the renormalization term coincides with the diagonal part of the pair interaction. Hence the formal derivation of the renormalization term from the Taylor expansion of $E_N(g^2)$ is valid by functional integral representation in Theorem 2.168 involving $\lim_{g\to 0} \frac{E_N(g^2)-E_N(0)}{g^2}$, which is due to [182]. In [180], it is proven that the semigroup $e^{-TH_{\mathrm{Nren}}(p)}$ is positivity improving for $p = 0$. The extension to the case $p \neq 0$ has been subsequently established in [250, 224, 162].

2.13

The study of the polaron model was initiated by Landau in 1933 to describe an electron moving in a crystal, where atoms move from their equilibrium positions to effectively screen the charge of an electron, known as a phonon cloud. This lowers the electron mobility and increases the so-called *effective mass* [223] of the electron. Fröhlich [103, 102] proposed a model for this polaron through which its dynamics are treated quantum mechanically. This model assumes that the wave-function describing the electron is spread out over many ions, which are all somewhat displaced from their equilibrium positions. Much of the theory focussed on solving this so-called Fröhlich polaron. Reference [120] is a useful summary of various versions of polaron models. Feynman introduced a variational principle for path integrals to study the polaron in [96], and a Feynman path integral approach to the ground state energy of polaron Hamiltonian is a remarkable result. In [294] the effective mass is studied by functional integration, where a double stochastic integral appears. Suppose that the strength of interaction between electron and phonon is expressed by a coupling constant α. In physics, despite the lack of an exact solution, some approximations of the polaron properties are known. When $|\alpha|$ is sufficiently small, the self-energy of the polaron can be approximated linearly in α. On the other hand, when $|\alpha|$ is large, a variational approach indicates that the self-energy of polaron is proportional to α^2. It was an open question for many years whether this expression was asymptotically exact as α tends to infinity. Applying *large deviation theory* to functional integration for the self-energy, [83] contributed to solving this problem. Later [228] gave a shorter proof using more conventional methods. We also refer to [101, 233, 247, 254] for the spectral analysis of polaron-type models.

2.14

The presentation here follows from Betz and Spohn [44]. The functional central limit theorem in Theorem 2.203 follows from [45, 153, 202]. The above theorem still contains the possibility that the diffusion matrix is zero, which would imply subdiffusive behavior of Brownian paths. There are two independent proofs of ruling this out, one in [44, 294] using ideas of [60], and another in [138]. Both require substantial additional techniques to the material of this book, and therefore they are omitted in our presentation.

Notes to Chapter 3

3.1

In 1938, the Pauli–Fierz model was introduced in [269], where the generation of infrared photons was studied. In particular, they considered the cross-section under assumption that a finite extent was ascribed to the electron, and thus the wavelength of photon shorter than the diameter of electron was cut, equivalently ultraviolet cutoff was introduced. For the model, they considered they took the *dipole approximation* and neglected $\hat{A}^2$ term for the sake of simplicity. Then it is of the form

$$-\frac{1}{2m}\Delta\otimes 1+\frac{1}{m}i\nabla\otimes\hat{A}(0)+V\otimes 1+1\otimes H_{\mathrm{rad}}$$

which can be diagonalized as

$$-\frac{1}{2(m+\delta m)}\Delta\otimes 1+V(\cdot-\phi)+1\otimes H_{\mathrm{rad}}+c.$$

Here, ϕ is some field operator and c and δm are constants. See also [52] for the infrared divergence. Bethe [41] and Welton [313] proposed a theoretical interpretation of the Lamb shift by a variant of the Pauli–Fierz model. We refer to [70, 245] for details, and to [300] for a summary of recent progress of the spectral analysis of the Pauli–Fierz and related models.

Consider a single classical particle in the configuration space $\mathbb{R}^3$, interacting with an electromagnetic field. The particle has mass $m = 1$ and a charge distribution $\hat{\varphi}$, and its dynamics is completely characterized by its momentum $p\in\mathbb{R}^3$ and position $q\in\mathbb{R}^3$. On the other hand, it is convenient for our purposes to describe the electromagnetic field in the Coulomb gauge by a complex vector field $\alpha=(\alpha_\lambda)_{\lambda=1,2}:\mathbb{R}^3\to\mathbb{C}$. The Hamiltonian of the Newton–Maxwell system takes the form

$$H(p,q,\alpha)=\frac{1}{2}(p-A(q,\alpha))^2+V(q)+\sum_{\lambda=1,2}\int_{\mathbb{R}^3}|\alpha_\lambda(k)|^2\,|k|\,dk.$$

This leads to the Newton–Maxwell equations

$$\begin{cases}\dot{p}=-\frac{\partial H}{\partial q},\\ \dot{q}=\frac{\partial H}{\partial p},\\ i\dot{\alpha}_\lambda(k)=\frac{\partial H}{\partial\overline{\alpha_\lambda(k)}}.\end{cases}$$

Hence we obtain

$$\text{(NM)}\quad \begin{cases} \dot{p} = (p - A(q,a)) \cdot \nabla_q A(q,a) - \nabla_q V(q), \\ \dot{q} = p - A(q,a), \\ i\dot{a}_\lambda(k) = |k| a_\lambda(k) - \hat{\varphi}(k) \frac{1}{\sqrt{2|k|}}(p - A(q,a)) \cdot e_\lambda(k) e^{-ik\cdot q}. \end{cases}$$

In the classical limit $\hbar \to 0$, the Pauli–Fierz model reduces to the Newton–Maxwell equations. We refer to [6] for further details.

3.2

In the dipole approximation, the quantized radiation field $\hat{A}(x)$ is replaced by $\hat{A}(0)$. In this case, collisions between electrons and photons are ignored, and there is no derivative coupling. Under specific conditions, the Pauli–Fierz model with dipole approximation and $V = 0$ can be diagonalized, i. e., $H_{\mathrm{PF}}^{\mathrm{dip}}$ can be approximated by the $(3 + 3L)$-dimensional harmonic oscillator

$$\frac{1}{2m}\Big(-i\nabla + \sum_{l=1}^{L} \hat{\varphi}(l) Q_l\Big)^2 + \frac{1}{2}\sum_{l=1}^{L}(P_l P_l + Q_l \omega^2(l) Q_l - \omega(l)).$$

Here, $[P_l, Q_{l'}] = -i\delta_{ll'}$ is satisfied. Then it can be shown that

$$H_{\mathrm{PF}}^{\mathrm{dip}} \cong -\frac{1}{2(m + e^2\frac{d-1}{d}\|\hat{\varphi}/\omega\|^2)}\Delta \otimes 1 + 1 \otimes H_{\mathrm{f}} + g,$$

where g is a constant. In the series of papers [13–16], Arai investigated the spectrum of the Pauli–Fierz Hamiltonian in the dipole approximation. In [194], the diagonalization of the Pauli–Fierz Hamiltonian with the dipole approximation by Bogoliubov transformation is discussed in detail. We also refer to [49, 297] for the scattering theory.

The spectral analysis of the Pauli–Fierz Hamiltonian without dipole approximation has seen much progress since the 1990s. The existence of a ground state has been proven by Bach, Fröhlich, and Sigal [32] for weak couplings and no infrared cutoff. Griesemer, Lieb, and Loss [130], as well as Lieb and Loss [227], successively removed the restriction on the coupling constant, and have also introduced a binding condition giving a criterion for the existence of a ground state; see also [35]. We emphasize that no infrared regularity is needed to show the existence of a ground state in Fock space, in contrast to the Nelson model. More precisely, to show the existence of the ground state of the Pauli–Fierz Hamiltonian the binding condition is a sufficient condition for massive cases. In massless cases, one can check (1) localizations of the number of bosons by pull-through formula and the exponential decay [129] of massive ground state Ψ_{g} and (2) regularities with respect to variables $x \in \mathbb{R}^3$ in position of nonrelativistic particle and $k = (k_1, \ldots, k_n) \in \mathbb{R}^{3n}$ of the quantized field of the nth sector of the massive ground state $\Psi_{\mathrm{g}}^{(n)}(x, k_1, \ldots, k_n)$. Then it can be shown (see [130]) that massive ground states strongly converge to a nonzero vector as the mass reduces to zero, which is the ground state of the massless Pauli–Fierz Hamiltonian. See [184] for a brief review.

A basic assumption to show the existence of the ground state is the existence of a ground state for the zero coupling Hamiltonian, i. e., that H_{p} has a ground state with positive spectral gap. Reference [197] shows that for sufficiently large coupling constants the Pauli–Fierz Hamiltonian in the dipole approximation has a ground state even when H_{p} has no ground state. This phenomenon is called *enhanced binding*. Enhanced binding for the Pauli–Fierz Hamiltonian is also studied in [25, 65, 69, 148, 200]; see [218] for the relativistic Pauli–Fierz Hamiltonian. The enhanced binding for the N-body Nelson Hamiltonian is due to [191], and in [193] a relativistic version is studied.

3.3

Functional integral representations for the Pauli–Fierz Hamiltonian are discussed in [92, 107, 143, 145, 174, 242, 300]. We emphasize that the exponent in the functional integral representation

$$-ie\hat{A}_{\mathrm{E}}(K_t) = -ie\int_0^t \hat{A}_{\mathrm{E}_s}(\tilde{\varphi}(\cdot - B_s))dB_s$$

depends on time s explicitly, in contrast to the classical case. Gross [135] discusses a $\mathscr{Q}$-space representation of a *Proca field*, which is a model of massive quantum electrodynamics. The diamagnetic inequality (Corollary 3.28) with a quantized radiation field is obtained in [27, 174].

3.4

This material is taken from [158]. In a similar manner to the Nelson Hamiltonian, the choice of Kato-class potentials is suitable to construct a path measure associated with the ground state of the Pauli–Fierz Hamiltonian.

3.5

Essential self-adjointness of $H + V$ such that e^{-tH} is positivity preserving is studied in [90, 216]. Self-adjointness of the Pauli–Fierz Hamiltonian is established in [32] for weak couplings, and in [176, 178] for arbitrary values of coupling constants. Theorem 3.47 is taken from [178]. Alternative proofs are given in [87, 150].

An infinite-dimensional version of the Perron–Frobenius theorem is due to [125, 132, 133]. Propositions 2.25 and 3.49 are taken from [89, 91]. The notion of a *positivity preserving* operator can be extended to *invariant cone* property. Let $\mathcal{H}$ be a real vector space. $\mathcal{K} \subset \mathcal{H}$ is a *cone* if and only if $u, v \in \mathcal{K} \to u + v \in \mathcal{K}$, $v \in \mathcal{K}$ and $a \geq 0 \to av \in \mathcal{K}$, and $\mathcal{K} \cap -\mathcal{K} = \emptyset$. The Perron–Frobenius theorem can then be extended to the invariant cone semigroup. One application is to show the uniqueness of the ground state of a fermion system, which was done by [91, 133]. Theorem 3.51 is due to [177]. We also refer to [29, 125, 179, 238, 290] for the uniqueness of the ground state of the Pauli–Fierz Hamiltonian; see [248, 249] for an abstract version. The spatial exponential decay of eigenvectors of the Pauli–Fierz model is discussed in [32, 130, 158]. By employing the Agmon metric, a lower bound on the ground state of the Pauli–Fierz model is established in [201]. In [158], it is shown that $e^{-tH_{\mathrm{PF}}^{\mathrm{Kato}}}$ is also positivity improving, and the ground state of $H_{\mathrm{PF}}^{\mathrm{Kato}}$ decays exponentially. These facts are proven in the same way as for H_{PF}. In the proof of Theorem 3.53, *log-convexity* is used. The general statement is as follows: Let $F \geq 0$ be a measurable function on a probability space. Then the function $0 < r \mapsto \log \mathbb{E}[F^r]$ is convex. The proof is simple. Let $r_0, r_1 > 0$ and $0 < \theta < 1$, and set $r = (1-\theta)r_0 + \theta r_1$. Then $F^r = (F^{r_0})^{1-\theta}(F^{r_1})^{\theta}$. Hence, by Hölder's inequality, $\mathbb{E}[F^r] = \mathbb{E}[(F^{r_0})^{1-\theta}(F^{r_1})^{\theta}] \leq (\mathbb{E}[F^{r_0}])^{1-\theta}(\mathbb{E}[F^{r_1}])^{\theta}$. Taking logarithms on both sides completes the proof.

3.6

This material is taken from [42]. The formal double stochastic integral of the type $\int_{-\infty}^{\infty} ds \int_{-\infty}^{\infty} e^{-i\omega|y-s|}\gamma(t)\gamma(s)dt$ appears in Feynman [95, p. 443]. The mathematical derivation is done in [177, 258, 300]; see also [72, 251].

3.7

This material is taken from [180]. The idea to construct the functional integral representation comes from [295]; see also [7, 237]. Chen [66] shows the existence of the ground state of $H_{\mathrm{PF}}(p)$ when $p = 0$. The *effective mass* m_{eff} of the Pauli–Fierz Hamiltonian is defined by

$$\frac{1}{m_{\mathrm{eff}}} = \frac{1}{3}\Delta_p E(p)\lceil_{p=0},$$

where $E(p)$ denotes the ground state energy of $H_{\mathrm{PF}}(p)$ [136, 137]. When a sharp infrared cutoff is introduced in H_{PF}, $E(p)$ is differentiable with respect to p_μ for sufficiently small coupling constants. It is, however, not clear that $E(\cdot)$ is twice differentiable in a neighborhood of $p = 0$, when no infrared cutoff is introduced. The infrared problem of the effective mass is studied in [28, 67, 68, 147]. The functional integral approach of the effective mass is [44, 294]. The asymptotic behavior of effective mass with respect to the ultraviolet cutoff is studied in [186, 199].

3.8

A basic reference for the material of this section is [187]. The degeneracy of ground states for weak enough coupling is proven in [179, 198]. The uniqueness of the ground state of the translation invariant spinless Pauli–Fierz Hamiltonian is also proven in [198]. Spohn [296] treats the model by functional integration.

The electronic part of the Pauli–Fierz Hamiltonian is nonrelativistic, given by a Schrödinger operator. Nelson's work on Euclidean field theory for spinless boson proved to be a very significant conceptual and technical stimulus in the program of constructive quantum field theory. A version for fermions has been also discussed. The Euclidean Fermi field was introduced in [266]. Reference [315] provides an axiomatic approach. See, furthermore, [38] for the Itô–Clifford integral, [12] for the fermionic Itô formula, [229] for fermion martingales, [277] for fermion path integrals, and [221] for functional integration of the Euclidean Dirac field.

3.9

Relativistic Pauli–Fierz Hamiltonian H_{RPF} is studied in many articles. In the functional integral representation, the subordinated Brownian motion $(B_{T_t})_{t\geq 0}$ is crucial. General *change of time* of processes is studied in [36]. In [106], the asymptotic field of the Pauli–Fierz Hamiltonian, as well as relativistic one, is investigated. In [181, 159], the self-adjointness and essential self-adjointness of the relativistic Pauli–Fierz Hamiltonian is shown by using a path measure, where an invariant domain of the semigroup $e^{-tH_{\mathrm{RPF}}}$ is found, and the uniqueness of the ground state is also derived by showing $e^{-tH_{\mathrm{RPF}}}$ is unitary equivalent to a positivity improving semigroup. The existence of the ground state of the relativistic Pauli–Fierz Hamiltonian is studied in, e. g., [217–220, 244] for the case of $H_{\mathrm{RPF}} = \sqrt{(-i\nabla - eA)^2 + m^2} + V + H_{\mathrm{rad}}$ with $m \neq 0$. Relativistic Pauli–Fierz Hamiltonian with zero particle mass, i. e.,

$$H_{\mathrm{RPF}} = |-i\nabla - eA| + V + H_{\mathrm{rad}},$$

is the most singular case, and the existence of the ground state of this is shown in [160] for massive case and, in [161] for massless case, where the functional integral representation and asymptotic field are applied. In [181], the Gibbs measure associated with the ground state of H_{RPF} is shown by using the existence of ground state, and the Gaussian domination of the ground state is proven. See Table 5.2. In [119, 192], binding conditions are discussed.

Table 5.2: Pauli–Fierz-type model.

	Gibbs measure	probability space	σ-field
H_{PF}	μ_∞	$\mathscr{X}$	$\mathcal{B}(\mathscr{X})$
$H_{\mathrm{PF}}^{\mathrm{rel}}$	$\mu_\infty^{\mathrm{rel}}$	$D(\mathbb{R};\mathbb{R}^{3+1})$	$\sigma(\mathscr{G})$

We introduce three classes V_1, V_2, and V_3 of external potentials V. We say that V is a relativistic Kato-class potential whenever $V \in V_3$, but $V_1 = V_2 = V_3$ if V is uniformly locally integrable. This is studied in [3]. In [232], not only relativistic Kato-class but also

Kato-class for Bernstein function of Laplacian is investigated. Let $\sigma = (\sigma^1, \sigma^2, \sigma^3)$ be the 2×2 Pauli matrices. We define the *relativistic Pauli–Fierz Hamiltonian with spin* 1/2 by

$$H_{\mathrm{RPF}}^{\mathrm{S}} = \sqrt{(\sigma \cdot (-i\nabla - eA))^2 + m^2} - m + V + H_{\mathrm{rad}}.$$

The functional integral representation of $e^{-tH_{\mathrm{RPF}}^{\mathrm{S}}}$ is given by an application of both of those of the Pauli–Fierz Hamiltonian with spin and the relativistic Pauli–Fierz Hamiltonian:

$$(F, e^{-tH_{\mathrm{RPF}}^{\mathrm{S}}} G) = e^{T_t} \sum_{\alpha=1,2} \int_{\mathbb{R}^3} \mathbb{E}^{x,0,\alpha} \left[e^{-\int_0^t V(B_{T_s})} (\mathrm{J}_0 F(q_{T_0}), e^{S_t} \mathrm{J}_t G(q_{T_t})) \right] dx,$$

where the exponent is given by

$$S_t = -ie\hat{A}_{\mathrm{E}}(K_{T_t}^{\mathrm{rel}}) - \int_0^{T_t} \hat{H}_{\mathrm{E,d}}(B_s, \theta_{N_s}, s) ds + \int_0^{T_t+} \log(-\hat{H}_{\mathrm{E,od}}(B_s, -\theta_{N_{s-}}, s)) dN_s.$$

3.10

Proposition 3.162 is proven by [16]. We also refer the reader to see [194, Theorem 3.28]. Proposition 3.168 referring to Ahiezer–Kac formula is taken from [2, p. 312]; see [210, (1.9)] for (3.10.7).

Notes to Chapter 4

4.1

The spin-boson model is a much studied variant of the Caldeira–Leggett model describing a two-state quantum system linearly coupled to a scalar quantum field. A general reference for the spin-boson model is [226]. Work on the spectral properties of the spin-boson and related models includes [1, 22, 23, 165, 171] for the ground state energy, [8, 151, 296, 298, 305] for the existence of ground state, and [17, 57, 88, 166, 205, 301] for the spectral analysis.

4.2

This is taken from [171]. In [40, 85, 171, 205, 296, 301], stochastic methods for spin-boson Hamiltonian are studied and applied. The functional integral representation of the spin-boson model follows directly from results in [187]; see also [40]. The shift invariance of spin processes is discussed in [232].

4.3

The existence and absence of the ground state is investigated in [22, 23, 74, 111, 112, 114, 171]. Arai and Hirokawa [22, 23] defined a generalized spin-boson model and proved the existence and uniqueness of a ground state for sufficiently small coupling constants. In [112, 171], the existence and uniqueness of the ground state is shown for arbitrary values of coupling constants. Theorem 4.12 is taken from [171]. In particular, in [151] it is shown that the spin-boson Hamiltonian has a ground state even when no infrared regularity condition is imposed. In [74] a general version of the spin-boson model is proposed, and the existence of a ground state of H_{SB}^{-} and the absence of a ground state of H_{SB}^{+} are shown by using the symmetries of the spin-boson model.

4.4

The Gibbs measure associated with the ground state and its applications have been investigated in [171]. See Table 5.3. Here the van Hove representation is also discussed and the ground state expectation $(\sigma_\alpha \otimes \Omega, e^{-tH_{\mathrm{SB}}}\sigma_\beta \otimes \Omega)$ is given in terms of a Poisson-driven process. This was also studied operator-theoretically in [164].

Table 5.3: Spin-boson model.

	Gibbs measure	probability space	σ-field
H_{SB}	μ_∞	$\mathscr{X}_D$	$\mathcal{B}(\mathscr{X}_D)$
H_{Rabi}	$\mu_\infty^{\mathrm{Rabi}}$	$\mathscr{X}_D$	$\mathcal{B}(\mathscr{X}_D)$

4.5

In 1936, Rabi introduced a model in [273] to discuss the effect of a rapidly varying weak magnetic field on an oriented atom possessing nuclear spin. The atom is treated as a two-level system and the field as a classically rotating field. Experimental physicists evidence the interaction by coupling a two-level atom with a one-mode light in a mirror cavity, which can be realized as the Rabi model. Jaynes and Cummings [209] introduced a similar quantum model in 1963 describing a two-level atom interacting with a quantized mode of an optical cavity. Traditionally, to solve the model, the so-called rotating wave approximation is taken. In this approximation, the counterrotating term of the model introduced by Jaynes and Cummings is neglected. Then it turned out to be a valid approximation for the near resonance and weak coupling regions; see [317] for the details and references therein. The Rabi model is studied from a mathematical point of view in, e. g., [58, 169, 168, 222]. In [170], a stochastic method is applied to estimate the ground state energy, and Theorem 4.34 is taken from here. A Gibbs measure can also be constructed for the Rabi model, and properties of its ground state have been investigated in [195], in a manner analogous to the spin-boson model.

Instead of $L^2(\mathbb{R})$, taking Segal–Bargmann space $\mathcal{K}$ is useful to study H_{Rabi}, see [284]. Let

$$\mathcal{B} = \left\{ f(z) \text{ is holomorphic on } \mathbb{C} \;\middle|\; \int_{\mathbb{C}} |f(z)|^2 e^{-|z|^2} dxdy < \infty \right\}.$$

Then $L^2(\mathbb{R})$ is unitarily equivalent to $\mathcal{B}$, and under this identification it follows that $a \cong d/dz$ and $a^* \cong z$. Then the eigenvalue equation

$$H_{\text{Rabi}} \begin{pmatrix} \psi_1 \\ \psi_2 \end{pmatrix} = E \begin{pmatrix} \psi_1 \\ \psi_2 \end{pmatrix}$$

can be reduced to find holomorphic solutions of the system

$$\begin{aligned} (z+g)\frac{d\psi_1}{dz} &= (E - gz)\psi_1 - \Delta\psi_2, \\ (z-g)\frac{d\psi_2}{dz} &= (E + gz)\psi_2 - \Delta\psi_1. \end{aligned}$$

Let $\kappa = 0$. Then the eigenvalue equation turns into

$$\left(\frac{d}{dz} - \frac{E+g^2}{z+g} + g \right)\phi = 0.$$

Thus $\phi(z) = Ce^{-gz}(z+g)^{E+g^2}$, and ϕ must be holomorphic, implying that $E + g^2$ is a nonnegative integer. Thus eigenvalues are of the form

$$E = n - g^2, \quad n = 0, 1, 2, \ldots$$

Similar arguments for H_{Rabi} with $\kappa \neq 0$ have been presented in, e. g., [222]. Let $\{E_n(g)\}_{n=0}^{\infty}$ denote the eigenvalues of the Rabi Hamiltonian $H_{\text{Rabi}} = H_{\text{Rabi}}(g)$. The associated spectral zeta function is defined by

$$\zeta_g(s) = \sum_{n=0}^{\infty} \frac{1}{E_n(g)^s}, \qquad \Re s > 1.$$

It admits the Mellin representation

$$\zeta_g(s) = \frac{1}{\Gamma(s)} \int_0^{\infty} t^{s-1} \operatorname{Tr}(e^{-tH_{\text{Rabi}}})\, dt.$$

It is known that $\lim_{g\to\infty} E_n(g) = \infty$ for each n. We define the Hurwitz-type spectral zeta function by

$$\zeta_g(s,\tau) = \sum_{n=0}^{\infty} \frac{1}{(E_n(g) + g^2 + \tau)^s}, \qquad \Re s > 1,$$

where $\tau > 0$. In [195, 196, 274], it is shown that this Hurwitz-type spectral zeta function converges to the classical Hurwitz zeta function:

$$\lim_{g\to\infty} \zeta_g(s;\tau) = 2\sum_{n=0}^{\infty} \frac{1}{(n+\tau)^s}.$$

The result mentioned above can be extended to $s \in \mathbb{C} \setminus \{1\}$. A similar analysis has been carried out for the two-photon Rabi Hamiltonian. A two-photon Rabi Hamiltonian is defined by

$$H_{\mathrm{Rabi},2p} = \kappa\sigma_z + \omega a^* a + g\sigma_x(a^2 + (a^*)^2)$$

which is bounded from below for $|g| < 1/2$. In [163], it is shown that the corresponding Hurwitz-type spectral zeta function,

$$\zeta_g(s;\tau) = \sum_{n=0}^{\infty} \frac{1}{\left(\frac{E_n(g)}{\sqrt{1-4g^2}} + \tau\right)^s}, \qquad \Re s > 1,$$

also converges to the classical Hurwitz zeta function:

$$\lim_{g\uparrow 1/2} \zeta_g(s;\tau) = 2\sum_{n=0}^{\infty} \frac{1}{(n+\tau)^s}.$$

The two-photon Rabi model is intimately connected with the representation theory of $\mathrm{SU}(1,1)$, whereas the Rabi model bears a profound relationship to $\mathrm{SU}(2)$.

Bibliography

[1] A. Abdesselam. The ground state energy of the massless spin-boson model. *Ann. Henri Poincaré*, 12:1321–1347, 2018.

[2] N. I. Ahiezer. The continuous analogue of some theorems on Toeplitz matrices. *Transl. from Ukr. Mat. Zh.*, 16:455–462, 1964. In Amer. Math. Soc. Transl. Ser. 2, 50, 295–316, 1966.

[3] M. Aizenman and B. Simon. Brownian motion and Harnack's inequality for Schrödinger operators. *Comm. Pure Appl. Math.*, 35:209–271, 1982.

[4] S. Albeverio, S. Paycha, and S. Scarlatti. A short overview of mathematical approaches to functional integration. Functional integration, geometry and strings. *Prog. Phys.*, 13:230–276, 1989.

[5] Z. Ammari. Asymptotic completeness for a renormalized non-relativistic Hamiltonian in quantum field theory: the Nelson model. *Math. Phys. Anal. Geom.*, 3:217–285, 2000.

[6] Z. Ammari, M. Falconi and F. Hiroshima. Towards a derivation of classical ElectroDynamics of charges and fields from QED. *Annales de l'Institut Fourier*, 76:701–787, 2026.

[7] L. Amour, B. Grebert, and J. C. Guillot. The dressed mobile atoms and ions. *J. Math. Pures Appl.*, 86:177–200, 2006.

[8] N. Angelescu, R. A. Minlos, J. Ruiz, and V. Zagrebnov. Lower spectral branches of spin-boson model. *J. Math. Phys.*, 49:102105, 2008.

[9] N. Angelescu, R. A. Minlos, and V. A. Zagrebnov. Lower spectral branches of a particle coupled to a Bose field. *Rev. Math. Phys.*, 17:1111–1142, 2005.

[10] D. Applebaum. Universal Malliavin calculus in Fock and Levy-Ito spaces. *Comm. Stoch. Anal.*, 3:119–141, 2009.

[11] D. Applebaum. *Lévy Processes and Stochastic Calculus*. Cambridge University Press, 2nd ed., 2009.

[12] D. Applebaum and R. L. Hudson. Fermion Ito's formula and stochastic evolutions. *Comm. Math. Phys.*, 96:473–496, 1984.

[13] A. Arai. On a model of a harmonic oscillator coupled to a quantized, massless, scalar field I. *J. Math. Phys.*, 22:2539–2548, 1981.

[14] A. Arai. On a model of a harmonic oscillator coupled to a quantized, massless, scalar field II. *J. Math. Phys.*, 22:2549–2552, 1981.

[15] A. Arai. Note on scattering theory in non-relativistic quantum electrodynamics. *J. Phys. A*, 16:49–70, 1983.

[16] A. Arai. Rigorous theory of spectra and radiation for a model in quantum electrodynamics. *J. Math. Phys.*, 24:1896–1910, 1983.

[17] A. Arai. An asymptotic analysis and its applications to the nonrelativistic limit of the Pauli–Fierz and a spin-boson model. *J. Math. Phys.*, 31:2653–2663, 1990.

[18] A. Arai. *Quantum Field Theory and Statistical Mechanics*. Nippon hyôronshya, in Japanese, 1990.

[19] A. Arai. Scaling limit for quantum systems of nonrelativistic particles interacting with a Bose field. *Hokkaido Univ. Preprint Series*, unpublished, 1992.

[20] A. Arai. Ground state of the massless Nelson model without infrared cutoff in a non-Fock representation. *Rev. Math. Phys.*, 13:1075–1094, 2001.

[21] A. Arai. *Analysis of Fock Spaces and Mathematical Theory of Quantum Fields*. World Scientific, 2018.

[22] A. Arai and M. Hirokawa. On the existence and uniqueness of ground states of a generalized spin-boson model. *J. Funct. Anal.*, 151:455–503, 1997.

[23] A. Arai and M. Hirokawa. Ground states of a general class of quantum field Hamiltonians. *Rev. Math. Phys.*, 12:1085–1135, 2000.

[24] A. Arai, M. Hirokawa, and F. Hiroshima. On the absence of eigenvectors of Hamiltonians in a class of massless quantum field models without infrared cutoff. *J. Funct. Anal.*, 168:470–497, 1999.

[25] A. Arai and H. Kawano. Enhanced binding in a general class of quantum field models. *Rev. Math. Phys.*, 15:387–423, 2003.

[26] A. Araki. *Mathematical Theory of Quantum Fields*. Oxford University Press, 2000.

https://doi.org/10.1515/9783112231494-006

[27] J. Avron, I. Herbst, and B. Simon. Schrödinger operators with magnetic fields. I. General interactions. *Duke Math.*, 45:847–883, 1978.

[28] V. Bach, T. Chen, J. Fröhlich, and I. M. Sigal. The renormalized electron mass in non-relativistic quantum electrodynamics. *J. Funct. Anal.*, 243:426–535, 2007.

[29] V. Bach, J. Fröhlich, and A. Pizzo. Infrared-finite algorithms in QED: the ground state of an atom interacting with the quantized radiation field. *Comm. Math. Phys.*, 264:145–165, 2006.

[30] V. Bach, J. Fröhlich, and I. M. Sigal. Quantum electrodynamics of confined nonrelativistic particles. *Adv. Math.*, 137:299–395, 1998.

[31] V. Bach, J. Fröhlich, and I. M. Sigal. Renormalization group analysis of spectral problems in quantum field theory. *Adv. Math.*, 137:205–298, 1998.

[32] V. Bach, J. Fröhlich, and I. M. Sigal. Spectral analysis for systems of atoms and molecules coupled to the quantized radiation field. *Comm. Math. Phys.*, 207:249–290, 1999.

[33] H. Bahouri, J. Y. Chemin, and R. Danchin. *Fourier Analysis and Nonlinear Partial Differential Equations*. Springer, 2011.

[34] C. Bär and K. Fredenhagen. *Quantum Field Theory on Curved Spacetimes*. Springer, 2009.

[35] J. M. Barbaroux, T. Chen, and S. Vugalter. Binding conditions for atomic N-electron systems in non-relativistic QED. *Ann. Henri Poincaré*, 4:1101–1136, 2003.

[36] O. E. Barndorff-Nielsen and A. Shiryaev. *Change of Time and Change of Measure*. World Scientific, 2015.

[37] V. Bargmann, On a Hilbert space of analytic functions and an associated integral transform. *Comm. Pure. Appl. Math.*, 14:187–214, 1961.

[38] C. Barnett, R. F. Streater, and I. F. Wilde. The Itô–Clliford integral. *J. Funct. Anal.*, 48:172–212, 1982.

[39] Y. M. Berezansky and Yu. G. Kondratiev. *Spectral Methods in Infinite-Dimensional Analysis I–II*. Kluwer Academic, 1995.

[40] J. Bertrand, B. Gaveau, and G. Rideau. Quantum fields and Poisson processes: interaction of a cut-off boson field with a quantum particle. *Lett. Math. Phys.*, 20:73–82, 1985.

[41] H. A. Bethe. The electromagnetic shift of energy levels. *Phys. Rev.*, 72:339–342, 1947.

[42] V. Betz and F. Hiroshima. Gibbs measures with double stochastic integrals on path space. *Infin. Dimens. Anal. Relat. Top.*, 12:135–152, 2009.

[43] V. Betz, F. Hiroshima, J. Lőrinczi, R. A. Minlos, and H. Spohn. Ground state properties of the Nelson Hamiltonian – a Gibbs measure-based approach. *Rev. Math. Phys.*, 14:173–198, 2002.

[44] V. Betz and H. Spohn. A central limit theorem for Gibbs measures relative to Brownian motion. *Probab. Theory Related Fields*, 131:459–478, 2005.

[45] R. Bhattacharya. On the functional central limit theorem and the law of the iterated logarithm for Markov processes. *Z. Wahrsch. verw. Gebiete*, 60:185–201, 1982.

[46] P. Billingsley. *Convergence of Probability Measures*. Wiley Interscience, 1968.

[47] N. D. Birrell and P. C. W. Davies. *Quantum Fields in Curved Space*. Cambridge University Press, 1999.

[48] J. D. Bjorken and S. D. Drell. *Relativistic Quantum Fields*. McGraw-Hill, 1965.

[49] Ph. Blanchard. Discussion mathématique du modèle de Pauli et Fierz relatif à la catastrophe infrarouge. *Comm. Math. Phys.*, 15:156–172, 1969.

[50] G. Bley. A lower bound on the renormalized Nelson model. *arXiv:1609.08590*, preprint, 2016.

[51] G. Bley and L. Thomas. Estimates on functional integrals of quantum mechanics and non-relativistic quantum field theory. *Comm. Math. Phys.*, 350:79–103, 2017.

[52] F. Bloch and M. Nordsieck. Notes on the radiation field of the electron. *Phys. Rev.*, 52:54–59, 1937.

[53] R. M. Blumenthal and R. L. Getoor. *Markov Processes and Potential Theory*. Academic Press, New York, 1968.

[54] K. Bogdan, T. Byczkowski, M. Ryznar, R. Song, and Z. Vondraček. *Potential Analysis of Stable Processes and Its Extensions*. Lecture Notes in Mathematics, Springer, 1980,2009.

[55] N. N. Bogoliubov, A. A. Logunov, and I. T. Todorov. *Introduction to Axiomatic Quantum Field Theory*. W. A. Benjamin, 1975.

[56] N. N. Bogoliubov and D. V. Shirkov. *Introduction to the Theory of Quantum Fields*. Interscience Publisher, 1959.

[57] A. Boutet de Monvel and J. Sahbani. On the spectral properties of the spin-boson Hamiltonians. *Lett. Math. Phys.*, 44:23–33, 1988.

[58] D. Braak. Integrability of Rabi model. *Phys. Rev. Lett.*, 107:100401, 7 pp., 2011.

[59] J. Cannon. Quantum field theoretic properties of a model of Nelson: domain and eigenvector stability for perturbed linear operators. *J. Funct. Anal.*, 8:101–152, 1971.

[60] H. J. Brascamp, J. L. Lebowitz, and E. Lieb. The statistical mechanics of anharmonic lattices. *Proceedings of the 40th Session of the International Statistics Institute*, 9:1–11, 1975.

[61] J. Cannon. Continuous sample paths in quantum field theory. *Comm. Math. Phys.*, 35:215–233, 1974.

[62] R. Carmona. Pointwise bounds for Schrödinger eigenstates. *Comm. Math. Phys.*, 62:97–106, 1978.

[63] R. Carmona, W. C. Masters, and B. Simon. Relativistic Schrödinger operators: asymptotic behavior of the eigenfunctions. *J. Funct. Anal.*, 91:117–142, 1990.

[64] P. Cartier and C. DeWitt-Morette. *Functional Integration: Action and Symmetries*. Cambridge University Press, 2006.

[65] I. Catto, P. Exner, and C. Hainzl. Enhanced binding revisited for a spinless particle in non-relativistic QED. *J. Math. Phys.*, 45:4174–4185, 2004.

[66] T. Chen. Operator theoretic infrared renormalization and construction of dressed 1-particle states in non-relativistic QED. *mp-arc 01-103*, preprint, 2001.

[67] T. Chen. Infrared renormalization in non-relativistic QED and scaling criticality. *J. Funct. Anal.*, 254:2555–2647, 2008.

[68] T. Chen and J. Fröhlich. Coherent infrared representations in non-relativistic QED. In *Proceedings Symp. Pure Math.*, Amer. Math. Soc., 2006.

[69] T. Chen, V. Vougalter, and S. A. Vugalter. The increase of binding energy and enhanced binding in nonrelativistic QED. *J. Math. Phys.*, 44:1961–1970, 2003.

[70] C. Cohen-Tannoudji, J. Dupont-Roc, and G. Grynberg. *Photons and Atoms*. Wiley Interscience, 1997.

[71] J. M. Cook. The mathematics of second quantization. *Trans. Amer. Math. Soc.*, 74:222–245, 1953.

[72] F. Cornu and Ph. A. Martin. Atom-wall dispersive forces: a microscopic approach. *J. Phys. A*, 42:495001, 2009.

[73] G. Da Prato. *An Introduction to Infinite-Dimensional Analysis*. Springer, 2006.

[74] T. N. Dam and J. S. Møller. Spin-boson type models analysed using symmetries. *arXiv:1803.05812*, preprint, 2018.

[75] A. Das. *Field Theory*. World Scientific, 1993.

[76] E. B. Davies. Asymptotic analysis of some abstract evolution equations. *J. Funct. Anal.*, 25:81–101, 1977.

[77] E. B. Davies. Particle-boson interactions and weak coupling limit. *J. Math. Phys.*, 20:345–351, 1979.

[78] J. Dereziński and C. Gérard. Asymptotic completeness in quantum field theory. Massive Pauli–Fierz Hamiltonians. *Rev. Math. Phys.*, 11:383–450, 1999.

[79] J. Dereziński and C. Gérard. Scattering theory of infrared divergent Pauli–Fierz Hamiltonians. *Ann. Henri Poincaré*, 5:523–578, 2004.

[80] J. Dereziński and V. Jakšić. Spectral theory of Pauli–Fierz operators. *J. Funct. Anal.*, 180:243–327, 2001.

[81] G. di Nunno, B. Øksendal and F. Proske. *Malliavin Calculus for Lévy Processes with Applications to Finance*. Springer, 2009.

[82] R. L. Dobrushin and R. A. Minlos. An investigation of the properties of generalized Gaussian random field. *Sel. Math. Sov.*, 1:215–263, 1981.

[83] M. D. Donsker and S. R. S. Varadhan. Asymptotics for the polaron. *Comm. Pure Appl. Math.*, 36:505–528, 1983.

[84] R. Dümcke. Convergence of multi time correlation functions in the weak and singular coupling limits. *J. Math. Phys.*, 24:311–315, 1983.

[85] R. Dümcke and H. Spohn. Quantum tunneling with dissipation and the Ising model over $\mathbb{R}$. *J. Stat. Phys.*, 41:389–423, 1985.

[86] S. N. Ethier and T. G. Kurtz. *Markov Processes*. Wiley Interscience, 1986.
[87] M. Falconi. Self-adjointness criterion for operators in Fock spaces. *Math. Phys. Anal. Geom.*, 18:18 pp., 2015.
[88] M. Fannes and B. Nachtergaele. Translating the spin-boson model into a classical system. *J. Math. Phys.*, 29:2288–2293, 1988.
[89] W. G. Faris. Invariant cones and uniqueness of the ground state for fermion systems. *J. Math. Phys.*, 13:1285–1290, 1972.
[90] W. G. Faris. Essential self-adjointness of operators in ordered Hilbert spaces. *Comm. Math. Phys.*, 30:23–34, 1973.
[91] W. G. Faris and B. Simon. Degenerate and nondegenerate ground states for Schrödinger operators. *Duke Math. J.*, 42:559–567, 1975.
[92] C. Fefferman, J. Fröhlich, and G. M. Graf. Stability of ultraviolet-cutoff quantum electrodynamics with non-relativistic matter. *Comm. Math. Phys.*, 190:309–330, 1997.
[93] J. Feldman, H. Knörrer, and E. Trubowitz. *Fermionic Functional Integrals and the Renormalization Group*. Amer. Math. Soc., 2002.
[94] B. Felsager. *Geometry, Particles, and Fields*. Springer, 1998.
[95] R. P. Feynman. Mathematical formulation of the quantum theory of electromagnetic interaction. *Phys. Rev.*, 80:440–457, 1950.
[96] R. P. Feynman. Slow electrons in a polar crystal. *Phys. Rev.*, 97:660–665, 1950.
[97] R. P. Feynman and A. R. Hibbs. *Quantum Mechanics and Path Integrals*. McGraw-Hill, 1965.
[98] V. Fock. Konfigurationsraum und zweite Quantelung. *Z. Phys.*, 75:622–647, 1932.
[99] G. B. Folland. *Quantum Field Theory: A Tourist Guide for Mathematicians*. Amer. Math. Soc., 2008.
[100] U. Franz, R. Léandre and R. Schott. Malliavin calculus and Skorohod integration for quantum stochastic processes. *Infin. Dim. Anal. Quant. Prob.*, 4:11–38, 2001.
[101] R. Frank, E. Lieb, R. Seiringer, and L. Thomas. Stability and absence of binding for multi-polaron systems. *arXiv:1004.4892*, preprint, 2010.
[102] H. Fröhlich. Electrons in lattice fields. *Adv. Phys.*, 3:325–361, 1954.
[103] H. Fröhlich, H. Pelzer, and S. Zienau. Properties of slow electrons in polar materials. *Philos. Mag.*, 41:221–242, 1950.
[104] J. Fröhlich. On the infrared problem in a model of scalar electrons and massless, scalar bosons. *Ann. Inst. H. Poincaré*, 19:1–103, 1973.
[105] J. Fröhlich. Existence of dressed one electron states in a class of persistent models. *Fortschr. Phys.*, 22:159–198, 1974.
[106] J. Fröhlich, M. Griesemer, and B. Schlein. Asymptotic electromagnetic fields in a mode of quantum-mechanical matter interacting with the quantum radiation field. *Adv. Math.*, 164:349–398, 2001.
[107] J. Fröhlich, and Y. M. Park. Correlation inequalities and thermodynamic limit for classical and quantum continuous systems II. Bose–Einstein and Fermi–Dirac statistics. *J. Stat. Phys.*, 23:701–753, 1980.
[108] S. A. Fulling. *Aspects of Quantum Field Theory in Curved Space-Time*. Cambridge University Press, 2003.
[109] V. Georgescu, C. Gérard, and J. S. Møller. Spectral theory of massless Pauli–Fierz models. *Comm. Math. Phys.*, 249:29–78, 2004.
[110] H. Georgi. *Lie Algebras in Particle and Physics*. Westview Press, 1999.
[111] C. Gérard. Asymptotic completeness for the spin-boson model with a particle number cutoff. *Rev. Math. Phys.*, 8:549–589, 1996.
[112] C. Gérard. On the existence of ground states for massless Pauli–Fierz Hamiltonians. *Ann. Henri Poincaré*, 1:443–459, 2000.
[113] C. Gérard. On the scattering theory of massless Nelson models. *Rev. Math. Phys.*, 14:1165–1280, 2002.

[114] C. Gérard. A remark on the paper: "On the existence of ground states for Hamiltonians". *mp-arc 06-146*, preprint, 2006.

[115] C. Gérard, F. Hiroshima, A. Panati, and A. Suzuki. Infrared divergence of a scalar quantum field model on a pseudo Riemannian manifold. *Interdiscip. Inform. Sci.*, 15:399–421, 2009.

[116] C. Gérard, F. Hiroshima, A. Panati, and A. Suzuki. Infrared problem for the Nelson model on static space-times. *Comm. Math. Phys.*, 308:543–566, 2011.

[117] C. Gérard, F. Hiroshima, A. Panati, and A. Suzuki. Absence of ground state for the Nelson model on static space-times. *J. Funct. Anal.*, 262:273–299, 2012.

[118] C. Gérard and A. Panati. Spectral and scattering theory for some abstract QFT Hamiltonians. *Rev. Math. Phys.*, 21:373–437, 2009.

[119] C. Gérard and I. Sasaki. Binding condition for a general class of quantum field Hamiltonians. *arXiv: 1206.4764*, preprint, 2012.

[120] B. Gerlach and H. Löwen. Analytic properties of polaron systems, or: do polaron phase transitions exist or not? *Rev. Modern Phys.*, 63:63–89, 1991.

[121] J. Glimm. Yukawa coupling of quantum fields in two dimensions. *Comm. Math. Phys.*, 5:343–386, 1967.

[122] J. Glimm and A. Jaffe. A $\lambda(\phi^4)_2$ quantum field theory without cutoffs. I. *Phys. Rev.*, 176:1945–1951, 1968.

[123] J. Glimm and A. Jaffe. The energy momentum spectrum and vacuum expectation values in quantum field theory. *J. Math. Phys.*, 11:3335–3338, 1970.

[124] J. Glimm and A. Jaffe. The $\lambda(\phi^4)_2$ quantum field theory without cutoffs III. The physical vacuum. *Acta Math.*, 125:203–267, 1970.

[125] J. Glimm and A. Jaffe. The $\lambda(\phi^4)_2$ quantum field theory without cutoffs: II. The field operators and the approximate vacuum. *Ann. Math.*, 91:362–401, 1970.

[126] J. Glimm and A. Jaffe. The energy momentum spectrum and vacuum expectation values in quantum field theory, II. *Comm. Math. Phys.*, 22:1–22, 1971.

[127] J. Glimm and A. Jaffe. The $\lambda(\phi^4)_2$ quantum field theory without cutoffs. IV. Perturbations of the Hamiltonian. *J. Math. Phys.*, 13:1568–1584, 1972.

[128] J. Glimm and A. Jaffe. *Quantum Physics: A Functional Integral Point of View*. Springer, 1981.

[129] M. Griesemer. Exponential decay and ionization thresholds in non-relativistic quantum electrodynamics. *J. Funct. Anal.*, 210:321–340, 2004.

[130] M. Griesemer, E. Lieb, and M. Loss. Ground states in non-relativistic quantum electrodynamics. *Invent. Math.*, 145:557–595, 2001.

[131] E. P. Gross. Particle-like solutions in field theory. *Ann. Phys.*, 19:219–233, 1962.

[132] L. Gross. A noncommutative extension of the Perron–Frobenius theorem. *Bull. Amer. Math. Soc.*, 77:343–347, 1971.

[133] L. Gross. Existence and uniqueness of physical ground states. *J. Funct. Anal.*, 10:52–109, 1972.

[134] L. Gross. The relativistic polaron without cutoffs. *Comm. Math. Phys.*, 31:25–73, 1973.

[135] L. Gross. The free Euclidean proca and electromagnetic fields, functional integral and its application. In *Proc. Int. Conf. at Cumberland Lodge, Windsor Great Park, London, 1974*, A. M. Arthurs (ed.), Clarendon Press, 1975.

[136] H. Grotch. Lamb shift in nonrelativistic quantum electrodynamics. *Ann. Phys.*, 49:48–51, 1981.

[137] H. Grotch and E. Kazes. Nonrelativistic quantum mechanics and the anomalous part of the electron g factor. *Amer. J. Phys.*, 45:618–623, 1977.

[138] M. Gubinelli. Gibbs measures for self-interacting Wiener paths. *Markov Process. Related Fields*, 12:747–766, 2006.

[139] M. Gubinelli, F. Hiroshima, and J. Lőrinczi. Ultraviolet renormalization of the Nelson Hamiltonian through functional integration. *J. Funct. Anal.*, 267:3125–3153, 2014.

[140] M. Gubinelli and J. Lőrinczi. Gibbs measures on Brownian currents. *Commun. Pure Appl. Math.*, 62:1–56, 2009.

[141] M. Gubinelli and J. Lőrinczi. Ultraviolet renormalization of Nelson's Hamiltonian through functional integration, unpublished manuscript, 2010.

[142] F. Guerra, L. Rosen, and B. Simon. The $P(\phi)_2$ Euclidean quantum field theory as classical statistical mechanics. *Ann. Math.*, 101:111–259, 1975.

[143] B. Güneysu, O. Matte, and J. S. Møller. Stochastic differential equations for models of non-relativistic matter interacting with quantized radiation fields. *Probab. Theory Relat. Fields*, 167:817–915, 2017.

[144] R. Haag. *Local Quantum Physics, Fields, Particles, Algebras*. Springer, 1992.

[145] Z. Haba. Feynman integral in regularized nonrelativistic quantum electrodynamics. *J. Math. Phys.*, 39:1766–1787, 1998.

[146] Z. Haba. *Feynman Integral and Random Dynamics in Quantum Physics: A Probabilistic Approach to Quantum Dynamics*. Kluwer Academic, 1999.

[147] C. Hainzl and R. Seiringer. Mass renormalization and energy level shift in non-relativistic QED. *Adv. Theor. Math. Phys.*, 6:847–881, 2002.

[148] C. Hainzl, V. Vougalter, and S. Vugalter. Enhanced binding in nonrelativistic QED. *Comm. Math. Phys.*, 233:13–26, 2003.

[149] E. Hamilton. The infinitely renormalized field in the scalar field model. *J. Math. Phys.*, 17:443–457, 1976.

[150] D. Hasler and I. Herbst. On the self-adjointness and domain of Pauli–Fierz type Hamiltonians. *Rev. Math. Phys.*, 20:787–800, 2008.

[151] D. Hasler and I. Herbst. Ground states in the spin boson model. *Ann. Henri Poincaré*, 12, 2011.

[152] W. Heitler. *The Quantum Theory of Radiation*. Oxford University Press, 1954.

[153] I. Helland. Central limit theorems for martingales with discrete or continuous time. *Scand. J. Stat.*, 9:79–94, 1982.

[154] E. M. Henley and W. Thirring. *Elementary Quantum Field Theory*. McGraw-Hill, 1962.

[155] T. Hida. *Brownian Motion*. Springer, 1980.

[156] T. Hida, H. H. Kuo, J. Potthoff and L. Streit. *White Noise*. Kluwer Academic, 1993.

[157] T. Hidaka. Existence of a ground state for the Nelson model with a singular perturbation. *J. Math. Phys.*, 52, 2011.

[158] T. Hidaka and F. Hiroshima. Pauli–Fierz model with Kato class potential and exponential decay. *Rev. Math. Phys.*, 22:1181–1208, 2010.

[159] T. Hidaka and F. Hiroshima. Self-adjointness of semi-relativistic Pauli–Fierz models. *Rev. Math. Phys.*, 27:1550015, 18 pp., 2015.

[160] T. Hidaka and F. Hiroshima. Spectrum of semi-relativistic Pauli–Fierz Hamiltonian I. *J. Math. Anal. Appl.*, 437:330–349, 2016.

[161] T. Hidaka, F. Hiroshima, and I. Sasaki. Spectrum of semi-relativistic Pauli–Fierz Hamiltonian II. *J. Spectr. Theory*, 11:1779–1830, 2021.

[162] B. Hinrichs and F. Hiroshima. On the ergodicity of renormalized translation-invariant Nelson-type semigroups. *arXiv:2412.09708*, 2024.

[163] B. Hinrichs and F. Hiroshima. On the spectrum of the two-photon quantum Rabi model at critical coupling. *preprint*, 2025.

[164] M. Hirokawa. An expression of the ground state energy of the spin-boson model. *J. Funct. Anal.*, 162:178–218, 1999.

[165] M. Hirokawa. Remarks on the ground state energy of the spin-boson model. An application of the Wigner–Weisskopf model. *Rev. Math. Phys.*, 13:221–251, 2001.

[166] M. Hirokawa. Ground state transition for two-level system coupled with Bose field. *Phys. Lett. A*, 294:13–18, 2002.

[167] M. Hirokawa. Infrared catastrophe for Nelson's model. *Publ. RIMS, Kyoto Univ.*, 42:897–922, 2003.

[168] M. Hirokawa. Dicke-type energy level crossings in cavity-induced atom cooling: another superradiant cooling. *Phys. Rev. A*, 79:043408, 2009.
[169] M. Hirokawa. The Dicke-type crossings among eigenvalues of differential operators in a class of non-commutative oscillators. *Indiana Univ. Math. J.*, 58:1493–1536, 2009.
[170] M. Hirokawa and F. Hiroshima. Absence of energy level crossing for the ground state energy of the Rabi model. *Commun. Stoch. Anal.*, 8:551–560, 2014.
[171] M. Hirokawa, F. Hiroshima, and J. Lőrinczi. Spin-boson model through a Poisson driven stochastic process. *Math. Z.*, 277:1165–1198, 2014.
[172] M. Hirokawa, F. Hiroshima, and H. Spohn. Ground state for point particles interacting through a massless scalar Bose field. *Adv. Math.*, 191:339–392, 2005.
[173] F. Hiroshima. Diamagnetic inequalities for systems of nonrelativistic particles with a quantized field. *Rev. Math. Phys.*, 8:185–203, 1996.
[174] F. Hiroshima. Functional integral representation of a model in quantum electrodynamics. *Rev. Math. Phys.*, 9:489–530, 1997.
[175] F. Hiroshima. Weak coupling limit removing an ultraviolet cut-off for a Hamiltonian of particles interacting with a quantized scalar field. *J. Math. Phys.*, 40:1215–1236, 1999.
[176] F. Hiroshima. Essential self-adjointness of translation-invariant quantum field models for arbitrary coupling constants. *Comm. Math. Phys.*, 211:585–613, 2000.
[177] F. Hiroshima. Ground states of a model in nonrelativistic quantum electrodynamics II. *J. Math. Phys.*, 41:661–674, 2000.
[178] F. Hiroshima. Self-adjointness of the Pauli–Fierz Hamiltonian for arbitrary values of coupling constants. *Ann. Henri Poincaré*, 3:171–201, 2002.
[179] F. Hiroshima. Multiplicity of ground states in quantum field models: applications of asymptotic fields. *J. Funct. Anal.*, 224:431–470, 2005.
[180] F. Hiroshima. Fiber Hamiltonians in the non-relativistic quantum electrodynamics. *J. Funct. Anal.*, 252:314–355, 2007.
[181] F. Hiroshima. Functional integral approach to semi-relativistic Pauli–Fierz models. *Adv. Math.*, 259:784–840, 2014.
[182] F. Hiroshima. Note on ultraviolet renormalization and ground state energy of the Nelson model. *arXiv:1507.05302*, preprint, 2015.
[183] F. Hiroshima. Translation invariant models in QFT without ultraviolet cutoffs. *arXiv:1506.07514*, preprint, 2015.
[184] F. Hiroshima. *Ground States of Quantum Field Models*. Springer, 2019.
[185] F. Hiroshima, T. Ichinose, and J. Lőrinczi. Probabilistic representation and fall-off of bound states of relativistic Schrödinger operators with spin 1/2. *Publ. RIMS Kyoto*, 49:189–214, 2013.
[186] F. Hiroshima and K. R. Ito. Mass renormalization in nonrelativistic quantum electrodynamics with spin 1/2. *Rev. Math. Phys.*, 19:405–454, 2007.
[187] F. Hiroshima and J. Lőrinczi. Functional integral representations of the Pauli–Fierz model with spin 1/2. *J. Funct. Anal.*, 254:2127–2185, 2008.
[188] F. Hiroshima, J. Lőrinczi, and T. Takaesu. A probabilistic representation of the ground state expectation of fractional powers of the boson number operator. *J. Math. Anal. Appl.*, 395:437–447, 2012.
[189] F. Hiroshima and O. Matte. Ground states and their associated Gibbs measures in the renormalized Nelson model. *Rev.Math. Phys.*, 22:2250002, 2022.
[190] F. Hiroshima and O. Matte, Two-sided bounds on the point-wise spatial decay of ground states in the renormalized Nelson model with confining potentials, *Math. Phys. Anal. Geom.* 29:19, (2026), https://doi.org/10.1007/s11040-026-09557-z.
[191] F. Hiroshima and I. Sasaki. Enhanced binding for *N* particle system interacting with a scalar field I. *Math. Z.*, 259:657–680, 2007.

[192] F. Hiroshima and I. Sasaki. Semi-relativistic Pauli–Fierz model. *RIMS Kôkyuroku Bessatsu*, 21:25–34, 2010.

[193] F. Hiroshima and I. Sasaki. Enhanced binding of an *N* particle system interacting with a scalar field II, relativistic version. *Publ. RIMS Kyoto*, 51:655–690, 2015.

[194] F. Hiroshima, I. Sasaki, H. Spohn, and A. Suzuki. *Enhanced Biding in Quantum Field Theory*. IMI Lecture Note, vol. 38, Kyushu University, 2012.

[195] F. Hiroshima, and T. Shirai. Fiber decomposition of non-commutative harmonic oscillators by two-photon quantum Rabi models. *J. Math. Phys.*, 66:032101, 2024.

[196] F. Hiroshima, and T. Shirai. Spectral zeta function and ground state of quantum Rabi model. *J. Funct. Anal.*, 289:110901, 2025.

[197] F. Hiroshima and H. Spohn. Enhanced binding through coupling to a quantum field. *Ann. Henri Poincaré*, 2:1159–1187, 2001.

[198] F. Hiroshima and H. Spohn. Ground state degeneracy of the Pauli–Fierz model with spin. *Adv. Theor. Math. Phys.*, 5:1091–1104, 2001.

[199] F. Hiroshima and H. Spohn. Mass renormalization in nonrelativistic quantum electrodynamics. *J. Math. Phys.*, 46:042302, 2005.

[200] F. Hiroshima, A. Suzuki, and H. Spohn. The non-binding regime of the Pauli–Fierz model. *J. Math. Phys.*, 52:062104, 12 pp., 2011.

[201] F. Hiroshima and Y. Tsujimoto. Lower bounds on the pointwise spatial decay of the ground state in the Pauli–Fierz model, preprint, 2026.

[202] H. Holzmann. Martingale approximations for continuous-time and discrete-time stationary Markov processes. *Stochastic Processes and their Applications*, 115:1518–1529, 2005.

[203] R. Høegh-Krohn. Asymptotic fields in some models of quantum field theory I. *J. Math. Phys.*, 9:2075–2080, 1968.

[204] R. L. Hudson and K. R. Parthasarathy. Quantum Ito's formula and stochastic evolutions. *Commun. Math. Phys.*, 93:301–323, 1984.

[205] M. Hübner and H. Spohn. Spectral properties of the spin-boson Hamiltonian. *Ann. Inst. Henri Poincaré*, 62:289–323, 1995.

[206] C. Itzykson and J. B. Zuber. *Quantum Field Theory*. McGraw-Hill, 1985.

[207] S. Janson. *Gaussian Hilbert Spaces*. Cambridge University Press, 1997.

[208] J. M. Jauch and F. Rohrlich. *The Theory of Photons and Electrons*. Springer, 1976.

[209] E. T. Jaynes and F. W. Cummings. On the process of space quantization. *Proc. IEEE*, 51:89–109, 1963.

[210] M. Kac. Toeplitz matrices, translation kernels and a related problem in probability theory. *Duke Mathematical Journal*, 21:501–509, 1954.

[211] G. Källén. *Quantum Electrodynamics*. Springer, 1972.

[212] G. Kallianpur and J. Xiong. *Stochastic Differential Equations in Infinite Dimensional Spaces*. Institute of Mathematical Statistics Lecture Notes – Monograph Series, 1995.

[213] I. Karatzas and S. E. Shreve. *Brownian Motion and Stochastic Calculus*. Springer, 1997.

[214] I. Karatzas and S. E. Shreve. *Methods of Mathematical Finance*. Springer, 2001.

[215] Y. Kato and N. Mugibayashi. Regular perturbation and asymptotic limits of operators in quantum field theory. *Prog. Theor. Phys.*, 30:103–133, 1963.

[216] A. Klein and L. J. Landau. Singular perturbations of positivity preserving semigroups via path space techniques. *J. Funct. Anal.*, 20:44–82, 1975.

[217] M. Könenberg and O. Matte. Ground states of semi-relativistic Pauli–Fierz and no-pair Hamiltonians in QED at critical Coulomb coupling. *J. Oper. Theory*, 70:211–237, 2013.

[218] M. Könenberg and O. Matte. On enhanced binding and related effects in the non- and semi-relativistic Pauli–Fierz models. *Commun. Math. Phys.*, 323:635–661, 2013.

[219] M. Könenberg, O. Matte, and E. Stockmeyer. Existence of ground states of hydrogen-like atoms in relativistic QED I: the semi-relativistic Pauli–Fierz operator. *Rev. Math. Phys.*, 23:635–661, 2011.

[220] M. Könenberg, O. Matte, and E. Stockmeyer. Existence of ground states of hydrogen-like atoms in relativistic quantum electrodynamics. II. The no-pair operator. *J. Math. Phys.*, 52:123501, 34 pp., 2011.
[221] L. Kupsch. Functional integration for Euclidean Dirac fields. *Ann. Henri Poincaré*, 50:143–160, 1989.
[222] M. Kuś. On the spectrum of a two-level system. *J. Math. Phys.*, 26:2792–2795, 1985.
[223] L. D. Landau and S. I. Pekar. Effective mass of a polaron. *Zh. Eksp. Teor. Fiz.*, 18:419–423, 1948.
[224] J. Lampart. The resolvent of the Nelson Hamiltonian improves positivity. *Math. Phys. Anal. Geom.*, 24:2021.
[225] T. D. Lee, F. E. Low, and D. Pines. The motion of slow electrons in a polar crystal. *Phys. Rev.*, 90:297–302, 1953.
[226] A. J. Leggett, S. Chakravarty, A. T. Dorsey, M. P. A. Fisher, A. Garg, and W. Zwerger. Dynamics of the dissipative two-state system. *Rev. Mod. Phys.*, 59:1–85, 1987.
[227] E. Lieb and M. Loss. Existence of atoms and molecules in non-relativistic quantum electrodynamics. *Adv. Theor. Math. Phys.*, 7:667–710, 2003.
[228] E. Lieb and L. E. Thomas. Exact ground state energy of the strong-coupling polaron. *Commun. Math. Phys.*, 183:511–519, 1997.
[229] J. M. Lindsay. Fermion martingale. *Probab. Theory Relat. Fields*, 71:307–320, 1986.
[230] J. Lőrinczi. Exponential integrability and DLR consistency of some rough functionals. In Seminar on Stochastic Analysis, Random Fields and Applications VI, Ascona, 2008, R. Dalang, M. Dozzi, F. Russo (eds.), pp. 191–208. Progress in Probability, Birkhäuser, 2011.
[231] J. Lőrinczi, F. Hiroshima, and V. Betz. *Feynman–Kac-type Theorems and Gibbs Measures on Path Space. With Applications to Rigorous Quantum Field Theory*. De Gruyter, 2011.
[232] J. Lőrinczi, F. Hiroshima, and V. Betz. *Feynman–Kac Type Theorems and Gibbs Measures on Path Space. Volume 1: Feynman–Kac-type Formulae and Gibbs Measures*, 2nd edition. De Gruyter, 2019.
[233] J. Lőrinczi and N. J. Mauser. A stochastic approach to the bipolaron model. *RIMS Kokyuroku*, 1482:1–10, 2006.
[234] J. Lőrinczi, R. A. Minlos. Gibbs measures for Brownian paths under the effect of an external and a small pair potential. *J. Stat. Phys.*, 105:605–647, 2001.
[235] J. Lőrinczi, R. A. Minlos, and H. Spohn. Infrared regular representation of the three dimensional massless Nelson model. *Lett. Math. Phys.*, 59:189–198, 2002.
[236] J. Lőrinczi, R. A. Minlos, and H. Spohn. The infrared behaviour in Nelson's model of a quantum particle coupled to a massless scalar field. *Ann. Henri Poincaré*, 3:1–28, 2002.
[237] M. Loss, T. Miyao, and H. Spohn. Lowest energy states in nonrelativistic QED: atoms and ions in motion. *J. Funct. Anal.*, 243:353–393, 2007.
[238] M. Loss, T. Miyao, and H. Spohn. Kramers degeneracy theorem in nonrelativistic QED. *Lett. Math. Phys.*, 89:21–31, 2009.
[239] E. Lukas. *Characteristic Functions*, 2nd edition, revised and enlarged. Hafner Publishing Co., New York, 1970.
[240] P. Malliavin and A. Thalmaier. *Stochastic Calculus of Variations in Mathematical Finance*. Springer, 2005.
[241] V. A. Malyshev and R. A. Minlos. *Linear Infinite-Particle Operators*. Amer. Math. Soc., 1995.
[242] O. Matte. Continuity properties of the semi-group and its integral kernel in non-relativistic QED. *Rev. Math. Phys.*, 28:1650011, 90 pp., 2016.
[243] O. Matte and J. S. Møller. Feynman–Kac formulas for the ultra-violet renormalized Nelson model. *arXiv:1701.02600*, preprint, 2017.
[244] O. Matte and E. Stockmeyer. Exponential localization of hydrogen-like atoms in relativistic quantum electrodynamics. *Commun. Math. Phys.*, 295:551–583, 2010.
[245] P. W. Milonni. *The Quantum Vacuum*. Academic Press, 1994.
[246] R. A. Minlos. Generalized random processes and their extension to a measure. *Tr. Mosk. Mat. Obŝĉ.*, 8:497–518, 1959.
[247] R. A. Minlos. Lower branch of the spectrum of a fermion interacting with a bosonic gas (polaron). *Theor. Math. Phys.*, 92:869–878, 1992.

[248] T. Miyao. Nondegeneracy of ground states in nonrelativistic quantum field theory. *J. Oper. Theory*, 64:207–241, 2010.

[249] T. Miyao. Self-dual cone analysis in condensed matter physics. *Rev. Math. Phys.*, 23:749–822, 2010.

[250] T. Miyao. On the semigroup generated by the renormalized Nelson Hamiltonian. *J. Funct. Anal.*, 276:19481977, 2019.

[251] T. Miyao and H. Spohn. The retarded van der Waals potential: revisited. *J. Math. Phys.*, 50:072103, 2009.

[252] O. Miyatake. On the non-existence of solution of field equations in quantum mechanics. *J. Inst. Polytech. Osaka City Univ.*, 2:89–99, 1952.

[253] J. S. Møller. The translation invariant massive Nelson model: I. The bottom of the spectrum. *Ann. Henri Poincaré*, 6:1091–1135, 2005.

[254] J. S. Møller. The polaron revisited. *Rev. Math. Phys.*, 18:485–517, 2006.

[255] V. P. Nair. *Quantum Field Theory*. Springer, 2005.

[256] E. Nelson. Analytic vectors. *Ann. Math.*, 70:572–615, 1959.

[257] E. Nelson. Interaction of nonrelativistic particles with a quantized scalar field. *J. Math. Phys.*, 5:1190–1197, 1964.

[258] E. Nelson. Schrödinger particles interacting with a quantized scalar field. In *Proc. Conf. on Analysis in Function Space*, W. T. Martin and I. Segal (eds.), p. 87. MIT Press, 1964.

[259] E. Nelson. Construction of quantum fields from Markoff fields. *J. Funct. Anal.*, 12:97–112, 1973.

[260] E. Nelson. Quantum fields and Markoff fields. In *Proc. of Sympos. Pure Math XXIII, Berkeley*, D. C. Spencer (ed.), pp. 413–420. Amer. Math. Soc., Providence, RI, 1973.

[261] E. Nelson. The free Markoff field. *J. Funct. Anal.*, 12:211–227, 1973.

[262] D. Nualart. *The Malliavin Calculus and Related Topics*. Springer, 1995.

[263] D. Nualart and E. Nualart. *Introduction to Malliavin Calculus*. Cambridge University Press, 2018.

[264] N. Obata. *White Noise Calculus and Fock Space*. Lecture Notes in Mathematics 1557, Springer, 1994.

[265] H. Osada and H. Spohn. Gibbs measures relative to Brownian motion. *Ann. Probab.*, 27:1183–1207, 1999.

[266] K. Osterwalder and R. Schrader. Feynman–Kac formula for Euclidean Fermi and Bose fields. *Phys. Rev. Lett.*, 29:1423–1425, 1972.

[267] P. F. Palmer. The singular coupling and weak coupling limit. *J. Math. Phys.*, 18:527–529, 1977.

[268] K. R. Parthasarathy. *An Introduction to Quantum Stochastic Calculus*. Springer, 1992.

[269] W. Pauli and M. Fierz. Zur Theorie der Emission langwelliger Lichtquanten. *Nuovo Cimento*, 15:167–188, 1938.

[270] M. E. Peskin and D. V. Schröder. *An Introduction to Quantum Field Theory*. Westview Press, 1995.

[271] M. A. Piech. The Ornstein–Uhlenbeck semigroup in an infinite dimensional L^2 setting. *J. Funct. Anal.*, 18:271–285, 1975.

[272] C. Prévôt and M. Röckner. *A Concise Course on Stochastic Partial Differential Equations*. Springer, 2007.

[273] I. I. Rabi. On the process of space quantization. *Phys. Rev.*, 49:324–328, 1936.

[274] C. Reyes Bustos and M. Wakayama. Zeta limits for the spectrum of quantum Rabi models. *J. Math. Phys.*, 65, 2024.

[275] M. Reed and B. Simon. *Methods of Modern Mathematical Physics IV*. Academic Press, 1978.

[276] G. Roepstorff. *Path Integral Approach to Quantum Physics*. Springer, 1996.

[277] A. Rogers. Fermion path integration and Grassmann Brownian motion. *Commun. Math. Phys.*, 113:353–368, 1987.

[278] L. H. Ryder. *Quantum Field Theory*. Cambridge University Press, 1985.

[279] I. Sasaki. Ground state of the massless Nelson model in a non-Fock representation. *J. Math. Phys.*, 46:102107, 2005.

[280] G. Scharf. *Finite Quantum Electrodynamics*. Springer, 1995.

[281] L. S. Schulman. *Techniques and Applications of Path Integration*. Wiley-Interscience, 1981.

[282] A. S. Schwarz. *Quantum Field Theory and Topology*. Springer, 1993.

[283] S. Schweber. *An Introduction to Relativistic Quantum Field Theory*. Harper & Row, 1961.

[284] S. Schweber. On the application of Bargmann Hilbert spaces to dynamical problems. *Ann. Phys.*, 41:205–229, 1967.

[285] I. Segal. Tensor algebras over Hilbert spaces, I. *Trans. Amer. Math. Soc.*, 81:106–134, 1956.

[286] I. Segal. Tensor algebras over Hilbert spaces, II. *Ann. Math.*, 63:160–175, 1956.

[287] D. Shale. A note on the scattering of boson fields. *J. Math. Phys.*, 5:915–921, 1962.

[288] B. Simon. *The $P(\phi)_2$ Euclidean (Quantum) Field Theory*. Princeton University Press, 1974.

[289] B. Simon and R. Høegh-Krohn. Hypercontractive semigroup and two dimensional self-coupled Bose fields. *J. Funct. Anal.*, 9:121–180, 1972.

[290] E. Skibsted. Spectral analysis of N-body systems coupled to a bosonic field. *Rev. Math. Phys.*, 10:989–1026, 1998.

[291] A. D. Sloan. A nonperturbative approach to nondegeneracy of ground states in quantum field theory: polaron models. *J. Funct. Anal.*, 16:161–191, 1974.

[292] A. D. Sloan. Analytic domination with quadratic form type estimates and nondegeneracy of ground states in quantum field theory. *Trans. Amer. Math. Soc.*, 194:325–336, 1974.

[293] A. D. Sloan. The polaron without cutoffs in two space dimensions. *J. Math. Phys.*, 15:190–201, 1974.

[294] H. Spohn. Effective mass of the polaron: a functional integral approach. *Ann. Phys.*, 175:278–318, 1987.

[295] H. Spohn. The polaron at large total momentum. *J. Phys. A*, 21:1199–1211, 1988.

[296] H. Spohn. Ground state(s) of the spin-boson Hamiltonian. *Commun. Math. Phys.*, 123:277–304, 1989.

[297] H. Spohn. Asymptotic completeness for Rayleigh scattering. *J. Math. Phys.*, 38:2281, 1997.

[298] H. Spohn. Ground state of a quantum particle coupled to a scalar boson field. *Lett. Math. Phys.*, 44:9–16, 1998.

[299] H. Spohn. Ground state of quantum particle coupled to a scalar boson field. *Lett. Math. Phys.*, 44:9–16, 1998.

[300] H. Spohn. *Dynamics of Charged Particles and Their Radiation Field*. Cambridge University Press, 2004.

[301] H. Spohn, R. Stück, and W. Wreszińki. Localisation for the spin J-boson Hamiltonian. *Ann. Inst. Henri Poincaré*, 53:225–244, 1990.

[302] M. Stone. *The Physics of Quantum Fields*. Springer, 2000.

[303] R. F. Streater and A. S. Wightman. *PCT, Spin and Statistics, and All That*. W. A. Benjamin, 1980.

[304] M. S. Swanson. *Path Integrals and Quantum Processes*. Academic Press, 1992.

[305] N. Teranishi. Absence of ground states of generalized spin-boson models. *Kyushu J. Math.*, 72:1–14, 2018.

[306] S. Teufel. *Adiabatic Perturbation Theory in Quantum Dynamics*. Springer, 2003.

[307] L. van Hove. Les difficultés de divergences pour un modèle particulier de champ quantifié. *Physica*, 18:145–159, 1952.

[308] V. S. Varadarajan. *Geometry of Quantum Theory*. Springer, 2007.

[309] M. Veraar. The stochastic Fubini theorem revisited. *Stochastics*, 84:543–551, 2012.

[310] S. Weinberg. *Quantum Physics I*. Cambridge University Press, 1995.

[311] S. Weinberg. *Quantum Physics II*. Cambridge University Press, 1996.

[312] S. Weinberg. *Quantum Physics III*. Cambridge University Press, 2000.

[313] T. Welton. Some observable effects of the quantum-mechanical fluctuations of the electromagnetic field. *Phys. Rev.*, 74:1157–1167, 1948.

[314] G. C. Wick. The evaluation of the collision matrix. *Phys. Rev.*, 80:267–272, 1950.

[315] D. N. Williams. Euclidean fermion fields with Hermitian Feynman–Kac–Nelson formula I. *Commun. Math. Phys.*, 38:65–80, 1974.

[316] L. M. Wu. Feynman–Kac semigroups, ground state diffusions, and large deviations. *J. Funct. Anal.*, 123:202–231, 1994.

[317] Q. Xie, H. Zhong, M. T. Batchelor, and C. Lee. The quantum Rabi model: solution and dynamics. *J. Math. Phys.*, 50:113001, 2017.

[318] Y. Yamasaki. *Measures on Infinite Dimensional Spaces*. World Scientific, 1983.

[319] E. Zeidler. *Quantum Field Theory II*. Springer, 2009.

Index

https://doi.org/10.1515/9783112231494-007

De Gruyter Studies in Mathematics

Volume 37
René L. Schilling, Renming Song, Zoran Vondraček
Bernstein Functions. Theory and Applications, 3rd Edition, 2026
ISBN 978-3-11-129427-8, e-ISBN 978-3-11-129512-1, e-ISBN (ePUB) 978-3-11-129580-0

Volume 91
Johnny Henderson, Rodica Luca
Boundary Value Problems for Second-Order Finite Difference Equations and Systems, 2nd Edition, 2025
ISBN 978-3-11-222482-3, e-ISBN 978-3-11-222483-0, e-ISBN (ePUB) 978-3-11-222484-7

Volume 103
David Yetter, Michael Nicholas Gurski, Juan Orendain
Compact Closed 2-Categories. Duality, Enrichment, and Strictification, 2025
ISBN 978-3-11-914815-3, e-ISBN 978-3-11-218854-5, e-ISBN (ePUB) 978-3-11-220458-0

Volume 102
Changxing Miao, Bo Zhang, Jiqiang Zheng
Harmonic Analysis Methods in Partial Differential Equations, 2025
ISBN 978-3-11-138451-1, e-ISBN 978-3-11-138472-6, e-ISBN (ePUB) 978-3-11-138486-3

Volume 101
Marko Kostić
Almost Periodic Type Solutions. to Integro-Differential-Difference Equations, 2025
ISBN 978-3-11-168728-5, e-ISBN 978-3-11-168974-6, e-ISBN (ePUB) 978-3-11-169021-6

Volume 64
Dorina Mitrea, Irina Mitrea, Marius Mitrea, Michael Taylor
The Hodge–Laplacian. Boundary Value Problems on Riemannian Manifolds, 2nd Edition, 2025
ISBN 978-3-11-148098-5, e-ISBN 978-3-11-148140-1, e-ISBN (ePUB) 978-3-11-148389-4

Volume 100
Changxing Miao, Ruipeng Shen
Regularity and Scattering of Dispersive Wave Equation. Multiplier Method and Morawetz Estimate, 2025
ISBN 978-3-11-148754-0, e-ISBN 978-3-11-148835-6, e-ISBN (ePUB) 978-3-11-148940-7

Volume 99
Marcus Laurel, Marius Mitrea
Weighted Morrey Spaces. Calderón-Zygmund Theory and Boundary Problems, 2024
ISBN 978-3-11-145816-8, e-ISBN 978-3-11-145827-4, e-ISBN (ePUB) 978-3-11-146145-8

www.degruyterbrill.com